On Certain *L*-Functions

Clay Mathematics Proceedings
Volume 13

On Certain L-Functions

Conference on Certain L-Functions in honor of Freydoon Shahidi
July 23–27, 2007
Purdue University, West Lafayette, Indiana

James Arthur
James W. Cogdell
Steve Gelbart
David Goldberg
Dinakar Ramakrishnan
Jiu-Kang Yu
Editors

American Mathematical Society
Clay Mathematics Institute

2000 *Mathematics Subject Classification.* Primary 11F30, 11F70, 11F80, 11S37, 20G05, 20G25, 22E35, 22E50, 22E55.

Front cover graphic: Shahidi, Freydoon. On Certain L-Functions. American Journal of Mathematics 103:2 (1981), p.297. ©1981 The Johns Hopkins University Press. Reprinted with permission of The Johns Hopkins University Press.

The back cover photograph of Freydoon Shahidi is used with the permission of Purdue University and Andrew Hancock.

Library of Congress Cataloging-in-Publication Data

On certain L-functions : conference in honor of Freydoon Shahidi on certain L-functions, Purdue University, West Lafayette, Indiana, July 23–27, 2007 / James Arthur . . . [et al.], editors.
p. cm. — (Clay mathematics proceedings ; v. 13)
Includes bibliographical references.
ISBN 978-0-8218-5204-0 (alk. paper)
1. L-functions—Congresses. I. Shahidi, Freydoon. II. Arthur, James, 1944–

QA246.O53 2011
512.7′3—dc22

2011003659

Visit the AMS home page at `http://www.ams.org/`
Visit the Clay Mathematics Institute home page at `http://www.claymath.org/`

10 9 8 7 6 5 4 3 2 1   16 15 14 13 12 11

Contents

Preface vii

Shahidi's Work "On Certain L-functions": A Short History of Langlands-Shahidi Theory 1
STEVE GELBART

The Embedded Eigenvalue Problem for Classical Groups 19
JAMES ARTHUR

A Cuspidality Criterion for the Exterior Square Transfer of Cusp Forms on GL(4) 33
MAHDI ASGARI AND A. RAGHURAM

Types and Explicit Plancherel Formulæ for Reductive p-adic Groups 55
COLIN J. BUSHNELL, GUY HENNIART AND PHILIP C. KUTZKO

Jacquet Modules and the Asymptotic Behaviour of Matrix Coefficients 81
BILL CASSELMAN

The ABS Principle : Consequences for $L^2(G/H)$ 99
LAURENT CLOZEL

Orbital Integrals and Distributions 107
L. CLOZEL AND J.-P. LABESSE

Functoriality for the Quasisplit Classical Groups 117
J.W. COGDELL, I.I. PIATETSKI-SHAPIRO AND F. SHAHIDI

Poles of L-Functions and Theta Liftings for Orthogonal Groups, II 141
DAVID GINZBURG, DIHUA JIANG AND DAVID SOUDRY

On Dual R–groups for Classical Groups 159
DAVID GOLDBERG

Irreducibility of the Igusa Tower over Unitary Shimura Varieties 187
HARUZO HIDA

On the Gross-Prasad Conjecture for Unitary Groups 205
HERVÉ JACQUET and STEPHEN RALLIS

On Local L-Functions and Normalized Intertwining Operators II; Quasi-Split Groups 265
HENRY H. KIM AND WOOK KIM

Reflexions on Receiving the Shaw Prize 297
ROBERT P. LANGLANDS

On Arthur's Asymptotic Inner Product Formula of Truncated Eisenstein Series 309
EREZ LAPID

Multiplicité 1 dans les paquets d'Arthur aux places p-adiques 333
C. MŒGLIN

Unramified Unitary Duals for Split Classical p–adic Groups; The Topology and Isolated Representations 375
GORAN MUIĆ AND MARKO TADIĆ

Parametrization of Tame Supercuspidal Representations 439
FIONA MURNAGHAN

On the Sato-Tate Conjecture, II 471
V. KUMAR MURTY

Icosahedral Fibres of the Symmetric Cube and Algebraicity 483
DINAKAR RAMAKRISHNAN

Pseudo Eisenstein Forms and the Cohomology of Arithmetic Groups III: Residual Cohomology Classes 501
JÜRGEN ROHLFS AND BIRGIT SPEH

A Functor from Smooth o-Torsion Representations to (φ, Γ)-Modules 525
PETER SCHNEIDER AND MARIE-FRANCE VIGNERAS

Motivic Galois Groups and L-Groups 603
HIROYUKI YOSHIDA

Preface

For over three decades Freydoon Shahidi has been making significant contributions to number theory, automorphic forms, and harmonic analysis. Shortly after receiving his Ph.D. in 1975 from Johns Hopkins, under the direction of Joseph Shalika, Shahidi laid out a program to address several open problems, as stated by Langlands, by a novel method now known as the Langlands-Shahidi method. In particular, Shahidi sought to exploit the Fourier coefficients of Eisenstein series and their local analogs to establish cases of Langlands functoriality. Key to this idea was the understanding of generic forms and their local components, from which he developed the theory of local coefficients. Shahidi believed this theory, combined with other methods (including converse theorems) could yield elusive examples of functoriality, such as the symmetric power transfers for GL_2 and functoriality from classical groups to general linear groups. It was well known that establishing such results would yield significant progress in number theory.

Through the first 10 years (or so) of this pursuit, Shahidi produced several important results, and another 20 years of results of such stature would, alone, be fitting of a 60th birthday conference and accompanying volume. However, the power and stature of his results grew significantly and continues through (and beyond) the publication of this volume. Simply put, in the last 15 to 20 years, Shahidi and the Langlands-Shahidi method have helped produce a series of significant results. Rather than a list, we will say that this record speaks for itself. That a number of events within and outside of the Langlands-Shahidi method transpired simultaneously has, no doubt, changed the face of the Langlands program in significant ways. That Shahidi's approach is crucial to so much recent progress is a testament to his persistence and perseverance. The influence of Shahidi's work continues to grow, and the breadth of the applications by Shahidi, his collaborators, and others is undeniable.

One of Shahidi's contributions that should not be overlooked is his service in mentoring young mathematicians within his chosen field. To date, Shahidi has produced eight Ph.D.'s and has at least six more in progress. In addition, he has sponsored roughly 15 postdoctoral appointments at Purdue. Further, several outstanding figures have noted his more informal, but just as crucial, role as a mentor. The list of these mathematicians includes several major contributors to the field, some of whom have contributed articles for this volume.

Freydoon Shahidi reached the age of 60 on June 19, 2007, and a conference was held to commemorate this occasion July 29-August 3, 2007, at Purdue University, West Lafayette, Indiana. As Shahidi has been a member of the Mathematics Department at Purdue since 1977 and has been designated by Purdue as a Distinguished Professor of Mathematics, this seemed a fitting location for such a conference. Funding for this conference was provided by Purdue University's Mathematics Department and College of Science, the National Science Foundation, The Clay Mathematics Institute, The Institute for Mathematics and its Applications, and the Number Theory Foundation. Over 100 mathematicians attended, and there were 23 one-hour lectures. The conference focused on several aspects of the Langlands program, including some exposition of Shahidi's work, recent progress, and future avenues of investigation. Far from being a retrospective, the conference emphasized the vast array of significant problems ahead. All lecturers were invited to contribute material for this volume. In addition, some important figures who were unable to attend or deferred on speaking at the conference were invited to submit articles as well. We hope this resulting volume will serve as a modest tribute to Shahidi's legacy to date, but should not be considered the final word on this subject.

The editors wish to thank all of the authors for their willingness to contribute manuscripts of such high quality in honor of our colleague. We also wish to thank the anonymous referees for their conscientious reading of these manuscripts and their helpful comments to authors which have improved the contents. We wish to express our deep gratitude to Purdue University's Mathematics Department and College of Science, the National Science Foundation, The Clay Mathematics Institute, The Institute for Mathematics and its Applications, and the Number Theory Foundation, for their sponsorship of the conference. We also wish to thank all of the conferees who made the conference so successful, and the staff of the Mathematics Department at Purdue University, particularly Julie Morris, for their help with organizational matters. We thank the Clay Mathematics Institute and the American Mathematical Society for agreeing to publish this work, and a special thanks goes to Vida Salahi for all her efforts in helping us through this process.

Clay Mathematics Proceedings
Volume 13, 2011

Shahidi's Work "On Certain L-functions": A Short History of Langlands-Shahidi Theory

Steve Gelbart

For Freydoon Shahidi, on his 60th birthday

Some History

The precursor of **Langlands-Shahidi** theory is **Selberg's** earlier relation between **real analytic Eisenstein series** and the **Riemann zeta function**.

More precisely, let

$$E(z,s) = \frac{1}{2}\sum \frac{y^s}{|cz+d|^{2s}}, \quad (c,d)=1,$$

where (c,d) means the greatest common divisor of c and d. This is the simplest real analytic Eisenstein series for $SL(2,\mathbf{R})$. Its zeroth Fourier coefficient is

$$a_0(y,s) = \int_0^1 E(z,s)dx = y^s + M(s)y^{1-s}$$

where

$$M(s) = \frac{\sqrt{\pi}\Gamma(s-\frac{1}{2})\zeta(2s-1)}{\Gamma(s)\zeta(2s)}; \tag{1}$$

its first Fourier coefficient is

$$a_1(y,s) = \int_0^1 E(z,s)exp(-2\pi ix)dx = \frac{2\pi^s y^{\frac{1}{2}} K_{s-\frac{1}{2}}(2\pi y)}{\Gamma(s)\zeta(2s)} \tag{2}$$

which implies what we want (see p. 46 of [**Kub**]). That is, the meromorphic continuation of $M(s)$ already gives the (in this case known) meromorphic continuability of $\zeta(s)$, and the functional equation of $E(z,s)$ gives the functional equation of $\zeta(s)$.

The generalization of these ideas would have to wait many more years, when **R.P.Langlands** began to publish his miraculous conjectures.

Langlands (1967 Whittemore Lectures)

In January 1967, **Langlands** wrote his famous letter to **Weil**. It contains precursors of most of the startling conjectures that today make up most of the broad "**Langlands Program**". In April 1967, **Langlands** delivered the Yale

2010 *Mathematics Subject Classification.* Primary 11F66 ; Secondary 11F70, 22E55.

Whittemore Lectures, wherein he investigated the (general) **Eisenstein** series of type

$$E(g,\varphi,s)=\sum_{\gamma\in P\setminus G}\varphi_s(\gamma g) \tag{3}$$

with G a reductive group (more about the terminology below). *Euler Products* [**Lan1**] is a monograph based on these lectures. In it, one finds the first (published) version of a crucial and key concept – that of an **L-group**. It comes out of the expression of the zeroth Fourier coefficient of the general $E(g,\varphi,s)$, in a surprisingly new way. Without going into too much detail, let's try to explain why. Complete details of everything can be found in many places; the basic source is [**Lan2**]. We also refer to Bill **Casselman's** paper "The L-Group" [**Cas1**] for a leisurely account of these and other important computations.

Our Eisenstein series are induced from a representation π and a complex parameter s on the Levi component M of P. Here $P = MU$ is a maximal parabolic, δ_P its modulus function, ϖ the fundamental weight corresponding to P, and K a good maximal compact subgroup. Fix a minimal parabolic $P_0 = M_0U_0$. Let T_M be the intersection of the maximal split torus of the center of M with the derived group of G. Then $T_M \simeq \mathbf{G_m}$ (the multiplicative group) and we let A_M be the subgroup $\mathbf{R}_+$ imbedded in $I_F \simeq T_M(\mathbf{A})$. If π is a cuspidal automorphic representation of $M(\mathbf{A})$, we assume that the central character of π is trivial on A_M, and consider π as a subspace of $L^2(A_M M(F)\backslash M(\mathbf{A}))$. Let $\mathcal{A}_P^\pi$ denote the space of automorphic forms φ on $U(\mathbf{A})M(F)\backslash G(\mathbf{A})$ such that for all $k \in K$ the function $m \mapsto \delta_P(m)^{-\frac{1}{2}}\varphi(mk)$ belongs to the space of π. The automorphic realization of π gives rise to an identification of $\mathcal{A}_P^\pi$ with (the K_∞-finite part of) the induced space $I(\pi) = \mathrm{Ind}_{P(\mathbf{A})}^{G(\mathbf{A})}\pi$. Set $\varphi_s(g) = \varphi(g)|\varpi|^s(g)$ for any $\varphi \in \mathcal{A}_P^\pi, s \in \mathbf{C}$. (Here ϖ is the fundamental weight corresponding to P in the vector space spanned by the rational characters of M and $|\varpi|$ is the continuation to $G(\mathbf{A})$ of the function $|\varpi|(m) = \prod |\varpi(m_v)|_v$ with $m = (m_v)$ inside $M(\mathbf{A})$.) The map $\varphi \mapsto \varphi_s$ identifies $I(\pi)$ (as a K-module) with any $I(\pi,s) = I_P(\pi,s) = \mathrm{Ind}_{P(\mathbf{A})}^{G(\mathbf{A})}\pi|\varpi|^s$. For any $\varphi \in \mathcal{A}_P^\pi$, we consider its Eisenstein series (3) which converges for $Re(s) \gg 0$. It defines an intertwining map from $I(\pi,s)$ into the space of automorphic forms on $G(\mathbf{A})$. Let $\overline{P}$ be the parabolic opposite to P containing M and let $P' = M'U'$ be the standard parabolic conjugate to $\overline{P}$ with $M' \supset M_0$. Thus, if w_0 is the longest Weyl element, then $M' = w_0 M w_0^{-1}$.

Langlands computed that the "zeroth Fourier coefficient"

$$E_{P'}(g,\varphi,s)=\int E(u'g,\varphi,s)\psi_0(u')du'$$

should equal

$$\epsilon_{P,P'}\varphi(g)+M(s)(\varphi)(g),$$

where $\epsilon_{P,P'} = 1$ if $P' = P$ and 0 otherwise, and

$$M(s)(\varphi)=\prod_{v\in S}M_v(\pi_v,s)\varphi_v\times\prod_{j=1}^{m}\frac{L^S(js,\pi,\tilde{r}_j)}{L^S(1+js,\pi,\tilde{r}_j)}.$$

Here $M(s)$ is the intertwining operator acting on the induced space $I(\pi,s)$, the space of $\mathrm{Ind}(\pi|\varpi|^s)$, and S is the finite set of primes v which includes the archimedean and ramified places. $L^S(js,\pi,\tilde{r}_j)$ is the **Langlands L-function** attached to S,π

and the contragredient of the irreducible representation r_j of the **L-group** ${}^L M$ on the Lie algebra of ${}^L N$. This formula reduces to (1) when G is $SL(2)$, M is the diagonal subgroup isomorphic to $\mathbf{A}^*$, and $S = \infty$. The point is that the analytic continuation of these $E(g, \varphi, s)$ (which Langlands accomplished a few years earlier, see [**Lan2**]) gives the meromorphic continuation of these Langlands L-functions. This setup contains a list of (M, G) pairs, which includes:

(a) **Rankin-Selberg** L-functions for

$$M = GL_n \times GL_m, \ G = GL_{n+m}$$

and

$$r(g_1, g_2)(x_{ij}) = g_1(x_{ij})g_2^{-1}.$$

L-functions for these $L(s, \pi_n \times \pi_m, r)$ were studied later on by **Jacquet, Piatetski-Shapiro, Shalika, Shahidi, Mœglin-Waldspurger, etc.**. More about these soon.

(b) G is the exceptional group of type G_2 and M is the group GL_2. Also $r_1 = Sym^3$ and $r_2 = \det$. This example Langlands called "particularly striking".

Where do we stand (1975)?

As is well known, with [**Lan1**], and more generally [**Lan3**], Langlands changed the course of modern number theory. On a smaller scale, for us the intimate relationship between general Eisenstein series and automorphic L-functions mapped out a theory of the latter. In particular, the **meromorphy** of the functions $L(s, \pi, r_j)$ already followed from analytically continuing the Eisenstein series.

But what about an exact functional equation? Is there a nice generalization of (2) for first Fourier coefficients (as opposed to just zeroth Fourier coefficients) in terms of L-functions which appear only in the denominator? And, more significantly, what about the **holomorphy** of $L(s, \pi, r_j)$, or at least meromorphy with a (prescribed) finite number of poles? (Think of Artin's Conjecture.)

Shahidi enters the picture (1975-)

Freydoon came to the Institute for Advanced Study in the fall of 1975, right after his doctoral studies under Joe **Shalika**. Shalika was sending Freydoon to learn from the master – Robert **Langlands**. What **Langlands** suggested he study was exactly the nonconstant Fourier coefficients of $E(g, \varphi, s)$, along the lines predicted in a letter to **Godement**. What resulted was the birth of "**Langlands-Shahidi** theory".

I first met Freydoon while he was still a graduate student of Shalika's, and feel honored to have had him as a friend ever since. Studying his work for over 30 years has been a rare treat. Indeed, very few mathematicians have worked singlehandedly towards such a difficult yet simple goal: to prove that the L-function $L(s, \pi, r)$ has a finite number of poles and satisfies the functional equation $L(s, \pi, r) = \epsilon(s, \pi, r)L(1 - s, \pi, \tilde{r})$. Amazingly, each successive paper of Freydoon's builds upon the ones that appeared before it, as we shall now see.

Acknowledgment

I thank Jim Cogdell for a large amount of help with this survey.

Ten Special Papers

We shall survey Freydoon's program (of Langlands-Shahidi theory) by looking at ten papers which we henceforth denote by numerals 1 through 10, putting a star next to 4 of those as being particularly important.

1*. **Functional Equation Satisfied by Certain L-functions**, Compositio Math., 1978, the predecessor of **On Certain L-functions**, Amer. J. Math, 1981. (See the backdrop of the Poster for Shahidi's 60th Birthday Conference.) These papers set the stage for (almost) all that follows: first for the GL_2 L-function $L(s, \pi, Sym^3)$ (see example (b) above), then for arbitrary Langlands-Shahidi $L(s, \pi, r_j)$.

For our purposes, let ψ be a non-degenerate character of U_0 and denote by ψ_M its restriction to $U_0 \cap M$ (a maximal unipotent of M).

In what follows, Shahidi assumes that π is generic with respect to ψ_M, i.e., π is globally generic. Then $E(g, \varphi, s)$ will be ψ generic.

In this case, one can compute non-trivial Whittaker coefficients ("Fourier coefficients") of the resulting Eisenstein series and relate them directly to the L-functions $L(s, \pi, r_j)$ (without the appearance of quotients of them). More exactly, **Shahidi** computes the *ψ-th* Fourier coefficient to satisfy

$$E^{\psi}(e, \varphi, s) = \prod_{v \in S} W_v(e_v) \prod_{j=1}^{m} L^S(1 + js, \pi, \tilde{r}_j)^{-1}.$$

Here $W_v(g) = \lambda_v(I_v(g_v)\varphi_v)$, where λ_v denotes a ψ-Whittaker functional on the space of $\mathrm{Ind}\, \pi_v |\varpi|^s$. Using this, he obtains a **crude functional equation** for the Langlands's L-functions:

$$\prod_{j=1}^{m} L^S(js, \pi, r_j) = \prod_{v \in S} C_{\overline{\psi}_v}(s, \tilde{\pi}_v) \prod_{j=1}^{m} L^S(1 - js, \pi, \tilde{r}_j),$$

where $C_{\overline{\psi}_v}(s, \tilde{\pi}_v)$ is Shahidi's (soon to be analyzed) **local coefficient**; it comes directly from the local uniqueness of Whittaker models.

Let us give an outline of the proof. The first thing to do is to write the expression $E^{\psi}(e, \varphi, s)$ as

$$\prod_{v \in S} W_v(e_v) \prod_{\notin S} W_v(e_v).$$

To do this, we apply Bruhat's decomposition. To make matters simpler, we consider the case of $G = SL(2)$. Here $P = P_0 = B = NU = P'$, π is a character χ, and we have the familiar Whittaker functional defined on the space of $\mathrm{Ind}\, \chi_p |\varpi|_p^s$; on pages 80-81 of [**Ge-Sh**], we showed precisely that

$$E^{\psi}(e, \varphi, s) = \prod_{v \in S} W_v(e_v) L^S(1 + s, \chi)^{-1}.$$

In general, one can proceed the same way, since only the "longest" Weyl group element will contribute non-trivially, and then the computation of W is precisely the subject matter of [**C-Shal**]: for v not in S,

$$W_v(e_v) = \prod_{j=1}^{m} L(1 + js, \pi_v, \tilde{r}_j)^{-1}.$$

For the *crude functional equation*, one needs to apply some results due to **Casselman** [**Cas2**] and **Wallach** [**Wal**] at the archimedean places (to show that $W_v's$ can be chosen so that $W_v(e_v) \neq 0$ if $v = \infty$). Then one uses **Shahidi's** theory of *local coefficients* by applying *intertwining integrals* (**Harish-Chandra, Silberger**). That is, one makes sense of a Whittaker functional λ'_v on the image of $I(s, \pi_v)$ by $M_v(s, \pi_v)$ by using a formula similar to the one defining λ_v; then one proves (by uniqueness of Whittaker functionals) that there is a complex function $C_{\psi_v}(s, \pi_v)$ relating λ_v to λ'_v acting on $M_v(\pi_v, s)$. The functional equation for $E(s)$ then gives the (crude) functional equation for $L^S(s)$. Along the way, we have that the holomorphy of the local coefficients determines the irreducibility of certain induced representations, and furthermore, that they can be used to normalize the intertwining operators.

Remark 1. The simpler argument for

$$\prod_{j=1}^{m} L^S(js, \pi, r_j) = \prod_{v \in S} C_{\overline{\psi}_v}(s, \tilde{\pi}_v) \prod_{j=1}^{m} L^S(1 - js, \pi, \tilde{r}_j).$$

for $G = SL(2)$ is carried out on pages 81-82 of [**Ge-Sh**].

Remark 2. Recall from above that there exists φ_v such that $W_v(e_v) \neq 0$. Consequently, the zeros of $\prod_{j=1}^{m} L^S(1 + js, \pi, \tilde{r}_j)$ are among the poles of $E(e, \varphi, s)$. From the fact that $E(e, \varphi, s)$ is known to have no poles on the imaginary axis $i\mathbf{R}$, it follows that

$$\prod_{j=1}^{m} L^S(1 + ijt, \pi, \tilde{r}_j) \neq 0$$

for all t in $\mathbf{R}$. In particular, for $M = GL(n) \times GL(m)$ and $G = GL(n+m)$ (example (a) of Langlands), and π and π' cusp forms on $GL(n, \mathbf{A_F})$ and $GL(m, \mathbf{A_F})$, we get

$$L(1, \pi \times \pi') \neq 0,$$

where the local L-functions at every place are the corresponding Artin factors (see [**Ha-Ta, Hen**] and more of the discussions below).

2. **Fourier Transforms of Intertwining Operators and Plancherel Measures for GL(n)**, Amer. J. Math, 1984.

Here is where Shahidi begins to work out some details of his general theory for the well known case $(M, G) = (GL_n \times GL_m,\ GL_{n+m})$ over a local non-archimedean field. This paper says mainly that (as Langlands had predicted) the local Shahidi coefficient may be expressed in terms of the usual gamma factor computed by **Jacquet, Piatetski-Shapiro,** and **Shalika** in [**J-PS-S**]. An important consequence of this equality is a formula for the Plancherel measure attached to s and $\pi_1 \otimes \pi_2$ by **Harish-Chandra** (see [**Harish**] and [**Sil**] for an introductory general theory). This formula was used by **Bushnell-Kutzko-Henniart** in [**B-H-K**]] to express the conductor for the pair (M, G) in terms of **types** of representations.

3. **Local Coefficients as Artin Factors for Real Groups**, Duke Math. Journal, 1985.

The purpose of this paper is to prove the equality of local Shahidi coefficients in the real case with their corresponding Artin factors of local class field theory. The beautiful result reads as follows:

$$C_\psi(s,\pi) = (*) \prod_{j=1} \epsilon(js, r_j \cdot \varphi_\pi, \psi) \frac{L(1-js, \tilde{r}_j \cdot \varphi_\pi)}{L(js, r_j \cdot \varphi_\pi)}$$

where $\psi = \psi_{\mathbf{R}}$, π is an irreducible admissible representation of $M = M_{\mathbf{R}}$, and $\varphi_\pi : W \to {}^L M$ is the homomorphism attached to π by Langlands' local class field theory at infinity (cf. [**Lan5**]); the Artin factors are those which **Langlands** attached to irreducible admissible representations. Since the statement is also true when v is unramified, the main theorem of the 1981 paper establishes the functional equation

$$\prod_{j=1}^{m} L(js, \pi, r_j) = \prod_{j=1}^{m} \epsilon(js, \pi, r_j) L(1-js, \pi, \tilde{r}_j)$$

whenever ∞ is the only ramification of π; in particular, this is true for every cusp form on $SL(2, \mathbf{Z})$. To define the C_ψ, one needs the analytic continuation of Whittaker functionals. Hence one needs the work of **Jacquet's** thesis [**Jac**], **Schiffmann** [**Schi**] and **Wallach** [**Wal**], **Vogan** [**Vog**], **Kostant** [**Kos**], etc. One also needs to apply **Knapp-Zuckernan** [**Kn-Zu**] and **Langlands** [**Lan5**] to show that the local coefficient is basically a product of Artin factors.

Remark

Around this time, about the middle of the 1980's, Freydoon wrote several interesting papers related to $L(s, \pi, Sym^m)$ with Carlos **Moreno** (see, for example, [**Mo-Sh**]). Shahidi also has a nice 1988 paper [**Ke-Sh**] with David **Keys** where they (among other things) studied Artin L-functions in the case of unitary principle series of a quasi-split p-adic group. These were about the only joint works Freydoon did between the dates 1975-1995, i.e., although Shahidi would have liked to do more work with others, the first 20 years he worked virtually alone. What came later (that is, after 1995) will be discussed after paper 5*.

4. **On the Ramanujan Conjecture and Finiteness of Poles for Certain L-functions**, Annals of Math., 1988.

First Freydoon completed the list of (M, G) pairs after **Langlands**; these included:

(c) $G = (E_7 - 1)$, the derived group of M is $SL_2 \times SL_3 \times SL_4$, and r_1 the tensor product representation (page 560 of paper 4);

(d) $G = (E_8 - 2)$, $M = SL_4 \times SL_5$, and r_1 the standard times exterior square (page 561); and

(e) $G = (F_4 - 2)$, with M given by $\alpha_1, \alpha_2, \alpha_4$ (page 563).

Then Freydoon established a **uniform** line ($Re(s) = 2$) of absolute convergence for all his "certain" L-functions. This is proved using a crucial lemma: For almost every local component of a cuspidal generic representation, every unramified L-function (obtained from this method) is holomorphic for $\mathrm{Re}(s) \geq 1$; in other words,

$$\det(I - r_i(A_v)q_v^{-s})^{-1}$$

is holomorphic for $\mathrm{Re}(s) \geq 1$ (where $A_v \in {}^LM$ is a semisimple element in the conjugacy class of ${}^LM$ attached to π_v).

To see how this is related to the Ramanujan-Petersson Conjecture (the up-to-date subject matter of paper 10), take π to be a cuspidal representation of $PGL_2(\mathbf{A}_F)$, and at each unramified v, let $A_v = \mathrm{diag}(\alpha_v, \alpha_v^{-1})$ be the corresponding conjugacy class in $SL_2(\mathbf{C})$; the conclusion is that

$$q_v{}^{-\frac{1}{5}} < |\alpha_v| < q_v{}^{\frac{1}{5}}. \tag{4}$$

Indeed, suppose we apply this "crucial lemma" in case (e) of (G,M) above. Let Π be the **Gelbart-Jacquet** lift of π. (We may assume Π is a *cuspidal* representation of $PGL_3(\mathbf{A})$ since **Ramanujan-Petersson** is known for monomial cusp forms.) The exact sequence

$$0 \to A \to M \to PGL_3 \times PGL_2 \to 0$$

(A is the split component of M) leads to a surjection

$$M \to PGL_3(\mathbf{A}) \times PGL_2(\mathbf{A}) \to 0$$

(by page 36 of [**Lan1**]), and the cuspidal representation $\Pi \times \pi$ of $PGL_3(\mathbf{A}) \times PGL_2(\mathbf{A})$ then defines a cuspidal representation ρ of M. Thus, among the factors dividing $L(s, \rho, r_1)^{-1}$ is

$$(1 - \alpha_v^5 q_v^{-s})(1 - \alpha_v^{-5} q_v^{-s}),$$

and by the "crucial lemma", this must be non-zero for $\mathrm{Re}(s) \geq 1$, i.e., (4) holds!

In the last part of this paper, a first response to the functional equation and entirety of these L-functions is addressed. However, the serious definitions of $L(s, \pi, r_j)$ and $\epsilon(s, \pi, r_j, \psi_F)$ will wait until paper number 5*.

5*. **A Proof of Langlands' Conjecture on Plancherel measures; Complementary Series for p-adic Groups.** Annals of Math. 1990.

In this important paper, several long awaited results follow from a (detailed and) non-trivial Theorem proving general results about the local Langlands root numbers ϵ and L-functions L, namely, if they are equal to the corresponding Artin factors, they are supposed to be "inductive", and (for the local component of a global cusp form) they must be the corresponding factor appearing in the functional equation satisfied by the cusp form. A nice corollary of this is Langlands' Conjecture expressing the Plancherel measure in terms of these ϵ and L.

The above are far-reaching results generalizing the case $(G, M) = [(GL(n+m), GL(n) \times GL(m))]$; see paper number 2 and the work of **Jacquet, Piatetski-Shapiro, and Shalika** [**J-PS-S**]. It also generalizes the theory of local coefficients as Artin factors for real groups (paper number 3 above).

One of the main applications here is the desired functional equation of the L-function of a generic (global) cusp form π: locally there exist gamma factors γ so that L and ϵ factors can be defined at *every* place, and globally

$$L(s, \pi, r_i) = \epsilon(s, \pi, r_i) L(1 - s, \pi, \tilde{r}_i).$$

This is obviously a generalization of what is true for the third symmetric power L-function for $M = GL(2)$ inside G_2; see the paper numbered 1.

To prove the individual functional equations, i.e., to get each $L(s, \pi, r_i)$ with precise root numbers, we need to appeal to the following "induction" statement: given $1 < i \leq m$, there exists a split group G_i over F, a maximal F-parabolic subgroup $P_i = M_i N_i$ and a cuspidal automorphic representation π' of $M_i = M_i(\mathbf{A}_F)$, unramified for every $v \notin S$, such that if the adjoint action of ${}^L M_i$ on ${}^L \mathbf{n}_i$ decomposes as $r' = \otimes r'_j$, then

$$L^S(s, \pi, r_i) = L^S(s, \pi', r'_1);$$

moreover $m' \leq m$.

Next, we need a generalization of paper number 3 to all places that have a vector fixed by an Iwahori subgroup. This, together with a local-global argument, can then be applied "inductively" as above. All in all, using the **Crude Functional Equation**, we arrive at one of the main results of the paper:

Theorem. Given a local field F, groups $G, P = MU$, representations r_j on ${}^L M$ as before, and an irreducible admissible *generic* representation σ of M, there exist complex functions $\gamma(s, \sigma, r_i, \psi_F), 1 \leq i \leq m$, such that

(1) whenever F is archimedean or σ has a vector fixed by an Iwahori subgroup,

$$\gamma(s, \sigma, r_i, \psi_F) = \epsilon(s, r_i \cdot \varphi, \psi_F) L(1 - s, \tilde{r}_i \cdot \varphi)/L(s, r_i \cdot \varphi);$$

(2) in general,

$$C_\psi(s, \sigma) = \prod_{i=1}^{m} \gamma(s, \sigma, r_i, \psi_F);$$

(3) $\gamma(s, \sigma, r_i, \psi_F)$ is *multiplicative* under induction, and

(4) whenever σ becomes a local component of a globally generic cusp form, then the γ's become the local factors needed in their functional equations. Moreover, (1), (3) and (4) determine the γ-factors uniquely.

Shahidi uses this Theorem to define L and ϵ factors for *all* irreducible admissible generic representations. Let us just look at the case of *tempered* such representations. For such representations, we define $L(s, \sigma, r_j)$ to be the inverse of the normalized polynomial $P_{\sigma,j}(q^{-s})$ which is the numerator of $\gamma(s, \sigma, r_j, \psi)$. Part (2) of the Theorem above implies

$$\gamma(s, \sigma, r_j, \psi_F) L(s, \sigma, r_j)/L(1 - s, \sigma, \tilde{r_j})$$

is a monomial in q^{-s}, which is denoted by $\epsilon(s, \sigma, r_j, \psi_F)$, the *root number* attached to σ and r_j. Thus,

$$\gamma(s, \sigma, r_j, \psi_F) = \epsilon(s, \sigma, r_j, \psi_F) L(1 - s, \sigma, \tilde{r_j})/L(s, \sigma, r_j).$$

For tempered σ it follows that $L(s, \sigma, r_j)$ is independent of ψ.

Back to the local theory, Shahidi derives Langlands' Conjecture (for generic representations) which concerns the normalization of local intertwining operators by means of local Langlands root numbers and L-functions (this is expressed in terms of Plancherel measures and some results from paper 3).

Finally, Freydoon obtains all the complementary series and special representations coming from (generic) **supercuspidal** representations of the Levi factors of **maximal** parabolics.

To tackle Langlands' Conjecture in general, i.e., for arbitrary π, Shahidi reduces it to two natural conjectures in harmonic analysis: one is the basis for stabilization of the trace formula, the other is about stable distributions (the Conjecture on genericity of tempered L-packets).

All in all, Shahidi's paper is very powerful, and a fitting 1990 end to paper number 1* – the functional equation of "all" automorphic L-functions is proved. The outstanding work Shahidi did since then will be discussed in papers numbered 6 through 10.

Here we must mention, especially for the non-generic case, a host of other names. For induced representations in the p-adic case, see **Lapid, Muic, Tadic** [**LMT**], and **Silberger** [**Sil**], etc. For interactions with the trace formula, see **Arthur** [**Art**], **Ngo**, **Clozel** (see, for example, [**Clo**]), **Langlands** [**Lan3**], and **Kottwitz-Shelstad** [**Ko-Sh**].

Recently, following up on Shahidi's work, **Hiraga-Ichino-Ikeda** [**H-I-I**] have conjectured a generalization of Langlands' Conjecture to formal degree and discrete series representations, and they have checked that their Conjecture agrees with Shahidi on GL_n and other groups (via twisted endoscopy.....see below).

6. **Twisted Endoscopy and Reducibility of Induced Representations for p-adic Groups**, Duke Math. Journal, 1992.

A major reason for studying (twisted) endoscopy is to detect representations coming from the corresponding (twisted) endoscopic groups. The theory goes back to **Langlands** and **Shelstad** [**Lan-Sh**] and is continued in [**Ko-Sh**]. Recall that there are no interesting endoscopic groups for $GL(n)$, but that symplectic groups and special orthogonal groups appear as **twisted** endoscopic groups of $GL(n)$.

Here Freydoon goes on to use the last paper to relate the poles of (local non-normalized) intertwining operators to the parametrization problem via twisted endoscopy. A typical example is the following:

Theorem. Suppose $G = Sp_{2n}$ and σ is a fixed irreducible unitary self-dual supercuspidal representation of $M = GL_n(F)$. Then $I(\sigma)$ is irreducible if and only if $L(s, \sigma, \Lambda^2)$ has a pole (at $s = 0$) and if and only if σ comes by twisted endoscopic transfer from $SO_{n+1}(F)$.

Closely related works are his 2000 (Appendix with **Shelstad** [**Shah1**]) paper and his 1998 paper with **Casselmen** [**C-Shah**]. In a number of papers with **Goldberg** (cf. [**Go-Sh**]), Freydoon later extended the work of paper number 6 to arbitrary parabolic subgroups of classical groups.

The last 10 years

Quite simply, Freydoon's research has had a dramatic impact on Langlands **functoriality**. We shall describe 4 principal papers under this heading, each carrying out a different role towards automorphic lifting.

7. With S. Gelbart, **Boundedness of Automorphic L-functions in Vertical Strips**, Journal of the Amer. Math. Society, 2001.

8*. With J. Cogdell, H. Kim, and I. Piatetski-Shapiro, **On Lifting from Classical Groups to GL_2**, Publ. Math. de l'IHES, 2001, along with **Functoriality for the Classical Groups**, Publ. Math. de l'IHES, 2004.

9*. With H. Kim, **Functorial Products for $GL_2 \times GL_3$ and the Symmetric Cube for GL_2**, Annals of Math., 2002.

10. With H.Kim, **Cuspidality of Symmetric Powers with Applications**, Duke Math. Journal, 2002.

Before looking at papers 7 through 10, let's see how far one got towards proving the analytic properties of, say $L(s, \pi, Sym^3)$, in about 1998. The functional equation was known (in fact, from paper 1 for $L(s, \pi, Sym^3)$, but from paper 5* for general Langlands-Shahidi L-functions). What about its entirety? For π generated by Ramanujan's τ-function the L-function for the symmetric cube had already been proved holomorphic by **Moreno** and **Shahidi** in 1976 [**Mo-Sh**]. But what about general cusp forms? In around 1998 Kim and Shahidi made a breakthrough in proving the (long awaited) holomorphicity of $L(s, \pi, Sym^3)$ [**Ki-Sh1**]. In particular, their method used the appearance of this function in the constant term of Eisenstein series on G_2, and some brand new results of **Kim** [**Kim1**] (more about this later), **Muic** [**Mui**], and **Ramakrishnan** [**Ram2**]. This marked the beginning of deeper analytic properties of L-functions like $L(s, \pi, Sym^3)$ being proved. For example, given **Cogdell** and **PS's** Converse Theorem, this meant showing that twisting by forms of $GL_2(\mathbf{A})$ was *nice*. All these things, like proving $L(s, \pi, Sym^3)$ was a standard GL_4 L-function, had to wait for the exciting Princeton gathering of **Cogdell-Kim-PS** and **Shahidi** in 1999.

More on the Background

Working in parallel to the theory of **Langlands-Shahidi** was the ongoing project of **Piatetski-Shapiro** on L-functions.

Anyone working with **PS**, from 1976 onwards, knew that one of Piatetski's dreams was to complete the Converse Theorem from $\mathbf{GL_2}$ and $\mathbf{GL_3}$ to $\mathbf{GL_n}$, then to use it to establish cases of functoriality.

On the other hand, also from 1976 onwards, Freydoon had been developing the **Langlands-Shahidi** theory, especially for $L(s, \pi, Sym^3)$.

Everything came together in 1999 - 2000, when a special year at the **IAS** was co-organized by **Bombieri, Iwaniec, Langlands** and **Sarnak**. Present at the same place, **Cogdell** and **PS** took their most recent version of the recent Converse Theorem and together with **Kim** and **Shahidi** saw how to apply it to several of the L-functions obtained by the **Langlands-Shahidi** method.

The Work Done

More precisely, this work was the happy marriage of the language of **Cogdell-PS** Converse Theorems (together with **stability**) with **Langlands-Shahidi** theory (together with **Kim's** observation), as we shall now see.

First, the purpose of paper 7 was to prove the boundedness in vertical strips of finite width for all the (completed) **Langlands-Shahidi** L-functions that appear in constant terms of Eisenstein series under a certain local assumption. In particular, we prove the boundedness of a number of important L-functions, among them the symmetric cube for GL_2 and several Rankin-Selberg product L-functions, where the local assumption was already proved. Our main theorem plays a fundamental role in establishing new and striking cases of functoriality such as for classical groups to GL_n and for $L(s,\pi,Sym^3)$ for GL_2 (both to be discussed soon).

Gelbart, Lapid and **Sarnak** then found a simpler proof of this boundedness on their way [**Ge-La-Sa**] to proving the stronger Conjecture stated in the Introduction to paper 7. Both papers rely heavily on the work [**Mül**] of **Müller** on the finiteness of order of Eisenstein series.

Next, we turn to papers 8 through 10. The papers numbered 8* prove global functorial lifting for the split classical groups $G_n = SO_{2n+1}$, SO_{2n}, or Sp_n to an appropriate general linear group GL_N, for **generic** cuspidal representations. Let's explain the lifting for SO_{2n+1} alone (which we shall denote by H). We follow fairly exactly the exposition given in [**Cog**] (see also **Cogdell's** paper in these Proceedings).

According to [**C-PS2**], the *Converse Theorem for* GL_N states the following: Let $\Pi = \otimes\Pi_v$ be an irreducible admissible representation of $GL_N(\mathbf{A})$ whose central character is invariant under k^* and whose L-function $L(s,\Pi)$ is absolutely convergent in some right half-plane. Let S be any finite set of finite places and η a continuous character of $\mathbf{A}^*$ mod k^*. Suppose also that for every $\tau \in T^S(N)\otimes\eta$ (where $T^S(N)$ denotes any cuspidal representation of $GL_d(\mathbf{A}), 1 \le d \le N-2$, *unramified* at all $v \in S$) the twisted L-function $L(s,\Pi\times\tau)$ is *nice*, i.e.,

(1) $L(s,\Pi\times\tau)$ and $L(s,\tilde{\Pi}\times\tilde{\tau})$ extend to entire functions of $s\in\mathbf{C}$,

(2) $L(s,\Pi\times\tau)$ and $L(s,\tilde{\Pi}\times\tilde{\tau})$ are bounded in vertical strips,

(3) $L(s,\Pi\times\tau)$ satisfies the functional equation $L(s,\Pi\times\tau) = \epsilon(s,\Pi\times\tau)L(1-s,\tilde{\Pi}\times\tilde{\tau})$.

Then there exists an automorphic representation Π' of $GL_N(\mathbf{A})$ such that $\Pi_v = \Pi'_v$ for all $v \notin S$.

To prove a global functorial lifting from H to GL_{2n}, a first step is to attach to a globally *generic* cuspidal representation

$$\pi = \otimes\pi_v$$

of $H(\mathbf{A})$ a *candidate* lifting $\Pi = \otimes\Pi_v$ of $GL_{2n}(\mathbf{A})$. (Now N equals $2n$.)

For $v \notin S$ (the finite set of finite places at which the local component π_v is ramified) and Π_v the local Langlands lift of π_v, let π'_v be the irreducible representation of $GL_m(k_v)$ with $m < 2n$; then

$$L(s,\pi_v\times\pi'_v) = L(s,\Pi_v\times\pi'_v)$$

and

$$\epsilon(s,\pi_v\times\pi'_v,\psi_v) = \epsilon(s,\Pi_v\times\pi'_v,\psi_v).$$

Now what about $v\in S$? As for paper 5*, we have a local twisted γ-factor $\gamma(s,\pi_v\times\pi'_v,\psi_v)$ where π'_v is a generic representation of $GL_m(k_v)$, $m<2n$; it is related to the local L and ϵ factors by

$$\gamma(s,\pi_v\times\pi'_v,\psi_v) = \frac{\epsilon(s,\pi_v\times\pi'_v,\psi_v)L(1-s,\tilde{\pi}_v\times\tilde{\pi}'_v)}{L(s,\pi_v\times\pi'_v)}.$$

By "multiplicativity of γ-factors", if $\pi'_v = \mathrm{Ind}(\pi'_{1,v}\otimes\pi'_{2,v})$ with $\pi'_{i,v}$ a representation of GL_{r_i}, then

$$\gamma(s,\pi_v\times\pi'_v,\psi_v) = \gamma(s,\pi_v\times\pi'_{1,v},\psi_v)\gamma(s,\pi_v\times\pi'_{2,v},\psi_v).$$

On the other hand, if $\pi_{1,v}$ and $\pi_{2,v}$ are two irreducible admissible representations of $H(k_v)$, then by "**stability of γ-factors**",

$$\gamma(s,\pi_{1,v}\times\eta_v,\psi_v)=\gamma(s,\pi_{2,v}\times\eta_v,\psi_v)$$

and

$$L(s,\pi_{1,v}\times\eta_v)=L(s,\pi_{2,v}\times\eta_v)=1$$

for every sufficiently highly ramified character η_v (see [**C-PS**] for the original work on "stability" by **Cogdell** and **PS**). As we shall soon see, this is enough to produce a replacement for the local Langlands Conjecture at $v\in S$.

Recall that the local multiplicativity and stability results *for* $GL_{2n}(k_v)$ are known by **Jacquet, Piatetski-Shapiro** and **Shalika**. This allows for a comparison of the stable forms for $GL_{2n}(k_v)$ *and* $H(k_v)$ and establishes the following analogue of a local Langlands lift for $v\in S$: let Π_v denote any $GL_{2n}(k_v)$ having trivial central character, and π'_v a generic irreducible admissible representation of $GL_m(k_v)$ with $m<2n$ of the form $\pi'_v=\pi'_{0,v}\otimes\eta_v$ with $\pi'_{0,v}$ unramified and η_v a fixed sufficiently highly ramified character of k_v^*; then

$$L(s,\pi_v\times\pi'_v)=L(s,\Pi\times\pi'_v)$$

and

$$\epsilon(s,\pi_v\times\pi'_v,\psi_v)=\epsilon(s,\Pi_v\times\pi'_v,\psi_v).$$

The end result is:

Proposition. For any fixed character η for which η_v is sufficiently ramified at the places $v\in S$ we have

$$L(s,\pi\times\pi')=L(s,\Pi\times\pi')$$

and

$$\epsilon(s,\pi\times\pi')=\epsilon(s,\Pi\times\pi').$$

for all $\pi'\in T^S(2n-1)\otimes\eta$.

The next step is to control the analytic properties of this twisted L-function $L(s,\pi\times\pi')$, i.e., to prove that $L(s,\pi\times\pi')$ is *nice*. The functional equations of course have been established in paper 5*, and the boundedness in vertical strips in 7; however, for $L(s,\pi\times\pi')$ to be entire for "sufficiently ramified" π' now requires **Kim's crucial observation**: the relevant Eisenstein series, hence the L-function $L(s,\pi\times\pi')$, can have poles only if the representation π' satisfies $\pi'=\tilde{\pi}'\otimes|\det|^t$ for some t. Since this can never happen for sufficiently ramified π', we have finally:

Proposition. Let π be a globally generic cuspidal representation of $H(\mathbf{A})$. Let S' be a non-empty set of finite places and suppose that η is an idèle class character such that at some place v of S' we have both η_v and η_v^2 ramified. Then the twisted L-functions $L(s,\pi\times\pi')$ are *nice* for all $\pi'\in T^{S'}(2n-1)\otimes\eta$.

Now [**C-K-PS-S**] is ready to prove that π has a global Langlands lift to $GL_{2n}(\mathbf{A})$.

Theorem. There is a finite set of places S and an automorphic representation Π' of $GL_{2n}(\mathbf{A})$ such that for all $v\notin S$ we have that Π'_v is the local Langlands lift of π_v.

Proof. Let π, S, and Π be as above. If S is non-empty, let $S' = S$ and if π is unramified at all finite places take $S' = v_0$ to be any chosen finite place. Choose a fixed idèle class character η which is suitably ramified for all $v \in S$ such that both the last two Propositions are valid. Then for all $\pi' \in T^{S'}(2n-1) \otimes \eta$ we have

$$L(\pi \times \pi') = L(s, \Pi \times \pi'), \epsilon(s, \pi \times \pi') = \epsilon(s, \Pi \times \pi')$$

and the $L(s, \Pi \times \pi')$ are thus *nice*. Then by the *Converse Theorem for* GL_{2n}, we conclude that there is an automorphic representation $\Pi' = \otimes \Pi_v^{'}$ of $GL_{2n}(\mathbf{A})$ such that $\Pi_v^{'} = \Pi_v$ is the local Langlands lift of π_v for all $v \notin S'$.

In short, this paper numbered 8* was a complete breakthrough.

Related Work of Others

Ginzburg, Rallis and **Soudry** had used their "descent technique" to characterize the image of global functoriality for generic representations of the split classical groups [**G-R-S**] (see also [**Sou**]); in particular, they show that the image of functoriality consists of isobaric sums of certain self-dual cuspidal representations of $GL_d(\mathbf{A})$ satisfying an appropriate L-function criterion.

In [**Ji-So**], **Jiang** and **Soudry** establish local Langlands reciprocity law, associating with each Langlands parameter of SO_{2n+1} an irreducible representation of $SO_{2n+1}(F)$ (F p-adic), preserving local twisted L and ϵ factors. Together with this, they establish the local Langlands functorial lift from generic representations of $SO_{2n+1}(F)$ to irreducible representations of $GL_{2n}(F)$. (Their main tool is the theory of "local" descent.) The γ factors were also handled by **Rallis-Soudry** [**Ra-So**], based on the doubling method of **PS-Rallis** [**PS-Ra**] and fine tuned by **Lapid-Rallis** [**La-Ra**].

Shahidi and **C-PS** [**C-PS-S1, C-PS-S2**] come back to handle stability for γ factors in the **Rallis** Volume and **Jussieu** Publication. An entirely different approach is by **Arthur** ([**Art**]); he uses the trace formula to develop transfer for the **whole** (not just generic) spectrum.

The Symmetric Cube Lift

In 9*, instead of working directly with $L(s, \pi, Sym^3)$, Kim and Shahidi prove the stronger result that $GL_2 \times GL_3$ is itself functorial! Let's see how that was done.

Keeping in mind the Converse Theorem of **Cogdell-PS**, **Kim** and **Shahidi** study the L-functions $L(s, \pi \times (\pi' \otimes \chi) \times \sigma)$, where σ is a cusp form on GL_r, $r = 1, 2, 3, 4$. For example, for $r = 4$, **Langlands-Shahidi** is used with (G, M) taken to be the case (c) of paper 4, namely $(E_7 - 1)$, with derived group of M equal to $SL_2 \times SL_3 \times SL_4$. Again, it is not a simple matter to prove that all these properties of $L(s, \pi \times (\pi' \otimes \chi) \times \sigma)$ are satisfied: in fact, the full power and subtlety of Shahidi's theory (the content of papers 1 through 5) is used. By the now familiar Cogdell-PS Converse Theorem, this $L(s, \pi \times (\pi' \otimes \chi))$ equals the L-function for $GL(6)$. Actually, much more is used, including paper 7, and the idea again of [**Kim1**]: Langlands' inner product formula for Eisenstein series implies that **all** of the automorphic L-functions appearing in the constant terms are entire if twisted by a Grössencharakter which is sufficiently ramified at one place.

To argue for $L(s, \pi, Sym^3)$, let $\pi' = \mathrm{Ad}(\pi)$ again denote the corresponding automorphic representation of $GL(3, \mathbf{A}_F)$. For each v, finite or infinite, let $\pi_v \boxtimes \pi_v'$

be the irreducible admissible representation of $GL(6, F_v)$ attached to v through the local Langlands correspondence by **Harris-Taylor** [**Ha-Ta**] and **Henniart** [**Hen**]. Then what **Kim** and **Shahidi** prove above is that $\pi \boxtimes \pi'$ is an irreducible admissible **automorphic** representation of $GL(6, \mathbf{A_F})$ and we argue as follows concerning its isobaric character.

Let τ be a cuspidal representation of $GL(2)$. Then (mostly by arguments of **Ramakrishnan** [**Ram1**]), $L(s, \mathrm{Ad}(\pi) \times \pi \times \sigma)$ has a pole at $s = 1$ iff σ and $\mathrm{Ad}(\pi)$ are self χ dual. The result is that $\pi \boxtimes \mathrm{Ad}(\pi)$ has $(GL(2) \times GL(4))$ isobaric data. Since it is easy to see that for almost every place $\pi_v \boxtimes \mathrm{Ad}(\pi_v) = \mathrm{Ad}^3(\pi_v) \boxplus \pi_v$, we conclude from the (generalized "multiplicity one") classification theorem of **Jacquet-Shalika** [**Ja-Sh**] that

$$\pi \boxtimes \mathrm{Ad}(\pi) = \mathrm{Ad}^3(\pi) \boxplus \pi,$$

exactly what we want.

We stress that the proof of the stronger result on $GL_2 \times GL_3$ uses the observation of Kim, the boundedness discussed on paper 7, and techniques developed in paper 5.

We must also mention **Kim's** important paper **Functoriality for the Exterior Square of GL_4 and the Symmetric Fourth of GL_2** [**Kim2**], with its interesting Corollaries for Ramanujan's Conjecture. (We'll get to that very soon with our discussion of paper 10.) The paper of [**As-Sh**], which appeared in Duke 2006 proves Langlands (generic) functoriality for general spin groups.

Paper number 10: Symmetry versus Ramanujan

Here, for a number field F, **Kim** and **Shahidi** produce the best known result on **Ramanujan-Petersson**: if π_v is an unramified local component of a cuspidal representation of $GL(2)$, then

$$q_v^{-1/9} < |\alpha_v|, |\beta_v| < q_v^{1/9}.$$

This uses the meromorphy of Sym^9 for GL_2 which is proved using case (d) of the (G, M) list.

When $F = \mathbf{Q}$, a stronger estimate of 7/64 is available by **Kim-Sarnak** [**Ki-Sa**]; it uses techniques of analytic number theory, as opposed to the estimates of 1/9 (which uses the same techniques as 1/5 did in 1988).

Concerning the failure of Ramanujan for groups other than $GL(2)$, **Howe** and **PS** found it to be false for Sp_4 and U_3 [**Ho-PS**]). Later, it was assumed to hold for **generic** cuspidal representations of quasi-split reductive groups (see **Shahidi** 5*).

In [**Lan4**], **Langlands** [**Lan4**] suggests that Ramanujan should hold for cuspidal representations of quasi-split groups which functorially lift to isobaric representations of $GL_N(\mathbf{A})$ (which is the case for globally generic representations of classical groups G_n by **C-K-PS-S**). Thus, with either formulation, we would expect Ramanujan to be true for $G_n(\mathbf{A})$. The best general bounds towards Ramanujan for $GL_N(\mathbf{A_F})$ are those of **Luo, Rudnick,** and **Sarnak** [**L-R-S**], and via functoriality, **C-K-PS-S** are able to link the **RC** for classical groups to the **RC** for GL_N. There is another approach to the problem due to **Burger, Li** and **Sarnak** [**B-L-S**]; this involves the **automorphic dual** of a group.

Where do we stand now (2008)?

Let's look at two major themes of Langlands which are developed in Shahidi's work:

1) the analytic continuation and functional equation of the Langlands automorphic L-functions $L(s, \pi, r)$, and

2) the principle of functoriality for automorphic forms.

Concerning the first, Shahidi proved the analytic continuation and functional equation for *all* L-functions $L(s, \pi, r_j)$ of **Langlands-Shahidi** type. (Further properties of $L(s, \pi, r_j)$ follow from his proof of a local conjecture on intertwining operators now known in *almost* all cases; see [**Shah1, Asg, C-Shah, Kim3**].) As regards *arbitrary* Langlands L-functions, our knowledge of automorphic representations is still too murky to ensure all the right methods to be followed. Certainly the trace formula is a central tool, and Freydoon has touched on this in various works not discussed here.

As for the second theme in Shahidi's work, the proof of the principle of general functoriality, we again know that this proof is far from complete. However, the surprising successes that Freydoon has had in proving various instances of functoriality, along with his other advances in the theory of L-functions, will always stand out for their depth and originality.

References

[Art] J. Arthur, *The Endoscopic Classification of Representations: Orthogonal and Symplectic Groups*, Colloquium Publication Series, AMS (to appear).

[Asg] M. Asgari, *Local L-functions for Split Spinor Groups*, Canad. J. Math., **154** (2002), pp. 673-693.

[As-Sh] ——— and F. Shahidi, *Generic Transfer for General Spin Groups*, Duke Math. J., **132** (2006), no. 1, pp. 137-190.

[B-H-K] C. Bushnell, P. Kutzko, and G. Henniart, *Correspondance de Langlands locale pour GL(n) et Conducteurs de pairs*, Ann. Sci. Ecole Norm. Sup., (4) **31** (1998), no. 4, pp. 537-560.

[B-L-S] M. Burger, J. Li, and P. Sarnak, *Ramanujan Duals and Automorphic Spectrum*, Bulletin of the A.M.S., **26** (1992), pp. 253-257.

[Cas1] W. Casselman, *The L-Group*, Advanced Studies in Pure Mathematics **30**, Class Field Theory – Its Centenary and Prospect, pp. 217-258.

[Cas2] ———, *Canonical Extensions of Harish-Chandra Modules*, Canadian J. of Math., **41** (1989), pp. 315-438.

[C-K-PS-S] J. Cogdell, H. Kim, I. Piatetski-Shapiro, and F. Shahidi, *On Lifting from Classical Groups of* GL_N, Pub. Math. Inst. des Hautes Etude Scientiques, **93** (2001), pp. 5-30.

[Clo] L. Clozel, *Motifs et Formes Automorphes: Applications du Principle de Fonctorialitè*, Automorphic Forms, Shimura Varieties, and L-functions, Volume 1, Perspectives in Mathematics, J. Coates and S. Helgason, Editors, Academic Press, 1990.

[Cog] J. Cogdell, *L-functions and Converse Theorems for* GL_n, in Automorphic Forms and Applications, Editors P. Sarnak and F. Shahidi, IAS/Park City Mathematics Series, Volume 12, 2007, pp. 95-177.

[C-PS] J. Cogdell, and I. Piatetski-Shapiro, *Stability of Gamma Factors for* $SO(2n+1)$, Manuscripta Math. **95** (1998), pp. 437-461.

[C-PS2] ———, *Converse Theorems for* $GL(n)$ *and Their Applications to Lifting*, in Proc. Internat. Conf. on Cohomology of Arithmetic Groups, L-functions and Automorphic Forms, Tata Inst. of Fund. Research, Ed. T.N. Venkataramana, 2001, pp. 1-34.

[C-PS-S1] J. Cogdell, I. Piatetski-Shapiro, and F. Shahidi, *Partial Bessel Functions for Quasi-split Groups; Progress and Prospects*, in Automorphic Representations, L-functions and Applications, de Gruyter, Berlin, New York, 2005, pp. 95-128; see

[C-PS-S2] J. Cogdell, I. Piatetski-Shapiro, and F. Shahidi *Stability of γ-factors for Quasi-split Groups*, J. Int. Math. Jussieu **7** (2008), no. 1, pp. 27-66.

[C-Shah] W. Casselman and F.Shahidi, *On the Irreducibility of Standard Modules for Generic Representations*, Ann. Scient. Éc. Norm. Sup., **31** (1998), pp. 561-588.

[C-Shal] W. Casselman and J. Shalika, *The Unramified Principal Series of p-adic Groups II: The Whittaker Function*, Comp. Math., **41** (1980), pp.207-231.

[Ge-Ja] S. Gelbart and H. Jacquet, *A Relation Between Automorphic Representations of $GL(2)$ and $GL(3)$*, Ann. Sci. Éc. Norm. Sup. (4) **11** (1978), pp. 471-552.

[Ge-Sh] ——— and F. Shahidi, *Analytic Properties of Automorphic L-functions*, Perspectives in Mathematics, No. 6, Academic Press Inc., Boston, Mass. 1988.

[Ge-La-Sa] S. Gelbart, E. Lapid, and P. Sarnak, *A New Method for Lower Bounds for L-series*, C.R. Math. Acad. Sci. Paris, **339** (2004), pp. 91-94; see also [Ge-La] S. Gelbart and E. Lapid, *Lower Bounds for L-functions at the Edge of the Critical Strip*, Amer. J. of Math., **128**, Number 3 (2006), pp. 619-638.

[Go-Sh] D. Goldberg and F. Shahidi, *On the Tempered Spectrum of Quasi-split Classical Groups, II*, Canadian J. of Math., **53** (2001), no. 2, pp. 244-277.

[G-R-S] D. Ginzburg, S. Rallis, and D. Soudry, *On explicit lifts of cusp forms from GL_m to classical groups*, Annals of Math., (2) **150** (1999), no. 3, pp. 807-866.

[Ha-Ta] M. Harris and R. Taylor, *The Geometry and Cohomology of Some Simple Shimura Varieties*, Annals of Math. Studies **151**, Princeton University Press, Princeton, 2001.

[Harish] Harish-Chandra, *Automorphic Forms on Semisimple Lie Groups*, Lecture Notes in Mathematics, **62** (1968), Springer, Berlin-Heidelberg-New York.

[Hen] G. Henniart, *Une preuve simple des conjectures de Langlands pour $GL(n)$ sur un corps p-adique*, Invent. Math. **139** (2000), pp.439-455.

[H-I-I] K. Hiraga, A. Ichino, and T. Ikeda, *Formal Degrees and Adjoint $\gamma-$ Factors*, J. Amer. Math. Soc. **21** (2008), no. 1, pp. 283-304.

[Ho-PS] R. Howe and I. Piatetski-Shapiro, *A Counterexample to the "General Ramanujan Conjecture" for (Quasi-) Split Groups*, Proceedings of Symposia in Pure Mathematics, Vol. **33** (1979), part 1, pp. 315-322.

[J-PS-S] H. Jacquet, I. Piatetski-Shapiro, J. Shalika, *Rankin-Selberg Convolutions*, Amer. J. Math., **105** (1983), pp. 367-464.

[Jac] —H. Jacquet, *Fonctions de Whittaker associées aux groupes de Chevalley*, Bull. Soc. Math. France, **95** (1967), pp.243-309.

[Ja-Sh] H. Jacquet and J. A. Shalika, *On Euler products and the classification of automorphic representations. I and II*, J. Math. **103** (1981), no.3, pp.499-558, no.4, pp.777-815.

[Ji-So] D. Jiang and D. Soudry, *The local converse theorem for $SO(2n+1)$ and applications*, Ann. of Math. **157** (2003), pp.743-806; see also *Generic representations and local Langlands reciprocity law for p-adic SO_{2n+1})*, Contributions to Automorphic Forms, Geometry and Number Theory (Shalikafest 2002) (H. Hilda, D. Ramakrishnan, and F. Shahidi, eds.), Johns Hopkins University Press, Baltimore (2007).

[Ke-Sh] D. Keys and F. Shahidi, *L-functions and Normalizations of Intertwining Operators*, Ann. Sci. Éc. Norm. Sup., (4) **21** (1988) no. 1, pp. 67-89.

[Kim1] H. Kim, *Langlands-Shahidi method and poles of automorphic L-functions: Application to exterior square L-functions*, Can. J. Math. **51** (1999), pp.835-849.

[Kim2] ———, *Functoriality for the exterior square of GL_4 and symmetric fourth of GL_2*, Journal of AMS **16** (2002), pp.139-183.

[Kim3] ———, *On Local L-functions and Normalized Intertwining Operators*, Canad. J. Math. **57** (2005), no. 3, pp. 535-597.

[Ki-Sa] ——— and P. Sarnak, *Refined estimates towards the Ramanujan and Selberg conjectures*, Appendix 2 to Kim2].

[Ki-Sh1] ——— and F. Shahidi *Symmetric cube L-functions for GL_2 are entire*, Annals of Mathematics, 150 (1999), pp. 645-662.

[Ko-Sh] Kottwitz, R. and Shelstad, D., *Foundations of twisted endoscopy*, Asterisque, No. 255, 1999.

[Kn-St] A. W. Knapp and E. M. Stein, *Intertwining operators for semisimple groups*, Ann. of Math. (2) **60** (1971), pp.489-578; see also Invent. Math. **60** (1980), no. 1, pp. 9-84.

[Kn-Zu] A. W. Knapp and Gregg J. Zuckerman, *Classification of irreducible tempered representations of semisimple groups*, Ann. of Math. (2) **116** (1982), no.2, pp.389-455.

[Kos] B. Kostant, *On Whitaker vectors and representation theory*, Invent. Math. **48** (1978), no. 2, pp.101-184.

[Kub] Kubota, T., *Elementary Theory of Eisenstein Series*, Kodansha Ltd. and John Wiley and Sons (Halsted Press), 1973.

[Lan1] Langlands, R., *Euler Products*, Yale University Press, 1971.

[Lan2] ———, *On the Functional Equation Satisfied by Eisenstein Series*, Springer Lecture Notes in Mathematics, Vol. 544, 1976.

[Lan3] ———, *Problems in the Theory of Automorphic Forms*, Lectures in Modern Analysis and Applications III, Vol. 120, Springer-Verlag, Berlin, 1970, pp. 18-61.

[Lan4] ———, *Automorphic Representations, Shimura Varieties, and Motives. Ein Marchen*, in Proceedings of Symposia in Pure Mathematics, Vol. **33** (1979), Part 2, pp. 205-246.

[Lan5] ———, *On the classification of irreducible representations of real algebraic groups*, Mimeographed notes, Institute for Advanced Study, 1973.

[Lan-Sh] ——— and D. Shelstad, *On the definition of transfer factors*, Math. Ann. **278** (1987), pp. 278-271.

[LMT] E. Lapid, G. Muić and M. Tadić, *On the Generic Unitary Dual of Quasi-split Classical Groups*, Int. Math. Res. Notes (2004), no. 26, pp. 1335-1354.

[La-Ra] Lapid, E. and Rallis, S., *On the local factors of representations of classical groups*, pp. 309-360, Automorphic Representations, L-functions and Applications: Progress and Prospects, de Gruyter, Berlin and New York

[L-R-S] W. Luo, Z. Rudnick and P. Sarnak, *On the generalized Ramanujan Conjecture for $GL(n)$*, Proc. Sympos. Pure Math. **66** (1999), pp. 301-310.

[Mo-Wa] C. Mœglin and J.-P. Waldspurger, *Le spectre résiduel de $GL(n)$*, Ann. Sci. Éc. Norm. Sup. (4) **22** (1999), pp. 605-674.

[Mo-Sh] C. J. Moreno and F. Shahidi, *The L-function $L_3(s,\Delta)$ is entire*, Invent. Math. **79** (1985), no. 2, pp. 247-251.

[Mui] G. Muic, *The Unitary Dual of G_2*, Duke Math. J., **90** (1997), pp. 465-493.

[Mül] W. Müller, *The trace class conjecture in the theory of automorphic forms*, Ann. of Math. **130** (1989), pp.473-529.

[PS-Ra] I. Piatetski-Shapiro and S. Rallis, *L-functions for the Classical Groups*, in "Explicit Constructions of Automorphic L-functions", Lecture Notes in Mathematics Volume 1254, Springer-Verlag, pp. 1-52.

[Ra-So] S. Rallis and D. Soudry, *Stability and the local gamma factor arising from the double method*, Math. Ann. **333** (2005), no. 5. pp. 291-313.

[Ram1] D. Ramakrishman, *On the Coefficients of Cusp Forms*, Math. Res. Lett., **4** (1997), pp. 295-307.

[Ram2] ———, *Modularity of the Rankin-Selberg L-series, and Multiplicity one for SL_2*, Annals of Math. **152** (2000), pp.45-111.

[Shah1] Shahidi, F., *Poles of intertwining operators via endoscopy; with an Appendix by Diana Shelstad*, Compositio Math. **120** (2000), no. 3, pp. 291-325.

[Schi] G. Schiffmann *Intégrales d'entrelacement et fonctions de Whittaker*, Bull. Soc. Math. France **99** (1971) , pp. 3-72.

[Sil] A. Silberger, Introduction to harmonic analysis on reductive p-adic groups. Based on lectures by Harish-Chandra at the Institute for Advanced Study, 1971–1973. Mathematical Notes, 23. Princeton University Press, Princeton, N.J.; University of Tokyo Press, Tokyo, 1979.

[Sou] Soudry, D., *Rankin-Selberg integrals, the descent method, and Langlands functoriality*, 2006 International Congress of Mathematicians, Vol. II, Eur. Math. Soc., Zurich, 1311-1325.

[Vog] D. A. Vogan, Jr., *Gelfand-Kirillov dimension for Harish-Chandra modules*, Invent. Math. **48** (1978), no.1, pp.75-98.

[Wal] N. Wallach, *Asymptotic Expansions of Generalized Matrix Coefficients of Representations of Real Reductive Groups*, Lie Group Representations I, Springer Lecture Notes **1024** (1983), pp. 287-369.

Nicki and J. Ira Harris Professorial Chair, Department of Mathematics, Weizmann Institute of Science, Rehovot 76100 Israel

E-mail address: steve.gelbart@weizmann.ac.il

Clay Mathematics Proceedings
Volume 13, 2011

The Embedded Eigenvalue Problem for Classical Groups

James Arthur

This paper is dedicated to Freydoon Shahidi on the occasion of his sixtieth birthday.

ABSTRACT. We report briefly on an endoscopic classification of representations by focusing on one aspect of the problem, the question of embedded Hecke eigenvalues.

1. The problem for G

By "eigenvalue", we mean the family of unramified Hecke eigenvalues of an automorphic representation. The question is whether there are any eigenvalues for the discrete spectrum that are also eigenvalues for the continuous spectrum. The answer for classical groups has to be part of any general classification of their automorphic representations.

The continuous spectrum is to be understood narrowly in the sense of the spectral theorem. It corresponds to representations in which the continuous induction parameter is unitary. For example, the trivial one-dimensional automorphic representation of the group $SL(2)$ does not represent an embedded eigenvalue. This is because it corresponds to a value of the one-dimensional induction parameter at a nonunitary point in the complex domain. For general linear groups, the absence of embedded eigenvalues has been known for some time. It is a consequence of the classification of Jacquet-Shalika [**JS**] and Moeglin-Waldspurger [**MW**]. For other classical groups, the problem leads to interesting combinatorial questions related to the endoscopic comparison of trace formulas.

We shall consider the case that G is a (simple) quasisplit symplectic or special orthogonal group over a number field F. Suppose for example that G is split and of rank n. The continuous spectrum of maximal dimension is then parametrized by n-tuples of (unitary) idele class characters. Is there any n-tuple whose unramified Hecke eigenvalue family matches that of an automorphic representation π in the discrete spectrum of G? The answer is no if π is required to have a global Whittaker model. This follows from the work of Cogdell, Kim, Piatetskii-Shapiro and Shahidi

2010 *Mathematics Subject Classification.* Primary 22E55, 22E50; Secondary 20G35, 58C40.
Key words and phrases. automorphic representations, unramified Hecke eigenvalues, symplectic and orthogonal groups.
The author was supported in part by NSERC Grant A3483.

[**CKPS**]. Any such π will automatically have a local Whittaker model at each place. However, it is by no means clear that π must also have a global Whittaker model. In fact, the general existence of global Whittaker models appears to be dependent on some a priori classification of the full discrete spectrum of G.

Our discussion of the embedded eigenvalue problem can therefore be regarded as a short introduction to the larger question of the endoscopic classification of representations. It represents an attempt to isolate a manageable part of a broader topic, which at the same time illustrates some of the basic techniques. These techniques rest on a comparison of trace formulas on different groups.

2. A distribution and its stabilization

It is the *discrete part* of the trace formula that carries the information about automorphic representations. This is by definition the linear form

$$(1)\qquad I_{\text{disc}}^G(f) = \sum_M |W(M)|^{-1} \sum_{w\in W(M)_{\text{reg}}} |\det(w-1)|^{-1}\text{tr}\big(M_P(w)\mathcal{I}_P(f)\big),$$

for a test function $f \in C_c^\infty\big(G(\mathbb{A})\big)$ on $G(\mathbb{A})$. We recall that M ranges over the finite set of conjugacy classes of Levi subgroups of G, that

$$W(M) = \text{Norm}_G(A_M)/M$$

is the Weyl group of M over F, and that $W(M)_{\text{reg}}$ is the set of elements $w \in W(M)$ such that the determinant of the associated linear operator

$$(w-1) = (w-1)_{\mathfrak{a}_M^G}$$

is nonzero. As usual,

$$\mathcal{I}_P(f) = \mathcal{I}_P(0, f), \qquad P \in \mathcal{P}(M),$$

is the representation of $G(\mathbb{A})$ on the Hilbert space

$$\mathcal{H}_P = L^2_{\text{disc}}\big(N_P(\mathbb{A})M(\mathbb{Q})A^+_{M,\infty}\backslash G(\mathbb{A})\big)$$

induced parabolically from the discrete spectrum of M, while

$$M_P(w):\ \mathcal{H}_P \longrightarrow \mathcal{H}_P$$

is the global intertwining operator attached to w. Recall that

$$A^+_{M,\infty} = (R_{F/\mathbb{Q}}A_M)(\mathbb{R})^0$$

is a central subgroup of $M(\mathbb{A})$ such that the quotient

$$M(F)A^+_{M,\infty}\backslash M(\mathbb{A})$$

has finite invariant volume.

This is the core of the trace formula. It includes what one hopes ultimately to understand, the automorphic discrete spectrum

$$\mathcal{H}_G = L^2_{\text{disc}}\big(G(\mathbb{Q})\backslash G(\mathbb{A})\big)$$

of G. Indeed, the term with $M = G$ is simply the trace of the right convolution operator of f on this space. The summands for smaller M represent contributions of Eisenstein series to the trace formula. They are boundary terms, which arise from the truncation methods required to deal with the noncompactness of the quotient $G(\mathbb{Q})\backslash G(\mathbb{A})$. The operators $M_P(w)$ are of special interest, being at the heart of the theory of Eisenstein series. It was their study that led to the Langlands-Shahidi method, and much recent progress in the theory of automorphic L-functions.

With its classical ingredients, the expression for $I_{\mathrm{disc}}^G(f)$ is remarkably simple. There are of course other terms in the trace formula, some of which are quite complex. We shall not discuss them here. Our purpose will be rather to see what can be established for the spectral information in $I_{\mathrm{disc}}^G(f)$, knowing that the complementary terms have already been taken care of.

To have a chance of understanding the terms in the formula for $I_{\mathrm{disc}}^G(f)$, we really need something to compare them with. A solution of sorts is provided by the *stabilization* of $I_{\mathrm{disc}}^G(f)$. This is an innocuous looking expansion

$$I_{\mathrm{disc}}^G(f) = \sum_{G'} \iota(G,G')\widehat{S}_{\mathrm{disc}}^{G'}(f^{G'}) \tag{2}$$

of $I_{\mathrm{disc}}^G(f)$ into stable distributions $S_{\mathrm{disc}}^{G'}$ on endoscopic groups G', with coefficients $\iota(G,G')$ that are defined by simple formulas. The sum is actually over the isomorphism classes of elliptic endoscopic data G' of G. For example, if G is the split adjoint group $SO(2n+1)$, the dual group $\widehat{G}$ equals $Sp(2n,\mathbb{C})$. We then have

$$\widehat{G}' = Sp(2m,\mathbb{C}) \times Sp(2n-2m,\mathbb{C})$$

and

$$G' = SO(2m+1) \times SO(2n-2m+1).$$

In particular, the sum in (2) is parametrized in this case by integers that range from 0 to the greatest integer in $\frac{1}{2}n$.

The mapping

$$f \longrightarrow f^{G'}, \qquad\qquad f \in C_c^\infty\big(G(\mathbb{A})\big),$$

in (2) is the Langlands-Shelstad transfer of functions. With Ngo's recent proof of the fundamental lemma [**N**], it is now known that this correspondence takes $C_c^\infty\big(G(\mathbb{A})\big)$ to the space $C_c^\infty\big(G'(\mathbb{A})\big)$ of test functions on $G'(\mathbb{A})$, as originally conjectured by Langlands. The general resolution of the problem is a culmination of work by many people, including Langlands [**L**], Shelstad [**S**], Langlands-Shelstad [**LS**], Goresky-Kottwitz-MacPherson [**GKM**], Waldspurger [**W1**], [**W3**], and Lauman-Ngo [**LN**], as well as Ngo. We recall that it is a local question, which has to be formulated for each completion F_v of F. It was first treated for archimedean v, in [**S**]. The fundamental lemma is required explicitly for the places v that are unramified (relative to f), and implicitly as a hypothesis in the solution [**W1**] for general p-adic v.

The formula (2) was established in [**A3**], following partial results [**L**] and [**K**] obtained earlier. It was predicated on a generalization of the fundamental lemma that applies to unramified *weighted* orbital integrals. This has now been established by Chaudouard and Laumon [**CL**], building on the techniques of Ngo. The stabilization formula (2) is therefore unconditionally valid.

We note that the proof of (2) is indirect. It is a consequence of a stabilization that must be established directly for all of the other terms in the trace formula. For example, the papers [**L**] and [**K**] can be regarded as stabilizations of, respectively, the regular elliptic and the singular elliptic terms. In general, the terms that are complementary to those in $I_{\mathrm{disc}}^G(f)$ each come with their own individual set of problems, all of which must be taken care of. This accounts for the difficulty of the proof of (2).

As we have said, the linear forms $S_{\mathrm{disc}}^{G'}$ in (2) are stable distributions on the groups $G'(\mathbb{A})$. (The symbol $\widehat{S}'$ is understood to be the pullback of S' to the space of stable orbital integrals on $C_c^\infty\big(G'(\mathbb{A})\big)$, a space in which the correspondence

$f \to f^{G'}$ takes values.) However, there is nothing in the formula (2) that tells us anything concrete about these objects. We can regard (2) as simply an inductive definition

$$S^G_{\text{disc}}(f) = I^G_{\text{disc}}(f) - \sum_{G' \neq G} \iota(G, G') \widehat{S}^{G'}_{\text{disc}}(f^{G'})$$

of a stable distribution on $G(\mathbb{A})$ in terms of its analogues for groups G' of smaller dimension. It does tell us that the right hand side, defined inductively on the dimension of G in terms of the right side of (1), is stable in f. This is an interesting fact, to be sure. But it is not something that by itself will give us concrete information about the automorphic discrete spectrum of G. To use (2) effectively, we must combine it with something further.

3. Its twisted analogue for $GL(N)$

The extra ingredient is the twisted trace formula for $GL(N)$, and its corresponding stablization. To describe what we need, we write

$$\widetilde{G} = GL(N) \rtimes \theta,$$

for the standard outer automorphism

$$\theta(x) = {}^t x^{-1}, \qquad x \in GL(N),$$

of $GL(N)$. Then $\widetilde{G}$ is the nonidentity component of the semidirect product

$$\widetilde{G}^+ = \widetilde{G}^0 \rtimes \langle \theta \rangle = GL(N) \rtimes (\mathbb{Z}/2\mathbb{Z}).$$

With this understanding, the twisted trace formula requires little change in notation. Its discrete part can be written in a form

$$(\tilde{1}) \qquad I^{\widetilde{G}}_{\text{disc}}(\widetilde{f}) = \sum_{M} |\widetilde{W}(M)|^{-1} \sum_{w \in \widetilde{W}(M)_{\text{reg}}} |\det(w-1)|^{-1} \text{tr}\big(M_P(w) \mathcal{I}_P(\widetilde{f})\big)$$

that matches (1). In particular, M ranges over the set of conjugacy classes of Levi subgroups in the connected group $\widetilde{G}^0 = GL(N)$, and P represents a parabolic subgroup of $\widetilde{G}^0$ with Levi component M. The only changes from (1) are that the test function $\widetilde{f} \in C^\infty_c\big(\widetilde{G}(\mathbb{A})\big)$ and the Weyl set

$$\widetilde{W}(M) = \text{Norm}_{\widetilde{G}}(M)/M$$

are taken relative to the component $\widetilde{G}$, and that $\mathcal{I}_P$ stands for a representation induced from P to $\widetilde{G}^+$. As before, $M_P(w)$ is the global intertwining operator attached to w. (See [**CLL**] and [**A1**].) The last step in the proof of the general (invariant) twisted trace formula has been the archimedean twisted trace Paley-Wiener theorem, established recently by Delorme and Mezo [**DM**].

The stabilization of $I^{\widetilde{G}}_{\text{disc}}(\widetilde{f})$ takes the form

$$(\tilde{2}) \qquad I^{\widetilde{G}}_{\text{disc}}(\widetilde{f}) = \sum_{G} \iota(\widetilde{G}, G) \widetilde{S}^G_{\text{disc}}(\widetilde{f}^G),$$

where the symbols S^G_{disc} represent stable distributions defined inductively by (2), and $\iota(\widetilde{G}, G)$ are again explicit coefficients. The sum is over isomorphism classes of elliptic twisted endoscopic data G for $\widetilde{G}$. For example, if $N = 2n+1$ is odd, the component

$$\widetilde{G} = GL(2n+1) \rtimes \theta$$

has a "dual set"

$$\widehat{G} = GL(2n+1, \mathbb{C}) \rtimes \widehat{\theta}.$$

We then have

$$\widehat{G} = Sp(2m, \mathbb{C}) \times SO(2n-2m+1, \mathbb{C})$$

and

$$G = SO(2m+1) \times Sp(2n-2m).$$

In general, a twisted endoscopic datum G entails a further choice, that of a suitable L-embedding

$$\xi_G : \ {}^L G \longrightarrow GL(N, \mathbb{C})$$

of the appropriate form of the L-group of G into $GL(N, \mathbb{C})$. However, if we forget this extra structure, we see in this case that G is just a group parametrized by an integer that ranges from 0 to n.

The mapping

$$\widetilde{f} \longrightarrow \widetilde{f}^G, \qquad \widetilde{f} \in C_c^\infty(\widetilde{G}(\mathbb{A})),$$

in $(\tilde{2})$ is the Kottwitz-Langlands-Shelstad correspondence of functions. The long-standing conjecture has been that it takes $C_c^\infty(\widetilde{G}(\mathbb{A}))$ to $C_c^\infty(G(\mathbb{A}))$. With the recent work of Ngo [N] and Waldspurger [W1]–[W3], this conjecture has now been resolved. The resulting transfer of functions becomes the fundamental starting point for a general stabilization of the twisted trace formula.

The actual identity $(\tilde{2})$ is less firmly in place. The twisted generalization of the weighted fundamental lemma does follow from the work of Chaudouard and Laumon, and of Waldspurger. However, the techniques of [**A3**] have not been established in the twisted case. Some of these techniques will no doubt carry over without much change. However, there will be others that call for serious refinement, and perhaps also new ideas. Still, there is again reason to be hopeful that a general version of $(\tilde{2})$ can be established in the not too distant future. We shall assume its stated version for $GL(N)$ in what follows.

Taken together, the stabilizations (2) and $(\tilde{2})$ offer us the possibility of relating automorphic representations of a classical group G with those of a twisted general linear group $\widetilde{G}$. As we have noted, the identity (2) represents an inductive definition of a stable distribution on $G(\mathbb{A})$ in terms of unknown spectral automorphic data (1) for G. The identity $(\tilde{2})$ provides a relation among the distributions in terms of known spectral automorphic data $(\tilde{1})$ for $GL(N)$.

This is not to say that the subsequent analysis is without further difficulty. It in fact contains many subtleties. For example, there is often more than one unknown stable distribution S^G_{disc} on the right hand side of the identity $(\tilde{2})$. The problem is more serious in case $N = 2n$ is even, where there are data G with dual groups $Sp(2n, \mathbb{C})$ and $SO(2n, \mathbb{C})$ that are both distinct and simple. This particular difficulty arises again and again in the analysis. Its constant presence requires a sustained effort finally to overcome.

4. Makeshift parameters

The comparison of (2) and $(\tilde{2})$ requires a suitable description of the automorphic discrete spectrum of the group $\widetilde{G}^0 = GL(N)$. Let $\Psi_2(N)$ be the set of formal tensor products

$$\psi = \mu \boxtimes \nu, \qquad N = mn,$$

where μ is a unitary cuspidal automorphic representation of $GL(m)$ and ν is the irreducible representation of the group $SL(2,\mathbb{C})$ of dimension n. The cuspidal representation μ comes with what we are calling an "eigenvalue". This, we recall, is the Hecke family

$$c(\mu) = \{c_v(\mu) = c(\mu_v) : v \notin S\}$$

of semisimple conjugacy classes in $GL(m,\mathbb{C})$ attached to the unramified constituents μ_v of v. To the tensor product ψ, we attach the "eigenvalue"

$$c(\psi) = c(\mu) \otimes c(\nu).$$

This is the family of semisimple conjugacy classes

$$c_v(\mu) \otimes \nu \begin{pmatrix} q_v^{\frac{1}{2}} & 0 \\ 0 & q_v^{-\frac{1}{2}} \end{pmatrix} = c_v(\mu) q_v^{\frac{n-1}{2}} \oplus \cdots \oplus c_v(\mu) q_v^{-\frac{n-1}{2}}, \qquad v \notin S,$$

in $GL(N,\mathbb{C})$. It follows from [**JS**] and [**MW**] that there is a bijection $\psi \to \pi_\psi$ from $\Psi_2(N)$ onto the set of unitary automorphic representations π_ψ in the discrete spectrum of $GL(N)$ (taken modulo the center) such that

$$c(\psi) = c(\pi_\psi).$$

More generally, one can index representations in the broader automorphic spectrum by sums of elements in $\Psi_2(N_i)$. Let $\Psi(N)$ be the set of formal direct sums

$$\psi = \ell_1 \psi_1 \boxplus \cdots \boxplus \ell_r \psi_r, \tag{3}$$

for positive integers ℓ_i and distinct elements $\psi_i = \mu_i \boxtimes \nu_i$ in $\Psi_2(N_i)$, whose ranks $N_i = m_i n_i$ satisfy

$$N = \ell_1 N_1 + \cdots + \ell_r N_r = \ell_1 m_1 n_1 + \cdots \ell_r m_r n_r.$$

For any ψ, we attach the "eigenvalue"

$$c(\psi) = \ell_1 c(\psi_1) \oplus \cdots \oplus \ell_r c(\psi_r),$$

of semisimple conjugacy classes

$$c_v(\psi) = \underbrace{c_v(\psi_1) \oplus \cdots \oplus c_v(\psi_1)}_{\ell_1} \oplus \cdots \oplus \underbrace{c_v(\psi_r) \oplus \cdots \oplus c_v(\psi_r)}_{\ell_r}$$

in $GL(N,\mathbb{C})$. It then follows from Langlands' theory of Eisenstein series that there is a bijection $\psi \to \pi_\psi$ from $\Psi(N)$ to the set of unitary representations π_ψ in the full automorphic spectrum of $GL(N)$ such that

$$c(\psi) = c(\pi_\psi).$$

The elements in $\Psi(N)$ are to be regarded as makeshift parameters. They are basically forced on us in the absence of the hypothetical automorphic Langlands group L_F. Recall that L_F is supposed to be a locally compact group whose irreducible, unitary, N-dimensional representations parametrize the unitary, cuspidal automorphic representation of $GL(N)$.

If we had the group L_F at our disposal, we could identify elements in our set $\Psi(N)$ with (equivalence classes of) N-dimensional representations

$$\psi : L_F \times SL(2,\mathbb{C}) \longrightarrow GL(N,\mathbb{C})$$

whose restrictions to L_F are unitary. This interpretation plays a conjectural role in the representation theory of the quasisplit group G. Regarding G as an elliptic twisted endoscopic datum for $GL(N)$, and $\Psi(N)$ as the set of N-dimensional representations of $L_F \times SL(2,\mathbb{C})$, we would be able to introduce the subset of mappings ψ in $\Psi(N)$ that factor through the embedded L-group

$$\xi_G : \ {}^LG \longrightarrow GL(N,\mathbb{C}).$$

Any such ψ would then give rise to a complex reductive group, namely the centralizer

$$S_\psi = S_\psi^G = \mathrm{Cent}\big(\mathrm{Im}(\psi), \widehat{G}\big)$$

in $\widehat{G} \subset {}^LG$ of its image in ${}^LG$. The finite quotient

$$\mathcal{S}_\psi = S_\psi / S_\psi^0 Z(\widehat{G})^{\Gamma_F}, \qquad \Gamma_F = \mathrm{Gal}(\overline{F}/F), \tag{4}$$

of S_ψ is expected to play a critical role in the automorphic representation theory of G.

5. The groups $\mathcal{L}_\psi$

The first challenge is to define the centralizers S_ψ and their quotients $\mathcal{S}_\psi$ without having the group L_F. For any makeshift parameter ψ as in (3), we can certainly form the contragredient parameter

$$\begin{aligned}\psi^\vee &= \ell_1\psi_1^\vee \boxplus \cdots \boxplus \ell_r\psi_r^\vee \\ &= \ell_1(\mu_1^\vee \boxtimes \nu_1) \boxplus \cdots \boxplus \ell_r(\mu_r^\vee \boxtimes \nu_r).\end{aligned}$$

The subset

$$\widetilde{\Psi}(N) = \{\psi \in \Psi(N) : \ \psi^\vee = \psi\}.$$

of self-dual parameters in $\Psi(N)$ consists of those ψ for which the corresponding automorphic representation π_ψ is θ-stable. The idea is to attach a makeshift group $\mathcal{L}_\psi$ to any ψ. The group $\mathcal{L}_\psi$ will then be our substitute for L_F. We shall formulate it as an extension of the Galois group Γ_F by a complex connected reductive group.

The main problem in the construction of $\mathcal{L}_\psi$ is to deal with the basic case that $\psi = \mu$ is cuspidal. Since ψ is assumed to lie in $\widetilde{\Psi}(N)$, μ equals $\mu^\vee$. It therefore represents a self dual cuspidal automorphic representation of $GL(N)$. At this point we have to rely on the following theorem.

THEOREM 1. *Suppose that μ is a self-dual, unitary, cuspidal automorphic representation of $GL(N)$. Then there is a unique elliptic, twisted endoscopic datum $G = G_\mu$ for $GL(N)$ that is simple, and such that*

$$c(\mu) = \xi_{G_\mu}\big(c(\pi)\big),$$

for a cuspidal automorphic representation π of $G(\mathbb{A})$.

The theorem asserts that there is exactly one G for which there is a cuspidal "eigenvalue" that maps to the "eigenvalue" of μ in $GL(N)$. Its proof is deep. In working on the general classification, one assumes inductively that the theorem holds for the proper self-dual components μ_i of a general parameter ψ. The resolution of this (and other) induction hypotheses then comes only at the end of the entire argument. However, we shall assume for the discussion here that the theorem is valid without restriction. In the case that $\psi = \mu$, this allows us to define

$$\mathcal{L}_\psi = {}^LG_\mu.$$

We then write $\widetilde{\psi}$ for the L-homomorphism $\xi_\mu = \xi_{G_\mu}$ of this group into $GL(N,\mathbb{C})$.

Consider now an arbitrary parameter $\psi \in \widetilde{\Psi}(N)$ of the general form (3). Since ψ is self-dual, the operation $\mu \to \mu^\vee$ acts as an involution on the cuspidal components μ_i of ψ. If i is an index with $\mu_i^\vee = \mu_i$, we introduce the group $G_i = G_{\mu_i}$ provided by the theorem, as well as the L-homomorphism

$$\xi_i = \xi_{\mu_i} : \ {}^LG_i \longrightarrow GL(m_i,\mathbb{C}).$$

If j parametrizes an orbit $\{\mu_j, \mu_j^\vee\}$ of order two, we set $G_j = GL(m_j)$, and we take

$$\xi_j : \ {}^L(GL(m_j)) \longrightarrow GL(2m_j,\mathbb{C})$$

to be the homomorphism that is trivial on Γ_F, and that restricts to the embedding

$$g \longrightarrow \begin{pmatrix} g & 0 \\ 0 & {}^tg^{-1} \end{pmatrix}$$

of $GL(m_j,\mathbb{C})$ into $GL(2m_j,\mathbb{C})$. We define our general makeshift group $\mathcal{L}_\psi$ to be the fibre product

$$\mathcal{L}_\psi = \prod_{k\in\{i,j\}} \left({}^LG_k \longrightarrow \Gamma_F\right)$$

of these L-groups over Γ_F. The various homomorphisms ξ_k can then be combined in the natural way with the corresponding representations

$$\nu_k : \ SL(2,\mathbb{C}) \longrightarrow GL(n_k,\mathbb{C})$$

to give a homomorphism

$$\widetilde{\psi} : \ \mathcal{L}_\psi \times SL(2,\mathbb{C}) \longrightarrow GL(N,\mathbb{C}).$$

We regard $\widetilde{\psi}$ as an equivalence class of N-dimensional representations of the group $\mathcal{L}_\psi \times SL(2,\mathbb{C})$.

Suppose that G represents a simple twisted endoscopic datum for $GL(N)$. We define $\widetilde{\Psi}(G)$ to be the subset of parameters $\psi \in \widetilde{\Psi}(N)$ such that $\widetilde{\psi}$ factors through the image of ${}^LG$ in $GL(N,\mathbb{C})$. For any $\psi \in \widetilde{\Psi}(G)$, we then have an L-embedding

$$\widetilde{\psi}_G : \ \mathcal{L}_\psi \times SL(2,\mathbb{C}) \longrightarrow {}^LG$$

such that

$$\xi_G \circ \widetilde{\psi}_G = \widetilde{\psi}.$$

We are treating $\widetilde{\psi}$ as an equivalence class of N-dimensional representations. This means that $\widetilde{\psi}_G$ is determined only up to the group $\mathrm{Aut}_{\widetilde{G}}(G)$ of L-automorphisms of ${}^LG$ induced by the stabilizer in $GL(N,\mathbb{C})$ of its image. Nevertheless, we can still write

$$S_\psi = S_\psi^G = \mathrm{Cent}\big(\mathrm{Im}(\widetilde{\psi}_G), \widehat{G}\big)$$

and

$$\mathcal{S}_\psi = S_\psi / S_\psi^0 Z(\widehat{G})^{\Gamma_F},$$

where $\widetilde{\psi}_G$ stands for some L-homomorphism in the associated $\mathrm{Aut}_{\widetilde{G}}(G)$-orbit. Since $\mathcal{S}_\psi$ is a finite abelian group (a 2-group actually), it is uniquely determined by ψ up a unique isomorphism.

The parameters $\psi \in \widetilde{\Psi}(G)$, along with the groups $\mathcal{L}_\psi$ and the associated centralizer groups S_ψ and $\mathcal{S}_\psi$, were described in §30 of [**A4**]. They will be discussed in greater detail in Chapter 1 of [**A5**]. The deeper properties of the hypothetical

Langlands group L_F probably mean that its existence will be one of the last theorems to be proved in the subject. However, if L_F does exist, its expected properties imply that the family of $\mathrm{Aut}_{\widetilde{G}}(G)$-orbits of homomorphisms

$$L_F \times SL(2,\mathbb{C}) \longrightarrow {}^L G$$

is in natural bijection with the set $\widetilde{\Psi}(G)$ we have just defined. Moreover, this bijection identifies the corresponding centralizers S_ψ and their quotients $\mathcal{S}_\psi$. It is also compatible with the localization

$$\psi \longrightarrow \psi_v$$

of parameters, something we will not discuss here.

This all means that our makeshift groups $\mathcal{L}_\psi$ capture the information from L_F that is relevant to the endoscopic classification of representations of G. In other words, the groups $\mathcal{L}_\psi$ are as good as the Langlands group for the purposes at hand, even though they vary with ψ. They are used in [**A5**] to formulate the classification of automorphic representations of G.

6. The ψ-components of distributions

The next step is to isolate the ψ-components of the terms in the expansions (1), (2), $(\tilde{1})$ and $(\tilde{2})$. Recall that a parameter $\psi \in \widetilde{\Psi}(N)$ comes with an "eigenvalue" $c(\psi)$. If D is a distribution that occurs in one of these expansions, its ψ-component D_ψ is a "ψ-eigendistribution", relative to the convolution action of the unramified Hecke algebra on the test function f (or $\widetilde{f}$). We thus obtain two expansions

$$(1)_\psi \qquad I^G_{\mathrm{disc},\psi}(f) = \sum_M |W(M)|^{-1} \sum_{w \in W(M)_{\mathrm{reg}}} |\det(w-1)| \mathrm{tr}\big(M_{P,\psi}(w)\mathcal{I}_{P,\psi}(f)\big)$$

and

$$(2)_\psi \qquad I^G_{\mathrm{disc},\psi}(f) = \sum_{G'} \iota(G,G') \widehat{S}^{G'}_{\mathrm{disc},\psi}(f^{G'})$$

of the ψ-component $I^G_{\mathrm{disc},\psi}(f)$. Similarly, we obtain two expansions $(\tilde{1})_\psi$ and $(\tilde{2})_\psi$ for the ψ-component $I^{\tilde{G}}_{\mathrm{disc},\psi}(\widetilde{f})$ of $I^{\tilde{G}}_{\mathrm{disc}}(\widetilde{f})$. The problem is to compare explicitly the terms in these two identities.

We are trying to describe these matters in the context of the embedded eigenvalue problem. According to general conjecture, a parameter $\psi \in \widetilde{\Psi}(G)$ would be expected to contribute to the discrete spectrum of G if and only if the group

$$\bar{S}_\psi = S_\psi / Z(\widehat{G})^{\Gamma_F}$$

is finite. In other words, the component group

$$\mathcal{S}_\psi = \pi_0(\bar{S}_\psi)$$

that is supposed to govern spectral multiplicities is actually equal to $\bar{S}_\psi$. If we apply this inductively to a Levi subgroup M of G, we see that ψ contributes Eisenstein series of rank k to the spectrum of G if and only if the rank of $\bar{S}_\psi$ equals k. The problem then is to show that if $\bar{S}_\psi$ is *not* finite, it does not contribute to the discrete spectrum of G. That is, there is no automorphic representation π of G in the discrete spectrum with $c(\psi) = c(\pi)$.

One has thus to show that if $\bar{S}_\psi$ is infinite, the term in $(1)_\psi$ with $M = G$ vanishes. However, we know nothing about this term. We can say (by induction) that ψ contributes to the term corresponding to a *unique* proper M. We would first try to express this term as concretely as possible. We would then want to express the terms on the right hand side of $(2)_\psi$ in such a way that their sum could be seen to cancel the term of M in $(1)_\psi$. This would tell us that the term of G in $(1)_\psi$ vanishes, as desired. But the distributions in $(2)_\psi$ are by no means explicit. They consist of the stable linear form $S^G_{\mathrm{disc},\psi}(f)$, about which we know very little, and its analogues for proper endoscopic groups G', which are at least amenable to induction. To deal with $S^G_{\mathrm{disc},\psi}(f)$, we have to compare the right hand side of $(2)_\psi$ (as G varies) with the right hand side of $(\tilde{2})_\psi$. We would then have to compare $(\tilde{2})_\psi$ with the expression on the right hand side of $(\tilde{1})_\psi$, about which we do know something (because it pertains to $GL(N)$).

7. Statement of theorems

It is a rather elaborate process. We shall describe the theorems that lead to a resolution of the problem. Our statements of these theorems will have to be somewhat impressionistic, since we will not take the time to describe all their ingredients precisely. We refer the reader to the forthcoming volume [**A5**] for a full account.

THEOREM 2 (Stable Multiplicity Formula). *Suppose that $\psi \in \widetilde{\Psi}(G)$. Then the term in $(2)_\psi$ corresponding to $G' = G$ satisfies an explicit formula*

$$S^G_{\mathrm{disc},\psi}(f) = m_\psi |\mathcal{S}_\psi|^{-1}\sigma(\bar{S}^0_\psi)\varepsilon_\psi(s_\psi) f^G(\psi),$$

where $m_\psi \in \{1, 2\}$ equals the number of $\widehat{G}$-orbits in the $\mathrm{Aut}_{\widetilde{G}}(G)$-orbit of embeddings $\widetilde{\psi}_G$, $\varepsilon(s_\psi) = \pm 1$ is a sign defined in terms of values at $s = \frac{1}{2}$ of global ε-factors attached to ψ, and $\sigma(\bar{S}^0_\psi)$ is the number attached to the complex connected group $\bar{S}^0_\psi$ in Theorem 4 below.

The last term $f^G(\psi)$ in the formula is harder to construct. It represents the pullback to $G(\mathbb{A})$ of the twisted character

$$\mathrm{tr}\big(\pi_\psi(\widetilde{f})\big), \qquad \widetilde{f} \in C^\infty_c\big(\widetilde{G}(\mathbb{A})\big),$$

on $GL(N, \mathbb{A})$. (We use the theory of Whittaker models for $GL(N)$ to extend the θ-stable representation π_ψ to the component

$$\widetilde{G}(\mathbb{A}) = GL(N, \mathbb{A}) \rtimes \theta$$

on which $\widetilde{f}$ is defined.) The construction is essentially local. Since the criterion of Theorem 1 that determines the subset $\widetilde{\Psi}(G)$ of $\widetilde{\Psi}(N)$ to which ψ belongs is global, the definition of $f^G(\psi)$ requires effort. It is an important part of the proof of Theorem 2.

The formula of Theorem 2 is easily specialized to the other summands in $(2)_\psi$. For any G', it gives rise to a sum over the subset $\widetilde{\Psi}(G', \psi)$ of parameters $\psi' \in \widetilde{\Psi}(G')$ that map to ψ. The formulas so obtained can then be combined in the sum over G'. The end result is an explicit expression for the right hand side of $(2)_\psi$ in terms of the distributions

$$f^{G'}(\psi'), \qquad f \in C^\infty_c\big(G(\mathbb{A})\big),\ \psi' \in \widetilde{\Psi}(G', \psi),$$

and combinatorial data attached to the (nonconnected) complex reductive group $\overline{S}_\psi$.

THEOREM 3. *Suppose that $\psi \in \widetilde{\Psi}(G)$ contributes to the induced discrete spectrum of a proper Levi subgroup M of G, and that w lies in $W(M)_{\mathrm{reg}}$. Then there is a natural formula for the corresponding distribution*

$$\mathrm{tr}\big(M_{P,\psi}(w)\mathcal{I}_{P,\psi}(f)\big)$$

in $(1)_\psi$ in terms of

(i) *the distributions*

$$f^{G'}(\psi'), \qquad\qquad \psi' \in \widetilde{\Psi}(G',\psi),$$

(ii) *the order of poles of global L-functions at $s = 1$, and*

(iii) *the values of global ε-factors at $s = \frac{1}{2}$.*

In this case, we have not tried to state even a semblance of a formula. However, the resulting expression for the sum in $(1)_\psi$ will evidently have ingredients in common with its counterpart for $(2)_\psi$ discussed above. It will also have two points of distinction. In $(1)_\psi$ there will be only one vanishing summand (other than the summand of G we are trying to show also vanishes). Furthermore, the summand of M contains something interesting beyond the distribution above, the coefficient

$$|\det(1-w)|^{-1}.$$

One sees easily that the distribution of Theorem 3 vanishes unless w has a representative in the subgroup $\overline{S}_\psi$ of $\widehat{G}/Z(\widehat{G})^{\Gamma_F}$. We can therefore analyze the combinatorial properties of the coefficients in the context of this group.

Suppose for a moment that S is any connected component of a general (nonconnected) complex, reductive algebraic group S^+. Let T be a maximal torus in the identity component $S^0 = (S^+)^0$ of this group. We can then form the Weyl set

$$W = W(S) = \mathrm{Norm}_S(T)/T,$$

induced by the conjugation action of elements in S on T. Let W_{reg} be the set of elements w in W that are regular, in the sense that as a linear operator on the real vector space

$$\mathfrak{a}_T = \mathrm{Hom}(X(T),\mathbb{R}),$$

the difference $(1-w)$ is nonsingular. We define the sign $\varepsilon^0(w) = \pm 1$ of an element $w \in W$ to be the parity of the number of positive roots of (S^0, T) mapped by w to negative roots. Given these objects, we attach a real number

$$i(S) = |W|^{-1} \sum_{w \in W_{\mathrm{reg}}} \varepsilon^0(w)|\det(w-1)|^{-1}$$

to S.

As is often customary, we write S_s for the centralizer in S^0 of a semisimple element $s \in S$. This is of course a complex reductive group, whose identity component we denote by S_s^0. We then introduce the subset

$$S_{\mathrm{ell}} = \big\{s : \ |Z(S_s^0)| < \infty\big\},$$

where $Z(S_1)$ denotes the center of any given complex connected group S_1. The set $\mathrm{Orb}(S_{\mathrm{fin}}, S^0)$ of orbits in S_{ell} under conjugation by S^0 is finite.

THEOREM 4. *There are unique constants $\sigma(S_1)$, defined whenever S_1 is a complex connected reductive group, such that for any S the number*

$$e(S) = \sum_{s \in \mathrm{Orb}(S_{\mathrm{ell}}, S^0)} |\pi_0(S_s)|^{-1}\sigma(S_s^0)$$

equals $i(S)$, *and such that*

$$\sigma(S_1) = \sigma(S_1/Z_1)|Z_1|^{-1},$$

for any central subgroup $Z_1 \subset Z(S_1)$ *of* S_1.

The numbers $i(S)$ and $e(S)$ of the theorem are elementary. However, they bear an interesting formal resemblance to the deeper expansions on the right hand sides of $(1)_\psi$ and $(2)_\psi$ respectively. In particular, the data in $(2)_\psi$ are vaguely endoscopic. I have sometimes wondered whether Theorem 4 represents some kind of broader theory of endoscopy for Weyl groups.

The proof of the theorem is also elementary. It was established in §8 of [**A2**]. We have displayed the result prominently here because of the link it provides between Theorems 2 and 3, or rather between the expressions for the right hand sides of $(1)_\psi$ and $(2)_\psi$ that these theorems ultimately yield. We have discussed these expressions in only the most fragmentary of terms. We add here only the following one-line summary. If the summand of G in $(1)_\psi$ is put aside, the two expressions are seen to match, up to coefficients that reduce respectively to the numbers $i(S)$ and $e(S)$ attached to the components S of the group $\bar{S}_\psi$. Theorem 4 then tells us that the right hand of $(2)_\psi$ equals the difference between the right side of $(1)_\psi$ and the summand of G in $(1)_\psi$. Since the left hand sides of $(1)_\psi$ and $(2)_\psi$ are equal, the summand of G does vanish for any $\psi \in \widetilde{\Psi}(G)$ with $\bar{S}_\psi$ infinite, as required. We thus obtain the following theorem.

THEOREM 5. *The automorphic discrete spectrum of G has no embedded eigenvalues.*

This is the result we set out to describe. As we have said, it is part of a general classification of the automorphic representations of G. The reader will have to refer to [A4, §30] and [A5] for a description of the classification. However, the theorems discussed here are at the heart of its proof.

References

[A1] J. Arthur, *The invariant trace formula II. Global theory*, J. Amer. Math. Soc. 1 (1988), 501–554.

[A2] ________ , *Unipotent automorphic representations: Global motivation*, in *Automorphic Forms, Shimura Varieties and L-functions*, Vol. I, 1–75, Academic Press, 1990.

[A3] ________ , *A stable trace formula III. Proof of the main theorems*, Annals of Math. **158** (2003), 769–873.

[A4] ________ , *An introduction to the trace formula*, in *Harmonic Analysis, the Trace Formula and Shimura Varieties*, 1–263, Clay Math. Proc. 4, Amer. Math. Soc., Providence RI, 2005.

[A5] ________ , *The Endoscopic Classification of Representations: Orthogonal and Symplectic Groups*, in preparation.

[CL] P.-H. Chaudouard and G. Laumon, *Le lemme fondamental pondéré* I and II, preprints.

[CLL] L. Clozel, J.-P. Labesse, and R.P. Langlands, *Morning Seminar on the Trace Formula*, Lecture Notes, Institute for Advanced Study, Princeton, N.J., 1984.

[CKPS] J. Cogdell, H. Kim, I. Piatetski-Shapiro, and F. Shahidi, *Functoriality for the classical groups*, Publ. Math. Inst. Hautes Études Sci. **99** (2004), 163–233.

[DM] P. Delorme and P. Mezo, *A twisted Paley-Wiener theorem for real reductive groups*, preprint.

[GKM] M. Goresky, R. Kottwitz and R. MacPherson, *Homology of affine Springer fibres in the unramified case*, Duke Math. J. **121** (2004), 509–561.

[JS] H. Jacquet and J. Shalika, *On Euler products and the classification of automorphic representations* II, Amer. J. Math. **103** (1981), 777-815.

[K] R. Kottwitz, *Stable trace formula: elliptic singular terms*, Math. Ann. **275** (1986), 365–399.

[L] R. Langlands, *Les débuts d'une formule stable des traces stables,* Publ. Math. Univ. Paris VII **13**, 1983.

[LN] G. Laumon and B.C. Ngo, *Le lemme fondamental pour les groupes unitaires*, to appear in Annals of Math.

[LS] R. Langlands and D. Shelstad, *On the definition of transfer factors,* Math. Ann **278** (1987), 219–271.

[MW] C. Moeglin and J.-L. Waldspurger, *Le spectre résiduel de $GL(n)$,* Ann. Scient. Ec. Norm. Sup. 4e série **22** (1989), 605–674.

[N] B.C. Ngo, *Le lemme fondamental pour les algebres de Lie*, Publ. Math. I.H.E.S. **111** (2010), 1–169.

[S] D. Shelstad, *L-indistinguishability for real groups,* Math. Ann. **259** (1982), 385–430.

[W1] J.-L. Waldspurger, *Le lemme fondamental implique le transfert*, Compositio Math. **105** (1997), 153–236.

[W2] ________ , *Endoscopie et changement de caractéristique,* J. of the Inst. of Math. Jussieu **5** (2006), 423–525.

[W3] ________ , *L'endoscopie tordue n'est pas si tordue*, Mem. Amer. Math. Soc. **908** (2008).

Department of Mathematics, University of Toronto, Toronto, Ontario M5S 2E1
E-mail address: arthur@math.toronto.edu

Clay Mathematics Proceedings
Volume 13, 2011

A Cuspidality Criterion for the Exterior Square Transfer of Cusp Forms on GL(4)

Mahdi Asgari and A. Raghuram

Dedicated to Freydoon Shahidi on the occasion of his sixtieth birthday

ABSTRACT. For a cuspidal automorphic representation Π of $\mathrm{GL}(4, \mathbb{A})$, H. Kim proved that the exterior square transfer $\wedge^2\Pi$ is nearly an isobaric automorphic representation of $\mathrm{GL}(6, \mathbb{A})$. In this paper we characterize those representations Π for which $\wedge^2\Pi$ is cuspidal.

1. Introduction and statement of the main theorem

Let F be a number field whose adèle ring we denote by $\mathbb{A}_F$. Let G_1 and G_2 be two connected reductive linear algebraic groups over F, with G_2 quasi-split over F, and let ${}^LG_1$ and ${}^LG_2$ be the corresponding L-groups. Given an L-homomorphism $r : {}^LG_1 \to {}^LG_2$, Langlands principle of functoriality predicts the existence of a transfer $\Pi \mapsto r(\Pi)$ of the L-packet of an automorphic representation Π of $G_1(\mathbb{A}_F)$ to an L-packet $r(\Pi)$ of automorphic representations of $G_2(\mathbb{A}_F)$. Now assume that G_2 is a general linear group. We note that an L-packet for a general linear group is a singleton set. For applications of functoriality one needs to understand the image and fibers of the correspondence $\Pi \mapsto r(\Pi)$. In particular, it is necessary to understand what conditions on Π ensure that the transfer $r(\Pi)$ is cuspidal.

The main aim of this paper is to describe a cuspidality criterion for the transfer of automorphic representations from $\mathrm{GL}(4, \mathbb{A}_F)$ to $\mathrm{GL}(6, \mathbb{A}_F)$ corresponding to the exterior square map $\wedge^2 : \mathrm{GL}(4, \mathbb{C}) \to \mathrm{GL}(6, \mathbb{C})$. Langlands functoriality in this case is a deep theorem due to H. Kim [**13**].

Let $\Pi = \otimes_v \Pi_v$ and $\Sigma = \otimes_v \Sigma_v$ be irreducible isobaric automorphic representations of $\mathrm{GL}(4, \mathbb{A}_F)$ and $\mathrm{GL}(6, \mathbb{A}_F)$, respectively. Assume that S is a finite set of places of F, including all the archimedean ones, outside of which both of the representations are unramified. We say Σ is an exterior square transfer of Π if for all $v \notin S$ we have $\Sigma_v = \wedge^2(\Pi_v)$, i.e., the semi-simple conjugacy class in $\mathrm{GL}(6, \mathbb{C})$ determining Σ_v is generated by the image under $\wedge^2$ of the semi-simple conjugacy class in $\mathrm{GL}(4, \mathbb{C})$ determining Π_v. By the strong multiplicity one theorem (see Theorem 2.1 below) such a Σ would be unique. We will denote it by $\wedge^2\Pi$. The existence of $\wedge^2\Pi$

2010 *Mathematics Subject Classification.* Primary: 11F70. Secondary: 11F55, 11F80, 22E55.

was established by H. Kim [**13**, Theorem A]. Furthermore, he showed that if Π_v is not supercuspidal for the places v dividing 2 or 3, then the local component Π_v and $(\wedge^2\Pi)_v$ are compatible via the local Langlands correspondence for $\mathrm{GL}(4, F_v)$ and $\mathrm{GL}(6, F_v)$. The assumption at $v|2, 3$ was made because of complications posed by supercuspidal representations, especially of $\mathrm{GL}(4, F_v)$. In any event, it has no bearing on our result as we do not need the fact that the local components of $\wedge^2\Pi$ and Π are compatible via the local Langlands correspondence at all places. We now state the main theorem of this article.

THEOREM 1.1. *Let F be a number field and let Π be a cuspidal automorphic representation of $\mathrm{GL}(4, \mathbb{A}_F)$. The following are equivalent:*

(i) $\wedge^2\Pi$ *is not cuspidal.*

(ii) Π *is one of the following:*

- (a) $\pi_1 \boxtimes \pi_2$, *the transfer from $\mathrm{GL}(2, \mathbb{A}_F) \times \mathrm{GL}(2, \mathbb{A}_F)$ to $\mathrm{GL}(4, \mathbb{A}_F)$ via the automorphic tensor product $\boxtimes$. (This may also be viewed as the transfer from split* GSpin(4) *to* GL(4).*)*
- (b) As(π), *the Asai transfer of a dihedral cuspidal automorphic representation π of $\mathrm{GL}(2, \mathbb{A}_E)$ where E/F is a quadratic extension. (This may also be viewed as the transfer to* GL(4) *from the quasi-split non-split* $\mathrm{GSpin}^*(4)$ *over F which splits over E.)*
- (c) *The functorial transfer of a cuspidal representation π of $\mathrm{GSp}(4, \mathbb{A}_F)$ associated with the natural embedding of the dual group $\mathrm{GSp}(4, \mathbb{C})$ into $\mathrm{GL}(4, \mathbb{C})$. The representation π may be taken to be globally generic.*
- (d) $\mathrm{I}_E^F(\pi)$, *the automorphic induction of a cuspidal automorphic representation π of $\mathrm{GL}(2, \mathbb{A}_E)$, where E/F is a quadratic extension.*

(iii) Π *satisfies one of the following:*

- (α) $\Pi \cong \Pi^\vee \otimes \chi$ *for some Hecke character χ of F, and Π is not the Asai transfer of a nondihedral cuspidal representation.*
- (β) $\Pi \cong \Pi \otimes \chi$ *for a nontrivial Hecke character χ of F.*

We observe that the groups in **(ii)**(a)–**(ii)**(d) are some of the twisted endoscopic groups for $\mathrm{GL}(4) \times \mathrm{GL}(1)$ which have the property that the image under $\wedge^2$ of the connected component of their dual groups are contained in proper Levi subgroups of $\mathrm{GL}(6, \mathbb{C})$. Recall from [**2**, §3.7] that the exterior square transfer we consider here is a special case of the more general transfer from $\mathrm{GSpin}(2n)$ to $\mathrm{GL}(2n)$. One would expect that the above theorem admits a generalization to that setting through the theory of twisted endoscopy.

We now briefly sketch the proof of the theorem. It is easy to verify that **(ii)** implies both **(i)** and **(iii)**. In Section 3 we explicitly write down the isobaric decomposition of $\wedge^2\Pi$ for each of the cases **(ii)**(a)–**(ii)**(d). In order to check that two isobaric representations are isomorphic we repeatedly use a strong multiplicity one theorem, due to Jacquet and Shalika, recalled in Section 2.1.

The proof of **(i)** $\Longrightarrow$ **(iii)**, described in Section 4, uses some details from the Langlands–Shahidi machinery. We show in Proposition 4.1 that if Π is not essentially self-dual, then $\wedge^2\Pi$ cannot have a degree 1 or degree 3 isobaric summand. In Proposition 4.2 we verify that if Π does not admit a nontrivial self-twist, then $\wedge^2\Pi$ cannot have a degree 2 isobaric summand. For a cuspidal representation σ of $\mathrm{GL}(m, \mathbb{A}_F)$ consider the Langlands L-function $L(s, \Pi \times \sigma, \wedge^2 \otimes \rho_m)$, where ρ_m is the standard representation of $GL(m, \mathbb{C})$. These L-functions appear in the Langlands–Shahidi machinery for a particular choice of Levi subgroup when the ambient group is $GSpin(2m+6)$. We use Kim [**13**, §3] to show that, under the above mentioned hypothesis on Π, these (partial) L-functions are holomorphic at $s = 1$. The implication then follows from a well-known result of Jacquet, Piatetski-Shapiro, and Shalika recalled in Section 2.1. We summarize some of the preliminaries we need from the Langlands–Shahidi machinery in Section 2.3.

Finally, the proof of **(iii)** $\Longrightarrow$ **(ii)** is given in Section 5. It relies on the so-called 'descent theory' for classical groups. The version of descent theory for GSpin groups that we need has now been announced by J. Hundley and E. Sayag [**9**]. As a general reference for descent theory we refer to Soudry's exposition [**31**].

In Sections 6.1 and 6.2 we present a few examples, some only conjectural, illustrating the above theorem. In Section 6.3 we comment on possible intersections among the cases in **(ii)**. In Section 6.4 we ask whether it is possible to see the cuspidality criterion from the 'Galois side'. The question can be made precise based on the philosophy that there is a correspondence between automorphic representations π of $GL(n, \mathbb{A}_F)$ and ℓ-adic n-dimensional representations σ of the absolute Galois group of F, or n-dimensional complex representations of the conjectural Langlands group $\mathcal{L}_F$. Let us denote this correspondence by $\pi \mapsto \sigma(\pi)$. Part of this philosophy is that π is supposed to be cuspidal if and only if $\sigma(\pi)$ is irreducible. We refer to Ramakrishnan [**27**] for the state of the art on this issue. In view of the above theorem one can ask the following question. Let σ be a four-dimensional irreducible Galois representation; what condition on σ will ensure that $\wedge^2\sigma$ is irreducible? Upon posing this question in a talk at the Oklahoma Joint Automorphic Forms Seminar, A. Kable came up with a very elegant theorem which reflects the equivalence of **(i)** and **(iii)** in Theorem 1.1. We are grateful to him for allowing us to include his theorem and its proof in Section 6.4. Recall that in **(iii)**(α) of Theorem 1.1 above, we had to exclude the Asai transfer of a nondihedral cuspidal representation if Π is essentially self-dual. On the Galois side, this is reflected in the fact that if a four-dimensional irreducible representation σ is essentially self-dual of orthogonal type, then for $\wedge^2\sigma$ to be reducible the image of σ should lie in the connected component of the identity in the algebraic group $GO(4)$; see Theorem 6.5.

Cuspidality criteria are important not only for their intrinsic value in helping us better understand a given instance of functoriality but also because they have important arithmetic applications. D. Ramakrishnan and S. Wang [**29**] proved a cuspidality criterion for the transfer from $GL(2) \times GL(3)$ to $GL(6)$ and used it to construct new cuspidal cohomology classes for $GL(6)$. We refer to [**22**] for a brief survey of cohomological applications of Langlands functoriality. H. Kim and F. Shahidi [**18**] proved a cuspidality criterion for the symmetric fourth transfer from $GL(2)$ to $GL(5)$, which has been used in the study of special values of symmetric power L-functions by the second author and F. Shahidi [**23**]. Such a potential arithmetic application was indeed our original motivation to seek a cuspidality criterion for the exterior square transfer.

We thank H. Kim, A. Kable, J.-P. Labesse, F. Shahidi, and R. Zierau for helpful discussions and comments. We also thank the referee for helpful comments. The first author is partially supported by an Alexander von Humboldt Research Fellowship and NSA Young Investigator grant H98230-09-1-0049. The second author is partially supported by an ASR+1 grant from the College of Arts and Science of Oklahoma State University and NSF grant DMS-0856113.

2. Some preliminaries

In this section we collect some results that we repeatedly use in later sections. To begin, we recall a theorem due to Jacquet and Shalika concerning strong multiplicity one for isobaric automorphic representations. Then we recall an analytic criterion in terms of Rankin–Selberg L-functions, due to Jacquet, Piatetski-Shapiro and Shalika, that characterizes when two cuspidal automorphic representations are equivalent. Next, we note that the natural transfer of automorphic representations of a quasi-split non-split general spin group $\mathrm{GSpin}^*(4)$ to $\mathrm{GL}(4)$ is in fact the Asai transfer. Finally, we recall some details from the Langlands–Shahidi machinery that will be of use to us, particularly when the ambient group is $\mathrm{GSpin}(m)$ with $m = 8, 10$ or 12.

2.1. Some results of Jacquet, Piatetski-Shapiro, and Shalika. The following strong multiplicity one theorem for isobaric representations is due to Jacquet and Shalika [**11, 12**].

THEOREM 2.1. *Let π_1 and π_2 be two isobaric automorphic representations of $\mathrm{GL}(n, \mathbb{A}_F)$. Let S be a finite set of places of F, containing the archimedean places, such that both π_1 and π_2 are unramified outside S. If $\pi_{1,v} \cong \pi_{2,v}$ for all $v \notin S$, then $\pi_1 \cong \pi_2$.*

Another useful technical tool for us is the following theorem, due to Jacquet, Piatetski-Shapiro and Shalika [**10**], concerning Rankin–Selberg L-functions.

THEOREM 2.2. *Let π_1 (resp., π_2) be a cuspidal automorphic representation of $\mathrm{GL}(n_1, \mathbb{A}_F)$ (resp., $\mathrm{GL}(n_2, \mathbb{A}_F)$). Let S be a finite set of places containing the archimedean places of F and the ramified places of π_1 and π_2. The partial Rankin–Selberg L-function $L^S(s, \pi_1 \times \pi_2)$ is holomorphic at $s = 1$ unless $n_1 = n_2$ and $\pi_2 \cong \pi_1^\vee$, in which case it has a simple pole at $s = 1$.*

2.2. The Asai transfer and the quasi-split non-split $\mathrm{GSpin}^*(4)$**.** Let E/F be a quadratic extension of number fields and let $\Gamma = \Gamma_F$ denote the absolute Galois group of F. In this section we let G denote the group $\mathrm{GSpin}^*(4)$, a quasi-split non-split linear algebraic group over F, which is isomorphic to the split $\mathrm{GSpin}(4)$ over E. The L-group of G can be written as ${}^LG = \mathrm{GSO}(4, \mathbb{C}) \rtimes \Gamma$, where the Galois action, which factors through $\mathrm{Gal}(E/F)$, is described below. We note that $\mathrm{GSO}(4, \mathbb{C})$ denotes the special orthogonal similitude group; one can identify it as a quotient of $\mathrm{GL}(2, \mathbb{C}) \times \mathrm{GL}(2, \mathbb{C})$ given by

$$\mathrm{GSO}(4, \mathbb{C}) = \beta(\mathrm{GL}(2, \mathbb{C}) \times \mathrm{GL}(2, \mathbb{C})),$$

where β is the map on the right of the exact sequence

$$1 \longrightarrow \mathbb{C}^* \longrightarrow \mathrm{GL}(2, \mathbb{C}) \times \mathrm{GL}(2, \mathbb{C}) \stackrel{\beta}{\longrightarrow} \mathrm{GO}(4, \mathbb{C}).$$

For details see [**26**, §2]. Furthermore, the Γ-action on $\mathrm{GSO}(4, \mathbb{C})$ is as follows. If $\gamma \in \Gamma$ and $g = \beta(g_1, g_2)$ with $g_i \in \mathrm{GL}(2, \mathbb{C})$, then

$$\gamma \cdot g = \begin{cases} \beta(g_1, g_2) & \text{if } \gamma|_E = 1, \\ \beta(g_2, g_1) & \text{if } \gamma|_E \neq 1. \end{cases}$$

We also need to recall the Asai transfer. Consider the group $H = \mathrm{Res}_{E/F}\mathrm{GL}(2)$ as a group over F. Its L-group is given by

$$^{L}H = (\mathrm{GL}(2,\mathbb{C}) \times \mathrm{GL}(2,\mathbb{C})) \rtimes \Gamma,$$

with the Galois action given by

$$\gamma \cdot (g_1, g_2) = \begin{cases} (g_1, g_2) & \text{if } \gamma|_E = 1, \\ (g_2, g_1) & \text{if } \gamma|_E \neq 1. \end{cases} \tag{2.3}$$

Let W be a 2-dimensional $\mathbb{C}$-vector space, and let $V = W \otimes W$. After fixing a basis for W, we identify $\mathrm{GL}(W)$ with $\mathrm{GL}(2,\mathbb{C})$. Consider the map

$$\mathrm{As} : (\mathrm{GL}(W) \times \mathrm{GL}(W)) \rtimes \Gamma \longrightarrow \mathrm{GL}(V) \cong \mathrm{GL}(4,\mathbb{C}) \tag{2.4}$$

given, on pure tensors, by

$$\mathrm{As}(g_1, g_2; \gamma)(\xi_1 \otimes \xi_2) = \begin{cases} g_1\xi_1 \otimes g_2\xi_2 & \text{if } \gamma|_E = 1, \\ g_1\xi_2 \otimes g_2\xi_1 & \text{if } \gamma|_E \neq 1, \end{cases}$$

for all $\xi_i \in W$ and all $g_i \in \mathrm{GL}(W)$. It is straightforward to check that this map is indeed a homomorphism. It is called the Asai (or 'twisted tensor') homomorphism. (Alternatively, one could take the map satisfying $\mathrm{As}(g_1, g_2; \gamma)(\xi_1 \otimes \xi_2) = -g_1\xi_2 \otimes g_2\xi_1$ when $\gamma|_E \neq 1$. This choice would lead to a quadratic twist of the above map.)

Further, let $\iota : {}^{L}G \longrightarrow \mathrm{GL}(V) \cong \mathrm{GL}(4,\mathbb{C})$ be the map defined via

$$\iota(\beta(g_1, g_2); \gamma)(\xi_1 \otimes \xi_2) = \begin{cases} g_1\xi_1 \otimes g_2\xi_2 & \text{if } \gamma|_E = 1, \\ g_1\xi_2 \otimes g_2\xi_1 & \text{if } \gamma|_E \neq 1. \end{cases}$$

Again, it is straightforward to check that this map is an L-homomorphism. It is now clear that $\iota \circ (\beta, \mathrm{id}) = \mathrm{As}$. In other words, the following diagram commutes:

$$\begin{array}{ccc} \mathrm{GL}(2,\mathbb{C}) \times \mathrm{GL}(2,\mathbb{C}) \rtimes \Gamma & \xrightarrow{\quad (\beta,\mathrm{id}) \quad} & \mathrm{GSO}(4,\mathbb{C}) \rtimes \Gamma \\ & {\scriptstyle \mathrm{As}} \searrow \qquad \swarrow {\scriptstyle \iota} & \\ & \mathrm{GL}(4,\mathbb{C}) & \end{array}$$

Assume that π is a cuspidal automorphic representation of $\mathrm{GL}(2,\mathbb{A}_E)$ and let $\Pi = \mathrm{As}(\pi)$ be its Asai transfer to $\mathrm{GL}(4,\mathbb{A}_F)$. (See Krishnamurthy [**19**] or Ramakrishnan [**26**].) Then $\Pi = \iota(\beta(\pi))$, where $\beta(\pi)$ denotes the transfer of π to the group $\mathrm{GSpin}^*(4,\mathbb{A}_F)$, and $\iota(\beta(\pi))$ denotes the transfer of $\beta(\pi)$ from $\mathrm{GSpin}^*(4,\mathbb{A}_F)$ to $\mathrm{GL}(4,\mathbb{A}_F)$. The transfer corresponding to β exists for formal reasons and the existence of the transfer corresponding to ι (for generic representations) is part of a joint work of the first author with F. Shahidi [**2, 3**].

2.3. The Langlands–Shahidi L-functions. Let $P = MN$ be a maximal proper parabolic subgroup of a connected reductive quasi-split linear algebraic group G, where M denotes a Levi subgroup and N denotes the unipotent radical of P. Let σ be a *generic* automorphic representation of $M(\mathbb{A}_F)$. Let r denote the adjoint action of the complex Langlands dual group $\widehat{M}$ on the Lie algebra of the dual of N. Write $r = r_1 \oplus \cdots \oplus r_m$, where the r_i's denote the irreducible constituents of r and the ordering is according to the eigenvalue of the adjoint action as in, for example, [**30**, p.278]. The Langlands–Shahidi method then constructs the L-functions $L(s, \sigma, r_i)$ for $1 \leq i \leq m$.

We need the following cases of the Langlands–Shahidi method. Let $G = \mathrm{GSpin}(2n+6)$ with $n = 1, 2, 3$, and consider a maximal parabolic subgroup of G with Levi subgroup $M = \mathrm{GL}(n) \times \mathrm{GSpin}(6)$. (One could also work with split spin groups as in [**13**, §3]; however, we find it more convenient to work with the similitude version of the groups.) The algebraic group GSpin(6) is isomorphic to a quotient of $\mathrm{GL}(1) \times \mathrm{Spin}(6)$ by a central subgroup $A = \{1, (-1, c)\}$, where c is the nontrivial element in the center of Spin(6) of order 2; see [**2**, Proposition 2.2]. The algebraic group GL(4) is isomorphic to a quotient of $\mathrm{GL}(1) \times \mathrm{SL}(4)$ by a cyclic central subgroup B of order 4. We identify Spin(6) with SL(4) such that B contains A. This way we get a natural map, defined over F, from GSpin(6) to GL(4), which in turn induces a map

$$f : M \longrightarrow \mathrm{GL}(n) \times \mathrm{GL}(4). \tag{2.5}$$

Let Π be an irreducible cuspidal representation of $\mathrm{GL}(4, \mathbb{A}_F)$ and let σ be an irreducible cuspidal automorphic representation of $\mathrm{GL}(n, \mathbb{A}_F)$, $n = 1, 2, 3$. Choose any irreducible constituent Σ of $\sigma \otimes \Pi|_{f(M(\mathbb{A}_F))}$ and let Σ also denote the corresponding representation of $M(\mathbb{A}_F)$. The Langlands–Shahidi method then gives

$$L(s, \Sigma, r_1) = L(s, \sigma \otimes \Pi, \rho_n \otimes \wedge^2 \rho_4), \tag{2.6}$$

where ρ_k denotes the standard representation of $\mathrm{GL}(k, \mathbb{C})$ and the L-function on the right-hand side is a Langlands L-function; see [**13**, §3]. We record a general fact that we need from the Langlands–Shahidi method.

PROPOSITION 2.7. *Let w_G and w_M denote the longest elements of the Weyl group of G and M, respectively. Let $w_0 = w_G w_M$. If $w_0(\Sigma) \not\cong \Sigma$, then $L(s, \Sigma, r_1)$ is entire.*

PROOF. This is a standard fact in the Langlands–Shahidi method. For example, see the proof of [**13**, Proposition 3.4]. □

In order to apply the above proposition one needs to know the action of w_0 on a representation of $M(\mathbb{A}_F)$.

PROPOSITION 2.8. *Let $G = \mathrm{GSpin}(2n+6)$ with n a positive integer and let $\sigma \otimes \Pi$ be a representation of $M(\mathbb{A}_F)$ as above. Moreover, let w_0 be as above and denote its image under the map (2.5) by w_0 again. Then we have*

$$w_0(\sigma \otimes \Pi) = \begin{cases} \sigma^\vee \otimes (\Pi^\vee \otimes \omega_\sigma) & \text{if } n \text{ is odd,} \\ \sigma^\vee \otimes (\Pi \otimes \omega_\sigma) & \text{if } n \text{ is even.} \end{cases}$$

Here, ω_σ denotes the central character of σ.

PROOF. Recall that the nontrivial automorphism of the Dynkin diagram of type A_m corresponds to an outer automorphism of $\mathrm{GL}(m+1)$ and it conjugates an irreducible representation to its dual representation. The proof of the proposition will follow from a description of how w_0 acts on the root system of type D_r.

We use the Bourbaki notation for the simple roots:

$$\alpha_1 = \epsilon_1 - \epsilon_2, \ldots, \alpha_{r-2} = \epsilon_{r-2} - \epsilon_{r-1}, \alpha_{r-1} = \epsilon_{r-1} - \epsilon_r, \alpha_r = \epsilon_{r-1} + \epsilon_r.$$

The Weyl group of G is isomorphic to $\{\pm 1\}^{r-1} \rtimes S_r$ which we identify with a subgroup of signed $r \times r$ permutation matrices acting on $\mathbb{R}^r = \mathbb{R}\epsilon_1 \oplus \cdots \oplus \mathbb{R}\epsilon_r$.

With this identification observe that

$$w_G = \begin{cases} -I_r & \text{if } r \text{ is even,} \\ \begin{pmatrix} -I_{r-1} & \\ & 1 \end{pmatrix} & \text{if } r \text{ is odd.} \end{cases}$$

Let $r = n + 3$ and let M be a maximal Levi subgroup of type $A_{n-1} \times A_3$ in $G = \mathrm{GSpin}(2r)$. The simple roots in the A_3-factor are $\alpha_{r-1}, \alpha_{r-2}, \alpha_r$. The proposition follows by observing that

$$\begin{aligned} w_0(\alpha_{r-1}) &= w_G w_M(\alpha_{r-1}) = w_G(-\alpha_r) \\ &= w_G(-\epsilon_{r-1} - \epsilon_r) = \begin{cases} \epsilon_{r-1} + \epsilon_r = \alpha_r & \text{if } r \text{ is even,} \\ \epsilon_{r-1} - \epsilon_r = \alpha_{r-1} & \text{if } r \text{ is odd,} \end{cases} \end{aligned}$$

and

$$\begin{aligned} w_0(\alpha_r) &= w_G w_M(\alpha_r) = w_G(-\alpha_{r-1}) \\ &= w_G(-\epsilon_{r-1} + \epsilon_r) = \begin{cases} \epsilon_{r-1} - \epsilon_r = \alpha_{r-1} & \text{if } r \text{ is even,} \\ \epsilon_{r-1} + \epsilon_r = \alpha_r & \text{if } r \text{ is odd,} \end{cases} \end{aligned}$$

while

$$w_0(\alpha_{r-2}) = w_G w_M(\alpha_{r-2}) = w_G(-\alpha_{r-2}) = w_G(-\epsilon_{r-2} - \epsilon_{r-1}) = \alpha_{r-2}$$

in either case. Moreover, for $1 \le j \le n-1$ we have

$$w_0(\alpha_j) = w_G w_M(\alpha_j) = w_G(-\alpha_{n-j}) = w_G(-\epsilon_{n-j} + \epsilon_{n-j+1}) = \alpha_{n-j}.$$

This means that w_0 induces the nontrivial automorphism of the Dynkin diagram of the A_{n-1}-factor of M, and on the A_3-factor it induces the nontrivial automorphism of the Dynkin diagram if and only if r is even.

Let $m = m(g, h)$ be an arbitrary element in the Levi subgroup M identified with $\mathrm{GL}(n) \times \mathrm{GSpin}(6)$, in $G = \mathrm{GSpin}(2n+6)$, where $g \in \mathrm{GL}(n)$ and $h \in \mathrm{GSpin}(6)$, and let $\nu = \nu(m)$ denote its similitude character value. Then

$$w_0 m(g, h) w_0^{-1} = m({}^t g^{-1} \nu(m), h^*),$$

where

$$h^* = \begin{cases} {}^t h^{-1} & \text{if } r \text{ is even,} \\ h & \text{if } r \text{ is odd.} \end{cases}$$

We conclude that

$$\begin{aligned} w_0(\sigma \otimes \Pi)(m(g,h)) &= (\sigma \otimes \Pi)(m({}^t g^{-1} \nu(m), h^*)) \\ &= \sigma({}^t g^{-1}) \Pi(h^*) \omega_\sigma(\nu(m)) \\ &= \sigma^\vee(g) \Pi^*(h) \omega_\sigma(\nu(h)) \\ &= (\sigma^\vee \otimes (\Pi^* \otimes \omega_\sigma))\,(m(g,h)), \end{aligned}$$

where

$$\Pi^* = \begin{cases} \Pi^\vee & \text{if } r \text{ is even,} \\ \Pi & \text{if } r \text{ is odd.} \end{cases}$$

Note that $r = n + 3$ is even if and only if n is odd. This completes the proof. $\square$

3. The proof of (ii)⇒(i)

We verify that for each of **(ii)(a)** through **(ii)(d)** the exterior square transfer $\wedge^2\Pi$ is not cuspidal. Indeed, it is not difficult to write down the isobaric decomposition for $\wedge^2\Pi$ in each case.

3.1. (ii)(a)⇒(i).

PROPOSITION 3.1. *Let π_1 and π_2 be two cuspidal automorphic representations of $\mathrm{GL}(2,\mathbb{A}_F)$. Let $\pi_1 \boxtimes \pi_2$ be the transfer to an automorphic representation of $\mathrm{GL}(4,\mathbb{A}_F)$, whose existence was established in* [**24**]. *For brevity, we let $\Pi = \pi_1 \boxtimes \pi_2$ and $\omega = \omega_{\pi_1}\omega_{\pi_2}$. We have*

(a) $\wedge^2(\pi_1 \boxtimes \pi_2) = \left(\mathrm{Sym}^2(\pi_1) \otimes \omega_{\pi_2}\right) \boxplus \left(\mathrm{Sym}^2(\pi_2) \otimes \omega_{\pi_1}\right).$

(b) *Assuming Langlands functoriality one should expect*

$$\mathrm{Sym}^2(\pi_1 \boxtimes \pi_2) = \left(\mathrm{Sym}^2(\pi_1) \boxtimes \mathrm{Sym}^2(\pi_2)\right) \boxplus \omega_{\pi_1}\omega_{\pi_2}.$$

(c) *The partial L-function $L^S(s, \wedge^2(\Pi) \otimes \omega^{-1})$ is entire while the partial L-function $L^S(s, \Pi, \mathrm{Sym}^2 \otimes \omega^{-1})$ has a pole at $s = 1$.*

PROOF. The proof of (a) and (b), using Theorem 2.1, is an easy calculation using Satake parameters on both sides. More precisely, for a finite place v at which both π_1 and π_2 are unramified we let $\pi_{1,v}$ and $\pi_{2,v}$ have Frobenius-Hecke eigenvalues $t_1 = \mathrm{diag}(a_1, b_1)$ and $t_2 = \mathrm{diag}(a_2, b_2)$, respectively. Then

$$\wedge^2(t_1 \otimes t_2) = \left(\mathrm{diag}(a_1^2, a_1b_1, b_1^2) \cdot a_2b_2\right) \boxplus \left(\mathrm{diag}(a_2^2, a_2b_2, b_2^2) \cdot a_1b_1\right)$$

and

$$\mathrm{Sym}^2(t_1 \otimes t_2) = \left(\mathrm{diag}(a_1^2, a_1b_1, b_1^2) \otimes \mathrm{diag}(a_2^2, a_2b_2, b_2^2)\right) \boxplus (a_1b_1 \cdot a_2b_2).$$

Part (a) has also been observed by others; see [**19**, (7.27)] and [**28**, (2.6)], for example. For (b) to make sense one has to assume the symmetric square transfer from GL(4) to GL(10) and the automorphic tensor product from GL(3) × GL(3) to GL(9), both particular instances of functoriality.

To prove (c), observe that $\Pi^\vee \cong \Pi \otimes \omega^{-1}$, which implies

$$L^S(s, \Pi \times \Pi^\vee) = L^S(s, \wedge^2(\Pi) \otimes \omega^{-1}) L^S(s, \Pi, \mathrm{Sym}^2 \otimes \omega^{-1}),$$

where S is a finite set of places including all the archimedean ones such that Π is unramified outside of S. From (a) we have $\wedge^2(\Pi) \otimes \omega^{-1} = \mathrm{Ad}(\pi_1) \boxplus \mathrm{Ad}(\pi_2)$. (Here $\mathrm{Ad}(\pi_i) = \mathrm{Sym}^2(\pi_i) \otimes \omega_{\pi_i}^{-1}$.) If π_i is not dihedral, then $\mathrm{Ad}(\pi_i)$ is cuspidal (by Gelbart-Jacquet [**7**]) and hence its partial L-function is entire. If π_i is dihedral, say $\pi_i = I_E^F(\chi)$, then it is easy to see that $\mathrm{Ad}(\pi_i) = \omega_{E/F} \boxplus I_E^F(\chi'\chi^{-1})$, where $\omega_{E/F}$ is the quadratic character of F associated to E by class field theory, and χ' is the nontrivial $\mathrm{Gal}(E/F)$-conjugate of χ. Since π_i is cuspidal, the inducing character χ is Galois regular, i.e., $\chi' \neq \chi$, or equivalently $\chi'\chi^{-1}$ is a nontrivial character, whence $L^S(s, \mathrm{Ad}(\pi_i)) = L^S(s, \omega_{E/F}) L^S(s, I_E^F(\chi'\chi^{-1}))$ is entire. (In particular, it does not have a pole at $s = 1$.) Therefore $L^S(s, \wedge^2(\Pi) \otimes \omega^{-1}) = L^S(s, \mathrm{Ad}(\pi_1)) L^S(s, \mathrm{Ad}(\pi_2))$ does not have a pole at $s = 1$. However, $L^S(s, \Pi \times \Pi^\vee)$ has a pole at $s = 1$, which implies that $L^S(s, \Pi, \mathrm{Sym}^2 \otimes \omega^{-1})$ has a pole at $s = 1$.

Note that (c), unlike (b), is unconditional and does not depend on assuming unproven instances of functoriality. □

3.2. (ii)(b)⇒(i).

PROPOSITION 3.2. *Let E/F be a quadratic extension. Let π be a cuspidal automorphic representation of $\mathrm{GL}(2, \mathbb{A}_E)$ and let $\Pi = \mathrm{As}(\pi)$ be its Asai transfer. Assume that Π is a cuspidal automorphic representation of $\mathrm{GL}(4, \mathbb{A}_F)$. Then $\wedge^2\Pi$ is cuspidal if and only if π is not dihedral.*

PROOF. The proof depends on the following identity:

$$\wedge^2(\mathrm{As}(\pi)) = \mathrm{I}_F^E(\mathrm{Sym}^2\pi \otimes \omega_\pi'),$$

where $'$ means the nontrivial $\mathrm{Gal}(E/F)$-conjugate. (See [**19**, §7].) To begin, assume that π is not dihedral. By [**26**, Theorem 1.4] we know that $\mathrm{As}(\pi)$ is cuspidal if and only if $\pi' \not\cong \pi \otimes \mu$ for any μ. If $\wedge^2\Pi$ is not cuspidal, then

$$\left(\mathrm{Sym}^2\pi \otimes \omega_\pi'\right)' \cong \mathrm{Sym}^2\pi \otimes \omega_\pi'.$$

This implies that $\mathrm{Sym}^2\pi' \otimes \omega_\pi \cong \mathrm{Sym}^2\pi \otimes \omega_\pi'$, i.e., $\mathrm{Ad}(\pi) \cong \mathrm{Ad}(\pi')$. This, in turn, implies that $\pi' \cong \pi \otimes \mu$ (by [**24**, Theorem 4.1.2]), contradicting the fact that there is no such twist. Hence $\wedge^2\Pi$ is cuspidal.

Next, assume that π is dihedral. In this case $\mathrm{Sym}^2(\pi)$ is not cuspidal, and therefore, $\mathrm{I}_F^E(\mathrm{Sym}^2\pi \otimes \omega_\pi')$ cannot possibly be cuspidal. □

3.3. (ii)(c)⇒(i).

PROPOSITION 3.3. *Let Π be a cuspidal automorphic representation of $\mathrm{GL}(4, \mathbb{A}_F)$ and assume that Π is a transfer from a cuspidal (generic) automorphic representation π of $\mathrm{GSp}(4, \mathbb{A}_F)$. Then*

$$\wedge^2\Pi = \tilde{r}_5(\pi) \boxplus \omega_\pi,$$

where $\tilde{r}_5$ is a degree 5 representation of $\mathrm{GSp}(4, \mathbb{C})$ defined below. In particular, $\wedge^2\Pi$ is not cuspidal.

PROOF. We use Kim [**14**, p. 2793]. As observed there, one has

$$\mathrm{Sp}(4, \mathbb{C}) \overset{\iota}{\hookrightarrow} \mathrm{GL}(4, \mathbb{C}) \overset{\wedge^2}{\longrightarrow} \mathrm{GL}(6, \mathbb{C})$$

and $\wedge^2 \circ \iota = r_5 \oplus \mathbb{1}$ decomposes into a direct sum of the trivial representation and a five-dimensional representation r_5. Similarly,

$$\mathrm{GSp}(4, \mathbb{C}) \overset{\tilde{\iota}}{\hookrightarrow} \mathrm{GL}(4, \mathbb{C}) \overset{\wedge^2}{\longrightarrow} \mathrm{GL}(6, \mathbb{C})$$

and $\wedge^2 \circ \tilde{\iota} = \tilde{r}_5 \oplus \nu$, where ν is the similitude character of $\mathrm{GSp}(4, \mathbb{C})$ and $\tilde{r}_5$ is a five-dimensional representation of $\mathrm{GSp}(4, \mathbb{C})$. This implies the desired equality of automorphic representations. □

REMARK 3.4. Embedded in the above proof is the assertion that a cuspidal (generic) automorphic representation π of $\mathrm{GSp}(4, \mathbb{A}_F)$ admits a transfer to an automorphic representation $\tilde{r}_5(\pi)$ of $\mathrm{GL}_5(\mathbb{A}_F)$ corresponding to the representation $\tilde{r}_5$. This depends on the generic transfer from GSp(4) to GL(4) (see Asgari-Shahidi [**4**]), and the exterior square transfer from GL(4) to GL(6) due to Kim [**13**].

3.4. (ii)(d)⇒(i).

PROPOSITION 3.5. *If π is a cuspidal automorphic representation of* $\mathrm{GL}(2,\mathbb{A}_E)$, *where E/F is a quadratic extension, and $\Pi = \mathrm{I}_F^E(\pi)$ is the automorphic induction of π to an automorphic representation of* $\mathrm{GL}(4,\mathbb{A}_F)$, *then $\wedge^2\Pi$ is not cuspidal.*

PROOF. It is known that

$$\wedge^2(\mathrm{I}_F^E(\pi)) = \mathrm{As}(\pi)\otimes\omega_{E/F} \boxplus \mathrm{I}_E^F(\omega_\pi),$$

where $\omega_{E/F}$ is the quadratic Hecke character of F associated to E/F by class field theory. See, for example, Kim [**16**, §3]. □

4. The proof of (i)⇒(iii)

It is equivalent to prove that, if Π is a cuspidal automorphic representation of $\mathrm{GL}(4,\mathbb{A}_F)$ which neither has a nontrivial self-twist nor is essentially self-dual, then $\wedge^2(\Pi)$ is cuspidal. Observe that if an isobaric automorphic representation ρ of $\mathrm{GL}(6,\mathbb{A}_F)$ is not cuspidal, then it must have an isobaric summand of degree 1, 2, or 3, i.e., there exists a cuspidal representation σ of $\mathrm{GL}(n,\mathbb{A}_F)$, with $1\le n\le 3$, such that $L^S(s,\rho\times\sigma)$ has a pole at $s=1$. Again, S denotes a finite set of places of F, including all the archimedean ones, such that all the representations involved are unramified at places outside S. Now with Π as above, the cuspidality of $\wedge^2\Pi$ follows from the following two propositions.

PROPOSITION 4.1. *If $\Pi \not\cong \Pi^\vee\otimes\chi$ for all χ, then $L^S(s,\Pi\otimes\sigma,\wedge^2\otimes\rho_2)$ is holomorphic at $s=1$ for every cuspidal representation σ of* $\mathrm{GL}(n,\mathbb{A}_F)$, $n=1,3$.

PROOF. Let Σ be as in (2.5). By Proposition 2.7, it is enough to show that $w_0(\Sigma)\not\cong\Sigma$. If we have $w_0(\Sigma)\cong\Sigma$, then $w_0(\sigma\otimes\Pi)\cong\sigma\otimes\Pi$. On the other hand, by Proposition 2.8 we have $w_0(\sigma\otimes\Pi)\cong\sigma^\vee\otimes(\Pi^\vee\otimes\omega_\sigma)$. In particular, we must have $\Pi\cong\Pi^\vee\otimes\omega_\sigma$ contradicting the hypothesis. □

PROPOSITION 4.2. *If $\Pi\not\cong\Pi\otimes\chi$ for all nontrivial χ, then $L^S(s,\Pi\otimes\sigma,\wedge^2\otimes\rho_2)$ is holomorphic at $s=1$ for every cuspidal representation σ of* $\mathrm{GL}(2,\mathbb{A}_F)$.

PROOF. The same argument as in the above proof works as long as $\omega_\sigma\neq\mathbb{1}$ because by Proposition 2.8 we have $w_0(\sigma\otimes\Pi)\cong\sigma^\vee\otimes(\Pi\otimes\omega_\sigma)\not\cong\sigma\otimes\Pi$. This means that if σ is a cuspidal representation of $\mathrm{GL}(2,\mathbb{A}_F)$ with nontrivial central character, then σ cannot occur as an isobaric summand of $\wedge^2(\Pi)$.

Now suppose that σ is a cuspidal representation of $\mathrm{GL}(2,\mathbb{A}_F)$ with trivial central character and that σ is an isobaric summand of $\wedge^2(\Pi)$. Then the representation $\sigma\otimes\theta^2$ occurs in $\wedge^2(\Pi\otimes\theta)$ for any Hecke character θ. Note that $\Pi\otimes\theta$ also satisfies the hypothesis that it has no nontrivial self-twists. Choose θ such that

$$\omega_{\sigma\otimes\theta^2}=\omega_\sigma\theta^4=\theta^4\neq\mathbb{1}$$

to get a contradiction. □

5. The proof of (iii)⇒(ii)

We prove that if Π satisfies **(iii)**(β), then it is of the form **(ii)**(d), and if it satisfies **(iii)**(α), then it is one of **(ii)**(a)–(c).

5.1. (iii)(β) $\Longrightarrow$ (ii)(d). Assume that

$$\Pi \cong \Pi \otimes \chi \tag{5.1}$$

for some nontrivial χ. Taking central characters we have $\chi^4 = \mathbb{1}$. If $\chi^2 \neq \mathbb{1}$, then we may replace χ with χ^2 in (5.1), which means we may assume that the character χ in (5.1) is quadratic. We want to show that Π is induced from a quadratic extension. If Π is a representation of GL(2), then the analogous statement is a well-known result due to Labesse-Langlands [**20**]. In our case it follows from the work of Arthur-Clozel [**1**] and some L-function arguments as we explain below.

LEMMA 5.2. *Let Π be a cuspidal representation of $\mathrm{GL}(2n, \mathbb{A}_F)$ satisfying $\Pi \cong \Pi \otimes \chi$ for a nontrivial quadratic character χ. Then $\Pi = \mathrm{I}_E^F(\pi)$, where E/F is the quadratic extension associated with χ and π is a cuspidal representation of $\mathrm{GL}(n, \mathbb{A}_E)$.*

PROOF. We first claim that the base change Π_E is not cuspidal. To see this, assume that it is cuspidal. For a finite set S of places of F and the corresponding set T of places of E lying above those in S, as before, we have

$$\begin{aligned} L^T(s, \Pi_E \times \Pi_E^\vee) &= L^S(s, \Pi \times \Pi^\vee) L^S(s, \Pi \times \Pi^\vee \otimes \chi) \\ &= L^S(s, \Pi \times \Pi^\vee)^2. \end{aligned} \tag{5.3}$$

For sufficiently large S, the left hand side of (5.3) has a simple pole at $s = 1$ while the right hand side has a double pole at $s = 1$. This contradiction shows that Π_E is not cuspidal. This means that $\Pi_E = \pi_1 \boxplus \pi_2$, where π_i are cuspidal representations of $\mathrm{GL}(n, \mathbb{A}_E)$. Further, $\pi_1 \not\cong \pi_2$, because if they are equivalent, then

$$L^S(s, \Pi \times \Pi^\vee)^2 = L^T(s, \Pi_E \times \Pi_E^\vee) = L^T(s, \pi_1 \times \pi_1^\vee)^4,$$

but $L^S(s, \Pi \times \Pi^\vee)^2$ has a double pole and $L^T(s, \pi_1 \times \pi_1^\vee)^4$ has a pole of order 4 at $s = 1$.

Next, we claim that $\Pi \cong \mathrm{I}_E^F(\pi_1)$. To show this it is enough to prove that the partial L-function $L^S(s, \mathrm{I}_E^F(\pi_1) \times \Pi^\vee)$ has a simple pole at $s = 1$. This follows from

$$\begin{aligned} L^S(s, \mathrm{I}_E^F(\pi_1) \times \Pi^\vee) &= L^S(s, \mathrm{I}_E^F(\pi_1 \times \Pi_E^\vee)) \\ &= L^T(s, \pi_1 \times \Pi_E^\vee) \\ &= L^T(s, \pi_1 \times \pi_1^\vee) L^T(s, \pi_1 \times \pi_2^\vee). \end{aligned}$$

Since $\pi_2 \not\cong \pi_1$ we know that $L^T(s, \pi_1 \times \pi_1^\vee) L^T(s, \pi_1 \times \pi_2^\vee)$ has a simple pole at $s = 1$. □

5.2. (iii)(α) $\Longrightarrow$ (ii)(a)–(c). Now assume that

$$\Pi \cong \Pi^\vee \otimes \chi \tag{5.4}$$

for some χ. For a finite set S of places of F, as before, we have

$$\begin{aligned} L^S(s, \Pi \times \Pi^\vee) &= L^S(s, \Pi \times (\Pi \otimes \chi^{-1})) \\ &= L^S(s, (\Pi \times \Pi) \otimes \chi^{-1}) \\ &= L^S(s, \Pi, \wedge^2 \otimes \chi^{-1}) L^S(s, \Pi, \mathrm{Sym}^2 \otimes \chi^{-1}). \end{aligned}$$

The last two L-functions are the standard twisted exterior square and twisted symmetric square L-functions of Π. If S is a sufficiently large set, then $L^S(s, \Pi \times \Pi^\vee)$ has a simple pole at $s = 1$. Therefore one and exactly one of the partial twisted exterior or symmetric square L-functions has a simple pole at $s = 1$.

First, assume that $L^S(s, \Pi, \wedge^2 \otimes \chi^{-1})$ has a pole at $s = 1$. Then there exists a cuspidal representation π, which may be taken to be globally generic, of $\mathrm{GSp}(4, \mathbb{A}_F)$ such that Π is the functorial transfer of Π, i.e., Π is a representation as in **(ii)**(c). This result has been known for a long time and, we believe, is originally due to Jacquet, Piatetski-Shapiro, and Shalika. See Gan-Takeda [**6**] for a proof. It would also follow from the more general method of "descent" as we explain below.

Next, assume that $L^S(s, \Pi, \mathrm{Sym}^2 \otimes \chi^{-1})$ has a pole at $s = 1$. Taking central characters in (5.4) we have $\omega_\Pi = \omega_\Pi^{-1}\chi^4$. In other words, $\mu = \omega_\Pi\chi^{-2}$ is a quadratic character. If μ is trivial, then Π is a transfer from a cuspidal representation π of $\mathrm{GSpin}(4)$, a split connected reductive group of type D_4 whose derived group is $\mathrm{Spin}(4)$. If μ is nontrivial, then Π is a functorial transfer from the quasi-split non-split group $\mathrm{GSpin}^*(4)$ associated with the quadratic extension E/F attached to μ. These facts can be proved using the "descent" method of Ginzburg-Rallis-Soudry. If χ is trivial, then Π would be a transfer from a special orthogonal or symplectic group. We refer to Ginzburg-Rallis-Soudry [**8**, Theorem A] and Soudry [**31**, Theorem 4 and 12] for the proofs for classical groups, and Hundley-Sayag [**9**, Corollary 3.2.1] for the case of GSpin groups.

With the above notation, if μ is trivial, then Π is as in **(ii)**(a), and if it is nontrivial then Π is as in **(ii)**(b).

6. Examples and Complements

In this section we give a few examples of our main result. In some of them the proposed representation Π of $\mathrm{GL}(4, \mathbb{A}_F)$ is not yet proved to be automorphic, but it is conjecturally so. We also comment on possible intersection among the four cases in part **(ii)** of Theorem 1.1. Finally, we present a theorem due to A. Kable on when the exterior square of an irreducible four-dimensional representation is reducible.

6.1. K. Martin's $\mathbf{G_{192}}$. The matrices

$$a = \begin{bmatrix} -1 & & & \\ & -1 & & \\ & & 1 & \\ & & & 1 \end{bmatrix}, \quad b = \begin{bmatrix} -1 & & & \\ & 1 & & \\ & & -1 & \\ & & & 1 \end{bmatrix},$$

$$c = \begin{bmatrix} & & -1 & \\ & & & 1 \\ -i & & & \\ & -i & & \end{bmatrix}, \quad d = \begin{bmatrix} & -1 & & \\ & & -1 & \\ 1 & & & \\ & & & 1 \end{bmatrix}$$

in $\mathrm{GL}(4, \mathbb{C})$ generate a group G_{192} of order 192. Let ρ be the four-dimensional representation of the group G_{192} given by inclusion. Then ρ is an irreducible representation. In [**21**] K. Martin showed that ρ is modular, i.e., there exists a (cuspidal) automorphic representation $\Pi(\rho)$ that corresponds to ρ.

EXAMPLE 6.1. Let $\Pi_1 = \Pi(\rho)$. Then Π_1 is a cuspidal automorphic representation of $\mathrm{GL}(4, \mathbb{A}_F)$. Moreover, it is neither essentially self-dual nor does it have a nontrivial self-twist. It is not on the list of possibilities of **(ii)**. Furthermore, $\wedge^2\Pi_1$ is cuspidal. In other words, Π_1 is an example of a cuspidal representation which does not satisfy any of **(i)**–**(iii)** of Theorem 1.1.

PROOF. First we check that $\wedge^2(\rho)$ is an irreducible representation. To see this consider the standard basis $\langle e_1, e_2, e_3, e_4\rangle$ for $\mathbb{C}^4$ and fix the ordered basis $\langle w_1, w_2, \dots, w_6\rangle$ of $\mathbb{C}^6 = \wedge^2\mathbb{C}^4$ given by

$$\begin{aligned} w_1 &= e_1 \otimes e_2 - e_2 \otimes e_1, \ w_2 = e_1 \otimes e_3 - e_3 \otimes e_1, \ w_3 = e_1 \otimes e_4 - e_4 \otimes e_1, \\ w_4 &= e_2 \otimes e_3 - e_3 \otimes e_2, \ w_5 = e_2 \otimes e_4 - e_4 \otimes e_2, \ w_6 = e_3 \otimes e_4 - e_4 \otimes e_3. \end{aligned}$$

Let A, B, C, D be the images of a, b, c, d under $\wedge^2$, respectively. Then, with respect to the above basis, we have

$$A = \begin{bmatrix} 1 & & & & & \\ & -1 & & & & \\ & & -1 & & & \\ & & & -1 & & \\ & & & & -1 & \\ & & & & & 1 \end{bmatrix}, \quad B = \begin{bmatrix} -1 & & & & & \\ & 1 & & & & \\ & & -1 & & & \\ & & & -1 & & \\ & & & & 1 & \\ & & & & & -1 \end{bmatrix},$$

$$C = \begin{bmatrix} & & & & & -1 \\ & -i & & & & \\ & & & -i & & \\ & & i & & & \\ & & & & i & \\ 1 & & & & & \end{bmatrix}, \quad D = \begin{bmatrix} & & & 1 & & \\ 1 & & & & & \\ & & & & -1 & \\ & 1 & & & & \\ & & & & & -1 \\ & & 1 & & & \end{bmatrix}.$$

It is easy to check that if a 6×6 matrix X commutes with A, B, and C, then it has to be of the form

$$X = \begin{bmatrix} a & & & & & b \\ & c & & & & \\ & & e & f & & \\ & & -f & e & & \\ & & & & d & \\ -b & & & & & a \end{bmatrix}.$$

Further, if $XD = DX$, then $a = c = d = e$ and $b = f = 0$, i.e., X is a scalar matrix. Therefore, $\mathrm{Hom}_{G_{192}}(\wedge^2\rho, \wedge^2\rho) = \mathbb{C}$ and, by Schur's lemma, the representation $\wedge^2\rho$ is irreducible. This implies that $\wedge^2\Pi_1$ is cuspidal. (This is because, if a complex Galois representation σ is modular, i.e., corresponds to an automorphic representation $\pi = \pi(\sigma)$, then σ is irreducible if and only if π is cuspidal. This fact follows from $L(s, \sigma \otimes \sigma^\vee) = L(s, \pi \times \pi^\vee)$; see Ramakrishnan [**27**, Introduction].) Martin observes that ρ, and hence Π_1, is not essentially self-dual. Clearly Π_1 is not on the list of possibilities in **(ii)** of the main theorem because if it were, then $\wedge^2\Pi_1$ would not be cuspidal (see Section 3). Hence Π_1 does not satisfy any of the equivalent statements of Theorem 1.1. □

In [**21**] Martin considers a four-dimensional irreducible representation ρ of the absolute Galois group of $\mathbb{Q}$ whose image in $\mathrm{PGL}(4, \mathbb{C})$, denoted $\bar{G}$, is an extension of A_4 by V_4. In this situation $\bar{G}$ is either $V_4 \rtimes A_4$ or $V_4 \cdot A_4$. In the former case ρ is of GO(4)-type and $\Pi = \Pi(\rho)$ is a transfer from $\mathrm{GL}(2, \mathbb{A}_F) \times \mathrm{GL}(2, \mathbb{A}_F)$, which is contained in our **(ii)**(a). The example of G_{192} is an instance of the latter situation. In either case, $\Pi(\rho)$ may also be thought of as being obtained by automorphic induction across a non-normal quartic extension with no quadratic subextension.

6.2. The standard representation of S_5. Consider a tower of number fields $\tilde{J}/J/E/F$. Here, E/F and $\tilde{J}/J$ are quadratic extensions, J/E is an A_5-extension, J/F is an S_5-extension, $\tilde{J}/E$ is an $\mathrm{SL}(2,\mathbb{F}_5)$-extension, and $\tilde{J}/F$ is Galois. (Recall that $A_5 \cong \mathrm{PSL}(2,\mathbb{F}_5)$.) In what follows we identify A_5, S_5 and $\mathrm{SL}(2,\mathbb{F}_5)$ with these Galois groups.

Let σ be the standard four-dimensional irreducible representation of S_5,

$$\sigma : S_5 \longrightarrow \mathrm{GL}(4,\mathbb{C}).$$

Here are some properties of σ:

(1) $\wedge^2\sigma$ is a six-dimensional irreducible representation of S_5 (see [**5**, §3.2]).
(2) $\sigma|_{A_5}$ is irreducible (because $\sigma \not\cong \sigma\otimes\epsilon$, where ϵ is the sign character of S_5).
(3) $\wedge^2(\sigma|_{A_5}) = (\wedge^2\sigma)|_{A_5}$ is reducible (because $\wedge^2(\sigma) \cong \wedge^2(\sigma)\otimes\epsilon$), and its irreducible constituents are both of degree 3.
(4) σ is self-dual (because the character of σ has integer values).
(5) $\sigma|_{A_5} = \rho_1\otimes\rho_2$, where ρ_1 and ρ_2 are the two-dimensional irreducible representations of $\mathrm{SL}(2,\mathbb{F}_5)$ (see [**15**, Lemma 5.1]).

We need some details about the ρ_i. Let $\tilde{\rho}$ be the unique (up to twists) cuspidal representation of $\mathrm{GL}(2,\mathbb{F}_5)$ whose restriction to $\mathrm{SL}(2,\mathbb{F}_5)$ is reducible. In this case, $\tilde{\rho}|_{\mathrm{SL}(2,\mathbb{F}_5)} = \rho_1\oplus\rho_2$. If $g\in \mathrm{GL}(2,\mathbb{F}_5) - Z(\mathbb{F}_5)\mathrm{SL}(2,\mathbb{F}_5)$, then $\rho_1^g = \rho_2$. Here Z is the center of $\mathrm{GL}(2)$. Conjugating $\mathrm{SL}(2,\mathbb{F}_5)$ by such an element g induces the nontrivial outer automorphism of $\mathrm{SL}(2,\mathbb{F}_5)$ because, if it were an inner automorphism, then we would have $\rho_1\cong\rho_2$, which contradicts the fact that the restriction from GL_2 to SL_2 is multiplicity-free.

Let ρ denote either ρ_1 or ρ_2. In constructing examples (to illustrate our main theorem), we make the following *assumption: ρ is modular*, i.e., there exists a cuspidal automorphic representation $\pi(\rho)$ of $\mathrm{GL}(2,\mathbb{A}_E)$ with $\rho\leftrightarrow\pi(\rho)$. In this situation, it is expected [**17**] that there exists an automorphic representation $\pi(\sigma)$ of $\mathrm{GL}(4,\mathbb{A}_F)$ with $\pi(\sigma)\leftrightarrow\sigma$ and $\pi(\sigma)$ is the Asai transfer of $\pi(\rho)$, i.e., $\pi(\sigma) = \mathrm{As}\,(\pi(\rho))$.

EXAMPLE 6.2. Let $\Pi_2 = \pi(\sigma) = \mathrm{As}(\pi(\rho))$. Then Π_2 is a cuspidal automorphic representation of $\mathrm{GL}(4,\mathbb{A}_F)$ which is self-dual. However, it is the Asai transfer of a nondihedral cuspidal representation. Moreover, it has no nontrivial self-twists and it is not on the list of possibilities in **(ii)**. Furthermore, its exterior square transfer $\wedge^2\Pi_2$ is cuspidal. In other words, Π_2 is an example of a cuspidal representation which does not satisfy any of **(i)**–**(iii)** of Theorem 1.1.

PROOF. Since $\Pi_2 = \pi(\sigma)$ and σ is irreducible, we conclude that Π_2 is cuspidal. (Cuspidality of Π_2 may also be seen by appealing to the cuspidality criterion for the Asai transfer due to Ramakrishnan [**26**, Theorem 1.4].)

Next, we note that Π_2 is self-dual and is the Asai transfer of a cuspidal representation, namely $\pi(\rho)$. Note that $\pi(\rho)$ is not dihedral, as ρ is not induced from a character of an index two subgroup because there is no such subgroup in $\mathrm{SL}(2,\mathbb{F}_5)$. Also, Π_2 has no nontrivial self-twists because σ has no nontrivial self-twists.

Finally, note that $\wedge^2\Pi_2$ is cuspidal since $\wedge^2\Pi_2 = \wedge^2\pi(\sigma) = \pi(\wedge^2\sigma)$ and $\wedge^2\sigma$ is an irreducible representation implying that $\pi(\wedge^2\sigma)$ is cuspidal. (See, for example, Ramakrishnan [**27**, Introduction].) □

EXAMPLE 6.3. Let $\Pi_3 = (\Pi_2)_E$ be the base change of Π_2 to an automorphic representation of $\mathrm{GL}(4,\mathbb{A}_E)$. Then Π_3 is a cuspidal automorphic representation of

$\mathrm{GL}(4, \mathbb{A}_E)$ which is self-dual and is not the Asai transfer of a nondihedral representation. It is contained in **(ii)**(a) and its exterior square transfer $\wedge^2\Pi_3$ is not cuspidal. In other words, Π_3 is an example of a cuspidal representation which satisfies **(i)**–**(iii)** of Theorem 1.1.

PROOF. To see cuspidality of Π_3, as well as the fact that it is contained in **(ii)**(a), note that

$$\Pi_3 = (\Pi_2)_E = \pi(\sigma)_E = \pi(\sigma|_{A_5}) = \pi(\rho_1 \otimes \rho_2) = \pi(\rho_1) \boxtimes \pi(\rho_2).$$

Neither ρ_i is monomial since $\mathrm{SL}(2, \mathbb{F}_5)$ does not have an index two subgroup. Applying the cuspidality criterion for $\pi(\rho_1) \boxtimes \pi(\rho_2)$ due to Ramakrishnan [**25**, Theorem 11.1], we see that Π_3 is not cuspidal if and only if $\pi(\rho_1) \cong \pi(\rho_2) \otimes \mu$ for some Hecke character μ of E. On the other hand, $\pi(\rho_2) \otimes \mu = \pi(\rho_2 \otimes \mu)$, where we identify the Hecke character μ with a character of the absolute Galois group of E via global class field theory. Hence, we have $\pi(\rho_1) = \pi(\rho_2 \otimes \mu)$. This implies that $\rho_1 \cong \rho_2 \otimes \mu$ (since, for any two Galois representations τ_1 and τ_2, one has $\pi(\tau_1) \cong \pi(\tau_2)$ if and only if $\tau_1 \cong \tau_2$; one can see this by considering the equality $L^S(s, \pi(\tau_1) \times \pi(\tau_2)^\vee) = L^S(s, \tau_1 \otimes \tau_2^\vee))$. Therefore, μ is a character of $\mathrm{Gal}(\tilde{J}/E) = \mathrm{SL}(2, \mathbb{F}_5)$, a perfect group, hence μ is trivial. Whence $\rho_1 = \rho_2$, which contradicts the fact that they are inequivalent, as was observed earlier.

Next, observe that $\wedge^2\Pi_3$ is not cuspidal because

$$\wedge^2\Pi_3 = \wedge^2(\pi(\rho_1) \boxtimes \pi(\rho_2)) = (\mathrm{Sym}^2(\pi(\rho_1)) \otimes \omega_{\pi(\rho_2)}) \oplus (\mathrm{Sym}^2(\pi(\rho_2)) \otimes \omega_{\pi(\rho_1)}),$$

which is of isobaric type $(3, 3)$. (See Proposition 3.1.)

Finally, we observe that Π_3 is self-dual because σ, and hence $\sigma|_{A_5}$, is self-dual and that Π_3 could not be an Asai transfer of a nondihedral representation because if it were, then $\wedge^2(\Pi_3)$ would be cuspidal by Proposition 3.2. □

6.3. On possible intersections between representations in (ii). The purpose of this subsection is to show that the cases **(ii)**(a) through **(ii)**(d) are not mutually exclusive.

EXAMPLE 6.4. Let $\pi = \mathrm{I}_E^F(\chi)$ be a cuspidal automorphic representation of $\mathrm{GL}(2, \mathbb{A}_F)$ which is automorphically induced from a Hecke character χ of E, where E/F is a quadratic extension. Let τ be a nondihedral cuspidal automorphic representation of $\mathrm{GL}(2, \mathbb{A}_F)$. Let $\Pi_4 = \pi \boxtimes \tau$. Then Π_4 is a representation that is common to **(ii)**(a), **(ii)**(c) and **(ii)**(d).

PROOF. From Ramakrishnan's cuspidality criterion [**25**, Theorem 11.1] we know that Π_4 is a cuspidal representation of $\mathrm{GL}(4, \mathbb{A}_F)$. By construction, Π_4 is in **(ii)**(a).

We observe that

$$\Pi_4 = \mathrm{I}_E^F(\chi) \boxtimes \tau = \mathrm{I}_E^F(\chi \otimes \tau_E).$$

Since the induced representation $\mathrm{I}_E^F(\chi \otimes \tau_E)$ is cuspidal, the inducing representation $\chi \otimes \tau_E$ is, *a fortiori*, cuspidal. Hence Π_4 is in **(ii)**(d).

Now we claim that Π is also a transfer from a (generic) cuspidal representation of $\mathrm{GSp}(4, \mathbb{A}_F)$. To see this we recall the following well known identities:

$$\begin{aligned} \mathrm{Sym}^2(\mathrm{I}_E^F(\chi)) &= \mathrm{I}_E^F(\chi^2) \boxplus \chi|_{\mathbb{A}_F^\times}, \\ \wedge^2(\mathrm{I}_E^F(\chi)) &= \chi|_{\mathbb{A}_F^\times} \cdot \omega_{E/F}, \end{aligned}$$

where $\omega_{E/F}$ is the quadratic Hecke character of F associated to E/F by class field theory. In particular, the central character of π is given by $\omega_\pi = \chi|_{\mathbb{A}_F^\times} \cdot \omega_{E/F}$. From Proposition 3.1 we have

$$\wedge^2(\pi \boxtimes \tau) = \left(\mathrm{Sym}^2(\pi) \otimes \omega_\tau\right) \boxplus \left(\mathrm{Sym}^2(\tau) \otimes \omega_\pi\right).$$

For brevity write $\omega = \omega_\pi \omega_\tau$. We deduce that

$$\wedge^2(\Pi_4) = \left(\mathrm{I}_E^F(\chi^2) \otimes \omega_\tau\right) \boxplus \omega\, \omega_{E/F} \boxplus \left(\mathrm{Sym}^2(\tau) \otimes \omega_\pi\right).$$

Hence, the partial L-function $L^S(s, \Pi, \wedge^2 \otimes (\omega\, \omega_{E/F})^{-1})$ has a pole at $s = 1$. Applying a recent result of Gan and Takeda [**6**] we conclude that Π is a transfer from GSp(4), i.e., Π is in **(ii)**(c). □

6.4. A calculation on the Galois side. As mentioned in the introduction, one may ask for an irreducibility criterion on the *Galois side*, i.e., for ℓ-adic Galois representations or complex representations of the Langlands group $\mathcal{L}_F$, which reflects the cuspidality criterion one is looking for. In this section we present such a theorem due to A. Kable. We are grateful to him for the permission to include this material here. Theorem 6.5 below is the analogue of the equivalence of **(i)** and **(iii)** in Theorem 1.1. We begin by reviewing some preliminaries.

Let k be an algebraically closed field whose characteristic is not two and let V be a four-dimensional k-vector space. Fix a nonzero element $\eta \in \wedge^4 V$. There is a nondegenerate symmetric bilinear form B on $\wedge^2 V$ defined by $\omega_1 \wedge \omega_2 = B(\omega_1, \omega_2)\eta$ for all $\omega_1, \omega_2 \in \wedge^2 V$. The bilinear space $(\wedge^2 V, B)$ is isomorphic to the orthogonal sum of three hyperbolic planes. Let $\mathrm{GO}(B)$ be the group of similitudes of B, $\lambda : \mathrm{GO}(B) \to k^\times$ the similitude character, and $\mathrm{GSO}(B)$ the subgroup of proper similitudes. This subgroup consists of those $T \in \mathrm{GO}(B)$ such that $\det(T) = \lambda(T)^3$, and it coincides with the connected component of the identity in the algebraic group $\mathrm{GO}(B)$. (In the literature, the group $\mathrm{GSO}(B)$ is also denoted by $\mathrm{SGO}(B)$ or $\mathrm{GO}^+(B)$.) We use similar notation also for the similitude groups of forms on V itself. Let $\rho : \mathrm{GL}(V) \to \mathrm{GL}(\wedge^2 V)$ be the homomorphism $\rho(S) = \wedge^2 S$.

Let G be a group and let σ be an irreducible representation of G on V. Recall that σ is *essentially self-dual* if there is a character χ of G such that $\sigma^\vee \cong \chi \otimes \sigma$. In this case, χ^{-1} is a subrepresentation of $\sigma \otimes \sigma$. We say that σ has *symplectic type* if χ^{-1} occurs in $\wedge^2\sigma$ and *orthogonal type* if χ^{-1} occurs in $\mathrm{Sym}^2\sigma$. If σ is essentially self-dual of orthogonal type, then there is a nonzero symmetric bilinear form C on V such that G acts on V by similitudes of C. The kernel of C is a G-invariant proper subspace of V and hence trivial. Thus, C is nondegenerate and $\sigma(G) \subset \mathrm{GO}(C)$. If $\sigma(G) \subset \mathrm{GSO}(C)$, then we say that σ is of *proper orthogonal type*; otherwise, we say that σ is of *improper orthogonal type*. Finally, we say that σ has a *nontrivial quadratic self-twist* if there is a nontrivial $\{\pm 1\}$-valued character χ of G such that $\sigma \cong \chi \otimes \sigma$.

THEOREM 6.5 (A. Kable). *Let σ be an irreducible 4-dimensional representation of a group G over an algebraically closed field whose characteristic is not two. Then the following two conditions on σ are equivalent:*

(1) *$\wedge^2\sigma$ is reducible.*
(2) *σ satisfies at least one of the following:*
   (a) *is essentially self-dual of symplectic type,*
   (b) *has a nontrivial quadratic self-twist, or*
   (c) *is essentially self-dual of proper orthogonal type.*

Toward the proof of the above theorem, we begin with a lemma.

LEMMA 6.6. *The bilinear space* $(\wedge^2 V, B)$ *has the following properties:*

(1) *The image of* ρ *is* $\mathrm{GSO}(B)$.
(2) *An isotropic line in* $\wedge^2 V$ *has the form* $\wedge^2 Q$, *where* $Q < V$ *is a uniquely determined* 2*-dimensional subspace of* V.
(3) *An isotropic* 3*-space in* $\wedge^2 V$ *either has the form* $L \wedge V$, *where* $L < V$ *is a uniquely determined line, or the form* $\wedge^2 U$, *where* $U < V$ *is a uniquely determined* 3*-space.*
(4) *Let* $W < \wedge^2 V$ *be a* 3*-space on which* B *is nondegenerate. Then there is a nondegenerate symmetric bilinear form* C *on* V *such that*
$$\{g \in \mathrm{GL}(V) \mid \rho(g)(W) = W\} = \mathrm{GSO}(C).$$
The form C *is determined by* W *up to scalars. Every nondegenerate symmetric bilinear form on* V *occurs in this way for a suitable choice of* W.

PROOF. We omit the proofs of (1), (2), and (3), as they are easy exercises, and briefly sketch the proof of (4). Let Q be the quadric hypersurface in $\mathbb{P}(\wedge^2 V)$ consisting of null vectors for B. By (2), we may identify Q with the Grassmannian of lines in $\mathbb{P}(V)$. Let W be a 3-dimensional subspace of $\wedge^2 V$ on which B is nondegenerate, and $Y = \mathbb{P}(W) \cap Q$ be the smooth plane conic defined by $B|_W$. Let T be the subvariety of $\mathbb{P}(V)$ obtained by taking the union of the lines in $\mathbb{P}(V)$ corresponding to points of Y. It is easily verified that T is a smooth quadric hypersurface, and so there is a nondegenerate symmetric bilinear form C on V, unique up to scalars, such that T has equation $C(v, v) = 0$. An element $g \in \mathrm{GL}(V)$ preserves T together with the ruling $T \to Y$ sending a line in T to the corresponding point of Y if and only if $g \in \mathrm{GSO}(C)$. From the construction, the set of g with this property is the same as the set of all g such that $\rho(g)(W) = W$. The last claim in (4) follows from the fact that $\mathrm{GL}(V)$ acts transitively on the set of all nondegenerate symmetric bilinear forms on V. □

PROPOSITION 6.7. *The representation* (σ, V) *is essentially self-dual of symplectic type if and only if* $\wedge^2 V$ *contains a* G*-invariant line.*

PROOF. This follows immediately from the definitions. □

LEMMA 6.8. *There is no* G*-invariant isotropic* 3*-space in* $\wedge^2 V$.

PROOF. If there is such a 3-space, then, by Lemma 6.6, it is of the form $L \wedge V$ or of the form $\wedge^2 U$. Note that G-invariance of the 3-space combined with the uniqueness statements in Lemma 6.6 imply that L or U is also G-invariant, which contradicts irreducibility of σ. □

The proof of the following proposition would be substantially simpler if the action of G on $\wedge^2 V$ were completely reducible. However, in the current generality, this need not be true.

PROPOSITION 6.9. *Suppose that* σ *is not essentially self-dual of symplectic type. Then* σ *has a nontrivial quadratic self-twist if and only if* $\wedge^2 V$ *contains a* G*-invariant* 2*-space.*

PROOF. Suppose first that σ has a nontrivial quadratic self-twist, say by the character χ. Let H be the kernel of χ and recall that, by Clifford theory, $\sigma|_H$ is the sum of two 2-dimensional subrepresentations. Let $W < V$ be the H-invariant 2-space on which one of these subrepresentations is realized. Then it is easy to see that the G-translates of $\wedge^2 W$ span a G-invariant 2-space in $\wedge^2 V$. (The reader should compare this with the proof of Proposition 3.5. Indeed, the 2-dimensional G-invariant subspace is the induction to G of the determinant character of the representation of H on W. Recall that if σ is a Galois representation that corresponds to an automorphic representation π, then the determinant character of σ corresponds to the central character of π.)

Now suppose that $\wedge^2 V$ contains a G-invariant 2-space P. The kernel of $B|_P$ is G-invariant and thus is either $\{0\}$ or P, for the first hypothesis implies that there can be no G-invariant line in $\wedge^2 V$. Suppose that the kernel is P. Then the 4-space $P^\perp$ contains P and the form B and the action of G pass down to $P^\perp/P$. Suppose that B has a nontrivial kernel in $P^\perp/P$. This kernel cannot be all of $P^\perp/P$, for then $P^\perp$ would be an isotropic 4-space in $\wedge^2 V$. Thus the kernel must be a line in $P^\perp/P$ and this kernel is necessarily G-invariant. The preimage of this line in $P^\perp$ is an isotropic G-invariant 3-space in $\wedge^2 V$, contrary to Lemma 6.8. We conclude that $(P^\perp/P, B)$ is a nondegenerate quadratic 2-space. Such a space is isomorphic to a hyperbolic plane and hence contains exactly two isotropic lines. The action of G on $P^\perp/P$ is by similitudes, hence it permutes these lines. Taking the preimage in $P^\perp$, we obtain two isotropic 3-spaces Λ_1 and Λ_2 in $\wedge^2 V$ that are permuted by G. By Lemma 6.8, these isotropic G-spaces cannot be fixed by G and we conclude that the stabilizer of each is a subgroup H of index two in G. By repeating the argument of Lemma 6.8 with H in place of G, we conclude that there is either a line $L < V$ or a 3-space $U < V$ that is H-invariant. By replacing σ by $\sigma^\vee$ if necessary, we may assume that the former possibility holds. Let $g_0 \in G - H$. Then the 2-space $L + \sigma(g_0)L$ is easily seen to be G-invariant, contrary to the irreducibility of σ and $\sigma^\vee$. This contradiction finally allows us to conclude that the restriction of B to P is nondegenerate.

We now repeat the argument of the previous paragraph with the space P in place of the space $P^\perp/P$. It yields an index two subgroup H of G and two isotropic lines in $\wedge^2 V$ that are fixed by H. By Lemma 6.6, each of these lines has the form $\wedge^2 Q$ with $Q < V$ a 2-space. By the uniqueness assertion from Lemma 6.6, each of these 2-spaces is H-invariant. It now follows from Clifford theory that if χ is the nontrivial $\{\pm 1\}$-valued character on G whose kernel is H, then $\chi \otimes \sigma \cong \sigma$. Thus σ has a nontrivial quadratic self-twist, as required. □

PROPOSITION 6.10. *Suppose that σ is neither essentially self-dual of symplectic type nor has a nontrivial quadratic self-twist. Then σ is essentially self-dual of proper orthogonal type if and only if $\wedge^2 V$ contains a G-invariant 3-space.*

PROOF. Let $W < \wedge^2 V$ be a G-invariant 3-space. By Lemma 6.8, W cannot be isotropic. The group G acts on W by similitudes and so the kernel of $B|_W$ is G-invariant. We have just observed that this kernel cannot be W and, by the hypotheses and the preceding results, it cannot be of dimension 1 or 2. Thus the restriction of B to W is nondegenerate. It follows from Lemma 6.6 that there is a nondegenerate symmetric bilinear form C on V such that $\sigma(G) \subset \mathrm{GSO}(C)$. This implies that σ is essentially self-dual of proper orthogonal type.

Now suppose that σ is essentially self-dual of proper orthogonal type, so that there is a nondegenerate bilinear form C on V such that $\sigma(G) \subset \mathrm{GSO}(C)$. By

Lemma 6.6, there is a 3-space $W \subset \wedge^2 V$ such that $\rho(\mathrm{GSO}(C))$ preserves W. In particular, W is G-invariant, and the reverse implication is proved. $\square$

PROOF OF THEOREM 6.5. We know that the representation $\wedge^2\sigma$ is essentially self-dual. Thus, if it has any proper nonzero G-invariant subspace, it necessarily has such a subspace of dimension at most 3. The proof follows from Propositions 6.7, 6.9 and 6.10. $\square$

6.5. Exception in (iii)(α). Using Theorem 6.5 it is possible to explain the seemingly strange exception in **(iii)**(α) of Theorem 1.1. For this we first set up some notation.

Let G be a group and let H be a subgroup of index two in G. If (τ, W) is a 2-dimensional representation of H, then $\mathrm{As}_{G/H}(\tau)$, the Asai lift of τ, which is a 4-dimensional representation of G, is defined as follows. Fix $g \in G - H$. Define the representation (τ', W) of H via

$$\tau'(h) = \tau(ghg^{-1}).$$

We have a homomorphism

$$\tau \times \tau' : H \longrightarrow \mathrm{GL}(W) \times \mathrm{GL}(W) \hookrightarrow (\mathrm{GL}(W) \times \mathrm{GL}(W)) \rtimes (G/H).$$

Here the action of G/H on $(\mathrm{GL}(W) \times \mathrm{GL}(W))$ is as in (2.3), i.e, via switching the two factors. We can extend the map $\tau \times \tau'$ from H to a map $\tilde{\tau}$ from G to $(\mathrm{GL}(W) \times \mathrm{GL}(W)) \rtimes (G/H)$ by setting $\tilde{\tau}(g) = (1_W, \tau(g^2), \gamma)$, where γ denotes the nontrivial element of G/H. In other words

$$\tilde{\tau}(x) = \begin{cases} (\tau(h), \tau(ghg^{-1}), 1) & \text{if } x = h \in H, \\ (\tau(h), \tau(ghg), \gamma) & \text{if } x = hg \in Hg. \end{cases}$$

It is easy to check that $\tilde{\tau}$ is a homomorphism. We define $\mathrm{As}_{G/H}(\tau) = \mathrm{As} \circ \tilde{\tau}$, where the map "As" is as in (2.4). This gives a representation of G on the space $V = W \otimes W$. To summarize, we have the following commutative diagram:

$$\begin{array}{ccc} H & \xrightarrow{\ \tau \times \tau'\ } & (\mathrm{GL}(W) \times \mathrm{GL}(W)) \rtimes (G/H) \\ \downarrow & \overset{\tilde{\tau}}{\nearrow} & \downarrow \mathrm{As} \\ G & \xrightarrow[\mathrm{As}_{G/H}(\tau)]{} & GL(V) \end{array}$$

PROPOSITION 6.11. *Assume that τ is an irreducible 2-dimensional representation of H and $\sigma = \mathrm{As}_{G/H}(\tau)$ is irreducible. Then τ has a nontrivial self-twist if and only if σ has a nontrivial self-twist.*

PROOF. First, assume that τ has a nontrivial self-twist, i.e., there is a nontrivial quadratic character χ of H and a nonzero $T \in \mathrm{Hom}_H(\tau, \tau \otimes \chi)$. Set $K = \ker(\chi)$. Since σ is irreducible, by [**26**, Theorem 1.4] we know that K is not normal in G. Hence $gKg^{-1} \neq K$. Let χ^* be the character of G obtained via composing χ with the transfer homomorphism from G^{ab} to H^{ab}. Observe that $\chi^*|_H$ is nontrivial because otherwise we would have $\chi^*(h) = \chi(h)\chi(ghg^{-1}) = 1$ for all $h \in H$, which implies that $K = gKg^{-1}$. Therefore, χ^* is nontrivial. It is now easy to see that $T \otimes T$ is a nonzero element of $\mathrm{Hom}_G(\sigma, \sigma \otimes \chi^*)$, i.e., σ has a nontrivial self-twist.

Next, assume that σ has a nontrivial self-twist. This implies that $\sigma \cong \mathrm{I}_M^G(\rho)$, where M is a subgroup of index 2 in G and ρ is a 2-dimensional representation of M. Hence, we have

$$\mathrm{As}_{G/H}(\tau) \cong \mathrm{I}_M^G(\rho).$$

Take $\wedge^2$ of both sides and restrict back to H. The semisimplification of the left hand side then gives

$$\begin{aligned}\mathrm{Res}_H(\wedge^2(\mathrm{As}_{G/H}(\tau)))_{ss} &\cong \wedge^2(\mathrm{Res}_H(\mathrm{As}_{G/H}(\tau)))_{ss} \\ &\cong \wedge^2(\tau \otimes \tau')_{ss} \\ &\cong \left((\mathrm{Sym}^2(\tau) \otimes \det(\tau')\right) \oplus \left((\mathrm{Sym}^2(\tau') \otimes \det(\tau)\right),\end{aligned}$$

i.e., a direct sum of two 3-dimensional representations of H. However, the right hand side, by [**16**, §3], gives

$$\mathrm{Res}_H(\wedge^2(\mathrm{I}_M^G(\rho)))_{ss} \cong \mathrm{Res}_H(\mathrm{As}_{G/M}(\rho) \otimes \omega_{G/M}) \oplus \mathrm{Res}_H(\mathrm{I}_M^G(\det(\rho))),$$

i.e., a direct sum of a 4-dimensional and a 2-dimensional representation. Therefore, at least one of the two 3-dimensional representations on the left hand side should be reducible. This implies that $\mathrm{Sym}^2(\tau)$ should be a reducible representation. Replacing τ by $\tau^\vee$ if necessary, we may assume that there is a character χ occurring as a quotient of $\mathrm{Sym}^2(\tau)$. Hence,

$$\tau \cong \tau^\vee \otimes \chi \cong (\tau \otimes \det(\tau)^{-1}) \otimes \chi \cong \tau \otimes (\chi \det(\tau)^{-1}),$$

i.e., τ has a self-twist by $\chi \det(\tau)^{-1}$. We claim that $\chi \det(\tau)^{-1}$ is nontrivial, for otherwise we would have

$$\begin{aligned}2 \leq \dim \mathrm{Hom}_H(\tau \otimes \tau, \det(\tau)) &= \dim \mathrm{Hom}_H(\tau, \tau^\vee \otimes \det(\tau)) \\ &= \dim \mathrm{Hom}_H(\tau, \tau) = 1\end{aligned}$$

which is a contradiction. □

The above proposition explains the strange exception in **(iii)**(α) of Theorem 1.1. Assume that (τ, W) is the *parameter* of a cuspidal representation π of GL(2) over a quadratic extension E/F of number fields. Then $\sigma = \mathrm{As}(\tau)$ is the *parameter* of $\Pi = \mathrm{As}_{E/F}(\pi)$. Assume that Π is cuspidal, i.e., σ is irreducible. Since τ is 2-dimensional, there is a symplectic form S on W which τ preserves up to similitudes. It is easy to see that σ preserves $S \otimes S$ on $W \otimes W$ up to similitudes. In fact, σ is an essentially self-dual representation of improper orthogonal type. Further, by the above proposition, we see that τ is dihedral if and only if σ has a nontrivial (quadratic) self-twist. By Theorem 6.5, one concludes that $\wedge^2\sigma$ is reducible if and only if τ is dihedral.

References

[1] Arthur J. and Clozel, L. Simple algebras, base change, and the advanced theory of the trace formula. Annals of Math. Studies 120 (1989), Princeton University Press, Princeton, NJ.

[2] Asgari, M. and Shahidi, F. *Generic transfer for general spin groups.* Duke Math. J., 132 (2006), no 1, 137–190.

[3] Asgari, M. and Shahidi, F. *Generic transfer for general spin groups II.* In Preparation.

[4] Asgari, M. and Shahidi, F. *Generic Transfer from* GSp*(4) to* GL*(4).* Compos. Math., 142 (2006), no. 3, 541–550.

[5] Fulton, W. and Harris, J. *Representation theory. A first course.* Graduate Texts in Mathematics, 129. Readings in Mathematics. Springer-Verlag, New York, 1991.

[6] Gan, W. and Takeda, S. *The local Langlands conjecture for* GSp(4). Preprint 2007. Available at http://arxiv.org/abs/0706.0952.

[7] Gelbart, S. and Jacquet, H. *A relation between automorphic representations of* GL(2) *and* GL(3). Ann. Sci. École Norm. Sup. (4) 11 (1978), 471–542.

[8] Ginzburg, D. and Rallis, S. and Soudry, D. *Generic Automorphic Forms on* $\mathrm{SO}(2n+1)$*: Functorial Lift to* $\mathrm{GL}(2n)$*, Endoscopy, and Base Change.* Internat. Math. Res. Notices, 2001, no. 14, 729-764.

[9] Hundley, J. and Sayag, E. *Descent construction For Gspin groups: main results and applications,* Electron. Res. Announc. Math. Sci. 16 (2009), 30–36.

[10] Jacquet, H., Piatetski-Shapiro, I. and Shalika, J. *Rankin–Selberg convolutions,* Amer. J. Math., 105 (1983), no. 2, 367–464.

[11] Jacquet, H. and Shalika, J. *On Euler products and the classification of automorphic forms, I.* Amer. J. Math., 103 (1981), no. 4, 499-558.

[12] Jacquet, H. and Shalika, J. *On Euler products and the classification of automorphic forms, II.* Amer. J. Math., 103 (1981), no. 4, 777-815.

[13] Kim, H. *Functoriality for the exterior square of* GL_4 *and the symmetric fourth of* GL_2. With appendix 1 by Dinakar Ramakrishnan and appendix 2 by Kim and Peter Sarnak. J. Amer. Math. Soc. 16 (2003), no. 1, 139–183.

[14] Kim, H. *Applications of Langlands' functorial lift of odd orthogonal groups.* Trans. Amer. Math. Soc. 354 (2002), no. 7, 2775–2796.

[15] Kim, H. *An example of non-normal quintic automorphic induction and modularity of symmetric powers of cusp forms of icosahedral type.* Invent. Math. 156 (2004), no. 3, 495–502.

[16] Kim, H. *An application of exterior square functoriality of* GL_4*; Asai lift.* Number theory, 197–202, CRM Proc. Lecture Notes, 36, Amer. Math. Soc., Providence, RI, 2004.

[17] Kim, H. *Private communication.*

[18] Kim, H., and Shahidi, F. *Cuspidality of symmetric powers with applications.* Duke Math. J. 112 (2002), no. 1, 177–197.

[19] Krishnamurthy, M. *The Asai transfer to* GL_4 *via the Langlands–Shahidi method.* Int. Math. Res. Not. 2003, no. 41, 2221–2254.

[20] Labesse, J.-P. and Langlands, R. *L-indistinguishability for* SL(2). Canad. J. Math., 31 (1979) 726–785.

[21] Martin, K. *Modularity of hypertetrahedral representations.* C. R. Math. Acad. Sci. Paris 339 (2004), no. 2, 99–102.

[22] Raghuram, A. and Shahidi, F. *Functoriality and special values of L-functions,* Eisenstein series and applications, 271–293, Progr. Math., 258, Birkhäuser Boston, Boston, MA, 2008.

[23] Raghuram, A. and Shahidi, F. *On certain period relations for cusp forms on* $\mathrm{GL}(n)$, Int. Math. Res. Not., 2008, Art. ID rnn 077, 23 pages.

[24] Ramakrishnan, D. *Modularity of the Rankin–Selberg L-series, and multiplicity one for* SL(2). Ann. of Math. (2) 152 (2000), no. 1, 45–111.

[25] Ramakrishnan, D. *Algebraic cycles on Hilbert modular fourfolds and poles of L-functions.* Algebraic groups and arithmetic, 221–274, Tata Inst. Fund. Res., Mumbai, 2004.

[26] Ramakrishnan, D. *Modularity of solvable Artin representations of* GO(4)*-type.* Int. Math. Res. Not. 2002, no. 1, 1–54.

[27] Ramakrishnan, D. *Irreducibility and cuspidality.* Representation theory and automorphic forms, 1–27, Progr. Math., 255, Birkhuser Boston, Boston, MA, 2008.

[28] Ramakrishnan, D. and Shahidi, F. *Siegel modular forms of genus 2 attached to elliptic curves.* Math. Res. Lett. 14 (2007), no. 2, 315–332.

[29] Ramakrishnan, D. and Wang, S. *A cuspidality criterion for the functorial product on* $\mathrm{GL}(2) \times \mathrm{GL}(3)$ *with a cohomological application.* Int. Math. Res. Not. 2004, no. 27, 1355–1394.

[30] Shahidi, F. *A proof of Langlands' conjecture on Plancherel measures; Complementary series for p-adic groups.* Ann. of Math., 132, 273–330 (1990).

[31] Soudry, D. *On Langlands functoriality from classical groups to* GL_n. Formes automorphes. I. Astérisque No. 298 (2005), 335–390.

Dept. of Mathematics, Oklahoma State University, Stillwater, OK 74078, USA
E-mail address: `asgari@math.okstate.edu`

Dept. of Mathematics, Oklahoma State University, Stillwater, OK 74078, USA
E-mail address: `araghur@math.okstate.edu`

Clay Mathematics Proceedings
Volume 13, 2011

Types and Explicit Plancherel Formulæ for Reductive p-adic Groups

Colin J. Bushnell, Guy Henniart, and Philip C. Kutzko

To Freydoon Shahidi, on his sixtieth birthday

0.1. Let F be a non-Archimedean local field with finite residue field, and let G be a connected, reductive F-group. That is, G is the group $\mathbf{G}(F)$ of F-rational points of a connected, reductive algebraic group $\mathbf{G}$ defined over F, equipped with its natural (locally profinite) topology inherited from F. This paper is concerned with the complex representation theory of G. The aim is to give a framework in which algebraic aspects of the smooth representation theory, expressed in the theory of types, can be connected via explicit formulæ with analytic aspects of the unitary representation theory, encapsulated in the Plancherel measure. In this introductory essay, we assume the reader to have some facility with the basic concepts of the two subjects involved, but more detail will be given in the body of the paper.

0.2. Let $\mathcal{H}(G)$ be the space of locally constant, compactly supported functions $f : G \to \mathbb{C}$. We fix a Haar measure μ on G. We endow $\mathcal{H}(G)$ with the operation $(a, b) \mapsto a\star b$ of μ-convolution, relative to which it becomes an associative $\mathbb{C}$-algebra. If (π, V) is a smooth representation of G, a standard construction extends π to an algebra homomorphism $\pi : \mathcal{H}(G) \to \mathrm{End}_{\mathbb{C}}(V)$. If e is an idempotent element of $\mathcal{H}(G)$ and (π, V) a smooth representation of G, the space $\pi(e)V$ is a module over the $\mathbb{C}$-algebra $e\mathcal{H}e = e \star \mathcal{H}(G) \star e$. If $\mathrm{Irr}_e\, G$ denotes the set of isomorphism classes of irreducible smooth representations (π, V) for which $\pi(e) \neq 0$, and if $\mathrm{Irr}(e\mathcal{H}e)$ is the set of isomorphism classes of simple $e\mathcal{H}e$-modules, the map $(\pi, V) \mapsto \pi(e)V$ induces a canonical bijection

$$(*) \qquad \mathrm{Irr}_e\, G \xrightarrow{\ \approx\ } \mathrm{Irr}(e\mathcal{H}e).$$

The central part of this paper studies the analogue of this bijection for *unitary* representations of G.

2010 *Mathematics Subject Classification.* Primary 22E50; Secondary 22D25.

Key words and phrases. Reductive p-adic group, Plancherel measure, type, Hecke algebra.

Research of the first- and second-named authors was partially supported by EU Network "Arithmetical Algebraic Geometry". Research of the third-named author was partially supported by NSF grant DMS-9003213. This author also thanks l'Institut des Hautes Études Scientifiques for support while working on an earlier draft of this paper.

0.3. So, let $\widehat{G}$ denote the topological space of isomorphism classes of irreducible, continuous, unitary representations of G on Hilbert spaces. This space carries the *Plancherel measure* $\hat{\mu}$ dual to μ; the support of $\hat{\mu}$ is the *reduced dual* ${}_r\widehat{G}$ of G.

We let $e \in \mathcal{H}(G)$ be idempotent and self-adjoint, relative to the canonical (anti-linear) involution on $\mathcal{H}(G)$. The algebra $e\mathcal{H}e$ carries an involution inherited from $\mathcal{H}$ and a related inner product. It thereby acquires the structure of a *normalized Hilbert algebra.* Taking an appropriate completion, we obtain a C^* algebra ${}_rC^*(e\mathcal{H}e)$ containing $e\mathcal{H}e$ as a dense subalgebra. The dual of ${}_rC^*(e\mathcal{H}e)$, denoted ${}_r\widehat{C}^*(e\mathcal{H}e)$, is a topological space which can be identified with a subset of $\mathrm{Irr}(e\mathcal{H}e)$. It carries a canonical Borel measure $\hat{\mu}_{e\mathcal{H}e}$, with a definition analogous to that of the Plancherel measure (see 3.2 below).

If $(\pi, V) \in \widehat{G}$, we may again extend π to a homomorphism $\pi : \mathcal{H}(G) \to \mathrm{End}_{\mathbb{C}}(V)$. Our first result is a direct C^*-analogue of the bijection $(*)$ above.

Theorem A. *Let $e \in \mathcal{H}(G)$ be self-adjoint and idempotent. Define ${}_r\widehat{G}(e)$ to be the set of $(\pi, V) \in {}_r\widehat{G}$ for which $\pi(e) \neq 0$.*

(a) *The set ${}_r\widehat{G}(e)$ is open in ${}_r\widehat{G}$.*

(b) *If $(\pi, V) \in {}_r\widehat{G}(e)$, there is a unique $\widehat{m}_e(V) \in {}_r\widehat{C}^*(e\mathcal{H}e)$ isomorphic to $\pi(e)V$ as $e\mathcal{H}e$-module. The map*

$$\widehat{m}_e : {}_r\widehat{G}(e) \longrightarrow {}_r\widehat{C}^*(e\mathcal{H}e)$$

is a homeomorphism.

(c) *If S is a Borel subset of ${}_r\widehat{G}(e)$, then*

$$\hat{\mu}(S) = e(1_G)\, \hat{\mu}_{e\mathcal{H}e}\big(\widehat{m}_e(S)\big).$$

0.4. The primary source for self-adjoint idempotents in $\mathcal{H}(G)$ is the representation theory of compact open subgroups of G: if K is a compact open subgroup and ρ is an irreducible smooth representation of K, then ρ gives a self-adjoint idempotent $e_\rho \in \mathcal{H}(G)$ such that, for any smooth representation (π, V) of G, the space $\pi(e_\rho)V$ is the ρ-isotypic subspace V^ρ of V. Thus ${}_r\widehat{G}(e_\rho)$ is the set of $(\pi, V) \in {}_r\widehat{G}$ for which $\mathrm{Hom}_K(\rho, \pi) \neq 0$: it is easier to denote this set by ${}_r\widehat{G}(\rho)$.

Theorem A determines the structure of the topological space ${}_r\widehat{G}(\rho)$, along with the measure $\hat{\mu}|{}_r\widehat{G}(\rho)$, purely in terms of the Hilbert algebra $e_\rho \star \mathcal{H}(G) \star e_\rho$. In practice, however, it is better to introduce a second stage. Let $\mathcal{H}(G, \rho)$ be the convolution algebra of compactly supported ρ-spherical functions on G. One often refers to $\mathcal{H}(G, \rho)$ as the ρ-spherical Hecke algebra of G. It carries a canonical Hilbert algebra structure, with associated C^*-algebra ${}_rC^*(G, \rho)$ and Plancherel measure $\hat{\mu}_{\mathcal{H}(G,\rho)} = \hat{\mu}_\rho$. Again, the C^*-algebra dual ${}_r\widehat{C}^*(G, \rho)$ is canonically identified with a set of isomorphism classes of simple $\mathcal{H}(G, \rho)$-modules.

If (π, V) is a smooth or continuous unitary representation of G, the space $V_\rho = \mathrm{Hom}_K(\rho, \pi)$ carries the structure of $\mathcal{H}(G, \rho)$-module. The algebras $\mathcal{H}(G, \rho)$, $e_\rho \star \mathcal{H}(G) \star e_\rho$ are Morita equivalent (in the algebraic sense). This implies that the C^*-algebras ${}_rC^*(G, \rho)$, ${}_rC^*(e_\rho\mathcal{H}e_\rho)$ are strongly Morita equivalent. Stitching these facts together, we get:

Theorem B. *Let K be a compact open subgroup of G and ρ an irreducible smooth representation of K. The map $(\pi, V) \mapsto V_\rho$ induces a homeomorphism*

$$\widehat{m}_\rho : {}_r\widehat{G}(\rho) \xrightarrow{\approx} {}_r\widehat{C}^*(G,\rho).$$

If S is a Borel subset of ${}_r\widehat{G}(\rho)$, then

$$\hat{\mu}(S) = \frac{\dim \rho}{\mu(K)}\, \hat{\mu}_\rho(\widehat{m}_\rho(S)).$$

The sole dependence of Plancherel measure on an abstract algebraic structure gives us a method of transferring information between groups. In its most direct form, we get:

Corollary C. *For $i = 1, 2$, let G_i be a connected reductive F-group, let K_i be a compact open subgroup of G_i, and let ρ_i be an irreducible smooth representation of K_i. Let μ_i be a Haar measure on G_i and let $\hat{\mu}_i$ be the corresponding Plancherel measure on ${}_r\widehat{G}_i$. Let*

$$j : \mathcal{H}(G_1, \rho_1) \xrightarrow{\approx} \mathcal{H}(G_2, \rho_2)$$

be an isomorphism of Hilbert algebras. The map j then induces a homeomorphism

$$\widehat{\jmath} : {}_r\widehat{G}_2(\rho_2) \xrightarrow{\approx} {}_r\widehat{G}_1(\rho_1)$$

such that

$$\frac{\mu_1(K_1)}{\dim \rho_1}\, \hat{\mu}_1\big(\widehat{\jmath}(S)\big) = \frac{\mu_2(K_2)}{\dim \rho_2}\, \hat{\mu}_2(S),$$

for any Borel subset S of ${}_r\widehat{G}_2(\rho_2)$.

0.5. Theorem A and Corollary C acquire interest from the fact that their hypotheses are often observed in concrete situations of some importance. To describe these, we need to recall some algebraic structures. Let $\mathfrak{R}(G)$ denote the category of smooth (complex) representations of G. The theory of the Bernstein Centre [**4**] gives a decomposition of $\mathfrak{R}(G)$ as the direct product of a family of full subcategories

$$\mathfrak{R}(G) = \prod_{\mathfrak{s} \in \mathfrak{B}(G)} \mathfrak{R}_\mathfrak{s}(G).$$

The index set $\mathfrak{B}(G)$, sometimes called the *Bernstein spectrum,* consists of classes of pairs (L, σ) in which L is an F-Levi subgroup of G and σ is an irreducible cuspidal[1] representation of L, modulo the relation of *inertial equivalence* (recalled in 1.3 below). The decomposition of $\mathfrak{R}(G)$ leads to a decomposition of ${}_r\widehat{G}$. If (π, V) is an irreducible unitary representation of G, the space V^∞ of G-smooth vectors in V carries an irreducible smooth representation π^∞ of G. We define ${}_r\widehat{G}(\mathfrak{s})$ to be the set of $\pi \in {}_r\widehat{G}$ for which $\pi^\infty \in \mathfrak{R}_\mathfrak{s}(G)$. We accordingly get a decomposition

$${}_r\widehat{G} = \bigcup_{\mathfrak{s} \in \mathfrak{B}(G)} {}_r\widehat{G}(\mathfrak{s})$$

in which the union is disjoint. We can place this structure within the context of Theorem A via the following result, proved in 3.6 below.

[1]We use "cuspidal" as synonymous with the standard usages "supercuspidal", "absolutely cuspidal". Experience with representations over fields of positive characteristic suggests that such a simplification is called for.

Proposition D. *Let $\mathfrak{s} \in \mathfrak{B}(G)$. There exists a self-adjoint idempotent $e_\mathfrak{s} \in \mathcal{H}(G)$ such that ${}_r\widehat{G}(\mathfrak{s}) = {}_r\widehat{G}(e_\mathfrak{s})$.*

The sets ${}_r\widehat{G}(\mathfrak{s})$ are all therefore open, and closed, in ${}_r\widehat{G}$, and the Plancherel measure $\hat{\mu}$ is determined by its restrictions to the components ${}_r\widehat{G}(\mathfrak{s})$.

0.6. Proposition D shows that, in principle at least, the analytic information embedded in the Plancherel meaure is already detectable in the basic algebraic structure of $\mathfrak{R}(G)$. However, one has little information concerning the idempotent $e_\mathfrak{s}$ or the structure of the Hilbert algebra $e_\mathfrak{s}\mathcal{H}e_\mathfrak{s}$ unless one can invoke the theory of types. The definition of $\mathfrak{s}$-type will be recalled later; here, it is sufficient to note that, if an irreducible representation λ of a compact open subgroup J of G is an $\mathfrak{s}$-type in G, then ${}_r\widehat{G}(\lambda) = {}_r\widehat{G}(\mathfrak{s})$.

The simplest example is that where $G = \mathrm{GL}_n(F)$. Here, any $\mathfrak{s} \in \mathfrak{B}(G)$ admits an $\mathfrak{s}$-type (J, λ). The Hecke algebra $\mathcal{H}(G, \lambda)$ is isomorphic to a tensor product of affine Hecke algebras. Refining mildly the techniques of [**11**], [**13**], one sees that this isomorphism is one of Hilbert algebras. The determination of Plancherel measure for G is thus reduced to one basic case, that of representations with Iwahori-fixed vector.

0.7. Explicit constructions of types, and the descriptions of their Hecke algebras, are now known for many groups. If G is $\mathrm{GL}_n(F)$ or $\mathrm{SL}_n(F)$, $\mathfrak{s}$-types have been constructed and their Hecke algebras calculated for all $\mathfrak{s} \in \mathfrak{B}(G)$ [**11**], [**12**], [**14**], [**17**], [**18**]. When $\mathbf{G}$ is an inner form of $\mathbf{GL}_n$ or a classical group, the most important kinds of type are covered in two series of papers culminating in [**27**] or [**29**] respectively. The paper [**6**] contains a very useful technique for constructing types in groups from types in Levi subgroups. For principal series representations of split groups G, see [**25**]; for "level zero" representations of arbitrary G, see [**22**] or [**23**] (and [**21**] for the Hecke algebras); for cuspidal representations of arbitrary G (in large residual characteristic) see [**31**].

The determination of the Hecke algebra of a type (K, ρ) follows a fairly standard pattern: one exhibits an explicit isomorphism of $\mathcal{H}(G, \rho)$ with a combinatorially defined algebra, closely related to an affine Hecke algebra. It seems intrinsic to the approach that the isomorphism may be normalized to preserve the Hilbert algebra structure. We do not pursue any examples beyond $\mathrm{GL}_n(F)$, but this background suggests that the methods of this paper will be widely applicable to computing Plancherel measure.

0.8. This paper is a refreshed version of a preprint [**10**] which has been in circulation for some ten years. Some of the ideas go back further, for example to the influential [**19**]. Various editions of [**10**] have been cited several times, for example in [**1**], [**2**], and its ideas have started to diffuse into the common consciousness. Formal publication is therefore somewhat overdue.

1. Smooth representation theory

We review the basic concepts. Throughout, $G = \mathbf{G}(F)$ is a connected reductive F-group, as in the Introduction.

1.1. For this subsection, however, it would be enough to assume only that G is a unimodular, locally profinite group. In the spirit of fixing our notation, we recall the elementary parts of the smooth representation theory of such a group G. The reader may consult the first four sections of [**8**] for a complete account.

Let (π, V) be a representation of G. Thus V is a complex vector space and π is a group homomorphism $G \to \mathrm{Aut}_{\mathbb{C}}(V)$. It is (algebraically) *irreducible* if V is nonzero and admits no G-invariant subspace other than $\{0\}$ and V.

If K is a compact open subgroup of G, we denote by V^K the space of $v \in V$ such that $\pi(k)v = v$, $k \in K$. We recall that (π, V) is called *smooth* if V is the union of its subspaces V^K, as K ranges over the compact open subgroups of G. It is called *admissible* if it is smooth and V^K has finite dimension, for all K.

If (π, V) is a (not necessarily smooth) representation of G, we write

$$V^\infty = \bigcup_K V^K,$$

where, again, K ranges over all compact open subgroups of G. The set V^∞ is indeed a subspace of V, and it is stable under $\pi(G)$. We denote by π^∞ the implied homomorphism $G \to \mathrm{Aut}_{\mathbb{C}}(V^\infty)$. Certainly, (π^∞, V^∞) is a smooth representation of G. We refer to the elements of V^∞ as the *G-smooth vectors in V*.

Let $\mathcal{H}(G)$ denote the space of functions $G \to \mathbb{C}$ which are both locally constant and compactly supported. Thus $\mathcal{H}(G)$ is spanned by the characteristic functions of double cosets KgK, as K ranges over the compact open subgroups of G and g over G. We fix a Haar measure μ on G and define

$$a \star b(g) = \int_G a(x)\, b(x^{-1}g)\, d\mu(x), \qquad \begin{cases} a, b \in \mathcal{H}(G), \\ g \in G. \end{cases}$$

The function $a \star b$ then lies in $\mathcal{H}(G)$. The binary operation $\star$, called *μ-convolution,* endows $\mathcal{H}(G)$ with the structure of associative $\mathbb{C}$-algebra. When there is no fear of confusion, we abbreviate $\mathcal{H}(G) = \mathcal{H}$ and $a \star b = ab$.

Let M be a left $\mathcal{H}$-module; one says that M is *nondegenerate*[2] if $\mathcal{H}M = M$. If (π, V) is a smooth representation of G, one may extend the homomorphism π to an algebra homomorphism $\pi : \mathcal{H} \to \mathrm{End}_{\mathbb{C}}(V)$ by setting

$$\pi(a)v = \int_G a(g)\, \pi(g)\, v\, d\mu(g), \quad a \in \mathcal{H},\ v \in V.$$

In this way, V becomes a nondegenerate $\mathcal{H}$-module. The categories of smooth representations of G and of nondegenerate $\mathcal{H}$-modules are then effectively identical: see [**8**] 4.2 for a full account.

1.2. We now make significant use of the fact that $G = \mathbf{G}(F)$ is a *connected reductive F-group.* The reader may consult [**15**], [**4**] or [**5**] for proofs of the results recalled in this subsection. A fundamental property of such groups is:

(1.2.1) *Any irreducible smooth representation of G is admissible.*

Let (π, V) be a smooth representation of G, and let $(\check{\pi}, \check{V})$ be the *smooth dual* of (π, V). By definition, $\check{\pi}$ denotes the natural action of G on the space $\check{V} = (V^*)^\infty$ of G-smooth vectors in the linear dual space $V^* = \mathrm{Hom}_{\mathbb{C}}(V, \mathbb{C})$ (carrying the obvious

---

[2]In the terminology of [**8**], M would be called a *smooth* module, but that usage is unhelpful here.

G-action). There is a canonical bilinear pairing $\check{V}\times V\to\mathbb{C}$, denoted $(\check{v},v)\mapsto\langle\check{v},v\rangle$, and satisfying

$$\langle\check{\pi}(g)\check{v},\pi(g)v\rangle=\langle\check{v},v\rangle,$$

for $\check{v}\in\check{V}$, $v\in V$, $g\in G$.

Suppose now that the smooth representation (π,V) is irreducible. We say (π,V) is *cuspidal* if every function on G of the form

$$g\longmapsto\langle\check{v},\pi(g)v\rangle,\quad \check{v}\in\check{V},\ v\in V,$$

is compactly supported modulo the centre of G.

1.3. We recall some ideas and results from [**4**]. Let G° denote the subgroup of G generated by the compact subgroups of G. Certainly, G° is an open normal subgroup of G, and one knows that G/G° is free abelian of finite rank. An *unramified quasicharacter* of G is a homomorphism $G\to\mathbb{C}^\times$ which is trivial on G°. We write $X(G)$ for the group of unramified quasicharacters of G.

Let L be an F-Levi subgroup of G. That is to say, $L=\mathbf{L}(F)$, where $\mathbf{L}$ is a Levi component of a parabolic subgroup $\mathbf{P}$ of $\mathbf{G}$, with both $\mathbf{L}$ and $\mathbf{P}$ defined over F. In particular, L is a connected reductive F-group, and we can form the group $X(L)$.

We consider the set of pairs (L,σ) consisting of an F-Levi subgroup L of G and an irreducible cuspidal representation σ of L: we call such a pair a *cuspidal datum in* G. Two cuspidal data (L_i,σ_i), $i=1,2$, are deemed *inertially equivalent in* G if there exist $g\in G$ and $\chi\in X(L_2)$ such that $L_2=L_1^g=g^{-1}L_1g$ and σ_2 is equivalent to the representation $\sigma_1^g\otimes\chi : x\mapsto\chi(x)\sigma_1(gxg^{-1})$ of L_2. We denote by $\mathfrak{B}(G)$ the set of inertial equivalence classes of cuspidal data in G: this is the *Bernstein spectrum* mentioned in the Introduction. We denote by $[L,\sigma]_G$ the G-inertial equivalence class of the cuspidal datum (L,σ) in G.

We recall one of the major building blocks of the theory. Let ι denote the functor of normalized smooth induction.

(1.3.1) *Let π be an irreducible smooth representation of G.*

(1) *There is a cuspidal datum (L,σ) in G and an F-parabolic subgroup P of G, with Levi component L, such that π is equivalent to a subquotient of $\iota_P^G\,\sigma$.*
(2) *The datum (L,σ) is uniquely determined by π, up to G-conjugacy.*

The result implies, in particular, that the inertial equivalence class $[L,\sigma]_G$ is determined by π: it is called the *inertial support* of π.

Let $\mathfrak{R}(G)$ denote the category of smooth representations of G, and let $\mathfrak{s}\in\mathfrak{B}(G)$. We define a full subcategory $\mathfrak{R}_\mathfrak{s}(G)$ of $\mathfrak{R}(G)$ by deeming that its object class shall consist of all smooth representations π of G with the property that every irreducible subquotient of π has inertial support $\mathfrak{s}$.

(1.3.2) *The abelian category $\mathfrak{R}(G)$ is the direct product of its subcategories $\mathfrak{R}_\mathfrak{s}(G)$, $\mathfrak{s}\in\mathfrak{B}(G)$.*

That is to say:

(1.3.3) *Let (π,V) be a smooth representation of G.*

(1) *The space V has a unique maximal G-subspace $(\pi_\mathfrak{s},V_\mathfrak{s})$ which is an object of $\mathfrak{R}_\mathfrak{s}(G)$.*

(2) *The space V is the direct sum of its subspaces $V_{\mathfrak{s}}$, $\mathfrak{s} \in \mathfrak{B}(G)$.*
(3) *If (π', V') is a further smooth representation of G, then*

$$\mathrm{Hom}_G(V, V') = \prod_{\mathfrak{s} \in \mathfrak{B}(G)} \mathrm{Hom}_G(V_{\mathfrak{s}}, V'_{\mathfrak{s}}).$$

Let $\mathrm{Irr}\, G$ denote the set of equivalence classes of irreducible smooth representations of G. For each $\mathfrak{s} \in \mathfrak{B}(G)$, let $\mathrm{Irr}_{\mathfrak{s}}\, G$ be the set of equivalence classes of irreducible smooth representations of G with inertial support $\mathfrak{s}$. As an instance of (1.3.3) we have

$$\mathrm{Irr}\, G = \bigcup_{\mathfrak{s} \in \mathfrak{B}(G)} \mathrm{Irr}_{\mathfrak{s}}\, G,$$

the union being disjoint.

1.4. Let $e \in \mathcal{H}(G)$ be a nonzero idempotent. Thus $e \star \mathcal{H}(G) \star e = e\mathcal{H}e$ is a subalgebra of $\mathcal{H}$ with unit element e. Further, $\mathcal{H}e\mathcal{H}$ is a subalgebra (and two-sided ideal) of $\mathcal{H}$.

Let (π, V) be a smooth representation of G. The space $\pi(e)V$ is an $e\mathcal{H}e$-module, on which e acts as the identity. The G-subspace V_e of V, generated by $\pi(e)V$, is $\mathcal{H}e\mathcal{H}V = \mathcal{H}eV$. By definition, V_e is a *nondegenerate* $\mathcal{H}e\mathcal{H}$-module, in that $\mathcal{H}e\mathcal{H}V_e = V_e$. Moreover, $(V_e)_e = V_e$ and $eV_e = eV$.

Let $\mathrm{Irr}_e\, G$ denote the set of isomorphism classes of irreducible smooth representations (π, V) of G for which $\pi(e) \neq 0$. Let $\mathrm{Irr}(e\mathcal{H}e)$ be the set of isomorphism classes of simple $e\mathcal{H}e$-modules. Using the same argument as in [**8**] 4.3, we get:

(1.4.1) *Let (π, V) be an irreducible smooth representation of G.*

(1) *If $\pi(e)V \neq 0$, then $\pi(e)V$ is a simple $e\mathcal{H}e$-module.*
(2) *The map $(\pi, V) \mapsto \pi(e)V$ induces a bijection*

$$\mathrm{Irr}_e\, G \xrightarrow{\approx} \mathrm{Irr}(e\mathcal{H}e).$$

(3) *Consequently, any simple $e\mathcal{H}e$-module has finite complex dimension.*

(Part (3) here follows from (1.2.1).)

More generally, let $\mathfrak{R}_e(G)$ denote the full subcategory of $\mathfrak{R}(G)$ with object class consisting of those representations (π, V) for which $V = V_e$. We have a functor

$$\mathbf{m}_e : \mathfrak{R}_e(G) \longrightarrow e\mathcal{H}e\text{-Mod},$$
$$(\pi, V) \longmapsto \pi(e)V.$$

One says that the idempotent e is *special* if $\mathbf{m}_e$ is an equivalence of categories $\mathfrak{R}_e(G) \cong e\mathcal{H}e$-Mod. Special idempotents relate to the considerations of 1.3 as follows [**13**] 3.12.

(1.4.2) *Let $e \in \mathcal{H}(G)$ be idempotent. The following conditions are equivalent:*

(1) *e is special;*
(2) *the category $\mathfrak{R}_e(G)$ is closed relative to the formation of G-subquotients;*
(3) *there is a finite subset $\mathfrak{S}(e)$ of $\mathfrak{B}(G)$ such that*

$$\mathfrak{R}_e(G) = \prod_{\mathfrak{s} \in \mathfrak{S}(e)} \mathfrak{R}_{\mathfrak{s}}(G).$$

There is a strong converse in [**13**] 3.13:

(1.4.3) *Let $\mathfrak{s} \in \mathfrak{B}(G)$. There exists a special idempotent $e \in \mathcal{H}$ such that $\mathfrak{S}(e) = \{\mathfrak{s}\}$. In particular, the functor $\mathbf{m}_e$ induces an equivalence of categories $\mathfrak{R}_\mathfrak{s}(G) \cong$* $e\mathcal{H}e$-Mod.

We review the proof of this result (and mildly strengthen it) in 3.6 below.

1.5. A fruitful source of interesting idempotents is the representation theory of the compact open subgroups of G.

Let K be a compact open subgroup of G, and let (ρ, W) be an irreducible smooth representation of K. The group K is profinite, whence $\dim W$ is finite. We define a function $e_\rho : G \to \mathbb{C}$ by

$$e_\rho(x) = \frac{\dim W}{\mu(K)}\ \mathrm{tr}(\rho(x^{-1})),$$

if $x \in K$, and $e_\rho(x) = 0$ otherwise. Surely $e_\rho \in \mathcal{H}(G)$, and it is idempotent. If (π, V) is a smooth representation of G, then $\pi(e_\rho)V$ is the sum V^ρ of all irreducible K-subspaces of V isomorphic to ρ. (See [**8**] 4.4 for these elementary facts.)

There is a slightly different algebra related to the irreducible representation (ρ, W) of K. Consider the space $\mathcal{H}(G, \rho)$ of compactly supported functions $\phi : G \to \mathrm{End}_\mathbb{C}(\check{W})$ which satisfy

(1.5.1) $$\phi(k_1 g k_2) = \check{\rho}(k_1)\,\phi(g)\,\check{\rho}(k_2),$$

for $k_i \in K$ and $g \in G$. This space carries an operation of μ-convolution, making it into an associative $\mathbb{C}$-algebra. The function $\mathbf{e}_\rho \in \mathcal{H}(G, \rho)$, defined by

(1.5.2) $$\mathbf{e}_\rho(x) = \begin{cases} \mu(K)^{-1}\,\check{\rho}(x) & \text{if } x \in K, \\ 0 & \text{otherwise,} \end{cases}$$

provides a unit element.

A function ϕ on G satisfying (1.5.1) is called *ρ-spherical.* Thus the elements of $\mathcal{H}(G, \rho)$ are the compactly supported ρ-spherical functions on G. The algebra $\mathcal{H}(G, \rho)$ is often called the *ρ-spherical Hecke algebra of G.*

1.6. With the same notation as in 1.5, the pair (K, ρ) is called a *type in G* if the idempotent e_ρ is special. For $\mathfrak{s} \in \mathfrak{B}(G)$, one says that (K, ρ) is an $\mathfrak{s}$-type in G if it is a type and $\mathfrak{S}(e_\rho) = \{\mathfrak{s}\}$.

2. Unitary representations

We continue with the same connected, reductive F-group G, to give a sketch of the standard theory of unitary representations and Plancherel measure. We follow [**16**] and [**30**].

2.1. A *unitary representation* of G is a pair (π, V), where V is a Hilbert space and π is a homomorphism from G to the group of unitary operators on V, such that the map

$$G \times V \longrightarrow V,$$
$$(g, v) \longmapsto \pi(g)v,$$

is continuous. We say that (π, V) is *(topologically) irreducible* if V is nonzero and admits no proper, G-invariant, *closed* subspace.

Starting from a unitary representation (π, V), we can form the smooth representation (π^∞, V^∞), as in 1.1. A cornerstone of the theory is the following sequence of results [**3**]:

(2.1.1) *If (π, V) is a topologically irreducible unitary representation of G, then*

(1) *the space V^∞ is nonzero;*
(2) *the smooth representation (π^∞, V^∞) is irreducible and hence admissible.*

In the opposite direction, let (π, V) be an irreducible admissible representation of G. One says that (π, V) is *pre-unitary* if there is a positive definite Hermitian form $[\,,\,]$ on V satisfying

$$[\pi(g)v, \pi(g)w] = [v, w], \quad g \in G,\ v, w \in V.$$

The irreducibility of V implies readily that such a Hermitian form is uniquely determined, up to positive scale. We may therefore unambiguously define the completion $\widetilde{V}$ of V relative to the norm $v \mapsto [v, v]^{\frac{1}{2}}$. This is a Hilbert space, carrying a unitary representation $\tilde{\pi}$ of G.

(2.1.2) *If (π, V) is an irreducible, smooth pre-unitary representation of G, then the unitary representation $(\tilde{\pi}, \widetilde{V})$ is topologically irreducible.*

Let $\mathrm{Irr}^u\, G$ denote the set of isomorphism classes of irreducible, smooth, pre-unitary representations of G. Let $\widehat{G}$ denote the set of isomorphism classes of irreducible unitary representations of G. The preceding discussion yields:

(2.1.3) *The map $(\pi, V) \mapsto (\pi^\infty, V^\infty)$ induces a bijection*

$$\widehat{G} \xrightarrow{\approx} \mathrm{Irr}^u\, G.$$

2.2. We recall a standard construction: see [**30**] 14.2 for details.

If $f : G \to \mathbb{C}$ is a function, we define another function $f^* : G \to \mathbb{C}$ by $f^*(g) = \overline{f(g^{-1})}$, where the bar denotes complex conjugation.

As usual, let $L^1(G)$ denote the space of μ-measurable functions $f : G \to \mathbb{C}$ such that

$$\|f\|_1 = \int_G |f(x)|\, d\mu(x) < \infty.$$

The space $L^1(G)$ is stable under the operation $f \mapsto f^*$. It admits the binary operation of μ-convolution (as in 1.1), relative to which it is an associative $\mathbb{C}$-algebra. We have the properties

$$(a \star b)^* = b^* \star a^*, \quad \|a \star b\|_1 \leqslant \|a\|_1\, \|b\|_1, \quad \|a^*\|_1 = \|a\|_1,$$

for $a, b \in L^1(G)$. Indeed, $L^1(G)$ is a Banach $*$-algebra in which $\mathcal{H}(G)$ is dense.

Let (π, V) be a (not necessarily irreducible) unitary representation of G. For $a \in L^1(G)$, we define a bounded linear operator $\pi(a)$ on V by

$$\pi(a)v = \int_G a(x)\pi(x)v\, d\mu(x), \quad v \in V.$$

The map $a \mapsto \pi(a)$ is a Banach $*$-representation of $L^1(G)$. The algebra $\mathcal{B}(V)$ of bounded linear operators on V carries the canonical operator norm, which we denote $x \mapsto \|x\|_{\mathcal{B}(V)}$. The norm $a \mapsto \|\pi(a)\|_{\mathcal{B}(V)}$ on $L^1(G)$ then depends only on the isomorphism class of (π, V). One defines

$$\|a\| = \sup_{(\pi,V)} \|\pi(a)\|_{\mathcal{B}(V)}, \quad a \in L^1(G),$$

where (π, V) ranges over the set of isomorphism classes of unitary representations of G. We then have $\|a\| \leqslant \|a\|_1$, while $\|a\| = 0$ if and only if $a = 0$. The C^* *algebra* $C^*(G)$ *of* G is defined to be the completion of $L^1(G)$ relative to the norm $a \mapsto \|a\|$. The algebra structure of $L^1(G)$ and its involution extend to $C^*(G)$. It is indeed a C^* algebra, containing $\mathcal{H}(G)$ as a dense subalgebra.

We recall that a *representation*, or *$*$-representation*, of a C^* algebra C is an algebra homomorphism π of C in the algebra $\mathcal{B}(H)$ of bounded linear operators on a Hilbert space H, such that $\pi(a^*) = \pi(a)^*$. Any such representation is continuous. A nonzero representation of C is topologically irreducible (that is, has no proper, C-invariant *closed* subspace) if and only if it is algebraically irreducible. A representation (π, H) is called *nondegenerate* if $\pi(C)H = H$.

In parallel to 1.1, we have [**30**] 14.2.5:

(2.2.1) *Let (π, V) be a unitary representation of G. The map $a \mapsto \pi(a)$, $a \in L^1(G)$, extends uniquely to a $*$-representation of $C^*(G)$ on V. This representation of $C^*(G)$ is nondegenerate. Every nondegenerate representation of $C^*(G)$ is of this form, and this extension process induces an equivalence of categories between unitary representations of G and nondegenerate representations of $C^*(G)$.*

The dual $\widehat{C}^*(G)$ of $C^*(G)$ is defined to be the set of isomorphism classes of irreducible $*$-representations of $C^*(G)$. The process described in (2.2.1) identifies the unitary dual $\widehat{G}$ of G with $\widehat{C}^*(G)$. Each of these spaces carries a natural topology, and the identification $\widehat{C}^*(G) \cong \widehat{G}$ is a homeomorphism. (These topologies, and the relation between them, are discussed in [**30**] 14.6, 14.7. It is not necessary for us to recall the details.)

2.3. Let (π, V) be an irreducible unitary representation of G. Let $a \in \mathcal{H}(G)$. There exists a compact open subgroup K of G such that $a(kg) = a(g)$, for all $g \in G$ and $k \in K$. It follows that $\pi(a)V \subset V^K \subset V^\infty$, and (1.2.1), (2.1.1) together imply that $\dim V^K < \infty$. The operator $\pi(a)$ has finite-dimensional range, therefore, whence its *trace* $\operatorname{tr}\pi(a)$ is defined. It follows that the image $\pi(C^*(G))$ is contained in the algebra of *compact* operators on V. This says exactly that $C^*(G)$ (or G) is *liminal* (translated from [**16**]) or *CCR*, in the terminology of [**30**].

As a consequence, we have the following fundamental result [**16**] 18.8.1, [**30**] 14.11.2:

Plancherel Theorem. *There is a unique positive Borel measure $\hat{\mu}$ on $\widehat{G}$ such that*

$$f(1_G) = \int_{\widehat{G}} \operatorname{tr}\pi(f)\, d\hat{\mu}(\pi), \quad f \in \mathcal{H}(G). \tag{2.3.1}$$

The measure $\hat{\mu}$ is called the *Plancherel measure* on G, relative to μ.

Remarks.

(1) The formula (2.3.1) is initially valid for functions $f = a^* \star a$, $a \in \mathcal{H}(G)$. The version we have given follows from standard manipulations.

(2) The uniqueness of $\hat{\mu}$ follows as in [**16**] 18.8.2, on replacing $L^1(G) \cap L^2(G)$ by $\mathcal{H}(G)$.
(3) The uniqueness property implies immediately the relation $\widehat{c\mu} = c^{-1}\hat{\mu}$, $c > 0$.

2.4. We let G act by left translation on the space $L^2(G)$:

$$g : a \longmapsto \lambda_g a, \quad \lambda_g a(x) = a(g^{-1}x),$$

for $a \in L^2(G)$, $g, x \in G$. Thus $(\lambda, L^2(G))$ is a unitary representation of G. The corresponding action of $\mathcal{H}(G)$ (*cf.* 1.1) is given by

$$\lambda(f)\, a = f \star a, \quad f \in \mathcal{H}(G),\ a \in L^2(G).$$

As in 2.2, we view $L^2(G)$ as a representation of $C^*(G)$ and set

(2.4.1) $$_rC^*(G) = \lambda(C^*(G)).$$

Thus $_rC^*(G)$ is a C^* algebra, known as the *reduced* C^* algebra of G. Since $\mathcal{H}(G)$ acts faithfully on $L^2(G)$, it is a dense subalgebra of $_rC^*(G)$. We may equivalently define $_rC^*(G)$ as the completion of $\mathcal{H}$ relative to the operator norm given by its canonical action on $L^2(G)$.

The definition (2.4.1) identifies the dual $_r\widehat{C}^*(G)$ of $_rC^*(G)$ with a closed subset of $\widehat{C}^*(G)$. Under the identification (2.2) of $\widehat{C}^*(G)$ with $\widehat{G}$, the set $_r\widehat{C}^*(G)$ corresponds to a closed subset $_r\widehat{G}$ of $\widehat{G}$. The set $_r\widehat{G}$ is the *reduced* unitary dual of G.

The support of the Plancherel measure $\hat{\mu}$ is precisely $_r\widehat{G}$ [**16**] 18.8.4. Consequently:

(2.4.2) *The Plancherel measure $\hat{\mu}$ is the unique positive Borel measure on $_r\widehat{G}$ such that*

$$f(1_G) = \int_{_r\widehat{G}} \operatorname{tr} \pi(f)\, d\hat{\mu}(\pi),$$

for all $f \in \mathcal{H}(G)$.

3. Idempotents and Hilbert algebras

In this section, we consider the analogue of (1.4.1) for unitary representations. This requires an excursion into the basic theory of Hilbert algebras, as in [**16**].

3.1. We recall the basic definition.

Definition. *A* Hilbert algebra *is a $\mathbb{C}$-algebra A, with an anti-linear involution $*$ and carrying a positive definite Hermitian form $[\ ,\ ] = [\ ,\ ]_A$, such that*
(1) *$[x, y] = [y^*, x^*]$, $x, y \in A$;*
(2) *$[xy, z] = [y, x^*z]$, $x, y, z \in A$;*
(3) *for every $x \in A$, the mapping $y \mapsto xy$ of A into A is continuous with respect to the topology induced by $[\ ,\ ]$;*
(4) *the set of elements xy, $x, y \in A$, is dense in A.*

Observe that condition (4) of the definition is automatically satisfied when A has a unit. A Hilbert algebra A will be called *normalized* if it has a unit $\mathbf{e}$ and the inner product satisfies $[\mathbf{e}, \mathbf{e}] = 1$.

Let A be a normalized Hilbert algebra, and let $\widetilde{A}$ be the Hilbert space obtained by completing A with respect to $[\ ,\ ]$. The action of A on itself by left multiplication induces an injection of A into the C^*-algebra $\mathcal{B}(\widetilde{A})$ of bounded linear operators on $\widetilde{A}$. We denote by ${}_rC^*(A)$ the closure of the image of A in $\mathcal{B}(\widetilde{A})$.

In particular, ${}_rC^*(A)$ is a C^* algebra, the dual of which we denote ${}_r\widehat{C}^*(A)$.

3.2. Let A be a normalized Hilbert algebra, and let $(\pi, V) \in {}_r\widehat{C}^*(A)$. The irreducibility of V implies that the unit element $\mathbf{e}$ of A acts on V as the identity. If ${}_rC^*(A)$ is *liminal*, $\pi(\mathbf{e})$ is a compact operator and has a trace. We deduce that V is finite-dimensional. It follows that ${}_rC^*(A)$ is liminal if and only if all its irreducible representations have finite dimension.

Supposing this condition holds, let (π, V) be an irreducible representation of ${}_rC^*(A)$. Since $\dim V$ is finite, any A-stable subspace is closed, and hence is ${}_rC^*(A)$-stable. It follows that V is simple as algebraic ${}_rC^*(A)$-module, whence $\pi({}_rC^*(A)) = \mathrm{End}_{\mathbb{C}}(V)$. Since $\pi(A)$ is dense in $\pi({}_rC^*(A))$, we conclude that $\pi(A) = \mathrm{End}_{\mathbb{C}}(V)$. This implies that V is simple as an A-module. Moreover, any A-isomorphism between irreducible representations of ${}_rC^*(A)$ is a ${}_rC^*(A)$-isomorphism. We have proved:

Lemma. *Let A be a normalized Hilbert algebra such that ${}_rC^*(A)$ is liminal. Let* $\mathrm{Irr}\,A$ *denote the set of isomorphism classes of irreducible (algebraic) representations of A. The map $(\pi, V) \mapsto (\pi|A, V)$ induces an injection ${}_r\widehat{C}^*(A) \to \mathrm{Irr}\,A$.*

Notation. *If A is a normalized Hilbert algebra such that ${}_rC^*(A)$ is liminal, then* ${}_r\mathrm{Irr}\,A$ *denotes the canonical image of ${}_r\widehat{C}^*(A)$ in* $\mathrm{Irr}\,A$.

In this situation, we have an analogue of the Plancherel Theorem:

Proposition. *Let A be a normalized Hilbert algebra, with unit element* $\mathbf{e}$, *such that ${}_rC^*(A)$ is liminal. There is then a unique positive Borel measure $\hat{\mu}_A$ on ${}_r\widehat{C}^*(A)$ such that*

$$[a, \mathbf{e}]_A = \int_{{}_r\widehat{C}^*(A)} \mathrm{tr}\,\pi(a)\, d\hat{\mu}_A(\pi), \quad a \in A.$$

Proof. By 17.2.1 of [**16**], the map $(x, y) \mapsto [x, y]_A$ extends to a maximal bi-trace on ${}_rC^*(A)$. Let f be the associated trace. We apply [**16**] 8.8.5 to obtain a positive Borel measure $\hat{\mu}_A$ on the dual space ${}_r\widehat{C}^*(A)$ with the property that

$$f(y^*y) = \int_{{}_r\widehat{C}^*(A)} \mathrm{tr}\,\pi(y^*y)\, d\hat{\mu}_A(\pi), \quad y \in {}_rC^*(A).$$

By definition, $f(b^*b) = [b, b]^{\frac{1}{2}}$, $b \in A$. Applying this in the cases $b = a \pm \mathbf{e}$, $b = a \pm i\cdot\mathbf{e}$ and making the usual manipulations, we see that $\hat{\mu}_A$ has the required property. To show that it is thereby uniquely determined, we argue exactly as in [**16**] 18.8.2, replacing $L^1(G) \cap L^2(G)$ by A. $\square$

We refer to this measure $\hat{\mu}_A$ as the *Plancherel measure* for A.

3.3. We return to the connected reductive F-group G. Let $e \in \mathcal{H}(G)$ be idempotent, and assume that e is *self-adjoint*, that is, $e^* = e$. The algebra $e\mathcal{H}e$ is therefore stable under the standard involution $x \mapsto x^*$ on $\mathcal{H}$. Imposing the inner product

(3.3.1) $$[a, b] = e(1_G)^{-1}\, a^* \star b(1_G), \quad a, b \in e\mathcal{H}e,$$

$e\mathcal{H}e$ becomes a normalized Hilbert algebra. We may form the C^* algebra ${}_rC^*(e\mathcal{H}e)$ of $e\mathcal{H}e$, as in 3.1.

The metric on $e\mathcal{H}e$, given by the inner product, is equivalent to the restriction of the L^2-metric on $\mathcal{H}$. So, in the notation of 3.1, the Hilbert space $(e\mathcal{H}e)^\sim$ is the closed subspace $e \star L^2(G) \star e$ of $L^2(G)$. The algebra norm on $e\mathcal{H}e$ is therefore given by the operator norm from its natural action on $e \star L^2(G) \star e$.

Remark. One can equally form a C^* algebra by taking the closure of $e\mathcal{H}e$ in ${}_rC^*(G)$: the resulting algebra is $e\,{}_rC^*(G)e$. This is given by the norm on $e\mathcal{H}e$ coming from its natural action on the closed subspace $e \star L^2(G)$ of $L^2(G)$. Since $e\mathcal{H}e$ does not annihilate the complement of $e \star L^2(G) \star e$ in $e \star L^2(G)$, the norm on $e\mathcal{H}e$ defining $e\,{}_rC^*(G)e$ is not, *a priori,* the same as that defining ${}_rC^*(e\mathcal{H}e)$. We will, however, show ((3.4.1), (3.5.3) below) that ${}_rC^*(e\mathcal{H}e) \cong e\,{}_rC^*(G)e$.

3.4. Let ${}_rC^*(e, G)$ denote the closure of $\mathcal{H}e\mathcal{H}$ in ${}_rC^*(G)$. Thus ${}_rC^*(e, G)$ is a closed two-sided ideal in ${}_rC^*(G)$ and a liminal C^* algebra. Its dual ${}_r\widehat{C}^*(G, e)$ is naturally identified with a subspace of ${}_r\widehat{C}^*(G)$. Observe that

$$e\,{}_rC^*(e, G)e = e\,{}_rC^*(G)e, \tag{3.4.1}$$

since each side is a closed subalgebra of ${}_rC^*(G)$ containing $e\mathcal{H}e = e\mathcal{H}e\mathcal{H}e$ as a dense subalgebra.

Following [**16**] 2.11, we put

$${}_r\widehat{G}(e) = \{\pi \in {}_r\widehat{G} : \pi(e) \neq 0\}.$$

The set ${}_r\widehat{G}(e)$ is open in ${}_r\widehat{G}$ ([**16**] 3.2.1), and the canonical homeomorphism ${}_r\widehat{G} \cong {}_r\widehat{C}^*(G)$ induces a homeomorphism

$${}_r\widehat{G}(e) \cong {}_r\widehat{C}^*(e, G). \tag{3.4.2}$$

3.5. Let $e \in \mathcal{H}$ be a self-adjoint idempotent as before, and let $(\pi, V) \in {}_r\widehat{G}(e)$. The space $\pi(e)V = \pi(e)V^\infty$ is therefore nonzero, it has finite dimension, and is an irreducible $e\mathcal{H}e$-module (1.4.1). We now give our central result.

Theorem. *Let e be a self-adjoint, idempotent element of $\mathcal{H}(G)$.*

(1) *The algebra ${}_rC^*(e\mathcal{H}e)$ is liminal.*
(2) *The map $(\pi, V) \mapsto \pi(e)V$ induces a bijection ${}_r\widehat{G}(e) \cong {}_r\mathrm{Irr}(e\mathcal{H}e)$, and a homeomorphism*

$$\widehat{\mathbf{m}}_e : {}_r\widehat{G}(e) \xrightarrow{\;\approx\;} {}_r\widehat{C}^*(e\mathcal{H}e). \tag{3.5.1}$$

(3) *If S is a Borel subset of ${}_r\widehat{G}(e)$, then*

$$\hat{\mu}(S) = e(1_G)\,\hat{\mu}_{e\mathcal{H}e}\big(\widehat{\mathbf{m}}_e(S)\big). \tag{3.5.2}$$

Preliminary remark. Take $(\pi, V) \in {}_r\widehat{G}(e)$. Obviously, $\pi(e)V$ is a module (in the algebraic sense) over the C^* algebra $e\,{}_rC^*(G)e$: indeed, it provides a $*$-representation of $e\,{}_rC^*(G)e$. This algebra, however, is not directly accessible in the manner of $e\mathcal{H}e$ or its associated C^* algebra ${}_rC^*(e\mathcal{H}e)$. We therefore need to work indirectly, showing *en route* that there is an isomorphism ${}_rC^*(e\mathcal{H}e) \cong e\,{}_rC^*(G)e$. Moreover, this isomorphism can be chosen to extend the identity map on $e\mathcal{H}e$, embedded canonically in either factor.

Proof. We consider ${}_rC^*(G)$ and its natural action on $L^2(G)$: recall that ${}_rC^*(G)$ is defined as the closure of the algebra $\mathcal{H}$ in $\mathcal{B}(L^2(G))$. The subspace $e \star L^2(G) \star e$ is stable under the subalgebra $e\,{}_rC^*(G)e$. Thus $e \star L^2(G) \star e$ affords a representation of $e\,{}_rC^*(G)e$. We denote by

$$\eta : e\,{}_rC^*(G)e \longrightarrow \mathcal{B}(e \star L^2(G) \star e)$$

the implied homomorphism. We identify $\mathcal{B}(e \star L^2(G) \star e)$ with the subalgebra of $\mathcal{B}(L^2(G))$ consisting of operators which annihilate the orthogonal complement of $e \star L^2(G) \star e$ in $L^2(G)$.

Lemma. *The homomorphism* $\eta : e\,{}_rC^*(G)e \to \mathcal{B}(e \star L^2(G) \star e)$ *is injective.*

Proof. We continue to denote multiplication in ${}_rC^*(G)$ by $(x, y) \mapsto xy$. On the other hand, we denote the natural action of $x \in \mathcal{B}(L^2(G))$ on $L^2(G)$ by

$$x : y \longmapsto x \cdot y, \quad y \in L^2(G).$$

We use the $\star$ notation when we wish to emphasize that one factor in the product belongs to $\mathcal{H}(G)$.

Let $a \in \operatorname{Ker}\eta$: in particular, $a \cdot e = 0$. Since ${}_rC^*(G)$ is the closure of $\mathcal{H}$ in $\mathcal{B}(L^2(G))$, the left action of ${}_rC^*(G)$ on $L^2(G)$ commutes with the right action of $\mathcal{H}$ by convolution:

$$x \cdot (y \star f) = (x \cdot y) \star f, \quad x \in {}_rC^*(G),\ f \in \mathcal{H},\ y \in L^2(G).$$

Therefore

$$a \cdot (e \star f) = (a \cdot e) \star f = 0,$$

for all $f \in \mathcal{H}$.

For $b, f \in \mathcal{H}$, we have $(b \cdot e) \star f = b \star e \star f = (be) \star f$. By continuity, this relation holds for $b \in {}_rC^*(G)$. Above, therefore, we can rewrite $(a \cdot e) \star f = (ae) \cdot f$ and so, by continuity, $(ae) \cdot f = 0$ for all $f \in L^2(G)$. However, $ae = a$, so $a \cdot f = 0$ for $f \in L^2(G)$. As $a \in {}_rC^*(G) \subset \mathcal{B}(L^2(G))$, we deduce that $a = 0$, as required. □

We return to the Hilbert algebra $e\mathcal{H}e$. As remarked in 3.3, ${}_rC^*(e\mathcal{H}e)$ is the closure in $\mathcal{B}(e \star L^2(G) \star e)$ of the image $\eta(e\mathcal{H}e)$ of $e\mathcal{H}e$. The algebra $\eta(e\,{}_rC^*(e, G)e)$ is closed, being the image of a morphism of C^* algebras. Therefore (recalling (3.4.1))

$${}_rC^*(e\mathcal{H}e) = \eta(e\,{}_rC^*(e, G)e).$$

The lemma thus implies that η induces a bijective, continuous homomorphism $e\,{}_rC^*(e, G)e \to {}_rC^*(e\mathcal{H}e)$ whence (*cf.* [**30**] 14.1.13):

Proposition. *The map* η *induces an isomorphism*

$$e\,{}_rC^*(G)e = e\,{}_rC^*(e, G)e \cong {}_rC^*(e\mathcal{H}e) \tag{3.5.3}$$

of C^* *algebras, which is the identity on* $e\mathcal{H}e$ *(embedded canonically in each factor).*

For $(\pi, V) \in {}_r\widehat{G}(e)$, we may now view $\pi(e)V$ as providing a representation of ${}_rC^*(e\mathcal{H}e)$. As $e\mathcal{H}e$-module, it is irreducible (1.4.1), so $\pi(e)V$ determines an element of ${}_r\widehat{C}^*(e\mathcal{H}e)$.

Next, we observe that since $\mathcal{H}e\mathcal{H} = \mathcal{H}e\mathcal{H}e\mathcal{H} \subset {}_rC^*(e, G)e\,{}_rC^*(e, G)$, the ideal ${}_rC^*(e, G)e\,{}_rC^*(e, G)$ is dense in ${}_rC^*(e, G)$. Thus $e\,{}_rC^*(G)e = e\,{}_rC^*(e, G)e$ is a "full corner" [**7**] in the algebra ${}_rC^*(e, G)$. Appealing to [**24**] Example 6.7, the C^* algebras ${}_rC^*(e, G)$, $e\,{}_rC^*(e, G)e$ are strongly Morita equivalent. Moreover, on categories of representations, this equivalence is the functor $(\pi, V) \mapsto \pi(e)V$. Consequently, the map

$$\widehat{\mathbf{m}}_e : {}_r\widehat{G}(e) \longrightarrow {}_r\widehat{C}^*(e\mathcal{H}e), \qquad (\pi, V) \longmapsto \pi(e)V, \tag{3.5.4}$$

is a homeomorphism.

In particular, every element of ${}_r\widehat{C}^*(e\mathcal{H}e)$ arises as $\pi(e)V$, for some $(\pi, V) \in {}_r\widehat{G}(e)$. As $\pi(e)V = \pi(e)V^\infty = \pi^\infty(e)V^\infty$, it follows that $\dim \pi(e)V < \infty$. Therefore every element of ${}_r\widehat{C}^*(e\mathcal{H}e)$ has finite dimension, whence ${}_rC^*(e\mathcal{H}e)$ is liminal. This proves part (1) of the theorem; part (2) follows from (3.5.4) and 3.2 Lemma.

As for part (3), let $f \in e\mathcal{H}e$ and let $(\pi, V) \in {}_r\widehat{G}(e)$. For $v \in \pi(e)V$, $v' \in (\pi(e)V)^\perp$, we have

$$[v, \pi(f)v'] = [\pi(f^*)v, v'] = [\pi(e)\pi(f^*)v, v'] = 0.$$

It follows that $\pi(f)v' \in \pi(e)V \cap (\pi(e)V)^\perp$, or $\pi(f)v' = 0$. Therefore

$$\operatorname{tr} \pi(f) = \operatorname{tr}\big(\pi(f)|\pi(e)V\big),$$

whence

$$e(1_G)\,[f, e] = f(1_G) = \int_{{}_r\widehat{G}(e)} \operatorname{tr} \pi(f)\, d\hat{\mu}_{e\mathcal{H}e}(\pi).$$

The result now follows from 3.2 Proposition and part (2) of the theorem. □

3.6. We make a connection with the considerations of 1.4. Let $\mathfrak{s} \in \mathfrak{B}(G)$, and define ${}_r\widehat{G}(\mathfrak{s})$ to be the set of $\pi \in {}_r\widehat{G}$ such that π^∞ has inertial support $\mathfrak{s}$. We write $\operatorname{Irr}_\mathfrak{s} G$ for the set of isomorphism classes of irreducible representations in $\mathfrak{R}_\mathfrak{s}(G)$, and use the other notation of 1.3, 1.4.

Proposition. *Let* $\mathfrak{s} \in \mathfrak{B}(G)$.

(1) *There exists a self-adjoint special idempotent* $e_\mathfrak{s} \in \mathcal{H}(G)$ *with* $\mathfrak{S}(e_\mathfrak{s}) = \{\mathfrak{s}\}$. *In particular,* ${}_r\widehat{G}(\mathfrak{s}) = {}_r\widehat{G}(e_\mathfrak{s}) \neq \emptyset$.

(2) *The set* ${}_r\widehat{G}(\mathfrak{s})$ *is open in* ${}_r\widehat{G}$ *and the map* $(\pi, V) \mapsto \pi(e_\mathfrak{s})V$ *induces a homeomorphism*

$$ {}_r\widehat{G}(\mathfrak{s}) \cong {}_r\widehat{C}^*\big(e_\mathfrak{s} \star \mathcal{H} \star e_\mathfrak{s}\big).$$

(3) *The space* ${}_r\widehat{G}$ *is the disjoint union of the open sets* ${}_r\widehat{G}(\mathfrak{s})$, $\mathfrak{s} \in \mathfrak{B}(G)$.

Proof. Assertion (2) follows from (1) and 3.5 Theorem, while (3) is implied by (1), (2.1.3), and the remark in 3.4.

From (1.4.3), we know that there exists a special idempotent e such that $\mathfrak{S}(e) = \{\mathfrak{s}\}$. We review the construction of e in [**13**] 3.13 to show e is self-adjoint.

Let K be a compact open subgroup of G, and take the idempotent e_K in $\mathcal{H}$ so that $\mu(K)e_K$ is the characteristic function of K. As in [**13**], we may choose K so that e_K is special and $\mathfrak{s} \in \mathfrak{S}(e_K)$. We let G act on $\mathcal{H}$ by left translation, and take the corresponding decomposition

$$\mathcal{H} = \bigoplus_{\mathfrak{t} \in \mathfrak{B}(G)} \mathcal{H}_\mathfrak{t},$$

with $\mathcal{H}_t \in \mathfrak{R}_\mathfrak{t}(G)$. The spaces $\mathcal{H}_\mathfrak{t}$ are two-sided ideals of $\mathcal{H}$ with the consequence that $\mathcal{H}_\mathfrak{t}\mathcal{H}_\mathfrak{u} = 0$ when $\mathfrak{t} \neq \mathfrak{u}$, $\mathfrak{t}, \mathfrak{u} \in \mathfrak{B}(G)$. We accordingly write $e_K = e + e'$, with

$e \in \mathcal{H}_{\mathfrak{s}}$ and $e' \in \bigoplus_{\mathfrak{t}\neq\mathfrak{s}} \mathcal{H}_{\mathfrak{t}}$. The functions e, e' are idempotent and $ee' = e'e = 0$. Moreover, $\mathcal{H}_{\mathfrak{s}} = \mathcal{H}e\mathcal{H}$, e is special, and $\mathfrak{S}(e) = \{\mathfrak{s}\}$. In particular, $\operatorname{Irr}_{\mathfrak{s}} G = \operatorname{Irr}_e G$.

We have to show that e is self-adjoint. If (π, V) is an irreducible smooth representation of G, let (π^*, V^*) denote the complex conjugate of its smooth dual $(\check\pi, \check V)$. Likewise, if $\mathfrak{t} \in \mathfrak{B}(G)$, $\mathfrak{t} = [L, \sigma]$ say, we can define $\mathfrak{t}^* = [L, \sigma^*]$. The class $\mathfrak{t}$ only determines the cuspidal representation σ up to twisting with some $\chi \in X(L)$, so we may assume that σ is pre-unitary. This implies $\sigma^* \cong \sigma$, so $\mathfrak{t}^* = \mathfrak{t}$, for all $\mathfrak{t}$. That is, $(\operatorname{Irr}_{\mathfrak{t}} G)^* = \operatorname{Irr}_{\mathfrak{t}} G$, for all $\mathfrak{t}$. Therefore

$$\operatorname{Irr}_{e^*} G = (\operatorname{Irr}_e G)^* = (\operatorname{Irr}_{\mathfrak{s}} G)^* = \operatorname{Irr}_{\mathfrak{s}} G.$$

In particular, $e^* \in \mathcal{H}_{\mathfrak{s}}$ and $(e')^* \in \bigoplus_{\mathfrak{t}\neq\mathfrak{s}} \mathcal{H}_{\mathfrak{t}}$. It follows that

$$e+e' = e_K = e_K^* = e^*+(e')^*,$$

whence $e^* = e$, as required. □

4. The ρ-spherical algebras

Let K be a compact open subgroup of G, and let $\rho : K \to \operatorname{Aut}_{\mathbb{C}}(W)$ be an irreducible smooth representation of K. The representation ρ defines a self-adjoint idempotent $e_\rho \in \mathcal{H}(G)$, as in 1.5. We henceforward abbreviate

$${}_r\widehat{G}(\rho) = {}_r\widehat{G}(e_\rho).$$

We consider the ρ-spherical Hecke algebra $\mathcal{H}(G, \rho)$ introduced in 1.5, first showing that it carries a canonical structure of *Hilbert algebra*. Taking a completion as in 3.1, we obtain a C^* algebra ${}_rC^*(G, \rho)$. The main result of this section compares the duals ${}_r\widehat{G}(\rho)$ and ${}_r\widehat{C}^*(G, \rho) = ({}_rC^*(G, \rho))^\wedge$, along with their Plancherel measures.

4.1. Let $(\check\rho, \check W)$ be the contragredient of (ρ, W). We write

$$E_\rho = \operatorname{End}_{\mathbb{C}}(\check W).$$

We first define a canonical Hilbert algebra structure on E_ρ. There is a K-invariant scalar product $[\,,\,]_W$ on $\check W$, uniquely determined up to a positive constant. The algebra E_ρ then carries the adjoint involution $a \mapsto a^*$ given by

$$[a^* v, w]_W = [v, aw]_W, \quad a \in E_\rho,\ v, w \in \check W.$$

We define a scalar product $[\,,\,] = [\,,\,]_{E_\rho}$ on E_ρ by

$$[a, b] = \frac{\operatorname{tr}(a^* b)}{\dim W}, \quad a, b \in E_\rho.$$

With this structure, E_ρ is a normalized Hilbert algebra.

As in 1.5, let $\mathcal{H}(G, \rho)$ be the μ-convolution algebra of compactly supported E_ρ-valued functions h on G which satisfy

$$h(k_1 x k_2) = \check\rho(k_1) h(x) \check\rho(k_2), \quad x \in G,\ k_i \in K.$$

This algebra $\mathcal{H}(G, \rho)$ has identity $\mathbf{e}_\rho$ (1.5.2). Further, it is a normalized Hilbert algebra, relative to the involution $h \mapsto h^*$ defined by

$$h^*(x) = \big(h(x^{-1})\big)^*, \quad h \in \mathcal{H}(G, \rho),\ x \in G,$$

and scalar product

$$[h_1, h_2] = \frac{\mu(K)}{\dim W} \operatorname{tr}\big(h_1^* \star h_2(1_G)\big).$$

We abbreviate ${}_rC^*(G,\rho) = {}_rC^*(\mathcal{H}(G,\rho))$.

4.2. We denote by $a \mapsto a^t$ the linear anti-isomorphism $E_\rho \to E_{\check\rho} = \mathrm{End}_{\mathbb{C}}(W)$ given by

$$\langle a^t w, \check w\rangle = \langle w, a\check w\rangle, \quad w \in W,\ \check w \in \check W.$$

Here, $\langle\,,\rangle$ is the canonical bilinear pairing $W \times \check W \to \mathbb{C}$. Observe that $\check\rho(k)^t = \rho(k^{-1})$, $k \in K$. We also use the notation $x \mapsto x^t$ for the inverse map $E_{\check\rho} \to E_\rho$.

This transposition operation gives rise to a linear anti-isomorphism of $\mathbb{C}$-algebras

$$\mathcal{H}(G,\rho) \longrightarrow \mathcal{H}(G,\check\rho),$$
$$h \longmapsto h^t,$$

where h^t denotes the function $g \mapsto h(g^{-1})^t$ on G.

We consider the G-representation $c\text{-Ind}_K^G\,\rho$ *compactly induced by* ρ. The underlying vector space consists of the compactly supported functions $\phi : G \to W$ such that

$$\phi(kg) = \rho(k)\phi(g), \quad k \in K,\ g \in G,$$

and G acts by right translation.

We define a *right* action of $\mathcal{H}(G,\rho)$ on $c\text{-Ind}\,\rho$ by

$$h : \phi \longmapsto \phi\, h = h^t \star \phi, \quad \phi \in c\text{-Ind}\,\rho,\ h \in \mathcal{H}(G,\rho).$$

This action induces an algebra isomorphism (see, for example, [**13**] 2.5)

$$\mathcal{H}(G,\rho) \cong \mathrm{End}_G(c\text{-Ind}\,\rho). \tag{4.2.1}$$

Frobenius Reciprocity [**8**] 2.5 gives an isomorphism

$$\mathrm{Hom}_G(c\text{-Ind}\,\rho, \pi) \cong \mathrm{Hom}_K(\rho,\pi),$$

for any smooth representation (π, V) of G. Set $V_\rho = \mathrm{Hom}_K(\rho,\pi)$. The right action of $\mathcal{H}(G,\rho)$ on $c\text{-Ind}\,\rho$ gives a left action on $\mathrm{Hom}_G(c\text{-Ind}\,\rho,\pi)$, and hence a representation π_ρ of $\mathcal{H}(G,\rho)$ on the space V_ρ.

Remark. Explicitly, the action of $\mathcal{H}(G,\rho)$ on V_ρ is given by

$$\pi_\rho(h)\phi : w \longrightarrow \int_G \pi(x)\,\phi\big((h(x))^t w\big)\,d\mu(x), \tag{4.2.2}$$

for $\phi \in V_\rho$, $h \in \mathcal{H}(G,\rho)$, $w \in W$.

4.3. Let (π, V) be an irreducible *unitary* representation of G. We can again form the space $V_\rho = \mathrm{Hom}_K(\rho,\pi)$. Since the kernel of ρ is open in G, we have $V_\rho = (V^\infty)_\rho$, whence V_ρ carries the structure of an $\mathcal{H}(G,\rho)$-module. Moreover,

$${}_r\widehat G(\rho) = {}_r\widehat G(e_\rho) = \{(\pi, V) \in {}_r\widehat G : V_\rho \neq 0\}.$$

We state our second result.

Theorem. *Let K be a compact open subgroup of G and (ρ, W) an irreducible smooth representation of K.*

(1) *The algebra ${}_rC^*(G,\rho)$ is liminal.*

(2) *Let $(\pi, V) \in {}_r\widehat G(\rho)$. The space V_ρ is nonzero and simple as $\mathcal{H}(G,\rho)$-module. The natural action of $\mathcal{H}(G,\rho)$ on V_ρ extends uniquely to a representation π_ρ of ${}_rC^*(G,\rho)$.*

(3) *The map* $(\pi, V) \mapsto (\pi_\rho, V_\rho)$ *induces a homeomorphism*

$$\widehat{m}_\rho : {}_r\widehat{G}(\rho) \xrightarrow{\approx} {}_r\widehat{C}^*(G, \rho).$$

(4) *If* S *is a Borel subset of* ${}_r\widehat{G}(\rho)$, *then*

$$\hat{\mu}(S) = \frac{\dim \rho}{\mu(K)}\, \hat{\mu}_\rho(\widehat{m}_\rho(S)),$$

where we abbreviate $\hat{\mu}_\rho = \hat{\mu}_{\mathcal{H}(G,\rho)}$.

The proof occupies the rest of the section. In outline, 3.5 gives ${}_r\widehat{G}(\rho) \cong {}_r\widehat{C}^*(e_\rho \mathcal{H} e_\rho)$ and the relation between Plancherel measures $\hat{\mu}|{}_r\widehat{G}(\rho)$, $\hat{\mu}_{e_\rho \mathcal{H} e_\rho}$. We therefore need to clarify the relation between ${}_r\widehat{C}^*(e_\rho \mathcal{H} e_\rho)$ and ${}_r\widehat{C}^*(G, \rho)$.

4.4. As the first step, we consider the algebra

$$\mathcal{H}_W(G, \rho) = \mathcal{H}(G, \rho) \otimes_{\mathbb{C}} E_{\check{\rho}}.$$

We may make $\mathcal{H}_W(G, \rho)$ into a normalized Hilbert algebra by setting

$$(h \otimes a)^* = h^* \otimes a^*,$$

and

$$[h \otimes a, h' \otimes a'] = [h, h'][a, a'],$$

for $h, h' \in \mathcal{H}(G, \rho)$, $a, a' \in E_{\check{\rho}}$ (*cf.* [**11**] 5.6.16, 5.6.17).

Proposition. *For* $h \in \mathcal{H}(G, \rho)$, $a \in E_{\check{\rho}}$, *define a function* $f = f_{(h,a)}$ *on* G *by*

$$f(x) = \dim \rho \cdot \mathrm{tr}(h(x) a^t), \quad x \in G.$$

The map $(h, a) \mapsto f_{(h,a)}$ *induces an isomorphism of Hilbert algebras:*

$$\Upsilon : \mathcal{H}_W(G, \rho) \xrightarrow{\approx} e_\rho \mathcal{H}(G) e_\rho. \tag{4.4.1}$$

Proof. This is essentially Proposition 4.3.3 of [**11**]. All that needs to be checked is that, for $w \in W$, $\check{w} \in \check{W}$, $b \in E_\rho$, we have $b \circ (w \otimes \check{w})^t = (w \otimes b\check{w})^t$, which is immediate. This same observation, together with the fact that $h(1_G)$, $h \in \mathcal{H}(G, \rho)$, is a scalar matrix, serves to verify that the scalar product given above is the one defined in §4.3 of [**11**]. □

We write ${}_rC^*_W(G, \rho) = {}_rC^*(\mathcal{H}_W(G, \rho))$. Immediately we get:

Corollary.

(1) *The map* Υ *extends to an isomorphism* ${}_rC^*_W(G, \rho) \cong {}_rC^*(e_\rho \mathcal{H} e_\rho)$, *which induces a homeomorphism* $\widehat{\Upsilon} : {}_r\widehat{C}^*_W(G, \rho) \cong {}_r\widehat{C}^*(e_\rho \mathcal{H} e_\rho)$.
(2) *The algebra* ${}_rC^*_W(G, \rho)$ *is liminal.*
(3) *If* S *is a Borel subset of* ${}_r\widehat{C}^*_W(G, \rho)$, *then*

$$\hat{\mu}_{\mathcal{H}_W(G,\rho)}(S) = \hat{\mu}_{e_\rho \mathcal{H} e_\rho}(\widehat{\Upsilon}(S)).$$

4.5. Let M be an $\mathcal{H}(G,\rho)$-module; we define an $\mathcal{H}_W(G,\rho)$-module M_W by setting $M_W = M \otimes_{\mathbb{C}} W$ and letting $\mathcal{H}_W(G,\rho)$ act in the obvious way. The map $M \mapsto M_W$ then induces an equivalence of categories

$$\mathcal{F}_W : \mathcal{H}(G,\rho)\text{-Mod} \xrightarrow{\approx} \mathcal{H}_W(G,\rho)\text{-Mod}. \tag{4.5.1}$$

Lemma. *The functor $\mathcal{F}_W$ induces a homeomorphism*

$$\widehat{\mathcal{F}}_W : {}_r\widehat{C}^*(G,\rho) \xrightarrow{\approx} {}_r\widehat{C}^*_W(G,\rho).$$

Proof. We note first that the natural embedding

$$\mathcal{H}_W(G,\rho) \cong \mathcal{H}(G,\rho) \otimes_{\mathbb{C}} E_{\check{\rho}} \longrightarrow {}_rC^*(G,\rho) \otimes_{\mathbb{C}} E_{\check{\rho}}$$

extends to an isomorphism of C^* algebras:

$${}_rC^*_W(G,\rho) \cong {}_rC^*(G,\rho) \otimes_{\mathbb{C}} E_{\check{\rho}}.$$

Thus we have

$${}_r\widehat{C}^*_W(G,\rho) \cong \left({}_rC^*(G,\rho) \otimes_{\mathbb{C}} E_{\check{\rho}}\right)^{\wedge}.$$

The algebras ${}_rC^*(G,\rho)$ and ${}_rC^*(G,\rho)\otimes_{\mathbb{C}} E_{\check{\rho}}$ are strongly Morita equivalent, relative to the "imprimitivity bimodule" ${}_rC^*(G,\rho) \otimes W$. The corresponding homeomorphism of dual spaces is induced by $H \mapsto H_W = H \otimes W$. □

We noted in 4.4 Corollary that the algebra ${}_rC_W(G,\rho)$ is liminal. The last step in the proof shows that ${}_rC^*(G,\rho)$ is also liminal, as required for 4.3 Theorem (1).

Given a subset S of ${}_r\widehat{C}^*(G,\rho)$, we denote by $S_W = \widehat{\mathcal{F}}_W(S)$ the set of equivalence classes of representations $H_W \in {}_r\widehat{C}^*_W(G,\rho)$ for which $H \in S$.

Proposition. *Write $\hat{\mu}_\rho = \hat{\mu}_{\mathcal{H}(G,\rho)}$, and let S be a Borel subset of ${}_r\widehat{C}^*(G,\rho)$. We then have*

$$\hat{\mu}_\rho(S) = \dim\rho \cdot \hat{\mu}_{\mathcal{H}_W(G,\rho)}(S_W).$$

Proof. Let $h \in \mathcal{H}(G,\rho)$. We have

$$[h \otimes \mathbf{1}_W, \mathbf{1}_{\mathcal{H}_W(G,\rho)}]_{\mathcal{H}_W(G,\rho)} = [h \otimes \mathbf{1}_W, \mathbf{e}_\rho \otimes \mathbf{1}_W] = [h, \mathbf{e}_\rho]_{\mathcal{H}(G,\rho)}.$$

If, on the other hand, (π, H) is an irreducible representation of ${}_rC^*(G,\rho)$ then

$$\operatorname{tr}\pi_W(h \otimes \mathbf{1}_W) = \dim\rho \cdot \operatorname{tr}\pi(h).$$

Therefore

$$[h, \mathbf{e}_\rho]_{\mathcal{H}(G,\rho)} = \dim\rho \cdot \int_{{}_r\widehat{C}^*_W(G,\rho)} \operatorname{tr}\pi(h)\, d\hat{\mu}_{\mathcal{H}_W(G,\rho)}(\pi_W).$$

Our result now follows from the uniqueness of the measure $\hat{\mu}_\rho$ given by 3.2 Proposition. □

4.6. We prove 4.3 Theorem. We abbreviate $e = e_\rho$ and return to the category $\mathfrak{R}_e(G)$ of 1.4. We have the functor

$$\mathbf{m}_\rho = \mathbf{m}_e : \mathfrak{R}_e(G) \longrightarrow e\mathcal{H}e\text{-Mod}, \quad (\pi, V) \longmapsto \pi(e)V = V^\rho,$$

and also a functor

$$m_\rho : \mathfrak{R}_e(G) \longrightarrow \mathcal{H}(G,\rho)\text{-Mod}, \quad (\pi, V) \longmapsto (\pi_\rho, V_\rho),$$

where π_ρ denotes the natural action of $\mathcal{H}(G,\rho)$ on V_ρ. Using the notation (4.4.1), (4.5.1), we appeal to [**13**] 2.13 to see that the functor $\mathbf{m}_\rho$ factors as

$$\mathbf{m}_\rho = \Upsilon_* \circ \mathcal{F}_W \circ m_\rho,$$

where $\Upsilon_* : \mathcal{H}_W(G,\rho)\text{-Mod} \to e_\rho\mathcal{H}(G)e_\rho\text{-Mod}$ is the equivalence of categories induced by the algebra isomorphism Υ.

The functor $\mathbf{m}_\rho$ takes irreducible objects to irreducible objects (1.4.1). The equivalences Υ_*, $\mathcal{F}_W$ certainly share this property, so m_ρ also preserves irreducibility. If $\mathcal{E}$ denotes any of these functors, we write $\mathcal{E}^0$ for the induced bijection on sets of isomorphism classes of irreducible objects. Thus

$$\mathbf{m}_\rho^0 = \Upsilon_*^0 \circ \mathcal{F}_W^0 \circ m_\rho^0 : \operatorname{Irr}_e G \longrightarrow \operatorname{Irr} e\mathcal{H}e.$$

Restricting to ${}_r\widehat{G}(\rho)$, viewed as a subset of $\operatorname{Irr}_e G$ (as in (2.1.3)), 3.5 Theorem (2) shows that $\mathbf{m}_\rho^0$ induces the homeomorphism $\widehat{\mathbf{m}}_\rho : {}_r\widehat{G}(\rho) \cong {}_r\widehat{C}^*(e\mathcal{H}e)$. Similarly, Υ_*^0 gives the homeomorphism $\widehat{\Upsilon} : {}_r\widehat{C}^*_W(G,\rho) \cong {}_r\widehat{C}^*(e\mathcal{H}e)$ of 4.4 Proposition and $\mathcal{F}_W^0$ the homeomorphism $\widehat{\mathcal{F}}_W : {}_r\widehat{C}^*(G,\rho) \cong {}_r\widehat{C}^*_W(G,\rho)$ of 4.5 Lemma. We deduce that m_ρ^0 induces a homeomorphism $\widehat{m}_\rho : {}_r\widehat{G}(\rho) \cong {}_r\widehat{C}^*(G,\rho)$. It then follows from 3.5 Theorem (3), 4.4 Proposition and 4.5 Proposition that, for any Borel subset S of ${}_r\widehat{G}(\rho)$, we have

$$\hat{\mu}(S) = \frac{e_\rho(1_G)}{\dim \rho}\, \hat{\mu}_\rho(\widehat{m}_\rho(S)).$$

However, $e_\rho(1_G) = (\dim\rho)^2/\mu(K)$, which completes the proof of the theorem. $\square$

5. Transfer theorem

We outline a framework within which the results above can be applied.

5.1. For $i = 1, 2$, let G_i be a connected reductive F-group, let K_i be a compact open subgroup of G_i, and let ρ_i be an irreducible smooth representation of K_i. We fix a Haar measure μ_i on G_i and denote by $\hat{\mu}_i$ the corresponding Plancherel measure on ${}_r\widehat{G}_i$.

We assume given an isomorphism of Hilbert algebras

$$j : \mathcal{H}(G_1,\rho_1) \xrightarrow{\approx} \mathcal{H}(G_2,\rho_2). \tag{5.1.1}$$

The map j then extends to an isomorphism ${}_rC^*(G_1,\rho_1) \cong {}_rC^*(G_2,\rho_2)$ of C^* algebras, which we continue to denote j. It induces a homeomorphism

$$\widehat{\jmath} : {}_r\widehat{C}^*(G_2,\rho_2) \longrightarrow {}_r\widehat{C}^*(G_1,\rho_1), \quad (\pi, H) \longmapsto (\pi \circ j, H).$$

As before, we let ${}_r\widehat{G}_i(\rho_i)$ denote the set of $(\pi, H) \in {}_r\widehat{G}_i$ for which $\pi(e_{\rho_i}) \neq 0$. Theorem 4.3 gives homeomorphisms

$$\widehat{m}_{\rho_i} : {}_r\widehat{G}_i(\rho_i) \xrightarrow{\approx} {}_r\widehat{C}^*(G_i, \rho_i), \quad i = 1, 2.$$

We define

$$\mathcal{J} = \widehat{m}_{\rho_1}^{-1} \circ \widehat{\jmath} \circ \widehat{m}_{\rho_2} : {}_r\widehat{G}_2(\rho_2) \longrightarrow {}_r\widehat{G}_1(\rho_1). \tag{5.1.2}$$

As an immediate consequence of 4.3 Theorem, we get:

Corollary.

(1) *The map $\mathcal{J}$ of (5.1.2) is a homeomorphism.*

(2) *If S is a Borel subset of ${}_r\widehat{G}_2(\rho_2)$, then*

$$\frac{\mu_1(K_1)}{\dim \rho_1}\, \hat{\mu}_1(\mathcal{J}(S)) = \frac{\mu_2(K_2)}{\dim \rho_2}\, \hat{\mu}_2(S). \tag{5.1.3}$$

5.2. We record a simple result, useful when applying 5.1 Corollary.

We take G_i, ρ_i as in 5.1, and let $\mathcal{H}_0(G_i, \rho_i)$ denote the space of functions $f \in \mathcal{H}(G_i, \rho_i)$ for which $f(1_{G_i}) = 0$. Thus $\mathcal{H}_0(G_i, \rho_i)$ is the orthogonal complement in $\mathcal{H}(G_i, \rho_i)$ of the space $\mathbb{C}\mathbf{e}_{\rho_i}$ spanned by the unit element $\mathbf{e}_{\rho_i}$ of $\mathcal{H}(G_i, \rho_i)$.

Proposition. *Let*

$$k : \mathcal{H}(G_1, \rho_1) \xrightarrow{\approx} \mathcal{H}(G_2, \rho_2)$$

be an isomorphism of $\mathbb{C}$-algebras with involution. Suppose that

$$k(\mathcal{H}_0(G_1, \rho_1)) \subset \mathcal{H}_0(G_2, \rho_2).$$

The map k is then an isomorphism of Hilbert algebras.

Proof. Define a linear functional Λ_i on $\mathcal{H}_0(G_i, \rho_i)$ by

$$\Lambda_i(x) = \langle x, \mathbf{e}_{\rho_i} \rangle, \quad x \in \mathcal{H}(G_i, \rho_i).$$

This is the unique linear functional on $\mathcal{H}(G_i, \rho_i)$ with kernel $\mathcal{H}_0(G_i, \rho_i)$ such that $\Lambda_i(\mathbf{e}_{\rho_i}) = 1$. We therefore have $\Lambda_1 = \Lambda_2 \circ k$. However, the functional Λ_i satisfies

$$\langle x, y \rangle = \Lambda_i(x^* y), \quad x, y \in \mathcal{H}(G_i, \rho_i),$$

whence the result follows. □

5.3. Example. A prime example of 5.1 Corollary is given by the Main Theorem 5.6.6 of [**11**]. There, $G = \mathrm{GL}_n(F)$ and (J, λ) is a *simple type* in G. In particular, (J, λ) is an $\mathfrak{s}$-type in G, where $\mathfrak{s} = [L, \sigma]_G$ is of the form

$$\begin{aligned} L &= \mathrm{GL}_m(F) \times \mathrm{GL}_m(F) \times \cdots \times \mathrm{GL}_m(F), \\ \sigma &= \tau \otimes \tau \otimes \cdots \otimes \tau, \end{aligned} \tag{5.3.1}$$

for a divisor m of n and an irreducible cuspidal representation τ of $\mathrm{GL}_m(F)$. Any $\mathfrak{s} \in \mathfrak{B}(G)$, of the form (5.3.1), admits an $\mathfrak{s}$-type which is also a simple type.

Consider the case of (5.3.1) in which $m = 1$ and τ is trivial. One then refers to $\mathfrak{s}$ as the *trivial class* in $\mathfrak{B}(G)$. An $\mathfrak{s}$-type is provided by the trivial character 1_I of an Iwahori subgroup I of G. The corresponding Hecke algebra $\mathcal{H}(G, 1_I)$ is an affine Hecke algebra of type A_{n-1}, with parameter q (the cardinality of the residue field of F). The standard presentation ([**11**] (5.4.6)) shows that $\mathcal{H}(G, 1_I)$ depends, up to isomorphism of *Hilbert algebras,* only on n and q. We accordingly denote it $\mathcal{H}(n, q)$.

Returning to an arbitrary inertial class $\mathfrak{s}$ of the form (5.3.1), Theorem 5.6.6 of [**11**] gives a canonical family of algebra isomorphisms $\mathcal{H}(G,\lambda) \cong \mathcal{H}(e, q^f)$, for a pair of integers e, f depending on $\mathfrak{s}$ (or λ). Corollary 5.6.17 of [**11**] and 5.2 Proposition above combine to show that, among this family, there are isomorphisms of Hilbert algebras. Corollary 5.1 thus gives $\hat{\mu}|_r\widehat{G}(\mathfrak{s})$ in terms of the basic case $\hat{\mu}_{\mathcal{H}(e,q^f)}$. It follows that $\hat{\mu}|_r\widehat{G}(\mathfrak{s})$ is determined by the numerical invariants e, q^f attached to $\mathfrak{s}$ via (J,λ).

5.4. We return to the general case of a connected reductive F-group G. Let Z_G be the centre of G and let Z be a closed subgroup of Z_G such that Z_G/Z is compact. We fix a *unitary* character χ of Z, and consider the closed subset ${}_r\widehat{G}_\chi$ of ${}_r\widehat{G}$ consisting of classes of representations (π, V) such that $\pi(z)v = \chi(z)v$, $z \in Z$, $v \in V$. This can be analyzed in exactly the same way, via pairs (K,ρ), where K is open, containing Z such that K/Z is compact, and ρ is an irreducible representation of K such that $\rho|Z$ is a multiple of χ. The formal degree calculations [**11**] 7.7.11, [**12**] 8.2 follow exactly this course.

6. Split covers

There is a specific family of applications of 5.1 Corollary within the theory of types. Greater generality is possible in the following arguments, but we concentrate on the main case.

6.1. We recall, with necessary detail, a basic construction from [**13**] §6 *et seq.*

As before, let G be a connected reductive F-group. We fix an F-Levi subgroup M of G and a parabolic subgroup P of G with Levi component M. Thus $P = MN$, where N is the unipotent radical of P. We let $\overline{P}$ be the M-opposite of P, so that $\overline{P} = M\overline{N}$, where $\overline{N}$ is the unipotent radical of $\overline{P}$.

We fix Haar measures μ_M, μ_G on M, G respectively. Let $\mathfrak{t} \in \mathfrak{B}(M)$. We make the following:

Hypotheses.

(1) *There exists a* $\mathfrak{t}$*-type* (K_M, ρ_M) *in* M.
(2) *The pair* (K_M, ρ_M) *admits a* G*-cover* (K,ρ).
(3) *Every function* $\phi \in \mathcal{H}(G,\rho)$ *has support contained in* KMK.

Remark. A G-cover (K,ρ) of (K_M,ρ_M) satisfying hypothesis (3) is called a *split cover.*

The definition of cover [**13**] 8.1 requires that $K \cap M = K_M$ and

$$K = K \cap \overline{N} \cdot K_M \cdot K \cap N.$$

Moreover, $\rho|K_M \cong \rho_M$, while $\operatorname{Ker}\rho$ contains both $K \cap \overline{N}$ and $K \cap N$.

We may write $\mathfrak{t} = [L,\sigma]_M$, for a cuspidal datum (L,σ) in G with $L \subset M$. We set

$$\mathfrak{s} = \mathfrak{t}^G = [L,\sigma]_G \in \mathfrak{B}(G).$$

The pair (K,ρ) is an $\mathfrak{s}$-type in G [**13**] 8.3.

6.2. Under the hypotheses of 6.1, there is a close relation between the Hecke algebras $\mathcal{H}(M,\rho_M)$, $\mathcal{H}(G,\rho)$. We may view ρ_M, ρ as sharing the same representation space W. We first recall [**13**] 6.3:

Lemma 1. *Let $\phi \in \mathcal{H}(M,\rho_M)$ have support $K_M m K_M$, for some $m \in M$.*

(1) *There exists a unique function $T\phi \in \mathcal{H}(G,\rho)$, with support KmK, such that $T\phi(m) = \phi(m)$ (as elements of $\operatorname{End}_{\mathbb{C}}(\check{W})$).*
(2) *The map $T : \mathcal{H}(M,\rho_M) \to \mathcal{H}(G,\rho)$ is an isomorphism of vector spaces.*

Let $m \in M$; one says that m is *K-positive* if

$$m(K \cap N)m^{-1} \subset K \cap N, \quad \text{and} \quad m(K \cap \overline{N})m^{-1} \supset K \cap \overline{N}.$$

Part (ii) of [**13**] Theorem 7.2 yields:

Lemma 2. *There is a unique algebra isomorphism $t_0 : \mathcal{H}(M,\rho_M) \to \mathcal{H}(G,\rho)$ such that, if $\phi \in \mathcal{H}(M,\rho_M)$ has positive support, then $t_0\phi = T\phi$. Moreover,*

$$\operatorname{supp} t_0\theta = K \cdot \operatorname{supp}\theta \cdot K,$$

for any $\theta \in \mathcal{H}(M,\rho_M)$.

Let δ_N denote the module of M (or P) acting on N. That is, if μ_N is a Haar measure on N and $m \in M$, then

$$\delta_N(m) = \mu_N(mSm^{-1})/\mu_N(S),$$

for any measurable subset S of N.

For $\phi \in \mathcal{H}(M,\rho_M)$, define a function ϕ' by

$$\phi' : x \longmapsto \phi(x)\,\delta_N^{1/2}(x), \quad x \in M.$$

The map $\phi \mapsto \phi'$ is then an algebra automorphism of $\mathcal{H}(M,\rho_M)$. Moreover:

Proposition. *The map*

$$\begin{aligned} j : \mathcal{H}(M,\rho_M) &\longrightarrow \mathcal{H}(G,\rho), \\ \phi &\longmapsto t_0(\phi') \end{aligned}$$

is an isomorphism of Hilbert algebras.

Proof. The map j is certainly an isomorphism of algebras, and it satisfies

$$\operatorname{supp} j\phi = K \cdot \operatorname{supp}\phi \cdot K,$$

for any $\phi \in \mathcal{H}(M,\rho_M)$. The proposition will therefore follow from 5.2 Proposition when we verify that j is a homomorphism of algebras with involution. That, however, follows from [**13**] 7.4. □

6.3. Let us now abbreviate $\mathcal{A} = \mathcal{H}(G,\rho)$, $\mathcal{B} = \mathcal{H}(M,\rho_M)$. We have equivalences of categories

$$\begin{aligned} \mathfrak{R}_{\mathfrak{t}}(M) &\longrightarrow \mathcal{B}\text{-Mod}, & \mathfrak{R}_{\mathfrak{s}}(G) &\longrightarrow \mathcal{A}\text{-Mod}, \\ (\sigma, W) &\longmapsto W_{\rho_M}, & (\pi, V) &\longmapsto V_\rho. \end{aligned}$$

The isomorphism j of 6.2 gives a functor, indeed an equivalence of categories,

$$\begin{aligned} j_\star : \mathcal{B}\text{-Mod} &\longrightarrow \mathcal{A}\text{-Mod}, \\ X &\longmapsto \operatorname{Hom}_{\mathcal{B}}(\mathcal{A}, X). \end{aligned}$$

This has the property [**13**] 8.4:

Proposition. *The diagram*

$$\begin{array}{ccc} \mathfrak{R}_{\mathfrak{t}}(M) & \xrightarrow{\iota_P^G} & \mathfrak{R}_{\mathfrak{s}}(G) \\ \approx\downarrow & & \downarrow\approx \\ \mathcal{B}\text{-Mod} & \xrightarrow[j_\star]{} & \mathcal{A}\text{-Mod} \end{array}$$

commutes.

Remarks. Assume for the moment only that (K_M, ρ_M) is a $\mathfrak{t}$-type admitting a G-cover (K, ρ). There is then an algebra homomorphism j relative to which the preceding Proposition remains valid. When (K, ρ) is a split cover of (K_M, ρ_M)), then $j_\star$ and ι_P^G are equivalences of categories. Indeed, $j_\star$ is an equivalence of categories if and only if (K, ρ) is a split cover of (K_M, ρ_M). A more general phenomenon of this kind holds, without assuming the existence of types and covers. The group $N_G(M)$ acts on $\mathfrak{B}(M)$ by conjugation: let $N_G(\mathfrak{t})$ denote the stabilizer of $\mathfrak{t}$ for this action. According to [**26**], the functor $\iota_P^G : \mathfrak{R}_{\mathfrak{t}}(M) \to \mathfrak{R}_{\mathfrak{s}}(G)$ is an equivalence of categories if and only if $N_G(\mathfrak{t}) = M$.

We now translate to the context of unitary representations. Directly from 5.1 Corollary, we obtain:

Theorem. *With the hypotheses of* 6.1, *there is a unique map* $\mathcal{I} : {}_r\widehat{M}(\rho_M) \to {}_r\widehat{G}(\rho)$ *such that*

$$(\mathcal{I}V)_\rho = j_\star(V_{\rho_M}) = (\iota_P^G \pi^\infty)_\rho, \quad (\pi, V) \in {}_r\widehat{M}(\rho_M).$$

The map $\mathcal{I}$ *is a homeomorphism and, if* S *is a Borel subset of* ${}_r\widehat{M}(\rho_M)$, *then*

$$\mu_M(K_M)\, \hat{\mu}_M(S) = \mu_G(K)\, \hat{\mu}_G(\mathcal{J}S).$$

We remark that the map $\mathcal{I}$ here is the inverse of the map $\mathcal{J}$ given by 5.1. The map $\mathcal{J}$, in this situation, corresponds to taking the $\mathfrak{t}$-component of the normalized Jacquet module at N.

6.4. Example. Take $G = \mathrm{GL}_n(F)$, let $\mathfrak{s} = [L, \sigma]_G \in \mathfrak{B}(G)$. Write $\mathfrak{s}_L = [L, \sigma]_L \in \mathfrak{B}(L)$. Let M be an F-Levi subgroup of G containing $N_G(\mathfrak{s}_L)$ and minimal for this property. Write $\mathfrak{s}_M = [L, \sigma]_M \in \mathfrak{B}(M)$. In the obvious notation, we then have $N_G(\mathfrak{s}_M) = M$.

By the choice of M, there is an $\mathfrak{s}_M$-type (K'_M, ρ'_M) in M which is a tensor product of simple types. The Plancherel measure $\hat{\mu}_M | {}_r\widehat{M}(\mathfrak{s}_M)$ is therefore determined as in 5.3.

The main construction in [**14**] shows:

(6.4.1) *There is an open subgroup* K_M *of* K'_M *and a smooth representation* ρ_M *of* K_M *such that*

(1) *the representation of* K'_M *induced by* ρ_M *is equivalent to* ρ'_M *and*
(2) *the pair* (K_M, ρ_M) *admits a split* G*-cover* (K, ρ).

In particular, (K_M, ρ_M) is an $\mathfrak{s}_M$-type in M and (K, ρ) is an $\mathfrak{s}$-type in G. Proposition 6.2 gives an isomorphism $\mathcal{H}(M, \rho_M) \cong \mathcal{H}(G, \rho)$ of Hilbert algebras.

The canonical algebra isomorphism $\mathcal{H}(M,\rho_M) \cong \mathcal{H}(M,\rho'_M)$ of [**11**] (4.1.3) is an isomorphism of Hilbert algebras, so $\hat{\mu}_G|_r\widehat{G}(\mathfrak{s})$ is given by $\hat{\mu}_M|_r\widehat{M}(\mathfrak{s}_M)$ and 6.2.

6.5. Comment. Take integers $n_1, n_2 \geqslant 1$, set $n = n_1+n_2$, $G = \mathrm{GL}_n(F)$, $G_i = \mathrm{GL}_{n_i}(F)$, and let M denote the maximal Levi subgroup $G_1 \times G_2$ of G. We take an irreducible cuspidal representation π_i of G_i and form the representation $\sigma = \pi_1 \otimes \pi_2$ of M. For simplicity, let us assume that (in the case $n_1 = n_2$) the pairs (G_i, π_i) are inertially inequivalent. Setting $\mathfrak{s}_M = [M,\sigma]_M$, we have $N_G(\mathfrak{s}_M) = M$. We take types (K_M, ρ_M), (K,ρ) as in 6.4.

Let ψ be a nontrivial character of F. In [**9**], we calculated the *conductor* $f(\pi_1 \times \check{\pi}_2, \psi)$ of the pair $(\pi_1, \check{\pi}_2)$, in the sense of [**20**]. We followed the approach of [**28**], which obtains the local constant $\varepsilon(\pi_1 \times \check{\pi}_2, s, \psi)$ by comparing a standard intertwining operator with uniqueness of Whittaker model. Taking the composite of two suitable operators, one obtains a scalar operator with eigenvalue $q^{-f(\pi_1\times\check{\pi}_2,\psi)}$, where q is the size of the residue field of F. We calculated this as the quotient of volumes $\mu_G(K)/\mu_M(K_M)$. As remarked in [**28**], this composite of intertwining operators is indeed the quotient of Plancherel measures, as shown here directly. Note, however, that in [**9**] there is a relation between the Haar measures μ_G, μ_M dictated, in a subtle way, by the character ψ.

References

[1] A-M. Aubert and R.J. Plymen, *Explicit Plancherel formula for the p-adic group* GL(n) , C. R. Math. Acad. Sci. Paris **338** (2004), 843–848.

[2] ———, *Plancherel measure for* GL(n, F) *and* GL(m, D)*: explicit formulas and Bernstein decomposition*, J. Number Theory **112** (2005), 26–66 .

[3] I.N. Bernstein, *All reductive p-adic groups are tame*, Funct. Anal. Appl. **8** (1974), 91–93.

[4] J.-N. Bernstein (rédigé par P. Deligne), *Le "centre" de Bernstein*, Représentations des groupes réductifs sur un corps local, Hermann, Paris, 1984, 1–32.

[5] I.N. Bernstein and A.V. Zelevinsky, *Representations of the group $GL(n,F)$ where F is a local non-archimedean field*, Uspekhi Mat. Nauk **31** (1976), 5–70 .

[6] C. Blondel, *Quelques propriétés des paires couvrantes*, Math. Ann. **331** (2005), 243–257.

[7] L.G. Brown, *Stable isomorphism of hereditary subalgebras of C^*-algebras*, Pacific J. Math. **71** (1977), 335–348 .

[8] C.J. Bushnell and G. Henniart, *The local Langlands Conjecture for* GL(2), Grundlehren der Math. Wiss., vol. 335, Springer, 2006.

[9] C.J. Bushnell, G. Henniart and P. C. Kutzko, *Local Rankin-Selberg convolutions for* GL_n*: explicit conductor formula*, J. Amer. Math. Soc. **11** (1998), 703–730 .

[10] ———, *Towards an explicit Plancherel theorem for reductive p-adic groups*, Preprint (King's College London) (1998).

[11] C.J. Bushnell and P. C. Kutzko, *The admissible dual of $GL(N)$ via compact open subgroups*, Annals of Math. Studies, vol. 129, Princeton University Press, Princeton, 1993.

[12] ———, *The admissible dual of $SL(N)$ II*, Proc. London Math. Soc. (3) **68** (1994), 317–379 .

[13] ———, *Smooth representations of p-adic reductive groups: Structure theory via types*, Proc. London Math. Soc. (3) **77** (1998), 582–634 .

[14] ———, *Semisimple types for $GL(N)$*, Compositio Math. **119** (1999), 53-97.

[15] W. Casselman, *Introduction to the theory of admissible representations of* $\mathfrak{p}$*-adic reductive groups*, (1974) Preprint, `http://www.math.ubc.ca/~cass/research.html`.

[16] J. Dixmier, *Les C^*-algèbres et leurs représentations*, Gauthiers-Villars, Paris, 1969.

[17] D. Goldberg and A. Roche, *Types in* SL_n, Proc. London Math. Soc. (3) **85** (2002), 119–138 .

[18] ———, *Hecke algebras and* SL_n*-types*, Proc. London Math. Soc. (3) **90** (2005), 87–131 .

[19] R.E. Howe and A. Moy, *Harish-Chandra homomorphisms for p-adic groups*, CBMS Regional Conf. Series in Math., vol. 59, Amer. Math. Soc., Providence RI, 1985.

[20] H. Jacquet, I.I. Piatetskii-Shapiro and J.A. Shalika, *Rankin-Selberg convolutions*, Amer. J. Math. **105** (1983), 367–483 .

[21] L.E. Morris, *Tamely ramified intertwining algebras*, Invent. Math. **114** (1993), 1–54 .
[22] ———, *Level zero G-types*, Compositio Math. **118** (1999), 135–157.
[23] A. Moy and G. Prasad, *Jacquet functors and unrefined minimal K-types*, Comment. Math. Helv. **71** (1996), 98–121 .
[24] M.A. Rieffel, *Induced representations of C^*-algebras*, Adv. Math. **13** (1974), 176–257.
[25] A. Roche, *Types and Hecke algebras for principal series representations of split reductive p-adic groups*, Ann. Scient. École Norm. Sup. (4) **31** (1998), 361–413 .
[26] ———, *Parabolic induction and the Bernstein decomposition*, Compositio Math. **134** (2002), 113–133 .
[27] V. Sécherre and S.A.R. Stevens, *Représentations lisses de* $\mathrm{GL}_m(D)$ *IV: représentations supercuspidales*, J. Inst. Math. Jussieu **7** (2008), 527–574 .
[28] F. Shahidi, *Fourier transforms of intertwining operators and Plancherel measures for* $\mathrm{GL}(n)$, Amer. J. Math. **106** (1984), 67–111.
[29] S.A.R. Stevens, *The supercuspidal representations of p-adic classical groups*, Invent. Math. **172** (2008), 289–352 .
[30] N.R. Wallach, *Real Reductive Groups II*, Academic Press, New York, 1992.
[31] J.-K. Yu, *Construction of tame supercuspidal representations*, J. Amer. Math. Soc. **14** (2001), 579–622 .

King's College London, Department of Mathematics, Strand, London WC2R 2LS, UK.
E-mail address: colin.bushnell@kcl.ac.uk

Université de Paris-Sud, Laboratoire de Mathématiques d'Orsay, Orsay cedex F-91405 France; CNRS, Orsay cedex F-91405 France
E-mail address: Guy.Henniart@math.u-psud.fr

Mathematics Department, University of Iowa, Iowa City, IA52242, USA.
E-mail address: philip-kutzko@uiowa.edu

Clay Mathematics Proceedings
Volume 13, 2011

Jacquet Modules and the Asymptotic Behaviour of Matrix Coefficients

Bill Casselman

There is an intimate relationship between the asymptotic behaviour at infinity of matrix coefficients of admissible representations of both real and $\mathfrak{p}$-adic reductive groups and the way in which these representations embed into representations induced from parabolic subgroups. Weak versions of this were known for a long time for real groups but, until work of Jacquet on $\mathfrak{p}$-adic groups around 1970, one didn't really understand very well what was going on. Starting with Jacquet's observations, something now called the **Jacquet module** was constructed, first and most easily for $\mathfrak{p}$-adic groups, and then, with somewhat more difficulty, for real groups (see [**Casselman: 1974**] and [**Casselman: 1979**]). More or less by definition, the Jacquet module of a representation controls its embeddings into induced representations, and following another hint by Jacquet it was established without a lot of difficulty that algebraic properties of the Jacquet module also controlled the asymptotic behaviour of matrix coefficients. What characterizes this best is something called the **canonical pairing** between Jacquet modules associated to a representation and its contragredient. It is not difficult to define the canonical pairing abstractly and to relate it to matrix coefficients, but it is not so easy to determine it in cases where one knows the Jacquet modules explicitly. The formula of [**Macdonald: 1971**] for spherical functions is a particular example that has been known for a long time, but I'm not aware that this has been generalized in the literature in the way that I'll do it. In this paper I'll sketch very roughly how things ought to go.

For $\mathfrak{p}$-adic groups, there exists also a relationship between the asymptotic behaviour of Whittaker functions and the Whittaker analogue of the Jacquet module. The best known example here is the formula found in [**Casselman-Shalika: 1980**] for unramified principal series. I think it likely that a similar relationship exists for real groups, and that it will explain to some extent the recent work of [**Hirano-Oda: 2007**] on Whittaker functions for $SL_3(\mathbb{C})$. I'll make a few comments on this at the end of the paper.

The results discussed in this paper were originally commissioned, in a sense, by Jim Arthur many years ago. He subsequently used them, at least the ones concerned with real groups, in [**Arthur: 1983**], to prove the Paley-Wiener theorem. His argument depended on Harish-Chandra's Plancherel formula for real reductive

2010 *Mathematics Subject Classification.* Primary 22E50; Secondary 22E45.

groups but in fact, with a little thought and a few observations about Plancherel measures, one can deduce that formula at the same time as following Jim's proof.

There is one intriguing question raised by the results I sketch here. One trend in representation theory over the past few years has been to replace analysis by algebraic geometry. This is particularly striking in the theory of unramified representations of $\mathfrak{p}$-adic groups, where sheaves replace functions, which are related to them by Grothendieck's dictionary. I have in mind the version of Macdonald's formula as a consequence of the 'geometric' Satake isomorphism of [**Mirkovic-Vilonen: 2000**], for example. (There is an efficient survey of results about unramified representations in [**Haines et al.2003**].) What do these ideas have to say in the presence of ramification? Or about representations of real groups (which, according to [**Manin: 1991**], ought to be considered infinitely ramified)?

As I have said, I shall give few details here. My principal purpose, rather, is to exhibit plainly the astonishing parallels between the real and $\mathfrak{p}$-adic cases.

Throughout this paper, G will be a reductive group defined over a local field. In addition:

$$\begin{aligned} P &= \text{ a parabolic subgroup} \\ N = N_P &= \text{ its unipotent radical} \\ M = M_P &= \text{ a subgroup of } P \text{ isomorphic to } P/N \\ A = A_P &= \text{ maximal split torus in } M \\ \Sigma_P &= \text{ eigencharacters of } \mathrm{Ad}_{\mathfrak{n}} \,\big|\, A \\ A^{--} &= \big\{a \in A \,\big|\, |\alpha(a)| \le 1 \text{ for all } \alpha \in \Sigma_P\big\} \\ \overline{P} &= \text{ opposite of } P \quad (\text{i.e. } P \cap \overline{P} = M) \\ \delta_P(p) &= |\det \mathrm{Ad}_{\mathfrak{n}}(p)| \\ W &= \text{ the Weyl group with respect to } A\,. \end{aligned}$$

Thus δ_P is the **modulus character** of P.

Part I: What happens for $\mathfrak{p}$-adic groups

1. Notation. Suppose $\mathfrak{k}$ to be a $\mathfrak{p}$-adic field, G the $\mathfrak{k}$-rational points on an unramified reductive group defined over $\mathfrak{k}$. In addition to basic notation:

$$\begin{aligned} A^{--}(\varepsilon) &= \big\{a \in A \,\big|\, |\alpha(a)| < \varepsilon \text{ for all } \alpha \in \Sigma_P\big\} \\ K = K_{\mathfrak{o}} &= \text{ what [\textbf{Bruhat-Tits: 1966}] call a `good' maximal compact.} \end{aligned}$$

Thus $G = PK_{\mathfrak{o}}$ and if P is minimal we have the Cartan decomposition $G = K_{\mathfrak{o}}A^{--}K_{\mathfrak{o}}$.

I write $a \to_P 0$ for a in A_P if $|\alpha(a)| \to 0$ for all α in Σ_P. Because of the Cartan decomposition, this is one way points on G travel off to infinity. I'll say that a is near 0 if all of those same $|\alpha(a)|$ are small.

2. Admissible representations. In these notes a **smooth** representation (π, V) of G will be a representation of G on a complex vector space V with the property that the subgroup of G fixing any v in V is open. It is **admissible** if in addition the dimension of the subspace fixed by any open subgroup is finite.

The simplest examples are the principal series. If (σ, U) is an admissible representation of M, hence of P, the normalized induced representation is the right regular representation of G on

$$\operatorname{Ind}(\sigma \mid P, G) = \{f \in C^\infty(G, U) \mid f(pg) = \delta_P^{1/2}(p)\sigma(p)f(g)\}.$$

If (π, V) is an admissible representation of G, its **Jacquet module** V_N is the quotient of V by the linear span $V(N)$ of the vectors $v - \pi(n)v$, the universal N-trivial quotient of V. It is in a natural way a smooth representation of M, which turns out in fact to be admissible (see [**Casselman: 1974**]). The **normalized** Jacquet module π_N is this twisted by $\delta_P^{-1/2}$.

The point of the normalization of the Jacquet module is that Frobenius reciprocity becomes

$$\operatorname{Hom}_G\big(\pi, \operatorname{Ind}(\sigma \mid P, G)\big) = \operatorname{Hom}_M(\pi_N, \sigma).$$

3. Matrix coefficients. The contragredient $(\widetilde{\pi}, \widetilde{V})$ of an admissible representation (π, V) is the subspace of smooth vectors in its linear dual. The **matrix coefficient** associated to $\widetilde{v}$ in $\widetilde{V}$, v in V is the function

$$\Phi_{\widetilde{v},v}(g) = \langle \widetilde{v}, \pi(g)v \rangle.$$

The asymptotic behaviour of matrix coefficients at infinity on the $\mathfrak{p}$-adic group G is fairly simple, at least qualitatively. Jacquet first observed that if v lies in $V(N)$ then $\langle \pi(a)v, \widetilde{v} \rangle = 0$ for a near 0—i.e. if $|\alpha(a)|$ is small for all α in Σ_P. This implied, for example, that the matrix coefficients of a cuspidal representation had compact support modulo the centre of G. A refinement of Jacquet's observation is this:

There exists a unique pairing $\langle \widetilde{u}, u \rangle_{\mathrm{can}}$ of $\widetilde{V}_{\overline{N}}$ and V_N with this property: for each $\widetilde{v}$, v with images $\widetilde{u}$ in $\widetilde{V}_{\overline{N}}$, u in V_N there exists $\varepsilon > 0$ such that

$$\langle \widetilde{v}, \pi(a)v \rangle = \delta_P^{1/2}(a)\langle \widetilde{u}, \pi_N(a)u \rangle_{\mathrm{can}}$$

for a in $A^{--}(\varepsilon)$.

This canonical pairing induces an isomorphism of $\widetilde{V}_{\overline{N}}$ with the admissible dual of V_N. It has a geometric interpretation. For example, if $G = \mathrm{SL}_2(\mathbb{Q}_p)$ and v is fixed on the right by $\mathrm{SL}_2(\mathbb{Z}_p)$, then the matrix coefficient becomes a function on certain vertices of the Bruhat-Tits tree of G, and their asymptotic behaviour is related to the embedding of π via boundary behaviour into a representation induced from P, a kind of complex line bundle over the copy of $\mathbb{P}^1(\mathbb{Q}_p)$ that compactifies the tree.

4. Principal series. One can describe rather explicitly the Jacquet modules of representations induced from parabolic subgroups. Can one then describe the canonical pairing explicitly?

I'll explain this problem by an example. Suppose P to be minimal and π to be the principal series $\operatorname{Ind}(\chi \mid P, G)$ with χ a generic character of M. Its contragredient may be identified with

$$\operatorname{Ind}(\chi^{-1} \mid P, G)$$

where the pairing is

$$\langle \varphi, f \rangle = \int_{P\backslash G} \varphi(x)f(x)\,dx = \int_K \varphi(k)f(k)\,dk.$$

Can we find an explicit formula for $\langle \varphi, R_a f \rangle$ as $a \to 0$?

Let me recall what we know about the Jacquet module in this situation. The Bruhat decomposition tells us that $G = \bigsqcup_{w \in W} PwN$, with $PxN \subseteq \overline{PyN}$ if $x \le y$ in

W. As explained in [**Casselman: 1974**], the space $\mathrm{Ind}(\chi)$ is filtered by subspaces I_w of f with support on the closure of PwN. Similarly $G = \bigsqcup P w \overline{N}$, and $\mathrm{Ind}(\chi^{-1})$ is filtered by the opposite order on W. These two double coset decompositions are transversal to one another. For each w in the Weyl group W we have a map

$$\Omega_w(f) = \int_{N \cap wNw^{-1} \backslash N} f(w^{-1}n)\, dn$$

well defined on $I_{w^{-1}}$. It extends generically (that is to say for generic χ) to all of $\mathrm{Ind}(\chi)$, and determines an M-covariant map from π_N to $\mathbb{C}_{w\chi}$. All together, as long as χ is generic, these induce an isomorphism of the (normalized) Jacquet module π_N with $\oplus\, w\chi$. Similar functionals

$$\widetilde{\Omega}_w(f) = \int_{\overline{N} \cap wNw^{-1} \backslash \overline{N}} f(w^{-1}\overline{n})\, d\overline{n}$$

determine an isomorphism of $\widetilde{\pi}_{\overline{N}}$ with $\oplus\, w\chi^{-1}$.

The agreement of these formulas with the canonical pairing is clear—the two Jacquet modules are dual, piece by piece, but the duality is only determined up to scalar multiplication. We have therefore an asymptotic equality of the form

$$\langle \varphi, R_a f \rangle = \sum_w c_{w,\chi}\, \delta_P^{1/2}(a)\, w\chi(a) \cdot \widetilde{\Omega}_w(\varphi)\, \Omega_w(f)$$

for a near 0, with suitable constants $c_{w,\chi}$.

The problem we now pose is this: *What are those constants?*

There is a classical formula found in [**Langlands: 1988**] that gives the leading term of the asymptotic behaviour for χ in a positive chamber. It involves an analytic estimate of an integral. (I'll present a simple case of Langlands' calculation later on, that of $\mathrm{SL}_2(\mathbb{R})$.) The result stated here is a stronger and more precise version of that result. What makes the new version possible is the apparently abstract result relating asymptotic behaviour to Jacquet modules. One point is that we don't have to find the asymptotic behaviour of all matrix coefficients, just enough to cover all the different components of the Jacquet modules. Another is that we just have to look at one component at a time.

5. Integration. The question about the constants $c_{w,\chi}$ is not quite precise, because we have to be more careful about what the integrals mean. The first point is that it is not functions on $P\backslash G$ that one integrates, but densities. These may be identified with functions in $\mathrm{Ind}(\delta^{1/2})$, but *the identification depends on a choice of measures, and is definitely not canonical.* There are two integral formulas commonly used to make the identification of densities with functions in $\mathrm{Ind}(\delta^{1/2})$, and I shall introduce a third.

The first formula depends on the factorization $G = PK$. The integral on densities must be K-invariant, so we must have

$$\int_{P\backslash G} f(x)\, dx = \mathit{constant} \cdot \int_K f(k)\, dk\,,$$

where we take the total measure of K to be 1. Indeed, we can just define the integral by this formula, with this choice (and the constant equal to 1).

Since $P\overline{N}$ is open in G and the integral must be $\overline{N}$-invariant, we may also set

$$\int_{P\backslash G} f(x)\, dx = \mathit{constant} \cdot \int_{\overline{N}} f(n)\, dn\,.$$

where now we can choose the measure on $\overline{N}$ so that $K \cap \overline{N}$ has measure equal to 1. Understanding this second formula requires some work to show that the integral always converges.

The two formulas can only differ by a scalar. So we have

$$\int_K f(k)\,dk = \mu \int_{\overline{N}} f(\overline{n})\,d\overline{n}$$

for some constant μ, easy enough to determine explicitly in all cases:

Let B be an Iwahori subgroup of K and w_ℓ in K represent the longest element of the Weyl group. Then $\mu = \mathrm{meas}(Bw_\ell B)/\mathrm{meas}(K)$.

This is because $Bw_\ell B$ is completely contained in the single Bruhat double coset $Pw_\ell P$.

Integration over $\overline{N}$ (or, in a mild variation, over $w_\ell N$) is in many ways the more natural choice. It is, for example, the one that arises in dealing with Tamagawa measures (implicit in [**Langlands: 1966**]). But it has a serious problem, and that is the question of convergence. Convergence shouldn't really arise here. The theory of admissible representations of $\mathfrak{p}$-adic groups is essentially algebraic, and one should be able to work with an arbitrary coefficient field, for which analysis is not in the toolbox. We would therefore like to modify the formula

$$\int_{P\backslash G} f(x)\,dx = \int_{\overline{N}} f(\overline{n})\,d\overline{n}\,.$$

so as to make all integrals into sums, and avoid all convergence considerations.

This is easy, and a very similar idea will reduce the analytical difficulties for real groups to elementary calculus. Choosing representatives of W in K, we get also measures and similar formulas on the translates $P\overline{N}w^{-1}$. The variety $P\backslash G$ is covered by these open translates, and we can express

$$f = \sum_w f_w \quad (\text{support of } f_w \text{ on } P\overline{N}w^{-1})$$

as a sum of functions f_w, each with compact support on one of them. Then

$$\int_{P\backslash G} f(x)\,dx = \sum_w \int_{\overline{N}} f_w(xw^{-1})\,dx = \sum_w \int_{w\overline{N}w^{-1}} f_w(w^{-1}x)\,dx\,.$$

All these integrals are now finite sums. This in turn gives explicit measures to choose for evaluating Ω_w and $\widetilde{\Omega}_w$ because if $N_w = w\overline{N}w^{-1}$ then

$$N_w = (N_w \cap N)(N_w \cap \overline{N})\,.$$

One can also choose measures on each one-dimensional unipotent root group compatibly with the action of w in K.

6. A sample calculation. To evaluate the canonical pairing, we may deal with each summand of the Jacquet module by itself. We can therefore choose both f and φ with support on $P\overline{N}w^{-1}$, in which case it is easy to see what the asymptotic behaviour of $\langle \varphi, R_a f\rangle$ is.

For example, if f, φ have support on $P\overline{N}$ and a near 0 we have

$$\begin{aligned}\int_{\overline{N}} f(\overline{n}a)\varphi(\overline{n})\,d\overline{n} &= \int_{\overline{N}} f(a\cdot a^{-1}\overline{n}a)\varphi(\overline{n})\,d\overline{n} \\ &= \delta_P^{1/2}(a)\chi(a)\int_{\overline{N}} f(a^{-1}\overline{n}a)\varphi(\overline{n})\,d\overline{n} \\ &= \delta_P^{1/2}(a)\chi(a)f(1)\int_{\overline{N}}\varphi(\overline{n})\,d\overline{n} \text{ (if } a \text{ is near 0)} \\ &= \delta_P^{1/2}(a)\chi(a)\cdot\Omega_1(f)\cdot\widetilde{\Omega}_1(\varphi)\end{aligned}$$

Similar calculations work for all principal series, and as this suggests the canonical pairing turns out to be that with $c_{\chi,w} = \mu$ for all w.

The most general result of this sort is that if

$$\pi = \mathrm{Ind}(\sigma\,|\,Q, G)$$

then the canonical pairing for the Jacquet module π_{N_P} can be expressed explicitly in terms of the canonical pairing for the Jacquet modules of σ and certain N_P-invariant functionals on π determined by integration over pieces of the Bruhat filtration, together with analogues for $\overline{P}$ for $\widetilde{\pi}$. This formula, an explicit formula for the canonical pairing, is too elaborate to present here.

7. Range of equality. *For what range of a does the 'asymptotic' equation hold?* The answer depends on the ramification of χ as well as on the particular f and φ. The most important result is that if χ is unramified and both f and φ are fixed by an Iwahori subgroup, then the equation is good on all of A^{--}. Macdonald's formula for the unramified spherical function is neither more nor less than the main formula together with this observation about the Iwahori-fixed case.

8. Whittaker functions. Let $P = MN$ be a Borel subgroup of a quasi-split group G, and let ψ be a non-degenerate character of the maximal unipotent subgroup N. One may define an analogue $V_{\psi,N}$ of the Jacquet module to be the quotient of V by the span of all $\big(\pi(n) - \psi(n)\big)v$.

The Whittaker functional on $V = \mathrm{Ind}(\sigma\,|\,P, G)$ is effectively in the dual of this, and is defined formally by

$$\langle W_\psi, f\rangle = \int_N f(w_\ell n)\psi^{-1}(n)\,dn\,.$$

and the Whittaker function as $W_\psi(g) = \langle W_\psi, R_g f\rangle$. Finding the asymptotic behaviour of Whittaker functions at infinity, or equivalently finding $W_\psi(a)$ for $a \to 0$, is similar to that of finding the asymptotic behaviour of matrix coefficients, with the Whittaker functional W_ψ replacing integration against φ. This is explained, maybe a bit hurriedly, first of all in [**Casselman-Shalika: 1980**] and then in more detail in [**Casselman-Shahidi: 1998**]. The approach given in that last paper was in fact motivated by the approach to matrix coefficients that I have used here. There exists in this situation a canonical map $V_N \to V_{\psi,N} \cong \mathbb{C}$ describing the asymptotic behaviour, roughly because as $a \to 0$ the value of $\psi(ana^{-1})$ becomes 1, and each component of this map is determined by the effect of the standard intertwining operators $T_w\colon \mathrm{Ind}(\chi) \to \mathrm{Ind}(w\chi)$ on the Whittaker functional W_ψ.

Part II. Real groups

9. Introduction. Let now G be the group of real points on a Zariski-connected reductive group defined over $\mathbb{R}$. In addition, let

$$\begin{aligned} K &= \text{ a maximal compact subgroup} \\ \mathfrak{g}, \text{ etc.} &= \text{ complex Lie algebra of } G \text{ etc.} \end{aligned}$$

In the first part, the simple nature of Jacquet modules as well as the phenomenon that 'asymptotic' expansions are asymptotic equalities made our task easy. For real groups, there are both algebraic and analytical complications:

- the behaviour of matrix coefficients at infinity on G is truly asymptotic, expressed in terms of Taylor series;
- the Jacquet module is, as it consequently has to be, more complicated;
- there is no Bruhat filtration for the usual representation of $(\mathfrak{g}, K)$ on the K-finite principal series. Instead, one has to consider certain smooth representations of G itself, for example the C^∞ principal series;
- there are now two Jacquet modules to be considered, one for K-finite, one for smooth spaces.

You can get a rough idea of what happens in general by looking at the case of harmonic functions on the unit disk D, a space on which $\mathrm{SL}_2(\mathbb{R})$ acts (since the Cayley transform $z \mapsto (z-i)/(z+i)$ takes the upper half plane $\mathcal{H}$ to D). There are two spaces of interest: (1) the finite sums of polynomials in z and their conjugates, a representation of $(\mathfrak{g}, K)$; (2) the space of harmonic functions which extend smoothly to $\overline{D}$, on which $\mathrm{SL}_2(\mathbb{R})$ itself acts. In either case, the constant functions are a stable subspace, and the quotient is the sum of two discrete series, holomorphic and anti-holomorphic.

The group P fixes the point 1 (corresponding to ∞ on the upper half-plane), and $\mathfrak{n}$ acts trivially on the tangent space there. This means that if I is the ideal of functions in the local ring $\mathcal{O}$ vanishing at 1, then $\mathfrak{n}$ takes $\mathcal{O}$ to I, and in general I^n to I^{n+1}. The asymptotic behaviour of a harmonic function at 1 is controlled by its Taylor series. The space of all harmonic Taylor series at 1 is a representation of P as well as the Lie algebra $\mathfrak{g}$. This space of formal power series is the correct analogue of the Jacquet module here. One thing that is deceptive here is that the K-finite harmonic functions are polynomials. This is unique to that case.

One feature seen here, a feature characteristic of real groups, is that the analogue of the Jacquet module has a simpler relationship to the representation on C^∞ functions than that on K-finite ones—the map from the first onto harmonic Taylor series is actually surjective, while the second is not.

What I, and presumably everyone who works with both real and $\mathfrak{p}$-adic groups, find so remarkable is that in spite of great differences in technique required to deal with the two cases, the results themselves are uncannily parallel. It might incline some to believe that there is some supernatural being at work in this business.

10. The real Jacquet module. If V is a finitely generated Harish-Chandra module over $(\mathfrak{g}, K)$ it is finitely generated as a module over $U(\mathfrak{n})$, and its contragredient $\widetilde{V}$ is finitely generated over $U(\overline{\mathfrak{n}})$. Its **Jacquet module** is the completion $V_{[\mathfrak{n}]}$—the projective limit of the quotients $V/\mathfrak{n}^k V$—with respect to powers of $\mathfrak{n}$ (introduced in [**Casselman: 1979**]). It is obviously a representation of $(\mathfrak{p}, K \cap P)$ and in fact one of P, even though V itself is not. Slightly more surprising is that it is a representation of all of $\mathfrak{g}$, although it is easy to verify. Not so surprising if you think about the example of harmonic functions, where this completion ts the space of harmonic Taylor series at 1 and the enveloping algebra $U(\mathfrak{g})$ acts by differential operators.

This Jacquet module is easily related to homomorphisms *via* Frobenius reciprocity from V to representations induced from finite-dimensional representations of P, since it is universal with respect to $\mathfrak{n}$-nilpotent modules.

The projective limit is a kind of non-abelian formal power series construction. As in the p-adic case:

10.1. Proposition. *The functor $V \to V_{[\mathfrak{n}]}$ is exact.*

Since a similar question will arise later in different circumstances, I recall how this goes. As proved in [**McConnell: 1967**], the Artin-Rees Lemma holds for the augmentation ideal $(\mathfrak{n})$ of $U(\mathfrak{n})$. This is one example of the fact that much of the theory of commutative Noetherian rings remains valid for $U(\mathfrak{n})$. If

$$0 \to A \to B \to C \to 0$$

is exact then we have a right exact sequence

$$\cdots \to A/\mathfrak{n}^n A \to B/\mathfrak{n}^n B \to C/\mathfrak{n}^n C \to 0\,.$$

The left inclusion is not necessarily injective. But by Artin-Rees, there exists $k \geq 0$ such that then $A \cap \mathfrak{n}^n B \subseteq \mathfrak{n}^{n-k} A$ for $n \gg 0$. Suppose (a_n) lies in the projective limit of the quotients $A/\mathfrak{n}^n A$ with image 0 in $B/\mathfrak{n}^n B$. Then for n large, a_{n+k} lies in $\mathfrak{n}^n A$, so $a_n = 0$. □

This argument will fail in later circumstances, but something close to it will succeed. For the moment, let R be the ring $U(\mathfrak{n})$, I the ideal generated by $\mathfrak{n}$. The long exact sequence above fits into

$$\cdots \to \mathrm{Tor}_1^R(R/I^n, B) \to \mathrm{Tor}_1^R(R/I^n, C) \to A/I^n A \to B/I^n B \to C/I^n C \to 0\,.$$

The following is equivalent to Artin-Rees.

10.2. Proposition. *If C is a finitely generated module over $U(\mathfrak{n})$, then for some k and $n \gg 0$, the canonical map from $\mathrm{Tor}_1^R(R/I^n, C)$ to $\mathrm{Tor}_1^R(R/I^{n-k}, C)$ is identically 0.*

Proof. Suppose

$$0 \to E \to F \to C \to 0$$

to be an exact sequence of finitely generated modules over $U(\mathfrak{n})$, where F is free. Choose k so that $E \cap I^n F \subseteq I^{n-k} E$ for $n \gg 0$. Since Tor of a free module vanishes, the proof follows from diagram chasing in:

$$\begin{array}{ccccccccc} 0 & \to & \mathrm{Tor}_1^R(R/I^n, C) & \to & E/I^n E & \to & F/I^n F & \ldots & \\ \downarrow & & \downarrow & & \downarrow & & \downarrow & & \\ 0 & \to & \mathrm{Tor}_1^R(R/I^{n-k}, C) & \to & E/I^{n-k} E & \to & F/I^{n-k} F & \ldots & \square \end{array}$$

10.3. Corollary. *If*

$$0 \to A \to B \to C \to 0$$

is an exact sequence of $U(\mathfrak{n})$ modules and C is finitely generated, then

$$0 \to A_{[\mathfrak{n}]} \to B_{[\mathfrak{n}]} \to C_{[\mathfrak{n}]} \to 0$$

is also exact.

Exactness for real groups is thus much more sophisticated than it is for p-adic ones. Still, that one can define a Jacquet module and that it again defines an exact functor seems almost miraculous. The one common feature in both cases is the connection with geometry—here with compactifications of symmetric spaces, in the other with compactifications of the building. But then the analogy between symmetric spaces and buildings is another miracle.

11. Verma modules. The Jacquet modules for real groups are closely related to the more familiar Verma modules.

Traditionally, a Verma module is a representation of $\mathfrak{g}$ on a space

$$U(\mathfrak{g}) \otimes_{U(\mathfrak{p})} V$$

where V is an irreducible finite-dimensional representation of $\mathfrak{p}$. Since any finite-dimensional representation of $\mathfrak{p}$ is necessarily annihilated by some power of $\mathfrak{n}$, every vector in such a space is also annihilated by some power of $\mathfrak{n}$. I shall therefore introduce a slightly more general notion with the same name—what I shall call a Verma module is a compatible pair of representations of $U(\mathfrak{g})$ and P on a space V which is finitely generated and has the property that every vector in V is annihilated by some power of $\mathfrak{n}$. In other words, V is the union $V^{[\mathfrak{n}]}$ of its $\mathfrak{n}$-torsion subspaces $V(\mathfrak{n}^n)$. The compatibility means that the representation of P agrees with that of its Lie algebra $\mathfrak{p}$ as a subalgebra of $\mathfrak{g}$.

Every Verma module is the quotient of one of the form $U(\mathfrak{g}) \otimes_{U(\mathfrak{p})} V$, where V is a finite-dimensional representation of P. A Verma module will always have finite length as a module over $U(\mathfrak{g})$, and will be annihilated by some ideal of $Z(\mathfrak{g})$ of finite codimension.

How do Verma modules relate to Jacquet modules? If V is a Verma module, its linear dual $\widehat{V}$ is the projective limit of the duals of its finite-dimensional $\mathfrak{n}$-stable subspaces. In other words we know that V is the direct limit of finite-dimensional subspaces:

$$V = \varinjlim V(\mathfrak{n}^k)$$

which means that

$$\widehat{V} = \text{ the projective limit of the duals of the } V(\mathfrak{n}^k)\,.$$

Furthermore, we have an exact sequence

$$0 \to V(\mathfrak{n}^k) \to V \to (\text{ dual of } \mathfrak{n}^k) \otimes V$$

and we deduce the exact sequence

$$\mathfrak{n}^k \otimes \widehat{V} \to \widehat{V} \to \text{ dual of } V(\mathfrak{n}^k) \to 0\,.$$

so that $\widehat{V}$ is the projective limit of the finite-dimensional quotients $\widehat{V}/\mathfrak{n}^k\widehat{V}$. It is a finitely generated module over the completion $U_{[\mathfrak{n}]}$ of $U(\mathfrak{n})$ with respect to powers of $\mathfrak{n}^n$.

Conversely, the topological dual of this completion —i.e. the space of linear functionals vanishing on some $\mathfrak{n}^k\widehat{V}$—is the original Verma module. Thus, a natural and straightforward duality exhibits a close relationship between Verma modules and $\mathfrak{g}$-modules finitely generated over $U_{[\mathfrak{n}]}$. In particular, the Jacquet module of V is the linear dual of the space of $\mathfrak{n}$-torsion in the linear dual $\widehat{V}$ of V, which is a Verma module in the sense defined above.

There is another duality relationship between Verma modules for P and those for its opposite $\overline{P}$. Something like this is to be expected in view of the duality between Jacquet modules in the $\mathfrak{p}$-adic case, where N and $\overline{N}$ both occur in the description of asymptotic behaviour of matrix coefficients. It is easy to see that a Verma module V is always finitely generated over $U(\overline{\mathfrak{n}})$. It is in fact the submodule of $\mathfrak{n}$-finite vectors in its completion $\overline{V} = V_{[\overline{\mathfrak{n}}]}$. The continuous dual U of $\overline{V}$ is then in turn a Verma module for $\overline{\mathfrak{p}}$. If we now perform the same construction for U we get V back again. So the categories of Verma modules for $\mathfrak{p}$ and $\overline{\mathfrak{p}}$ are naturally dual to each other. This is crucial, as we shall see, in understanding the relationship between Jacquet modules and matrix coefficients.

12. Jacquet modules and matrix coefficients. Matrix coefficients satisfy certain differential equations which have regular singularities at infinity on G. This implies that we have a convergent expansion

$$\langle \widetilde{v}, \pi(a)v \rangle = \sum_{\varphi\in\Phi, n\geq 0} c_{\varphi,n}\, \varphi(a)\alpha^n(a)$$

where Φ is a finite collection of A-finite functions on A. If $A \cong \mathbb{R}^\times$, for example, functions in Φ will be of the form $|x|^s \log^m |x|$.

An analogue of Jacquet's observation holds:

For v in $\mathfrak{n}^k V$ or $\widetilde{v}$ in $\overline{\mathfrak{n}}^k \widetilde{V}$ the coefficient $c_{\varphi,n}$ vanishes for $n < k$.

This is easy to see for harmonic functions, since $\mathfrak{n}^k$ takes $\mathcal{O}$ to I^k.

In the limit we therefore get a pairing of $V_{[\mathfrak{n}]}$ with $\widetilde{V}_{[\overline{\mathfrak{n}}]}$ taking values in a space of formal series. The pairing of $V_{[\mathfrak{n}]}$ with $\widetilde{V}_{[\overline{\mathfrak{n}}]}$ is best expressed in terms of the duality explained in the previous section. If $\widetilde{v}$ is annihilated by $\overline{\mathfrak{n}}$ then the series associated to v and $\widetilde{v}$ will be finite, hence defining an A-finite function. So we are now in a situation much like that for p-adic groups. This A-finite function may be evaluated at 1, and in this we we get a 'canonical pairing' between $V_{[\mathfrak{n}]}$ and the $\mathfrak{n}$-torsion in $\widetilde{V}_{[\overline{\mathfrak{n}}]}$ ([**Casselman: 1979**]).

That the pairing is in some strong sense non-degenerate is highly non-trivial, first proven in [**Miličič : 1977**]. His argument was rather indirect. It will also be a corollary of the computation of the canonical pairing for induced representations, which is what this paper is all about.

13. Langlands' calculation for $\mathbf{SL_2(R)}$. The results for arbitrary reductive groups are quite complicated, even to state (and remain so far unpublished). To give you at least some idea of what goes on I'll look just at the principal series of $\mathrm{SL}_2(\mathbb{R})$. But in order to offer some contrast to what is to come later, I'll begin with a 'classical' argument to be found in [**Langlands: 1988**], which is itself presumably based on earlier results of Harish-Chandra.

Suppose that f lies in $\mathrm{Ind}^\infty(\chi)$, φ in $\mathrm{Ind}^\infty(\chi^{-1})$. The associated matrix coefficient is

$$\langle \varphi, R_g f \rangle = \int_{P\backslash G} \varphi(x) f(xg)\, dx\,.$$

In the rest of this paper, let

$$w = \begin{bmatrix} 0 & -1 \\ 1 & 0 \end{bmatrix}.$$

In the following result, let

$$a_t = \begin{bmatrix} t & 0 \\ 0 & 1/t \end{bmatrix}.$$

13.1. Theorem. (Langlands) *If* $\chi(x) = |x|^s$ *with* $\Re(s) > 0$ *then*

$$\langle \varphi, R_{a_t} f\rangle \sim \delta^{1/2}(a_t)\chi^{-1}(a_t)\varphi(w)\int_N f(wn)\,dn$$

as $t \to 0$.

This is an asymptotic equation, with the interpretation that the limit of

$$\lim_{t\to 0}\frac{\langle \varphi, R_{a_t} f\rangle}{\delta^{1/2}(a_t)\chi^{-1}(a_t)} = \varphi(w)\int_N f(wn)\,dn\,.$$

It is well known (and we'll see in a moment) that for $\Re(s) > 0$ the integral

$$\Omega_w(f) = \int_N f(wn)\,dn$$

is absolutely convergent and defines an N-invariant functional on $\mathrm{Ind}(\chi)$.

Proof. We have

$$\begin{aligned}
\langle \varphi, R_{a_t} f\rangle &= \int_N f(wna)\varphi(wn)\,dn \\
&= \int_N f(wa_tw^{-1}\cdot w\cdot a_t^{-1}na_t)\varphi(wn)\,dn \\
&= \delta^{-1/2}(a_t)\chi^{-1}(a_t)\int_N f(w\cdot a_t^{-1}na_t)\varphi(wn)\,dn \\
&= \delta^{-1/2}(a_t)\chi^{-1}(a_t)\delta(a_t)\int_N f(wn)\varphi(w\cdot a_tna_t^{-1})\,dn \\
&= \delta^{1/2}(a_t)\chi^{-1}(a_t)\int_N f(wn)\varphi(w\cdot a_tna_t^{-1})\,dn\,.
\end{aligned}$$

We'll be through if I show that

$$\lim_{t\to 0}\int_N f(wn)\varphi(w\cdot a_tna_t^{-1})\,dn = \varphi(w)\cdot\int_N f(wn)\,dn\,.$$

First I recall the explicit Iwasawa factorization for $\mathrm{SL}_2(\mathbb{R})$:

$$\begin{bmatrix} a & b \\ c & d\end{bmatrix} = \begin{bmatrix} 1 & (ac+bd)/r^2 \\ 0 & 1\end{bmatrix}\begin{bmatrix} 1/r & 0 \\ 0 & r\end{bmatrix}\begin{bmatrix} d/r & -c/r \\ c/r & d/r\end{bmatrix}$$

where $r = \sqrt{c^2+d^2}$. Thus we can write

$$wn = \begin{bmatrix} 0 & -1 \\ 1 & 0\end{bmatrix}\begin{bmatrix} 1 & x \\ 0 & 1\end{bmatrix} = \begin{bmatrix} 0 & -1 \\ 1 & x\end{bmatrix} = n_x\begin{bmatrix} (1+x^2)^{-1/2} & 0 \\ 0 & (1+x^2)^{1/2}\end{bmatrix}k_x$$

where n_x, k_x are continuous functions of x.

As a consequence, the integral is

$$\int_{\mathbb{R}}(x^2+1)^{-(s+1)/2}f(k_x)\big((t^2x)^2+1\big)^{(s-1)/2}\varphi(k_{t^2x})\,dx\,.$$

The integrand converges to

$$(x^2+1)^{-(s+1)/2}f(k_x)\varphi(w)$$

as $t \to 0$. According to the dominated convergence theorem (elementary in this case, and justified in a moment) the integral converges to $\varphi(w)\cdot\Omega_w(f)$.

In later sections we'll see a more precise description of the behaviour of matrix coefficients at infinity on G.

I include here a Lemma needed to apply the dominated convergence theorem, a pleasant exercise in calculus.

13.2. Lemma. *For all* $0 \le |t| \le 1$ *and* $0 < s$ *the product*

$$(x^2+1)^{-(s+1)/2}\big((t^2x)^2+1\big)^{(s-1)/2}$$

lies between $(x^2+1)^{-(s+1)/2}$ *and* $(x^2+1)^{-1}$.

Proof. These are what you get for $t=0$, $t=1$, and the derivative with respect to t is always of constant sign in between.

We shall see later that this result gives only the leading term in an infinite asymptotic series.

14. The Bruhat filtration. Now let's begin a new analysis, following the p-adic case as closely as possible. We need first to say something about the Jacquet module for principal series. Here, as in the p-adic case, this depends on the Bruhat decomposition $G = P \bigsqcup PwN$.

Let V be $\mathrm{Ind}^\infty(\chi)$. For f in this space let Ω_f be the map from $U(\mathfrak{g})$ taking X to $Xf(1)$. This lies in a kind of **infinitesimal principal series**

$$V_1 = \mathrm{Hom}_{U(\mathfrak{p})}(U(\mathfrak{g}), \mathbb{C}_{\chi\delta_P^{1/2}}).$$

In this way, we get a $\mathfrak{p}$-covariant map

$$\Omega\colon\ f \longmapsto \Omega_f .$$

Let V_w be the subspace of functions in V vanishing of infinite order along P, which are the closure in $\mathrm{Ind}^\infty(\chi)$ of those functions with compact support on PwN. By a theorem of E. Borel this fits into a short exact sequence

$$0 \to V_w \to V \overset{\Omega}{\to} V_1 \to 0 .$$

I call this the **Bruhat filtration** of V.

It is important to realize—originally, it took me quite a while to fully absorb—that:

There does not exist such a sequence for the K-finite principal series.

After all, the K-finite functions are analytic, and an analytic function cannot vanish of infinite order anywhere. In other words, we do not have a good Bruhat filtration for the representation of $(\mathfrak{g}, K)$ on the K-finite principal series. In spite of this, we *do* have however a filtration of the Jacquet module of the K-finite principal series, because according to the main result of [**Casselman: 1989**], *the Jacquet modules of a K-finite principal series and its C^∞ version are the same*, in a very strong sense. The way in which this is phrased in [**Casselman: 1989**] is that any $(\mathfrak{g}, K)$-covariant homomorphism from one K-finite principal series to another extends continuously to a map between the associated smooth principal series (the phenomenon called in [**Wallach: 1983**] 'automatic continuity'). These two assertions are essentially equivalent because of Frobenius reciprocity. Another miracle to put in the pot.

As before, I define the Jacquet module of any representation U of $\mathfrak{g}$ to be the projective limit of the quotients $U/\mathfrak{n}^k U$. As in the p-adic case, the Bruhat filtration of V gives rise to a filtration of its Jacquet module.

14.1. Proposition. *The exact sequence defining the Bruhat filtration gives rise to an exact sequence of Jacquet modules*

$$0 \to (V_w)_{[\mathfrak{n}]} \to V_{[\mathfrak{n}]} \to (V_1)_{[\mathfrak{n}]} \to 0 \,.$$

As we have seen, if the terms in the original exact sequence were finitely generated over $U(\mathfrak{n})$, this would be a consequence of the Artin-Rees Lemma. This would still be true if they were finitely generated over $U(\mathfrak{n})_{[\mathfrak{n}]}$. However, I have proved a variation of Artin-Rees which applies here, because of:

14.2. Lemma. *The space V_1 is the linear dual of the Verma module $U(\mathfrak{g}) \otimes_{U(\mathfrak{p})} \mathbb{C}_{\chi^{-1}\delta_P^{1/2}}$.*

This is straightforward to verify, and well known. In particular, the space V_1 is finitely generated over $U(\mathfrak{n})_{[\mathfrak{n}]}$, and with the help of a few standard results comparing Tor for $U(\mathfrak{n})$ and its completion, we can deduce the exactness we want. □

Incidentally, in the case at hand we can prove everything directly. Again let $R = U(\mathfrak{n})$, and let I be the ideal generated by $\mathfrak{n}$. If ν is a generator of $\mathfrak{n}$, the ring R is just a polynomial algebra in ν. The group $\mathrm{Tor}_1^R(I^n, A)$ is the subspace of A annihilated by I^n. The completion of R with respect to I is still a principal ideal domain, and the torsion in V_1 is finite-dimensional. In particular, there exists some $k \geq 0$ annihilating all its torsion. The canonical projection from $\mathrm{Tor}_1^R(R/I^n, V_1)$ to $\mathrm{Tor}_1^R(R/I^{n-k}, V_1)$ may be identified with multiplication by ν^k, which annihilates all torsion. This is a special case of what I called the variant of Artin-Rees.

One immediate corollary of the Lemma is this:

14.3. Corollary. *The subspace $\mathfrak{n}^k V_1$ is closed in V_1 of finite codimension.*

In order to fully understand the Bruhat filtration of the Jacquet module, we must figure out what $(V_w)_{[\mathfrak{n}]}$ is. The group N is isomorphic to the additive group $\mathbb{R}$. Its Schwartz space $\mathcal{S}(N)$ is defined by this identification. An application of the same calculations we made for Langlands' Theorem, using the Iwasawa decomposition, tells us:

14.4. Proposition. *The restriction of V_w to N is isomorphic to the Schwartz space $\mathcal{S}(N)$.*

This is a slight generalization of Schwartz' identification of $\mathcal{S}(\mathbb{R})$ with the space of those smooth functions on the projective line that vanish of infinite order at ∞ (which is in fact a special case).

14.5. Proposition. *A function f in V_w lies in $\mathfrak{n}^k V_w$ if and only if*

$$\int_N P(n) f(n)\, dn = 0$$

for every polynomial P of degree $< k$.

This identifies the $\mathfrak{n}$-torsion in the dual of V_w. It is a simple exercise in calculus.

14.6. Corollary. *The space $\mathfrak{n}^k V_w$ is closed in V_w, and every quotient $V_w/\mathfrak{n}^k V_w$ is free of rank one over $U(\mathfrak{n})/\mathfrak{n}^k U(\mathfrak{n})$.*

This gives us:

14.7. Proposition. *Each space $\mathfrak{n}^k V$ is closed in V and has finite codimension.*

Proof. If $U \to V$ is a continuous map of Fréchet spaces with image of finite codimension, then its image is closed. This is because if F is a finite-dimensional complement, the associated map from $U \oplus F$ to V is continuous and surjective, so that the induced map from $\big(U/\ker(f)\big) \oplus F$ is an isomorphism of topological vector spaces, by the open mapping theorem. The long exact sequence

$$\cdots \to V_w/\mathfrak{n}^k V_w \to V/\mathfrak{n}^k V \to V_1/\mathfrak{n}^k V_1 \to 0$$

tells us that $\mathfrak{n}^k V$ has finite codimension, but it is also the image of a map from $\otimes^{(k)}\mathfrak{n} \otimes V$ to V. □

Now, let's try to understand what we have on hand. Dual to the Bruhat filtration of $\mathrm{Ind}(\chi)$ by double cosets PwN is that of its contragredient $\mathrm{Ind}(\chi^{-1})$ by cosets $Pw\overline{N}$. Let the corresponding exact sequence be

$$0 \to \widetilde{V}_w \to \widetilde{V} \to \widetilde{V}_1 \to 0\,,$$

giving rise to

$$0 \to (\widetilde{V}_w)_{[\overline{\mathfrak{n}}]} \to \widetilde{V}_{[\overline{\mathfrak{n}}]} \to (\widetilde{V}_1)_{[\overline{\mathfrak{n}}]} \to 0\,.$$

These two coset decompositions are transversal to one another–the coset $P\overline{N}$ is open in G and contains $P = PN$, while PwN is open in G and contains $Pw\overline{N}$. Every smooth function on PwN therefore determines a Taylor series along $Pw = Pw\overline{N}$, and in particular every polynomial $P(n)$ on PwN determines one. As we have seen these are all annihilated by some power of $\mathfrak{n}$, so it should not be too surprising to see:

14.8. Proposition. *As a module over $\mathfrak{g}$, the $\mathfrak{n}$-completion of V_w is isomorphic to the $\mathfrak{n}$-completion of the continuous dual $(\widetilde{V}_w)_{[\overline{\mathfrak{n}}]}$.*

Similarly for $\widetilde{V}_1$ and V_1.

So now we find ourselves in exactly the same situation we saw in the p-adic case—Bruhat filtrations of $\mathrm{Ind}(\chi)$ and $\mathrm{Ind}(\chi^{-1})$ with corresponding terms in the associated graded spaces dual to one another. In the next section we shall see that this duality matches with the asymptotic expansion of matrix coefficients.

Generically, the Bruhat filtration of Jacquet modules will split, but for isolated values of χ it will not.

There is one final remark to make. Of course the K-finite principal series $V_{(K)}$ embeds into the smooth one V, inducing a map of their Jacquet modules. As I have already mentioned, it is a consequence of the 'automatic continuity' theorem in [**Wallach: 1983**] that this is an isomorphism. Thus, although there is no Bruhat filtration of $V_{(K)}$, there is one of its Jacquet module. It's a curious fact, and presumably a fundamental one.

15. The explicit formula.

Let me recall where we are in the discussion. We want to calculate $\langle \varphi, R_a f\rangle$ for φ in $\mathrm{Ind}^\infty(\chi^{-1})$, f in $\mathrm{Ind}^\infty(\chi)$. As with p-adic groups, since the asymptotic expansion of matrix coefficients factors through Jacquet modules, but here I do not see how to use that to simplify calculations. The problem is that we must look at what happens for smooth functions.

I'll look here at a principal series representation of $\mathrm{SL}_2(\mathbb{R})$. Let

$$a = \begin{bmatrix} t & 0 \\ 0 & 1/t \end{bmatrix}\,.$$

Set $\chi(a) = |t|^s$ as in the discussion of Langlands' formula.

We express f in $\mathrm{Ind}(\chi)$ and φ in $\mathrm{Ind}(\chi^{-1})$ as sums of functions with support on the open Bruhat double cosets:

$$f = f_w + f_1, \quad \varphi = \varphi_w + \varphi_1$$

where $*_w$ has support on PwN, $*_1$ on $P\overline{N}$. We will see what happens to each term in

$$\langle\varphi, R_a f\rangle = \langle\varphi_1, R_a f_1\rangle + \langle\varphi_1, R_a f_w\rangle + \langle\varphi_w, R_a f_1\rangle + \langle\varphi_w, R_a f_w\rangle$$

as $t \to 0$.

I'll look first at $\langle\varphi_w, R_a f_w\rangle$ because it offers an interesting comparison with the verification of Langlands' formula. To simplify notation I'll set $\varphi = \varphi_w$, $f = f_w$. We get here as before that

$$\langle\varphi, R_a f\rangle = \delta^{1/2}\chi^{-1}(a)\int_N f(wn)\varphi(w \cdot ana^{-1})\, dn\,,$$

We must look at the integral, also as before. But here things are somewhat simpler analytically—the functions $f(wn)$ and $\varphi(wn)$ are both of compact support as functions of n. Set

$$n = \begin{bmatrix} 1 & x \\ 0 & 0 \end{bmatrix}$$

and change both the variable n and the functions $f(wn)$ and $\varphi(w \cdot ana^{-1})$ of n to functions of x. So we are now considering the integral

$$\int_{\mathbb{R}} f(x)\varphi(t^2 x)\, dx$$

where f and φ are both of compact support on $\mathbb{R}$. You can see immediately and roughly what is going to happen—as $t \to 0$ $\varphi(t^2x)$ will be more or less determined by its behaviour near 0, or in other words by its Taylor series at 0. We get formally

$$\begin{aligned}\int_{\mathbb{R}} f(x)\varphi(t^2x)\, dx &= \int_{\mathbb{R}} f(x) \sum_{m\geq 0} t^{2m}\frac{x^m}{m!}\varphi^{(m)}(0)\, dx \\ &= \sum_{m\geq 0} t^{2m}\frac{\varphi^{(m)}(0)}{m!}\int_{\mathbb{R}} x^m f(x)\, dx\,.\end{aligned}$$

Because f and φ have compact support, it is not hard to justify this as an asymptotic expansion.

We can find a more enlightening interpretation of this. Let

$$\overline{\nu} = \begin{bmatrix} 0 & 0 \\ 1 & 0 \end{bmatrix}$$

be a generator of $\overline{\mathfrak{n}}$. In terms of our choice of coordinates in $\overline{N}$ the functional

$$\widetilde{\Omega}_{w,m}\colon\ \mathrm{Ind}(\chi^{-1}) \to \mathbb{C},\ \varphi \mapsto R_{\overline{\nu}^m} f(w)$$

is the same as that taking it to $\varphi^{(m)}(0)$. It is annihilated by $\overline{\mathfrak{n}}^m$. Let

$$\Omega_{w,m}\colon\ \mathrm{Ind}(\chi) \to \mathbb{C},\ f \mapsto \int_N x^m f(wn)\, dn\,,$$

which is a meromorphic function of χ. In these terms, the formula becomes

$$\langle\varphi, R_a f\rangle \sim \delta^{1/2}(a)\chi^{-1}(a)\Big(\sum_{m\geq 0}\widetilde{\Omega}_{w,m}(\varphi)\,\Omega_{w,m}(f)\Big)\,.$$

Actually, this whole expansion can be deduced from the algebra of Verma modules, if one knows just the leading term. The expansion expresses, essentially, the unique pairing between generic terms in Jacquet module of the principal series. After all, generic Verma modules are irreducible, so the pairing is unique up to scalar multiplication.

The term $\langle \varphi_1, R_a f_1 \rangle$ has a similar asymptotic expression in terms of the Jacquet modules of $\widehat{V}_1$ and $\widehat{V}$, and the cross terms vanish asymptotically. Define functionals $\Omega_{1,m}$ $\widetilde{\Omega}_{1,m}$ as I did $\Omega_{w,m}$ and $\widetilde{\Omega}_{w,m}$ In the end we get as asymptotic expansion a sum of two infinite series

$$\begin{aligned}\langle \varphi, R_a f \rangle \sim \delta^{1/2}(a)\chi^{-1}(a)\Big(\sum_{m \geq 0} \widetilde{\Omega}_{w,m}(\varphi)\, \Omega_{w,m}(f) \Big) \\ + \delta^{1/2}(a)\chi(a)\Big(\sum_{m \geq 0} \widetilde{\Omega}_{1,m}(\varphi)\Omega_{1,m}(f) \Big) .\end{aligned}$$

There is one series for each component in the Bruhat filtration. This is therefore the analogue for real groups of Macdonald's formula for spherical functions on $\mathrm{SL}_2(\mathbb{Q}_p)$. Of course, Macdonald's formula is an exact formula, but here we are given an asymptotic expansion. But, in fact, K-finite matrix coefficients satisfy an analytic ordinary differential equation, and the formula for them becomes one involving convergent series valid everywhere except for $a = I$, where the differential equation has a (regular) singularity.

Macdonald's formula, incidentally, is proven along similar lines in [**Casselman: 2009**].

16. Whittaker functions. Whittaker functions for real groups also satisfy a differential equation with regular singularities along the walls $\alpha = 0$ for simple roots α (although irregular at infinity in other directions). Suppose ψ to be a non-degenerate character of $\mathfrak{n}$. If V is a finitely generated $(\mathfrak{g}, K)$-module, then its **Kostant module** (following [**Kostant: 1978**]) is the space $\widehat{V}^{[\mathfrak{n},\psi]}$ of $\mathfrak{n}_\psi$-torsion in the continuous dual of its canonical G-representation—that is to say continuous linear functionals annihilated by some power of the $U(\mathfrak{n})$-ideal generated by the $x - \psi(x)$ for x in $\mathfrak{n}$. We have a map from $\widehat{V}^{[\mathfrak{n},\psi]} \otimes V$, taking $W \otimes v$ to the series expansion of $\langle W, \pi(a)v \rangle$ at $a = 0$. How can we fit this into a scheme such as we have seen above?

I have only a rough idea of what to propose. In the p-adic case we have a canonical map from V_N to $V_{\psi,N}$. In the real case, both the Jacquet module and the Kostant module are very different as modules over $U(\mathfrak{n})$, but have similar structures as modules over $\overline{\mathfrak{n}}$. A Kostant module is finitely generated over $\overline{\mathfrak{n}}$, and if $U = \widehat{V}^{[\mathfrak{n},\psi]}$ we get a map from the Jacquet module of V to the $\overline{\mathfrak{n}}$-completion. An explicit formula for the expansion of Whittaker functions on the smooth principal series would then follow from a calculation of the scalars defined by intertwining operators on Whittaker models. This would not be all that different, conceptually, from what happens for p-adic groups.

References

[Arthur: 1983] J. Arthur, 'A Paley-Wiener theorem for real groups', *Acta Mathematica* **150** (1983), 1–89. Available at `http://www.claymath.org/cw/arthur/`

[Bruhat-Tits: 1966] F. Bruhat and J. Tits, 'Groupes algébriques simples sur un corps local: cohomologie galoisienne, décompositions d'Iwasawa et de Cartan', *C. R. Acad. Sc. Paris* **263** (1966), 867–869.

[Casselman: 1974] W. Casselman, **Introduction to the theory of admissible representations of p-adic reductive groups**. This preprint was written originally in 1974. It is now out of date and full of minor errors, but available at http://www.math.ubc.ca/~cass/research/pdf/casselman.pdf

[Casselman: 1979] —, 'Jacquet modules for real reductive groups', Proceedings of the ICM in Helsinki, 1979, 557–563.

[Casselman: 1989] —, 'Canonical extensions of Harish-Chandra modules to representations of G', *Canadian Journal of Mathematics* **XLI** (1989), 385–438.

[Casselman: 2009] —, 'Jacquet modules', one of several revisions of material in my old notes, available at http://www.math.ubc.ca/~cass/research/pdf/Jacquet.pdf

[Casselman: 2009] —, 'A short account of subsequent developments concerning $\mathfrak{p}$-adic spherical functions', preprint, 2009. This is macdonald.pdf, available at the same site.

[Casselman et al.: 2000] —, H. Hecht, and D. Miličić, 'Bruhat filtrations and Whittaker vectors for real groups', *Proceedings of Symposia in Pure Mathematics* **68** (2000), 151–189.

[Casselman-Shahidi: 1998] — and F. Shahidi, 'On irreducibility of standard modules for generic representations', *Annales scientifiques de l'École Normale Supérieure* (1998), 561–589. Also at: http://www.numdam.org

[Casselman-Shalika: 1980] — and J. Shalika, 'The unramified principal series of p-adic groups II. The Whittaker function', *Compositio Mathematicae* **41** (1980), 207–231.

[Haines et al.2003] T. Haines, R. Kottwitz, and A. Prasad, 'Iwahori-Hecke algebras', *J. Ramanujan Math. Soc.* **25** (2010), 113–145.

[Hirano-Oda: 2007] M. Hirano and T. Oda, 'Calculus of principal series Whittaker functions on GL$(3,\mathbb{C})$', preprint, University of Tokyo, 2007.

[Kostant: 1978] B.Kostant, 'On Whittaker vectors and representation theory', *Inventiones Mathematicae* **48** (1978), 101–184.

[Langlands: 1966] R. P. Langlands, 'The volume of the fundamental domain for some arithmetical subgroups of Chevalley groups', 143–148 in **Algebraic groups and discontinuous subgroups**, *Proc. Symp. Pure Math.*, A.M.S., 1966.

[Langlands: 1988] —, 'The classification of representations of real reductive groups', **Mathematical Surveys and Monographs 31**, A. M. S., 1988. This and the previous item also at http://sunsite.ubc.ca/DigitalMathArchive/Langlands/intro.html

[McConnell: 1967] J. McConnell, 'The intersection theorem for a class of non-commutative rings', *Proceedings of the London Mathematical Society* **17** (1967), 487–498.

[Macdonald: 1971] I. G. Macdonald, **Spherical functions on a group of p-adic type**, University of Madras, 1971.

[Manin: 1991] Y. Manin, 'Three-dimensional geometry as ∞-adic Arakelov geometry', *Inventiones Mathematicae* **104** (1991), 223–244.

[Miličić : 1977] D. Miličić, 'Asymptotic behaviour of matrix coefficients of the discrete series', *Duke Math. J.* **44** (1977), 59–88.

[Mirkovic-Vilonen: 2000] I. Mirkovic and K. Vilonen, 'Perverse sheaves on affine Grassmannians and Langlands duality', *Math. Res. Lett.* **7** (2000) 13–24.

[Wallach: 1983] N. Wallach, 'Asymptotic expansions of generalized matrix entries of representations of real reductive groups', 287–389 in **Lie group representations I**, *Lecture Notes in Mathematics* **1024**, Springer, 1983.

DEPARTMENT OF MATHEMATICS, UNIVERSITY OF BRITISH COLUMBIA, VANCOUVER V6T 1Z2, CANADA

E-mail address: cass@math.ubc.ca

Clay Mathematics Proceedings
Volume 13, 2011

The ABS Principle : Consequences for $L^2(G/H)$

Laurent Clozel

Pour Freydoon Shahidi, à l'occasion de son soixantième anniversaire

1. The purpose of this note is exclusively propagandistic. As we have shown in [**8**], the principle in the title has remarkable consequences for the harmonic analysis of (real or p–adic) semi–simple groups. Precisely, when $H \subset G$ are two such groups, this principle imposes very strong constraints

(i) on the representations of H weakly contained in an irreducible representation π_G of G

(ii) on the representations of G weakly contained in $\mathrm{ind}_H^G \pi_H$, where π_H is an irreducible representation of H.

(All the representations considered are unitary).

In both cases, the given representation π_H (or π_G) has to be assumed to belong to the set of representations singled out by Arthur [**2**], and which should be the constituents of the spaces of automorphic forms (for the action of the local, p–adic group). The Burger–Sarnak principle ([**5**], [**6**]) states that these sets are preserved by induction or restriction. Thus "ABS" stands for Arthur, Burger and Sarnak. The rigidity argument, which implies the constraints, was introduced in [**8**], § 3. It is strongest when we consider **unramified** representations of G, or H. It involves a trick called the (p, q)–trick in [**9**].

When (ii) is applied to $\pi_H = \mathbb{1}_H$= trivial representation, the induced representation is simply

$$L^2(G/H)$$

and the ABS principle gives significant **a priori** information on its spectral decomposition. It is not complete, because Arthur's conjectures are not proven. However approximations of the conjectures are known (and rather precise ones will be, for classical groups, when Arthur's program ([**2**], Ch. 30) is completed). It is quite possible that these approximations may be useful to determine the spectrum, much like an argument of Bernstein [**4**] was successfully used by Delorme [**10**]. Meanwhile, the principle gives much (conjectural) **a priori** information which should be useful. We have tried to illustrate, by examples, how it sheds light on some known or unknown cases, in particular when (G, H) is a symmetric pair as in the work of

2010 *Mathematics Subject Classification.* Primary 22E30 ; Secondary 22D30, 22E35.

Delorme and others. We will not describe the consequences of the ABS principle in case (i) as this was done in [**9**], and made explicit by Lapid, Rogawski [**13**] and Venkatesh [**16**].

2. Arthur's formalism can be studied in Arthur [**2**], and has been described, in the terms needed here, in [**8**], Ch. 3 and [**9**]. We refer the reader to these papers for more details.

Assume for simplicity G, H split over a local field F (which may be real, p–adic or a function field). If $\widehat{G}$ is the dual group Arthur considers homomorphisms ("Arthur parameters")

$$\psi : \quad W'_F \times SL(2,\mathbb{C}) \longrightarrow \widehat{G}$$

where

$$W'_F = \text{ the Weil group } W_F \quad \text{if } F = \mathbb{R} \text{ or } \mathbb{C}$$
$$W'_F = W_F \times SL(2,\mathbb{C}) \quad \text{if } F \text{ is } p\text{–adic}.$$

The restriction of ψ to W_F is supposed to be tempered. Unramified parameters are of the form

$$\psi : W_F \times SL(2,\mathbb{C}) \longrightarrow \widehat{G}$$

(thus, trivial on the "first" $SL(2)$ in the p–adic case), and unramified as maps $W_F \longrightarrow \widehat{G}$. In the real or complex case, this means that the parameters factor as sums of unramified characters of $\mathbb{R}^\times$ or $\mathbb{C}^\times$.

The **type** of ψ is the map $SL(2,\mathbb{C}) \longrightarrow \widehat{G}$, which defines uniquely a unipotent conjugacy class U in $\widehat{G}$. (Parameters are, of course, considered up to conjugacy). These parameters are expected to parametrize representations of $G = G(F)$ occurring in $L^2(G(k)\backslash G(\mathbb{A}))$ where k is a global field of which F is a completion, and $\mathbb{A} = \mathbb{A}_k$. There will be some ambiguity (ψ will determine a **set** $\Pi(\psi)$ of representations, expected however to be finite) and some overlap ($\Pi(\psi)$ and $\Pi(\psi')$ will not, in general, be disjoint). Let us denote by $\mathcal{U}(\widehat{G})$ the set of unipotent orbits in the dual group.

To each ψ we associate a unipotent orbit $U \in \mathcal{U}(\widehat{G})$, determined by $\psi|_{SL(2,\mathbb{C})}$. (Recall that unipotent orbits correspond uniquely to maps $SL(2,\mathbb{C}) \longrightarrow \widehat{G}$ by the Jacobson–Morozov theorem). In particular :

- $U = \{1\}$ if the parameter ψ is **tempered**
- U is unipotent regular if $\psi|_{SL(2,\mathbb{C})}$ is the "maximal" representation $SL(2,\mathbb{C}) \longrightarrow \widehat{G}$. In this case, ψ sends W'_F into $Z(\widehat{G})$ (the center) and the associated packet $\Pi(\psi)$ is an Abelian character of G – the trivial representation if $\psi = 1$ on W'_F.

Recall that a Langlands parameter is simply

$$\varphi : W'_F \longrightarrow \widehat{G}$$

(with semi–simple image). It is expected to parametrize an L–packet $\Pi(\varphi)$ of representations; this is known for $F = \mathbb{R}$ or $\mathbb{C}$, and for any F if $G = GL(n)$. To an Arthur parameter we associate a Langlands parameter

$$\text{(2.1)} \qquad \varphi_\psi : w \longmapsto \psi\left(w, \begin{pmatrix} |w|^{1/2} & \\ & |w|^{-1/2} \end{pmatrix}\right) \in \widehat{G}$$

where $w \in W'_F$ and $|w|$ is simply $|w_1|$ if $w = (w_1, s) \in W_F \times SL(2,\mathbb{C})$. (The absolute value is given by the reciprocity map $W_F \longrightarrow F^\times$). The L–packet $\Pi(\varphi_\psi)$ should be a subset of $\Pi(\psi)$.

Suppose ψ is unramified. Then φ_ψ defines, by the known Langlands functoriality in the unramified case, a unique unramified representation $\pi(\psi)$, with Hecke matrix

$$t(\pi(\psi)) = \psi\left(\mathrm{Frob}, \begin{pmatrix} q^{1/2} & \\ & q^{-1/2} \end{pmatrix}\right) \in \widehat{G}\,;$$

q is the cardinality of the residue field, and Frob $\in \mathrm{Gal}(\overline{k}/k)$ a Frobenius element. (In the Archimedean case, $\pi(\psi)$ is defined by the homomorphism (2.1) : $W_F \longrightarrow F^\times \longrightarrow \widehat{G}$ and the Langlands parametrization). As in [**8**], [**9**] we will assume :

ASSUMPTION 2.1.— For ψ unramified, $\pi(\psi)$ is the unique unramified element of $\Pi(\psi)$.

This is true for $SL(n)$. For F real or complex, it should be included in the results of [**1**]. For G classical (i.e., orthogonal or symplectic) it follows from the results announced by Arthur in [**3**], with some ambiguity if $G = SO(2n)$ (split).

Now return to the situation (ii) of § 1. We have assumed G, H split, but we will allow ourselves to relax this condition in the examples. In the split case we take models of G, H over $\mathbb{Z}$ and we have maximal compact subgroups $G(\mathfrak{o}_F)$, $H(\mathfrak{o}_F)$ — in the p–adic case. (This was implicit in Assumption 2.1 : "unramified" means unramified w.r. to $G(\mathfrak{o}_F)$). We can consider the unramified part $L^2_{nr}(G/H)$ of $L^2(G/H)$, i.e., the subspace generated under G by

$$L^2(K_G\backslash G/H)$$

where $K_G = G(\mathfrak{o}_F)$. In the real case, K_G is the maximal compact subgroup.

By the Burger–Sarnak principle for induction (due to Burger, Li and Sarnak [**5**] ; and **proved** in [**8**], § 3) the support of this representation in $\widehat{G}$ belongs to the automorphic spectrum [**5**] and therefore, conjecturally, to the set of representations parametrized by (unramified) Arthur parameters. In particular, to each constituent (possibly continuous, of course) of L^2_{nr} corresponds a unipotent orbit.

PRINCIPLE 2.2.— *The constituents of L^2_{nr} are associated to a unique unipotent orbit $U = \mathrm{Ind}_H^G(U_{H,reg}) \subset \widehat{G}$.*

Here $U_{H,reg} \subset \widehat{H}$ is the regular orbit, and the symbol Ind_H^G will be explained below.

Note that this is a rather strong contraint. For instance, if **one** tempered representation belongs to the support of L^2_{nr}, this whole representation is tempered.

We sketch a "proof" assuming (the global variant of) Arthur's conjectures. (For more details see [**9**]). We can choose a global field k such that F, F' are two completions of k, say k_v and k_w.

By Burger–Li–Sarnak (extended in [**8**], § 3 to the S–arithmetic case), the support of

$$\mathrm{ind}_{H(k_v)\times H(k_w)}^{G(k_v)\times G(k_w)} \mathbb{1} = \mathrm{ind}_{H(k_v}^{G(k_v)} \mathbb{1} \otimes \mathrm{ind}_{H(k_w)}^{G(k_w)} \mathbb{1}$$

is composed of automorphic representations. Assume $\pi_v \otimes \pi_w$ is irreducible in the support. If they are unramified, they are associated to the same orbit. But this applies to $\pi_v \otimes \pi_w$ and $\pi'_v \otimes \pi_w$. Thus, if π_v and π'_v occur, they belong to the same orbit. (This is the "p, q–trick").

We note that in many cases the Principle is (approximately) true. For instance, if $G = SL(n)$, Arthur's partition of the automorphic spectrum is a theorem, due to Mœglin–Waldspurger and Luo, Rudnick and Sarnak [**8**], § 3. (It is not known that automorphic representations are of the form (2.1) with $\psi|_{W'_F}$ tempered, but it is true modulo the "approximation to the Ramanujan Conjecture". At any rate, the orbit U is well–defined). Interpreted in this fashion, the Principle is a theorem – with the proof given above. When Arthur's results announced in [**3**], Chapter 30 are complete, this will apply to classical groups.

What happens if we induce another representation of H? (Of course, this is less interesting for the analysis on homogeneous spaces...). If π_H is unramified, and of Arthur's type, consider

$$I = \mathrm{ind}_H^G \pi_H\,, \quad \text{and} \quad I_{nr} \subset I\,.$$

PRINCIPLE 2.3.— *The constituents of I_{nr} are associated to a unique unipotent orbit $U := \mathrm{Ind}_H^G(U_H)$ where U_H is the orbit associated to π_H.*

In [**9**] we shew that these considerations, for restriction, gave a natural map Res : $\mathcal{U}(\widehat{G}) \longrightarrow \mathcal{U}(\widehat{H})$ between unipotent orbits. Here we get an induction map Ind : $\mathcal{U}(\widehat{H}) \longrightarrow \mathcal{U}(\widehat{G})$. It will have similar properties, e.g.

$$\mathrm{Ind}_H^G \circ \mathrm{Ind}_{H'}^H = \mathrm{Ind}_{H'}^G\,.$$

However, one of the most interesting parts of [**9**], i.e. the multiplication

$$\mathcal{U}(\widehat{G}) \times \mathcal{U}(\widehat{G}) \longrightarrow \mathcal{U}(\widehat{G})$$

associated to the diagonal embedding $G \hookrightarrow G \times G$, is here lacking : for $H = G \subset G \times G$ we should consider

$$\mathrm{ind}_G^{G\times G}(\pi_G)\,.$$

If $\pi_G = \mathbb{1}_G$ – the "largest" representation from the spectral point of view – this is tempered, thus

$$\mathrm{Ind}(U_{reg,G}) = 1_{G\times G} \qquad \text{(trivial orbit)}\,.$$

By Lemma 1 of [**16**], this implies that the induced representation from any unitary representation is tempered. Thus the "comultiplication" is trivial.

An explicit description of the (conjectural) map Ind_H^G is unknown. For $G = GL(n)$ and $H = M$ a Levi subgroup, the map has been computed by Lapid–Rogawski [**13**] and [**16**]. In this case $\widehat{M}$ is a Levi subgroup of $\widehat{G}$.

For instance, assume $M = GL(n_1) \times GL(n_2) \subset G$, so $\widehat{M} = GL(n_1, \mathbb{C}) \times GL(n_2, \mathbb{C}) \subset \widehat{G}$. A unipotent orbit in $GL(n, \mathbb{C})$ is given by a partition

$$n = m_1 + \cdots + m_k\,,$$

the trivial representation (minimal orbit) being associated to the partition $n = n$. If (m_i), (m'_j) are partitions of n_1 and n_2, the orbit associated to

$$\mathrm{ind}_M^G(\pi),$$

where π is a representation of $GL(n_1) \times GL(n_2)$ of types (m_i, m'_j), has type

$$< m_i + n_1 - n\,, \;\; m'_j + n_2 - n >_n$$

where, for a set of integers a_k (here indexed by $\{i\} \cup \{j\}$), the partition $< a_k >_n$ is obtained by keeping the integers ≥ 2 and completing by 1's so as to get a partition of n.

For instance, if π is trivial (so we are decomposing $L^2(G/M)$) we get the type

$$< 2n_1 - n\,, \;\; 2n_2 - n >_n$$

which is tempered (equal to $(1,1,\ldots,1)$) if $n_1 = n_2 = \frac{n}{2}$. Similarly, if there are r blocks we get $< 2n_i - n >$ which is the orbit $\{1\}$ unless one of the n_i's is $> \frac{n}{2}$.

In particular this induction is **not** the Lusztig–Spaltenstein induction of orbits, which sends the regular orbit (associated to the trivial representation) to the regular orbit [**14**]. Note that if $r = 2$, G/M is a "reductive symmetric space" in the sense of [**10**], [**15**], i.e., M is the set of fixed points of an involution.

What can we say about the whole spectrum of $I = \mathrm{ind}_H^G(\mathbb{1}_H)$? Assume we have determined the orbit $U \subset \widehat{G}$ in Principle 2.3. Thus we have a representation $SL(2,\mathbb{C}) \longrightarrow \widehat{G}$. (Of course there **are** unramified representations in I since there are K–invariant functions on G/H). If we apply the (p,q)- trick again (taking π_w unramified) we see that :

PRINCIPLE 2.4.— *Any representation π in the support of $L^2(G/H)$ belongs to an Arthur packet $\Pi(\psi)$, of type U.*

In terms of Langlands parameters, it is expected that $\Pi(\psi)$ contains $\Pi(\varphi_\psi)$ (Langlands packet) as well as other representations that are "smaller" from the spectral point of view, that is, **more tempered**. If they belong to a $\Pi(\varphi_{\psi'})$, for instance, the unipotent orbit of ψ' should be smaller. (I do not know if, for instance, the orbit $\mathcal{U}(\psi')$ should be in the closure of $\mathcal{U}(\psi)$).

3. Queries and examples

We first recall that the arguments in section 2 assumed G split, or at least unramified, at the local prime used to determine the type. If we consider, for instance, a group over $\mathbb{R}$ that is not quasi-split, the spherical representations of G occurring in $L^2(G/H)$ can be quite different, cf. Faraut [**11**] for $SO(n,1)$. They can probably belong to different Arthur packets $\Pi(\psi)$. This restriction should be borne in mind. (I thank N. Bergeron for pointing this out to me.)

3.1. If $H \subset G$ is the set of fixed points of an involution, $L^2(G/H)$ has been decomposed for F archimedean by Delorme, and van den Ban and Schlichtkrull [**10**], [**15**]. Moreover the Arthur packets are known [**1**]. It would be interesting to check Principles 2.2 and 2.4 in this case.

3.2. We consider now a case which has been treated by P. Harinck [**12**]. Here G is a real group, and she considers $L^2(G(\mathbb{C})/G(\mathbb{R}))$. She determined the spectral decomposition – of course this is now a special case of the results of Delorme and van den Ban and Schlichtkrull. In particular the spectrum is tempered.

From our point of view this result is obvious – assuming the conjectural arguments, of course. For we can choose a real quadratic extension $F/\mathbb{Q}$ and a quadratic extension E/F, complex at one Archimedean prime v of F and split at the other real prime v', and then a group $\mathbf{G}$ over F giving G by extension of scalars at the prime v, and split at the prime v'. Call G' the split form of G.

Then $\mathbf{G}(E_\infty) = G(\mathbb{C}) \times (G'(\mathbb{R}) \times G'(\mathbb{R}))$, with subgroup $H = \mathbf{G}(F_\infty) = G(\mathbb{R}) \times G'(\mathbb{R})$, the embedding being the obvious one. The (v,v')–trick says that the types of $L^2(G(\mathbb{C})/G(\mathbb{R}))$, and of $L^2(G'(\mathbb{R}) \times G'(\mathbb{R})/G'(\mathbb{R}))$ – diagonal embedding – coincide. But the second space has tempered support (Harish–Chandra) ; therefore so does the first. (Cf. [**16**], § 3.7.)

In fact this last result also follows from these principles. Assume G is a group over $\mathbb{R}$. Choose F, $\mathbf{G}$ as before such that $\mathbf{G}(F_v) = G(\mathbb{R})$ and $U = \mathbf{G}(F_{v'})$ is compact. (There may be an obstruction to doing this ; if necessary use more primes

v'). Then choose E/F as above, inverting the roles of v and v'. At the prime v, we have $L^2(G(\mathbb{R}) \times G(\mathbb{R})/G(\mathbb{R}))$; at the prime v', $L^2(G(\mathbb{C})/U)$: by the **spherical** theory this is tempered. In both cases, we have ensured that the group is split at the controlling prime.

The argument of 3.2 will of course apply in the p–adic case. There is no need to consider a quadratic extension :

CONJECTURE 3.1.— *Assume E/F is a finite extension of local fields, and G a group over F. Then*

$$L^2(G(E)/G(F))$$

is tempered.

Indeed, the (p,q)- trick reduces us to $L^2(G(F)^d/G(F))$ (diagonal embedding), isomorphic under $G(F)^{d-1}$ (any $(d-1)$ factors) to $\bigotimes^{d-1} L^2(G(F)$, and this is tempered. This implies that the representation of $G(F)^d$ is tempered.

3.3. We now consider the case of symplectic groups, first over $\mathbb{R}$. Assume $G = Sp(g)$, the symplectic group given by the alternating form of matrix

$$J = \begin{pmatrix} & -1_g \\ 1_g & \end{pmatrix}.$$

The centralizer of the matrix

$$I_- = \begin{pmatrix} & 1_g \\ -1_g & \end{pmatrix} \in G$$

is the maximal compact subgroup $U(g)$, embedded in G by

$$A + i\,B \longmapsto \begin{pmatrix} A & B \\ -B & A \end{pmatrix} \qquad (A + i\,B \in U(g)).$$

The centralizer of the matrix

$$I_+ = \begin{pmatrix} & 1_g \\ 1_g & \end{pmatrix} \in GSp(g)$$

is $GL(n,\mathbb{R})$ embedded in G by

$$C \in GL(n,\mathbb{R}) \longmapsto \begin{pmatrix} A & B \\ B & A \end{pmatrix}, \quad \text{where} \quad \begin{cases} A - B = C \\ A + B = {}^t C^{-1} \end{cases}.$$

It is a Levi subgroup for a maximal parabolic subgroup of G.

Assume that k is a real quadratic field and that $\varepsilon \in k$ is an element that is positive at one Archimedean prime ∞_1 and negative at the other prime ∞_2. Then $\begin{pmatrix} & 1 \\ -\varepsilon & \end{pmatrix}$ (blocks of type n) belongs to $GSp(g,k)$; its square is central, so it defines an involution of $Sp(g,k)$, with fixed points $U(g) \subset Sp(g,\mathbb{R})$ at the prime ∞_1 and $GL(g,\mathbb{R})$ at the prime ∞_2. Therefore we see – and this is confirmed by the results in [**10**], [**15**] – that $L^2(G/M)$ is tempered for $G = Sp(g,\mathbb{R})$ and M the Levi subgroup of type $GL(g)$.

If v is a finite prime of k, we can now use the "p,q"–trick (here "v,∞_2"). So we are naturally led to[1] :

1. This was pointed out by Akshay Venkatesh.

CONJECTURE 3.2.— *If F is a p–adic field,*

$$L^2(Sp(g,F)/GL(g,F)) \text{ is tempered}.$$

The same argument will apply - at least for G quasi-split - when a Levi subgroup of a maximal parabolic subgroup has a "form" – similarly to the previous construction for $Sp(g)$ – that is a maximal compact subgroup. For instance, it will apply to the Levi subgroup M of a maximal parabolic subgroup of G/F such that some form of G **over** $\mathbb{R}$ is Hermitian symmetric, the maximal compact subgroup being M after some algebraic extension of the ground field k.

3.4. We conclude with more musings.

3.4.1. We refer to [**9**] for the similar questions concerning **restriction**. For $G \subset G \times G$ (diagonal embedding), a very geometric construction of the restriction map

$$\mathrm{Res} : \mathcal{U}(\widehat{G}) \times \mathcal{U}(\widehat{G}) \longrightarrow \mathcal{U}(\widehat{G})$$

(called product in [**9**]) was provided by Waldspurger [**9**] : Conjecture 4.2. Of course, this is conjectural. In many cases his construction can be extended to

$$\mathrm{Res} : \mathcal{U}(\widehat{G}) \longrightarrow \mathcal{U}(\widehat{H})$$

for $H \subset G$. It would be interesting, of course, to obtain a conjectural, **geometric** description of the induction map.

3.4.2. In the case of restriction, very strong constraints are given by the exponents of representations in the sense of Harish–Chandra and Casselman [**9**], § 4. The behaviour of exponents is, in fact, encoded in Waldspurger's geometric "projection". How can one control the exponents of induced representations ? Of course, partial answers are given by [**13**], [**16**].

3.4.3. We end this survey with, perhaps, the must interesting question. Assume (G, H) is a symmetric pair (G/H is a "reductive" symmetric space, in the accepted terminology). In the real case, a crucial element in Delorme's determination of $L^2(G/H)$ is an **a priori** estimate on the H–invariant distributions on a unitary representation π of G. This is given by a theorem of Bernstein [**4**], which implies that the distribution α on the space of π must be "w–tempered", for a weight w on G/H which is explicit ([**7**], Appendix C).

No doubt such a weight can be defined in the p–adic case ; Delorme and his co–workers (Ph. Blanc, V. Sécherre) are making impressive progress. On the other hand, from our point of view, we have the unipotent orbit $U_G = \mathrm{Ind}_H^G(U_H^{reg})$ – of course, unknown in general, but which can actually be computed by the "p, q–trick" and comparison with the known real case. There should be a natural relation between the weight and the unipotent orbit.

Références

[1] J. ADAMS, D. BARBASCH, D. VOGAN, *The Langlands classification and irreducible characters for real reductive groups,* Birkhaüser, 1992.

[2] J. ARTHUR, *Unipotent automorphic representations,* in *Orbites unipotentes et représentations II : Groupes p–adiques et réels,* Astérisque **171–172** (1981), 13–71.

[3] J. ARTHUR, *An introduction to the trace formula,* in *Harmonic analysis, the trace formula and Shimura varieties,* Clay Math. Proceedings **4** (2005), 1-263.

[4] J. BERNSTEIN, *On the support of the Plancherel measure,* J. Geom. Phys. **5** (1988), 663–710.

[5] M. BURGER, J.S. LI, P. SARNAK, *Ramanujan duals and automorphic spectrum,* Bull. A.M.S. **26** (1992), 253–257.

[6] M. BURGER, P. SARNAK, *Ramanujan duals II,* Inv. Math. **106** (1991), 1–11.

[7] J. CARMONA, P. DELORME, *Base méromorphe de vecteurs distributions H-invariants dans les séries principales généralisées d'espaces symétriques réductifs : Equation fonctionnelle,* J. Funct. Anal. **122** (1994), 152–221.

[8] L. CLOZEL, *Spectral theory of automorphic forms,* in Automorphic forms and applications, IAS/Park City Math. ser. 12, AMS, Providence, 2007, 43–93.

[9] L. CLOZEL, *Combinatorial consequences of Arthur's conjectures and the Burger–Sarnak method,* Int. Math. Res. Notices **11** (2004), 511–523.

[10] P. DELORME, *Formule de Plancherel pour les espaces symétriques réductifs,* Ann. of Math. 147 (1998), 417–452.

[11] J. FARAUT, *Distributions sphériques sur les espaces hyperboliques,* J. Math. Pures Appl. **58** (1979), 369–444.

[12] P. HARINCK, *Functions orbitales sur $G_{\mathbb{C}}/G_{\mathbb{R}}$. Formule d'inversion des intégrales orbitales et formule de Plancherel,* J. Funct. Analysis **153** (1998), 52–107.

[13] E. LAPID, J. ROGAWSKI , *On a result of Venkatesh on Clozel's conjecture*, Contemporary Math. **489** (2009), 173-178.

[14] G. LUSZTIG, N.S. SPALTENSTEIN, *Induced unipotent classes,* J. London Math. Soc. **19** (1979), 41–52.

[15] E. VAN DEN BAN, H. SCHLICHTKRULL, *The Plancherel decomposition for a reductive symmetric space II. Representation theory,* Inv. Math. **161** (2005), 567–628.

[16] A. VENKATESH, *The Burger–Sarnak method and operations on the unitary dual of $GL(n)$,* Represent. Theory **9** (2005), 268–286 (electronic).

L.C. MATHÉMATIQUES BAT. 425, UNIVERSITÉ PARIS-SUD, 91405 ORSAY CEDEX - FRANCE
E-mail address: laurent.clozel@math.u-psud.fr

Clay Mathematics Proceedings
Volume 13, 2011

Orbital Integrals and Distributions

L. Clozel and J.-P. Labesse

A Freydoon Shahidi, en témoignage d'amitié

The aim of this note is to give a detailed proof of a generalization of a result in [CL]. We take this opportunity to correct the statement and the proof of theorem A.1.1 in [CL].

1. Twisted spaces and centralizers

When dealing with the twisted case we shall use the language of twisted spaces instead of twisted orbital integrals; for the basic properties of twisted spaces we refer the reader to [Lab3] sections I and II.

Consider a real connected reductive group $\mathbf{G}$ and θ an automorphism of $\mathbf{G}$ of finite order defined over $\mathbb{R}$. We identify $\mathbf{G}$ with the group of its complex points. We introduce the twisted algebraic space

$$\mathbf{L} = \mathbf{G} \rtimes \theta \,.$$

Recall that semisimple elements in $\mathbf{L}$ are elements that induce semisimple automorphisms of $\mathbf{G}$. Regular semisimple elements are the $\delta \in \mathbf{L}$ whose connected centralizer $\mathbf{G}_\delta$ – i.e. the neutral connected component of the centralizer $\mathbf{G}^\delta$ – is a torus. For such an element the stable centralizer $\mathbf{I}_\delta$ (in the sense of [Lab3] section II.1) is an abelian group. More precisely $\mathbf{G}_\delta$ is a torus whose centralizer is a maximal torus $\mathbf{T}$ and

$$\mathbf{I}_\delta = \mathbf{T}^\delta$$

the centralizer of δ in $\mathbf{T}$. Strongly regular elements are semisimple elements δ whose centralizer $\mathbf{G}^\delta$ is commutative; in such a case $\mathbf{I}_\delta = \mathbf{T}^\delta = \mathbf{G}^\delta$.

In the non-twisted case or in the base change situation $\mathbf{T}^\delta$ is connected but this is not the case in general. Nevertheless the possible non-connectivity is mild: in fact the stable centralizer of a semi-simple element δ is quasi-connected (in the sense of [Lab2]) and in particular

$$\mathbf{I}_\delta = \mathbf{I}_\delta^{\mathbf{0}}.\mathbf{Z}_\delta$$

where $\mathbf{I}_\delta^{\mathbf{0}} = \mathbf{G}_\delta$ is the neutral component of $\mathbf{I}_\delta$ and $\mathbf{Z}_\delta$ is the centralizer of δ in the center Z_G of G. This is lemma II.1.4 of [Lab3].

2010 *Mathematics Subject Classification.* Primary 22E45; Secondary 22F30.

LEMMA 1.1. *Let δ be a regular semisimple element in $\mathbf{L}$. Then $\mathbf{I}_\delta = \mathbf{I}_{t\delta}$ for $t \in \mathbf{I}_\delta$ small enough.*

Proof: This follows from the above characterization of the stable centralizer for regular elements.

□

2. Orbital integrals and measures

Let G (resp. L, etc.) be the set of real points of $\mathbf{G}$ (resp. $\mathbf{L}$, etc.) and let K be a maximal compact subgroup in G which is θ-invariant. Let $\mathfrak{g} = \mathfrak{k} \oplus \mathfrak{p}$ be the corresponding Cartan decomposition of the Lie algebra $\mathfrak{g}$ of G. We introduce the twisted real spaces

$$L = G \rtimes \theta \quad \text{and} \quad M = K \rtimes \theta$$

We fix a Haar measure on G and on K. This defines invariant measures on L and M. This gives us an isomorphism between the space of distributions and the space of generalized functions on L.

Let μ be the measure supported on M and defined by the product of our invariant measure and a smooth function χ on M. By abuse of notation we shall again denote by μ the corresponding generalized function. For any smooth function φ on L we have

$$\mu(\varphi) = \int_L \varphi(x)\, d\mu(x) = \int_M \varphi(m)\chi(m)\, dm$$

Viewed as a distribution on L it has a wave front set $W(\mu) \subset T^*(L)$ (cf. [Hör] and [GS]) whose fiber $W_\delta(\mu)$ above $\delta \in L$ is either the empty set or, if it is non-empty, then δ belongs to $M \subset L$ and the fiber is the set of non-zero cotangent vectors $\xi \in T^*_\delta(L)$ orthogonal to the tangent space of M at δ.

We shall make the following assumption:

ASSUMPTION 2.1. *Any $\delta \in M$ regular semisimple in L has a stable centralizer I_δ which is a compact abelian group contained in K.*

When $\theta = 1$ this is equivalent to the condition rank G = rank K and more generally this is a generalization of Harish-Chandra's condition that guarantees the existence of L-discrete series (also called θ-discrete series).

Let $\delta \in L$ be a regular semisimple element. Denote by $\mathfrak{i}_\delta$ the Lie algebra of I_δ. If $\delta \in M$ and the assumption 2.1 is satisfied we introduce

$$J^G(\delta) = \det(1 - \mathrm{Ad}\,(\delta)|\mathfrak{g}/\mathfrak{i}_\delta)$$

and

$$J^K(\delta) = \det(1 - \mathrm{Ad}\,(\delta)|\mathfrak{k}/\mathfrak{i}_\delta)$$

We denote by $d\dot{g}$ (resp. $d\dot{k}$) the quotient measure on $I_\delta\backslash G$ (resp. $I_\delta\backslash K$) for some choice of a Haar measure on I_δ. The orbital integral of a smooth compactly supported function φ is defined by

$$\mathcal{O}_\delta(\varphi) = \int_{I_\delta\backslash G} \varphi(g^{-1}\delta\, g)\, d\dot{g}$$

THEOREM 2.2. *Under assumption 2.1 the orbital integral $\mathcal{O}_\delta(\mu)$ of μ for δ is well-defined and is given by*

$$\mathcal{O}_\delta(\mu) = \frac{J^K(\delta_1)}{J^G(\delta_1)} \int_{I_\delta \backslash K} \chi(k^{-1}\delta_1 k)\, d\dot{k}$$

if δ has a conjugate δ_1 in M. When χ is invariant by K-conjugacy we have simply

$$\mathcal{O}_\delta(\mu) = \frac{J^K(\delta_1)}{J^G(\delta_1)}\, \chi(\delta_1)\ \mathrm{vol}(I_{\delta_1} \backslash K)$$

The orbital integral vanishes otherwise.

Proof: Let φ be a smooth compactly supported function and let dx be the measure on L defined by the Haar measure on G. The product is a smooth compactly supported density. Let $\delta \in L$ be regular semisimple. The orbital integral of φ on the orbit of δ can be seen as the integral of the generalized function σ_δ defined by the orbit S_δ of δ against the smooth density $\varphi(x)dx$. We denote by $W(\sigma_\delta)$ its wave front set. Now observe that $W(\sigma_\delta)$ is transverse to $W(\mu)$. By this we mean that for any $\xi_1 \in W(\mu)$ and any $\xi_2 \in W(\sigma_\delta)$ above the same point one has

$$\xi_1 + \xi_2 \neq 0\ .$$

In fact if the intersection of the orbit S_δ with M is non-empty we may assume $\delta \in M$; we shall show that the tangent spaces to the two submanifolds M and S_δ at δ generate the full tangent space and hence their orthogonal spaces minus zero, which are the wave front sets, are transverse. For $\delta \in M$ the tangent space to S_δ at δ, is the right translation by δ of

$$\mathfrak{s}_\delta = (1 - \mathrm{Ad}\,(\delta))\mathfrak{g}$$

and by assumption 2.1

$$\mathfrak{i}_\delta = \ker(1 - \mathrm{Ad}\,(\delta))$$

is contained in $\mathfrak{k}$. Since δ is semisimple we then have

$$\mathfrak{g} = \mathfrak{i}_\delta \oplus \mathfrak{s}_\delta = \mathfrak{k} + \mathfrak{s}_\delta\ .$$

Recall that a generalized function can be integrated against a compactly supported distribution provided their wave front sets are transverse (see [Hör] Th. 8.2.10 or [GS] Chap. VI, Proposition 3.10, p.335). This shows that the orbital integral $\mathcal{O}_\delta(\mu)$ makes sense. We now want to compute it. Let α be a smooth positive and compactly supported function supported on a small neighbourhood of the identity in G with integral 1. By convolution μ becomes a smooth density on L:

$$\alpha * \mu(x) = f_\alpha(x)dx\ .$$

The above given references state that the product of distributions is continuous and hence

$$(*) \qquad \mathcal{O}_\delta(\mu) = \lim_\alpha \int_{I_\delta \backslash G} f_\alpha(x^{-1}\delta\, x)\, d\dot{x}$$

when α converges to the Dirac measure at the origin in G. Let $\varphi = \varphi_0\psi$ be a function on L that is the product of a smooth function φ_0 invariant under conjugacy and a function ψ smooth with compact support equal to 1 on a neighbourhood of K. Provided the support of α is small enough, $\psi = 1$ on the support of f_α. The map

$$I_\delta \backslash G \times I_\delta \to L : (x, t) \mapsto x^{-1} t\delta\, x$$

is a local diffeomorphism near the identity in I_δ with Jacobian $J^G(t\delta)$. Finally, if δ is strongly regular and if t is restricted to a sufficiently small neighbourhood of 1 in I_δ, the map is one to one onto its image: in fact for $t \in I_\delta$ small enough $t\delta$ is also strongly regular and the centralizer G^δ of a strongly regular element equals its stable centralizer I_δ. Therefore, provided φ_0 has a small enough support near the orbit of δ, assumed for a while to be strongly regular,

$$\int_L \varphi(x)\, f_\alpha(x)\, dx = \int_{I_\delta} J^G(t\delta)\varphi_0(t\delta) \left(\int_{I_\delta\backslash G} f_\alpha(x^{-1}t\delta x)\, dx \right) dt\ .$$

But, similarly,

$$\int_L \varphi(x)\, d\mu(x) =$$

$$\int_M \varphi(m)\chi(m)\, dm = \int_{I_\delta} J^K(t\delta)\varphi_0(t\delta) \left(\int_{I_\delta\backslash K} \chi(k^{-1}t\delta k)\, dk \right) dt\ .$$

Now return to formula $(*)$. It remains true for δ replaced by $t\delta$ with t sufficently small in I_δ. Denote the corresponding generalized function by $\sigma_{t\delta}$. Then f_α (for α converging to the Dirac measure at the origin) and $\sigma_{t\delta}$ (for varying t) are both continuous (as follows from [Hör, Thm. 8.2.10] and [S, p. 110]), and since we are integrating over a compact set in t we see that

$$\lim_\alpha \int_{I_\delta} J^G(t\delta)\varphi_0(t\delta)\mathcal{O}_{t\delta}(f_\alpha)\, dt = \int_{I_\delta} J^G(t\delta)\varphi_0(t\delta) \lim_\alpha \mathcal{O}_{t\delta}(f_\alpha)\, dt$$

This implies

$$J^G(\delta) \lim_\alpha \int_{I_\delta\backslash G} f_\alpha(x^{-1}\delta x)\, dx = J^K(\delta) \int_{I_\delta\backslash K} \chi(k^{-1}t\delta k)\, dk$$

which proves the theorem for strongly regular elements. Now, using 1.1, we see that $t \mapsto \mathcal{O}_{t\delta}$ is smooth for t small enough and by continuity the result extends to all regular elements.

$\square$

3. Lefschetz function

Let $G^+ = G \rtimes \langle\theta\rangle$, where $\langle\theta\rangle$ denotes the finite group generated by θ. Let F be a finite dimensional irreducible representation supposed θ-stable with a chosen extension, again denoted by F, to the semidirect product G^+. Consider π, an admissible irreducible representation of G^+; the Lefschetz number is by definition

$$\mathrm{Lef}(\pi, F) = \sum (-1)^i \,\mathrm{trace}\,(\theta \mid H^i(\mathfrak{g}, \mathfrak{k}; \pi \otimes F)\ .$$

Recall that $\mathrm{Lef}(\pi, F) = 0$ unless the restriction of π to G remains irrreducible and $\lambda_\pi = \lambda_{\check{F}}$, where λ_π and $\lambda_{\check{F}}$ denote the infinitesimal characters of π and $\check{F}$ respectively.

Following [Lab1], consider μ_F, the measure supported on M that gives the Lefschetz number for F:

$$\mathrm{trace}\,\pi(\mu_F) = \mathrm{Lef}(\pi, F)\ .$$

This measure is defined by the function on M

$$\chi_F(\delta) = \sum_i (-1)^i \operatorname{trace}(\delta \mid \wedge^i \mathfrak{p} \otimes F) = \det(1 - \delta \mid \mathfrak{p}) \operatorname{trace} F(\delta)$$

and the normalized Haar measure on K.

LEMMA 3.1. *Either the measure μ_F is zero or the assumption 2.1 is satisfied.*

Proof: Assume δ is regular. Then $\chi_F(\delta)$ is zero whenever the Lie algebra $\mathfrak{i}_\delta$ of the centralizer of δ projects on $\mathfrak{p}$ non-trivially. This shows that either the measure μ_F is zero or the assumption 2.1 is satisfied at the level of Lie algebras: $\mathfrak{i}_\delta \subset \mathfrak{k}$. But this implies that I_δ is a compact abelian group. Recall that the centralizer of the compact torus I_δ^0 in G is a maximal torus T and hence $U = T \cap K$ is a maximal torus in K. Clearly T is stable by the Cartan involution; this implies that its maximal compact subgroup is contained in K. In particular $I_\delta \subset K$. □

By definition Lefschetz functions are smooth compactly supported densities on L such that for any representation π of G^+ one has

$$\operatorname{trace} \pi(\phi_F) = \operatorname{Lef}(\pi, F) .$$

As shown in [Lab1] one obtains Lefschetz functions ϕ_F for F by regularization via Arthur's multipliers of the measure μ_F.

LEMMA 3.2. *Let α be an Arthur multiplier such that $\hat{\alpha}(\lambda_{\check{F}}) = 1$. The Lefschetz function defined by μ_F and α,*

$$\phi_F = \mu_{F,\alpha}$$

depends continuously, as a distribution, on α and has limit μ_F, in the sense of distributions, when α tends to the Dirac measure at the origin. Moreover its orbital integrals are independent of α.

Proof: The value of ϕ_F against a test function φ is given by the integral over G of the product of the functions $\phi'_F(x) = \phi_F(x \rtimes \theta)$ and $\varphi'(x) = \varphi(x \rtimes \theta)$; it can be evaluated by the Plancherel formula

$$\int_L \phi_F(y)\varphi(y)\, dy = \int_G \phi'_F(x)\varphi'(x)\, dx = \int_{\hat{G}} \operatorname{trace}(\pi(\phi'_F)\pi(\overline{\varphi}')^*)\, d\pi$$

but since, by definition of ϕ_F,

$$\pi(\phi'_F) = \hat{\alpha}(\lambda_\pi)\pi(\mu'_F)$$

the continuity follows by dominated convergence. We can view ϕ_F as a function on G^+; as such its trace is non-zero only on the representations of G^+ irreducible under G; on such a representation its trace is

$$\hat{\alpha}(\lambda_\pi) \operatorname{trace} \pi(\mu_F) .$$

This is non-zero only if $\lambda_\pi = \lambda_{\check{F}}$ (by the cohomological property of μ_F) and then it equals $\operatorname{trace}\pi(\mu_F)$. Hence, by the density theorem of Kottwitz and Rogawski [KR], the orbital integrals of ϕ_F are independent of α. (Note that the assumptions of [KR] are now satisfied, thanks to Delorme and Mezo [DM]). □

PROPOSITION 3.3. *Assume that the Lefschetz numbers are not identically zero. The orbital integrals of a Lefschetz function are given for regular semisimple elements by*

$$\int_{I_\delta\backslash G} \phi_F(x^{-1}\delta x)\,dx = \mathrm{vol}(I_{\delta_1}\backslash K)\ \mathrm{trace}\,F(\delta_1)$$

if δ has a conjugate δ_1 in M; the orbital integral vanishes otherwise.

Proof: According to 3.2, its orbital integrals are independent of the regularization via Arthur's multipliers, and when the multiplier tends to the Dirac measure at the origin the distribution defined by the Lefschetz function tends to μ_F. We have

$$\chi_F(\delta) = \frac{J^G(\delta)}{J^K(\delta)}\ \mathrm{trace}\,F(\delta)$$

Thanks to 3.1 we may appeal to 2.2 and we get

$$\mathcal{O}_\delta(\mu_F) = \mathrm{vol}(I_\delta\backslash K)\ \mathrm{trace}\,F(\delta)\ .$$

Now the continuity of the product of distributions yields the proposition. □

4. Erratum and complements to [CL]

We computed in [CL, theorem A.1.1] the semi-simple orbital integrals of Lefschetz functions when F is the trivial representation of G^+. Moreover we assumed that the stable centralizers were connected. The proof was hasty and the result incorrect: there is a sign error in the formula given for singular elements. Here we shall give the corrected statement, in its natural generality: we make no connectedness assumption and we allow arbitrary coefficients.

Recall that, for a quasi-connected reductive group $\mathbf{I}$ over $\mathbb{R}$ [Lab2] with maximal compact subgroup K_I

$$q(I) = \frac{1}{2}\dim(I/K_I)$$

is an integer when I has discrete series. In such a case we choose the Haar measure di such that the formal degree of the discrete series with trivial infinitesimal character is 1. Following section A.1 of [CL], this defines an Euler-Poincaré function (i.e. a Lefschetz function for the trivial automorphism) on I which we denote by f^I_{ep}. We have

$$f^I_{\mathrm{ep}}(1) = (-1)^{q(I)}\,d(I)$$

where $d(I)$ is the order of the quotient $W_{\mathbb{C}}/W_{\mathbb{R}}$ of the complex / real Weyl group for a compact Cartan subgroup of I^0. One can use Kottwitz signs [Kot] to express $(-1)^{q(I)}$. Denote by $\overline{I}^0$ the compact inner form of I^0. Then

$$(-1)^{q(I)} = e(I^0)e(\overline{I}^0)\ .$$

We will write $e(I) = e(I^0)$ and $e(\overline{I}) = e(\overline{I}^0)$. We refer to [LBC,§ 2.7] for the definition of the stable orbital integrals. Let $T \subset K \subset G$ be a maximal compact torus and let

$$d(G) = \#\ker[H^1(\mathbb{R},T) \to H^1(\mathbb{R},G)]\ .$$

If G has discrete series, this coincides with the previous definition.

THEOREM 4.1. *Assume δ is semisimple with stable centralizer I_δ. Then*

(*i*)
$$\mathcal{O}_\delta(\phi_F) = 0$$
if I_δ does not have a discrete series;

(*ii*)
$$\mathcal{O}_\delta(\phi_F) = e(I_\delta)e(\overline{I_\delta})\, d(I_\delta)\,\mathrm{trace}\, F(\delta) = f_{\mathrm{ep}}^{I_\delta}(1)\,\mathrm{trace}\, F(\delta)$$
if I_δ has discrete series and is endowed with the measure di. The stable orbital integral of ϕ_F is given by

(*iii*)
$$\mathcal{SO}_\delta(\phi_F) = 0$$
if I_δ does not have discrete series;

(*iv*)
$$\mathcal{SO}_\delta(\phi_F) = e(\overline{I_\delta})\, d(G)\,\mathrm{trace}\, F(\delta)$$
otherwise.

Proof: For δ regular in M we have seen in 3.3 that
$$\mathcal{O}_\delta(\phi_F) = \mathrm{trace}\, F(\delta)$$
where we use the normalized measure on the compact group I_δ. The descent argument in [CL, p. 122] can be copied; for t close to 1 in I_δ, δ being semi-simple and $t\delta$ regular, we have
$$\mathcal{O}_t^{I_\delta}(f_{\mathrm{ep}}^{I_\delta}) = 1$$
(for instance, by theorem 2.2 applied to I_δ, with a simple extension to the quasi-connected case), whence
$$\mathcal{O}_{t\delta}(\phi_F) = \mathrm{trace}\, F(t\delta) = \mathcal{O}_t^{I_\delta}(f_{\mathrm{ep}}^{I_\delta}\ \Theta_{\delta,F})$$
where
$$\Theta_{\delta,F}(t) = \mathrm{trace}\, F(t\delta)$$
is invariant on I_δ. We deduce that
$$\mathcal{O}_\delta(\phi_F) = f_{\mathrm{ep}}^{I_\delta}(1)\,\Theta_{\delta,F}(1)\ .$$
The same argument applies to the stabilization, the rational character trace F being invariant under conjugation by $\mathbf{G}$. The vanishing statements follow similarly from 3.3.

□

Our theorem 4.1 above is essentially theorem A.1.1 in [CL]. But there are sign errors in [CL]. First, on page 120 line 14 and 16 the sign $e(\overline{I_\delta})$ is omitted; this has no bearing on the considerations preceding the theorem. However Kottwitz's sign $e(\overline{I_\delta})$ is again repeatedly omitted on page 122 line −8 and −6 and 123 line 6, which introduces the sign mistakes in the final statement.

5. The unitary case

Finally we will give an explicit statement in the case of base change for unitary groups, the case used in [CL]. Let $H = U(p,q)$, $G = GL(n,\mathbb{C})$ $(n = p+q)$ viewed as a real Lie group and let θ be the automorphism of G whose group of fixed points is H. Let E be an irreducible algebraic representation of the complex group G and consider $F = E \otimes E$, where G acts by
$$g(e_1 \otimes e_2) = ge_1 \otimes \theta(g)e_2$$

One can extend F to G^+ by letting θ act by $e_1 \otimes e_2 \mapsto e_2 \otimes e_1$. For $g \in G$ define $Ng = g\theta(g)$. Then, for $\delta = g \rtimes \theta \in L$ one has

$$\delta^2 = \gamma \rtimes 1 \quad \text{with} \quad \gamma = Ng$$

and

$$\operatorname{trace} F(\delta) = \operatorname{trace} E(\gamma)$$

as follows from [Clo1]. Take $K = U(n) \subset G$; then $\delta \in M = K \rtimes \theta$ is regular semi-simple if and only if γ is regular semi-simple in K, and the stable centralizer of δ is the connected component of the centralizer of γ in K.

THEOREM 5.1. *Let $\delta = g \rtimes \theta$ be semi-simple with stable centralizer I_δ. Then:*

(i) $$\mathcal{O}_\delta(\phi_F) = 0$$

if I_δ does not have a discrete series;

(ii) $$\mathcal{O}_\delta(\phi_F) = e(I_\delta)e(\overline{I_\delta})\, d(I_\delta) \operatorname{trace} E(\gamma)$$

if I_δ has discrete series and is endowed with the measure di. The stable orbital integral of ϕ_F is given by

(iii) $$\mathcal{SO}_\delta(\phi_F) = 0$$

if I_δ does not have discrete series

(iv) $$\mathcal{SO}_\delta(\phi_F) = e(\overline{I_\delta})\, d(G) \operatorname{trace} E(\gamma)$$

otherwise. In this case

$$d(G) = 2^n$$

This shows that the results of [CL], in particular the Theorem A.3.1, extend with similar proof to the case of an arbitrary local system.

Acknowledgement. – We would like to thank Richard Taylor for pointing out the inadequacy of our proof in [CL] of the main result of this note.

References

[Clo1] L. Clozel, *Changement de base pour les représentations tempérées des groupes réductifs réels*, Ann. Sci. E.N.S. 4e série **15**, (1982), pp. 45-115.

[CL] L. Clozel, J.-P. Labesse, *Changement de base pour les représentations cohomologiques de certains groupes unitaires* Astérisque **257**, (1999) pp. 121-136.

[DM] P. Delorme, P. Mezo, *A twisted invariant Paley-Wiener theorem for real reductive groups* Duke Math. J. Volume **144**, Number 2 (2008), pp. 341-380.

[GS] V. Guillemin, S. Sternberg, *Geometric Asymptotics*, Amer. Math. Soc. Math. Surveys **14**, (1977).

[H] L. Hörmander, *The Analysis of Linear Partial Differential Operators I* Springer-Verlag, Grundlehren der Math. Wissensch. **256**, (1983).

[Kot] R. Kottwitz, *Sign changes in harmonic analysis on reductive groups* Trans. A.M.S. **278**, (1983), pp. 289-297.

[KR] R. Kottwitz, J. Rogawski, *The distributions in the invariant trace formula are supported on characters*, Canadian. J. Math. **52**, (2000) pp. 804-814.

[Lab1] J.-P. Labesse, *Pseudo-coefficients très cuspidaux et K-théorie*, Math. Ann. **291**, (1991), pp. 607-616.

[Lab2] J.-P. Labesse, *Nombres de Tamagawa des groupes réductifs quasi-connexes*, Manuscripta Math. **104**, (2001) pp. 407-430.

[Lab3] J.-P. Labesse, *Stable twisted trace formula: Elliptic terms* J.I.M.J. **3**, (2004), pp. 473-530.

[LBC] J.-P. Labesse, *Cohomologie, stabilisation et changement de base*, Asterisque **257**, (1999) pp. 1-116.

[S] L. Schwartz, *Théorie des distributions*, Publication Univ. Strasb. IX-X Hermann (1966).

L.C. MATHÉMATIQUES BAT. 425, UNIVERSITÉ PARIS-SUD, 91405 ORSAY CEDEX - FRANCE
E-mail address: laurent.clozel@math.u-psud.fr

INSTITUT DE MATHÉMATIQUES DE LUMINY (UMR 6206), CAMPUS DE LUMINY, CASE 907, 13288 MARSEILLE CEDEX 9, FRANCE
E-mail address: labesse@iml.univ-mrs.fr

Clay Mathematics Proceedings
Volume 13, 2011

Functoriality for the Quasisplit Classical Groups

J.W. Cogdell, I.I. Piatetski-Shapiro, and F. Shahidi

Functoriality is one of the most central questions in the theory of automorphic forms and representations [**3, 6, 31, 32**]. Locally and globally, it is a manifestation of Langlands' formulation of a non-abelian class field theory. Now known as the Langlands correspondence, this formulation of class field theory can be viewed as giving an arithmetic parametrization of local or automorphic representations in terms of admissible homomorphisms of (an appropriate analogue) of the Weil-Deligne group into the L-group. When this conjectural parametrization is combined with natural homomorphisms of the L-groups it predicts a transfer or lifting of local or automorphic representations of two reductive algebraic groups. As a purely automorphic expression of a global non-abelian class field theory, global functoriality is inherently an arithmetic process.

Global functoriality from a quasisplit classical group G to GL_N associated to a natural map on the L-groups has been established in many cases. We recall the main cases:

(i) For G a split classical group with the natural embedding of the L-groups, this was established in [**10**] and [**11**].
(ii) For G a quasisplit unitary group with the L-homomorphism associated to stable base change on the L-groups, this was established in [**29**],[**26**], and [**27**].
(iii) For G a split general spin group, this was established in [**5**].

In this paper we consider simultaneously the cases of quasisplit classical groups G. This includes all the cases mentioned in (i) and (ii) above as well as the new case of the quasisplit even special orthogonal groups. Similar methods should work for the quasisplit GSpin groups, and this will be pursued by Asgari and Shahidi as a sequel to [**5**].

As with the previous results above, our method combines the Converse Theorem for GL_N with the Langlands-Shahidi method for controlling the L-functions of the quasisplit classical groups. One of the crucial ingredients in this method is the use of the "stability of local γ–factors" to finesse the lack of the Local Langlands Conjecture at the ramified non-archimedean places. The advance that lets us now

2010 *Mathematics Subject Classification.* 22E55, 11F70, 11R39.

During the work on this project J.W.C. was partially supported by the NSF through grant DMS-0654017, I.I.P-S. by the NSF through grant DMS-0072450, and F.S. by the NSF through grant DMS-0700280.

handle the quasisplit orthogonal groups is our general stability result in [**14**]. In the past, the stability results were established on a case-by-case basis as needed. The general stability result in [**14**] now lets us give a uniform treatment of all quasisplit classical groups. As before, once we have established the existence of functoriality for the quasisplit classical groups, the descent results of Ginzburg, Rallis, and Soudry [**16, 42**] then give the complete characterization of the image of functoriality in these cases.

This paper can be considered as a survey of past results, an exposition of how to apply the general stability result of [**14**], and the first proof of global functoriality for the quasisplit even orthogonal groups. We have included an appendix containing specific calculations for the even quasisplit orthogonal groups. We will return to local applications of these liftings in a subsequent paper.

Finally, we would like to thank the referee who helped us improve the readability of the paper.

Dedication. The first two authors would like to take this opportunity to dedicate their contributions to this paper to their friend and coauthor, Freydoon Shahidi. The collaboration with Freydoon has been a high point in our careers and we feel it is fitting for a paper reflecting this to appear in this volume in his honor.

1. Functoriality for quasisplit classical groups

Let k be a number field and let $\mathbb{A}_k$ be its ring of adeles. We fix a non-trivial continuous additive character ψ of $\mathbb{A}_k$ which is trivial on the principal adeles k. We will let G_n denote a quasisplit classical group of rank n defined over k. More specifically, we will consider the following cases.

(i) *Odd orthogonal groups.* In this case $G_n = \mathrm{SO}_{2n+1}$, the split special orthogonal group in $2n+1$ variables defined over k, i.e., type B_n. The connected component of the L-group of G_n is ${}^LG_n^0 = \widehat{G}_n = \mathrm{Sp}_{2n}(\mathbb{C})$ while the L-group is the direct product ${}^LG_n = \mathrm{Sp}_{2n}(\mathbb{C}) \times W_k$.

(ii) *Even orthogonal groups.* In this case either (a) $G_n = \mathrm{SO}_{2n}$, the split special orthogonal group in $2n$ variables defined over k, type D_n, or (b) $G_n = \mathrm{SO}_{2n}^*$ is the quasisplit special orthogonal group associated to a quadratic extension E/k, i.e, type ${}^2D_n$. In either case, the connected component of the L-group of G_n is ${}^LG_n^0 = \widehat{G}_n = \mathrm{SO}_{2n}(\mathbb{C})$. In the split case (a), the L-group of the product ${}^LG_n = \mathrm{SO}_{2n}(\mathbb{C}) \times W_k$, while in the quasisplit case (b), the L-group is the semi-direct product ${}^LG_n = \mathrm{SO}_{2n}(\mathbb{C}) \rtimes W_k$ where the Weil group acts through the quotient $W_k/W_E \simeq \mathrm{Gal}(E/k)$ which gives the Galois structure of SO_{2n}^*. We will need to make this Galois action more explicit. Let $\mathrm{O}_{2n}(\mathbb{C})$ denote the even orthogonal group of size $2n$. Then we have $\mathrm{Gal}(E/k) \simeq \mathrm{O}_{2n}(\mathbb{C})/\mathrm{SO}_{2n}(\mathbb{C})$. Conjugation by an element of $\mathrm{O}_{2n}(\mathbb{C})$ of negative determinant gives an outer automorphism of $\mathrm{SO}_{2n}(\mathbb{C})$ corresponding to the diagram automorphism which exchanges the roots α_n and α_{n-1} in Bourbaki's numbering [**7**] or the numbering in Shahidi [**38**]. So if we let $h' \in \mathrm{O}_{2n}(\mathbb{C})$ be any element of negative determinant then for $\sigma \in \mathrm{Gal}(E/k)$

the non-trivial element of the Galois group the action of σ on ${}^L G_n^0$ is

$$\sigma(g) = (h')^{-1} g h'.$$

We will discuss this case in more detail below when we discuss the relevant L-homomorphism and in the appendix (Section 7.1). (Note that when $n = 4$ except for the SO_8^* defined by a quadratic extension the other non-split quasisplit forms of D_4 are not considered to be classical groups.)

(iii) *Symplectic groups.* In this case $G_n = \mathrm{Sp}_{2n}$, the symplectic group in $2n$ variables defined over k, type C_n. The connected component of the L-group of G_n is ${}^L G_n^0 = \widehat{G}_n = \mathrm{SO}_{2n+1}(\mathbb{C})$ and the L-group is the product ${}^L G = \mathrm{SO}_{2n+1}(\mathbb{C}) \times W_k$.

(iv) *Unitary groups.* In this case either (a) $G_n = \mathrm{U}_{2n}$ is the even quasisplit unitary group defined with respect to a quadratic extension E/k or (b) $G_n = \mathrm{U}_{2n+1}$ is the odd quasisplit unitary group defined with respect to a quadratic extension E/k. Both are of type ${}^2A_n$. In case (a) the connected component of the L-group is ${}^L G_n^0 = \widehat{G}_n = \mathrm{GL}_{2n}(\mathbb{C})$ and the L-group is the semi-direct product ${}^L G_n = \mathrm{GL}_{2n}(\mathbb{C}) \rtimes W_k$ where the Weil group acts through the quotient $W_k/W_E \simeq \mathrm{Gal}(E/k)$ which gives the Galois structure of U_{2n}. In case (b) the connected component of the Langlands dual group is ${}^L G_n^0 = \mathrm{GL}_{2n+1}(\mathbb{C})$ and the L-group is the semi-direct product ${}^L G_n = \mathrm{GL}_{2n+1}(\mathbb{C}) \rtimes W_k$ where the Weil groups acts through the quotient $W_k/W_E \simeq \mathrm{Gal}(E/k)$ which gives the Galois structure of U_{2n+1}. We will need to make precise the Galois action. Following [**26, 27**] we let

$$J_n = \begin{pmatrix} & & 1 \\ & \cdot^{\cdot^{\cdot}} & \\ 1 & & \end{pmatrix} \quad \text{and set} \quad J_n' = \begin{cases} \begin{pmatrix} & J_n \\ -J_n & \end{pmatrix} & \text{in case (a)} \\ \begin{pmatrix} & & J_n \\ & 1 & \\ -J_n & & \end{pmatrix} & \text{in case (b)} \end{cases}$$

so that $G_n = \mathrm{U}(J_n')$. Then if σ is the non-trivial element in $\mathrm{Gal}(E/k)$ then the action of σ on ${}^L G_n^0$ is

$$\sigma(g) = (J_n')^{-1}\,{}^t g^{-1} J_n',$$

the outer automorphism of ${}^L G_n$ conjugated by the form.

In each of these cases let δ_1 denote the first fundamental representation, or standard representation, of the connected component of the L-group. This is the defining representation of $\widehat{G}_n$ on the appropriate $\mathbb{C}^N$. As can be seen from the description of ${}^L G_n^0$ above, in each case either $N = 2n$ or $N = 2n+1$. Associated to this representation is a natural embedding of ${}^L G_n^0$ into $\mathrm{GL}_N(\mathbb{C}) = {}^L\mathrm{GL}_N^0$. There is an associated standard representation of the L-group ${}^L G_n$ on either $\mathbb{C}^N$ or $\mathbb{C}^N \times \mathbb{C}^N$ which gives rise to a natural L-homomorphism ι which we now describe.

In the case of the split classical groups, the standard representation of the L-group is still on $\mathbb{C}^N$ and is obtained by extending δ_1 to be trivial on the Weil group. This representation then determines an L-homomorphism $\iota : {}^L G_n \hookrightarrow {}^L\mathrm{GL}_N$. By Langlands' principle of functoriality [**3, 6, 9**], associated to these L-homomorphisms there should be a *transfer* or *lift* of automorphic representations from $G_n(\mathbb{A}_k)$ to $\mathrm{GL}_N(\mathbb{A}_k)$. These were the cases treated in [**10, 11**].

In the case of the quasisplit even orthogonal group, we extend the first fundamental representation of ${}^LG^0$ to an embedding of the L-group in the natural way, to obtain $\iota : {}^LG_n \hookrightarrow {}^L\mathrm{GL}_N$. So in this case we again expect a transfer from $\mathrm{SO}^*_{2n}(\mathbb{A}_k)$ to $\mathrm{GL}_{2n}(\mathbb{A}_k)$. Let us elaborate on this embedding, since although well known it is not all together straightforward. It is related to the theory of twisted endoscopy and can be found in [**1, 2, 4**]. The first fundamental representation gives an embedding of $\mathrm{SO}_{2n}(\mathbb{C}) \hookrightarrow \mathrm{GL}_N(\mathbb{C})$ with $N = 2n$. In fact this extends to an embedding of $\mathrm{O}_{2n}(\mathbb{C}) \hookrightarrow \mathrm{GL}_N(\mathbb{C})$. We then choose an L-homomorphism

$$\xi : W_k \to \mathrm{O}_{2n}(\mathbb{C}) \times W_k \subset \mathrm{GL}_N(\mathbb{C}) \times W_k$$

which induces the isomorphism

$$W_k/W_E \to \mathrm{Gal}(E/k) \simeq \mathrm{O}_{2n}(\mathbb{C})/\mathrm{SO}_{2n}(\mathbb{C})$$

that is, such that ξ factors through $W_k/W_E \simeq \mathrm{Gal}(E/k)$ and sends the non-trivial Galois automorphism σ to an element of negative determinant in $\mathrm{O}_{2n}(\mathbb{C})$ times σ. Let us write this as $\xi(w) = \xi'(w) \times w$ with $\xi'(w) \in \mathrm{O}_{2n}(\mathbb{C})$. Then in the construction of ${}^L\mathrm{SO}^*_{2n} = \mathrm{SO}_{2n}(\mathbb{C}) \rtimes W_k$ the Weil group acts on $\mathrm{SO}_{2n}(\mathbb{C})$ through conjugation by $\xi'(w)$. We now turn to the embedding of the L-group. If we represent elements of ${}^L\mathrm{SO}^*_{2n}$ as products $h \times w = (h \times 1)(1 \times w)$ with $h \in \mathrm{SO}_{2n}(\mathbb{C})$ and $w \in W_k$ then $\iota : \mathrm{SO}_{2n}(\mathbb{C}) \rtimes W_k \hookrightarrow \mathrm{GL}_N(\mathbb{C}) \times W_k$ is given by $\iota(h \times 1) = h \times 1 \in \mathrm{GL}_N(\mathbb{C}) \times W_k$ and $\iota(1 \times w) = \xi(w) = \xi'(w) \times w$. One can find a more detailed description of the embedding in the appendix (Section 7.1).

In the case of unitary groups we follow the description in [**26, 27**], to which the reader can refer for more details. The standard representation of ${}^LG_n$ is now on $\mathbb{C}^N \times \mathbb{C}^N$. The action of the connected component ${}^LG^0_n$ is by

$$[g \times 1](v_1, v_2) = (gv_1, \sigma(g)v_2)$$

while the Weil group acts through the quotient $W_k/W_E \simeq \mathrm{Gal}(E/k)$ with the non-trivial Galois element acting by

$$[1 \times \sigma](v_1, v_2) = (v_2, v_1).$$

It determines an embedding ι of ${}^LG_n \simeq {}^LG^0_n \rtimes W_k$ into $(\mathrm{GL}_N(\mathbb{C}) \times \mathrm{GL}_N(\mathbb{C})) \rtimes W_k$ given by $\iota(g \times w) = (g \times \sigma(g)) \times w$, where on the right hand side, W_k acts on $\mathrm{GL}_N(\mathbb{C}) \times \mathrm{GL}_N(\mathbb{C})$ through the quotient $W_k/W_E \simeq \mathrm{Gal}(E/k)$ with $\sigma(g_1 \times g_2) = g_2 \times g_1$. The group $(\mathrm{GL}_N(\mathbb{C}) \times \mathrm{GL}_N(\mathbb{C})) \rtimes W_k$ defined in this way is the L-group of the restriction of scalars $\mathrm{Res}_{E/k}\mathrm{GL}_N$. Hence the map on L-groups we consider is that associated to stable base change $\iota : {}^LG_n \hookrightarrow {}^L(\mathrm{Res}_{E/k}\mathrm{GL}_N)$.

To give a unified presentation of these functorialities, we let

$$H_N = \begin{cases} \mathrm{GL}_N & \text{if } G_n \text{ is orthogonal or symplectic} \\ \mathrm{Res}_{E/k}\mathrm{GL}_N & \text{if } G_n \text{ is unitary} \end{cases}$$

where $N = 2n$ or $2n+1$ as described above. Then the functorialities that we will establish are from G_n to H_N given in the following table. The embedding ι of L-groups is that described above.

G_n	$\iota : {}^L G_n \hookrightarrow {}^L H_N$	H_N
SO_{2n+1}	$\mathrm{Sp}_{2n}(\mathbb{C}) \times W_k \hookrightarrow \mathrm{GL}_{2n}(\mathbb{C}) \times W_k$	GL_{2n}
SO_{2n}	$\mathrm{SO}_{2n}(\mathbb{C}) \times W_k \hookrightarrow \mathrm{GL}_{2n}(\mathbb{C}) \times W_k$	GL_{2n}
SO^*_{2n}	$\mathrm{SO}_{2n}(\mathbb{C}) \rtimes W_k \hookrightarrow \mathrm{GL}_{2n}(\mathbb{C}) \times W_k$	GL_{2n}
Sp_{2n}	$\mathrm{SO}_{2n+1}(\mathbb{C}) \times W_k \hookrightarrow \mathrm{GL}_{2n+1}(\mathbb{C}) \times W_k$	GL_{2n+1}
U_{2n}	$\mathrm{GL}_{2n}(\mathbb{C}) \rtimes W_k \hookrightarrow (\mathrm{GL}_{2n}(\mathbb{C}) \times \mathrm{GL}_{2n}(\mathbb{C})) \rtimes W_k$	$\mathrm{Res}_{E/k}\mathrm{GL}_{2n}$
U_{2n+1}	$\mathrm{GL}_{2n+1}(\mathbb{C}) \rtimes W_k \hookrightarrow (\mathrm{GL}_{2n+1}(\mathbb{C}) \times \mathrm{GL}_{2n+1}(\mathbb{C})) \rtimes W_k$	$\mathrm{Res}_{E/k}\mathrm{GL}_{2n+1}$

By Langlands' principle of functoriality, as explicated in [**3, 6, 9**], associated to these L-homomorphisms there should be a *transfer* or *lift* of automorphic representations from $G_n(\mathbb{A}_k)$ to $H_N(\mathbb{A}_k)$. To be more precise, for each place v of k we have the local versions of the L-groups, obtained by replacing the Weil group W_k with the local Weil group W_{k_v}. The natural maps $W_{k_v} \to W_k$ make the global and local L-groups compatible. We will not distinguish between our local and global L-groups notationally. Our global L-homomorphism ι then induces a local L-homomorphism, which we will denote by $\iota_v : {}^L G_n \to {}^L H_N$.

Let $\pi = \otimes' \pi_v$ be an irreducible automorphic representation of $G_n(\mathbb{A}_k)$. For v a finite place of k where π_v is unramified, and if necessary the local quadratic extension E_w/k_v is also unramified, the unramified arithmetic Langlands classification or the Satake classification [**6, 35**] implies that π_v is parametrized by an unramified admissible homomorphism $\phi_v : W_{k_v} \to {}^L G_n$ where W_{k_v} is the Weil group of k_v. By composing with $\iota_v : {}^L G_n \hookrightarrow {}^L H_N$ we have an unramified admissible homomorphism $\Phi_v = \iota_v \circ \phi_v : W_{k_v} \longrightarrow {}^L H_N$ and this defines an irreducible admissible unramified representation Π_v of $H_N(k_v)$ [**17, 18**]. Then Π_v is the *local functorial lift* of π_v. The process is outlined in the following *local functoriality diagram.*

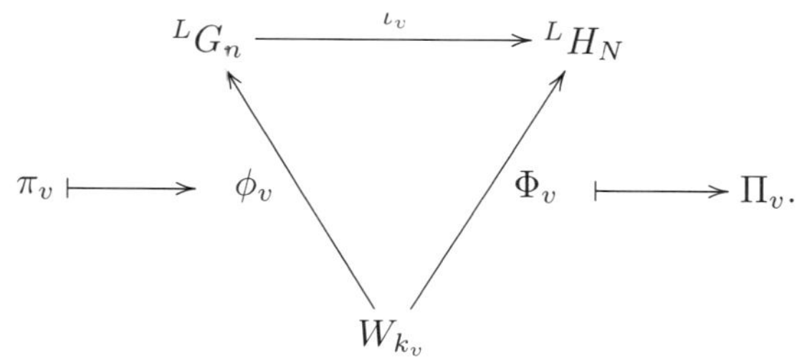

Similarly, if v is an archimedean place, then by the arithmetic Langlands classification π_v is determined by an admissible homomorphism $\phi_v : W_v \longrightarrow {}^L G_n$ where

W_v is the local Weil group of k_v [**6, 30**]. The composition $\iota_v \circ \phi_v$ is an admissible homomorphism of W_v into ${}^L H_N$ and hence determines a representation Π_v of $H_n(k_v)$ via the same diagram. This is again the *local functorial lift* of π_v. Note that in either case we have an equality of local L-functions

$$L(s, \Pi_v) = L(s, \Phi_v) = L(s, \iota_v \circ \phi_v) = L(s, \pi_v, \iota_v)$$

as well as equalities for the associated ε–factors (if ψ_v is unramified as well at the finite place in question).

An irreducible automorphic representation $\Pi = \otimes' \Pi_v$ of $H_N(\mathbb{A}_k)$ is called a *functorial lift* of π if for every archimedean place v and for almost all non-archimedean places v for which π_v is unramified we have that Π_v is a local functorial lift of π_v. In particular this entails an equality of (partial) Langlands L-functions

$$L^S(s, \Pi) = \prod_{v \notin S} L(s, \Pi_v) = \prod_{v \notin S} L(s, \pi_v, \iota_v) = L^S(s, \pi, \iota),$$

where S is the (finite) complement of the places where we know the local Langlands classification, so the ramified places.

We will let B_n denote a Borel subgroup of G_n and let U_n denote the unipotent radical of B_n. The abelianization of U_n is a direct sum of copies of k and we may use ψ to define a non-degenerate character of $U_n(\mathbb{A}_k)$ which is trivial on $U_n(k)$. By abuse of notation we continue to call this character ψ.

Let π be an irreducible cuspidal representation of $G_n(\mathbb{A}_k)$. We say that π is *globally generic* if there is a cusp form $\varphi \in V_\pi$ such that φ has a non-vanishing ψ-Fourier coefficient along U_n, i.e., such that

$$\int_{U_n(k) \backslash U_n(\mathbb{A}_k)} \varphi(ug) \psi^{-1}(u)\, du \neq 0.$$

Cuspidal automorphic representations of GL_n are always globally generic in this sense. For cuspidal automorphic representations of the classical groups this is a condition. In general the notion of being globally generic may depend on the choice of splitting of the group. However, as is shown in the Appendix to [**11**], given a π which is globally generic with respect to some splitting there is always an "outer twist" which is globally generic with respect to a fixed splitting. This outer twist provides an abstract isomorphism between globally generic cuspidal representations and will not effect the L- or ε-factors nor the notion of the functorial lift. Hence we lose no generality in considering cuspidal representations that are globally generic with respect to our fixed splitting.

The principal result that we will prove in this paper is the following.

THEOREM 1.1. *Let k be a number field and let π be an irreducible globally generic cuspidal automorphic representation of a quasisplit classical group $G_n(\mathbb{A}_k)$ as above. Then π has a functorial lift to $H_N(\mathbb{A}_k)$ associated to the embedding ι of L-groups above.*

The low-dimensional cases of this theorem are already well understood. In the split cases, they were discussed in [**11**]. Thus we will concentrate primarily on the cases where $n \geq 2$, except for the quasisplit orthogonal groups where we restrict to $n \geq 4$.

2. The Converse Theorem

In order to effect the functorial lifting from G_n to H_N we will use the Converse Theorem for GL_N [**12, 13**] as we did in [**10, 11**]. Let us fix a number field K and a finite set S of finite places of K. For the case of G_n orthogonal or symplectic, the target for functoriality is $\mathrm{GL}_N(\mathbb{A}_k)$ and we will need $K = k$. However, in the case of unitary G_n, the target of functoriality is $\mathrm{Res}_{E/k}\mathrm{GL}_N(\mathbb{A}_k) \simeq \mathrm{GL}_N(\mathbb{A}_E)$ and we will need to apply the converse theorem for $K = E$.

For each integer m, let

$$\mathcal{A}_0(m) = \{\tau \mid \tau \text{ is a cuspidal representation of } \mathrm{GL}_m(\mathbb{A}_K)\}$$

and

$$\mathcal{A}_0^S(m) = \{\tau \in \mathcal{A}_0(m) \mid \tau_v \text{ is unramified for all } v \in S\}.$$

We set

$$\mathcal{T}(N-1) = \coprod_{m=1}^{N-1} \mathcal{A}_0(m) \quad \text{and} \quad \mathcal{T}^S(N-1) = \coprod_{m=1}^{N-1} \mathcal{A}_0^S(m).$$

If η is a continuous character of $K^\times \backslash \mathbb{A}_K^\times$, let us set

$$\mathcal{T}(S;\eta) = \mathcal{T}^S(N-1) \otimes \eta = \{\tau = \tau' \otimes \eta \; : \; \tau' \in \mathcal{T}^S(N-1)\}.$$

THEOREM 2.1 (Converse Theorem). *Let $\Pi = \otimes' \Pi_v$ be an irreducible admissible representation of $\mathrm{GL}_N(\mathbb{A}_K)$ whose central character ω_Π is invariant under $K^\times$ and whose L-function $L(s,\Pi) = \prod_v L(s,\Pi_v)$ is absolutely convergent in some right half-plane. Let S be a finite set of finite places of K and let η be a continuous character of $K^\times \backslash \mathbb{A}_K^\times$. Suppose that for every $\tau \in \mathcal{T}(S;\eta)$ the L-function $L(s,\Pi \times \tau)$ is* nice, *that is, it satisfies*

(1) *$L(s,\Pi \times \tau)$ and $L(s,\widetilde{\Pi} \times \tilde{\tau})$ extend to entire functions of $s \in \mathbb{C}$,*
(2) *$L(s,\Pi \times \tau)$ and $L(s,\widetilde{\Pi} \times \tilde{\tau})$ are bounded in vertical strips, and*
(3) *$L(s,\Pi \times \tau)$ satisfies the functional equation*

$$L(s,\Pi \times \tau) = \varepsilon(s,\Pi \times \tau) L(1-s,\widetilde{\Pi} \times \tilde{\tau}).$$

Then there exists an automorphic representation Π' of $\mathrm{GL}_N(\mathbb{A}_K)$ such that $\Pi_v \simeq \Pi'_v$ for almost all v. More precisely, $\Pi_v \simeq \Pi'_v$ for all $v \notin S$.

In the statement of the theorem, the twisted L- and ϵ-factors are defined by the products

$$L(s,\Pi \times \tau) = \prod_v L(s,\Pi_v \times \tau_v) \qquad \varepsilon(s,\Pi \times \tau) = \prod_v \varepsilon(s,\Pi_v \times \tau_v,\psi_v)$$

of local factors as in [**12, 10**].

To motivate the next few sections, let us describe how we will apply this theorem to the problem of Langlands lifting from G_n to H_N. We begin with our globally generic cuspidal automorphic representation $\pi = \otimes' \pi_v$ of $G_n(\mathbb{A}_k)$.

If G_n is an orthogonal or symplectic group, then for each place v we need to associate to π_v an irreducible admissible representation Π_v of $H_N(k_v) = \mathrm{GL}_N(k_v)$ such that for every $\tau \in \mathcal{T}(S;\eta)$ we have

$$L(s, \pi_v \otimes \tau_v, \iota_v \otimes \iota'_v) = L(s, \Pi_v \times \tau_v)$$
$$\varepsilon(s, \pi_v \otimes \tau_v, \iota_v \otimes \iota'_v, \psi_v) = \varepsilon(s, \Pi_v \times \tau_v, \psi_v)$$

where ι' is the identity map on $\mathrm{GL}_m(\mathbb{C})$, or more accurately from the L-group ${}^L\mathrm{GL}_m = \mathrm{GL}_m(\mathbb{C}) \times W_k$ to $\mathrm{GL}_m(\mathbb{C})$ given by projection on the first factor, and similarly ι now represents the representation of ${}^LG_n$ given by ι followed by the projection onto the first factor of ${}^LH$, or the connected component of the identity, i.e., the associated map to $\mathrm{GL}_N(\mathbb{C})$.

If G_n is a unitary group, then for each place v we need to associate to π_v an irreducible admissible representation Π_v of $H_N(k_v) = \mathrm{GL}_N(E_v)$, where $E_v = E \otimes k_v$ is either an honest quadratic extension or the split quadratic algebra over k_v, such that for every $\tau \in \mathcal{T}(S;\eta)$ we have

$$L(s, \pi_v \otimes \tau_v, \iota_v \otimes \iota'_v) = L(s, \Pi_v \times \tau_v)$$
$$\varepsilon(s, \pi_v \otimes \tau_v, \iota_v \otimes \iota'_v, \psi_v) = \varepsilon(s, \Pi_v \times \tau_v, \psi_v).$$

Now τ_v must be viewed as a representation of $\mathrm{GL}_m(E_v)$, i.e., of $\mathrm{Res}_{E/k}\mathrm{GL}_m(k_v)$. If E_v/k_v is an honest quadratic extension, then $G_n(k_v)$ is an honest local unitary group and $H_N(k_v) \simeq \mathrm{GL}_N(E_v)$. If v splits in E, so $E_v \simeq E_{w_1} \oplus E_{w_2}$ with $E_{w_i} \simeq k_v$, then $G_n(k_v) \simeq \mathrm{GL}_N(k_v)$ and $H_N(k_v) \simeq \mathrm{GL}_N(E_v) \simeq \mathrm{GL}_N(k_v) \times \mathrm{GL}_N(k_v)$. In this case $\Pi_v = \Pi_{1,v} \otimes \Pi_{2,v}$, an outer tensor product, and similarly $\tau_v = \tau_{1,v} \otimes \tau_{2,v}$ and we have a product of two factors on each side. A more detailed description for this case can be found in [**26, 27**].

For archimedean places v and those non-archimedean v where π_v is unramified, we take Π_v to be the local functorial lift of π_v described above. For those places v where π_v is ramified, we will finesse the lack of a local functorial lift using the stability of γ–factors as described in Section 4 below. This will allow us to associate to π_v a representation Π_v of $H_N(k_v)$ at these places as well. The process involves the choice of a highly ramified character η_v of $H_1(k_v)$. If we then take $\Pi = \otimes' \Pi_v$, this is an irreducible representation of $H_N(\mathbb{A}_k)$. With the choices above we will have

$$L(s, \pi \otimes \tau, \iota \otimes \iota') = L(s, \Pi \times \tau)$$
$$\varepsilon(s, \pi \otimes \tau, \iota \otimes \iota') = \varepsilon(s, \Pi \times \tau)$$

for $\mathrm{Re}(s) >> 0$ and all $\tau \in \mathcal{T}(S;\eta)$ for a suitable fixed character η of $H_1(\mathbb{A}_k)$. This is our candidate lift. The theory of L-functions for $G_n \times H_m$, which we address in the next section, will then guarantee that the twisted L-functions $L(s, \pi \otimes \tau, \iota \otimes \iota')$ are nice for all $\tau \in \mathcal{T}(S;\eta)$. Then the $L(s, \Pi \times \tau)$ will also be nice and Π satisfies the hypotheses of the Converse Theorem. Hence there exists an irreducible automorphic representation Π' of $H_N(\mathbb{A}_k)$ such that $\Pi_v \simeq \Pi'_v$ for all archimedean v and almost all finite v where π_v is unramified. Hence Π' is a functorial lift of π.

3. L-functions for $G_n \times H_m$

Let π be a globally generic cuspidal representation of $G_n(\mathbb{A}_k)$ and τ a cuspidal representation of $H_m(\mathbb{A}_k)$, with $m \geq 1$. We let ι and ι' be the representations of the L–groups defined in Sections 1 and 2 respectively. To effect our lifting, we must control the analytic properties of the twisted L-functions $L(s, \pi \otimes \tau, \iota \otimes \iota')$. This we do by the method of Langlands and Shahidi, as we outline here.

The L-functions $L(s, \pi \otimes \tau, \iota \otimes \iota')$ are the completed L-functions as defined in [**39**] via the theory of Eisenstein series. If we let $M_{n,m}$ denote $G_n \times H_m$ with $m \geq 1$, then this appears as the Levi factor of a maximal self-associate parabolic subgroup $P_{n,m} = M_{n,m}N_{n,m}$ of G_{n+m} associated to the root α_m as in [**38**]. The representation $\iota \otimes \iota'$ then occurs in the adjoint action of ${}^LM_{n,m}$ on the Lie algebra ${}^L\mathfrak{n}_{n,m}$ as the representation $\tilde{r}_1$ of [**38**]. Then these L-functions can be defined and controlled by considering the induced representation $I(s, \pi \otimes \tau)$ described in [**38, 39**] since $\pi \otimes \tau$ is a cuspidal representation of $M_{n,m}(\mathbb{A}_k)$. The local factors are then defined in [**39**] via the arithmetic Langlands classification for archimedean places, through the Satake parameters for finite unramified places, as given by the poles of the associated γ–factor (or local coefficient) if π_v and τ_v are tempered, by analytic extension if π_v and τ_v are quasi-tempered, and via the representation theoretic Langlands classification otherwise. Since it is only these representations that we will be considering, we will abbreviate our notation by suppressing the L–homomorphism, so for example

$$L(s, \pi \times \tau) = \prod_v L(s, \pi_v \times \tau_v) = \prod_v L(s, \pi_v \otimes \tau_v, \iota_v \otimes \iota'_v) = L(s, \pi \otimes \tau, \iota \otimes \iota')$$

with similar conventions for the ε– and γ–factors.

The global theory of these twisted L-functions is now quite well understood.

THEOREM 3.1. *Let S be a non-empty set of finite places of k. Let $K = k$ when G_n is orthogonal or symplectic or $K = E$ if G_n is a unitary group associated to the quadratic extension E/k and continue to let S denote the corresponding set of places of K. Let η be a character of $K^\times \backslash \mathbb{A}_K^\times$ such that, for some $v \in S$, either the square η_v^2 is ramified if $K = k$, or if $K = E$ then for the places w of E above v we have both η_w and $\eta_w \overline{\eta}_w$ are ramified. Then for all $\tau \in \mathcal{T}(S; \eta)$ the L-function $L(s, \pi \times \tau)$ is* nice, *that is,*

(1) *$L(s, \pi \times \tau)$ is an entire function of s,*
(2) *$L(s, \pi \times \tau)$ is bounded in vertical strips of finite width, and*
(3) *we have the functional equation*

$$L(s, \pi \times \tau) = \varepsilon(s, \pi \times \tau) L(1 - s, \tilde{\pi} \times \tilde{\tau}).$$

Proof: (1) In all cases this follows from the more general Proposition 2.1 of [**28**]. Note that in view of the results of Muić [**34**] and of [**8**], the necessary result on normalized intertwining operators, Assumption 1.1 of [**28**], usually referred to as Assumption A [**24**], is valid in all cases as proved in [**24, 25**]. Note that this is the only part of the theorem where the twisting by η is needed. The specific ramification stated comes from [**11**] or [**27**].

(2) The boundedness in vertical strips of these L-functions is known in wide generality, which includes the cases of interest to us. It follows from Corollary 4.5 of [**15**] and is valid for all $\tau \in \mathcal{T}(N-1)$, provided one removes neighborhoods of the finite number of possible poles of the L-function.

(3) The functional equation is also known in wide generality and is a consequence of Theorem 7.7 of [**39**]. It is again valid for all $\tau \in \mathcal{T}(N-1)$. □

4. Stability of γ-factors

This section is devoted to the formulation of the stability of the local γ-factors for generic representations of the quasisplit groups under consideration. This result is necessary for defining a suitable local lift at the non-archimedean places where we do not have the local Langlands conjecture at our disposal.

For this section, let k denote a p-adic local field, that is, a non-archimedean local field of characteristic zero. Let G_n now be a quasisplit classical group of the types defined in Section 1, but now over k. These will correspond to the local situations that arise in our global problem, with the exception of the global unitary groups at a place which splits in the defining quadratic extension (see Remark 4.1 below).

4.1. Stability. Let π be a generic irreducible admissible representation of $G_n(k)$ and let η be a continuous character of $H_1(k) \simeq k^\times$ (resp. $E^\times$ in the local unitary case). Let ψ be a fixed non-trivial additive character of k. Let $\gamma(s, \pi\times\eta, \psi)$ be the associated γ-factor as defined in Theorem 3.5 of [**39**]. These are defined inductively through the local coefficients $C_\psi(s, \pi\otimes\eta)$ of the local induced representations analogous to those given above. They are related to the local L- and ε-factors by

$$\gamma(s, \pi\times\eta, \psi) = \frac{\varepsilon(s, \pi\times\eta, \psi)L(1-s, \tilde{\pi}\times\eta^{-1})}{L(s, \pi\times\eta)}.$$

We begin by recalling the main result of [**14**], with a slight shift in notation for consistency.

THEOREM 4.1. *Let G be a quasisplit connected reductive algebraic group over k such that the Γ–diagram of G_D is of either type $B_{n+1}, C_{n+1}, D_{n+1}, {}^2A_{n+1}$ or ${}^2D_{n+1} (n+1 \geq 4)$. Let $P = MN$ be a self-associate maximal parabolic subgroup of G over k such that the unique simple root in N is the root α_1 in Bourbaki's numbering [**7**]. Let π be an irreducible admissible generic representation of $M(k)$.*

Then $C_\psi(s,\pi)$ is stable, that is, if ν is a character of $K^\times$, realized as a character $\tilde{\nu}$ of $M(k)$ by

$$\tilde{\nu}(m) = \nu(\det(\mathrm{Ad}_{\mathfrak{n}}(m))),$$

then

$$C_\psi(s, \pi_1 \otimes \tilde{\nu}) = C_\psi(s, \pi_2 \otimes \tilde{\nu})$$

for any two such representations π_1 and π_2 with the same central characters and all sufficiently highly ramified ν. Here $\mathfrak{n}$ is the Lie algebra of $N(k)$.

Note that this covers the local quasisplit classical groups that are under consideration here if we take $G = G_{n+1}$. The splitting field K is then k itself except in the ${}^2A_n$ and ${}^2D_n$ cases, where it is the associated quadratic extension E as in Section 1. According to the tables in Section 4 of [**38**], the Levi subgroups M in Theorem 4.1 are of the form $M \simeq H_1 \times G_n$. To use the stability result in the application to functoriality we need the following elementary lemma.

LEMMA 4.1. *Let $m \in M(k)$ and write $m = a \times m'$ with $a \in H_1(k)$ and $m' \in G_n(k)$. Then $\det(\mathrm{Ad}_{\mathfrak{n}}(m')) = 1$, i.e.*

$$\det(\mathrm{Ad}_{\mathfrak{n}}(m)) = \det(\mathrm{Ad}_{\mathfrak{n}}(a)).$$

Proof: An elementary matrix calculation shows that in the symplectic and special orthogonal cases we have $\det(\mathrm{Ad}_{\mathfrak{n}}(m')) = \det(m') = 1$. In the unitary case, $\det(\mathrm{Ad}_{\mathfrak{n}}(m')) = N_{E/k}\det(m') = 1$. □

Thus we see that in Theorem 4.1 the twisting character $\tilde{\nu}$ of $M(k)$ factors to a character of the GL_1 factor in M. We will call a character $\tilde{\nu}$ of $H_1(k)$, *suitable* if it arises from a character ν of $K^\times$ via a composition of the embedding $\mathrm{GL}_1 \hookrightarrow M$ followed by the character $m \mapsto \nu(\det(\mathrm{Ad}_{\mathfrak{n}}(m)))$ of M. Now Theorem 4.1 has the following corollary.

COROLLARY 4.1.1. *Let G_n be a quasisplit classical group over k as in Section 1, so that it satisfies the hypotheses of Theorem 4.1. Let π_1 and π_2 be two irreducible admissible representations of $G_n(k)$ having the same central character. Then for every sufficiently highly ramified suitable character $\tilde{\nu}$ of $H_1(k)$ we have*

$$\gamma(s, \pi_1 \times \tilde{\nu}, \psi) = \gamma(s, \pi_2 \times \tilde{\nu}, \psi).$$

Proof: Let G_{n+1} be the quasisplit connected reductive algebraic group over k of rank one larger such such that the parabolic subgroup P in Theorem 4.1 has Levi subgroup $M \simeq H_1 \times G_n$. Then $\mathbf{1} \otimes \pi_1$ and $\mathbf{1} \otimes \pi_2$ determine irreducible admissible representations of $M(k)$ with the same central character. By Theorem 4.1 we know that for every sufficiently highly ramified character ν of $K^\times$, determining a character of $M(k)$ by $\tilde{\nu}(m) = \nu(\det(\mathrm{Ad}_{\mathfrak{n}}(m))$, we have $C_\psi(s, (\mathbf{1}\otimes\pi_1)\otimes\tilde{\nu}) = C_\psi(s, (\mathbf{1}\otimes\pi_2)\otimes\tilde{\nu})$. By our lemma $\tilde{\nu}$ factors to only the H_1 variable in M and $(\mathbf{1} \otimes \pi_i) \otimes \tilde{\nu} = \tilde{\nu} \otimes \pi_i$ as representations of $M(k)$. Then the statement of the corollary follows from the definition of $\gamma(s, \pi_i \times \tilde{\nu}, \psi)$ given in [**38**]. □

REMARK 4.1. Theorem 4.1 and its corollary cover the possible local situations that arise in our global problem *except* for the case of unitary groups at a place that splits in the global quadratic extension. At these places, locally the hypotheses of Theorem 4.1 are not satisfied since the parabolic subgroup P in question is no longer maximal. In this case, $G_n \simeq \mathrm{GL}_N$, $H_1 \simeq \mathrm{GL}_1 \times \mathrm{GL}_1$, and, remembering the implicit L-homomorphism $\iota \otimes \iota'$, both sides are a pair of local $\mathrm{GL}_N \times \mathrm{GL}_1$ γ-factors. In this case the stability result is due to Jacquet and Shalika [**22**]. However, as we shall see, in our application we will not need stability in this situation since the local Langlands correspondence is known for GL_N.

4.2. Stability and parametrization. Let π be an irreducible admissible representation of one of our local G_n. Assume that π is ramified, so we may not know how to parametrize π by an admissible homomorphism of the Weil-Deligne group W'_k into ${}^LG_n$. We wish to replace π by a second representation π' for which we have an arithmetic Langlands parameter and for which we still have a modicum of control over its L- and ε-factors.

We replace π with an induced representation having the same central character. To this end, let T_n be the maximal torus of G_n, take a character λ of T_n, and let $I(\lambda)$ be the associated induced representation. By appropriate choice of λ we can guarantee that π and $I(\lambda)$ have the same central character and that $I(\lambda)$ is irreducible. Let $\phi_\lambda : W_k \to {}^LT$ be the Langlands parameter for λ so that the composition $\phi_\lambda : W_k \to {}^LT_n \to {}^LG_n$ is the arithmetic Langlands parameter for $I(\lambda)$. We can take for π' any of the so constructed $I(\lambda)$. We fix one. By the corollary above, for sufficiently highly ramified suitable $\tilde{\nu}$, depending on π and our choice of $\pi' = I(\lambda)$, we have

$$\gamma(s, \pi \times \tilde{\nu}, \psi) = \gamma(s, \pi' \times \tilde{\nu}, \psi).$$

Once we have the stable γ–factor is expressed in terms of a principal series representations that we can arithmetically parametrize, then we can express the analytic γ–factor as one from arithmetic, the Artin γ–factor associated to the Galois representation $\iota \circ \phi_\lambda$.

PROPOSITION 4.1. *With notation as above,*

$$\gamma(s, I(\lambda) \times \tilde{\nu}, \psi) = \gamma(s, (\iota \circ \phi_\lambda) \otimes \tilde{\nu}, \psi).$$

Proof: The embedding of L-groups $\iota : {}^LG_n \hookrightarrow {}^LH_N$ is defined so that ι is the map coming from the restriction of this adjoint action on ${}^L\mathfrak{n}_{n+1}$ to G_n and similarly for ι' as a representation of H_1. By our convention, $\gamma(s, I(\lambda) \times \tilde{\nu}, \psi)$ is the γ-factor associated to this representation of the L-group, i.e.,

$$\gamma(s, I(\lambda) \times \tilde{\nu}, \psi) = \gamma(s, I(\lambda) \otimes \tilde{\nu}, \iota \otimes \iota', \psi).$$

To relate this analytic γ-factor to that from the parametrization, we embed $G_n \hookrightarrow G_{n+1}$ as part of the Levi subgroup M_{n+1} of the self-associate parabolic subgroup $P_{n+1} = M_{n+1}N_{n+1} \subset G_{n+1}$ such that the unique simple root in N_{n+1} is α_1 as above. Using the product formula (or "cocycle relation") for the local coefficients from Proposition 3.2.1 of [**36**], the local coefficient $C_\psi(s, I(\lambda)\otimes\tilde{\nu})$ factors into a product over the roots appearing in the adjoint representation of ${}^LT_{n+1} \subset {}^LM_{n+1}$ on ${}^L\mathfrak{n}_{n+1}$. Sections 2 and 3 of [**23**] give a computation of the contribution of an individual root space to $C_\psi(s, I(\lambda)\otimes\tilde{\nu})$ in terms of rank one Artin factors coming from the co-roots composed with λ. The resulting expression (at $s = 0$) is found in Proposition 3.4 of [**23**]. If one takes the expression for a general s and then extracts the γ-factor from the local coefficient, one arrives at

$$\gamma(s, I(\lambda)\otimes\tilde{\nu}, \iota\otimes\iota', \psi) = \gamma(s, (\iota\circ\phi_\lambda)\otimes\tilde{\nu}, \psi).$$

This proves the proposition. □

5. The candidate lift

We now return to k denoting a number field. Let $\pi = \otimes'\pi_v$ be a globally generic cuspidal representation of $G_n(\mathbb{A}_k)$. In this section we will construct our candidate $\Pi = \otimes'\Pi_v$ for the functorial lift of π as an irreducible admissible representation of $\mathrm{GL}_N(\mathbb{A}_k)$. We will construct Π by constructing each local component, or local lift, Π_v. There will be three cases: (i) the archimedean lift, (ii) the non-archimedean unramified lift, and finally (iii) the non-archimedean ramified lift.

5.1. The archimedean lift. Let v be an archimedean place of k. By the arithmetic Langlands classification [**30, 6**], π_v is parametrized by an admissible homomorphism $\phi_v : W_{k_v} \to {}^LG_n^0$ where W_{k_v} is the Weil group of k_v. By composing with $\iota_v : {}^LG_n \hookrightarrow {}^LH_N$ we have an admissible homomorphism $\Phi_v = \iota_v\circ\phi_v : W_{k_v} \longrightarrow {}^LH_N$ and this defines an irreducible admissible representation Π_v of $H_N(k_v)$. Then Π_v is the local functorial lift of π_v. We take Π_v as our local lift of π_v. (See the local functoriality diagram in Section 1.)

The local archimedean L– and ε–factors defined via the theory of Eisenstein series we are using are the same as the Artin factors defined through the arithmetic Langlands classification [**37**]. Since the embedding $\iota_v : {}^LG_n \hookrightarrow {}^LH_N$ is the standard representation of the L–group of $G_n(k_v)$ then by the definition of the local L- and ε–factors given in [**6**] we have

$$L(s, \pi_v) = L(s, \iota_v\circ\phi_v) = L(s, \Pi_v)$$

and

$$\varepsilon(s, \pi_v, \psi_v) = \varepsilon(s, \iota_v\circ\phi_v, \psi_v) = \varepsilon(s, \Pi_v, \psi_v)$$

where in both instances the middle factor is the local Artin-Weil L– and ε–factor attached to representations of the Weil group as in [**43**].

If τ_v is an irreducible admissible representation of $H_m(k_v)$ then it is in turn parametrized by an admissible homomorphism $\phi'_v : W_{k_v} \longrightarrow {}^LH_m$. Then the tensor product homomorphism $(\iota_v\circ\phi_v)\otimes(\iota'_v\circ\phi'_v) : W_{k_v} \longrightarrow {}^LH_{mN}$ is admissible and again we have by definition

$$L(s, \pi_v\times\tau_v) = L(s, (\iota_v\circ\phi_v)\otimes(\iota'_v\circ\phi'_v)) = L(s, \Pi_v\times\tau_v)$$

and

$$\varepsilon(s,\pi_v\times\tau_v,\psi_v)=\varepsilon(s,(\iota_v\circ\phi_v)\otimes(\iota'_v\circ\phi'_v),\psi_v)=\varepsilon(s,\Pi_v\times\tau_v,\psi_v).$$

This then gives the following matching of the twisted local L- and ε-factors.

PROPOSITION 5.1. *Let v be an archimedean place of k and let π_v be an irreducible admissible generic representation of $G_n(k_v)$, Π_v its local functorial lift to $H_N(k_v)$, and τ_v an irreducible admissible generic representation of $H_m(k_v)$ with $m<N$. Then*

$$L(s,\pi_v\times\tau_v)=L(s,\Pi_v\times\tau_v)\quad\textit{and}\quad\varepsilon(s,\pi_v\times\tau_v,\psi_v)=\varepsilon(s,\Pi_v\times\tau_v,\psi_v).$$

5.2. The non-archimedean unramified lift. Now let v be a place of k which is non-archimedean and assume that π_v is an unramified representation. By the unramified arithmetic Langlands classification or the Satake classification [**6, 35**], π_v is parametrized by an unramified admissible homomorphism $\phi_v:W_{k_v}\to {}^LG_n^0$ where W_{k_v} is the Weil group of k_v. By composing with $\iota_v:{}^LG_n\hookrightarrow {}^LH_N$ we have an unramified admissible homomorphism $\Phi_v=\iota_v\circ\phi_v:W_{k_v}\longrightarrow {}^LH_N$ and this defines an irreducible admissible unramified representation Π_v of $H_N(k_v)$ [**17, 18**]. Then Π_v is again the local functorial lift of π_v and we take it as our local lift. (Again, see the local functoriality diagram in Section 1.)

We will again need to know that the twisted L- and ε-factors agree for π_v and Π_v.

PROPOSITION 5.2. *Let v be a non-archimedean place of k and let π_v be an irreducible admissible generic unramified representation of $G_n(k_v)$. Let Π_v be its functorial local lift to $H_N(k_v)$ as above, and τ_v an irreducible admissible generic representation of $H_m(k_v)$ with $m<N$. Then*

$$L(s,\pi_v\times\tau_v)=L(s,\Pi_v\times\tau_v)\quad\textit{and}\quad\varepsilon(s,\pi_v\times\tau_v,\psi_v)=\varepsilon(s,\Pi_v\times\tau_v,\psi_v).$$

Proof: Since π_v is unramified its parameter ϕ_v factors through an unramified homomorphism into the maximal torus ${}^LT_n\hookrightarrow {}^LG_n$. The composition $\iota\circ\phi_v=\Phi_v$ then has image in a torus ${}^LT'_N\hookrightarrow {}^LH_N$, which necessarily splits, and Π_v is the corresponding unramified (isobaric) representation. Then the functoriality diagram gives that $L(s,\pi_v,\iota_v)=L(s,\Pi_v)$ and $\varepsilon(s,\pi_v,\iota_v,\psi_v)=\varepsilon(s,\Pi_v,\psi_v)$ and both can be expressed as products of one dimensional abelian Artin L-functions and ε–factors. This is the multiplicativity of the local L- and ε–factors in this case. For twisting by τ_v one appeals to the general multiplicativity of local factors from [**19, 40**] with respect to the preceding data. This is done in detail for the split groups in [**11**] and the calculation here is the same. □

5.3. The non-archimedean ramified lift. Now consider a non-archimedean place v of k where the local component π_v of π is ramified. Assume for now that we are not in the situation where G_n is a unitary group associated to a quadratic extension E/k in which the place v splits; we will return to this at the end of the section. Now we do not have the local Langlands correspondence to give us a natural local functorial lift. Instead we will use the results of Section 4.

Given π_v we choose an induced representation $\pi'_v = I(\lambda_v)$ as in Section 4.1 which has the same central character as π_v and which we do know how to parametrize. Let $\phi_{\lambda_v} : W_{k_v} \to {}^LT_n \to {}^LG_n$ be the associated parameter. By composing with $\iota_v : {}^LG_n \hookrightarrow {}^LH_N$ we have an admissible homomorphism $\Phi_v = \iota_v \circ \phi_{\lambda_v} : W_{k_v} \longrightarrow {}^LH_N$ and this defines an irreducible admissible representation Π_v of $H_N(k_v)$. We now use the local functoriality diagram in the following form:

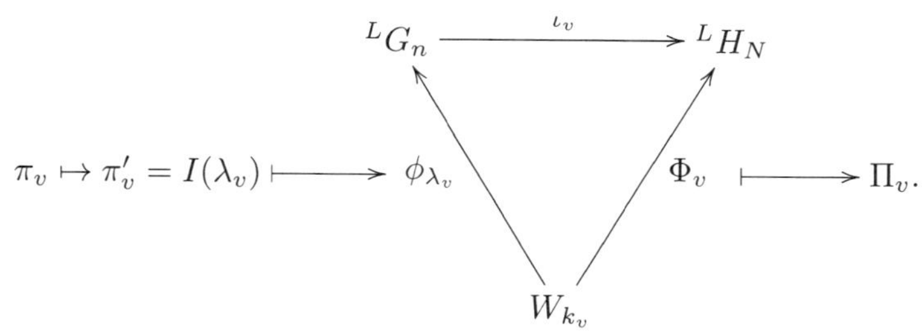

Then Π_v is the local functorial lift of $\pi'_v = I(\lambda_v)$. We take Π_v as our local lift of π_v.

Now let $\tilde{\nu}_v$ be a sufficiently ramified suitable character of $H_1(k_v)$ as in Section 4. Then by Corollary 4.1.1 we know that

$$\gamma(s, \pi_v \times \tilde{\nu}_v, \psi_v) = \gamma(s, \pi'_v \times \tilde{\nu}_v, \psi_v)$$

and by Proposition 4.1 we have

$$\gamma(s, \pi'_v \times \tilde{\nu}_v, \psi_v) = \gamma(s, I(\lambda_v) \times \tilde{\nu}_v, \psi_v) = \gamma(s, (\iota_v \circ \phi_{\lambda_v}) \otimes \tilde{\nu}_v, \psi_v).$$

On the other hand, by the functoriality diagram above

$$\gamma(s, (\iota_v \circ \phi_{\lambda_v}) \otimes \tilde{\nu}_v, \psi_v) = \gamma(s, \Phi_v \otimes \tilde{\nu}_v, \psi_v)$$

and the work of Harris-Taylor and Henniart establishing the local Langlands conjecture for GL_n gives

$$\gamma(s, \Phi_v \otimes \tilde{\nu}_v, \psi_v) = \gamma(s, \Pi_v \times \tilde{\nu}_v, \psi_v).$$

Thus finally

$$\gamma(s, \pi_v \times \tilde{\nu}_v, \psi_v) = \gamma(s, \Pi_v \times \tilde{\nu}_v, \psi_v).$$

For sufficiently ramified $\tilde{\nu}$ the local L-functions $L(s, \pi_v \times \tilde{\nu}_v)$ and $L(s, \Pi_v \times \tilde{\nu}_v)$ both stabilize to 1 [**41, 22**] and so the stability of local γ-factors is essentially the stability of local ε–factors.

PROPOSITION 5.3. *Let π_v be an irreducible admissible generic representation of $G_n(k_v)$ and let Π_v be the irreducible admissible representation of $H_N(k_v)$ as above. Then for sufficiently ramified suitable characters $\tilde{\nu}$ of $H_1(k_v)$ we have*

$$L(s, \pi_v \times \tilde{\nu}_v) = L(s, \Pi_v \times \tilde{\nu}_v) \quad \textit{and} \quad \varepsilon(s, \pi_v \times \tilde{\nu}_v, \psi_v) = \varepsilon(s, \Pi_v \times \tilde{\nu}_v, \psi_v).$$

There is a natural extension of this to the class of representations of $H_m(k_v)$ that we require for the application of the Converse Theorem given in the following proposition.

PROPOSITION 5.4. *Let v be a non-archimedean place of k. Let π_v be an irreducible admissible generic representation of $G_n(k_v)$ and let Π_v be the irreducible admissible representation of $H_N(k_v)$ as above. Let τ_v be an irreducible admissible generic representation of $H_m(k_v)$ with $m < N$ of the form $\tau_v \simeq \tau_{0,v} \otimes \tilde{\nu}_v$ with $\tau_{0,v}$ unramified and $\tilde{\nu}_v$ suitable and sufficiently ramified as above. Then*

$$L(s, \pi_v \times \tau_v) = L(s, \Pi_v \times \tau_v) \quad \text{and} \quad \varepsilon(s, \pi_v \times \tau_v, \psi_v) = \varepsilon(s, \Pi_v \times \tau_v, \psi_v).$$

Proof: The proof of this proposition is by the use of multiplicativity of the local factors with respect to the H_m-variable [**40**]. Since $\tau_{0,v}$ is unramified and generic we can write it as a full induced representation from characters [**21**]

$$\tau_{0,v} \simeq \mathrm{Ind}_{B'_m(k_v)}^{H_m(k_v)}(\chi_{1,v} \otimes \cdots \otimes \chi_{m,v})$$

with each $\chi_{i,v}$ unramified. If we let $\chi_{i,v}(x) = |x|_v^{b_i}$ and let $\mu(x) = |x|_v$, then we may write τ_v as

$$\tau_v \simeq \mathrm{Ind}_{B'_m(k_v)}^{H_m(k_v)}(\tilde{\nu}_v \mu^{b_1} \otimes \cdots \otimes \tilde{\nu}_v \mu^{b_m}).$$

By the multiplicativity of the local factors [**40**] we find

$$L(s, \pi_v \times \tau_v) = \prod_{i=1}^{m} L(s + b_i, \pi_v \times \tilde{\nu}_v)$$

and

$$\varepsilon(s, \pi_v \times \tau_v, \psi_v) = \prod_{i=1}^{m} \varepsilon(s + b_i, \pi_v \times \tilde{\nu}_v, \psi_v).$$

On the other hand, by the same results of [**19**] we also have

$$L(s, \Pi_v \times \tau_v) = \prod_{i=1}^{m} L(s + b_i, \Pi_v \times \tilde{\nu}_v)$$

and

$$\varepsilon(s, \Pi_v \times \tau_v, \psi_v) = \prod_{i=1}^{m} \varepsilon(s + b_i, \Pi_v \times \tilde{\nu}_v, \psi_v).$$

By Proposition 5.3 above we see that after factoring the L- and ε–factors for π_v and Π_v twisted by such τ_v the factors are term by term equal for $\tilde{\nu}_v$ a suitable sufficiently ramified character. This establishes the proposition. □

In the appendix (Section 7.2) we explicitly calculate the local lift of a principal series representation for the case of $G_n = \mathrm{SO}_{2n}^*$. For the other cases (at least in the unramified situation) these explicit calculations are in [**11**] and [**26, 27**].

Now let us return to the situation where G_n is a unitary group associated to a quadratic extension E/k in which the place v splits. Then $G_n(k_v) \simeq \mathrm{GL}_N(k_v)$ and $H_N(k_v) \simeq \mathrm{GL}_N(k_v) \times \mathrm{GL}_N(k_v)$. This situation is analyzed in the beginning of Section 6 of [**26**]. The L-homomorphism is simply understood in this case. If π_v is an irreducible admissible representation of $G_n(k_v)$ then, ramified or not, we take the representation Π_v of $H_N(k_v)$ to be $\pi_v \otimes \tilde{\pi}_v$. The twisting representation τ_v it a representation of $H_m(k_v) \simeq \mathrm{GL}_m(k_v) \times \mathrm{GL}_m(k_v)$ and hence of the form $\tau_v \simeq \tau_{1,v} \otimes \tau_{2,v}$. Then we have

$$\gamma(s, \pi_v \otimes \tau_v, \iota \otimes \iota', \psi_v) = \gamma(s, \pi_v \times \tau_{1,v}, \psi_v)\gamma(s, \tilde{\pi}_v \times \tau_{2,v}, \psi_v) = \gamma(s, \Pi_v \times \tau_v, \psi_v)$$

and

$$L(s, \pi_v \otimes \tau_v, \iota \otimes \iota') = L(s, \pi_v \times \tau_{1,v})L(s, \tilde{\pi}_v \times \tau_{2,v}) = L(s, \Pi_v \times \tau_v).$$

So we are in the same situation as in the unramified case, i.e., the stronger Proposition 5.2 holds in this case.

5.4. The global candidate lift. Return now to the global situation. Let $\pi \simeq \otimes' \pi_v$ be a globally generic cuspidal representation of $G_n(\mathbb{A}_k)$. Let S be a finite set of finite places such that for all non-archimedean places $v \notin S$ we have π_v and ψ_v are unramified (and if necessary the local extension E_w/k_v unramified as well). For each $v \notin S$ let Π_v be the local functorial lift of π_v as in Section 5.1 or 5.2. For the places $v \in S$ we take Π_v to be the irreducible admissible representation of $H_N(k_v)$ obtained in Section 5.3. Then the restricted tensor product $\Pi \simeq \otimes' \Pi_v$ is an irreducible admissible representation of $H_N(\mathbb{A}_k)$. It is self-dual except in the case of unitary groups, where it is self-conjugate-dual. This is our candidate lift.

For each place $v \in S$ choose a suitable sufficiently ramified character $\eta_v = \tilde{\nu}_v$ of $H_1(k_v)$ so that Proposition 5.4 is valid. Let η be any idele class character of $H_1(\mathbb{A}_k)$ which has local component η_v at those $v \in S$. Then combining Propositions 5.1 – 5.4 we obtain the following result on our candidate lift.

PROPOSITION 5.5. *Let π be a globally generic cuspidal representation of $G_n(\mathbb{A}_k)$ and let Π be the candidate lift constructed above as a representation of $H_N(\mathbb{A}_k)$. Then for every representation $\tau \in \mathcal{T}(S;\eta) = \mathcal{T}^S(N-1) \otimes \eta$ we have*

$$L(s, \pi \times \tau) = L(s, \Pi \times \tau) \quad \text{and} \quad \varepsilon(s, \pi \times \tau) = \varepsilon(s, \Pi \times \tau).$$

6. Global functoriality

6.1. Functoriality. Let us now prove Theorem 1.1. The proof is the usual one [**11, 27**], but it is short and we repeat it for completeness.

We begin with our globally generic cuspidal representation of $G_n(\mathbb{A}_k)$. Decompose $\pi \simeq \otimes' \pi_v$ into its local components and let S be a non-empty set of non-archimedean places such that for all non-archimedean places $v \notin S$ we have that π_v and ψ_v (and if necessary E_w/k_v) are unramified. Let $\Pi \simeq \otimes' \Pi_v$ be the irreducible admissible representation of $H_N(\mathbb{A}_k)$ constructed in Section 5 as our candidate lift. By construction Π is self-dual or self-conjugate-dual and is the local functorial lift of π at all places $v \notin S$. Choose η, an idele class character of $H_1(\mathbb{A}_k)$, such that its local components η_v are suitable and sufficiently ramified at those $v \in S$ so that Proposition 5.5 is valid. Furthermore, since we have taken S non-empty, we may choose η so that for at least one place $v_0 \in S$ we have that η_{v_0} is sufficiently ramified so that Theorem 3.1 is also valid. Fix this character.

We are now ready to apply the Converse Theorem to Π. Consider $\tau \in \mathcal{T}(S;\eta)$. By Proposition 5.5 we have that

$$L(s, \pi \times \tau) = L(s, \Pi \times \tau) \qquad \text{and} \qquad \varepsilon(s, \pi \times \tau) = \varepsilon(s, \Pi \times \tau).$$

On the other hand, by Theorem 3.1 we know that each $L(s, \pi \times \tau)$ and hence $L(s, \Pi \times \tau)$ is nice. Thus Π satisfies the hypotheses of the Converse Theorem, Theorem 2.1. Hence there is an automorphic representation $\Pi' \simeq \otimes' \Pi'_v$ of $H_N(\mathbb{A}_k)$ such that $\Pi'_v \simeq \Pi_v$ for all $v \notin S$. But for $v \notin S$, by construction Π_v is the local functorial lift of π_v. Hence Π' is a functorial lift of π as required in the statement of Theorem 1.1. The lift is then uniquely determined, independent of the choice of S and η, by the classification theorem of Jacquet and Shalika [**20**]. □

6.2. The image of functoriality. In this section we would like to record the image of functoriality. Assuming the existence of functoriality, the global image has been analyzed in the papers of Ginzburg, Rallis, and Soudry using their method of descent [**16, 42**]. From their method of descent of automorphic representations from $H_N(\mathbb{A}_k)$ to the classical groups $G_n(\mathbb{A}_k)$ and its local analogue, Ginzburg, Rallis, and Soudry were able to characterize the image of functoriality from generic representations.

There is a central character condition that must be satisfied by the lift. For each classical group we associate a quadratic idele class character of $\mathbb{A}_k^\times$ as follows. If G_n is of type B_n, C_n, D_n, or ${}^2A_n$ we simply take χ_{G_n} to be the trivial character $\mathbf{1}$. If G_n is of type ${}^2D_n$, so a quasisplit even special orthogonal group SO_{2n}^*, then the two-dimensional anisotropic kernel of the associated orthogonal space is given by the norm form of a quadratic extension E/k; in this case we set $\chi_{G_n} = \eta_{E/k}$ the quadratic character coming from class field theory.

The arithmetic part of their characterization relies on a certain L-function $L(s, \Pi_i, R)$ for a H_{N_i} having a pole at $s = 1$. The corresponding representation R of the L-group depends on the G_n from which we are lifting. If the dual group ${}^LG_n$ is of orthogonal type, then $R = \mathrm{Sym}^2$, if it is of symplectic type then $R = \Lambda^2$, and in the unitary case it is either the Asai representation $R = \mathrm{Asai}_{E/k}$ for $G_n = \mathrm{U}_{2n+1}$ or the twist by the quadratic character $\eta_{E/k}$ of the associated quadratic extension $R = \mathrm{Asai}_{E/k} \otimes \eta_{E/k}$ for $G_n = \mathrm{U}_{2n}$. For the definition of the Asai representation, if it is not familiar, see [**26, 27**].

The image of the lifting then has the following characterization [**42**].

THEOREM 6.1. *Let π be a globally generic cuspidal representation of $G_n(\mathbb{A}_k)$. Then any functorial lift of π to an automorphic representation Π of $H_N(\mathbb{A}_k)$ is self-dual (respectively self-conjugate-dual in the unitary case) with central character $\omega_\Pi = \chi_{G_n}$ (resp. $\omega_\Pi|_{\mathbb{A}^\times} = \chi_{G_n}$) and is of the form*

$$\Pi = \mathrm{Ind}(\Pi_1 \otimes \cdots \otimes \Pi_d) = \Pi_1 \boxplus \cdots \boxplus \Pi_d,$$

where each Π_i is a unitary self-dual (resp. self-conjugate-dual) cuspidal representation of $H_{N_i}(\mathbb{A}_k)$ such that the partial L-function $L^T(s, \Pi_i, R)$, with any sufficiently large finite set of places T containing all archimedean places, has a pole at $s = 1$ and $\Pi_i \not\simeq \Pi_j$ for $i \neq j$. Moreover, any such Π is the functorial lift of some π.

Let us make a few elementary observations on the distinguishing characterizations of the various lifts. We summarize the image characterization data in the following table:

G_n	R	χ_{G_n}
SO_{2n+1}	Λ^2	$\mathbf{1}$
SO_{2n}	Sym^2	$\mathbf{1}$
SO^*_{2n}	Sym^2	$\eta_{E/k}$
Sp_{2n}	Sym^2	$\mathbf{1}$
U_{2n}	$\mathrm{Asai}_{E/k} \otimes \eta_{E/k}$	$\mathbf{1}$
U_{2n+1}	$\mathrm{Asai}_{E/k}$	$\mathbf{1}$

We first consider lifts from orthogonal and symplectic groups. Suppose for simplicity that Π is a self-dual cuspidal representation of $\mathrm{GL}_N(\mathbb{A}_k)$. If $N = 2n+1$ is odd, then Π can only be a lift from Sp_{2n} and will be iff $L^T(s, \Pi, \mathrm{Sym}^2)$ has a pole at $s = 1$ and ω_Π is trivial. If $N = 2n$ is even, then Π can only be a lift from an orthogonal group. If $L^T(s, \Pi, \Lambda^2)$ has a pole at $s = 1$ and ω_Π is trivial then it is a lift from the split SO_{2n+1}. On the other hand, if it is a lift from an even orthogonal group, then necessarily $L^T(s, \Pi, \mathrm{Sym}^2)$ has a pole at $s = 1$. Since the central character ω_Π is necessarily quadratic, this character will distinguish between the various even orthogonal groups. If ω_Π is trivial, then Π is a lift from the split SO_{2n} while if $\omega_\Pi = \eta_{E/k}$ for some quadratic extension E/k, then Π is a lift from the quasisplit SO^*_{2n} associated to this extension. (If Π is isobaric, one applies the same conditions to the summands.)

We next consider lifts from unitary groups and we begin with a cuspidal representation Π of $\mathrm{GL}_N(\mathbb{A}_k)$ for some number field k. If Π is to be a lift from a unitary group U_N, then we must have a quadratic sub-field $k_0 \subset k$ with non-trivial Galois automorphism σ such that both $\omega_\Pi|_{\mathbb{A}_{k_0}^\times} = \mathbf{1}$ and $\Pi \simeq \widetilde{\Pi}^\sigma$. Then we can realize $\mathrm{GL}_N(\mathbb{A}_k) = H_N(\mathbb{A}_{k_0})$ with $H_N = \mathrm{Res}_{k/k_0}\mathrm{GL}_N$. If $N = 2n+1$ is odd, then for Π to be a transfer from $\mathrm{U}_N(\mathbb{A}_{k_0})$ we would need $L^T(s, \Pi, \mathrm{Asai}_{k/k_0})$ to have a pole at $s = 1$ and if $N = 2n$ is even then for Π to be a transfer from $\mathrm{U}_{2n}(\mathbb{A}_{k_0})$ we would need $L^T(s, \Pi, \mathrm{Asai}_{k/k_0} \otimes \eta_{k/k_0})$ to have a pole at $s = 1$.

From these descriptions, it is clear that there is no intersection between lifts from different orthogonal groups nor between them and symplectic groups since the Λ^2 and Sym^2 L-functions can never share poles. On the other hand, there seems to be much room for overlap in the images from orthogonal/symplectic and unitary groups as well as potential overlap in the images from different unitary groups. It would be interesting to understand these.

7. Appendix: Quasisplit orthogonal groups

In this section we present some explicit computation for the case of $G_n = \mathrm{SO}^*_{2n}$. While these are not necessary for what preceded, they can be quite helpful in understanding this case.

7.1. The L-homomorphism. Let us start with some L-group generalities. Let k be a local or global field and let $\Gamma_k = \mathrm{Gal}(\overline{k}/k)$. Let G be a connected reductive group over k. Let B be a Borel subgroup and $T \subset B$ a Cartan subgroup. Let $\Psi_0(G) = (X, \Delta, X^\vee, \Delta^\vee)$ be the based root datum for G associated to (B, T).

Since $\mathrm{Out}(G) \simeq \mathrm{Aut}(\Psi_0(G))$ we have the short exact sequence of automorphisms

$$1 \to \mathrm{Int}(G) \to \mathrm{Aut}(G) \to \mathrm{Aut}(\Psi_0(G)) \to 1.$$

We fix a splitting of this sequence as follows. For each $\alpha \in \Delta$ we fix $x_\alpha \in G_\alpha$. Then

$$\mathrm{Aut}(\Psi_0(G)) \simeq \mathrm{Aut}(G, B, T, \{x_\alpha\})$$

realizes $\mathrm{Aut}(\Psi_0(G))$ as a subgroup of $\mathrm{Aut}(G)$.

The cocycle

$$[\sigma \mapsto f(f^\sigma)^{-1}] \in H^1(\Gamma_k, \mathrm{Aut}(G))$$

then lands in $\mathrm{Aut}(\Psi_0(G))$ and becomes a homomorphism

$$\mu_G : \Gamma_k \to \mathrm{Aut}(\Psi_0(G)).$$

We then have the dual action

$$\mu_G^\vee : \mathrm{Gal}(\overline{k}/k) \to \mathrm{Aut}(\Psi_0(G)^\vee) = \mathrm{Aut}(\Psi_0(\widehat{G})).$$

In the case of SO^*_{2n} which splits over E, with $(E : k) = 2$, let $\mathrm{Gal}(E/k) = \{1, \sigma\}$. As before, we realize $\mathrm{Aut}(\Psi_0(\widehat{G})) \simeq \mathrm{Aut}(\widehat{G}, \widehat{B}, \widehat{T}, \{x_{\alpha^\vee}\})$. If $\tau \in \sigma\mathrm{Gal}(\overline{k}/E)$ then $\mu_G^\vee(\tau)$ must send T to itself, each $x_{\alpha_i^\vee}$ to itself for $1 \le i \le n-2$, and interchange $x_{\alpha_{n-1}^\vee}$ and $x_{\alpha_n^\vee}$. In particular, it must send $e_n^\vee$ to $-e_n^\vee$. So the element $\mu_G^\vee(\tau)$ can be represented by an element $[\hat{w}]$ representing a coset of T in the normalizer of T in $\mathrm{O}_{2n}(\mathbb{C})$. In fact

$$[\hat{w}] = \begin{pmatrix} 1 & & & & & & \\ & \ddots & & & & & \\ & & 1 & & & & \\ & & & \begin{pmatrix} 0 & 1 \\ 1 & 0 \end{pmatrix} & & & \\ & & & & 1 & & \\ & & & & & \ddots & \\ & & & & & & 1 \end{pmatrix} T.$$

Let

$$\hat{w} = \begin{pmatrix} 1 & & & & & & \\ & \ddots & & & & & \\ & & 1 & & & & \\ & & & \begin{pmatrix} 0 & 1 \\ 1 & 0 \end{pmatrix} & & & \\ & & & & 1 & & \\ & & & & & \ddots & \\ & & & & & & 1 \end{pmatrix} \hat{t}$$

be an element of this coset that fixes the splitting $\{x_{\alpha^\vee}\}$. Then $\hat{t} \in Z(\widehat{G}) = Z(\mathrm{O}_{2n}(\mathbb{C})) = \{\pm 1\}$. Thus

$$\hat{w} = \pm \begin{pmatrix} 1 & & & & & & \\ & \ddots & & & & & \\ & & 1 & & & & \\ & & & \begin{pmatrix} 0 & 1 \\ 1 & 0 \end{pmatrix} & & & \\ & & & & 1 & & \\ & & & & & \ddots & \\ & & & & & & 1 \end{pmatrix}$$

are the only possibilities for an element in $\mathrm{O}_{2n}(\mathbb{C})$ representing $\mu_G^\vee(\tau)$ by conjugation. The choice of $\pm$ is irrelevant. So we set

$$\hat{w} = \begin{pmatrix} 1 & & & & & & \\ & \ddots & & & & & \\ & & 1 & & & & \\ & & & \begin{pmatrix} 0 & 1 \\ 1 & 0 \end{pmatrix} & & & \\ & & & & 1 & & \\ & & & & & \ddots & \\ & & & & & & 1 \end{pmatrix}.$$

Then our embedding $\iota : \mathrm{SO}_{2n}(\mathbb{C}) \rtimes \Gamma_k \hookrightarrow \mathrm{GL}_{2n}(\mathbb{C}) \times \Gamma_k$ must send

$$(1, \tau) \mapsto (\hat{w}, \tau) \quad \text{if} \quad \tau \in \sigma \mathrm{Gal}(\overline{k}/E)$$

while

$$(1, \tau) \mapsto (1, \tau) \quad \text{it} \quad \tau \in \mathrm{Gal}(\overline{k}/E).$$

This follows from the fact that $\mu_G^\vee(\tau)(g) = \hat{w} g \hat{w}^{-1}$ and

$$(1, \tau)(g, 1) = (\tau(g), \tau) = (\mu_G^\vee(\tau)(g), \tau) = (\hat{w} g \hat{w}^{-1}, \tau).$$

In particular

$$(1, \tau)(g, 1)(1, \tau)^{-1} = (\hat{w} g \hat{w}^{-1}, 1).$$

Note that by the matrix representation given for $\hat{w}$ we are clearly fixing

$$\widehat{T} = \mathrm{GL}_1(\mathbb{C})^{n-1} \times (\mathrm{GL}_1(\mathbb{C}) \times \mathrm{GL}_1(\mathbb{C}))/\mathbb{C}^\times$$

the latter being the L-group of SO_2^*. Moreover note that $\hat{w}$ basically represents only one sign change ($e_n^\vee \mapsto -e_n^\vee$) and thus cannot be in the Weyl group of $\widehat{T}$ in

$SO_{2n}(\mathbb{C})$. It represents the outer automorphism (the graph automorphism) of SO^*_{2n} and $(\hat{w}, \tau)$ gives the embedding of $\tau \in \sigma\mathrm{Gal}(\overline{k}/E)$ in $GL_{2n}(\mathbb{C}) \times \Gamma_k$.

7.2. Computation of the local lift and its central character: SO^*_{2n}. In the local lifts of Section 5, both the unramified lift and the ramified lift relied on lifting a principal series representation of G_n to a representation of GL_{2n}. We will analyze this in a bit more detail here, computing the local lift in the quasisplit case and the central character of the local lift in both cases.

Let k be a non-archimedean local field. Let $T' \subset GL_{2n}$ be the standard maximal split torus. Let

$$\varphi : W_k \to {}^LT' \qquad \text{and} \qquad \chi : T'(k) \to \mathbb{C}^\times.$$

Write

$$\varphi((x, w)) = (\varphi_0(x), w) \qquad \text{with} \qquad \varphi_0 \in H^1(k, \widehat{T}').$$

Let $\rho^\vee \in X_*(T') = X^*(\widehat{T}')$ be such that for $x \in \overline{k}^\times$ we have $\rho^\vee(x) = \mathrm{diag}(x, \dots, x) \in GL_{2n}(\overline{k})$. Then $\chi(\rho^\vee(x)) = \omega_\pi(x)$ if $x \in k^\times$ and $\pi = \mathrm{Ind}(\chi)$.

Suppose we are in the case of a split SO_{2n} and $\varphi = \iota \circ \phi'$ with $\iota : SO_{2n}(\mathbb{C}) \hookrightarrow GL_{2n}(\mathbb{C})$ and $\phi' : W_k \to {}^LT$. Then

$$\chi(\rho^\vee(x)) = \det(\varphi(x)) = \det(\iota(\phi'(x))) = 1.$$

Now let us look at the non-split case SO^*_{2n} associated to the quadratic extension E/k as above. In this case

$$T(k) = (k^\times)^{n-1} \times E^1 \qquad \text{or} \qquad T = \mathbb{G}_m^{n-1} \times SO^*_2.$$

Thus

$${}^LT = GL_1(\mathbb{C})^{n-1} \times (GL_1(\mathbb{C}) \times GL_1(\mathbb{C}))/\mathbb{C}^\times.$$

and

$$\varphi = \iota \circ \phi' : W_k \to {}^LT \to {}^L\widetilde{T}'$$

where $\widetilde{T}'(k) \simeq (k^\times)^{n-2} \times E^\times \times (k^\times)^{n-2}$ is a torus of $GL_{2n}(k)$ with $E^\times$ embedded in $GL_2(k)$ as in Langlands-Labesse [**33**]. Let $\mu = (\mu_1, \dots, \mu_{n-1}, \chi_n)$, with $\mu_i \in \widehat{k^\times}$ and $\chi_n \in \widehat{E^1}$, be a character of $T(k)$. Then, by Hilbert's Theorem 90, $E^\times/k^\times \simeq E^1$ through the map $x \mapsto x/x^\sigma$. Thus we can extend χ_n to a character $\widetilde{\chi}_n$ of $E^\times$. We can then consider the character

$$\widetilde{\mu} = (\mu_1, \dots, \mu_{n-1}, \widetilde{\chi}_n, \mu_{n-1}^{-1}, \dots, \mu_1^{-1})$$

of $\widetilde{T}'(k)$.

To get a principal series on $GL_{2n}(k)$, $\widetilde{\chi}_n$ must factor through the norm map, so write $\widetilde{\chi}_n = \mu_n \circ N_{E/k}$ with $\mu_n \in \widehat{k^\times}$. Since $\widetilde{\chi}_n$ is trivial on restriction to $k^\times$, then $\mu_n^2 = 1$. Since endoscopy then gives the Weil representation of $GL_2(k)$ defined by $\mathrm{Ind}_{W_E}^{W_k} \widetilde{\chi}_n$, it gives the principal series representation $I(\mu_n \eta_{E/k}, \mu_n) = I(\mu_n \eta_{E/k}, \mu_n^{-1})$, where $\eta_{E/k}$ is the quadratic character of $k^\times$ associated to the quadratic extension E/k by local class field theory. So if the principal series representation $I(\mu)$ of $SO^*_{2n}(k)$ transfers to a principal series representation Π_v of $GL_{2n}(k)$, it will be induced from the character

$$(\mu_1, \dots, \mu_{n-1}, \mu_n \eta_{E/k}, \mu_n^{-1}, \mu_{n-1}^{-1}, \dots, \mu_1^{-1})$$

of $T'(k)$. Its central character is then simply $\eta_{E/k}$.

Even if $\widetilde{\chi}_n$ does not factor through the norm, the lift Π_v will be the representation of $\mathrm{GL}_{2n}(k)$ induced from

$$(\mu_1, \ldots, \mu_{n-1}, \pi(\mathrm{Ind}_{W_E}^{W_k} \widetilde{\chi}_n), \mu_{n-1}^{-1}, \ldots, \mu_1^{-1})$$

and its central character is still $\eta_{E/k}$. In fact, the central character of $\pi(\mathrm{Ind}_{W_E}^{W_k} \widetilde{\chi}_n)$ is the restriction of $\eta_{E/k}\widetilde{\chi}_n$ to $k^\times$, which is simply $\eta_{E/k}$. Note that now the transfer is tempered, but not necessarily a principal series.

References

[1] J. Arthur, Unipotent automorphic representations: global motivation. *Automorphic forms, Shimura varieties, and L-functions*, Vol. I (Ann Arbor, MI, 1988), 1–75, Perspect. Math., 10, Academic Press, Boston, MA, 1990.

[2] J. Arthur, The problem of classifying automorphic representations of classical groups. *Advances in mathematical sciences: CRM's 25 years* (Montreal, PQ, 1994), 1–12, *CRM Proc. Lecture Notes*, **11**, Amer. Math. Soc., Providence, RI, 1997.

[3] J. Arthur, The principle of functoriality. *Bull. Amer. Math. Soc.* **40** (2002), 39–53.

[4] J. Arthur, Automorphic representations of GSp(4). *Contributions to automorphic forms, geometry, and number theory*, 65–81, Johns Hopkins Univ. Press, Baltimore, MD, 2004.

[5] M. Asgari and F. Shahidi, Generic transfer for general spin groups. *Duke Math. J.* **132** (2006), no. 1, 137–190.

[6] A. Borel, Automorphic L-functions. *Proc. Sympos. Pure Math.* **33**, part 2, (1979), 27–61.

[7] N. Bourbaki, Lie Groups and Lie Algebras, Chapters 4–6. Elements of Mathematics, Springer-Verlag, Berlin, 2002.

[8] W. Casselman and F. Shahidi, On reducibility of standard modules for generic representations. *Ann. Scient. Éc. Norm. Sup.* **31** (1998), 561–589.

[9] J.W. Cogdell, Dual groups and Langlands Functoriality. *An Introduction to the Langlands Program* (J. Bernstein and S. Gelbart, eds.). Birkhäuser, Boston, 2003, 251–268.

[10] J.W. Cogdell, H.H. Kim, I.I. Piatetski-Shapiro, and F. Shahidi, On lifting from classical groups to GL_N. *Publ. Math. IHES* **93** (2001), 5–30.

[11] J.W. Cogdell, H.H. Kim, I.I. Piatetski-Shapiro, and F. Shahidi, Functoriality for the classical groups. *Publ. Math. IHES* **99** (2004), 163–233.

[12] J.W. Cogdell and I.I. Piatetski-Shapiro, Converse Theorems for GL_n. *Publ. Math. IHES* **79** (1994), 157–214.

[13] J.W. Cogdell and I.I. Piatetski-Shapiro, Converse Theorems for GL_n, II. *J. reine angew. Math.* **507** (1999), 165–188.

[14] J.W. Cogdell, I.I. Piatetski-Shapiro, and F. Shahidi, Stability of γ-factors for quasisplit groups. *J. Inst. Math. Jussieu* **7** (2008), 27-66.

[15] S. Gelbart and F. Shahidi, Boundedness of automorphic L-functions in vertical strips. *J. Amer. Math. Soc.* **14** (2001), 79–107.

[16] D. Ginzburg, S. Rallis, and D. Soudry, Generic automorphic forms on $\mathrm{SO}(2n+1)$: functorial lift to $\mathrm{GL}(2n)$, endoscopy, and base change. *Internat. Math. Res. Notices* **2001**, no. 14, 729–764.

[17] M. Harris and R. Taylor, *The Geometry and Cohomology of Some Simple Shimura Varieties*, Annals of Math. Studies No. 151, Princeton University Press, Princeton, 2001.

[18] G. Henniart, Une preuve simple des conjectures de Langlands pour $\mathrm{GL}(n)$ sur un corps p-adique. *Invent. math.* **139** (2000), 439–455.

[19] H. Jacquet, I.I. Piatetski-Shapiro, and J. Shalika, Rankin-Selberg convolutions. *Amer. J. Math.* **105** (1983), 367–464.

[20] H. Jacquet and J. Shalika, On Euler products and the classification of automorphic representations. *Amer. J. Math.* **103** (1981), 499–558 and 777–815.

[21] H. Jacquet and J. Shalika, The Whittaker models for induced representations. *Pacific J. Math.* **109** (1983). 107–120.

[22] H. Jacquet and J. Shalika, A lemma on highly ramified ϵ–factors. *Math. Ann.* **271** (1985), 319–332.

[23] D. Keys and F. Shahidi, Artin L-functions and normalization of intertwining operators. *Ann. Sci. École Norm. Sup.* (4) **21** (1988), no. 1, 67–89.

[24] H. Kim, Langlands–Shahidi method and poles of automorphic L-functions, II. *Israel J. Math.* **117** (2000), 261–284.

[25] H. Kim, On local L-functions and normalized intertwining operators. *Canad. J. Math.*, **57** (2005), 535–597.

[26] H. Kim and M. Krishnamurthy, Base change lift for odd unitary groups. *Functional analysis VIII*, 116–125, Various Publ. Ser. (Aarhus), **47**, Aarhus Univ., Aarhus, 2004.

[27] H. Kim and M. Krishnamurthy, Stable base change lift from unitary groups to GL_n. *IMRP Int. Math. Res. Pap.* **2005**, no. 1, 1–52.

[28] H. Kim and F. Shahidi, Functorial products for $\mathrm{GL}_2 \times \mathrm{GL}_3$ and the symmetric cube for GL_2. *Ann. of Math.* **155** (2002), 837–893.

[29] M. Krishnamurthy, he Asai transfer to GL_4 via the Langlands-Shahidi method. *Int. Math. Res. Not.* **2003**, no. 41, 2221–2254.

[30] R.P. Langlands, On the classification of irreducible representations of real algebraic groups. *Representation Theory and Harmonic Analysis on Semisimple Lie Groups*, Math. Surveys Monographs, 31, Amer. Math. Soc., Providence, RI, 1989, 101–170

[31] R.P. Langlands, Automorphic representations, Shimura varieties, and motives. Ein Märchen. *Proc. Sympos. Pure Math.* **33**, Part 2, (1979), 205–246.

[32] R.P. Langlands, Where stands functoriality today. *Proc. Sympos. Pure Math.* **61** (1997), 457–471.

[33] J-P. Labesse snd R.P. Langlands, L-indistinguishability for SL(2). *Canad. J. Math.* **31** (1979), no. 4, 726–785.

[34] G. Muić, A proof of Casselman-Shahidi's conjecture for quasisplit classical groups. *Canad. Math. Bull.* **44** (2001), 298-312.

[35] I. Satake, Theory of spherical functions on reductive algebraic groups over $\mathfrak{p}$-adic fields. *Publ. Math. IHES* **18** (1963), 5–69.

[36] F. Shahidi, On certain L-functions. *Amer. J. Math.* **103** (1981), 297–355.

[37] F. Shahidi, Local coefficients as Artin factors for real groups *Duke Math. J.* **52** (1985), 973–1007.

[38] F. Shahidi, On the Ramanujan conjecture and finiteness of poles for certain L-functions. *Ann. of Math.* **127** (1988), 547–584.

[39] F. Shahidi, A proof of Langlands' conjecture on Plancherel measures; complementary series for p-adic groups. *Ann. of Math.* **132** (1990), 273–330.

[40] F. Shahidi, On multiplicativity of local factors. *Festschrift in honor of I. I. Piatetski-Shapiro on the occasion of his sixtieth birthday*, Part II (Ramat Aviv, 1989), Israel Math. Conf. Proc., 3, Weizmann, Jerusalem, 1990, 279–289.

[41] F. Shahidi, Twists of a general class of L-functions by highly ramified characters. *Canad. Math. Bull.* **43** (2000), 380–384.

[42] D. Soudry, On Langlands functoriality from classical groups to GL_n. *Formes Automorphes (I), Asterisque* **298** (2005), 335–390.

[43] J. Tate, Number theoretic background. *Proc. Sympos. Pure Math.* **33**, Part 2, (1979), 3–26.

Department of Mathematics, Ohio State university, Columbus, OH 43210 USA
E-mail address: cogdell@math.ohio-state.edu

Department of Mathematics, Yale University, New Haven, CT 06520 USA, and School of Mathematics, Tel Aviv University, Tel Aviv 69978 Israel
E-mail address: ilya@math.yale.edu

Department of Mathematics, Purdue University, West Lafayette, IN 47907 USA
E-mail address: shahidi@math.purdue.edu

Clay Mathematics Proceedings
Volume 13, 2011

Poles of L-functions and Theta Liftings for Orthogonal Groups, II

David Ginzburg, Dihua Jiang, and David Soudry

Dedicated to Freydoon Shahidi

ABSTRACT. We bound the first occurrence in the theta correspondence of irreducible cuspidal automorphic representations σ of orthogonal groups, in terms of their generalized Gelfand-Graev periods. We also obtain a local analog at a finite place. As a result, we determine a range of holomorphy of $L^S(s,\sigma)$ in the right half-plane in terms of the local generalized Gelfand-Graev models of σ at one finite place.

1. Introduction

In [**GJS09**], we characterized the first occurrence of irreducible cuspidal automorphic representations of $\mathrm{O}_m(\mathbb{A})$ under the theta correspondence to $\mathrm{Mp}_{2n}(\mathbb{A})$, where $\mathrm{Mp}_{2n}(\mathbb{A})$ is either $\widetilde{\mathrm{Sp}}_{2n}(\mathbb{A})$ (when m is odd) or $\mathrm{Sp}_{2n}(\mathbb{A})$ (when m is even) in terms of the existence of poles of certain Eisenstein series (Theorem 1.3, [**GJS09**]). Here, $\mathbb{A}$ is the ring of adèles of a number field k. As a consequence, we determined a range of holomorphy in the right half-plane for the standard partial L-functions $L^S(s,\sigma)$ of irreducible cuspidal automorphic representations σ of $\mathrm{O}_m(\mathbb{A})$ (Theorem 1.1 in [**GJS09**]). These results can be viewed as a natural extension to orthogonal groups of the work of Kudla and Rallis on symplectic groups ([**KR94**]) and as a completion to Mœglin's work ([**M97a**] and [**M97b**]).

In this paper, we discuss the relations between the global or local theta correspondence and the generalized Gelfand-Graev periods or models. As a consequence, we determine a range of holomorphy in the right half-plane of the standard partial L-functions $L^S(s,\sigma)$ in terms of local generalized Gelfand-Graev models supported by a local component σ_v at one finite place v. A preliminary version of such a result was given in [**GJS09**] (Theorem 1.7), and some related very interesting applications were discussed in §7 of [**GJS09**].

2010 *Mathematics Subject Classification.* Primary 11F70; Secondary 11F66, 22E55.

All three authors are partially supported by the USA-Israel Binational Science Foundation, and the second named author is also supported in part by NSF grant DMS-0653742 and DMS-1001672.

For an irreducible cuspidal automorphic representation (σ, V_σ) of $\mathrm{O}_m(\mathbb{A})$, we define in §2.2 $\psi_{t,\alpha}$-Fourier coefficients of $\phi_\sigma \in V_\sigma$ in (2.14). The characters $\psi_{t,\alpha}$ are parametrized by integers t and square classes α in k. Let r be the Witt index of the quadratic space defining O_m. We assume that r is positive, $1 \leq t \leq r$, and $2t < m$. Similarly, we define in (2.17) the notion of a $\psi_{t,\alpha}$-functional in the local setting. The main result of this paper can be formulated as follows.

THEOREM 1.1 (Main). *Let σ be an irreducible cuspidal automorphic representation of $\mathrm{O}_m(\mathbb{A})$ and t as above.*

1. *If there exists one finite local place v of k such that the local component σ_v of σ has a nonzero $\psi_{t,\alpha}$-functional, then the partial L-function $L^S(s,\sigma)$ is holomorphic for* $\mathrm{Re}(s) > \frac{m}{2} - t$. *In particular, if σ has a nonzero $\psi_{t,\alpha}$-Fourier coefficient, then the partial L-function $L^S(s,\sigma)$ is holomorphic for* $\mathrm{Re}(s) > \frac{m}{2} - t$.
2. *Assume that σ has a nonzero $\psi_{t,\alpha}$-Fourier coefficient. If either $t < r-1$, or $t < r$ and α is represented by the quadratic form corresponding to the anisotropic kernel of the quadratic space defining O_m, then $L^S(s,\sigma)$ is holomorphic for* $\mathrm{Re}(s) > \frac{m}{2} - t - 1$.

REMARK 1.2. *Write $m = m_0 + 2r$. When $m_0 = 1$, O_m is the split orthogonal group in $2r+1$ variables. When $t = r$, $\psi_{t,\alpha}$ is a Whittaker character, and the assertion of the first part of the main theorem is that $L^S(s,\sigma)$ is holomorphic when* $\mathrm{Re}(s) > \frac{1}{2}$. *This is Theorem 1.5 in* [**GJS09**]. *If $t = r-1$, then σ has a nonzero $\psi_{r-1,\alpha}$-Fourier coefficient. This case was discussed in § 7 of* [**GJS09**].

When $m_0 = 0$, O_m is the split orthogonal group in $2r$ variables. If $t = r-1$, then $\psi_{r-1,\alpha}$ is a Whittaker character, and the assertion of the first part of the main theorem is that $L^S(s,\sigma)$ is holomorphic when $\mathrm{Re}(s) > 1$. *This is Theorem 1.5 in* [**GJS09**].

We first prove in §4 that the nonvanishing of $\psi_{t,\alpha}$-Fourier coefficients of σ determines a range of the lowest occurrence $\mathrm{LO}_\psi(\sigma)$ (defined in §3.1) of ψ-theta lifts of σ. Then we establish the corresponding local version of this global result. This is done by an explicit calculation of the $\psi_{t,\alpha}$-Fourier coefficient of theta lifts of cuspidal automorphic representations from $\mathrm{Mp}_{2n}(\mathbb{A})$ to $\mathrm{O}_m(\mathbb{A})$, and an analogous calculation in the local setting. At the first occurrence, we get a relation between these $\psi_{t,\alpha}$- Fourier coefficients (respectively, functionals in the local setting) and Whittaker coefficients (resp. models), corresponding to ψ and α, on the symplectic or metaplectic side. Finally, we use Theorem 1.1 in [**GJS09**]. Since we quote this theorem several times in this paper, we state it here for convenience.

THEOREM 1.3 (Theorem 1.1 in [**GJS09**]). *Let σ be an irreducible cuspidal automorphic representation of $\mathrm{O}_m(\mathbb{A})$.*

1. *If $L^S(s,\sigma)$ has a pole at $s_0 = \frac{m}{2} - j > 0$, or if m is odd and $L^S(s,\sigma)$ does not vanish at $s = \frac{1}{2}$, and we let $j = 2[\frac{m}{2}]$, then there is an automorphic sign character ϵ of $\mathrm{O}_m(\mathbb{A})$ such that the ψ-theta lift of $\sigma \otimes \epsilon$ to $\mathrm{Mp}_{2j}(\mathbb{A})$ does not vanish, i.e. $\mathrm{LO}_\psi(\sigma) \leq 2j$.*
2. *If $\mathrm{LO}_\psi(\sigma) = 2j_0 < m$, then $L^S(s,\sigma)$ is holomorphic for* $\mathrm{Re}(s) > \frac{m}{2} - j_0$.
3. *If $\mathrm{LO}_\psi(\sigma) = 2j_0 \geq m$, then $L^S(s,\sigma)$ is holomorphic for* $\mathrm{Re}(s) \geq \frac{1}{2}$.

The authors thank the referee for his comments and suggestions.

2. The Generalized Gelfand-Graev Periods

Let k be a number field and $\mathbb{A}$ be the ring of adèles of k. Let $(X_m, (\cdot,\cdot))$ be a non-degenerate quadratic vector space over k of dimension m and Witt index r. We assume that $r \geq 1$. For any nonnegative integer a, we denote by

$$\mathcal{H}_a = \ell_a^+ \oplus \ell_a^- \tag{2.1}$$

the polarization of the $2a$-dimensional quadratic k-vector space $\mathcal{H}_a$, which is the direct sum of a copies of the hyperbolic plane. Then X_m can be written as

$$X_m := X_{m_0} \perp (\ell_r^+ \oplus \ell_r^-) = X_{m_0} \perp \mathcal{H}_r, \tag{2.2}$$

where X_{m_0} ($m_0 = m-2r$) is the m_0-dimensional anisotropic quadratic vector space, which is called the anisotropic kernel of X_m.

We may choose a basis for X_m

$$\{e_1, \cdots, e_r; \epsilon_1, \cdots, \epsilon_{m_0}; e_{-r}, \cdots, e_{-1}\} \tag{2.3}$$

such that

$$(e_i, e_j) = \begin{cases} 1 & \text{if } j = -i; \\ 0 & \text{if } j \neq -i, \end{cases}$$

and $(e_i, \epsilon_j) = 0$ for all $i \in \{\pm 1, \cdots, \pm r\}$ and $j \in \{1, \cdots, m_0\}$, where $\{e_1, \cdots, e_r\}$ is a basis for ℓ_r^+, $\{e_{-r}, \cdots, e_{-1}\}$ is a basis for ℓ_r^- and $\{\epsilon_1, \cdots, \epsilon_{m_0}\}$ is a basis for X_{m_0}.

Denote the Gram matrix of $\{\epsilon_1, \cdots, \epsilon_{m_0}\}$ by T_{m_0}. Then the Gram matrix of the basis (2.3) is

$$T_m = \begin{pmatrix} & & \omega_r \\ & T_{m_0} & \\ \omega_r & & \end{pmatrix},$$

where ω_r is the $r \times r$ permutation matrix with 1 in its second main diagonal, i.e. $(\omega_r)_{i,j} = \delta_{i,r+1-j}$.

For each $t \in \{1, 2, \cdots, r\}$, we have the following partial polarization

$$X_m = \ell_t^+ \oplus X_{m-2t} \oplus \ell_t^- \tag{2.4}$$

where ℓ_t^+ (resp. ℓ_t^-) is the totally isotropic subspace of dimension t of ℓ_r^+ (resp. ℓ_r^-), generated by $\{e_1, \cdots, e_t\}$ (resp. by $\{e_{-t}, \cdots, e_{-1}\}$). We will write the elements of O_m as matrices according to (2.4) and (2.3). Denote by T_{m-2t} the Gram matrix of the basis $\{e_{t+1}, \cdots, e_r, \epsilon_1, \cdots, \epsilon_{m_0}, e_{-r}, \cdots, e_{-t-1}\}$ of X_{m-2t};

$$T_{m-2t} = \begin{pmatrix} & & \omega_{r-t} \\ & T_0 & \\ \omega_{r-t} & & \end{pmatrix}.$$

Assume that $m - 2t \geq 1$. Let $Q_t = L_t V_t$ be the standard parabolic subgroup of O_m such that

$$L_t = \mathrm{GL}_1^t \times \mathrm{O}_{m-2t} \subset \mathrm{O}_m \tag{2.5}$$

and

$$V_t = \{v = v'(u, x, z) = \begin{pmatrix} u & x^* & z \\ & \mathrm{I}_{m-2t} & x \\ & & u^* \end{pmatrix} \in \mathrm{O}_m\}, \tag{2.6}$$

where $u \in U_t$, the maximal standard (upper-triangular) unipotent subgroup of GL_t. Denote the first column of x by x_1. Then the vectors

$$x_1 = {}^t(x_{1,1}, \cdots, x_{m-2t,1}) \tag{2.7}$$

form a k-vector space which is isomorphic to X_{m-2t}. Consider the action of $\mathrm{GL}_1 \times \mathrm{O}_{m-2t}$ on X_{m-2t} by

$$(a, h) \circ x_1 := ahx_1. \tag{2.8}$$

By the Witt theorem, the space X_{m-2t} decomposes into the following disjoint union of k-rational $\mathrm{GL}_1 \times \mathrm{O}_{m-2t}$-orbits:

$$X_{m-2t} = \{0\} \cup \mathcal{O}_0 \cup (\cup_{\alpha \in k^\times/(k^\times)^2} \mathcal{O}_\alpha), \tag{2.9}$$

where $\mathcal{O}_0$ consists of all (nonzero) isotropic vectors in X_{m-2t} and $\mathcal{O}_\alpha$ consists of all vectors x_1 in X_{m-2t} with $(x_1, x_1) \equiv \alpha \mod (k^\times)^2$. It is clear that the disjoint union $\cup_{\alpha \in k^\times/(k^\times)^2} \mathcal{O}_\alpha$ is k-stable.

2.1. Global periods. Let ψ be a nontrivial character of $\mathbb{A}/k$. Take μ_α in $\mathcal{O}_\alpha$ and define a character $\psi_{t,\alpha}$ of $V_t(\mathbb{A})$ as follows. For $v = v'(u, x, z) \in V_t(\mathbb{A})$, we define

$$\psi_{t,\alpha}(v) := \psi(u_{1,2} + \cdots + u_{t-1,1t})\psi^{-1}((\mu_\alpha, x_1)). \tag{2.10}$$

It is clear that the character $\psi_{t,\alpha}$ is trivial when restricted to $V_t(k)$. Since the Levi subgroup $L_t = \mathrm{GL}_1^t \times \mathrm{O}_{m-2t}$ normalizes V_t, the group of k-rational points $L_t(k)$ also acts on the characters $\psi_{t,\alpha}$, as α runs through a square class in $k^\times$. Consider the following decomposition

$$X_{m-2t} = (k \cdot \mu_\alpha) \perp (k \cdot \mu_\alpha)^\perp. \tag{2.11}$$

Since μ_α is anisotropic, the orthogonal complement $(k \cdot \mu_\alpha)^\perp$ is a non-degenerate quadratic k-vector space of dimension $m - 2t - 1$ with respect to the restriction of the bilinear form $(\cdot, \cdot)$ on X_m. The stabilizer of $\psi_{t,\alpha}$ in O_{m-2t} is

$$\mathrm{D}_{t,\alpha} := \mathrm{O}((k \cdot \mu_\alpha)^\perp) \tag{2.12}$$

We want to calculate the Witt index of $(k \cdot \mu_\alpha)^\perp$. Recall that $m - 2t \geq 1$. If $t = r$, then $X_{m-2t} = X_{m_0}$ is anisotropic, and hence the Witt index of $(k \cdot \mu_\alpha)^\perp$ is zero. If $t < r$, then the Witt index of X_{m-2t} is $r - t$ and we have

$$X_{m-2t} = \ell^+_{r-t} \oplus X_{m_0} \oplus \ell^-_{r-t}. \tag{2.13}$$

If α is representable by X_{m_0}, then the Witt index of $(k \cdot \mu_\alpha)^\perp$ is $r - t$. If α is not representable by X_{m_0}, then the Witt index of $(k \cdot \mu_\alpha)^\perp$ is $r - t - 1$.

For an automorphic form ϕ on $\mathrm{O}_m(\mathbb{A})$, we define the $\psi_{t,\alpha}$-Fourier coefficient of φ by the following integral:

$$\mathcal{F}^{\psi_{t,\alpha}}(\phi)(g) := \int_{V_t(k)\backslash V_t(\mathbb{A})} \phi(vg)\psi_{t,\alpha}^{-1}(v)dv. \tag{2.14}$$

It is clear that the restriction of $\mathcal{F}^{\psi_{t,\alpha}}(\phi)$ to $\mathrm{D}_{t,\alpha}(\mathbb{A})$ is left $\mathrm{D}_{t,\alpha}(k)$-invariant. We note that when O_m is quasi-split, or split over k, and $t = [\frac{m-1}{2}]$, then the $\psi_{t,\alpha}$-Fourier coefficient is a Whittaker-Fourier coefficient.

Let ϕ' be an automorphic form on $\mathrm{D}_{t,\alpha}(\mathbb{A})$. Then we define the generalized Gelfand-Graev period (or Bessel period) of ϕ of type $(\mathrm{D}_{t,\alpha}, \psi_{t,\alpha}, \phi')$, or simply the $(\mathrm{D}_{t,\alpha}, \psi_{t,\alpha}, \phi)$-period of ϕ, by the following integral:

$$\mathcal{P}_{\mathrm{D}_{t,\alpha};\psi_{t,\alpha}}(\phi, \phi') = \mathcal{P}_{\mathrm{D}_{t,\alpha}}(\phi, \phi') := \int_{\mathrm{D}_{t,\alpha}(k)\backslash \mathrm{D}_{t,\alpha}(\mathbb{A})} \mathcal{F}^{\psi_{t,\alpha}}(\phi)(h)\phi'(h)dh, \tag{2.15}$$

if the last integral converges. We refer to [**GPSR97**] for applications of such periods to the theory of automorphic L-functions.

2.2. Local models. Let v be a finite local place of k and k_v be the local field of k at v. Let ψ_v be a nontrivial character of k_v. We define the v-analogue of $\psi_{t,\alpha}$ for $V_t(k_v)$ by

$$\psi_{t,\alpha;v}(v'(u, x, z)) = \psi_v(u_{1,2} + \cdots + u_{t-1,1t})\psi^{-1}((\mu_{\alpha,v}, x_1)) \tag{2.16}$$

where $\alpha \in k_v^\times$ and $\mu_{\alpha,v} \in X_{m-2t}(k_v)$ is such that $(\mu_{\alpha,v}, \mu_{\alpha,v}) = \alpha$. Let (σ_v, V_{σ_v}) be an irreducible admissible representation of $\mathrm{O}_m(k_v)$. We say that σ_v has a nontrivial $\psi_{t,\alpha;v}$-functional if the space

$$\mathrm{Hom}_{V_t(k_v)}(V_{\sigma_v}, \psi_{t,\alpha;v}) \tag{2.17}$$

is nonzero. It is clear that in case $\mathrm{O}_m(k_v)$ is quasi-split or split over k_v, and $t = [\frac{m-1}{2}]$, then a $\psi_{t,\alpha}$-functional is a Whittaker functional.

Let τ_v be an irreducible admissible representation of $\mathrm{D}_{t,\alpha}(k_v)$. Then $\tau_v \otimes \psi_{t,\alpha;v}$ is a representation of the semi-direct product

$$\mathcal{J}_{t,\alpha}(k_v) = \mathrm{D}_{t,\alpha}(k_v) \ltimes V_t(k_v). \tag{2.18}$$

We say that σ_v has a nontrivial generalized Gelfand-Graev model (or Bessel model) of type $(\mathcal{J}_{t,\alpha}, \tau_v \otimes \psi_{t,\alpha;v})$, or a nontrivial $(\mathcal{J}_{t,\alpha}, \tau_v \otimes \psi_{t,\alpha;v})$-model if the space

$$\mathrm{Hom}_{\mathcal{J}_{t,\alpha}(k_v)}(V_{\sigma_v}, \tau_v \otimes \psi_{t,\alpha;v}) \tag{2.19}$$

is nonzero. In this case, take $0 \neq \ell_v \in \mathrm{Hom}_{\mathcal{J}_{t,\alpha}(k_v)}(V_{\sigma_v}, \tau_v \otimes \psi_{t,\alpha;v})$. Then the corresponding $(\mathcal{J}_{t,\alpha}, \tau_v \otimes \psi_{t,\alpha;v})$-model is the space consisting of all functions of the following type:

$$\mathcal{B}_x^{\psi_{t,\alpha;v}}(g) := \ell_v(\sigma_v(g)(x)), \quad g \in \mathrm{O}_m(k_v) \tag{2.20}$$

when x runs through V_{σ_v}.

3. Global and Local Theta Correspondences

In this section we recall the global and local theta correspondences for O_m and then study the global and local first occurrences of theta correspondences in terms of the periods or models defined in the previous sections.

3.1. Global and local theta liftings. Let Sp_{2l} be the symplectic group of k-rank l. Then $(\mathrm{O}_m, \mathrm{Sp}_{2l})$ forms a reductive dual pair in Sp_{2lm} in the sense of R. Howe ([**H79**]). We denote by $\mathrm{Mp}_{2l}(\mathbb{A})$ the metaplectic double cover $\widetilde{\mathrm{Sp}}_{2l}(\mathbb{A})$ of $\mathrm{Sp}_{2l}(\mathbb{A})$ if $m = 2n+1$ or the $\mathbb{A}$-rational points $\mathrm{Sp}_{2l}(\mathbb{A})$ of Sp_{2l} if $m = 2n$. Similarly, we denote by $\mathrm{Mp}_{2l}(k_v)$ the metaplectic double cover $\widetilde{\mathrm{Sp}}_{2l}(k_v)$ of $\mathrm{Sp}_{2l}(k_v)$ if $m = 2n+1$ or the k_v-rational points $\mathrm{Sp}_{2l}(k_v)$ of Sp_{2l} if $m = 2n$. Details about $\mathrm{Mp}_{2l}(k_v)$ and $\mathrm{Mp}_{2l}(\mathbb{A})$ and their splitting properties can be found in many references. See, for instance, [**K94**] or [**JngS07b**].

For a non-trivial character ψ of $\mathbb{A}/k$, there exists the Weil representation ω_ψ of $\widetilde{\mathrm{Sp}}_{2lm}(\mathbb{A})$, which is realized in the Schrödinger model $\mathcal{S}(\mathbb{A}^{ml})$, where $\mathcal{S}(\mathbb{A}^{ml})$ is the space of $\mathbb{C}$-valued Schwartz-Bruhat functions on $\mathbb{A}^{ml}$.

For $\varphi \in \mathcal{S}(\mathbb{A}^{ml})$, we form the theta function

$$\theta_{\psi,\varphi}(x) := \sum_{\xi \in k^{ml}} \omega_\psi(x)(\varphi)(\xi),$$

on $\widetilde{\mathrm{Sp}}_{2ml}(\mathbb{A})$. This series is absolutely convergent and defines a function of moderate growth on $\widetilde{\mathrm{Sp}}_{2ml}(\mathbb{A})$. There is a natural homomorphism

$$\mathrm{O}_m(\mathbb{A}) \times \mathrm{Mp}_{2l}(\mathbb{A}) \to \widetilde{\mathrm{Sp}}_{2ml}(\mathbb{A})$$

with kernel $C_2 = \{\pm 1\}$, and the center of $\mathrm{O}_m(\mathbb{A})$ diagonally embedded. We pull the Weil representation ω_ψ back to $\mathrm{O}_m(\mathbb{A}) \times \mathrm{Mp}_{2l}(\mathbb{A})$. This allows us to restrict $\theta_{\psi,\varphi}$ to $\mathrm{O}_m(\mathbb{A}) \times \mathrm{Mp}_{2l}(\mathbb{A})$. See [**JngS07b**], for instance.

For an irreducible cuspidal automorphic representation (σ, V_σ) of $\mathrm{O}_m(\mathbb{A})$, the integral

$$\theta^{2l}_{\psi,m}(h;\phi_\sigma,\varphi) := \int_{\mathrm{O}_m(k)\backslash \mathrm{O}_m(\mathbb{A})} \phi_\sigma(g)\theta_{\psi^{-1},\varphi}(g,h)dg \tag{3.1}$$

with $\phi_\sigma \in V_\sigma$, defines an automorphic form on $\mathrm{Mp}_{2l}(\mathbb{A})$. We denote by $\theta^{2l}_{\psi,m}(\sigma)$ the space generated by all $\theta^{2l}_{\psi,m}(g;\phi_\sigma,\varphi)$ as φ and ϕ_σ vary. This defines a genuine automorphic representation of $\mathrm{Mp}_{2l}(\mathbb{A})$, which we denote by $\theta^{2l}_{\psi,m}(\sigma)$. We call this representation the ψ-theta lifting of σ to $\mathrm{Mp}_{2l}(\mathbb{A})$. Similarly, for a genuine irreducible cuspidal automorphic representation $(\tilde{\pi}, V_{\tilde{\pi}})$ of $\mathrm{Mp}_{2l}(\mathbb{A})$, we get the automorphic representation $\theta^m_{\psi,2l}(\tilde{\pi})$ of $\mathrm{O}_m(\mathbb{A})$. Its space is generated by the automorphic forms

$$\theta^m_{\psi,2l}(g;\phi_{\tilde{\pi}},\varphi) := \int_{\mathrm{Mp}_{2l}(k)\backslash \mathrm{Mp}_{2l}(\mathbb{A})} \phi_{\tilde{\pi}}(h)\theta_{\psi,\varphi}(g,h)dh \tag{3.2}$$

as φ and $\phi_{\tilde{\pi}}$ vary. We say that $\theta^m_{\psi,2l}(\tilde{\pi})$ is the ψ-theta lifting of $\tilde{\pi}$ to $\mathrm{O}_m(\mathbb{A})$. In this paper, all representations of the metaplectic group (global or local) are assumed to be genuine.

Recall that a basic problem in the theory of the theta correspondence is to determine when the ψ-theta lifting $\theta^{2l}_{\psi,m}(\sigma)$ is nonzero for a given irreducible cuspidal automorphic representation σ of $\mathrm{O}_m(\mathbb{A})$ (similarly for $\theta^m_{\psi,2l}(\tilde{\pi})$). In [**GJS09**], we introduced the notion of the lowest occurrence $\mathrm{LO}_\psi(\sigma)$ of σ, with respect to all twists by automorphic sign characters of $\mathrm{O}_m(\mathbb{A})$, in the tower $\mathrm{Mp}_{2l}(\mathbb{A})$, via the ψ-theta correspondence, namely

$$\mathrm{LO}_\psi(\sigma) := \min_\epsilon\{\mathrm{FO}_\psi(\sigma\otimes\epsilon)\}, \tag{3.3}$$

where ϵ runs through all automorphic sign characters of $\mathrm{O}_m(\mathbb{A})$. As the notation suggests, $\mathrm{FO}_\psi(\sigma\otimes\epsilon)$ denotes the first occurrence of $\sigma\otimes\epsilon$ in the tower $\mathrm{Mp}_{2l}(\mathbb{A})$ via the ψ-theta correspondence.

Next, we recall briefly from [**MVW87**] the local theta correspondence over the local field k_v, where v is a finite local place of k.

For a nontrivial character ψ_v of k_v, let ω_{ψ_v} be the Weil representation of the reductive dual pair $\mathrm{O}_m(k_v) \times \mathrm{Mp}_{2l}(k_v)$ acting on the local Schrödinger model $\mathcal{S}(k_v^{ml})$, where $\mathcal{S}(k_v^{ml})$ is the space of local k_v-valued Schwartz-Bruhat functions on k_v^{ml}. A detailed discussion of the splitting of the double cover and the related cocycles can be found in [**JngS07b**], for example. See [**K94**] for general reductive dual pairs.

Let (σ_v, V_{σ_v}) ($(\tilde{\pi}_v, V_{\tilde{\pi}_v})$, resp.) be an irreducible admissible representation of $\mathrm{O}_m(k_v)$ ($\mathrm{Mp}_{2l}(k_v)$, resp.). If

$$\mathrm{Hom}_{\mathrm{O}_m(k_v)\times \mathrm{Mp}_{2l}(k_v)}(\mathcal{S}(k_v^{ml}), V_{\sigma_v}\otimes V_{\tilde{\pi}_v}) \neq 0, \tag{3.4}$$

then we say that $\tilde{\pi}_v$ is a local ψ_v-theta lift of σ_v, and σ_v is a local ψ_v-theta lift of $\tilde{\pi}_v$. We do not assume that the local Howe duality conjecture holds for the case we are discussing here. The local Howe duality conjecture was proved by J.-L. Waldspurger [**W90**], when the residual characteristic of k is odd. In such a circumstance, the local ψ_v-theta lift is the same as the local ψ_v-Howe lift. We refer to [**MVW87**] for more detailed discussions.

We define the first occurrence for the local ψ_v-theta liftings based on (3.4). More precisely, we say that the first occurrence of σ_v is $\mathrm{FO}_{\psi_v}(\sigma_v) = 2l_0$ if

$$\mathrm{Hom}_{\mathrm{O}_m(k_v)\times \mathrm{Mp}_{2l_1}(k_v)}(\mathcal{S}(k_v^{ml_1}), V_{\sigma_v}\otimes V_{\tilde{\pi}_{v,l_1}}) = 0,$$

for all $l_1 < l_0$ and for all irreducible admissible representations $\tilde{\pi}_{v,l_1}$ of $\mathrm{Mp}_{2l_1}(k_v)$, but there exists at least one irreducible admissible representation $\tilde{\pi}_{v,l_0}$ of $\mathrm{Mp}_{2l_0}(k_v)$ such that

$$\mathrm{Hom}_{\mathrm{O}_m(k_v)\times \mathrm{Mp}_{2l_0}(k_v)}(\mathcal{S}(k_v^{ml_0}), V_{\sigma_v}\otimes V_{\tilde{\pi}_{v,l_0}}) \neq 0.$$

By the local tower property of ([**K96**], for instance), if the first occurrence of σ_v is $\mathrm{FO}_{\psi_v}(\sigma_v) = 2l_0$, then for any $l > l_0$, there always exists an irreducible admissible representation $\tilde{\pi}_{v,l}$ of $\mathrm{Mp}_{2l}(k_v)$ such that the space

$$\mathrm{Hom}_{\mathrm{O}_m(k_v)\times \mathrm{Mp}_{2l}(k_v)}(\mathcal{S}(k_v^{ml}), V_{\sigma_v}\otimes V_{\tilde{\pi}_{v,l}}) \neq 0.$$

We define the local lowest occurrence of σ_v by

$$\mathrm{LO}_{\psi_v}(\sigma_v) := \min\{\mathrm{FO}_{\psi_v}(\sigma_v), \mathrm{FO}_{\psi_v}(\sigma_v\otimes \det)\}.$$

We mention here the conservation relation conjectured by Kudla and Rallis, namely that $\mathrm{FO}_{\psi_v}(\sigma_v) + \mathrm{FO}_{\psi_v}(\sigma_v\otimes\det) = m$. See [**KR05**].
The local first occurrence for $\tilde{\pi}_v$ can be defined in the same way.

3.2. Vanishing of theta liftings. For an irreducible cuspidal automorphic representation σ of $\mathrm{O}_m(\mathbb{A})$, we are going to relate, by doing some explicit calculations, the nonvanishing of the $\psi_{t,\alpha}$-Fourier coefficient on σ to the first occurrence $\mathrm{FO}_\psi(\sigma)$ of σ.

Following [**MW87**], [**M96**] and [**GRS03**], we say that σ has $\psi_{t,\alpha}$ as a top Fourier coefficient, for given $t \in \{1, 2, \cdots, r\}$ and $\alpha \in k^\times \mod (k^\times)^2$, if there is some $\phi_\sigma \in V_\sigma$ such that the $\psi_{t,\alpha}$-Fourier coefficient $\mathcal{F}^{\psi_{t,\alpha}}(\phi_\sigma)$ is not identically zero, but the $\psi_{t',\alpha'}$-Fourier coefficients $\mathcal{F}^{\psi_{t',\alpha'}}(\phi_\sigma)$ are all identically zero, for all $\phi_\sigma \in V_\sigma$, $\alpha' \in k^\times \mod (k^\times)^2$, and $t' > t$. Recall again that we assume that $m - 2t \geq 1$. Note that if r, the k-rank of O_m, is zero, i.e. O_m is k-anisotropic, then σ has no such Fourier coefficients at all.

The first result in this paper is

THEOREM 3.1. *Let σ be an irreducible cuspidal automorphic representation of $\mathrm{O}_m(\mathbb{A})$. If σ has $\psi_{t,\alpha}$ as a top Fourier coefficient, for some $t \in \{1, 2, \cdots, r\}$, with $r = \frac{m-m_0}{2} \geq 1$, $m - 2t \geq 1$, and some $\alpha \in k^\times \mod (k^\times)^2$, then the lowest occurrence of σ, $\mathrm{LO}_\psi(\sigma)$ is greater than or equal to $2t$, i.e. for any automorphic* *sign character ϵ of $\mathrm{O}_m(\mathbb{A})$, the first occurrence $\mathrm{FO}_\psi(\sigma\otimes\epsilon)$ is greater than or equal to $2t$.*

Note that when $m_0 \leq 1$ and $t = [\frac{m-1}{2}]$, $\psi_{t,\alpha}$ is a Whittaker character. In these cases the theorem is well known (at least the version with special orthogonal groups). See [**W80**], for $m=3$, [**PSS87**], for $m=5$, [**F95**], for m odd, in general, and [**GRS97**], for m even.

Here is an outline of the proof of Theorem 3.1. It suffices to show that for any $l < t$, the ψ-theta lifting $\theta^{2l}_{\psi,m}(\sigma\otimes\epsilon)$ is zero as an automorphic representation of $\mathrm{Mp}_{2l}(\mathbb{A})$, for all automorphic sign characters ϵ of $\mathrm{O}_m(\mathbb{A})$.

If this is not the case, then by the Rallis tower property of theta liftings ([**R84**]), there is an integer $l<t$ and an automorphic sign character ϵ of $\mathrm{O}_m(\mathbb{A})$ such that the ψ-theta lifting $\theta^{2l}_{\psi,m}(\sigma\otimes\epsilon)$ is nonzero and cuspidal. Clearly, $\sigma\otimes\epsilon$ has a nontrivial $\psi_{t,\alpha}$-Fourier coefficient (and this is its top Fourier coefficient). Thus, we may assume that ϵ is trivial, and hence that $\theta^{2l}_{\psi,m}(\sigma)$ is nonzero and cuspidal.

By the main theorem of [**M97b**] and Theorem 1.2 of [**JngS07b**], the ψ-theta lifting $\tilde\pi_{2l} := \theta^{2l}_{\psi,m}(\sigma)$ is a nonzero irreducible cuspidal automorphic representation of $\mathrm{Mp}_{2l}(\mathbb{A})$ and we have

$$\sigma = \theta^m_{\psi,2l}(\tilde\pi_{2l}) = \theta^m_{\psi,2l}(\theta^{2l}_{\psi,m}(\sigma)). \tag{3.5}$$

We consider the following polarizations for X_m and W_{2l}:

$$X_m = \ell_t^+ \oplus X_{m-2t} \oplus \ell_t^-, \tag{3.6}$$

$$W_{2l} = Y_l^+ \oplus Y_l^-, \tag{3.7}$$

where W_{2l} is the non-degenerate symplectic k-vector space defining Sp_{2l}, and hence Mp_{2l}. We assume that O_m acts from the left on X_m and Sp_{2l} acts from the right on W_{2l}. We may take a canonical basis

$$\{f_1,\cdots,f_l;f_{-l},\cdots,f_{-1}\} \tag{3.8}$$

for W_{2l}, such that Y_l^+ is generated by $\{f_1,\cdots,f_l\}$, Y_l^- is generated by $\{f_{-l},\cdots,f_{-1}\}$, and $(f_i,f_{-j})_{W_{2l}} = \delta_{ij}$.

We consider the Weil representation ω_ψ on the mixed Schrödinger model

$$\mathcal{S}_{m\otimes 2l} := \mathcal{S}(\ell_t^-(\mathbb{A})\otimes W_{2l}(\mathbb{A}) \oplus X_{m-2t}(\mathbb{A})\otimes Y_l^+(\mathbb{A})). \tag{3.9}$$

The Schwartz-Bruhat function φ in $\mathcal{S}_{m\otimes 2l}$ is written as

$$\varphi(w_1,\cdots,w_t;y_1,\cdots,y_{m-2t}) \tag{3.10}$$

where $w_i\in W_{2l}(\mathbb{A})$ and $y_j\in Y_l^+(\mathbb{A})$ for $i=1,\cdots,t$ and $j=1,\cdots,m-2t$.

By assumption, σ has a nonzero $\psi_{t,\alpha}$-Fourier coefficient for $t\leq r$ and for some $\alpha\in k^\times$ (see (2.14)), i.e.

$$\mathcal{F}^{\psi_{t,\alpha}}(\phi_\sigma)(g) := \int_{V_t(k)\backslash V_t(\mathbb{A})} \phi_\sigma(vg)\psi^{-1}_{t,\alpha}(v)dv \tag{3.11}$$

is nonzero, for some $\phi_\sigma\in V_\sigma$ and some $g\in\mathrm{O}_m(\mathbb{A})$. By (3.5), we may take ϕ_σ to be

$$\phi_\sigma(g) = \int_{\mathrm{Mp}_{2l}(k)\backslash\mathrm{Mp}_{2l}(\mathbb{A})} \phi_{\tilde\pi}(h)\theta_{\psi,\varphi}(g,h)dh \tag{3.12}$$

for some $\phi_{\tilde{\pi}} \in V_{\tilde{\pi}}$. Then the $\psi_{t,\alpha}$-Fourier coefficient $\mathcal{F}^{\psi_{t,\alpha}}(\phi_\sigma)$ can be written as

$$\begin{aligned}\mathcal{F}^{\psi_{t,\alpha}}(\phi_\sigma)(e) &= \int_{V_t(k)\backslash V_t(\mathbb{A})}\int_{\mathrm{Mp}_{2l}(k)\backslash \mathrm{Mp}_{2l}(\mathbb{A})} \phi_{\tilde{\pi}}(h)\theta_{\psi,\varphi}(v,h)dh\psi_{t,\alpha}^{-1}(v)dv \\ &= \int_{\mathrm{Mp}_{2l}(k)\backslash \mathrm{Mp}_{2l}(\mathbb{A})} \phi_{\tilde{\pi}}(h)\int_{V_t(k)\backslash V_t(\mathbb{A})} \theta_{\psi,\varphi}(v,h)\psi_{t,\alpha}^{-1}(v)dvdh.\end{aligned} \tag{3.13}$$

The switch of order of integrations is easily justified, since $V_t(k)\backslash V_t(\mathbb{A})$ is compact, $\phi_{\tilde{\pi}}$ is rapidly decreasing and $\theta_{\psi,\varphi}(v,h)$ is of moderate growth. The inner integral is the $\psi_{t,\alpha}$-Fourier coefficient of the theta function

$$\mathcal{F}^{\psi_{t,\alpha}}(\theta_{\psi,\varphi}(\cdot,h)) := \int_{V_t(k)\backslash V_t(\mathbb{A})} \theta_{\psi,\varphi}(v,h)\psi_{t,\alpha}^{-1}(v)dv \tag{3.14}$$

Then we have

PROPOSITION 3.2. *The $\psi_{t,\alpha}$-Fourier coefficient of the theta function $\theta_{\psi,\varphi}(g,h)$, $\mathcal{F}^{\psi_{t,\alpha}}(\theta_{\psi,\varphi}(\cdot,h))$, is zero for all $\varphi \in \mathcal{S}(\mathbb{A}^{ml})$, if $l < t$.*

We postpone the proof of Proposition 3.2 to §4.1.

By Proposition 3.2, if $l < t$, the $\psi_{t,\alpha}$-Fourier coefficient of the theta function $\theta_{\psi,\varphi}(g,h)$, $\mathcal{F}^{\psi_{t,\alpha}}(\theta_{\psi,\varphi}(\cdot,h))$, is zero for all $\varphi \in \mathcal{S}(\mathbb{A}^{ml})$. It follows that the $\psi_{t,\alpha}$-Fourier coefficient $\mathcal{F}^{\psi_{t,\alpha}}(\phi_\sigma)$ as in (3.11) is zero for all $\phi_\sigma \in V_\sigma$. This contradicts our assumption. This will prove Theorem 3.1.

By applying Theorem 3.1 above to Theorem 1.1 in [**GJS09**], we obtain

COROLLARY 3.3. *Let σ be an irreducible cuspidal automorphic representation of $\mathrm{O}_m(\mathbb{A})$. If σ has $\psi_{t,\alpha}$ as a top Fourier coefficient, for some $t \in \{1,2,\cdots,r\}$, with $r = \frac{m-m_0}{2} \geq 1$, $m-2t \geq 1$, and some $\alpha \in k^\times \mod (k^\times)^2$, then the partial L-function $L^S(s,\sigma)$ is holomorphic for $\mathrm{Re}(s) > \frac{m}{2} - t$.*

4. Fourier Coefficients of Theta Functions

We shall prove Proposition 3.2 first and then develop its local version afterwards.

4.1. Proof of Proposition 3.2. We shall use the notation in §3 for the calculation of the $\psi_{t,\alpha}$-Fourier coefficient of the theta function $\theta_{\psi,\varphi}(g,h)$ as in Proposition 3.2,

$$\mathcal{F}^{\psi_{t,\alpha}}(\theta_{\psi,\varphi}(\cdot,h)) := \int_{V_t(k)\backslash V_t(\mathbb{A})} \theta_{\psi,\varphi}(v,h)\psi_{t,\alpha}^{-1}(v)dv. \tag{4.1}$$

Let us rewrite the elements (2.6) in V_t in the form

$$v = v(u,x,z) = \begin{pmatrix} u & x & z \\ & \mathrm{I}_{m-2t} & x^* \\ & & u^* \end{pmatrix}$$

The subgroup $Z_t = \{v(z) = v(\mathrm{I}_t,0,z) \in V_t\}$ is the center of $N_t = \{v(x,z) = v(\mathrm{I}_t,x,z) \in V_t\}$, and the subgroup $U_t = \{v(u) = v(u,0,0) \in V_t\}$ normalizes N_t. We may write the elements of $Z_t\backslash N_t$ as $v(x) = v(\mathrm{I}_t,x,z)Z_t$, for any z, such that $v(\mathrm{I}_t,x,z) \in V_t$. Note that

$$\psi_{t,\alpha}(v(\mathrm{I}_t,x,z)) = \psi(x_t \cdot \mu_\alpha),$$

where x_t is the last row of x. We have

$$\mathcal{F}^{\psi_{t,\alpha}}(\theta_{\psi,\varphi}(\cdot,h)) = \int_{U_t(k)\backslash U_t(\mathbb{A})}\int_{M_{t\times(m-2t)}(k)\backslash M_{t\times(m-2t)}(\mathbb{A})} \cdot\int_{Z_t(k)\backslash Z_t(\mathbb{A})} \theta_{\psi,\varphi}(v(z)v(x)v(u),h)\psi_{t,\alpha}^{-1}(v(x)v(u))dzdxdu. \tag{4.2}$$

By the definition of the mixed Schrödinger model as in (3.9) and (3.10), the theta function $\theta_{\psi,\varphi}(v(z)g,h)$ can be written as

$$\sum_{w_i\in W_{2l}(k),y_j\in Y_l^+(k)} \omega_\psi(v(z)g,h)\varphi(w_1,\cdots,w_t;y_1,\cdots,y_{m-2t}). \tag{4.3}$$

We have the following formula for the action of $\omega_\psi(v(z),1)$ on the mixed Schrödinger model:

$$\begin{aligned}&\omega_\psi(v(z),1)\varphi(w_1,\cdots,w_t;y_1,\cdots,y_{m-2t})\\ =\ &\varphi(w_1,\cdots,w_t;y_1,\cdots,y_{m-2t})\psi(\frac{1}{2}\mathrm{tr}(\mathrm{Gr}(w_1,\cdots,w_t)\omega_t z)),\end{aligned}$$

where $\mathrm{Gr}(w_1,\cdots,w_t)$ is the Gram matrix of $(w_1,\cdots,w_t)$ (see [**K96**], p. 37, and also [**JngS07b**], p. 727).

Hence the dz-integration in (4.2) can be expressed as

$$\begin{aligned}\int_{Z_t(k)\backslash Z_t(\mathbb{A})}\theta_{\psi,\varphi}(v(z)g,h)dz &= \sum_{w_i,y_j}\omega_\psi(g,h)\varphi(w_1,\cdots,w_t;y_1,\cdots,y_{m-2t})\\ &\cdot\int_{Z_t(k)\backslash Z_t(\mathbb{A})}\psi^{-1}(\frac{1}{2}\mathrm{tr}(\mathrm{Gr}(w_1,\cdots,w_t)\omega_t z))dz,\end{aligned}$$

where the summation over w_i, y_j is the same as in (4.3). The order switch of integral and sum is easily justified, since $Z_t(k)\backslash Z_t(\mathbb{A})$ is compact and the summation over w_i, y_j is absolutely convergent. Note that

$$\int_{Z_t(k)\backslash Z_t(\mathbb{A})}\psi(\frac{1}{2}\mathrm{tr}(\mathrm{Gr}(w_1,\cdots,w_t)\omega_t z))dz$$

must be zero unless the Gram matrix $\mathrm{Gr}(w_1,\cdots,w_t)$ is zero, i.e. $(w_i,w_j)_{W_{2l}} = 0$ for all $i,j = 1,2,\cdots,t$. This means that the subspace of W_{2l} generated by $w_1,w_2,\cdots,w_t$ is totally isotropic. Since we assume that $l < t$, we deduce that $w_1,w_2,\cdots,w_t$ must be linearly dependent in W_{2l}. When $\mathrm{Gr}(w_1,\cdots,w_t)$ is zero,

$$\int_{Z_t(k)\backslash Z_t(\mathbb{A})}\psi(\frac{1}{2}\mathrm{tr}(\mathrm{Gr}(w_1,\cdots,w_t)\omega_t z))dz = 1$$

by the choice of the Haar measure on $Z_t(k)\backslash Z_t(\mathbb{A})$. Therefore we have

$$\int_{Z_t(k)\backslash Z_t(\mathbb{A})}\theta_{\psi,\varphi}(v(z)g,h)dz = \sum_{w_i,y_j}\omega_\psi(g,h)\varphi(w_1,\cdots,w_t;y_1,\cdots,y_{m-2t}) \tag{4.4}$$

where the summation is over all $y_1,\cdots,y_{m-2t}\in Y_l^+(k)$, and all $w_1,\cdots,w_t\in W_{2l}(k)$ with the property that $w_1,w_2,\cdots,w_t$ generate a totally isotropic subspace of $W_{2l}(k)$. Again, since $\dim_k \mathrm{Span}_k(w_1,\cdots,w_t)\le l < t$, $w_1,\cdots,w_t$ are automatically linearly dependent. Hence we obtain

$$\mathcal{F}^{\psi_{t,\alpha}}(\theta_{\psi,\varphi}(\cdot,h)) = \int_{U_t(k)\backslash U_t(\mathbb{A})}\int_{M_{t\times(m-2t)}(k)\backslash M_{t\times(m-2t)}(\mathbb{A})}\sum_{w_i,y_j}\omega_\psi(v(x)v(u),h) \tag{4.5}$$
$$\cdot\varphi(w_1,\cdots,w_t;y_1,\cdots,y_{m-2t})\psi_{t,\alpha}^{-1}(v(x)v(u))dxdu,$$

with summation as above. Denote

$$\begin{aligned} d &= \dim_k \operatorname{Span}_k(w_1,\cdots,w_t), \\ E_d &= \operatorname{Span}_k(w_1,\cdots,w_t). \end{aligned}$$

Then we can split the last integral as a sum over $0 \leq d \leq l$, where in each summand we compute the last integral with $w_1,\cdots,w_t \in E_d$ and E_d varies over all d-dimensional totally isotropic subspaces of W_{2l}. Let P_d be the standard parabolic subgroup of Sp_{2l}, which preserves the totally isotropic subspace Y_d^- generated by $\{f_{-d},\cdots,f_{-1}\}$. Then we may write $E_d = Y_d^-\gamma$, where $\gamma \in P_d(k)\backslash\mathrm{Sp}_{2l}(k)$. Thus,

$$\mathcal{F}^{\psi_{t,\alpha}}(\theta_{\psi,\varphi}(\cdot,h)) = \sum_{d=0}^{l}\int_{U_t(k)\backslash U_t(\mathbb{A})}\int_{M_{t\times(m-2t)}(k)\backslash M_{t\times(m-2t)}(\mathbb{A})}\sum_{\gamma\in P_d(k)\backslash\mathrm{Sp}_{2l}(k)}$$
$$\sum_{w_i\in Y_d^-}\sum_{y_j\in Y_l^+}\omega_\psi(v(x)v(u),\gamma h)\cdot\varphi(w_1,\cdots,w_t;y_1,\cdots,y_{m-2t})$$
$$\cdot\psi_{t,\alpha}^{-1}(v(x)v(u))dxdu. \tag{4.6}$$

Here, we used the automorphy of theta series. More explicitly, if we write in (4.5), $w_i = v_i\gamma$, where $w_i \in E_d$ and $v_i \in Y_d^-$, then

$$\begin{aligned} &\sum_{y_j\in Y_l^+}\omega_\psi(g,h)\cdot\varphi(v_1\gamma,\cdots,v_t\gamma;y_1,\cdots,y_{m-2t}) \\ =\ &\sum_{y_j\in Y_l^+}\omega_\psi(g,\gamma h)\cdot\varphi(v_1,\cdots,v_t;y_1,\cdots,y_{m-2t}). \end{aligned}$$

The point is that the summation over $y_j \in Y_l^+$ defines the theta series on $\mathrm{O}_{m-2t}(\mathbb{A})\times \mathrm{Mp}_{2l}(\mathbb{A})$. To explain this, we may assume that $g=1$, and that $\varphi = \varphi_1\otimes\varphi_2$, where

$$\varphi(w_1,\cdots,w_t;y_1,\cdots,y_{m-2t}) = \varphi_1(w_1,\cdots,w_t)\varphi_2(y_1,\cdots,y_{m-2t}).$$

Then

$$\sum_{y_j\in Y_l^+}\omega_\psi(1,h)\cdot\varphi(w_1,\cdots,w_t;y_1,\cdots,y_{m-2t}) = \varphi_1(w_1h,\cdots,w_th)\theta_{\psi,\varphi_2}(1,h),$$

where θ_{ψ,φ_2} is the corresponding theta series for $\mathrm{O}_{m-2t}(\mathbb{A})\times\mathrm{Mp}_{2l}(\mathbb{A})$. Let us use now the action of $v(x)$, which follows from the formulae of the Weil representation on the mixed model. For $w_1,\cdots,w_t \in Y_l^-$ and $y_1,\cdots,y_{m-2t}\in Y_l^+$,

$$\begin{aligned} &\omega_\psi(v(x),1)\cdot\varphi(w_1,\cdots,w_t;y_1,\cdots,y_{m-2t}) \\ =\ &\psi\Big(\sum_{i=1}^{t}\sum_{j=1}^{m-2t}x_{t+1-i,j}(w_i,y_j)\Big)\varphi(w_1,\cdots,w_t;y_1,\cdots,y_{m-2t}). \end{aligned}$$

As before, we may switch the order of summations and the dx-integration and get

$$\begin{aligned}\mathcal{F}^{\psi_{t,\alpha}}(\theta_{\psi,\varphi}(\cdot,h)) &= \sum_{d=0}^{l}\int_{U_t(k)\backslash U_t(\mathbb{A})}\sum_{\gamma\in P_d(k)\backslash \mathrm{Sp}_{2l}(k)}\sum_{W_{\alpha,d}(k)} \\ (4.7)\qquad & \omega_\psi(v(u),\gamma h)\cdot\varphi(w_1,\cdots,w_t;y_1,\cdots,y_{m-2t})\psi_{t,\alpha}^{-1}(v(u))du,\end{aligned}$$

where $W_{\alpha,d}$ is the variety of all $(w_1,\cdots,w_t;y_1,\cdots,y_{m-2t})$, such that the w_i lie in Y_d^-, the y_j lie in Y_l^+, $(w_i,y_j)=0$, for all $2\le i\le t$, $1\le j\le m-2t$, and similarly,

$$(w_1,y_j)=(\mu_\alpha)_j.$$

Recall that μ_α is the (column) vector in $X_{m-2t}(k)$, such that $(\mu_\alpha,\mu_\alpha)=\alpha$, which enters into the definition of $\psi_{t,\alpha}$. The action of $\omega_\psi(v(u),1)$ is linear on $\ell_{-t}(\mathbb{A})\otimes W_{2l}(\mathbb{A})$ and trivial on $X_{m-2t}(\mathbb{A})\otimes Y_l^+(\mathbb{A})$. The precise form is

$$\omega_\psi(v(u),1)\cdot\varphi(w_1,\cdots,w_t;y_1,\cdots,y_{m-2t})=\varphi((w_1,\cdots,w_t)\cdot\omega_t u\omega_t;y_1,\cdots,y_{m-2t}).$$

Note that

$$(4.8)\qquad (w_1,\cdots,w_t;y_1,\cdots,y_{m-2t})\mapsto((w_1,\cdots,w_t)\cdot\omega_t u\omega_t;y_1,\cdots,y_{m-2t})$$

defines a k-rational action of U_t on $W_{\alpha,d}$. The $U_t(k)$-orbits in $W_{\alpha,d}(k)$ are given by elements $(w_1,\cdots,w_t;y_1,\cdots,y_{m-2t})\in W_{\alpha,d}(k)$, such that $(w_1,\cdots,w_t)$ is of the following form

$$(4.9)\qquad (w_1,\cdots,0,w_{i_2},0,\cdots,0,w_{i_3},0,\cdots,0,\cdots,0,w_{i_d},0,\cdots,0),$$

where $w_1,w_{i_1},w_{i_2},\cdots,w_{i_d}$ are linearly independent elements in Y_d^-. Note that by definition of $W_{\alpha,d}$, we must have $d\ge 1$ and $w_1\neq 0$. Denote by $w'_{(t:d)}$ the element in (4.9), and let $w_{(t:d)}=(w'_{(t:d)};y_1,\cdots,y_{m-2t})$. Denote its $U_t(k)$-orbit by $\mathcal{O}_{w_{(t:d)}}$ and its stabilizer in U_t, via the action (4.8), by $\mathcal{L}_{w_{(t:d)}}$, i.e.

$$\mathcal{L}_{w_{(t:d)}}:=\{u\in U_t(k)\mid w'_{(t:d)}\cdot\omega_t u\omega_t=w'_{(t:d)}\}.$$

Again, in (4.7), we may switch order of summations and the du-integration. Then, the contribution of the $U_t(k)$-orbit $\mathcal{O}_{w_{(t:d)}}$ is

$$\begin{aligned}&\int_{U_t(k)\backslash U_t(\mathbb{A})}\sum_{\mathcal{O}_{w_{(t:d)}}}\omega_\psi(v(u),\gamma h)\varphi(w_1,\cdots,w_t;y_1,\cdots,y_{m-2t})\psi_{t,\alpha}^{-1}(v(u))du\\ =&\int_{U_t(k)\backslash U_t(\mathbb{A})}\sum_{\eta\in\mathcal{L}_{w_{(t:d)}}(k)\backslash U_t(k)}\omega_\psi(v(\eta u),\gamma h)\varphi(w_{(t:d)})\psi_{t,\alpha}^{-1}(v(u))du\\ =&\int_{\mathcal{L}_{w_{(t:d)}}(k)\backslash U_t(\mathbb{A})}\omega_\psi(v(u),\gamma h)\varphi(w_{(t:d)})\psi_{t,\alpha}^{-1}(v(u))du\\ =&\int_{\mathcal{L}_{w_{(t:d)}}(\mathbb{A})\backslash U_t(\mathbb{A})}\omega_\psi(v(u),\gamma h)\varphi(w_{(t:d)})\psi_{t,\alpha}^{-1}(v(u))du\\ (4.10)\qquad &\cdot\int_{\mathcal{L}_{w_{(t:d)}}(k)\backslash\mathcal{L}_{w_{(t:d)}}(\mathbb{A})}\psi_{t,\alpha}^{-1}(v(a))da.\end{aligned}$$

Hence, if the restriction of the character $\psi_{t,\alpha}$ to the stabilizer $\mathcal{L}_{w_{(t:d)}}(\mathbb{A})$ is nontrivial, then we must have

$$\int_{\mathcal{L}_{w_{(t:d)}}(k)\backslash\mathcal{L}_{w_{(t:d)}}(\mathbb{A})} \psi_{t,\alpha}^{-1}(v(a))da = 0.$$

This implies that for such a $U_t(k)$-orbit $\mathcal{O}_{w_{(t:d)}}$, we have

(4.11)

$$\int_{U_t(k)\backslash U_t(\mathbb{A})} \sum_{\mathcal{O}_{w_{(t:d)}}} \omega_\psi(v(u),\gamma h)\varphi(w_1,\cdots,w_t;y_1,\cdots,y_{m-2t})\psi_{t,\alpha}^{-1}(v(u))du = 0.$$

Note again that in the orbit $\mathcal{O}_{w_{(t:d)}}$, $y_1,\cdots,y_{m-2t}$ are fixed. Now for a $U_t(k)$-orbit $\mathcal{O}_{w_{(t:d)}}$ with representative of the form $(w'_{(t:d)};y_1,\cdots,y_{m-2t})$, where $w'_{(t:d)}$ is as in (4.9), since $r\le l<t$, it is easy to check that the stabilizer $\mathcal{L}_{w_{(t:d)}}$ contains at least one simple root of GL_t in U_t. Recall again, that in $W_{\alpha,d}$, we must have $w_1\neq 0$. Hence we must have that the restriction of the character $\psi_{t,\alpha}$ to the stabilizer $\mathcal{L}_{w_{(t:d)}}(\mathbb{A})$ is nontrivial. This proves Proposition 3.2.

4.2. Genericity of theta liftings. We are going to calculate the $\psi_{t,\alpha}$-Fourier coefficient $\mathcal{F}^{\psi_{t,\alpha}}(\phi_\sigma)$ of $\phi_\sigma\in V_\sigma$, as defined in (3.12) when $l=t$. In other words, we will calculate explicitly the following integral:

$$\mathcal{F}^{\psi_{t,\alpha}}(\phi_\sigma) = \int_{\mathrm{Mp}_{2t}(k)\backslash\mathrm{Mp}_{2t}(\mathbb{A})} \phi_{\tilde{\pi}}(h)\int_{V_t(k)\backslash V_t(\mathbb{A})} \theta_{\psi,\varphi}(v,h)\psi_{t,\alpha}^{-1}(v)dvdh \tag{4.12}$$

for $\phi_\sigma\in V_\sigma$ and $\varphi\in\mathcal{S}(\mathbb{A}^{mt})$. For this, we simply continue our calculation in §4.1, with $l=t$. What we did shows that in (4.7) only $d=t$ may contribute a nonzero summand. Note also that for $(w_1,\cdots,w_t;y_1,\cdots,y_{m-2t})\in W_{\alpha,t}(k)$, $w_1,\cdots,w_t$ form a basis of Y_t^-. Denote by S the unipotent radical of the Siegel parabolic subgroup P_t. Then we may replace in (4.7) ($l=t$) the summation over $\gamma\in P_t(k)\backslash\mathrm{Sp}_{2t}(k)$ by the summation over $\gamma\in S(k)\backslash\mathrm{Sp}_{2t}(k)$, but now $(w_1,\cdots,w_t)=(f_{-t},\cdots,f_{-1})$ are fixed to be the standard basis of Y_t^-. We get

$$\begin{aligned}\mathcal{F}^{\psi_{t,\alpha}}(\theta_{\psi,\varphi}(\cdot,h)) &= \int_{U_t(k)\backslash U_t(\mathbb{A})}\sum_{\gamma\in S(k)\backslash\mathrm{Sp}_{2t}(k)}\sum_{Y_\alpha(k)} \\ &\qquad \omega_\psi(v(u),\gamma h)\varphi(f_{-t},\cdots,f_{-1};y_1,\cdots,y_{m-2t})\psi_{t,\alpha}^{-1}(v(u))du,\end{aligned} \tag{4.13}$$

where Y_α is the set of all $(y_1,\cdots,y_{m-2t})$, such that the y_j lie in Y_t^+, satisfy $(f_{-i},y_j)=0$, for all $1\le i<t$ and $1\le j\le m-2t$, and similarly

$$(f_{-t},y_j)=(\mu_\alpha)_j.$$

This implies that Y_α is a single point. Indeed, we must have $y_j=a_jf_t$, where $a_j\in k$, and

$${}^t(a_1,\cdots,a_{m-2t})=\mu_\alpha.$$

In terms of our notation (3.9), (3.10), it is now more convenient to re-denote the vector $(a_1f_t,\cdots,a_{m-2t}f_t)$ by $\mu_\alpha\otimes f_t\in X_{m-2t}(k)\otimes Y_t^+(k)$. We conclude that

$$\begin{aligned}\mathcal{F}^{\psi_{t,\alpha}}(\theta_{\psi,\varphi}(\cdot,h)) &= \sum_{\gamma\in S(k)\backslash\mathrm{Sp}_{2t}(k)}\int_{U_t(k)\backslash U_t(\mathbb{A})} \\ &\qquad \omega_\psi(v(u),\gamma h)\cdot\varphi(f_{-t},\cdots,f_{-1};\mu_\alpha\otimes f_t)\psi_{t,\alpha}^{-1}(v(u))du,\end{aligned} \tag{4.14}$$

Substitute this in (3.13). We get

$$\mathcal{F}^{\psi_{t,\alpha}}(\phi_\sigma)(e) = \int_{S(k)\backslash \mathrm{Mp}_{2t}(\mathbb{A})} \phi_{\tilde{\pi}}(h) \int_{U_t(k)\backslash U_t(\mathbb{A})} \omega_\psi(v(u),h)\cdot\varphi(f_{-t},\cdots,f_{-1};\mu_\alpha\otimes f_t)\psi_{t,\alpha}^{-1}(v(u))dudh, \tag{4.15}$$

In the mixed Schrödinger model (3.9), we have, for $s = \begin{pmatrix} I_t & b \\ & I_t \end{pmatrix} \in S(\mathbb{A})$,

$$\omega_\psi(1,s)\cdot\varphi(f_{-t},\cdots,f_{-1};\mu_\alpha\otimes f_t) = \psi(\frac{1}{2}\alpha b_{t,1})\varphi(f_{-t},\cdots,f_{-1};\mu_\alpha\otimes f_t).$$

Factoring integration in the last integral through $S(\mathbb{A})$, we get

$$\mathcal{F}^{\psi_{t,\alpha}}(\phi_\sigma)(e) = \int_{S(\mathbb{A})\backslash \mathrm{Mp}_{2t}(\mathbb{A})} \int_{S(k)\backslash S(\mathbb{A})} \phi_{\tilde{\pi}}(sh)\psi(\frac{1}{2}\alpha s_{t,t+1})ds \cdot\int_{U_t(k)\backslash U_t(\mathbb{A})} \omega_\psi(v(u),h)\cdot\varphi(f_{-t},\cdots,f_{-1};\mu_\alpha\otimes f_t)\psi_{t,\alpha}^{-1}(v(u))dudh, \tag{4.16}$$

Next, we have the following formula, for $u\in U_t(\mathbb{A})$,

$$\omega_\psi(u,1)\cdot\varphi(f_{-t},\cdots,f_{-1};\mu_\alpha\otimes f_t) = \omega_\psi(1,u^{-1}h)\varphi(f_{-t},\cdots,f_{-1};\mu_\alpha\otimes f_t),$$

Then by changing the variable $h\mapsto uh$ in the last integral, we get that

$$\mathcal{F}^{\psi_{t,\alpha}}(\phi_\sigma)(e) = \int_{S(\mathbb{A})\backslash \mathrm{Mp}_{2t}(\mathbb{A})} \omega_\psi(1,h)\cdot\varphi(f_{-t},\cdots,f_{-1};\mu_\alpha\otimes f_t) \cdot\int_{U_t(k)\backslash U_t(\mathbb{A})}\int_{S(k)\backslash S(\mathbb{A})} \phi_{\tilde{\pi}}(suh)\psi(\frac{1}{2}\alpha s_{t,t+1})\psi_{t,\alpha}^{-1}(u)dudsdh. \tag{4.17}$$

It is clear that the semi-direct product $U_t \ltimes S$ is the unipotent radical R_t of the standard Borel subgroup of Sp_{2t} and the product of the two characters $\psi_{t,\alpha}(u)$ and $\psi(-\frac{1}{2}\alpha\cdot s_{t,t+1})$ is a generic character $\psi_{R_t,\alpha}$ of R_t. Hence the inner integrations ds and du give a Whittaker-Fourier coefficient of $\phi_{\tilde{\pi}}$, which is denoted by $\mathcal{W}_{\phi_{\tilde{\pi}}}^{\psi_{R_t,\alpha}}(h)$. Hence $\mathcal{F}^{\psi_{t,\alpha}}(\phi_\sigma)(e)$ is equal to

$$\int_{S(\mathbb{A})\backslash \mathrm{Mp}_{2t}(\mathbb{A})} \omega_\psi(1,h)\varphi(f_{-t},\cdots,f_{-1};\mu_\alpha\otimes f_t)\mathcal{W}_{\phi_{\tilde{\pi}}}^{\psi_{R_t,\alpha}}(h)dh. \tag{4.18}$$

We record the calculation above in the following proposition.

PROPOSITION 4.1. *Let $\tilde{\pi}$ be an irreducible cuspidal automorphic representation of $\mathrm{Mp}_{2t}(\mathbb{A})$. Let $\sigma = \theta_{\psi,2t}^m(\tilde{\pi})$ be the theta lift of $\tilde{\pi}$ to $\mathrm{O}_m(\mathbb{A})$. We assume that $r\geq t$ and $m>2t$. Let ϕ_σ be the element of V_σ given by (3.12), namely*

$$\phi_\sigma(g) = \int_{\mathrm{Mp}_{2l}(k)\backslash \mathrm{Mp}_{2l}(\mathbb{A})} \phi_{\tilde{\pi}}(h)\theta_{\psi,\varphi}(g,h)dh.$$

Then the $\psi_{t,\alpha}$- Fourier coefficient of σ is related to the $\psi_{R_t,\alpha}$-Whittaker -Fourier coefficient of $\tilde{\pi}$ by

$$\mathcal{F}^{\psi_{t,\alpha}}(\phi_\sigma)(e) = \int_{S(\mathbb{A})\backslash \mathrm{Mp}_{2t}(\mathbb{A})} \omega_\psi(1,h)\varphi(f_{-t},\cdots,f_{-1};\mu_\alpha\otimes f_t)\mathcal{W}_{\phi_{\tilde{\pi}}}^{\psi_{R_t,\alpha}}(h)dh. \tag{4.19}$$

In particular, if $\tilde{\pi}$ is not generic, or if α is not represented by X_{m-2t}, then σ has zero $\psi_{t,\alpha}$-Fourier coefficients.

As a corollary, we get

PROPOSITION 4.2. *Let σ be an irreducible cuspidal automorphic representation of $\mathrm{O}_m(\mathbb{A})$. Assume that σ has a nonzero $\psi_{t,\alpha}$-Fourier coefficient ($1 \le t \le r$, $m - 2t \ge 1$). Assume that the first occurrence of σ, $\mathrm{FO}_\psi(\sigma)$ is $2t$. Let $\tilde{\pi} = \theta^{2t}_{\psi,m}(\sigma)$ be the ψ-theta lift of σ to $\mathrm{Mp}_{2t}(\mathbb{A})$. Then $\tilde{\pi}$ is globally generic, with respect to the Whittaker character $\psi_{R_t,\alpha}$, as above. Moreover, the formula relating the $\psi_{t,\alpha}$-Fourier coefficient of σ and the $\psi_{R_t,\alpha}$-Whittaker-Fourier coefficient of $\tilde{\pi}$ is given (in the above notation) by* (4.19).

Conversely, start with an irreducible, cuspidal automorphic representation $\tilde{\pi}$ of $\mathrm{Mp}_{2t}(\mathbb{A})$, which is globally generic with respect to a character of the form $\psi_{R_t,\alpha}$, where $\alpha \in k^\times$. Assume that $t \le r$ and $2t < m$. We use the same notation pertaining to X_m (X_{m-2t}, the symmetric non-degenerate matrices T_{m-2t} etc.). Since the quadratic form defined by T_{m_0+2} is not anisotropic, α is represented by T_{m_0+2}. Let $\mu_\alpha \in X_{m_0+2}(k)$ be such that $(\mu_\alpha, \mu_\alpha) = \alpha$, and consider the r.h.s. of (4.19), where we take ω_ψ to be the Weil representation of the dual pair $\mathrm{O}_{m_0+2+2t}(\mathbb{A}) \times \mathrm{Mp}_{2t}(\mathbb{A})$. It is easy to see that the r.h.s is not identically zero. By Proposition 4.1, we conclude that the $\psi_{t,\alpha}$-Fourier coefficient of the ψ theta lift of $\tilde{\pi}$ to $\mathrm{O}_{m_0+2+2t}(\mathbb{A})$ is nontrivial. In particular, the ψ-theta lift of $\tilde{\pi}$ to $\mathrm{O}_{m_0+2+2t}(\mathbb{A})$ is nonzero. Now, let σ be an irreducible cuspidal automorphic representation of $\mathrm{O}_m(\mathbb{A})$, which has a nontrivial $\psi_{t,\alpha}$-Fourier coefficient. We already proved that $\mathrm{FO}_\psi(\sigma) \ge 2t$. If $\mathrm{FO}_\psi(\sigma) = 2t$, then, by what we just explained, we must have that $r \le t + 1$. Thus, if we assume that $t < r - 1$, then we get that $\mathrm{FO}_\psi(\sigma) \ge 2t + 2$, and hence, by Theorem 1.1 in [**GJS09**], $L^S(s, \sigma)$ is holomorphic at $\mathrm{Re}(s) > \frac{m}{2} - t - 1$.

We can repeat the same considerations if α is represented by T_{m_0}. Let $\mu_\alpha \in X_{m_0}(k)$ represent α. Now we repeat the same argument with X_{m_0} replacing X_{m_0+2} and obtain that if $\mathrm{FO}_\psi(\sigma) = 2t$ and $\tilde{\pi}$ is the ψ-theta lift of σ to $\mathrm{Mp}_{2t}(\mathbb{A})$, then since $\tilde{\pi}$ is globally $\psi_{R_t,\alpha}$-generic, it has a nontrivial ψ-theta lift to $\mathrm{O}_{m_0+2t}(\mathbb{A})$, and we conclude that $t \ge r$. Thus, if we assume, in this case, that $t < r$, then we get, as before, that $\mathrm{FO}_\psi(\sigma) \ge 2t+2$, and that $L^S(s, \sigma)$ is holomorphic for $\mathrm{Re}(s) > \frac{m}{2} - t - 1$. Let us summarize this.

THEOREM 4.3. *Let σ be an irreducible cuspidal automorphic representation of $\mathrm{O}_m(\mathbb{A})$. Assume that σ has a nonzero $\psi_{t,\alpha}$-Fourier coefficient ($1 \le t \le r$, $m - 2t \ge 1$).*

1. *Assume that $t < r-1$. Then the partial L-function $L^S(s, \sigma)$ is holomorphic for $\mathrm{Re}(s) > \frac{m}{2} - t - 1$.*
2. *Assume that α is represented by the quadratic form corresponding to T_{m_0}, and that $t < r$. Then the partial L-function $L^S(s, \sigma)$ is holomorphic for $\mathrm{Re}(s) > \frac{m}{2} - t - 1$.*

5. Completion of the Proof of Theorem 1.1

The proof of Part (1) of Theorem 1.1 is completely analogous to the one in §4.1. We will use the same notation as before, adapted to the local setting.

5.1. Local models and theta lifts. We determine the vanishing of local theta lifts in terms of the local $\psi_{t,\alpha}$-functional. Here is the result.

THEOREM 5.1. *Let F be a finite extension of the p-adic field $\mathbb{Q}_p$. Let σ be an irreducible admissible representation of $\mathrm{O}_m(F)$. Assume that σ has a nonzero*

$\psi_{t,\alpha}$-functional as defined in (2.17) *for some $t \le r$, the Witt index of the quadratic space X_m defining $\mathrm{O}_m(F)$. Then $\mathrm{LO}_\psi(\sigma) \ge 2t$.*

This is the local analogue of Theorem 3.1. The proof uses the local version of the global arguments used in the proof of Theorem 3.1 and is modeled after the proof of Proposition 2.1, [**JngS03**] and of Theorem 4.2, [**JngS07a**]. Here is a sketch.

It is enough to show that the local ψ-theta lift $\theta^{2l}_{\psi,m}(\sigma)$ of σ to $\mathrm{Mp}_{2l}(F)$ is zero for all $l < t$. Assume that this is not the case. Then there is a integer $l < t$ such that the ψ-theta lift $\theta^{2l}_{\psi,m}(\sigma)$ of σ to $\mathrm{Mp}_{2l}(F)$ is nonzero. This means by (3.4) that there is an irreducible admissible representation $\tilde{\pi}$ of $\mathrm{Mp}_{2l}(F)$ such that

$$\mathrm{Hom}_{\mathrm{O}_m(F)\times \mathrm{Mp}_{2l}(F)}(\omega_\psi, \sigma \otimes \tilde{\pi}) \neq 0 \tag{5.1}$$

or equivalently,

$$\mathrm{Hom}_{\mathrm{O}_m(F)}(\omega_\psi \otimes \tilde{\pi}^\vee, \sigma) \neq 0 \tag{5.2}$$

where ω_ψ is the local Weil representation of $\mathrm{Mp}_{2lm}(F)$, restricted to the dual pair $(\mathrm{O}_m(F), \mathrm{Mp}_{2l}(F))$.

Following §3, we consider the analogous polarizations for X_m and W_{2l}:

$$X_m = \ell_t^+ \oplus X_{m-2t} \oplus \ell_t^-, \tag{5.3}$$

$$W_{2l} = Y_l^+ \oplus Y_l^-. \tag{5.4}$$

We consider the local Weil representation ω_ψ on the mixed Schrödinger model

$$\mathcal{S}_{m\otimes 2l} := \mathcal{S}(\ell_t^- \otimes W_{2l} \oplus X_{m-2t} \otimes Y_l^+). \tag{5.5}$$

Using similar bases as in §3, we write a local Schwartz-Bruhat function φ in $\mathcal{S}_{m\otimes 2l}$ as

$$\varphi(w_1, \cdots, w_t; y_1, \cdots, y_{m-2t}) \tag{5.6}$$

where $w_i \in W_{2l}$ and $y_j \in Y_l^+$ for $i = 1, \cdots, t$ and $j = 1, \cdots, m-2t$.

By hypothesis, σ has a nonzero $\psi_{t,\alpha}$-functional ℓ, i.e. a nonzero element in

$$\mathrm{Hom}_{V_t(F)}(V_\sigma, \psi_{t,\alpha}).$$

By (5.2), the functional ℓ induces a nonzero functional β over $\mathcal{S}_{m\otimes 2l} \otimes V_{\tilde{\pi}^\vee}$, such that

$$\beta(\omega_\psi(v,h)\varphi, \xi) = \psi_{t,\alpha}(v)\beta(\varphi, \xi) \tag{5.7}$$

for $v \in V_t(F)$, $h \in \mathrm{Mp}_{2l}(F)$, $\xi \in V_{\tilde{\pi}^\vee}$, and φ is a function in the mixed model.

We consider the local version of the dz-integration as in (4.4), and obtain as in [**JngS03**], p. 755, that for each fixed ξ, β is supported on

$$C_0 = \{(w_1, \cdots, w_t; y_1, \cdots, y_{m-2t}))|(w_i, w_j) = 0, \forall\ 1 \le i, j \le t\}.$$

Indeed, let i be the restriction map from $\mathcal{S}(W_{2l}^t \oplus (Y_l^+)^{m-2t})$ to $\mathcal{S}(C_0)$. It is surjective. Let i^* be the corresponding map on Jacquet modules with respect to Z_t and the trivial character. Then i^* is an isomorphism, i.e.

$$J_{Z_t}(\mathcal{S}(W_{2l}^t \oplus (Y_l^+)^{m-2t})) \cong J_{Z_t}(\mathcal{S}(C_0)).$$

Let C be the complement of C_0 in $W_{2l}^t \oplus (Y_l^+)^{m-2t}$. Then it is easy to see that $J_{Z_t}(\mathcal{S}(C)) \cong 0$ and $J_{Z_t}(\mathcal{S}(C_0)) \cong \mathcal{S}(C_0)$. We regard $\mathcal{S}(C_0)$ as a module over $(Z_t(F)\backslash V_t(F)) \times \mathrm{Mp}_{2l}(F)$. Denote $U'_t(F) = Z_t(F)\backslash V_t(F)$. We identify $U'_t(F)$ with $U_t(F)M_{t\times(m-2t)}(F)$ and regard $\psi_{t,\alpha}$ as a character of $U'_t(F)$. Thus, we have to prove that $J_{U_t'(F),\psi_{t,\alpha}}(\mathcal{S}(C_0)) = 0$, when $l < t$. Write C_0 as the disjoint union, over $0 \leq d \leq l$, of the varieties

$$C_0^d = \{(w_1, \cdots, w_t; y_1, \cdots, y_{m-2t}) \in C_0 \mid \dim_F \mathrm{Span}\{w_1, \cdots, w_t\} = d\}.$$

Then it is enough to prove that $J_{U_t'(F),\psi_{t,\alpha}}(\mathcal{S}(C_0^d)) = 0$, for all $0 \leq d \leq l$. We can embed $\mathcal{S}(C_0^d)$ inside $ind_{U_t'(F)\times \bar{P}_d(F)}^{U_t'(F)\times \mathrm{Mp}_{2l}(F)} \mathcal{S}(C_0^{d,-})$, where

$$C_0^{d,-} = \{(w_1, \cdots, w_t; y_1, \cdots, y_{m-2t}) \in C_0^d \mid w_1, \cdots, w_d \in Y_d^-\},$$

and $\bar{P}_d(F)$ is the inverse image of $P_d(F)$ inside $\mathrm{Mp}_{2l}(F)$. Thus, we have an embedding

$$J_{U_t'(F),\psi_{t,\alpha}}(\mathcal{S}(C_0^d)) \hookrightarrow \mathrm{ind}_{\bar{P}_d(F)}^{\mathrm{Mp}_{2l}(F)} J_{U_t'(F),\psi_{t,\alpha}}(\mathcal{S}(C_0^{d,-})),$$

and so it is enough to show that $J_{U_t'(F),\psi_{t,\alpha}}(\mathcal{S}(C_0^{d,-})) = 0$, when $l < t$. Using the action of $M_{t\times(m-2t)}(F)$ on $\mathcal{S}(C_0^{d,-})$ through the formulae of the Weil representation, we conclude, as above, and as in §4.1 that

$$J_{U_t'(F),\psi_{t,\alpha}}(\mathcal{S}(C_0^{d,-})) \cong J_{U_t(F),\psi_{t,\alpha}}(\mathcal{S}(W_{\alpha,d})),$$

where $W_{\alpha,d}$ is defined exactly by the same relations as in the global case. Finally, it remains to show that $J_{U_t(F),\psi_{t,\alpha}}(\mathcal{O}_{w_{(t:d)}}) = 0$, for every $U_t(F)$-orbit $\mathcal{O}_{w_{(t:d)}}$ of $W_{\alpha,d}$ (same definition as in the global case). This follows, as in the global case, from the fact that since $d \leq l < t$, there is a simple root subgroup in $U_t(F)$, which lies in the stabilizer of the representative $w_{t:d}$. This proves Theorem 5.1.

5.2. Proof of part (1) of Theorem 1.1. Let σ be an irreducible cuspidal automorphic representation of $\mathrm{O}_m(\mathbb{A})$. Assume that there is a finite local place v of the number field k such that the local v-component σ_v of σ has a nonzero local $\psi_{t,\alpha}$-functional. Then, by Theorem 5.1, the local ψ-theta lift of σ_v to $\mathrm{Mp}_{2l}(k_v)$ is zero for all $l < t$. Hence the global ψ-theta lift of σ to $\mathrm{Mp}_{2l}(\mathbb{A})$ must be zero for all $l < t$. This property holds also for all twists of σ by automorphic sign characters, since the twist of σ_v by any sign character also has a nonzero local $\psi_{t,\alpha}$-functional. Hence the lowest occurrence of σ in the global ψ-theta liftings, $\mathrm{LO}_\psi(\sigma)$, must be greater than or equal to $2t$. By Theorem 1.1 of **[GJS09]**, the partial L-function $L^S(s,\sigma)$ must be holomorphic for $\mathrm{Re}(s) > \frac{m}{2} - t$. This completes the proof of Theorem 1.1.

References

[F95] Furusawa, Masaaki *On the theta lift from SO_{2n+1} to $\widetilde{\mathrm{Sp}}_n$*. J. Reine Angew. Math. 466 (1995), 87–110.

[GJS09] Ginzburg, David; Jiang, Dihua; Soudry, David *Poles of L-functions and theta liftings for orthogonal groups.* Journal of the Institute of Math. of Jussieu (2009) 8(4), 693–741.

[GPSR97] Ginzburg, D.; Piatetski-Shapiro, I.; Rallis, S. *L functions for the orthogonal group.* Mem. Amer. Math. Soc. 128 (1997), no. 611.

[GRS97] Ginzburg, David; Rallis, Stephen; Soudry, David *Periods, poles of L-functions and symplectic-orthogonal theta lifts.* J. Reine Angew. Math. 487 (1997), 85–114.

[GRS03] Ginzburg, D.; Rallis, S.; Soudry, D. *On Fourier coefficients of automorphic forms of symplectic groups.* Manuscripta Math. 111 (2003), no. 1, 1–16.

[H79] Howe, Roger *θ-series and invariant theory.* Automorphic forms, representations and L-functions (Proc. Sympos. Pure Math., Oregon State Univ., Corvallis, Ore., 1977), Part 1, pp. 275–285, Proc. Sympos. Pure Math., XXXIII, Amer. Math. Soc., Providence, R.I., 1979.

[JngS03] Jiang, Dihua; Soudry, David *The local converse theorem for* SO$(2n+1)$ *and applications.* Ann. of Math. (2) 157 (2003), no. 3, 743–806.

[JngS07a] Jiang, Dihua; Soudry, David *On the genericity of cuspidal automorphic forms of* SO_{2n+1}. J. Reine Angew. Math. 604 (2007), 187–209.

[JngS07b] Jiang, Dihua; Soudry, David *On the genericity of cuspidal automorphic forms of* SO_{2n+1}*, II.* Compositio Math. 143 (2007), 721–748.

[K94] Kudla, Stephen S. *Splitting metaplectic covers of dual reductive pairs.* Israel J. Math.,Vol. 87 (1994), 361-401.

[K96] Kudla, Stephen S. *Notes on the local theta correspondence.* Lecture series in the European School in Group Theory, at Schloss Hirschberg, September 1996 (http://www.math.umd.edu/ ssk/ssk.research.html).

[KR94] Kudla, Stephen; Rallis, Stephen. *A regularized Siegel-Weil formula: the first term identity.* Ann. of Math.(2) 140 (1994), no. 1, 1-80.

[KR05] Kudla, Stephen; Rallis, Stephen. *On first occurrence in the local theta correspondence.* Automorphic Representations, L-functions and Applications: Progress and Prospects. Ohio-State University MRI Publications 11, de Gruyter (2005), 273–308.

[M96] Mœglin, Colette *Front d'onde des représentations des groupes classiques p-adiques.* (French) [Wave front of representations of p-adic classical groups] Amer. J. Math. 118 (1996), no. 6, 1313–1346.

[M97a] Mœglin, Colette *Non nullité de certaines relêvements par séries théta.* Journal of Lie Theory, 7(1997) 201–229.

[M97b] Mœglin, Colette *Quelques propriétés de base des séries théta.* Journal of Lie Theory, 7(1997) 231–238.

[MVW87] Mœglin, Colette; Vignéras, Marie-France; Waldspurger, Jean-Loup *Correspondances de Howe sur un corps p-adique.* (French) [Howe correspondences over a p-adic field] Lecture Notes in Mathematics, 1291. Springer-Verlag, Berlin, 1987

[MW87] Mœglin, Colette; Waldspurger, Jean-Loup *Modéles de Whittaker dégénérés pour des groupes p-adiques.* (French) [Degenerate Whittaker models for p-adic groups] Math. Z. 196 (1987), no. 3, 427–452.

[PSS87] Piatetski-Shapiro, Ilya; Soudry David *On a correspondence of automorphic forms of orthogonal groups of order five.* J. Math. Pures Appl. 66 (1987), 407–436.

[R84] Rallis, Stephen *On the Howe duality conjecture.* Compositio Math. 51 (1984), no. 3, 333–399.

[W80] Waldspurger, J.-L. *Correspondance de Shimura.* (French) [Shimura Correspondence] J. Math Pures Appl. 59 (1980), 1–132.

[W90] Waldspurger, J.-L. *Démonstration d'une conjecture de dualité de Howe dans le cas p-adique,* $p \neq 2$. (French) [Proof of a duality conjecture of Howe in the p-adic case, $p \neq 2$] Festschrift in honor of I. I. Piatetski-Shapiro on the occasion of his sixtieth birthday, Part I (Ramat Aviv, 1989), 267–324, Israel Math. Conf. Proc., 2, Weizmann, Jerusalem, 1990.

DAVID GINZBURG: SCHOOL OF MATHEMATICAL SCIENCES, THE RAYMOND AND BEVERLY SACKLER FACULTY OF EXACT SCIENCES, TEL AVIV UNIVERSITY, RAMAT-AVIV, 69978 ISRAEL

E-mail address: ginzburg@post.tau.ac.il

DIHUA JIANG: SCHOOL OF MATHEMATICS, UNIVERSITY OF MINNESOTA, MINNEAPOLIS, MN 55455, USA

E-mail address: dhjiang@math.umn.edu

DAVID SOUDRY: SCHOOL OF MATHEMATICAL SCIENCES, THE RAYMOND AND BEVERLY SACKLER FACULTY OF EXACT SCIENCES, TEL AVIV UNIVERSITY, RAMAT-AVIV, 69978 ISRAEL

E-mail address: soudry@post.tau.ac.il

Clay Mathematics Proceedings
Volume 13, 2011

On Dual R–Groups for Classical Groups

David Goldberg

To Freydoon Shahidi on the occasion of his 60th birthday

Contents

1. Preliminaries 160
2. Preliminary results on R–groups: The maximal case 165
3. The case of Levi factors contained in the Siegel 170
4. The case of non-Siegel parabolic subgroups 174
5. Duality implements the isomorphism 182
References 184

Introduction

One major facet of the Langlands program is the problem of connecting results on local harmonic analysis with the arithmetic of Artin L–functions. One such connection should come through the determination of reducibility points of induced representations via local Langlands L–functions. This was established (in the generic case) by Shahidi [**25**]. Problems on reducibility of induced representations of groups over local fields are an important aspect of the global theory of automorphic forms, and particularly crucial in the theory of Eisenstein series. The Langlands-Shahidi method has been a powerful approach, and has yielded remarkable results in the past decade. Here we consider the more modest problem of describing the structure of induced representations of classical p-adic groups via the local Langlands correspondence.

Knapp and Stein developed the theory of intertwining operators and R–groups in the case of archimedean fields [**21**] and Silberger extended this theory to the case of p-adic groups [**27, 28**]. The R-group gives a combinatorial description of the intertwining algebra of parabolically induced representations. For the archimedean case Shelstad [**26**] showed the R-group could be determined from the representation of the Weil-Deligne group parametrizing the inducing representation. In [**1**] Arthur proposed a generalization of this result to arbitrary local fields, refining

2000 *Mathematics Subject Classification.* Primary 22E50; Secondary 11F70.

ideas proposed earlier by Langlands [22]. Keys determined R-groups for the principal series of Chevalley groups, quasi-split unitary groups, and special unitary groups, [18, 19, 20], and, in many cases, showed the isomorphism of the Knapp-Stein R–group with the R–group proposed by Langlands, Shelstad, and Arthur. The R-groups for reperesentations of classical groups parabolically induced from arbitrary discrete series were determined in [9]. In [3] Ban and Zhang showed the isomorphism of the Knapp-Stein and Arthur R–groups for all such representations of $SO_{2n+1}(F)$. Here, we study induced from unitary supercuspidal representations of the classical groups $SO_{2n+1}(F)$, $Sp_{2n}(F)$ and $SO_{2n}(F)$. In the case of a non-Siegel maximal parabolic subgroup we impose the condition that the representation of the Weil-Deligne group which parametrizes the inducing representation is irreducible. We also must assume that the Rankin product L–functions defined by the Langlands-Shahidi method [25] are Artin. We show, by direct computation, that the Knapp-Stein and Arthur R–groups are isomorphic. Of course, in the case of $SO_{2n+1}(F)$ these results are covered by [3]; however, our proof is different. We also show the isomorphism can be realized by the map from roots to coroots.

Let F be a nonarchimedean local field of characteristic zero, and suppose $\mathbf{G}$ is a connected reductive group defined over F. Suppose $\mathbf{P} = \mathbf{MN}$ is a proper parabolic subgroup. We assume σ is an irreducible discrete series representation of $M = \mathbf{M}(F)$. The local Langlands conjecture predicts σ is a member of an L–packet $\Pi_\varphi(M)$ with parameter $\varphi : W_F' \to {}^LM$, with W_F' the Weil-Deligne group of F and ${}^LM$ the L–group of M. Then composition with the inclusion ${}^LM \hookrightarrow {}^LG$ gives a parameter for an L–packet $\Pi_\varphi(G)$ of $G = \mathbf{G}(F)$. The conjecture is $\Pi_G(\varphi)$ consists of the irreducible constituents of $\mathrm{Ind}_P^G(\sigma')$ for $\sigma' \in \Pi_\varphi(M)$. Arthur defines the R-group, R_φ, attached to the packet $\Pi_\varphi(M)$, and a subgroup $R_{\varphi,\sigma}$. This is accomplished by looking at the centralizer, S_φ, of the image of φ in $\hat{G} = {}^LG^\circ$, and identifying certain subgroups of the Weyl group of S_φ. Let $R(\sigma)$ be the Knapp–Stein R–group attached to σ. Then, our main result is (under the assumptions we imposed above) $R_{\varphi,\sigma} \simeq R(\sigma)$.

In Section 1 we review the theory of intertwining operators, the Knapp-Stein R-group, the Arthur R–group, and recall the computation of the Knapp-Stein R–groups for the classical groups. In Section 2 we discuss the isomorphism of $R(\sigma)$ and $R_{\varphi,\sigma}$ in the case where $\mathbf{M}$ is a maximal proper Levi subgroup. In Section 3 we discuss the case where $\mathbf{M}$ is a subgroup of the Siegel Levi subgroup, and in Section 4 we address the case where $\mathbf{M}$ is not a subgroup of the Siegel Levi subgroup. Finally, in Section 5 we show the isomorphism can be realized by the map $\alpha \mapsto \alpha^\vee$ from roots to coroots.

The author thanks Jim Arthur for originally suggesting this problem many years ago, and Dubravka Ban and Freydoon Shahidi for several encouraging and informative conversations, as well as pointing out some carelessness in earlier versions. We wish to acknowledge several conversations with Alan Roche and thank him for several suggestions. Finally, we thank the referee for pointing out several inconsistencies in the preliminary version, significantly improving the exposition.

1. Preliminaries

Let F be a local nonarchimedean field of characteristic zero. We let $\mathbf{G}$ be a quasi–split connected reductive group defined over F, and $G = \mathbf{G}(F)$, and we use similar notation for other groups defined over F. Fix a Borel subgroup $\mathbf{B} = \mathbf{TU}$ of

$\mathbf{G}$,with $\mathbf{T}$ a maximal torus of $\mathbf{G}$ and $\mathbf{U}$ the unipotent radical of $\mathbf{B}$. Let $\Phi = \Phi(\mathbf{G},\mathbf{T})$ be the roots of $\mathbf{T}$ in $\mathbf{G}$, let $\Phi^+ = \Phi^+(\mathbf{G},\mathbf{T})$ be the system of positive roots given by $\mathbf{B}$, and Δ the simple roots. For $\theta \subset \Delta$ we denote by $\mathbf{A}_\theta$ the corresponding subtorus of $\mathbf{T}$, namely

$$\mathbf{A}_\theta = \left(\bigcap_{\alpha\in\theta} \operatorname{Ker}\alpha\right)^\circ.$$

Set $\mathbf{P}_\theta = \mathbf{M}_\theta\mathbf{N}_\theta$ to be the standard parabolic subgroup of $\mathbf{G}$ defined by θ. Then $\mathbf{M}_\theta = Z_{\mathbf{G}}(\mathbf{A}_\theta)$, and $\mathbf{N}_\theta = \prod_{\alpha\in\Phi^+\setminus\Sigma^+(\theta)} \mathbf{U}_\alpha$, with $\mathbf{U}_\alpha$ the root subgroup defined by α. Here $\Sigma^+(\theta) = \Phi^+ \cap \operatorname{span}_{\mathbb{Z}}\theta$. (See [**6**] for more details.) If θ is fixed, then we denote the above groups by $\mathbf{A} = \mathbf{A}_\theta, \mathbf{P} = \mathbf{P}_\theta, \mathbf{M} = \mathbf{M}_\theta$, and $\mathbf{N} = \mathbf{N}_\theta$. As $\mathbf{A}$ is the maximal spit torus in the center of $\mathbf{M}$, we may denote it by $\mathbf{A}_{\mathbf{M}}$.

We denote by $\mathfrak{a}_{\mathbb{C}}^*$ the complexified dual of the real Lie algebra of $\mathbf{A}$. If $\nu \in \mathfrak{a}_{\mathbb{C}}^*$ and σ is a discrete series representation of M, then we denote the parabolically induced representation associated to P, σ, and ν by

$$I(\nu,\sigma) = \operatorname{Ind}_P^G(\sigma\otimes q^{\langle\nu,H_P(\cdot)\rangle}).$$

The induction here is normalized induction, and the function H_P is defined as in [**25**]. We let $i_{G,M}(\sigma) = I(0,\sigma)$. Let $w \in W(\mathbf{G},\mathbf{A}) = N_G(\mathbf{A})/\mathbf{M}$. We fix a representative $\tilde{w}$ for w. Let $A(\nu,\sigma,\tilde{w})\colon I(\nu,\sigma) \to I(w\nu,w\sigma)$ be the standard intertwining operator [**21, 27, 28**]. Harish–Chandra proved that there is a meromorphic function $\mu(\nu,\sigma,\tilde{w})$ on $\mathfrak{a}_{\mathbb{C}}^*$ so that

$$A(\tilde{w}\nu, w\sigma, \tilde{w}^{-1})A(\nu,\sigma,\tilde{w}) = \mu(\nu,\sigma,\tilde{w})^{-2}\,\mathrm{Id}\,.$$

THEOREM 1.1. *(Harish-Chandra [**14**]). Suppose $\mathbf{P}$ is a maximal proper parabolic subgroup of $\mathbf{G}$. Then $i_{G,M}(\sigma)$ is reducible if and only if there is a non–trivial element w of $W(\mathbf{G},\mathbf{A})$, with $w\sigma\simeq\sigma$, and $\mu(\sigma) = \mu(0,\sigma,w) \neq 0$.*

If $\tilde{w}_0$ represents the longest element of $W(\mathbf{G},\mathbf{A})$, then we let $\mu(\sigma) = \mu(0,\sigma,\tilde{w}_0)$. We call $\mu(\sigma)$ the Plancherel measure of σ. Note, Theorem 1.1 says, in the case of a maximal proper Levi subgroup, $\mu(\sigma) = 0$ if and only if $w\sigma \simeq \sigma$ and $\nu \mapsto A(\nu,\sigma,\tilde{w})f(g)$ has a pole at $\nu = 0$, for some $f \in I(\nu,\sigma)$ and $g \in G$. (See [**29**] for the details of this). We note the existence (or not) of a pole of the intertwining operator does not depend on the choice of $\tilde{w}$.

For an arbitrary standard parabolic subgroup $\mathbf{P} = \mathbf{P}_\theta$ of $\mathbf{G}$, we let $\Phi(\mathbf{P},\mathbf{A})$ be the roots of $\mathbf{A}$ in $\mathbf{P}$. For $\alpha \in \Phi(\mathbf{P},\mathbf{A})$, we define $\mathbf{M}_\alpha = \mathbf{M}_{\theta\cup\{\alpha\}}$, and $\mathbf{N}_\alpha^* = \mathbf{N}\cap\mathbf{M}_\alpha$. Then $\mathbf{P}_\alpha^* = \mathbf{M}\mathbf{N}_\alpha$ is a maximal proper parabolic subgroup of $\mathbf{M}_\alpha$. Thus, we have a Plancherel measure $\mu_\alpha(\sigma)$ attached to $i_{M_\alpha,M}(\sigma)$ as defined above. We denote $W(\sigma) = \{w \in W(\mathbf{G},\mathbf{A}) | w\sigma \simeq \sigma\}$. We let $\Delta' = \{\alpha \in \Phi(\mathbf{P},\mathbf{A}) | \mu_\alpha(\sigma) = 0\}$. For $\alpha \in \Phi(\mathbf{P},\mathbf{A})$ we let $w_\alpha \in W(\mathbf{G},\mathbf{A})$ be the associated (relative) reflection. Let $W' = \langle w_\alpha | \alpha \in \Delta'\rangle$. Note, Theorem 1.1 implies $W' \subseteq W(\sigma)$. Let

$$R(\sigma) = \{w \in W(\sigma) | w\Delta' = \Delta'\} = \{w \in W(\sigma) | w\alpha > 0 \text{ for all } \alpha \in \Delta'\}.$$

For $w \in W(\sigma)$, one can define normalized intertwining operators $\mathcal{A}'(\sigma,\tilde{w})$: $i_{G,M}(\sigma) \to i_{G,M}(\sigma)$, i.e., an element in $\mathcal{C}(\sigma) = \operatorname{End}_G(i_{G,M}(\sigma))$. One can see [**20**] for an explicit description of $\mathcal{A}'(\sigma,w)$. There is a two cocycle

$$\eta\colon R(\sigma)\times R(\sigma) \to \mathbb{C}$$

so that $\mathcal{A}'(\sigma,w_1)\mathcal{A}'(\sigma,w_2) = \eta(w_1,w_2)\mathcal{A}'(\sigma,w_1w_2)$.

THEOREM 1.2. *(Knapp–Stein, Silberger). The above groups satisfy the following conditions:*

(i): $W' \triangleleft W(\sigma)$;
(ii): $R(\sigma) \simeq W(\sigma)/W'$;
(iii): $W(\sigma) = R(\sigma) \ltimes W'$.

Further, W' is the set of $w \in W(\sigma)$, so that the normalized self intertwining operators $\mathcal{A}'(\sigma, w)\colon i_{G,M}(\sigma) \mapsto i_{G,M}(\sigma)$ are scalar. □

Then the intertwining algebra $\mathcal{C}(\sigma)$ of $i_{G,M}(\sigma)$ is isomorphic to $\mathbb{C}[R(\sigma)]_\eta$, the group algebra of $R(\sigma)$ twisted by η. Thus, $R(\sigma)$ (and its representation theory) determines the structure of $i_{G,M}(\sigma)$. This can be made more explicit [**2, 20**].

Let W_F be the Weil group of F and $W_F' = W_F \times SL_2(\mathbb{C})$ the Weil-Deligne group [**30**]. For a quasi-split connected reductive group $\mathbf{G}$ defined over F, we let $\hat{G}$ be the complex algebraic group with root datum dual to that of $\mathbf{G}$. The L-group ${}^LG$ is given by ${}^LG = \hat{G} \rtimes W_F$, where W_F acts on $\hat{G}$ by the Galois action on the root datum of $\hat{G}$. (Note, when $\mathbf{G}$ is split, W_F acts trivially on $\hat{G}$, so the semidirect product is, in fact, a direct product.) The local Langlands conjecture says, in part, that the irreducible tempered representations of $G = \mathbf{G}(F)$ are partitioned into finite subsets (called L–packets), and there is a correspondence between L–packets and admissible homomorphisms (as defined in [**30**]) $\varphi : W_F' \to {}^LG$. We denote the L-packet corresponding to φ by $\Pi_G(\varphi)$, and may denote $\pi \in \Pi_G(\varphi)$ by π_φ. One fundamental property of this correspondence is the equality of Langlands L-functions and Artin L–functions. That is, the conjecture asserts the correspondence can be chosen such that, for any complex representation $r : {}^LG \to GL_m(\mathbb{C})$, we have $L(s, \pi_\varphi, r) = L(s, r \circ \varphi)$, for any $\pi_\varphi \in \Pi_G(\varphi)$.

The local Langlands correspondence also gives rise to a notion of R–group. Suppose $\varphi\colon W_F' \to {}^LM$ is an admissible homomorphism parametrizing the L–packet $\prod_\varphi(M)$ containing σ. Since ${}^LM \hookrightarrow {}^LG$, $\varphi\colon W_F \to {}^LG$ parametrizes an L–packet $\Pi_\varphi(G)$ of G, expected to be the set of irreducible components of $i_{G,M}(\sigma')$ for $\sigma' \in \prod_\varphi(M)$. We write $S_\varphi = Z_{\hat{G}}(\operatorname{Im}\varphi)$, and S_φ° for its connected component. The quotient $\mathbf{S}_\varphi = S_\varphi/S_\varphi^\circ$ should contain information on reducibility, as described below.

Fix a maximal torus T_φ in S_φ°, and let $W_\varphi^\circ = W(S_\varphi^\circ, T_\varphi) = N_{S_\varphi^\circ}(T_\varphi)/Z_{S_\varphi^\circ}(T_\varphi)$. Similarly let $W_\varphi = W(S_\varphi, T_\varphi)$. Then we denote by R_φ the group $W_\varphi/W_\varphi^\circ$. By duality, one can identify W_φ as the subgroup of $W(G, A_M)$ for which $w\sigma \in \prod_\varphi(M)$. Thus, there is a an identification $W_{\varphi,\sigma} = \{w \in W_\varphi | w\sigma \simeq \sigma\} \simeq W(\sigma)$. We let $W_{\varphi,\sigma}^\circ = W_\varphi^\circ \cap W_{\varphi,\sigma}$. Then we define $R_{\varphi,\sigma} = W_{\varphi,\sigma}/W_{\varphi,\sigma}^\circ$. Arthur conjectures $R_{\varphi,\sigma} \simeq R(\sigma)$. We will prove this claim in many cases. Note, if $\mathbf{G}$ is split, we may consider homomorphisms $\varphi : W_F' \to \hat{M}$, since W_F centralizes all of $\hat{G}$. For our computations, we only consider supercuspidal representations of M, and therefore need only consider the image of the Weil group, W_F. We use the notation $\hat{A} = A_{\hat{M}}$ for the split component of $\hat{M}$.

We will recall the computation of R–groups for the classical groups $\mathbf{G} = \mathbf{G}(n) = SO_{2n+1}$, Sp_{2n}, SO_{2n}. This description can be made explicit, in that [**9**] shows exactly which elements of $W(\sigma)$ lie in $R(\sigma)$.

We fix forms for the groups $\mathbf{G}(n)$. Let

$$s_n = \begin{pmatrix} & & & & & 1 \\ & & & & \cdot & \\ & & & \cdot & & \\ & & \cdot & & & \\ & 1 & & & & \\ 1 & & & & & \end{pmatrix} \in GL_n,$$

and

$$u_{2n} = \begin{pmatrix} & & & & & 1 \\ & & & & \cdot & \\ & & & \cdot & & \\ & & \cdot & & & \\ & -1 & & & & \\ 1 & & & & & \\ -1 & & & & & \end{pmatrix} \in GL_{2n}.$$

Then, for any $n > 1$,

$$SO_n = \left\{ g \in SL_n | \, {}^t g s_n g = s_n \right\},$$

and

$$Sp_{2n} = \left\{ g \in GL_{2n} | \, {}^t g u_{2n} g = u_{2n} \right\}.$$

We also use s_n and u_{2n} to denote the complex matrices with the same entries, and fix these as our our forms for $\hat{G}$. When computing the centralizers S_φ we will often find an isomorphism with a complex orthogonal (or special orthogonal) group of some rank r. In each case we are studying, the orthogonal groups arising are split forms, and we emphasize this by using the standard notation $O_{r,r}(\mathbb{C})$ or $O_{r+1,r}(\mathbb{C})$ for the full orthogonal groups. In order to avoid confusion with other forms, we also use this notation when we refer to the dual groups $\hat{G}$. Fix $\mathbf{T}$ to be the maximal torus of diagonal elements in $\mathbf{G}$. Then

$$\mathbf{T} = \{\text{diag}\{x_1, x_2, \ldots, x_n, x_n^{-1}, \ldots, x_1^{-1}\} | x_i \in \mathbf{G}_m\}$$

if $\mathbf{G} = Sp_{2n}$, or SO_{2n}, and

$$\mathbf{T} = \{\text{diag}\{x_1, \ldots, x_n, 1, x_n^{-1}, \ldots, x_1^{-1}\} | x_i \in \mathbf{G}_m\}$$

if $\mathbf{G} = SO_{2n+1}$, with $\mathbf{G}_m$ the algebraic multiplicative group over F. We fix $\mathbf{B}$ to be the Borel subgroup of upper triangular matrices in $\mathbf{G}$. The root system $\Phi(\mathbf{G}, \mathbf{T})$ is of type B_n, C_n, or D_n and has simple roots

$$\Delta = \{e_1 - e_2, e_2 - e_3, \ldots, e_{n-1} - e_n, \alpha_n\},$$

with $\alpha_n = e_n, 2e_n$, or $e_{n-1} + e_n$, respectively. If $\theta \subseteq \Delta$, with $\alpha_n \notin \theta$, then

$$\mathbf{M} = \mathbf{M}_\theta \simeq GL_{n_1} \times GL_{n_2} \times \ldots \times GL_{n_r}, \tag{1.1}$$

with $n_1 + n_2 + \cdots + n_r = n$. If, on the other hand $\alpha_n \in \theta$, then

$$\mathbf{M} = \mathbf{M}_\theta = GL_{n_1} \times GL_{n_2} \times \cdots \times GL_{n_r} \times \mathbf{G}(m), \tag{1.2}$$

with $n_1 + \cdots + n_r + m = n$. If we take $\mathbf{G}(0) = \{I\}$, then we can include (1.1) into (1.2). However, for many of our computations, we work with (1.1) and (1.2) separately.

We now recall results of Shahidi.

THEOREM 1.3. (Shahidi [**24**]) *Let* $\mathbf{G} = Sp_{2n}$, $\mathbf{G}' = SO_{2n+1}$, *and* $\mathbf{G}'' = SO_{2n}$. *Consider* $\mathbf{M} \simeq GL_n$ *as the Siegel Levi subgroup of all these groups. We fix an irreducible unitary supercuspidal representation of* M. *We denote the induced representations as* $\pi = i_{G,M}(\sigma)$, $\pi' = i_{G',M}(\sigma)$, *and* $\pi'' = i_{G'',M}(\sigma)$.

(i): *If* $\sigma \not\simeq \tilde{\sigma}$, *then* π, π', *and* π'' *are all irreducible.*

(ii): *If* $\sigma \simeq \tilde{\sigma}$ *then precisely one of the two Langlands L–functions,* $L(s, \sigma, \mathrm{Sym}^2)$ *or* $L(s, \sigma, \wedge^2)$ *has a pole at* $s = 0$. *When* n *is even, and* $L(s, \sigma, \mathrm{Sym}^2)$ *has a pole at* $s = 0$, *then* π' *is irreducible and* π, π'' *are both reducible. Again, if* n *is even ,and* $L(s, \sigma, \wedge^2)$ *has the pole, then* π *and* π'' *are irreducible, while* π' *is reducible. If* n *is odd, then* $L(s, \sigma, \wedge^2)$ *is always holomorphic at* $s = 0$, *and thus* $L(s, \mathrm{Sym}^2, \sigma)$ *has a pole at* $s = 0$. *In this case* π' *and* π'' *are both irreducible, and* π *is reducible.*

In terms of the parametrization, this can be viewed as follows. Let $\varphi\colon W_F \to GL_n(\mathbb{C}) \simeq \hat{M}$ be a parameter for σ. We further, by composing with the obvious injections, consider φ as a parameter with image in $\hat{G} = SO_{n+1,n}(\mathbb{C}), \hat{G}' = Sp_{2n}(\mathbb{C})$, and $\hat{G}'' = SO_{n,n}(\mathbb{C})$, respectively. Since σ is supercuspidal, $\varphi(W_F)$ is an irreducible subgroup of $GL_n(\mathbb{C})$. If φ does not fix a non-degenerate bilinear form, then $L(s, \mathrm{Sym}^2 \varphi)$ and $L(s, \wedge^2\varphi)$ are both entire. If, on the other hand φ fixes a form, then the type of the form determines which of these two Artin L–functions has a pole. That is, $L(s, \wedge^2\varphi)$ has a pole if and only if φ fixes an alternating form, and $L(s, \mathrm{Sym}^2 \varphi)$ has a pole if and only if φ fixes a symmetric form. The first of these means φ factors through $Sp_n(\mathbb{C})$, i.e., n is even and φ is symplectic. In the second case φ is orthogonal, i.e. φ factors through $SO_n(\mathbb{C})$. The connection of Shahidi's results with Artin L–functions is made explicit by Henniart.

THEOREM 1.4. (Henniart [**15**]). *If* $\varphi\colon W_F \to GL_n(\mathbb{C})$ *is a tempered parameter, and* $\sigma = \sigma_\varphi$ *is the corresponding irreducible admissible representation of* $GL_n(F)$, *then* $L(s, \wedge^2\varphi) = L(s, \sigma, \wedge^2)$ *and* $L(s, \mathrm{Sym}^2 \varphi) = L(s, \sigma, \mathrm{Sym}^2)$. *In each of these equalities, the left hand side is an Artin L–function, while the right hand side is the Langlands L–function, as defined by Shahidi* [**25**].

We can summarize the results of [**9**] as follows:

THEOREM 1.5. *Let* $\mathbf{G} = \mathbf{G}(n) = SO_{2n+1}, Sp_{2n}$, *or* SO_{2n}. *Let* $\mathbf{M} \simeq GL_{n_1} \times GL_{n_2} \times \cdots \times GL_{n_r} \times \mathbf{G}(m)$, *with* $n_1 + \cdots + n_r + m = n$. *Let* $\sigma \simeq \sigma_1 \otimes \sigma_2 \otimes \cdots \otimes \sigma_r \otimes \tau$ *be an irreducible discrete series representation of* M, *with each* σ_i *an irreducible discrete series representation of* $GL_{n_i}(F)$ *and* τ *one such of* $G(m)$. *In the case of* SO_{2n}, *we assume* $n_1, \ldots, n_t$ *are even, while* $n_{t+1}, \ldots, n_r$ *are odd, and set*

$$C_0 = \begin{pmatrix} I_{m-1} & & & \\ & 0 & 1 & \\ & 1 & 0 & \\ & & & I_{m-1} \end{pmatrix} \in O_{2m}(F) \setminus SO_{2m}(F).$$

(i): *If* $\mathbf{G} = SO_{2n+1}$, *or* Sp_{2n}, *then* $R(\sigma) \simeq \mathbb{Z}_2^d$, *where* d *is the number of equivalence classes among the* σ_i *for which* $i_{G(n_i+m_i),GL_{n_i}(F)\times G(m)}(\sigma_i \otimes \tau)$ *is reducible.*

(ii): *Suppose* $\mathbf{G} = SO_{2n}$:

a): *Suppose* $m = 0$. *Then* $R(\sigma) \simeq \mathbb{Z}_2^{d_1+d_2-1}$, *where* d_1 *is the number of equivalence classes among* $\sigma_1, \ldots, \sigma_t$ *such that* $i_{G(n_i),GL_{n_i}(F)}(\sigma_i)$ *is reducible,* *and d_2 is the number of equivalence classes among $\sigma_{t+1}, \dots, \sigma_r$ for which $\tilde{\sigma}_i \simeq \sigma_i$.*

b): *Suppose $m > 0$ and $C_0\tau \not\simeq \tau$. Then $R(\sigma) \simeq \mathbb{Z}_2^{d_1+d_2-1}$, where d_1 is the number of equivalence classes among $\sigma_1, \dots, \sigma_t$ such that*

$$i_{G(n_i+m),GL_{n_i}(F)\times G(m)}(\sigma_i \otimes \tau)$$

is reducible, and d_2 is the number of equivalence classes among $\sigma_{t+1}, \dots, \sigma_r$ for which $\tilde{\sigma}_i \simeq \sigma_i$.

c): *Suppose $m > 0$ and $C_0\tau \simeq \tau$. Then $R(\sigma) \simeq \mathbb{Z}_2^d$, where d is the number of equivalence classes among the σ_i for which $i_{G(n_i+m_i),GL_{n_i}(F)\times G(m)}(\sigma_i \otimes \tau)$ is reducible.* □

Of course these results can be phrased in terms of L–functions in the case where the inducing representation is supercuspidal, using Theorems 1.3 or 1.4. Furthermore, Theorems 1.3 and 1.4 have extensions to discrete series using Shahidi's multiplicativity principle and the work of Jacquet, Piatetskii-Shapiro, and Shalika [**16**], as detailed in [**24**]. Thus, in fact, all of Theorem 1.5 can be phrased in terms of L–functions.

2. Preliminary results on R–groups: The maximal case

We now give some preliminary results on the equality of R–groups on the group and dual side. These results for maximal parabolic subgroups will be used to compute $R_{\varphi,\sigma}$ in the general case. We begin with a well known result (see, for example, [**4**]).

LEMMA 2.1. *If $\mathbf{G} = GL_n$, and $\mathbf{M} = GL_m \times GL_k$, then for any irreducible parameter $\varphi\colon W_F \to \hat{M}$, we have $R_\varphi = \{1\} = R(\sigma)$.* □

Recall for $GL_n(F)$ it has long been known (see, for example [**5**]) every L–packet is a singleton set. Thus, for the Siegel parabolic subgroup $R_{\varphi,\sigma} = R_\varphi$.

LEMMA 2.2. *Let $\mathbf{G} = Sp_{2n}, \mathbf{G}' = SO_{2n}$, or $\mathbf{G}'' = SO_{2n+1}$, and $\mathbf{M} \simeq GL_n$ the Siegel Levi subgroup of $\mathbf{G}, \mathbf{G}'$, or $\mathbf{G}''$. Suppose $\varphi\colon W_F' \to GL_n(\mathbb{C})$ is an irreducible representation parametrizing an irreducible unitary supercuspidal representation σ of M. Let $R(\sigma), R'(\sigma)$, and $R''(\sigma)$ be the Knapp–Stein R–groups attached to $i_{G,M}(\sigma)$, $i_{G',M}(\sigma)$, and $i_{G'',M}(\sigma)$, respectively. Let R_φ, R'_φ, and R''_φ be the Arthur R–groups attached to φ as a parameter for G, G', and G'', respectively. Then $R(\sigma) \simeq R_\varphi$, $R'(\sigma) \simeq R'_\varphi$, and $R''(\sigma) = R''_\varphi$.*

PROOF. We first consider $\mathbf{G} = Sp_{2n}$. Then $\hat{G} = SO_{n+1,n}(\mathbb{C})$, and so

$$\hat{M} = \left\{ \begin{pmatrix} g & 0 & 0 \\ 0 & 1 & 0 \\ 0 & 0 & \theta(g) \end{pmatrix} \middle| g \in GL_n(\mathbb{C}) \right\},$$

with $\theta(g) = s_n {}^t g^{-1} s_n^{-1}$. By abuse of notation we think of φ, an M–parameter, as

$$w \mapsto \begin{pmatrix} \varphi(w) & & \\ & 1 & \\ & & \theta(\varphi(w)) \end{pmatrix},$$

for a $GL_n(F)$ parameter, also labeled φ. Consider elements of S_φ in block matrix form. Thus, if $A \in S_\varphi$, write

$$A = \begin{pmatrix} A_{11} & A_{12} & A_{13} \\ A_{21} & A_{22} & A_{23} \\ A_{31} & A_{32} & A_{33} \end{pmatrix}$$

with $A_{11}, A_{13}, A_{31}, A_{33} \in M_n(\mathbb{C})$, $A_{12}, A_{32} \in M_{n\times 1}(\mathbb{C})$, $A_{21}, A_{23} \in M_{1\times n}(\mathbb{C})$, and $A_{22} \in \mathbb{C}$. Since φ is irreducible, we quickly see, for $i = 1$ or 3, $A_{ii} = \lambda_i I_n$ for some $\lambda_i \in \mathbb{C}$. Also, $A_{21} = A_{23} = 0$, $A_{12} = A_{32} = 0$. Further A_{13}, if it is non–zero, is an equivalence between φ and $\theta \circ \varphi$, while A_{31} must be an equivalence in the other direction. Thus, if $\varphi \not\simeq \theta\varphi$, we easily see $S_\varphi = \left\{ \begin{pmatrix} \lambda I & & \\ & 1 & \\ & & \lambda^{-1} I \end{pmatrix} \middle| \lambda \in \mathbb{C}^\times \right\}$ and $R_\varphi = 1 = R(\sigma)$, by Theorem 1.3. Now suppose $\varphi \simeq \theta \circ \varphi$. Then we fix $B \neq 0$ with $B^{-1}\varphi B = \theta \circ \varphi$. Then, by Schur's Lemma, $B\theta(B) = \varepsilon I$, and this implies $Bs_n = \varepsilon s_n {}^t B$. We denote $Bs_n = J$, and see ${}^t J = \varepsilon J$, so $\varepsilon = \pm 1$, and Bs_n is a symmetric or symplectic form fixed by φ. Now, we set $J' = s_n {}^t B^{-1} = B^{-1} J^t B^{-1}$, which is a form of the same type as J. Since $A \in SO_{n+1,n}(\mathbb{C})$, direct computation shows

$$A = \begin{pmatrix} \lambda_{11} I_n & 0 & \lambda_{12} B \\ 0 & w & 0 \\ \lambda_{21} B^{-1} & 0 & \lambda_{22} I_n \end{pmatrix}$$

must satisfy $\lambda_{11}\lambda_{12}(1+\varepsilon)J = 0 = \lambda_{21}\lambda_{22}(1+\varepsilon)J'$ and $(\lambda_{12}\lambda_{21}\varepsilon + \lambda_{11}\lambda_{22})s_n = s_n$. So if $\varepsilon = 1$, then $A \mapsto \begin{pmatrix} \lambda_{11} & \lambda_{12} \\ \lambda_{21} & \lambda_{21} \end{pmatrix}$ is an isomorphism with $O_{1,1}(\mathbb{C})$, while if $\varepsilon = -1$, $A \mapsto \begin{pmatrix} \lambda_{11} & \lambda_{12} \\ \lambda_{21} & \lambda_{22} \end{pmatrix}$ is an isomorphism with $SL_2(\mathbb{C})$. Thus,

$$R_\varphi = \begin{cases} \mathbb{Z}_2 & \text{if } \varphi \text{ is orthogonal;} \\ 1 & \text{otherwise.} \end{cases}$$

Therefore, $R_\varphi \simeq R(\sigma)$ by Theorem 1.3, as claimed. Note, if φ is orthogonal, the non–trivial element of R_φ is the standard (block) sign change, C, and thus under the isomorphism

$$s_\alpha \mapsto s_{\alpha^\vee} \text{ of } W(G, A) \text{ with } W(\hat{G}, \hat{A}),$$

we have $R(\sigma) \tilde{\to} R_\varphi$. The computation for $\mathbf{G} = SO_{2n}$ is similar. Here $\hat{G} = SO_{n,n}(\mathbb{C})$. Then $\hat{M} = \left\{ \begin{pmatrix} g & 0 \\ 0 & \theta(g) \end{pmatrix} \middle| g \in GL_n(\mathbb{C}) \right\}$. We let $A = \begin{pmatrix} A_{11} & A_{12} \\ A_{21} & A_{22} \end{pmatrix} \in S_\varphi$. If n is even essentially the same calculation as in the case $G = Sp_{2n}$ shows

$$S_\varphi \simeq \begin{cases} O_{1,1}(\mathbb{C}) & \text{if } \varphi \text{ is orthogonal;} \\ SL_2(\mathbb{C}) & \text{if } \varphi \text{ is symplectic;} \\ \mathbb{C}^\times & \text{otherwise.} \end{cases}$$

If n is odd and φ is orthogonal, then $A = \begin{pmatrix} \lambda_{11} I_n & \lambda_{12} B \\ \lambda_{21} B^{-1} & \lambda_{22} I_n \end{pmatrix} \in SO_{n,n}(\mathbb{C})$ if and only if

$$\det \begin{pmatrix} \lambda_{11} & \lambda_{12} \\ \lambda_{21} & \lambda_{22} \end{pmatrix} = 1.$$

Thus, we have $S_\varphi \simeq SO_{1,1}(\mathbb{C}) \simeq \mathbb{C}^\times$ in this case. Since φ cannot be symplectic, the only other possibility is $\theta \circ \varphi \not\simeq \varphi$, in which case $S_\varphi \simeq \mathbb{C}^\times$, as before. Thus, when n is odd, $R_\varphi = \{1\}$. Then, by Theorems 1.3 and 1.4, $R'_\varphi \simeq R'(\sigma)$ as claimed, and the remark about this isomorphism being implemented by $\alpha \mapsto \alpha^\vee$ is valid here as well.

Finally, consider $\mathbf{G}'' = SO_{2n+1}$. Then $\hat{G} = Sp_{2n}(\mathbb{C})$. Setting

$$\theta^*(g) = u_n\, {}^t g^{-1} u_n^{-1},$$

for $g \in GL_n(\mathbb{C})$, we have

$$\hat{M} = \left\{ \begin{pmatrix} g & 0 \\ 0 & \theta^*(g) \end{pmatrix} \middle| g \in GL_n(\mathbb{C}) \right\}.$$

Now, again by abuse of notation, we consider φ as the map $w \mapsto \begin{pmatrix} \varphi(w) & 0 \\ 0 & \theta^*(\varphi(w)) \end{pmatrix}$. Then, for $A = \begin{pmatrix} A_{11} & A_{12} \\ A_{21} & A_{22} \end{pmatrix} \in S_\varphi$, we have $A_{ii} = \lambda_i I_n$. Further, A_{12}, if it is non–zero, is an equivalence $\varphi \simeq \theta^* \circ \varphi$. Similarly, A_{21}, when non–zero, is an equivalence in the opposite direction. Thus, as in the other cases, when $\varphi \not\simeq \theta^*\varphi$, we have $S_\varphi \simeq \mathbb{C}^\times$. Now suppose $\varphi \simeq \theta^*\varphi$, and fix B with $B^{-1}\varphi B = \theta^* \circ \varphi$. Then $B\theta^*(B) = \varepsilon I$, and now set $J = Bu_n = \varepsilon u_n^t B$. In this case ${}^tJ = -\varepsilon J$ and thus J is a symplectic form if φ is orthogonal, and is symmetric if φ is symplectic. Now, the requirement that $A = \begin{pmatrix} \lambda_{11} I_n & \lambda_{12} B \\ \lambda_{21} B^{-1} & \lambda_{22} I_n \end{pmatrix} \in Sp_{2n}(\mathbb{C})$ gives

$$\begin{aligned} &\lambda_{12}\lambda_{11}(1+\varepsilon) = 0 = \lambda_{21}\lambda_{22}(1+\varepsilon), \text{ and} \\ &\lambda_{11}\lambda_{22} + \varepsilon\lambda_{12}\lambda_{22} = 1. \end{aligned}$$

So in this case we have

$$S_\varphi = \begin{cases} O_{1,1}(\mathbb{C}) & \text{if } \varphi \text{ is symplectic;} \\ SL_2(\mathbb{C}) & \text{if } \varphi \text{ is orthogonal.} \end{cases}$$

This gives $R''_\varphi \cong R''(\sigma)$, as claimed. □

Now we consider the case of non-Siegel maximal parabolic subgroups. Thus, $\mathbf{P} = \mathbf{MN}$, with $\mathbf{M} \simeq GL_k \times \mathbf{G}(m)$, for some $m + k = n$, and $m > 0$. We let $\varphi = \varphi_1 \oplus \psi$, with both $\varphi_1 \colon W_F \to GL_k(\mathbb{C})$, and $\psi \colon W_F \to \hat{G}(m)$ irreducible. We emphasize that this latter asumption is significant, as, in general, ψ need not be irreducible, even for supercuspidal L–packets. We let $\sigma = \sigma_{\varphi_1}$ be the supercuspidal representation attached to φ_1. We let $\prod_\psi(G(m))$ be the L–packet of $G(m)$ parametrized by ψ. We fix a member $\tau \in \prod_\psi(G(m))$. Then $\pi = \sigma \otimes \tau$ is an element of $\prod_\varphi(M)$. When $\mathbf{G} = SO_{2n+1}$, the work of Jiang-Soudry [**17**] gives the generic member of the L-packet $\prod_\psi(G(m))$. We expect that, based on the work of Cogdell-Kim-Piatetski-Shapiro-Shahidi [**8**] these results will eventually be extended to other classical groups. For the purposes of comparing the Arthur R–group with the Knapp-Stein R–group, we make the following assumption throughout the rest of this exposition.

Assumption A: We assume the local Langlands conjecture for Rankin product L–functions. Thus, we assume the local correspondence $\psi \leftrightarrow \Pi_{G(m)}(\psi)$ is established. Further, we assume if σ is an irreducible supercuspidal representation of $GL_k(F)$ corresponding to φ_1, as above, then $L(s, \varphi_1 \oplus \psi) = L(s, \sigma \times \tau)$, where this latter is the local Rankin product L–function, as defined by Shahidi [**25**], for generic representations τ. For non-generic representations we assume Shahidi's conjecture, every L–packet contains a generic element [**25**]. Thus, $L(s,\sigma\times\tau')=L(s,\sigma\times\tau)=L(s,\varphi_1\oplus\psi)$, for all $\tau,\tau'\in\Pi_\psi(\hat{G}(m))$.

Note, in [**12**] the pole of the intertwining operator $A(s,\sigma\otimes\tau)$ is determined in terms of the theory of twisted endoscopy. The normalization factor for the intertwining operator is given by a product $L(s,\sigma\otimes\tau)L(2s,\sigma,r_2)$, with $r_2=\mathrm{Sym}^2\rho_k$ or $\wedge^2\rho_k$. A reasonable assumption is that the pole of $L(s,\sigma\times\tau)$ is controlled by a certain *regular* term appearing in the residue as described in [**10, 12**]. In [**12**] it is shown, for $\mathbf{G}=SO_{2n+1}$, and with this assumption, $L(s,\sigma\times\tau)$ has a pole at $s=0$ if and only if σ is the local transfer of τ, i.e., σ is the local component of an automorphic transfer [**7, 8**] to the general linear group of a cusp form on the special orthogonal group, and the corresponding local component of the cusp form on the special orthogonal group is τ. This result will be extended to the other classical groups by work in progress of Asgari, Cogdell, and Shahidi.

We first consider the case where $\mathbf{G}=SO_{2n+1}$. Thus, $G=SO_{2n+1}(F)$ and $\hat{G}=Sp_{2n}(\mathbb{C})$. Since

$$\hat{M}=\left\{\begin{pmatrix} g & & \\ & h & \\ & & \theta^*(g)\end{pmatrix}\middle|\begin{matrix} g\in GL_k(\mathbb{C}),\\ h\in Sp_{2m}(\mathbb{C})\end{matrix}\right\},$$

we see that

$$\varphi(w)=\begin{pmatrix}\varphi_1(w) & & \\ & \psi(w) & \\ & & \theta^*\varphi_1(w)\end{pmatrix},$$

with θ^* as in the proof of Lemma 2.2. If $A\in S_\varphi$, and we write $A=\begin{pmatrix}A_{11} & A_{12} & A_{13}\\ A_{21} & A_{22} & A_{23}\\ A_{31} & A_{32} & A_{33}\end{pmatrix}$, with the block decomposition given by $\hat{M}$. Then we have

$$\begin{aligned}&A_{11}\varphi_1=\varphi_1A_{11},\ A_{12}\psi=\varphi_1A_{12},\ A_{13}\theta^*\varphi_1=\varphi_1A_{13},\\ &A_{21}\varphi_1=\psi A_{21},\ A_{22}\psi=\psi A_{22},\ A_{23}\theta^*\varphi_1=\psi A_{23},\\ &A_{31}\varphi_1=\theta^*\varphi_1A_{31},\ A_{32}\psi=\theta^*\varphi_1A_{32},\text{ and }A_{33}\theta^*\varphi_1=\theta^*\varphi_1A_{33}.\end{aligned}$$

Thus, $A_{ii}=\lambda_iI$. Further as φ_1 and ψ are both irreducible, each A_{ij}, for $i\neq j$ is either zero or an equivalence. Note, since $\psi\colon W_F\to Sp_{2n}(\mathbb{C})$, $\varphi_1\simeq\psi$ only if $\varphi_1\simeq\theta^*\varphi_1$. Thus, if $\varphi_1\not\simeq\psi$ then $\theta^*\varphi_1\not\simeq\psi$ and in this case $A_{21}=A_{23}=0$, $A_{12}=A_{32}=0$, and with $A_{22}=\lambda_2I\in Sp_{2m}(\mathbb{C})$ we have $\lambda_{22}=\pm1$.

So, if $\varphi_1\not\simeq\theta^*\varphi_1$, then

$$S_\varphi=\left\{\begin{pmatrix}\lambda_{11}I_k & & \\ & I_{2m} & \\ & & \lambda_{11}^{-1}I_k\end{pmatrix}\middle|\lambda_{11}\in\mathbb{C}^\times\right\}\simeq\mathbb{C}^\times.$$

Now, if $\varphi_1\simeq\theta^*\varphi_1$, but $\varphi_1\not\simeq\psi$, then $A\in S_\varphi$ implies A is of the form

$$\begin{pmatrix}\lambda_{11}I_k & 0 & \lambda_{12}B\\ 0 & \omega(A)I_{2m} & 0\\ \lambda_{21}B^{-1} & 0 & \lambda_{22}I_k\end{pmatrix}.$$

Note $\omega(A) = \pm 1$, and in fact $\omega(A) = \det \begin{pmatrix} \lambda_{11} I_k & \lambda_{12} B \\ \lambda_{21} B & \lambda_{22} I_k \end{pmatrix}$. Now, as in the Siegel case, $B\theta^*(B) = \varepsilon I$ and $J = Bu_n$ is a form fixed by φ_1. Since ${}^tJ = -\varepsilon J$, we get

$$S_\varphi \simeq \begin{cases} SL_2(\mathbb{C}) & \text{if } \varphi_1 \text{ is orthogonal;} \\ O_{1,1}(\mathbb{C}) & \text{if } \varphi_1 \text{ is symplectic.} \end{cases}$$

Now suppose $\varphi_1 \simeq \psi$. Since φ is only given up to $\hat{G}$–conjugacy, we may take $\varphi_1 = \psi \simeq \theta^* \varphi_1$. We still take $J = Bu_n$, a form fixed by φ_1. Since φ is symplectic, $\varepsilon = 1$. Let $\langle\ ,\ \rangle$ be the form on $\mathbb{C}^k = \mathbb{C}^{2m} = V_1$ given by $\langle v,\ v' \rangle = {}^tvJv'$. So $\langle \varphi_1(w)v,\ \varphi_1(w)v' \rangle = \langle v, v' \rangle$ for all $w \in W_F$, v, $v' \in \mathbb{C}^\times$. Then $\mathbb{C}^{3k} \simeq V_1 \otimes \mathbb{C}^3$. There is then a unique orthogonal form $\langle\ ,\ \rangle'$ on $\mathbb{C}^3$ so that $\langle\ ,\ \rangle \otimes \langle\ ,\ \rangle'$ is the symplectic form on $\mathbb{C}^{3k}$ giving rise to $\hat{G}$. So, by the dual pair construction, $S_\varphi \simeq SO_{2,1}(\mathbb{C})$. Note, in each case, $W_{\varphi,\pi} \simeq W_\varphi$, for even if $\prod_\psi(G(m))$ is not a singleton set, the Weyl group $W(\hat{G}, A_{\hat{M}})$ acts trivially on the $G(m)$ component τ. With Assumption A, we have $R_\varphi \simeq R(\pi)$, as predicted.

Suppose now that $\mathbf{G} = Sp_{2n}$. Then $G = Sp_{2n}(F)$ and $\hat{G} = SO_{n+1,n}(\mathbb{C})$. The computations are essentially the same as above. Thus, if $\varphi_1 \not\simeq \psi$ we have

$$S_\varphi \simeq \begin{cases} \mathbb{C}^\times & \text{if } \varphi_1 \not\simeq \theta \circ \varphi_1; \\ SL_2(\mathbb{C}) & \text{if } \varphi_1 \text{ is symplectic;} \\ O_{1,1}(\mathbb{C}) & \text{if } \varphi_1 \text{ is orthogonal.} \end{cases}$$

If $\varphi_1 \simeq \psi$, then φ_1 is orthogonal, and $k = 2m+1$. We set $\langle\ ,\ \rangle$ to be the form on $V_1 = \mathbb{C}^k$ given by $\langle v, v' \rangle = {}^tvJv'$. So, $\langle \varphi_1(w)v,\ \varphi_1(w)v' \rangle = \langle v, v' \rangle$. Here $J = Bs_n$, with $B\varphi_1 = \theta\varphi_1 B$. Choose $\langle\ ,\ \rangle'$ on $\mathbb{C}^3$ so that the form on $\mathbb{C}^{3k} = V_1 \otimes \mathbb{C}^3$, given by $\langle , \rangle \otimes \langle\ ,\ \rangle'$. Then we see, again, $S_\varphi \simeq SO_{2,1}(\mathbb{C})$ in this case. Thus far we have proved the following.

Lemma 2.3. *If* $\mathbf{G} = Sp_{2n}$ *or* SO_{2n+1} *and* $\mathbf{M} \simeq GL_k \times G(m)$*, then, under Assumption A,* $R_{\varphi,\pi} \simeq R(\pi)$ *for any* $\pi \in \Pi_\varphi(M)$. □

Now suppose $\mathbf{G} = SO_{2n}$, so $G = SO_{2n}(F)$ and $\hat{G} = SO_{n,n}(\mathbb{C})$. Here we consider 2 cases. Let $c_0\colon SO_{m,m}(\mathbb{C}) \to SO_{m,m}(\mathbb{C})$ be the outer automorphism given by conjugation by an element of $O_{m,m}(\mathbb{C})$. The two cases we consider are $c_0\psi \not\simeq \psi$ and $c_0\psi \simeq \psi$.

In the first case, the computations are essentially as before. If $\varphi_1 \not\simeq \theta\varphi_1$ then $S_\varphi \simeq \mathbb{C}^\times$ as before. If $\varphi_1 \simeq \theta\varphi_1$, but $\varphi_1 \not\simeq \psi$, then

$$S_\varphi \simeq \begin{cases} SL_2(\mathbb{C}) & \text{if } \varphi_1 \text{ is symplectic;} \\ O_{1,1}(\mathbb{C}) & \text{if } \varphi_1 \text{ is orthogonal.} \end{cases}$$

Now suppose $\varphi_1 \simeq \psi$. Then we may take $\varphi_1 = \psi \simeq \theta\varphi_1^*$. Let $\langle\ ,\ \rangle$ be the orthogonal form on $V_1 = \mathbb{C}^k = \mathbb{C}^{2m}$ fixed by φ_1, and consider $V = \mathbb{C}^{6m} \simeq V_1 \otimes \mathbb{C}^3$. Take $\langle\ ,\ \rangle'$ to be a symmetric form on $\mathbb{C}^3$ so that $\langle , \rangle \otimes \langle\ ,\ \rangle'$ is the ambient symmetric form. Then dual pairs gives $S_\varphi \simeq \operatorname{stab}_{GL_3(\mathbb{C})}(\langle\ ,\ \rangle') \cap \hat{G} \simeq SO_{2,1}(\mathbb{C})$ as before. Now suppose $c_0\psi \simeq \psi$. The computations for $\varphi_1 \not\simeq \theta\varphi_1$ are as before. So assume $\varphi_1 \simeq \theta\varphi_1$. First assume that $\varphi_1 \not\simeq \psi$. Then, in block form $A \in S_\varphi$,

$$A = \begin{pmatrix} \lambda_{11} I & 0 & B\lambda_{12} \\ 0 & \delta & 0 \\ B^{-1}\lambda & 0 & \lambda_{22} I \end{pmatrix}, \text{ with } \delta\psi \simeq \psi.$$

If k is even, then $\delta = \pm 1$, so

$$S_\varphi \simeq \begin{cases} O_{1,1}(\mathbb{C}) & \text{if } \varphi_1 \text{ is orthogonal,} \\ SL_2(\mathbb{C}) & \text{if } \varphi_1 \text{ is symplectic,} \end{cases}$$

and further $W_\varphi = W_{\varphi,\pi}$. If k is odd, then $\delta = \pm c_0$. Note, in this case, $S_\varphi \cong O_{1,1}(\mathbb{C})$ and $W_\varphi \cong \mathbb{Z}_2$. However,

$$W_{\varphi,\pi} \cong \begin{cases} \mathbb{Z}_2 & \text{if } c\tau \simeq \tau, \\ 1 & \text{if } c\tau \not\simeq \tau, \end{cases}$$

where c is the automorphism dual to c_0. Now, if $W_{\varphi,\pi} = \mathbb{Z}_2$, then as $c \notin SO_{m,m}(\mathbb{C})$, $W^\circ_{\varphi,\pi} = \{1\}$, and $R_{\varphi,\pi} \simeq R(\pi)$, as predicted. Finally, suppose $c_0\psi \simeq \psi$ and $\varphi_1 = \psi$. Then $\varphi_1 \simeq \theta\varphi_1$, and $J = Bs_k$ is our form fixed by φ_1. Then the computation is, in fact, the same as in the case $c_0\psi \not\simeq \psi$, and thus, $S_\varphi \simeq SO_{2,1}(\mathbb{C})$. Now comparison with [**9**] shows we have the result we wish.

PROPOSITION 2.4. *If* $\mathbf{G} = SO_{2n}$, *and* $\mathbf{M} = GL_k \times SO_{2m}$, *then, under Assumption A, for any* $\pi = \sigma \otimes \tau$, *supercuspidal, with the property that the parameter* ψ *for* τ *is irreducible, we have* $R_{\varphi,\pi} \simeq R(\pi)$. □

3. The case of Levi factors contained in the Siegel

We now consider the case where $\mathbf{M} \hookrightarrow GL_n$. Thus, let $\mathbf{M} \simeq GL_{n_1} \times GL_{n_2} \times \cdots \times GL_{n_r}$, with $n_1 + n_2 + \cdots + n_r = n$. In this case each L–packet $\Pi_\varphi(M)$ is a singleton, so $R_{\varphi,\sigma} = R_\varphi$. Then $\hat{M} \simeq GL_{n_1}(\mathbb{C}) \times GL_{n_2}(\mathbb{C}) \times \cdots \times GL_{n_r}(\mathbb{C})$. We consider $\varphi = \varphi_1 \oplus \varphi_2 \oplus \ldots \oplus \varphi_r$, with each φ_i irreducible as in Section 2. We now reduce to the simplest cases. First assume, if $\mathbf{G} = SO(2n)$, then each n_i is even. Recall, in all of these cases, $W = W(\hat{G}, A_{\hat{M}}) = \langle (ij) \rangle \ltimes \langle C_i \rangle$, where $(ij) \in W$ if and only if $n_i = n_j$, and $C_i\colon g_i \mapsto \Theta(g_i)$, where $\Theta(g_i) = \theta(g_i)$ if $\mathbf{G} = Sp_{2n}$ or SO_{2n}, and $\Theta(g_i) = \theta^*(g_i)$ if $\mathbf{G} = SO_{2n+1}$ (where θ and θ^* are as in Section 2). Let $w = sc \in W$. We suppose $w = (1\ 2\ \cdots\ j)$ and C acts only on the indices $1, 2, \ldots, j$. We assume $c = C_{k_1}C_{k_2}\ldots C_{k_\ell}$, and set $\Omega = \{k_1, k_2, \ldots, k_\ell\}$. We write

$$w\varphi = \bigoplus_{i=1}^{r} (w\varphi)_i.$$

If $i > j$, then $(w\varphi)_i = \varphi_i$. On the other hand if $1 \le i \le j$, then

$$(w\varphi)_i = \begin{cases} \varphi_{i-1} & \text{if } i \notin \Omega, \\ \Theta \circ \varphi_{i-1} & \text{if } i \in \Omega, \end{cases}$$

where we must take $i-1$ modulo j. Then

$$w\varphi \simeq \Theta^{\gamma_1}\varphi_2 \oplus \Theta^{\gamma_2}\varphi_3 \oplus \cdots \oplus \Theta^{\gamma_{j-1}}\varphi_j \oplus \Theta^{\gamma_j}\varphi_1 \oplus \varphi_{j+1} \oplus \cdots \oplus \varphi_r,$$

with $\gamma_i = 1$ if $i \in \Omega$, and $\gamma_i = 0$, otherwise. Thus, $w\varphi = \varphi$ if and only if

$$\varphi_1 \simeq \Theta^{\gamma_1}\varphi_2 \simeq \Theta^{\gamma_1+\gamma_2}\varphi_3 \simeq \cdots \simeq \Theta^{\gamma_1+\ldots+\gamma_{j-1}}\varphi_j \simeq \Theta^{\gamma_1+\ldots+\gamma_j}\varphi_1.$$

We may choose a conjugate of φ so that

$$\varphi = m_1\varphi_1 \oplus m_1'\Theta\varphi_1 \oplus m_2\varphi_2 \oplus m_2'\Theta\varphi_2 \oplus \\ \cdots \oplus m_\ell\varphi_\ell \oplus m_\ell'\Theta\varphi_\ell,$$

with $\varphi_i \not\simeq \varphi_j$, and $\varphi_i \not\simeq \Theta\varphi_j$, for $i \neq j$. Let $\mathbf{M}_i = GL_{n_i}^{m_i+m_i'}$, and $\mathbf{G}_i = \mathbf{G}(m_i+m_i')$. Consider $\Phi_i = m_i\varphi_i \oplus m_i'\varphi_i'$, as a parameter with image in $\hat{G}_i$. Then

$$S_\varphi = \prod_{i=1}^{\ell} S_{\Phi_i},\ S_\varphi^\circ = \prod_{i=1}^{\ell} S_{\Phi_i}^\circ,\ W_\varphi = \prod_{i=1}^{\ell} W_{\Phi_i},$$

and $W_\varphi^\circ = \prod_{i=1}^{\ell} W_{\Phi_i}^\circ$. This is the same analysis as in §4 and §5 of [**9**]. Thus, we reduce to the case $n_1 = n_2 = \cdots = n_r$, and

$$\varphi_i = \Theta^{\gamma_i}\varphi_1 \text{ for } i = 1, 2, 3, \ldots, r.$$

Further, by conjugation we may assume $\varphi_i = \varphi_1$ for $i = 1, 2, \ldots, r$. The following is due to Gross–Prasad when $\mathbf{G} = SO_{2n}$ or SO_{2n+1}, and we adopt their proof for $\mathbf{G} = Sp_{2n}$. Note, this generalizes Lemma 2.2.

PROPOSITION 3.1. *Let* $\mathbf{G} = SO_{2n+1}, Sp_{2n}$, *or* SO_{2n}. *Let* $n = kr$, *and let* $\mathbf{M} \simeq GL_k^r$. *If* $\mathbf{G} = SO_{2n}$, *we assume* k *is even. Let* σ_1 *be an irreducible unitary supercuspidal representation of* $GL_k(F)$ *and take* $\sigma = \otimes^r\sigma_1$ *as a representation of* M. *Let* $\varphi = \oplus^r\varphi_1$ *be the parameter associated to* σ.

(a): *If* $\varphi_1 \not\simeq \Theta\varphi_1$, *then* $S_\varphi \simeq GL_r(\mathbb{C})$.

(b): *If* $\mathbf{G} = SO_{2n+1}$, *and* $\varphi_1 \simeq \theta^*\varphi_1$, *then*

$$S_\varphi \simeq \begin{cases} Sp_{2r}(\mathbb{C}) & \textit{if } \varphi_1 \textit{ is orthogonal;} \\ O_{r,r}(\mathbb{C}) & \textit{if } \varphi_1 \textit{ is symplectic.} \end{cases}$$

(c): *If* $\mathbf{G} \simeq SO_{2n}$, *and* $\varphi_1 \simeq \theta\varphi_1$, *then*

$$S_\varphi \simeq \begin{cases} Sp_{2r}(\mathbb{C}) & \textit{if } \varphi_1 \textit{ is symplectic;} \\ O_{r,r}(\mathbb{C}) & \textit{if } \varphi_1 \textit{ is orthogonal.} \end{cases}$$

(d): *If* $\mathbf{G} = Sp_{2n}$, *and* $\varphi_1 \simeq \theta\varphi_1$, *then*

$$S_\varphi \simeq \begin{cases} Sp_{2r}(\mathbb{C}) & \textit{if } \varphi \textit{ is symplectic;} \\ O_{r,r}(\mathbb{C}) & \textit{if } \varphi \textit{ is orthogonal.} \end{cases}$$

PROOF. Due to Sections 6 and 7 of [**13**], we only need to address the case of $\mathbf{G} = Sp_{2n}$. Let $k = n_1$. First consider case (a). If $g_0 = (\lambda_{ij}) \in GL_r(\mathbb{C})$, we set

$$h(g_0) = \begin{pmatrix} \lambda_{11}I_k & \lambda_{12}I_k & \cdots & \lambda_{1r}I_k \\ \vdots & & \ddots & \vdots \\ \lambda_{r_1}I_k & \lambda_{r_2}I_k & \cdots & \lambda_{rr}I_k \end{pmatrix} \in GL_n(\mathbb{C}). \text{ Then}$$

$$S_\varphi = \left\{ \begin{pmatrix} h(g_0) & 0 & 0 \\ 0 & 1 & 0 \\ 0 & 0 & \theta(h(g_0)) \end{pmatrix} \middle| g_0 \in GL_r(\mathbb{C}) \right\} \simeq GL_r(\mathbb{C}).$$

For (d) we adopt the argument of [**13**]. Let V_1 be the space of φ_1. Then, identifying V_1 with $V_1^\vee$ through the form fixed by φ_1, we have $V = V_1 \otimes \mathbb{C}^{2r} \otimes \mathbb{C}^\times$ is the ambient space for $\hat{G}$. We let $\langle\ ,\ \rangle_1$, be the form fixed by φ_1, and $\langle\ ,\ \rangle_0$ the standard orthogonal form given by multiplication on $\mathbb{C}^\times$. Then we can choose a (unique up to scalars) form $\langle\ ,\ \rangle'$ on $\mathbb{C}^{2r}$ so that $\langle\ ,\ \rangle = \langle\ ,\ \rangle_1 \otimes \langle\ ,\ \rangle' \otimes \langle,\ \rangle_0$, is the form given by s_{2n+1}. Thus $S_\varphi = O(\mathbb{C}^{2r}, \langle\ ,\ \rangle') \cap \hat{G}$

$$\simeq \begin{cases} Sp_{2r}(\mathbb{C}) & \text{if } \varphi_1 \text{ is symplectic;} \\ O_{2r}(\mathbb{C}) & \text{if } \varphi_1 \text{ is orthogonal.} \end{cases}$$

□

THEOREM 3.2. *Let* $\mathbf{G} = Sp_{2n}$, SO_{2n}, *or* SO_{2n+1}, *and let* $\mathbf{M} = GL_{n_1} \times \ldots \times GL_{n_r}$. *If* $\mathbf{G} = SO_{2n}$, *we assume each* n_i *is even. Let* $\varphi\colon W_F' \to \hat{M}$ *be an irreducible representation, and suppose* $\varphi = \varphi_1 \oplus \varphi_2 \oplus \ldots \oplus \varphi_r$. *Then* $R_\varphi \simeq R(\sigma)$, *where* φ *parametrizes* σ.

PROOF. As per the discussion preceding Proposition 3.1, it is enough to show this in the case $n_1 = n_2 = \cdots = n_r = k$, for some k, and $\sigma \simeq \otimes^r \sigma_1$. Let $\varphi = \oplus^r \varphi_1$. We take T_φ to be the maximal torus of block diagonal elements in $\hat{M}$. Note, in each case, $T_\varphi \subset S_\varphi$. Then $W_\varphi = W(S_\varphi, T_\varphi) \simeq S_r \ltimes \mathbb{Z}_2^r$, in the standard way. If $\varphi_1 \not\simeq \Theta\varphi_1$, then $W_\varphi = W_\varphi^\circ$ by Prop. 3.1, and hence $R_\varphi = \{1\} = R(\sigma)$, by [**9**]. If φ_1 is orthogonal, then

$$W_\varphi = \begin{cases} W_\varphi^\circ & \text{if } \mathbf{G} = SO_{2n+1}; \\ W_\varphi^\circ \ltimes \mathbb{Z}_2, & \text{if } \mathbf{G} = Sp_{2n}.SO_{2n}, \end{cases}$$

Therefore,

$$R_\varphi \simeq \begin{cases} \mathbb{Z}_2 & \text{if } \mathbf{G} = Sp_{2n}, SO_{2n}, \\ 1 & \text{if } \mathbf{G} = SO_{2n+1}, \end{cases}$$

and, again by [**9**] $R_\varphi \simeq R(\sigma)$. If φ_1 is symplectic, the argument is similar, with the roles of the groups reversed. □

Now suppose $\mathbf{G} = SO_{2n}$, $\mathbf{M} = GL_{n_1} \times \cdots \times GL_{n_r}$, and at least one n_i is odd. We may assume $n_1, n_2, \ldots, n_t$ are even, and $n_{t+1}, n_{t+2}, \ldots, n_r$ are odd. Consider $m_1 = n_1 + n_2 + \cdots + n_t$, $m_2 = n_{t+1} + \cdots + n_r$. Set $\mathbf{G}_i = SO_{2m_i}$. We let

$$\begin{aligned} \mathbf{M}_1 &= GL_{n_1} \times \cdots \times GL_{n_t} \subseteq \mathbf{G}_1, \text{ and} \\ \mathbf{M}_2 &= GL_{n_{t+1}} \times \cdots \times GL_{n_r} \subseteq \mathbf{G}_2. \end{aligned}$$

Let $\varphi = \varphi_1 \oplus \cdots \oplus \varphi_r$, and define $\psi_1\colon W_F \to \hat{M}_1$ by $\psi_1 = \varphi_1 \oplus \cdots \oplus \varphi_t$, and $\psi_2 = \varphi_{t+1} \oplus \cdots \oplus \varphi_r$. Let π_i be the corresponding irreducible unitary supercuspidal representation of M_i. Let $W_i = W(\hat{G}_i, A_{\hat{M}_i})$.

LEMMA 3.3. *With notation as above;*

(a): $\hat{M} \simeq \hat{M}_1 \times \hat{M}_2$;
(b): $W(\hat{G}, A_{\hat{M}}) = W_1 \times W_2$;
(c): $S_\varphi = S_{\psi_1} \times S_{\psi_2}$;
(d): $S_\varphi^\circ = S_{\psi_1}^\circ \times S_{\psi_2}^\circ$.

PROOF. Parts (a) and (b) are obvious, and are also parts of lemma 5.1 of [**9**]. Part (d) will follow from (c). For (c), we consider an element $A \in S_\varphi$ as block matrices relative to $m_1 + m_2 = n$, i.e.,

$$A = \begin{pmatrix} A_{11} & A_{12} & A_{13} & A_{14} \\ A_{21} & A_{22} & A_{23} & A_{24} \\ A_{31} & A_{32} & A_{33} & A_{34} \\ A_{41} & A_{42} & A_{43} & A_{44} \end{pmatrix},$$

and each $A_{ij} = \left(\lambda^{ij}_{k\ell}\right)$ is scalar in the appropriate blocks. We note that $A \in S_\varphi$ implies

$$\begin{aligned} A_{12}\psi_2 &= \psi_1 A_{12}, & A_{13}\theta\psi_2 &= \psi_1 A_{13}; \\ A_{21}\psi_1 &= \psi_2 A_{21}, & A_{24}\theta\psi_1 &= \psi_2 A_{24}; \\ A_{31}\psi_1 &= \theta\psi_2 A_{32}, & A_{34}\theta\psi_1 &= \theta\psi_2\ A_{34}; \\ A_{42}\ \psi_2 &= \theta\psi_1 A_{42}, & A_{43}\theta\psi_2 &= \theta\psi_1 A_{43}. \end{aligned}$$

Since each of the blocks for ψ_1 is even, and those for ψ_2 are odd, the above equations show $A_{ij} = 0$, for

$$(i,j) = (1,2),\ (1,3),\ (2,1),\ (2,4),\ (3,1),\ (3,4),\ (4,2),\ (4.3).$$

Then

$$S_\varphi = \left\{ \begin{pmatrix} A_{11} & 0 & 0 & A_{12} \\ 0 & A_{22} & A_{23} & 0 \\ 0 & A_{32} & A_{33} & 0 \\ A_{41} & 0 & 0 & A_{44} \end{pmatrix} \right\} \simeq S_{\psi_1} \times S_{\psi_2},$$

in the obvious way. □

COROLLARY 3.4. *With the notation of Lemma 3.2,* $R_\varphi \simeq R_{\psi_1} \times R_{\psi_2}$. □

Since we have computed R_{ψ_1}, we are reduced to the case $\mathbf{M} = \mathbf{M}_2$, i.e., $n_1, n_2, \ldots, n_r$ are all odd. Letting S_r act on the blocks of M, we have $(ij) \in W(\hat{G}, A_{\hat{M}})$ if and only if $n_i = n_j$. Let $W_0 = \langle (ij) | n_i = n_j \rangle$. We let $\mathcal{C} = \langle C_i C_{i+1} | i = 1, 2, \ldots, r-1 \rangle$ where C_i is the ith block sign change. Then we have

$$W = W(\hat{G}, A_{\hat{M}}) = W_0 \ltimes \mathcal{C} \simeq W_0 \ltimes \mathbb{Z}_2^{r-1}.$$

Consider $\varphi = \varphi_1 \oplus \varphi_2 \oplus \ldots \oplus \varphi_r$. Note, since each n_i is odd, $\theta\varphi_i \simeq \varphi_i$ if and only if φ_i is orthogonal.

We now, briefly, change notation so

$$\begin{aligned} \mathbf{M} &= GL_{n_1}^{m_1} \times GL_{n_2}^{m_2} \times \ldots \times GL_{n_r}^{m_r}, \text{ with} \\ \varphi &= \bigoplus_i \oplus^{m_i} \varphi_i, \end{aligned}$$

with n_i odd, and $\varphi_1, \varphi_2, \ldots, \varphi_t$ not orthogonal, while $\varphi_{t+1}, \varphi_{t+2}, \ldots \varphi_r$ are orthogonal. Further assume $\varphi_i \not\simeq \varphi_j$ for $i \neq j$.

LEMMA 3.5. *We let* $k_1 = m_1 n_1 + \cdots + m_t n_t$, $k_2 = m_{t+1} n_{t+1} + \cdots + m_r n_r$; $\mathbf{G}_i = SO_{2k_i}$, *and* $\mathbf{M}_i$ *the obvious Levi subgroup of* $\mathbf{G}_i$, *as in Lemma 3.3. Let* $\varphi = \Psi_1 \oplus \Psi_2$, *with*

$$\begin{aligned} \Psi_1 &= \bigoplus_{i=1}^{t} \oplus^{m_i} \varphi_i, \text{ and} \\ \Psi_2 &= \bigoplus_{i=t+1}^{r} \oplus^{m_i} \varphi_i. \end{aligned}$$

Then

$$S_\varphi = S_{\Psi_1} \times S_{\Psi_2}.$$

PROOF. This is precisely the same argument as lemma 3.3(c), noting that the appropriate blocks are now zero owing to φ_i being orthogonal if and only if $i \geq t+1$. □

An easy block matrix computation shows

$$S_{\Psi_1} \simeq GL_{m_1}(\mathbb{C}) \times \ GL_{m_2}(\mathbb{C}) \times \cdots \times GL_{m_t}(\mathbb{C}).$$

Thus, we may assume each φ_i is orthogonal.

PROPOSITION 3.6. *Let* $\mathbf{G} = SO_{2n}$, $\mathbf{M} = GL_{n_1}^{m_1} \times \cdots \times GL_{n_r}^{m_r}$, *with each* n_i *odd. Suppose* $\varphi = \varphi_1^{m_1} \oplus \cdots \oplus \varphi_r^{m_r}$ *with each* φ_i *orthogonal, and* $\varphi_i \not\simeq \varphi_j$, $i \neq j$. *Then* $S_\varphi \simeq S(O_{m_1,m_1}(\mathbb{C}) \times \cdots \times O_{m_r,m_r}(\mathbb{C}))$.

PROOF. For each i fix an equivalence B_i of φ_i with $\theta\varphi_i$, and take the form $\langle\ ,\ \rangle_i$ on the space V_i of φ_i given by $B_i s_{n_i}$. Then there is a unique, up to scalars, choice of forms $\langle\ ,\ \rangle_i'$ on $\mathbb{C}^{m_i}$ for each i so that on $\mathbb{C}^{2m_i n_i}$ the form $\langle\ ,\ \rangle$ on the ambient space of $\hat{G} = SO_{n,n}(\mathbb{C})$ is $\bigotimes_{i=1}^{r} \langle\ ,\rangle_i \otimes \langle\ ,\ \rangle_i'$. Then each $\langle\ ,\ \rangle_i'$ is orthogonal, and by duality

$$\begin{aligned} S_\varphi &= (O_{m_1,m_1}(\mathbb{C}) \times O_{m_2,m_2}(\mathbb{C}) \times \cdots \times O_{m_r,m_r}(\mathbb{C})) \cap \hat{G} \\ &= S(O_{m_1,m_1}(\mathbb{C}) \times O_{2m_2}(\mathbb{C}) \times \cdots \times O_{2m_r}(\mathbb{C})), \text{ as claimed.} \end{aligned}$$

□

THEOREM 3.7. *If* $\mathbf{G} = SO_{2n+1}, Sp_{2n}$, *or* SO_{2n}, *and* $\mathbf{M} = GL_{n_1} \times GL_{n_2} \times \cdots \times GL_{n_r}$, *then for any irreducible* $\varphi : W_F \to \hat{M}$, *we have* , $R_\varphi \simeq R(\sigma)$, *with* σ *parametrized by* φ.

PROOF. Combining Proposition 3.1(c), Theorem 3.2, Corollary 3.4, and Lemma 3.5, it is enough to prove the statement when each n_i is odd and φ_i is orthogonal. By Proposition 3.6, we have

$$W_\varphi = W(S_\varphi, T_\varphi) = \prod_{i=1}^{r} S_{m_i} \ltimes \mathcal{C},$$

where $\mathcal{C} \simeq \mathbb{Z}_2^{m_1+\cdots+m_r-1}$ is as in Corollary 3.4. Further, since

$$\begin{aligned} S_\varphi^\circ &= [S(O_{m_1,m_1}(\mathbb{C}) \times \cdots \times O_{m_r,m_r}(\mathbb{C}))]^\circ \\ &= SO_{m_1,m_1}(\mathbb{C}) \times SO_{m_2,m_2}(\mathbb{C}) \times \cdots \times SO_{m_r,m_r}(\mathbb{C}), \text{ we have} \end{aligned}$$

$$W(S_\varphi^\circ, T_\varphi) \simeq \prod_{i=1}^{r} \left(S_{m_i} \ltimes \mathbb{Z}_2^{m_i-1}\right).$$

Thus, $R_\varphi \simeq W_\varphi / W_\varphi^\circ \simeq \mathbb{Z}_2^{r-1}$, and by Theorem 6.8 of [**9**], $R_\varphi \simeq R(\sigma)$. □

4. The case of non-Siegel parabolic subgroups

We now turn to the case where

$$\mathbf{M} \cong GL_{n_1} \times GL_{n_2} \times \cdots \times GL_{n_r} \times \mathbf{G}(m), \text{ with}$$

$n_1 + n_2 + \cdots + n_r + m = n$, and $m > 0$.

We fix $\varphi = \varphi_1 \oplus \varphi_2 \oplus \cdots \oplus \varphi_r \oplus \psi$, with φ_i an (irreducible) parameter for an irreducible unitary supercuspidal representation σ_i of GL_{n_i}, and ψ a parameter for a supercuspidal L–packet $\{\tau\}$ of $G(m)$. We make the assumption that ψ is irreducible. We will first prove some results in general, but eventually we will need to consider the case of $\mathbf{G} = SO_{2n}$ separately. Recall here we are working under Assumption A.

By changing $\mathbf{M}$ and φ by conjugates, we may assume $\mathbf{M} \simeq GL_{n_1}^{m_1} \times GL_{n_2}^{m_2} \times \cdots \times GL_{n_t}^{m_t} \times \mathbf{G}(m)$, and

$$\sigma \simeq \otimes^{m_1}\sigma_1 \bigotimes \otimes^{m_2}\sigma_2 \bigotimes \cdots \bigotimes \otimes^{m_t}\sigma_t \otimes \tau,$$

with $\sigma_i \not\simeq \sigma_j$ and $\sigma_i \not\simeq \tilde{\sigma}_j$ for $i \neq j$. Thus, we may take

$$\varphi = \oplus^{m_1}\varphi_1 \bigoplus \oplus^{m_2}\varphi_2 \bigoplus \cdots \bigoplus \oplus^{m_t}\varphi_t \oplus \psi,$$

with $\varphi_i \not\simeq \varphi_j$, and $\varphi_i \not\simeq \Theta \circ \varphi_j$, for $i \neq j$. For $i = 1, 2, \ldots, t$, set $\mathbf{G}_i = \mathbf{G}(n_i m_i + m)$, and $\mathbf{M}_i = GL_{n_i}^{m_i} \times \mathbf{G}(m)$, and consider $\mathbf{M}_i$ as a Levi subgroup of $\mathbf{G}_i$. Denote by Φ_i the parameter $m_i\varphi_i \oplus \psi$, and let $\pi_i = \otimes^{m_i}\sigma_i \oplus \tau$. Then π_i is in the L–packet $\prod_{\Phi_i}(M_i)$. We denote $S_i = S_{\Phi_i}$, the $\hat{G}_i$ centralizer of $\Phi_i(W_F)$. Since $\mathrm{Hom}(\varphi_i, \psi) \neq 0$ for at most one i, we may assume (by replacing the pair $(\mathbf{M}, \sigma)$ with a conjugate) $\mathrm{Hom}(\varphi_i, \psi) = 0$ for $i = 1, 2, \ldots, t-1$.

We fix

$$C_0 = \begin{pmatrix} I_{m-1} & 0 & 0 & 0 \\ 0 & 0 & 1 & 0 \\ 0 & 1 & 0 & 0 \\ 0 & 0 & 0 & I_{m-1} \end{pmatrix} \in O_{m,m}(\mathbb{C}) \setminus SO_{m,m}(\mathbb{C}).$$

LEMMA 4.1. *We use the above notation.*
(a) Suppose $\mathbf{G} = SO_{2n}$ *and* $C_0 \cdot \psi \not\simeq \psi$. *Then*

(i) $S_\varphi \simeq S_1/\mathbb{Z}_2 \times S_2/\mathbb{Z}_2 \times \cdots \times S_{t-1}/\mathbb{Z}_2 \times S_t$;
(ii) $S_\varphi^\circ \simeq S_1^\circ \times S_2^\circ \times \cdots \times S_{t-1}^\circ \times S_t^\circ$.

(b) Suppose $\mathbf{G} = SO_{2n}$ *and* $C_0 \cdot \psi \simeq \psi$. *Then*

(i) $S_\varphi \simeq (S_1/\mathbb{Z}_2 \times \mathbb{Z}_2) \times (S_2/\mathbb{Z}_2 \times \mathbb{Z}_2) \times \cdots \times (S_{t-1}/\mathbb{Z}_2 \times \mathbb{Z}_2) \times S_t$;
(ii) $S_\varphi^\circ \simeq S_1^\circ \times S_2^\circ \times \cdots \times S_{t-1}^\circ \times S_t^\circ$.

(c) Suppose $\mathbf{G} = SO_{2n+1}$ *or* Sp_{2n}, *then*

(i)$S_\varphi \simeq S_1 \times S_2 \times \cdots \times S_t$;
(ii)$S_\varphi^\circ \simeq S_1^\circ \times S_2^\circ \times \cdots \times S_{t-1}^\circ \times S_t^\circ$.

PROOF.

(a) Let $A \in S_\varphi$. Then, relative to the decomposition $\hat{M} \simeq GL_{n_1}(\mathbb{C})^{m_1} \times \cdots \times GL_{n_t}(\mathbb{C})^{m_t} \times \hat{G}(m)$, we have

$$A = \begin{pmatrix} A_{11} & A_{12} & \dots & A_{1t} & A_{1\,t+1} & A_{1\,t+2} & \dots & A_{1\,2t+1} \\ A_{21} & A_{22} & \dots & A_{2t} & A_{2\,t+1} & A_{2\,t+2} & \dots & A_{2\,2t+1} \\ \vdots & & \ddots & \dots & & & & \vdots \\ \vdots & & \dots & A_{tt} & & & \vdots & \\ \vdots & & & & A_{t+1\,t+1} & \dots & & \vdots \\ & & & & & \ddots & \vdots & \\ A_{2t+1\,1} & \dots & & & \dots & & \dots & A_{2t+1\,2t+1} \end{pmatrix}.$$

For $1 \le i \le t$, we set $i' = 2t+2-i$. By our assumption that $\varphi_i \not\simeq \varphi_j$, and $\varphi_i \not\simeq \Theta\varphi_j$, for $i \ne j$, we see $A_{ij} = 0$ for $1 \le i < j \le t, t+2 \le i < j \le 2t+1$. Similarly, $A_{ij} = 0$ if $1 \le i \le t, t+2 \le j \le 2t+1$, unless $j = i'$. or $1 \le j \le t, t+2 \le i \le 2t+1$, unless $j = i'$. Further, by our assumption on $\mathrm{Hom}(\varphi_i, \psi)$ for $1 \le i \le t$, we have $A_{i\,t+1} = 0$ and $A_{t+1\,i} = 0$ for $i = 1, 2, \dots, t-1, i = t+3, \dots, 2t+1$. Finally, by Schur's Lemma, and our assumption that $C_0 \cdot \psi \not\simeq \psi$, we see $A_{t+1\,t+1}$ is scalar. Thus,

$$A = \begin{pmatrix} A_{11} & 0 & 0 & \dots & & & \dots & 0 & A_{11'} \\ 0 & A_{22} & 0 & \dots & & & \dots & A_{22'} & 0 \\ \vdots & & \ddots & & \vdots & \vdots & & & \\ \vdots & & & A_{tt} & A_{t\,t+1} & A_{tt'} & & \vdots & \\ & & & A_{t+1\,t} & c_{t+1} & A_{t+1\,t'} & & & \\ 0 & \dots & & A_{t't} & A_{t'\,t+1} & A_{t't'} & 0\dots & & \vdots \\ \vdots & & & & & & \ddots & & \\ A_{1'1} & 0 & \dots & & & & \dots & 0 & A_{1'1'} \end{pmatrix},$$

with c_{t+1} a scalar.

We also see that if $1 \le i \le t-1$, then

$$S_i = \left\{ \begin{pmatrix} A_{ii} & 0 & A_{ii'} \\ 0 & c_i & 0 \\ A_{i'i} & 0 & A_{i'i'} \end{pmatrix} \right\},$$

with c_i a scalar and the obvious block decomposition. But, as this element must be in $SO_{m_in_i,m_in_i}(\mathbb{C})$, we have $c_i = \pm I_{2m}$. We let the elements of S_i be denoted by $[A_i, c_i]$. We also have

$$S_t = \left\{ \begin{pmatrix} A_{tt} & A_{t\,t+1} & A_{tt'} \\ A_{t+1\,t} & c_{t+1} & A_{t+1\,t'} \\ A_{t't} & A_{t'\,t+1} & A_{t't'} \end{pmatrix} \right\},$$

with c_t a scalar, and we denote these elements by A_t. Consider the map $\eta : S_1 \times S_2 \times \cdots \times S_t \to S_\varphi$ given by

$$\eta\left(\left(\prod_{i=1}^{t-1} [A_i, c_i] \right), A_t \right) =$$

$$
(4.1)\qquad \begin{pmatrix}
A_{11} & 0 & 0 & \dots & & & \dots & 0 & A_{11'}\\
0 & A_{22} & 0 & \dots & & & \dots & A_{22'} & 0\\
\vdots & & \ddots & & \vdots & \vdots & & & \\
\vdots & & & A_{t-1\,t-1} & 0 & A_{t-1(t-1)'} & & \vdots & \\
 & & & 0 & A_t & 0 & & & \\
0 & \dots & & A_{(t-1)'\,t-1} & 0 & A_{(t-1)'(t-1)'} & \dots & & \vdots\\
\vdots & & & & & & \ddots & & \\
A_{1'1} & 0 & \dots & & & & \dots & 0 & A_{1'1'}
\end{pmatrix}.
$$

Then η is a surjective homomorphism with kernel

$$\left\{\left(\prod_{i=1}^{t-1}[I, c_i], I\right)\right\},$$

which gives (i). Part (ii) then follows immediately. This completes (a).

(b) The proof above need only be modified by replacing each c_i with $c_iC_0^{\varepsilon_i}$, with $\varepsilon_i \in \{0,1\}$. With this modification, the proof of (i) follows the same steps as (a) and again (ii) is immediate. Finally (c) follows from simple block matrix computations as in Section 3, taking into account each element of W_φ acts trivially on ψ. □

Continuing with the notation above, and that of the proof, we see that we may take the maximal torus

$$(4.2)\qquad T_\varphi = \{\mathrm{diag}\,\{A_{11}, A_{22}\dots, A_{tt}, I_{m'}, A'_{tt}, \dots, A'_{11}\}\} \subset S_\varphi,$$

with blocks of the form

$$\begin{aligned}
A_{ii} &= \mathrm{diag}\,\{a_{i1}I_{n_i}, a_{i2}I_{n_i}, \dots, a_{im_i}I_{n_i}\},\\
A'_{ii} &= \mathrm{diag}\,\{a_{im_i}^{-1}I_{n_i}, a_{im_i-1}^{-1}I_{n_i}, \dots, a_{i1}^{-1}I_{n_i}\} = \Theta(A_{ii}),
\end{aligned}$$

with $m' = 2m$ or $2m+1$, as appropriate, and Θ as in Section 3. Also note we may take the torus $T_i = \{\mathrm{diag}\,\{A_{ii}, I_{m'}, A'_{ii}\}\} \subset S_i$, and so $T_\varphi \simeq T_1 \times \cdots \times T_t$. If $\mathbf{G} = Sp_{2n}$, or SO_{2n+1}, or if $\mathbf{G} = SO_{2m}$ and each n_i is even, then each element of $W(S_\varphi, T_\varphi)$ acts trivially on the middle $\mathbf{G}(m)$ block, and hence trivially on ψ. Thus we have the following result.

Lemma 4.2. *With the notation of the previous lemma, suppose* $\mathbf{G} = Sp_{2n}$*, or* SO_{2n+1}*, or* $\mathbf{G} = SO_{2n}$ *and each* n_i *is even. Then*

(i) $W_\varphi \cong W_{\Phi_1} \times W_{\Phi_2} \times \cdots \times W_{\Phi_t}$;

(ii) $R_\varphi \simeq R_{\Phi_1} \times R_{\Phi_2} \times \cdots \times R_{\Phi_t}$;

(iii) $R_{\varphi,\sigma} \simeq R_{\Phi_1,\pi_1} \times R_{\Phi_2,\pi_2} \times \cdots \times R_{\Phi_t,\pi_t}$.

□

We now restrict ourselves to the cases of Lemma 4.2, namely, $\mathbf{G} = Sp_{2n}, SO_{2n+1}$, or $\mathbf{G} = SO_{2n}$ and each n_i is even. Then, by Lemma 4.2, we may reduce to the case where $\mathbf{M} \simeq GL_k^r \times G(m)$, and $\sigma = \bigotimes^r \sigma_0 \otimes \tau$. Assume φ_0 and ψ are parameters for σ_0 and τ, respectively, and we continue to assume ψ is irreducible.

LEMMA 4.3. *Suppose* $\varphi_0 \simeq \Theta\varphi_0$, *but* $\varphi_0 \not\simeq \psi$.

(i): *If* $\mathbf{G} = Sp_{2n}$, *or* SO_{2n}, *(with* k *even in the latter case), then*

$$S_\varphi = \begin{cases} O_{r,r}(\mathbb{C}) & \text{if } \varphi_0 \text{ is orthogonal;} \\ Sp_{2r}(\mathbb{C}) & \text{if } \varphi_0 \text{ is symplectic.} \end{cases}$$

(ii): *If* $\mathbf{G} = SO_{2n+1}$, *then*

$$S_\varphi = \begin{cases} O_{r,r}(\mathbb{C}) & \text{if } \varphi_0 \text{ is symplectic;} \\ Sp_{2r}(\mathbb{C}) & \text{if } \varphi_0 \text{ is orthogonal.} \end{cases}$$

PROOF. (i) Let V_0 be the space of φ_0 (so $V_0 \simeq \mathbb{C}^k$). Let $\langle\ ,\ \rangle_0$ be the form on V_0 given by $J_0 = Bs_k$, with $B\varphi_0 \simeq \theta\varphi_0$, and s_k as in Sections 2 and 3. Let $\langle\ ,\ \rangle_\psi$ be the form on $\mathbb{C}^{m'}$ fixed by ψ, with $m' = 2m+1$, or $2m$, as appropriate. Then there is a unique, up to scalars, non–degenerate form $\langle\ ,\ \rangle'$ on $\mathbb{C}^{2r}$ so that $\langle\ ,\ \rangle_0 \otimes \langle\ ,\ \rangle' \otimes \langle\ ,\ \rangle_\psi$ is the form defined by s_{2n} if $\mathbf{G} = SO_{2n}$, and and s_{2n+1}, if $\mathbf{G} = Sp_{2n}$. As this is the form which defines $\hat{G}$ we have $S_\varphi \simeq \operatorname{Stab}_{\langle\ ,\ \rangle'} \cap \hat{G}$. In each case we see that $S_\varphi \cong O_{r,r}(\mathbb{C})$ if φ_0 is orthogonal, and $S_\varphi \cong Sp_{2r}(\mathbb{C})$ if φ_0 is symplectic. This completes (i). The proof of (ii) is similar, with $J_0 = Bu_k$, and noting J_0 is orthogonal if φ_0 is symplectic and vice-versa. □

LEMMA 4.4. *We continue with the assumption* $\mathbf{M} \simeq GL_k^r \times \mathbf{G}(m)$, *with the additional assumption that* k *is even if* $\mathbf{G} = SO_{2n}$. *We also assume* $\sigma \simeq \bigotimes^r \sigma_0 \otimes \tau$, *and* φ_0 *and* ψ *are the parameters of* σ_0 *and* τ, *respectively. (Of course we still assume* ψ *is irreducible). If* $\varphi_0 \simeq \psi$ *then* $S_\varphi \simeq \begin{cases} SO_{r+1,r}(\mathbb{C}) & \text{if } k \text{ is odd,} \\ O_{r+1,r}(\mathbb{C}) & \text{if } k \text{ is even.} \end{cases}$

PROOF. As in the proof of the last lemma, let V_0 be the space of φ_0. Since $\varphi_0 \simeq \psi$ we have $V \simeq V_0 \otimes \mathbb{C}^{2r+1}$ is the ambient space for $\hat{G}$. If $\langle\ ,\ \rangle_0$ is the form fixed by φ_0, then there is a unique, up to scalars, non–degenerate form $\langle\ ,\ \rangle'$ on $\mathbb{C}^{2r+1}$ so that $\langle\ ,\ \rangle = \langle\ ,\ \rangle_0 \otimes \langle\ ,\ \rangle'$ is the form defining $\hat{G}$. Clearly $\langle\ ,\ \rangle'$ must be orthogonal, so by this dual pair construction $S_\varphi = O_{r+1,r} \cap \hat{G}$. The determinant condition shows this intersection is $SO_{r+1,r}$, if k is odd. However, when k is even, this copy of $O_{r+1,r}(\mathbb{C})$ is a subgroup of $\mathbf{G}$. Thus, S_φ is as claimed. □

We note, in the last lemma, when k is even, the difference in $N_{S_\varphi}(T_\varphi)$ and $N_{S_\varphi^\circ}(T_\varphi)$ can be realized as $-I_{2r+1}$, which lies in $Z_{S_\varphi}(T_\varphi)$. Thus, regardless of the parity of k, we have $W_\varphi \simeq W_\varphi^\circ$, so $R_\varphi = \{1\}$.

THEOREM 4.5. *Let* $\mathbf{M} \simeq GL_k^r \times \mathbf{G}(m)$, *with the additional assumption that* k *is even if* $\mathbf{G} = SO_{2m}$. *Assume* σ *is as in Lemmas 4.3 and 4.4.*

(i): $R_\varphi \simeq \{1\}$ *if* $\begin{cases} \varphi_0 \simeq \psi, \text{ or} \\ \varphi_0 \not\simeq \Theta\varphi_0. \end{cases}$

(ii): $R_\varphi \simeq \mathbb{Z}_2$ *if* $\begin{cases} \mathbf{G} = Sp_{2n},\ \varphi_0 \text{ is orthogonal}, \varphi_0 \not\simeq \psi; \\ \mathbf{G} = SO_{2n+1},\ \varphi_0 \text{ is symplectic}, \varphi_0 \not\simeq \psi; \text{ or} \\ \mathbf{G} = SO_{2n},\ k \text{ is even},\ \varphi_0 \text{ is orthogonal, and } \varphi_0 \not\simeq \psi. \end{cases}$

□

Since $L(s, \varphi_0 \oplus \psi)$ has a pole at $s = 0$ if and only if $\varphi_0 \simeq \psi$, this can be phrased as follows. If $\varphi_0 \not\simeq \Theta\varphi_0$, then $R_\varphi = \{1\}$. On the other hand, if $\varphi_0 \simeq \Theta\varphi_0$, then

$R_\varphi \simeq \mathbb{Z}_2$, unless the product $L(s, \varphi_0 \oplus \psi)L(2s, r_2 \circ \varphi_0)$ has a pole at $s = 0$. Here $r_2 = \mathrm{Sym}^2$ if $\mathbf{G} = SO_{2n+1}$, and is $\wedge^2$ otherwise. This is as expected.

THEOREM 4.6. *Suppose $\mathbf{G} = Sp_{2n}$, SO_{2n+1}, or SO_{2n} and $\mathbf{M} \simeq GL_{n_1} \times \cdots \times GL_{n_r} \times \mathbf{G}(m)$, with each n_i even in the case $\mathbf{G} = SO_{2n}$. Assume $\varphi\colon W_F \to \hat{M}$ is an irreducible parameter for an irreducible unitary supercuspidal representation σ of M. Under Assumption A, $R_{\varphi,\sigma} = R_\varphi$ and $R_\varphi \simeq R(\sigma)$.*

PROOF. That $R_\varphi = R_{\varphi,\sigma}$ was already observed in these cases, owing to W_φ acting trivially on the summand ψ. The R-groups $R(\sigma)$ were computed in [**9**], and there the computation in these cases was reduced to the special cases of Theorem 4.5. Of course, if $\sigma \not\simeq \tilde{\sigma}$, then $R(\sigma) = \{1\}$. Otherwise, if $\sigma \simeq \tilde{\sigma}$, then $R(\sigma) \simeq \mathbb{Z}_2$, unless $L(s, \sigma_0 \times \tau)L(2s, \sigma_0, r)$ has a pole at $s = 0$ [**25, 10, 12**]. Here r_2 is Sym^2 if $\mathbf{G} = SO_{2n+1}$ and is $\wedge^2$, otherwise. Using Assumption A, Theorem 1.4, Theorem 4.5, and the remarks following Theorem 4.5, we have $R_\varphi \simeq R(\sigma)$. □

Now let $\mathbf{G} = SO_{2n}$, so $\hat{G} = SO_{n,n}(\mathbb{C})$. We suppose $\mathbf{M} \simeq GL_{n_1} \times GL_{n_2} \times \cdots \times GL_{n_r} \times \mathbf{G}(m)$, with $m \geq 1$. We assume $n_1, n_2, \ldots, n_t$ are odd, and $n_{t+1}, \ldots, n_r$ are even. Let $\varphi : W_F \to \hat{M}$, and we assume $\varphi = \varphi_1 \oplus \varphi_2 \oplus \cdots \oplus \varphi_r \oplus \psi$. We let $\mathbf{G}_1 = \mathbf{G}(n_1 + n_2 + \cdots + n_t + m)$, and $\mathbf{G}_2 = \mathbf{G}(n_{t+1} + \cdots + n_r + m)$. Let $\mathbf{M}_1 = GL_{n_1} \times GL_{n_2} \times \cdots \times GL_{n_t} \times \mathbf{G}(m) \simeq \mathbf{M}_1' \times \mathbf{G}(m)$, and $\mathbf{M}_2 = GL_{n_{t+1}} \times \cdots \times GL_{n_r} \times \mathbf{G}(m) = \mathbf{M}_2' \times \mathbf{G}(m)$. We let $\Psi_1 : W_F \to \hat{M}_1$ be given by

$$\Psi_1 = \oplus_{i=1}^t \varphi_i \oplus \psi = \Psi_1' \oplus \psi,$$

and $\Psi_2 : W_F \to \hat{M}_2$, be given by

$$\Psi_2 = \oplus_{i=t+1}^r \varphi_i \oplus \psi = \Psi_2' \oplus \psi.$$

Denote the Weyl groups by $W_i = W(\hat{G}_i, A_{\hat{M}_i})$. We also denote $m_1 = n_1 + n_2 + \cdots + n_t$, and $m_2 = n_{t+1} + \cdots + n_r$.

LEMMA 4.7. *With he notation above we have the following identites:*

(a): $\hat{M} \simeq \hat{M}_1' \times \hat{M}_2' \times SO_{m,m}(\mathbb{C})$;
(b): $W(\hat{G}, A_{\hat{M}}) \simeq W_1 \times W_2$;
(c): $S_\varphi \simeq (S_{\Psi_1}/\mathbb{Z}_2^{\varepsilon_1}) \times S_{\Psi_2}$, *with* $\varepsilon_1 = 1$, *or* 2;
(d): $S_\varphi^\circ \simeq S_{\Psi_1}^\circ \times S_{\Psi_2}^\circ$;
(e): $W_\varphi \simeq W_{\Psi_1} \times W_{\Psi_2}$;
(f): $W_\varphi^\circ \simeq W_{\Psi_1}^\circ \times W_{\Psi_2}^\circ$;
(g): $R_\varphi \simeq R_{\Psi_1} \times R_{\Psi_2}$.

In part (c), $\varepsilon_1 = 1$ if $C_0 \cdot \psi \not\simeq \psi$, and $\varepsilon_1 = 2$ if $C_0\psi \simeq \psi$.

PROOF. As in Lemma 3.3, (a) and (b) are straightforward computations, and the proofs are as in Lemma 5.10 of [**9**] We note (c) is a consequence of Lemma 4.1a(i) or b(i) and (d) is one of Lemma 4.1 a(ii) or b(ii). The last three parts now follow immediately. □

By Theorem 4.6 we are now reduced to considering the case where $\mathbf{G} = SO_{2n}$ and each n_i is odd. Let's suppose, as before $\mathbf{M} \simeq GL_{n_1}^{m_1} \times GL_{n_2}^{m_2} \times \ldots GL_{n_t}^{m_t} \times SO_{2m}$,

$$\sigma \simeq \oplus^{m_1}\sigma_1 \bigoplus \oplus^{m_2}\sigma_2 \bigoplus \cdots \oplus^{m_t} \sigma_t \oplus \tau,$$

and $\sigma_i \not\simeq \sigma_j, \tilde{\sigma}_j$, for $i \neq j$. Thus, take $\varphi = \oplus^{m_1}\varphi_1 \bigoplus \cdots \oplus^{m_t} \varphi_t \oplus \psi$. Assume $\sigma_1, \sigma_2, \ldots, \sigma_s$ are not orthogonal, and $\sigma_{s+1}, \ldots, \sigma_t$ are orthogonal. Then, using the notation of (4.1), we see $A_{ii'} = A_{i'i} = 0$, for $1 \leq i \leq s$. Further, for $1 \leq i \leq s$,

$$A_{ii} = \begin{pmatrix} a_{11}^{(i)} I_{n_i} & a_{12}^{(i)} I_{n_i} & \cdots & a_{1m_i}^{(i)} I_{n_i} \\ a_{21}^{(i)} I_{n_i} & \ddots & \cdots & a_{2m_i}^{(i)} I_{n_i} \\ \vdots & & \ddots & \vdots \\ a_{m_i1}^{(i)} I_{n_i} & \cdots & & \ldots a_{m_im_i}^{(i)} I_{n_i} \end{pmatrix},$$

with each $a_{jk}^{(i)} \in \mathbb{C}$. Also note (again for $1 \leq i \leq s$) $A_{i'i'} = \Theta(A_{ii})$. As each φ_i and ψ are irreducible, and each n_i is odd, we have $\operatorname{Hom}(\varphi_i, \psi) = 0$, for each i. Thus, again using the notation of (4.1), we have $A_t = \begin{pmatrix} A_{tt} & 0 & A_{tt'} \\ 0 & \pm C_0^{\varepsilon_0} & 0 \\ A_{t't} & 0 & A_{t't'} \end{pmatrix}$, and $\varepsilon_0 \in \{0,1\}$. We have the possibility of $\varepsilon_0 = 1$ only if $C_0 \cdot \psi \simeq \psi$. We now describe the centralizer S_φ.

Proposition 4.8. *Let $\mathbf{G} = SO_{2n}$, $\mathbf{M} \simeq GL_{n_1}^{m_1} \times \cdots \times GL_{n_t}^{m_t} \times \mathbf{G}(m)$, with each n_i odd,*

$$\sigma \simeq \otimes^{m_1}\sigma_1 \bigotimes \otimes^{m_2}\sigma_2 \bigotimes \cdots \bigotimes \otimes^{m_t}\sigma_t \otimes \tau,$$

and $\sigma_i \not\simeq \sigma_j$, $\sigma_i \not\simeq \tilde{\sigma}_j$, for $i \neq j$. Take

$$\varphi = \oplus^{m_1}\varphi_1 \bigoplus \oplus^{m_2}\varphi_2 \bigoplus \cdots \bigoplus \oplus^{m_t}\varphi_t \oplus \psi.$$

Assume $\sigma_1, \sigma_2, \ldots, \sigma_s$ are not orthogonal and $\sigma_{s+1}, \ldots, \sigma_t$ are orthogonal.
(a) If $C_0 \cdot \psi \not\simeq \psi$, then

$$S_\varphi \simeq GL_{m_1}(\mathbb{C}) \times GL_{m_2}(\mathbb{C}) \times \cdots \times GL_{m_s}(\mathbb{C}) \times \\ S\left(O_{m_{s+1},m_{s+1}}(\mathbb{C}) \times \cdots \times O_{m_t,m_t}(\mathbb{C})\right) \times \mathbb{Z}_2;$$

(b) If $C_0 \cdot \psi \simeq \psi$, then

$$S_\varphi \simeq GL_{m_1}(\mathbb{C}) \times GL_{m_2}(\mathbb{C}) \times \cdots \times GL_{m_s}(\mathbb{C}) \times O_{m_{s+1},m_{s+1}}(\mathbb{C}) \times \cdots \times O_{m_t,m_t}(\mathbb{C}) \times \mathbb{Z}_2.$$

Proof. The proofs follow the reasoning used in earlier results and, in particular the argument of Proposition 3.6. For $1 \leq i \leq s$, the map

$$\begin{pmatrix} a_{11}^{(i)} I_{n_i} & a_{12}^{(i)} I_{n_i} & \cdots & a_{1m_i}^{(i)} I_{n_i} \\ a_{21}^{(i)} I_{n_i} & \cdots & \cdots & a_{2m_i}^{(i)} I_{n_i} \\ \vdots & & & \vdots \\ a_{m_i1}^{(i)} I_{n_i} & \cdots & & \ldots a_{m_im_i}^{(i)} I_{n_i} \end{pmatrix} \mapsto \begin{pmatrix} a_{11}^{(i)} & a_{12}^{(i)} & \cdots & a_{1m_i}^{(i)} \\ a_{21}^{(i)} & \cdots & \cdots & a_{2m_i}^{(i)} \\ \vdots & & & \vdots \\ a_{m_i1}^{(i)} & \cdots & & \ldots a_{m_im_i}^{(i)} \end{pmatrix},$$

is an isomorphism of

$$\left\{ \begin{pmatrix} A_{ii} & 0 \\ 0 & \Theta(A_i i) \end{pmatrix} \right\}$$

with $GL_{m_i}(\mathbb{C})$. Also for these indices, let $\langle\ ,\ \rangle_i$ be the form on $\mathbb{C}^{2m_in_i}$ given by $s_{2m_in_i}$. Let $\langle\ ,\rangle_0$ be the form on $\mathbb{C}^{2m}$ given by s_{2m}.

For each $s+1 \leq i \leq t$ fix an equivalence B_i of φ_i with $\theta\varphi_i$. Fix the form $\langle\ ,\ \rangle_i$ on the space V_i of φ_i given by $B_is_{n_i}$. Then there is a unique, up to scalars, choice of forms, $\langle\ ,\ \rangle_i'$ on $\mathbb{C}^{m_i}$ for each $i \geq s+1$ so the form $\langle\ ,\ \rangle$ on $\mathbb{C}^{2n}$ is $\bigotimes_{i=1}^{r}\langle\ ,\rangle_i \otimes$

$\langle\ ,\ \rangle_i' \bigotimes \langle\ ,\ \rangle_0$. Then each $\langle\ ,\ \rangle_i'$ is orthogonal. We now, as in previous cases, use the dual pair construction to identify S_φ. For (a), since the middle $2m \times 2m$ block must be $\pm I_{2m}$, we see, as in the case of Proposition 3.6, that the determinant condition forces S_φ to have the form claimed in (a). In case (b), we see the determinant forces the choice of ε_0, and gives S_φ the form as claimed □

We now compute the R-group $R_{\varphi,\sigma}$. We take the maximal torus $T_\varphi \subset S_\varphi$ given by (4.2).

THEOREM 4.9. *Let* $\mathbf{G} = SO_{2n}$ *and* $\mathbf{M} \simeq GL_{n_1}^{m_1} \times \cdots \times GL_{n_t}^{m_t} \times SO_{2m}$, *with each* n_i *odd. Assume* $\sigma \simeq \otimes^{m_1}\sigma_1 \bigotimes \otimes^{m_2} \bigotimes \cdots \bigotimes \otimes^{m_t}\sigma_t \otimes \tau$ *is an irreducible supercuspidal representation of* M. *We assume* $\sigma_i \not\simeq \sigma_j$, *and* $\sigma_i \not\simeq \tilde{\sigma}_j$, *for* $i \neq j$. *We let* φ_i *be a parameter for* σ_i, *and* ψ *a (conjectural) parameter for* τ. *We assume* ψ *is irreducible. Finally, assume* $\sigma_1, \sigma_2, \ldots, \sigma_s$ *are not orthogonal, while* $\sigma_{s+1}, \ldots, \sigma_t$ *are orthogonal. Let* $d = t - s$.
(a) If $C_0 \cdot \psi \not\simeq \psi$, *then* $R_{\varphi,\sigma} \simeq \mathbb{Z}_2^{d-1}$;
(b) If $C_0 \cdot \psi \simeq \psi$, *then* $R_{\varphi,\sigma} \simeq \mathbb{Z}_2^{d}$.

PROOF. We first describe $W_\varphi = W(S_\varphi, T_\varphi)$, irrespective of whether $C_0\psi \simeq \psi$ or not. Let the permutation group S_t act on $\{1, 2, \ldots, t\}$, and let $W_0 = \langle (ij) | n_i = n_j \rangle$. For each i, the permutation group S_{m_i} acts on the m_i blocks giving rise to σ_i. Further, since $m > 0$, we have the subgroup of block sign changes in W_φ is

$$\mathcal{C} = \left\langle C_{ij} C_0 \middle| 1 \leq i \leq t, 1 \leq j \leq m_i \right\rangle,$$

where C_{ij} is the block sign change on the j-th block of the blocks corresponding to σ_i. Now, we have

$$W_\varphi = \left(W_0 \cdot \prod_{i=1}^{t} S_{m_i} \right) \ltimes \mathcal{C}.$$

Let

$$\mathcal{C}_0 = \left\langle C_{ij} C_{kl} \middle| s+1 \leq i, k \leq t, 1 \leq j \leq m_i, 1 \leq l \leq m_k \right\rangle,$$

and

$$\mathcal{C}_1 = \left\langle C_{ij} C_0 \middle| s+1 \leq i \leq t, 1 \leq j \leq m_i \right\rangle,$$

We note, with our assumptions

$$W_{\varphi,\sigma} = \prod_{i=1}^{t} S_{m_i} \ltimes \mathcal{C}_0$$

if $C_0 \cdot \psi \not\simeq \psi$, and

$$W_{\varphi,\sigma} = \prod_{i=1}^{t} S_{m_i} \ltimes \mathcal{C}_1$$

if $C_0 \cdot \psi \simeq \psi$. Note $\mathcal{C}_0 \simeq \mathbb{Z}_2^{m_{s+1}+m_2+\cdots+m_t-1}$, and $\mathcal{C}_1 \simeq \mathbb{Z}_2^{m_{s+1}+\cdots+m_t}$. Now

$$S_\varphi^\circ \simeq GL_{m_1}(\mathbb{C}) \times \cdots \times GL_{m_s}(\mathbb{C}) \times SO_{m_{s+1},m_{s+1}}(\mathbb{C}) \times \cdots \times SO_{m_t,m_t}(\mathbb{C}).$$

So in either case

$$W_{\varphi,\sigma}^\circ \simeq \prod_{i=1}^{s} S_{m_i} \times \prod_{i=s+1}^{t} \left(S_{m_i} \ltimes \mathbb{Z}_2^{m_i-1} \right).$$

Thus, in case (a) $R_{\varphi,\sigma} \simeq \mathbb{Z}_2^{d-1}$, while in case (b), $R_{\varphi,\sigma} \simeq \mathbb{Z}_2^{d}$. □

THEOREM 4.10. *Let* $\mathbf{G} = SO_{2n+1}$, Sp_{2n}, *or* SO_{2n} *and* $\mathbf{M} \simeq GL_{n_1} \times GL_{n_2} \times \cdots \times GL_{n_r} \times SO_{2m}$, *Assume* $\sigma \simeq \sigma_1 \otimes \sigma_2 \otimes \cdots \otimes \sigma_r \otimes \tau$ *is an irreducible supercuspidal representation of* M. *We let* φ_i *be a parameter for* σ_i, *and* ψ *a (conjectural) parameter for* τ. *We assume* ψ *is irreducible. Then, under Assumption* A, $R(\sigma) \simeq R_{\varphi,\sigma}$.

PROOF. From Theorem 4.6, we can reduce to the case covered by Theorem 4.9. Then we only need compare the results of Sections 5 and 6 of [**9**] (See Theorem 1.5) with those of Theorem 4.9. This comparison gives $R(\sigma) \simeq R_{\varphi,\sigma}$ in each case. □

REMARK 4.11. *We believe the process of extending these results to the case where* ψ *is reducible should follow from some combinatorics and the multiplicity one results of Cogdell, Kim, Piatetski-Shapiro, and Shahidi* [**8**]. *Further extension to the case where* σ_i *and* τ *are discrete series should be tractable using the classifications of Zelevinsky* [**31**] *and Mœglin and Tadić* [**23**]. *We leave these considerations to future work. We also emphasize, again, in the case of* $\mathbf{G} = SO_{2n+1}$ *our results are subsumed by* [**3**], *and it seems possible an approach along the lines of that work may be generalized.*

5. Duality implements the isomorphism

We continue with the previous notation. So, $\varphi : W_F \to \hat{M} \hookrightarrow \hat{G}$, is a parameter for an L–packet $\Pi_\varphi(M)$, with $\sigma \in \Pi_\varphi(M)$ an irreducible unitary supercuspidal representation. We have shown that $R_{\varphi,\sigma} \simeq R(\sigma)$, for all cases we have considered. Now, we claim that this isomorphism can be realized by the dual map $\alpha \mapsto \alpha^\vee$ sending roots of G, to coroots of G, i.e., to roots of $\hat{G}$.

LEMMA 5.1. *Let* $\mathbf{G} = G(n) = Sp_{2n}, SO_{2n}$, *or* SO_{2n+1}, *and* $\mathbf{M} = GL_{n_1} \times \ldots \times GL_{n_r} \times \mathbf{G}(m)$. *Then, for any* M*–parameter* φ: $W_F \to \hat{M}$, *we have* $W_\varphi \simeq W(\sigma_\varphi)$, *and this map arises from the duality* $\alpha \to \alpha^\vee$.

PROOF. By Lemma 2.2 of [**3**], $W_\varphi \subset W(\hat{G}, A_{\hat{M}})$. We abuse notation and identify $W(G, A_M)$ and $W(\hat{G}, A_{\hat{M}})$. Comparing the computations preceding Proposition 3.1, those preceding Lemmas 4.2, 4.7, Theorem 4.9, and those of [**9**], the lemma is clear. □

In order to complete the goal of this section, it is enough to show that $W_\varphi^\circ \cong W' = W(\Delta')$. Recall Δ' is the positive subroot system of $\Phi(P, A)$ generated by those $\alpha > 0$ with $\mu_\alpha(\sigma) = 0$. Thus, it is enough to show $W_\varphi^\circ = W(\Delta')^\vee$. Since $|W_\varphi / W_\varphi^\circ| = |W(\sigma)/W'|$, it is enough to show the simple reflection $s_{\alpha^\vee}$ associated with $\alpha^\vee$ belongs to W_φ° for every $\alpha \in \Delta'$. We write

$$\Phi(P, A) = \{E_i \pm E_j |\ 1 \leq i < j \leq r_{-1}\} \cup \{\beta_i\}_{i=1}^r ,$$

where

$$\beta_i = \begin{cases} 2E_i & G = Sp_{2n}, \\ E_i & G = SO_{2n+1}, \text{ and} \\ E_i^\circ + E_i', & G = SO_{2n}, \end{cases}$$

where these are the obvious elements, as described in [**9**]. Then

$$\Phi(\hat{P}, \hat{A}) = \{E_i^\vee \pm E_j^\vee\} \cup \{\beta_i^\vee\}$$

and $\beta_i^\vee$ is $E_i^\vee, 2E_i^\vee$, or $E_i^{\vee_0} + E_i^{\vee_1}$, as above.

Now if $\alpha^\vee = E_i^\vee - E_j^\vee$, then $s_{\alpha^\vee} = (ij) \in W_\varphi$ if and only if $\varphi_i \simeq \varphi_j$ which is equivalent to $\sigma_i \simeq \sigma_j$, which happens if and only if $s_\alpha \in W'$, by [**4, 9**]. Note, by Lemma 2.1, $s_\alpha^\vee \in W_\varphi^\circ$, as claimed. If $\alpha^\vee = E_i^\vee + E_j^\vee$, then the result holds by conjugation.

PROPOSITION 5.2. *If $\alpha^\vee = \beta_i^\vee$, then $s_\alpha^\vee \in W_\varphi^\circ$ if and only if $s_\alpha \in W'$.*

PROOF. We first assume $m = 0$. Suppose $\mathbf{G} = Sp_{2n}$. Then $s_{\alpha^\vee} = s_{\beta_i^\vee} = C_i \in W_\varphi$ if and only if $\varphi_i \simeq \theta\varphi_i$, and is equivalent to $\sigma_i \simeq \tilde{\sigma}_i$. Now, by the proof of Lemma 2.2, and Hennniart's theorem [**15**] (Theorem 1.4), $s_{\alpha^\vee} \in W_\varphi^\circ$ if and only φ is symplectic, i.e., if and only if $L(s, \wedge^2\varphi_i) = L(s, \sigma_i, \wedge^2)$ has a pole at $s = 0$, and this is equivalent to $\alpha \in \Delta'$ by Shahidi [**24**] (Theorem 1.3). If $\mathbf{G} = SO_{2n+1}$, and $\alpha^\vee = E_i^\vee$, then, $s_{\alpha^\vee}$ is again C_i, and $C_i \in W_\varphi$ if and only if $\varphi_i \simeq \theta^*\varphi_i$. Again, by Lemma 2.2, $C_i \in W_\varphi^\circ$ if and only if φ is orthogonal and this is equivalent to $L(s, \mathrm{Sym}^2 \varphi_i) = L(s, \sigma_i, \mathrm{Sym}^2)$ having a pole (using Theorem 1.4 again) and equivalent to $\mu_\alpha(\sigma) = 0$, i.e. to $\alpha \in \Delta'$ (Theorem 1.3). Finally, suppose $\mathbf{G} = SO_{2n}$. Then $\beta_i^\vee = E_i^{\vee_0} + E_i^{\vee'} = e_{k_{i-1}} + e_{k_i}$ for some k_i (see [**9**]). Then

$$s_{\alpha^\vee} = \begin{cases} C_i & \text{if } n_i \text{ is even,} \\ 1 & \text{if } n_i \text{ is odd.} \end{cases}$$

Thus, we may assume n_i is even. In this case, by Lemma 2.2, $C_i \in W_\varphi$ if and only if $\varphi_i \simeq \theta\varphi_i$, which is equivalent to $\sigma_i \simeq \tilde{\sigma}_i$. We also see from the proof of Lemma 2.2 that $C_i \in W_\varphi^\circ$ if and only if φ_i is symplectic, which is equivalent to $C_i \in W'$, i.e., $\alpha \in \Delta'$ (Theorem 1.3).

Now assume $m > 0$. If $\mathbf{G} = SO_{2n+1}$, then $\hat{G} = Sp_{2n}(\mathbb{C})$, and $\beta_i^\vee = 2\breve{E}_i$. In this case $s_{\beta_i^\vee} = C_i$. We see $s_{\beta_i^\vee} \in W_\varphi$ if and only if $\varphi_i \simeq \theta^*\varphi_i$. By the proof of Lemma 2.3 (which precedes its statement) we have $C_i \in W_\varphi^\circ$ only if φ_i is orthogonal, or $\varphi_i \simeq \psi$. This is equivalent to poles at $s = 0$ for $L(2s, \mathrm{Sym}^2 \varphi_i)$ or $L(s, \varphi_i \oplus \psi)$. We know $L(2s, \sigma, \mathrm{Sym}^2) = L(2s, \mathrm{Sym}^2 \varphi_i)$ by Henniart [**15**]. Under Assumption A, $L(s, \sigma_i \times \tau) = L(s, \varphi_i \oplus \psi)$, and the latter has a pole if and only if $\varphi_i \simeq \psi$. In [**12**] it is shown $L(s, \sigma_i \times \tau)$ having a pole means σ_i comes from τ by twisted endoscopy (cf. the discussion following Assumption A). Now we see $C_i \in W_{\varphi,\sigma}^\circ$ if and only if $\beta_i^\vee \in \Delta'$. If $\mathbf{G} = Sp_{2n}$, then $\beta_i^\vee = E_i^\vee$. Again, $s_{\beta_i^\vee} = C_i$. Here $C_i \in W_{\varphi,\sigma}$ if and only if $\varphi_i \simeq \theta\varphi_i$. By the above observation on L–functions, we see $C_i \in W_{\varphi,\sigma}^\circ$ if and only if one of $L(2s, \wedge^2\varphi_i)$ or $L(s, \varphi_i \oplus \psi)$ has a pole at $s = 0$, i.e., φ_i is symplectic or $\varphi_i \simeq \psi$. By Theorems 1.3 and 1.4 we have $s_{\beta_i^\vee} \in W_{\varphi,\sigma}^\circ$ if and only if $\beta_i \in \Delta'$. Finally, suppose $\mathbf{G} = SO_{2n}$. Then, $\beta_i = E_i^\circ + E_i'$. Then $C_i \in W_{\varphi,\sigma}$ if and only if either φ_i is symplectic or φ_i is orthogonal and n_i s even. The discussion preceding the proof of Lemma 2.4 shows $C_i \in W_{\varphi,\sigma}^\circ$ if and only if φ_i is symplectic or $\varphi_i \simeq \psi$ (necessitating n_i be even). Now, we again appeal to [**15, 25, 24**] to see $s_{\beta_i^\vee} \in W_{\varphi,\sigma}^\circ$ if and only if $\beta_i \in \Delta'$. This completes the proof. □

We have now proved the following.

THEOREM 5.3. *Under the assumptions of Lemma 5.1, Assumption A, and the assumption ψ is irreducible, $W_\varphi^\circ = [W(\Delta')]^\vee = W((\Delta')^\vee)$. Thus, the map $s_\alpha \mapsto s_{\alpha^\vee}$ induces an isomorphism $R(\sigma) \simeq R_{\varphi,\sigma}$.* □

COROLLARY 5.4. *Suppose* $\mathbf{G} = Sp_{2n}, SO_{2n}$, *or* SO_{2n+1}, *and suppose* $\mathbf{M}$ *is a Levi subgroup of* $\mathbf{G}$. *Under the assumptions of Theorem 5.3 the sequence*

$$1 \to W_\varphi^\circ \to W_\varphi \to R_\varphi \to 1 \text{ splits},$$

$W_\varphi = R_\varphi \ltimes W_\varphi^\circ$, *and* $R_\varphi = \{w \in W_\varphi | w\Delta^\vee = \Delta^\vee\} = \{w \in W_\varphi | w\alpha^\vee > 0\ \forall \alpha^\vee \in \Delta^\vee\}$.

PROOF. This is a restatement of Theorems 3.7, and 4.10, along with Theorem 5.3 and the Knapp–Stein Theorem[**27, 28**] □

References

[1] J. Arthur, *Unipotent automorphic representations, conjectures* in Orbites unipotentes et représentations II, Astérisque 171-172, (1989), pp. 13-72

[2] J. Arthur, *On elliptic tempered characters*, Acta Math. 171 (1993), pp. 73–138

[3] D. Ban and Y. Zhang, *Arthur R-groups, classical R-groups, and Aubert involutions for SO(2n+1)*, Compo. Math. 141 (2005), pp. 323-343

[4] Bernstein, I. N.; Zelevinskii, A. V. Induced representations of the group $GL(n)$ over a p-adic field. (Russian) Funkcional. Anal. i Prilozen. 10 (1976), pp 74–75, (English) Functional Analysis and Its Applications, 1976, 10, pp 225-227

[5] A. Borel, *Automorphic L-functions* in Automorphic forms, representations and L-functions (Proc. Sympos. Pure Math., Oregon State Univ., Corvallis, Ore., 1977), Part 2, pp. 27–61, Proc. Sympos. Pure Math., XXXIII, Amer. Math. Soc., Providence, R.I., 1979

[6] W. Casselman *Introduction to the theory of admissible representations of p–adic reductive groups* preprint

[7] J. Cogdell, H. Kim, I. Piatetski–Shapiro and F. Shahidi, *On lifting from classical groups to GL(n)*, Inst. Haut. Études Sci. Publ. Math., No. 93 (2001), 5–30.

[8] J.W. Cogdell, H.H. Kim, I.I. Piatetski-Shapiro, and F. Shahidi, *Functoriality for the classical groups* Publ. Math. Inst. Hautes Études Sci. No. 99 (2004), pp. 163–233

[9] D. Goldberg *Reducibility of induced representations for* $SP(2N)$ *and* $SO(N)$, Amer. J. Math, 116 (1994), pp 1101-1151

[10] D. Goldberg and F. Shahidi *On the tempered spectrum of quasi-split classical groups*, Duke Math. J. 92 (1998), pp 255–294

[11] D. Goldberg and F. Shahidi *On the tempered spectrum of quasi-split classical groups II* Canad. J. Math. 53 (2001), pp. 244–277

[12] D. Goldberg and F. Shahidi *On the tempered spectrum of quasi-split classical groups III: The odd orthogonal groups* Forum Math., to appear

[13] B. Gross and D. Praasad,*On the decomposition fo a representation of* SO_n *when restricted to* SO_{n-1}, Canad. J. Math, 44 (1992), pp. 974-1002

[14] Harish-Chandra, *Harmonic Analyss on reductive p–adic groups*, Proc. Sympos. Pure Math. (AMS), 26 (1973), pp. 167-192

[15] G. Henniart, *Correspondance de Langlands et fonctions L des carrés extérieur et symétrique*, Int. Math. Res. Not. (IMRN) 2010, no. 4, pp. 633-673

[16] H. Jacquet, I. I. Piatetskii-Shapiro,; J. A. Shalika, *Rankin–Selberg convolutions* Amer. J. Math. 105 (1983), pp. 367–464

[17] D. Jiang and D. Soudry *The local converse theorem for* SO$(2n+1)$ *and applications*, Ann. of Math. (2) 157 (2003), 743–806

[18] C.D. Keys, *On the decomposition of reducible principal series of p–adic Chevalley groups*, Pac. J. Math., 101 (1982) pp. 351-388

[19] C.D. Keys, *Principal series representations of special unitary groups over local fields*, Compo. Math., 51 (1984), pp. 115-130

[20] C.D. Keys, *L-indistinguishability and R-groups for quasi split groups: unitary groups in even dimension*, Ann. Sci. Ec. Norm. Sup., 20 (1987), pp. 31-64

[21] A.W. Knapp and A E.M. Stein, *Irreducibility theorems for the principal series* in Conference on Harmonic Analysis, Lecture Notes in Mathematics, No. 266, pp 197-214, Springer, (1972)

[22] R. P. Langlands, *Problems in the theory of automorphic forms* , Lectures in Modern Analysis and Applications, III, Springer, Berlin, Lecture Notes in Math., Vol. 170 (1970), pp. 18–61. .

[23] C.Mœglin and M. Tadić, *Construction of discrete series for classical p-adic groups*, J. Amer. Math. Soc. 15 (2002), pp.715–786

[24] F. Shahidi, *Twisted endoscopy and reducibility of induced representations for p–adic groups*, Duke Math J., 66 (1992) pp 1-41

[25] Shahidi. F., *A proof of Langlands' conjecture on Plancherel measures; complementary series for p-adic groups* Ann. of Math. , 132 (1990), pp. 273-330

[26] D. Shelstad, *L–indistinguishability for real groups*, Math., Ann., 259 (1982), pp 385-430

[27] A.J. Silberger *The Knapp-Stein dimension theorem for p–adic groups* Proc. Amer. Math. Soc. 68 (1978), 243–246

[28] A.J. Silberger . *Correction: The Knapp-Stein dimension theorem for p–adic groups* Proc. Amer. Math. Soc. 76 (1979), pp. 169-170

[29] A.J. Silberger, **Introduction to Harmonic analysis on Reductive p-adic Groups**, Mathematical Notes, 23. Princeton University Press, (1979)

[30] J. Tate *Number theoretic background* Proc. Sympos. Pure Math. (AMS), 33, II (1979), pp. 3-26

[31] A.V.Zelevinsky, *Induced representations of reductive p-adic groups. II. On irreducible representations of* GL(n), Ann. Sci. cole Norm. Sup. (4) 13 (1980), pp.165–210

Department of Mathematics, Purdue University, West Lafayette, IN 47907 USA
E-mail address: goldberg@math.purdue.edu

Clay Mathematics Proceedings
Volume 13, 2011

Irreducibility of the Igusa Tower over Unitary Shimura Varieties

Haruzo Hida

To Freydoon Shahidi on the occasion of his sixtieth anniversary

CONTENTS

1. Elliptic modular Igusa tower	189
2. Shimura varieties of unitary groups	193
3. Igusa tower over unitary Shimura variety	198
References	202

In this paper, we describe a proof (more concise than the treatments in [**PAF**] Chapter 8 and [**H06**]) of irreducibility of the modulo p Igusa tower over a (unitary) Shimura variety. We study the decomposition group of the mixed characteristic valuation associated to each irreducible component of the Igusa tower (so the argument is closer to [**PAF**] Chapter 8 than the purely characteristic p argument in [**H06**]). The author hopes that the account here is easier to follow than the technical but more general treatment in [**H06**] and [**PAF**].

There are at least two ways of showing irreducibility: (i) the use of the automorphism group of the function field of the Shimura variety of characteristic 0 (cf. [**PAF**] Sections 6.4.3 and 8.4.4), which uses characteristic 0 results to prove the characteristic p assertion, and (ii) a purely characteristic p proof following a line close to (i) (see [**H06**]). There are some other arguments (purely in characteristic p) to prove the same result (covering different families of reductive groups giving the Shimura variety) as sketched in [**C1**] for the Siegel modular variety.

Here is an axiomatic approach to prove irreducibility of an étale covering $\pi : I \to S$ of a smooth irreducible variety S over the algebraic closure $\mathbb{F}$ of $\mathbb{F}_p$. Write $\pi_0(I)$ for the set of connected components of I. We start with the following two axioms:

2010 *Mathematics Subject Classification.* Primary: 11G18.

This note is based on the lecture delivered at a Luminy conference in July 2006 and four lectures at UCLA in Fall 2007. The author is partially supported by the NSF grant: DMS 0244401, DMS 0456252 and DMS 0753991.

(A1) A group $\mathcal{G} = \mathcal{M} \times \mathcal{G}_1$ acts on I and S compatibly so that $\mathcal{M} \subset \mathrm{Aut}(I/S)$, $\mathcal{G}_1 \subset \mathrm{Aut}(S)$ and $\mathcal{G}_1$ acts trivially on $\pi_0(I)$.
(A2) $\mathcal{M}$ acts on each fiber of I/S **transitively**; so, $\mathcal{M}$ acts transitively on $\pi_0(I)$.

Under (A1–2), we study the stabilizer subgroup T_x in $\mathcal{G}$ of a point x in a connected component I° of I and try to prove the following conclusion:

(C) $\{T_x\}_x$ for a good choice of a collection of points x and $\mathcal{G}_1$ generate a dense subgroup of $\mathcal{G}$.

Once we reach the conclusion (C), by the transitivity (A2), we obtain $I^\circ = I$ getting the irreducibility of I.

In the setting of Shimura variety Sh of PEL type (of level away from a given finite set Σ of places), assuming that we have a smooth integral compactification of Sh over a p-adic discrete valuation ring $\mathcal{W}$ (see [**ACS**] 6.4.1), we can easily verify the axioms (A1–2) for the following reasons: compatibility of the action in (A1) and and the transitivity in (A2) follow from the definition. In this case of a Shimura variety, S is the ordinary locus of the modulo p Shimura variety $Sh_{/\mathbb{F}}$ of level away from a given finite set Σ of places including p and ∞. Then, for the adéle ring $\mathbb{A}^{(\Sigma)}$ away from Σ, $\mathcal{G}_1$ is the adéle group $G(\mathbb{A}^{(\Sigma)})$ for the semi-simple group $G_{/\mathbb{Q}}$ (which is the derived group of the starting reductive group in Shimura's data), and $\mathcal{M}$ is the $\mathbb{Z}_p$-points $M(\mathbb{Z}_p)$ of the reductive part M of a parabolic subgroup of G. If we choose Σ so that $G(\mathbb{Q}_\ell)$ is generated by unipotent elements for all $\ell \notin \Sigma$, $\mathcal{G}_1$ has no nontrivial finite quotient group (because unipotent groups over a characteristic 0 field are uniquely divisible). For any finite subcovering I'/S of I, $\mathcal{G}_1$ acts on the finite set $\pi_0(I')$ through a finite quotient of $\mathcal{G}_1$; thus, the action is trivial, proving (A1).

In the above discussion of how to verify (A1), a key ingredient is that $\mathcal{G}_1$ is large enough not to have finite (nontrivial) quotient. As we will do in this paper, this is deduced from the existence of a smooth toroidal compactification (if the Shimura variety is not projective) and a characteristic 0 determination of the automorphism group of the Shimura variety. Alternatively, one can prove that $\mathcal{G}_1$ is large by showing that the ℓ-adic monodromy homomorphism for primes $\ell \neq p$ has large open image in $G(\mathbb{A}^{(\Sigma)})$. Indeed, C.-L. Chai [**C**] (in the symplectic case) has deduced the open image result via group theory from the semi-simplicity theorem of Grothendieck-Deligne of the ℓ-adic representation. The method in [**C**] should also work for $\ell \notin \Sigma$ (for an appropriate Σ) in our setting.

Let $I^\circ_{/\mathbb{F}}$ be an irreducible component of $I_{/\mathbb{F}}$. We want to prove $I^\circ = \pi^{-1}(S) = I$ (irreducibility). Then $\mathrm{Gal}(I^\circ/S) \subset \mathcal{M}$, and if $\mathcal{M} = \mathrm{Gal}(I^\circ/S)$, we get $I^\circ = \pi^{-1}(S)$. Let D be the stabilizer of $I^\circ \in \pi_0(I)$ in $\mathcal{G}$. Pick a point $x \in I$ (which can be a generic point), and look at the stabilizer $T_x \subset \mathcal{G}$ of x. Since $g_x(x) \in I^\circ$ ($g_x \in \mathcal{M}$) by the transitivity of the action, we have $g_x T_x g_x^{-1} \subset D$. Then we show that $\mathcal{M} = \mathcal{G}/\mathcal{G}_1$ is generated topologically by $\{g_x T_x g_x^{-1} | x \in I\}$, which implies $\mathcal{M} = \mathrm{Gal}(I^\circ/S)$ and the conclusion (C).

In the setting of the Igusa tower of a Shimura variety, we can have at least three choices of the points $x \in I$:

(∞) A cusp, assuming that the group $G = \mathrm{Res}_{F/\mathbb{Q}} G_0$ for a quasi-split group G_0 over a number field F (acting on a tube domain). This is the proof given for $GSp(2n)$ in [**DAV**].

(cm) A closed point $x \in I(\mathbb{F})$ is fixed by a maximal torus T_x of G anisotropic at ∞; thus, $g_x T_x(\mathbb{Z}_{(p)}) g_x^{-1} \subset D$ for $\mathbb{Z}_{(p)} = \mathbb{Q} \cap \mathbb{Z}_p$). A well chosen finite set of closed points $X := \{x\}$ is enough to generate a dense subgroup of D by $\{g_x T_x(\mathbb{Z}_{(p)}) g_x^{-1}\}_{x \in X}$. This is Ribet's choice for Hilbert modular varieties and is also taken in [**H06**]. If one uses a CM point (the so-called "hyper-symmetric point") carrying a product of copies of CM elliptic curves, often one such point is sufficient (see Section 3.5);

(gn) Take a coordinate system $T = (T_1, \ldots, T_d)$ around $x \in Sh(W)$ with (x mod p) $\in I^\circ(\mathbb{F})$ (so that $\widehat{\mathcal{O}}_{Sh,x} \cong W[[T_1, \ldots, T_d]]$) and take the valuation

$$v_x(\sum_\alpha c(\alpha, f) T^\alpha) = \mathrm{Inf}_\alpha \, \mathrm{ord}_p(c(\alpha, f)).$$

Then the decomposition group D of v_x contains T_x (for all $x \in I^\circ$), and D is the stabilizer of the generic point of I° containing x (this choice is taken in [**PAF**] 8.4.4). The valuation v_x corresponds to the generic point of I°. The point x can be a cusp as in (∞), and in the case of the modular curve (see Section 1.3), the Hilbert-Siegel modular variety and $U(n, n)$ Shimura variety, the choice of the infinity cusp works as well (cf. [**PAF**] 6.4.3).

Actually there is (at least) one more choice. Igusa completed his tower over modular curves adding super singular points and used such points to prove his irreducibility theorem in the 1950s. Here we describe the method (gn), but the base point x we use is the infinity cusp in the elliptic modular case and a hyper symmetric point in the unitary case.

Fix a prime p and an algebraic closure $\mathbb{F}$ of $\mathbb{F}_p$. We fix an algebraic closure $\overline{\mathbb{Q}}$ (resp. $\overline{\mathbb{Q}}_p$) of $\mathbb{Q}$ (resp. $\mathbb{Q}_p$), respectively. We fix field embeddings $i_p : \overline{\mathbb{Q}} \hookrightarrow \overline{\mathbb{Q}}_p$ and $i_\infty : \overline{\mathbb{Q}} \hookrightarrow \mathbb{C}$. Throughout this paper, proofs of the results claimed are given assuming $p > 2$ (just for simplicity; see [**H06**] for the treatment in the case $p = 2$).

1. Elliptic modular Igusa tower

As an introduction to the subject, we first describe the simplest case: the modular curves by the method (gn).

1.1. Elliptic modular function fields. We consider a field $\mathcal{K}$ given by $\bigcup_{p \nmid N} \mathbb{Q}(\mu_N)$ inside $\overline{\mathbb{Q}}$; so $\mathcal{K} \overset{i_p}{\hookrightarrow} \overline{\mathbb{Q}}_p$. Take a p-adic place $\mathfrak{P}$ of $\mathcal{K}$ given by i_p and write $\mathcal{W} \subset \mathcal{K}$ for the *discrete* valuation ring of $\mathfrak{P}$. We thus have a continuous embedding $i_p : \mathcal{W} \hookrightarrow \overline{\mathbb{Q}}_p$, and for the maximal ideal $\mathfrak{m}$ of $\mathcal{W}$, $\mathbb{F} = \mathcal{W}/\mathfrak{m}$ is an algebraic closure of $\mathbb{F}_p$. Put $\mathcal{G} = GL_1(\mathbb{Z}_p) \times SL_2(\mathbb{A}^{(p\infty)})$ and we embed diagonally $\mathbb{Z}_{(p)}$-points of the standard diagonal torus $M \subset SL(2)$ (of the upper triangular Borel subgroup $P = \left\{\left(\begin{smallmatrix} a & * \\ 0 & a^{-1} \end{smallmatrix}\right) \middle| a \in GL(1)\right\}$ of $SL(2)$) into $\mathcal{G}$ so that $\left(\begin{smallmatrix} a & 0 \\ 0 & a^{-1} \end{smallmatrix}\right)$ is sent to $a \in GL_1(\mathbb{Z}_p)$ at p and $\left(\begin{smallmatrix} a & 0 \\ 0 & a^{-1} \end{smallmatrix}\right) \in SL_2(\mathbb{Q}_\ell)$ at all primes $\ell \nmid p$.

We consider the modular curve $X(N)_{/\mathbb{Z}[\frac{1}{N}]}$ for an integer N prime to p which classifies pairs $(E, \phi_N)_{/A}$, where E is an elliptic over A and $\phi_N : (\mathbb{Z}/N\mathbb{Z})^2 \cong A[N] = \mathrm{Ker}(N : A \to A)$ is an isomorphism of finite flat group schemes over A. The level structure ϕ_N specifies a primitive root of unity $\zeta_N \in \mu_N$ via the Weil pairing

$$\zeta_N := \langle \phi_N(1, 0), \phi_N(0, 1) \rangle.$$

Thus $X(N)$ has a scheme structure over $\mathbb{Z}[\mu_N, \frac{1}{N}]$, but we may consider it defined over $\mathbb{Z}[\frac{1}{N}]$, composing with the morphism $\mathrm{Spec}(\mathbb{Z}[\mu_N, \frac{1}{N}]) \to \mathrm{Spec}(\mathbb{Z}[\frac{1}{N}])$. If we consider level p^m-structure ϕ_p of type $\Gamma = \Gamma_?(p^m)$ ($? = 0, 1$) given as follows: ϕ_p is a subgroup isomorphic to μ_{p^m} étale locally if $\Gamma = \Gamma_0(p^m)$ and $\phi_p : \mu_{p^m} \hookrightarrow E[p^m]$ (a closed immersion of finite flat group schemes) if $\Gamma = \Gamma_1(p^m)$ (and $Np^m \geq 4$), we can think of the fine moduli space $X(N,\Gamma)_{/B}$ over the base ring $B_{/\mathbb{Z}[\frac{1}{N}]}$ which classifies triples $(E, \phi_N, \phi_p)_{/A}$ over B-algebras A. As the ring B, we take one of $\mathcal{W}$, $\mathbb{F}$ or $\mathcal{K}$. As we observed, the open curves $X(N)$ (resp. $X(N,\Gamma)$) can be regarded as schemes over $\mathrm{Spec}(\mathbb{Z}[\frac{1}{N}, \mu_N])$ (resp. over $\mathrm{Spec}(\mathbb{Q}[\mu_N])$). For N prime to p, $X(N)_{/\mathbb{Q}[\mu_N]}$ is geometrically irreducible.

We can think of the p-integral Shimura curve

$$Sh_{/\mathbb{Z}_{(p)}} = \varprojlim_{p\nmid N} X(N)_{/\mathbb{Z}_{(p)}},$$

and more generally over $\mathbb{Q}$,

$$Sh_{\Gamma/\mathbb{Q}} = \varprojlim_{p\nmid N} X(N,\Gamma)_{/\mathbb{Q}}$$

(regarding these schemes as $\mathbb{Z}_{(p)}$-schemes or $\mathbb{Q}$-schemes). Let $X(N,\Gamma)_{/B} = X(N,\Gamma)_{/\mathbb{Z}[\mu_N,\frac{1}{N}]} \times_{\mathbb{Z}[\mu_N]} B$ and $X(N)_{/\mathcal{W}} = X(N)_{/\mathbb{Z}[\mu_N,\frac{1}{N}]} \times_{\mathbb{Z}[\mu_N,\frac{1}{N}]} B$. The pro-schemes

$$X_{\Gamma/B} = \varprojlim_{N} X(N,\Gamma)_{/B} \quad \text{for } B = \mathcal{K} \quad \text{and} \quad X^{(p)}_{/\mathcal{W}} = \varprojlim_{p\nmid N} X(N)_{/\mathcal{W}}$$

give geometrically irreducible components of $Sh_{\Gamma/\mathbb{Q}} \times_{\mathbb{Q}} \mathcal{K}$ and $Sh^{(p)}_{/\mathbb{Z}_{(p)}} \times_{\mathbb{Z}_{(p)}} \mathcal{W}$ (the neutral components). If convenient, we write $Sh_{\Gamma_1(p^0)/\mathbb{Z}_{(p)}}$ for $Sh_{/\mathbb{Z}_{(p)}}$ (abusing the notation). By the interpretation of Deligne–Kottwitz, we have

$$Sh_\Gamma(A) \cong \frac{\{(E, \eta : (\mathbb{A}^{(p\infty)})^2 \cong V(E), \phi_p)_{/A}\}}{\text{prime-to-}p \text{ isogenies}}, \tag{1.1}$$

where A runs over $\mathbb{Z}_{(p)}$-algebras if $\Gamma = \Gamma_1(p^0)$ and B-algebras ($B = \mathbb{F}$ or $\mathbb{Q}$) if $\Gamma = \Gamma_?(p^m)$ with $m > 0$ ($? = 0, 1$), $V(E) = \mathbb{A}^{(p\infty)} \otimes \varprojlim_{p\nmid N} E[N]$. Thus $(a, g) \in \mathcal{G}$ ($a \in GL_1(\mathbb{Z}_p)$ and $g \in SL_2(\mathbb{A}^{(p\infty)})$) acts on Sh_Γ by

$$(E, \eta, \phi_p) \mapsto (E, \eta \circ g, \phi_p \circ a),$$

where $a \in GL_1(\mathbb{Z}_p) \cong M(\mathbb{Z}_p)$. Write $\mathfrak{F}_\Gamma$ for the function field $\mathcal{K}(X_\Gamma)$ and $\mathfrak{F}^{(p)}$ for $\mathcal{K}(X^{(p)})$ (the arithmetic automorphic function fields). This action produces an embedding

$$\tau : \mathcal{G}/\{\pm 1\} \hookrightarrow \mathrm{Aut}(\mathfrak{F}_{\Gamma_1(p^\infty)}/\mathcal{K}) = \mathrm{Aut}(X_{\Gamma_1(p^\infty)/\mathcal{K}}).$$

The action of $\tau(a, g)$ on the function field $\mathfrak{F}_\Gamma$ is on the left and has the following property (by Shimura; e.g., [**IAT**] Theorem 6.23 or [**PAF**] Theorem 4.14): For $a \in GL_1(\mathbb{Z}_{(p)})$ (corresponding to $\left(\begin{smallmatrix} a & 0 \\ 0 & a^{-1} \end{smallmatrix}\right)$ in $M(\mathbb{Z}_{(p)})$ diagonally embedded in $SL_2(\mathbb{A}^{(\infty)})$), we have for $f \in \mathcal{F}_\Gamma$

$$\tau(a)(f)(z) = f(a^{-2} z); \tag{1.2}$$

so, we have $\tau(\alpha)(f) = f(\alpha^{-1}(z))$ for $\alpha = \left(\begin{smallmatrix} a & 0 \\ 0 & a^{-1} \end{smallmatrix}\right)$. This formula is valid for general $\alpha \in GL_2(\mathbb{Z}_{(p)})$ if $f \in \mathfrak{F}^{(p)}$ (thus, our normalization is different form Shimura's).

We define a valuation

$$v_\Gamma(f) = \inf_\xi \mathrm{ord}_p(c(\xi, f))$$

of modular functions $f = \sum_\xi c(\xi, f)q^\xi \in \mathfrak{F}_\Gamma$. We write v_m for v_Γ if $\Gamma = \Gamma_1(p^m)$. Thus the valuation $v_0 : \mathfrak{F}^{(p)} \to \mathbb{Z} \cup \{\infty\}$ has a standard unramified extension $v_\Gamma : \mathfrak{F}_\Gamma \twoheadrightarrow \mathbb{Z} \cup \{\infty\}$. Here are some easy facts:

LEMMA 1.1. (1) *If $a \in GL_1(\mathbb{Z}_{(p)}) \cong M(\mathbb{Z}_{(p)})$, then*

$$c(\xi, \tau(a)(f)) = c(a^2\xi, f).$$

In particular, the diagonally embedded $M(\mathbb{Z}_{(p)}) \subset \mathcal{G}$ preserves the valuation v_Γ;

(2) *The vertical divisor $X^{(p)}_{/\mathbb{F}} := X^{(p)}_{/\mathcal{W}} \otimes_{\mathcal{W}} \mathbb{F}$ of $X^{(p)}_{/\mathcal{W}}$ is a prime divisor (geometrically irreducible) and gives rise to a* **unique** *valuation of $\mathfrak{F}^{(p)}$, whose explicit form is given by the valuation v_0.*

PROOF. The first assertion follows directly from (1.2). By the existence of a smooth compactification of $X^{(p)}$ over $\mathcal{W}$, Zariski's connectedness theorem tells us that $X^{(p)}_{/\mathbb{F}} = X^{(p)} \times_{\mathcal{W}} \mathbb{F}$ is irreducible. Thus the vertical Weil prime divisor $X^{(p)}_{/\mathbb{F}}$ on the smooth arithmetic surface $X^{(p)}_{/\mathcal{W}}$ gives rise to a unique valuation. By the irreducibility of $X^{(p)}_{/\mathbb{F}}$, a $\mathcal{W}$-integral modular form of level away from p vanishes on the divisor $X^{(p)}_{/\mathbb{F}}$ if and only if its q-expansion vanishes modulo p. Thus the valuation v_0 is the one corresponding to the vertical prime divisor $X^{(p)}_{/\mathbb{F}} \subset X^{(p)}_{/\mathcal{W}}$. □

1.2. mod p connected components and the valuation v_m. Let S be the ordinary locus $X^{(p)}[\frac{1}{H}]_{/\mathbb{F}}$ for the Hasse invariant H. Then S is an irreducible variety over $\mathbb{F}$, because H is a global section of the ample modular line bundle $\underline{\omega}^{\otimes(p-1)}$ of the compactification of $X^{(p)}_{/\mathbb{F}}$. Consider the valuation ring V of $\mathfrak{F}^{(p)}$ of the valuation v_0. Thus the residue field $V/\mathfrak{m}_V$ is the function field $\mathbb{F}(S)$ of S. Let $\mathbb{E}_{/X^{(p)}}$ be the universal elliptic curve. Then we consider the Cartesian diagram for $\mathbb{E}_V = \mathbb{E} \times_{X^{(p)}} \mathrm{Spec}(V)$:

$$\begin{array}{ccc} \mathbb{E}_V & \xrightarrow{\hookrightarrow} & \mathbb{E} \\ \downarrow & & \downarrow \\ \mathrm{Spec}(V) & \underset{\hookrightarrow}{\longrightarrow} & X^{(p)}. \end{array}$$

Since any lift of a power of H is inverted in V, $\mathbb{E}_{\widehat{V}} = \mathbb{E}_V \times_V \widehat{V}$ is an ordinary abelian scheme for the completed valuation ring $\widehat{V} = \varprojlim_n V/p^nV$. Thus we can think of the functor $I_{\widehat{V},m} = \mathrm{Isom}_{\widehat{V}}(\mu_{p^m}, \mathbb{E}_{\widehat{V}}[p^m])$ which assigns to each p-adic $\widehat{V}$-algebra $R = \varprojlim_n R/p^nR$ the set of closed immersions: $\mu_{p^m/R} \to \mathbb{E}_{\widehat{V}}[p^m]_{/R}$ defined over R.

Since $\mathbb{E}_{\widehat{V}}[p^m]$ has a well defined connected component over $\widehat{V}$ isomorphic to μ_{p^m} étale locally ($\widehat{V}$ is a henselian local ring), we have canonical isomorphisms of formal schemes:

$$\begin{aligned} I_{\widehat{V},m} = \mathrm{Isom}_{\widehat{V}}(\mu_{p^m/\widehat{V}}, \mathbb{E}_{\widehat{V}}[p^m]^\circ) &\overset{(*)}{\cong} \mathrm{Isom}_{\widehat{V}}((\mathbb{Z}/p^m\mathbb{Z})_{/\widehat{V}}, \mathbb{E}_{\widehat{V}}[p^m]^{\text{ét}}) \\ &\overset{(**)}{\cong} \mathbb{E}_{\widehat{V}}[p^m]^{\text{ét}} - \mathbb{E}_{\widehat{V}}[p^{m-1}]^{\text{ét}}, \end{aligned}$$

where the identity $(*)$ is given by taking the inverse of the Cartier dual map and $(**)$ is given by $\phi \mapsto \phi(1)$ for $1 \in \mathbb{Z}/p^m\mathbb{Z}$ and $\phi \in \mathrm{Isom}_{\widehat{V}}((\mathbb{Z}/p^m\mathbb{Z})_{/\widehat{V}}, \mathbb{E}_{\widehat{V}}[p^m]^{\text{ét}})$.

Thus $I_{\widehat{V},m}/\mathrm{Spf}(\widehat{V})$ is étale finite. Note here $\mathbb{E}_{\widehat{V}}[p^m]^\circ$ is isomorphic to $\mu_{p^n/\widehat{V}}$ étale locally. By the second expression, $I_{\widehat{V},m}/\mathrm{Spf}(\widehat{V})$ is an étale finite covering over $\widehat{V}$, and $GL_1(\mathbb{Z}_p)$ naturally acts on $I_{\widehat{V},m}$. Since $I_{\widehat{V},m}$ is étale faithfully flat over $\mathrm{Spf}(\widehat{V})$, it is affine, and we may write $I_{\widehat{V},m} = \mathrm{Spf}(\widehat{V}_m)$. Then $\widehat{V}_m$ is a semi-local normal $\widehat{V}$-algebra étale finite over $\widehat{V}$; so, it is a product of complete discrete valuation rings whose maximal ideal is generated by the rational prime p. Write $W = \varprojlim_n \mathcal{W}/p^n\mathcal{W}$, and take a modular form E on $X^{(p)}_{/\mathcal{W}}$ lifting a positive power of the Hasse invariant H. Let $\widehat{X}^{(p)}$ be a formal completion of $X^{(p)}[\frac{1}{E}]_{/W}$ along S (the ordinary locus). The p-adic formal scheme $\widehat{X}^{(p)}$ does not depend on the choice of the lift E. Then we define a p-adic formal scheme $\widehat{X}_{\Gamma/W} = \mathrm{Isom}_{\widehat{X}^{(p)}}(\mu_{p^n}, \mathbb{E}) \cong \mathbb{E}[p^m]^{\text{ét}} - \mathbb{E}[p^{m-1}]^{\text{ét}}$ over $\widehat{X}^{(p)}$, which is étale finite over $\widehat{X}^{(p)}$. We may regard $\widehat{X}_{\Gamma/W}$ as the formal completion of $X_{\Gamma/\mathcal{W}}$ along $X_{\Gamma/\mathbb{F}}$. By definition, we have an open immersion

$$I_{\widehat{V},m} \hookrightarrow \widehat{X}_{\Gamma_1(p^m)/W} \times_{\widehat{X}^{(p)}_{/\mathcal{W}}} \mathrm{Spf}(\widehat{V}),$$

and $\widehat{V}_m$ is the product of the completions of valuation rings of $\mathfrak{F}_{\Gamma_1(p^m)}$ unramified over V. Thus $V_m = \widehat{V}_m \cap \mathfrak{F}_{\Gamma_1(p^m)}$ inside $\mathfrak{F}_{\Gamma_1(p^m)} \otimes_V \widehat{V}$ is a semi-local ring V_m with $\widehat{V}_m = \varprojlim_n V_m/p^nV_m = V_m \otimes_V \widehat{V}$.

We put $I_{V,m} = \mathrm{Spec}(V_m)$ and $X_{\Gamma/\mathbb{F}} = \varprojlim_{p\nmid N} X(N,\Gamma)_{/\mathbb{F}}$. Then

$$X_{\Gamma_1(p^m)/\mathbb{F}} = \mathrm{Isom}_S(\mu_{p^m}, \mathbb{E}[p^m]^\circ) =: I_m$$

gives rise to the Igusa tower $I \twoheadrightarrow \cdots \twoheadrightarrow I_m \twoheadrightarrow \cdots \twoheadrightarrow I_1 \twoheadrightarrow S$ over S. We may regard the moduli scheme $X(N,\Gamma)_{/\mathbb{F}}$ as a scheme over $X(N)[\frac{1}{H}]$ (forgetting the level p-structure). The set of generic points $\{\eta_{I^\circ_m} \in I^\circ_{m/\mathbb{F}} | I^\circ_{m/\mathbb{F}} \in \pi_0(I_{m/\mathbb{F}})\}$ is in bijection with $\pi_0(I_m)$, and

$$\widehat{V}_m \otimes_{\mathbb{Z}_p} \mathbb{F} = V_m \otimes_{\mathbb{Z}_{(p)}} \mathbb{F} = \prod_{I^\circ_m \in \pi_0(I_m)} \mathbb{F}(I^\circ_m) \ (\Leftrightarrow I_{V,m} \otimes_{\mathbb{Z}_{(p)}} \mathbb{F} = \bigsqcup_{I^\circ \in \pi_0(I_m)} \{\eta_{I^\circ_m}\}).$$

By the definition of the action of $(a,g) \in \mathcal{G}$:

$$(E, \eta^{(p)}, \phi_p) \mapsto (E, \eta^{(p)} \circ g, \phi_p \circ a),$$

$\mathcal{G} := GL_1(\mathbb{Z}_p) \times SL_2(\mathbb{A}^{(p\infty)})$ acts on $I_{\widehat{V},m}$ and hence on $I_{V,m}$ ($m = 1, 2, \ldots, \infty$), $\mathrm{Spec}(V)$ (by Lemma 1.1 (2)), $\mathfrak{F}_\Gamma$, I_m, $X_{\Gamma/\mathbb{F}}$ and $X_{\Gamma/\mathcal{K}}$. Thus we can form the étale quotient $I_{\Gamma_0(p^m)} := I_{V,m}/GL_1(\mathbb{Z}/p^m\mathbb{Z})$. Again we have $I_{\Gamma_0(p^m)} = \mathrm{Spec}(V_{\Gamma_0(p^m)})$, and $V_{\Gamma_0(p^m)}$ is a valuation ring finite flat over V sharing the same residue field. Indeed, there is a unique connected subgroup of E (isomorphic to μ_{p^m} étale locally) if $(E, \phi_N)_{/A}$ gives rise to a unique A-point of $X(N, \Gamma_0(p^m))_{/\mathbb{F}}$. Thus for any $m > 0$, $S_{/\mathbb{F}} = \varprojlim_{p\nmid N} X(N, \Gamma_0(p^m))_{/\mathbb{F}}$. This shows that the residue field of $V_{\Gamma_0(p^m)}$ is the function field of S and that the quotient field of $V_{\Gamma_0(p^m)}$ is $\mathfrak{F}_{\Gamma_0(p^m)}$. Since V_m/V is étale, we have

$$\widehat{V}_{\Gamma_0(p^m)} = \varprojlim_m V_{\Gamma_0(p^m)}/p^m V_{\Gamma_0(p^m)} = \varprojlim_m V/p^mV = \widehat{V},$$

and V_m is étale finite over $V_{\Gamma_0(p^m)}$. This shows

LEMMA 1.2. *We have the following one-to-one onto correspondences:*

$$\left\{v : \mathfrak{F}_{\Gamma_1(p^m)} \to \mathbb{Z} \middle| v|_{\mathfrak{F}_{\Gamma_0(p^m)}} = v_{\Gamma_0(p^m)} \text{ unramified over } v_0\right\}$$
$$\leftrightarrow \mathrm{Max}(V_m) \leftrightarrow \pi_0(I_m) \leftrightarrow \{\eta_{I_m^\circ}\},$$

where v is a p-adic valuation of $\mathfrak{F}_{\Gamma_1(p^m)}$ unramified (of degree 1) over v_0 and $\mathrm{Max}(V_m)$ *is the set of maximal ideals of V_m.*

The correspondence is given by

$$v \leftrightarrow \mathfrak{m}_v = \{x \in V_m | v(x) > 0\} \leftrightarrow I_m^\circ \quad \text{with } \mathbb{F}(I_m^\circ) = V_m/\mathfrak{m}_v.$$

LEMMA 1.3. *The action of $\mathcal{G}_1 := SL_2(\mathbb{A}^{(p\infty)})$ fixes $v_m = v_{\Gamma_1(p^m)}$ and each element of $\pi_0(I_m)$.*

PROOF. Since $\mathfrak{F}_{\Gamma_1(p^m)}/\mathfrak{F}_{\Gamma_0(p^m)}$ is a finite Galois extension, the set of extensions of $v_{\Gamma_0(p^m)}$ to $\mathfrak{F}_{\Gamma_1(p^m)}$ is a finite set, and by the above lemma, it is in bijection with $\pi_0(I_m)$. Thus the action of $SL_2(\mathbb{A}^{(p\infty)})$ on $\pi_0(I_m)$ gives a finite permutation representation of $SL_2(\mathbb{A}^{(p\infty)})$. Since $SL_2(k)$ of any field k of characteristic 0 does not have a nontrivial finite quotient group (because it is generated by divisible unipotent subgroups), the action of $SL_2(\mathbb{A}^{(p\infty)})$ fixes every irreducible component of $\pi_0(I_m)$. □

1.3. Proof of irreducibility of elliptic Igusa tower. Let $v_\infty = v_{\Gamma_1(p^\infty)}$, and define

$$D = \left\{x \in (GL_1(\mathbb{Z}_p) \times SL_2(\mathbb{A}^{(p\infty)})) \middle| v_\infty \circ \tau(x) = v_\infty\right\}.$$

Since $M(\mathbb{Z}_{(p)})$ and $SL_2(\mathbb{A}^{(p\infty)})$ fixes v_∞ (Lemmas 1.1 and 1.3) and the subgroup $(M(\mathbb{Z}_{(p)})SL_2(\mathbb{A}^{(p\infty)}))$ is dense in $\mathcal{G} = GL_1(\mathbb{Z}_p) \times SL_2(\mathbb{A}^{(p\infty)})$, we conclude (C):

THEOREM 1.4. *We have $D = \mathcal{G}$.*

Let $K^{(p)}$ be a compact open subgroup of $SL_2(\mathbb{A}^{(p\infty)})$ and $K = K^{(p)} \times GL_2(\mathbb{Z}_p)$. Put $X_K = X^{(p)}/K^{(p)}$ (which is the level K modular curve). Let $I_K = I/K^{(p)}$, which is the Igusa tower over X_K. Since I is irreducible by

$$\mathrm{Aut}(I^\circ/S) = GL_1(\mathbb{Z}_p) \cong M(\mathbb{Z}_p) \quad \text{(the above theorem)},$$

I_K is irreducible. Thus we have reproved

COROLLARY 1.5 (Igusa). *The Igusa tower I_K over $X_{K/\mathbb{F}}$ is irreducible for $K = GL_2(\mathbb{Z}_p) \times K^{(p)}$ for each compact open subgroup $K^{(p)}$ of $SL_2(\mathbb{A}^{(p\infty)})$.*

2. Shimura varieties of unitary groups

We give an example S of smooth Shimura varieties for which irreducibility of the full Igusa tower is false but one can study the irreducible components explicitly. In other words, we construct a partial tower I°/S for which the axioms (A1–2) can be proved. Write W for the ring of Witt vectors of the algebraic closure $\mathbb{F}$ of $\mathbb{F}_p$ and embed W inside $\mathbb{C}_p$ (the p-adic completion of $\overline{\mathbb{Q}}_p$). Hereafter, we write $\mathcal{W}$ for the valuation ring $i_p^{-1}(W)$ and $\mathcal{K}$ for the field of fractions of $\mathcal{W}$. The (additive) valuation of W and $\mathcal{W}$ is written as ord_p; so, $\mathrm{ord}_p(p) = 1$. As before, we prove that $S_{/\mathcal{W}}$ is irreducible and smooth and that the Igusa tower $I_{/\mathbb{F}}$ is étale over $S_{/\mathbb{F}}$. Then for each point $x \in I(\mathcal{W})$, we take a coordinate system $X_1, \ldots, X_d$ of I and define a valuation v_x of the function field of I by $v_x(\sum_\alpha c(\alpha)X^\alpha) = \mathrm{Inf}_\alpha \mathrm{ord}_p(c(\alpha))$ $(X^\alpha = X_1^{\alpha_1}X_2^{\alpha_2}\cdots X_d^{\alpha_d})$. For any automorphism σ of $I_{/\mathcal{W}}$ fixing x, plainly $v_x \circ \sigma = v_x$. Then we conclude the irreducibility by showing that the stabilizers $\{T_x\}_{x\in I(\mathcal{W})}$ inside $\mathrm{Aut}(I_{/\mathcal{W}})$ of $x \in I(\mathcal{W})$ cover sufficiently many conjugacy classes of tori to prove (A1–2). Actually, in the simple case we study, a well chosen single point $x_0 \in I(\mathcal{W})$ is sufficient.

We first recall briefly the definition of unitary groups over an imaginary quadratic field F and the construction of the Shimura variety for the unitary groups. The main source of the information for this part is [**PAF**] Chapter 7. Then we prove the irreducibility of the Igusa tower.

Suppose that the imaginary quadratic field F is sitting inside $\overline{\mathbb{Q}}$, and write $1 : F \hookrightarrow \overline{\mathbb{Q}}$ for the identity embedding. Suppose for simplicity that the fixed prime p is *split* in F and that the embedding $1 : F \hookrightarrow \overline{\mathbb{Q}}$ composed with $i_p : \overline{\mathbb{Q}} \hookrightarrow \overline{\mathbb{Q}}_p$ gives the standard p-adic place $\mathfrak{p}$ of F. Write O for the integer ring of F.

2.1. Unitary groups. Write c for the generator of $\mathrm{Gal}(F/\mathbb{Q})$ (the complex conjugation on F). We fix a vector space V over F with c-Hermitian alternating form $\langle\ ,\ \rangle : V \times V \to \mathbb{Q}$. We assume we have an O-submodule $L \subset V$ of finite type such that

(L1) $L \otimes_{\mathbb{Z}} \mathbb{Q} = V$;
(L2) $\langle\ ,\ \rangle$ induces $\mathrm{Hom}_{\mathbb{Z}_p}(L_p, \mathbb{Z}_p) \cong L_p$, where $L_p = L \otimes_{\mathbb{Z}} \mathbb{Z}_p$.

We fix an O-lattice L of V as above.

We identify V with the column vector space F^r by fixing a basis of V over F. Let $C = \mathrm{End}_F(V) = M_r(F)$. There exists an invertible matrix $s \in M_r(F)$ with ${}^ts^c = -s$ such that $\langle v, w\rangle = \mathrm{Tr}_{F/\mathbb{Q}}({}^tvs \cdot w^c)$, where $\mathrm{Tr}_{F/\mathbb{Q}}$ is the trace map: $F \to \mathbb{Q}$. On C, we have the involution ι given by $x^\iota = s^{-1}{}^tx^cs$. Define algebraic groups defined over $\mathbb{Q}$ by the following group functors from $\mathbb{Q}$-algebras R to groups:

$$\begin{aligned}(2.1)\qquad GU(R) &= \left\{x \in C \otimes_{\mathbb{Q}} R \middle| x^\iota x \in R^\times\right\}\\ &= \left\{x \in C \otimes_{\mathbb{Q}} R \middle| {}^tx^cs \cdot x = \nu(x)s \ \text{ for }\ \nu(x) = x^\iota x \in R^\times\right\},\\ U(R) &= \left\{x \in GU(R) \middle| x^\iota x = 1\right\},\ SU(R) = \left\{x \in U(R) \middle| \det(x) = 1\right\},\end{aligned}$$

where $\det(x)$ is the determinant of x as an F-linear automorphism of V. Then SU is the derived group of GU and U. Let $Z \subset GU$ be the center; so $Z(R) = (R \otimes_{\mathbb{Q}} F)^\times$ as a group functor. Since $F_{\mathbb{R}} = F \otimes_{\mathbb{Q}} \mathbb{R} = \mathbb{C}$ with $b^c = \overline{b}$ for complex conjugation $b \mapsto \overline{b}$, $S = \sqrt{-1}s \in M_r(F_{\mathbb{R}}) = M_r(\mathbb{C})$ is a Hermitian matrix. Thus $U(\mathbb{R})$ is the unitary group of S. We have $\mathrm{Hom}_{\mathrm{field}}(F, \mathbb{C}) = \{1, c\}$ for the identity inclusion 1. Writing the signature of S as (m_1, m_c), we find $U(\mathbb{R}) \cong U_{m_1,m_c}(\mathbb{R}) = \left\{x \in GL_r(\mathbb{C}) \middle| {}^t\overline{x}I_{m_1,m_c}x = I_{m_1,m_c}\right\}$ for $I_{m_1,m_c} = \left(\begin{smallmatrix}1_{m_1} & 0\\ 0 & -1_{m_c}\end{smallmatrix}\right)$.

EXAMPLE 2.1. For a $\mathbb{Q}$-algebra R,

(1) if $s = \left(\begin{smallmatrix}0 & -1\\ 1 & 0\end{smallmatrix}\right)$, then $\left(\begin{smallmatrix}a & b\\ c & d\end{smallmatrix}\right)^\iota = \left(\begin{smallmatrix}\overline{d} & -\overline{b}\\ -\overline{c} & \overline{a}\end{smallmatrix}\right)$ and $SU(R) = SL_2(R)$,

$$GL_2(R) = \left\{x \in GU(R) \middle| \det(x) = \nu(x)\right\};$$

(2) $GU(\mathbb{Q}) = GL_2(\mathbb{Q})Z(\mathbb{Q})^\times$ and $GU(R) = GL_2(R)Z(R)$.

2.2. Abelian schemes of hermitian type. To put a complex structure on the real vector space $V_\infty = V \otimes_{\mathbb{Q}} \mathbb{R}$, we use an $\mathbb{R}$-algebra homomorphism $h : \mathbb{C} \hookrightarrow C_\infty = C \otimes_{\mathbb{Q}} \mathbb{R}$ with $h(\overline{z}) = h(z)^\iota$. We call such an algebra homomorphism an *ι-homomorphism.* Then $h(i)^\iota = -h(i)$ for $i = \sqrt{-1}$ and hence $x^\rho = h(i)^{-1}x^\iota h(i)$ is an involution of C_∞.

EXAMPLE 2.2. If $s = \left(\begin{smallmatrix} 0 & -1 \\ 1 & 0 \end{smallmatrix}\right)$, the morphism $a + bi \mapsto h(a + bi) = \left(\begin{smallmatrix} a & -b \\ b & a \end{smallmatrix}\right) \in M_2(\mathbb{R}) \subset C_\infty$ is an ι-homomorphism.

We suppose

(pos) The symmetric real bilinear form $(v, w) \mapsto \langle v, h(i)w\rangle$ on V_∞ is positive definite.

It is easy to check that h in Example 2.2 satisfies (pos).

By (pos), we have $0 < (xv, xv) = (v, (x^\rho x)v)$ for all $0 \neq v \in V_\infty$ and $x \in C_\infty$, and hence $x^\rho x$ only has positive eigenvalues; therefore, ρ is a positive involution of C (i.e., $\mathrm{Tr}_{C/\mathbb{Q}}(x^\rho x) > 0$ unless $x = 0$).

Fix one such $h := h_{\mathbf{0}} : \mathbb{C} \to C_\infty$, and define $\mathfrak{X}$ (resp. $\mathfrak{X}^+$) by the collection of all conjugates of $h_{\mathbf{0}}$ under $GU(\mathbb{R})$ (resp. under $SU(\mathbb{R})$). Any two homomorphisms satisfying (pos) are conjugates under $SU(\mathbb{R})$ (see [**PAF**] Lemma 7.3). Thus $\mathfrak{X}^+ = SU(\mathbb{R})/C_{\mathbf{0}}$ for the stabilizer $C_{\mathbf{0}}$ of $h_{\mathbf{0}}$ in $SU(\mathbb{R})$ is connected and is a connected component of $\mathfrak{X}$. On $\mathfrak{X}$, $GU(\mathbb{R})$ acts by conjugation (from the left), and by (pos) the stabilizer $C_{\mathbf{0}} \subset GU(\mathbb{R})$ of $h_{\mathbf{0}}$ is a maximal compact subgroup of $GU(\mathbb{R})$ modulo center.

EXAMPLE 2.3. Assume that $s = \left(\begin{smallmatrix} 0 & -1 \\ 1 & 0 \end{smallmatrix}\right)$ and take $h_{\mathbf{0}}(a + bi) = \left(\begin{smallmatrix} a & -b \\ b & a \end{smallmatrix}\right)$. Since $h_{\mathbf{0}}(\mathbb{C}^\times)$ gives the stabilizer of $i \in \mathfrak{H} = \{z \in \mathbb{C} | \operatorname{Im}(z) > 0\}$, we have $\mathfrak{X}^+ \cong \mathfrak{H}$ by sending $gh_{\mathbf{0}}g^{-1}$ to $g(i)$. We also have $\mathfrak{X} \cong \mathfrak{H} \sqcup \overline{\mathfrak{H}} = (\mathbb{C} - \mathbb{R})$ in the same way.

Since $h : \mathbb{C} \to C_\infty$ is an $\mathbb{R}$-algebra homomorphism, we can split $V_{\mathbb{C}} = V \otimes_{\mathbb{Q}} \mathbb{C}$ into the direct sum of eigenspaces $V_{\mathbb{C}} = V_1 \oplus V_2$ so that $h(z)$ acts on V_1 (resp. V_2) through multiplication by z (resp. $\overline{z}$); thereby, we get a complex vector space structure on V_∞ by the projection $V_\infty \cong V_1$. Since $h(\mathbb{C}) \subset C_\infty$, $h(z)$ commutes with the action of F; so, V_j is stable under the action of $F_{\mathbb{C}} = F \otimes_{\mathbb{Q}} \mathbb{C}$. We get the representation $\rho_1 : F \hookrightarrow \mathrm{End}_{\mathbb{C}}(V_1)$. We define E to be the subfield of $\mathbb{C}$ fixed by the open subgroup $\left\{\sigma \in \mathrm{Aut}(\mathbb{C}) \middle| \rho_1^\sigma \cong \rho_1\right\}$. If $h'(z) = g \cdot h(z)g^{-1}$ for $g \in GU(\mathbb{R})$, h' induces a similar decomposition $V_{\mathbb{C}} = V_1' \oplus V_2'$, and g induces an F-linear isomorphism between V_1 and V_1'; thus, E is independent of the choice of h' in the $GU(\mathbb{R})$-conjugacy class of h. This field E is called the *reflex field* of $(GU, \mathfrak{X})$ (and is a canonical field of definition of our canonical models of the Shimura variety).

By the positivity (pos), the quotient complex torus $V_\infty/L = V_1/L$ has a Riemann form induced by $\langle\cdot,\cdot\rangle$. The theta functions with respect to the Hermitian form $\langle\cdot,\cdot\rangle$ give rise to global sections of an ample line bundle (e.g., [**ABV**] Chapter I) on V_1/L and hence embed V_1/L into a projective space over $\mathbb{C}$. The embedded image is the analytic space $A_h(\mathbb{C})$ associated with an abelian variety $A_{h/\mathbb{C}}$ by Chow's theorem (see [**ABV**] page 33). Multiplication by $b \in O$ on V_1/L induces an embedding $i : O \hookrightarrow \mathrm{End}(A_{h/\mathbb{C}})$ and $i : F \hookrightarrow \mathrm{End}^{\mathbb{Q}}(A_{h/\mathbb{C}}) = \mathrm{End}(A_{h/\mathbb{C}}) \otimes_{\mathbb{Z}} \mathbb{Q}$.

The representation ρ_1 is given by the action of F on the Lie algebra $Lie(A_h) = V_1$ at the origin of $A_h(\mathbb{C})$. Since A_h is projective, the field of definition of the abelian variety A_h is a field of finite type over $\mathbb{Q}$.

The reflex field E is the field of rationality of the representation of F on $Lie(A_h)$; therefore, the field of definition of (A_h, ι) always contains this field E. It would then be natural to expect that the moduli variety of triples (A, λ, ι) for an abelian variety A with F-linear isomorphism $Lie(A) \cong V_1$ is defined over E.

Since the isomorphism class of ρ_1 is determined by $\mathrm{Tr}(\rho_1)$ (see [**MFG**] Proposition 2.9), E is generated over $\mathbb{Q}$ by $\mathrm{Tr}(\rho_1(b))$ for all $b \in F$. Thus we have $E = F$ or $\mathbb{Q}$ and that $E = \mathbb{Q}$ implies $m_1 = m_c$, because $\mathrm{Tr}(\rho_1(\xi)) = m_1\xi + m_c\xi^c$ for $\xi \in F$. We write O_E for the integer ring of E. Let $\mathbb{Z}_{(p)} = \mathbb{Z}_p \cap \mathbb{Q}$, put $O_{(p)} = O \otimes_{\mathbb{Z}} \mathbb{Z}_{(p)}$, $O_{E,(p)} = O_E \otimes_{\mathbb{Z}} \mathbb{Z}_{(p)}$, and write $\mathcal{V}$ for the valuation ring $\mathcal{W} \cap E \supset O_{E,(p)}$ ($\mathcal{V}$ is the localization of O_E at $\mathfrak{p}$). More generally, for a finite set of places Σ, we write $\mathbb{Z}_\Sigma$ for the product of $\mathbb{Z}_\ell$ over finite places $\ell \in \Sigma$, and we put $\mathbb{Z}_{(\Sigma)} = \mathbb{Q} \cap \mathbb{Z}_\Sigma$ and $O_{(\Sigma)} = O \otimes_{\mathbb{Z}} \mathbb{Z}_{(\Sigma)}$. The ring $\mathcal{V}$ has residue field $\mathbb{F}_p$ since p is split in E because $E \subset F$.

2.3. Shimura variety for GU. We study the classification problem of quadruples $(A, \lambda, i, \overline{\eta}^{(p)})_{/R}$: A is a (projective) abelian scheme over a base R, ${}^tA = \mathrm{Pic}^0_{A/R}(A)$ is the dual abelian scheme of A, $\lambda : A \to {}^tA$ is a prime-to-p polarization (that is, an isogeny with degree prime to p fiber-by-fiber geometrically induced from an ample divisor), $i : O_{(p)} \hookrightarrow \mathrm{End}_R^{\mathbb{Z}_{(p)}}(A) = \mathrm{End}_R(A) \otimes_{\mathbb{Z}} \mathbb{Z}_{(p)}$ is a $\mathbb{Z}_{(p)}$-algebra embedding (taking 1 to the identity of A) with $\lambda \circ i(\alpha^c) = {}^t i(\alpha) \circ \lambda$ for all $\alpha \in O$, and $\eta^{(p)}$ is a level structure. Regarding ${}^tA$ as a left O-module by $O \ni b \mapsto {}^t i(b^c) \in \mathrm{End}({}^tA)$, λ is F-linear. Hereafter we call λ *F-linear* in this sense. The base scheme R is assumed to be a scheme over $\mathrm{Spec}(\mathcal{V})$.

We clarify the meaning of the level structure $\eta^{(p)}$. Fix a base (geometric) point $s \in R$ and write A_s for the fiber of A at s. We consider the Tate module $\mathcal{T}(A_s) = \varprojlim_N A[N](k(s))$ and $V^{(p)}(A_s) = \mathcal{T}(A_s) \otimes_{\mathbb{Z}} \mathbb{A}^{(p\infty)}$, where N runs over all positive integers ordered by divisibility. The prime-to-p level structure $\eta^{(p)} : V(\mathbb{A}^{(p\infty)}) = V \otimes_{\mathbb{Q}} \mathbb{A}^{(p\infty)} \cong V^{(p)}(A_s)$ is an O-linear isomorphism. The duality pairing $e_N : A[N] \times {}^tA[N] \to \mu_N$ composed with λ gives, after taking the limit with respect to N, an alternating form $(\cdot,\cdot)_\lambda : V^{(p)}(A_s) \times V^{(p)}(A_s) \to \mathbb{A}^{(p\infty)}(1) := \lim_{p\nmid N} \mu_N$ satisfying the following conditions:

(P1) $(\alpha(x), y)_\lambda = (x, \alpha^c(y))_\lambda$ for $\alpha \in \mathrm{End}(A_{/B})$;

(P2) The pairing induces the self-duality: $A[p^n] \cong \mathrm{Hom}(A[p^n], \mu_{p^n})$ if $N = p^n$.

We require that $\eta^{(p)}$ send the alternating form $\langle\cdot,\cdot\rangle$ to $(\cdot,\cdot)_\lambda$ up to multiple of scalars in $(\mathbb{A}^{(p\infty)})^\times$. This is possible, because $\mathbb{A}^{(p\infty)}(1) \cong \mathbb{A}^{(p\infty)}$ up to scalars in $(\mathbb{A}^{(p\infty)})^\times$. Then $\eta^{(p)}$ is required to be an isomorphism of skew Hermitian F-modules with respect to the pairing $\langle\cdot,\cdot\rangle_\lambda$ on $V^{(p)}(A_s)$.

The algebraic fundamental group $\pi_1(R, s)$ acts on $V^{(p)}(A_s)$ preserving the skew Hermitian form $\langle\cdot,\cdot\rangle_\lambda$ up to scalars in $(\mathbb{A}^{(p\infty)})^\times$ (because it preserves the Weil e_N-pairing; see [**ABV**] Section 20). Take a closed subgroup $K^{(p)} \subset GU(\mathbb{A}^{(p\infty)})$. We write $\overline{\eta}^{(p)}$ for the orbit $\eta^{(p)} \circ K^{(p)}$. If $\sigma \circ \overline{\eta}^{(p)} = \overline{\eta}^{(p)}$ for all $\sigma \in \pi_1(R, s)$, we say the level structure $\overline{\eta}^{(p)}$ is defined over R. Even if we change the point $s \in R$, everything will be conjugated by an isomorphism; therefore, the definition does not depend on the choice of s as long as R is connected. For nonconnected R, we choose one geometric point at each connected component.

A quadruple $\underline{A}_{/R} = (A, \lambda, i, \overline{\eta}^{(p)})$ is *isomorphic* to $\underline{A}'_{/R} = (A', \lambda', i', \overline{\eta}'^{(p)})$ if we have an O-linear isogeny $\phi : A \to A'$ defined over R such that $p \nmid \deg(\phi)$,

$\phi^*\lambda' = {}^t\phi \circ \lambda' \circ \phi = \nu\lambda$ with $\nu \in \mathbb{Z}^\times_{(p)+}$, $\phi \circ i \circ \phi^{-1} = i'$, and $\overline{\eta}'^{(p)} = \phi \circ \overline{\eta}^{(p)}$. Here $\mathbb{Z}^\times_{(p)+}$ is the collection of all positive elements in $\mathbb{Z}^\times_{(p)}$. Thus ϕ associates the prime-to-p polarization class $\overline{\lambda}' = \{\nu\lambda' | \nu \in \mathbb{Z}^\times_{(p)+}\}$ of λ' to the class $\overline{\lambda}$ of λ: $\phi^*\overline{\lambda}' = \overline{\lambda}$. In this case, we write $A \approx A'$. We write $A \cong A'$ if the isogeny is an isomorphism of abelian schemes; that is, $\deg(\phi) = 1$.

We take the fibered category $\mathcal{C} = \mathcal{C}_{F,V}$ of the quadruples $(A, \lambda, i, \eta^{(p)})_{/R}$ over the category $\mathcal{V}$-SCH of $\mathcal{V}$-schemes and define

$$(2.2)\quad \mathrm{Hom}_{\mathcal{C}_{/R}}((A,\lambda,i,\eta^{(p)})_{/R},(A',\lambda',i',\eta'^{(p)})_{/R}) \\ = \left\{ \phi \in \mathrm{Hom}_R(A,A') \otimes_{\mathbb{Z}} \mathbb{Z}_{(p)} \middle| \begin{matrix} {}^t\phi \circ \lambda' \circ \phi = \nu\lambda \text{ with } 0 < \nu \in \mathbb{Z}^\times_{(p)+}, \\ \phi \circ i = i' \circ \phi \text{ and } \eta'^{(p)} = \phi \circ \eta^{(p)} \end{matrix} \right\}.$$

The representation ρ_1 is well defined over $\mathcal{V}$, since p splits in F; thus, it is well defined over $\mathcal{O}_R$ for any $\mathcal{V}$-scheme R. We consider the functor $\mathcal{E}^{(p)} : \mathcal{V}$-$SCH \to SETS$ given by

$$\mathcal{E}^{(p)}(R) = \left\{ \underline{A}_{/R} = (A, \overline{\lambda}, i, \eta^{(p)})_{/R} \middle| Lie(A) \cong \rho_1 \text{ over } \mathcal{O}_R \right\} / \approx .$$

Since $A_{/R}$ is a group scheme, its tangent space at the zero section has a Lie algebra structure over $\mathcal{O}_R$. We write $Lie(A)$ for this Lie algebra. Since A is smooth over R, $Lie(A)$ is a locally free $\mathcal{O}_R$-module of rank $\dim_R A$. In our case, for a given quadruple $\underline{A} = (A, \lambda, i, \overline{\eta}^{(p)})_{/R}$, the Lie algebra $Lie(A)$ of A over $\mathcal{O}_R$ is an $O_{(p)}$-module via i. Since $Lie(A)$ is locally free of rank $\dim_R A$ over $\mathcal{O}_R$, we can think of an isomorphism $Lie(A) \cong \rho_1$ of $\mathcal{O}_R$-representations of $O_{(p)}$. One can find in [**PAF**] Chapter 7 a proof of the following theorem due to Shimura, Deligne and Kottwitz.

THEOREM 2.1. *The functor $\mathcal{E}^{(p)}$ is representable by a quasi-projective smooth pro-scheme $Sh^{(p)}$ over $\mathcal{V}$. Letting $g \in GU(\mathbb{A}^{(p\infty)})$ act on $Sh^{(p)}$ by $\eta^{(p)} \mapsto \eta^{(p)} \circ g$, for each compact open subgroup $K \subset G(\mathbb{A}^{(p\infty)})$, the quotient scheme $Sh^{(p)}_K = Sh^{(p)}/K$ exists as a quasi-projective scheme of finite type over $\mathcal{V}$, and $Sh^{(p)} = \varprojlim_K Sh^{(p)}_K$. The Shimura variety $Sh^{(p)}_K$ is projective over $\mathcal{V}$ if the Hermitian pairing $\langle\cdot,\cdot\rangle$ is anisotropic.*

For a finite set of primes Σ containing p and ∞, we can think of the Shimura variety away from Σ as follows. Write $\Sigma = \{p, \infty\} \sqcup \Sigma'$. If $\Sigma' \neq \emptyset$, let $GU(\mathbb{Z}_{\Sigma'}) = \{g \in GU(\mathbb{Q}_{\Sigma'}) | gL_{\Sigma'} = L_{\Sigma'}\}$, and put $Sh^{(\Sigma)} = Sh^{(p)}/GU(\mathbb{Z}_{\Sigma'})$. It is known that $Sh^{(\Sigma)}_{/\mathcal{V}}$ is a smooth (quasi-projective) pro-scheme.

Recall the embedding $i_p : \overline{\mathbb{Q}} \hookrightarrow \overline{\mathbb{Q}}_p$ and the valuation ring $\mathcal{W}$ which is the pull-back by i_p of the p-adic integer ring of the maximal unramified extension of $\mathbb{Q}_p$. By our choice, $1 : F \hookrightarrow \overline{\mathbb{Q}} \overset{i_p}{\hookrightarrow} \overline{\mathbb{Q}}_p$ induces the valuation ring $\mathcal{V}$. Write $\mathcal{K}$ be the filed of fraction of $\mathcal{W}$. Let $Sh^{(\Sigma)}_{/\mathcal{W}} = Sh^{(\Sigma)} \times_{\mathrm{Spec}(\mathcal{V})} \mathrm{Spec}(\mathcal{W})$ and put $W = \varprojlim_n \mathcal{W}/p^n\mathcal{W}$. By the reduction map (see [**ACS**] Corollary 6.4.1.3), we have $\pi_0(Sh^{(\Sigma)}_{/\mathcal{K}}) \cong \pi_0(Sh^{(\Sigma)}_{/\mathbb{F}})$ for $Sh^{(\Sigma)}_{/\mathbb{F}} = Sh^{(\Sigma)}_{/\mathcal{W}} \times_{\mathcal{W}} \mathbb{F}$ by Zariski's connectedness theorem and the existence of a smooth toroidal compactification of $Sh^{(p)}_{K/W}$, and $SU(\mathbb{A}^{(\Sigma)})$ leaves stable each irreducible component in $\pi_0(Sh^{(\Sigma)}_{/\mathcal{K}})$ because $\mathfrak{X}^+$ is a quotient of $SU(\mathbb{R})$. A proof of the existence of a smooth toroidal compactification of $Sh^{(p)}_{K/W}$ can also be found in [**ACS**] 6.4.1. Thus, by the existence of a smooth toroidal compactification of $Sh^{(\Sigma)}_{/\mathcal{W}}$ (and Zariski's connectedness theorem), we get

PROPOSITION 2.2. *Geometrically irreducible components of $Sh^{(\Sigma)}_{/\mathcal{K}}$ are generic fibers of irreducible components of $Sh^{(\Sigma)}_{/\mathcal{W}}$. Each irreducible component of $Sh^{(\Sigma)}_{/\mathcal{W}}$ has irreducible special fiber over $\mathbb{F}$, and the group $SU(\mathbb{A}^{(\Sigma)})$ leaves stable each irreducible component of the Shimura variety $Sh^{(\Sigma)}_{/\mathbb{F}} = Sh^{(\Sigma)}_{/\mathcal{W}} \times_{\mathcal{W}} \mathbb{F}$.*

We can compute the stabilizer in $GU(\mathbb{A}^{(\Sigma)})$ of each point of $\pi_0(Sh^{(\Sigma)}_{/\mathbb{F}})$ explicitly ([**H06**] Lemma 1.1).

3. Igusa tower over unitary Shimura variety

We first define the Igusa tower over the GU Shimura variety and prove that the tower is not irreducible. Then we prove the irreducibility of the partial SU-tower. Let $G(\mathbb{Z}_p) = \{g \in G(\mathbb{Q}_p) | gL_p = L_p\}$ for $G = GU, U$ and SU. Let Σ be a finite set of rational places including p and ∞.

3.1. Unitary group over $\mathbb{Z}_p$. Recall our simplifying assumption: $p = \mathfrak{p}\overline{\mathfrak{p}}$ ($\mathfrak{p} \neq \overline{\mathfrak{p}}$) in O so that $\mathfrak{p}$ is induced by i_p. Since $O_p = O_{\mathfrak{p}} \times O_{\overline{\mathfrak{p}}} = \mathbb{Z}_p \times \mathbb{Z}_p$ on which c acts by interchanging the coordinates, $(x, y)^c = (y, x)$ and $\xi \in O$ is sent to $(\xi, \xi^c) \in \mathbb{Z}_p \times \mathbb{Z}_p$, we thus have $GL_r(O_p) = GL_r(O_{\mathfrak{p}}) \times GL_r(O_{\overline{\mathfrak{p}}}) = GL_r(\mathbb{Z}_p) \times GL_r(\mathbb{Z}_p)$. Since $x^\iota = s^{-1}{}^t x^c s$ for the skew-hermitian matrix $s = -{}^t s^c$, if $(x, y) \in U(\mathbb{Z}_p)$, we have

$$(x^{-1}, y^{-1}) = (x, y)^{-1} = x^\iota = (s, s^c)^{-1}({}^t y, {}^t x)(s, s^c) = (s^{-1}{}^t y s, s^{-c}{}^t x s^c)$$

and $y = {}^t s^{-1} x^{-1} {}^t s$. Thus, choosing a basis of L_p over O_p, we have $U(\mathbb{Z}_p) \cong GL_r(\mathbb{Z}_p)$ by sending $(x, y) \in U(\mathbb{Z}_p)$ to $x \in GL_r(\mathbb{Z}_p)$. Similarly, $SU(\mathbb{Z}_p) \cong SL_r(\mathbb{Z}_p)$ and $GU(\mathbb{Z}_p) \cong GL_r(\mathbb{Z}_p) \times GL_1(\mathbb{Z}_p)$ by $g = (x, y) \mapsto (x\nu(x, y)^{-1}, \nu(x, y))$.

3.2. The Igusa tower. Let $S_{/\mathcal{W}} = S^{(\Sigma)}_{/\mathcal{W}}$ be an irreducible component of the ordinary locus of $Sh^{(\Sigma)}_{/\mathcal{W}}$. Thus S is the subscheme obtained from $Sh^{(\Sigma)}_{/\mathcal{W}}$ by removing the closed subscheme of non-ordinary locus at the special fiber at p. By $\langle \cdot, \cdot \rangle$, L_p is self-dual. Since $O_p = O_{\mathfrak{p}} \oplus O_{\overline{\mathfrak{p}}}$, we have the corresponding decomposition $L_p = L_{\mathfrak{p}} \oplus L_{\overline{\mathfrak{p}}}$.

Let $\mathbf{A}_{/S}$ be the universal ordinary abelian scheme over S with its fiber A_x at $x \in S$. Pick a base point x_0 of $S(W)$ ($W = \varprojlim_n \mathcal{W}/p^n\mathcal{W}$) with reduction $\overline{x}_0 \in S(\mathbb{F})$ modulo p. We fix an identification: $L_{\mathfrak{p}} \cong T_p A_{x_0}[\mathfrak{p}^\infty]$ for the p-adic Tate module $T_p A_{x_0}[\mathfrak{p}^\infty]$ of the Barsotti-Tate group $A_{x_0}[\mathfrak{p}^\infty]$. Then over the formal completion $\widehat{S}$ along the special fiber, we have the reduction map $T_p A_{x_0}[\mathfrak{p}^\infty] \to T_p A_{\overline{x}_0}[\mathfrak{p}^\infty]^{\text{ét}}$. The kernel of the reduction map gives rise to an $O_{\mathfrak{p}}$-direct summand $L_1 \subset L_{\mathfrak{p}}$. Since O acts on the tangent space at 0 via the identity inclusion into $\mathbb{Z}_p$ by multiplicity m_1 and the tangent space of $A[\mathfrak{p}]^\circ_{/x_0}$ is equal to this eigenspace in the tangent space of $A_{\overline{x}_0}$, we find that $L_1 \otimes_{O_{\mathfrak{p}}} \mathbb{F}_p \cong \mathbb{F}_p^{m_1}$; thus, $L_1 \cong O_{\mathfrak{p}}^{m_1}$. Similarly, we define $L_c \subset L_{\overline{\mathfrak{p}}}$ using the reduction map on $\overline{\mathfrak{p}}$-torsion points of A_{x_0}. Then $L_c \cong O_{\overline{\mathfrak{p}}}^{m_c}$. Note that $L_{\mathfrak{p}}/L_1 \cong \mathrm{Hom}_{\mathbb{Z}_p}(L_c, \mathbb{Z}_p)$ and $L_{\overline{\mathfrak{p}}}/L_c \cong \mathrm{Hom}_{\mathbb{Z}_p}(L_1, \mathbb{Z}_p)$ by $\langle \cdot, \cdot \rangle$. Let $\mathcal{L} = L_1 \oplus L_c$ as O-modules.

We consider the functor $I_n = I_n^{(\Sigma)}$ from the category of $S_{/\mathcal{W}}$-schemes R into the category of sets taking R to the set of O-linear closed immersions of $\mathcal{L}\otimes_{\mathbb{Z}}\mu_{p^n/R}$ into $\mathbf{A}_{/R}[p^n]$, where $\mathbf{A}_{/R} = \mathbf{A}\times_S R$. Since the two schemes $\mathcal{L}\otimes_{\mathbb{Z}}\mu_{p^n}$ and $\mathbf{A}[p^n]$ are finite flat over S, by the theory of Hilbert schemes, this functor is representable by a scheme I_n. Then $I_{n/\mathcal{W}}$ classifies quintuples $(A, i, \overline{\lambda}, \eta^{(\Sigma)}, \phi_p)$ for an O-linear closed immersion $\phi_p : \mathcal{L}\otimes_{\mathbb{Z}}\mu_{p^n} \hookrightarrow A[p^n]$.

The formal completion $\widehat{S}$ along the special fiber $S_{/\mathbb{F}} = S\times_{\mathcal{W}}\mathbb{F}$ is a formal W-scheme. The connected component $\mathbf{A}[p^n]^\circ$ of $\mathbf{A}[p^n]$ is well defined over $\widehat{S}$, and hence the formal completion $\widehat{I}_{n/W}$ of I_n along its special fiber $I_{n/\mathbb{F}} = I_n\times_{\mathcal{W}}\mathbb{F}$ can be written as $\mathrm{Isom}_{\widehat{S}}(\mathcal{L}\otimes_{\mathbb{Z}}\mu_{p^n}, \mathbf{A}[p^n]^\circ)$. Then $\widehat{I}_{n/\widehat{S}}$ is isomorphic to the scheme $\mathrm{Isom}_{\widehat{S}}(\mathcal{L}^\vee/p^n\mathcal{L}^\vee, \mathbf{A}[p^n]^{\text{ét}})$ étale finite over $\widehat{S}$, since by duality, $\phi_p : \mathcal{L}\otimes_{\mathbb{Z}}\mu_{p^n} \cong \mathbf{A}[p^n]^\circ$ gives rise to ${}^t\phi_p^{-1} : \mathcal{L}^\vee/p^n\mathcal{L}^\vee \cong \mathbf{A}[p^n]^{\text{ét}}$ for $\mathcal{L}^\vee = L_p/\mathcal{L}$.

Let $I_{/\mathcal{W}} = \varprojlim_n I_{n/\mathcal{W}}$. Its special fiber

$$I_{/\mathcal{W}}\times_{\mathcal{W}}\mathbb{F} = \varprojlim_n I_{n/\mathbb{F}}$$

is called the Igusa tower over $S_{/\mathbb{F}}$. By the projection $L_p \twoheadrightarrow L_{\mathfrak{p}}$, we have $U(\mathbb{Z}_p) \cong GL(L_{\mathfrak{p}}) \cong GL_r(\mathbb{Z}_p)$. Consider the universal level structure $\boldsymbol{\phi}_p : \mathcal{L}\otimes\mu_{p^\infty} \hookrightarrow \mathbf{A}[p^\infty]$ over I. The group $GU(\mathbb{Z}_p)$ acts on L. Let

$$P(\mathbb{Z}_p) = \left\{g\in U(\mathbb{Z}_p) = GL(L_{\mathfrak{p}})\middle| gL_1 = L_1\right\}.$$

Then, identifying $GL(L_1) = GL_{m_1}(\mathbb{Z}_p)$ and $GL(L_{\mathfrak{p}}/L_1) = GL_{m_c}(\mathbb{Z}_p)$, $P(\mathbb{Z}_p)$ is a parabolic subgroup of $U(\mathbb{Z}_p) = GL_r(\mathbb{Z}_p)$ of the following form,

$$\left\{\left(\begin{smallmatrix} a & b\\ 0 & d\end{smallmatrix}\right)\middle|(a,d)\in GL(L_1)\times GL(L_c^\vee)\right\} \cong \left\{\left(\begin{smallmatrix} a & b\\ 0 & d\end{smallmatrix}\right)\middle|(a,d)\in GL_{m_1}(\mathbb{Z}_p)\times GL_{m_c}(\mathbb{Z}_p)\right\}.$$

Here the action of $d\in GL(L_c^\vee)$ on $L_c^\vee = L_{\mathfrak{p}}/L_1$ is given by the matrix d and hence it acts on $L_c = \mathrm{Hom}(L_{\mathfrak{p}}/L_1, \mathbb{Z}_p)$ by the dual action (induced by $\langle\cdot,\cdot\rangle$) written as d^{-*}. Define $M(\mathbb{Z}_p) = GL(L_1)\times GL(L_c^\vee)$ for the reductive part of P. Put $M_1(\mathbb{Z}_p) = M(\mathbb{Z}_p)\cap SU(\mathbb{Z}_p)$. Then $M(\mathbb{Z}_p)$ acts on each fiber of I transitively, since $I/S_{/\mathbb{F}}$ is an $M(\mathbb{Z}_p)$-torsor by the action

$$(\phi_{\mathfrak{p}}, \phi_{\overline{\mathfrak{p}}})\circ(a,d) = (\phi_{\mathfrak{p}}\circ a, \phi_{\overline{\mathfrak{p}}}\circ d^{-*}),$$

where the original action of d on $L_{\mathfrak{p}}/L_1$ is dualized by the polarization pairing

$$\langle\cdot,\cdot\rangle_\lambda : \mathbf{A}[\mathfrak{p}^\infty]^{\text{ét}}\times\varprojlim_n \mathbf{A}[\overline{\mathfrak{p}}^n]^\circ \to \mu_{p^\infty}.$$

3.3. Reducibility and irreducibility. First, we may assume that $S(\mathbb{C})$ is the image of $SU(\mathbb{A}^{(\Sigma)})\times\mathfrak{X}^+$ in $Sh^{(\Sigma)}(\mathbb{C}) = GU(\mathbb{Q})\backslash(GU(\mathbb{A}^{(\infty)})\times\mathfrak{X})/GU(\mathbb{Z}_\Sigma)\overline{Z(\mathbb{Q})}$, where $\mathbb{Z}_\Sigma = \prod_{\ell\in\Sigma-\{\infty\}}\mathbb{Z}_\ell$, $\mathbb{Q}_\Sigma = \prod_{\ell\in\Sigma-\{\infty\}}\mathbb{Q}_\ell$, $\mathbb{Z}_{(\Sigma)} = \mathbb{Q}\cap\mathbb{Z}_\Sigma$ in $\mathbb{Q}_\Sigma$ and $GU(\mathbb{Z}_\Sigma) = \{x\in GU(\mathbb{Q}_\Sigma)|xL_\Sigma = L_\Sigma\}$ for $L_\Sigma = L\otimes_{\mathbb{Z}}\mathbb{Z}_\Sigma$.

On S, the universal level structure $\boldsymbol{\eta}^{(\Sigma)} : V(\mathbb{A}^{(\Sigma)}) \cong V^{(\Sigma)}(\mathbf{A})$ induces the trivialization of the étale $\mathbb{A}^{(\Sigma)}$-sheaf:

$$\det(\boldsymbol{\eta}^{(\Sigma)}) : \mathbb{A}^{(\Sigma)} \cong \bigwedge^r_{F_{\mathbb{A}^{(\Sigma)}}} V(\mathbb{A}^{(\Sigma)}) \cong \bigwedge^r_{F_{\mathbb{A}^{(\Sigma)}}} V^{(\Sigma)}(\mathbf{A}).$$

For any prime ℓ outside Σ, take a compact open subgroup K of $GU(\mathbb{A}^{(\Sigma)})$ such that $K = K_\ell\times K^{(\ell)}$ with $K_\ell = \{x\in GU(\mathbb{Z}_\ell)|xL_\ell = L_\ell\}$ and such that $Sh^{(\Sigma)}/Sh_K^{(\Sigma)}$

($Sh_K^{(\Sigma)} = Sh^{(\Sigma)}/K$) is an étale covering. Then for the principal congruence subgroup $K(\ell^n) \subset K$ modulo ℓ^n, $Sh_{K(\ell^n)/\mathcal{W}}$ is constructed as $\mathrm{Isom}_{Sh_K^{(\Sigma)}}(L/\ell^n L, \mathbf{A}_K[\ell^n])$ for the universal abelian scheme $\mathbf{A}_K$ over $Sh_K^{(\Sigma)}$. Let S_K be the image of S in $Sh_K^{(\Sigma)}$ and write again as x_0 the image of x_0 in S_K. By this expression, the action of $\pi_1(S_K, x_0)$ on the étale sheaf $\mathbf{A}_K[\ell^n]_{/S_K}$ factors through the action of $K_\ell \cap SU(\mathbb{Z}_\ell)$. In particular, its action on $\bigwedge^r_{O_\ell} \mathbf{A}_K[\ell^n]$ factors through $\det : K_\ell \cap SU(\mathbb{Z}_\ell) \to O_\ell^\times$ which is the trivial character by the definition of SU. Thus $\bigwedge^r_{O_\ell} \mathbf{A}_K[\ell^n]$ is a constant étale sheaf over $S_{K/\mathcal{W}}$. In other words, the action of $GU(\mathbb{A}^{(\Sigma)})$ on $\bigwedge^r_{F_{\mathbb{A}^{(\Sigma)}}} V(\mathbb{A}^{(\Sigma)})$ factors through the determinant map, it is trivial on $SU(\mathbb{A}^{(\Sigma)})$, and $V^{(\Sigma)}(\mathbf{A})$ over the irreducible component $S_{/\mathcal{W}}$ is constant; thus, the ℓ-adic sheaf $\bigwedge^r_{O_\ell} \mathcal{T}_\ell \mathbf{A}$ ($\mathcal{T}_\ell \mathbf{A} = \varprojlim_n \mathbf{A}[\ell^n]$) is identical to $\bigwedge^r_{O_\ell} \mathcal{T}_\ell A_x$ for the fiber of $\mathbf{A}$ at any closed point $x \in S(\mathcal{K})$.

For any exact sequence of free $\mathbb{Z}_p$-modules $X_1 \hookrightarrow X \twoheadrightarrow X_2$ with ranks r_1, r and r_2 respectively, we have a natural direct summand $\bigwedge^{r_1} X_1 \otimes \bigwedge^{r_2} X_2$ in $\bigwedge^r X$, because the ambiguity of lifting $x_2 \in X_2$ to $x \in X$ is killed by wedge product with $\bigwedge^{r_1} X_1$.

As for the fppf abelian sheaf $\bigwedge^{m_1}_{O_\mathfrak{p}} \mathbf{A}[\mathfrak{p}^n]^\circ_{/\widehat{S}_{/\mathcal{W}}}$ over $\widehat{S}_{/W}$, it is isomorphic to $\bigwedge^{m_1}_{\mathbb{Z}_p}(O_\mathfrak{p} \otimes \mu_{p^n})^{m_1}$; thus, its dual étale sheaf $\bigwedge^{m_1}_{O_{\overline{\mathfrak{p}}}} \mathbf{A}[\overline{\mathfrak{p}}^n]^{\text{ét}}_{/\widehat{S}_{/\mathcal{W}}}$ is constant over $\widehat{S}_{/W}$. Similarly $\bigwedge^{m_c}_{O_\mathfrak{p}} \mathbf{A}[\mathfrak{p}^n]^{\text{ét}}_{/\widehat{S}_{/\mathcal{W}}}$ is constant. Thus

$$\mathbb{E}[p^n] = \bigwedge_{\mathbb{Z}_p}^{m_c} \mathbf{A}[\mathfrak{p}^n]^{\text{ét}} \otimes_{\mathbb{Z}_p} \bigwedge_{\mathbb{Z}_p}^{m_1} \mathbf{A}[\overline{\mathfrak{p}}^n]^{\text{ét}}_{/\widehat{S}}$$

is isomorphic to the constant sheaf $\mathbb{Z}/p^n\mathbb{Z}$ over $\widehat{S}_{/W}$. Thus we have a morphism

$$\det : \widehat{I}_{n/\widehat{S}} = \mathrm{Isom}_{\widehat{S}}\Big(\frac{\mathcal{L}^\vee}{p^n \mathcal{L}^\vee}, \mathbf{A}[p^n]^{\text{ét}}\Big) \to \mathrm{Isom}\Big(\bigwedge^{m_c} \frac{L_1^\vee}{p^n L_1^\vee} \otimes_{\mathbb{Z}_p} \bigwedge^{m_1} \frac{L_c^\vee}{p^n L_c^\vee}, \mathbb{E}[p^n]\Big) \cong (\mathbb{Z}/p^n\mathbb{Z})^\times$$

over $\widehat{S}$ taking ${}^t\phi_p^{-1} : \mathcal{L}^\vee/p^n\mathcal{L}^\vee \cong \mathbf{A}[p^n]^{\text{ét}}$ to

$$\left(\bigwedge^{m_1} ({}^t\phi_p^{-1}|_{L_c^\vee/p^n L_c^\vee}) \otimes \bigwedge^{m_c} ({}^t\phi_p^{-1}|_{L_1^\vee/p^n L_1^\vee})\right).$$

Pick a generator

$$v \in \varprojlim_n \mathrm{Isom}\Big(\bigwedge^{m_c} \frac{L_1^\vee}{p^n L_1^\vee} \otimes_{\mathbb{Z}_p} \bigwedge^{m_1} \frac{L_c^\vee}{p^n L_c^\vee}, \mathbb{E}[p^n]\Big)$$

over $\mathbb{Z}_p$, and define $I_n^{SU} = I_n^{SU,(\Sigma)} = \det^{-1}(v \bmod p^n)$ and $I^{SU} = I^{SU,(\Sigma)} = \varprojlim_n I_n^{SU,(\Sigma)}$. We claim

THEOREM 3.1. *For each finite set Σ of rational places containing p and ∞, $I_n^{SU,(\Sigma)}/S$ is a geometrically irreducible component of $I_{n/S}$.*

3.4. Proof. By construction, $I_n^{SU,(\Sigma)}$ contains an irreducible component of $I^{(\Sigma)}_{n/S}$. Thus we need to prove irreducibility of $I_n^{SU,(\Sigma)}/S$ showing axioms (A1–2). For a point $x \in I^{SU,(\Sigma)}(\mathbb{F})$, consider the formal completion $\widehat{I}^{SU,(\Sigma)}_{x/W}$ along x. Then $\mathcal{O}_{\widehat{I}^{SU,(\Sigma)}_{x/W}} \cong W[[X_1,\dots,X_d]]$ for $d = \dim_{\mathcal{W}} S$ ($\Leftrightarrow \widehat{I}^{SU,(\Sigma)}_{x/W} \cong \mathrm{Spf}(W[[X_1,\dots,X_d]])$). Define the valuation $v_x : \mathcal{O}_{\widehat{I}^{SU,(\Sigma)}_{x/W}} \to \mathbb{Z}\cup\{\infty\}$ as already mentioned:

$$v_x(\sum_\alpha c(\alpha)X^\alpha) = \mathrm{Inf}_\alpha\, \mathrm{ord}_p(c(\alpha))$$

where $X^\alpha = X_1^{\alpha_1}X_2^{\alpha_2}\cdots X_d^{\alpha_d}$. Then the stalk $\mathcal{O}_{I^{SU,(\Sigma)},x} \subset \mathcal{O}_{\widehat{I}^{SU,(\Sigma)}_{x/W}}$ inherits the valuation v_x and hence its function field $\mathfrak{F} = \mathcal{K}(I^{SU,(\Sigma)})$ gets the valuation v_x. The valuation v_x is unramified over the function field $\mathfrak{F}_{S^{(\Sigma)}} = \mathcal{K}(S^{(\Sigma)})$. Let D (resp. T_x) be the stabilizer of v_x (resp. x) in $M_1(\mathbb{Z}_p)\times SU(\mathbb{A}^{(\Sigma)})$. Then $T_x \subset D$.

First take Σ to be Σ_0 given by $\{p,\infty\}\cup\{\ell | SU \text{ is not quasi-split at } \ell\}$. Then $SU(\mathbb{A}^{(\Sigma)})$ does not have any finite quotient. In particular, $SU(\mathbb{A}^{(\Sigma)})$ fixes each connected component of I_n^{SU}, and $SU(\mathbb{A}^{(\Sigma)}) \subset D$. As will be seen in the following section, we can find one base point $x = x_0$ such that T_{x_0} has p-adically dense image in $M_1(\mathbb{Z}_p)$ under the projection: $SU(\mathbb{A}^{(\Sigma)})\times M_1(\mathbb{Z}_p) \to M_1(\mathbb{Z}_p)$. Thus $T_{x_0}\cdot SU(\mathbb{A}^{(\Sigma)})$ is dense in $SU(\mathbb{A}^{(\Sigma)})\times M_1(\mathbb{Z}_p)$. Since $D \supset T_{x_0}\cdot SU(\mathbb{A}^{(\Sigma)})$, D contains $SU(\mathbb{A}^{(\Sigma)})\times M_1(\mathbb{Z}_p)$ and in particular contains $M_1(\mathbb{Z}_p)$. This shows the irreducibility of $I^{SU,(\Sigma)}$.

If Σ_0 as above is bigger than the minimal choice $\sigma = \{p,\infty\}$, we note that $\mathbb{F}(S^{(\sigma)})$ and $\mathbb{F}(I^{SU,(\Sigma_0)})$ are linearly disjoint over $\mathbb{F}(S^{(\Sigma_0)})$. Indeed, we have

$$\mathbb{F}(S^{(\sigma)})\cap\mathbb{F}(I^{SU,(\Sigma_0)}) = \mathbb{F}(S^{(\Sigma_0)})$$

by construction, and the two extensions are Galois extensions over $\mathbb{F}(S^{(\Sigma_0)})$. The quotient field $\mathbb{K}$ of the integral domain $\mathbb{F}(S^{(\sigma)})\otimes_{\mathbb{F}(S^{(\Sigma_0)})}\mathbb{F}(I_n^{SU,(\Sigma_0)})$ has degree equal to the covering degree $[I_n^{SU,(\sigma)} : S^{(\sigma)}]$ and $\mathbb{K}$ is an intermediate field of $\mathbb{F}(I_n^{SU,(\sigma)})/\mathbb{F}(S^{(\sigma)})$; therefore, $\mathbb{K}$ is the function field of the full Igusa tower $I^{SU,(\sigma)}_{n/S^{(\sigma)}}$. This shows that $I^{SU,(\sigma)}_{/S^{(\sigma)}}$ is still irreducible.

For an arbitrary $\Sigma \supsetneq \sigma$, the natural projection $I^{SU,(\sigma)} \to I^{SU,(\Sigma)}$ is surjective dominant; therefore, the irreducibility of $I^{SU,(\sigma)}$ implies the irreducibility of $I^{SU,(\Sigma)}$.

3.5. Finding the base point x_0. Here is how to find the point x_0 with p-adically dense image in $M_1(\mathbb{Z}_p)$. For simplicity, we assume that $p > 2$. The unitary group $GU_{/\mathbb{Q}}$ depends only on the hermitian vector space V not the lattice L. The unitary group $GU_{/\mathbb{Z}_{(p)}}$ depends on the hermitian form on $L_{(p)} = L\otimes_{\mathbb{Z}}\mathbb{Z}_{(p)}$, and $Sh^{(p)}_{/\mathcal{V}}$ only depends on $GU_{/\mathbb{Z}_{(p)}}$; thus, we may change the lattice L without changing $L_{(p)}$. In particular, if necessary, replacing L keeping $L_{(p)}$ intact, we may assume that the hermitian matrix s is diagonalizable over L (if $p > 2$).

Since $g \in M(\mathbb{Z}_p)$ acts transitively on $\mathbb{E}[p^n] - \mathbb{E}[p^{n-1}] \cong (\mathbb{Z}/p^n\mathbb{Z})^\times$ by multiplication by $\det(g)$, we can change the element

$$v \in \varprojlim_n \mathrm{Isom}(\bigwedge^{m_c}\frac{L_1^\vee}{p^nL_1^\vee}\otimes_{\mathbb{Z}_p}\bigwedge^{m_1}\frac{L_c^\vee}{p^nL_c^\vee}, \mathbb{E}[p^n])$$

(appearing in the definition of $I^{SU,(\Sigma_0)}$) at our will. Thus, changing v if necessary, we only need to find a hyper symmetric point $x_0 \in I^{(\Sigma_0)}$ with $T_{x_0}SU(\mathbb{A}^{(\Sigma_0)})$ dense in $SU(\mathbb{A}^{(\Sigma_0)}) \times M_1(\mathbb{Z}_p)$. We may assume that $m_1 m_c \neq 0$. Diagonalize the hermitian matrix s over L. By the self-duality of L_p, s has p-adic unit diagonal entries $s_1, \dots, s_r \in F$ and $\mathrm{Im}(s_j) > 0 \Leftrightarrow j \leq m_1$. Note that $|s_j|\sqrt{-1} = \pm s_j$ has positive imaginary part. Take an elliptic curve $E_{j/\mathcal{W}}$ with complex multiplication by F with Riemann form given by $F \times F \ni (v, w) \mapsto \mathrm{Tr}_{F/\mathbb{Q}}(v|s_j|\sqrt{-1}w^c)$. Since s_j is a p-adic unit, we may assume that $E_j(\mathbb{C}) \cong \mathbb{C}/\mathfrak{a}_j$ for a lattice $\mathfrak{a}_j$ in F with $\mathfrak{a}_{j,p} = O_p$. We identify $\mathrm{End}^{\mathbb{Q}}(E_j) := \mathrm{End}(E_j) \otimes_{\mathbb{Z}} \mathbb{Q}$ with F by sending $\xi \in F$ to the multiplication by ξ on $\mathbb{C}$. Take $A = E_1 \oplus E_2 \oplus \cdots \oplus E_r$. Embed F into $\mathrm{End}^{\mathbb{Q}}(A)$ so that $F \to \mathrm{End}^{\mathbb{Q}}(A) \twoheadrightarrow \mathrm{End}^{\mathbb{Q}}(E_j)$ is 1 if and only if $j \leq m_1$ (thus, $F \to \mathrm{End}^{\mathbb{Q}}(A) \twoheadrightarrow \mathrm{End}^{\mathbb{Q}}(E_j)$ is complex conjugation c if and only if $j > m_1$). By our construction, we have an isomorphism $H_1(A(\mathbb{C}), \mathbb{Z}) \cong L$ which takes the Riemann form on $H_1(A(\mathbb{C}), \mathbb{Z})$ to $\langle \cdot, \cdot \rangle$ on L. The Hodge decomposition $H_1(A(\mathbb{C}), \mathbb{C}) = H^{-1,0} \oplus H^{0,-1}$ gives the decomposition $V \otimes_{\mathbb{Q}} \mathbb{C} = V_1 \oplus V_2$ and hence a point in $h_A \in \mathfrak{X}^+$.

Since p splits in F, E_j is ordinary; so $h_A \in \mathfrak{X}^+$ projects down to a point $S(\mathcal{W})$. We have

$$\mathrm{End}_O^{(\Sigma_0)}(A_{/F}) := \mathrm{End}_O(A_{/F}) \otimes \mathbb{Z}_{(\Sigma_0)} = M_{m_1}(O_{(\Sigma_0)}) \times M_{m_c}(O_{(\Sigma_0)}).$$

Over the place $\mathfrak{p}$, $E_j[\mathfrak{p}^\infty]_{/\mathcal{W}} \cong \mu_{p^\infty/\mathcal{W}}$ if and only if $j \leq m_1$. We may identify $T_pE_j[\mathfrak{p}^\infty] \cong T_p(\mathfrak{a}_{j,\mathfrak{p}} \otimes \mu_{p^\infty}) = \mathbb{Z}_p(1)$ if $j \leq m_1$ and $T_pE_j[\overline{\mathfrak{p}}^\infty] \cong T_p(\mathfrak{a}_{j,\overline{\mathfrak{p}}} \otimes \mu_{p^\infty}) = \mathbb{Z}_p(1)$ if $j > m_1$. In this way we get $\phi_p : \mathcal{L} \cong T_pA[p^\infty]^\circ$. By duality, we get

$$\left(\bigoplus_{j=1}^{m_1} \mathfrak{a}_{j,\overline{\mathfrak{p}}}\right) \oplus \left(\bigoplus_{j=m_1+1}^{r} \mathfrak{a}_{j,\mathfrak{p}}\right) \overset{{}^t\phi_p^{-1}}{\cong} \left(\bigoplus_{j=1}^{m_1} T_pE_j[\overline{\mathfrak{p}}^\infty]^{\text{ét}}_{/W}\right) \oplus \left(\bigoplus_{j=m_1+1}^{r} T_pE_j[\mathfrak{p}^\infty]^{\text{ét}}_{/W}\right).$$

Then we put $\eta_p = \phi \oplus {}^t\phi_p^{-1} : L_p = \mathcal{L} \oplus \mathcal{L}^\vee \cong T_pA[p^\infty]^\circ \oplus T_pA[p^\infty]^{\text{ét}} = T_pA[p^\infty]$. We choose $\eta^{(\Sigma_0)}$ of A defined over $\mathcal{W}$ so that $(A, \phi_p, \eta^{(\Sigma_0)})$ is over $x_0 \in I(\mathbb{F})$, and write $\eta = (\eta_p, \eta^{(\Sigma_0)})$. For each isogeny $\alpha \in \mathrm{End}_O^{(\Sigma_0)}(A_{/F})$ preserving polarization up to scalars and fixing the generator $v \in \varprojlim_n \mathrm{Isom}(\bigwedge^{m_c} \frac{L_1^\vee}{p^n L_1^\vee} \otimes_{\mathbb{Z}_p} \bigwedge^{m_1} \frac{L_c^\vee}{p^n L_c^\vee}, \mathbb{E}[p^n])$, we can define $\rho^{(\Sigma_0)}(\alpha) \in SU(\mathbb{A}^{(\Sigma_0)})$ by $\alpha \circ \eta^{(\Sigma_0)} = \eta^{(\Sigma_0)} \circ \rho^{(\Sigma_0)}(\alpha)$ and $\rho_p(\alpha) \in M(\mathbb{Z}_p)$ by $\alpha \circ \eta_p = \eta_p \circ \rho_p(\alpha)$. Then we embed α in $SU(\mathbb{A}^{(\Sigma_0)}) \times M(\mathbb{Z}_p)$ diagonally by $\alpha \mapsto (\rho^{(\Sigma_0)}(\alpha) \times \rho_p(\alpha))$. Note that $\alpha \circ v = v \Leftrightarrow \rho_p(\alpha) \in M_1(\mathbb{Z}_p)$. Since the abelian scheme above $\rho(\alpha)(x_0)$ is

$$(A, \eta \circ \rho(\alpha)) = (\mathrm{Im}(\alpha), \alpha \circ \eta) \overset{\alpha^{-1}}{\cong} (A, \eta),$$

we find that $\rho(\alpha)(x_0) = x_0$. By construction, the stabilizer of $x_0 \in I^{(\Sigma_0)}(\mathbb{F})$ contains the image $\mathrm{Im}(\rho)$ whose projection to $M_1(\mathbb{Z}_p)$ is the p-adically dense subgroup

$$(GL_{m_1}(O_{(\Sigma_0)}) \times GL_{m_c}(O_{(\Sigma_0)})) \cap SU(\mathbb{Q})$$

as desired.

References

[ABV] D. Mumford, *Abelian Varieties*, TIFR Studies in Mathematics, Oxford University Press, New York, 1994.

[ACS] Kai-Wen Lan, *Arithmetic compactifications of PEL-type Shimura varieties*, Ph. D. thesis at Harvard, 2008 (717 pages; version of March 18, 2008)

[C] C.-L. Chai, Monodromy of Hecke-invariant Subvarieties, Pure and Applied Mathematics Quarterly **1** (2005), (Special Issue: In memory of Armand Borel), 291–303

[C1] C.-L. Chai, Methods for p-adic monodromy, Journal of the Inst. of Math. Jussieu **7** (2008), 247–268

[DAV] G. Faltings and C.-L. Chai, *Degeneration of Abelian Varieties*, Springer, New York, 1990

[GIT] D. Mumford, *Geometric Invariant Theory*, Ergebnisse **34**, Springer, New York, 1965.

[H05] H. Hida, p-adic automorphic forms on reductive groups, Astérisque **298** (2005), 147–254, SMF

[H06] H. Hida, Irreducibility of the Igusa tower, Acta Math. Sin. (Engl. Ser.) **25** (2009), 1–20

[IAT] G. Shimura, *Introduction to the Arithmetic Theory of Automorphic Functions*, Princeton University Press, Princeton, NJ, and Iwanami Shoten, Tokyo, 1971.

[MFG] H. Hida, *Modular Forms and Galois Cohomology*, Cambridge Studies in Advanced Mathematics **69**, Cambridge University Press, Cambridge, England, 2000.

[PAF] H. Hida, *p-adic Automorphic Forms on Shimura Varieties*, Springer Monographs in Mathematics, 2004, Springer

Department of Mathematics, UCLA, Los Angeles, CA 90095-1555
E-mail address: hida@math.ucla.edu

Clay Mathematics Proceedings
Volume 13, 2011

On the Gross-Prasad Conjecture for Unitary Groups

Hervé Jacquet and Stephen Rallis

This paper is dedicated to Freydoon Shahidi

Abstract. We propose a new approach to the Gross-Prasad conjecture for unitary groups. It is based on a relative trace formula. As evidence for the soundness of this approach, we prove the infinitesimal form of the relevant fundamental lemma in the case of unitary groups in three variables.

Contents

1. Introduction 206
2. Orbits of $Gl_{n-1}(E)$ 210
3. Orbits of $Gl_{n-1}(F)$ 214
4. Orbits of U_{n-1} 215
5. Comparison of the orbits, the fundamental lemma 216
6. Smooth matching and the fundamental Lemma for $n = 2$ 217
7. The trace formula for $n = 2$ 220
8. Orbits of $Gl_2(E)$ 221
9. Orbits of $Gl_2(F)$ 224
10. Orbits of the unitary group 224
11. Comparison of orbits 226
12. The fundamental lemma for $n = 3$ 226
13. Orbital integrals for $Sl_2(F)$ 230
14. Proof of the fundamental lemma for $n = 3$ 239
15. Proof of the fundamental Lemma: $a^2 + b$ is not a square 240
16. Proof of the fundamental Lemma: $a^2 + b$ is a square 246
17. Proof of the fundamental Lemma: $a^2 + b = 0$ 255
18. Other regular elements 256
19. Orbital integrals for Sl_2 259
20. Orbital integrals for $Gl_2(F)$ 260
21. Verifcation of $\Omega_{Sl_2}(X) = \eta(b_2)\Omega_{Gl_2}(Y)$ 263
References 263

2010 *Mathematics Subject Classification.* Primary 22E55; Secondary 11F72.
First named author partially supported by NSF grant DMS 9988611.
Second named author partially supported by NSF grant DMS 9970342.

1. Introduction

Consider a quadratic extension of number fields E/F. Let η be the corresponding quadratic idèle-class character of F. Denote by σ the non-trivial element of $\mathrm{Gal}(E/F)$. We often write $\sigma(z) = \overline{z}$ and $N_r(z) = z\overline{z}$. Let U_n be a unitary group in n variables and U_{n-1} a unitary group in $(n-1)$ variables. Suppose that $\iota : U_{n-1} \to U_n$ is an embedding. In a precise way, let β be a Hermitian non-degenerate form on an E vector space V_n and let $e_n \in V_n$ be a vector such that $\beta(e_n, e_n) = 1$. Let V_{n-1} be the orthogonal complement of e_n. Then let U_n be the automorphism group of β_n and let U_{n-1} be the automorphism group of $\beta|_{V_{n-1}}$. Then ι is defined by the conditions $\iota(h)e_n = e_n$ and $\iota(h)v = hv$ for $v \in V_{n-1}$.

Let π be an automorphic cuspidal representation of U_n and σ an automorphic cuspidal representation of U_{n-1}. For ϕ_π in the space of π and ϕ_σ in the space of σ set

$$A_U(\phi_\pi, \phi_\sigma) := \int_{U_{n-1}(F)\backslash U_{n-1}(F_{\mathbb{A}})} \phi_\pi(\iota(h))\phi_\sigma(h)dh\,. \tag{1}$$

Suppose that this bilinear form does not vanish identically. Let Π be the standard base change of π to $Gl_n(E)$ and let Σ be the standard base change of σ to $Gl_{n-1}(E)$. For simplicity, assume that Π and Σ are themselves cuspidal. The conjecture of Gross-Prasad for orthogonal groups extends to the present set-up of unitary groups and predicts that the central value of the $L-$function $L(s, \Pi \times \Sigma)$ does not vanish. Cases of this conjecture have been proved by Jiang, Ginzburg and Rallis, at least in the context of orthogonal groups ([**16**] and [**17**]). The conjecture has to be made much more precise. One must ask to what extent the converse is true. One must specify which forms of the unitary group and which elements of the packets corresponding to Π and Σ are to be used in the formulation of the converse. Finally, the relation between A_U (or rather $A_U\overline{A_U}$) and the $L-$value should be made more precise.

We will not discuss the general case, where there is no restriction on the representations. We remark however that the case where σ is trivial or one-dimensional is already very interesting even in the case $n = 2$ (See [**11**]) and $n = 3$ (See [**19**], [**20**], [**21**], also [**4**], [**5**]).

In this note we propose an approach based on a relative trace formula. The results of this note are quite modest. We only prove the infinitesimal form of the fundamental lemma for the case $n = 3$. We do not claim that this implies the fundamental lemma itself or the smooth matching of functions. We hope, however, this will interest other mathematicians. In particular, we feel that the fundamental lemma itself is an interesting problem.

We now describe in rough form the relative trace formula at hand. Let f_n and f_{n-1} be smooth functions of compact support on $U_n(F_{\mathbb{A}})$ and $U_{n-1}(F_{\mathbb{A}})$ respectively. We introduce the distribution

$$A_{\pi,\sigma}(f_n \otimes f_{n-1}) := \sum A_U(\pi(f_n)\phi_\pi, \sigma(f_{n-1})\phi_\sigma)\overline{A_U(\phi_\pi, \phi_\sigma)}\,, \tag{2}$$

where the sum is over orthonormal bases for each representation.

Let $\iota : Gl_{n-1} \to Gl_n$ be the obvious embedding. For ϕ_Π in the space of Π and ϕ_Σ in the space of Σ, we define

$$A_G(\phi_\Pi, \phi_\Sigma) := \int_{Gl_{n-1}(E)\backslash Gl_{n-1}(E_{\mathbb{A}})} \phi_\Pi(\iota(g))\phi_\Sigma(g)dg \tag{3}$$

Thus the bilinear form A_G is non-zero if and only if $L(\frac{1}{2}, \Pi \times \Sigma) \neq 0$. In fact we understand completely the relation between the special value and the bilinear form A_G.

Say that n is odd. Let us also set

$$P_n(\phi_\Pi) = \int_{Gl_n(F)\backslash Gl_n(F_\mathbb{A})} \phi_\Pi(g_0) dg_0 \tag{4}$$

$$P_{n-1}(\phi_\Sigma) = \int_{Gl_{n-1}(F)\backslash Gl_{n-1}(F_\mathbb{A})} \eta(\det g_0)\phi_\Sigma(g_0) dg_0 \tag{5}$$

Strictly speaking, the first integral should be over the quotient of

$$\{g \in Gl_n(F_\mathbb{A}) : |\det g| = 1\}$$

by $Gl_n(F)$. Similarly for the other integral. The study of the poles of the Asai L–function and its integral representation (see [**3**] and [**4**], also [**10**]) predict that P_n and P_{n-1} are not identically 0. If n is even, then η must appear in the definition of P_n and not appear in the definition of P_{n-1}. This will change somewhat the following discussion but will lead to the same infinitesimal analog.

Let f'_n and f'_{n-1} be smooth functions of compact support on $Gl_n(E_\mathbb{A})$ and $Gl_{n-1}(E_\mathbb{A})$ respectively. Consider the distribution

$$A_{\Pi,\Sigma}(f'_n \otimes f'_{n-1}) := \tag{6}$$

$$\sum A_G(\Pi(f'_n)\phi_\Pi, \sigma(f'_{n-1})\phi_\Sigma)\overline{P_n(\phi_\Pi)P_{n-1}(\phi_\Sigma)}\,,$$

where the sum is over an orthonormal basis of the representations.

One should have an equality

$$A_{\pi,\sigma}(f_n \otimes f_{n-1}) = A_{\Pi,\Sigma}(f'_n \otimes f'_{n-1})\,, \tag{7}$$

for pairs (f_n, f_{n-1}) and (f'_n, f'_{n-1}) satisfying an appropriate condition of **matching orbital integrals**. In turn, the equality should be used to understand the precise relation between the L value and the bilinear form A_U.

To continue, we associate to the function $f_n \otimes f_{n-1}$ in the usual way a kernel $K_{f_n \otimes f_{n-1}}(g_1 : g_2, h_1 : h_2)$ on

$$(U_n(F_\mathbb{A}) \times U_{n-1}(F_\mathbb{A})) \times (U_n(F_\mathbb{A}) \times U_{n-1}(F_\mathbb{A}))\,.$$

The kernel is invariant on the left by the group of rational points. We consider the (regularized) integral

$$\int_{(U_{n-1}(F)\backslash U_{n-1}(F_\mathbb{A}))^2} K_{f_n \otimes f_{n-1}}(\iota(g_2) : g_2, \iota(h_2) : h_2) dg_2 dh_2\,. \tag{8}$$

Likewise, we associate to the function $f'_n \otimes f'_{n-1}$ a kernel $K'_{f'_n \otimes f'_{n-1}}(g_1 : g_2, h_1 : h_2)$ on

$$(Gl_n(E_\mathbb{A}) \times Gl_{n-1}(E_\mathbb{A})) \times (Gl_n(E_\mathbb{A}) \times Gl_{n-1}(E_\mathbb{A}))$$

and we consider the (regularized) integral

$$\int K'_{f'_n \otimes f'_{n-1}}(\iota(g_2) : g_2, h_1 : h_2) dg_2 dh_1 \eta(\det h_2) dh_2 \tag{9}$$

where

$$g_2 \in Gl_{n-1}(E)\backslash Gl_{n-1}(E_\mathbb{A})\,,\; h_1 \in Gl_n(F)\backslash Gl_n(F_\mathbb{A})\,,\; h_2 \in Gl_{n-1}(F)\backslash Gl_{n-1}(F_\mathbb{A})\,.$$

The conditions of matching orbital integrals should guarantee that (8) and (9) are equal. In turn this should imply (7).

In more detail, (8) is equal to

$$\int \sum_{\gamma\in U_n(F)} f_n\left(\iota(g_2)^{-1}\gamma\,\iota(h_2)\right) \sum_{\xi\in U_{n-1}(F)} f_{n-1}\left(g_2^{-1}\xi h_2\right))dg_2dh_2$$

or

$$\int \sum_{\gamma\in U_n(F)} f_n\left(\iota(g_2)\gamma\,\iota(h_2)\right) \sum_{\xi\in U_{n-1}(F)} f_{n-1}\left(g_2\xi h_2\right))dg_2dh_2\,.$$

In the sum over γ we may replace γ by $\iota(\xi)\gamma$. Then $\iota(g_2\xi)$ appears. Now we combine the sum over ξ and the integral over $g_2 \in U_{n-1}(F)\backslash U_{n-1}(E_{\mathbb{A}})$ into an integral for $g_2 \in U_{n-1}(E_{\mathbb{A}})$ to get

$$\int \sum_{\gamma} f_n\left(\iota(g_2)\gamma\iota(h_2)\right) f_{n-1}\left(g_2h_2\right))dg_2dh_2\,.$$

After a change of variables, this becomes

$$\int \sum_{\gamma} f_n\left(\iota(g_2)\iota(h_2)^{-1}\gamma\,\iota(h_2)\right) f_{n-1}\left(g_2\right) dg_2dh_2\,.$$

At this point, we introduce a new function $f_{n,n-1}$ on $U_n(F_{\mathbb{A}})$ defined by

$$f_{n,n-1}(g) := \int_{U_{n-1}(F_{\mathbb{A}})} f_n(\iota(g_2)g)f_{n-1}(g_2)dg_2\,. \tag{10}$$

Then we can rewrite the previous expression as

$$\int_{U_{n-1}(F)\backslash U_{n-1}(F_{\mathbb{A}})} \sum_{\gamma} f_{n,n-1}\left(\iota(h_2)^{-1}\gamma\,\iota(h_2)\right) dh_2\,.$$

The group U_{n-1} operates on U_n by conjugation:

$$\gamma \mapsto \iota(h)^{-1}\gamma\iota(h)$$

For **regular** elements of $U_n(F)$ the stabilizer is trivial. Thus, ignoring terms which are not regular, the above expression can be rewritten

$$\sum_{\gamma} \int_{U_{n-1}(F_{\mathbb{A}})} f_n\left(\iota(h)^{-1}\gamma\iota(h)\right) dh\,, \tag{11}$$

where the sum is now over a set of representatives for the regular orbits of $U_{n-1}(F)$ in $U_n(F)$.

Likewise, we can write (9) in the form

$$\int \sum_{\gamma\in Gl_n(E)} f'_n(\iota(g_2)^{-1}\gamma h_1) \sum_{\xi\in Gl_{n-1}(F)} f'_{n-1}(g_2^{-1}\xi h_2)\eta(\det h_2)dg_2dh_1dh_2\,.$$

The same kind of manipulation as before gives

$$= \int \sum_{\gamma\in Gl_n(E)} f'_n(\iota(g_2)\gamma h_1)f'_{n-1}(g_2h_2)dg_2dh_1\eta(\det h_2)dh_2$$

where now g_2 is in $Gl_{n-1}(E_{\mathbb{A}})$. If we change variables, this becomes

$$= \int \sum_{\gamma\in Gl_n(E)} f'_n(\iota(g_2)\iota(h_2)^{-1}\gamma h_1)f'_{n-1}(g_2)dg_2dh_1\eta(\det h_2)dh_2\,.$$

We introduce a new function $f'_{n,n-1}$ on $Gl_n(E_{\mathbb{A}})$ defined by

$$f'_{n,n-1}(g) := \int_{Gl_{n-1}(E_{\mathbb{A}})} f'_n(\iota(g_2)g) f'_{n-1}(g_2) dg_2 \,.$$

The above expression can be rewritten

$$\int \sum_{\gamma \in Gl_n(E)} f'_{n,n-1}(\iota(h_2)^{-1}\gamma h_1) dh_1 \eta(\det h_2) dh_2 \,,$$

where h_1 is in $Gl_n(F)\backslash Gl_n(F_{\mathbb{A}})$ and h_2 is in $Gl_{n-1}(F)\backslash Gl_{n-1}(F_{\mathbb{A}})$. We also write this as

$$\int \sum_{\gamma \in Gl_n(E)/Gl_n(F)} \left(\int f'_{n,n-1}(\iota(h_2)^{-1}\gamma h_1) dh_1 \right) \eta(\det h_2) dh_2 \tag{12}$$

with $h_1 \in Gl_n(F_{\mathbb{A}})$.

At this point we introduce the symmetric space S_n defined by the equation $ss^\sigma = 1$. Thus

$$S_n(F) := \{ s \in Gl_n(E) : s\overline{s} = 1 . \} \tag{13}$$

Let $\Phi_{n,n-1}$ be the function on $S_n(F_{\mathbb{A}})$ defined by

$$\Phi_{n,n-1}(g\overline{g}^{-1}) = \int_{Gl_n(F_{\mathbb{A}})} f'_{n,n-1}(gh_1) dh_1 \,.$$

The expression (12) can be written as

$$\int_{Gl_{n-1}(F_{\mathbb{A}})/Gl_{n-1}(F)} \sum_{\xi \in S_n(F)} \Phi_{n,n-1} \left[\iota(h_2)^{-1} \xi \iota(h_2) \right] \eta(\det h_2) dh_2 \,.$$

The group $Gl_n(F)$ operates on $S_n(F)$ by

$$s \mapsto \iota(g)^{-1} s \iota(g) \,.$$

Again, for **regular** elements of $S_n(F)$ the stabilizer under $Gl_{n-1}(F)$ is trivial. Thus, at the cost of ignoring non-regular elements, we get

$$\sum_{\xi} \int_{Gl_{n-1}(F_{\mathbb{A}})} \Phi_{n,n-1} \left(\iota(h)^{-1} \xi \iota(h) \right) \eta(\det h) dh \,, \tag{14}$$

where the sum is over a set of representatives for the regular orbits of $Gl_{n-1}(F)$ in $S_n(F)$.

To carry through our trace formula we need to find a way to match regular orbits of $U_{n-1}(F)$ in $U_n(F)$ with regular orbits of $Gl_{n-1}(F)$ in $S_n(F)$. We will use the notation $\xi \to \xi'$ for such a matching. The global condition of **matching orbital integrals** is then

$$\int_{U_{n-1}(F_{\mathbb{A}})} f_{n,n-1}(\iota(h)^{-1}\xi\iota(h)) dh =$$

$$\int_{Gl_{n-1}(F_{\mathbb{A}})} \Phi_{n,n-1}(\iota(h)^{-1}\xi'\iota(h)) \eta(\det h) dh$$

if $\xi \to \xi'$. If ξ' does not correspond to any ξ then

$$\int \Phi_{n,n-1}(\iota(h)^{-1}\xi'\iota(h)) \eta(\det h) dh = 0 \,.$$

A formula of this type is discussed in [7], [8], [9] for $n = 2$. Or rather, the results of these papers could be modified to recover a trace formula of the above type.

As a first step, we consider the infinitesimal analog of the above trace formula. Now n needs not be odd. We set $\mathfrak{G}_n = M(n \times n, E)$. We often drop the index n if this does not create confusion. We let $\mathfrak{U}_n \subset \mathfrak{G}_n$ be the Lie algebra of the group U_n. Then U_{n-1} operates on $\mathfrak{U}_n$ by conjugation. Likewise, we consider the vector space $\mathfrak{S}_n$ tangent to S_n at the origin. This is the vector space of matrices $X \in \mathfrak{G}_n$ such that $X + \overline{X} = 0$. Again the group $Gl_{n-1}(F)$ operates by conjugation on $\mathfrak{S}_n$. The trace formula we have in mind is

$$\int_{U_{n-1}(F)\backslash U_{n-1}(F_{\mathbb{A}})} \sum_{\xi \in \mathfrak{U}_n(F)} f\left(\iota(h)^{-1}\xi\iota(h)\right) dh = \tag{15}$$

$$\int_{Gl_{n-1}(F)\backslash Gl_{n-1}(F_{\mathbb{A}})} \sum_{\xi' \in \mathfrak{S}_n(F)} \Phi\left(\iota(h)^{-1}\xi'\iota(h)\right) \eta(\det h) dh\,,$$

where f is a smooth function of compact support on $\mathfrak{U}_n(F_{\mathbb{A}})$ and Φ a smooth function of compact support on $\mathfrak{S}_n(F_{\mathbb{A}})$. Once more, the integrals on both sides are not convergent and need to be regularized. The equality occurs if the functions satisfy a certain matching orbital integral condition. We will define a notion of strongly regular elements and a condition of matching of strongly regular elements denoted by

$$\xi \to \xi'\,.$$

Then the global condition of matching between functions is as before: if $\xi \to \xi'$ then

$$\int_{U_{n-1}(F_{\mathbb{A}})} f\left(\iota(h)\xi\iota(h)^{-1}\right) dh$$

$$= \int_{Gl_{n-1}(F_{\mathbb{A}})} \Phi\left(\iota(h)\xi'\iota(h)^{-1}\right) \eta(\det h) dh\,;$$

if ξ' does not correspond to a ξ then

$$\int_{Gl_{n-1}(F_{\mathbb{A}})} \Phi\left(\iota(h)\xi\iota(h)^{-1}\right) \eta(\det h) dh = 0\,.$$

We now investigate in detail the matching of orbits announced above.

2. Orbits of $Gl_{n-1}(E)$

Let E be an arbitrary field. We first introduce a convenient definition. Let P_n, P_{n-1} be two polynomials of degree n and $n-1$ respectively in $E[X]$. We will say that they are **strongly relatively prime** if the following condition is satisfied. There exists a sequence of polynomials P_i of degree i, $n \geq i \geq 0$, where P_n and P_{n-1} are the given polynomials, and the P_i are defined inductively by the relation

$$P_{i+2} = Q_i P_{i+1} + P_i\,.$$

In particular, P_0 is a non-zero constant. In other words, we demand that the P_n and P_{n-1} be relatively prime and the Euclidean algorithm which gives the (constant) G.C.D. of P_n and P_{n-1} have exactly $n-1$ steps. Of course the sequence, if it exists, is unique. Moreover, for each i, the polynomials P_{i+1}, P_i are strongly relatively prime.

Let V_n be a vector space of dimension n over the field E. We often write $V_n(E)$ for V_n. We set $\mathfrak{G} = \mathrm{Hom}_E(V_n, V_n)$. Let $e_n \in V_n$ and $e_n^* \in V_n^*$ (dual vector space). Assume $\langle e_n^*, e_n\rangle \neq 0$. Let V_{n-1} be the kernel of e_n^*. Thus

$$V_n = V_{n-1} \oplus Ee_n\,.$$

We define an embedding $\iota : Gl(V_{n-1}(E)) \to Gl(V_n(E))$ by

$$\begin{aligned}\iota(g)v_{n-1} &= gv_{n-1} \text{ for } v_{n-1} \in V_{n-1}\,,\\ \iota(g)e_n &= e_n\,.\end{aligned}$$

We let $Gl(V_n(E))$ act on V_n^* on the right by

$$\langle v^*g, v\rangle = \langle v^*, gv\rangle\,.$$

Then $\iota(Gl(V_{n-1}(E)))$ is the subgroup of $Gl(V_n(E))$ which fixes e_n^* and e_n.

Suppose $A_n \in \mathfrak{G}$. We can represent A_n by a matrix

$$\begin{pmatrix} A_{n-1} & e_{n-1} \\ e_{n-1}^* & a_n \end{pmatrix},$$

with $A_{n-1} \in \mathrm{Hom}(V_{n-1}, V_{n-1})$, $e_{n-1} \in V_{n-1}$, $e_{n-1}^* \in V_{n-1}^*$, $a_n \in E$. This means that, for all $v_{n-1} \in V_{n-1}(E)$,

$$A_n(v_{n-1}) = A_{n-1}(v_{n-1}) + \langle e_{n-1}^*, v_{n-1}\rangle e_n$$

and

$$A_n(e_n) = e_{n-1} + a_n e_n\,.$$

In particular

$$A_n(e_{n-1}) = A_{n-1}(e_{n-1}) + \langle e_{n-1}^*, e_{n-1}\rangle e_n\,.$$

The group $Gl(V_{n-1}(E))$ acts on $\mathfrak{G}$ by

$$A \mapsto \iota(g)A\iota(g)^{-1}\,.$$

The operator $\iota(g)A\iota(g)^{-1}$ is represented by the matrix

$$\begin{pmatrix} gA_{n-1}g^{-1} & ge_{n-1} \\ e_{n-1}^*g^{-1} & a_n \end{pmatrix}.$$

Thus the scalar product $\langle e_{n-1}^*, e_{n-1}\rangle$ is an invariant of this action. We often call it the first invariant of this action. Moreover, if we replace e_n and e_n^* by scalar multiples, the spaces V_{n-1}, Ee_n and the scalar product $\langle e_{n-1}^*, e_{n-1}\rangle$ do not change. We will say that A_n is **strongly regular with respect to the pair** (e_n, e_n^*) (or with respect to the pair (V_{n-1}, e_n)) if the polynomials

$$\det(A_n - \lambda) \text{ and } \det(A_{n-1} - \lambda)$$

are strongly relatively prime.

Now assume that A_n is strongly regular with respect to (e_n, e_n^*). We have

$$\det(A_n - \lambda) = (a_n - \lambda)\det(A_{n-1} - \lambda) + R(\lambda)$$

with R of degree $n-2$. The leading term of R is $-\langle e_{n-1}^*, e_n\rangle(-\lambda)^{n-2}$. Thus $\langle e_{n-1}^*, e_n\rangle$ is non-zero. Thus we can write

$$V_{n-1} = V_{n-2} \oplus Ee_{n-1}$$

where V_{n-2} is the kernel of e_{n-1}^* and represent A_{n-1} by a matrix

$$\begin{pmatrix} A_{n-2} & e_{n-2} \\ e_{n-2}^* & a_{n-1} \end{pmatrix},$$

with $A_{n-2} \in \mathrm{Hom}(V_{n-2}, V_{n-2})$, $e_{n-2} \in V_{n-2}$, $e^*_{n-1} \in V^*_{n-2}$, $a_{n-1} \in E$. As before, this means that

$$A_{n-1}(v_{n-2}) = A_{n-2}(v_{n-2}) + \langle e^*_{n-2}, v_{n-2} \rangle e_{n-1}$$

$$A_{n-1}(e_{n-1}) = e_{n-2} + a_{n-1}e_{n-1}\,.$$

Choose a basis ϵ_i, $1 \le i \le n-2$, of V_{n-2}. Since $\langle e^*_{n-1}, \epsilon_i \rangle = 0$ we have

$$A_n(\epsilon_i) = A_{n-1}(\epsilon_i) + \langle e^*_{n-1}, \epsilon_i \rangle e_n = A_{n-1}(\epsilon_i) = A_{n-2}(\epsilon_i) + \langle e^*_{n-2}, \epsilon_i \rangle e_{n-1}\,.$$

On the other hand,

$$A_n(e_{n-1}) = e_{n-2} + a_{n-1}e_{n-1} + \langle e^*_{n-1}, e_{n-1} \rangle e_n\,.$$

Thus the matrix of A_n with respect to the basis

$$(\epsilon_1, \epsilon_2, \ldots, \epsilon_{n-2}, e_{n-1}, e_n)$$

has the form

$$\begin{pmatrix} \mathrm{Mat}(A_{n-2}) & *_{n-2} & 0_{n-2} \\ *^{n-2} & a_{n-1} & 1 \\ 0^{n-2} & \langle e^*_{n-1}, e_{n-1} \rangle e_n & a_n \end{pmatrix} \tag{16}$$

where $\mathrm{Mat}(A_{n-2})$ is the matrix of A_{n-2} with respect to the basis $(\epsilon_1, \epsilon_2, \ldots, \epsilon_{n-2})$. The index $n-2$ indicates a column of size $n-2$ and the exponent $n-2$ a row of size $n-2$. Likewise the matrix of A_{n-1} with respect to the basis $(\epsilon_1, \epsilon_2, \ldots, \epsilon_{n-2}, e_{n-1})$ has the form

$$\begin{pmatrix} \mathrm{Mat}(A_{n-2}) & *_{n-2} \\ *^{n-2} & a_{n-1} \end{pmatrix}.$$

It follows that

$$\det(A_n - \lambda) = \det(A_{n-1} - \lambda)(a_n - \lambda) - \langle e^*_{n-1}, e_{n-1} \rangle \det(A_{n-2} - \lambda)\,.$$

Thus the polynomials $\det(A_{n-1} - \lambda)$ and $\det(A_{n-2} - \lambda)$ are strongly relatively prime and the operator A_{n-1} is strongly regular with respect to (e_{n-1}, e^*_{n-1}). At this point we proceed inductively. We construct a sequence of subspaces

$$V_1 \subset V_2 \subset \cdots \subset V_{n-1} \subset V_n$$

with $\dim(V_i) = i$, vectors $e_i \in V_i$, and linear forms $e^*_i \in V^*_i$ such that V_{i-1} is the kernel of e^*_i. The matrix of A_n with respect to the basis

$$(e_1, e_2, \ldots, e_{n-1}, e_n)$$

is the tridiagonal matrix

$$\begin{pmatrix} a_1 & 1 & 0 & 0 & \cdots & 0 & 0 & 0 & 0 \\ c_1 & a_2 & 1 & 0 & \cdots & 0 & 0 & 0 & 0 \\ 0 & c_2 & a_3 & 1 & \cdots & 0 & 0 & 0 & 0 \\ \cdots & \cdots & \cdots & \cdots & \cdots & \cdots & \cdots & \cdots & \cdots \\ \cdots & \cdots & \cdots & \cdots & \cdots & \cdots & \cdots & \cdots & \cdots \\ 0 & 0 & 0 & 0 & \cdots & c_{n-3} & a_{n-2} & 1 & 0 \\ 0 & 0 & 0 & 0 & \cdots & 0 & c_{n-2} & a_{n-1} & 1 \\ 0 & 0 & 0 & 0 & \cdots & 0 & 0 & c_{n-1} & a_n \end{pmatrix} \tag{17}$$

where $c_i = \langle e^*_i, e_i \rangle \neq 0$. We note the relations

$$\det(A_i - \lambda) = \det(A_{i-1} - \lambda) - c_{i-1} \det(A_{i-2} - \lambda)\,,\ i \ge 2\,.$$

Now suppoe

$$(e'_1, e'_2, \ldots, e'_{n-1})$$

is a basis of V_{n-1} and the matrix of A_n with respect to the basis

$$(e'_1, e'_2, \ldots, e'_{n-1}, e_n)$$

has the form

$$\begin{pmatrix} a'_1 & 1 & 0 & 0 & \cdots & 0 & 0 & 0 & 0 \\ c'_1 & a'_2 & 1 & 0 & \cdots & 0 & 0 & 0 & 0 \\ 0 & c'_2 & a'_3 & 1 & \cdots & 0 & 0 & 0 & 0 \\ \cdots & \cdots & \cdots & \cdots & \cdots & \cdots & \cdots & \cdots & \cdots \\ \cdots & \cdots & \cdots & \cdots & \cdots & \cdots & \cdots & \cdots & \cdots \\ 0 & 0 & 0 & 0 & \cdots & c'_{n-3} & a'_{n-2} & 1 & 0 \\ 0 & 0 & 0 & 0 & \cdots & 0 & c'_{n-2} & a'_{n-1} & 1 \\ 0 & 0 & 0 & 0 & \cdots & 0 & 0 & c'_{n-1} & a'_n \end{pmatrix}.$$

Thus, for $i \geq 1$

$$A_n e'_i = e'_{i-1} + a'_i e'_i + c_{i-1} e_{i+1}$$

(where $e'_n = e_n$, $e_{-1} = 0$ and $e'_{n+1} = 0$) Call A'_i the sub square matrix obtained by deleting the last $n - i$ rows and the last $n - i$ columns. Then we have

$$\det(A'_i - \lambda) = \det(A'_{i-1} - \lambda) - c'_{i-1} \det(A'_{i-2} - \lambda)\,,\ i \geq 2\,.$$

Also

$$\det(A_n - \lambda) = \det(A'_n - \lambda)\,,\ \det(A_{n-1} - \lambda) = \det(A'_{n-1} - \lambda)\,.$$

It follows inductively that $a_i = a'_i$, $c_j = c'_j$, $e'_i = e_i$.

We have proved the following Proposition.

PROPOSITION 1. *If A is strongly regular with respect to the pair (V_{n-1}, e_n) there is a unique basis*

$$(e_1, e_2, \ldots, e_{n-1})$$

of V_{n-1} such that the matrix of A with respect to the basis

$$(e_1, e_2, \ldots, e_{n-1}, e_n)$$

has the form (17). In particular, the a_i, $1 \leq i \leq n$, and the c_j, $1 \leq j \leq n-1$, are uniquely determined.

REMARK. If we demand that the matrix have the form

$$\begin{pmatrix} a'_1 & b'_1 & 0 & 0 & \cdots & 0 & 0 & 0 & 0 \\ c'_1 & a'_2 & b'_2 & 0 & \cdots & 0 & 0 & 0 & 0 \\ 0 & c'_2 & a'_3 & b'_3 & \cdots & 0 & 0 & 0 & 0 \\ \cdots & \cdots & \cdots & \cdots & \cdots & \cdots & \cdots & \cdots & \cdots \\ \cdots & \cdots & \cdots & \cdots & \cdots & \cdots & \cdots & \cdots & \cdots \\ 0 & 0 & 0 & 0 & \cdots & c'_{n-3} & a'_{n-2} & b'_{n-2} & 0 \\ 0 & 0 & 0 & 0 & \cdots & 0 & c'_{n-2} & a'_{n-1} & b'_{n-1} \\ 0 & 0 & 0 & 0 & \cdots & 0 & 0 & c'_{n-1} & a'_n \end{pmatrix},$$

with respect to a basis of the form

$$(e'_1, e'_2, \ldots, e'_{n-1}, e_n)\,,$$

where $(e'_1, e'_2, \ldots, e'_{n-1})$ is a basis of V_{n-1}, then $a'_i = a_i$, $1 \leq i \leq n$, $b'_j c'_j = c_j$, $1 \leq i \leq n-1$, and the e'_i are scalar multiples of the e_i.

According to [**1**], an element $A_n \in \mathfrak{G}$ is **regular** if the vectors

$$A^i_{n-1}e_{n-1}\,,\, 0 \le i \le n-2$$

are linearly independent and the linear forms

$$e^*_i A^i_{n-1}\,,\, 0 \le i \le n-2$$

are linearly independent. This is equivalent to the condition that the stabilizer of A_n in $Gl(V_n(E))$ be trivial and the orbit of A_n under $Gl(V_n(\overline{E}))$ be Zariski closed. A strongly regular element is regular. The above and forthcoming discussion concerning strongly regular elements should apply to regular elements as well. However, we have verified it is so only in the case $n = 2, 3$.

3. Orbits of $Gl_{n-1}(F)$

Now suppose that E is a quadratic extension of F. Let σ be the non trivial element of the Galois-group of E/F.

Suppose that V_n is given an F form. For clarity we often write $V_n(E)$ for V_n and $V_n(F)$ for the $F-$form. We denote by $v \mapsto v^\sigma$ the corresponding action of σ on $V_n(E)$. Then $V_n(F)$ is the space of $v \in V_n(E)$ such that $v^\sigma = v$. We assume $e^\sigma_n = e_n$ and $V^\sigma_{n-1} = V_{n-1}$. We have an action of σ on $\mathrm{Hon}_E(V_n, V_n)$ denoted by $A \mapsto A^\sigma$ and defined by

$$A^\sigma(v) = A(v^\sigma)^\sigma\,.$$

We denote by $\mathfrak{S}$ the space of $A \in \mathrm{Hon}_E(V_n, V_n)$ such that

$$A^\sigma = -A\,.$$

The group $Gl(V_{n-1}(F))$ can be identified with the group of $g \in Gl(V_{n-1}(E))$ fixed by σ. It operates on $\mathfrak{S}$.

We say that an element of $\mathfrak{S}_n$ is **strongly regular** if it is strongly regular as an element of $\mathrm{Hon}_E(V_n, V_n)$. We study the orbits of $Gl(V_n(F))$ in the set of strongly regular elements of $\mathfrak{S}$.

We fix $\sqrt{\tau}$ such that $E = F(\sqrt{\tau})$. If A is strongly regular, there is a unique basis $(e_1, e_2, \ldots, e_{n-1})$ of $V_n(F)$ such that the matrix of A with respect to the basis

$$(e_1, e_2, \ldots, e_{n-1}, e_n)$$

has the form

$$\begin{pmatrix}
a_1 & \sqrt{\tau} & 0 & 0 & \cdots & 0 & 0 & 0 & 0 \\
\frac{c_1}{\sqrt{\tau}} & a_2 & \sqrt{\tau} & 0 & \cdots & 0 & 0 & 0 & 0 \\
0 & \frac{c_2}{\sqrt{\tau}} & a_3 & \sqrt{\tau} & \cdots & 0 & 0 & 0 & 0 \\
\cdots & \cdots & \cdots & \cdots & \cdots & \cdots & \cdots & \cdots & \cdots \\
\cdots & \cdots & \cdots & \cdots & \cdots & \cdots & \cdots & \cdots & \cdots \\
0 & 0 & 0 & 0 & \cdots & \frac{c_{n-3}}{\sqrt{\tau}} & a_{n-2} & \sqrt{\tau} & 0 \\
0 & 0 & 0 & 0 & \cdots & 0 & \frac{c_{n-2}}{\sqrt{\tau}} & a_{n-1} & \sqrt{\tau} \\
0 & 0 & 0 & 0 & \cdots & 0 & 0 & \frac{c_{n-1}}{\sqrt{\tau}} & a_n
\end{pmatrix}. \tag{18}$$

Then the a_i and the c_j are the invariants of A. Furthermore, $a_i \in F\sqrt{\tau}$ and $c_j \in F^\times$. Two strongly regular elements A and A' of $\mathfrak{S}_n$ are conjugate under $Gl(V_{n-1}(F))$ if and only they are conjugate under $Gl(V_{n-1}(E))$, or, equivalently, if and only if they have the same invariants. Finally, given $a_i \in F\sqrt{\tau}$, $1 \le i \le n$, and $c_j \in F^\times$, $1 \le j \le n-1$, there is a strongly regular element of $\mathfrak{S}_n$ with those invariants.

4. Orbits of U_{n-1}

Let V_n be a $E-$vector space of dimension n and β a non-degenerate Hermitian form on V_n. Let e_n be an anisotropic vector, that is,

$$\beta(e_n, e_n) \neq 0 .$$

Usually, we will scale β by demanding that $\beta(e_n, e_n) = 1$.

Let V_{n-1} be the subspace orthogonal to e_n. Thus

$$V_n = V_{n-1} \oplus E e_n .$$

Let $U(\beta)$ be the unitary group of β. Let θ be the restriction of β to V_{n-1}. and $U(\theta)$ the unitary group of θ. Thus we have an injection $\iota : U(\theta) \to U(\beta)$. We have the adjoint action of $U(\beta)$ on $\mathrm{Lie}(U(\beta))$ and thus an action of $U(\theta)$ on $\mathrm{Lie}(U(\beta))$. We have an embedding of $\mathrm{Lie}(U(\beta))$ into $\mathrm{Hom}(V_n, V_n)$. We say that an element of $\mathrm{Lie}(U(\beta))$ is **strongly regular** if it is strongly regular as an element of $\mathrm{Hom}_E(V_n, V_n)$. As before, to $A_n \in \mathrm{Hom}_E(V_n, V_n)$ we associate a matrix

$$\begin{pmatrix} A_{n-1} & e_{n-1} \\ e^*_{n-1} & a_n \end{pmatrix} .$$

The condition that A_n be in $\mathrm{Lie}(U(\beta))$ is

$$A_{n-1} \in \mathrm{Lie}(U(\theta)), a_n + \overline{a_n} = 0$$

and

$$\langle e^*_{n-1}, v \rangle = -\frac{\beta(v, e_{n-1})}{\beta(e_n, e_n)} ,$$

for all $v \in V_{n-1}$. Thus the first invariant of the matrix is

$$\langle e^*_n, e_n \rangle = -\frac{\beta(e_{n-1}, e_{n-1})}{\beta(e_n, e_n)} .$$

Assume that A_n is strongly regular. Then $\beta(e_{n-1}, e_{n-1}) \neq 0$ and V_{n-1} is an orthogonal direct sum

$$V_{n-1} = V_{n-2} \oplus E e_{n-1} .$$

We can then repeat the process and obtain in this way an orthogonal basis

$$(e_1, e_2, \dots, e_{n-1}, e_{n-1})$$

such that $\beta(e_i, e_i) \neq 0$ and the matrix of A_n with respect to the basis

$$(e_1, e_2, \dots, e_{n-1}, e_n)$$

has the form (17). Moreover, it is the only orthogonal basis with this property. In addition, for $1 \leq i \leq n-1$,

$$c_i = -\frac{\beta(e_i, e_i)}{\beta(e_{i+1}, e_{i+1})} .$$

Finally, $a_i \in F\sqrt{\tau}$ for $1 \leq i \leq n$ and $c_j \in F^\times$ for $1 \leq j \leq n-1$. Two strongly regular elements of $\mathrm{Lie}(U(\beta))$ are conjugate under $U(\theta)$ if and only if they are conjugate under $Gl(V_{n-1})$, or, what amounts to the same thing, have the same invariants.

From now on let us scale β by demanding that $\beta(e_n, e_n) = 1$. Then θ determine β and we write $\beta = \theta^e$.

Given $a_i \in F\sqrt{\tau}$, $1 \le i \le n$, $c_j \in F^\times$, $1 \le j \le n-1$ there is a non degenerate Hermitian form θ on V_{n-1}, a strongly regular element A of $\mathrm{Lie}(U(\theta^e))$ whose invariants are the a_i and the c_j. The isomorphism class of θ is uniquely determined and for any choice of θ the conjugacy class of A under $U(\theta)$ is uniquely determined.

The determinant of θ is equal to

$$(-1)^{\frac{(n-1)n}{2}} c_1 c_2^2 \cdots c_{n-1}^{n-1} .$$

5. Comparison of the orbits, the fundamental lemma

We now consider an $E-$vector space V_n, a vector $e_n \neq 0$, and a linear complement V_{n-1} of e_n. We are also given an $F-$form of V_n or, what amounts to the same thing, an action of σ on V_n. We assume that $e_n^\sigma = e_n$ and $V_{n-1}^\sigma = V_{n-1}$. For a Hermitian form θ on V_{n-1} we denote by θ^e the Hermitian form on V_n such that V_{n-1} and E_n are orthogonal, $\theta^e|V_{n-1} = \theta$, $\theta^e(e_n, e_n) = 1$. Then $U(\theta) \subset Gl(V_{n-1}(E))$ and $Gl(V_{n-1}(F)) \subset Gl(V_{n-1}(E))$. Let ξ be a strongly regular element of $\mathrm{Lie}(U(\theta^e))$ and ξ' a strongly regular element of $\mathfrak{S}$. We say that ξ' matches ξ and we write

$$\xi \to \xi'$$

if ξ and ξ' have the same invariants, or, what amounts to the same thing, are conjugate under $Gl(V_n(E))$. Every ξ matches a ξ'. The converse is not true. However, given ξ' there is a θ and a strongly regular element ξ of $\mathrm{Lie}(U(\theta^e))$ such that $\xi \to \xi'$. The form θ is unique, up to equivalence, and the element ξ is unique, up to conjugation by $U(\theta)$.

For instance, suppose that E is a quadratic extension of F, a local, non-Archimedean field. Up to equivalence, there are only two choices for θ. Let θ_0 be a form whose determinant is a norm and θ_1 a form whose determinant is not a norm. Let ξ' be a strongly regular element of $\mathfrak{S}(F)$ and c_i, $1 \le i \le n-1$ the corresponding invariants. If

$$(-1)^{\frac{(n-1)n}{2}} c_1 c_2^2 \cdots c_{n-1}^{n-1}$$

is a norm then ξ' matches an element $\mathrm{Lie}(U(\theta_0^e))$. Otherwise it matches an element of $\mathrm{Lie}(U(\theta_1^e))$.

We have a conjecture of **smooth matching**. If Φ is a smooth function of compact support on $\mathfrak{S}(F)$ and ξ' is strongly regular, we define the orbital integral

$$\Omega_G(\xi', \Phi) = \int_{Gl(V_{n-1}(F))} \Phi\left(\iota(g)\xi'\iota(g)^{-1}\right) \eta(\det g) dg .$$

Likewise, if f_i, $i = 0, 1$, is a smooth function of compact support on $\mathrm{Lie}(U(\theta_i^e)(F)$, ξ_i a strongly regular element, we define the orbital integral

$$\Omega_{U_i}(\xi_i, f_i) = \int_{U(\theta_i^e)(F)} f_i(\iota(g)\xi_i\iota(g)^{-1}) dg .$$

Conjecture 1 (Smooth matching). *There is a factor $\tau(\xi')$, defined for ξ' strongly regular with the property. Given Φ, there is a pair (f_0, f_1), and conversely, such that*

$$\Omega_G(\xi', \Phi) = \tau(\xi')\Omega_{U_i}(\xi_i, f_i)$$

if $\xi_i \to \xi'$.

We have a conjectural **fundamental lemma**. Assume that E/F is an unramified quadratic extension and the residual characteristic is odd. Thus -1 is a norm in E. To be specific let us take $V_n = E^n$, $V_n(F) = F^n$,

$$e_n = \begin{pmatrix} 0 \\ 0 \\ * \\ 0 \\ 1 \end{pmatrix},$$

$V_{n-1}(E) \simeq E^{n-1}$ the space of column vectors whose last entry is 0. Finally, let θ_0 be the form whose matrix is the identity matrix. Thus $\mathrm{Lie}(U(\theta_0^e))$ is the space of matrices $A \in M(n \times n, E)$ such that $A + {}^t\overline{A} = 0$. On the other hand $\mathfrak{S}(F)$ is the space of matrices A such that $A + \overline{A} = 0$.

Let f_0 (resp. Φ_0) be the characteristic function of the matrices with integral entries in $\mathrm{Lie}(U(\theta_0^e))$ (resp. $\mathfrak{S}(F)$). Choose the Haar measures so that the standard maximal compact subgroups have mass 1.

CONJECTURE 2 (fundamental lemma). *Let ξ' be a strongly regular element of $\mathfrak{S}(F)$ and a_i, c_j the corresponding invariants. If*

$$c_1 c_2^2 \cdots c_{n-1}^{n-1}$$

has even valuation, then

$$\Omega_G(\xi', \Phi_0) = \tau(\xi')\Omega_{U_0}(\xi, f_0),$$

where $\xi \in \mathrm{Lie}(U(\theta_0^e))$ matches ξ' and $\tau(\xi') = \pm 1$. Otherwise

$$\Omega_G(\xi', \Phi_0) = 0.$$

Before we proceed we remark that in the general setting the linear forms

$$A_n \mapsto \mathrm{Tr}(A_n), \mapsto \mathrm{Tr}(A_{n-1})$$

are invariant under $Gl(V_{n-1}(E))$. Thus in the above discussion and conjectures we may replace $\mathfrak{G} := \mathrm{Hom}(V_n, V_n)$ by the space

$$\mathfrak{g} := \{A_n : \mathrm{Tr}(A_n) = 0, \mathrm{Tr}(A_{n-1}) = 0\}.$$

Then $\mathrm{Lie}(U(\theta_0)^e)$ is replaced by

$$\mathfrak{u}_{\theta_0} := \mathrm{Lie}(U(\theta_0^e)) \cap \mathfrak{g}$$

and $\mathfrak{S}$ by

$$\mathfrak{s} := \mathfrak{S} \cap \mathfrak{g}.$$

6. Smooth matching and the fundamental Lemma for $n = 2$

Let E/F be an arbitrary quadratic extension. We choose τ such that $E = F\sqrt{\tau}$. For $n = 2$ we take $V_2 = E^2$ and $V_1 = E$. Then

$$\mathfrak{g} = \left\{ \begin{pmatrix} 0 & b \\ c & 0 \end{pmatrix} : b, c \in E \right\}.$$

The only invariant is the determinant. There is no difference between between regular and strongly regular. The above element is regular if and only if $bc \neq 0$.

Similarly,

$$\mathfrak{s} = \left\{ \begin{pmatrix} 0 & b' \\ c' & 0 \end{pmatrix} : b' + \overline{b'} = 0, c' + \overline{c'} = 0 \right\}.$$

The matrix of β has the form

$$\begin{pmatrix} \theta & 0 \\ 0 & 1 \end{pmatrix}$$

with $\theta \in F^\times$. The isomorphism class of β depends on the class of θ modulo the subgroup $N_r(E^\times)$ of norms. The corresponding vector space $\mathfrak{u}_\theta(F)$ is the space of matrices of the form

$$\begin{pmatrix} 0 & b \\ -\bar{b}\theta & 0 \end{pmatrix}.$$

Such an element is regular if $b \neq 0$. The group $U_1(F) = \{t : t\bar{t} = 1\}$ operates by conjugation. The action of t is given by:

$$\begin{pmatrix} 0 & b \\ -\bar{b}\theta & 0 \end{pmatrix} \mapsto \begin{pmatrix} 0 & bt \\ -\overline{bt}\theta & 0 \end{pmatrix}.$$

The only invariant of this action is the determinant. Two regular elements

$$\begin{pmatrix} 0 & b_1 \\ -\overline{b_1}\theta & 0 \end{pmatrix}, \begin{pmatrix} 0 & b_2 \\ -\overline{b_2}\theta & 0 \end{pmatrix}$$

are in the same orbit if and only if $b_1\overline{b_1} = b_2\overline{b_2}$. The only non-regular element is the 0 matrix.

On the other hand $\mathfrak{s}(F)$ is the space of matrices of the form

$$\begin{pmatrix} 0 & b\sqrt{\tau} \\ \frac{c}{\sqrt{\tau}} & 0 \end{pmatrix}, \; b, c \in F.$$

Such an element is regular if and only if $bc \neq 0$. The group $F^\times$ operates by conjugation. The action of $t \in F^\times$ is given by

$$\begin{pmatrix} 0 & b\sqrt{\tau} \\ \frac{c}{\sqrt{\tau}} & 0 \end{pmatrix} \mapsto \begin{pmatrix} 0 & bt\sqrt{\tau} \\ \frac{t^{-1}c}{\sqrt{\tau}} & 0 \end{pmatrix}.$$

The orbits of non-regular elements are the 0 matrix and the orbit of the following elements:

$$\begin{pmatrix} 0 & \sqrt{\tau} \\ 0 & 0 \end{pmatrix}, \begin{pmatrix} 0 & 0 \\ \frac{1}{\sqrt{\tau}} & 0 \end{pmatrix}.$$

The only invariant of this action is the determinant. Two regular elements

$$\begin{pmatrix} 0 & b_1\sqrt{\tau} \\ \frac{c_1}{\sqrt{\tau}} & 0 \end{pmatrix}, \begin{pmatrix} 0 & b_2\sqrt{\tau} \\ \frac{c_2}{\sqrt{\tau}} & 0 \end{pmatrix}$$

are conjugate if and only if $b_1c_1 = b_2c_2$.

The correspondence between regular elements is as follows:

$$\begin{pmatrix} 0 & b \\ -\bar{b}\theta & 0 \end{pmatrix} \to \begin{pmatrix} 0 & b'\sqrt{\tau} \\ \frac{c'}{\sqrt{\tau}} & 0 \end{pmatrix}$$

if $b\bar{b}\theta = -b'c'$. Thus we have a bijection between the disjoint union of the regular orbits of the spaces $\mathfrak{u}_\theta(F)$, $\theta \in E^\times/N_rF^\times)$, and the regular orbits in $\mathfrak{s}(F)$.

Now suppose that E/F is a local extension. Modulo the group of norms we have two choices θ_0 and θ_1 for θ. For f_i smooth of compact support on $\mathfrak{u}_i := \mathfrak{u}_{\theta_i}$ the orbital integral evaluated on

$$\xi_i = \begin{pmatrix} 0 & b \\ -\theta_i\bar{b} & 0 \end{pmatrix}$$

has the form

$$\Omega_U(f_i, \xi_i) = \int_{U_1} f_i \begin{pmatrix} 0 & bu \\ -\theta_i \overline{bu} & 0 \end{pmatrix} du\,.$$

The integral depends only on $b\bar{b}$ and can be written as

$$\Omega_U(f_i, -\theta_i b\bar{b})\,.$$

For Φ smooth of compact support on $\int$ the orbital integral evaluated on

$$\xi' = \begin{pmatrix} 0 & a\sqrt{\tau} \\ \frac{1}{\sqrt{\tau}} & 0 \end{pmatrix}$$

takes the form

$$\Omega(\Phi, a) := \Omega_G(f, \xi') = \int_{F^\times} \begin{pmatrix} 0 & a\sqrt{\tau}t \\ \frac{1}{\sqrt{t\tau}} & 0 \end{pmatrix} \eta(t) d^\times t\,.$$

We appeal to the following Lemma

LEMMA 1. *Let E/F be a quadratic extension of local fields and η the corresponding quadratic character. Given a smooth function of compact support ϕ on F^2, there are two smooth functions of compact support on F ϕ_1, ϕ_2 such that*

$$\int \phi(t^{-1}, at)\eta(t) d^\times t = \phi_1(a) + \eta(a)\phi_2(a)$$

and

$$\phi_1(0) = \int \phi(x, 0)\eta(x) d^\times x\,,\ \phi_2(0) = \int \phi(0, x)\eta(x) d^\times x\,.$$

Conversely, given ϕ_1, ϕ_2 there is ϕ such that the above conditions are satisfied.

Here we recall that the local Tate integral

$$\int \phi(x)\eta(x)|x|^s d^\times x$$

converges absolutely for $\Re s > 0$ and extends to a meromorphic function of s which is holomorphic at $s = 0$. The improper integral

$$\int \phi(x)\eta(x) d^\times x$$

is the value at $s = 0$.

The lemma implies that

$$\Omega_G(\Phi, a) = \phi_1(a) + \eta(a)\phi_2(a)$$

where ϕ_1, ϕ_2 are smooth functions of compact support on F. Then the condition that the pair (f_0, f_1) matches Φ becomes

$$\Omega_U(f_i, -b\bar{b}\theta_i) = \phi_1(-b\bar{b}\theta_i) + \eta(-\theta_i)\phi_2(-b\bar{b}\theta_i)\,.$$

It is then clear that given Φ there is a matching pair (f_0, f_1) and conversely.

We pass to the fundamental lemma. We assume the fields are non-Archimedean, the residual characteristic is odd, and the extension is unramified. We take τ to be a unit. We also take $\theta_0 = 1$. On the other hand θ_1 is any element with odd valuation. Let f_0 be the characteristic function of the integral elements of $\mathfrak{u}_0$. Then, with the previous notations,

$$\Omega(f_0, -b\bar{b}) = \Omega(f_0, \xi_0) = f_0 \begin{pmatrix} 0 & b \\ -\bar{b} & 0 \end{pmatrix}.$$

This is zero unless $|b\bar{b}| \leq 1$ in which case it is 1. On the other hand, let Φ_0 be the characteristic function of the integrals elements of $\mathfrak{s}$. Then

$$\Omega_G(\Phi_0, a) = \int_{1 \leq |t| \leq |a|^{-1}} \eta(t) d^\times t\,.$$

This is zero unless $|a| \leq 1$. Then it is zero unless a is a norm in which case it is one.

Thus if $\xi_0 \to \xi'$, that is, $a = -b\bar{b}$, we get

$$\Omega(f_0, \xi) = \Omega(\Phi_0, \xi')\,.$$

Otherwise, we get

$$\Omega(\Phi_0, \xi') = 0\,.$$

The fundamental lemma is established.

7. The trace formula for $n = 2$

In general, it will be convenient to consider all pairs (U_n, U_{n-1}) simultaneously. We illustrate this idea for the case $n = 2$. Let E/F be a quadratic extension of number fields.

The trace formula we want to consider has the following shape:

$$\sum_{\theta \in E^\times / N_\tau E^\times} \int_{U_1(F)\backslash U_1(F_\mathbb{A})} \sum_{\xi \in \mathfrak{U}_\theta(F)} f_\theta \left(\iota(h)^{-1} \xi \iota(h)\right) dh = \tag{19}$$

$$\int_{Gl_2(F)\backslash Gl_2(F_\mathbb{A})} \sum_{\xi' \in \mathfrak{s}(F)} \Phi\left(\iota(h)^{-1} \xi' \iota(h)\right) \eta(\det h) dh\,.$$

The left hand side converges and is equal to

$$\sum_\theta \left[f_\theta(0) \mathrm{Vol}(U_1(F)\backslash U_1(F_\mathbb{A})) + \sum_{\beta \in E^\times / N_\tau E^\times} \int_{U_1(F_\mathbb{A})} f_\theta \begin{pmatrix} 0 & t\beta \\ -\bar{\beta} t \theta & 0 \end{pmatrix} dt \right].$$

The right hand side must be interpreted as an improper integral. It is equal to

$$\int_{F^\times} \Phi \begin{pmatrix} 0 & t\sqrt{\tau} \\ 0 & 0 \end{pmatrix} \eta(t) d^\times t + \int_{F_\mathbb{A}^\times} \Phi \begin{pmatrix} 0 & 0 \\ \frac{t}{\sqrt{\tau}} & 0 \end{pmatrix} \eta(t) d^\times t$$

$$+ \sum_{\alpha \in F^\times} \int \Phi \begin{pmatrix} 0 & \alpha t \sqrt{\tau} \\ \frac{1}{t\sqrt{\tau}} & 0 \end{pmatrix} \eta(t) d^\times t\,.$$

For the first two terms, we recall that if ϕ is a Schwartz-Bruhat function on $F_\mathbb{A}$ then the global Tate integral

$$\int \phi(t) |t|^s \eta(t) d^\times t$$

converges for $\Re s > 1$ and has analytic continuation to an entire function of s. The improper integral

$$\int \phi(t) \eta(t) d^\times t$$

is the value of this function at $s = 0$. The remaining terms are absolutely convergent.

The matching condition is between a family (f_θ) and a function Φ. The global matching condition has the following form:

$$\int_{U_1(F_{\mathbb{A}})} f_\theta \begin{pmatrix} 0 & t\beta \\ -\overline{\beta}t\theta & 0 \end{pmatrix} dt = \int_{F_{\mathbb{A}}^\times} \Phi \begin{pmatrix} 0 & \alpha t\sqrt{\tau} \\ \frac{1}{t\sqrt{\tau}} & 0 \end{pmatrix} \eta(t) d^\times t$$

if $-\beta\overline{\beta}\theta = \alpha$. At a place of F inert in E, the corresponding local matching condition is described in the previous section. At a place which splits in E, it is elementary. The local matching conditions imply

$$\sum_\theta f_\theta(0)\mathrm{Vol}(U_1(F)\backslash U_1(F_{\mathbb{A}})) =$$

$$\int_{F^\times} \Phi \begin{pmatrix} 0 & t\sqrt{\tau} \\ 0 & 0 \end{pmatrix} \eta(t) d^\times t + \int_{F_{\mathbb{A}}^\times} \Phi \begin{pmatrix} 0 & 0 \\ \frac{t}{\sqrt{\tau}} & 0 \end{pmatrix} \eta(t) d^\times t\,.$$

We will not give the proof. It can be derived from [**9**].

8. Orbits of $Gl_2(E)$

We take $V_3(E) = E^3$ (column vectors). We set

$$e_3 = \begin{pmatrix} 0 \\ 0 \\ 1 \end{pmatrix}.$$

We identify V_3^* with the space of row vectors with 3 entries. We take $e_3^* = (0,0,1)$. Then $V_2(E) = E^2$ is the space of row vectors whose last component is 0. We denote by $\mathfrak{G}$ the space $\mathrm{Hom}_E(V_3, V_3)$ and by $\mathfrak{g}$ the subspace of A such that $\mathrm{Tr}(A) = 0$ and $\mathrm{Tr}(A|V_2) = 0$. Thus $\mathfrak{g}(E)$ is the space of 3×3 matrices X with entries in E of the form

$$X = \begin{pmatrix} a & b & x_1 \\ c & -a & x_2 \\ y_1 & y_2 & 0 \end{pmatrix}$$

The group $Gl_2(E)$ operates on $\mathfrak{g}(E)$. We introduce several invariants of this action:

$$A_1(X) = \det \begin{pmatrix} a & b \\ c & -a \end{pmatrix}, \tag{20}$$

$$A_2(X) = (y_1, y_2) \begin{pmatrix} x_1 \\ x_2 \end{pmatrix}, \tag{21}$$

$$B_1(X) = \det X\,. \tag{22}$$

We denote by $R(X)$ the resultant of the following polynomials in λ:

$$\det\left[\begin{pmatrix} a & b \\ c & -a \end{pmatrix} - \lambda\right], -\det[X - \lambda]\,.$$

It is also an invariant. More explicitly,

$$A_1(X) = -a^2 - bc \tag{23}$$

$$A_2(X) = x_1 y_1 + x_2 y_2 \tag{24}$$

$$B_1(X) = (x_1 y_1 - x_2 y_2)a + x_1 y_2 c + x_2 y_1 b \tag{25}$$

$$R(X) = A_1(X) A_2(X)^2 + B_1(X)^2 \tag{26}$$

Clearly, X is strongly regular if and only if $A_2(X) \neq 0$ and $R(X) \neq 0$. If X is strongly regular the invariants c_1, c_2 and a_1, a_2, a_3 introduced earlier can be computed in terms of the new invariants as follows:

$$c_2 = A_2(X) \tag{27}$$
$$-c_1 c_2^2 = R(X) \tag{28}$$
$$a_1 = -B_1(X) A_2^{-1}(X) \tag{29}$$
$$a_2 = -a_1 \tag{30}$$
$$a_3 = 0 \tag{31}$$

We also introduce

$$B_2(X) := \begin{pmatrix} -x_2 & x_1 \end{pmatrix} \begin{pmatrix} a & b \\ c & -a \end{pmatrix} \begin{pmatrix} x_1 \\ x_2 \end{pmatrix} \tag{32}$$

$$B_3(X) := \begin{pmatrix} y_1 & y_2 \end{pmatrix} \begin{pmatrix} a & b \\ c & -a \end{pmatrix} \begin{pmatrix} -y_2 \\ y_1 \end{pmatrix} \tag{33}$$

Explicitly,

$$\begin{aligned} B_2(X) &= -2x_1x_2a + x_1^2c - x_2^2b \\ B_3(X) &= -2y_1y_2a + y_1^2b - y_2^2c \end{aligned}$$

We remark that if we replace $\begin{pmatrix} x_1 \\ x_2 \end{pmatrix}$ by $h \begin{pmatrix} x_1 \\ x_2 \end{pmatrix}$ with $h \in Sl(2, F)$ then $(-x_2, x_1)$ is replaced by $(-x_2, x_1)h^{-1}$. It follows that B_2 is $Sl_2(E)$ invariant. Likewise for B_3.

We let $\mathfrak{g}(E)'$ be the set of X such that $A_2(X) \neq 0$ and $\mathfrak{g}(E)^s$ the set of $X \in \mathfrak{g}(E)'$ such that $R(X) \neq 0$. Thus $\mathfrak{g}(E)^s$ is the set of strongly regular elements.

LEMMA 2. *Every $Sl_2(E)$ orbit in $\mathfrak{g}(E)'$ contains a unique element of the form*

$$X = \begin{pmatrix} a & b & 0 \\ c & -a & 1 \\ 0 & t & 0 \end{pmatrix}$$

and then $A_1(X) = -a^2 - bc$, $A_2(X) = t \neq 0$, $B_1(X) = -at$, $B_2(X) = -b$, $B_3(X) = -t^2c$, $R(X) = -t^2bc$. In particular, A_2, B_1, B_2, B_3 form a complete set of invariants for the orbits of $Sl_2(E)$ in $\mathfrak{g}(E)'$.

PROOF: If $A_2(X) \neq 0$ then a fortiori $\begin{pmatrix} x_1 \\ x_2 \end{pmatrix} \neq 0$. Since $Sl_2(F)$ is transitive on the space of non-zero vectors in F^2, we may as well assume

$$X = \begin{pmatrix} a & b & 0 \\ c & -a & 1 \\ y_1 & y_2 & 0 \end{pmatrix}$$

Then $y_2 = A_2(X) \neq 0$. We now conjugate X by

$$\iota \begin{pmatrix} 1 & 0 \\ -\frac{y_1}{y_2} & 1 \end{pmatrix}$$

and obtain a matrix like the one in the lemma. In $Gl_2(E)$ the stabilizer of the column $\left(\begin{smallmatrix} 0 \\ 1 \end{smallmatrix}\right)$ and the row $\left(\begin{smallmatrix} 0 & t \end{smallmatrix}\right)$ (where $t \neq 0$) is the group

$$H = \left\{ \begin{pmatrix} \alpha & 0 \\ 0 & 1 \end{pmatrix}, \alpha \in E^\times \right\}$$

Thus the stabilizer in $Sl_2(E)$ of a matrix like the one in the lemma is indeed trivial. The remaining assertions of the lemma are easy. □

LEMMA 3. *If X is in $\mathfrak{g}(E)'$ then X is strongly regular if and only if it is regular.*

PROOF: We may assume that

$$X = \begin{pmatrix} a & b & 0 \\ c & -a & 1 \\ 0 & t & 0 \end{pmatrix},$$

with $t \neq 0$. Then X is strongly regular if and only $R(X) = -t^2bc \neq 0$. On the other hand, it is regular if and only if the column vectors

$$\begin{pmatrix} 0 \\ 1 \end{pmatrix}, \begin{pmatrix} b \\ -a \end{pmatrix}$$

are linearly independent and the row vectors

$$(0, t), (ct, -ta)$$

are linearly independent. Then it is so if and only if $b \neq 0$ and $c \neq 0$. Our assertion follows. □

LEMMA 4. *Every orbit of $Gl_2(E)$ in $\mathfrak{g}(E)^s$ contains a unique element of the form*

$$X = \begin{pmatrix} a & b & 0 \\ 1 & -a & 1 \\ 0 & t & 0 \end{pmatrix},$$

where $b \neq 0$ and $t \neq 0$. Then

$$\begin{aligned} A_1(X) &= -a^2 - b \\ A_2(X) &= t \\ B_1(X) &= -at \\ R(X) &= -bt^2 \end{aligned}$$

If the invariants A_1, A_2, B_1 take the same values on two matrices in $\mathfrak{g}(E)^s$, then they are in the same orbit of $Gl_2(E)$. Finally, given a_1, a_2, b_1 in E with $a_2 \neq 0$ and $a_1a_2^2 + b_1^2 \neq 0$ there is $X \in \mathfrak{g}(E)^s$ such that $A_1(X) = a_1$, $A_2(X) = a_2$ and $B_1(X) = b_1$.

PROOF: The first assertion follows from the general case, or more simply, from the previous Lemma. Indeed, by the previous lemma, every orbit contains an element of the form

$$X = \begin{pmatrix} a & b & 0 \\ c & -a & 1 \\ 0 & t & 0 \end{pmatrix}$$

and then $-bct^2 = R(X)$. Thus $bc \neq 0$. Conjugating by

$$\iota\begin{pmatrix} c & 0 \\ 0 & 1 \end{pmatrix}$$

we obtain an element of the required form. The stabilizer of this element in $Gl_2(E)$ is trivial. The remaining assertions are obvious. □

9. Orbits of $Gl_2(F)$

Now we consider the orbits of $Gl_2(F)$ in $\mathfrak{s}$. Of course, $\mathfrak{s} = \sqrt{\tau}\mathfrak{g}(F)$. We define $\mathfrak{s}' = \mathfrak{s} \cap \mathfrak{g}(E)'$ and $\mathfrak{s}^s = \mathfrak{s} \cap \mathfrak{g}(E)^s$. For $Y \in \mathfrak{g}(F)$, we have

$$\begin{aligned} A_1(\sqrt{\tau}Y) &= \tau A_1(Y) \\ A_2(\sqrt{\tau}Y) &= \tau A_2(Y) \\ B_1(\sqrt{\tau}Y) &= \tau\sqrt{\tau}B_1(Y)\,. \end{aligned}$$

Also

$$R(\sqrt{\tau}Y) = \tau^3 R(Y)\,.$$

Thus, on $\mathfrak{s}^s$, the functions A_1, A_2 (with values in F) together with the function B_1 (with values in $F\sqrt{\tau}$) form a complete set of invariants for the action of $Gl_2(F)$. Conversely, given $a_1 \in F$, $a_2 \in F^\times$ and $b_1 \in F\sqrt{\tau}$ such that $a_1a_2^2 + b_1^2 \neq 0$ there is $X \in \mathfrak{s}^s$ with those numbers for invariants.

10. Orbits of the unitary group

We formulate the fundamental lemma in terms of the Hermitian matrix

$$\theta_0 = \begin{pmatrix} 0 & 1 \\ 1 & 0 \end{pmatrix},$$

rather than in terms of the Hermitian unit matrix. Then

$$\theta_0^e = \begin{pmatrix} 0 & 1 & 0 \\ 1 & 0 & 0 \\ 0 & 0 & 1 \end{pmatrix}.$$

We let $U_{2,1}$ be the unitary group for the Hermitian matrix θ_0^e. Thus the Lie algebra of $U_{2,1}$ is the space $\mathfrak{U}(F)$ of matrices Ξ of the form

$$\Xi = \begin{pmatrix} a & b & z_1 \\ c & d & z_2 \\ -\overline{z_2} & \overline{z_1} & e \end{pmatrix}$$

with $a + \overline{d} = 0$, $b \in F\sqrt{\tau}$, $c \in F\sqrt{\tau}$, $e \in F\sqrt{\tau}$. We let $U_{1,1}$ be the unitary group for the Hermitian matrix θ_0. The corresponding Hermitian form is

$$Q(z_1, z_2) = z_1\overline{z_2} + z_2\overline{z_1}$$

We embed $U_{1,1}$ into $U_{2,1}$ by

$$\iota(u) = \begin{pmatrix} u & 0 \\ 0 & 1 \end{pmatrix}.$$

We obtain an action of $U_{1,1}(F)$ by conjugation. As before, we set $\mathfrak{u} = \mathfrak{U} \cap \mathfrak{g}$. Thus $\mathfrak{u}$ is the space of matrices Ξ of the form

$$\Xi = \begin{pmatrix} a & b & z_1 \\ c & -a & z_2 \\ -\overline{z_2} & -\overline{z_1} & 0 \end{pmatrix},\ a \in F, b \in F\sqrt{\tau}\,,\ c \in F\sqrt{\tau}\,. \tag{34}$$

Then

$$\begin{aligned} A_1(\Xi) &= -a^2 - bc \\ A_2(\Xi) &= -Q(z_1, z_2) \\ B_1(\Xi) &= a(\overline{z_1}z_2 - \overline{z_2}z_1) - bz_2\overline{z_2} - cz_1\overline{z_1} \end{aligned}$$

We set $\mathfrak{u}' = \mathfrak{u} \cap \mathfrak{g}'$ and $\mathfrak{u}^s = \mathfrak{u} \cap \mathfrak{g}^s$. We study directly the orbits of $U_{1,1}$ on $\mathfrak{u}^s$.

LEMMA 5. *For $t \in F^\times$ choose $(z_{1,0}, z_{2,0})$ such that $Q(z_{1,0}, z_{2,0}) = -t$. Any orbit of $SU_{1,1}$ in $\mathfrak{u}'$ on which A_1 takes the value t contains a unique element of the form*

$$\begin{pmatrix} a & b & z_{1,0} \\ c & -a & z_{2,0} \\ -\overline{z_{2,0}} & -\overline{z_{1,0}} & 0 \end{pmatrix}$$

PROOF: Since $SU_{1,1}$ acting on E^2 is transitive on the sphere $S_{-t} = \{v \in E^2 | Q(v) = -t\}$ and each point of the sphere has a trivial stabilizer in $SU_{1,1}$, our assertion is trivial. □

LEMMA 6. *For $t \in F^\times$ choose $(z_{1,0}, z_{2,0})$ such that $Q(z_{1,0}, z_{2,0}) = -t$. Any orbit of $U_{1,1}$ in $\mathfrak{u}^s$ on which A_1 takes the value t contains an element of the form*

$$\Xi = \begin{pmatrix} a & b & z_{1,0} \\ c & -a & z_{2,0} \\ -\overline{z_{2,0}} & -\overline{z_{1,0}} & 0 \end{pmatrix}$$

The stabilizer in $U_{1,1}$ of such an element is trivial. Moreover $A_1(\Xi) \in F$, $A_2(\Xi) \in F$, $B_1(\Xi) \in F\sqrt{\tau}$ and $-R(\Xi)$ is a non-zero norm. $A_1(\Xi)$, $A_2(\Xi)$, $B_1(\Xi)$ completely determine the orbit of Ξ. Finally, if $a_1 \in F$, $a_2 \in F$ and $b_1 \in F\sqrt{\tau}$ are such that $a_2 \neq 0$, $a_1 a_2^2 + b_1^2 \neq 0$ and $-(a_1 a_2^2 + b_1^2)$ is a norm, then there is Ξ in $\mathfrak{u}^s$ such that $A_1(\Xi) = a_1$, $A_2(\Xi) = a_2$ and $B_1(\Xi) = b_1$.

PROOF: As before, the orbit in question contains at least one element of this type, say Ξ_0. To prove the remaining assertions we introduce the matrix

$$M = \begin{pmatrix} -\overline{z_{1,0}}t^{-1} & z_{1,0} \\ \overline{z_{2,0}}t^{-1} & z_{2,0} \end{pmatrix} \in Sl_2(E)\,.$$

Then

$${}^t\overline{M} \begin{pmatrix} 0 & 1 \\ 1 & 0 \end{pmatrix} M = \begin{pmatrix} t^{-1} & 0 \\ 0 & -t \end{pmatrix}.$$

It follows that $\iota(M)^{-1}\mathfrak{U}\iota(M)$ is the Lie algebra of the unitary group for the Hermitian matrix

$$\begin{pmatrix} t^{-1} & 0 & 0 \\ 0 & -t & 0 \\ 0 & 0 & 1 \end{pmatrix}$$

Then $\iota(M)^{-1}\mathfrak{u}\iota(M)$ becomes the space of matrices of the form

$$\begin{pmatrix} \alpha & \beta & z_1 \\ \overline{\beta}t^{-2} & -\alpha & z_2 \\ -\overline{z_1}t^{-1} & \overline{z_2}t & 0 \end{pmatrix},\ \alpha \in F\sqrt{\tau}\,.$$

and $\Xi_1 = \iota(M)^{-1}\Xi_0\iota(M)$ is a matrix of the form

$$\Xi_1 = \begin{pmatrix} \alpha_1 & \beta_1 & 0 \\ \overline{\beta_1}t^{-2} & -\alpha_1 & 1 \\ 0 & t & 0 \end{pmatrix}.$$

We have

$$\begin{aligned} A_1(\Xi_0) = A_1(\Xi_1) &= -\alpha_1^2 - \beta_1\overline{\beta_1}t^{-2} \\ A_2(\Xi_0) = A_2(\Xi_1) &= t \\ B_1(\Xi_0) = B_1(\Xi_1) &= -\alpha_1 t \\ R(\Xi_0) = R(\Xi_1) &= -\beta_1\overline{\beta_1} \end{aligned}$$

The stabilizer H of the column $\left(\begin{smallmatrix} 0 \\ 1 \end{smallmatrix}\right)$ and the row $\left(\begin{smallmatrix} 0 & t \end{smallmatrix}\right)$ in the group $\iota(M)^{-1}U_{1,1}\iota(M)$ is the group

$$\begin{pmatrix} u & 0 \\ 0 & 1 \end{pmatrix}, \; u \in U_1 .$$

Since Ξ_1 is in $\mathfrak{g}(E)^s$ we have $\beta_1 \neq 0$. Thus the stabilizer of Ξ_1 of Ξ_1 in H or in $\iota(M)^{-1}U_{1,1}\iota(M)$ is trivial. If the invariants A_1, A_2, B_1 take the same value on two such elements Ξ_1 and Ξ_2 of $\iota(M)^{-1}\mathfrak{u}\iota(M)$, then we have $t_1 = t_2$, $\alpha_1 = \alpha_2$ and $\beta_1\overline{\beta_1} = \beta_2\overline{\beta_2}$. Then $\beta_1 = \beta_2 u$ with $u \in U_1$. Then Ξ_1 and Ξ_2 are conjugate by an element of H. □

11. Comparison of orbits

In accordance with our general discussion, we match the orbit of $\Xi \in \mathfrak{u}^s$ with the orbit of $X \in \mathfrak{s}^s$ and we write $\Xi \to X$ if the matrices are conjugate by $Gl_2(E)$, or, what amounts to the same thing, if they have the same invariants A_1, A_2, B_1. In particular, we have the following Proposition.

PROPOSITION 2. *Given $X \in \mathfrak{s}^s$, there is a matrix Ξ in $\mathfrak{u}^s$ which matches X if and only if $-R(X)$ is a (non-zero) norm.*

12. The fundamental lemma for $n = 3$

We now let E/F be an unramified quadratic extension of non-Archimedean fields. We assume the residual characteristic is not 2. We let $f_{\mathfrak{u}}$ be the characteristic function of the matrices with integral entries in $\mathfrak{u}$ and $\Phi_{\mathfrak{s}}$ be similarly the characteristic function of the set of matrices with integral entries in $\mathfrak{s}$. For $\Xi \in \mathfrak{u}^s$ we set

$$\Omega_U(\Xi) = \int_{U_{1,1}} f_{\mathfrak{u}}(u\Xi u^{-1})du \tag{35}$$

Likewise, for $X \in \mathfrak{s}^s$ we set

$$\Omega_G(X) = \int_{Gl_2(F)} \Phi_0(gXg^{-1})\eta(\det g)dg \tag{36}$$

The **fundamental lemma** asserts that if Ξ matches X then

$$\Omega_U(\Xi) = \tau(X)\Omega_G(X) \tag{37}$$

where $\tau(X) = \pm 1$ is the transfer factor. If, on the contrary, X matches no Ξ then

$$\Omega_G(X) = 0 .$$

To prove the fundamental lemma we exploit the isomorphism between $U_{1,1}$ and $Sl(2, F)$. Now $U_{1,1}$ is the product of the normal subgroup $SU_{1,1}$ and the torus

$$T = \left\{ t = \begin{pmatrix} z & 0 \\ 0 & \overline{z}^{-1} \end{pmatrix}, z \in E^\times \right\} .$$

with intersection

$$T \cap SU_{1,1} = \left\{ t = \begin{pmatrix} a & 0 \\ 0 & a^{-1} \end{pmatrix}, a \in F^\times \right\}.$$

Let T_0 be the subgroup of $t \in T$ with $|z| = 1$. Then $U_{1,1} = SU_{1,1}T_0$.

The function $f_\mathfrak{u}$ is invariant under T_0. Thus, in fact,

$$\Omega_U(\Xi) = \int_{SU_{1,1}} f_\mathfrak{u}(u\Xi u^{-1})du\,.$$

To establish the fundamental lemma we will use the isomorphism $\theta : SU_{1,1} \to Sl_2(F)$ defined by

$$\theta(g) = \begin{pmatrix} \sqrt{\tau} & 0 \\ 0 & 1 \end{pmatrix} g \begin{pmatrix} \frac{1}{\sqrt{\tau}} & 0 \\ 0 & 1 \end{pmatrix} \tag{38}$$

and a compatible $F-$linear bijective map $\Theta : \mathfrak{u} \to \mathfrak{g}(F)$ defined as follows. If

$$\Xi = \begin{pmatrix} \alpha & \beta & z_1 \\ \gamma & -\alpha & z_2 \\ -\overline{z_2} & -\overline{z_1} & 0 \end{pmatrix}, \alpha \in F, \beta \in \sqrt{\tau}F, \gamma \in \sqrt{\tau}F$$

then

$$\Theta(\Xi) = X\,,\; X = \begin{pmatrix} a & b & x_1 \\ c & -a & x_2 \\ y_1 & y_2 & 0 \end{pmatrix} \tag{39}$$

where

$$\begin{array}{lll} a = \alpha & b = \beta\sqrt{\tau} & c = \frac{\gamma}{\sqrt{\tau}} \\ x_1 = \frac{z_1+\overline{z_1}}{2} & y_1 = \frac{z_2+\overline{z_2}}{2} & \\ x_2 = \frac{z_2-\overline{z_2}}{2\sqrt{\tau}} & y_2 = \frac{-\sqrt{\tau}(z_1-\overline{z_1})}{2} & \end{array}$$

The inverse formulas for z_1, z_2 read

$$z_1 = x_1 - \frac{y_2}{\sqrt{\tau}}\,,\; z_2 = y_1 + x_2\sqrt{\tau}\,.$$

Note that

$$\begin{pmatrix} a & b \\ c & -a \end{pmatrix} = \begin{pmatrix} \sqrt{\tau} & 0 \\ 0 & 1 \end{pmatrix} \begin{pmatrix} \alpha & \beta \\ \gamma & -\alpha \end{pmatrix} \begin{pmatrix} \frac{1}{\sqrt{\tau}} & 0 \\ 0 & 1 \end{pmatrix}.$$

The linear bijection Θ has the following property of compatibility with the isomorphism θ:

$$\Theta(\iota(g)\Xi\iota(g)^{-1}) = \iota(\theta(g))\Theta(\Xi)\iota(\theta(g))^{-1}$$

for $g \in SU(1,1)$.

We can use Θ to define an action μ of T on $\mathfrak{g}$. It is defined by

$$\Theta\left(\iota(t)\Xi\iota(t)^{-1}\right) = \mu(t)\left(\Theta(\Xi)\right)\,.$$

Explicitly, if $t = \mathrm{diag}(z, \overline{z}^{-1})$, $z = p + \sqrt{\tau}$, then

$$\mu(t)\left[\begin{pmatrix} a & b & x_1 \\ c & -a & x_2 \\ y_1 & y_2 & 0 \end{pmatrix}\right] =$$

$$\begin{pmatrix} a & bz\overline{z} & px_1 - qy_2 \\ c(z\overline{z})^{-1} & -a & \frac{px_2+qy_1}{p^2-q^2\tau} \\ \frac{py_1+q\tau x_2}{p^2-q^2\tau} & py_2 - q\tau x_1 & 0 \end{pmatrix}$$

For $t \in T \cap SU_{1,1} = T \cap Sl_2(F)$, $\mu(t)$ is the conjugation by $\iota(t)$. Again, $T = T_0(T \cap Sl_2(F))$.

We compare the invariants of Ξ and $X = \Theta(\Xi)$. From

$$-\overline{z_2}z_1 - \overline{z_1}z_2 = -2(x_1y_1 + x_2y_2)$$

and

$$\alpha(\overline{z_1}z_2 - \overline{z_2}z_1) - \beta z_2\overline{z_2} - \gamma z_1\overline{z_1} =$$
$$\sqrt{\tau}(2ax_1x_2 + bx_2^2 - cx_1^2) + \frac{1}{\sqrt{\tau}}(2ay_1y_2 - by_1^2 + cy_2^2)$$

we get

$$A_1(\Xi) = A_1(\Theta(\Xi)) \tag{40}$$
$$A_2(\Xi) = -2A_2(\Theta(\Xi)) \tag{41}$$
$$B_1(\Xi) = -\sqrt{\tau}B_2(\Theta(\Xi)) - \frac{1}{\sqrt{\tau}}B_3(\Theta(\Xi)) \tag{42}$$

Also

$$R(\Xi) = 4A_1(X)A_2(X)^2 + \tau B_2(X)^2 + \frac{1}{\tau}B_3(X)^2 + 2B_2(X)B_3(X).$$

We let $\tilde{\mathfrak{g}}(F)$ be the image of $\mathfrak{u}^s$ under Θ. Thus $\tilde{\mathfrak{g}}(F)$ is contained in $\mathfrak{g}(F)'$. The functions A_1, A_2 and $-\sqrt{\tau}B_2 - \frac{1}{\sqrt{\tau}}B_3$ form a complete set of invariants for the action of $Sl_2(F)$ and T_0 on $\tilde{\mathfrak{g}}$.

We will let Φ_0 be the characteristic function of the set of integers in $\mathfrak{g}(F)$. For $X \in \mathfrak{g}'$ we set

$$\Omega_{Sl_2}(X) = \int_{Sl_2(F)} \Phi_0(\iota(g)X\iota(g)^{-1})dg. \tag{43}$$

Thus $\Omega_U(\Xi) = \Omega_{Sl_2}(\Theta(\Xi))$.

We match the orbits of $U_{2,1}$ in $\mathfrak{u}^s$ with the orbits of $Gl_2(F)$ in $\mathfrak{s}^s$ by matching the invariants: for Ξ in $\mathfrak{u}^s$ and $Y \in \mathfrak{g}(F)^s$, $\Xi \to \sqrt{\tau}Y$ if

$$\begin{aligned} A_1(\Xi) &= A_1(\sqrt{\tau}Y) \\ A_2(\Xi) &= A_2(\sqrt{\tau}Y) \\ B_1(\Xi) &= B_1(\sqrt{\tau}Y) \end{aligned}$$

This leads to the following relation in terms of $X = \Theta(\Xi)$ and Y:

$$A_1(X) = \tau A_1(Y)$$
$$-2A_2(X) = \tau A_2(Y)$$
$$-\sqrt{\tau}B_2(X) - \frac{1}{\sqrt{\tau}}B_3(X) = \tau\sqrt{\tau}B_1(Y)$$

The last relation can be simplified:

$$-\tau B_2(X) - B_3(X) = \tau^2 B_1(Y)$$

To make this relation explicit, we may replace $X \in \tilde{\mathfrak{g}}(F)$ by a conjugate under $Sl_2(F)$ and thus assume

$$X = \begin{pmatrix} a_1 & b_1 & 0 \\ c_1 & -a_1 & 1 \\ 0 & t_1 & 0 \end{pmatrix} \tag{44}$$

The condition that X be in $\tilde{\mathfrak{g}}(F)$ reads

$$t_1 \neq 0\,,\ \tau b_1^2 + \frac{t_1^4 c_1^2}{\tau} - 2b_1c_1t_1^2 - 4a_1^2t_1^2 \neq 0\,.$$

The second condition can also be written as

$$(\sqrt{\tau}b_1 - \frac{t_1^2c_1}{\sqrt{\tau}})^2 - 4a_1^2t_1^2 \neq 0\,.$$

As a matter of fact, assuming $t_1 \neq 0$, the second condition fails only if $a_1 = 0$ and $\tau b_1 = t_1^2 c_1$.

Likewise, we may assume:

$$Y = \begin{pmatrix} a & b & 0 \\ c & -a & 1 \\ 0 & t & 0 \end{pmatrix} \tag{45}$$

Then

$$\begin{aligned} A_1(Y) &= -a^2 - bc \\ A_2(Y) &= t \\ B_1(Y) &= -ta \end{aligned}$$

Moreover $R(\sqrt{Y}) = \tau^3 R(Y) = -bc\tau^3 t^2$. This matrix is in $\mathfrak{g}(F)^s$ if and only if $t \neq 0$ and $bc \neq 0$. It matches some X if and only if $-R(\sqrt{Y})$ is a norm. Since $-\tau$ is a norm this is equivalent to $-bc$ being a norm.

The condition of matching of orbits becomes: $X \to Y$ if

$$a_1^2 + b_1c_1 = \tau(a^2 + bc) \tag{46}$$

$$-2t_1 = \tau t \tag{47}$$

$$\tau b_1 + t_1^2 c_1 = -\tau^2 ta \tag{48}$$

In a precise way, this system of equations for (a_1, b_1, c_1, t_1) has a solution if and only if $-bc$ is a norm. If we write

$$-\tau^2 bc = y^2 - \tau a_1^2 \tag{49}$$

then we can take a_1 for the first entry of X, and then take $t_1 = -\frac{\tau t}{2}$,

$$b_1 = -\frac{t}{2}(y + \tau a)\,,\ c_1 = \frac{2}{t\tau}(y - \tau a)\,. \tag{50}$$

Note that $a_1 = 0$ and $\tau b_1 = t_1^2 c_1$ would imply $y = 0$ and thus $bc = 0$. Thus X is indeed in $\tilde{\mathfrak{g}}(F)$.

The fundamental lemma then takes the following form.

THEOREM 1 (The fundamental lemma for $n = 3$). *For $Y \in \mathfrak{g}(F)^s$ of the form (45) define*

$$\Omega_{Gl_2}(Y) = \int_{Gl_2(F)} \Phi_0(gYg^{-1})\eta(\det g)dg\,. \tag{51}$$

If $-bc$ is not a norm then $\Omega_{Gl_2}(Y) = 0$. If $-bc$ is a norm, let (a_1, b_1, c_1, t_1) satisfy the conditions (46) and let X be the element of $\tilde{\mathfrak{g}}(F)$ defined by (44). Then

$$\Omega_{Gl_2}(Y) = \eta(c)\Omega_{Sl_2}(X)$$

We now prove the fundamental lemma.

13. Orbital integrals for $Sl_2(F)$

In this section we compute the orbital integral $\Omega_{Sl_2}(X)$, where

$$X = \begin{pmatrix} a & b & 0 \\ c & -a & 1 \\ 0 & t & 0 \end{pmatrix}. \tag{52}$$

Suppose $\Omega_{Sl_2}(X) \neq 0$. This implies that the orbit of X intersects the support of Φ_0, and we get that the invariants of X are integral. In particular a^2+bc, t, at, b, t^2c are all integers.

We set

$$g = k \begin{pmatrix} m & 0 \\ 0 & m^{-1} \end{pmatrix} \begin{pmatrix} 1 & u \\ 0 & 1 \end{pmatrix}, \; k \in Gl_2(\mathcal{O}_F),$$

$$dg = dk|m|^2 d^\times m dk$$

The integration over k is superfluous. Thus we get

$$\Omega_{Sl_2}(X) =$$

$$\int\int \Phi_0 \left[\begin{pmatrix} a+cu & m^2(b-2au-u^2c) & mu \\ cm^{-2} & -a-cu & m^{-1} \\ 0 & tm & 0 \end{pmatrix} \right] du|m|^2 d^\times m.$$

LEMMA 7. *The integral converges absolutely, provided $t \neq 0$.*

PROOF: Indeed, the ranges of u and m are limited by

$$|u| \leq |m|^{-1}, \; 1 \leq |m| \leq |t|^{-1}.$$

Thus the integral is less than the integral

$$\int\int_{|u|\leq|m|^{-1}, 1\leq|m|\leq|t|^{-1}} du|m|^2 d^\times m$$

$$= \int_{1\leq|m|\leq|t|^{-1}} |m| d^\times m$$

which is finite. □

Explicitly, the integral is equal to

$$\int\int du|m|^2 d^\times m$$

over

$$\begin{cases} |a+cu| \leq 1 & |u| \leq |m|^{-1} \\ |c| \leq |m|^2 & 1 \leq |m| \leq |t|^{-1} \\ |b-2au-u^2c| \leq |m|^{-2} \end{cases}$$

We first compute the integral for $c \neq 0$. We may change u to uc^{-1} to get

$$|c|^{-1} \int\int du|m|^2 d^\times m$$

$$\begin{cases} |a+u| \leq 1 & |u| \leq |cm^{-1}| \\ |c| \leq |m|^2 & 1 \leq |m| \leq |t|^{-1} \\ |a^2+bc-(a+u)^2| \leq |cm^{-2}| \end{cases}$$

Since $|a^2+bc| \leq 1$ and $|cm^{-2}| \leq 1$ we see that the condition $|a+u| \leq 1$ is superfluous. We may then change u to $u-a$ to obtain

$$\Omega_{Sl_2}(X) = |c|^{-1} \int\int du|m|^2 d^\times m \tag{53}$$

$$\begin{cases} |u-a| \leq |cm^{-1}| & |a^2+bc-u^2| \leq |cm^{-2}| \\ |c| \leq |m|^2 & 1 \leq |m| \leq |t^{-1}| \end{cases}$$

Before embarking on the computation, we prove a lemma which will show that the orbital integral Ω_{Gl_2} converges absolutely.

LEMMA 8. *Let ω be a compact set of $F^\times$. Then, with the previous notations, the relations $A_2(X) \in \omega$, $R(X) \in \omega$ and $\Omega_{Sl_2}(X) \neq 0$ imply that c is in a compact set of $F^\times$.*

PROOF: Indeed, both t and bc are then in compact sets of $F^\times$. If $\Omega_{Sl_2}(X) \neq 0$ then there are m and u satisfying the above conditions. We have then $|c| \leq |t^{-2}|$ so that $|c|$ is bounded above. If $|bc| \leq |cm^{-2}|$ then, since $|m^{-1}| \leq 1$ we have $|c| \geq |bc|$ and $|c|$ is bounded below. If $|cm^{-2}| < |bc|$ then $|a^2 - u^2| = |bc|$. Now $|a^2 + bc| \leq 1$ so $|a|$ is bounded above. Thus $|u|$ is also bounded above. Hence $|a + u|$ is bounded above by A say. Then $|bc| \leq A|a - u| \leq |cm^{-1}|A \leq |c|A$. Hence $|c| \geq |bc|A^{-1}$. Thus $|c|$ is bounded below, away from zero, in all cases. □

We have now to distinguish various cases depending on the square class of $-A_1(X) = a^2 + bc$.

13.1. Some notations. To formulate the result of our computations in a convenient way, we will introduce some notations.

For $A \in F^\times$ we set

$$\mu(A) := \int_{1 \leq |m| \leq |A|} |m| d^\times m \tag{54}$$

Thus $\mu(A) = 0$ if $|A| < 1$. Otherwise $\mu(A) = \frac{|A| - q^{-1}}{1 - q^{-1}}$. In particular, if $|A| = 1$, then $\mu(A) = 1$. Note that the above integral can be written as a sum

$$\sum_{1 \leq |m| \leq |A|} |m|$$

where the sum is over powers of a uniformizer satisfying the required inequalities.

If $A, B, C, \ldots,$ are given then we set

$$\mu(A, B, C, \ldots) := \mu(D) \text{ where } |D| = \inf(|A|, |B|, |C|, \ldots) \tag{55}$$

We also define

$$\mu(A : B) := \int_{|B| \leq |m| \leq |A|} |m| d^\times m\,.$$

Thus $\mu(A : 1) = \mu(A)$. We also define

$$\mu(A, B, C, \cdots : P, Q, R, \ldots) = \mu(D : S)$$

where $|D| = \inf(|A|, |B|, |C|, \ldots)$ while $|S| = \sup(|P|, |Q|, |R|, \ldots)$. Then

$$\mu(A, B, C \cdots : D) = |D| \mu(AD^{-1}, BD^{-1}, CD^{-1} \ldots)\,.$$

Clearly, if $1 \leq |C| \leq \inf(|A|, |B|)$, then

$$\mu(A, B : C\varpi^{-1}) + \mu(C) = \mu(A, B)\,. \tag{56}$$

We will frequently use the following elementary lemma.

LEMMA 9. *The difference*

$$\mu(A, B, C) - \mu(A\varpi, B, C)$$

is 0 unless $1 \leq |A| \leq \inf(|B|, |C|)$, in which case the difference is $|A|$.

For $A \in F^\times$ we set

$$\nu(A) := \int_{1 \le |m| \le |A|} d^\times m \tag{57}$$

Thus $\nu(A) = 0$ if $|A| < 1$. Otherwise $\nu(A) = 1 - v(A)$. In particular, if $|A| = 1$, then $\nu(A) = 1$. If $A, B, C, \ldots$, are given then we set

$$\nu(A, B, C, \ldots) = \nu(D)\,,\ |D| = \inf\left(|A|, |B|, |C|, \ldots\right) \tag{58}$$

We also define

$$\nu(A : B) = \int_{|B| \le |m| \le |A|} d^\times m$$

Thus $\nu(A : 1) = \nu(A)$. We also define

$$\nu(A, B, C, \cdots : P, Q, R, \ldots) = \nu(D : S)$$

where

$$|D| = \inf\left(|A|, |B|, |C|, \ldots\right)\ ,\ |S| = \sup\left(|P|, |Q|, |R|, \ldots\right)\,.$$

Clearly,

$$\nu(A, B, C \cdots : D) = \nu(AD^{-1}, BD^{-1}, CD^{-1} \ldots)\,. \tag{59}$$

We will use frequently the following elementary lemma:

LEMMA 10. *The difference*

$$\nu(A, B, C) - \nu(A\varpi, B, C)$$

is zero unless $1 \le |A| \le \inf(|B|, |C|)$ *in which case it is* 1.

If $x \in F^\times$ is an element of even valuation, then we denote by $\sqrt[v]{x}$ **any** element of $F^\times$ whose valuation is one-half the valuation of x. If x has odd valuation then $\sqrt[v]{x\varpi}$ is defined but not $\sqrt[v]{x}$. With this convention, the condition

$$|a| \le |x^2| \le |b|$$

is equivalent to

$$\left|\left\{\begin{matrix} \sqrt[v]{a} \\ \sqrt[v]{a\varpi^{-1}} \end{matrix}\right\}\right| \le |x| \le \left|\left\{\begin{matrix} \sqrt[v]{b} \\ \sqrt[v]{b\varpi} \end{matrix}\right\}\right|\,. \tag{60}$$

If $|a| \le |b|$ then

$$|a| \le \left|\left\{\begin{matrix} \sqrt[v]{ab} \\ \sqrt[v]{ab\varpi} \end{matrix}\right\}\right| \le \left|\left\{\begin{matrix} \sqrt[v]{ab} \\ \sqrt[v]{ab\varpi^{-1}} \end{matrix}\right\}\right| \le |b|\,. \tag{61}$$

13.2. Case where $a^2 + bc$ is odd. Suppose first $a^2 + bc$ has odd valuation, or, as we shall say, is **odd**. Then there is a uniformizer ϖ such that $a^2 + bc = \delta^2\varpi$. In the range (53) for the integral the quadratic condition becomes $|\delta^2\varpi - u^2| \le |cm^{-2}|$ and, in turn, this is equivalent to $|\delta^2\varpi| \le |cm^{-2}|$ and $|u^2| \le |cm^{-2}|$. Thus the integral is equal to

$$|c|^{-1} \int\int du |m|^2 d^\times m \tag{62}$$

over

$$\left\{\begin{matrix} |u| \le \left|\left\{\begin{matrix} \sqrt[v]{c} \\ \sqrt[v]{c\varpi} \end{matrix}\right\}\right| |m^{-1}| & |u - a| \le |cm^{-1}| \\ 1 \le |m| \le |t^{-1}| & |c| \le |m^2| \le |c\delta^{-2}\varpi^{-1}| \end{matrix}\right.$$

If $|c| \le 1$ then the condition $|c| \le |m^2|$ is superfluous. Moreover

$$|c| \le \left|\begin{Bmatrix} \sqrt[v]{c} \\ \sqrt[v]{c\varpi} \end{Bmatrix}\right| .$$

Thus the two conditions on u can be rewritten

$$|u-a| \le |cm^{-1}|, \, |a| \le \left|\begin{Bmatrix} \sqrt[v]{c} \\ \sqrt[v]{c\varpi} \end{Bmatrix}\right| |m^{-1}|$$

The integral over u is then equal to $|cm^{-1}|$ and so we are left with

$$\int |m| d^\times m \tag{63}$$

over the domain

$$1 \le |m|$$

$$|m| \le |t^{-1}|, \, |m| \le \left|\begin{Bmatrix} \sqrt[v]{c} \\ \sqrt[v]{c\varpi} \end{Bmatrix}\right| |a^{-1}|, \, |m| \le \left|\begin{Bmatrix} \sqrt[v]{c} \\ \sqrt[v]{c\varpi^{-1}} \end{Bmatrix}\right| |\delta^{-1}| .$$

With the notation (55) we have, for $|c| \le 1$,

$$\Omega_{Sl_2}(X) = \mu\left(t^{-1}, \delta^{-1}\begin{Bmatrix} \sqrt[v]{c} \\ \sqrt[v]{c\varpi^{-1}} \end{Bmatrix}, a^{-1}\begin{Bmatrix} \sqrt[v]{c} \\ \sqrt[v]{c\varpi} \end{Bmatrix}\right) .$$

We pass to the case $|c| > 1$. Then the condition $|c| \le |m^2|$ implies the condition $1 \le |m|$. On the other hand, since

$$\left|\begin{Bmatrix} \sqrt[v]{c} \\ \sqrt[v]{c\varpi} \end{Bmatrix}\right| \le |c| .$$

the conditions on u become

$$|u| \le \left|\begin{matrix} \sqrt[v]{c} \\ \sqrt[v]{c\varpi} \end{matrix}\right| |m^{-1}|, \, |a| \le cm^{-1}| .$$

The integral over u is then equal to

$$\left|\begin{matrix} \sqrt[v]{c} \\ \sqrt[v]{c\varpi} \end{matrix}\right| |m^{-1}|$$

and so we are left with

$$\left|\begin{matrix} \frac{1}{\sqrt[v]{c}} \\ \frac{1}{\sqrt[v]{c\varpi^{-1}}} \end{matrix}\right| \int |m| d^\times m \tag{64}$$

over

$$\left|\begin{matrix} \sqrt[v]{c} \\ \sqrt[v]{c\varpi^{-1}} \end{matrix}\right| \le |m|$$

$$|m| \le |ca^{-1}|, \, |m| \le |t^{-1}|, \, |m| \le \left|\begin{matrix} \sqrt[v]{c} \\ \sqrt[v]{c\varpi^{-1}} \end{matrix}\right| |\delta^{-1}|$$

We change m to

$$m\begin{Bmatrix} \sqrt[v]{c} \\ \sqrt[v]{c\varpi^{-1}} \end{Bmatrix}$$

and we get

$$\int |m| d^\times m$$

over

$$1 \le |m|$$

$$|m| \leq \left| \begin{array}{c} \sqrt[v]{c} \\ \sqrt[v]{c\varpi} \end{array} \right| |a^{-1}|\,,\ |m| \leq \left| \begin{array}{c} \frac{1}{\sqrt[v]{c}} \\ \frac{1}{\sqrt[v]{c\varpi^{-1}}} \end{array} \right| |t^{-1}|\,,\ |m| \leq |\delta^{-1}|$$

Thus, for $|c| > 1$, we find

$$\Omega_{Sl_2}(X) = \mu\left(t^{-1} \left\{ \begin{array}{c} \frac{1}{\sqrt[v]{c}} \\ \frac{1}{\sqrt[v]{c\varpi^{-1}}} \end{array} \right\}, \delta^{-1}, a^{-1} \left\{ \begin{array}{c} \sqrt[v]{c} \\ \sqrt[v]{c\varpi} \end{array} \right\} \right)$$

PROPOSITION 3. *In summary, if* $a^2 + bc = \delta^2\varpi$, *(or more generally if* $a^2 + bc = \delta^2\varpi\epsilon$ *where* ϵ *is a unit and* ϖ *a uniformizer), then*

$$(65)\quad \Omega_{Sl_2}(X) = \begin{cases} \mu\left(t^{-1}, \delta^{-1} \left\{ \begin{array}{c} \sqrt[v]{c} \\ \sqrt[v]{c\varpi^{-1}} \end{array} \right\}, a^{-1} \left\{ \begin{array}{c} \sqrt[v]{c} \\ \sqrt[v]{c\varpi} \end{array} \right\} \right) & \text{if } |c| \leq 1 \\ \mu\left(t^{-1} \left\{ \begin{array}{c} \frac{1}{\sqrt[v]{c}} \\ \frac{1}{\sqrt[v]{c\varpi^{-1}}} \end{array} \right\}, \delta^{-1}, a^{-1} \left\{ \begin{array}{c} \sqrt[v]{c} \\ \sqrt[v]{c\varpi} \end{array} \right\} \right) & \text{if } |c| > 1 \end{cases}.$$

We note that if $a = 0$ the identity is to be interpreted as

$$\Omega_{Sl_2}(X) = \begin{cases} \mu\left(t^{-1}, \delta^{-1} \left\{ \begin{array}{c} \sqrt[v]{c} \\ \sqrt[v]{c\varpi^{-1}} \end{array} \right\} \right) & \text{if } |c| \leq 1 \\ \mu\left(t^{-1} \left\{ \begin{array}{c} \frac{1}{\sqrt[v]{c}} \\ \frac{1}{\sqrt[v]{c\varpi^{-1}}} \end{array} \right\}, \delta^{-1} \right) & \text{if } |c| > 1 \end{cases}.$$

13.3. Case where $a^2 + bc$ is even but not a square. We now assume that $a^2 + bc$ has even valuation but is not a square. Thus $a^2 + bc = \delta^2\tau$ where τ is a unit and a non-square. In the range for the integral (53) the quadratic condition on u becomes $|\delta^2\tau - u^2| \leq |cm^{-2}|$. In turn this is equivalent to $|\delta^2| \leq |cm^{-2}|$ and $|u^2| \leq |cm^{-2}|$. Thus the integral is equal to

$$(66)\qquad |c|^{-1} \int du |m|^2 d^\times m$$

over

$$\begin{cases} |u| \leq \left| \left\{ \begin{array}{c} \sqrt[v]{c} \\ \sqrt[v]{c\varpi} \end{array} \right\} \right| |m^{-1}| & |u - a| \leq |cm^{-1}| \\ 1 \leq |m| \leq |t^{-1}| & |c| \leq |m^2| \leq |c\delta^{-2}| \end{cases}$$

If $|c| \leq 1$ then the condition $|c| \leq |m^2|$ is superfluous. The conditions on u can be rewritten

$$|u - a| \leq |cm^{-1}|\,,\ |a| \leq \left| \left\{ \left\{ \begin{array}{c} \sqrt[v]{c} \\ \sqrt[v]{c\varpi} \end{array} \right\} \right\} \right| |m^{-1}|$$

After integrating over u we find

$$(67)\qquad \int |m| d^\times m$$

over

$$1 \leq |m|$$

$$|m| \leq |t^{-1}|\,,\ |m| \leq \left| \left\{ \begin{array}{c} \sqrt[v]{c} \\ \sqrt[v]{c\varpi} \end{array} \right\} \right| |a^{-1}|\,,\ |m| \leq \left| \left\{ \begin{array}{c} \sqrt[v]{c} \\ \sqrt[v]{c\varpi} \end{array} \right\} \right| |\delta^{-1}|$$

Thus, for $|c| \leq 1$,

$$\Omega_{Sl_2}(X) = \mu\left(t^{-1}, \delta^{-1} \left\{ \begin{array}{c} \sqrt[v]{c} \\ \sqrt[v]{c\varpi} \end{array} \right\}, a^{-1} \left\{ \begin{array}{c} \sqrt[v]{c} \\ \sqrt[v]{c\varpi} \end{array} \right\} \right)$$

If $|c| > 1$, then the condition $1 \leq |m|$ is superfluous. On the other hand, the conditions on u become

$$|u| \leq \left|\left\{ \begin{array}{c} \sqrt[v]{c} \\ \sqrt[v]{c\varpi} \end{array} \right\}\right| |m^{-1}|, \; |a| \leq |cm^{-1}|$$

After integrating over u we find

$$\left|\left\{ \begin{array}{c} \frac{1}{\sqrt[v]{c}} \\ \frac{1}{\sqrt[v]{c\varpi^{-1}}} \end{array} \right\}\right| \int |m| d^\times m$$

over

$$\left|\left\{ \begin{array}{c} \sqrt[v]{c} \\ \sqrt[v]{c\varpi^{-1}} \end{array} \right\}\right| \leq |m|$$

$$|m| \leq |t^{-1}|, \; |m| \leq |ca^{-1}|, \; |m| \leq |\delta^{-1}| \left|\left\{ \begin{array}{c} \sqrt[v]{c} \\ \sqrt[v]{c\varpi} \end{array} \right\}\right|$$

We change m to

$$m \left\{ \begin{array}{c} \sqrt[v]{c} \\ \sqrt[v]{c\varpi^{-1}} \end{array} \right\}$$

to get

$$\int |m| d^\times m$$

over

$$1 \leq |m|$$

$$|m| \leq |a^{-1}| \left|\left\{ \begin{array}{c} \sqrt[v]{c} \\ \sqrt[v]{c\varpi} \end{array} \right\}\right|, \; |m| \leq |\delta^{-1}| \left|\left\{ \begin{array}{c} 1 \\ \varpi \end{array} \right\}\right|, \; |m| \leq |t^{-1}| \left| \begin{array}{c} \frac{1}{\sqrt[v]{c}} \\ \frac{1}{\sqrt[v]{c\varpi^{-1}}} \end{array} \right|$$

Thus, for $|c| > 1$ we get

$$\Omega_{sL_2}(X) = \mu\left(t^{-1}\left\{ \begin{array}{c} \frac{1}{\sqrt[v]{c}} \\ \frac{1}{\sqrt[v]{c\varpi^{-1}}} \end{array} \right\}, \delta^{-1}\left\{ \begin{array}{c} 1 \\ \varpi \end{array} \right\}, a^{-1}\left\{ \begin{array}{c} \sqrt[v]{c} \\ \sqrt[v]{c\varpi} \end{array} \right\}\right).$$

We have proved the following Proposition.

PROPOSITION 4. *If $a^2 + bc = \delta^2\tau$ where τ is a non-square unit and $\delta \neq 0$, then*

(68)
$$\Omega_{Sl_2}(X) = \begin{cases} \mu\left(t^{-1}, \delta^{-1}\left\{ \begin{array}{c} \sqrt[v]{c} \\ \sqrt[v]{c\varpi} \end{array} \right\}, a^{-1}\left\{ \begin{array}{c} \sqrt[v]{c} \\ \sqrt[v]{c\varpi} \end{array} \right\}\right) & \text{if } |c| \leq 1 \\ \mu\left(t^{-1}\left\{ \begin{array}{c} \frac{1}{\sqrt[v]{c}} \\ \frac{1}{\sqrt[v]{c\varpi^{-1}}} \end{array} \right\}, \delta^{-1}\left\{ \begin{array}{c} 1 \\ \varpi \end{array} \right\}, a^{-1}\left\{ \begin{array}{c} \sqrt[v]{c} \\ \sqrt[v]{c\varpi} \end{array} \right\}\right) & \text{if } |c| > 1 \end{cases}$$

The meaning of the notations is that if c is even, then the formula is true with the top element of each column $\left\{ \begin{array}{c} \bullet \\ \bullet \end{array} \right\}$. On the contrary, if c is odd, then the formula is true with the bottom element of each column $\left\{ \begin{array}{c} \bullet \\ \bullet \end{array} \right\}$.

13.4. Case where a^2+bc is a square and $c \neq 0$. We now assume that $a^2+bc=\delta^2$ with $\delta \in F^\times$ and $c \neq 0$. Then $a \pm \delta \neq 0$. In (53), the quadratic condition on u becomes $|\delta^2-u^2| \leq |cm^{-2}|$. This condition is satisfied if and only if one of the three following conditions is satisfied:

$$\begin{array}{llll} & I & |\delta^2| \leq |cm^{-2}| & |u^2| \leq |cm^{-2}| \\ (69) & II & |cm^{-2}| < |\delta^2| & |u-\delta| \leq |cm^{-2}\delta^{-1}| \\ & III & |cm^{-2}| < |\delta^2| & |u+\delta| \leq |cm^{-2}\delta^{-1}| \end{array}$$

Accordingly, we write the integral as a sum of three terms $\Omega^I_{Sl_2}$, $\Omega^{II}_{Sl_2}$, $\Omega^{III}_{Sl_2}$.

The term $\Omega^I_{Sl_2}$ is given by the same expression as before, namely (68).

It clear that the term $\Omega^{III}_{Sl_2}$ is obtained from the term $\Omega^{II}_{Sl_2}$ by exchanging δ and $-\delta$. Thus we have only to compute $\Omega^{II}_{Sl_2}$:

$$\Omega^{II}_{Sl_2} = |c|^{-1} \int |m|^2 d^\times m \tag{70}$$

over

$$\begin{cases} |u-a| \leq |cm^{-1}| & |u-\delta| \leq |cm^{-2}\delta^{-1}| \\ |c\delta^{-2}| < |m^2| & |c| \leq |m^2| \\ 1 \leq |m| & |m| \leq |t^{-1}| \end{cases}$$

We remark that $|a^2+bc| \leq 1$ implies $|\delta| \leq 1$ and so the condition $|c\delta^{-2}| < |m^2|$ implies $|c| \leq |m^2|$. We further divide the domain of integration into two subdomains defined by $|m| \leq |\delta^{-1}|$ and $|\delta^{-1}| < |m|$ respectively. The last condition implies $1 \leq |m|$. Correspondingly, we write $\Omega^{II}_{Sl_2}$ as the sum of two terms $\Omega^{II.1}_{Sl_2}$ and $\Omega^{II.2}_{Sl_2}$ defined respectively by

$$\Omega^{II.1}_{Sl_2} = |c|^{-1} \int |m|^2 d^\times m \tag{71}$$

over

$$\begin{cases} |u-a| \leq |cm^{-1}| & |u-\delta| \leq |cm^{-2}\delta^{-1}| \\ |c\delta^{-2}| < |m^2| & 1 \leq |m| \\ |m| \leq |\delta^{-1}| & |m| \leq |t^{-1}| \end{cases}$$

and

$$\Omega^{II.2}_{Sl_2} = |c|^{-1} \int |m|^2 d^\times m \tag{72}$$

over

$$\begin{cases} |u-a| \leq |cm^{-1}| & |u-\delta| \leq |cm^{-2}\delta^{-1}| \\ |c\delta^{-2}| < |m^2| & |\delta^{-1}| < |m| \\ |m| \leq |t^{-1}| & \end{cases}$$

In $\Omega^{II.1}_{Sl_2}$ the conditions on u are equivalent to

$$|u-a| \leq |cm^{-1}|,\ |a-\delta| \leq |cm^{-2}\delta^{-1}|$$

The second condition can be written

$$|m| \leq \left| \begin{Bmatrix} \delta^{-1}\sqrt[v]{c\delta(a-\delta)^{-1}} \\ \delta^{-1}\sqrt[v]{c\delta(a-\delta)^{-1}\varpi} \end{Bmatrix} \right| .$$

After integrating over u, we find:

$$\Omega^{II.1}_{Sl_2} = \int |m| d^\times m \tag{73}$$

over

$$\begin{cases} |m| \leq |\delta^{-1}| & |m| \leq |t^{-1}| \\ 1 \leq |m| & |c\delta^{-2}| < |m|^2 \\ |m| \leq \left| \begin{Bmatrix} \delta^{-1} \sqrt[v]{c\delta(a-\delta)^{-1}} \\ \delta^{-1} \sqrt[v]{c\delta(a-\delta)^{-1}\varpi} \end{Bmatrix} \right| & \end{cases}$$

If $|c\delta^{-2}| < 1$ then the condition $|c\delta^{-2}| < |m^2|$ is implied by $1 \leq |m|$. Thus we find, for $|c\delta^{-2}| < 1$,

$$\Omega_{Sl_2}^{II.1} = \mu\left(t^{-1}, \delta^{-1}, \delta^{-1} \begin{Bmatrix} \sqrt[v]{c\delta(a-\delta)^{-1}} \\ \sqrt[v]{c\delta(a-\delta)^{-1}\varpi} \end{Bmatrix}\right)$$

If $|c\delta^{-2}| \geq 1$ then the condition $|c\delta^{-2}| < |m^2|$ implies the condition $1 \leq |m|$. On the other hand, the condition $|c\delta^{-2}| < |m^2|$ is equivalent to

$$\left| \begin{Bmatrix} \delta^{-1}\varpi^{-1} \sqrt[v]{c} \\ \delta^{-1} \sqrt[v]{c\varpi^{-1}} \end{Bmatrix} \right| \leq |m| \, .$$

Thus we find, for $|c\delta^{-2}| \geq 1$,

$$\Omega_{Sl_2}^{II.1} = \mu\left(t^{-1}, \delta^{-1}, \delta^{-1} \begin{Bmatrix} \sqrt[v]{c\delta(a-\delta)^{-1}} \\ \sqrt[v]{c\delta(a-\delta)^{-1}\varpi} \end{Bmatrix} : \begin{Bmatrix} \delta^{-1}\varpi^{-1} \sqrt[v]{c} \\ \delta^{-1} \sqrt[v]{c\varpi^{-1}} \end{Bmatrix}\right)$$

We pass to the computation of $\Omega_{Sl_2}^{II.2}$. The conditions on u read

$$|u-\delta| \leq |cm^{-2}\delta^{-2}|, |a-\delta| \leq |cm^{-1}| .$$

Thus, after integrating over u, we find

$$\Omega_{Sl_2}^{II.2} = |\delta^{-1}| \int d^\times m \tag{74}$$

over

$$\begin{cases} |\delta^{-1}| < |m| & |c\delta^{-2}| < |m^2| \\ |m| \leq |t^{-1}| & |m| \leq |c(a-\delta)^{-1}| \end{cases}$$

If $|c| \leq 1$ then the condition $|c\delta^{-2}| < |m^2|$ is already implied by $|\delta^{-1}| < |m|$. Thus we find that the domain of integration is

$$|\delta^{-1}\varpi^{-1}| \leq |m| , \ |m| \leq |t^{-1}| , \ |m| \leq |c(a-\delta)^{-1}| .$$

Thus, after a change of variables, we get

$$|\delta^{-1}| \int d^\times m$$

over

$$1 \leq |m| , \ |m| \leq \delta\varpi|t^{-1}| , \ |m| \leq |\delta\varpi c(a-\delta)^{-1}|$$

or

$$|\delta^{-1}|\nu\left(c\delta\varpi(a-\delta)^{-1}, \delta\varpi t^{-1}\right) .$$

If $|c| > 1$ then the relation $|\delta^{-1}| < |m|$ is implied by $|c\delta^{-2}| < |m^2|$. This relation is equivalent to

$$\left| \begin{Bmatrix} \sqrt[v]{c}\delta^{-1}\varpi^{-1} \\ \sqrt[v]{c\varpi}\varpi^{-1}\delta^{-1} \end{Bmatrix} \right| .$$

After a change of variables, we find, for $|c| > 1$,

$$\Omega_{Sl_2}^{II.2} = |\delta^{-1}|\nu\left(\begin{Bmatrix} \sqrt[v]{c}\delta(a-\delta)^{-1}\varpi \\ \sqrt[v]{c\varpi}\delta(a-\delta)^{-1} \end{Bmatrix}, \begin{Bmatrix} \frac{\delta\varpi t^{-1}}{\sqrt[v]{c}} \\ \frac{\delta\varpi t^{-1}}{\sqrt[v]{c\varpi}} \end{Bmatrix}\right) . \tag{75}$$

In summary, we have proved:

PROPOSITION 5. *If $a^2+bc=\delta^2$ with $\delta \neq 0$ and $c \neq 0$ then $\Omega_{Sl_2}(X)$ is the sum of*
(76)

$$\Omega^I_{Sl_2}(X)=\begin{cases}\mu\left(t^{-1},\delta^{-1}\left\{\begin{array}{c}\sqrt[v]{c}\\ \sqrt[v]{c\varpi}\end{array}\right\},a^{-1}\left\{\begin{array}{c}\sqrt[v]{c}\\ \sqrt[v]{c\varpi}\end{array}\right\}\right) & |c|\leq 1\\ \mu\left(t^{-1}\left\{\begin{array}{c}\frac{1}{\sqrt[v]{c}}\\ \frac{1}{\sqrt[v]{c\varpi^{-1}}}\end{array}\right\},\delta^{-1}\left\{\begin{array}{c}1\\ \varpi\end{array}\right\},a^{-1}\left\{\begin{array}{c}\sqrt[v]{c}\\ \sqrt[v]{c\varpi}\end{array}\right\}\right) & |c|>1\end{cases}$$

(77)
$$\Omega^{II.1}_{Sl_2}=$$
$$\begin{cases}\mu\left(t^{-1},\delta^{-1},\delta^{-1}\left\{\begin{array}{c}\sqrt[v]{c\delta(a-\delta)^{-1}}\\ \sqrt[v]{c\delta(a-\delta)^{-1}\varpi}\end{array}\right\}\right) & |c\delta^{-2}|<1\\ \mu\left(t^{-1},\delta^{-1},\delta^{-1}\left\{\begin{array}{c}\sqrt[v]{c\delta(a-\delta)^{-1}}\\ \sqrt[v]{c\delta(a-\delta)^{-1}\varpi}\end{array}\right\}:\left\{\begin{array}{c}\delta^{-1}\varpi^{-1}\sqrt[v]{c}\\ \delta^{-1}\sqrt[v]{c\varpi^{-1}}\end{array}\right\}\right) & |c\delta^{-2}|\geq 1\end{cases}$$

(78)
$$\Omega^{II.2}_{Sl_2}=\begin{cases}|\delta^{-1}|\nu\left(c\delta\varpi(a-\delta)^{-1},\delta\varpi t^{-1}\right) & |c|\leq 1\\ |\delta^{-1}|\nu\left(\left\{\begin{array}{c}\sqrt[v]{c\delta(a-\delta)^{-1}\varpi}\\ \sqrt[v]{c\varpi}\delta(a-\delta)^{-1}\end{array}\right\},\left\{\begin{array}{c}\frac{\delta\varpi t^{-1}}{\sqrt[v]{c}}\\ \frac{\delta\varpi t^{-1}}{\sqrt[v]{c\varpi}}\end{array}\right\}\right) & |c|>1\end{cases}$$

plus the terms $\Omega^{III.1}_{Sl_2}$ and $\Omega^{III.2}_{Sl_2}$ obtained by changing δ into $-\delta$.

We also note that if $\delta=0$ but $c\neq 0$ then the conditions (69) become $|u^2|\leq |cm^{-2}|$ so that $\Omega_{Sl_2}=\Omega^I_{Sl_2}$ with $|\delta^{-1}|=\infty$. We record this as a Proposition.

PROPOSITION 6. *If $a^2+bc=0$ but $c\neq 0$ then*

(79)
$$\Omega_{Sl_2}(X)=\begin{cases}\mu\left(t^{-1},a^{-1}\left\{\begin{array}{c}\sqrt[v]{c}\\ \sqrt[v]{c\varpi}\end{array}\right\}\right) & \text{if } |c|\leq 1\\ \mu\left(t^{-1}\left\{\begin{array}{c}\sqrt[v]{c^{-1}}\\ \sqrt[v]{c^{-1}\varpi}\end{array}\right\},a^{-1}\left\{\begin{array}{c}\sqrt[v]{c}\\ \sqrt[v]{c\varpi}\end{array}\right\}\right) & \text{if } |c|>1\end{cases}$$

In particular if $a=0$, $b=0$ but $c\neq 0$ then

(80)
$$\Omega_{Sl_2}(X)=\begin{cases}\mu\left(t^{-1}\right) & \text{if } |c|\leq 1\\ \mu\left(t^{-1}\left\{\begin{array}{c}\sqrt[v]{c^{-1}}\\ \sqrt[v]{c^{-1}\varpi}\end{array}\right\}\right) & \text{if } |c|>1\end{cases}$$

13.5. Case where $c=0$. We will need the corresponding result when $c=0$ (and $a=\delta$).

PROPOSITION 7. *If $c=0$ then*

$$\Omega_{Sl_2}(X)=$$
$$\mu\left(t^{-1},a^{-1},\left\{\begin{array}{c}\frac{1}{\sqrt[v]{b}}\\ \frac{1}{\sqrt[v]{b\varpi^{-1}}}\end{array}\right\}\right)+|a^{-1}|\nu(at^{-1}\varpi,a^2\varpi b^{-1})$$

PROOF:

$$\Omega_{Sl_2}(X)=\int\int du|m|^2d^\times m$$

over

$$|u|,\leq|m^{-1}| \quad , \quad |\frac{b}{2a}-u|\leq|m^{-2}a^{-1}|$$
$$1\leq|m| \quad , \quad |m|\leq|t^{-1}|$$

Since $A_1(X)$ is an integer we have $|a| \leq 1$.

We first consider the contribution of the terms for which $|m| \leq |a^{-1}|$. Then the condition on u become

$$|u| \leq |m^{-1}|\,,\ |\frac{b}{2a}| \leq |m^{-2}a^{-1}|\,.$$

After integrating over u we find

$$\int |m| d^{\times} m$$

over

$$1 \leq |m|\ |m| \leq |t^{-1}|\,,\ |m^2| \leq |b^{-1}|$$

that is,

$$\mu\left(t^{-1}, a^{-1}, \left\{\begin{array}{c} \frac{1}{\sqrt[v]{b}} \\ \frac{1}{\sqrt[v]{b\varpi^{-1}}} \end{array}\right\}\right).$$

Next, we consider the contributions of the terms for which $|a^{-1}\varpi^{-1}| \leq |m|$. Then the conditions on u become

$$|u| \leq |m^{-2}a^{-1}|\,,\ |\frac{b}{2a}| \leq |m^{-1}|\,.$$

After integrating over u we find

$$|a^{-1}| \int d^{\times} m$$

over

$$\begin{aligned} &1 \leq |m|\,,\ |a^{-1}\varpi^{-1}| \leq |m|\,, \\ &|m| \leq |t^{-1}|\,,\ |m| \leq |ab^{-1}|\,. \end{aligned}$$

However, $|a| \leq 1$. Thus the condition $1 \leq |m|$ is superfluous. Thus this is

$$\nu(t^{-1}, ab^{-1} : a^{-1}\varpi^{-1}) = \nu(at^{-1}\varpi, a^2\varpi b^{-1})\,.$$

The Proposition follows. □

14. Proof of the fundamental lemma for $n = 3$

We let

(81) $$Y = \begin{pmatrix} a & b & 0 \\ 1 & -a & 1 \\ 0 & t & 0 \end{pmatrix}$$

with $t \neq 0$ and $b \neq 0$. Then

$$\Omega_{Gl_2}(Y) = \int_{F^{\times}} \Omega_{Sl_2} \begin{pmatrix} a & bs^{-1} & 0 \\ s & -a & 1 \\ 0 & t & 0 \end{pmatrix} \eta(s) d^{\times} s$$

Since the integrand depends only on the absolute value of s, this integral can be computed as a sum:

$$\sum_s \Omega_{Sl_2} \begin{pmatrix} a & bs^{-1} & 0 \\ s & -a & 1 \\ 0 & t & 0 \end{pmatrix} \eta(s)\,,$$

where s is summed over the powers of a uniformizer ϖ. It follows from lemma (8) that the sum is finite, that is, the integral converges absolutely, provided Y is in $\mathfrak{g}(F)^s$. In the two next sections, we compute this integral and check Theorem (1).

That is, if $-b$ is not a norm we show that $\Omega_{Gl_2}(Y) = 0$. Otherwise we solve the equations (46), define X by (44) and check that

$$\Omega_{Sl_2}(X) = \Omega_{Gl_2}(Y)\,. \tag{82}$$

Before we proceed we remark that $\Omega_{Gl_2}(Y) \neq 0$ implies $|A_1(Y)| \leq 1$ and $|A_2(Y)| \leq 1$. Likewise, if X is defined, $\Omega_{Sl_2}(X) \neq 0$ implies $|A_1(X)| \leq 1$ and $|A_2(X)| \leq 1$. Finally, if X is defined then $|A_1(X)| = |A_1(Y)|$ and $|A_2(X)| = |A_2(Y)|$. Thus if $|A_1(Y)| > 1$ or $|A_2(Y)| > 1$ our assertions are trivially true. Thus we may assume $|A_1(Y)| \leq 1$ and $|A_2(Y)| \leq 1$, that is, $|a^2 + b| \leq 1$ and $|t| \leq 1$.

As before, the discussion depends on the square class of $a^2 + b = -A_1(Y)$.

15. Proof of the fundamental Lemma: $a^2 + b$ is not a square

15.1. Case where $a^2 + b$ is odd. We consider the case where $a^2 + b = -A_1(Y)$ is odd (that is, has odd valuation) and we write $a^2 + b = \delta^2 \varpi$ where ϖ is a uniformizer. The integral Ω_{Gl_2} is then the sum of two terms $\Omega^A_{Gl_2}$ and $\Omega^B_{Gl_2}$ corresponding to the contributions of $|s| \leq 1$ and $|s| > 1$ respectively. If $|s| \leq 1$ we write $s = r^2$ or $s = r^2 \varpi$ with $|r| \leq 1$. Then

$$\Omega^A_{Gl_2} = \sum_{|r| \leq 1} \left[\mu(t^{-1}, \delta^{-1} r, a^{-1} r) - \mu(t^{-1}, \delta^{-1} r, a^{-1} r \varpi)\right]\,. \tag{83}$$

By Lemma 9, the expression $\Omega^A_{Gl_2}$ is equal to

$$\sum |a^{-1} r|$$

over

$$|r| \leq 1\,,\, 1 \leq |a^{-1} r| \leq \inf(|t^{-1}|, |\delta^{-1} r|)\,.$$

This is zero unless $|\delta| \leq |a|$. If $|\delta| \leq |a|$, after changing r to ra, we find

$$\sum_{1 \leq |r| \leq \inf(|a^{-1}|, |t^{-1}|)} |r|\,.$$

In other words, we find

$$\Omega^A_{Gl_2} = \begin{cases} \mu(a^{-1}, t^{-1}) & \text{if } |\delta| \leq |a| \\ 0 & \text{if } |\delta| > |a| \end{cases} \tag{84}$$

We pass to the contribution of $|s| > 1$. We write $s = r^2$ or $s = r^2 \varpi$ with $|r| > 1$. Then

$$\Omega^B_{Gl_2} = \sum_{1 < |r|} \left[\mu(t^{-1} r^{-1}, \delta^{-1}, a^{-1} r) - \mu(t^{-1} r^{-1}, \delta^{-1}, a^{-1} r \varpi)\right]\,. \tag{85}$$

Applying lemma (9) we get

$$\sum |a^{-1} r|$$

over

$$1 < |r|\,,\, 1 \leq |a^{-1} r| \leq \inf(|\delta^{-1}|, |t^{-1} r^{-1}|)$$

This is zero unless $|\delta| < |a|$. If $|\delta| < |a|$, after changing r to ra, we find that this is

$$\sum |r|$$

over

$$\sup(|a^{-1} \varpi^{-1}|, 1) \leq |r|\,,\, |r| \leq |\delta^{-1}|\,,\, |r^2| \leq |t^{-1} a^{-1}|$$

Thus we find

$$\Omega^B_{Gl_2} = \begin{cases} \mu\left(\delta^{-1}, \left\{ \begin{array}{c} \sqrt[v]{t^{-1}a^{-1}} \\ \sqrt[v]{t^{-1}a^{-1}\varpi} \end{array} \right\} : 1, a^{-1}\varpi^{-1}\right) & \text{if } |\delta| < |a| \\ 0 & \text{if } |\delta| \geq |a| \end{cases} \tag{86}$$

We can combine both results to obtain

PROPOSITION 8. *If $a^2 + b = \delta^2\varpi$ then*

$$\Omega_{Gl_2}(Y) = \begin{cases} \mu\left(t^{-1}, \delta^{-1}, \left\{ \begin{array}{c} \sqrt[v]{t^{-1}a^{-1}} \\ \sqrt[v]{t^{-1}a^{-1}\varpi} \end{array} \right\}\right) & \text{if } |\delta| \leq |a| \\ 0 & \text{if } |\delta| > |a| \end{cases}$$

PROOF: Clearly, our integral is 0 if $|\delta| > |a|$. If $|\delta| = |a|$ then the integral reduces to $\mu(t^{-1}, \delta^{-1})$. However,

$$\left|\left\{ \begin{array}{c} \sqrt[v]{t^{-1}\delta^{-1}} \\ \sqrt[v]{t^{-1}\delta^{-1}\varpi} \end{array} \right\}\right|$$

belongs to the interval determined by $|t^{-1}|$ and $|\delta^{-1}|$ and so the integral can be written in the stated form.

Assume now that $|\delta| < |a|$. If $|a| > 1$ then $\mu(a^{-1}, t^{-1}) = 0$ and $|a^{-1}\varpi^{-1}| \leq 1$. Thus $\Omega^A_{Gl_2} = 0$ and $\Omega^B_{Gl_2}$ reduces to

$$\mu\left(\delta^{-1}, \left\{ \begin{array}{c} \sqrt[v]{t^{-1}a^{-1}} \\ \sqrt[v]{t^{-1}a^{-1}\varpi} \end{array} \right\}\right).$$

Since $|t| \leq 1$ we have $|at| > |t^2|$ or

$$|t^{-1}| > \left|\left\{ \begin{array}{c} \sqrt[v]{t^{-1}a^{-1}} \\ \sqrt[v]{t^{-1}a^{-1}\varpi} \end{array} \right\}\right|$$

so that the result can again being written in the required form.

Finally, assume $|\delta| < |a| \leq 1$. Then $|a^{-1}\omega^{-1}| > 1$ and

$$\Omega_{Gl_2} = \mu\left(\delta^{-1}, \left\{ \begin{array}{c} \sqrt[v]{t^{-1}a^{-1}} \\ \sqrt[v]{t^{-1}a^{-1}\varpi} \end{array} \right\} : a^{-1}\varpi^{-1}\right) + \mu(a^{-1}, t^{-1}).$$

Suppose first $|t| \leq |a|$. Then $\mu(a^{-1}, t^{-1}) = \mu(a^{-1})$. Then $|a^{-1}\varpi^{-1}| \leq |\delta^{-1}|$ and

$$|a^{-1}| \leq \left|\left\{ \begin{array}{c} \sqrt[v]{t^{-1}a^{-1}} \\ \sqrt[v]{t^{-1}a^{-1}\varpi} \end{array} \right\}\right|$$

The sum for Ω_{Gl_2} is then by (56) equal to

$$\mu\left(\delta^{-1}, \left\{ \begin{array}{c} \sqrt[v]{t^{-1}a^{-1}} \\ \sqrt[v]{t^{-1}a^{-1}\varpi} \end{array} \right\}\right)$$

Since

$$\left|\left\{ \begin{array}{c} \sqrt[v]{t^{-1}a^{-1}} \\ \sqrt[v]{t^{-1}a^{-1}\varpi} \end{array} \right\}\right| \leq |t^{-1}|$$

this can be written in the required form.

Suppose now $|t| > |a|$. Then $\mu(a^{-1}, t^{-1}) = \mu(t^{-1})$. On the other hand,

$$\left|\left\{ \begin{array}{c} \sqrt[v]{t^{-1}a^{-1}} \\ \sqrt[v]{t^{-1}a^{-1}\varpi} \end{array} \right\}\right| < |a^{-1}\varpi^{-1}|$$

so that $\Omega^B_{Gl_2}$ vanishes. On the other hand, since $|\delta|^{-1} \geq |t^{-1}|$ and

$$\left|\left\{ \begin{array}{c} \sqrt[v]{t^{-1}a^{-1}} \\ \sqrt[v]{t^{-1}a^{-1}\varpi} \end{array} \right\}\right| \geq |t^{-1}|$$

the expression given in the Proposition is indeed equal to $\mu(t^{-1})$. $\square$.

We now check the fundamental lemma in the case at hand. If $-b = a^2 - \delta^2\varpi$ is not a norm, then the valuation of b is odd and $|\delta| > |a|$. Then $\Omega_{Gl_2}(Y) = 0$. Now suppose that $-b$ is a norm, that is, $|a| \geq |\delta|$. Then $-b$ is in fact a square. Thus we may solve the equations of matching (46) in the following way. If $|u| < 1$ we denote by $\sqrt{1+u}$ the square root of $1+u$ which is congruent to one modulo $\varpi\mathcal{O}_F$. Recall that τ is a non-square unit. Then we write

$$-\tau^2 b = y^2\,,\; y = -\tau a\sqrt{1 - \delta^2 a^{-2}\varpi}\,;$$

Then we take

$$a_1 = 0\,,\; b_1 = -\frac{t}{2}(y + \tau a)\,,\; c_1 = \frac{2}{\tau t}(y - \tau a)\,,\; t_1 = -\frac{\tau t}{2}\,.$$

We then have $a_1^2 + b_1c_1 = \tau(a^2 + b) = \delta^2\varpi\tau$. Thus $a_1^2 + b_1c_1$ is odd. We have also $|c_1| = |at^{-1}|$ and $|t_1| = |t|$. Let X be as in (44). We then have by Proposition 3,

$$\Omega_{Sl_2}(X) = \begin{cases} \mu\left(t^{-1}, \delta^{-1}\left\{ \begin{array}{c} \sqrt[v]{at^{-1}} \\ \sqrt[v]{at^{-1}\varpi^{-1}} \end{array} \right\}\right) & \text{if } |a| \leq |t| \\ \mu\left(t^{-1}\left\{ \begin{array}{c} \frac{1}{\sqrt[v]{at^{-1}}} \\ \frac{1}{\sqrt[v]{at^{-1}\varpi^{-1}}} \end{array} \right\}, \delta^{-1}\right) & \text{if } |a| > |t| \end{cases}$$

Suppose first that $|a| \leq |t|$. Since $|\delta| \leq |a|$ we easily get

$$|t^{-1}| \leq \left|\delta^{-1}\left\{ \begin{array}{c} \sqrt[v]{at^{-1}} \\ \sqrt[v]{at^{-1}\varpi^{-1}} \end{array} \right\}\right|$$

and so the expression for $\Omega_{Sl_2}(X)$ reduces to $\mu(t^{-1})$. But the same is true of the expression for $\Omega_{Gl_2}(Y)$.

Now suppose $|a| > |t|$. Then the expression for $\Omega_{SL_2}(X)$ becomes

$$\mu\left(\left\{ \begin{array}{c} \sqrt[v]{t^{-1}a^{-1}} \\ \sqrt[v]{t^{-1}a^{-1}\varpi} \end{array} \right\}, \delta^{-1}\right)\,.$$

Since

$$|t^{-1}| \geq \left|\left\{ \begin{array}{c} \sqrt[v]{t^{-1}a^{-1}} \\ \sqrt[v]{t^{-1}a^{-1}\varpi} \end{array} \right\}\right|$$

this is also the expression for $\Omega_{Gl_2}(Y)$ and we are done. $\square$

15.2. Case where $a^2 + b$ is even and not a square. Suppose now that $a^2 + b = \delta^2\tau$ where τ is, as before, a non-square unit.

PROPOSITION 9. *Suppose $a^2 + b = \delta^2\tau$. Then $\Omega_{Gl_2}(Y)$ is the sum of*

$$|\delta^{-1}|\nu\left(\delta t^{-1}, \varpi\delta^2 t^{-1}a^{-1}\right)$$

and

$$\begin{cases} \mu\left(\delta^{-1}, \left\{ \begin{array}{c} \sqrt[v]{t^{-1}a^{-1}} \\ \sqrt[v]{t^{-1}a^{-1}\varpi} \end{array} \right\},\right) & \text{if } |a| \geq \sup(|\delta|, |t|) \\ \mu(t^{-1}, \delta^{-1}\varpi) & \text{if } |a| < \sup(|\delta|, |t|) \end{cases}$$

PROOF: We proceed as before and write $\Omega_{Gl_2}(Y)$ as the sum of $\Omega^A_{Gl_2}$ and $\Omega^B_{Gl_2}$, these being respectively the contributions of the terms corresponding to $|s| \leq 1$ and $|s| > 1$. For $|s| \leq 1$, we set aside the term $|s| = 1$ and we write $s = r^2\varpi^2$ or $s = r^2\varpi$ with $|r| \leq 1$. We find

$$\begin{aligned}\Omega^A_{Gl_2} &= \mu(t^{-1}, \delta^{-1}, a^{-1}) \\ &+ \sum_{|r|\leq 1} \left[\mu(t^{-1}, \delta^{-1}r\varpi, a^{-1}r\varpi) - \mu(t^{-1}, \delta^{-1}r\varpi, a^{-1}r\varpi)\right] \\ &= \mu(t^{-1}, \delta^{-1}, a^{-1})\end{aligned}$$

For $|s| > 1$ we write $s = r^2$ or $s = r^2\varpi$ with $|r| > 1$. We find

$$\Omega^B_{Gl_2} = \sum_{|r|>1} \left[\mu(t^{-1}r^{-1}, \delta^{-1}, a^{-1}r) - \mu(t^{-1}r^{-1}, \delta^{-1}\varpi, a^{-1}r\varpi)\right] \tag{87}$$

If we add to this $\Omega^A_{Gl_2}$ we find

$$\Omega_{Gl_2} = \mu(t^{-1}, \delta^{-1}\varpi, a^{-1}\varpi) \tag{88}$$

$$+ \sum_{|r|\geq 1} \left[\mu(t^{-1}r^{-1}, \delta^{-1}, a^{-1}r) - \mu(t^{-1}r^{-1}, \delta^{-1}\varpi, a^{-1}r\varpi)\right] \tag{89}$$

Applying lemma (9), the second sum can be computed as

$$\sum \inf\left(|\delta^{-1}|, |a^{-1}r|\right) \tag{90}$$

the sum over

$$|r| \geq 1\,,\ 1 \leq \inf\left(|\delta^{-1}|, |a^{-1}r|\right) \leq |t^{-1}r^{-1}|$$

We first consider the contribution of the terms with $|a^{-1}r| \leq |\delta^{-1}|$:

$$\sum |a^{-1}r| \tag{91}$$

over

$$\begin{gathered} 1 \leq |r|\,,\ |a| \leq |r| \\ |r| \leq |a\delta^{-1}|\,,\ |r^2| \leq |at^{-1}| \end{gathered}$$

If we change r to ra this becomes

$$\mu\left(\delta^{-1}, \left\{\begin{array}{c} \sqrt[v]{t^{-1}a^{-1}} \\ \sqrt[v]{t^{-1}a^{-1}\varpi} \end{array}\right\} : 1, a^{-1}\right) \tag{92}$$

Next, we consider the contribution of the terms with $|\delta^{-1}| < |a^{-1}r|$:

$$\sum |\delta^{-1}|$$

over

$$\begin{gathered} 1 \leq |r|\,,\ |\delta^{-1}a| < |r| \\ |r| \leq |\delta t^{-1}| \end{gathered}$$

After a change of variables, this can be written as

$$|\delta^{-1}| \sum 1$$

over

$$1 \leq |r| \leq \inf\left(|\delta t^{-1}|, |\varpi\delta^2 t^{-1}a^{-1}|\right)$$

so that this is

$$|\delta^{-1}|\nu\left(\delta t^{-1}, \varpi\delta^2 t^{-1}a^{-1}\right)$$

In summary we have found that Ω_{Gl_2} is the sum of

$$\mu(t^{-1}, \delta^{-1}\varpi, a^{-1}\varpi) \tag{93}$$

$$\mu\left(\delta^{-1}, \left\{ \begin{array}{c} \sqrt[v]{t^{-1}a^{-1}} \\ \sqrt[v]{t^{-1}a^{-1}\varpi} \end{array} \right\} : 1, a^{-1}\right) \tag{94}$$

$$|\delta^{-1}|\nu\left(\delta t^{-1}, \varpi\delta^2 t^{-1}a^{-1}\right) \tag{95}$$

If $|a| < |\delta|$ then the second term is zero and the first can be written as $\mu(t^{-1}, \delta^{-1}\varpi)$.

If $|a| < |t|$ then

$$|a^{-1}| > \left|\left\{ \begin{array}{c} \sqrt[v]{t^{-1}a^{-1}} \\ \sqrt[v]{t^{-1}a^{-1}\varpi} \end{array} \right\}\right|$$

so that the second term is 0 and the first can be again written as $\mu(t^{-1}, \delta^{-1}\varpi)$.

Now assume that $|a| \geq \sup(|\delta|, |t|)$. Then $\mu(t^{-1}, \delta^{-1}\varpi, a^{-1}\varpi) = \mu(a^{-1}\varpi)$. If $|a| \geq 1$ then $\mu(a^{-1}\varpi) = 0$ while the second term reduces to

$$\mu\left(\delta^{-1}, \left\{ \begin{array}{c} \sqrt[v]{t^{-1}a^{-1}} \\ \sqrt[v]{t^{-1}a^{-1}\varpi} \end{array} \right\}\right)$$

and we obtain the Proposition. If $|a| < 1$ then the second term is in fact

$$\mu\left(\delta^{-1}, \left\{ \begin{array}{c} \sqrt[v]{t^{-1}a^{-1}} \\ \sqrt[v]{t^{-1}a^{-1}\varpi} \end{array} \right\} : a^{-1}\right).$$

Adding $\mu(a^{-1}\varpi)$ to this and using (56) we obtain the Proposition. □

We now check the fundamental lemma for the case at hand. Of course $-b = a^2 - \delta^2\tau$ is a norm. Thus we may solve the conditions of matching (46) as follows:

$$a_1 = \delta\tau\,,\ c_1 = 0\,,\ b_1 = -\tau ta\,,\ t_1 = -\frac{\tau t}{2}\,.$$

Then $a_1^2 + b_1c_1 = a_1^2 = \delta_1^2$ where $\delta_1 = \delta\tau$. Thus by section 6.3,

$$\Omega_{Sl_2}(X) =$$

$$\mu\left(t^{-1}, \delta^{-1}, \left\{ \begin{array}{c} \sqrt[v]{t^{-1}a^{-1}} \\ \sqrt[v]{t^{-1}a^{-1}\varpi} \end{array} \right\}\right) + |\delta^{-1}|\nu(\delta t^{-1}\varpi, \delta^2 t^{-1}a^{-1}\varpi)\,.$$

If $|a| \geq \sup(|\delta|, |t|)$ then

$$|t^{-1}| \geq \left|\left\{ \begin{array}{c} \sqrt[v]{t^{-1}a^{-1}} \\ \sqrt[v]{t^{-1}a^{-1}\varpi} \end{array} \right\}\right|,$$

$$|\delta^2 t^{-1}a^{-1}\varpi| \leq |\delta t^{-1}\varpi| < |\delta t^{-1}|\,.$$

Hence Ω_{SL_2} is equal to

$$\mu\left(\delta^{-1}, \left\{ \begin{array}{c} \sqrt[v]{t^{-1}a^{-1}} \\ \sqrt[v]{t^{-1}a^{-1}\varpi} \end{array} \right\}\right) + |\delta^{-1}|\nu(\delta t^{-1}, \delta^2 t^{-1}a^{-1}\varpi)$$

which is Ω_{Gl_2} in this case.

Now assume $|a| < \sup(|\delta|, |t|)$. Suppose first that $|t| \leq |a| < |\delta|$. Then $|\delta a^{-1}| > 1$, $|\delta t^{-1}| > 1$ and $|\delta^2| > |ta|$. Thus

$$|\delta^{-1}| \leq \left|\left\{ \begin{array}{c} \sqrt[v]{t^{-1}a^{-1}} \\ \sqrt[v]{t^{-1}a^{-1}\varpi} \end{array} \right\}\right|.$$

Recall taht $|\delta| \leq 1$. Hence

$$\begin{aligned}\Omega_{Sl_2} &= \mu(\delta^{-1}) + |\delta^{-1}|\nu(\delta t^{-1}\varpi) \\ &= \frac{|\delta^{-1}| - q^{-1}}{1-q^{-1}} + |\delta^{-1}|(-v(\delta t^{-1}))\end{aligned}$$

while

$$\Omega_{Gl_2} = \mu(\delta^{-1}\varpi) + |\delta^{-1}|\nu(\delta t^{-1})$$

If $|\delta| < 1$ then we find

$$\Omega_{Gl_2} = \frac{|\delta^{-1}|q^{-1} - q^{-1}}{1-q^{-1}} + |\delta^{-1}|(1 - v(\delta t^{-1}))$$

If $|\delta| = 1$ then we find

$$\Omega_{Gl_2} = 1 - v(\delta t^{-1})$$

In any case the two expressions are indeed equal.

Now assume that $|\delta| \leq |a| < |t|$. Then

$$|t^{-1}| \leq \left| \begin{Bmatrix} \sqrt[v]{t^{-1}a^{-1}} \\ \sqrt[v]{t^{-1}a^{-1}\varpi} \end{Bmatrix} \right|$$

and both orbital integrals are equal to

$$\mu(t^{-1}) + |\delta^{-1}|\nu(\delta^2 t^{-1}a^{-1}\varpi)\,.$$

Finally assume $|a| < |\delta|$ and $|a| < |t|$. Then again

$$|t^{-1}| \leq \left| \begin{Bmatrix} \sqrt[v]{t^{-1}a^{-1}} \\ \sqrt[v]{t^{-1}a^{-1}\varpi} \end{Bmatrix} \right|$$

and Ω_{Sl_2} is equal to

$$\mu(t^{-1}, \delta^{-1}) + |\delta^{-1}|\nu(\delta t^{-1}\varpi)$$

while Ω_{Gl_2} is equal to

$$\mu(t^{-1}, \delta^{-1}\varpi) + |\delta^{-1}|\nu(\delta t^{-1})\,.$$

If $1 > |\delta| > |t|$ then

$$\Omega_{Sl_2} = \mu(\delta^{-1}) + |\delta^{-1}|\nu(\delta t^{-1}\varpi) = \frac{|\delta^{-1}| - q^{-1}}{1-q^{-1}} + |\delta^{-1}|(-v(\delta t^{-1})$$

while

$$\Omega_{Gl_2} = \mu(\delta^{-1}\varpi) + |\delta^{-1}|\nu(\delta t^{-1}) = \frac{|\delta^{-1}|q^{-1} - q^{-1}}{1-q^{-1}} + |\delta^{-1}|(1 - v(\delta t^{-1}))$$

and those two expressions are indeed equal.

If $1 = |\delta| > |t|$ then

$$\Omega_{Sl_2} = \mu(\delta^{-1}) + |\delta^{-1}|\nu(\delta t^{-1}\varpi) = 1 - v(t^{-1})$$

while

$$\Omega_{Gl_2} = |\delta^{-1}|\nu(\delta t^{-1}) = 1 - v(t^{-1})$$

and the two expressions are indeed equal.

Now suppose $|\delta| = |t|$. Recall that $|\delta| \leq 1$. Then

$$\Omega_{Sl_2} = \mu(\delta^{-1}) = \frac{|\delta|^{-1} - q^{-1}}{1-q^{-1}}$$

while

$$\Omega_{Gl_2} = \mu(\delta^{-1}\varpi) + |\delta|^{-1}\nu(1) = \frac{|\delta|^{-1}q^{-1} - q^{-1}}{1-q^{-1}} + |\delta|^{-1}$$

and the two expressions are indeed equal.

If $|\delta| < |t|$ then both orbital integrals are equal to $\mu(t^{-1})$. So the fundamental lemma has been completely checked in this case. □

16. Proof of the fundamental Lemma: $a^2 + b$ is a square

Finally we consider the case where $a^2 + b = \delta^2$, $\delta \neq 0$. Recall that we compute $\Omega_{Gl_2}(Y)$ as the sum

$$\sum_s \Omega_{Sl_2}\begin{pmatrix} a & bs^{-1} & 0 \\ s & -a & 1 \\ 0 & t & 0 \end{pmatrix}\eta(s)$$

and $a^2 + bs^{-1}s = a^2 + b = \delta^2$. Recall that we have written the orbital integral Ω_{SL_2} as a sum of terms labeled $\Omega^{I}_{Sl_2}$, $\Omega^{II.1}_{Sl_2}$, $\Omega^{II.2}_{Sl_2}$, $\Omega^{III.1}_{Sl_2}$, $\Omega^{III.2}_{Sl_2}$ respectively. Correspondingly, we write $\Omega_{Gl_2}(Y)$ as the sum of terms labeled $\Omega^{I}_{Gl_2}$, $\Omega^{II.1}_{Gl_2}$ and so on. For instance,

$$\Omega^{I}_{Gl_2} = \sum_s \Omega^{I}_{Sl_2}\begin{pmatrix} a & bs^{-1} & 0 \\ s & -a & 1 \\ 0 & t & 0 \end{pmatrix}\eta(s).$$

16.1. Computation of $\Omega^{I}_{Gl_2}$. The term $\Omega^{I}_{Gl_2}$ can be computed as Ω_{Gl_2} in the previous case (where $a^2 + b$ is even and not a square). We write it as a sum

$$\Omega^{I}_{Gl_2} = \Omega^{I.1}_{Gl_2} + \Omega^{I.2}_{Gl_2} \tag{96}$$

where

$$\Omega^{I.1}_{Gl_2} = \begin{cases} \mu\left(\delta^{-1}, \left\{ \begin{matrix} \sqrt[v]{t^{-1}a^{-1}} \\ \sqrt[v]{t^{-1}a^{-1}\varpi} \end{matrix} \right\}\right) & \text{if } |a| \geq \sup(|\delta|, |t|) \\ \mu(t^{-1}, \delta^{-1}\varpi) & \text{if } |a| < \sup(|\delta|, |t|) \end{cases} \tag{97}$$

and

$$\Omega^{I.2}_{Gl_2} = |\delta^{-1}|\nu(\delta t^{-1}, \delta^2 t^{-1}a^{-1}\varpi) \tag{98}$$

16.2. Computation of $\Omega^{II.1}_{Gl_2}$. After changing s into $s\delta^2$ we see that

$$\Omega^{II.1}_{Gl_2} = \sum_s \Omega^{II.1}_{Sl_2}\begin{pmatrix} a & bs^{-1}\delta^{-2} & 0 \\ s\delta^2 & -a & 1 \\ 0 & t & 0 \end{pmatrix}\eta(s)$$

and so, by Proposition 5, we get $\Omega^{II.1}_{Gl_2} = \Omega^{II.1.1}_{Gl_2} + \Omega^{II.1.2}_{Gl_2}$ where

$$\Omega^{II.1.1}_{Gl_2} = \sum_{|s|<1} \eta(s)\mu\left(t^{-1}, \delta^{-1}, \left\{ \begin{matrix} \sqrt[v]{s\delta(a-\delta)^{-1}} \\ \sqrt[v]{s\delta(a-\delta)^{-1}\varpi} \end{matrix} \right\}\right) \tag{99}$$

and

$$\Omega^{II.1.2}_{Gl_2} = \sum_{|s|\geq 1} \eta(s)\mu\left(t^{-1}, \delta^{-1}, \left\{ \begin{matrix} \sqrt[v]{s\delta(a-\delta)^{-1}} \\ \sqrt[v]{s\delta(a-\delta)^{-1}\varpi} \end{matrix} \right\} : \left\{ \begin{matrix} \varpi^{-1}\sqrt[v]{s} \\ \sqrt[v]{s\varpi^{-1}} \end{matrix} \right\}\right) \tag{100}$$

Suppose first that $\delta(a-\delta)^{-1}$ is even. For $\Omega_{Gl_2}^{II.1.1}$ we write $s=r^2\varpi^2$ or $s=r^2\varpi$ with $|r|\leq 1$. We find, for $|r|\leq 1$, each term

$$\mu(t^{-1},\delta^{-1},\varpi r\sqrt[v]{\delta(a-\delta)^{-1}})$$

once with a + sign and once with a − sign. So we get zero. For $\Omega_{Gl_2}^{II.1.2}$ we write $s=r^2$ or $s=r^2\varpi^{-1}$ with $|r|\geq 1$. We find, for $|r|\geq 1$, each term

$$\mu(t^{-1},\delta^{-1},r\sqrt[v]{\delta(a-\delta)^{-1}}):\varpi^{-1}r)$$

one with a + sign and once with a − sign. So we get 0. Thus $\Omega_{Gl_2}^{II.1}=0$ if $\delta(a-\delta)^{-1}$ is even.

Now we assume $\delta(a-\delta)^{-1}$ is odd. For $\Omega_{Gl_2}^{II.1.1}$ we write $s=r^2$ or $s=r^2\varpi$ with $|r|\leq 1$. We have then added a term corresponding to $s=r^2$ with $|r|=1$ that we must subtract. We find

$$-\mu\left(t^{-1},\delta^{-1},\sqrt[v]{\delta(a-\delta)^{-1}\varpi}\right)+$$

$$\sum_{|r|\leq 1}\mu\left(t^{-1},\delta^{-1},r\sqrt[v]{\delta(a-\delta)^{-1}\varpi}\right)-\sum_{|r|\leq 1}\mu\left(t^{-1},\delta^{-1},r\sqrt[v]{\delta(a-\delta)^{-1}\varpi}\right)$$

or

$$\Omega_{Gl_2}^{II.1.1}=-\mu\left(t^{-1},\delta^{-1},\sqrt[v]{\delta(a-\delta)^{-1}\varpi}\right).$$

In particular, this is 0 unless $|\delta(a-\delta)^{-1}\varpi|\geq 1$. For $\Omega_{Gl_2}^{II.1.2}$ we write $s=r^2$ or $s=r^2\varpi^{-1}$ with $|r|\geq 1$. We find

$$\sum_{|r|\geq 1}\left(\mu\left(t^{-1},\delta^{-1},r\sqrt[v]{\delta(a-\delta)^{-1}\varpi}:\varpi^{-1}r\right)-\right.$$

$$\left.\mu\left(t^{-1},\delta^{-1},r\sqrt[v]{\delta(a-\delta)^{-1}\varpi^{-1}}:\varpi^{-1}r\right)\right)$$

$$=|\varpi^{-1}|\sum_{|r|\geq 1}|r|\left(\mu\left(t^{-1}r^{-1}\varpi,\delta^{-1}r^{-1}\varpi,\varpi\sqrt[v]{\delta(a-\delta)^{-1}\varpi}\right)-\right.$$

$$\left.\mu\left(t^{-1}r^{-1}\varpi,\delta^{-1}r^{-1}\varpi,\sqrt[v]{\delta(a-\delta)^{-1}\varpi}\right)\right).$$

Once more we apply Lemma 9. We find that this is zero unless $|\delta(a-\delta)^{-1}\varpi|\geq 1$. Then this is equal to

$$=-|\varpi^{-1}|\left|\sqrt[v]{\delta(a-\delta)^{-1}\varpi}\right|\sum_r|r|$$

where the sum is for

$$1\leq|r|,\ |r|\leq\left|\frac{t^{-1}\varpi}{\sqrt[v]{\delta(a-\delta)^{-1}\varpi}}\right|,\ |r|\leq\left|\frac{\delta^{-1}\varpi}{\sqrt[v]{\delta(a-\delta)^{-1}\varpi}}\right|$$

Thus

$$\Omega_{Gl_2}^{II.1.2}=$$

$$-|\varpi^{-1}|\left|\sqrt[v]{\delta(a-\delta)^{-1}\varpi}\right|\mu\left(\frac{t^{-1}\varpi}{\sqrt[v]{\delta(a-\delta)^{-1}\varpi}},\frac{\delta^{-1}\varpi}{\sqrt[v]{\delta(a-\delta)^{-1}\varpi}}\right).$$

Hence we find that $\Omega^{II.1}_{Gl_2}$ is zero unless $\delta(a-\delta)^{-1}$ is odd and $|\delta(a-\delta)^{-1}\varpi| \geq 1$. It is then given by

$$-|\varpi^{-1}|\left|\sqrt[v]{\delta(a-\delta)^{-1}\varpi}\right|\mu\left(\frac{t^{-1}\varpi}{\sqrt[v]{\delta(a-\delta)^{-1}\varpi}},\frac{\delta^{-1}\varpi}{\sqrt[v]{\delta(a-\delta)^{-1}\varpi}}\right)$$
$$-\mu\left(t^{-1},\delta^{-1},\sqrt[v]{\delta(a-\delta)^{-1}\varpi}\right).$$

We claim that this is $-\mu(t^{-1},\delta^{-1})$. Indeed, this is clear if

$$\left|\sqrt[v]{\delta(a-\delta)^{-1}\varpi}\right| \geq \inf(|t^{-1}|,|\delta^{-1}|)$$

because the first term is then 0 and the second term equal to $-\mu(t^{-1},\delta^{-1})$. Now assume that $\left|\sqrt[v]{\delta(a-\delta)^{-1}\varpi}\right| < \inf(|t^{-1}|,|\delta^{-1}|)$. Recall $|\delta| \leq 1$ and $|t| \leq 1$. To be definite assume $|t^{-1}| \leq |\delta^{-1}|$. Then our sum is

$$-|\varpi^{-1}|\left|\sqrt[v]{\delta(a-\delta)^{-1}\varpi}\right|\mu\left(\frac{t^{-1}\varpi}{\sqrt[v]{\delta(a-\delta)^{-1}\varpi}}\right)$$
$$-\mu\left(\sqrt[v]{\delta(a-\delta)^{-1}\varpi}\right)$$
$$=\frac{q^{-1}-|t^{-1}|}{1-q^{-1}}=-\mu(t^{-1})$$

as was claimed. We have proved:

PROPOSITION 10. *$\Omega^{II.1}_{Gl_2}(Y)=0$ unless $\delta(a-\delta)^{-1}$ is odd and $|(a-\delta)| \leq |\delta\varpi|$. Then*

$$\Omega^{II.1}_{Gl_2}(Y)=-\mu(t^{-1},\delta^{-1}).$$

16.3. Computation of $\Omega^{II.2}_{Gl_2}$. As before

$$\Omega^{II.2}_{Gl_2}(Y)=\sum_s \Omega^{II.2}_{Sl_2}\begin{pmatrix} a & bs^{-1} & 0\\ s & -a & 1\\ 0 & t & 0\end{pmatrix}\eta(s)$$

and we denote by $\Omega^{II.2.1}_{Gl_2}$ and $\Omega^{II.2.2}_{Gl_2}$ the respective contributions of the terms $|s| \leq 1$ and $|s| > 1$. Then

$$\Omega^{II.2}_{Gl_2}(Y)=\Omega^{II.2.1}_{Gl_2}+\Omega^{II.2.2}_{Gl_2}.$$

We now appeal to Proposition 5. To compute $\Omega^{II.2.1}_{Gl_2}$ we write $s=r^2$ or $s=r^2\varpi$ with $|r| \leq 1$. We find:

$$\Omega^{II.2.1}_{Gl_2}=|\delta^{-1}|\sum_{|r|\leq 1}\left[\nu\left(r^2\varpi\delta(a-\delta)^{-1},\delta t^{-1}\varpi\right)-\nu\left(r^2\varpi^2\delta(a-\delta)^{-1},\delta t^{-1}\varpi\right)\right]$$

By Lemma 10 this is

$$|\delta^{-1}|\sum 1$$

over

$$|r| \leq 1\,,\ 1 \leq |r^2\varpi\delta(a-\delta)^{-1}| \leq |\delta t^{-1}\varpi|\,.$$

This is 0 unless $|a-\delta| \leq |\varpi\delta|$ and $|t\varpi^{-1}| \leq |\delta|$. It can then be written as $|\delta^{-1}|$ times

$$\nu\left(1,\left\{\begin{array}{c}\sqrt[v]{(a-\delta)t^{-1}}\\ \sqrt[v]{(a-\delta)t^{-1}\varpi}\end{array}\right\}:\left\{\begin{array}{c}\varpi^{-1}\sqrt[v]{(a-\delta)\delta^{-1}}\\ \sqrt[v]{\varpi^{-1}(a-\delta)\delta^{-1}}\end{array}\right\}\right)$$

or

$$\nu\left(\left\{\begin{array}{c}\varpi\sqrt[v]{(a-\delta)^{-1}\delta}\\ \sqrt[v]{\varpi(a-\delta)^{-1}\delta}\end{array}\right\},\frac{\left\{\begin{array}{c}\sqrt[v]{(a-\delta)t^{-1}}\\ \sqrt[v]{(a-\delta)t^{-1}\varpi}\end{array}\right\}}{\left\{\begin{array}{c}\varpi^{-1}\sqrt[v]{(a-\delta)\delta^{-1}}\\ \sqrt[v]{\varpi^{-1}(a-\delta)\delta^{-1}}\end{array}\right\}}\right).$$

This can be further simplified

(101) $$\Omega_{Gl_2}^{II.2.1}=|\delta^{-1}|\times$$

$$\begin{array}{ll}\nu\left(\varpi\sqrt[v]{\delta(a-\delta)^{-1}},\left\{\begin{array}{c}\varpi\sqrt[v]{\delta t^{-1}}\\ \varpi\sqrt[v]{\delta t^{-1}\varpi}\end{array}\right\}\right) & \text{if } \delta(a-\delta) \text{ is even}\\ \nu\left(\sqrt[v]{\varpi\delta(a-\delta)^{-1}},\left\{\begin{array}{c}\varpi\sqrt[v]{\delta t^{-1}}\\ \sqrt[v]{\delta t^{-1}\varpi}\end{array}\right\}\right) & \text{if } \delta(a-\delta) \text{ is odd}\end{array}.$$

To compute $\Omega_{Gl_2}^{II.2.2}$ we write $s=r^2$ or $s=r^2\varpi$ with $|r|>1$. We find

$$\Omega_{Gl_2}^{II.2.2}=|\delta^{-1}|\sum_{|r|>1}\left[\nu\left(\varpi r\delta(a-\delta)^{-1},\delta r^{-1}t^{-1}\varpi\right)-\nu\left(\varpi r\delta(a-\delta)^{-1},\delta r^{-1}t^{-1}\right)\right].$$

By Lemma 10 this is

$$-|\delta^{-1}|\sum 1$$

over

$$|\varpi^{-1}|\le|r|,\ |\varpi^{-1}(a-\delta)t^{-1}|\le|r^2|,\ |r|\le|\delta t^{-1}|.$$

This is 0 unless

$$|a-\delta|\le|\delta^2t^{-1}\varpi|,\ |t\varpi^{-1}|\le|\delta|$$

and can then be written then as

$$-|\delta^{-1}|\nu\left(\delta t^{-1}:\varpi^{-1},\left\{\begin{array}{c}\varpi^{-1}\sqrt[v]{(a-\delta)t^{-1}}\\ \sqrt[v]{\varpi^{-1}(a-\delta)t^{-1}}\end{array}\right\}\right)$$

or

(102) $$\Omega_{Gl_2}^{II.2.2}=-|\delta^{-1}|\nu\left(\varpi\delta t^{-1},\left\{\begin{array}{c}\varpi\delta t^{-1}\sqrt[v]{t(a-\delta)^{-1}}\\ \delta t^{-1}\sqrt[v]{\varpi t(a-\delta)^{-1}}\end{array}\right\}\right)$$

We can simplify our result:

PROPOSITION 11. *Suppose*

$$|a-\delta|\le|\varpi\delta|,\ |t\varpi^{-1}|\le|\delta|.$$

Then

$$\Omega_{Gl_2}^{II.2}(Y)=2^{-1}|\delta^{-1}|\left\{v(\delta t^{-1})+\begin{array}{|c|c|c|}\hline \delta t\text{ even} & \delta t\text{ odd} & \\ \hline 0 & -1 & \delta(a-\delta)\text{ even}\\ \hline 0 & 1 & \delta(a-\delta)\text{ odd}\\ \hline\end{array}\right\}$$

Suppose

$$|\delta|\le|a-\delta|\le|\varpi\delta^2t^{-1}|,\ |t\varpi^{-1}|\le|\delta|.$$

Then

$$\Omega_{Gl_2}^{II.2}(Y)=$$

$$2^{-1}|\delta^{-1}|\left\{v\left(\delta t^{-1}\right)-v\left((a-\delta)\delta^{-1}\right)+\begin{array}{|c|c|c|}\hline \delta t\text{ even} & \delta t\text{ odd} & \\ \hline 0 & -1 & \delta(a-\delta)\text{ even}\\ \hline -1 & 0 & \delta(a-\delta)\text{ odd}\\ \hline\end{array}\right\}$$

In all other cases $\Omega_{Gl_2}^{II.2}(Y)=0$.

PROOF: In any case both $\Omega_{Gl_2}^{II.2.1}(Y)$ and $\Omega_{Gl_2}^{II.2.2}(Y)$ vanish unless $|t\varpi^{-1}| \le |\delta|$. So we assume that this is the case. Suppose $|a-\delta| \le |\varpi\delta|$. Then $\Omega_{Gl_2}^{II.2.1}(Y)$ is non-zero. Since $|\delta t^{-1}\varpi| \ge 1$ we have also $|a-\delta| < |\delta^2 t^{-1}\varpi|$ so $\Omega_{Gl_2}^{II.2.2}(Y)$ is non-zero as well. We have then to consider 4 cases depending on the parity of $(a-\delta)\delta$ and $t\delta$. Suppose for instance that both are even. Then $\Omega_{Gl_2}^{II.2}(Y)$ is $|\delta^{-1}|$ times

$$\nu\left(\varpi\sqrt[v]{\delta(a-\delta)^{-1}}, \varpi\sqrt[v]{\delta t^{-1}}\right) - \nu\left(\varpi\delta t^{-1}, \varpi\delta t^{-1}\sqrt[v]{t(a-\delta)^{-1}}\right)$$

If $|a-\delta| \le |t|$ then this is

$$\begin{aligned} & \nu\left(\varpi\sqrt[v]{\delta t^{-1}}\right) - \nu\left(\varpi\delta t^{-1}\right) \\ = & \left(1 - v\left(\varpi\sqrt[v]{\delta t^{-1}}\right)\right) - \left(1 - v\left(\varpi\delta t^{-1}\right)\right) \\ = & \frac{1}{2}v(\delta t^{-1}). \end{aligned}$$

If, on the contrary, $|t| < |a-\delta|$ then this is

$$\begin{aligned} & \nu\left(\varpi\sqrt[v]{\delta(a-\delta)^{-1}}\right) - \nu\left(\varpi\delta t^{-1}\sqrt[v]{t(a-\delta)^{-1}}\right) \\ = & \left(1 - v\left(\varpi\sqrt[v]{\delta(a-\delta)^{-1}}\right)\right) - \left(1 - v\left(\varpi\delta t^{-1}\sqrt[v]{t(a-\delta)^{-1}}\right)\right) \\ = & \frac{1}{2}v(\delta t^{-1}). \end{aligned}$$

The other cases are treated in a similar way and we have proved the first assertion of the Proposition.

Now assume $|\delta| \le |a-\delta|$. Then $\Omega_{Gl_2}^{II.2.1} = 0$ and $\Omega_{Gl_2}^{II.2.2} \ne 0$ if and only if $|a-\delta| \le |\delta^2 t^{-1}\varpi|$. Note that these conditions imply $|(a-\delta)\varpi| \ge |t|$. Assume $t(a-\delta)$ even. Then $\Omega_{Gl_2}^{II.2.2}$ is equal to $|\delta^{-1}|$ times

$$-\nu\left(\varpi\delta t^{-1}, \varpi\delta t^{-1}\sqrt[v]{t(a-\delta)^{-1}}\right).$$

Since $|(a-\delta)\varpi| \ge |t|$, this is in fact

$$-\nu\left(\varpi\delta t^{-1}\sqrt[v]{t(a-\delta)^{-1}}\right) = v(\delta) - \frac{1}{2}v(t) - \frac{1}{2}v(a-\delta).$$

Assume now $t(a-\delta)$ odd. Then $\Omega_{Gl_2}^{II.2.2}$ is equal to $|\delta^{-1}|$ times

$$-\nu\left(\varpi\delta t^{-1}, \delta t^{-1}\sqrt[v]{\varpi t(a-\delta)^{-1}}\right).$$

Since $|(a-\delta)\varpi| \ge |t|$ this is

$$-\nu\left(\delta t^{-1}\sqrt[v]{\varpi t(a-\delta)^{-1}}\right) = v(\delta) - \frac{1}{2}v(t) - \frac{1}{2}v(a-\delta) - \frac{1}{2}.$$

Thus we have completely proved the Proposition. □

16.4. Case where $-b$ is odd. We are now ready to compute Ω_{Gl_2} completely.

PROPOSITION 12. *If a^2+b is a square but $-b$ is not a norm then $\Omega_{Gl_2}(Y) = 0$.*

PROOF: Assume that $-b$ is not a norm, that is, has odd valuation. Recall that $-b = (a+\delta)(a-\delta)$. Thus $a+\delta$ and $a-\delta$ have different parities. In particular they have different absolute values. Thus, choosing the sign $\pm$ suitably, we must have $|a+\delta| = |a| = |\delta|$ and $|a-\delta| \le |\varpi\delta|$. In particular $(a-\delta)\delta$ is odd and $(a+\delta)\delta$

even. At this point we recall that the terms $\Omega^{III.1}$ and $\Omega^{III.2}$ are obtained from $\Omega^{II.1}$ and $\Omega^{II.2}$ by changing δ into $-\delta$. If $|a| = |\delta| \geq |t|$ then

$$\Omega^{I.1}_{Gl_2} = \mu\left(\delta^{-1}, \left\{ \begin{array}{c} \sqrt[v]{\delta^{-1}t^{-1}} \\ \sqrt[v]{\delta^{-1}t^{-1}\varpi} \end{array} \right\}\right) = \mu(\delta^{-1})\,.$$

If $|a| = |\delta| < |t|$ then

$$\Omega^{I.1}_{Gl_2} = \mu(t^{-1}, \delta^{-1}\varpi) = \mu(t^{-1})\,.$$

Thus, in any case,

$$\Omega^{I.1}_{Gl_2} = \mu(t^{-1}, \delta^{-1})\,.$$

On the other hand,

$$\Omega^{II.1}_{Gl_2} = -\mu(t^{-1}, \delta^{-1})\,,\ \Omega^{III.1}_{Gl_2} = 0\,.$$

Thus

$$\Omega^{I.1}_{Gl_2} + \Omega^{II.1}_{Gl_2} + \Omega^{III.1}_{Gl_2} = 0\,.$$

We study the remaining terms. We have

$$\Omega^{I.2}_{Gl_2} = |\delta^{-1}|\nu(\delta t^{-1}, \delta t^{-1}\varpi) = |\delta^{-1}|\nu(\delta t^{-1}\varpi) = \,.$$

This is 0 unless $|\delta| \geq |\varpi^{-1}t|$. Similarly, the terms $\Omega^{II.2}_{Gl_2}$ and $\Omega^{III.2}_{Gl_2}$ vanish unless $|\delta| \geq |\varpi^{-1}t|$. Thus we may assume $|\delta| \geq |\varpi^{-1}t|$. Then

$$\Omega^{I.2}_{Gl_2} = -|\delta^{-1}|v(\delta t^{-1})\,.$$

Since $|a-\delta| \leq |\varpi\delta|$ and $(a-\delta)\delta$ is odd, we have

$$\Omega^{II.2}_{Gl_2} = 2^{-1}|\delta^{-1}|\left\{v(\delta t^{-1}) + \begin{array}{|c|c|}\hline \delta t \text{ even} & \delta t \text{ odd} \\ \hline 0 & 1 \\ \hline \end{array}\right\}.$$

On the other hand since $|a+\delta| = |\delta|$ and $|\delta| \leq |\delta^2 t^{-1}\varpi|$ we get

$$\Omega^{III.2}_{Gl_2} = 2^{-1}|\delta^{-1}|\left\{v(\delta t^{-1}) + \begin{array}{|c|c|}\hline \delta t \text{ even} & \delta t \text{ odd} \\ \hline 0 & -1 \\ \hline \end{array}\right\}.$$

Thus we do get

$$\Omega^{I.2}_{Gl_2} + \Omega^{II.2}_{Gl_2} + \Omega^{III.2}_{Gl_2} = 0\,.$$

This concludes the proof. □

16.5. Case where b is even. We compute $\Omega_{Gl_2}(Y)$ when $a^2 + b = \delta^2$, $\delta \neq 0$ and b is even. Then $a+\delta$ and $a-\delta$ have the same parity. The result is as follows:

PROPOSITION 13. *Suppose $a^2 + b = \delta^2$, $\delta \neq 0$ and b is even. Then*

$$\Omega_{Gl_2}(Y) = \mu\left(t^{-1}, \left\{ \begin{array}{c} \sqrt[v]{a^{-1}t^{-1}} \\ \sqrt[v]{a^{-1}t^{-1}\varpi} \end{array} \right\}\right) \text{ if } |t| \geq |\delta| \tag{103}$$

$$\Omega_{Gl_2}(Y) = \mu\left(\delta^{-1}, \left\{ \begin{array}{c} \sqrt[v]{a^{-1}t^{-1}} \\ \sqrt[v]{a^{-1}t^{-1}\varpi} \end{array} \right\}\right) - \epsilon|\delta^{-1}| \text{ if } |\delta| > |t| \tag{104}$$

where

$$\epsilon = \left\{ \begin{array}{ll} 1 & \text{if } |a| \leq |\varpi\delta^2 t^{-1}|,\ (a \pm \delta)t \text{ odd} \\ 0 & \text{otherwise} \end{array} \right. \tag{105}$$

PROOF: First we claim that $\Omega_{Gl_2}^{II.1}$ and $\Omega_{Gl_2}^{III.1}$ are both zero. Indeed, if $\Omega_{Gl_2}^{II.1} \neq 0$ then $|a-\delta| \leq |\varpi\delta|$ and $(a-\delta)\delta$ is odd. Then $(a+\delta)\delta$ is also odd. However $|a+\delta| = |\delta|$ and so we get a contradiction and $\Omega_{Gl_2}^{II.1} = 0$. Likewise $\Omega_{Gl_2}^{III.1} = 0$. We compute the other terms.

We first consider the case $|\delta| < |t|$. Then the terms $\Omega_{Gl_2}^{I.2}$, $\Omega_{Gl_2}^{II.2}$, and $\Omega_{Gl_2}^{III.2}$ all vanish. Thus

$$\Omega_{Gl_2}(Y) = \Omega_{GL_2}^{I.1}\,.$$

We use the formula for $\Omega_{GL_2}^{I.1}$. If $|a| \geq |t| > |\delta|$ we find

$$\Omega_{Gl_2}(Y) = \mu\left(\delta^{-1}, \left\{ \begin{array}{c} \sqrt[v]{a^{-1}t^{-1}} \\ \sqrt[v]{a^{-1}t^{-1}\varpi} \end{array} \right\}\right) = \mu\left(\left\{ \begin{array}{c} \sqrt[v]{a^{-1}t^{-1}} \\ \sqrt[v]{a^{-1}t^{-1}\varpi} \end{array} \right\}\right).$$

If $|t| > |a|$ then

$$\Omega_{Gl_2}(Y) = \mu(t^{-1}, \delta^{-1}\varpi) = \mu(t^{-1})$$

Now assume $|\delta| = |t|$. Then $\Omega_{Gl_2}^{II.2} = \Omega_{Gl_2}^{III.2} = 0$. On the other hand,

$$\Omega_{Gl_2}^{I.2} = |\delta^{-1}|\nu(1, \delta a^{-1}\varpi)\,.$$

This is zero unless $|\delta| > |a|$ in which case this is $|\delta^{-1}|$. Thus, if $|a| \geq |\delta| = |t|$, we find

$$\Omega_{Gl_2} = \Omega_{Gl_2}^{I.1} = \mu\left(\delta^{-1}, \left\{ \begin{array}{c} \sqrt[v]{a^{-1}t^{-1}} \\ \sqrt[v]{a^{-1}t^{-1}\varpi} \end{array} \right\}\right) = \mu\left(t^{-1}, \left\{ \begin{array}{c} \sqrt[v]{a^{-1}t^{-1}} \\ \sqrt[v]{a^{-1}t^{-1}\varpi} \end{array} \right\}\right).$$

If $|a| < |\delta| = |t|$, then

$$\Omega_{Gl_2} = \Omega_{Gl_2}^{I.1} + \Omega_{Gl_2}^{I.2} = \mu(\delta^{-1}\varpi) + |\delta^{-1}| = \mu(\delta^{-1})$$

Thus if $|t| \geq |\delta|$ we find the first formula of the Proposition.

From now on, we assume $|\delta| > |t|$. Then we find

$$\Omega_{Gl_2}^{I.1} = \left\{ \begin{array}{ll} \mu\left(\delta^{-1}, \left\{ \begin{array}{c} \sqrt[v]{a^{-1}t^{-1}} \\ \sqrt[v]{a^{-1}t^{-1}\varpi} \end{array} \right\}\right) & \text{if } |a| \geq |\delta| \\ \mu(\delta^{-1}\varpi) & \text{if } |a| < |\delta| \end{array} \right.$$

This can also be written

$$\Omega_{Gl_2}^{I.1} = \mu\left(\delta^{-1}, \left\{ \begin{array}{c} \sqrt[v]{a^{-1}t^{-1}} \\ \sqrt[v]{a^{-1}t^{-1}\varpi} \end{array} \right\}\right) + \left\{ \begin{array}{ll} 0 & \text{if } |a| \geq |\delta| \\ -|\delta^{-1}| & \text{if } |a| < |\delta| \end{array} \right. . \tag{106}$$

Similarly,

$$\Omega_{Gl_2}^{I.2} = \left\{ \begin{array}{ll} |\delta^{-1}|\nu(\delta^2 t^{-1}a^{-1}\varpi) & \text{if } |a| \geq |\delta| \\ |\delta^{-1}|\nu(\delta t^{-1}) & \text{if } |a| < |\delta| \end{array} \right.$$

Adding up these results we find

$$\Omega_{Gl_2}^{I} = \mu\left(\delta^{-1}, \left\{ \begin{array}{c} \sqrt[v]{a^{-1}t^{-1}} \\ \sqrt[v]{a^{-1}t^{-1}\varpi} \end{array} \right\}\right) +$$

$$\left\{ \begin{array}{ll} 0 & \text{if } |a| \geq |\delta|\,,\ |a| \geq |\delta^2 t^{-1}| \\ -|\delta^{-1}|v(\delta^2 t^{-1}a^{-1}) & \text{if } |a| \geq |\delta|\,,\ |a| \leq |\delta^2 t^{-1}\varpi| \\ -|\delta^{-1}|v(\delta t^{-1}) & \text{if } |a| < |\delta| \end{array} \right. .$$

We compute the remaining terms.

Suppose $|a| \geq |\delta|$. Suppose first that $|a+\delta| = |\delta - a| = |a|$ (or for short, $|\delta \pm a| = |a|$). Of course, this is always the case if $|a| > |\delta|$. Both $\Omega^{II.2}_{Gl_2}$ and $\Omega^{III.2}_{Gl_2}$ are 0 unless $|a| \leq |\varpi\delta^2 t^{-1}|$; then they are equal and

$$\Omega^{II.2}_{Gl_2} + \Omega^{III.2}_{Gl_2} = |\delta^{-1}| \left\{ v(\delta^2 t^{-1} a^{-1}) + \begin{array}{|c|c|} \hline (a\pm\delta)t \text{ even} & (a\pm\delta)t \text{ odd} \\ \hline 0 & -1 \\ \hline \end{array} \right\}.$$

Now suppose $|\delta| = |a|$ but $|\delta \pm a|$ is not equal to $|a| = |\delta|$ for both choices of $\pm$. Say $|\delta - a| \leq |\varpi\delta|$ and $|\delta + a| = |\delta|$. Both $\Omega^{II.2}_{Gl_2}$ and $\Omega^{III.2}_{Gl_2}$ are non-zero. In addition we remark that $\delta(\delta \pm a)$ have the same parity and are thus even. Thus we again fidn the same result. Note that here $|a| = |\delta| \leq |\varpi\delta^2 t^{-1}|$. We conclude that if $|a| \geq |\delta|$ then $\Omega^{II.2}_{Gl_2} + \Omega^{III.2}_{Gl_2} = 0$ unless $|a| \leq |\varpi\delta^2 t^{-1}|$. Then

$$\Omega^{II.2}_{Gl_2} + \Omega^{III.2}_{Gl_2} = |\delta^{-1}| \left\{ v(\delta^2 t^{-1} a^{-1}) + \begin{array}{|c|c|} \hline (a\pm\delta)t \text{ even} & (a\pm\delta)t \text{ odd} \\ \hline 0 & -1 \\ \hline \end{array} \right\}.$$

Finally, suppose $|a| < |\delta|$. Then $|a \pm \delta| = |\delta|$ so $(a \pm \delta)\delta$ is even and both $\Omega^{II.2}_{Gl_2}$ and $\Omega^{III.2}_{Gl_2}$ are non-zero with the same value. Then

$$\Omega^{II.2}_{Gl_2} + \Omega^{III.2}_{Gl_2} = |\delta^{-1}| \left\{ v(\delta t^{-1}) + \begin{array}{|c|c|} \hline (a\pm\delta)t \text{ even} & (a\pm\delta)t \text{ odd} \\ \hline 0 & -1 \\ \hline \end{array} \right\}.$$

Summing up, we find the second formula of the Proposition.

16.6. Verification of $\Omega_{Gl_2}(Y) = \Omega_{Sl_2}(X)$. We verify the identity of the fundamental lemma when $a^2 + b = \delta^2$, $\delta \neq 0$ and b is even. We solve the equations of matching (46) as before. We write

$$-\tau^2 b = y^2 - \tau a_1^2$$

and then we take

$$t_1 = -\frac{\tau t}{2}, \; c_1 = \frac{2}{t\tau}(y - \tau a), \; b_1 = -\frac{t}{2}(y + \tau a).$$

Then

$$a_1^2 + b_1 c_1 = \tau(a^2 + b) = \tau\delta^2.$$

Thus $a_1^2 + b_1 c_1$ is even but not a square. We need to compute $|c_1|$. We have

$$-\tau^2 b = y^2 - \tau a_1^2 = \tau^2 a^2 - \tau^2 \delta^2.$$

Suppose $|a| \geq |\delta|$. If $|a| = |\delta|$ we choose δ in such a way that $|\delta - a| = |a|$. We have $|b| = |a^2 - \delta^2| \leq |a|^2$. From $-\tau^2 b = y^2 - \tau a_1^2$ we conclude that $|y| \leq |a|$ and $|a_1| \leq |a|$. From

$$y^2 - \tau^2 a^2 = \tau(a_1^2 - \tau\delta^2)$$

we conclude that

$$|(y - \tau a)(y + \tau a)| \leq |a|^2.$$

Hence either $|y - \tau a| = |a|$ or $|y + \tau a| = |a|$. Thus we can choose y in such a way that $|y - \tau a| = |a|$. Then

$$|c_1| = |a t^{-1}| = |(\delta - a) t^{-1}|.$$

Now suppose $|\delta| > |a|$. Then $|b| = |\delta|^2$. From $-\tau^2 b = y^2 - \tau a_1^2$ we conclude that $|y| \leq |\delta|$ and $|a_1| \leq |\delta|$. Suppose $|y| < |\delta|$. Then $|a_1| = |\delta|$. From $y^2 - \tau a_1^2 = \tau^2 a^2 - \tau^2 \delta^2$ we get

$$\tau = \left(1 - \frac{a^2}{\delta^2}\right) \frac{\tau^2\delta^2}{a_1^2} + \frac{y^2}{a_1^2}.$$

Thus τ is congruent to a square unit modulo $\varpi\mathcal{O}_F$ hence is a square, a contradiction. Thus $|y| = |\delta|$ and we again find

$$|c_1| = |\delta t^{-1}| = |(\delta - a)t^{-1}|.$$

Now we can write down the formula for $\Omega_{Sl_2}(X)$. It reads as follows.

If $|(\delta - a)t^{-1}| \leq 1$,

$$\Omega_{Sl_2}(X) =$$

$$\mu\left(t^{-1}, \delta^{-1}\left\{\begin{array}{c}\sqrt[v]{(\delta-a)t^{-1}} \\ \sqrt[v]{(\delta-a)t^{-1}\varpi}\end{array}\right\}, a^{-1}\left\{\begin{array}{c}\sqrt[v]{(\delta-a)t^{-1}} \\ \sqrt[v]{(\delta-a)t^{-1}\varpi}\end{array}\right\}\right).$$

If $|(\delta - a)t^{-1}| > 1$,

$$\Omega_{Sl_2}(X) =$$

$$\mu\left(t^{-1}\left\{\begin{array}{c}\frac{1}{\sqrt[v]{(\delta-a)t^{-1}}} \\ \frac{1}{\sqrt[v]{(\delta-a)t^{-1}\varpi^{-1}}}\end{array}\right\}, \delta^{-1}\left\{\begin{array}{c}1 \\ \varpi\end{array}\right\}, a^{-1}\left\{\begin{array}{c}\sqrt[v]{(\delta-a)t^{-1}} \\ \sqrt[v]{(\delta-a)t^{-1}\varpi}\end{array}\right\}\right)$$

Suppose first that $|a| \geq |\delta|$. Recall that if $|a| = |\delta|$ then we choose δ in such a way that $|\delta - a| = |a|$. Thus $|\delta - a| = |a|$ in all cases. Then we find

$$\Omega_{Sl_2}(X) =$$

$$\begin{cases} \mu\left(t^{-1}, \delta^{-1}\left\{\begin{array}{c}\sqrt[v]{at^{-1}} \\ \sqrt[v]{at^{-1}\varpi}\end{array}\right\}, \left\{\begin{array}{c}\sqrt[v]{a^{-1}t^{-1}} \\ \sqrt[v]{a^{-1}t^{-1}\varpi}\end{array}\right\}\right) & \text{if } |a| \leq |t| \\ \mu\left(\delta^{-1}\left\{\begin{array}{c}1 \\ \varpi\end{array}\right\}, \left\{\begin{array}{c}\sqrt[v]{a^{-1}t^{-1}} \\ \sqrt[v]{a^{-1}t^{-1}\varpi}\end{array}\right\}\right) & \text{if } |t| < |a| \end{cases}$$

Consider first the case $|a| \leq |t|$ so that $|\delta| \leq |a| \leq |t|$. This is

$$\Omega_{Sl_2}(X) = \mu\left(t^{-1}, \left\{\begin{array}{c}\sqrt[v]{a^{-1}t^{-1}} \\ \sqrt[v]{a^{-1}t^{-1}\varpi}\end{array}\right\}\right) = \Omega_{Gl_2}(Y).$$

Consider now the case $|t| < |a|$. If $|\delta| \leq |t|$ this is

$$\Omega_{Sl_2}(X) = \mu\left(\left\{\begin{array}{c}\sqrt[v]{a^{-1}t^{-1}} \\ \sqrt[v]{a^{-1}t^{-1}\varpi}\end{array}\right\}\right) = \Omega_{Gl_2}(Y).$$

If $|\delta| > |t|$ then we have to distinguish two cases. If $|a| > |\varpi\delta^2 t^{-1}|$ we find

$$\Omega_{Sl_2} = \mu\left(\left\{\begin{array}{c}\sqrt[v]{a^{-1}t^{-1}} \\ \sqrt[v]{a^{-1}t^{-1}\varpi}\end{array}\right\}\right) = \mu\left(\delta^{-1}, \left\{\begin{array}{c}\sqrt[v]{a^{-1}t^{-1}} \\ \sqrt[v]{a^{-1}t^{-1}\varpi}\end{array}\right\}\right)$$

which is again equal to Ω_{Gl_2} since $\epsilon = 0$ in this case. If $|a| \leq |\varpi\delta^2 t^{-1}|$ and at (or equivalently $(a - \delta)t$) is even we find

$$\Omega_{Sl_2}(X) = \mu(\delta^{-1}).$$

Since $\epsilon = 0$ in this case, this is again Ω_{Gl_2}. If $|a| \leq |\varpi\delta^2 t^{-1}|$ and at (or equivalently $(a - \delta)t$) is odd we find

$$\Omega_{Sl_2}(X) = \mu(\delta^{-1}\varpi) = \mu(\delta^{-1}) - |\delta^{-1}|.$$

This is again equal to Ω_{Gl_2}, since $\epsilon = 1$ in this case.

We now discuss the case where $|a| < |\delta|$. Then $|a - \delta| = |\delta|$ and our expression for Ω_{Sl_2} simplifies:

$$\begin{cases} \mu\left(t^{-1}, \left\{ \begin{array}{c} \sqrt[v]{\delta^{-1}t^{-1}} \\ \sqrt[v]{\delta^{-1}t^{-1}\varpi} \end{array} \right\}\right) & \text{if } |\delta| \leq |t| \\ \mu\left(\left\{ \begin{array}{c} \sqrt[v]{\delta^{-1}t^{-1}} \\ \sqrt[v]{\delta^{-1}t^{-1}\varpi} \end{array} \right\}, \delta^{-1}\left\{ \begin{array}{c} 1 \\ \varpi \end{array} \right\}, a^{-1}\left\{ \begin{array}{c} \sqrt[v]{\delta t^{-1}} \\ \sqrt[v]{\delta t^{-1}\varpi} \end{array} \right\}\right) & \text{if } |t| < |\delta| \end{cases}$$

This simplifies further as follows:

$$\Omega_{Sl_2}(X) = \begin{cases} \mu\left(t^{-1}\right) & \text{if } |\delta| \leq |t| \\ \mu(\delta^{-1}) & \text{if } |t| < |\delta|,\ \delta t \text{ even} \\ \mu(\delta^{-1}\varpi) & \text{if } |t| < |\delta|,\ \delta t \text{ odd} \end{cases}.$$

Likewise, the expression for $\Omega_{Gl_2}(Y)$ simplifies as follows:

$$\Omega_{Gl_2}(Y) = \begin{cases} \mu\left(t^{-1}\right) & \text{if } |\delta| \leq |t| \\ \mu(\delta^{-1}) & \text{if } |t| < |\delta|,\ (a \pm \delta)t \text{ even} \\ \mu(\delta^{-1}) - |\delta^{-1}| & \text{if } |t| < |\delta|,\ (a \pm \delta t) \text{ odd} \end{cases}.$$

Again δt and $(\delta - a)t$ have the same parity and $\mu(\delta^{-1}\varpi) = \mu(\delta^{-1}) - |\delta^{-1}|$. Thus $\Omega_{Sl_2}(X) = \Omega_{GL_2}(Y)$ in all cases.

17. Proof of the fundamental Lemma: $a^2 + b = 0$

It remains to treat the case where $a^2 + b = 0$. THen $-b = a^2$ is a norm. We proceed as before. We write the integral for Ω_{Gl_2} as the sum of $\Omega^A_{GL_2}$ and $G^B_{Gl_2}$ corresponding respectively to the contributions of $|s| \leq 1$ and $|s| > 1$. We use Proposition 6. For $|s| \leq 1$ we write $s = r^2$ or $s = r^2\varpi$ with $|r| \leq 1$. We obtain

$$\begin{aligned} \Omega^A_{Gl_2} &= \sum_{|r|\leq 1} \left(\mu(t^{-1}, a^{-1}r) - \mu(t^{-1}, a^{-1}r\varpi)\right) \\ &= \mu(t^{-1}, a^{-1}). \end{aligned}$$

For $|s| > 1$ we write $s = r^2$ or $s = r^2\varpi$ with $|r > |1$. We find

$$\Omega^B_{Gl_2} = \sum_{|r|>1} \left(\mu(t^{-1}r^{-1}, a^{-1}r) - \mu(t^{-1}r^{-1}, a^{-1}r\varpi)\right)$$

Applying Lemma 9 we find that this is

$$\sum |a^{-1}r|$$

over

$$|\varpi^{-1}| \leq |r|,\ |a| \leq |r|,\ |r^2| \leq |at^{-1}|.$$

This is

$$\mu\left(\left\{ \begin{array}{c} \sqrt[v]{a^{-1}t^{-1}} \\ \sqrt[v]{a^{-1}t^{-1}\varpi} \end{array} \right\} : a^{-1}\varpi^{-1}, 1\right).$$

If $|a| \leq |t|$ then $\mu(t^{-1}, a^{-1}) = \mu(t^{-1})$ and $\mu\left(\left\{ \begin{array}{c} \sqrt[v]{a^{-1}t^{-1}} \\ \sqrt[v]{a^{-1}t^{-1}\varpi} \end{array} \right\} : a^{-1}\varpi^{-1}, 1\right) = 0.$
If $|a| \geq |t|$ then $\mu(t^{-1}, a^{-1}) = \mu(a^{-1})$. Moreover, if $|a| \leq 1$ then

$$\mu(a^{-1}) + \mu\left(\left\{ \begin{array}{c} \sqrt[v]{a^{-1}t^{-1}} \\ \sqrt[v]{a^{-1}t^{-1}\varpi} \end{array} \right\} : a^{-1}\varpi^{-1}, 1\right) = \mu\left(\left\{ \begin{array}{c} \sqrt[v]{a^{-1}t^{-1}} \\ \sqrt[v]{a^{-1}t^{-1}\varpi} \end{array} \right\}\right).$$

If $|a| > 1$ then $\mu(a^{-1}) = 0$ and

$$\mu\left(\left\{ \begin{array}{c} \sqrt[v]{a^{-1}t^{-1}} \\ \sqrt[v]{a^{-1}t^{-1}\varpi} \end{array} \right\} : a^{-1}\varpi^{-1}, 1\right) = \mu\left(\left\{ \begin{array}{c} \sqrt[v]{a^{-1}t^{-1}} \\ \sqrt[v]{a^{-1}t^{-1}\varpi} \end{array} \right\}\right)$$

Thus the above equality remains true. In summary,

$$\Omega_{Gl_2}(Y) = \begin{cases} \mu(t^{-1}) & \text{if } |a| \leq |t| \\ \mu\left(\left\{ \begin{array}{c} \sqrt[v]{a^{-1}t^{-1}} \\ \sqrt[v]{a^{-1}t^{-1}\varpi} \end{array} \right\}\right) & \text{if } |a| > |t| \end{cases}$$

On the other hand, the conditions of matching (46) can be solved with

$$a_1 = 0\,,\, b_1 = 0\,,\, c_1 = \frac{-4a}{t}\,,\, t_1 = -\frac{\tau t}{2}\,.$$

For the corresponding element X we find

$$\Omega_{Sl_2}(X) = \begin{cases} \mu(t^{-1}) & \text{if } |a| \leq |t| \\ \mu\left(t^{-1}\left\{ \begin{array}{c} \sqrt[v]{a^{-1}t} \\ \sqrt[v]{a^{-1}t\varpi} \end{array} \right\}\right) & \text{if } |a| > |t| \end{cases}$$

Clearly $\Omega_{Sl_2}(X) = \Omega_{Gl_2}(Y)$.

We have now completely proved the fundamental lemma for strongly regular elements.

18. Other regular elements

Recall the definition of a regular element. A matrix $X \in M(3\times 3, E)$ is **regular** if writing X in the form

$$\begin{pmatrix} A & B \\ C & d \end{pmatrix}$$

the column vectors B, AB are linearly independent and the row vectors C, CA are linearly independent. We have seen that if X is in $\mathfrak{g}(E)'$ then it is regular if and only if it is strongly regular. We consider now the elements X which are regular but not strongly regular. For such an element we have necessarily $A_2(X) = CB = 0$.

LEMMA 11. *Any element $X \in \mathfrak{g}(E)$ which is regular but not strongly regular is conjugate under $\iota Gl_2(E)$ to a unique matrix of the form*

$$\begin{pmatrix} 0 & b & 0 \\ c & 0 & 1 \\ 1 & 0 & 0 \end{pmatrix}$$

with $b \neq 0$. In addition

$$\begin{aligned} A_1(X) &= -bc \\ B_1(X) &= b \end{aligned}$$

PROOF: First B and C are not 0. After conjugation we may assume $B = \left(\begin{smallmatrix} 0 \\ 1 \end{smallmatrix}\right)$. Since $CB = 0$ we have

$$C = \begin{pmatrix} t & 0 \end{pmatrix}, t \neq 0\,.$$

Conjugating by a diagonal matrix in $Gl_2(E)$ we may assume $t = 1$. Thus we are reduced to the case of a matrix of the form

$$\begin{pmatrix} a & b & 0 \\ c & -a & 1 \\ 1 & 0 & 0 \end{pmatrix}.$$

If we conjugate by the matrix $\iota\begin{pmatrix} 1 & 0 \\ \frac{a}{b} & 1 \end{pmatrix}$ we arrive at a matrix of the prescribed form. The other assertions are obvious. □.

REMARK: Similarly, the element is conjugate to a unique matrix of the form

$$\begin{pmatrix} 0 & b & 0 \\ c & 0 & 1 \\ -1 & 0 & 0 \end{pmatrix}.$$

Any element X of $\mathfrak{s}(F)$ which is regular but not strongly regular is conjugate under $Gl_2(F)$ to a unique element of the form

$$\xi = \begin{pmatrix} 0 & b & 0 \\ c & 0 & \sqrt{\tau} \\ \sqrt{\tau} & 0 & 0 \end{pmatrix}$$

with $b, c \in F\sqrt{\tau}$ and $b \neq 0$. Then

$$\begin{aligned} A_1(X) &= -bc \\ A_2(X) &= b\tau \end{aligned}$$

Two such elements are conjugate under $Gl_2(F)$ if and only if they are conjugate under $Gl_2(E)$.

LEMMA 12. *Any element X of $\mathfrak{u}(F)$ which is regular but not strongly regular is conjugate under $\iota U_{1,1}$ to a unique element of the form*

$$\begin{pmatrix} 0 & b & 0 \\ c & 0 & 1 \\ -1 & 0 & 0 \end{pmatrix},$$

with $b, c \in F\sqrt{\tau}$ and $b \neq 0$. In addition

$$\begin{aligned} A_1(X) &= -bc \\ B_1(X) &= -b \end{aligned}$$

Two such elements are conjugate under $U_{1,1}$ if and only if they are conjugate under $Gl_2(E)$.

PROOF: Write

$$X = \begin{pmatrix} a & b & z_1 \\ c & -a & z_2 \\ -\overline{z_2} & -\overline{z_1} & 0 \end{pmatrix}.$$

By assumption we have $\overline{z_2}z_1 + \overline{z_1}z_2 = 0$. Conjugating by a diagonal matrix in $U_{1,1}$ we may assume $z_2 = 1$. Then $z_1 + \overline{z_1} = 0$. Conjugating by the matrix $\begin{pmatrix} 1 & z_1 \\ 0 & 1 \end{pmatrix}$ we are reduced to the case where the matrix has the form

$$\begin{pmatrix} a & b & 0 \\ c & -a & 1 \\ -1 & 0 & 0 \end{pmatrix}.$$

We finish the proof as before. □

We see now that any element ξ' of $\mathfrak{s}(F)$ which is regular but not strongly regular matches an element ξ of $\mathfrak{u}(F)$. Explicitly

$$\xi = \begin{pmatrix} 0 & b & 0 \\ c & 0 & 1 \\ -1 & 0 & 0 \end{pmatrix}$$

matches

$$\xi' = \begin{pmatrix} 0 & b' & 0 \\ c' & 0 & \sqrt{\tau} \\ \sqrt{\tau} & 0 & 0 \end{pmatrix}$$

if and only

$$bc = b'c' \,,\; -b = b'\tau \,.$$

As before, we set

$$\begin{aligned} \Omega_U(\xi) &= \int_U f_0(\iota(u)\xi\iota(u)^{-1})du \\ \Omega_{Gl_2}(\xi') &= \int_{Gl_2(F)} \Phi_0(\iota(g)\xi'\iota(g)^{-1})\eta(\det g)dg \end{aligned}$$

The fundamental lemma asserts that if $\xi \to \xi'$ then

$$\Omega_U(\xi) = \tau(\xi')\Omega_{Gl_2}(\xi') \,.$$

To prove the lemma we proceed as before. We set

$$X = \Theta(\xi) \,,\; \xi' = \sqrt{\tau}Y \,.$$

Then

$$X = \begin{pmatrix} 0 & b_1 & 0 \\ c_1 & 0 & 1 \\ -1 & 0 & 0 \end{pmatrix}$$

with

$$b_1 = b\sqrt{\tau} \,,\; c_1 = \frac{c}{\sqrt{\tau}} \,.$$

On the other hand

$$Y = \begin{pmatrix} 0 & b_2 & 0 \\ c_2 & 0 & 1 \\ 1 & 0 & 0 \end{pmatrix}$$

with

$$b_2 = \frac{b'}{\sqrt{\tau}} \,,\; c_2 = \frac{c'}{\sqrt{\tau}} \,.$$

Thus in terms of X and Y the matching conditions become

$$c_2 = -c_1\tau \,,\; b_2 = -\frac{b_1}{\tau^2} \,.$$

We have

$$|b_1| = |b_2| \,,\; |b_2| = |c_2| \,.$$

Moreover, if b_1c_1 (and thus b_2c_2) is even, then b_1c_1 is a square if and only if b_2c_2 is not a square.

THEOREM 2 (Remaining case of the fundamental Lemma). *If X and Y are as above and*

$$c_2 = -c_1\tau\,,\; b_2 = -\frac{b_1}{\tau^2}\,,$$

then

$$\Omega_{Sl_2}(X) = \eta(b_2)\Omega_{Gl_2}(Y)\,.$$

19. Orbital integrals for Sl_2

We compute the orbital integral under $SL_2(F)$ of

$$X = \begin{pmatrix} 0 & b & 0 \\ c & 0 & 1 \\ -1 & 0 & 0 \end{pmatrix},$$

where $b \neq 0$, $c \neq 0$. We also write $\Omega_{Sl_2}(X) = \Omega_{Sl_2}(b, c)$.

We have

$$\Omega_{Sl_2}(X) = \int \Phi \begin{pmatrix} -bu & bm^2 & 0 \\ m^{-2}(c-u^2b) & ub & m^{-1} \\ -m^{-1} & 0 & 0 \end{pmatrix} du|m|^{-2}d^\times m\,.$$

If the integral is non-zero then $|b| \leq 1$ and $|bc| \leq 1$. Explicitly, the domain of integration is

$$1 \leq |m|\,,\; |bu| \leq 1\,,\; |bm^2| \leq 1\,,$$
$$|bc - u^2b^2| \leq |m^2b| \leq 1\,.$$

Under the assumption $|bc| \leq 1$ the condition $|ub| \leq 1$ is superfluous. After a change of variables, we can rewrite the integral as

$$|b|^{-1}\int du|m|^{-2}d^\times m$$

over

$$|bc - u^2| \leq |m^2b| \leq 1\,,\; 1 \leq |m|\,.$$

We divide the integral into the sum of the contribution $\Omega^1_{Sl_2}(X)$ of $|c| \leq |m^2|$ and the contribution $\Omega^2_{Sl_2}(X)$ of $|m^2| < |c|$.

We have

$$\Omega^1_{Sl_2}(X) = |b|^{-1}\int du|m|^{-2}d^\times m$$

over

$$|u^2| \leq |m^2b|\,,\; \sup(1,|c|) \leq |m^2| \leq |b|^{-1}\,.$$

This integral can be computed as follows:

$$\Omega^1_{Sl_2}(X) =$$

$\frac{|b|^{-1/2}-q^{-1}}{1-q^{-1}}$	$|c| \leq 1$	b even	
$\frac{|\varpi^{-1}b|^{-1/2}-q^{-1}}{1-q^{-1}}$	$|c| \leq 1$	b odd	
$\frac{|\varpi^{-1}bc|^{-1/2}-q^{-1}}{1-q^{-1}}$	$|c| > 1$		bc odd
$\frac{|bc|^{-1/2}-q^{-1}}{1-q^{-1}}$	$|c| > 1$	b even	bc even
$\frac{q^{-1}|bc|^{-1/2}-q^{-1}}{1-q^{-1}}$	$|c| > 1$	b odd	bc even

For $\Omega^2_{Sl_2}(X)$ we first compute the integral

$$\int_{|bc-u^2|\leq|m^2b|} du\,.$$

It is 0 unless bc is a square and then it is equal to $2|bc|^{-1/2}|bm^2|$. We thus have

$$\Omega^2_{Sl_2}(X)=|bc|^{-1/2}2\int_{1\leq|m^2|<|c|} d^\times m\,.$$

This is 0 unless $|c|>1$. Then it is equal to

$$\Omega^2_{Sl_2}(X)=|bc|^{-1/2}\begin{cases} c \text{ even} & -v(c)\\ c \text{ odd} & 1-v(c)\end{cases}$$

Adding our two results we arrive at the following Proposition.

PROPOSITION 14. *$\Omega_{Sl_2}(b,c)$ is given by the following formula.*

$\frac{\lvert b\rvert^{-1/2}-q^{-1}}{1-q^{-1}}$	$\lvert c\rvert\leq 1$	b even	
$\frac{\lvert\varpi^{-1}b\rvert^{-1/2}-q^{-1}}{1-q^{-1}}$	$\lvert c\rvert\leq 1$	b odd	
$\frac{\lvert\varpi^{-1}bc\rvert^{-1/2}-q^{-1}}{1-q^{-1}}$	$\lvert c\rvert> 1$		bc odd
$\frac{\lvert bc\rvert^{-1/2}-q^{-1}}{1-q^{-1}}$	$\lvert c\rvert> 1$	b even	bc even non square
$\frac{q^{-1}\lvert bc\rvert^{-1/2}-q^{-1}}{1-q^{-1}}$	$\lvert c\rvert> 1$	b odd	bc even non square
$\frac{\lvert bc\rvert^{-1/2}-q^{-1}}{1-q^{-1}}-v(c)\lvert bc\rvert^{-1/2}$	$\lvert c\rvert> 1$		bc square

20. Orbital integrals for $Gl_2(F)$

We let

$$Y=\begin{pmatrix} 0 & b & 0\\ c & 0 & 1\\ 1 & 0 & 0\end{pmatrix},$$

and we write $\Omega_{Gl_2}(Y)=\Omega_{Gl_2}(b,c)$. We have

$$\Omega_{Gl_2}(Y)=\int_{Gl_2(F)}\Phi(\iota(g)Y\iota(g)^{-1})\eta(\det g)dg$$

Explicitly, this is

$$\int\Phi\begin{pmatrix} -b\alpha u & b\alpha m^2 & 0\\ m^{-2}(c\alpha^{-1}-u^2b\alpha) & b\alpha u & m^{-1}\\ \alpha^{-1}m^{-1} & 0 & 0\end{pmatrix}\eta(\alpha)d^\times\alpha du|m|^{-2}d^\times m\,.$$

or

$$\int\eta(\alpha)d^\times\alpha du|m|^{-2}d^\times m$$

over

$$|m^{-1}|\leq 1\,,\ |\alpha^{-1}m^{-1}|\leq 1$$
$$|b\alpha u|\leq 1\,,\ |b\alpha m^2|\leq 1$$
$$|cb-u^2b^2\alpha^2|\leq|m^2b\alpha|\,.$$

As before, if the integral is non-zero then $|b|\leq 1$ and $|bc|\leq 1$. Under these assumptions the condition $|b\alpha u|\leq 1$ is superfluous. After a change of variables this becomes

$$|b|^{-1}\int\int\eta(\alpha)|\alpha|^{-1}d^\times\alpha du|m|^{-2}d^\times m$$

over

$$1 \leq |m|, |\alpha|^{-1} \leq |m|,$$
$$|cb - u^2| \leq |m^2 b\alpha| \leq 1.$$

After a new change of variables, we get

$$|b|^{-1} \int\int \eta(\alpha)|\alpha|^{-1} d^\times \alpha du d^\times m$$

over

$$1 \leq |m| \leq |\alpha| \leq |b|^{-1},$$
$$|bc - u^2| \leq |\alpha b|.$$

Now, if $|\alpha| \geq 1$ then

$$\int_{1\leq|m|\leq|\alpha|} d^\times m = 1 - v(\alpha).$$

Thus we get

$$|b|^{-1} \int \eta(\alpha)|\alpha|^{-1}(1 - v(\alpha)) d^\times \alpha du$$

over

$$1 \leq |\alpha| \leq |b|^{-1}, \ |bc - u^2| \leq |\alpha b|$$

or, after a new change of variables,

$$\eta(b) \int \eta(\alpha)|\alpha|^{-1}(1 - v(\alpha) + v(b)) d^\times \alpha du$$

over

$$|b| \leq |\alpha| \leq 1, \ |bc - u^2| \leq |\alpha|,$$

We divide the integral into the sum of the contribution $\Omega^1_{Gl}(Y)$ of $|bc| \leq |\alpha|$ and the contribution $\Omega^2_{Gl}(Y)$ of $|bc| > |\alpha|$.

To compute $\Omega^1_{Gl}(Y)$ we may write $\alpha = \omega^{2s}$ or $\alpha = \omega^{2s+1}$ with $s \geq 0$ and sum over s. We set $A = b$ or $A = bc$ in such a way that

$$|A| = \sup(|b|, |bc|).$$

We get

$$\Omega^1_{Gl}(\xi) =$$
$$\eta(b) \sum_{s\geq 0, |A|\leq|\varpi^{2s}|} (1 - 2s + v(b))q^s$$
$$-\eta(b) \sum_{s\geq 0, |A|\leq|\varpi^{2s+1}|} (v(b) - 2s)q^s.$$

If $|A| = |\varpi^{2r}|$ the first sum is for $0 \leq s \leq r$ and the second sum is for $0 \leq s \leq r-1$. We find

$$\eta(b)\left(\sum_{0\leq s\leq r} q^s + (v(b) - 2r)q^r\right) =$$
$$\eta(b)\left(\frac{|A|^{-1/2} - q^{-1}}{1 - q^{-1}} + (v(b) - 2r)|A|^{-1/2}\right).$$

If $|c| \leq 1$, then $A = b$, b is even, and we are left with

$$\eta(b)\frac{|b^{-1}|^{1/2} - q^{-1}}{1 - q^{-1}}.$$

If $|c| > 1$ then $A = bc$, bc is even, and we are left with

$$\eta(b)\left(\frac{|bc|^{-1/2} - q^{-1}}{1 - q^{-1}} - v(c)|bc|^{-1/2}\right).$$

If $|A| = |\varpi^{2r+1}|$ then both sums are for $0 \le s \le r$. We are left with

$$\eta(b)(\sum_{0\le s\le r} q^s) = \eta(b)\frac{|\varpi|^{1/2}|A|^{-1/2} - q^{-1}}{1 - q^{-1}}.$$

Now we compute $\Omega^2_{Gl}(Y)$. Now $|b| \le |\alpha| < |bc|$. Thus in order to have a non-zero result we need $|c| > 1$. The integral

$$\int_{|bc-u^2|\le|\alpha|} du$$

is 0 unless bc is a square. Then it is equal to $2|\alpha||bc|^{-1/2}$. Thus we find

$$2\eta(b)|bc|^{-1/2}\int_{|b|\le|\alpha|<|bc|}(1 - v(\alpha) + v(b))\eta(\alpha)d^\times\alpha$$

or

$$2|bc|^{-1/2}\int_{1\le|\alpha|<|c|}(1 - v(\alpha))\eta(\alpha)d^\times\alpha$$

$$= 2|bc|^{-1/2}\int_{1\le|\alpha|<|c|}\eta(\alpha)d^\times\alpha + 2|bc|^{-1/2}\int_{|c|^{-1}<|\alpha|\le1} v(\alpha)\eta(\alpha)d^\times\alpha.$$

Let us write $|c^{-1}| = |\varpi^r|$ and use the formula

$$\sum_{n=0}^{r-1} n(-1)^n = \frac{1}{4}(-1 + (-1)^r - 2(-1)^r r).$$

The first integral is 0 unless r is odd in which case it is 1. We find

$$\Omega_{Gl_2}(Y) = \begin{cases} c \text{ even} & |bc|^{-1/2}v(c) \\ c \text{ odd} & |bc|^{-1/2}(1 - v(c)) \end{cases}$$

Adding our two results we arrive at the following Proposition.

PROPOSITION 15. *$\Omega_{Gl_2}(b, c)$ is given by the following formula.*

$\eta(b)\frac{\lvert b\rvert^{-1/2}-q^{-1}}{1-q^{-1}}$	$\lvert c\rvert \le 1$	b even	
$\eta(b)\frac{\lvert\varpi^{-1}b\rvert^{-1/2}-q^{-1}}{1-q^{-1}}$	$\lvert c\rvert \le 1$	b odd	
$\eta(b)\frac{\lvert\varpi^{-1}bc\rvert^{-1/2}-q^{-1}}{1-q^{-1}}$	$\lvert c\rvert > 1$		bc odd
$\eta(b)\left(\frac{\lvert bc\rvert^{-1/2}-q^{-1}}{1-q^{-1}} - v(c)\lvert bc\rvert^{-1/2}\right)$	$\lvert c\rvert > 1$		bc even non square
$\eta(b)\frac{\lvert bc\rvert^{-1/2}-q^{-1}}{1-q^{-1}}$	$\lvert c\rvert > 1$	b even	bc square
$\eta(b)\frac{q^{-1}\lvert bc\rvert^{-1/2}-q^{-1}}{1-q^{-1}}$	$\lvert c\rvert > 1$	b odd	bc square

21. Verifcation of $\Omega_{Sl_2}(X) = \eta(b_2)\Omega_{Gl_2}(Y)$

Under our condition of matching we have

$$|b_1| = |b_2|, \; |c_1| = |c_2|.$$

In addition if b_1c_1 and b_2c_2 are even then b_1c_1 is a square if and only b_2c_2 is not a square. By direct inspection we find

$$\Omega_{Sl_2}(b_1, c_1) = \eta(b_2)\Omega_{Gl_2}(b_2, c_2).$$

This concludes the proof of the fundamental Lemma.

References

[1] Aizenbud, Avraham ; Gourevitch, Dmitry ; Rallis, Stephen ; Schiffmann, Grard *Multiplicity one theorems.* Ann. of Math. (2) 172 (2010), no. 2, 1407–1434.

[2] Flicker, Yuval Z. *Twisted tensors and Euler products.* Bull. Soc. Math. France 116 (1988), no. 3, 295–313.

[3] Flicker, Yuval Z. *On distinguished representations.* J. Reine Angew. Math. 418 (1991), 139–172.

[4] Flicker, Yuval Z. *Automorphic forms with anisotropic periods on a unitary group.* J. Algebra 220 (1999), no. 2, 636–663.

[5] Flicker, Yuval Z. *Cyclic automorphic forms on a unitary group.* J. Math. Kyoto Univ. 37 (1997), no. 3, 367–439.

[6] Jacquet, Hervé *A guide to the relative trace formula.* Automorphic representations, L-functions and applications: progress and prospects, 257–272, Ohio State Univ. Math. Res. Inst. Publ., 11, de Gruyter, Berlin, 2005.

[7] Jacquet, Hervé; Chen, Nan *Positivity of quadratic base change L-functions.* Bull. Soc. Math. France 129 (2001), no. 1, 33–90.

[8] Jacquet, Hervé *Sur un résultat de Waldspurger. II.* (French) [On a result of Waldspurger. II] Compositio Math. 63 (1987), no. 3, 315–389.

[9] Jacquet, Hervé *Sur un résultat de Waldspurger.* (French) [On a result of Waldspurger] Ann. Sci. École Norm. Sup. (4) 19 (1986), no. 2, 185–229.

[10] Gelbart, Stephen; Jacquet, Hervé; Rogawski, Jonathan *Generic representations for the unitary group in three variables.* Israel J. Math. 126 (2001), 173–237.

[11] Harder, G.; Langlands, R. P.; Rapoport, M. *Algebraische Zyklen auf Hilbert-Blumenthal-Flächen.* (German) [Algebraic cycles on Hilbert-Blumenthal surfaces] J. Reine Angew. Math. 366 (1986), 53–120.

[12] Gross, Benedict H. *L-functions at the central critical point.* Motives (Seattle, WA, 1991), 527–535, Proc. Sympos. Pure Math., 55, Part 1, Amer. Math. Soc., Providence, RI, 1994.

[13] Gross, Benedict H. *Some applications of Gelfand pairs to number theory.* Bull. Amer. Math. Soc. (N.S.) 24 (1991), no. 2, 277–301.

[14] Gross, Benedict H.; Prasad, Dipendra *On irreducible representations of* $SO_{2n+1} \times SO_{2m}$. Canad. J. Math. 46 (1994), no. 5, 930–950.

[15] Gross, Benedict H.; Prasad, Dipendra *On the decomposition of a representation of* SO_n *when restricted to* SO_{n-1}. Canad. J. Math. 44 (1992), no. 5, 974–1002.

[16] Ginzburg, David; Jiang, Dihua; Rallis, Stephen *On the nonvanishing of the central value of the Rankin-Selberg L-functions.* II. Automorphic representations, L-functions and applications: progress and prospects, 157–191, Ohio State Univ. Math. Res. Inst. Publ., 11, de Gruyter, Berlin, 2005.

[17] Ginzburg, David; Jiang, Dihua; Rallis, Stephen *On the nonvanishing of the central value of the Rankin-Selberg L-functions.* J. Amer. Math. Soc. 17 (2004), no. 3, 679–722 .

[18] Oda, Takayuki *Distinguished cycles and Shimura varieties.* Automorphic forms of several variables (Katata, 1983), 298–332, Progr. Math., 46, Birkhäuser Boston, Boston, MA, 1984.

[19] Gelbart, Stephen; Rogawski, Jonathan; Soudry,David *Endoscopy, theta-liftings, and period integrals for the unitary group in three variables.* Ann. of Math. (2) 145 (1997), no. 3, 419–476.

[20] Gelbart, S.; Rogawski, J.; Soudry, D. *On periods of cusp forms and algebraic cycles for* U(3). Israel J. Math. 83 (1993), no. 1-2, 213–252.

[21] Gelbart, Stephen; Rogawski, Jonathan; Soudry, David *Periods of cusp forms and L-packets.* C. R. Acad. Sci. Paris Sr. I Math. 317 (1993), no. 8, 717–722.

DEPARTMENT OF MATHEMTICS, COLUMBIA UNIVERSITY, MC 4408, NEW YORK , NY 10027
E-mail address: hj@math.columbia.edu

MATHEMATICS DEPARTMENT, THE OHIO STATE UNIVERSITY, 231 W. 18TH AVENUE, COLUMBUS, OHIO 43210
E-mail address: haar@math.ohio-state.edu

Clay Mathematics Proceedings
Volume 13, 2011

On Local L-Functions and Normalized Intertwining Operators II; Quasi-Split Groups

Henry H. Kim and Wook Kim

To Freydoon Shahidi on the occasion of his 60th birthday

ABSTRACT. Continuing our earlier paper, we make explicit the L-functions obtained by the Langlands-Shahidi method for quasi-split groups. We also study the conjecture of Shahidi regarding the holomorphy of the local L-functions, and holomorphy of the normalized local intertwining operators for $\mathrm{Re}(s) \geq \frac{1}{2}$. The recent result by Heiermann and Muić, which says that Shahidi's conjecture implies the standard module conjecture, settles several exceptional cases left open earlier.

CONTENTS

Introduction 265
1. Preliminaries 267
2. Local L-functions made explicit 268
3. Proof of a Conjecture of Shahidi 285
4. Proof of Assumption (A) 289
5. Correction to [**14**] 293
References 294

Introduction

This paper is a continuation of [**14**]. As in [**14**], we make explicit the L-functions obtained by the Langlands-Shahidi method [**27**] for quasi-split reductive groups. Then we study Conjecture 7.1 of [**27**] (Shahidi's conjecture) and Assumption (A) (cf. Section 4).

More precisely, let $\mathbf{G}$ be a connected reductive quasi-split algebraic group, and let $\mathbf{M}$ be a maximal Levi subgroup of $\mathbf{G}$. If $\mathbf{G}$ is simply connected, then the derived

2010 *Mathematics Subject Classification.* Primary 11F70; Secondary 11F66, 22E55.
The first named author is partially supported by NSERC grant.
The second named author is partially supported by BK21 postdoctoral fellowship at Yonsei University.

group $\mathbf{M}_D$ is also simply connected. This fact allows us to compute Levi groups explicitly for simply connected exceptional groups. To compute L-functions for a representation π of $\mathbf{M}(F_v)$, we follow the general theory of [**25, 27**] and use the table of [**25**] for the quasi-split cases as we did for split cases in [**14**]. Of course, we only deal with new cases not covered in [**14**]. For example, there is no Galois action in B_n cases and so we do not obtain any new L-functions. Such cases will be omitted unless we want to complement the preceding paper [**14**] (e.g., Section 2.6). An interesting case is ${}^2E_6 - 3$, where we obtain L-functions for $HSpin_8^-$ (whose derived group is $Spin_8^-$). These L-functions agree with the L-functions of W.T. Gan and J. Hundley that are obtained by the Rankin-Selberg integral [**9**] (See Remark 2.6 for details). After giving case-by-case explicit computations for L-functions we study general spinor groups and their properties in Section 2.8. General spinor groups such as $GSpin_m$ and $HSpin_m$ naturally appear as Levi subgroups of an algebraic group $\mathbf{G}$ even if $\mathbf{G}$ is simply connected. For example, $HSpin_{10}$ is a maximal Levi subgroup of the simply connected split group E_6^{sc}. The result of Section 2.8 will be used in Section 5.

We consider Conjecture 7.1 of [**27**] (Shahidi's conjecture) in Section 3. Using the machinery of multiplicativity of γ-factors and the corresponding multiplicativity of L-functions we study Shahidi's conjecture. Another key ingredient is the base change developed by Langlands [**21**], Arthur and Clozel [**2**] which enables us to use the local-global argument. With these theories in hand we can prove Shahidi's conjecture for quasi-split cases. The result is summarized in Theorem 3.5.

In Section 4 we study Assumption (A) which concerns holomorphy of the normalized intertwining operators (defined in Section 1). The recent result by V. Heiermann and G. Muić [**10**] says that Shahidi's conjecture implies the standard module conjecture. This new ingredient settles several exceptional cases left out in [**14**]. More precisely, Assumption (A) holds for $B_n - 1, D_n - 1$, (xxx) in [**20**], and (xxxii) in [**20**] (See Theorem 4.11 of [**14**]). Other recent progress on Langlands' functoriality from classical groups to general linear groups by J. Cogdell, H. Kim, I. Piatetski-Shapiro and F. Shahidi [**11**] provides a new ingredient for Assumption (A) and settles (xviii),(xxii), and (xxiv) in [**20**] (See Theorem 4.11 of [**14**]). The results concerning Assumption (A) are summarized in Theorem 4.7.

In Section 5, we correct several minor mistakes of [**14**] where the L-groups of certain Levi subgroups of exceptional groups are incorrect. However, it does not affect the computation of L-functions and the subsequent results of [**14**].

Finally, we make several remarks on the notation. In general we follow the notation of [**14**]. For example, F denotes a field of characteristic zero, local or global, which will be specified in each case, and G denotes F-points of an algebraic group $\mathbf{G}$. Since we use the Langlands-Shahidi method as a main tool, we also use the standard notation from [**27**]. Hence, see [**14**] or [**27**] for any unexplained notation. When we need to use both restricted and non-restricted roots for explicit computations we use $\{\alpha_1, ..., \alpha_n\}$ for the set of non-restricted simple roots and $\{\beta_1, ..., \beta_m\}$ for the set of restricted simple roots. For example, $\{\beta_1, \beta_2, \beta_3, \beta_4\}$ is the set of restricted simple roots for ${}^2E_6$ while $\{\alpha_1, ..., \alpha_6\}$ is the set of non-restricted simple roots of E_6, where α_1 and α_5 restrict to β_1 and so on. We also need to keep in mind that we use [**30**] for the non-restricted root systems of exceptional groups rather than those of Bourbaki.

Acknowledgments. We would like to thank Prof. F. Shahidi for his constant help in explaining his results and for many discussions. We would like to also thank Prof. S. Kudla for many helpful discussions.

1. Preliminaries

Let $\mathbf{G}$ be a connected reductive quasi-split algebraic group over a p-adic field F of characteristic zero. Let $\mathbf{A}_0$ be the split component in the center of $\mathbf{G}$. Let π be a representation of $\mathbf{G}(F)$. Then the central character of π is defined to be $\omega_\pi(a) = \pi(a), a \in \mathbf{A}_0(F)$. Hence we will talk about the central character only when $\mathbf{G}$ has a split central torus.

For a positive reduced root β, let $\widetilde{\mathbf{G}}_{\beta,D}$ be the simply connected covering of the derived group of the rank-one subgroup attached to β. There are only two possibilities; either $\widetilde{\mathbf{G}}_{\beta,D} = \mathrm{Res}_{E/F} SL_2$ or $SU(2,1)_{E/F}$ where E/F is a finite separable extension in the first case and E/F is a quadratic extension defining $SU(2,1)_{E/F}$ in the second case.

Let $\mathbf{P} = \mathbf{MN}$ be a maximal parabolic subgroup and let α be the unique simple root in $\mathbf{N}$ so that $\mathbf{P} = \mathbf{P}_\theta$, where $\theta = \Delta - \{\alpha\}$. As in [**27**], let $\tilde{\alpha} = \langle \rho, \alpha\rangle^{-1}\rho$, where ρ is half the sum of roots in $\mathbf{N}$. The pairing $\langle \alpha, \beta^\vee\rangle = \langle \alpha, \beta\rangle$ is defined as follows: Let $\tilde{\psi}^+$ be the set of non-restricted roots of $\mathbf{T}$ in $\mathbf{U}$, restricting to ψ^+. Here we choose a Borel subgroup $\mathbf{B}$ of $\mathbf{G}$ over F, and let $\mathbf{B} = \mathbf{TU}$, where $\mathbf{T}$ is a maximal torus and $\mathbf{U}$ is the unipotent radical of $\mathbf{B}$. For $\alpha, \beta \in \psi^+$, we identify α, β with roots of $\tilde{\psi}^+$, and then set $\langle \alpha, \beta\rangle = \frac{2(\alpha,\beta)}{(\beta,\beta)}$.

Let π be an unramified representation of $M = \mathbf{M}(F)$ and χ be the inducing character of the torus. Denote by ${}^L M$ the L-group of $\mathbf{M}$ and let ${}^L\mathfrak{n}$ be the Lie algebra of the L-group of $\mathbf{N}$. The adjoint action r of ${}^L M$ on ${}^L\mathfrak{n}$ decomposes as $r = \oplus_{i=1}^m r_i$, with ordering as in [**27**]. For each i, $1 \le i \le m$, let $L(s, \pi, r_i)$ be the local L-function defined in [**27**]. Let $\hat{t}$ be the semi-simple conjugacy class in ${}^L M^0$ corresponding to π. Note the relationship

$$\chi \circ \beta^\vee(\varpi) = \beta^\vee(\hat{t}),$$

where $\beta^\vee$ on the right is considered as a root of ${}^L M^0$. Then we have

$$L(s, \pi, r_i) = \prod_{\beta>0, \langle\tilde{\alpha},\beta\rangle = i} L(s, \chi\circ\beta^\vee)$$

where

$$L(s,\chi\circ\beta^\vee) = \begin{cases} L_E(s,\chi\circ\beta^\vee) & \text{if } \widetilde{\mathbf{G}}_{\beta,D} = Res_{E/F}SL_2, \\ L_E(s,\chi\circ\beta^\vee)L_F(2s, \omega_{E/F}(\chi\circ\beta^\vee|_{F^\times})) & \text{if } \widetilde{\mathbf{G}}_{\beta,D} = SU(2,1)_{E/F}. \end{cases}$$

and $\omega_{E/F}$ is defined as follows: Let E/F be an unramified quadratic extension of p-adic fields and let ϖ_E, ϖ_F be uniformizers of E, F, resp. Then $\varpi_E = \varpi_F$, and $|\varpi_E|_E = q_E^{-1}$, and $|\varpi_F|_F = q_F^{-1}$. Hence for any $x \in E$, $|x|_E = |N_{E/F}(x)|_F$, and $F^\times/N_{E/F}(E^\times)$ has order 2. Let $\omega_{E/F}$ be the character of $F^\times$ defined by local class field theory, i.e.,

$$\omega_{E/F} = \begin{cases} 1 & \text{on } N_{E/F}(E^\times), \\ -1 & \text{otherwise.} \end{cases}$$

We identify $s \in \mathbb{C}$ with $s\tilde{\alpha} \in \mathfrak{a}_{\mathbb{C}}^*$ and denote by

$$I(s,\pi) = I(s\tilde{\alpha},\pi) = \mathrm{Ind}_P^G\, \pi \otimes \exp(\langle s\tilde{\alpha}, H_P(\cdot)\rangle),$$

the induced representation of π from $P = \mathbf{P}(F)$ to $G = \mathbf{G}(F)$. Let $A(s\tilde{\alpha}, \pi, w_0)$ be the standard intertwining operator from $I(s\tilde{\alpha}, \pi)$ to $I(w_0(s\tilde{\alpha}), w_0(\pi))$. We define the normalized intertwining operator $N(s, \pi, w_0)$ by the relation

$$A(s, \pi, w_0) = r(s, \pi, w_0) N(s, \pi, w_0),$$

where

$$r(s, \pi, w_0) = \prod_{i=1}^{m} \frac{L(is, \pi, r_i)}{L(1 + is, \pi, r_i)\epsilon(s, \pi, r_i, \psi_v)}.$$

2. Local L-functions made explicit

In this section we make explicit the L-functions which appear in the constant term of Eisenstein series. Using the table in Section 4 of [**25**] we look at it case by case.

Let F be a number field and $\mathbb{A}_F$ its ring of adeles. In each case we choose a quasi-split reductive algebraic group $\mathbf{G}$ and then compute explicitly each maximal Levi subgroup $\mathbf{M}$ and $L(s, \pi, r_i)$, $i = 1, ..., m$, for any (generic) cuspidal representation π of $\mathbf{M}(\mathbb{A}_F)$.

Let χ be a character of $\mathbf{M}$. We let $\pi_\chi = \pi \otimes \chi$ be the representation of $\mathbf{M}(\mathbb{A}_F)$ given by $(\pi \otimes \chi)(m) = \pi(m)\chi(m)$. In the following, we will consider the twisted L-function only when it gives rise to a new L-function.

2.1. $^2A_{2n-1}$ **case.** Let $\{\alpha_1, ..., \alpha_{2n-1}\}$ be simple roots of type A_{2n-1}. Let $\beta_1 = \frac{1}{2}(\alpha_1 + \alpha_{2n-1}), ..., \beta_{n-1} = \frac{1}{2}(\alpha_{n-1} + \alpha_{n+1}), \beta_n = \alpha_n$. Then $\Delta = \{\beta_1, ..., \beta_n\}$ forms simple roots of type C_n.

More explicitly, we use the unitary group: Let E/F be a quadratic extension of number fields and $\mathbf{G} = U(n, n)$ be the quasi-split unitary group in $2n$ variables defined with respect to E/F. Let $G = \mathbf{G}(F)$. It is given as follows: Let J_n be the $n \times n$ matrix given by

$$J_n = \begin{pmatrix} & & & & 1 \\ & & & 1 & \\ & & \cdot & & \\ & \cdot & & & \\ 1 & & & & \end{pmatrix}.$$

Let $J'_{2n} = \begin{pmatrix} & J_n \\ -J_n & \end{pmatrix}$. Then

$$U(n, n) = \left\{ g \in GL(2n) \middle| \, {}^t\bar{g} J'_{2n} g = J'_{2n} \right\},$$

where $x \longmapsto \bar{x}$ is the Galois automorphism of E/F. We note that $\mathbf{G}(E) = GL_{2n}(E)$. When v splits in E, then $\mathbf{G}(F_v) = GL_{2n}(F_v)$. The derived group of $\mathbf{G}$ is $\mathbf{G}' = SU(n, n) = U(n, n) \cap SL(2n)$. The maximal Levi subgroups are of the form $\mathbf{M} = \mathrm{Res}_{E/F} GL_k \times U(l, l)$. When v splits, $\mathbf{M}(F_v) = \mathrm{diag}(A, B, J_k {}^tA^{-1} J_k) \simeq GL_k(F_v) \times GL_{2l}(F_v) \times GL_k(F_v)$.

We need to deal also with similitude groups: Let

$$\tilde{\mathbf{G}} = GU(n, n) = \left\{ g \in GL(2n) \middle| \, {}^t\bar{g} J'_{2n} g = x J'_{2n} \text{for some } x = \bar{x} \right\}.$$

Then $\tilde{\mathbf{G}}(E) = GL(2n, E) \times E^\times$.

Let $\mathbf{T}_d$ be the maximal F-split torus consisting of diagonal elements in $\mathbf{G}$. Then

$$\mathbf{T}_d(F) = T_d = \{t(\lambda_1, ..., \lambda_n) = \mathrm{diag}(\lambda_1, ..., \lambda_n, \lambda_n^{-1}, ..., \lambda_1^{-1}) | \lambda_i \in F^\times\},$$

The centralizer of $\mathbf{T}_d$ in $\mathbf{G}$ is the maximal torus $\mathbf{T}$ of diagonal elements:

$$\mathbf{T}(F) = T = \{t(\lambda_1, ..., \lambda_n) = \mathrm{diag}(\lambda_1, ..., \lambda_n, \bar{\lambda}_n^{-1}, ..., \bar{\lambda}_1^{-1} | \lambda_i \in E^\times\}.$$

Then the root system $\Phi(\mathbf{G}, \mathbf{T})$ is of type A_{2n-1}. But the restricted root system $\Phi(\mathbf{G}, \mathbf{T}_d)$ is of type C_n. Let $\mathrm{Gal}(E/F) = \{1, \tau\}$. The maximal torus $\tilde{\mathbf{T}}$ in $\tilde{\mathbf{G}}$ is given by

$$\tilde{\mathbf{T}} = \{t(\lambda_1, ..., \lambda_n, x) = \mathrm{diag}(\lambda_1, ..., \lambda_n, x\bar{\lambda}_n^{-1}, ..., x\bar{\lambda}_1^{-1})\},$$

where $x = \bar{x}$. We extend the coroots $\alpha^\vee : F^\times \longmapsto T_d$ to $\alpha^\vee : F^\times \longmapsto T$ as follows. For $\alpha = e_i - e_j$, $\alpha^\vee(\lambda) = t(1, ..., \underset{i}{\lambda}, ..., \underset{j}{\lambda^{-1}}, ..., 1) \in T$ for $1 \le i < j \le n$. For $\alpha = e_i + e_j$, $\alpha^\vee(\lambda) = t(1, ..., \underset{i}{\lambda}, ..., \underset{j}{\bar{\lambda}}, ..., 1)$, for $1 \le i < j \le n$. For $\alpha = 2e_i$, $\alpha^\vee(\lambda) = t(1, ..., \underset{i}{\lambda}, ..., 1)$ for $1 \le i \le n$. Here dots represent 1.

2.1.1. ${}^2A_{2n-1} - 2$. Let $\theta = \Delta - \{\beta_n\}$. In this case, $\mathbf{M} = \mathbf{M}_\theta = \mathrm{Res}_{E/F} GL_n$. Then ${}^LM = GL_n(\mathbb{C}) \times GL_n(\mathbb{C}) \rtimes \mathrm{Gal}(E/F)$. Let r_A be the Asai representation, i.e.,

$$r_A : {}^LM \longrightarrow GL_{n^2}(\mathbb{C}); \quad r_A(g_1, g_2, 1) = g_1 \otimes g_2, \quad r_A(g_1, g_2, \tau) = g_2 \otimes g_1.$$

Let $\sigma = \otimes_v \sigma_v$ be a cuspidal representation of $\mathbf{M}(\mathbb{A}_F) = GL_n(\mathbb{A}_E)$. Let $\mathbf{M}(F_v) = GL_n(E_v)$, where $E_v = E \otimes F_v$; then

$$\mathbf{M}(F_v) = \begin{cases} GL_n(E_v) & \text{if } v \text{ inert}, \\ GL_n(E_{w_1}) \times GL_n(E_{w_2}) & \text{if } v \text{ splits}, v = w_1 w_2. \end{cases}$$

Let χ be a grössencharacter of E with restriction χ_0 to F. Recall that χ_0 corresponds to the transfer from E to F of the character of the Weil group W_E associated to χ by class field theory. Let $\pi = \sigma \otimes \chi$. Suppose σ_v is unramified and E_v/F_v is unramified. Then if v is inert, σ_v is an unramified representation of $GL_n(E_v)$, and we denote it by $\sigma_v = \pi(\mu_1, ..., \mu_n)$, where $\mu_1, ..., \mu_n$ are unramified quasi-characters of $E_v^\times$. If v splits, then $\sigma_v = \sigma_{w_1} \otimes \sigma_{w_2}$, where σ_{w_i} is an unramified representation of $GL_n(E_{w_i})$ for $i = 1, 2$. Then we see that $m = 1$, and

$$\begin{aligned} L(s, \pi_v, r_1) &= L(s, \sigma_v, r_A \otimes \chi_{0v}) \\ &= \begin{cases} \prod_{i=1}^n L_{F_v}(s, \chi_{0v}\mu_i|_{F_v^\times}) \prod_{1 \le i < j \le n} L_{E_v}(s, \mu_i \mu_j \chi_v) & \text{if } v \text{ is inert} \\ L(s, \sigma_{w_1} \times \sigma_{w_2} \otimes \chi_{0v}) & \text{if } v \text{ splits}, \end{cases} \end{aligned}$$

where $L_{F_v}(s, \chi_{0v}\mu_1|_{F_v^\times})$ is the local Hecke L-function over F_v, and $L(s, \sigma_{w_1} \times \sigma_{w_2} \otimes \chi_{0v})$ is the local Rankin-Selberg L-function. Here $L(s, \sigma, r_A \otimes \chi_0)$ is called the Asai L-function (twisted tensor L-function).

2.1.2. ${}^2A_{2n-1} - 1$. Let $\theta = \Delta - \{\beta_1\}$. We separate this case, because it gives the well-known standard L-function of the unitary group. In this case, $\mathbf{M} = \mathrm{Res}_{E/F} GL_1 \times U(n, n)$. Let $\Sigma = \chi \otimes \pi$ be a cuspidal representation of $\mathbf{M}(\mathbb{A}_F)$. Let Σ_v be an unramified representation. If v splits, let $w_1 w_2 = v$. Then π_v is an unramified representation of $GL_{2n}(F_v)$, and

$$L(s, \Sigma_v, r_1) = L(s, \chi_{w_1} \otimes \pi_v) L(s, \chi_{w_2} \otimes \tilde{\pi}_v)$$
$$L(s, \Sigma_v, r_2) = L(s, \chi_{w_1} \chi_{w_2}).$$

If v is inert, we write π_v as $\pi_v = \pi_v(\mu_1, ..., \mu_n)$, where μ_i's are unramified quasi-characters of $E_v^\times$. Then

$$L(s, \Sigma_v, r_1) = L(s, \chi_v\mu_1^{-1}) \cdots L(s, \chi_v\mu_n^{-1})L(s, \chi_v\bar{\mu}_1) \cdots L(s, \chi_v\bar{\mu}_n)$$
$$L(s, \Sigma_v, r_2) = L_{F_v}(s, \chi_v|_{F_v^\times}).$$

Therefore, the global L-function is $L(s, \Sigma, r_2) = L(s, \chi|_{\mathbb{A}_F^\times})$ (Hecke L-function). When $\chi = 1$, $L(s, \Sigma_v, r_1)$ is the standard L-function of $U(n, n)$.

2.1.3. ${}^2A_{2n-1} - 4$. Let $\theta = \Delta - \{\beta_k\}$, and $k + l = n$. In this case, $\mathbf{M} = \mathbf{M}_\theta = \mathrm{Res}_{E/F} GL_k \times U(l, l)$. Let σ, τ be cuspidal representations of $GL_k(\mathbb{A}_E)$, $U(l, l)(\mathbb{A}_F)$, resp. Let $\pi = \sigma \otimes \tau$ be the cuspidal representation of $\mathbf{M}(\mathbb{A}_F)$.

If v splits ($v = w_1w_2$), then τ_v is a representation of $GL_{2l}(F_v)$, and $\pi_v = \sigma_{w_1} \otimes \tau_v \otimes \sigma_{w_2}$ is a representation of $GL_k(F_v) \times GL_{2l}(F_v) \times GL_k(F_v) \subset GL_{2n}(F_v)$. In this case, we need to consider the normalizing factors attached to the non-maximal parabolic subgroup:

$$L(s, \pi_v, r_1) = L(s, \sigma_{w_1} \times \tau_v)L(s, \sigma_{w_2} \times \tilde{\tau}_v), \quad L(s, \pi_v, r_2) = L(s, \sigma_{w_1} \times \sigma_{w_2}).$$

If v is inert, let $\sigma_v = \pi(\mu_1, ..., \mu_k)$, $\tau_v = \pi(\nu_1, ..., \nu_l)$, where μ_i, ν_j's are unramified quasi-characters of $E_v^\times$. Then

$$L(s, \pi_v, r_1) = \prod_{i=1}^{k} \prod_{j=1}^{l} L(s, \mu_i\nu_j^{-1})L(s, \mu_i\bar{\nu}_j), \quad L(s, \pi_v, r_2) = L(s, r_A(\sigma_v)).$$

In this case, we obtain the Rankin-Selberg L-function $L(s, \sigma \times \tau)$.

REMARK 2.1. *The lift from $U(n, n)$ to GL_{2n}/E (called the base change) is given by $\pi = \otimes_v \pi_v \longmapsto \Pi = \otimes_w \Pi_w$, where*

$$\Pi_w = \begin{cases} \Pi_w = \pi(\mu_1, ..., \mu_n, \bar{\mu}_1^{-1}, ..., \bar{\mu}_n^{-1}) & \text{if } v \text{ is inert and } \pi_v = \pi(\mu_1, ..., \mu_n), \\ \Pi_{w_1} = \pi_v, \Pi_{w_2} = \tilde{\pi}_v & \text{if } v \text{ splits and } w_1w_2 = v. \end{cases}$$

REMARK 2.2. *We note the difference between split group and quasi-split group: When $G = Sp_{2n}$, $M = GL_{n-1} \times SL_2$, we obtain the L-function $L(s, \sigma \times \mathrm{Ad}(\tau))$, where σ, τ are cuspidal representations of $GL_{n-1}(F), GL_2(F)$, resp. However, when $G = U(n, n)$, $M = \mathrm{Res}_{E/F} GL_{n-1} \times U(1, 1)$, even though $\theta = \Delta - \{\beta_{n-1}\}$ as an F-root system, two L-functions show up in the constant term of the Eisenstein series.*

2.2. ${}^2A_{2n}$ case. Let $\{\alpha_1, ..., \alpha_{2n}\}$ be simple roots of type A_{2n}. Let $\beta_1 = \frac{1}{2}(\alpha_1 + \alpha_{2n}), ..., \beta_n = \frac{1}{2}(\alpha_n + \alpha_{n+1})$. Then $\Delta = \{\beta_1, ..., \beta_n\}$ forms simple roots of type BC_n. More explicitly, we use the unitary group $\mathbf{G} = U(n, n+1)$: It is given by

$$U(n, n+1) = \left\{ g \in GL(2n+1) \middle| \ {}^t\bar{g}J'_{2n}g = J'_{2n} \right\},$$

where $J'_{2n} = \begin{pmatrix} & & J_n \\ & 1 & \\ -J_n & & \end{pmatrix}$. We note that $\mathbf{G}(E) = GL(2n+1, E)$. Let $\mathbf{T}_d$ be the maximal F-split torus consisting of diagonal elements in $\mathbf{G}$. Then

$$\mathbf{T}_d(F) = T_d = \left\{ t(\lambda_1, ..., \lambda_n, 1) = \mathrm{diag}(\lambda_1, ..., \lambda_n, 1, \lambda_n^{-1}, ..., \lambda_1^{-1}) | \lambda_i \in F^\times \right\},$$

The centralizer of T_d in G is the maximal torus T of diagonal elements:

$$\mathbf{T}(F) = T = \left\{ t(\lambda_1, ..., \lambda_n, y) = \mathrm{diag}(\lambda_1, ..., \lambda_n, y, \bar{\lambda}_n^{-1}, ..., \bar{\lambda}_1^{-1} | \lambda_i \in E^\times, y \in E^1 \right\}.$$

Then the root system $\Phi(\mathbf{G},\mathbf{T})$ is of type A_{2n}. But the root system $\Phi(\mathbf{G},\mathbf{T}_d)$ is of type BC_n. We extend the coroots $\beta^\vee : F^\times \longmapsto T_d$ to $\alpha^\vee : F^\times \longmapsto T$ as follows. For $\beta = e_i - e_j$, $\beta^\vee(\lambda) = t(1,...,\underset{i}{\lambda},...,\underset{j}{\lambda^{-1}},...,1) \in T$ for $1 \le i < j \le n$. For $\alpha = e_i + e_j$, $\beta^\vee(\lambda) = t(1,...,\underset{i}{\lambda},...,\underset{j}{\bar\lambda},...,1)$, for $1 \le i < j \le n$. For $\alpha = 2e_i$, $\beta^\vee(\lambda) = t(1,...,\underset{i}{\lambda},...,1,\bar\lambda\lambda^{-1})$ for $1 \le i \le n$. Here dots represent 1.

2.2.1. ${}^2A_{2n}-1$. Let $\theta = \Delta - \{\beta_1\}$. In this case, $\mathbf{M} = \mathrm{Res}_{E/F}GL_1 \times U(n,n+1)$. Let $\Sigma = \chi \otimes \pi$ be a cuspidal representation of $\mathbf{M}(\mathbb{A}_F)$. Let Σ_v be an unramified representation. If v splits, let $w_1w_2 = v$. Then π_v is an unramified representation of $GL_{2n+1}(F_v)$ and

$$L(s,\Sigma_v,r_1) = L(s,\chi_{w_1}\otimes\pi_v)L(s,\chi_{w_2}\otimes\tilde\pi_v),$$
$$L(s,\Sigma_v,r_2) = L(s,\chi_{w_1}\chi_{w_2}).$$

If v is inert, π_v can be written as $\pi_v = \pi_v(\mu_1,...,\mu_n,\nu)$, where μ_i's are unramified quasi-characters of $E_v^\times$ and ν is a character of E_v^1. It acts as

$$\pi_v(\mu_1,...,\mu_n,\nu)(t(\lambda_1,...,\lambda_n,y)) = \mu_1(\lambda_1)\cdots\mu_n(\lambda_n)\nu(y(\lambda_1\cdots\lambda_n)(\bar\lambda_1\cdots\bar\lambda_n)^{-1}).$$

Then

$$L(s,\Sigma_v,r_1) = L(s,\chi_v\mu_1^{-1})\cdots L(s,\chi_v\mu_n^{-1})L(s,\chi_v)L(s,\chi_v\bar\mu_1)\cdots L(s,\chi_v\bar\mu_n),$$
$$L(s,\Sigma_v,r_2) = L_{F_v}(s,\chi_v|_{F_v^\times}).$$

2.2.2. ${}^2A_{2n}-3$. Let $\theta = \Delta - \{\beta_n\}$. Then $\mathbf{M} = \mathrm{Res}_{E/F}GL_n \times U(1)$ and $\tilde\alpha = e_1 + \cdots + e_n - (e_{n+2} + \cdots + e_{2n+1})$. Let $\Sigma = \pi\otimes\chi$ be a cuspidal representation of $GL_n(\mathbb{A}_E) \times U(1,\mathbb{A}_F)$.

If v splits ($v = w_1w_2$), let $\pi_v = \pi_{w_1}\otimes\pi_{w_2}$, where π_{w_i} is an unramified representation of $GL_n(E_{w_i})$ for $i = 1,2$, and χ_v is a character of $F_v^\times$. Then $m = 2$ and

$$L(s,\Sigma_v,r_1) = L(s,\pi_{w_1})L(s,\pi_{w_2}), \quad L(s,\Sigma_v,r_2) = L(s,\pi_{w_1}\times\pi_{w_2}).$$

If v is inert, π_v is an unramified representation of $GL_n(E_v)$ and χ_v is a character of E_v^1, and we denote it by $\pi_v = \pi(\mu_1,...,\mu_n)$, where $\mu_1,...,\mu_n$ are unramified quasi-characters of $E_v^\times$. Then

$$L(s,\Sigma_v,r_1) = L(s,\pi_v), \quad L(s,\Sigma_v,r_2) = L(s,\pi_v,r_A\otimes\omega_{E/F}).$$

2.2.3. ${}^2A_{2n}-4$. Then $\mathbf{M} = \mathrm{Res}_{E/F}GL_k \times U(l,l+1)$. Let $\pi = \sigma\otimes\tau$ be a cuspidal representation of $\mathbf{M}(\mathbb{A}_F)$.

If v splits ($v = w_1w_2$), then τ_v is a representation of $GL_{2l+1}(F_v)$, and $\pi_v = \sigma_{w_1}\otimes\tau_v\otimes\sigma_{w_2}$ is a representation of $GL_k(F_v)\times GL_{2l+1}(F_v)\times GL_k(F_v) \subset GL_{2n}(F_v)$. In this case, we need to consider the normalizing factors attached to the non-maximal parabolic subgroup:

$$L(s,\pi_v,r_1) = L(s,\sigma_{w_1}\times\tau_v)L(s,\sigma_{w_2}\times\tilde\tau_v), \quad L(s,\pi_v,r_2) = L(s,\sigma_{w_1}\times\sigma_{w_2}).$$

If v is inert, let $\sigma_v = \pi(\mu_1,...,\mu_k)$, $\tau_v = \pi(\nu_1,...,\nu_l,\nu_0)$, where μ_i,ν_j's are unramified quasi-characters of $E_v^\times$ and ν_0 is a character of E_v^1. Then

$$L(s,\pi_v,r_1) = \prod_{i=1}^k\prod_{j=1}^l L(s,\mu_i\nu_j^{-1})L(s,\mu_i\bar\nu_j)\prod_{i=1}^k L(s,\mu_i),$$
$$L(s,\pi_v,r_2) = L(s,r_A(\sigma_v)\otimes\omega_{E/F}).$$

In this case, we obtain the Rankin-Selberg L-function $L(s,\sigma\times\tau)$.

REMARK 2.3. *The lift from $U(n,n+1)$ to GL_{2n+1}/E (the base change) is given by $\pi=\otimes_v\pi_v\longmapsto\Pi=\otimes_w\Pi_w$, where*

$$\Pi_w=\begin{cases}\Pi_w=\pi(\mu_1,...,\mu_n,1,\bar\mu_1^{-1},...,\bar\mu_n^{-1}) & \text{if } v \text{ is inert and } \pi_v=\pi(\mu_1,...,\mu_n,\nu),\\ \Pi_{w_1}=\pi_v,\Pi_{w_2}=\tilde\pi_v & \text{if } v \text{ splits and } w_1w_2=v.\end{cases}$$

2.3. ^2D_n case. Let $\mathbf{G}=Spin_{2n}^-$ be a quasi-split spin group over a quadratic extension E/F. It is a simply connected 2-fold covering group of the quasi-split orthogonal group SO_{2n}^- corresponding to a quadratic form of index $n-1$ relative to F but index n relative to E.Let $\{\alpha_1=e_1-e_2,...,\alpha_{n-2}=e_{n-2}-e_{n-1},\alpha_{n-1}=e_{n-1}-e_n,\alpha_n=e_{n-1}+e_n\}$ be the non-restricted simple roots. Let $\beta_1=\alpha_1=e_1-e_2,...,\beta_{n-2}=\alpha_{n-2}=e_{n-2}-e_{n-1},\beta_{n-1}=\frac{1}{2}(\alpha_{n-1}+\alpha_n)=e_{n-1}$. Then $\Delta=\{\beta_1,...,\beta_{n-1}\}$ form simple roots of type B_{n-1}. Any element in the F-points $\mathbf{T}(F)$ of the maximal torus $\mathbf{T}$ can be written as

$$H_{\alpha_1}(t_1)\cdots H_{\alpha_{n-2}}(t_{n-2})H_{\alpha_{n-1}}(t_{n-1})H_{\alpha_n}(\bar t_{n-1}),$$

where $t_i=\bar t_i$ for $i=1,...,n-2$ and $t_i\in E^\times$. On the other hand an element in the F-points of the maximal F-split torus is

$$H_{\alpha_1}(t_1)\cdots H_{\alpha_{n-2}}(t_{n-2})H_{\alpha_{n-1}}(t_{n-1})H_{\alpha_n}(t_{n-1}),$$

where $t_i=\bar t_i$ for all i. There is, up to isomorphism, a unique non simply-connected, non-adjoint group of type ^2D_n, namely, SO_{2n}^-.

We define $GSpin_{2n}^-$ to be the maximal Levi subgroup of $Spin_{2n+2}^-$, which has $Spin_{2n}^-$ as its derived group. More precisely, we add $\beta_0=e_0-e_1$ in the root system and consider the Levi subgroup attached to $\theta=\Delta-\{\beta_0\}$. Then

$$\mathbf{A}=\{H_{\alpha_0}(t^2)H_{\alpha_1}(t^2)\cdots H_{\alpha_{n-2}}(t^2)H_{\alpha_{n-1}}(t)H_{\alpha_n}(t)\},$$

and

$$\mathbf{M}_D=Spin_{2n}^-,\quad \mathbf{A}\cap\mathbf{M}_D=\{H_{\alpha_{n-1}}(t)H_{\alpha_n}(t):t^2=1\}.$$

We define

$$GSpin_{2n}^-=(GL_1\times Spin_{2n}^-)/(\mathbf{A}\cap\mathbf{M}_D).$$

Then $^LGSpin_{2n}^-=GSO_{2n}(\mathbb{C})\rtimes \mathrm{Gal}(E/F)$.

2.3.1. $^2D_n-1$. Let $\theta=\Delta-\{\beta_k\}$. Let $\mathbf{P}=\mathbf{MN}$. The derived group $\mathbf{M}_D$ of $\mathbf{M}$ is

$$\mathbf{M}_D=SL_k\times Spin_{2l}^-,$$

where $k+l=n$. We identify $\mathbf{A}$ with GL_1. We fix an identification of $\mathbf{M}_D$ and $SL_k\times Spin_{2l}$ under which the element $H_{\alpha_1}(t)H_{\alpha_2}(t^2)\cdots H_{\alpha_{k-1}}(t^{k-1})$ goes to the diagonal element $\mathrm{diag}(t,t,...,t,t^{-(k-1)})$ of SL_k, and

$$b(t)=H_{\alpha_{k+1}}(t^2)\cdots H_{\alpha_{n-2}}(t^2)H_{\alpha_{n-1}}(t)H_{\alpha_n}(t)$$

is the toral element in $Spin_{2l}$. We define a map $\bar f:\mathbf{A}\times\mathbf{M}_D\longrightarrow GL_1\times GL_1\times SL_k\times Spin_{2l}$ by

$$\bar f:(a(t),x,y)\longmapsto\begin{cases}(t,t^{\frac{k}{2}},x,y) & \text{if } k \text{ even},\\ (t^2,t^k,x,y) & \text{if } k \text{ odd}.\end{cases}$$

Now, $\mathbf{M} \simeq (GL_1 \times SL_k \times Spin_{2l})/S$, where

$$S = \begin{cases} \{(a(t), tI_k, b(t^{\frac{k}{2}})) : t^k = 1\} & \text{if } n \text{ even,} \\ \{(a(t), t^2 I_k, b(t^k)) : t^{2k} = 1\} & \text{if } n \text{ odd.} \end{cases}$$

We obtain a map $f : \mathbf{M} \longrightarrow GL_k \times GSpin_{2l}$ so that

$$f(H_{\alpha_k}(t)) = (\mathrm{diag}(1, ..., 1, t), c(t)),$$

where $c(t)$ is an element in $GSpin_{2l}$.

Let π_1, π_2 be cuspidal representations of $GL_k(\mathbb{A}_F), GSpin^-(2l, \mathbb{A}_F)$, resp. Let π be a cuspidal representation of $\mathbf{M}(\mathbb{A}_F)$, induced by f and π_1, π_2. Then the central character of π is

$$\omega_\pi = \begin{cases} \omega_1 \omega_2^{\frac{k}{2}}, & \text{if } k \text{ even} \\ \omega_1^2 \omega_2^k. & \text{if } k \text{ odd} \end{cases}$$

If v splits, π_{2v} is an unramified representation of $GSpin_{2l}(F_v)$. Hence by $D_n - 1$ case in [**14**], $L(s, \pi_v, r_1) = L(s, \pi_{1v} \times \pi_{2v})$, $\quad L(s, \pi_v, r_2) = L(s, \pi_{1v}, \wedge^2 \otimes \omega_{2v})$.

Suppose v is inert, and let $\hat{t}_1 = \mathrm{diag}(a_1, ..., a_k) \in GL_k(\mathbb{C}) = {}^L GL_k$ and $\hat{t}_2 = \mathrm{diag}(b_1, ..., b_l, b_l^{-1} b_0, ..., b_1^{-1} b_0) \rtimes \sigma \in GSO_{2l}(\mathbb{C}) \rtimes \mathrm{Gal}(E/F)$ be the Satake parameters attached to π_{1v}, π_{2v}, resp. Here we note that

$$\mathrm{diag}(b_1, ..., b_l, b_l^{-1} b_0, ..., b_1^{-1} b_0) \longmapsto b_0$$

generates the character group of GSO_{2l} and hence by Lemma 1.2 of [**14**], $b_0 = \omega_2(\varpi)$. Then

$$\chi \circ H_{\alpha_1} = a_1 a_2^{-1}, \ldots, \chi \circ H_{\alpha_{k-1}} = a_{k-1} a_k^{-1},$$
$$\chi \circ H_{\alpha_{k+1}} = b_1 b_2^{-1}, \ldots, \chi \circ H_{\alpha_{n-1}} = b_{l-1} b_l^{-1}, \chi \circ H_{\alpha_n} = b_{l-1} b_l b_0^{-1},$$
$$\chi(a(t)) = \omega_{\pi_v} = \begin{cases} (a_1 \cdots a_k)(b_0)^{\frac{k}{2}}, & \text{if } k \text{ even} \\ (a_1 \cdots a_k)^2 (b_0)^k, & \text{if } k \text{ odd} \end{cases}.$$

Since $f(H_{\alpha_k}(t)) = (\mathrm{diag}(1, ..., 1, t), b(t))$, we can see $\chi \circ H_{\alpha_k} = a_k b_1^{-1} b_0$. Hence, we see that $m = 2$,

$$L(s, \pi_v, r_1) = L(s, \pi_{1v} \times \pi_{2v})$$
$$L(s, \pi_v, r_2) = L(s, \pi_{1v}, \wedge^2 \otimes \omega_{2v}).$$

2.3.2. ${}^2D_n - 2$. Let $\theta = \Delta - \{\beta_{n-2}\}$. Then $\mathbf{P} = \mathbf{MN}$, and $\rho_P = e_1 + \cdots + e_{n-2}$. The connected component of the center of the torus is an F-split torus and is given by

$$\mathbf{A} = \begin{cases} \{H_{\alpha_1}(t) H_{\alpha_2}(t^2) \cdots H_{\alpha_{n-2}}(t^{n-2}) H_{\alpha_{n-1}}(t^{\frac{n-2}{2}}) H_{\alpha_n}(t^{\frac{n-2}{2}})\} & \text{if } n \text{ even,} \\ \{H_{\alpha_1}(t^2) H_{\alpha_2}(t^4) \cdots H_{\alpha_{n-2}}(t^{2(n-2)}) H_{\alpha_{n-1}}(t^{n-2}) H_{\alpha_n}(t^{n-2})\} & \text{if } n \text{ odd.} \end{cases}$$

Since $\mathbf{G}$ is simply connected, the derived group $\mathbf{M}_D$ of $\mathbf{M}$ is simply connected. Hence $\mathbf{M}_D = SL_{n-2} \times \mathrm{Res}_{E/F} SL_2$. There exists a map $f : \mathbf{M} \longrightarrow GL_{n-2} \times \mathrm{Res}_{E/F} GL_2$. Let π_1, π_2 be cuspidal representations of $GL_{n-2}(\mathbb{A}_F), GL_2(\mathbb{A}_E)$, resp. Let π be a cuspidal representation of $\mathbf{M}(\mathbb{A}_F)$, given by f, π_1, π_2. Let π_v be an unramified representation and let χ be the inducing character. If $v = w_1 w_2$ splits, then it is $D_n - 2$ case. So

$$L(s, \pi_v, r_1) = L(s, \pi_{1v} \times \pi_{2w_1} \times \pi_{2w_2}), \quad L(s, \pi_v, r_2) = L(s, \pi_{1v}, \wedge^2 \otimes \omega_{\pi_{2w_1}} \omega_{\pi_{2w_2}}).$$

Suppose v is inert and let $\pi_{1v} = \pi(\mu_1, ..., \mu_{n-2})$, and $\pi_{2v} = \pi(\nu_1, \nu_2)$, where $\mu_1, ..., \mu_{n-2}$ are characters of $F_v^\times$ and ν_1, ν_2 are characters of $E_v^\times$. Then

$$\chi \circ H_{\alpha_1} = \mu_1\mu_2^{-1}, \ldots, \chi \circ H_{\alpha_{n-3}} = \mu_{n-3}\mu_{n-2}^{-1}$$

$$\chi \circ H_{\alpha_{n-1}}(t)H_{\alpha_n}(t) = \nu_1\nu_2^{-1}(t^2), \quad \chi(a(t)) = \omega_1(t)\omega_2^{\frac{n-3}{2}}(u).$$

From this, we see that $\chi \circ H_{\alpha_{n-2}} = \mu_{n-2}\nu_2|_{F_v^\times}$. Hence $m = 2$ and

$$L(s, \pi_v, r_1) = L(s, \pi_{1v} \times r_A(\pi_{2v})), \quad L(s, \pi_v, r_2) = L(s, \pi_{1v}, \wedge^2 \otimes (\omega_{\pi_{2v}})|_{F_v^\times}),$$

where $r_A(\pi_v) = \pi(\nu_1|_{F_v^\times}, \nu_2|_{F_v^\times}, ((\nu_1\nu_2)|_{F_v^\times})^{\frac{1}{2}}, \omega_{E/F}((\nu_1\nu_2)|_{F_v^\times})^{\frac{1}{2}})$.

REMARK 2.4. *This case is important because we get the Asai lift from cuspidal representations of GL_2/E to GL_4/F in the following way. It corresponds to the L-group homomorphism*

$$r_A : {}^L\mathrm{Res}_{E/F}GL_2 = (GL_2(\mathbb{C}) \times GL_2(\mathbb{C})) \rtimes \mathrm{Gal}(E/F) \longrightarrow GL(\mathbb{C}^2 \otimes \mathbb{C}^2) = GL_4(\mathbb{C}),$$
$$r(g_1, g_2, 1) = g_1 \otimes g_2, \quad r(g_1, g_2, \theta) = g_2 \otimes g_1.$$

The details had been worked out by Krishnamurthy in his thesis [**18**]*: Let $\pi = \otimes_w \pi_w$ be a cuspidal representation of $GL_2(\mathbb{A}_E)$. Let $\pi_w = \pi(\mu, \nu)$ be an unramified representation. Then the local lift of π_w is given by*

$$\Pi_v = \begin{cases} \pi(\mu|_{F_v^\times}, \nu|_{F_v^\times}, ((\mu\nu)|_{F_v^\times})^{\frac{1}{2}}, \omega_{E/F}((\mu\nu)|_{F_v^\times})^{\frac{1}{2}}) & \text{if } v \text{ is inert}, \\ \pi_{w_1} \boxtimes \pi_{w_2} & \text{if } v = w_1w_2, \end{cases}$$

where $\omega = \omega_{E/F}$. This is different from automorphic induction; Automorphic induction also gives the lifting from cuspidal representations of GL_2/E to GL_4/F. It corresponds to the L-group homomorphism

$$r : {}^L\mathrm{Res}_{E/F}GL_2 = (GL_2(\mathbb{C}) \times GL_2(\mathbb{C})) \rtimes \mathrm{Gal}(E/F) \longrightarrow GL(\mathbb{C}^2 \oplus \mathbb{C}^2) = GL_4(\mathbb{C}),$$
$$r(g_1, g_2, 1) = g_1 \oplus g_2, \quad r(g_1, g_2, \theta) = g_2 \oplus g_1.$$

Let $\pi = \otimes_w \pi_w$ be a cuspidal representation of $GL_2(\mathbb{A}_E)$. Let $\pi_w = \pi(\mu, \nu)$ be an unramified representation, where the local lift of π_w is given by

$$\Pi_v = I_E^F(\pi_v) = \begin{cases} \pi(\mu|_{F_v^\times}^{\frac{1}{2}}, \omega(\mu|_{F_v^\times})^{\frac{1}{2}}, \nu|_{F_v^\times}^{\frac{1}{2}}, \omega(\nu|_{F_v^\times})^{\frac{1}{2}}) & \text{if } v \text{ is inert}, \\ \pi_{w_1} \boxplus \pi_{w_2} & \text{if } v = w_1w_2. \end{cases}$$

Hence it satisfies the adjoint relation: for any representation σ of $GL_2(F_v)$,

$$L(s, \sigma \times I_E^F(\pi_v)) = L(s, \sigma_E \times \pi_v).$$

We also note that if π is a cuspidal representation of $GL_2(\mathbb{A}_F)$ and π_E is the base change to E, then $As(\pi_E) = \mathrm{Sym}^2(\pi) \oplus (\omega_\pi\omega_{E/F})$.

2.3.3. ${}^2D_n - 3$. Let $\theta = \Delta - \{\beta_{n-1}\}$, and let $\mathbf{P} = \mathbf{MN}$. Then

$$\mathbf{A} = \begin{cases} \{H_{\alpha_1}(t)H_{\alpha_2}(t^2)\cdots H_{\alpha_{n-2}}(t^{n-2})H_{\alpha_{n-1}}(t^{\frac{n-1}{2}})H_{\alpha_n}(t^{\frac{n-1}{2}})\} & \text{if } n \text{ odd}, \\ \{H_{\alpha_1}(t^2)H_{\alpha_2}(t^4)\cdots H_{\alpha_{n-2}}(t^{2(n-2)})H_{\alpha_{n-1}}(t^{n-1})H_{\alpha_n}(t^{n-1})\} & \text{if } n \text{ even}. \end{cases}$$

In this case, the derived group $\mathbf{M}_D$ is simply connected, and hence $\mathbf{M}_D = SL_{n-2}$. We have a map $f : \mathbf{M} \longrightarrow GL_{n-2} \times \mathrm{Res}_{E/F}GL_1$. Let σ be a cuspidal representation of $GL_{n-2}(\mathbb{A}_F)$, η a grössencharacter of E, and π be a cuspidal representation of $\mathbf{M}(\mathbb{A}_F)$, given by f and σ, η.

If v splits, we need to consider the normalizing factor of the intertwining operator attached to the non-maximal parabolic subgroup P_θ, where $\theta = \Delta - \{e_{n-1} \pm e_n\}$. Then

$$L(s,\pi_v,r_1) = L(s,\sigma_v \otimes \eta_{w_1})L(s,\sigma_v \otimes \eta_{w_2}), \quad L(s,\pi_v,r_2) = L(s,\sigma_v,\wedge^2 \otimes \omega_{\sigma_v}\eta_{w_1}\eta_{w_2}).$$

If v is inert, let π_v be an unramified representation and let χ be the inducing character. Let $\pi_{1v} = \pi(\mu_1,...,\mu_{n-1})$, where $\mu_1,...,\mu_{n-1}$ are characters of $F_v^\times$. Then

$$\chi \circ H_{\alpha_1} = \mu_1\mu_2^{-1}, \ldots, \chi \circ H_{\alpha_{n-2}} = \mu_{n-2}\mu_{n-1}^{-1},$$
$$\chi(a(t)) = \omega_1(t)\eta(t).$$

From this, we see that $\chi \circ H_{\alpha_{n-1}}H_{\alpha_n} = \mu_{n-1} \circ N\eta_v$. Hence $m = 2$,

$$L(s,\pi_v,r_1) = L(s,\pi_{E,v} \otimes \eta_v), \quad L(s,\pi_v,r_2) = L(s,\pi_v,\wedge^2 \otimes \omega_{\sigma_v}(\eta_v|_{F_v^\times})),$$

where π_E is the base change of π to E.

2.3.4. ${}^2D_n - 4$. Let $\theta = \Delta - \{\beta_{n-3}\}$. We separate this case because this gives a twisted exterior square lift from $GU(2,2)_{E/F}$ to GL_6/F corresponding to the map $\wedge_t^2 : {}^LGU(2,2) = (GL_4(\mathbb{C}) \times GL_1(\mathbb{C})) \rtimes \mathrm{Gal}(E/F) \longrightarrow GL_6(\mathbb{C})$. (See **[15]** for the details.) For $g \in GL_4(\mathbb{C}), \lambda \in GL_1(\mathbb{C})$, we consider the six-dimensional representation

$$\wedge^2 : GL_4(\mathbb{C}) \times GL_1(\mathbb{C}) \longrightarrow GL_6(\mathbb{C})$$

given by $(g,\lambda) \longmapsto (\wedge^2 g)\lambda$, where $\wedge^2$ is the usual exterior square. (We abuse the notation by using $\wedge^2$ as 6-dimensional representations of both $GL_4(\mathbb{C})$ and $GL_4(\mathbb{C}) \times GL_1(\mathbb{C})$.) Let $\theta(g,\lambda) = (J_4'^{-1}{}^tg^{-1}J_4', \lambda\det(g))$. Then there is a matrix $A \in GL_6(\mathbb{C})$ such that

$$\wedge^2(\theta(g,\lambda)) = A^{-1} \wedge^2 (g,\lambda)A, \quad \forall g \in GL_4(\mathbb{C}).$$

Now, we can extend $\wedge^2$ to ${}^LGU(2,2)$ by mapping

$$(g,\lambda,1) \longmapsto \wedge^2(g)\lambda, \text{ and } (1,1,\sigma) \longmapsto A.$$

Let $\mathbf{P} = \mathbf{P}_\theta = \mathbf{MN}$, and let $\mathbf{A}$ be the connected component of the center of $\mathbf{M}$. Then $\mathbf{A}(\overline{F}) = \{a(t)|\, t \in \overline{F}\}$ where

$$a(t) = \begin{cases} H_{\alpha_1}(t)\cdots H_{\alpha_{n-3}}(t^{n-3})H_{\alpha_{n-2}}(t^{n-3})H_{\alpha_{n-1}}(t^{\frac{n-3}{2}})H_{\alpha_n}(t^{\frac{n-3}{2}}) & \text{if } n \text{ odd}, \\ H_{\alpha_1}(t^2)\cdots H_{\alpha_{n-3}}(t^{2(n-3)})H_{\alpha_{n-2}}(t^{2(n-3)})H_{\alpha_{n-1}}(t^{n-3})H_{\alpha_n}(t^{n-3}) & \text{if } n \text{ even}. \end{cases}$$

We note that $\mathbf{A}$ is a 1-dimensional torus that splits over F. Since $\mathbf{G}$ is simply connected, the derived group $\mathbf{M}_D$ of $\mathbf{M}$ is simply connected, and hence

$$\mathbf{M}_D = SL_{n-3} \times SU(2,2).$$

Note that $Spin_6^- \simeq SU(2,2)$. Furthermore,

$$\mathbf{A} \cap \mathbf{M}_D(\overline{F}) = \{H_{\alpha_1}(t)\cdots H_{\alpha_{n-4}}(t^{n-4})H_{\alpha_{n-1}}(t^{\frac{n-3}{2}})H_{\alpha_n}(t^{\frac{n-3}{2}})|\; t^{n-3} = 1\},$$

if n odd, and

$$\mathbf{A} \cap \mathbf{M}_D(\overline{F}) = \{H_{\alpha_1}(t^2)\cdots H_{\alpha_{n-4}}(t^{2(n-4)})H_{\alpha_{n-1}}(t^{n-3})H_{\alpha_n}(t^{n-3})|\, t^{2(n-3)} = 1\},$$

if n even. We obtain an injection $f : \mathbf{M} \longrightarrow GL_{n-3} \times GU(2,2)$ which is the identity map when restricted to the derived group $\mathbf{M}_D$. Let π_1, π_2 be a cuspidal representation of $GL_{n-3}(\mathbb{A}_F), GU(2,2)(\mathbb{A}_F)$, resp. Let π be a cuspidal representation of $\mathbf{M}(\mathbb{A}_F)$, induced by the map f and π_1, π_2. Let π_v be an unramified representation.

If v splits as (w_1, w_2) in E, then π_{2v} is an unramified representation of $GL_4(F_v)\times GL_1(F_v)$. Write $\pi_{2v} = \tau_v \otimes \chi_v$. In this case,

$$L(s, \pi_v, r_{1,v}) = L(s, \pi_{1v} \otimes \tau_v, \rho_{n-3} \otimes \wedge^2\rho_4 \otimes \chi_v) = L(s, \pi_{1v} \times \wedge_t^2(\pi_{2v}))$$
$$L(s, \pi_v, r_{2,v}) = L(s, \pi_{1v}, \wedge^2 \otimes (\omega_{\tau_v}\chi_v^2)).$$

Here note that $\omega_{\tau_v}\chi_v^2 = (\omega_\pi|_{\mathbb{A}_F^*})_v = (\omega_\pi)_{w_1}(\omega_\pi)_{w_2}$.

Suppose v is inert. Let $\pi_{1v} = \pi(\mu_1, ..., \mu_{n-3})$, and $\pi_{2v} = \pi(\nu_1, \nu_2, \nu_0)$, where $\mu_1, ..., \mu_{n-3}, \nu_0$ are characters of $F_v^\times$ and ν_1, ν_2 are characters of $E_v^\times$. We compute the local L-function for n odd. The "even" case is similar. Let π_v be induced from a character χ. Then

$$\chi \circ H_{\alpha_1} = \mu_1\mu_2^{-1}, \ldots, \chi \circ H_{\alpha_{n-4}} = \mu_{n-4}\mu_{n-3}^{-1}, \ \chi \circ H_{\alpha_{n-2}} = \nu_2|_{F_v^\times},$$
$$\chi \circ H_{\alpha_{n-1}}(\varpi)H_{\alpha_n}(\varpi) = \nu_1\nu_2^{-1}(\varpi), \quad \chi(a(\varpi)) = \omega_{\sigma_v}(\varpi)\omega_{\pi_v}(\varpi)^{\frac{n-3}{2}},$$

where ϖ is a uniformizing element in $F_v^\times$. From this, we see that $\chi \circ H_{\alpha_{n-3}} = \mu_{n-3}\nu_0$, and

$$L(s, \pi_v, r_{1,v}) = L(s, \pi_{1v} \otimes \pi_{2v}, \rho_{n-3} \otimes \wedge_t^2) = L(s, \pi_{1v} \times \wedge_t^2(\pi_{2v}))$$
$$L(s, \pi_v, r_{2,v}) = L(s, \pi_{1v}, \wedge^2 \otimes (\omega_{\pi_{2v}}|_{F_v^\times})),$$

where $\wedge_t^2(\pi_{2v})$ is the unramified representation of $GL_6(F_v)$ given by

$$\wedge_t^2(\pi_{2v}) = \pi(\nu_0(\nu_1\nu_2|_{F_v^\times})^{\frac12}, \nu_0\omega_{E_v/F_v}(\nu_1\nu_2|_{F_v^\times})^{\frac12}, \nu_0(\nu_1|_{F_v^\times}), \nu_0(\nu_2|_{F_v^\times}), \nu_0(\nu_1\nu_2|_{F_v^\times}), \nu_0).$$

2.4. $^3D_4$ **case.** Let E/F be a cubic Galois extension. Then there exists a simply connected, absolutely simple quasi-split group $\mathbf{G}$ defined over F, which splits over E, and whose non-restricted root system is of type D_4. We also note that the only other quasi-split group of type D_4 is of adjoint type.

Let $\beta_1 = \frac13(\alpha_1 + \alpha_3 + \alpha_4) = \frac13(e_1 - e_2 + 2e_3)$, $\beta_2 = \alpha_2 = e_2 - e_3$. Then $\Delta = \{\beta_1, \beta_2\}$ form simple roots of type G_2. For a reference, we record positive roots: Short roots are

$$\beta_1, \quad \beta_1+\beta_2 = \frac13(e_1-e_3+e_2-e_4+e_2+e_4), \quad 2\beta_1+\beta_2 = \frac13(e_1-e_4+e_1+e_4+e_2+e_3).$$

Long roots are

$$\beta_2, \quad 3\beta_1 + 2\beta_2 = e_1 + e_2, \quad 3\beta_1 + \beta_2 = e_1 + e_3.$$

Note that any element in the maximal torus in $\mathbf{G}(F)$ can be written as $H_{\alpha_1}(t_1)H_{\alpha_3}(\bar{t}_1)H_{\alpha_4}(\bar{\bar{t}}_1)H_{\alpha_2}(t_2)$, where $t_2 = \bar{t}_2 = \bar{\bar{t}}_2$ and $t_1, t_2 \in E^*$. Here $^-$ and $^=$ denote the Galois conjugate of $\mathrm{Gal}(E/F)$ corresponding to the graph automorphism. The element in the maximal F-split torus is $H_{\alpha_1}(t_1)H_{\alpha_3}(t_1)H_{\alpha_4}(t_1)H_{\alpha_2}(t_2)$, where $t_i = \bar{t}_i = \bar{\bar{t}}_i$ for $i = 1, 2$.

2.4.1. $^3D_4 - 1$. Let $\theta = \Delta - \{\beta_2\}$. Then $\tilde{\alpha} = e_1 + e_2$

$$\mathbf{A} = (\bigcap_{\beta\in\theta} \ker \beta)^0 = \{H_{\alpha_1}(t)H_{\alpha_3}(t)H_{\alpha_4}(t)H_{\alpha_2}(t^2)\}.$$

Let $\mathbf{P} = \mathbf{P}_\theta = \mathbf{MN}$, where $\mathbf{M}$ is the centralizer of $\mathbf{A}$. Since $\mathbf{G}$ is simply connected, the derived group $\mathbf{M}_D$ of $\mathbf{M}$ is simply connected, and hence

$$\mathbf{M}_D = \mathrm{Res}_{E/F}SL_2.$$

We obtain a map $f : \mathbf{M} \longrightarrow \mathrm{Res}_{E/F} GL_2$. Let σ be a cuspidal representation of $GL_2(\mathbb{A}_E)$ and π be the cuspidal representation of $\mathbf{M}$ obtained by f and σ.

If v splits ($v = w_1w_2w_3$), then by the result in [**16**], $m = 2$ and

$$L(s, \pi_v, r_1) = L(s, \sigma_{w_1} \times \sigma_{w_2} \times \sigma_{w_3}), \quad L(s, \pi_v, r_2) = L(s, \omega),$$

where $\omega = \omega_{\sigma_{w_1}} \omega_{\sigma_{w_2}} \omega_{\sigma_{w_3}}$.

If v is inert, let $\sigma_v = \pi(\mu, \nu)$, where μ, ν's are characters of $E_v^\times$. Let χ be the inducing character of the torus. Then

$$\chi \circ H_{\alpha_1}(t) H_{\alpha_3}(\bar{t}) H_{\alpha_4}(\bar{\bar{t}}) = \mu\nu^{-1}(t), \quad \chi(a(t)) = \mu\nu.$$

From this, we see that $\chi \circ H_{\alpha_2} = \nu|_{F_v^\times}$. Hence $m = 2$ and

$$L(s, \pi_v, r_1)^{-1} = (1 - \mu|_{F_v^\times} q_{F_v}^{-s})(1 - \nu|_{F_v^\times} q_{F_v}^{-s})(1 - \mu\nu^2 q_{E_v}^{-s})(1 - \mu^2\nu q_{E_v}^{-s}),$$
$$L(s, \pi_v, r_2) = L(s, \omega_{\sigma_v}|_{F_v^\times}).$$

2.4.2. ${}^3D_4 - 2$. Let $\theta = \Delta - \{\beta_1\}$. Then $\tilde{\alpha} = 2e_1 + e_2 + e_3$.

$$\mathbf{A} = (\bigcap_{\beta \in \theta} ker\beta)^0 = \{H_{\alpha_1}(t^2) H_{\alpha_3}(t^2) H_{\alpha_4}(t^2) H_{\alpha_2}(t^3)\}.$$

Let $\mathbf{P} = \mathbf{P}_\theta = \mathbf{MN}$, where $\mathbf{M}$ is the centralizer of $\mathbf{A}$. Since $\mathbf{G}$ is simply connected, the derived group $\mathbf{M}_D$ of $\mathbf{M}$ is simply connected, and hence

$$\mathbf{M}_D = SL_2.$$

We obtain a map $f : \mathbf{M} \longrightarrow \mathrm{Res}_{E/F} GL_1 \times GL_2$. Let σ be a cuspidal representation of $GL_2(\mathbb{A}_F)$ and η be a grössencharacter of E. Let π be a cuspidal representation of $\mathbf{M}(\mathbb{A}_F)$, obtained by f, σ, η.

If v splits ($v = w_1w_2w_3$), then we need to consider the non-maximal parabolic subgroup attached to $\theta = \{\alpha_2\}$ and normalizing factors attached to π_v. In this case,

$$\mathbf{A} = \{H_{\alpha_1}(t_1) H_{\alpha_3}(t_3) H_{\alpha_4}(t_4) H_{\alpha_2}(t_2)\},$$

where $t_2^2 = t_1t_3t_4$. Let $\mathbf{P} = \mathbf{P}_\theta = \mathbf{MN}$. In this case,

$$\chi \circ H_{\alpha_2} = \mu\nu^{-1}, \quad \chi(a(t_1, t_3, t_4)) = \omega_{\sigma_v}(t_2)(\eta_{w_1}(t_1)\eta_{w_2}(t_3)\eta_{w_3}(t_4))^2.$$

Then we have $\chi \circ H_{\alpha_1} = \nu\eta_{w_1}$, $\chi \circ H_{\alpha_3} = \nu\eta_{w_2}$, $\chi \circ H_{\alpha_4} = \nu\eta_{w_3}$. Hence $m = 3$ and

$$L(s, \pi_v, r_1) = L(s, \sigma_v \otimes \eta_{w_1}) L(s, \sigma_v \otimes \eta_{w_2}) L(s, \sigma_v \otimes \eta_{w_3}),$$
$$L(s, \pi_v, r_2) = \prod_{1 \le i \le 3} L(s, \eta_{w_i}\eta_{w_j}\omega_{\sigma_v}),$$
$$L(s, \pi_v, r_3) = L(s, \sigma_v \otimes \omega),$$

where $\omega = \omega_{\sigma_v}\eta_{w_1}\eta_{w_2}\eta_{w_3}$.

If v is inert, then

$$\chi \circ H_{\alpha_2}(t) = \mu\nu^{-1}(t), \quad \chi(a(t)) = \eta_v^2(\mu\nu) \circ N.$$

From this, we see that $\chi \circ H_{\alpha_1} H_{\alpha_3} H_{\alpha_4} = \eta_v \nu \circ N$. Hence $m = 3$ and

$$L(s, \pi_v, r_1) = L(s, \sigma_{E,v} \otimes \eta_v),$$
$$L(s, \pi_v, r_2) = L(s, \omega_{\sigma_{E,v}}\eta_v^2), \quad L(s, \pi_v, r_3) = L(s, \sigma_v \otimes \omega),$$

where $\omega = \omega_{\sigma_v}\eta_v|_{F_v^\times}$ and $\sigma_{E,v}$ is the base change of σ_v to E_v.

2.5. $^6D_4$ case. This is the case when E/F is not a Galois extension. This is the non-Galois version of the case $^3D_4$. The L-group is $PSO_8(\mathbb{C}) \rtimes S_3$.

The cases $^6D_4 - 1$ and $^6D_4 - 2$ are essentially the same as those of $^3D_4 - 1$ and $^3D_4 - 2$, resp. The only difference is that we need to consider the case when v is unramified, and $v = w_1w_2$, where $E_{w_1} = F_v$ and E_{w_2}/F_v is a quadratic extension. In this case, the local L-function is $L(s, \sigma_v \otimes \eta_v)L(s, \sigma_v \otimes \eta_v)$ for $\pi_{w_1} \otimes \pi_{w_2}$.

REMARK 2.5. *Let π be a cuspidal representation of GL_2/F, and let π_E be a base change to E. It exists whether E/F is Galois or not. But the L-functions $L(s,\pi,r_1)$ in the cases of $^3D_4 - 1$ and $^6D_4 - 1$ are different. In the case of $^3D_4 - 1$, $L(s,\pi,r_1) = L(s, \pi\times\pi\times\pi)$. On the other hand, in the case of $^6D_4 - 1$, $L(s,\pi,r_1) = L(s, \pi \times \pi \times (\pi \otimes \chi))$, where χ is the quadratic character attached to L/F. Here K/F is the Galois closure of E/F and L/F is the unique quadratic intermediate extension.*

2.6. $^2E_6$ case. Let $\mathbf{G} = {}^2E_6$ be the simply connected, absolutely simple, quasi-split group of type E_6. Let $\beta_1 = \frac{1}{2}(\alpha_1 + \alpha_5), \beta_2 = \frac{1}{2}(\alpha_2 + \alpha_4), \beta_3 = \alpha_3, \beta_4 = \alpha_6$. Then $\Delta = \{\beta_1, \beta_2, \beta_3, \beta_4\}$ are simple roots of type F_4. The Dynkin diagram is

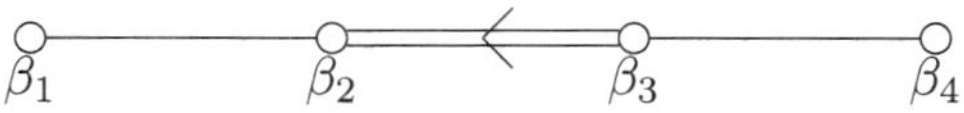

For a reference, we record positive roots as follows: Short roots are

$$\beta_1, \quad \beta_2, \quad \beta_1 + \beta_2 = \frac{1}{2}(e_1 - e_3 + e_4 - e_6), \quad \beta_1 + \beta_2 + \beta_3 = \frac{1}{2}(e_1 - e_4 + e_3 - e_6),$$
$$\beta_2 + \beta_3 = \frac{1}{2}(e_2 - e_4 + e_3 - e_5), \quad \beta_1 + 2\beta_2 + \beta_3 = \frac{1}{2}(e_1 - e_5 + e_2 - e_6),$$
$$\beta_1 + \beta_2 + \beta_3 + \beta_4 = \frac{1}{2}(e_1 + e_5 + e_6 + \epsilon + e_3 + e_4 + e_5 + \epsilon),$$
$$\beta_1 + 2\beta_2 + 2\beta_3 + \beta_4 = \frac{1}{2}(e_2 + e_3 + e_5 + \epsilon + e_1 + e_3 + e_6 + \epsilon),$$
$$\beta_1 + 3\beta_2 + 2\beta_3 + \beta_4 = \frac{1}{2}(e_2 + e_3 + e_4 + \epsilon + e_1 + e_2 + e_6 + \epsilon),$$
$$\beta_1 + 2\beta_2 + \beta_3 + \beta_4 = \frac{1}{2}(e_2 + e_4 + e_5 + \epsilon + e_1 + e_4 + e_6 + \epsilon),$$
$$2\beta_1 + 3\beta_2 + 2\beta_3 + \beta_4 = \frac{1}{2}(e_1 + e_3 + e_4 + \epsilon + e_1 + e_2 + e_5 + \epsilon),$$
$$\beta_2 + \beta_3 + \beta_4 = \frac{1}{2}(e_2 + e_5 + e_6 + \epsilon + e_3 + e_4 + e_6 + \epsilon),$$

and long roots are

$$\beta_3, \quad \beta_4, \quad \beta_3 + \beta_4 = e_3 + e_5 + e_6 + \epsilon, \quad 2\beta_2 + 2\beta_3 + \beta_4 = e_2 + e_3 + e_6 + \epsilon,$$
$$2\beta_2 + \beta_3 = e_2 - e_5, \quad 2\beta_1 + 2\beta_2 + \beta_3 = e_1 - e_6,$$
$$2\beta_2 + \beta_3 + \beta_4 = e_2 + e_4 + e_6 + \epsilon, \quad 2\beta_1 + 2\beta_2 + \beta_3 + \beta_4 = e_1 + e_4 + e_5 + \epsilon,$$
$$2\beta_1 + 4\beta_2 + 3\beta_3 + \beta_4 = e_1 + e_2 + e_3 + \epsilon, \quad 2\beta_1 + 4\beta_2 + 3\beta_3 + 2\beta_4 = 2\epsilon,$$
$$2\beta_1 + 4\beta_2 + 2\beta_3 + \beta_4 = e_1 + e_2 + e_4 + \epsilon, 2\beta_1 + 2\beta_2 + 2\beta_3 + \beta_4 = e_1 + e_3 + e_5 + \epsilon.$$

Note that any element in the F-points of a maximal torus $\mathbf{T}$ can be written as

$$H_{\alpha_1}(t_1)H_{\alpha_5}(\bar{t}_1)H_{\alpha_2}(t_2)H_{\alpha_4}(\bar{t}_2)H_{\alpha_3}(t_3)H_{\alpha_6}(t_4),$$

where $t_3 = \bar{t}_3, t_4 = \bar{t}_4$ and $t_i \in E^\times$, and an element in the F-points of the maximal F-split torus is

$$H_{\alpha_1}(t_1)H_{\alpha_5}(t_1)H_{\alpha_2}(t_2)H_{\alpha_4}(t_2)H_{\alpha_3}(t_3)H_{\alpha_6}(t_4),$$

where $t_i = \bar{t}_i$ for $i = 1, 2, 3, 4$.

2.6.1. ${}^2E_6 - 1$. Let $\theta = \Delta - \{\beta_3\}$. Then $\tilde{\alpha} = 2(e_1 + e_2 + e_3) + e_4 + e_5 + e_6 + 3\epsilon$.

$$\mathbf{A} = (\bigcap_{\beta \in \theta} ker\beta)^0 = \{H_{\alpha_1}(t^2)H_{\alpha_5}(t^2)H_{\alpha_2}(t^4)H_{\alpha_4}(t^4)H_{\alpha_3}(t^6)H_{\alpha_6}(t^3)\}.$$

Let $\mathbf{P} = \mathbf{P}_\theta = \mathbf{MN}$, where $\mathbf{M}$ is the centralizer of $\mathbf{A}$. Since $\mathbf{G}$ is simply connected, the derived group $\mathbf{M}_D$ of $\mathbf{M}$ is simply connected, and hence

$$\mathbf{M}_D = \text{Res}_{E/F} SL_3 \times SL_2.$$

And

$$\mathbf{A} \cap \mathbf{M}_D = \{H_{\alpha_1}(t^2)H_{\alpha_5}(t^2)H_{\alpha_2}(t^4)H_{\alpha_4}(t^4)H_{\alpha_3}(t^6)H_{\alpha_6}(t^3) | t^6 = 1\}.$$

Note that $\mathbf{A} \cap \mathbf{M}_D$ is finite.

We obtain a map $f : \mathbf{M} \longrightarrow \text{Res}_{E/F} GL_3 \times GL_2$. Let π_1, π_2 be a cuspidal representation of $GL_3(\mathbb{A}_E), GL_2(\mathbb{A}_F)$, resp. Let π be a cuspidal representation of $\mathbf{M}(\mathbb{A}_F)$, induced by the map f and π_1, π_2. Let π_v be an unramified representation. If v splits, then from $E_6 - 1$ case in [**14**],

$$\begin{aligned} L(s, \pi_v, r_1) &= L(s, \pi_{1w_1} \times \pi_{1w_2} \times \pi_{2v}), \\ L(s, \pi_v, r_2) &= L(s, (\tilde{\pi}_{1w_1} \otimes \omega) \times \tilde{\pi}_{1w_2}), \\ L(s, \pi_v, r_3) &= L(s, \pi_{2v} \otimes \omega), \end{aligned}$$

where $\omega = \omega_{\pi_{1w_1}} \omega_{\pi_{1w_2}} \omega_{\pi_{2v}}$.

If v is inert, let $\pi_{1v} = \pi(\mu_1, \mu_2, \mu_3)$, where μ_i's are unramified quasi-characters of $E_v^\times$. Let $\pi_{2v} = \pi(\nu_1, \nu_2)$, where ν_j's are unramified quasi-characters of $F_v^\times$. Then π_v is induced from the character χ of the torus. We have

$$\chi \circ H_{\alpha_1}(t)H_{\alpha_5}(\bar{t}) = \mu_1\mu_2^{-1}(t), \quad \chi \circ H_{\alpha_2}(t)H_{\alpha_4}(\bar{t}) = \mu_2\mu_3^{-1}(t)$$
$$\chi \circ H_{\alpha_6} = \nu_1\nu_2^{-1}, \chi(a(u)) = \omega_1(u^2)\omega_2(u^3),$$

where $u = \bar{u}$. From this, we see that $\chi \circ H_{\alpha_3} = (\mu_3|_{F_v^\times})\nu_2$. Then by direct computation, short roots $\{\beta_1 + \beta_2 + \beta_3, \beta_2 + \beta_3, \beta_1 + 2\beta_2 + \beta_3, \beta_1 + \beta_2 + \beta_3 + \beta_4, \beta_1 + 2\beta_2 + \beta_3 + \beta_4, \beta_2 + \beta_3 + \beta_4\}$ and long roots $\{\beta_3, \beta_3 + \beta_4, 2\beta_2 + \beta_3 + \beta_4, 2\beta_2 + \beta_3, 2\beta_1 + 2\beta_2 + \beta_3, 2\beta_1 + 2\beta_2 + \beta_3 + \beta_4\}$ contribute to $L(s, \pi_v, r_1)$ and so on. Hence

$$\begin{aligned} L(s, \pi_v, r_1) &= L(s, r_A(\pi_{1v}) \times \pi_{2v}), \\ L(s, \pi_v, r_2) &= L(s, r_A(\tilde{\pi}_{1v}) \otimes \omega) \\ L(s, \pi_v, r_3) &= L(s, \pi_{2v} \otimes \omega) \end{aligned}$$

where $\omega = \omega_{\pi_{1v}}|_{F_v^\times} \omega_{\pi_{2v}}$, and $r_A(\pi_{1v})$ is the unramified representation of $GL_9(F_v)$, given by

$$\begin{aligned} r_A(\pi_{1v}) = \pi(&\mu_1|_{F_v^\times}, \mu_2|_{F_v^\times}, \mu_3|_{F_v^\times}, ((\mu_1\mu_2)|_{F_v^\times})^{\frac{1}{2}}, ((\mu_2\mu_3)|_{F_v^\times})^{\frac{1}{2}}, ((\mu_1\mu_3)|_{F_v^\times})^{\frac{1}{2}}, \\ &\omega((\mu_1\mu_2)|_{F_v^\times})^{\frac{1}{2}}, \omega((\mu_2\mu_3)|_{F_v^\times})^{\frac{1}{2}}, \omega((\mu_1\mu_3)|_{F_v^\times})^{\frac{1}{2}}). \end{aligned}$$

where $\omega = \omega_{E_v/F_v}$. It is the Asai lift of π_{1v} from $GL_3(E_v)$ to $GL_9(F_v)$.

2.6.2. ${}^2E_6-2$. Let $\theta=\Delta-\{\beta_2\}$. Then $\tilde\alpha=3(e_1+e_2)+2(e_3+e_4)+e_5+e_6+\epsilon$.

$$\mathbf{A}=\{H_{\alpha_1}(t^3)H_{\alpha_5}(t^3)H_{\alpha_2}(t^6)H_{\alpha_4}(t^6)H_{\alpha_3}(t^8)H_{\alpha_6}(t^4)\}.$$

Let $\mathbf{P}=\mathbf{P}_\theta=\mathbf{MN}$. Since $\mathbf{G}$ is simply connected, the derived group $\mathbf{M}_D$ of $\mathbf{M}$ is simply connected, and hence

$$\mathbf{M}_D=\mathrm{Res}_{E/F}SL_2\times SL_3.$$

And

$$\mathbf{A}\cap\mathbf{M}_D=\{H_{\alpha_1}(t^3)H_{\alpha_5}(t^3)H_{\alpha_2}(t^6)H_{\alpha_4}(t^6)H_{\alpha_3}(t^8)H_{\alpha_6}(t^4)|t^6=1\}.$$

We obtain a map $f:\mathbf{M}\longrightarrow\mathrm{Res}_{E/F}GL_2\times GL_3$. Let π_1,π_2 be a cuspidal representation of $GL_2(\mathbb{A}_E),GL_3(\mathbb{A}_F)$, resp. Let π be a cuspidal representation of $\mathbf{M}(\mathbb{A}_F)$, induced by the map f and π_1,π_2. Let π_v be an unramified representation.

If v splits, then we need to consider the non-maximal parabolic subgroup attached to $\Delta-\{\alpha_2,\alpha_4\}$ and normalizing factors attached to $\pi_{1w_1},\pi_{2v},\pi_{1w_2}$. In this case,

$$\mathbf{A}=\{H_{\alpha_1}(t_1)H_{\alpha_2}(t_1^2)H_{\alpha_3}(u^4)H_{\alpha_4}(t_5^2)H_{\alpha_5}(t_5)H_{\alpha_6}(u^2)\},$$

where $t_1t_5=u^3$. Let $\pi_{1w_1}=\pi(\mu_1,\nu_1),\pi_{1w_2}=\pi(\mu_2,\nu_2),\pi_{2v}=\pi(\eta_1,\eta_2,\eta_3)$. Then we have

$$\chi\circ H_{\alpha_1}=\mu_1\nu_1^{-1},\quad H_{\alpha_5}=\mu_2\nu_2^{-1},\quad \chi\circ H_{\alpha_3}=\eta_2\eta_3^{-1},$$
$$\chi\circ H_{\alpha_6}=\eta_1\eta_2^{-1},\quad \chi(a(t_1,t_2))=\omega_{\pi_{1w_1}}(t_1)\omega_{\pi_{1w_2}}(t_2)\omega_{\pi_{2v}}(u^2).$$

Then we have $\chi\circ H_{\alpha_2}=\nu_1\eta_3$, $\chi\circ H_{\alpha_4}=\nu_2\eta_3$. Hence $m=4$ and

$$\begin{aligned}
L(s,\pi_v,r_1)&=L(s,\pi_{1w_1}\times\pi_{2v})L(s,\pi_{1w_2}\times\pi_{2v}),\\
L(s,\pi_v,r_2)&=L(s,\pi_{1w_1}\times\pi_{1w_2}\times\tilde\pi_{2v}\otimes\omega_{\pi_{2v}}),\\
L(s,\pi_v,r_3)&=L(s,\pi_{1w_1}\otimes\omega_{\pi_{1w_2}}\omega_{\pi_{2v}}),L(s,\pi_{1w_2}\otimes\omega_{\pi_{1w_1}}\omega_{\pi_{2v}}),\\
L(s,\pi_v,r_4)&=L(s,\pi_{2v}\otimes\omega),
\end{aligned}$$

where $\omega=\omega_{\pi_{1w_1}}\omega_{\pi_{1w_2}}\omega_{\pi_{2v}}$.

If v is inert, let $\pi_{1v}=\pi(\mu_1,\mu_2)$, where μ_i's are unramified quasi-characters of $E_v^\times$. Let $\pi_{2v}=\pi(\eta_1,\eta_2,\eta_3)$, where η_j's are unramified quasi-characters of $F_v^\times$, and π_v be induced from the character χ of the torus. We have

$$\chi\circ H_{\alpha_1}(t)H_{\alpha_5}(\bar t)=\mu_1\mu_2^{-1}(t),\quad \chi\circ H_{\alpha_3}=\eta_2\eta_3^{-1}$$
$$\chi\circ H_{\alpha_6}=\eta_1\eta_2^{-1},\chi(a(t))=\omega_1(t\bar t)\omega_2(u^3).$$

Then $\chi\circ H_{\alpha_2}(t)H_{\alpha_4}(\bar t)=\mu_2(\eta_3\circ N)$ and

$$\begin{aligned}
L(s,\pi_v,r_1)&=L(s,\pi_{1v}\times\pi_{2,E_v}),\\
L(s,\pi_v,r_2)&=L(s,r_A(\pi_{1v})\times\tilde\pi_{2v}\otimes\omega_{\pi_{2v}}),\\
L(s,\pi_v,r_3)&=L(s,\pi_{1v}\otimes\omega_{\pi_{1v}}(\omega_{\pi_{2v}}\circ N)),\\
L(s,\pi_v,r_4)&=L(s,\pi_{2v}\otimes(\omega_{\pi_{1v}}|_{F_v^\times})\omega_{\pi_{2v}}).
\end{aligned}$$

where π_{2,E_v} is the base change of π_{2v} to E_v and $r_A(\pi_{1v})$ is the Asai lift as in the ${}^2E_6-1$ case.

2.6.3. ${}^2E_6-3$. Let $\theta=\Delta-\{\beta_1\}$. Then $\tilde\alpha=2e_1+e_2+e_3+e_4+e_5+2\epsilon$ and

$$\mathbf{A}=\{H_{\alpha_1}(t^2)H_{\alpha_5}(t^2)H_{\alpha_2}(t^3)H_{\alpha_4}(t^3)H_{\alpha_3}(t^4)H_{\alpha_6}(t^2)\}.$$

Let $\mathbf{P}=\mathbf{P}_\theta=\mathbf{MN}$. Since $\mathbf{G}$ is simply connected, the derived group $\mathbf{M}_D$ of $\mathbf{M}$ is simply connected, and hence

$$\mathbf{M}_D=Spin_8^-.$$

There is an F-rational map $\mathbf{M}\longrightarrow \mathrm{Res}_{E/F}GL_1\times HSpin_8^-$. Let η be a grössencharacter of E and σ be a cuspidal representation of $HSpin_8^-(\mathbb{A}_F)$. Let π be a cuspidal representation of $\mathbf{M}(\mathbb{A}_F)$, given by f,σ,η. Let π_v be an unramified representation and χ be the inducing character.

If v splits, then we need to consider the non-maximal parabolic subgroup attached to $\Delta-\{\alpha_1,\alpha_5\}$ and normalizing factors attached to π_v. In this case,

$$\mathbf{A}=\{H_{\alpha_1}(u^2)H_{\alpha_5}(v^2)H_{\alpha_2}(u^2v)H_{\alpha_4}(uv^2)H_{\alpha_3}(u^2v^2)H_{\alpha_6}(uv)\}.$$

Hence we have a rational map $\mathbf{M}\longrightarrow GL_1\times HSpin_8\times GL_1$. Since ${}^LHSpin_8=HSpin_8(\mathbb{C})$, let $\phi: HSpin_8(\mathbb{C})\longrightarrow GSO_8(\mathbb{C})$ be the 2 to 1 map. We can write $\phi(\hat t|_{HSpin_8(\mathbb{C})})=\mathrm{diag}(b_1^2,...,b_4^2,b_0^2b_4^{-2},...,b_0^2b_1^{-2})\in GSO_8(\mathbb{C})$ and we have

$$\chi\circ H_{\alpha_2}=b_3^2b_4^{-2},\quad \chi\circ H_{\alpha_4}=b_3^2b_4^2b_0^{-2},\quad \chi\circ H_{\alpha_3}=b_2^2b_3^{-2},$$
$$\chi\circ H_{\alpha_6}=b_1^2b_2^{-2},\quad \chi(a(u,v))=\eta_{w_1}\eta_{w_2}b_0^2.$$

Then we have $\chi\circ H_{\alpha_1}=\eta_{w_1}b_0^2(b_1b_2b_3b_4^{-1})^{-1}$, $\chi\circ H_{\alpha_5}=\eta_{w_2}b_0^3(b_1b_2b_3b_4)^{-1}$. Hence $m=2$ and

$$\begin{aligned}L(s,\pi_v,r_1)^{-1}&=(1-\eta_{w_2}b_0^{-1}(b_1b_2b_3b_4)q_v^{-s})(1-\eta_{w_2}b_0^3(b_1b_2b_3b_4)^{-1}q_v^{-s})\\&\cdot\prod_{i=1}^4(1-\eta_{w_1}(b_1b_2b_3b_4)b_i^{-2}q_v^{-s})\prod_{i=1}^4(1-\eta_{w_1}b_0^2(b_1b_2b_3b_4)^{-1}b_i^2q_v^{-s})\\&\cdot\prod_{1\le i<j\le 4}(1-\eta_{w_2}b_0(b_1b_2b_3b_4)(b_ib_j)^{-2}q_v^{-s}),\\L(s,\pi_v,r_2)^{-1}&=\prod_{i=1}^4(1-\eta_{w_1}\eta_{w_2}b_0b_i^2q_v^{-s})(1-\eta_{w_1}\eta_{w_2}b_0b_i^{-2}b_0^2q_v^{-s}),\end{aligned}$$

where $q_v=q_{F_v}$.

Let v be inert. Using the map $\phi: HSpin_8(\mathbb{C})\longrightarrow GSO_8(\mathbb{C})$ we can write $\phi(\hat t|_{HSpin_8(\mathbb{C})})=\mathrm{diag}(b_1^2,...,b_4^2,b_0^2b_4^{-2},...,b_0^2b_1^{-2})\rtimes\tau$, where τ is the nontrivial element in $\mathrm{Gal}(E/F)$, and we have

$$\chi\circ H_{\alpha_2}H_{\alpha_4}=b_3^4b_0^{-2},\quad \chi\circ H_{\alpha_3}=b_2^2b_3^{-2},$$
$$\chi\circ H_{\alpha_6}=b_1^2b_2^{-2},\quad \chi(a(t))=\eta_v^2b_0^2.$$

Then $\chi\circ H_{\alpha_1}H_{\alpha_5}=\eta_vb_1^{-2}b_2^{-2}b_3^{-2}b_0^4$. Hence $m=2$ and

$$\begin{aligned}L(s,\pi_v,r_1)^{-1}&=(1-\eta_vb_0^4(b_1b_2b_3)^{-2}q_v^{-2s})(1-\eta_vb_0^{-2}(b_1b_2b_3)^2q_v^{-2s})\\&\cdot\prod_{i=1}^3(1-\eta_vb_0^2(b_1b_2b_3)^{-2}b_i^4q_v^{-2s})\prod_{i=1}^3(1-\eta_v(b_1b_2b_3)^2b_i^{-4}q_v^{-2s}),\\L(s,\pi_v,r_2)^{-1}&=(1-(\eta_v|_{F_v^\times})^2b_0^2q_v^{-2s})\prod_{i=1}^3(1-\eta_v|_{F_v^\times}b_i^2q_v^{-s})(1-\eta_v|_{F_v^\times}b_i^{-2}b_0^2q_v^{-s}).\end{aligned}$$

REMARK 2.6. *Using the Rankin-Selberg integral, W.T. Gan and J. Hundley obtained the L-functions for a cuspidal representation of quasi-split $PSO_8(\mathbb{A}_F)$. The L-functions of (i) and (ii) of Section 1.4 of* [**9**] *agree with the L-functions of this paper. However, the L-function of the case (iii) of Section 1.4 of* [**9**] *(attached to a cubic extension of F) cannot be obtained by the Langlands-Shahidi method.*

2.6.4. ${}^2E_6 - 4$. Let $\theta = \Delta - \{\beta_4\}$. Then $\tilde{\alpha} = e_1 + e_2 + e_3 + e_4 + e_5 + e_6 + 2\epsilon$ and

$$\mathbf{A} = \{H_{\alpha_1}(t)H_{\alpha_5}(t)H_{\alpha_2}(t^2)H_{\alpha_4}(t^2)H_{\alpha_3}(t^3)H_{\alpha_6}(t^2)\}.$$

Let $\mathbf{P} = \mathbf{MN}$. Since $\mathbf{G}$ is simply connected, the derived group $\mathbf{M}_D$ of $\mathbf{M}$ is simply connected, and hence

$$\mathbf{M}_D = SU(3,3)$$

We obtain a map $f : \mathbf{M} \longrightarrow GL_1 \times GU(3,3)$.

Let π be a cuspidal representation of $GU(3,3)(\mathbb{A}_F)$. Let π_v be an unramified representation. If v splits, then then π_v is a representation of $GL_6(F_v) \times F_v^\times$, namely, $\pi_v = \sigma_v \otimes \eta_v$. In this case, by (x) case in [**14**], $m = 2$ and

$$L(s, \pi_v, r_1) = L(s, \sigma_v, \wedge^3 \otimes \eta_v), \quad L(s, \pi_v, r_2) = L(s, \omega_v \eta_v^2).$$

If v is inert, let $\pi_v = \pi(\mu_1, \mu_2, \mu_3, \eta)$, where μ_i's are characters of $E_v^\times$ and η is a character of $F_v^\times$. The central character is $((\mu_1\mu_2\mu_3)|_{F_v^\times})\eta^2$. Let χ be the inducing character of the torus. Then

$$\chi \circ H_{\alpha_1}(t)H_{\alpha_5}(\bar{t}) = \mu_1\mu_2^{-1}(t), \quad \chi \circ H_{\alpha_2}(t)H_{\alpha_4}(\bar{t}) = \mu_2\mu_3^{-1}(t),$$
$$\chi \circ H_{\alpha_3} = \mu_3|_{F_v^\times}, \quad \chi(a(t)) = ((\mu_1\mu_2\mu_3)|_{F_v^\times})\eta^2.$$

From this, we see that $\chi \circ H_{\alpha_6} = \eta$. Hence $m = 2$ and

$$L(s, \pi_v, r_1)^{-1} = (1 - \eta q_v^{-1})(1 - \eta(\mu_1\mu_2\mu_3|_{F_v^\times})q_v^{-1}) \prod_{i=1}^{3}(1 - \eta(\mu_i|_{F_v^\times})q_v^{-1})$$
$$\cdot \prod_{1\le i<j\le 3} (1 - \eta(\mu_i\mu_j|_{F_v^\times})q_v^{-1}) \prod_{1\le i<j\le 3} (1 \pm \eta(\mu_i\mu_j|_{F_v^\times})^{\frac{1}{2}} q_v^{-1})$$
$$\cdot \prod_{i=1}^{3}(1 \pm \eta(\mu_1\mu_2\mu_3|_{F_v^\times})^{\frac{1}{2}} \mu_i|_{F_v^\times} q_v^{-1}),$$

and $L(s, \pi_v, r_2) = L(s, ((\mu_1\mu_2\mu_3)|_{F_v^\times})\eta^2)$. This gives the twisted exterior cube L-function of $U(3,3)$.

2.7. B_n **case.** In [**14**], we only dealt with the $B_n - 1$ case. Here we deal with the general case. Let $\mathbf{G} = Spin_{2n+1}$ be a split spin group as in [**14**]. We define $GSpin_{2n+1}$ to be the maximal Levi subgroup of $Spin_{2n+3}$ which has $Spin_{2n+1}$ as the derived group. More precisely, we add $\alpha_0 = e_0 - e_1$ in the root system and consider the Levi subgroup attached to $\theta = \Delta - \{\alpha_0\}$. Then

$$\mathbf{A} = \{H_{\alpha_0}(t^2)H_{\alpha_1}(t^2)\cdots H_{\alpha_{n-1}}(t^2)H_{\alpha_n}(t) : t \in \overline{F}^\times\},$$

and

$$\mathbf{M}_D = Spin_{2n+1}, \quad \mathbf{A} \cap \mathbf{M}_D = \{H_{\alpha_n}(t) : t^2 = 1\}.$$

We define

$$GSpin_{2n+1} = (GL_1 \times Spin_{2n+1})/(\mathbf{A} \cap \mathbf{M}_D).$$

Note that the center of $\mathbf{G}$ is

$$\mathbf{Z}(G) = \{H_{\alpha_n}(t) : t^2 = 1\}.$$

Since the center of $GSpin_{2n+1}$ is connected, the derived group of ${}^LGSpin_{2n+1}$ is simply connected, and is $Sp_{2n}(\mathbb{C})$. Therefore, ${}^LGSpin_{2n+1} = GSp_{2n}(\mathbb{C})$.

Let $\theta = \Delta - \{\alpha_k\}$. Let $n = k + l$. Let $\mathbf{P} = \mathbf{P}_\theta$ be the parabolic subgroup attached to θ and let $\mathbf{A}$ be the connected component of the center of $\mathbf{M}$. Then

$$\mathbf{A} = \begin{cases} \{H_{\alpha_1}(t)\cdots H_{\alpha_k}(t^k)H_{\alpha_{k+1}}(t^k)\cdots H_{\alpha_{n-1}}(t^k)H_{\alpha_n}(t^{\frac{k}{2}}) : t \in \overline{F}^*\} & \text{if } k \text{ even}, \\ \{H_{\alpha_1}(t^2)\cdots H_{\alpha_k}(t^{2k})H_{\alpha_{k+1}}(t^{2k})\cdots H_{\alpha_{n-2}}(t^{2k})H_{\alpha_n}(t^k) : t \in \overline{F}^*\} & \text{if } k \text{ odd}. \end{cases}$$

Since $\mathbf{G}$ is simply connected, the derived group $\mathbf{M}_D$ of $\mathbf{M}$ is simply connected, and hence $\mathbf{M} = SL_k \times Spin_{2l+1}$. Then

$$\mathbf{M} = GL_1 \times SL_k \times Spin_{2l+1}/(\mathbf{A} \cap \mathbf{M}_D).$$

We can define a map $f : \mathbf{M} \longrightarrow GL_k \times GSpin_{2l+1}$.

Let π_1, π_2 be cuspidal representations of $GL_k(\mathbb{A}_F), GSpin_{2l+1}(\mathbb{A}_F)$ with the central character ω_1, ω_2, resp. Let π be a cuspidal representation of $\mathbf{M}(\mathbb{A}_F)$, induced by f and π_1, π_2. Then the central character of π is

$$\omega_\pi = \begin{cases} \omega_1\omega_2^{\frac{k}{2}}, & \text{if } k \text{ even} \\ \omega_1^2\omega_2^k. & \text{if } k \text{ odd}. \end{cases}$$

Here ${}^LGSpin_{2n+1} = GSp_{2n}(\mathbb{C})$. Let $\hat{t}_1 = \text{diag}(a_1, ..., a_k) \in GL_k(\mathbb{C}) = {}^LGL_k$ and $\hat{t}_2 = \text{diag}(b_1, ..., b_l, b_l^{-1}b_0, ..., b_1^{-1}b_0) \in GSp_{2l}(\mathbb{C})$ be the Satake parameters attached to π_{1v}, π_{2v}, resp. Here we note that

$$\text{diag}(b_1, ..., b_l, b_l^{-1}b_0, ..., b_1^{-1}b_0) \longmapsto b_0$$

generates the character group of GSp_{2n} and hence by Lemma 1.2 of [**14**], $b_0 = \omega_2(\varpi)$. Then

$$\begin{aligned} \chi \circ H_{\alpha_1} &= a_1a_2^{-1}, \ldots, \chi \circ H_{\alpha_{k-1}} = a_{k-1}a_k^{-1}, \\ \chi \circ H_{\alpha_{k+1}} &= b_1b_2^{-1}, \ldots, \chi \circ H_{\alpha_{n-1}} = b_{l-1}b_l^{-1}, \chi \circ H_{\alpha_n} = b_{l-1}b_lb_0^{-1}, \\ \chi(a(t)) = \omega_{\pi_v} &= \begin{cases} (a_1\cdots a_k)(b_0)^{\frac{k}{2}}, & \text{if } k \text{ even} \\ (a_1\cdots a_k)^2(b_0)^k, & \text{if } k \text{ odd}. \end{cases} \end{aligned}$$

Since $f(H_{\alpha_k}(t)) = (\text{diag}(1, ..., 1, t), b(t))$, we can see $\chi \circ H_{\alpha_k} = a_kb_1^{-1}b_0$. Hence, we see that $m = 2$,

$$\begin{aligned} L(s, \pi, r_1) &= L(s, \pi_1 \times \pi_2), \\ L(s, \pi, r_2) &= L(s, \pi_1, Sym^2 \otimes \omega_2). \end{aligned}$$

REMARK 2.7. *By a low-dimensional accident, $GSpin_5 \simeq GSp_4$. Using this isomorphism, we obtain the spin L-function of cuspidal representations of GSp_4 and also the degree 8 L-function of cuspidal representations of $GL_2 \times GSp_4$. These L-functions have been studied extensively by the Rankin-Selberg method.*

2.8. C_n case. In order to demonstrate that we do not obtain any new L-functions by considering similitude groups, we calculate L-functions for $\mathbf{G} = GSp_{2n}$; The element of the torus is written as

$$t = t(u_1, ..., u_n, u_0) = \mathrm{diag}(u_1, ..., u_n, u_n^{-1}u_0, ..., u_1^{-1}u_0).$$

Let $e_i(t) = u_i$ for $i = 0, ..., n$. Then the simple roots are $\alpha_1 = e_1 - e_2, ..., \alpha_{n-1} = e_{n-1} - e_n, \alpha_n = 2e_n - e_0$. The corresponding coroots are

$$\alpha_1^\vee(u) = t(u, u^{-1}, 1, ..., 1, 1), ..., \alpha_{n-1}^\vee = t(1, ..., 1, u, u^{-1}, 1), \alpha_n^\vee = t(1, ..., 1, u, 1).$$

Any character χ of $T(F_v)$ is given by $\chi = \chi(\eta_1, ..., \eta_n, \eta_0)$ so that

$$\chi(\eta_1, ..., \eta_n, \eta_0)(t(u_1, ..., u_n, u_0)) = \eta_1(u_1) \cdots \eta_n(u_n)\eta_0(u_0).$$

Let $\mathbf{M} = GL_k \times GSp_{2l}$, $k + l = n$. Then $\tilde{\alpha} = e_1 + \cdots + e_k - e_0$. Let π_1, π_2 be cuspidal representations of GL_k, GSp_{2l} with the central characters ω_1, ω_2, resp. Let $\pi = \pi_1 \otimes \pi_2$. Let $\pi_{1v} = \pi(\mu_1, ..., \mu_k), \pi_{2v} = \pi(\nu_1, ..., \nu_l, \nu_0)$. Then $m = 2$ and

$$L(s, \pi_v, r_1) = L(s, \pi_{1v} \times \pi'_{2v}),$$
$$L(s, \pi_v, r_2) = L(s, \pi_{1v}, \wedge^2),$$

except when $k = 1$. When $k = 1$, the second L-function does not occur. Here π'_2 is any irreducible constituent of $\pi_2|_{Sp_{2n}(\mathbb{A})}$.

2.9. General spinor groups. We construct several maps between general spinor groups and GSO_{2n}. The center of $Spin_{2n}$ is $\{1, c, z, cz\}$ where

$$c = H_{\alpha_{n-1}}(-1)H_{\alpha_n}(-1),$$

and

$$z = \begin{cases} \prod_{i=1}^{n-2} H_{\alpha_i}((-1)^i)H_{\alpha_{n-1}}(-1) & \text{if n is even,} \\ \prod_{i=1}^{n-2} H_{\alpha_i}((-1)^i)H_{\alpha_{n-1}}(-\sqrt{-1})H_{\alpha_n}(\sqrt{-1}) & \text{if n is odd.} \end{cases}$$

Following [**14**], we define $HSpin_{2n}$ to be

$$HSpin_{2n} = \frac{\mathrm{GL}_1 \times Spin_{2n}}{\{(1, 1), (\sqrt{-1}, z), (-1, c), (-\sqrt{-1}, zc)\}}.$$

By the definition we see that there is a natural 2 to 1 map

$$GSpin_{2n} \to HSpin_{2n} \simeq \frac{GSpin_{2n}}{\{(1, 1), (\sqrt{-1}, z)\}}.$$

Note that

$$GSO_{2n} = \frac{GL_1 \times SO_{2n}}{\{(1, 1), (-1, z)\}} \simeq \frac{\mathrm{GL}_1 \times Spin_{2n}}{\{(1, 1), (-1, z), (1, c), (-1, zc)\}}.$$

The map $GL_1 \times Spin_{2n} \to GL_1 \times Spin_{2n}, (t, x) \mapsto (t^2, x)$ induces a 2 to 1 map $HSpin_{2n} \to GSO_{2n}$. The kernel of the homomorphism $GSpin_{2n} \to GSO_{2n}$ is easy to compute, and so we have the following proposition.

PROPOSITION 2.8. *There are natural 2 to 1 maps $GSpin_{2n} \to HSpin_{2n}$ and $HSpin_{2n} \to GSO_{2n}$ as described above. The composite of the two maps, $GSpin_{2n} \to GSO_{2n}$, is a 4 to 1 map and it has kernel $\{(1, 1), (1, c), (\sqrt{-1}, z), (\sqrt{-1}, cz)\}$.*

REMARK 2.9. *In the literature, $GSpin_{2n}$ is defined by Clifford algebras. It is usually referred as Clifford group, e.g., see §20.2 of* [**8**]. *Namely, let V be a vector space of dim $2n$ with a symmetric bilinear form Q. Then one can construct Clifford algebras $C(Q)$ and $C^+(Q)$. If $V = W \oplus W'$, where $\dim W = \dim W' = n$, then $C^+(Q) \simeq \mathrm{End}(\wedge^{\mathrm{even}} W) \oplus \mathrm{End}(\wedge^{\mathrm{odd}} W)$. Let $x \mapsto x^*$ be the anti-involution such that $(v_1 \cdots v_r)^* = (-1)^r v_r \cdots v_1$ for $v_1, ..., v_r \in V$. Then $GSpin(Q) = \{x \in C^+(Q) |\, xVx^* \subset V\}$. Then using the fact that $v \cdot v = -v \cdot v^* = Q(v, v)$ for $v \in V$, we can see that the map $\rho : GSpin(Q) \longrightarrow GSO(Q)$, $\rho(x)(v) = x \cdot v \cdot x^*$, is a homomorphism. If Q is attached to the identity matrix, the kernel is*

$$\{\pm 1, \pm\sqrt{-1} e_1 \cdot e_2 \cdots e_{2n-1} \cdot e_{2n}\}.$$

Note that $Spin(Q) = \{x \in C^+(Q) |\, xVx^* \subset V, x \cdot x^* = 1\}$. The fact that this is equivalent to our definition has been pointed out by S. Kudla.

3. Proof of a Conjecture of Shahidi

We recall Conjecture 7.1 of [**27**] which we call Shahidi's conjecture:

CONJECTURE (Shahidi's conjecture). *Let $\pi = \otimes_v \pi_v$ be a generic cuspidal representation of $\mathbf{M}(\mathbb{A}_F)$. If π_v is tempered then each $L(s, \pi_v, r_i)$ is holomorphic for* $\mathrm{Re}(s) > 0$.

In many cases this conjecture is proved to be true. First of all, if F_v is an archimedean local field, then Shahidi's conjecture is true since the L-functions and ϵ-factors are Artin factors. From now on we assume that F_v is nonarchimedean. Casselman and Shahidi prove the conjecture for quasi-split classical groups in [**7**]. For simply connected groups of type B_n or D_n the conjecture is proved by Asgari [**3**] in the split case and by the second named author [**17**] in the quasi-split case. The first named author settled Shahidi's conjecture for split exceptional groups except for a few cases (cf. Theorem 3.16 of [**14**]).

We want to prove Shahidi's conjecture for quasi-split groups. To prepare we state several general results concerning the holomorphy of L-functions [**27**].

PROPOSITION 3.1. *Let π_v be a generic tempered representation of $M = \mathbf{M}(F_v)$.*
(1) If $m = 1$ then $L(s, \pi_v, r_1)$ is holomorphic for $\mathrm{Re}(s) > 0$.
(2) If $m = 2$ and $L(s, \pi_v, r_2) = \prod_j (1 - \alpha_j q^{-s})^{-1}$ for $|\alpha_j| = 1, \alpha_j \in \mathbb{C}$, then $L(s, \pi_v, r_1)$ is holomorphic for $\mathrm{Re}(s) > 0$. *In particular, if r_2 is one-dimensional, then $L(s, \pi_v, r_1)$ is holomorphic for* $\mathrm{Re}(s) > 0$.

The multiplicativity of γ-factors plays an important role in studying the holomorphy of L-functions and the properties of intertwining operators. The general theory is explained in [**26, 27**] and we use the notation of [**14**] for consistency.

PROPOSITION 3.2. *Let π_v be an irreducible admissible generic representation of $M = \mathbf{M}(F_v)$ such that $\pi_v \subset \mathrm{Ind}_{M_\theta N_\theta}(\sigma \otimes 1)$, where σ is an irreducible generic admissible representation of $M_\theta = \mathbf{M}_\theta(F)$. For each $j \in S_i$, let $\gamma(s, \overline{w}_j(\sigma), r_{i(j)}, \psi)$, $1 \leq i \leq m$, be the corresponding γ-factors. Then*

$$\gamma(s, \pi_v, r_i, \psi) = \prod_{j \in S_i} \gamma(s, \overline{w}_j(\sigma), r_{i(j)}, \psi).$$

Furthermore, if Shahidi's conjecture is true for each $L(s, \overline{w}_j(\sigma), r_{i(j)})$, then we have the corresponding multiplicativity for L-functions.

Let $\phi_v : W_{F_v} \times SL_2(\mathbb{C}) \to {}^L M$ be the parametrization of π_v. Then there is a corresponding multiplicativity for Artin γ-factors of ϕ_v as explained in page 280 of [**26**]. Since Artin L-functions satisfy holomorphy we obtain the following result.

PROPOSITION 3.3. *Let π_v and σ be as in Proposition 3.2, and suppose π_v is tempered. If each $\gamma(s, \overline{w}_j(\sigma), r_{i(j)}, \psi)$ is an Artin factor then so are $\gamma(s, \pi_v, r_i, \psi)$ and $L(s, \pi_v, r_i)$. In this case, $L(s, \pi_v, r_i)$ is holomorphic for* $\mathrm{Re}(s) > 0$.

It is expected that $\gamma(s, \pi_v, r_i, \psi)$ and $L(s, \pi_v, r_i)$ are Artin factors for any tempered representation π_v. However, we do not know this yet except for a few cases. For example, Shahidi proves that his L-functions are Artin L-functions for $GL_n \times GL_m \subset GL_{n+m}$. It follows from Proposition 3.3 that Shahidi's conjecture is true for tempered representations of $GL_n(F_v) \times GL_m(F_v) \subset GL_{n+m}(F_v)$. Using the local-global argument and the base change we prove another case in the following lemma.

LEMMA 3.4. *Let E/F be a quadratic extension of number fields. Fix a finite place v of F, and suppose w is a unique place of E lying over v with $[E_w : F_v] = 2$. Let σ be a tempered representation of $GL_{n-1}(F_v)$ and μ be a unitary character of* $\mathrm{Res}_{E/F} GL_1(F_v) = GL_1(E_w) = E_w^\times$ *so that $\sigma \otimes \mu$ is a tempered representation of* $\mathbf{M}(F_v) = GL_{n-1}(F_v) \times \mathrm{Res}_{E/F} GL_1(F_v) (\subset GSpin_{2n}^-(F_v)$ *in* ${}^2D_n - 3$ *case). Then $L(s, \sigma \otimes \mu, r_1) = L(s, \sigma_{E,v} \otimes \mu)$ is an Artin L-function, where $\sigma_{E,v}$ is the base change of σ.*

PROOF. By multiplicativity (cf. Proposition 3.2), it is enough to prove it for a supercuspidal representation σ. Let π be a cuspidal representation of $GL_{n-1}(\mathbb{A}_F)$, and χ be a grössencharacter of E such that $\pi_v = \sigma$, $\chi_w = \mu$ and $\pi_{v'}, \chi_{w'}$ are unramified for all finite places $v' \neq v$ of F and finite places $w' \neq w$ of E. Let π_E be a cuspidal representation of $GL_{n-1}(\mathbb{A}_E)$ obtained by the base change of π (cf. See I.6 and III.1-5 of [**2**]). Now we have functional equations

$$\begin{aligned} L(s, \pi \otimes \chi, r_1) &= \epsilon(s, \pi \otimes \chi) L(1 - s, \pi \otimes \chi, r_1) \\ L(s, \pi_E \otimes \chi) &= \epsilon(s, \pi_E \otimes \chi) L(1 - s, \pi_E \otimes \chi) \end{aligned}$$

Since $L(s, \pi_{v'} \otimes \chi_{v'}, r_1) = L(s, (\pi_E)_{v'} \otimes \chi_{v'})$ for all $v' \neq v$ (cf. Section 2.3.3), we see that

$$\gamma(s, \sigma \otimes \mu, r_1, \psi) = \gamma(s, \sigma_{E,v} \otimes \mu, \psi).$$

The fact that $\sigma_{E,v}$ is tempered (Theorem 6.2, (a) in [**2**]) implies the corresponding equation of L-functions

$$L(s, \sigma \otimes \mu, r_1) = L(s, \sigma_{E,v} \otimes \mu)$$

and so the L-function is an Artin L-function. □

We now prove Shahidi's conjecture by case-by-case analysis for a p-adic field F_v. For simplicity of notation we drop the notation v. For example, we write F for F_v and π for a tempered representation π_v of $\mathbf{M}(F_v)$, and so F does not mean the number field but a p-adic field. Furthermore, we assume that v is non-split since the split cases are considered in [**14**].

3.1. ^2D_n case. Let $\mathbf{G} = GSpin_{2n}^-$, and let E/F be a quadratic extension over which $GSpin_{2n}^-$ splits. This case is proved in [**17**] where a detailed proof was given for $^2D_n - 1$ but the proof of $^2D_n - 2$ and $^2D_n - 3$ was sketchy there. So we treat these two cases in detail. First, note that $^2D_n - 3$ case follows from Lemma 3.4 and Proposition 3.3.

Now we consider the $^2D_n - 2$ case. Let π_1, π_2 be tempered representations of $GL_{n-2}(F), GL_2(E)$, respectively so that $\pi = \pi_1 \otimes \pi_2$ is a tempered representation of $\mathbf{M}(F) = GL_{n-2}(F) \times \mathrm{Res}_{E/F} GL_2(F)$. We may assume that π_2 is a discrete series. Then π_2 is given as a unique subrepresentation of $\mathrm{Ind}(\mu|\ |_E^{1/2} \otimes \mu|\ |_E^{-1/2})$, where μ is a character of $GL_1(E)$, and we have

$$\gamma(s, \pi_1 \otimes \pi_2, r_1, \psi) = \gamma(s+1, \pi_1 \otimes \mu, r_1, \psi)\gamma(s-1, \pi_1 \otimes \mu, r_1, \psi).$$

The γ-factors on the right are Artin factors and so $\gamma(s, \pi_1 \otimes \pi_2, r_1, \psi)$ is also an Artin factor. Now holomorphy of $L(s, \pi_1 \otimes \pi_2, r_1)$ for $\mathrm{Re}(s) > 0$ follows from by Proposition 3.3.

3.2. $^3D_4$ case. Let E/F be a cubic extension over which $Spin_8^-$ splits.

3.2.1. $^3D_4 - 1$. Since r_2 is one-dimensional we apply Proposition 3.1 (2).

3.2.2. $^3D_4 - 2$. Let σ be a tempered representation of $GL_2(F)$, and let η be a character of $GL_1(E) = E^\times$. Let π be a tempered representation of $\mathbf{M}(F)$ obtained from σ and η via the F-rational map f. The local-global argument (used in the proof of Lemma 3.4) with L-function computation in 2.4.2 shows that

$$L(s, \pi, r_1) = L(s, \sigma_E \otimes \eta),$$

where σ_E is the base change of σ to E. Now we note that $L(s, \sigma_E \otimes \eta)$ is holomorphic for $\mathrm{Re}(s) > 0$.

3.3. $^6D_4$. The proof is similar to that of $^3D_4$.

3.4. $^2E_6$ case. Let E/F be a quadratic extension over which $^2E_6$ splits.

3.4.1. $^2E_6 - 1$. Let π_1, π_2 be tempered representations of $GL_3(E)$,$GL_2(F)$ respectively. Let π be a tempered representation of $\mathbf{M}(F)$ obtained from π_1 and π_2 via the F-rational map f. By multiplicativity (Proposition 3.2) we assume that π is a discrete series. Since Shahidi's conjecture is true for supercuspidal representations we may assume one of π_1 or π_2 is not supercuspidal. If π_1 is given as a unique square integrable subrepresentation of $\mathrm{Ind}(|\ |_E^{1/2}\chi \otimes \chi \otimes |\ |_E^{-1/2}\chi)$, where χ is a character of $GL_1(E)$, and π_2 is supercuspidal, then

$$\gamma(s, \pi, r_1, \psi) = \prod_{i=0}^{2} \gamma(s+i-1, \chi \times \widetilde{\pi}_2, \psi) \prod_{i=0}^{2} \gamma(s+i-1, \chi^2 \times \widetilde{\pi}_2, \psi)$$

After cancellation we see that the corresponding L-function is

$$L(s, \pi, r_1) = L(s+1, \chi \times \widetilde{\pi}_2) L(s+1, \chi^2 \times \widetilde{\pi}_2)$$

which is holomorphic for $\mathrm{Re}(s) > 0$. If $\pi_1 \hookrightarrow \mathrm{Ind}(|\ |_E^{1/2}\chi \otimes \chi \otimes |\ |_E^{-1/2}\chi)$ and $\pi_2 \hookrightarrow \mathrm{Ind}(|\ |_F^{1/2}\mu \otimes |\ |_F^{-1/2}\mu)$, where μ is a character of $GL_1(F)$, then

$$L(s, \pi, r_1) = L(s+3/2, \chi \times \mu^{-1}) L(s+3/2, \chi^2 \times \mu^{-1})$$

which is holomorphic for $\mathrm{Re}(s) > 0$. Finally, if π_1 is supercuspidal and $\pi_2 \hookrightarrow \mathrm{Ind}(|\ |_F^{1/2}\mu \otimes |\ |_F^{-1/2}\mu)$, then

$$L(s, \pi, r_1) = L(s + 1/2, \pi_1 \times \mu^{-1})$$

which is holomorphic for $\mathrm{Re}(s) > 0$.

3.4.2. ${}^2E_6 - 2$. Let π_1, π_2 be tempered representations of $GL_2(E)$,$GL_3(F)$ respectively. Let π be a tempered representation of $\mathbf{M}(F)$ obtained from π_1 and π_2 via the F-rational map f. The local-global argument with unramified computation in 2.6.2 shows that

$$L(s, \pi, r_1) = L(s, \pi_1 \times \pi_{2,E}),$$

where $\pi_{2,E}$ is the base change of π_2 to E. The proof follows from the fact that $L(s, \pi_1 \times \pi_{2,E})$ is holomorphic for $\mathrm{Re}(s) > 0$.

3.4.3. ${}^2E_6 - 3$. Let σ be a tempered representation of $HSpin_8^-(F)$, and let η be a character of $GL_1(E)$. Let π be a tempered representation of $\mathbf{M}(F)$ obtained from σ and η via the F-rational map f as in 2.6.3. We may assume that π is a discrete series since any tempered representation is unitarily induced from a discrete series representation. If σ is supercuspidal, then Shahidi's conjecture is true by Proposition 7.3 of [**27**].

If σ is a unique square integrable subrepresentation of $\mathrm{Ind}(|\ |_F^e\rho \otimes \sigma_0)$, where ρ is a unitary character of $GL_1(F)$ and σ_0 is a unitary supercuspidal representation of $HSpin_6^-(F)$, then $\rho \simeq \widetilde{\rho}$ and $2e \in \mathbb{Z}, e > 0$ by Casselman's square integrability criterion [**6**]. Then

$$\gamma(s, \pi, r_1, \psi) = \gamma(s+e, \eta\times\widetilde{\rho}, \psi)\gamma(s, \eta\times\widetilde{\rho}, \psi)\gamma(s-e, \eta\times\widetilde{\rho}, \psi)\gamma(s+e, \rho\otimes(\sigma_0\otimes\eta), r_1, \psi)$$

where r_1 on the right is for ${}^2D_4 - 1$. By Corollary 7.6 of [**27**], we see that $e = 1/2, 1$ and

$$\gamma(s+e, \eta\times\widetilde{\rho}, \psi)\gamma(s, \eta\times\widetilde{\rho}, \psi)\gamma(s-e, \eta\times\widetilde{\rho}, \psi) = \begin{cases} \frac{L(2-s,\widetilde{\eta}\times\widetilde{\rho})}{L(s+1,\eta\times\rho)} & \text{if } e = 1 \\ \frac{L(1-s,\widetilde{\eta}\times\widetilde{\rho})}{L(s,\eta\times\rho)} \cdot \frac{L(\frac{3}{2}-s,\widetilde{\eta}\times\widetilde{\rho})}{L(s+\frac{1}{2},\eta\times\rho)} & \text{if } e = \frac{1}{2}. \end{cases}$$

From this observation we see that $L(s, \pi, r_1)$ is holomorphic for $\mathrm{Re}(s) > 0$.

If σ is a unique square integrable subrepresentation of $\mathrm{Ind}(|\det|_F^e\rho\otimes\sigma_0)$, where ρ is a unitary supercuspidal representation of $GL_2(F)$ and σ_0 is a unitary supercuspidal representation of $HSpin_4^-(F)$, then $\rho \simeq \widetilde{\rho}$ and $2e \in \mathbb{Z}$ by Casselman's square integrability criterion [**6**]. Then

$$\gamma(s, \pi, r_1, \psi) = \gamma(s + e, \rho \otimes (\sigma_0 \otimes \eta), r_1, \psi)\gamma(s - e, \rho \otimes (\sigma_0 \otimes \eta), r_1, \psi)$$

where r_1 on the right is for ${}^2D_4 - 2$. Note that $\gamma(s \pm e, \rho \otimes (\sigma_0 \otimes \eta), r_1, \psi)$ are Artin γ-factors (whose proof is similar to that of Lemma 3.4). Now apply Proposition 3.3.

If σ is a unique square integrable representation of $\mathrm{Ind}(|\det|_F^e\rho\otimes\sigma_0)$, where ρ is a unitary supercuspidal representation of $GL_3(F)$ and σ_0 is a unitary supercuspidal representation of $HSpin_2^-(F)$, then $\rho \simeq \widetilde{\rho}$ and $2e \in \mathbb{Z}$ by Casselman's square integrability criterion [**6**]. Then

$$\begin{aligned} \gamma(s, \pi, r_1, \psi) &= \gamma(s + e, \rho \otimes (\sigma_0 \otimes \eta), r_1, \psi)\gamma(s - e, \rho \otimes (\sigma_0 \otimes \eta), r_1, \psi) \\ &\times \gamma(s + e, \rho \times \widetilde{\eta}, \psi)\gamma(s - e, \rho \times \widetilde{\eta}, \psi) \end{aligned}$$

where r_1 on the right is for ${}^2D_4 - 3$. The factors on the right are Artin factors and so we get holomorphy of L-functions. The remaining cases can be handled similarly by applying multiplicativity in a suitable way and noting the factors are Artin factors.

3.4.4. ${}^2E_6 - 4$. Since r_2 is one-dimensional we apply Proposition 3.1 (2).

THEOREM 3.5. *Let* π *be a generic tempered representation of* $\mathbf{M}(F)$. *Then* $L(s, \pi, r_1)$ *is homomorphic for* $\mathrm{Re}(s) > 0$ *except possibly for the cases:* $E_7 - 3, E_8 - 3, E_8 - 4$ *and (xxviii) of* [**20**].

4. Proof of Assumption (A)

We recall the following assertion which is called Assumption (A), e.g., [**14**].

ASSUMPTION (A). *Let* $\pi = \otimes_v \pi_v$ *be a generic cuspidal representation of* $\mathbf{M}(\mathbb{A}_F)$. *Then the normalized intertwining operator* $N(s, \pi_v, w_0)$ *is holomorphic and non-zero for* $\mathrm{Re}(s) \geq \frac{1}{2}$ *for every* v.

The first named author has developed a quite general machinery of proving Assumption (A) through several papers [**12, 13, 14**]. For the sake of completeness, we recall several results in the following.

PROPOSITION 4.1. *Let* π_v *be a tempered, generic representation of* $\mathbf{M}(F_v)$ *for which Shahidi's conjecture is true. Then* $N(s, \pi_v, w_0)$ *is holomorphic and non-zero for* $\mathrm{Re}(s) \geq 0$.

LEMMA 4.2. *Let* π_v *be a generic tempered representation of* $\mathbf{M}(F_v)$ *which is a subrepresentation of* $I(\Lambda, \rho)$ *where* ρ *is a supercuspidal representation. If* $\langle \Lambda, \beta^\vee \rangle$ *is a half-integer for each positive root* β, *then* $N(s, \pi_v, w_0)$ *is holomorphic and non-zero for* $\mathrm{Re}(s) > -\frac{1}{2m}$, *where* m *is the number of irreducible constituents of the adjoint action of* ${}^LM$ *on* ${}^L\mathfrak{n}$ *as in Section 1.*

Proposition 4.1 and Lemma 4.2 are direct generalizations of Lemma 4.2 and Lemma 4.3 of [**14**] for quasi-split groups. In [**14**] the statements are given for split groups, but the proofs are general and work for quasi-split groups, too.

In the special rank-one case of $GL_k \times GL_l \subset GL_{k+l}$, we have the following well-known result (cf. Proposition I.10 of [**24**] or Lemma 2.10 of [**13**]):

LEMMA 4.3. *Let* σ *and* τ *be tempered representations of* $GL_k(F_v)$ *and* $GL_l(F_v)$, *respectively. Then* $N(s, \sigma \otimes \tau, w_0)$ *is holomorphic and non-zero for* $\mathrm{Re}(s) > -1$.

The next result concerns non-vanishing of the normalized intertwining operator $N(s, \pi_v, w_0)$ which is now known as Zhang's lemma (Lemma 1.7 of [**13**] or the proof of Theorem 3 of [**31**]).

LEMMA 4.4. *Let* π_0 *is an irreducible, generic tempered representation of* M *and let* $I(\Lambda, \pi_0)$ *be the induced representation. Assume Shahidi's conjecture for each rank-one situation. If* $N(\Lambda, \pi_0, w_0)$ *is holomorphic at* Λ_0 *then it is nonzero at* Λ_0.

In the study of the normalized intertwining operators it is important to know when a representation is a fully induced representation. To be specific, let π_v be an irreducible admissible representation of $\mathbf{M}(F_v)$. By Langlands' classification theorem (Chapter IV, Theorem 4.11 and Chapter XI, Theorem 2.10 of [**5**]), there exists Langlands' data $(\mathbf{P}_0, \Lambda_0, \pi_0)$ such that $\pi_v = J(\Lambda_0, \pi_0)$ where $\mathbf{P}_0 = \mathbf{M}_0\mathbf{N}_0$ is a parabolic subgroup of $\mathbf{M}$, Λ_0 is a complex parameter in the positive Weyl chamber, and π_0 is a tempered representation of $\mathbf{M}_0(F_v)$. In general, the Langlands' quotient

$J(\Lambda_0, \pi_0)$ is a quotient of $I(\Lambda_0, \pi_0)$. However, if π_v is generic then it is expected that $J(\Lambda_0, \pi_0) = I(\Lambda_0, \pi_0)$.

CONJECTURE (Standard module conjecture). *Let π_v be a non-tempered representation of $\mathbf{M}(F_v)$. If π_v is generic, then there is a tempered representation π_0 of $\mathbf{M}_0(F_v)$ and a complex parameter Λ_0 in the positive Weyl chamber such that*

$$\pi_v = I(\Lambda_0, \pi_0) = \mathrm{Ind}_{\mathbf{M}_0(F_v)}^{\mathbf{M}(F_v)}\big(\pi_0 \otimes q_v^{\langle \Lambda_0, H_{P_0}^M(\cdot)\rangle}\big).$$

This conjecture is proved for many cases. For archimedean places, it is due to Vogan [**29**]. For nonarchimedean cases, it is proved in [**7**] for supercuspidal representations, and G. Muić settled many cases for classical groups [**22, 23**]. Recently, V. Heiermann and G. Muić proved the following fundamental result [**10**].

THEOREM 4.5. *If the local coefficients $C_\psi(s, \pi, w_0)$ attached to (M, π) (cf. [**27**]) are regular in the negative Weyl chamber, then the standard module conjecture is true. In particular, Shahidi's conjecture implies the standard module conjecture.*

Next we recall that the analogue of Assumption (A) is true for $\mathrm{Re}(s) \geq 1$ provided that Shahidi's conjecture is true (Proposition 4.9 of [**14**]). Another fundamental result concerning normalized intertwining operators is the cocycle relation of intertwining operators (cf. [**1, 27**]). The following version is taken from Theorem 1 of [**31**].

PROPOSITION 4.6. *Let π be an irreducible admissible generic representation of $\mathbf{M}_\theta(F_v)$ where $\theta \subset \Delta$. If $w_1\theta, w_2w_1\theta \subset \Delta$, then*

$$N(\Lambda, \pi_v, w_2w_1) = N(w_1\Lambda, w_1\pi_v, w_2)N(\Lambda, \pi_v, w_1).$$

Now we assume that π_v is generic and non-tempered. We will show that $N(s, \pi_v, w_0)$ is holomorphic and non-zero for $\mathrm{Re}(s) \geq \frac{1}{2}$ under the following three assumptions (i),(ii) and (iii) to be specified below:

(i) Shahidi's conjecture and (hence) the standard module conjecture.
(ii) The assumption of Lemma 4.2 (half integer condition).

From now on we assume (i) and (ii). By the standard module conjecture, we may write $\pi_v = I(\Lambda_0, \pi_0)$ for a generic tempered representation π_0, and

$$I(s, \pi_v) = I(s\tilde{\alpha} + \Lambda_0, \pi_0).$$

Then we have

$$N(s, \pi_v, w_0) = N(s\tilde{\alpha} + \Lambda, \pi_0, w')|_{I(s,\pi_v)}.$$

Thus it is enough to prove that $N(s\tilde{\alpha} + \Lambda, \pi_0, w')$ is holomorphic and non-zero for $\mathrm{Re}(s) \geq \frac{1}{2}$. By Zhang's lemma (Lemma 4.4), all we have to do is to prove $N(s\tilde{\alpha}+\Lambda_0, \pi_0, w')$ is holomorphic for $\mathrm{Re}(s) \geq \frac{1}{2}$. By Proposition 4.6, we know that $N(s\tilde{\alpha} + \Lambda_0, \pi_0, w')$ is a product of rank-one operators whose complex parameters are of the form $\langle s\tilde{\alpha} + \Lambda_0, \beta^\vee\rangle$ for positive roots β by identifying roots $\psi(\mathbf{G}, \mathbf{A}_\theta)$ with roots $\psi(\mathbf{G}, \mathbf{A}_\emptyset)$ where $\mathbf{A}_\emptyset$ is the maximal split torus. Further we assume that

$$\text{(iii)} \begin{cases} \mathrm{Re}(\langle s\tilde{\alpha} + \Lambda_0, \beta^\vee\rangle) > -1 & \text{if the rank-one situation is } GL_k \times GL_l \subset GL_{k+l} \\ \mathrm{Re}(\langle s\tilde{\alpha} + \Lambda_0, \beta^\vee\rangle) > -\frac{1}{2m} & \text{other rank-one situation} \end{cases}$$

for all positive roots β. By Lemma 4.2 and 4.3, we see that $N(s\tilde{\alpha} + \Lambda_0, \pi_0, w')$ is a product of rank-one operators, each of which is holomorphic for $\mathrm{Re}(s) \geq \frac{1}{2}$. This completes the proof of Assumption (A) under the assumptions (i),(ii) and (iii).

We shall prove Assumption (A) for ${}^2A_n, {}^2D_n, {}^3D_4$ and ${}^2E_6$. First, we note that assumptions (i) and (ii) are fulfilled for these cases. In fact, we proved Shahidi's conjecture for these cases in Section 3, and (partial) classifications of tempered (or square integrable) representations for $GL_n(F), \mathbf{U}(F), GSpin_{2n}(F)$ (cf. [**17, 19, 28**]) imply assumption (ii). Therefore, the proof of assumption (iii) is the only issue for these cases. For simplicity of notation, we drop the notation v of place and write π for π_v and F for F_v. In view of the discussion above, we may assume that all the parameters are real, in particular, we assume that s is real. With this notational convention we prove that $N(s, \pi, w_0)$ is holomorphic and non-zero for $s \geq \frac{1}{2}$ by proving the assumption (iii).

4.1. Unitary groups. Let $\mathbf{U}$ be either $\mathbf{U}(n, n)$ or $\mathbf{U}(n, n+1)$, and let E/F be the quadratic extension which defines the unitary group $\mathbf{U}$. Denote by $\Delta = \{\beta_1, ..., \beta_n\}$ the set of simple roots of $\mathbf{U}$. Any Levi subgroup $\mathbf{M}$ of $\mathbf{U}$ is of the form

$$\mathbf{M} \simeq \mathrm{Res}_{E/F} GL_{n_1} \times \cdots \times \mathrm{Res}_{E/F} GL_{n_k} \times \mathbf{U}'$$

where $\mathbf{U}'$ is a unitary group of the same type of smaller rank.

According to [**19**], any generic unitary representation of π of $\mathbf{U}(F)$ is of the form

$$\pi = \mathrm{Ind} |\det|_E^{r_1} \sigma_1 \otimes \cdots \otimes |\det|_E^{r_k} \sigma_k \otimes \tau$$

where $0 < r_k \leq \cdots \leq r_1 < 1$, σ_i are discrete series of $GL_{n_i}(E)$ and τ is a generic tempered representation of $\mathbf{U}'(F)$. Then we may write

$$\Lambda_0 = s_1 E_1 + \cdots + s_n E_n$$

where $0 \leq s_n \leq \cdots \leq s_1 < 1$. Here $E_1, ..., E_n$ is the standard basis for $\mathbb{R}^n$ so that $\beta_i = E_i - E_{i+1}$ for $1 \leq i \leq n-1$ and

$$\beta_n = \begin{cases} 2E_n & \text{if } \mathbf{U} = U(n, n), \\ E_n & \text{if } \mathbf{U} = U(n, n+1). \end{cases}$$

Hence we can argue and prove Assumption (A) in this case as we do for classical groups. More precisely, if $\mathbf{U} = \mathbf{U}(n, n)$ then the (restricted) root system is of type C_n and thus the case is identical to that of Sp_{2n}. If $\mathbf{U} = \mathbf{U}(n, n+1)$ then the (restricted) root system is of type BC_n and thus the case is similar to that of SO_{2n+1} (cf. [**13**]).

4.2. D_n case. The root system of ${}^2D_{2n}$ is of type B_{n-1} and so the proof of Assumption (A) is similar to that of classical groups of type B_{n-1} (cf. [**13**]).

For completeness, we include a proof for the ${}^3D_4$ case which is similar to that of type G_2. In this case $[E : F] = 3$, and we consider parameters with respect to $|\ |_E$.

4.2.1. ${}^3D_4 - 1$. In this case,

$$\tilde{\alpha} = e_1 + e_2 = 3\beta_1 + 2\beta_2.$$

Let π be a representation of $GL_2(E) = \mathrm{Res}_{E/F} GL_2(F)$. Then we may write $\Lambda_0 = r\beta_1, 0 \leq r < \frac{3}{2}$. The rank one situations are those for $GL_l(E) \times GL_k(E) \subset GL_{l+k}(E)$ and the least value of $\langle s\tilde{\alpha} + \Lambda_0, \beta^\vee \rangle$ is $s - r > -1$ if $s \geq \frac{1}{2}$.

4.2.2. ${}^3D_4 - 2$. In this case,

$$\tilde{\alpha} = \frac{1}{3}(e_1 + e_2) = 2\beta_1 + \beta_2.$$

Let π be a representation of $GL_2(F)$. Then we may write $\Lambda_0 = r\beta_2, 0 \le r < \frac{1}{2}$. The rank one situations are those for $GL_l(E) \times GL_k(E) \subset GL_{l+k}(E)$ and the least value of $\langle s\tilde{\alpha} + \Lambda_0, \beta^\vee \rangle$ is $s - 3r > -1$ if $s \ge \frac{1}{2}$.

4.3. ${}^2E_6$ case. In this case $[E : F] = 2$, and we consider parameters with respect to $| \ |_E$. Since the (restricted) root system ${}^2E_6$ is of type F_4, the proof of Assumption (A) is similar to that of the split F_4 case.

4.3.1. ${}^2E_6 - 1$. In this case

$$\tilde{\alpha} = 2(e_1 + e_2 + e_3) + e_4 + e_5 + e_6 + 3\epsilon = 4\beta_1 + 8\beta_2 + 6\beta_3 + 3\beta_4$$

Let π_1 be a representation of $GL_3(E)$ and π_2 a representation of $GL_2(F)$. We write $\Lambda_0 = r_1\beta_1 + r_1\beta_2 + r_2\beta_4$, where $0 \le r_1 < 1$ and $0 \le r_2 < \frac{1}{2}$, and

$$s\tilde{\alpha} + \Lambda_0 = (4s + r_1)\beta_1 + (8s + r_2)\beta_2 + 6s\beta_3 + (3s + r_2)\beta_4.$$

The rank one operators are for $GL_l(E) \times GL_k(E) \subset GL_{l+k}(E)$ and the least value of $\langle s\tilde{\alpha} + \Lambda_0, \beta^\vee \rangle$ is $s - (r_1 + r_2) > -1$ if $s \ge \frac{1}{2}$.

4.3.2. ${}^2E_6 - 2$. In this case

$$\tilde{\alpha} = 3(e_1 + e_2) + 2(e_3 + e_4) + e_5 + e_6 + 3\epsilon = 3\beta_1 + 6\beta_2 + 4\beta_3 + 2\beta_4$$

Let π_1 be a representation of $GL_2(E)$ and π_2 a representation of $GL_3(F)$. We write $\Lambda_0 = r_1\beta_1 + r_2\beta_3 + r_2\beta_4$, where $0 \le r_1 < 1$ and $0 \le r_2 < \frac{1}{2}$, and

$$s\tilde{\alpha} + \Lambda_0 = (3s + r_1)\beta_1 + 6s\beta_2 + (4s + r_2)\beta_3 + (2s + r_2)\beta_4.$$

The rank one operators are for $GL_l(E) \times GL_k(E) \subset GL_{l+k}(E)$ and the least value of $\langle s\tilde{\alpha} + \Lambda_0, \beta^\vee \rangle$ is $s - (r_1 + r_2) > -1$ if $s \ge \frac{1}{2}$.

4.3.3. ${}^2E_6 - 3$. In this case

$$\tilde{\alpha} = 2e_1 + e_2 + e_3 + e_4 + e_5 + 2\epsilon = 2\beta_1 + 3\beta_2 + 2\beta_3 + \beta_4$$

Let π_1 be a character of $GL_1(E)$ and π_2 a representation of $HSpin_8^-(F)$. Since there is a map $GSpin_8^- \to HSpin_8^-$ we may consider π_2 as a representation of $GSpin_2^-(F)$. Let E_0, E_1, E_2, E_3 be the standard basis for $\mathbb{R}^4$ so that $\beta_2 = E_1 - E_2, \beta_3 = E_2 - E_3, \beta_4 = E_3$ are the simple roots of $GSpin_8^-$. Using the standard module conjecture for $GSpin_8^-$ (and hence for $HSpin_8^-$), we may write $\Lambda_0 = a_1E_1 + a_2E_2 + a_3E_3 = a_1\beta_2 + (a_1 + a_2)\beta_3 + (a_1 + a_2 + a_3)\beta_4$, where $0 \le a_3 \le a_2 \le a_1 < \frac{1}{2}$. In terms of β_i's,

$$s\tilde{\alpha} + \Lambda_0 = 2s\beta_1 + (3s + a_1)\beta_2 + (2s + a_1 + a_2)\beta_3 + (s + a_1 + a_2 + a_3)\beta_4.$$

The least value of $\langle s\tilde{\alpha} + \Lambda_0, \beta^\vee \rangle$ is $s - (a_1 + a_2 + a_3) > -1$ if $s \ge \frac{1}{2}$.

4.3.4. ${}^2E_6 - 4$. This case is dual to ${}^2E_6 - 3$, and the argument is similar.

We summarize the progress of Assumption (A). Since we have the standard module conjecture available for $B_n - 1, D_n - 1$, (xxx) in [**20**], and (xxxii) in [**20**], we have Assumption (A) for these cases. On the other hand, the long-sought functorial lift from classical groups to GL_N is proved in [**11**]. This result provides the necessary ingredient in the proof of Assumption (A) for (xviii)($SO_6 \to GL_6$),(xxii),(xxiv)($SO_{10} \to GL_{10}$) which was not available at the time of [**14**]. In conclusion, we have the following theorem.

THEOREM 4.7. *Assumption (A) holds except possibly for the cases E_7-3, (xxvi) of* [**20**], E_8-3, E_8-4, *and (xxviii) of* [**20**].

REMARK 4.8. *The difficulties of the unsettled cases of Assumption (A) lie in the complicated nature of Levi subgroups in the exceptional group cases, and the lack of the (partial) classifications of discrete representations of those Levi subgroups. For example, in the cases of* E_7-3, E_8-3, E_8-4, *and (xxviii) of* [**20**], *it is hard to apply multiplicativity to prove Shahidi's conjecture, which is a key ingredient of Assumption (A) due to the complicated nature of Levi subgroups.*

5. Correction to [14]

The calculations of the Langlands L-groups of certain Levi subgroups are incorrect. Sections 2.5.4, 2.6.3, 2.6.6, 2.7.3 and 2.7.7 need revision. However, this does not affect the results of the paper [**14**].

In Section 2.5.4, we have the maximal Levi subgroup $\mathbf{M}$, denoted by $HSpin_{10}$, which corresponds to the simple root $\{\alpha_1\}$ of the exceptional group of simply connected type E_6^{sc}. The Langlands L-group ${}^LM$ is not $GSpin_{10}(\mathbb{C})$. In fact, ${}^LM = \mathbf{M}(\mathbb{C})$. To see this, let E_6^{ad} be the exceptional group of the adjoint type. Then $E_6^{ad} = E_6^{sc}/S$, where S is the center of E_6^{sc}, and it has order 3. Note that ${}^LE_6^{sc} = E_6^{ad}(\mathbb{C})$, and ${}^LM$ is the maximal Levi subgroup of $E_6^{ad}(\mathbb{C})$. Since $\mathbf{A} \cap \mathbf{M}_D$ has order 4, we can see that ${}^LM = \mathbf{M}(\mathbb{C})$. Hence given a generic cuspidal representation π of $GSO_{10}(\mathbb{A}_F)$, we can consider it as a cuspidal representation of $HSpin_{10}(\mathbb{A}_F)$ via the 2 to 1 map $HSpin_{10}(\mathbb{A}_F) \to GSO_{10}(\mathbb{A}_F)$ of Proposition 2.8, and obtain the degree 16 spin L-function in Section 2.5.4.

In Section 2.6.2 and 2.7.3, $GSpin_{10}(\mathbb{C})$ should be replaced by ${}^LHSpin_{10}(\mathbb{C}) = HSpin_{10}(\mathbb{C})$ and the map $GSpin_{10} \to GSO_{10}$ is not a 2 to 1 map but a 4 to 1 map (cf. See Proposition 2.8). These corrections do not affect L-function computations there.

In Section 2.6.6, ${}^LM$ is not $GSpin_{12}(\mathbb{C})$. Let E_7^{ad} be the exceptional group of the adjoint type. Then $E_7^{ad} = E_7^{sc}/S$, where S is the center of E_7^{sc}, and it has order 2. More explicitly, $S = \{H_{\alpha_1}(t)H_{\alpha_3}(t)H_{\alpha_7}(t) : t^2 = 1\}$. Let $\mathbf{M}' \simeq \mathbf{M}/S$. Then ${}^LM = \mathbf{M}'(\mathbb{C})$. The derived group of ${}^LM$ is $HS(12, \mathbb{C})$ (the half-spin group in Section 2.3.4). Let $c = H_{\alpha_5}(-1)H_{\alpha_7}(-1)$, $z = H_{\alpha_1}(-1)H_{\alpha_3}(-1)H_{\alpha_5}(-1)$. Then $\mathbf{M} = GL_1 \times Spin_{12}/\{(1,1),(-1,z)\}$. Hence we have a 2 to 1 map $f : \mathbf{M} \longrightarrow GSO_{12}$. (cf. Proposition 2.8).

Let $\pi' = \otimes_v \pi'_v$ be a generic cuspidal representation of $GSO_{12}(\mathbb{A}_F)$. Let π'_v be a spherical representation of $GSO_{12}(F_v)$ with the corresponding semi-simple conjugacy class $\hat{t} = e_0^*(b_0^2)e_1^*(b_1^2)\cdots e_6^*(b_6^2)$ in $\hat{T}(\mathbb{C})$, the torus in ${}^LGSO_{12}(\mathbb{C}) = GSpin_{2n}(\mathbb{C})$, where $e_i^* : GL_1(\mathbb{C}) \to \hat{T}(\mathbb{C})$ are the standard cocharacters (cf. [**4**]). Now let π be a cuspidal representation of $\mathbf{M}(\mathbb{A}_F)$, induced by π' and the 2 to 1 map f. The Satake parameter of π_v is ${}^Lf(\hat{t})$ in ${}^LM$. The rest of the calculations are correct. Since ${}^LM$ is complicated, we are not able to write down explicitly the L-functions of cuspidal representations of $\mathbf{M}(\mathbb{A}_F)$ which do not come from $GSO_{12}(\mathbb{A}_F)$.

Finally, let E_8^{ad} be the exceptional group of the adjoint type. Then $E_8^{ad} = E_8^{sc}$. So ${}^LE_8^{sc} = E_8^{sc}(\mathbb{C})$. Hence in this case, for any maximal Levi subgroup $\mathbf{M}$, ${}^LM = \mathbf{M}(\mathbb{C})$. In particular, ${}^LM = M(\mathbb{C}) = HSpin_{14}(\mathbb{C})$ for $\mathbf{M} = HSpin_{14}$ in Section 2.7.7. The computation of L-functions remains true.

References

1. J. Arthur, *Intertwining operators and residues I, Weighted characters*, J. Funct. Anal. **84** (1989), 19–84.
2. J. Arthur and L. Clozel, *Simple algebras, base change, and the advanced theory of the trace formula*, Annals of Mathematics studies, Princeton University press, Princeton, 1989.
3. M. Asgari, *Local L-functions for split spinor groups*, Canad. J. Math. **54** (2002), no. 4, 673–693.
4. M. Asgari and F. Shahidi, *General transfer for general spin groups*, Duke Math J., **132** (2006), 137-190.
5. A. Borel and N. Wallach, *Continuous cohomology, discrete groups and representation of reductive groups*, Annals of Mathematics studies, Princeton University press, Princeton, 1980.
6. W. Casselman *Introduction to the theory of admissible representations of p-adic reductive groups*, Preprint.
7. W. Casselman and F. Shahidi, *On irreducibility of standard modules for generic representations*, Ann. Sci. École Norm. Sup. **31** (1998), 561–589.
8. W. Fulton and J. Harris, *Representation theory, a first course*, Springer-Verlag, 1991.
9. W. T. Gan and J. Hundley, *The spin L-function of quasi-split D_4*, IMRN (2006), 1–74.
10. V. Heiermann and G. Muić, *On the standard module conjecture*, Math. Z. **255** (2007), 847-853.
11. I. Piatetski-Shapiro J. Cogdell, H. Kim and F. Shahidi, *Functoriality for classical groups*, Publ. Math. Inst. Hautes Études Sci. **99** (2004), 163–233.
12. H. Kim, *Langlands-Shahidi method and poles of automorphic L-functions: application to exterior square L-functions*, Canad. J. Math. **51** (1999), 835–849.
13. H. Kim, *Langlands-Shahidi method and poles of automorphic L-functions II*, Israel. J. Math. **117** (2000), 261–284.
14. H. Kim, *On local L-functions and normalized intertwining operators*, Canad. J. Math. **57** (2005), no. 3, 535–597.
15. H. Kim and M. Krishnamurthy, *Twisted exterior square lift from $GU(2,2)_{E/F}$ to GL_6/F*, J. Ramanujan Math. Soc, **23** (2008), 381–412.
16. H. Kim and F. Shahidi, *Functorial products for $GL_2 \times GL_3$ and symmetric cube for GL_2*, Ann. of Math. **155** (2002), no. 3, 837–893.
17. W. Kim, *Square integrable representations and the standard module conjecture for general spin groups*, Canad. J. Math., **61** (2009), 617–640.
18. M. Krishnamurthy, *The Asai transfer to* GL_4 *via the Langlands-Shahidi method*, IMRN 2003, 2221-2254.
19. G. Muić, E. Lapid and M. Tadić, *On the generic unitary dual of quasisplit classical groups*, Ann. Sci. École Norm. Sup. **31** (1998), 561–589.
20. R. P. Langlands, *Euler products*, Yale University Press, 1971.
21. R. P. Langlands, *Base change for $GL(2)$*, Princeton University Press, 1980.
22. G. Muić, *Some results on square integrable representations; irreducibility of standard representations*, Internat. Math. Res. Notices **14** (1998), 705–726.
23. ———, *A proof of Casselman-Shahidi's conjecture for quasi-split classical groups*, Canad. Math. Bull. **44** (2001), no. 3, 298–312.
24. C. Moeglin and J.L. Waldspurger, *Le spectre résiduel de $GL(n)$*, Ann. Sci. École Norm. Sup. **22** (1989), 605–674.
25. F. Shahidi, *On the Ramanujan conjecture and finiteness of poles for certain L-functions*, Ann. of Math. **127** (1988), 547–584.
26. ———, *On the multiplicativity of local factors*, Festschrift in honor of I.I. Piatetski-Shapiro on the occasion of his sixtieth birthday (S. Gelbart, R. Howe and P. Sarnak, eds.), Israel Math. Conf. Proc., vol. 3, 1990, pp. 279–289.
27. ———, *A proof of Langlands' conjecture on Plancherel measures; complementary series for p-adic groups*, Ann. of Math **132(2)** (1990), 273–330.
28. M. Tadić, *Classification of unitary representations in irreducible representations of general linear group (non-Archimedean case)*, Ann. Sci. Ec. Norm. Sup. **19** (1986), 335–382.
29. D. Vogan, *Gelfand-Kirillov dimension for Harish-Chandra modules*, Invent. Math. **48** (1978), 75–98.
30. A.L. Onishshik V.V., Gorbatsevich and E.B. Vinberg, *Lie groups and lie algebras III*, Encyclopedia of Mathematical Sciences, vol. 41, Springer-Verlag, 1990.

31. Y. Zhang, *The holomorphy and nonvanishing of normalized local intertwining operators*, Pacific J. Math. **180** (1997), 3855–398.

Department of Mathematics, University of Toronto, Toronto, ON M5S 2E4, Canada, and Korea Institute for Advanced Study, Seoul, Korea
E-mail address: henrykim@math.toronto.edu

Department of Mathematics, Yonsei University, 134 Shinchondong, Seodaemungu, Seoul, Korea and Korea Institute for Advanced Study, Seoul, Korea
E-mail address: wkim@kias.re.kr

Clay Mathematics Proceedings
Volume 13, 2011

Reflexions on Receiving the Shaw Prize

Robert P. Langlands

ABSTRACT. As its title indicates this is the text for a lecture delivered in Hong Kong in September, 2007 on the occasion of the receipt of the Shaw Prize in Mathematics. It will be published together with an autobiographical essay by the Shaw Foundation in the Shaw Prize Book, but for the sake of wider circulation among specialists it is also reproduced here with the kind permission of the Shaw Foundation. There is a good deal to be said for further discussion of many of the points made in the text, but that will require a much more mature understanding of the mathematical issues. I hope that, for the moment, the lecture is of some value as it stands.

To receive the Shaw Prize is of course a great honor, but it was also an occasion to discover, or to be reminded, that a number of mathematicians have a perception of the development of the theory of automorphic forms over the last four decades that differs from mine if not in a radical, certainly in an essential way. Some of the differences are a result of misapprehensions that are a natural consequence of the variety of the theory's relations to fields practiced by mathematicians with many different temperaments and training. With a little explanation these misapprehensions can be dissipated. The prize is an opportunity to do so. Others are the result of conflicting methodological stances, mostly unrecognized and certainly unresolved. Their resolution will certainly demand a deeper understanding of the subject than is yet available. In this lecture I attempt to describe the current, unresolved situation. My emphasis will be on my own stance, although my purpose here is not to advocate but to explain it

My own views are best explained with reference to the accompanying diagram, in which there are five circles of different sizes, the sizes reflecting nothing more than the space the associated fields of mathematics occupy in my own mind. The upper left-hand corner is the analytic theory of automorphic forms, a theory that came into prominence in the fifties and sixties, as the legacy of mathematicians like Erich Hecke, C. L. Siegel, Atle Selberg and, as it became more and more appropriate to employ the language of infinite-dimensional representations, Harish-Chandra. It is an analytic theory. In the mid-sixties, as a young mathematician there were several serious questions that I tried to broach, not all in this area and for most of them with little success. With two I was lucky, simultaneously

2010 *Mathematics Subject Classification.* Primary: 22-53, Secondary: 17-53, 11F66, 11F70, 11F80, 22F55.

and as a result of my own earlier work on the general theory of Eisenstein series, basically the study of the spectra of specific commuting families of differential operators on certain noncompact Riemannian manifolds. The spectra are highly structured and their qualitative properties difficult to establish. To my surprise, their study ultimately led to a conjectural response to two of the questions or problems: the definition of a natural family of analytically – at least potentially – tractable L-functions associated to automorphic forms and the possible structure of a nonabelian class field theory. The second came immediately after the first, more the result of inspiration than of effort.

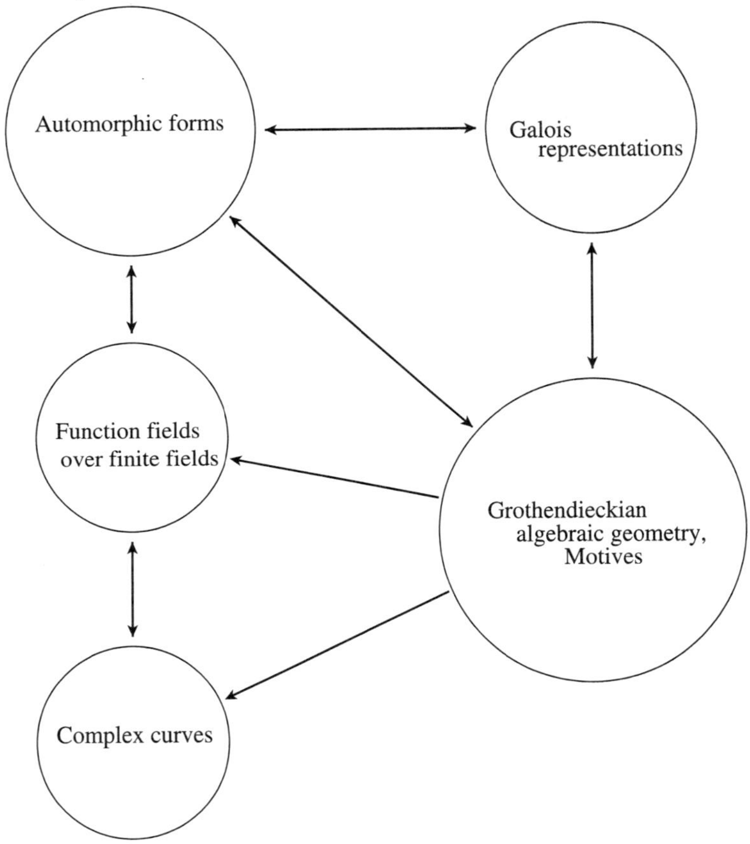

I recall here that not long before, in the proceedings of a mathematical conference celebrating the second centenary of Princeton University in 1956, Artin had suggested that such a theory might not exist, or at least might not contain any new elements. So I may well have been the only one who was searching for it in the 1960's.

The suggested answer took the form of a construction and a conjecture. The basic object in the theory of automorphic forms is, today, an automorphic representation of the adelic points $G(\mathbb{A}_F)$ of a reductive algebraic group G over the algebraic number field F, all objects that need not be defined here ([**1**]). For many expository purposes, the representation can be replaced by an element of the function space on which it acts. If, in addition, the group is taken to be $GL(2)$, and, for simplicity, the adeles replaced by the real numbers, this element is often just a classical elliptic modular form. This simplification entails, however, a real possibility of misunderstanding the import of the construction and the conjecture.

The first step in the construction is to attach to G a complex algebraic group ${}^LG$, or better ${}^LG_K$, usually referred to as the L-group. K is a sufficiently large finite Galois extension of the ground field F, itself a finite-dimensional extension of $\mathbb{Q}$ at the time of the group's initial introduction. The L-group has a connected component $\hat{G}$ of the same dimension as G and its group of connected components comes with an isomorphism with $\mathrm{Gal}(K/F)$. So there is an exact sequence

$$1 \to \hat{G} \to {}^LG_K \to \mathrm{Gal}(K/F) \to 1.$$

The second step is to attach to each automorphic representation π and to each finite-dimensional algebraic representation r of ${}^LG$ an L-function defined by an Euler product, at first partial,

$$L_S(s, \pi, r) = \prod_{v \notin S} L(s, \pi_v, r). \tag{1}$$

The set S is a finite set of places of F containing all infinite places, and $L(s, \pi, r)$ has the form

$$\frac{1}{\det\left(I - \frac{r(A(\pi_v))}{\mathrm{Nm}\,\mathfrak{p}_v^s}\right)},$$

where $\{A_{\pi_v} = A_v(\pi)\}$ is a conjugacy class in ${}^LG_K$ attached to π or its local representative π_v. These products converge in a half-plane. Of course, the L-functions introduced by Hecke, and more generally by H. Maaß, for $GL(2)$ were the source of the impulse to search for such general L-functions.

The definition of the L-functions (1) was inspired by the general theory of Eisenstein series, for it was there that a substantial number of them emerged and could be continued to the whole complex plane. The first problem that presents itself is the continuation of all of them, not just as meromorphic functions but as meromorphic functions with a very limited number of poles. If G is $GL(n)$ and $r = r_0$ the standard representation of $GL(n)$ it was pretty clear that this could be done, using ideas already proposed, as I recall, in their first form by T. Tamagawa ([**2**]). The final theory was developed by Godement-Jacquet.

Artin's proof of the analytic continuation of abelian Artin L-functions came quickly to my mind and a conjecture simple to state presented itself immediately with great force. Suppose H and G are two groups over F and ϕ is a homomorphism $\phi : {}^LH_K \to {}^LG_K$ compatible with the projections onto the Galois group. Then for any automorphic representation π_H of $H(\mathbb{A}_F)$ there is an automorphic representation $\pi_G = \phi(\pi_H)$ of $G(\mathbb{A}_F)$ such that $\{A_v(\pi_G)\} = \{\phi(A_v(\pi_H))\}$ for almost all v. The informed reader will notice that for simplicity all problems related to L-packets have been passed over in silence.

It is immediately clear that this conjecture is already deep and pregnant with consequence even for $H = \{1\}$ and $G = GL(n)$. For suppose ρ is a representation of the Galois group $\mathrm{Gal}(K/F)$ in $GL(n, \mathbb{C})$. Then taking advantage of the freedom in the choice of K – an inevitable consequence of the initial freedom in the choice of G – we take ${}^LH = \mathrm{Gal}(K/F)$, ${}^LG = GL(n) \times \mathrm{Gal}(K/F)$, $\phi(\sigma) = \sigma \times \rho(\sigma)$, π_H

the unique one-dimensional representation of the trivial group $H(\mathbb{A}_F) = \{1\}$ and $\pi_G = \phi(\pi_H)$, and conclude that

$$L(s,\rho) = L(s,\pi_H,\rho) = L(s,\pi_G,r_0'), \tag{2}$$

r_0' being the product of the standard representation of $GL(n,\mathbb{C})$ with the trivial representation of $\mathrm{Gal}(K/F)$. As a consequence of (2) and the Tamagawa-Godement-Jacquet theory for $GL(n)$, $L(s,\rho)$ can be extended to the entire complex plane.

The general conjecture that $\phi(\pi_G)$ always exists I began after some time to call *functoriality*. I was amazed by it at the time and remain so today. It has, I believe, to be regarded as a striking historical fact that the solution – still itself in large part conjectural, but no longer entirely – to the Artin conjecture (for the first of the very few available cases, see [**3**]) appeared as part of a much larger conjecture with implications of a much broader compass. To deny this context and this historical origin by referring to the conjectured existence of the π_G attached to ρ as in (2) as the strong Artin conjecture seems to me wrong-headed. It lends an unmerited legitimacy to clearly limited methods. The denial can charitably be ascribed to ignorance and a fear of the analytic theory of automorphic forms.

For number theorists in the 1960's and subsequent decades, Galois cohomology and elliptic curves were much more intensively cultivated than algebraic number theory as such. Legions of practitioners were produced in these domains for whom, by and large, the analytic theory of automorphic forms, especially nonabelian harmonic analysis, was anathema. The use by A. Wiles of some simple cases of functoriality that could be proven by such means in the proof of the Shimura-Taniyama conjecture and therefore of Fermat's theorem was at first simply overlooked ([**4**]). Even now that it has been generally noticed, there is among many number theorists a reluctance to accept the imbrication of number theory and other domains entailed by a systematic reference to functoriality and nonabelian harmonic analysis and a failure to recognize the possibilities that this offers.

Once the general conjecture was formulated, the first order of business was to examine its simpler consequences and to verify in so far as possible that they could be proved or were compatible with what was then known. There were also over the years some accretions to the original conjecture. I would now be inclined to add to the conjectured existence of $\phi(\pi_H)$ just described a second one and to label the two together *functoriality*. Functoriality as such applies to all automorphic representations, even to those that, like most of the representations associated to Maaß forms, probably have no strictly diophantine significance.[1]

There are some fine points concerning the second conjecture for which I would hesitate to lay my hand in the fire and that I pass over in silence here, but I describe it nonetheless because something like it has certainly to be proved in any theory that aspires to completeness. To describe it, I have to assume a notion adumbrated by Arthur ([**5**]) that would be a consequence of any complete theory of the trace formula, namely the notion of *Ramanujan type* for an automorphic representation π, essentially the type for which the Ramanujan conjecture would be true. Functoriality offers of course the possibility of proving the Ramanujan conjecture for these representations, which will be in the majority, and of disproving

[1]Peter Sarnak observed to me that this view is too narrow and referred me, in particular, to the work of Cogdell, Piatetski-Shapiro and himself on the number of representations of integers by ternary quadratic forms.

it for the rest. If π is of Ramanujan type, the critical strip for $L(s, \pi, r)$ will have the same significance as for Dirichlet L-functions, thus lie between $\Re s = 0$ and $\Re s = 1$. Moreover the order $m(\pi, r)$ of the pole of $L(s, \pi, r)$ at $s = 1$ will be greater than or equal to 0. Call π thick if $m(\pi, r)$ is always equal to the number of times the trivial representation of ${}^L G$ is contained in r. The second conjecture is that for any $\pi = \pi_G$ there always exists an H, a thick π_H and a $\phi : {}^L H \to {}^L G$ such that $\pi_G = \phi(\pi_H)$. For a thick π the distribution of the conjugacy classes $\{A_v(\pi)\}$ would, basically by definition, be given by the usual Weyl distribution on conjugacy classes of ${}^L H$.

So functoriality contains a very general form of the Sato-Tate conjecture. Here, in contrast to any work on the Artin conjecture, the Sato-Tate conjecture was formulated before functoriality. So there are historically sound reasons for singling it out. Its early formulation is, like that of the Taniyama-Shimura conjecture, no doubt a reflection of the strong early interest in elliptic curves and their zeta-functions.

The two conjectures of functoriality are in themselves related to Artin's conjecture, largely through their application to the trivial group $H = \{1\}$, but, as formulated here, their purely arithmetic content is otherwise still limited. Not only do they have a validity extending beyond those automorphic forms strictly related to diophantine problems but also there is not yet in them any reference to diophantine problems for varieties of dimension greater than zero, for example no reference to the Taniyama-Shimura conjecture.

A good deal of work has been done on functoriality by F. Shahidi, I. Piatetski-Shapiro, and others without any pretense that the methods would ever offer the ultimate insights, but which, in my view, was nevertheless of great importance because it persuaded many analytic number theorists of the relevance of functoriality to their problems ([**6**]). This is, in some sense, quite separate from any interest that functoriality may have as a tool for more purely diophantine problems. The trace formula was developed – in higher dimensions created – by J. Arthur and used as a tool by him and many others in the treatment of specific cases of functoriality, largely those accessible to endoscopy, especially twisted endoscopy. The book [**7**] will be a valuable introduction to the results of many years of effort.

The techniques referred to in the field as *endoscopy* had, however, from the beginning an obvious and important limitation. They could provide cases of functoriality that have been widely used and in quite different contexts, base change or the Jacquet-Langlands correspondence in various guises, but functoriality in general was not within their range. At the same time, there was a severe technical difficulty that caused me, and others, to despair: the *fundamental lemma*. It was a simply stated general combinatorial lemma and I expected that as such I would be able to prove it with time. Matters turned out quite differently.

Endoscopy, a feature of nonabelian harmonic analysis on reductive groups over local or global fields, arose implicitly in a number of contexts, in its twisted form both implicitly in the early work of Saito-Shintani on what was later called base change, and somewhat more explicitly in suggestions of – I believe – Jacquet for functoriality from orthogonal groups or symplectic groups to $GL(n)$. It arose for me in the context of the trace formula and Shimura varieties.

Over the years a number of my students were introduced to the fundamental lemma and its difficulties, especially, R. Kottwitz, J. Rogawski and T. Hales. Some went on, as is well known, to quite different things, but Kottwitz continued to reflect not only on it, but also on Shimura varieties and the number-theoretical difficulties attached to them and on applications of the trace formula. It was he, in the beginning alone and then later together with M. Goresky and R. MacPherson, who first had some genuine insight into the topological nature of the lemma.

In the hands of J-L. Waldspurger, G. Laumon and most recently B. C. Ngo, the lemma and the associated problems took on quite different features. Notice that, in the diagram under the large circle in the upper left-hand corner, there are two slightly smaller circles, the size reflecting, as I observed, my own predilections. These are theories that were inspired by the theory for automorphic forms over number fields: first of automorphic forms over the second examples of global fields, namely function fields over finite fields, and then in the very lowest circle over the complex numbers. By the time we arrive at this third circle the theory has quite a different flavor. Two names associated to the second circle are V. Drinfeld and L. Lafforgue ([**8**]). There is a whole school, strongly influenced by Drinfeld and largely a Russo-American school, associated to the third circle.

The fundamental lemma is a local lemma, over p-adic fields. The recognition informing recent work is, first, that to prove it over a p-adic field it is enough to prove it for the second type of local fields, fields of Laurent series over finite fields, and secondly that to prove it over such fields it is best to work not with local orbital integrals but with the corresponding global objects as they appear in the trace formula. The first step is far from easy but was taken by Waldspurger in an important paper; for the second we pass naturally from the first of the three circles on the left of the diagram to the second.

Before passing to the third, I have to indulge in a good deal of somewhat reckless speculation, but I am growing old and the need to correct false impressions is growing more urgent. I may no longer have enough time to pursue any insight slowly to the point of genuine understanding and conviction. So, in the face of what seem to me the serious misunderstandings that have emerged, I must take my chances and state my case without delay as clearly as I can. The reader is warned that prudence is expected of him. He will have to take a great deal of what follows with a grain of salt until he has reflected on it himself.

I have been troubled for years and often discouraged by my failure, indeed by the general failure, to broach functoriality in any decisive way. Not so long ago, I suggested a different approach to the question with which I began to amuse myself ([**9**],[**10**]) but it was all very tentative. At the same time, I resolved to learn more in general about the various researches referred to often in a blanket way by the catch phrase*Langlands program*, a phrase that can mean many things.

I also had occasion to listen to lectures of Ngo (supplemented by the report of J-F. Dat ([**11**])) and to try to understand them. In particular, I had to attach for myself some meaning to the notion of stack and algebraic stack. It was a revelation. I discovered that I had been thinking for decades of orbital integrals in an incorrect way. I had separated the local from a global part. With the notion of stack, with the suppleness of the etale cohomology, the two parts are, over global fields of the second kind, thus over function fields, to be fused and regarded as yielding the number of points on a stack, a number that can be calculated cohomologically. The problems encountered in [**9**],[**10**] suddenly appeared in an entirely new light. I shall try to explain this, although I am still dealing with concepts that I may have misunderstood.

In [**9**] a tentative method for approaching functoriality by taking limits in various trace formulas over an appropriate sequence of functions was introduced. The global field was taken to be $\mathbb{Q}$, the group $GL(2)$. The following difference, written in a notation that is not quite the same as that of [**9**], was encountered

$$\frac{\sum_{p<X} \ln(p)\theta_m(p)}{X^{m/2+1}} - c_m X^{m/2}. \tag{3}$$

I define neither the constant c_m nor the expression $\theta_m(p)$. The question of whether the limit of this difference exists as $X \to \infty$ was discussed, but inconclusively. In [**10**], I passed to the rational function field over a finite field with q elements. Then the sum (3) is replaced by

$$\frac{\sum_{\deg \mathfrak{p}=n} n\theta_m(\mathfrak{p})}{q^{nm/2+n}} - c'_m q^{nm/2}. \tag{4}$$

The limit is to be taken for a fixed m but with $n \to \infty$ and the constant has changed. There would be something similar for the function fields of curves of positive genus. The divisor $\mathfrak{p}$ is here prime.

We write (4) as

$$\frac{\sum_{\deg \mathfrak{p}=n} n\theta_m(\mathfrak{p}) - c'_m q^{mn+n}}{q^{nm/2+n}}. \tag{5}$$

We need to show that this expression has a limit as n approaches infinity. The first term of the numerator is a fused orbital integral, and thus can – I suppose – be calculated cohomologically. Thus, the dimension of the associated stack being $mn+n$, it will be of the form

$$\sum_{k=0}^{2(mn+n)} (-1)^k \sum_{j=1}^{d_k} \gamma_{j,k} q^{k/2},$$

where $|\gamma_{j,k}| = 1$. The kth term is the contribution of the cohomology with compact support in degree k, thus of the cohomology in degree $2(mn+n)-k$. So what is necessary is to show that after the cohomology in degree 0 or at least very small degrees, which will just contribute the term $c'_m q^{mn+n}$, there is no cohomology in positive dimension less than (approximately) the intermediate dimension $mn+n$ and that the dimension of the cohomology in all degrees can be bounded. Then the cohomology in degrees around $mn+n$ can contribute to the limit, and the cohomology in higher degrees will contribute 0 because of the denominator.

All this looks far-fetched. It is suggested by a simple phenomenon, first described to me by N. Katz, that is discussed in [**10**]. In the naive reflexions of that paper, the stack is replaced by the moduli space of hyperelliptic curves of some large genus, thus by the space of monic polynomials of a given degree with distinct roots. This space has cohomology over $\mathbb{Q}$ only in degrees 0 and 1. It is an Eilenberg-MacLane space for a braid group, itself fairly closely related to congruence subgroups. For congruence subgroups, the phenomenon of concentration of cohomology in only a few dimensions, in particular those around the middle dimension, seems to have presented itself in other contexts ([**12**]), but all this is still very new to me.

I have, of course, passed rather glibly from function fields over finite fields to ordinary topology. This is the passage from the second circle on the left to the third. For vanishing theorems, this is perfectly natural because there are comparison theorems between etale cohomology and other cohomologies or between etale cohomology for a variety (or for stacks!) and its reductions. Moreover, it is quite likely that as the theory over complex curves progresses, the stacks that appear for orbital integrals when we are examining the trace formula over function fields over finite fields will, as a variant of the stacks $\mathcal{H}ecke_\lambda$ of E. Frenkel's report on recent advances [**8**], also appear there.

If so, a gratifying unity will appear. Functoriality, as in the first circle, is to this point in this presentation largely analytic, the only link to algebraic number theory being the Artin L-functions. In both geometric forms of the theory, the reciprocity, both local and global, between Galois representations or representations of the fundamental group on the one hand and, on the other, automorphic forms for curves over finite fields, or $\mathcal{D}$-modules and perverse sheaves over the complex field, is the focus of attention. In these two cases, the functoriality is a consequence of the reciprocity. Over number fields, functoriality is, as I have stressed, also applicable to automorphic representations for which there is no reciprocal Galois representation and there is no real sign that it can be deduced in any generality except from the trace formula. The possibility that the topological study of the varieties (or stacks) appearing in the purely geometric theory will be pertinent to the trace formula is appealing.

There will be at least two major problems. The cohomology of braid groups is difficult and not well understood. That of the stacks $\mathcal{H}ecke_\lambda$ and their variants may be even more challenging. In addition, even if this strategy works, it is limited at first to global fields associated to curves over finite fields. On the other hand, a well-defined technique with a well-defined structure that was successful for the trace formula over function fields would certainly stimulate the search for related techniques over number fields. It is apparent from [**10**] that the difficulties, even for function fields, are related to the behaviour of class numbers, so that it is not impossible that questions like those raised by the heuristics of Cohen-Lenstra ([**13**]) will be relevant when we turn to number fields. I expect, however, that for number fields there will be very large, still unforeseeable difficulties that will make great demands on the inventive powers of analytic number theorists.

We could continue down the circles on the left to the last of the three and examine its relation to various aspects of ordinary differential equations or to conformal field theory, but that is not, so far as I know, where the misunderstandings lie. They lie largely in a failure to appreciate the autonomous merit of functoriality, but also in a misapprehension of its relation to motives and to Galois representations.

I myself am inclined to regard the Galois representations as instruments, and the central relation between the left and the right sides of the diagram as the diagonal arrow between automorphic forms and motives, not the horizontal and vertical arrow passing through Galois representations. What the diagonal arrow provides, as in the proof of Fermat's last theorem, is passage from a context where a given, critical assertion is difficult, even impossible, to one in which it is almost transparent. It may, for example, not be possible to prove directly that there is no free-standing elliptic curve with various constraints on its ramification, but when the curve or an isogenous curve is assumed to be contained in the jacobian of a modular variety the same conclusion can be immediate.

Grothendieck appears to have been grievously disappointed when his cherished notion of motives and the theorems needed to establish it turned out to be unnecessary for the proof of the last of the Weil conjectures. Perhaps he could have drawn a different conclusion. As explained in N. Katz's report ([**14**]), the last of the Weil conjectures was proven, by Deligne, in essence on the basis of a profound understanding of the etale cohomology theory accompanied by an observation arising in the theory of automorphic forms, namely that Ramanujan's conjecture, in its original or in its generalized forms, is an immediate consequence of functoriality and the resulting knowledge of the analytic properties of the family of all L-functions associated to the corresponding automorphic form or representation. In the context of the Weil conjectures, there are only the Galois representations, where functoriality is almost formal, and so no need for unproved assertions, just for a complete mastery of the etale cohomology theory. The conclusion to be drawn from this might have been that the theory of motives will have to be founded simultaneously with functoriality.

At the moment, I cannot make too much of this suggestion. There is, however, one point to which I shall return. Reflections on Shimura varieties led to the introduction of the Taniyama group ([**15**]). This Taniyama group was then shown ([**16**]) to be the motivic Galois group of a restricted family of motives, not in the sense of Grothendieck but in a different sense, that defined by absolute Hodge cycles, thus the family of motives of potentially CM-type. It is likely that the two senses will be shown ultimately to coincide. Since the Taniyama group was shown at its introduction to be closely related to automorphic forms on tori, this is a genuine connection between automorphic forms and motives – or Galois representations – whose interest should not be overlooked.

I had already observed that the Taniyama-Shimura conjecture, like the Sato-Tate conjecture, preceded the introduction of functoriality for automorphic forms. I myself only became aware of it after my letter to Weil, when he drew my attention to his paper on the Hecke theory in which he mentions it. With this conjecture and the large number of L-functions introduced in connection with functoriality at hand, it was natural to suppose that they would account for all the L-functions attached to algebraic varieties – in the sense associated in a general way to the pair of names Hasse-Weil.

Given the Eichler-Shimura theory and the extensive researches of Shimura on what I later referred to as Shimura varieties, these were the clear context in which to test the supposition. As I already observed, there were difficulties associated with endoscopy and therefore with the fundamental lemma. There were also – or so it seemed to me at first until I was enlightened by Kottwitz – independent combinatorial obstacles. Finally there was a serious problem connected with the action of the Galois group on abelian varieties over finite fields that was finally clarified by Kottwitz and by Reimann-Zink ([**17**]). At the time (1992), the general fundamental lemma still missing, Kottwitz was able to develop a reasonably complete theory only for a limited class of varieties ([**18**]), but these are in themselves of considerable importance.

We can now hope that, with the recent work of Laumon-Ngo, it can be established in general that the L-functions attached to Shimura varieties are automorphic. It is, however, not yet clear to me what pertinence this will ultimately have for the general reciprocity between motives and that large but special class of automorphic representations (sometimes called arithmetic) to which motives are thought to correspond. The final structure of the arguments can hardly be certain at this stage.

The proof of the Taniyama-Shimura conjecture, first for semi-stable curves by Wiles (with the help of R. Taylor) and then in general, introduced an entirely new element into the correspondence between automorphic representations π and Galois representations σ or, if one immediately passes to the diagonal arrow, motives M. Here there are many things with which I am completely unfamiliar and many more that I barely understand. So we are leaving the domains in which I have any claim whatsoever to authority. In particular, the theory of Galois representations as it developed in the hands of, say, B. Mazur and J-M. Fontaine is a subject that is not easily mastered and that I neglected in favor of other interests for too long a time. This makes it difficult to understand not only the work of Taylor but also the p-adic local reciprocity, which I am only beginning to learn.

Whether it is the horizontal arrow or the diagonal arrow from motives to automorphic forms that is being considered, there is also a necessity to establish an independent stance. My own first impressions were described in my review ([**19**]) of Hida's book ([**20**]). There is a seeding and there are deformations, apparently of two kinds: the first are moves from a $\mathbb{Q}_l$ representation to a $\mathbb{Q}_{l'}$ representation, but for the same motive; the others simultaneous deformations of automorphic representation and Galois representations. The change from l to l' is some wondrous phenomenon at the heart of the etale theory that I have not yet been able to internalize. There is nothing I can add at present to the comments of M. Harris and R. Taylor on deformations of both kinds and on the p-adic local reciprocity that are contained in the text supplementary to my review.

My view of the seeding is different from that of, say, Taylor, perhaps largely because I am so attached to functoriality, which has a wider scope than the arithmetic automorphic representations alone. This attachment suggests to me that the best seeding is that given by motives of potentially CM-type, a class that includes all motives of dimension 0, thus all Artin representations. As I observed, for motives of CM-type the correspondence can be established thanks to the Taniyama group and its properties.

It is, on the other hand, almost an explicit demand of the approach described here for establishing functoriality for function fields that motives whose cohomology consists of arbitrary Galois representations with finite image can be isolated. Something similar will have to be available for number fields, and at the moment it is not clear to me where to look. So it is best to keep an open mind.

Recall what the correspondence is to associate to what. We are trying to establish the isomorphism of two Tannakian categories, perhaps with a fibre functor. That for automorphic forms will be defined by its group, which will necessarily be over $\mathbb{C}$. Apart from some obscurities and difficulties caused by centres that I prefer to disregard at present, this will be essentially the product over all thick π_H of the groups ${}^LH$. There is, of course, a restriction to elements with the same image in $\mathrm{Gal}(K/F)$ and an inverse limit over K. Notice that H is freely varying.

Thus the analogue of a motive, better a motive with values in ${}^LG$, corresponds to a choice of a thick π_H and a homomorphism $\phi : {}^LH \to {}^LG$. Motives are of course defined quite differently and the associated group defined by a categorical construction that still presents severe problems. There is also a fibre functor to be introduced by an imbedding $\bar{Q}_l \to \mathbb{C}$. As a consequence the correspondence will be $M \to \{\pi_H, \phi : {}^LH \to GL(n)\}$. So the complete construction does not seem to be possible without functoriality. This is a point on which to reflect!

The simplest example of the seeding provided by the Taniyama group is of course that for the trivial group, thus for the trivial representation π_H, $H = \{1\}$. Supplemented by functoriality, this would mean that every motive of Artin type, thus essentially every linear representation of $\mathrm{Gal}(\bar{F}/F)$, would have its automorphic correspondent. It would mean as well that any base change was possible. It would also mean, I suppose, that, when the relation between automorphic representations of Ramanujan type and the remaining ones was taken into account, induction to include various motives whose Galois representation is not irreducible would be possible. This is the kind of information available to R. Taylor and his collaborators in their recent papers, except that the base change and, in general, the functoriality at hand are extremely limited, largely to solvable base change, some form of the Jacquet-Langlands correspondence, and the functoriality provided by the converse method.

So it is startling to me, initially even somewhat disturbing, that Taylor is able to deduce from their results the Sato-Tate conjecture. This conjecture is, as observed, just one case of a statement expected to be valid for all automorphic representations (of Ramanujan type of course — but these are typical and all others are deduced from them). Nevertheless, because it anticipated the general assertion and refers to one of the simplest and most studied classes of diophantine objects, elliptic curves, a proof of it, even if it turns out to be of limited import for the conjecture in general, is of special interest. Taylor's proof lies, in part, outside the strategy described in this lecture for it does not work with automorphic forms alone and does not rely solely on functoriality, but combines some special cases of functoriality already at hand with deformation.

The strategy of the lecture, in spite of a large conjectural element, is coherent and has a solid record of proved predictions. A major departure from it is at least a methodological challenge. Moreover, that two different strategies will succeed in such a highly structured subject seems to me unlikely. Perhaps that described here is correct and hidden somewhere in the arguments of Taylor is a method that, say, surmounts the analytic difficulties for number fields, about which I have been able to suggest very little. Maybe a way will be found to handle with the deformations not only other automorphic representations of arithmetic type but even all automorphic representations; on the other hand the Sato-Tate conjecture, even in its general form for all automorphic representations, may turn out to be only a weak consequence of functoriality and not lead back to it. The relation of the conjecture, in its original or in its general form, to functoriality appears, on reflection, to be like that of the Chebotarev theorem to the Artin conjecture. Although of importance in its own right, it is a weaker, more accessible assertion.

Until more insight into these questions is acquired, there will remain a serious intellectual, or methodological, gap between my stance and that of Richard Taylor. Although we have been yoked by the Shaw Prize, we are to some extent pulling in different directions. Perhaps that is not so bad. There is still a long way to go and the road uncertain.[2]

References

[1] A. Borel and W. Casselman, editors, *Automorphic forms, representations, and L-functions*, Proc. of Symp. in Pure Math., American Mathematical Society **33** (1977).

[2] T. Tamagawa, *On the ζ-function of a division algebra*, Ann. of Math. **77** (1963), 387–405.

[3] R. P. Langlands,*Base change for $GL(2)$*, Princeton Univ. Press, Ann. of Math. Study **96** (1980); J. Tunnell *Artin's conjecture for representations of octahedral type*, BAMS **5** (1981), 173–175.

[4] Alf van der Poorten, *Notes on Fermat's Last Theorem*, Wiley-Interscience.

[5] J. Arthur, *Unipotent automorphic representations: conjectures*, Astérisque 171–172 (1989), 13–71.

[6] Peter Sarnak and Freydoon Shahidi, editors,*Automorphic Forms and Applications*, Amer. Math. Soc., (2007).

[7] J. Arthur, *On the automorphic classification of representations: orthogonal and symplectic groups*, to appear

[8] E. Frenkel, *Recent advances in the Langlands program*, BAMS **41** (2004), 151–184.

[9] R. P. Langlands, *Beyond endoscopy*, Contributions to automorphic forms, geometry, and number theory, Johns Hopkins University Press, Baltimore, MD, (2004), 611–697.

[10] R. P. Langlands, *Un nouveau point de vue dans la théorie des formes automorphes*, Can. Math. Bull., Can. Math. Soc., **50** (2007) 234–267.

[11] J-F. Dat, *Lemme fondamental et endoscopie, une approche géométrique*, Sém. Bourbaki, 2004-05, Astérisque No. 38 (2006), 71–112.

[12] J-S. Li and J. Schwermer, On the cuspidal cohomology of arithmetic groups, Amer. J. Math. **131** (2009), 1431–1464.

[13] H. Cohen and H.W. Lenstra, *Heuristics on class groups*, Number theory (New York 1982), SLN **1052** Springer, Berlin (1984) 26–36; *Heuristics on class groups of number fields*, Number theory (Noordwijkerhout 1984), SLN **1068** Springer, Berlin (1984), 33–62.

[14] Nicholas M. Katz, *An overview of Deligne's proof of the Riemann hypothesis for varieties over finite fields*, Proc. of Symp. in Pure Mathematics, Developments arising from Hilbert problems, Amer. Math. Soc., (1976), 275–305.

[15] R. P. Langlands, *Automorphic representations, Shimura varieties and motives*, Automorphic forms, representations, and L-functions (A. Borel and W. Casselman, ed.), Proc. of Symp. in Pure Math., American Mathematical Society **33** (1977), 205–246.

[16] Pierre Deligne, James S. Milne, Arthur Ogus, Kuang-yen Shih *Hodge cycles, motives and Shimura varieties*, SLN **900** Springer, Berlin (1982).

[17] Robert Kottwitz, *Points on some Shimura varieties over finite fields*, Jour. of the AMS, **5** (1992), 373–444.

[18] Robert Kottwitz, *On the λ-adic representations associated to some simple Shimura varieties*, Inv. Math. **108** (1992), 653–665.

[19] R.P. Langlands, *Review of* [**20**], Bull. Amer. Math. Soc. **44** (2007) 291–308.

[20] Haruzo Hida, *p-adic automorphic forms on Shimura varieties*, Springer-Verlag, New York (2004).

INSTITUTE FOR ADVANCED STUDY, EINSTEIN DRIVE, PRINCETON NJ 08540
E-mail address: rpl@math.ias.edu

[2]Over the past few months, my reflexions on the fate of functoriality and the methods proposed here have profited from conversations and communications, both sometimes very brief, with several mathematicians: Nicholas Katz, Peter Sarnak, Mark Goresky, Dipendra Prasad, C. Rajan, Şahin Koçak and Joachim Schwermer. I am grateful to them all.

Clay Mathematics Proceedings
Volume 13, 2011

On Arthur's Asymptotic Inner Product Formula of Truncated Eisenstein Series

Erez Lapid

To Freydoon Shahidi, on the occasion of his 60th birthday

ABSTRACT. We give a new proof of Arthur's asymptotic formula for the inner product of truncated Eisenstein series, based on the idea of regularization developed in [**JLR99**]. We also rederive Langlands' formula for the inner product of wave packets of Eisenstein series.

CONTENTS

1. Introduction 309
2. Preliminaries 311
3. The main results 313
4. Polynomial exponential functions 314
5. The regularized integral 317
6. Regularized inner products of Eisenstein series 321
7. Proof of Proposition 1 323
8. Proof of Theorems 1 and 2 328
References 330

1. Introduction

Let G be a reductive group over a number field F, and let $\mathbb{A}$ be the ring of adeles of F. The theory of Eisenstein series provides a description of the continuous spectrum of $L^2(G(F)\backslash G(\mathbb{A}))$ in terms of the discrete spectrum of $L^2(M(F)\backslash M(\mathbb{A})^1)$ of Levi subgroups M of G ([**Lan76**], [**MW95**]). The role of Eisenstein series is analogous to the role of the exponential functions in the spectral theory of $L^2(\mathbb{R})$. Just as in Fourier analysis, the Eisenstein series are not themselves in $L^2(G(F)\backslash G(\mathbb{A}))$. Instead, one starts with a better-behaved class of functions (from an analytic point of view) called pseudo-Eisenstein series, which are smooth and rapidly decreasing. The

2010 *Mathematics Subject Classification.* Primary 22E55; Secondary 11F72, 11M36.

Partially sponsored by a grant from the Israel Science Foundation.

inner product of two pseudo-Eisenstein series is a formal computation ([**Lan66**]) and gives the coarse spectral decomposition of $L^2(G(F)\backslash G(\mathbb{A}))$. The finer spectral expansion is obtained by performing a shift of contour to the imaginary axis in the expression for the inner product.

A useful variant of pseudo-Eisenstein series is the analytically cruder truncated Eisenstein series, which show up in Langlands' work. Langlands obtained a formula for the inner product formula of truncated Eisenstein series. In the case of the upper half-plane this is a consequence of what is known as the Maass-Selberg relations (cf. [**Iwa02**]). The complete details of the proof of the formula in the higher rank case were given by Arthur, who also defined the truncation operator for all automorphic forms and studied its properties ([**Art80**]). The proof uses complex analysis (residue calculus) in a rather mysterious way.

The truncation operator plays a crucial role in the development and the analysis of the trace formula. In fact, Arthur had to consider the inner product of truncated Eisenstein series which are induced from square-integrable, but not necessarily cuspidal, automorphic forms on the Levi subgroup. As it turns out, the main term in the inner product is the one appearing in Langlands' formula, but there are additional terms which tend to zero exponentially as the truncation parameter grows [**Art82c**]. Arthur's formula is derived from Langlands' formula using the description of the discrete spectrum as residues of cuspidal Eisenstein series. It is a key step in Arthur's fine spectral expansion of the trace formula ([**Art82a**], [**Art82b**]). (A more explicit version of the fine spectral expansion, building on Arthur's work, was recently obtained in [**FLM09**].)

In [**JLR99**] the notion of regularized periods was developed and used to study periods of Eisenstein series in certain cases. (See also [**LR03**], [**LR01**].) As a by-product, a new and simple proof of Langlands' inner product formula was obtained by reducing it to the vanishing of the regularized inner product of Eisenstein series, which in turn, immediately follows for local reasons. The purpose of this note is to extend this argument to the more general case considered by Arthur, namely when the inducing data is not necessarily cuspidal. The proof is a little trickier than in the cuspidal case, but hopefully it is still reasonably conceptual and short. Perhaps more importantly, it is independent of Langlands' description of the discrete spectrum in terms of residues of cuspidal Eisenstein series.[1] As a bonus, we obtain "one-half" of Langlands' spectral decomposition of $L^2(G(F)\backslash G(\mathbb{A}))$ without appealing to the description of the discrete spectrum. This raises the question of whether the same can be done for the other half (exhaustion part) as well. While we do not answer this question here we point out that its plausibility is suggested by the existence of such an argument in the local case (cf. [**Wal03**]).

Another motivation is to generalize these results to the relative setup, for instance, the one considered in [**JLR99**]. We hope to consider this in a subsequent work.

Acknowledgement. I would like to thank Joseph Bernstein for useful discussions and the referee for carefully reading this manuscript.

[1] Of course, the analytic properties of Eisenstein series are used, but they too can be proved independently of the description above.

2. Preliminaries

Throughout, G will be a reductive group over a number field F and $\mathbb{A}$ the ring of adeles of F. We will freely use Arthur's notation and conventions from [**Art80**]. In particular, we fix a maximal split torus T_0 over F and let $M_0 = C_G(T_0)$ be a minimal Levi subgroup defined over F and $\Omega = N_G(T_0)/M_0$ the Weyl group. We also fix a maximal compact K of $G(\mathbb{A})$ which is in good position with respect to M_0, a minimal parabolic subgroup P_0 over F with Levi M_0, a Siegel domain $\mathfrak{S}$ in $G(\mathbb{A})$ and a height function $\|\cdot\|$ on $G(\mathbb{A})$. Until further notice the letters P and Q will be reserved for standard parabolic subgroups defined over F, and $M = M_P$ will denote the Levi subgroup of P containing M_0. The vector spaces $\mathfrak{a}_P$, $\mathfrak{a}_P^Q$, the sets Δ_P, $\hat{\Delta}_P$, $R(T_M, U)$, $\Omega(P, Q)$, the lattice L_P, the characteristic functions τ_P, $\hat{\tau}_P$, the decomposition

$$M(\mathbb{A}) = M(\mathbb{A})^1 \times A_M,$$

and the map

$$H_P : G(\mathbb{A}) \to \mathfrak{a}_P$$

are as in [ibid.]. We choose Haar measures on $G(\mathbb{A})$, $M(\mathbb{A})$, $U(\mathbb{A})$ K and $\mathfrak{a}_P$ compatibly with respect to the Iwasawa decomposition. By abuse of notation, we will often denote by X the F-points of a variety X defined over F.

Denote by $\mathcal{A}_G$ the space of automorphic forms on $G\backslash G(\mathbb{A})$. More generally, for any P we denote by $\mathcal{A}_P$ the space of automorphic forms on $U(\mathbb{A})P\backslash G(\mathbb{A})$. The constant term map $\varphi \mapsto \varphi_P$ from $\mathcal{A}_G$ to $\mathcal{A}_P$ is defined in the usual way. We also denote by $\mathcal{A}_P^n$ the subspace of $\mathcal{A}_P$ consisting of those φ such that $\varphi(ag) = \delta_P(a)^{\frac{1}{2}}\varphi(g)$ for any $a \in A_M$, and by $\mathcal{A}_P^2$ the subspace of $\mathcal{A}_P^n$ of those φ such that

$$\|\varphi\|_P^2 := \int_{A_M PU(\mathbb{A})\backslash G(\mathbb{A})} |\varphi(g)|^2 \, dg < \infty.$$

We can view $\mathcal{A}_P^2$ as a dense subspace of $\operatorname{Ind}_{P(\mathbb{A})}^{G(\mathbb{A})} L^2_{\text{disc}}(A_M M\backslash M(\mathbb{A}))$, where $L^2_{\text{disc}}(A_M M\backslash M(\mathbb{A}))$ is the direct sum of all irreducible subrepresentations of the regular representation of $M(\mathbb{A})$ on $L^2(A_M M\backslash M(\mathbb{A}))$.

Any $\varphi \in \mathcal{A}_P$ can be uniquely written as

$$\varphi(g) = \sum_i e^{\langle \lambda_i, H_P(g)\rangle} Q_i(H_P(g))\psi_i(g) \tag{1}$$

where $\lambda_i \in \mathfrak{a}^*_{P,\mathbb{C}}$ are distinct, $0 \neq Q_i \in \mathbb{C}[\mathfrak{a}_P]$ and $0 \neq \psi_i \in \mathcal{A}_P^n$. We denote by $\mathcal{E}_P(\varphi)$ the multiset $\{\lambda_1, \ldots, \lambda_n\}$ where λ_i appears $\deg Q_i + 1$ times. We also set $\mathcal{E}_P(\varphi) = \mathcal{E}_P(\varphi_P)$ for $\varphi \in \mathcal{A}$.

Given $\varphi \in \mathcal{A}_P$ and $\lambda \in \mathfrak{a}^*_{P,\mathbb{C}}$ let

$$\varphi_\lambda(g) = \varphi(g)e^{\langle \lambda, H_P(g)\rangle}.$$

The Eisenstein series

$$E_P(g, \varphi, \lambda) = \sum_{\gamma \in P\backslash G} \varphi_\lambda(\gamma g)$$

converges for $\operatorname{Re}\lambda$ sufficiently regular in the positive Weyl chamber. Similarly, for any $Q = M_Q V$ and $w \in \Omega(P, Q)$ the intertwining operators

$$M(w, \lambda) : \mathcal{A}_P \to \mathcal{A}_Q$$

defined by

$$M(w,\lambda)\varphi(g)=\int_{(V(\mathbb{A})\cap wU(\mathbb{A})w^{-1})\backslash V(\mathbb{A})}\varphi_\lambda(w^{-1}vg)\,dv$$

converge for $\operatorname{Re}\lambda$ sufficiently regular in the positive Weyl chamber.

Working hypothesis. For any $\varphi\in\mathcal{A}_P^2$ and $\lambda_0\in\mathfrak{a}_{P,\mathbb{C}}^*$ there exist an open neighborhood N of λ_0 and integers n, m such that

$$\Big[\prod_{\alpha\in R(T_M,U)}\langle\lambda-\lambda_0,\alpha^\vee\rangle\Big]^m E(\cdot,\varphi,\lambda)$$

is a holomorphic function from N to the Fréchet space of smooth functions on $G\backslash G(\mathbb{A})$ such that the norms

$$\sup_{g\in\mathfrak{S}}|[R(X)f](g)|\,\|g\|^{-n},\quad X\in\mathcal{U}(\mathfrak{g}_{\mathbb{C}}),$$

are finite; similarly, for any Q and $w\in\Omega(P,Q)$

$$\Big[\prod_{\alpha\in R(T_M,U)}\langle\lambda-\lambda_0,\alpha^\vee\rangle\Big]^m M(w,\lambda)\varphi$$

is a holomorphic function on N taking values in a finite-dimensional subspace of $\mathcal{A}_Q^2$ determined by the K- and $\mathfrak{z}$-types of φ. Moreover, we have functional equations

$$E_Q(M(w,\lambda)\varphi,w\lambda)=E(\varphi,\lambda),\quad \varphi\in\mathcal{A}_P^2,\ w\in\Omega(P,Q)$$
$$M(w_2,w_1\lambda)\circ M(w_1,\lambda)=M(w_2w_1,\lambda)\quad w_1\in\Omega(P,Q_1),\ w_2\in\Omega(Q_1,Q_2)$$

for $\lambda\in\mathfrak{a}_{P,\mathbb{C}}^*$.

As is well known, these results are a consequence of Langlands' theory (cf. [**Lan76**], [**MW95**], [**Lap08**]). However, they can also be proved independently (for any $\varphi\in\mathcal{A}_P$) by a method of Bernstein. Details will appear elsewhere.

The functional equations of the intertwining operators immediately imply that for $\lambda\in\mathrm{i}\mathfrak{a}_P^*$, $M(w,\lambda)$ is unitary, hence holomorphic, and therefore extends to a unitary operator on $\operatorname{Ind}_{P(\mathbb{A})}^{G(\mathbb{A})}L^2_{\mathrm{disc}}(A_M M\backslash M(\mathbb{A}))$. (Cf. [**MW95**, IV.3.12].) The Eisenstein series are also holomorphic near the imaginary axis, but we will not assume this a priori.

Arthur's truncation operator is defined for any left G-invariant locally bounded measurable functions φ by

$$\Lambda^T\varphi(g)=\sum_{P_0\subseteq P}(-1)^{\dim\mathfrak{a}_P^G}\sum_{\gamma\in P\backslash G}\varphi_P(\gamma g)\hat{\tau}_P(H_P(\gamma g)-T).$$

More generally, for any P the relative truncation operator is defined for functions on $P\backslash G(\mathbb{A})$ by

$$\Lambda^{T,P}\varphi(g)=\sum_{P_0\subseteq Q\subseteq P}(-1)^{\dim\mathfrak{a}_Q^G}\sum_{\gamma\in Q\backslash P}\varphi_Q(\gamma g)\hat{\tau}_Q^P(H_Q(\gamma g)-T).$$

Note that $\Lambda^{T,P}\varphi=\Lambda^{T,P}\varphi_P$.

3. The main results

Our goal is to give an alternative proof for the following result of Arthur.

THEOREM 1 ([**Art82c**]). *Let $\varphi \in \mathcal{A}^2_P$, and $\varphi' \in \mathcal{A}^2_{P'}$ and let $\mathcal{C}$ be a cone generated by $\dim \mathfrak{a}_0^G$ elements in the positive Weyl chamber of $\mathfrak{a}_0^G$. Then there exists $\delta > 0$ such that*

$$\left(\Lambda^T E_P(g,\varphi,\lambda), \Lambda^T E_{P'}(g,\varphi',\lambda')\right)_{G\backslash G(\mathbb{A})^1} = \sum_Q \sum_{w\in\Omega(P,Q)} \sum_{w'\in\Omega(P',Q)}$$
$$\operatorname{vol}(\mathfrak{a}_Q^G/L_Q)\frac{e^{\langle w\lambda+w'\overline{\lambda'},T\rangle}\left(M(w,\lambda)\varphi, M(w',\lambda')\varphi'\right)_Q}{\prod_{\alpha\in\Delta_Q}\langle w\lambda+w'\overline{\lambda'},\alpha^\vee\rangle} + O(e^{-\delta\|T\|})$$

for all $(\lambda,\lambda') \in \mathrm{i}(\mathfrak{a}_P^G)^* \times \mathrm{i}(\mathfrak{a}_{P'}^G)^*$ and $T \in \mathcal{C}$ with $\|T\| \gg 0$. The implied constant is independent of T and can be chosen uniformly for (λ,λ') in a compact set.

We recall that if φ and φ' are cuspidal then the asymptotic formula in Theorem 1 is exact for T sufficiently regular in the positive Weyl chamber ([**Art80**, §4]; cf. [**JLR99**] for an alternative proof). In the general case, Arthur's proof is rather involved. However, it contains two relatively easy ingredients. The first is the computation of the constant term of Eisenstein series in terms of the inducing data – a global (but easier) analogue of the geometric Lemma of [**BZ77**]. The second is the criterion of square-integrability in terms of the exponents ([**MW95**, I.4.11]).

In our approach we use these ingredients as well. However, to avoid some of the more technical parts of [**Art82c**] we will adopt the method of proof of [**JLR99**] utilizing the regularized inner products of automorphic forms. One can express the inner product of truncated automorphic forms in terms of the regularized inner products with respect to a Levi subgroup of the constant terms. In the cuspidal case considered in [ibid.], this immediately reduces the Theorem to the vanishing of the regularized inner product of Eisenstein series. In the general case, the vanishing is still a key ingredient, but the reduction is more subtle. (See Remark 1 below.)

Since we do not assume a priori that the Eisenstein series are holomorphic on the imaginary axis we will have to resort to the following expedient of Theorem 1.

PROPOSITION 1. *Let φ be a smooth map from $\mathrm{i}(\mathfrak{a}_P^G)^*$ to a finite-dimensional subspace of $\mathcal{A}^2_P$; similarly for φ'. Assume that $E_P(\cdot,\varphi(\lambda),\lambda)$ is smooth on $\mathrm{i}(\mathfrak{a}_P^G)^*$; similarly for $E_{P'}(\cdot,\varphi'(\lambda'),\lambda')$. Let $\mathcal{C}$ be as in Theorem 1. Then there exists $\delta > 0$ such that*

$$(2)\quad \left(\Lambda^T E_P(g,\varphi(\lambda),\lambda), \Lambda^T E_{P'}(g,\varphi'(\lambda'),\lambda')\right)_{G\backslash G(\mathbb{A})^1} = \sum_Q \sum_{w\in\Omega(P,Q)} \sum_{w'\in\Omega(P',Q)}$$
$$\operatorname{vol}(\mathfrak{a}_Q^G/L_Q)\frac{e^{\langle w\lambda+w'\overline{\lambda'},T\rangle}\left(M(w,\lambda)\varphi(\lambda), M(w',\lambda')\varphi'(\lambda')\right)_Q}{\prod_{\alpha\in\Delta_Q}\langle w\lambda+w'\overline{\lambda'},\alpha^\vee\rangle} + O(e^{-\delta\|T\|})$$

for all $(\lambda,\lambda') \in \mathrm{i}(\mathfrak{a}_P^G)^* \times \mathrm{i}(\mathfrak{a}_{P'}^G)^*$ and $T \in \mathcal{C}$ with $\|T\| \gg 0$. The implied constant is independent of T and can be chosen uniformly for (λ,λ') in a compact set.

Once again, the asymptotic formula is in fact an exact formula if $\varphi(\lambda)$, $\varphi'(\lambda')$ are cuspidal. Of course, Theorem 1 would follow from Proposition 1 once the holomorphy of $E_P(\cdot,\varphi,\lambda)$ on $\mathrm{i}\mathfrak{a}_P^*$ is established.

As a consequence of Proposition 1 we will compute the inner product of wave packets of Eisenstein series. To formulate this, denote by $\mathcal{W}_P$ the space of compactly supported smooth functions on $i\mathfrak{a}_P^*$ taking values in a finite-dimensional subspace of $\mathcal{A}_P^2$ such that $E_P(\cdot, \varphi(\lambda), \lambda)$ is smooth on $i\mathfrak{a}_P^*$. (Ultimately, the last condition is redundant.) Write

$$\|\varphi\|_*^2 = \int_{i\mathfrak{a}_P^*} \|\varphi(\lambda)\|_P^2 \, d\lambda. \tag{3}$$

For $\varphi \in \mathcal{W}_P$ let

$$\Theta_{P,\varphi}(g) = \Theta_\varphi(g) = \int_{i\mathfrak{a}_P^*} E_P(g, \varphi(\lambda), \lambda) \, d\lambda.$$

PROPOSITION 2. *For any $\varphi \in \mathcal{W}_P$ $\Theta_{P,\varphi} \in L^2(G\backslash G(\mathbb{A}))$. The inner product is given by*

$$(\Theta_{P,\varphi}, \Theta_{P,\varphi'})_{G\backslash G(\mathbb{A})} = \int_{i\mathfrak{a}_P^*} \sum_{w \in \Omega(P,P')} (M(w,\lambda)\varphi(\lambda), \varphi'(w\lambda))_{P'} \, d\lambda. \tag{4}$$

for any $\varphi' \in \mathcal{W}_{P'}$.

Consider the Hilbert space $\mathcal{L}$ consisting of families of functions

$$F_P : i\mathfrak{a}_P^* \to \operatorname{Ind}_{P(\mathbb{A})}^{G(\mathbb{A})} L^2_{\text{disc}}(A_M M\backslash M(\mathbb{A})), \quad P \supseteq P_0$$

satisfying

$$\|(F_P)_P\|^2 := \sum_{P \supseteq P_0} |\mathcal{P}(M_P)|^{-1} \|F_P\|_*^2 < \infty,$$

$$F_{P'}(w\lambda) = M(w,\lambda)F_P(\lambda) \text{ for all } w \in \Omega(P,P'), \ \lambda \in i\mathfrak{a}_P^*.$$

The subspace $\mathcal{L}'$ consisting of those families such that $F_P \in \mathcal{W}_P$ for all P is dense in $\mathcal{L}$. We conclude the following result which is "one-half" of Langlands' L^2 decomposition.

THEOREM 2 ([**Lan76**]; cf. [**MW95**]). *The map*

$$\mathcal{E} : (F_P)_{P\supseteq P_0} \mapsto \sum_P |\mathcal{P}(M_P)|^{-1} \Theta_{P,F_P}, \quad (F_P) \in \mathcal{L}'$$

extends to an isometry from $\mathcal{L}$ to a subspace *of $L^2(G\backslash G(\mathbb{A}))$.*

Of course, Langlands showed that $\mathcal{E}$ is *onto* as well. We will not discuss this aspect here.

Proposition 2 will also be used to show the holomorphy of the Eisenstein series on the imaginary axis, to conclude Theorem 1.

4. Polynomial exponential functions

Let V be a real vector space of dimension d. A function on V of the form

$$f(v) = \sum_{i=1}^n e^{\langle \lambda_i, v\rangle} P_i(v) \quad P_i \in \mathbb{C}[V], \ \lambda_i \in V_{\mathbb{C}}^*,$$

is called a *polynomial exponential.* Any polynomial exponential function is determined by its restriction to a non-empty open subset. The decomposition above is unique if the λ_i's are distinct and $P_i \neq 0$ for all i. We write $\mathcal{E}(f)$ for the multiset $\{\lambda_1, \dots, \lambda_n\}$, where each λ_i appears $\deg P_i + 1$ times.

We denote the space of polynomial exponential functions by $\mathcal{PE}^V$. For any multiset A of elements of $V_{\mathbb{C}}^*$ we define the subspace

$$\mathcal{PE}^V(A) = \{f \in \mathcal{PE}^V : \mathcal{E}(f) \subseteq A\}.$$

One can characterize the polynomial exponential functions as follows. For $v \in V$ and $\lambda \in V_{\mathbb{C}}^*$ let $D_{v,\lambda}$ be the (generalized) difference operator

$$D_{v,\lambda} f(u) = f(u+v) - e^{\langle \lambda, v\rangle} f(u).$$

Then $f \in \mathcal{PE}^V(\{\lambda_1, \dots, \lambda_m\})$ if and only if

$$D_{v_m,\lambda_m} \circ \cdots \circ D_{v_1,\lambda_1} f \equiv 0 \tag{5}$$

for all $v_1, \dots, v_m \in V$. (It suffices to take $v_1, \dots, v_m$ in a neighborhood of 0.)

Let Γ be a piecewise continuously differentiable Jordan curve contained in the strip $-\frac{\pi}{2} \leq \operatorname{Im} z \leq \frac{\pi}{2}$ and let D be the bounded domain surrounded by Γ. Let $R > 0$ and set

$$RD = \{R\lambda : \lambda \in D\}.$$

For any integer k we define the continuous function

$$a_k^{\Gamma,R}(\lambda_1, \dots, \lambda_k; x) = \frac{1}{2\pi \mathrm{i}} \int_\Gamma \frac{e^{Rzx}\, dz}{\prod_{l=1}^k (e^z - e^{\lambda_l/R})}, \quad \lambda_1, \dots, \lambda_k \in RD, \ x \in \mathbb{R}.$$

For any $\lambda_1, \dots, \lambda_k \in RD$ we have $a_k^{\Gamma,R}(\lambda_1, \dots, \lambda_k; \cdot) \in \mathcal{PE}^{\mathbb{R}}(\{\lambda_1, \dots, \lambda_k\})$. Moreover, for any compact subset C of RD we have

$$\sup_{\lambda_1, \dots, \lambda_k \in C, x \geq 0} \left| a_k^{\Gamma,R}(\lambda_1, \dots, \lambda_k; x) \right| e^{-R\delta x} < \infty$$

where

$$\delta = \sup_{\gamma \in \Gamma} \operatorname{Re} \gamma.$$

For any integer m and $\lambda_1, \dots, \lambda_m \in RD$ let

$$b_0^{\Gamma,R}(\lambda_1, \dots, \lambda_m; x), \dots, b_{m-1}^{\Gamma,R}(\lambda_1, \dots, \lambda_m; x)$$

be the coefficients of the polynomial (in t)

$$\sum_{k=1}^m a_k^{\Gamma,R}(\lambda_1, \dots, \lambda_k; x) \prod_{l=1}^{k-1} (t - e^{\lambda_l/R}).$$

Once again, for all $i = 0, \dots, m-1$, $b_i^{\Gamma,R}(\lambda_1, \dots, \lambda_m; x)$ is a continuous function on $(RD)^k \times \mathbb{R}$, $b_i^{\Gamma,R}(\lambda_1, \dots, \lambda_m; \cdot) \in \mathcal{PE}^{\mathbb{R}}(\{\lambda_1, \dots, \lambda_m\})$, and for any compact set C of RD we have

$$\sup_{\lambda_1, \dots, \lambda_m \in C, x \geq 0} \left| b_i^{\Gamma,R}(\lambda_1, \dots, \lambda_m; x) \right| e^{-R\delta x} < \infty. \tag{6}$$

Suppose that $f \in \mathcal{PE}^{\mathbb{R}}(\{\lambda_1, \dots, \lambda_m\})$ with $\lambda_i \in RD$, $i = 1, \dots, m$. Then by the argument of [**MW95**, Lemma I.4.2] (based on that of [**Wal74**, Lemma 6.3.1]) we have the following extrapolation formula:

$$f(x) = \sum_{k=1}^m f(\frac{k}{R}) b_{k-1}^{\Gamma,R}(\lambda_1, \dots, \lambda_m; x) \quad x \in \mathbb{R}. \tag{7}$$

For a subset $A \subseteq V_{\mathbb{C}}^*$ and an integer $l \geq 1$ we write

$$\mathcal{PE}^V_{\leq l}(A) = \{f \in \mathcal{PE}^V : \mathcal{E}(f) = \{\lambda_1, \dots, \lambda_m\}, m \leq l, \lambda_i \in A \text{ for all } i\}.$$

(Note that this is not a subspace, or even a subset, of $\mathcal{PE}^V(A)$.) We will need the following closedness property of $\mathcal{PE}^V_{\leq l}(A)$.

LEMMA 1. *Suppose that $A \subseteq V_{\mathbb{C}}^*$ is compact and $l \geq 1$ is an integer. Let f_n be a sequence in $\mathcal{PE}^V_{\leq l}(A)$ and U a non-empty open subset of V such that the limit*

$$f(v) := \lim_{n \to \infty} f_n(v) \qquad v \in U$$

exists pointwise. Then the limit exists for all $v \in V$ and $f \in \mathcal{PE}^V_{\leq l}(A)$.

PROOF. We can assume, upon translating f_n, that U contains 0. Upon passing to a subsequence, we can also assume that

$$\mathcal{E}(f_n) = \{\lambda_1(n), \dots, \lambda_m(n)\}$$

where $m \leq l$ is independent of n and $\lambda_i(n)$ converges (say to λ_i) for all $i = 1, \dots, d$.

Consider first the case $V = \mathbb{R}$. Take Γ to be the unit circle. Choose R sufficiently large so that A is contained in the open ball of radius R, and that $[0, d/R] \subseteq U$. By (7) we have

$$f_n(x) = \sum_{k=1}^{m} f_n(\frac{k}{R}) b_{k-1}^{\Gamma,R}(\lambda_1(n), \dots, \lambda_m(n); x).$$

Passing to the limit, we obtain

$$f(x) = \lim_{n \to \infty} f_n(x) = \sum_{k=1}^{m} f(\frac{k}{R}) b_{k-1}^{\Gamma,R}(\lambda_1, \dots, \lambda_m; x).$$

Thus, $f \in \mathcal{PE}^{\mathbb{R}}(\{\lambda_1, \dots, \lambda_m\})$.

Consider now the general case. By restricting f_n to any line we infer that $f(v) = \lim f_n(v)$ exists for all $v \in V$. Using the criterion (5) for f_n and passing to the limit, we obtain once again that $f \in \mathcal{PE}(\{\lambda_1, \dots, \lambda_m\})$. □

Suppose that $\mathcal{C}$ is a simplicial cone in V, generated by $e_1, \dots, e_d$. Let $\chi_{\mathcal{C}}$ be the characteristic function of $\mathcal{C}$. We say that $\lambda \in V_{\mathbb{C}}^*$ is negative (resp. non-degenerate) with respect to $\mathcal{C}$ if $\operatorname{Re}\langle \lambda, e_i \rangle < 0$ (resp. $\langle \lambda, e_i \rangle \neq 0$) for all $i = 1, \dots, d$.

The following Lemma is elementary, but it is basic to the regularization procedure. See [**JLR99**] for more details.

LEMMA 2. *For any $f \in \mathcal{PE}^V$ and $u \in V$ the integral*

$$I^V(f, u, \omega) = \int_V f(v) e^{\langle \omega, v \rangle} \chi_{\mathcal{C}}(v - u) \, dv$$

converges provided that $\omega + \lambda$ is negative with respect to $\mathcal{C}$ for all $\lambda \in \mathcal{E}(f)$. As a function of ω, $I^V(f, u, \omega)$ admits meromorphic continuation to $V_{\mathbb{C}}^*$ with hyperplane singularities contained in

$$\langle \omega + \lambda, e_j \rangle = 0 \quad \lambda \in \mathcal{E}(f), j = 1, \dots, d.$$

Outside the singular hyperplanes, $I^V(f, \cdot, \omega) \in \mathcal{PE}^V(\mathcal{E}(f) + \omega)$.

The following Lemma will be used for the uniformity statement in Theorem 1.

LEMMA 3. *Let A be a compact subset of V^* such that λ is negative with respect to $\mathcal{C}$ for all $\lambda \in A$. Then there exists $\delta > 0$ depending only on the set* $\operatorname{Re} A$ *such that for any integer $l \geq 1$, a neighborhood U of $u \in V$ and a family of functions $\mathcal{F} \subseteq \mathcal{PE}^V_{\leq l}(A)$ such that*

$$\sup_{f \in \mathcal{F}} \sup_{v \in U} |f(v)| < \infty$$

we have

$$\sup_{f \in \mathcal{F}} \sup_{v \in u + \mathcal{C}} |f(v)| \, e^{\delta \|v\|} < \infty.$$

PROOF. As before, considering the translates of $f \in \mathcal{F}$ by u we can assume that $u = 0$. We write $\mathcal{E}(f) = \{\lambda_1(f), \dots, \lambda_m(f)\}$ for $f \in \mathcal{F}$ with $m \leq k$. Without loss of generality we can assume that m does not depend on f. Choose R so that $|\langle \lambda, e_j \rangle| < R$ for all $\lambda \in A$ and that $x_1 e_1 + \dots + x_d e_d \in U$ if $0 \leq x_i \leq m/R$ for all i. By repeated application of (7) we can write

$$f(x_1 e_1 + \dots + x_d e_d) = \sum_{k_1, \dots, k_d = 1}^{m} f(\frac{k_1}{R} e_1 + \dots + \frac{k_d}{R} e_d)$$
$$b^{\Gamma_1, R}_{k_1 - 1}(\langle \lambda_1(f), e_1 \rangle, \dots, \langle \lambda_m(f), e_1 \rangle; x_1) \cdots b^{\Gamma_d, R}_{k_d - 1}(\langle \lambda_1(f), e_d \rangle, \dots, \langle \lambda_m(f), e_d \rangle; x_d)$$

where Γ_j is the boundary of the rectangle with vertices

$$\frac{\gamma_j}{2R} \pm \frac{\pi \mathrm{i}}{2}, -1 \pm \frac{\pi \mathrm{i}}{2}.$$

and $\gamma_j = \max_{\lambda \in A} \operatorname{Re} \langle \lambda, e_j \rangle < 0$. By (6) and our assumption we have

$$\sup_{f \in \mathcal{F}} \sup_{x_1, \dots, x_d \geq 0} \left| f(\sum_{j=1}^{d} x_j e_j) \right| e^{-\frac{1}{2} \sum_{j=1}^{d} \gamma_j x_j} < \infty$$

The lemma follows. □

5. The regularized integral

5.1. Definition and properties. The definition of the regularized integral is based on the identity

$$(8) \qquad \varphi(g) = \sum_{P} \sum_{\gamma \in P \backslash G} \Lambda^{T,P} \varphi(\gamma g) \tau_P(H_P(g) - T), \quad g \in G(\mathbb{A})$$

([**Art80**, Lemma 1.5]) which is a formal consequence of Langlands' combinatorial Lemma

$$(9) \qquad \sum_{P : R \subseteq P} (-1)^{\dim \mathfrak{a}_P^G} \tau_R^P \widehat{\tau}_P = \begin{cases} 1 & \text{if } R = G, \\ 0 & \text{otherwise.} \end{cases}$$

To define the regularized integral, consider

$$(10) \qquad I^G(\varphi, T, \omega) = \sum_{P} \int_{P \backslash G(\mathbb{A})^1} \Lambda^{T,P} \varphi(g) e^{\langle \omega, H_P(g) \rangle} \tau_P(H_P(g) - T \, dg).$$

(We often suppress the superscript G if it is clear from the context.) This integral is convergent for $\operatorname{Re} \omega$ sufficiently regular in the negative obtuse Weyl chamber $\mathfrak{a}^*_{0,-}$, more precisely, when $\mathrm{Re}(\omega_P + \mathcal{E}_P(\varphi) - \rho_P) \subseteq \mathfrak{a}^*_{P,-}$ for all P, where

$$\mathfrak{a}^*_{P,-} = \{ \sum_{\alpha \in \Delta_P} c_\alpha \alpha : c_\alpha < 0 \ \forall \alpha \in \Delta_P \}.$$

$I^G(\varphi, T, \omega)$ admits meromorphic continuation in ω with hyperplane singularities along the finitely many hyperplanes

$$\langle \omega + \lambda - \rho_P, \varpi^\vee \rangle = 0, \quad P \supseteq P_0, \lambda \in \mathcal{E}_P(\varphi), \varpi \in \hat{\Delta}_P.$$

In fact, if

$$\varphi_P(g) = \sum_i e^{\langle \lambda_i, H_P(g) \rangle} Q_i(H_P(g)) \psi_i(g)$$

where $Q_i \in \mathbb{C}[\mathfrak{a}_p]$ and $\psi_i \in \mathcal{A}_P^n$ then

$$\int_{P\backslash G(\mathbb{A})^1} \Lambda^{T,P} \varphi(g) e^{\langle \omega, H_P(g) \rangle} \tau_P(H_P(g) - T) \, dg$$
$$= \sum_i \int_K \int_{M\backslash M(\mathbb{A})^1} \Lambda^{T,M} \psi_i(mk) \int_{\mathfrak{a}_P^G} e^{\langle \omega + \lambda_i - \rho_P, X \rangle} Q_i(X) \tau_P(X - T) \, dX,$$

so we can appeal to Lemma 2. In particular, if the exponents of φ satisfy the regularity conditions

$$\langle \lambda - \rho_P, \varpi^\vee \rangle \neq 0 \ \ \forall P, \lambda \in \mathcal{E}_P(\varphi), \varpi \in \hat{\Delta}_P,$$

(in which case we say that φ is $*$-integrable) then $I(\varphi, T, 0)$ is well-defined. It was shown in [**JLR99**] that in this case $I(\varphi, T, 0)$ does not depend on T and is called the regularized integral of φ, denoted by

$$\int^*_{G\backslash G(\mathbb{A})^1} \varphi(g) \, dg.$$

Recall that if in fact $\varphi \in L^1(G\backslash G(\mathbb{A}))$ then for all $\lambda \in \mathcal{E}_P(\varphi)$, $\mathrm{Re}\,\lambda - \rho_P \in \mathfrak{a}^*_{P,-}$ (and conversely – cf. [**MW95**, I.4.11] for an analogous statement). Therefore, in that case, $\omega = 0$ is in the range of convergence of (10) and by (8), $\int^*_{G\backslash G(\mathbb{A})^1} \varphi(g) \, dg$ coincides with the usual integral of φ. Another crucial fact proved in [**JLR99**] is that $\int^*_{G\backslash G(\mathbb{A})^1}$ is a $G(\mathbb{A}_f)^1$-invariant functional on the space of $*$-integrable automorphic forms.

We can say a little bit more about $I(\varphi, T, \omega)$ if we take into account the following Proposition. *Henceforth T will always denote a sufficiently regular element in the positive Weyl chamber of $\mathfrak{a}_0^G$.*

PROPOSITION 3. *For all $\varphi \in \mathcal{A}$ we have*

$$T \mapsto \int_{G\backslash G(\mathbb{A})^1} \Lambda^T \varphi(g) \, dg \in \mathcal{PE}^{\mathfrak{a}_0^G} \left(\bigcup_P (\mathcal{E}_P(\varphi) - \rho_P) \right).$$

This is proved exactly as in [**LR03**, Proposition 8.4.1].

COROLLARY 1. *For any ω in general position*

$$I(\varphi, \cdot, \omega) \in \mathcal{PE}^{\mathfrak{a}_0^G} \left(\bigcup_P \bigcup_{Q \subseteq P} (\mathcal{E}_Q(\varphi) + \omega_P - \rho_Q) \right).$$

More generally, for $\varphi \in \mathcal{A}_P$ and a simplicial cone $\mathcal{C}$ of $\mathfrak{a}_P^G$ generated by $v_1, \dots, v_d$ where $d = \dim \mathfrak{a}_P^G$ we can define

$$I_P(\varphi, \chi_{\mathcal{C}}, T, \omega) = \sum_{P_0 \subseteq Q \subseteq P} \int_{Q\backslash G(\mathbb{A})^1} \Lambda^{T,Q} \varphi(g) e^{\langle \omega, H_Q(g) \rangle} \tau_Q^P(H_Q(g) - T) \chi_{\mathcal{C}}(H_P(g) - T).$$

Using the decomposition (1) this can be written as

$$\sum_i \int_K I^M(\psi_i(\cdot k), T, \omega)\, dk \int_{\mathfrak{a}_P^G} e^{\langle \omega + \lambda_i - \rho_P, X \rangle} Q_i(X) \chi_{\mathcal{C}}(X - T)\, dX.$$

Once again, the integral converges for $\operatorname{Re}\omega$ in an appropriate cone and admits a meromorphic continuation in $\omega \in \mathfrak{a}_{0,\mathbb{C}}^*$ with hyperplane singularities along

$$\langle \omega + \lambda - \rho_Q, \varpi^\vee \rangle = 0,\ Q \subseteq P,\ \varpi \in \hat{\Delta}_Q^P,\ \lambda \in \mathcal{E}_Q(\varphi)$$
$$\langle \omega + \lambda - \rho_P, v_i \rangle = 0,\ i = 1, \dots, d,\ \lambda \in \mathcal{E}_P(\varphi).$$

Outside the singular hyperplanes,

$$I_P(\varphi, \chi_{\mathcal{C}}, \cdot, \omega) \in \mathcal{PE}^{\mathfrak{a}_0^G}\left(\bigcup_{R \subseteq Q \subseteq P} (\omega_Q + \mathcal{E}_R(\varphi) - \rho_R) \right).$$

In particular, if I_P is holomorphic at $\omega = 0$ we denote the value by

$$\int^*_{P\backslash G(\mathbb{A})^1} \varphi(g) \chi_{\mathcal{C}}(H_P(g) - T)\, dg.$$

If $\varphi(g)\chi_{\mathcal{C}}(H_P(g) - T)$ belongs to $L^1(P\backslash G(\mathbb{A})^1)$ which happens precisely when

(1) $\operatorname{Re}\lambda^P - \rho_Q^P \in (\mathfrak{a}_Q^P)^*_-$ for all $Q \subseteq P$ and $\lambda \in \mathcal{E}_Q(\varphi)$, and
(2) $\lambda - \rho_P$ is negative with respect to $\mathcal{C}$ for all $\lambda \in \mathcal{E}_P(\varphi)$,

then the integral defining $I(\varphi, T, \omega)$ is convergent at $\omega = 0$ and

$$\int^*_{P\backslash G(\mathbb{A})^1} \varphi(g) \chi_{\mathcal{C}}(H_P(g) - T)\, dg = \int_{P\backslash G(\mathbb{A})^1} \varphi(g) \chi_{\mathcal{C}}(H_P(g) - T)\, dg.$$

LEMMA 4. *We have the following equality of meromorphic functions*

$$\sum_P (-1)^{\dim \mathfrak{a}_P^G} I_P(\varphi_P, \hat{\tau}_P, T, \omega) = \int_{G\backslash G(\mathbb{A})^1} \Lambda^T \varphi(g)\, dg.$$

In particular, the left-hand side does not depend on ω.

PROOF. When $\operatorname{Re}\omega$ is sufficiently regular in the negative Weyl chamber of $\mathfrak{a}_0^*$ the left-hand side is

$$\sum_P \sum_{R: R \subseteq P} (-1)^{\dim \mathfrak{a}_P^G} \int_{R\backslash G(\mathbb{A})^1} \Lambda^{T,R} \varphi(g)\, e^{\langle \omega, H_R(g) \rangle} \tau_R^P(H(g) - T) \hat{\tau}_P(H(g) - T)\, dg.$$

Interchanging the sum, and using the relation (9), we get the required statement. □

5.2. Inner product. ([**JLR99**]). In a similar vein we define, for any $\varphi, \psi \in \mathcal{A}$

$$I^G(\varphi, \psi, T, \omega) = \sum_P \int_{P\backslash G(\mathbb{A})^1} \Lambda^{T,P}\varphi(g)\overline{\Lambda^{T,P}\psi(g)}e^{\langle \omega, H_P(g)\rangle}\tau_P(H_P(g)-T)\,dg.$$

This integral converges if

$$\mathrm{Re}(\omega + \mathcal{E}_P(\varphi) + \mathcal{E}_P(\psi)) \subseteq \mathfrak{a}_{P,-}^*$$

for all P. Moreover, $I(\varphi, \psi, T, \omega)$ admits meromorphic continuation in ω with hyperplane singularities along

$$\langle \omega + \lambda + \overline{\mu}, \varpi^\vee\rangle = 0, \quad P \supseteq P_0, \lambda \in \mathcal{E}_P(\varphi), \mu \in \mathcal{E}_P(\psi), \varpi \in \hat{\Delta}_P.$$

Outside the singular hyperplanes

$$T \mapsto I(\varphi, \psi, T, \omega) \in \mathcal{PE}^{\mathfrak{a}_0^G}\left(\bigcup_{Q \subseteq P} (\mathcal{E}_Q(\varphi) + \overline{\mathcal{E}_Q(\psi)} + \omega_P)\right).$$

If $I(\varphi, \psi, T, \omega)$ is holomorphic near $\omega = 0$ then

$$\int^*_{G\backslash G(\mathbb{A})^1} \varphi(g)\overline{\psi(g)}\,dg := I(\varphi, \psi, T, 0)$$

is independent of T. In particular, this is the case if φ and ψ are square-integrable, and in this case,

$$\int^*_{G\backslash G(\mathbb{A})^1} \varphi(g)\overline{\psi(g)}\,dg = \int_{G\backslash G(\mathbb{A})^1} \varphi(g)\overline{\psi(g)}\,dg.$$

This follows from the decomposition (8) and the fact that

$$\int_{P\backslash G(\mathbb{A})^1} \Lambda^{T,P}\varphi(g)\overline{\psi(g)}\tau_P(H_P(g)-T)\,dg =$$
$$\int_{P\backslash G(\mathbb{A})^1} \Lambda^{T,P}\varphi(g)\overline{\Lambda^{T,P}\psi_P(g)}\tau_P(H_P(g)-T)\,dg.$$

The sesquilinear form $\int^*_{G\backslash G(\mathbb{A})^1}$ is $G(\mathbb{A}_f)^1$-invariant (whenever defined).

More generally for any P, a simplicial cone $\mathcal{C}$ of $\mathfrak{a}_P^G$ and $\varphi \in \mathcal{A}_P$ one defines

$$I_P(\varphi, \psi, \chi_{\mathcal{C}}, T, \omega)$$
$$= \sum_{Q \subseteq P} \int_{Q\backslash G(\mathbb{A})^1} \Lambda^{T,Q}\varphi(g)\overline{\Lambda^{T,Q}\psi(g)}e^{\langle \omega, H_Q(g)\rangle}\tau_Q^P(H_Q(g)-T)\chi_{\mathcal{C}}(H_P(g)-T)\,dg$$

and

$$\int^*_{P\backslash G(\mathbb{A})^1} \varphi(g)\overline{\psi(g)}\chi_{\mathcal{C}}(H_P(g)-T)\,dg = I_P(\varphi, \psi, \chi_{\mathcal{C}}, T, 0),$$

the latter, provided that

(1) for any $Q \subseteq P$ and any $\lambda \in \mathcal{E}_Q(\varphi)$ and $\mu \in \mathcal{E}_Q(\psi)$, $\lambda + \overline{\mu}$ is non-degenerate with respect to the Weyl chamber of $\mathfrak{a}_P^Q$, and,
(2) for any $\lambda \in \mathcal{E}_P(\varphi)$ and $\mu \in \mathcal{E}_P(\psi)$, $\lambda + \overline{\mu}$ is non-degenerate with respect to $\mathcal{C}$

As before, we have

LEMMA 5. (1) *For any* $\varphi, \psi \in \mathcal{A}_G$ *we have*

$$\sum_{P \subseteq G} (-1)^{\dim \mathfrak{a}_P^G} I_P(\varphi_P(g), \psi_P(g), \hat{\tau}_P, T, \omega) = \int_{G(F)\backslash G(\mathbb{A})^1} \Lambda^T \varphi(g) \overline{\Lambda^T \psi(g)} \, dg.$$

(2) *For any* $\varphi, \psi \in \mathcal{A}_P$ *we have*

$$I_P(\varphi, \psi, \chi_{\mathcal{C}}, \cdot, \omega) \in \mathcal{PE}^{\mathfrak{a}_0^G} \left(\bigcup_{R \subseteq Q \subseteq P} (\omega_Q + \mathcal{E}_R(\varphi) + \overline{\mathcal{E}_R(\psi)}) \right)$$

outside the singular hyperplanes.

This first part is proved exactly as Lemma 4 using (8), (9) and the fact that Λ^T is a projection. The second part follows from the fact that for any $\varphi, \psi \in \mathcal{A}$

$$(11) \qquad T \mapsto \int_{G \backslash G(\mathbb{A})^1} \Lambda^T \varphi(g) \overline{\Lambda^T \psi(g)} \, dg \in \mathcal{PE}^{\mathfrak{a}_0^G} \left(\bigcup_P (\mathcal{E}_P(\varphi) + \overline{\mathcal{E}_P(\psi)}) \right)$$

(cf. [**LR03**, Proposition 8.4.1]).

REMARK 1. The relation

$$\int_{G \backslash G(\mathbb{A})^1} \Lambda^T \varphi(g) \overline{\Lambda^T \psi(g)} \, dg = \sum_P (-1)^{\dim \mathfrak{a}_P^G} \int^*_{P \backslash G(\mathbb{A})^1} \varphi_P(g) \overline{\psi_P(g)} \hat{\tau}_P(H_P(g) - T) \, dg$$

holds if each term on the right-hand side is well-defined. Unfortunately, this is not necessarily the case if φ, ψ are square-integrable. For example, if we write the positive roots for the group $G = G_2$ as α (long), β (short), $\alpha + \beta$, $\alpha + 2\beta$, $\alpha + 3\beta$, $2\alpha + 3\beta$, then G admits a square-integrable automorphic form φ with $\mathcal{E}_{P_0}(\varphi) = \{\gamma_1 = -\alpha - \beta, \gamma_2 = -\alpha - 2\beta\}$ ([**Lan76**], [**MW95**, Appendix III]). Since $\langle \gamma_2, \beta^\vee \rangle = 0$ the term

$$\int^*_{P \backslash G(\mathbb{A})^1} |\varphi_P(g)|^2 \, \hat{\tau}_P(H_P(g) - T) \, dg$$

is not defined for the maximal parabolic subgroup P such that $\Delta_0^P = \{\beta\}$. This is why we have to introduce the parameter ω and it explains why the proof in the non-cuspidal case is more subtle.

6. Regularized inner products of Eisenstein series

For any P, Q let ${}_Q\Omega_P$ be the set of elements of Ω which are left-$\Omega(M_Q)$ and right-$\Omega(M_P)$ reduced. This is a set of representatives for $Q \backslash G / P$. For any $w \in {}_Q\Omega_P$ the group $M_P \cap w^{-1} M_Q w$ is the Levi subgroup of a parabolic subgroup P_w contained in P, and $M_Q \cap w M_P w^{-1}$ is the Levi subgroup of a parabolic subgroup Q_w contained in Q. We have $w \in \Omega(P_w, Q_w)$. Denote by $\Omega(P; Q)$ the subset of ${}_Q\Omega_P$ consisting of those w such that $P_w = P$, that is, $w M_P w^{-1} \subseteq M_Q$. Recall also that we set

$$\Omega(P, Q) = \{w \in {}_Q\Omega_P : w M_P w^{-1} = M_Q\} = \Omega(P; Q) \cap \Omega(Q; P)^{-1}.$$

PROPOSITION 4. *For any* $\varphi \in \mathcal{A}_P$ *the constant term of* $E_P(\cdot, \varphi, g)$ *along* Q *is given (at least for* $\operatorname{Re} \lambda$ *sufficiently regular in the positive Weyl chamber) by*

$$\sum_{w \in {}_Q\Omega_P} E^Q_{Q_w}(\cdot, M(w, \lambda) \varphi_{P_w}, w\lambda).$$

This is a straightforward generalization of the computation of [**MW95**, II.1.7]. The main difference is that unlike in [loc. cit.] we don't assume that φ is cuspidal, so that we get a contribution from all $w \in {}_Q\Omega_P$, and not merely from $w \in \Omega(P;Q)$.

It is also easy to see that the constant term of the summand

$$E^Q_{Q_w}(\cdot, M(w,\lambda)\varphi_{P_w}, w\lambda)$$

along a parabolic subgroup $R \subseteq Q$ is given by

$$\sum_{s \in \iota^{-1}_{Q,R}(w)} E^R_{R_s}(\cdot, M(s,\lambda)\varphi_{P_s}, s\lambda)$$

where $\iota_{Q,R} : {}_R\Omega_P \to {}_Q\Omega_P$ corresponds to the canonical map $R\backslash G/P \to Q\backslash G/P$. Note that

$$\iota_{Q,R}(\Omega(P;R)) \subseteq \Omega(P;Q).$$

We will write the constant term of $E_P(\varphi,\lambda)$ along Q as

$$\mathfrak{M}_Q(\cdot,\varphi,\lambda) + \mathfrak{E}_Q(\cdot,\varphi,\lambda) \tag{12}$$

where

$$\mathfrak{M}_Q(\cdot,\varphi,\lambda) = \sum_{w \in \Omega(P;Q)} E^Q_{Q_w}(\cdot, M(w,\lambda)\varphi, w\lambda).$$

(If φ is cuspidal then $\mathfrak{E}_Q \equiv 0$.) Thus,

$$\mathcal{E}_Q(\mathfrak{E}_Q(\varphi,\lambda)) \subseteq \bigcup_{w \in {}_Q\Omega_P \setminus \Omega(P;Q)} [w(\mathcal{E}_{P_w}(\varphi) + \lambda)]_Q$$

and more generally for any $R \subseteq Q$

$$\mathcal{E}_R(\mathfrak{E}_Q(\varphi,\lambda)) \subseteq \bigcup_{w \in \iota^{-1}_{Q,R}({}_Q\Omega_P \setminus \Omega(P;Q))} [w(\mathcal{E}_{P_w}(\varphi) + \lambda)]_R. \tag{13}$$

Recall that

$$\iota^{-1}_{Q,R}({}_Q\Omega_P \setminus \Omega(P;Q)) \subseteq {}_R\Omega_P \setminus \Omega(P;R).$$

By analytic continuation (13) continues to hold whenever $E_P(\cdot,\varphi,\lambda)$ and $M(w,\lambda)\varphi$, $w \in \Omega(P;Q)$ are regular.

LEMMA 6. *Let $\varphi \in \mathcal{A}^2_P$, $w \in {}_Q\Omega_P$, $\mu \in \mathcal{E}_{P_w}(\varphi)$ and $\varpi \in \hat{\Delta}_{Q_w}$. Then, $\operatorname{Re}\langle w\mu, \varpi^\vee\rangle \le 0$ with equality if and only if $\varpi^\vee \in w\mathfrak{a}_P$. In particular, if $\operatorname{Re} w\mu = 0$ then $w \in \Omega(P;Q)$.*

PROOF. (Cf. [**Art82c**, p. 61–62].) Clearly, $\langle w\mu, \varpi^\vee\rangle = 0$ if $\varpi^\vee \in w\mathfrak{a}_P$ since $\mu \in (\mathfrak{a}^P_0)^*_{\mathbb{C}}$. By [**MW95**, I.4.11] we have

$$\mu = \sum_{\alpha \in \Delta^P_{P_w}} c_\alpha \alpha$$

with $\operatorname{Re} c_\alpha < 0$ for all α. Thus,

$$w\mu = \sum_{\alpha \in \Delta^P_{P_w}} c_\alpha w\alpha.$$

The $w\alpha$'s are positive roots of $\mathfrak{a}_{Q_w}$. Therefore $\langle w\alpha, \varpi^\vee\rangle \ge 0$ for all $\varpi \in \hat{\Delta}_{Q_w}$. It follows that $\operatorname{Re}\langle w\mu, \varpi^\vee\rangle \le 0$. If equality holds then necessarily $\langle w\alpha, \varpi^\vee\rangle = 0$ for all $\alpha \in \Delta^P_{P_w}$. Then $w^{-1}\varpi^\vee \in \mathfrak{a}_{P_w}$ is orthogonal to $(\mathfrak{a}^P_{P_w})^*$. Therefore $w^{-1}\varpi^\vee \in \mathfrak{a}_P$ as required.

If $\operatorname{Re} w\mu = 0$ then by the above $\hat{\Delta}_Q \subseteq w\mathfrak{a}_P$. Thus, $\mathfrak{a}_Q \subseteq w\mathfrak{a}_P$, which means that $M_Q \supseteq wMw^{-1}$. □

COROLLARY 2. *Let $\varphi \in \mathcal{A}^2_P$. There exists a constant k such that for any $\lambda \in \mathfrak{a}^*_{P,\mathbb{C}}$ outside the singular hyperplanes of $E(\cdot,\lambda)$ and the intertwining operators, and all $R \subseteq Q \subseteq G$,*

$$\mathcal{E}_R(\mathfrak{M}_Q(\varphi,\lambda)) \cup \mathcal{E}_R(\mathfrak{E}_Q(\varphi,\lambda))$$

is of size $\leq k$ (as a multiset) and is contained in a ball of radius $\|\lambda\| + k$ around 0 in $\mathfrak{a}^*_{Q,\mathbb{C}}$. Moreover,

$$\bigcup_{\lambda \in i\mathfrak{a}^*_P} \operatorname{Re} \mathcal{E}_R(\mathfrak{M}_Q(\varphi,\lambda))$$

and

$$\bigcup_{\lambda \in i\mathfrak{a}^*_P} \operatorname{Re} \mathcal{E}_R(\mathfrak{E}_Q(\varphi,\lambda))$$

are finite subsets of the closure of $\mathfrak{a}^*_{R,-}$. The second set does not contain 0.

We can also infer another crucial property.

PROPOSITION 5. *Let $\varphi \in \mathcal{A}^2_P$, $\varphi' \in \mathcal{A}^2_{P'}$ and suppose that either P or P' is proper. Then for $\lambda \in (\mathfrak{a}^G_P)^*_{\mathbb{C}}$, $\lambda' \in (\mathfrak{a}^G_{P'})^*_{\mathbb{C}}$ in general position we have*

$$\int^*_{G\backslash G(\mathbb{A})^1} E_P(g,\varphi,\lambda)\overline{E_{P'}(g,\varphi',\lambda')}\, dg = 0.$$

PROOF. We can assume that $\varphi(\cdot g)$ (resp. $\varphi'(\cdot g)$) belongs to an irreducible subspace π (resp. π') of $L^2(A_M M\backslash M(\mathbb{A}))$ (resp. $L^2(A_{M'}M'\backslash M'(\mathbb{A}))$). The regularized integral, if defined, gives a $G(\mathbb{A}_f)^1$-invariant sesquilinear form on $\operatorname{Ind}_{P(\mathbb{A})}(\pi,\lambda) \times \operatorname{Ind}_{P'(\mathbb{A})}(\pi',\lambda')$. However, such a form does not exist (even locally) for λ, λ' in general position unless $P = P' = G$. It therefore remains to show that

$$\int^*_{G\backslash G(\mathbb{A})^1} E_P(g,\varphi,\lambda)\overline{E_{P'}(g,\varphi',\lambda')}\, dg$$

is well-defined for λ, λ' in general position. Let $\Lambda \in \mathcal{E}_Q(E_P(\cdot,\varphi,\lambda))$ and $\Lambda' \in \mathcal{E}_Q(E_{P'}(\cdot,\varphi',\lambda'))$. Then there exist $w \in {}_Q\Omega_P$, $w' \in {}_Q\Omega_{P'}$, $\mu \in \mathcal{E}_{P_w}(\varphi)$ and $\mu' \in \mathcal{E}_{P'_{w'}}(\varphi')$ such that $\Lambda = (w\lambda + \mu)_Q$ and $\Lambda' = (w'\lambda' + \mu')_Q$. Suppose that $\varpi \in \hat{\Delta}_Q$. Then $\langle \Lambda + \overline{\Lambda'}, \varpi^\vee \rangle$ is a non-constant affine functional of $(\lambda,\lambda') \in \mathfrak{a}^*_{P,\mathbb{C}} \times \mathfrak{a}^*_{P',\mathbb{C}}$ unless $\varpi^\vee \in w\mathfrak{a}^P_0 \cap w'\mathfrak{a}^{P'}_0$, in which case,

$$\operatorname{Re} \langle \Lambda + \overline{\Lambda'}, \varpi^\vee \rangle = \operatorname{Re} \langle w\mu, \varpi^\vee \rangle + \langle w'\mu, \varpi^\vee \rangle < 0$$

by Lemma 6. In any case $\Lambda + \overline{\Lambda'}$ is non-degenerate with respect to the Weyl chamber of $\mathfrak{a}_Q$ for λ, λ' in general position. □

7. Proof of Proposition 1

We are now ready to prove Proposition 1, our makeshift for Theorem 1. Denote by $J(T)$ the left-hand side of (2) and by $M(T)$ the main term on the right-hand side of (2). Also, let $R(T) = J(T) - M(T)$. For the moment, we suppress the dependence on λ, λ' (as well as φ, φ') from the notation. By (11) and Corollary 2 we have $\operatorname{Re} \mathcal{E}(J(T)) \subseteq \overline{\mathfrak{a}^*_{0,-}}$. Since $\mathcal{E}(M(T)) \subseteq i\mathfrak{a}^*_0$ we also have $\operatorname{Re} \mathcal{E}(R(T)) \subseteq \overline{\mathfrak{a}^*_{0,-}}$. To prove (2) we have to show that

$$\operatorname{Re} \mathcal{E}(R(T)) \subseteq \overline{\mathfrak{a}^*_{0,-}} \setminus \{0\}.$$

Using the first part of Lemma 5 and the decomposition (12) for the constant term of Eisenstein series, we can write

$$J(T) = J_1(T,\omega) + J_2(T,\omega)$$

for ω in general position, where

$$J_1(T,\omega) = \sum_Q (-1)^{\dim \mathfrak{a}_Q^G} I_Q(\mathfrak{M}_Q(\varphi(\lambda),\lambda), \mathfrak{M}_Q(\varphi'(\lambda'),\lambda'), \hat{\tau}_Q, T, \omega)$$

and

$$J_2(T,\omega) = \sum_Q (-1)^{\dim \mathfrak{a}_Q^G} [I_Q(\mathfrak{E}_Q(\varphi(\lambda),\lambda), \mathfrak{M}_Q(\varphi'(\lambda'),\lambda'), \hat{\tau}_Q, T, \omega) + I_Q(\mathfrak{M}_Q(\varphi(\lambda),\lambda), \mathfrak{E}_Q(\varphi'(\lambda'),\lambda'), \hat{\tau}_Q, T, \omega) + I_Q(\mathfrak{E}_Q(\varphi(\lambda),\lambda), \mathfrak{E}_Q(\varphi'(\lambda'),\lambda'), \hat{\tau}_Q, T, \omega)].$$

Assume that $\omega \in i\mathfrak{a}_0^*$ is in general position.

It follows from Corollary 2 and the second part of Lemma 5 that there exists a finite set $0 \notin A \subseteq \overline{\mathfrak{a}_{0,-}^*}$ and an integer l, both independent of λ, λ' and ω such that

$$J_2(T,\omega) \in \mathcal{PE}_{\leq l}^{\mathfrak{a}_0^G}(A + B_{\|\omega\|+\|\lambda\|+\|\lambda'\|+l}(i\mathfrak{a}_0^*)) \tag{14}$$

where $B_a(i\mathfrak{a}_0^*)$ denotes the ball of radius a around 0 in $i\mathfrak{a}_0^*$.

Assume first that (λ,λ') is in general position. The term $J_1(T,\omega)$ is the sum over Q, $w \in \Omega(P;Q)$ and $w' \in \Omega(P';Q)$, of

$$\int_K I^{M_Q}(E_{Q_w}^Q(\cdot k, M(w,\lambda)\varphi(\lambda), w\lambda), E_{Q_{w'}}^Q(\cdot k, M(w',\lambda')\varphi'(\lambda'), w'\lambda'), T, \omega)\, dk \times \operatorname{vol}(\mathfrak{a}_Q^G/L_Q) \frac{e^{\langle w\lambda + w'\overline{\lambda'} + \omega, T_Q\rangle}}{\prod_{\alpha\in\Delta_Q} \langle w\lambda + w'\overline{\lambda'} + \omega, \alpha^\vee\rangle}.$$

By Proposition 5 this is 0 at $\omega = 0$ unless $w \in \Omega(P,Q)$ and $w' \in \Omega(P',Q)$ in which case the value at $\omega = 0$ is equal to

$$\operatorname{vol}(\mathfrak{a}_Q^G/L_Q) \frac{e^{\langle w\lambda + w'\overline{\lambda'}, T\rangle} (M(w,\lambda)\varphi(\lambda), M(w',\lambda')\varphi'(\lambda'))_Q}{\prod_{\alpha\in\Delta_Q} \langle w\lambda + w'\overline{\lambda'}, \alpha^\vee\rangle}.$$

Thus, $J_1(T,\omega)$ is holomorphic at $\omega = 0$ and $J_1(T,0) = M(T)$. Let ω_n be a sequence of $i\mathfrak{a}_0^*$ which converges to 0 and which lies outside the singular hyperplanes of $J_1(T,\omega)$ and $J_2(T,\omega)$. Then

$$R(T) = J(T) - M(T) = J_2(T,0) = \lim_{n\to\infty} J_2(T,\omega_n).$$

Applying Lemma 1 for $f_n = J_2(T,\omega_n)$ and using (14) we conclude that

$$R(T) \in \mathcal{PE}_{\leq l}^{\mathfrak{a}_0^G}(A + B_{\|\lambda\|+\|\lambda'\|+l}(i\mathfrak{a}_0^*)).$$

This gives the relation (2) for λ, λ' in general position.

To prove (2) for arbitrary λ, λ' choose a sequence (λ_n, λ_n') in general position which converges for (λ, λ'). We will emphasize the dependence on $J(T)$, $M(T)$ and $R(T)$ on λ λ' by writing $J(\lambda,\lambda',T)$, $M(\lambda,\lambda',T)$ and $R(\lambda,\lambda',T)$ respectively. By the argument of [**Art82b**, top of p. 1299] (see also §8 below) $M(\lambda, \lambda', T)$ is smooth for $(\lambda, \lambda') \in \mathrm{i}(\mathfrak{a}_P^G)^* \times \mathrm{i}(\mathfrak{a}_{P'}^G)^*$.[2] Thus,

$$\lim_{n\to\infty} M(\lambda_n, \lambda_n', T) = M(\lambda, \lambda', T).$$

On the other hand, by the properties of the truncation operator, we also have

$$\lim_{n\to\infty} J(\lambda_n, \lambda_n', T) = J(\lambda, \lambda', T).$$

We infer that

$$R(\lambda, \lambda', T) = \lim_{n\to\infty} R(\lambda_n, \lambda_n', T).$$

Since we know that

$$R(\lambda_n, \lambda_n', T) \in \mathcal{PE}_{\leq l}^{\mathfrak{a}_0^G}(A + B_{\|\lambda_n\|+\|\lambda_n'\|+l}(\mathrm{i}\mathfrak{a}_0^*))$$

we infer once again from Lemma 1 that

$$R(\lambda, \lambda', T) \in \mathcal{PE}_{\leq l}^{\mathfrak{a}_0^G}(A + B_{\|\lambda\|+\|\lambda'\|+l}(\mathrm{i}\mathfrak{a}_0^*))$$

which implies (2).

The uniformity of the error term follows from Lemma 3 applied to the family $R(\lambda, \lambda', T)$ where λ, λ' range over a compact set of $\mathrm{i}(\mathfrak{a}_P^G)^* \times \mathrm{i}(\mathfrak{a}_{P'}^G)^*$.

This concludes the proof of Proposition 1.

More preparations. Recall that ultimately we want to prove Theorem 1. What remains to be shown is the holomorphy of the Eisenstein series on the imaginary axis. This will be done in the next section using Proposition 2 (which also implies Theorem 2). To that end we need some variations on a theme of Arthur. We recall the notion of (G, M)-families and switch to the notation of [**Art82b**], which we will freely use without further comment.

We start with the following elementary Lemma. Let V be a Euclidean space and denote by $\mathcal{S}(V)$ the Fréchet space of Schwartz functions on V.

LEMMA 7. *Let H be the hyperplane in V defined by $0 \neq \lambda \in V^*$. Then the map $f \mapsto f/\langle\lambda, \cdot\rangle$ defines a continuous linear map*

$$\{f \in \mathcal{S}(V) : f|_H \equiv 0\} \to \mathcal{S}(V).$$

PROOF. The statement immediately reduces to the case $V = \mathbb{R}$, in which case it follows from the formula

$$\frac{f(x)}{x} = \int_0^1 f'(tx)\, dt$$

valid for any smooth function f with $f(0) = 0$. □

COROLLARY 3. *Suppose that $(c_Q(\lambda))_{Q\in\mathcal{P}(M)}$ is a (G, M)-family and $c_Q \in \mathcal{S}(\mathrm{i}\mathfrak{a}_P^*)$. Then $c_M \in \mathcal{S}(\mathrm{i}\mathfrak{a}_P^*)$. Moreover c_M is a continuous linear map from the space of (G, M)-families (a closed subspace of $\mathcal{S}(\mathrm{i}\mathfrak{a}_P^*)^{\mathcal{P}(M)}$) to $\mathcal{S}(\mathrm{i}\mathfrak{a}_P^*)$.*

[2] This is based only on the holomorphy of the intertwining operators, and not on the results of [**Art82c**].

PROOF. Let

$$\Theta(\lambda) = \prod_{\alpha \in R(T_M, U)} \langle \lambda, \alpha^\vee \rangle .$$

Then

$$c_M = \frac{\sum_{Q \in \mathcal{P}(M)} c_Q(\lambda) \frac{\Theta(\lambda)}{\theta_Q(\lambda)}}{\Theta(\lambda)}.$$

Since $\theta_Q | \Theta$ for all $Q \in \mathcal{P}(M)$, the numerator is a continuous function of $(c_Q)_Q$ and by [**Art81**, Lemma 6.2] it vanishes on the root hyperplanes. Therefore by repeatedly applying the Lemma for each root, we obtain the result. □

The (G, M)-family $(c_Q(T; \lambda))_{Q \in \mathcal{P}(M)}$ is defined in [**Art82b**, §2] as

$$c_Q(T; \lambda) = e^{\langle \lambda, Y_Q(T) \rangle}$$

where $Y_Q(T)$ are certain affine transformations in T defined (for any $Q \in \mathcal{F}(M)$) in [ibid.]. Fix a cone $\mathcal{C}$ as in Theorem 1. A simple fact which is an immediate consequence of the definition is that there exists $c > 0$ such that for any $Q \in \mathcal{F}(M)$ and $\alpha \in \Delta_Q$ we have

$$\langle \alpha, Y_Q(T) \rangle \geq c \|T\| \tag{15}$$

for all $T \in \mathcal{C}$.

The following is essentially contained in [ibid., §4]. For completeness we give a proof.

LEMMA 8. *For any n there exists a continuous seminorm μ_n on the space of Schwartz functions on $\mathrm{i}\mathfrak{a}_P^*$ such that for any (G, M)-family $d_Q(\lambda)$ consisting of Schwartz functions and $T \in \mathcal{C}$ we have*

$$\left| \int_{\mathrm{i}(\mathfrak{a}_P^G)^*} \sum_{Q \in \mathcal{P}(M)} \frac{c_Q(T; \lambda) d_Q(\lambda)}{\theta_Q(\lambda)} \, d\lambda - d_P(0) \right| \leq (1 + \|T\|)^{-n} \sum_{Q \in \mathcal{P}(M)} \mu_n(d_Q).$$

PROOF. We use the product formula of [**Art88**, §7] to write the integrand as

$$\sum_{Q_1, Q_2} \alpha(Q_1, Q_2) c_M^{Q_1}(T; \lambda) d_M^{Q_2}(\lambda)$$

where the sum is over pairs $Q_1, Q_2 \in \mathcal{F}(M)$ such that $\mathfrak{a}_P^{Q_1} + \mathfrak{a}_P^{Q_2} = \mathfrak{a}_P^G$ and $\mathfrak{a}_P^{Q_1} \cap \mathfrak{a}_P^{Q_2} = 0$, and $\alpha(Q_1, Q_2)$ are certain constants, which we do not need to care about except that

$$\alpha(G, Q) = \begin{cases} 1 & Q = P, \\ 0 & \text{otherwise.} \end{cases}$$

By [**Art82b**, (3.1)]

$$c_M^{Q_1}(T; \mu) = \int_{Y_{Q_1}(T) + \mathfrak{a}_P^{Q_1}} \chi_M^{Q_1}(T, H) e^{\langle \mu, H \rangle} \, dH$$

where $\chi_M^{Q_1}(T, \cdot)$ is the characteristic function in $\mathfrak{a}_P$ of the convex hull of the set $\mathcal{Y}_M^{Q_1}(T)$. Thus,

$$\int_{\mathrm{i}(\mathfrak{a}_P^G)^*} c_M^{Q_1}(T; \lambda) d_M^{Q_2}(\lambda) \, d\lambda = \int_{Y_{Q_1}(T) + \mathfrak{a}_P^{Q_1}} \chi_M^{Q_1}(T, H) \phi_{Q_2}(H) \, dH$$

where

$$\phi_{Q_2}(H) = \int_{\mathrm{i}(\mathfrak{a}_P^G)^*} e^{\langle \mu, H\rangle} d_M^{Q_2}(\mu)\, d\mu.$$

Suppose first that $Q_1 \neq G$ and let $\alpha \in \Delta_{Q_1}$. We have $\langle \alpha, H\rangle = \langle \alpha, Y_{Q_1}(T)\rangle$ for all $H \in Y_{Q_1}(T) + \mathfrak{a}_P^{Q_1}$. It follows from (15) that there exists a constant C such that $1 + \|T\| \le C(1 + \|H\|)$ for all $H \in Y_{Q_1}(T) + \mathfrak{a}_P^{Q_1}$. Thus, for any n

$$\left| \int_{\mathrm{i}(\mathfrak{a}_P^G)^*} c_M^{Q_1}(T;\lambda) d_M^{Q_2}(\lambda) \right| \, d\lambda \le c_{n,Q_2}(1 + \|T\|)^{-n} \int_{Y_{Q_1}(T)+\mathfrak{a}_P^{Q_1}} \chi_M^{Q_1}(T,H)\, dH$$

where

$$c_{n,Q_2} = \sup_{X \in \mathfrak{a}_P^G} (C(1 + \|X\|))^n \left|\phi_{Q_2}(X)\right|.$$

Since $\int_{Y_{Q_1}(T)+\mathfrak{a}_P^{Q_1}} \chi_M^{Q_1}(T,H)\, dH$ is polynomial in T we obtain

$$\left| \int_{\mathrm{i}(\mathfrak{a}_P^G)^*} c_M^{Q_1}(T;\lambda) d_M^{Q_2}(\lambda) \right| \le c_{n',Q_2}(1 + \|T\|)^{-n} \tag{16}$$

for appropriate n'. By Corollary 3 c_{n,Q_2} is a continuous seminorm on the space of (G, M)-families.

On the other hand, the contribution from $Q_1 = G$ is

$$\int_{\mathfrak{a}_P^G} \chi_M(T,H)\phi_P(H)\, dH = \int_{\mathfrak{a}_P^G} \phi_P(H)\, dH + \int_{\mathfrak{a}_P^G} (1 - \chi_M(T,H))\phi_P(H)\, dH.$$

Since

$$\phi_P(H) = \int_{\mathrm{i}(\mathfrak{a}_P^G)^*} e^{\langle \mu, H\rangle} d_P(\mu)\, d\mu,$$

and there exists $C > 0$ such that $\chi_M(T,H) = 1$ unless $1 + \|T\| \le C(1 + \|H\|)$, we get

$$\left| \int_{\mathfrak{a}_P^G} \chi_M(T,H)\phi_P(H)\, dH - d_P(0) \right| \le c_{n,P}(1 + \|T\|)^{-n}. \tag{17}$$

Combining (16) and (17) we get the Lemma. □

We also need another simple property of the truncation operator.

LEMMA 9. *Suppose that F is a function of uniform moderate growth on $G\backslash G(\mathbb{A})$. Then $F \in L^2(G\backslash G(\mathbb{A}))$ if and only if $(\Lambda^T F, F)_{G\backslash G(\mathbb{A})}$ converges as $T \to \infty$ in $\mathcal{C}$, in which case it converges to $\|F\|^2$.*

PROOF. Recall that $(\Lambda^T F, F) = \|\Lambda^T F\|^2 \le \|F\|^2$ ([**Art80**, §1]). By [**Lap06**, Lemma 6.2] there exists a constant c such that $\Lambda^T F(g) = F(g)$ for all $g \in \mathfrak{S}_T$ where

$$\mathfrak{S}_T = \{g \in \mathfrak{S} : \langle \varpi, H_0(g) - T\rangle < c \text{ for all } \varpi \in \hat{\Delta}_0\}.$$

It follows that

$$\|\Lambda^T F\|^2 \ge \|F|_{X_T}\|^2$$

where X_T is the image of $\mathfrak{S}_T$ in $G\backslash G(\mathbb{A})$. The Lemma now follows from Lebesgue's monotone convergence theorem. □

8. Proof of Theorems 1 and 2

Next we prove Proposition 2, which immediately implies Theorem 2. We fix a cone $\mathcal{C}$ as in Theorem 1. By Lemma 9 it suffices to show that the inner product

$$(\Lambda^T \Theta_{P,\varphi}, \Lambda^T \Theta_{P',\varphi'})_{G\backslash G(\mathbb{A})}$$

converges to the right-hand side of (4) as $T \to \infty$ in $\mathcal{C}$. Using Fourier transform on $\mathfrak{a}_G$ and the properties of the truncation operator ([**Art80**]) we write this inner product as

$$\int_{\mathfrak{a}_G^*} \int_{\mathrm{i}(\mathfrak{a}_P^G)^*} \int_{\mathrm{i}(\mathfrak{a}_{P'}^G)^*} \left(\Lambda^T E_P(\cdot, \varphi(\lambda+\mu), \lambda), \Lambda^T E_{P'}(\cdot, \varphi'(\lambda'+\mu), \lambda')\right)_{G\backslash G(\mathbb{A})^1} \, d\lambda' \, d\lambda \, d\mu.$$

By Proposition 1 the limit of this expression as $T \to \infty$ in $\mathcal{C}$ is equal to the limit of

$$(18) \quad \int_{\mathfrak{a}_G^*} \int_{\mathrm{i}(\mathfrak{a}_P^G)^*} \int_{\mathrm{i}(\mathfrak{a}_{P'}^G)^*} \sum_Q \sum_{w\in\Omega(P,Q)} \sum_{w'\in\Omega(P',Q)} \mathrm{vol}(\mathfrak{a}_Q^G/L_Q)$$
$$\frac{e^{\langle w\lambda + w'\overline{\lambda'}, T\rangle} \left(M(w,\lambda)\varphi(\lambda+\mu), M(w',\lambda')\varphi'(\lambda'+\mu)\right)_Q}{\prod_{\alpha\in\Delta_Q} \langle w\lambda + w'\overline{\lambda'}, \alpha^\vee\rangle} \, d\lambda' \, d\lambda \, d\mu.$$

We have to show that this limit exists and is equal to the right-hand side of (4). This is clear if P and P' are not associate, in which case both sides are 0. Otherwise, by choosing $w \in \Omega(P, P')$ and changing φ' to $M(w,\lambda)\varphi'(w\lambda)$ we can assume without loss of generality that $P' = P$. Following [**Art82b**, §§1-2] we recast the integrand in terms of (G, M)-families by writing it as the sum over $s \in \Omega(P, P)$ of

$$\sum_{Q\in\mathcal{P}(M)} \frac{c_Q(T;\Lambda) d_Q(\lambda;\Lambda)}{\theta_Q(\Lambda)}$$

where $\Lambda = s\lambda' - \lambda$, $c_Q(T;\Lambda)$ is as in the previous section and

$$d_Q(\lambda;\Lambda) = \left(\varphi(\lambda+\mu), M_{Q|P}(\lambda)^{-1} M_{Q|P}(s, s^{-1}(\lambda+\Lambda))\varphi'(s^{-1}(\lambda+\Lambda)+\mu)\right)_P.$$

By Lemma 8 the inner integral in (18) approaches

$$\sum_{s\in\Omega(P,P)} \left(\varphi(\lambda+\mu), M(s, s^{-1}\lambda)\varphi'(s^{-1}\lambda+\mu)\right)_P$$

as $T \to \infty$ in $\mathcal{C}$ uniformly in λ and μ. Thus, the limit of (18) exists and is equal to

$$\int_{\mathfrak{a}_G^*} \int_{\mathrm{i}(\mathfrak{a}_P^G)^*} \sum_{s\in\Omega(P,P)} \left(\varphi(\lambda+\mu), M(s, s^{-1}\lambda)\varphi'(s^{-1}\lambda+\mu)\right)_P \, d\lambda \, d\mu.$$

This is equal to the right-hand side of (4) by adjointness and change of variable. This concludes the proof of Proposition 2.

Finally, we explain how to derive from Proposition 2 the holomorphy of the Eisenstein series on the imaginary axis, which is needed to infer Theorem 1. The prototype of the argument is the fact that a meromorphic function f on $\mathbb{C}^n$ with singularities along hyperplanes of the form $\langle \lambda, v - v_0\rangle = 0$, $\lambda \in \mathbb{R}^n$, $v_0 \in \mathbb{C}^n$, and whose restriction to $\mathbb{R}^n$ is in $L^2(\mathbb{R}^n)$, must be holomorphic near $\mathbb{R}^n$. Indeed, otherwise there exist $v_0, \lambda \in \mathbb{R}^n$ and an integer $k \geq 1$ such that

$$\langle \lambda, v - v_0\rangle^k f(v)$$

is holomorphic and non-zero near v_0. Thus, $|f(v)|$ grows at least like $|\langle \lambda, v - v_0 \rangle|^{-1}$ near v_0, and the latter in not square-integrable on any neighborhood of v_0 in $\mathbb{R}^n$.

Note that the conclusion is not true for arbitrary meromorphic functions as shown by the example

$$f(z_1, \ldots, z_n) = \frac{1}{e^{z_1^2 + \cdots + z_n^2}(z_1^2 + \cdots + z_n^2)}, \qquad n \geq 5.$$

For the case at hand, we first sharpen Proposition 2 as follows.

LEMMA 10. *Suppose that φ is a bounded measurable function from $i\mathfrak{a}_P^*$ into a finite-dimensional space V of $\mathcal{A}_P^2$. Suppose that φ is supported in a compact set C on which $E_P(\psi, \cdot)$ is regular for all $\psi \in V$. Similarly for φ'. Then Proposition 2 holds for φ, φ'. In particular, we have*

$$\|\Theta_\varphi\|_{L^2(G\backslash G(\mathbb{A}))}^2 \leq |\Omega| \, \|\varphi\|_*^2 \tag{19}$$

where $\|\varphi\|_*$ is as in (3).

PROOF. It is enough to consider the case $\varphi' = \varphi$. Let φ_n be a sequence in $\mathcal{W}_P$ such that

(1) φ_n take values in V;
(2) φ_n are uniformly bounded;
(3) φ_n are supported inside a fixed, compact neighborhood of C on which $E_P(\psi, \cdot)$ is regular for all $\psi \in V$.
(4) $\varphi_n \to \varphi$ with respect to $\|\cdot\|_*$.

We will show that

$$\Theta_{\varphi_n} \to \Theta_\varphi \text{ in } L^2(G\backslash G(\mathbb{A})). \tag{20}$$

Taking the limit in the identity (4) for φ_n we will obtain the Lemma. To show (20) observe that, once again by (4), Θ_{φ_n} is a Cauchy sequence in $L^2(G\backslash G(\mathbb{A}))$ and therefore it has a limit F. On the other hand, Θ_{φ_n} converges pointwise to Θ_φ by Lebesgue's dominated convergence Theorem. Thus $F = \Theta_\varphi$. □

PROPOSITION 6. *For any $\varphi \in \mathcal{A}_P^2$ the Eisenstein series $E_P(\cdot, \varphi, \lambda)$ is holomorphic for $\lambda \in i\mathfrak{a}_P^*$.*

PROOF. Suppose on the contrary that for some $\varphi_0 \in \mathcal{A}_P^2$ (say, with $\|\varphi_0\| = 1$) $E_P(\cdot, \varphi_0, \lambda)$ is not holomorphic near $i\mathfrak{a}_P^*$. Then there exists $\lambda_0 \in i\mathfrak{a}_P^*$, $r > 0$, $\alpha \in R(T_M, U)$ and an integer $k \geq 1$ such that

$$E^*(\cdot, \lambda) := l(\lambda)^k E_P(\cdot, \varphi_0, \lambda)$$

is holomorphic on $\|\lambda - \lambda_0\| < 2r$ and non-zero at λ_0 where $l(\lambda) = \langle \lambda - \lambda_0, \alpha^\vee \rangle$.

Fix $x_0 \in G(\mathbb{A})$ such that $E^*(x_0, \lambda_0) \neq 0$. For any $\eta > 0$ define

$$\varphi(\lambda) = \begin{cases} \overline{E(x_0, \varphi_0, \lambda)}\varphi_0 & \lambda \in A_\eta \\ 0 & \text{otherwise} \end{cases}$$

where

$$A_\eta = \{\lambda \in i\mathfrak{a}_P^* : \|\lambda - \lambda_0\| < r \text{ and } |l(\lambda)| > \eta\}.$$

We have

$$\Theta_\varphi(x_0) = \|\varphi\|_*^2 = \int_{A_\eta} \frac{|E^*(x_0, \lambda)|^2}{|l(\lambda)|^{2k}} \, d\lambda.$$

It follows that

$$c_1\eta^{1-2k} < \Theta_\varphi(x_0) = \|\varphi\|_*^2 < c_2\eta^{1-2k}$$

with $c_1 > 0$ where we will denote by c_i constants which are independent of η (and later on of ϵ). Using (19) we obtain,

$$\|\Theta_\varphi\|_{L^2(G\backslash G(\mathbb{A}))} \leq c_3\eta^{\frac{1}{2}-k}.$$

For any $\epsilon > 0$ let N_ϵ be a compact neighborhood of x_0 so that

$$|E^*(x,\lambda) - E^*(x_0,\lambda)| < \epsilon$$

for all $x \in N_\epsilon$ and $\|\lambda - \lambda_0\| < r$. Then for any $x \in N_\epsilon$

$$|\Theta_\varphi(x) - \Theta_\varphi(x_0)| \leq \epsilon \int_{A_\eta} |l(\lambda)|^{-2k}\, d\lambda < c_4\epsilon\eta^{1-2k}.$$

All in all,

$$c_1\eta^{1-2k}\operatorname{vol}(N_\epsilon) < \|\Theta_\varphi(x_0)\|_{L^1(N_\epsilon)} \leq \|\Theta_\varphi\|_{L^1(N_\epsilon)} + \|\Theta_\varphi - \Theta_\varphi(x_0)\|_{L^1(N_\epsilon)} \leq$$
$$\operatorname{vol}(N_\epsilon)^{\frac{1}{2}}\|\Theta_\varphi\|_{L^2(N_\epsilon)} + c_4\epsilon\eta^{1-2k}\operatorname{vol}(N_\epsilon) \leq c_3\operatorname{vol}(N_\epsilon)^{\frac{1}{2}}\eta^{\frac{1}{2}-k} + c_4\epsilon\eta^{1-2k}\operatorname{vol}(N_\epsilon).$$

Taking $\epsilon = c_1/2c_4$ we obtain

$$\eta^{1-2k} \leq c_5\eta^{\frac{1}{2}-k}$$

with c_5 independent of η. This is impossible if η is sufficiently small. □

References

[Art80] James Arthur, *A trace formula for reductive groups. II. Applications of a truncation operator*, Compositio Math. **40** (1980), no. 1, 87–121. MR558260 (81b:22018)

[Art81] ———, *The trace formula in invariant form*, Ann. of Math. (2) **114** (1981), no. 1, 1–74. MR625344 (84a:10031)

[Art82a] ———, *On a family of distributions obtained from Eisenstein series. I. Application of the Paley-Wiener theorem*, Amer. J. Math. **104** (1982), no. 6, 1243–1288. MR681737 (85k:22044)

[Art82b] ———, *On a family of distributions obtained from Eisenstein series. II. Explicit formulas*, Amer. J. Math. **104** (1982), no. 6, 1289–1336. MR681738 (85d:22033)

[Art82c] ———, *On the inner product of truncated Eisenstein series*, Duke Math. J. **49** (1982), no. 1, 35–70. MR650368 (83e:22023)

[Art88] ———, *The invariant trace formula. I. Local theory*, J. Amer. Math. Soc. **1** (1988), no. 2, 323–383. MR928262 (89e:22029)

[BZ77] I. N. Bernstein and A. V. Zelevinsky, *Induced representations of reductive $\mathfrak{p}$-adic groups. I*, Ann. Sci. École Norm. Sup. (4) **10** (1977), no. 4, 441–472. MR0579172 (58 #28310)

[FLM09] Tobias Finis, Erez Lapid, and Werner Müller, *The spectral side of Arthur's trace formula*, Proc. Natl. Acad. Sci. USA **106** (2009), no. 37, 15563–15566. MR2550527

[Iwa02] Henryk Iwaniec, *Spectral methods of automorphic forms*, second ed., Graduate Studies in Mathematics, vol. 53, American Mathematical Society, Providence, RI, 2002. MR1942691 (2003k:11085)

[JLR99] Hervé Jacquet, Erez Lapid, and Jonathan Rogawski, *Periods of automorphic forms*, J. Amer. Math. Soc. **12** (1999), no. 1, 173–240. MR1625060 (99c:11056)

[Lan66] R. P. Langlands, *Eisenstein series*, Algebraic Groups and Discontinuous Subgroups (Proc. Sympos. Pure Math., Boulder, Colo., 1965), Amer. Math. Soc., Providence, R.I., 1966, pp. 235–252. MR0249539 (40 #2784)

[Lan76] Robert P. Langlands, *On the functional equations satisfied by Eisenstein series*, Springer-Verlag, Berlin, 1976, Lecture Notes in Mathematics, Vol. 544. MR0579181 (58 #28319)

[Lap06] Erez M. Lapid, *On the fine spectral expansion of Jacquet's relative trace formula*, J. Inst. Math. Jussieu **5** (2006), no. 2, 263–308. MR2225043 (2007d:11059)

[Lap08] ———, *A remark on Eisenstein series*, Eisenstein series and applications, Progr. Math., vol. 258, Birkhäuser Boston, Boston, MA, 2008, pp. 239–249. MR2402686

[LR01] Erez Lapid and Jonathan Rogawski, *Periods of Eisenstein series*, C. R. Acad. Sci. Paris Sér. I Math. **333** (2001), no. 6, 513–516. MR1860921 (2002k:11072)

[LR03] Erez M. Lapid and Jonathan D. Rogawski, *Periods of Eisenstein series: the Galois case*, Duke Math. J. **120** (2003), no. 1, 153–226. MR2010737 (2004m:11077)

[MW95] C. Mœglin and J.-L. Waldspurger, *Spectral decomposition and Eisenstein series*, Cambridge Tracts in Mathematics, vol. 113, Cambridge University Press, Cambridge, 1995, Une paraphrase de l'Écriture [A paraphrase of Scripture]. MR1361168 (97d:11083)

[Wal74] Michel Waldschmidt, *Nombres transcendants*, Springer-Verlag, Berlin, 1974, Lecture Notes in Mathematics, Vol. 402. MR0360483 (50 #12931)

[Wal03] J.-L. Waldspurger, *La formule de Plancherel pour les groupes p-adiques (d'après Harish-Chandra)*, J. Inst. Math. Jussieu **2** (2003), no. 2, 235–333. MR1989693 (2004d:22009)

Institute of Mathematics, Hebrew University of Jerusalem, Jerusalem Israel, 91904
E-mail address: erezla@math.huji.ac.il

Clay Mathematics Proceedings
Volume 13, 2011

Multiplicité 1 dans les paquets d'Arthur aux places p-adiques

C. Mœglin

En l'honneur de Freydoon Shahidi, pour son 60e anniversaire

RÉSUMÉ. The main goal of this paper is to summarize the construction of Arthur's packet of representations and to finish the proofs of the fact that each representation occurs with multiplicity at most 1 in such a packet. This will complete the work of [**21**] and [**28**]. More precise results are in the unpublished paper [**22**].

TABLE DES MATIÈRES

1. Introduction 333
2. Notations 337
3. Le cas des groupes métaplectiques 350
4. Construction générale 355
5. Transfert 366
6. Paquets d'Arthur pour les groupes métaplectiques 369
7. Paquet de Langlands à l'intérieur d'un paquet d'Arthur 370
Références 372

1. Introduction

Le but de cet article est de résumer les grandes lignes qui permettent de montrer que les paquets d'Arthur locaux, aux places p-adiques n'ont pas de multiplicité. Ceci part comme hypothèse générale que l'on connaît l'existence de paquets de séries discrètes pour les groupes considérés via des formules de transfert de caractères, transfert endoscopique et transfert endoscopiques tordus et que l'on sait que ces paquets de séries discrètes ont multiplicité 1. Ces résultats sont annoncés au moins dans certains cas par Arthur (cf. en particulier le dernier chapitre de [**3**]) et les exposés faits par Arthur mais au moment où on écrit ces lignes, le texte définitif d'Arthur n'est pas disponible, nos résultats sont donc conditionnels.

2010 *Mathematics Subject Classification.* Primary 22E50 ; Secondary 22E35, 11F70.

On ne récrit pas les démonstrations techniques déjà disponibles sous forme de publications [**28**], [**21**] mais on ajoute la démonstration de l'irréductibilité dans la définition même des représentations dans le cas général (cf. 4.1.2, 4.2) qui n'était pas incluse dans ces publications ; c'est nécessaire pour avoir la multiplicité 1. On essaie aussi de considérer tous les groupes pour lesquels nos méthodes s'appliquent ; aux groupes classiques usuels, on ajoute les groupes $GSpin$ (en suivant les idées d'Arthur [**5**]) et on aborde aussi le cas des groupes métaplectiques ; on prouve ici la classification des séries discrètes des groupes métaplectiques en s'appuyant sur celles des groupes orthogonaux ; on doit supposer que la caractéristique résiduelle est différente de 2 pour ce résultat et on donne une description des paquets généraux en prenant une définition ad hoc à l'aide de la correspondance de Howe. Pour les groupes métaplectiques, nous n'écrivons pas de propriétés de transfert.

Décrivons un peu plus précisément l'article. Ici F est un corps p-adique. Shahidi et ses collaborateurs, en particulier Goldberg, Kim et Asgari ont montré comment les résultats d'Harish-Chandra sur les opérateurs d'entrelacement permettent de comprendre les fonctions L attachées à des représentations cuspidales génériques de certains groupes M, M étant vu comme sous-groupes de Levi de groupes H, la représentation du L-groupe de M donnant lieu à la fonction L est alors celle qui s'obtient naturellement dans la dualité de Langlands à l'aide du parabolique dual dans ${}^LH$ et de l'action de ${}^LM$ dans l'algèbre de Lie du radical unipotent de ce parabolique. Shahidi en interprétant les résultats d'Harish-Chandra a déduit de ces observations des résultats importants sur les points de réductibilité des induites de cuspidales : exprimons les sur l'exemple le plus simple. Ici $G = SO(2n+1, F)$, la forme déployée, σ est une représentation cuspidale générique de G et ρ est une représentation cuspidale unitaire d'un groupe linéaire $GL(d_\rho, F)$. On voit $GL(d_\rho, F) \times G$ comme un sous-groupe de Levi de $H = SO(2(n+d_\rho)+1, F)$ et on considère les induites de la représentation $\rho|\,|^s \otimes \sigma$ où $s \in \mathbb{R}_{\geq 0}$, notées simplement $\rho|\,|^s \times \sigma$. D'après les résultats d'Harish-Chandra une telle induite est irréductible pour tout s réel si $\rho \not\simeq \rho^*$ c'est la condition dite improprement de "ramification" ou encore il faut qu'un élément du groupe de Weyl de H stabilise la représentation $\rho \otimes \sigma$. De plus Silberger a montré que si la condition de "ramification" est satisfaite alors il existe exactement un réel $s_{\rho,\sigma}$ positif ou nul tel que l'induite $\rho|\,|^{s_{\rho,\sigma}} \times \sigma$ soit réductible. Shahidi a alors démontré que $s_{\rho,\sigma} = 0, 1/2$ ou 1 et le fait que $s_{\rho,\sigma}$ soit entier ou demi-entier ne dépend que de ρ et non de σ. En effet, d'après Shahidi, $s_{\rho,\sigma} = 1/2$ si et seulement si la fonction $L(\rho, Sym^2, s)$ a un pôle en $s = 0$. On peut maintenant généraliser ces résultats en enlevant l'hypothèse que σ est générique mais cela se fait au prix de l'utilisation de lemmes fondamentaux et de la formule des traces simplifiée (la simplification permet uniquement d'éviter les lemmes fondamentaux pondérés). Ceci est expliqué par les travaux d'Arthur (cf. par exemple [**4**] et [**6**]). La meilleure façon d'expliquer la situation pour le G fixé ici est de considérer les homomorphismes ψ de $W_F \times SL(2, \mathbb{C})$ dans $Sp(2n, \mathbb{C})$, pris évidemment à conjugaison près, dont le centralisateur dans $Sp(2n, \mathbb{C})$ est un groupe fini. A un tel ψ, Arthur associe (cf [**6**]) un ensemble sans multiplicité de séries discrtètes (ensemble noté $\Pi(\psi)$) tel que la somme des traces, $\sum_{\pi \in \Pi(\psi)} \operatorname{tr} \pi$ se calcule en fonction de la trace de la représentation tempérée de $GL(2n, F)$ associée par la correspondance de Langlands locale à Ψ ; plus exactement cette représentation $\pi^{GL}(\psi)$ est invariante par l'automorphisme extérieur de $GL(2n, F)$, noté θ_G et on prolonge cette représentation en une représentation du produit semi-direct de $GL(2n, F) \times \{1, \theta\}$;

ici on appelle θ_G l'automorphisme $h \mapsto J({}^t g^{-1})J^{-1}$ où J est tel que θ_G conserve un épinglage. Et alors, Arthur montre que pour un bon choix de ce prolongement, le choix qui est tel que θ_G induise une action triviale sur le modèle de Whittaker usuel, et pour tout $g \in G$ semi-simple suffisamment régulier

$$\sum_{\pi \in \Pi(\psi)} \operatorname{tr} \pi(g) = \delta(g, \theta, h_g) tr\, \pi^{GL}(\psi)(h_g \theta_G), \qquad (*)$$

où g est h_g sont simplement reliés par le fait que les valeurs propres de g différentes de 1 sont celles de $h_g\theta_G(h_g)$ et où $\delta(g, \theta, h_g)$ est un facteur de transfert géométrique qui vaut identiquement 1 dans un certain nombre de cas en particulier dans le cas que nous considérons ici. En d'autres termes, G est un groupe endoscopique de $GL(n).\theta_G$ et $\pi^{GL}(\psi)$ est un transfert de $\sum_{\pi \in \Pi(\psi)} \pi$ pour l'endoscopie tordue ; si G est un groupe endoscopique principal, les facteurs de transfert valent identiquement 1. Ce résultat d'Arthur n'est pas encore complètement rédigé mais Arthur a largement expliqué sa démonstration. Le transfert commute à 2 opérations clé, l'induction et la restriction (ceci se trouve déjà dans les travaux de Shelstad dans le cas de l'endoscopie ordinaire, cf. aussi [**28**] 4.2.1 pour la restriction). On a donc des renseignements sur les modules de Jacquet des éléments de $\Pi(\psi)$ en fonction des modules de Jacquet de $\pi^{GL}(\psi)$ on le reprend en 2.3 ci-dessous. Ainsi si on revient à ρ, σ ci-dessus sans supposer que σ est générique, on vérifie que si $s_{\rho,\sigma} > 0$ alors l'induite $\rho|\,|^{-s_{\rho,\sigma}} \times \sigma$ contient une sous-représentation qui est une série discrète irréductible ; cette série discrète appartient à un paquet paramétré par un morphisme noté ψ et le $s_{\rho,\sigma}$ se calcule à l'aide du module de Jacquet de $\pi^{GL}(\psi)$ et on retrouve alors les résultats de Shahidi, la fonction L qui intervient est maintenant $L({}^L\rho, Sym^2, s)$ où ${}^L\rho$ est la représentation de W_F qui correspond à ρ par la correspondance de Langlands ; dans tout l'article on notera ${}^L\rho$ tout simplement ρ. Henniart a montré dans un cadre très général que ces fonctions L coïncident et on a donc bien une généralisation du résultat de Shahidi. A partir de là on a une connaissance très précise des représentations intervenant dans $\Pi(\psi)$ et ceci est rappelé en 2.3.

On vient d'expliquer comment (*) permet de comprendre les points de réductibilité des induites de cuspidales mais pour établir la classification des séries discrètes, Arthur utilise toutes les propriétés de l'endoscopie. Jusqu'ici on avait supposé le groupe considéré quasi-déployé ; c'est l'endoscopie ordinaire qui permet de traiter aussi les groupes non quasi-déployés, en utilisant le transfert stable vers la forme quasi-déployée. Pour traiter les groupes métaplectiques, on utilise la correspondance de Howe donc on n'a plus de formule de caractères ; on suppose alors que la caractéristique résiduelle est différente de 2.

Pour les applications à la formule des traces le point est de généraliser (*) à tout morphisme ψ de $W_F \times SL(2,\mathbb{C}) \times SL(2,\mathbb{C})$ dans $Sp(2n,\mathbb{C})$ en modifiant les signes du membres de gauche. Pour un tel ψ Arthur sait démontrer l'existence du paquet de représentations $\Pi(\psi)$ tel que (*) soit satisfait mais avec éventuellement des coefficients dans le membre de gauche qui sont, à un signe près qu'Arthur sait calculer, des entiers ; on ne sait pas encore avec quelle généralité Arthur écrira ses résultats. Dans cet article on résume les points qui permettent de prouver que les coefficients sont ± 1 dès qu'on le sait pour les paquets de séries discrètes (nos méthodes ne permettent nullement de le démontrer pour les paquets de séries discrètes). La démonstration est technique et elle s'appuie sur 2 types de résultats ; d'une part du côté des groupes linéaires, pour ψ général il faut écrire la θ_G-trace de $\pi^{GL}(\psi)$ dans le groupe de Grothendieck de façon à pouvoir calculer de quelle représentation elle est un transfert (ceci est fait dans [**28**]). De façon analogue, on a une formule pour le groupe classique considéré dans le groupe de Grothendieck qui ne nécessite d'ailleurs pas que le groupe soit quasi-déployé. Le point est alors de décrire cette formule comme combinaison linéaire avec coefficient ± 1 (explicites) de traces de représentations irréductibles ; ceci est fait dans [**21**]. En fait la méthode est un peu plus fine, pour chaque caractère, ϵ, du centralisateur de ψ, on définit dans le groupe de Grothendieck de G une représentation $\pi(\psi, \epsilon)$ et on montre que cette représentation est une somme sans multiplicité de représentations irréductibles (donc à coefficient 1) puis on prouve qu'une bonne combinaison linéaire avec des coefficients ± 1 de ces représentations $\pi(\psi, \epsilon)$ a pour transfert la θ-trace de $\pi^{GL}(\psi)$. La description des représentations est complètement explicite sous l'hypothèse que la restriction de ψ à $W_F \times \Delta_{SL(2,\mathbb{C})}$ est sans multiplicité, où $\Delta_{SL(2,\mathbb{C})}$ est la diagonale de $SL(2, \mathbb{C})$; dans ce cas, on a une paramétrisation précise des représentations dans $\Pi(\psi)$. On passe au cas général sans perdre la multiplicité 1 (cf 4.2 ci-dessous) mais en perdant de la précision sur la paramétrisation ; dans cet article on donne une démonstration complète de l'irréductibilité de certaines induites (la démonstration se trouve dans une prépublication [**22**] qui restera prépublication, vue sa technicité) de façon à ce que toutes les étapes de la preuve soient disponibles sous forme de publication. Quand on travaille dans le groupe classique, on a exactement la même combinatoire que le groupe soit quasi-déployé ou non ; la différence porte sur la restriction du caractère ϵ (ci-dessus) au centre du groupe dual ; dans le cas de $GSpin$ ce n'est pas exactement le groupe dual, ceci est expliqué en toute généralité par Arthur en [**4**] et repris ici dans le cas particulier considéré ; par exemple si $G = GSpin(2n + 1, F)$ la composante neutre du groupe dual est $Gsp(2n, \mathbb{C})$ mais les centralisateurs sont calculés dans $Sp(2n, \mathbb{C})$ et c'est le centre de $Sp(2n, \mathbb{C})$ qui distingue la forme déployée de celle qui ne l'est pas.

Les groupes pour lesquels les méthodes développées ci-dessous fonctionnent bien sont les groupes classiques et leurs variantes ; c'est-à-dire ce sont les groupes qui peuvent être mis en famille indexée par le rang, $\{G(n); n \in \mathbb{N}\}$ ici n est le rang du groupe, tel que les classes de conjugaison des sous-groupes de Levi de $G(n)$ pour tout n sont indexés par les multiensembles $\{n_0; (n_1, \cdots, n_\ell)\}$, où $n = n_0 + n_1 + \cdots + n_\ell$, $n_0 \geq 0$ et les n_i pour $i \in [1, \ell]$ sont définis à l'ordre près, le sous-groupe de Levi étant isomorphe à $\times_{i \in [1,\ell]} GL(n_i) \times G(n_0)$ où $G(n_0)$ est le groupe de même type que $G(n)$ mais de rang n_0 ; ceci n'est pas suffisant, on veut aussi que la composante neutre du L groupe de G ait une représentation naturelle injective dans un groupe linéaire de rang $2n$ ou $2n + 1$, c'est-à-dire que l'on exclut encore les groupes $G = Gsp$ ou GO. On obtient finalement les groupes classiques usuels sans supposer nécessairement qu'ils sont quasi-déployés, ce qui sous entend qu'il faut considérer $O(2n, F)$ et non $SO(2n, F)$ (pour ces groupes il faut tenir compte du discriminant de la forme et de l'invariant de Hasse dans les paramétrisations ; ceci est amplement écrit dans la littérature, par exemple [**17**] et on ne revient pas là-dessus) ; les groupes unitaires sont aussi acceptables et les groupes $GSpin(2m + 1, F)$ (les deux formes, déployée ou non déployée) ; ceci est remarqué dans [**7**] et repris dans [**8**], [**14**]. Le groupe dual à considérer est alors $GSp(2m, \mathbb{C})$ que l'on considère en suivant Arthur (cf [**5**]) comme un sous-groupe de $GL(2m, \mathbb{C}) \times GL(1, \mathbb{C})$. On peut aussi traiter une variante non connexe de $GSpin(2m, F)$ avec comme groupe dual $GO(2m, \mathbb{C})$ encore inclus dans $GL(2m, \mathbb{C}) \times GL(1, \mathbb{C})$. Il faut préciser ici que nous n'avons pas fait toutes les vérifications nécessaires pour les groupes orthogonaux paires et leurs variantes $GSpin(2n, F)$; la difficulté vient de la non connexité. Pour les représentations, il suffit d'appliquer la théorie de Mackey et la difficulté est dans le cas tempéré et non dans le passage du cas tempéré au cas général. Il y a aussi une difficulté avec les facteurs de transfert, ces facteurs de transfert n'apparaissent pas pour les autres groupes qui sont des groupes endoscopiques principaux mais viennent pour les groupes orthogonaux pairs et la non connexité du groupe n'améliore pas leur définition (je ne sais même pas si elle existe) ; là aussi la difficulté est pour les représentations tempérées et je préfère attendre les résultats précis annoncés par Arthur pour utiliser son yoga et faire les vérifications. Le groupe métaplectique rentre aussi dans ce shéma, au moins si $p \neq 2$ à la différence de taille près que l'on n'a pas d'analogue de (*) ; on remplace cette caractérisation des paquets par la correspondance de Howe. Ce sont des idées dûes, me semble-t-il, à Adams et exploitées dans ses travaux en particulier communs avec Barbasch et Trapa et dans les travaux de Renard et Trappa, au moins dans le cas des groupes archimédiens. Pour donner la forme d'un paquet d'Arthur général de $Mp(2n, F)$, on a en plus l'hypothèse que le caractère d'une représentation est localement L^1, je ne sais pas si cela est démontré dans la littérature ; cette hypothèse vient du fait que [**26**] 4.II. 2 n'est pas algébrique. On caractérise un paquet d'Arthur par son image dans la correspondance de Howe pour la paire $Mp(2n, F), O(2n+2N+1, F)$ pour N grand en demandant que cette image soit l'ensemble des représentations de $O(2n+2N+1, F)$ dont la restriction à $SO(2n+2N+1, F)$ soit dans le paquet d'Arthur prédit par Adams et qui interviennent dans cette correspondance de Howe.

2. Notations

F est un corps p-adique, n est toujours un entier, G est l'un des groupes suivants, quasi-déployé ou non : $Sp(2n, F)$, $SO(2n+1, F)$, $O(2n, F)$, $GSpin(2n+1, F)$, $GSpin^{nc}(2n, F)$ (cf ci-dessous pour la définition), $Mp(2n, F)$ (notation abusive, puisque le groupe n'est pas le groupe des points sur F d'un groupe algébrique). On pose respectivement suivant les cas $G*$ le groupe complexe $SO(2n+1, \mathbb{C})$, $Sp(2n, \mathbb{C})$, $O(2n, \mathbb{C})$, $GSp(2n, \mathbb{C})$, $GO(2n, \mathbb{C})$. On considère la représentation naturelle de G^* dans le GL évident sauf dans le cas GSp et GO où on considère la représentation naturelle dans $GL(2n, \mathbb{C}) \times GL(1)$ comme expliquée ci-dessous (on suit Arthur). On note θ_G, ou θ l'automorphisme extérieur de GL qui respecte un épinglage dans tous les cas sauf quand G est un groupe $GSpin$ où l'automorphisme est décrit ci-dessous.

On considère aussi des morphismes ψ de $W_F \times SL(2, \mathbb{C}) \times SL(2, \mathbb{C})$ dans G^* vu grâce à la représentation de G^* décrite comme une représentation de $W_F \times SL(2, \mathbb{C}) \times SL(2, \mathbb{C})$. On suppose toujours que cette représentation est unitaire, continue au sens habituel sur W_F algébrique sur les copies de $SL(2, \mathbb{C})$. On décompose alors cette représentation en sous-représentations irréductibles et on note $Jord(\psi)$ l'ensemble de ces sous-représentations irréductibles comptées avec multiplicité ; dans le cas des groupes $GSpin$, on a en plus un caractère de W_F noté ν_ψ ; on l'oublie en général des notations. Un élément de $Jord(\psi)$ est donc un triplet (ρ, a, b) où ρ est une représentation irréductible unitaire de W_F et a, b sont des entiers déterminant des représentations irréductibles de $SL(2, \mathbb{C})$, représentation irréductible que l'on écrira parfois $\rho \otimes [a] \otimes [b]$.

On considérera aussi le centralisateur de ψ, il est à prendre dans G^* sauf dans le cas des groupes $G = GSpin$ où on le prend dans $Sp(2n, \mathbb{C})$ ou dans $GO(2n, \mathbb{C})$. On considérera ensuite des caractères de ce centralisateur ; dans tous les cas sauf celui des groupes métaplectiques, ces caractères ont pour restrictions au centre de G^* le caractère trivial si et seulement si G est quasi-déployé. Dans le cas des groupes métaplectiques, il n'y a pas de condition sur la restriction des caractères au centre de G^* ici $Sp(2n,\mathbb{C})$. Il est facile d'identifier un caractère du centralisateur de ψ à une application de $Jord(\psi)$ dans ± 1, où $Jord(\psi)$ est ici vu comme un ensemble sans multiplicité, application qui vaut $+1$ sur tous les triplets (ρ,a,b) tel que la représentation $\rho\otimes[a]\otimes[b]$ n'est pas à valeurs dans un groupe de même type que G^* ; plus précisément si (ρ,a,b) n'a pas la bonne parité (cf. 4.1)

La notation ρ qui revient dans tout cet article, signifie indifféremment une représentation irréductible unitaire de W_F ou une représentation cuspidale unitaire irréductible de $GL(d_\rho,F)$ (ce qui définit d_ρ).

On a essayé d'éviter le maximum de technicité dans cet article, mais on a par endroit besoin de la notation bien commode $Jac_{\rho|\,|^x}\pi$; elle signifie que π est une représentation lisse de longueur finie de G, x est un nombre réel (en général un demi-entier), ρ est comme ci-dessus et $Jac_{\rho|\,|^x}\pi$ est l'élément du groupe de Grothendieck du groupe G' de même type que G mais de rang d_ρ plus petit tel que le module de Jacquet de π pour le parabolique de Levi $GL(d_\rho,F)\times G'$ soit de la forme $\left(\rho|\,|^x\otimes Jac_{\rho|\,|^x}\pi\right)\oplus\left(\oplus_{\sigma',\sigma''}\sigma'\otimes\sigma''\right)$, où σ',σ'' décrive un ensemble de représentations irréductibles de $GL(d_\rho,F)$ et G' respectivement telles que σ' ne soit pas isomorphe à $\rho|\,|^x$. On peut composer ces applications $Jac_{\rho|\,|^x}$ en faisant attention à la notation

$$Jac_{\rho|\,|^y}\circ Jac_{\rho|\,|^x}\pi =: Jac_{\rho|\,|^x,\rho|\,|^y}\pi.$$

Soit ψ comme ci-dessus ; on note $\pi^{GL}(\psi)$ la représentation du GL convenable isomorphe à l'induite $\times_{(\rho,a,b)\in Jord(\psi)}Speh(St(\rho,a),b)$, où St signifie Steinberg et $Speh(St(\rho,a),b)$ est l'unique quotient irréductible de l'induite $St(\rho,a)|\,|^{(b-1)/2}\times\cdots\times St(\rho,a)|\,|^{-(b-1)/2}$. Pour donner la définition de $\pi^{GL}(\psi)$, il suffit bien évidemment d'avoir une représentation de $W_F\times SL(2,\mathbb{C})\times SL(2,\mathbb{C})$ (il n'est pas utile de savoir qu'elle est à valeurs dans G^* ; on utilisera par endroit cette généralisation.

Convention pour les ψ tempérés

Quand ψ est trivial sur la 2e copie de $SL(2,\mathbb{C})$, tout $(\rho,a,b)\in Jord(\psi)$ est tel que $b=1$ et on remplace donc les triplets par des couples.

2.1. Les groupes ***GSpin*****.** On suit ici [**7**] et [**8**] : $Spin(2m+1,F)$ est le revêtement d'ordre 2 de $SO(2m+1,F)$ non trivial ; ce qui suppose que l'on a fixé une forme orthogonale non nécessairement sans noyau anisotrope de dimension $2n+1$. Ce groupe a un centre isomorphe à $\mathbb{Z}/2\mathbb{Z}$ (cf. [**7**] proposition 2.2). On note c l'élément non trivial de ce centre et on considère $GSpin(2m+1,F):=GL(1,F)\times Spin(2m+1,F)/\{(1,1),(-1,c)\}$ (cf. [**7**] def. 2.3). On définit aussi $Spin(2m,F)$ comme le revêtement d'ordre 2 non trivial de la forme quasidéployée de $SO(2m,F)$; l'automorphisme extérieur de $SO(2m,F)$ se relève en un automorphisme , *aut*, de $Spin(2m,F)$. Le centre de $Spin(2m,F)$ est un groupe à 4 éléments et son sous-groupe formé par les éléments invariants sous *aut* est d'ordre 2 ; c'est le sous-groupe $\{1,c\}$ avec les notations de [**7**] proposition 2.2. On note $Spin^{nc}(2m,F)$ le produit semi-direct de $Spin(2m,F)$ avec $\{1,aut\}$ et on pose encore $GSpin^{nc}(2m,F)$ le quotient $GL(1,F)\times Spin^{nc}(2m,F)/\{(1,1),(-1,c)\}$. Toute représentation irréductible d'un de ces groupes a un caractère central. On supposera toujours que ce caractère central est unitaire. Si π est la représentation considérée, on note ν_π la restriction du caractère central de π au facteur $GL(1,F)$ qui apparaît ci-dessus. D'après [**7**] 2.7, les sous-groupes de Levi des groupes $GSpin(2m+1,F)$ déployés sont de la forme

$$GL(n_1,F)\times\cdots\times GL(n_\ell,F)\times GSpin(2(m-n_1-\cdots-n_\ell)+1,F);$$

son argument (qui utilise le L-groupe comme décrit ci-dessous) montre aussi que la composante connexe du centre de $GSpin(2m+1,F)$ s'identifie à celle de

$$GSpin(2(m-n_1-\cdots-n_\ell)+1,F).$$

Le cas de $GSpin(2m,F)$ déployé est fait dans [**14**] page 545. Kim y calcule effectivement les sous-groupes à un paramètre et montre ainsi qu'un sous-groupe parabolique maximal de $GSpin(2m,F)$ est de la forme $GL(1)\times GL(1)\times SL(k)\times Spin(2(m-k))/S$ où $k\leq m$ et le premier $GL(1)$ est celui qui sert à définir $GSpin$ et le 2e est le tore déployé maximal dans le centre du sous-groupe de Levi de $Spin$ obtenu par intersection ; ce sont des arguments sur les groupes dérivés puisque l'on sait que le groupe dérivé doit être simplement connexe ; le groupe S se calcule à l'aide des coracines et Kim montre qu'en plus de l'élément $\{-1,1,1,c\}$, il contient, si k est pair

$$\{(1,t,t\mathrm{Id}_k,1);t^{k/2}=1\}\cup\{(-1,t,tId,c);t^{k/2}=-1\}.$$

Si k est impair, ce groupe vaut $\{(1,t,t^2\mathrm{Id}_k,1);t^k=1\}\cup\{-1,t,t^2\mathrm{Id}_k,c);t^k=-1\}$. En tant que groupe algébrique ce Levi est isomorphe naturellement à $GL(k)\times GSpin(2(m-k))$. Avec cette présentation, il est clair que ceci reste vrai si $GSpin$ n'est pas quasidéployé et que ceci s'étend aussi à $GSpin^{nc}(2n)$.

Le groupe dual est $Gsp(2m,\mathbb{C})$ pour $GSpin(2m+1,F)$ et $GO(2m,\mathbb{C})$ pour $GSpin^{nc}(2m,F)$. En suivant [**5**] on voit ces groupes comme des sous-groupes de $GL(2m,\mathbb{C})\times GL(1,\mathbb{C})$ par le plongement $g\mapsto(g,\lambda_g)$ où λ_g est le scalaire tel que ${}^tgJ_Gg=\lambda_gJ_G$ où J_G est la matrice antidiagonale avec des 1 dans le 2e cas et la matrice antidiagonale mais avec des -1 comme entrées non nulles sur les m premières colonnes et des 1 pour les m dernières.

A la suite de [**5**] on note θ_{G^*} l'automorphisme de $GL(2m,\mathbb{C})\times GL(1,\mathbb{C})$ défini par $\theta_{G^*}(g,\lambda)=J({}^tg^{-1})J^{-1}\lambda,\lambda)$. On considère les homomorphismes de $W_F\times SL(2,\mathbb{C})\times SL(2,\mathbb{C})$ dans $GL(2m,\mathbb{C})\times GL(1,\mathbb{C})$ dont la classe de conjugaison est θ_{G^*} invariante. La projection sur le facteur $GL(1,\mathbb{C})$ définit un caractère de W_F, noté ω_ψ ; on considère ω_ψ comme un caractère de $W_F\times SL(2,\mathbb{C})$ trivial sur $SL(2,\mathbb{C})$. La projection sur le facteur $GL(2m,\mathbb{C})$ définit une représentation, ψ^{GL} de dimension $2m$ de $W_F\times SL(2,\mathbb{C})\times SL(2,\mathbb{C})$. On suppose toujours que cette représentation est semi-simple de restriction à W_F bornée. Supposons que la classe de conjugaison de ψ soit θ_{G^*}-invariante ; cette hypothèse se traduit exactement par le fait que $\psi^{GL,*}\otimes\omega_\psi\simeq\psi^{GL}$. On supposera toujours que ω_ψ est un caractère unitaire. En [**5**], Arthur a classifié les classes de conjugaison d'homomorphisme ψ. Dualement l'automorphisme θ_G de $GL(2m,G)\times GL(1,F)$ à considérer et l'automorphisme $\theta_G(g,x)=J({}^tg^{-1}))J^{-1},det(g)x)$ (cf. [**5**]) page 66.

2.2. Le groupe métaplectique et la filtration de Kudla. Ici on suppose que $p\neq 2$, p étant la caractéristique résiduelle du corps de base. On peut comprendre les représentations irréductibles du groupe métaplectique en utilisant la représentation métaplectique et tous les groupes orthogonaux d'une forme de dimension impaire ; comme on le verra ci-dessous, il est justifié de considérer $Sp(2n,\mathbb{C})$

comme groupe dual. C'est-à-dire que l'on considérera des homomorphismes de $W_F \times SL(2,\mathbb{C}) \times SL(2,\mathbb{C})$ dans $Sp(2n,\mathbb{C})$.

On aura besoin des résultats suivants sur la représentation métaplectique, essentiellement dûs à Kudla dans sa thèse ([**15**]). On fixe 2 tours de Witt d'espaces orthogonaux de dimension impaire, l'une correspondant aux formes orthogonales de noyau anisotrope de dimension 1 et l'autre à celles ayant un noyau anisotrope de dimension 3 ; on suppose que les discriminants valent 1 pour tous ces espaces. La première tour est formé d'espaces orthogonaux d'invariant de Hasse 1 et la deuxième d'espaces orthogonaux d'invariant de Hasse -1. Les objets associés dans ce qui suit via l'une ou l'autre des tours seront notés avec un indice $\zeta = \pm$ représentant l'invariant de Hasse attaché à la tour. On fixe le groupe métaplectique $Mp(2n,F)$ et on a donc 2 familles de représentations métaplectiques donnant des correspondances de Howe. Soit π une représentation cuspidale de $Mp(2n,F)$; on note $\tilde{\pi}_\zeta$ la première occurrence de π dans ces représentations ; Kudla a démontré que $\tilde{\pi}_\zeta$ est une représentation cuspidale. Notons m_ζ l'entier impair tel que $\tilde{\pi}_\zeta$ est une représentation de $O(m_\zeta,F)$. On sait d'après les travaux de Kudla et de Rallis ([**16**], theorem 3.8) que l'on a

$$(m_+ - 1) + (m_- - 1) \geq 4n + 2.$$

Une conjecture de Kudla est que cette inégalité est une égalité et l'égalité est démontré par Kudla et Rallis dans un certain nombre de cas qui incluent le cas où π est cuspidale ([**16**], theorem 3.9). On a une inégalité symétrique : on fixe m et l'invariant de Hasse ζ et on regarde la tour de Witt quand n varie. Soit $\tilde{\pi}$ une représentation cuspidale de $O(m,F)_\zeta$ et on note $n(\tilde{\pi})$ la première occurrence de $\tilde{\pi}$. Pour $sign$ le caractère signe de $O(m,F)_\zeta$, on définit de même $n(\tilde{\pi} \otimes sign)$ et on a $n(\tilde{\pi}) + n(\tilde{\pi} \otimes sign) \geq 2(m-1) + 2$.

Le résultat clé dû à Kudla que nous utiliserons est le suivant. Pour n comme ci-dessus, m un entier impair ζ un signe qui donne l'invariant de Hasse d'une forme orthogonale de dimension m et de déterminant 1, on note $\Omega_{n,m,\zeta}$ la représentation métaplectique pour la paire $Mp(2n,F), O(m,F)_\zeta$. Pour $d \leq n$, Kudla a calculé les modules de Jacquet de cette représentation pour le parabolique maximal isomorphe à $GL(d,F) \times Mp(2(n-d),F)$ de $Mp(2n,F)$; il a décrit une filtration à 2 termes de ce module de Jacquet, le quotient est isomorphe à $\Omega_{n-d,m,\zeta}$ est le sous-module est isomorphe à $ind_Q^{O(m,F)_\zeta} \omega_{d,d} \otimes \Omega_{n-d,m-2d,\zeta}$ où Q est un parabolique de $O(m,F)$ de Levi isomorphe à $GL(d,F) \times O(m-2d,F)$ (ce terme n'existe pas si Q n'existe pas), $\omega_{d,d}$ est à torsion près essentiellement la représentation régulière gauche et droite de $GL(d,F) \times GL(d,F)$ où le premier $GL(d,F)$ est dans $Mp(2n,F)$ et le deuxième est un facteur de Q. Le groupe $GL(d,F)$ sous-groupe de $Mp(2n,F)$ opère sur $\Omega_{n-d,m,\zeta}$ par le caractère $|det|^{-(n-(m-1)/2-1/2)}$. Soit n', m' et $\pi' \otimes \tilde{\pi}'$ un quotient irréductible de $\Omega_{n',m',\zeta}$ et on suppose qu'il existe une représentation cuspidale irréductible ρ' d'un groupe $GL(d_{\rho'},F)$ et une représentation irréductible σ' de $Mp(2(n-d_{\rho'}),F)$ tel que π' soit un sous-module de l'induite $\rho' \times \sigma'$. En utilisant la filtration ci-dessus et quelques dualités, on vérifie que 2 cas sont possibles : soit $\rho' = |\,|^{-(n-(m-1)/2-1/2}$ et $\sigma' \otimes \tilde{\pi}'$ est un quotient de $\Omega_{n-d_{\rho'},m,\zeta}$ soit il existe une représentation irréductible $\tilde{\sigma}'$ telle que $\sigma \otimes \tilde{\sigma}'$ soit un quotient de $\Omega_{n'-d_{\rho'},m'-2d_{\rho'},\zeta}$ et $\tilde{\pi}'$ est un sous-module de l'induite $\rho' \times \tilde{\sigma}'$; pour ce que l'on fait, le problème des torsions n'est pas grave, ce qui compte est la valeur du caractère exceptionnel et celle là est facile à calculer.

Kudla a aussi établi une filtration symétrique en échangeant les rôles des groupes métaplectiques et orthogonaux ; le caractère exceptionnel est alors $-((m-1)/2-n-1/2)$. On applique ce résultat de la façon suivante :

soit ρ une représentation cuspidale unitaire d'un $GL(d_\rho, F)$ et $x \in \mathbb{R}$. On considère les correspondances pour $Mp(2n+2d_\rho, F)$ avec les tours de Witt déjà décrites. On fixe un quotient irréductible de l'induite π' de $\rho|\,|^x \times \pi$. On suppose que soit ρ n'est pas le caractère trivial soit x n'est pas un demi-entier non entier. Ainsi π', par la filtration de Kudla, est un quotient de la représentation métaplectique $\Omega_{n+d_\rho, m_\zeta+d_\rho,\zeta}$. Et il existe une représentation irréductible $\tilde{\pi}'$ de $O(m_\zeta+2d_\rho, F)_\zeta$ telle que $\tilde{\pi}'$ soit l'image de π' dans la correspondance définie par $\Omega_{n+d_\rho, m_\zeta+2d_\rho,\zeta}$. On sait alors que $\tilde{\pi}'$ est un sous-quotient de l'induite $\rho|\,|^x \times \tilde{\pi}_\zeta$ (cf. [**34**]). De plus, par la filtration de Kudla, $Jac_{\rho|\,|^{-x}}\tilde{\pi}' \neq 0$ puisque ceci est vrai pour π et donc $\tilde{\pi}'$ est un quotient de l'induite $\rho|\,|^x \times \tilde{\pi}$; on vérifie symétriquement l'implication réciproque ; on a donc

$$Jac_{\rho|\,|^{-x}}\pi' \neq 0 \Leftrightarrow Jac_{\rho|\,|^{-x}}\tilde{\pi}' \neq 0.$$

Si l'une des induites est irréductible, on a aussi $Jac_{\rho|\,|^x}\pi' \neq 0$ et $Jac_{\rho|\,|^x}\tilde{\pi}' \neq 0$ et les 2 induites sont irréductibles si $x \neq 0$. Si $x = 0$, les induites $\rho \times \pi$ et $\rho \times \tilde{\pi}$ sont semi-simples et tout sous-module irréductible de l'une a pour image un sous-module irréductible de l'autre. On sait que pour le groupe orthogonal, une telle induite est sans multiplicité mais on ne le sait pas pour le groupe métaplectique puisque cela fait partie des résultats d'Harish-Chandra. Supposons donc que $\rho \times \pi$ soit de la forme $\sigma \oplus \sigma$; dans ce cas nécessairement $\rho \times \tilde{\pi}$ est irréductible. Le module de Jacquet de $\rho \times \pi$ est lui semi-simple de longueur 2. Mais alors l'induite déjà considérée $ind_Q^{O(m,F)_\zeta}\omega_{d,d} \otimes \Omega_{n,m_\zeta,\zeta}$ a 2 homomorphismes linérairement indépendants dans $\left(ind_Q^{O(m_\zeta+2d_\rho,F)}\rho \times \tilde{\pi}\right) \otimes \pi$ surjectifs par irréductibilité. Comme le module de Jacquet de $\rho \times \tilde{\pi}$ a un unique quotient irréductible, par réciprocité de Frobenius, cela donne 2 homomorphismes linéairement indépendants de $\Omega_{n,m_\zeta,\zeta}$ sur $\pi \otimes \tilde{\pi}$. Ceci est exclu et on a donc montré :

LEMME 2.2.1. *si ρ n'est pas la représentation triviale ou si x n'est pas un demi-entier non entier, on a l'équivalence pour ζ fixé, $\rho|\,|^x \times \pi$ est irréductible si et seulement si $\rho|\,|^x \times \tilde{\pi}_\zeta$ est irréductible.*

Supposons maintenant que ρ est le caractère trivial et que x est un demi-entier non entier ; on fixe encore un sous-quotient irréductible π' de l'induite $|\,|^x \times \pi$. Pour ζ fixé, la première occurrence de π' est soit m_ζ soit $m_\zeta+2$; le premier cas se produit exactement quand $x = x_0 := -(n+1-(m_\zeta-1)/2-1/2)$ comme on le voit en utilisant la filtration ci-dessus pour la représentation $\Omega_{n+1,m_\zeta,\zeta}$. Supposons donc d'abord que $x = -(n-(m_\zeta-1)/2+1/2)$; alors π' est l'image de $\tilde{\pi}_\zeta$ dans la correspondance $\Omega_{n+1,m_\zeta,\zeta}$ et on sait (cela se vérifie avec la filtration) que le module de Jacquet de π' est réduit à un terme. Ainsi l'induite $|\,|^{n-(m_\zeta-1)/2+1/2} \times \pi$ est réductible. On pose $x_1 := (n-(m_\zeta-1)/2-1/2)$; on remarque que $x_1 = -((m_\zeta+2-1)/2-n-1/2)$ et symétriquement, on démontre que l'induite $|\,|^{x_1} \times \tilde{\pi}_\zeta$ est réductible. Par le résultat de Silberger ([**33**]), on sait que les induites $|\,|^x \times \tilde{\pi}_\zeta$ sont irréductibles pour tout réel x différent de $\pm x_1$. On veut en déduire que les induites $|\,|^x \times \pi$ sont irréductibles pour tout $x \neq \pm x_0$. Pour $x \neq \pm x_0, \pm x_1$, le résultat est clair avec la filtration : π' à pour image toute l'induite $|\,|^x \times \tilde{\pi}_\zeta$ par irréductibilité. En utilisant la filtration symétrique de celle décrite on voit que le module de Jacquet de π' a 2 termes et l'induite est donc irréductible. Supposons donc que $x = x_1$. On vérifie alors que le sous-module irréductible de l'induite $\rho|\,|^{x_1} \times \tilde{\pi}_\zeta$ a sa première occurrence dans la correspondance $\Omega_{n,m_\zeta+2,\zeta}$ et a donc son image dans la correspondance $\Omega_{n+1,m_\zeta+2,\zeta}$ qui est un sous-quotient de l'induite $|\,|^{x_0} \times \pi$; on peut le voir en considérant d'abord la filtration de Kudla pour le parabolique $GL(1) \times O(m_\zeta, F)_\zeta$ de $O(m_\zeta+2, F)_\zeta$ puis en appliquant encore la filtration telle que décrite au quotient $\Omega_{n+1,m_\zeta,\zeta}$. Donc tout sous-quotient irréductible π' de l'induite $|\,|^{x_1} \times \pi$ correspond à un sous-quotient de l'induite $|\,|^{x_1} \times \tilde{\pi}_\zeta$ qui n'est pas le sous-module. Il n'y a plus qu'un choix et l'existence de la correspondance de Howe assure alors l'irréductibilité de l'induite $|\,|^{x_1} \times \pi$.

Remarquons que l'on a donc étendu le résultat de Silberger au cas des groupes métaplectiques et même un peu plus :

LEMME 2.2.2. *soit π une représentation cuspidale irréductible de $Mp(2n, F)$ et soit ρ une représentation cuspidale unitaire d'un groupe $GL(d_\rho, F)$. Si ρ n'est pas autoduale pour tout réel x l'induite $\rho|\,|^x \times \pi$ est irréductible. Si ρ est autoduale il existe exactement une valeur de $x \in \mathbb{R}_{\geq 0}$ tel que $\rho|\,|^x \times \pi$ soit réductible ; notons $x_{\rho,\pi}$ cette valeur de x et notons pour $\zeta = \pm$, $\tilde{\pi}_\zeta$ la première occurrence de π. On a $x_{\rho,\pi} = x_{\rho,\tilde{\pi}_\zeta}$ pour toute représentation ρ autoduale sauf la représentation triviale pour laquelle on a $x_{1,\pi} = |n-(m_\zeta-1)/2+1/2|$; en particulier $x_{1,\pi}$ est demi-entier non entier.*

2.3. Blocs de Jordan et classification des représentations cuspidales. On rappelle la définition bien commode ; soit π une série discrète irréductible d'un groupe G ; on note $Jord(\pi)$ l'ensemble des couples (ρ, a) formés d'une représentation cuspidale irréductible unitaire ρ d'un groupe $GL(d_\rho, F)$ et d'un entier $a \geq 1$ tel que l'induite $St(\rho, a) \times \pi$ soit irréductible alors qu'il existe un entier a' de même parité que a tel que l'induite $St(\rho, a') \times \pi$ soit réductible ; St signifie "Steinberg". Ce n'est pas la définition originale de [**19**] car on évite ici le recours aux fonctions L. L'interprétation en termes de fonctions L découle des résultats d'Arthur, précisément, on va montrer qu'avec cette définition, on retrouve la classification de Langlands telle que montrée par Arthur.

Via la correspondance de Langlands, on identifie une représentation cuspidale irréductible, ρ, d'un groupe $GL(d_\rho, F)$ à un morphisme, encore noté ρ, de W_F dans $GL(d_\rho, \mathbb{C})$. On pose $n_\pi := \sum_{(\rho,a)\in Jord(\pi)} a\, d_\rho$, l'infini si $Jord(\pi)$ n'était pas fini (cf. [**18**] où la finitude est démontrée a priori pour au moins certains de nos groupes). A un tel ensemble $Jord(\pi)$, on associe un morphisme semi-simple ψ_π de $W_F \times SL(2, \mathbb{C})$ dans $GL(n_\pi, \mathbb{C})$ dont la décomposition en sous-représentations irréductibles est précisément $\sum_{(\rho,a)\in Jord(\pi)} \rho \otimes [a]$ où $[a]$ est la représentation irréductible de $SL(2, \mathbb{C})$ de dimension a. Soit $(\rho, a) \in Jord(\pi)$ et fixons a' de même parité que a tel que l'induite $St(\rho, a') \times \pi$ soit réductible. Si G n'est pas un groupe métaplectique, on sait grâce aux travaux d'Harish-Chandra qu'il existe un élément du groupe de Weyl de G qui stabilise le sous-groupe de Levi $GL(a\, d_\rho, F) \times G$ et la représentation $St(\rho, a') \otimes \pi$ de ce sous-groupe de Levi. On vérifie que cela entraîne que $\rho \simeq \rho^*$ si G n'est pas un groupe $GSpin$. Si G est un groupe $GSpin$, le centre de G a pour composante neutre le groupe $GL(1, F)$ qui apparaît naturellement dans la construction et ce groupe agit par un caractère ν_π ; la condition est alors $\rho \simeq \rho^* \otimes \nu_\pi$. Dans les 2 cas la représentation ρ est nécessairement unitaire car l'on a supposé que π est unitaire.

On note $m^*(G)$ la dimension de la représentation naturelle du G^* (la composante neutre du L-groupe de G). Soit ψ un morphisme semi-simple de $W_F \times SL(2, \mathbb{C})$ dans $GL(m^*(G), \mathbb{C})$ ou, si $G = GSpin(m, F)$, $GL(m^*(G), \mathbb{C}) \times \mathbb{C}^*$. On peut alors définir $\pi^{GL}(\psi)$ la représentation de $GL(m^*(G), F)$ (ou de $GL(m^*(G), F) \times F^*$) qui correspond à ψ via la correspondance de Langlands locale. On a déjà défini θ_G et θ_{G^*}. On suppose que la classe de conjugaison de ψ est invariante sous θ_{G^*} ; on prolonge alors $\pi^{GL}(\psi)$ au produit semi-direct de $GL(m^*(G), F)$ (ou $GL(m^*(G), F) \times F^*$)avec $\{1, \theta_G\}$. On suppose ici que G est quasidéployé et n'est pas le groupe métaplectique et on dit qu'une série discrtèe π de G est dans $\Pi(\psi)$ s'il existe une combinaison linéaire de traces de série discrète contenant de façon non triviale la trace de π dont le transfert à $GL(m^*(G), F)$ (ou $GL(m^*(G), F) \times F^*$) est la θ_G trace de $\pi^{GL}(\psi)$. Pour simplifier on admet ici que si ψ est comme ci-dessus, la θ_G-trace de $\pi^{GL}(\psi)$ ne vit que sur un groupe endoscopique du produit semi-direct ci-dessus ; c'est annoncé par Arthur. En combinant l'article [**2**] et l'argument général de [**24**] on montre alors aisément qu'il existe au moins un morphisme ψ tel que $\pi \in \Pi(\psi)$ mais ceci fait aussi partie des annonces faites par Arthur. On admet aussi que les paquets de séries discrètes sont sans multiplicité c'est-à-dire que la somme (sans multiplicité) des traces des représentations dans $\Pi(\psi)$ est une distribution stable dont le transfert à $GL(m^*_G, F).\theta_G$ est la θ-trace de $\pi^{GL}(\psi)$; cela ne sert que pour avoir la multiplicité 1 pour tout paquet d'Arthur.

Le théorème ci-dessous fait le lien entre l'approche de [**19**] et [**25**] et celle évidemment plus efficace d'Arthur. Ce qui importe pour nous est que l'on a ainsi un calcul des points de réductibilité des induites de cuspidales dont on a besoin pour décrire les représentations. De plus, la définition des blocs de Jordan est une traduction des propriétés de certaines fonctions μ de Harish-Chandra et le résultat ci-dessous montre, comme on s'y attendait, que ces propriétés sont des invariants des paquets de Langlands de séries discrètes.

THÉORÈME 2.3.1. *Soit G un groupe classique quasi-déployé ce qui inclut $GSpin$ et exclut les groupes métaplectiques et soit π une série discrète de G. Alors $\pi \in \Pi(\psi)$ si et seulement si ψ est conjugué de ψ_π, ce qui sous-entend que n_π vaut nécessairement $m^*(G)$.*

Remarque

Soit π une représentation cuspidale de G et ρ une représentation cuspidale d'un GL telle que $\rho \simeq \rho^* \otimes \nu_\pi$ (cf. 2.1). On note $x_{\rho,\pi}$ le réel positif ou nul tel que l'induite $\rho|\,|^{x_{\rho,\pi}} \times \pi$ soit réductible. Alors $Jord(\pi)$ est l'ensemble des couples (ρ', a') tel que $x_{\rho',\pi} > 1/2$ est un demi-entier et $a \leq 2x_{\rho',\pi}$ avec $a \equiv 2x_{\rho',\pi'}$. Ceci est démontré dans [**18**] pour les groupes classiques usuels par des méthodes complètement élémentaires de théorie des représentations, qui se généralisent donc sans problème à tous les groupes que nous considérons ici. Cela donne une caractérisation facile du paquet contenant une représentation cuspidale en termes de points de réductibilité des induites.

Venons en aux preuves. Dans ce qui suit on suppose que le groupe G n'est pas le groupe métaplectique, on traitera ce groupe ultérieurement.

On suppose d'abord que π est une représentation cuspidale irréductible de G.

On fixe ψ un morphisme θ_{G^*} invariant. On décompose ψ en sous représentations irréductibles, c'est-à-dire un ensemble de couples (ρ, a) et on a la condition que $\rho^* \simeq \rho \otimes \nu_\psi$ où ν_ψ est trivial si $G \neq GSpin$ et s'identifie (via la théorie du corps de classes) au caractère de W_F décrit en 2.1 sinon. Fixons a priori ρ vérifiant $\rho^* \simeq \rho \otimes \nu_\psi$ et on note $a_{\rho,\psi}$ le plus grand entier tel que $(\rho, a_{\rho,\psi})$ intervient dans ψ et si cet entier n'existe pas on pose $a_{\rho,\psi} = 0$ si le morphisme $\psi \oplus \rho \otimes \sigma_2$ définit une représentation du produit semi-direct de $GL(m^*(G)+2d_\rho,F)$ avec $\{1,\theta_G\}$ dont la θ_G-trace est un transfert d'une distribution stable à support des séries discrètes d'un groupe de même type que G et $a_{\rho,\psi}=-1$ dans le cas restant. On pose alors $x_{\rho,\psi}:=(a_{\rho,\psi}+1)/2$. On appelle $Jord(\psi)$ l'ensemble des sous-représentations irréductibles incluses dans ψ, c'est un ensemble que l'on identifie à un ensemble de couple (ρ,a) où ρ a la propriété d'invariance ci-dessus et a est un entier strictement positif.

On suppose que $\pi\in\Pi(\psi)$. Dans le cas de $G=GSpin$, on a $\nu_\psi=\nu_\pi$.

On fixe ρ et on montre d'abord que si $x_{\rho,\psi}\geq 1/2$ alors $x_{\rho,\psi}=x_{\rho,\pi}$; en effet, on considère le morphisme, ψ_ρ^+, analogue à ψ mais où on a remplacé la sous-représentation associée à $(\rho,a_{\rho,\psi})$ par la sous-représentation associée à $(\rho,a_{\rho,\psi}+2)$. On réalise $\pi^{GL}(\psi_\rho^+)$ comme l'unique sous-module irréductible de l'induite $\rho|\,|^{x_{\rho,\psi}}\times\pi^{GL}(\psi)\times\rho|\,|^{-x_{\rho,\psi}}$ qui se prolonge aux actions de θ_G (au moins pour des bons choix qui n'importent pas). On écrit la θ_G-trace de $\pi^{GL}(\psi_\rho^+)$ comme transfert et on calcule les modules de Jacquet qui viennent de la réciprocité de Frobenius pour l'inclusion ci-dessus (cf. [**28**] 4.2.1) ; on montre ainsi que nécessairement la trace d'un sous-quotient de l'induite $\rho|\,|^{x_{\rho,\psi}}\times\pi$ intervient pour calculer ce transfert et que ce sous-quotient est nécessairement un sous-module de $\rho|\,|^{x_{\rho,\psi}}\times\pi$ mais n'est pas un sous-module de $\rho|\,|^{-x_{\rho,\psi}}\times\pi$ car son module de Jacquet ne peut contenir le terme $\rho|\,|^{-x_{\rho,\psi}}\otimes\pi$ (par positivité). Cela prouve la réductibilité de l'induite $\rho|\,|^{x_{\rho,\psi}}\times\pi$ et l'égalité $x_{\rho,\psi}=x_{\rho,\pi}$ pour cette représentation ρ. Ceci entraîne que $Jord(\psi)$ est un sous-ensemble de $Jord(\pi)$. Supposons momentanément que G soit un groupe orthogonal ou symplectique ; dans ce cas, on peut utiliser l'inégalité de [**18**]

$$\sum_{(\rho,a)\in Jord(\pi)} a\,d_\rho\leq m_G^*. \tag{1}$$

Comme ψ définit une représentation de dimension m_G^*, on a aussi avec ce que l'on vient de démontrer :

$$m_G^*=\sum_{(\rho,a)\in Jord(\psi)} a\,d_\rho\geq\sum_{(\rho,a)\in Jord(\pi)} a\,d_\rho.$$

On a donc l'égalité de $Jord(\psi)$ et de $Jord(\pi)$. Si $G=GSpin$ ou est un groupe unitaire, l'inégalité (1) n'a pas été démontrée même si les méthodes employées s'y prêteraient sans doute. Vus les résultats d'Arthur, on peut retrouver cette inégalité différemment, ce sont les arguments développés dans [**24**] en particulier 5.3, qui eux sont tout à fait généraux quand on sait a priori que toute série discrète appartient à un paquet dont on sait transférer la trace à un groupe linéaire tordu par un automorphisme extérieur : on montre que l'induite suivante :

$$\times_{(\rho,a)\in Jord(\pi)}\rho|\,|^{(a+1)/2}\times\pi,$$

où les (ρ,a) sont ordonnés tels que (ρ,a') arrive plus à gauche que (ρ,a) si $a'<a$, contient une série discrète π_{++} comme sous-module irréductible et que $Jord(\pi_{++})=\cup_{(\rho,a)\in Jord(\pi)}(\rho,a+2)$. On met π_{++} dans un paquet $\Pi(\psi_{++})$ pour un bon choix de morphisme ψ_{++}. Et on calcule les modules de Jacquet de $\pi^{GL}(\psi_{++})$ comme transfert convenable. On montre en particulier facilement que $Jord(\psi_{++})$ contient $Jord(\pi_{++})$; en écrivant que ψ_{++} est une représentation de dimension

$$m_G^*+2\sum_{(\rho,a)\in Jord(\pi)} d_\rho$$

on obtient alors l'inégalité (1).

On a donc démontré le théorème dans le cas des représentations cuspidales. Pour l'étendre à toutes les séries discrètes, on introduit la notion de support cuspidal étendu ; soit π une représentation irréductible de G de caractère central unitaire. On choisit un sous-groupe de Levi de G et une représentation cuspidale, σ, tels que π soit un sous-quotient de l'induite de σ ; on écrit σ sous la forme $\otimes_{i\in[1,\ell]}\sigma_i \otimes \pi_{cusp}$ en utilisant un isomorphisme du Levi comme produit de ℓ-groupes linéaires (pour ℓ convenable) avec un groupe de même type que G. Pour tout $i \in [1,\ell]$, on définit $\theta_G\sigma_i$ comme étant σ_i^* si G n'est pas un groupe $GSpin$ et $\sigma_i^* \otimes \nu_\pi$ si $G = GSpin$, où ν_π est le caractère par lequel $GL(1,F)$, vu comme composante neutre du centre de $GSpin$ agit sur π. On définit le support cuspidal étendu de π_{cusp} comme la collection des représentations $\cup_{(\rho,a)\in Jord(\pi_{cusp}} \cup_{x\in[-(a-1)/2,(a-1)/2]} \rho| \, |^x$ et le support cuspidal étendu de π comme l'union du support cuspidal étendu de π_{cusp} avec la collection $\cup_{i\in[1,\ell]}\sigma_i \cup_{i\in[1,\ell]} \theta_G\sigma_i$. On note $Supp_{cusp,e}(\pi)$ l'ensemble ainsi défini ; cet ensemble est considéré non ordonné mais avec multiplicité. On montre par des méthodes de théorie des représentations que si π est une série discrète,

$$Supp_{cusp,e}\pi = \cup_{(\rho,a)\in Jord(\pi)} \cup_{x\in[-(a-1)/2,(a-1)/2]} \rho| \, |^x.$$

Cela force une propriété très particulière de l'ensemble $Supp_{cusp,e}(\pi)$ on peut l'écrire comme union de segments symétriques en l'origine. De plus le résultat du théorème se transforme simplement en l'assertion que pour toute série discrète π si $\pi \in \Pi(\psi)$ pour ψ convenable, alors $Supp_{cusp,e}\pi$ est le support cuspidal usuel de la représentation $\pi^{GL}(\psi)$. On a donc démontré cette assertion pour π_{cusp} et on la déduit pour π en prenant des modules de Jacquet dans l'identité qui donne la θ_G trace de $\pi^{GL}(\psi)$ comme transfert d'une combinaison linéaire de trace de séries discrètes incluant celle de π ; c'est la démonstration faite dans [**24**] qui est totalement générale. On a ainsi le théorème 1 pour toutes les séries discrètes.

2.4. Le cas des groupes non quasi-déployés.

THÉORÈME 2.4.1. *Le théorème de 2.3 est aussi vrai pour les groupes algébriques classiques non quasi-déployés.*

On suppose ici que G n'est pas quasi-déployé et on note G_{dep} la forme intérieure quasi-déployée de G. Pour définir les paquets, on utilise le transfert stable entre G et G_{dep}. On fixe ψ comme précédemment, en particulier avec $Jord(\psi)$ sans multiplicité. Et on définit le paquet $\Pi_{dep}(\psi)$ pour le groupe G_{dep}. On transfert la distribution stable $\sum_{\pi\in\Pi_{dep}(\psi)} \operatorname{tr} \pi$ en une distribution stable sur G qui est une somme fini de caractère de représentations elliptiques de G (d'après [**28**]) avec des coefficients a priori dans $\mathbb{C}^*$. On note alors $\Pi(\psi)$ l'ensemble des représentations qui interviennent dans cette combinaison linéaire. On montre encore que toute série discrète π de G est dans un paquet de la forme $\Pi(\psi)$ et qu'alors $Jord(\pi) = Jord(\psi)$ pour un tel paquet (cf. [**23**]).

On montre aussi que tout élément de $\Pi(\psi)$ est une série discrète : si π est une représentation elliptique sans être une série discrète, il existe des couples (ρ_i, a_i) pour $i \in [1,\ell]$ et une série discrète π' tels que π soit une combinaison linéaire de sous-modules irréductibles de l'induite $\times_{i\in[1,\ell]} St(\rho_i, a_i) \times \pi'$. On pose $Jord(\pi) = Jord(\pi') \cup_{i\in[1,\ell]} (\rho_i, a_i) \cup_{i\in[1,\ell]} (\rho_i, a_i)$; c'est donc un ensemble avec multiplicité. Et étant donné ce que l'on a déjà démontré pour les séries discrètes, on a certainement l'égalité :

$$\sum_{(\rho,a)\in Jord(\pi)} a\, d_\rho = m_G^*.$$

On va montrer encore que $Jord(\pi) = Jord(\psi)$ ce qui donnera une contradiction. Pour cela, on utilise le morphisme ψ_+ qui consiste à changer tous les (ρ, a) de $Jord(\psi)$ en $(\rho, a+2)$ et on considère $\Pi_{dep}(\psi_+)$. On écrit pour des bons coefficients $\sum_{\pi_+} \alpha_{\pi_+} \operatorname{tr} \pi_+(h)$ comme transfert de $\sum_{\pi_{dep,+}\in\Pi(\psi_+)} \operatorname{tr} \pi_{dep,+}$. On calcule encore $\sum_{\pi_+\in\Pi_{dep}(\psi_+)} \circ_{(\rho,a)\in Jord(\psi)} Jac_{\rho|\,|^{(a+1)/2}} \pi_{dep,+}$ où on ordonne $Jord(\psi)$ pour que (ρ, a) soit plus à gauche que (ρ, a') si $a < a'$. On trouve nécessairement la distribution stable associée à Ψ et donc pour tout $\pi \in \Pi(\psi)$ il existe au moins une représentation $\pi_+ \in \Pi(\psi_+)$ tel que le module de Jacquet

$$\circ_{(\rho,a)\in Jord(\psi)} Jac_{\rho|\,|^{(a+1)/2}} \pi_+$$

contienne π. Mais on sait parfaitement calculé ces modules de Jacquet puisque l'on sait que π_+ est une combinaison linéaire de représentations tempérées. Et cela montre que nécessairement pour tout $(\rho, a) \in Jord(\psi)$, on a aussi $(\rho, a) \in Jord(\pi)$. Mais par définition de $Jord(\psi)$, on a l'égalité

$$\sum_{(\rho,a)\in Jord(\psi)} a\, d_\rho = m_G^*.$$

En comparant avec la même égalité pour $Jord(\pi)$ cela force $Jord(\pi) = Jord(\psi)$.

2.5. Classification des séries discrètes. Ici encore, on exclut les groupes métaplectiques. On a donc déjà réexpliqué comment Arthur associe un morphisme de $W_F \times SL(2,\mathbb{C})$ dans G^* à toute série discrète π ; notons ψ ce morphisme. De plus, supposons que G ne soit pas un groupe $GSpin$; en utilisant les propriété de transfert endoscopique, Arthur associe à π un caractère du groupe $Centr_{G^*}\psi$ trivial sur le centre Z_{G^*} exactement quand G est quasidéployé ; même si ce n'est écrit que pour les groupes quasi-déployé, il est facile de l'étendre au cas non quasidéployé en utilisant la stabilisation de [**28**] avec la modification de [**4**] pour l'action des automorphismes sur les données endoscopiques. Si G est un groupe $GSpin(m)$ les mêmes résultats restent vrais à condition d'utiliser le centralisateur de π dans $Sp(m-1,\mathbb{C})$ ou $O(m,\mathbb{C})$ suivant que m est pair ou impair et non dans $Gsp(m-1,\mathbb{C})$ ou $GO(m,\mathbb{C})$ et on tire cela de [**4**] qui explique la situation en général. Dans le cas quasi-déployé et au moins pour les groupes considérés ici, il n'y a pas de différence mais la différence apparaît dans le cas non quasidéployé ; l'application de restriction de l'ensemble des caractères du centralisateur de ψ dans G^* au sous-groupe qui est le centralisateur de ψ dans Sp ou O est surjective sur l'ensemble des caractères de ce sous-groupe trivial sur le centre du groupe. On note ϵ_π ce caractère.

On rappelle juste ici une propriété de ce caractère ϵ_π. Nous avions aussi défini en [**19**] pour toute série discrète un sous-groupe du groupe ci-dessus et un caractère de ce sous-groupe : à tout élément (ρ, a) de $Jord(\psi)$ est associé canoniquement un élément du centralisteur dans G^* de ψ que l'on note $z_{\rho,a}$; c'est l'élément qui agit par -1 sur l'espace de la représentation associé à ρ, a et par 1 "ailleurs". Notre sous-groupe est engendré par les éléments $z_{\rho,a}$ pour $(\rho, a) \in Jord(\psi)$ avec a pair et par les éléments $z_{\rho,a} z_{\rho,a'}$ pour $(\rho, a), (\rho, a') \in Jord(\psi)$ (le même ρ) sans hypothèse de parité sur a, a'. Cette définition avait été faite avec des propriétés de modules de Jacquet et on a vérifié aisément (cf. [**24**] 8.4.3) qu'elle coïncidait avec la restriction de ϵ_π défini par Arthur à notre sous-groupe. C'est comme cela que l'on obtient la propriété suivante : soit $(\rho, a), (\rho, a') \in Jord(\psi)$ le même ρ tel que $a' < a$ et il n'existe pas $b \in]a', a[$ avec $(\rho, b) \in Jord(\psi)$; on dit alors que (ρ, a) et (ρ, a') sont consécutifs. Alors $\epsilon_\pi(z_{\rho,a} z_{\rho,a'}) = 1$ si et seulement si π est un sous-module d'une représentation induite de la forme $\rho| |^{(a-1)/2} \times \cdots \times \rho| |^{(a'+1)/2} \times \sigma$, où σ est convenable. Si $(\rho, a) \in Jord(\psi)$ avec a pair on a la même interprétation pour $\epsilon_\pi(z_{\rho,a}) = 1$ en faisant $a' = 0$ ci-dessus. On note temporairement $a_{\rho,min}$ le plus petit entier tel que $(\rho, a_{\rho,min}) \in Jord(\psi)$ si cet entier existe.

On dit que le caractère ϵ du centralisateur de ψ dans G^* ou sa variante si $G = GSpin(m)$ est alterné si pour $(\rho, a), (\rho, a')$ consécutifs, $\epsilon_\pi(z_{\rho,a} z_{\rho,a'}) = -1$ et si pour tout $(\rho, a_{\rho,min}) \in Jord(\psi)$ tel que $a_{\rho,min}$ soit pair, $\epsilon_\pi(z_{\rho,a_{\rho,min}}) = -1$. On a déjà défini la notion $Jord(\pi)$ ou $Jord(\psi)$ est sans trou (rappelons : si (ρ, a) est dans l'ensemble avec $a > 2$ alors $(\rho, a-2)$ est aussi dans l'ensemble). Pour avoir un théorème plus joli et éviter de devoir faire intervenir explicitement la condition sur la restriction au centre, quand G est défini via l'utilisation d'une forme orthogonale, on note pour $\zeta = \pm$, G_ζ le groupe pour la forme orthogonale de même discriminant et d'invariant de Hasse ζ. On note alors G l'union de G_+ et G_-. Dans le cas où la forme orthogonale est de dimension paire, dans ses résultats annoncés, Arthur montre aussi que le discrimant de la forme orthogonale est relié au déterminant des morphismes $\psi_{|W_F}$ qui interviennent via la théorie du corps de classes. Ceci intervient de façon cruciale et naturelle quand on caractérise les groupes orthogonaux $O(2n, F)$ (ou $GSpin(2n, F)$) comme groupe endoscopique mais cela n'intervient pas explicitement ici et on l'oublie donc des énoncés pour les alléger.

Ceci amène à la classification des représentations cuspidales de G sous la forme suivante :

THÉORÈME 2.5.1. *La classification de Langlands des séries discrètes de G, telle qu'établie par Arthur, induit une bijection entre l'ensemble des représentations cuspidales de G (à isomorphisme près) et l'ensemble des couples ψ, ϵ tel que $Jord(\psi)$ est sans trou et ϵ est alterné ; la bijection, $\pi \mapsto (\psi, \epsilon)$ est définie par le fait que $Jord(\psi) = Jord(\pi)$ et $\epsilon = \epsilon_\pi$.*

Il résulte de ce qui précède que si π est cuspidal $Jord(\pi)$ est sans trou et ϵ_π est alterné ; comme $Jord(\psi) = Jord(\pi)$ si $\pi \in \Pi(\psi)$, on a un des sens du théorème. Réciproquement, soit ψ, ϵ tel que $Jord(\psi)$ soit sans trou et ϵ soit alterné. On sait par les travaux d'Arthur, comme expliqué dans les paragraphes précédents, qu'il existe une série discrète π de G tel que $Jord(\pi) = Jord(\psi)$ et $\epsilon_\pi = \epsilon$. Il faut montrer que π est cuspidal. Mais si π n'est pas cuspidal, il existe une représentation cuspidale unitaire irréductible ρ' d'un groupe linéaire et un réel positif (nécessairement) ainsi qu'une représentation irréductible σ tel que π soit un sous-module de l'induite $\rho'| |^x \times \sigma$. On a démontré (cf. [**19**]) qu'il existe nécessairement $(\rho, a) \in Jord(\psi)$ tel que $\rho' \simeq \rho$ et $x = (a-1)/2$ et c'est élémentaire : on montre que nécessairement σ est tempérée, on peut étendre la définition de $Jord(\pi)$ aux représentation tempérée (cela a été fait ci-dessus) donc définir $Jord(\sigma)$; on étend aussi le résultat qui calcule $Supp_{cusp,e}(\sigma)$ en fonction de $Jord(\sigma)$; en particulier $Supp_{cusp,e}(\sigma)$ est aussi une réunion de segments symétriques en 0. Il est clair que $Supp_{cusp,e}(\pi) = Supp_{cusp,e}(\sigma) \cup \rho'| |^x \cup \theta_G(\rho')| |^{-x}$; si on rajoute à des segments symétriques en 0, 2 éléments et que le résultat est encore une union de segments symétriques en 0, nécessairement $\rho' \simeq \theta_G \rho'$ et les éléments s'ajoutent aux extrêmité d'un segment de $Supp_{cusp,e}(\sigma)$, éventuellement le segment vide si $x = 1/2$. Cela donne l'assertion.

Remarque

Le théorème ne détermine pas complètement la bijection; Arthur la précise en utilisant toutes les propriétés de l'endoscopie. Il y a des choix à faire mais qui semble-t-il reviennent à utiliser les modèles de Whittaker pour les représentations des groupes GL. Il apparaît que quand tous les résultats d'Arthur seront disponibles, on devrait pouvoir donner une classification précise à la Shahidi de la façon suivante : soit donc π une représentation cuspidale et ψ_π le morphisme qui lui est associé simplement par $Jord(\pi) = Jord(\psi_\pi)$. Soit (ρ, a) un couple formé d'une représentation cuspidale unitaire d'un groupe linéaire et a un entier. On regarde l'opérateur d'entrelacement standard défini méromorphiquement pour $s \in \mathbb{C}$:

$$M(\rho, a, \pi, s) := St(\rho, a)|\,|^s \times \pi \to \theta_G(St(\rho, a))|\,|^{-s} \times \pi.$$

Il faut normaliser cet opérateur d'entrelacement en utilisant uniquement (ρ, a) et ψ_π (et non $\pi \in \Pi(\psi_\pi)$. Pour faire simple, on pourrait dire que l'on prend la normalisation de Langlands-Shahidi pour l'unique représentation ayant un modèle de Whittaker dans $\Pi(\psi)$ quand une telle représentation existe. Une autre façon de faire est d'utiliser les fonctions L et les facteurs ϵ de la représentation $\pi^{GL}(\psi)$; on prend la formule conjecturale de Langlands-Shahidi pour le groupe G mais on remplace les fonctions L et ϵ par leur expression en terme de $\pi^{GL}(\psi_\pi)$ conformément à la fonctorialité de Langlands; on note $r(\rho, a, \psi, s)$ cette fonction méromorphe; explicitement pour chaque groupe, s'introduit une représentation r_G d'un GL : si $G = Sp(2n, F)$, r_G est la représentation Sym^2, si G est un groupe orthogonal r_G est la représentation $\wedge^2$ si G est un groupe $GSpin$ c'est la représentation de [**14**] et si G est un groupe unitaire, $U(n)$ r_G est soit la fonction L d'Asai soit un twist de cette fonction suivant la parité de n (cf. [**11**]) et on a, si G est quasi-déployé,

$$r(\rho, a, \psi, s) =$$

$$\times_{(\rho', b') \in Jord(\psi)} \epsilon(St(\rho, a) \times St(\rho', b'), s)^{-1} \frac{L(St(\rho, a) \times St(\rho', b'), s)}{L(St(\rho, a) \times St(\rho', b'), 1 + s)}$$

$$\times \epsilon(St(\rho, a), r_G, 2s)^{-1} \frac{L(St(\rho, a), r_G, 2s)}{L(St(\rho, a), r_G, 1 + 2s)}$$

et dans le cas non quasi déployé, il faut sans doute rajouter des facteurs abéliens. Ce que l'on veut et qui pour le moment n'est qu'une conjecture : on pose $N(\rho, a, \psi, s) := r(\rho, a, \psi, s) M(\rho, a, \pi, s)$ et on veut, pour les couples ρ, a tel que $St(\rho, a) \simeq St(\rho, a)^* \otimes \nu_\pi$ la formule de produit $N(\rho, a, \psi, -s) \circ N(\rho, a, \psi, s) = 1$. Supposons alors que $(\rho, a) \in Jord(\psi)$ et on veut alors que le caractère ϵ associé à π soit tel que $\epsilon(\rho, a)$ soit le scalaire par lequel $N(\rho, a, \psi, 0)$ agit sur l'induite irréductible $St(\rho, a) \times \pi$.

Ceci est lié aux résultats sur l'endoscopie annoncés par Arthur et pourrait même en résulter directement. En effet une telle assertion est en fait le calcul de la distribution caractère

$$\sum_{\pi \in \Pi(\psi)} \operatorname{tr} N(\rho, a, \psi, 0) St(\rho, a) \times \pi)$$

comme transfert. Si (ρ, a) n'a pas la bonne parité c'est-à-dire si $St(\rho, a)$ ne provient pas par endoscopie d'un groupe de même type que G, cette distribution doit être stable et sinon elle doit être un transfert endoscopique à partir d'un groupe endoscopique; ici le groupe à considérer est un groupe de même type que G mais de rang $a\, d_\rho$ plus grand et le groupe endoscopique est la forme quasidéployée du groupe ayant pour L-groupe $G^*(a\,d_\rho,\mathbb{C}) \times G^*(a\,d_\rho + m^*_G,\mathbb{C})$. Et un tel calcul, doit, me semble-t-il, intervenir nécessairement dans la comparaison des formules de traces. La fonction $r(\rho,a,\psi,s)$ est essentiellement le facteur de normalisation pour l'entrelacement :

$$St(\rho,a)|\,|^s \times \pi^{GL}(\psi) \times St(\rho,a)|\,|^{-s},$$

la seule différence vient de la fontion $L(St(\rho,a) \times St(\rho,a), 2s)$ (et le facteur ϵ correspondant) qui intervient dans le cas du groupe GL et se factorise en

$$L(St(\rho,a), r_G, 2s)L(St(\rho,a), r'_G, 2s)$$

pour r'_G une représentation convenable de $GL(a\,d_\rho,\mathbb{C})$; pour le groupe G, on ne garde que le premier facteur, le 2e facteur a un pôle en $s=0$ et il faut donc de toute façon l'enlever ; dans le cas du groupe linéaire ce pôle compense un pôle de l'opérateur d'entrelacement standard puisque l'induite est irréductible alors que l'opérateur d'entrelacement standard n'a pas de pôle pour le groupe classique, l'induite étant réductible.

L'intérêt de cette description est qu'elle ne fait pas intervenir les facteurs de transfert et les choix sont clairs.

2.6. Paquets élémentaires, 1e version . Pour traiter les groupes métaplectiques, on utilise les groupes orthogonaux en se plaçant dans le domaine de petit rang, c'est-à-dire que le groupe orthogonal considéré est de rang beaucoup plus grand que le groupe métaplectique considéré ; cette notion est dûe à R. Howe et dans ce cas Adams a conjecturé le calcul de la correspondance de Howe. Suivant ces conjectures, on sait que l'image d'une série discrète ne reste pas une série discrète mais est dans un paquet d'Arthur simple. On décrit ici un peu plus généralement les paquets qui vont intervenir et on reviendra en 4.2 sur la construction générale.

Ici on fixe un morphisme ψ de $W_F \times SL(2,\mathbb{C}) \times SL(2,\mathbb{C})$ dans G^*, on décompose ce morphisme en sous-représentations irréductibles indexées par des triplets (ρ,a,b) $\in Jord(\psi)$ et on suppose que tous ces triplets ont la propriété que $inf(a,b)=1$. Ce cas a été traité complètement en [**20**] et on reprend ici quelques propriétés. On se limite ici au cas où la restriction de ψ à W_F fois la diagonale de $SL(2,\mathbb{C}) \times SL(2,\mathbb{C})$ est sans multiplicité. On a construit en [**20**] un paquet de représentations associées à ψ ; ce paquet est paramétré par les caractères du centralisateur de ψ avec la bonne restriction au centre de G^* et le bon déterminant de $\psi_{|W_F}$; on note $\pi(\psi,\epsilon)$ la représentation correspondant au caractère ϵ. On aura besoin des propriétés suivantes (qui sont celles qui permettent de construire par récurrence ces paquets de représentations) :

1e propriété : soit $(\rho,a,b),(\rho,a',b')$ 2 triplets distincts tels que $inf(a,b) = inf(a',b') = 1$ et pour tout triplet $(\rho,a'',b'') \in Jord(\psi)$

$$sup(a'',b'') \notin [sup(a,b), sup(a',b')];$$

on suppose que $(a-b)(a'-b') \geq 0$ et que $sup(a,b) > sup(a',b')$ et on note ζ le signe de $a-b$. On note ψ' le morphisme qui se déduit de ψ en ajoutant les 2 triplets et on suppose que ψ' se factorise par un groupe de même type que G^*. On note $\langle \rho|\,|^{\zeta(sup(a,b)-1)/2}, \cdots, \rho|\,|^{-\zeta(sup(a',b')-1)/2}\rangle$ l'unique sous-représentation irréductible de l'induite, pour un GL convenable

$$\rho|\,|^{\zeta(sup(a,b)-1)/2} \times \cdots \times \rho|\,|^{-\zeta(sup(a',b')-1)/2}.$$

Soit $\pi(\psi,\epsilon)$ une représentation du paquet associé à ψ ; alors l'induite

$$\langle\rho|\,|^{\zeta(sup(a,b)-1)/2},\cdots,\rho|\,|^{-\zeta(sup(a',b')-1)/2}\rangle\times\pi$$

a exactement 2 sous-modules irréductibles ; ces 2 sous-modules sont dans le paquet associé à ψ' et sont les représentations $\pi(\psi',\epsilon')$ où ϵ' est un prolongement de ϵ au centralisateur de ψ' vérifiant $\epsilon'(\rho,a,b)=\epsilon'(\rho,a',b')$.

2e propriété : soit π une représentation du paquet associé à ψ. Soit x un réel et ρ une représentation cuspidal unitaire d'un groupe $GL(d_\rho,F)$. On ne peut avoir $Jac_{\rho|\,|^x}\pi\neq 0$ que s'il existe $(\rho,a,b)\in Jord(\psi)$ avec $2|x|+1=sup(a,b)$ et $x(a-b)>0$ (en particulier nécessairement $x>0$). Réciproquement si cette condition est satisfaite et si $Jord(\psi)$ ne contient pas de triplet (ρ,a',b') avec $sup(a',b')=sup(a,b)-2$ alors $Jac_{\rho|\,|^x}\pi\neq 0$.

Une variante de cette propriété est la suivante : soit $(\rho,a,b)\in Jord(\psi)$; on note (ρ,a',b') un triplet tel que $sup(a',b')=sup(a,b)+2$, $inf(a',b')=1$ et $(a'-b')(a-b)\geq 0$. On note ζ le signe de $a'-b'$ et on note ψ' le morphisme qui se déduit de ψ en remplaçant (ρ,a,b) par (ρ,a',b'). On suppose encore que pour tout $(\rho,a'',b'')\in Jord(\psi)$, $sup(a'',b'')\neq sup(a',b')$ ou encore que la restriction de ψ' à W_F fois la diagonale de $SL(2,\mathbb{C})\times SL(2,\mathbb{C})$ est sans multiplicité. Soit $\pi(\psi,\epsilon)$ une représentation dans le paquet associé à ψ, alors l'induite $\rho|\,|^{\zeta(sup(a',b')-1)/2}\times\pi(\psi,\epsilon)$ a un unique sous-module irréductible et ce sous-module irréductible appartient au paquet associé à ψ' ; de plus il est associé au caractère ϵ (vu naturellement comme une application de $Jord(\psi')\simeq Jord(\psi)$ dans ± 1).

3e propriété : on suppose que $Jord(\psi)$ contient, pour ℓ un entier convenable, des éléments $(\rho,a_1,1)$, $\cdots$, $(\rho,a_\ell,1)$ avec $a_1\leq 2$ et tels que pour tout $i\in[1,\ell]$, $a_i=a_1+2(i-1)$. On note ψ' le morphisme obtenu à partir de ψ en changeant chaque triplet $(\rho,a_i,1)$ en $(\rho,1,a_i)$. Soit $\pi=\pi(\psi,\epsilon)$ une représentation du paquet associé à ψ correspondant au caractère ϵ. Alors, π est dans le paquet associé à ψ' si et seulement si pour tout $i\in[1,\ell]$, $\epsilon(\rho,a_i,1)=\eta(-1)^{i-1}$, où $\eta=\pm 1$ est indépendant de i et vaut nécessairement -1 si $a_1=2$.

En [**20**], on a développé cette construction et montré qu'elle est équivalente à une construction plus adaptée au calcul du transfert, construction que l'on rappellera en 4.1 ci-dessous.

3. Le cas des groupes métaplectiques

Le but de ce paragraphe est d'étendre la classification de Langlands aux séries discrètes des groupes métaplectiques. On ne veut utiliser que la notion de module de Jacquet et de réciprocité de Frobenius pour les groupes métaplectiques. On définit les séries discrètes par leurs propriétés sur les modules de Jacquet ; les exposants doivent être dans la chambre de Weyl obtuse positive ouverte ; on a quand même besoin de savoir qu'une représentation ayant cette propriété est unitaire (sinon les démonstrations sont autrement plus compliquées) ; on l'admet donc ici. Si on remplace la chambre de Weyl obtuse ouverte par fermée, on obtient les représentations dites tempérées. Le lemme combinatoire de Bernstein-Zelevinsky qui permet de calculer les modules de Jacquet d'une induite s'applique et on vérifie sans problème que l'induite d'une représentation tempérée, a tous ses sous-quotients qui sont des représentations tempérées. L'hypothèse d'unitarité entraîne qu'une telle induite est semi-simple.

3.1. Propriété des blocs de Jordan d'une série discrète du groupe métaplectique. Soit π une série discrète de $Mp(2n,F)$; on a déjà défini $Jord(\pi)$; on définit aussi $Jord^L(\pi)$ comme l'ensemble des couples (ρ,a) où ρ est une représentation cuspidale irréductible autoduale d'un groupe linéaire et a est un entier pair exactement quand $L(\rho,Sym^2,s)$ a un pôle en $s=0$ et tel que l'induite $St(\rho,a)$ soit irréductible ; la différence avec $Jord(\pi)$ et que l'on remplace le fait que la parité de a est déterminée par une propriété de fonction L et non par le fait qu'il existe un entier a' de même parité que a tel que l'induite $St(\rho,a')\times\pi$ soit réductible. On montrera que $Jord(\pi)=Jord^L(\pi)$ mais ce n'est pas immédiat.

On fixe un ensemble $\mathcal{J}$ de couples (ρ,a) comme ci-dessus avec $\rho\simeq\rho^*$ et a pair exactement quand $L(\rho,Sym^2,s)$ a un pôle en $s=0$. On note $\mathcal{D}_{\mathcal{J}}$ l'ensemble des séries discrètes, π, des groupes métaplectiques $Mp(2n',F)$ où n' peut varier mais telles que $Jord^L(\pi)=\{\mathcal{J}\}$. A priori $\mathcal{D}$ pourrait être un ensemble infini. On fixe N un entier pair grand et on note, $m_N:=1+N+\sum_{(\rho,a)\in\mathcal{J}}a\,d_\rho$ et on note $\mathcal{E}_{\mathcal{J},N}$, l'ensemble des représentations de $O(m_N,F)$ dont la restriction à $SO(m,F)$ est dans le paquet non tempéré associé à l'ensemble des triplets $\{(\rho,a,1);(\rho,a)\in\mathcal{J}\cup(1,1,N)\}$. On sait que pour N grand, (exactement pour tout N strictement supérieur à tout a tel que $(1,a)\in\mathcal{J}$) le cardinal de $\mathcal{E}_{\mathcal{J},N}$ est exactement $2\times 2^{|\mathcal{J}|}$; le premier 2 vient du fait que l'on considère des représentations de $O(m_N,F)$ et non $SO(m_N,F)$. On va montrer alors l'assertion :

PROPOSITION 3.1.1. *On fixe n. Soit $\pi\in\mathcal{D}_{\mathcal{J}}$ une représentation de $Mp(2n,F)$. Pour tout N grand, notion relative à n, l'image de π dans la correspondance de Howe $Mp(2n,F),O(m_N,F)$ est un élément de $\mathcal{E}_{\mathcal{J},N}$. De plus $2n=\sum_{(\rho,a)\in\mathcal{J}}a\,d_\rho$ et $Jord^L(\pi)=Jord(\pi)$.*

On démontre d'abord l'assertion sous l'hypothèse que π est cuspidale.

On note $\tilde\pi$ la première occurrence de π dans la tour de Witt des espaces orthogonaux impairs de noyau anisotrope de dimension 1 et de discriminant 1. On reprend 2.2 ; on connaît le demi-entier $x_{1,\tilde\pi}$ tel que l'induite $|\,|^{x_{1,\tilde\pi}}\times\tilde\pi$ soit réductible, il s'écrit sous la forme $(a_{max,\tilde\pi}+1)/2$ où a_{max} est le plus petit entier pair éventuellement 0 tel que si $(1,a)\in Jord(\tilde\pi)$, $a_{max}\geq a$. On a donc montré, avec les notations de loc. cit. :

$$x_1=(n-(m_{\tilde\pi}-1)/2-1/2)=\pm(a_{max,\tilde\pi}+1)/2.$$

On a aussi montré une égalité symétrique en remplaçant $\tilde\pi$ par π

$$x_0=(-n+(m_{\tilde\pi}-1)/2-1/2)=\pm(a_{max,\pi}+1)/2.$$

De plus pour tout ρ non la représentation triviale $(\rho,a)\in Jord(\pi)$ si et seulement si $(\rho,a)\in Jord(\tilde\pi)$.

Supposons d'abord que $(m_{\tilde\pi}-1)/2-n>0$. On a alors

$$a_{max,\tilde\pi}=(m_{\tilde\pi}-1)-2n; a_{max,\pi}=(m_{\tilde\pi}-1)-2n-2.$$

On rappelle que $Jord(\tilde\pi)$ et $Jord(\pi)$ sont sans trou ; on passe donc de $Jord(\tilde\pi)$ à $Jord(\pi)$ en enlevant simplement le couple $(\rho,a_{max,\tilde\pi})$. En particulier $Jord(\pi)\subset Jord^L(\pi)$ car $Jord(\tilde\pi)$ est formé de couple (ρ,a) avec la parité de a déterminé par les pôles des fonctions L de la représentation ρ vu comme représentation de W_F et donc de la fonction L de ρ vue comme représentation d'un GL d'après [**13**]. Réciproquement soit $(\rho,a)\in Jord^L(\pi)$; la seule possibilité pour que (ρ,a) ne soit pas dans $Jord(\pi)$ est que pour tout a' de même parité que a l'induite $St(\rho,a')\times\pi$ soit irréductible. On suppose d'abord que ρ n'est pas la représentation triviale ;

on sait qu'il existe a' comme ci-dessus tel que $St(\rho, a') \times \tilde{\pi}$ soit réductible ; on regarde l'image des 2 sous-représentations irréductibles de $St(\rho, a') \times \tilde{\pi}$ dans la correspondance de Howe pour cette fois la tour de Witt d'espaces symplectiques. On vérifie que chacune de ces représentations a une image qui est un sous-module de $St(\rho, a') \times \pi$; cela force la réductibilité de cette induite. On suppose maintenant que ρ est la représentation triviale ; ici on sait que si $(\rho, a) \in Jord^L(\pi)$, a est nécessairement paire et on a alors vu ci-dessus que l'induite $St(\rho, a) \times \pi$ est irréductible, pour a pair, exactement quand $a \leq (2x_{1,\pi} - 1)/2$; c'est-à-dire aussi $(\rho, a) \in Jord(\pi)$.

On vient donc de montrer que pour les représentations cuspidales π des groupes métaplectiques, on a $Jord(\pi) = Jord^L(\pi)$.

On montre maintenant que pour N grand (avec les notations de l'énoncé) l'image de π est dans $\mathcal{E}_N$; c'est une retraduction des résultats de Kudla. On sait a priori que l'image de π, $\tilde{\pi}_N$ est un sous-quotient de l'induite :

$$\times_{x \in [-(N-1)/2, n-(m_{\tilde{\pi}}-1)/2-1/2]} | \, |^x \times \tilde{\pi}.$$

On sait aussi, avec la filtration de Kudla que $Jac_x \tilde{\pi}_N = 0$ pour tout $x \neq -(N-1)/2$; on note $\langle | \, |^{-(N-1)/2}, \cdots, | \, |^{n-(m_{\tilde{\pi}}-1)/2-1/2} \rangle$ l'unique sous-module irréductible de l'induite, pour le GL convenable, $\times_{x \in [-(N-1)/2, n-(m_{\tilde{\pi}}-1)/2-1/2]} | \, |^x$ et $\tilde{\pi}_N$ est donc un sous-module irréductible de l'induite

$$\langle | \, |^{-(N-1)/2}, \cdots, | \, |^{n-(m_{\tilde{\pi}}-1)/2-1/2} \rangle \times \tilde{\pi}.$$

On distingue 2 cas ; le 1e cas est celui où $n \leq (m_{\tilde{\pi}} - 1)/2$ dans ce cas, l'induite ci-dessus a un unique sous-module irréductible car $(1, (m_{\tilde{\pi}} - 1) - 2n)$ est dans $Jord(\tilde{\pi})$ et pour tout $(1, a) \in Jord(\tilde{\pi})$, nécessairement $a \leq (m_{\tilde{\pi}} - 1) - 2n$. Ce sous-module irréductible est un élément de $\mathcal{E}_{\mathcal{J},N}$ (variante de la 2e propriété de 2.6). Dans le cas où $n > (m_{\tilde{\pi}} - 1)/2$, l'induite écrite a 2 sous-modules irréductibles car si $(1, a) \in Jord(\tilde{\pi})$ on a $a \leq 2n - (m_{\tilde{\pi}} - 1) - 2)$; (cf. 1e propriété de 2.6) ces 2 sous-modules irréductibles sont les 2 représentations associées au morphisme élémentaire ψ_N tel que $Jord(\psi_N) = \{(\rho, a, b)\}$ tels que pour tout ρ non la représentation triviale, $b = 1$ et $(\rho, a) \in Jord(\pi)$ et pour ρ la représentation triviale, on a soit, $b = 1$ et a est un entier pair vérifiant $a < a_{max,\pi}$, soit $a = 1$ et $b = 2n - (m_{\tilde{\pi}} - 1)$ ou $b = N$; une de ces 2 représentations exactement est dans $\mathcal{E}_{\mathcal{J},N}$ celle qui vérifie $Jac_{| \, |^{-n+(m_{\tilde{\pi}}-1)/2+1/2}} = 0$. Or la filtration de Kudla dit que c'est $\tilde{\pi}_N$ qui a cette propriété (cf. 3e propriété de 2.6). Cela termine la preuve de l'assertion.

On démontre le théorème par récurrence sur n. Pour $n = 0$, il n'y a rien à démontrer car il n'y a pas de représentation du côté du groupe métaplectique. On fixe donc n tel que l'énoncé soit vrai pour tout $n' < n$ et soit $\mathcal{J}$ un ensemble de couples (ρ, a) comme dans l'énoncé. Soit π une série discrète irréductible de $Mp(2n, F)$ telle que $Jord(\pi) = \mathcal{J}$. On fixe ρ et x tel que $Jac_{\rho| \, |^x} \pi \neq 0$; x est nécessairement un demi-entier, (parceque c'est vrai pour les cuspidales) strictement positif (parce que π est une série discrète). Par réciprocité de Frobenius on écrit π comme sous-module d'une induite de la forme $\rho| \, |^x \times \sigma$ pour σ une représentation irréductible convenable. Un calcul complétement élémentaire (cf. [**19**] 2.1) montre que soit σ est une série discrète soit il existe une série discrète σ' et une inclusion de σ dans l'induite :

$$\langle \rho| \, |^{x-1}, \cdots, \rho| \, |^{-(x-1)} \rangle \times \sigma'.$$

On fixe un tel couple ρ, x en supposant x minimum et on fixe alors σ' convenant ; on suppose que l'on est dans le 2e cas qui est le cas le plus difficile ; nécessairement

$x > 1/2$ et π est donc un sous-module irréductible de l'induite :

$$\langle \rho| \,|^x, \cdots, \rho| \,|^{-(x-1)}\rangle \times \sigma'. \tag{1}$$

On pose $\mathcal{J}' := Jord(\sigma')$. Pour tout N grand, on définit $\mathcal{E}_{\mathcal{J}',N'}$. Par récurrence, on sait que l'image de σ' dans la correspondance de Howe pour N grand est un élément, noté $\tilde{\sigma}'_N$ dans $\mathcal{E}_{\mathcal{J}',N}$. Pour les mêmes valeurs de N, on note $\tilde{\pi}_N$ l'image de π dans la correspondance de Howe et on vérifie en utilisant la filtration de Kudla que $\tilde{\pi}_N$ est un sous-module irréductible de l'induite :

$$\langle \rho| \,|^x, \cdots, \rho| \,|^{-(x-1)}\rangle \times \tilde{\sigma}'_N. \tag{2}$$

On considère d'abord le cas où soit ρ n'est pas autoduale soit $2x+1$ n'a pas la bonne parité ; dans ce cas, l'induite (2) a un unique sous-module irréductible qui n'est pas une série discrète. Il vérifie nécessairement $Jac_{\rho| \,|^{x-1}}\tilde{\pi}_N \neq 0$. On en déduit avec la filtration de Kudla que $Jac_{\rho| \,|^{x-1}}\pi \neq 0$ ce qui contredit la minimalité de x.

On suppose donc maintenant que ρ est autoduale et que $2x+1$ a la bonne parité.

On sait décrire les sous-modules irréductibles de (2) (cf. 1e propriété de 2.6) ; ce sont des représentations qui sont associées au morphisme élémentaire ψ_N tel que $(\rho', a', b') \in Jord(\psi_N)$ si et seulement si soit $b' = 1$ et $(\rho, a') \in \mathcal{J}'$, soit $\rho' = 1, a' = 1, b' = N$ soit $\rho' = \rho$, $b' = 1$ et $a' = 2x+1$ ou $2x-1$. Le point est donc de démontrer que $\mathcal{J} = \mathcal{J}' \cup \{(\rho, 2x+1), (\rho, 2x-1)\}$. On suppose d'abord que $(\rho, 2x-1) \in \mathcal{J}'$; on sait alors que $\tilde{\pi}_N$ est un sous-module d'une induite de la forme $St(\rho, 2x-1) \times \sigma''$ pour σ'' convenable et vérifie donc $Jac_{\rho| \,|^{x-1}}\tilde{\pi}_N \neq 0$. On aurait alors aussi $Jac_{\rho| \,|^{x-1}}\pi \neq 0$ ce qui contredirait la minimalité de x. Ainsi $(\rho, 2x-1) \notin \mathcal{J}'$. On montre aussi que $(\rho, 2x+1) \notin \mathcal{J}'$; sinon, $\tilde{\pi}_N$ serait un sous-module irréductible de l'induite $St(\rho, 2x) \times \pi''$ pour π'' convenable et on aurait $Jac_{\rho| \,|^{x-1}, \cdots, \rho| \,|^{-(x-1)}}\tilde{\pi}_N \neq 0$. Cela entraîne une assertion de non nullité du même type pour π mais ceci contredit le fait que π est une série discrète. Il reste à montrer que $\mathcal{J} = \mathcal{J}' \cup \{(\rho, 2x+1), (\rho, 2x-1)\}$. On vérifie d'abord que pour $\zeta = \pm 1$, l'induite $St(\rho, 2x+\zeta) \times \pi$ est irréductible ; supposons qu'il n'en soit pas ainsi et que l'induite est réductible. Par unitarité, cette induite est semi-simple et soit π_1 l'un de ses sous-modules irréductible. On vérifie que $Jac_{\rho| \,|^{x+(\zeta-1)/2}, \cdots, | \,|^{-x-(\zeta-1)/2}}St(\rho, 2x+\zeta) \times \pi$ est isomorphe à 2 copies de π. Cela prouve d'une part que la longueur de l'induite est 2 et d'autre part que $\mathrm{Hom}(\pi_1, St(\rho, 2x+\zeta) \times \pi)$ est de dimension 1, par réciprocité de Frobenius. Ainsi les 2 sous-représentations de l'induite sont non isomorphes. Les images de ces 2 sous-représentations par la correspondance de Howe, sont pour N grand des sous-modules irréductibles de l'induite $St(\rho, 2x+\zeta) \times \tilde{\pi}_N$. Il faut donc que cette induite soit réductible ce qui n'est pas le cas puisque $(\rho, 2x+\zeta, 1)$ intervient dans la décomposition en sous-représentation irréductible du morphisme associé à $\tilde{\pi}_N$. Il faut encore démontrer qu'il existe a' de même parité que $2x+1$ tel que l'induite $St(\rho, a') \times \pi$ soit réductible. Par récurrence on sait que $Jord(\sigma') = Jord^L(\sigma')$ et d'après la condition de parité déjà montrée on sait que pour a' grand $St(\rho, a') \times \sigma'$ est réductible. Or

$$St(\rho, a') \times \langle \rho| \,|^x, \cdots, \rho| \,|^{-(x-1)}\rangle \times \sigma' \simeq \langle \rho| \,|^x, \cdots, \rho| \,|^{-(x-1)}\rangle \times St(\rho, a') \times \sigma'$$

a 4 sous-modules irréductibles. Un calcul de module de Jacquet montre alors que $St(\rho, a') \times \pi$ ne peut être irréductible. On vient donc de montrer que l'image de π est un élément de $\mathcal{E}_{\mathcal{J},N}$; on a aussi l'égalité $2n = \sum_{(\rho,a)\in\mathcal{J}} a\, d_\rho$ en appliquant l'hypothèse de récurrence à σ' en tenant compte de l'inclusion (1). Il faut encore montrer que $\mathcal{J} = Jord^L(\pi)$ puisque l'on vient de démontrer que $\mathcal{J} = Jord(\pi)$. Puisque $\mathcal{J}$ intervient dans la décomposition en sous-représentations irréductibles du morphisme associé à $\tilde{\pi}_N$, on a certainement $Jord(\pi) = \mathcal{J} \subset Jord^L(\pi)$. Soit $(\rho, a) \in Jord^L(\pi)$; on sait donc que l'induite $St(\rho, a) \times \pi$ est irréductible et que la parité de a dépend des pôles de la fonction $L(\rho, Sym^2, s)$ en $s = 0$; on utilise le même argument que ci-dessus pour montrer que pour a' grand de même parité que a, l'induite $St(\rho, a') \times \pi$ est réductible.

Cela termine la preuve de la récurrence.

3.2. Caractère associé à une série discrète de $Mp(2n, F)$. Soit donc π une série discrète de $Mp(2n, F)$; on lui a associé $Jord(\pi)$ et on lui associe une application, ϵ_π, de $Jord(\pi)$ dans ± 1 en utilisant son image par la correspondance de Howe $\tilde{\pi}_N$ définie pour N grand ; cette application qui s'identifie à un caractère du centralisateur de ψ pourra définir soit le caractère trivial soit le caractère non trivial par restriction au centre de $Sp(2n, \mathbb{C})$. En effet on a vu que $\tilde{\pi}_N \in \mathcal{E}_{Jord(\pi),N}$; à la restriction de $\tilde{\pi}_N$ est associée une application de $Jord(\pi) \cup (1, 1, N)$ dans ± 1, notée $\epsilon_{\pi,N}$; on vérifie aisément que cette définition ne dépend pas du choix de N grand. On note simplement ϵ_π la restriction de $\epsilon_{\pi,N}$ à $Jord(\pi)$. Remarquons que si $\tilde{\pi}_N$ et $\tilde{\pi}'_N$ sont les images de 2 séries discrètes distincts π et π' alors nécessairement $\tilde{\pi}_N \not\simeq \tilde{\pi}_N \otimes sign$ et l'application qui à π associe $Jord(\pi), \epsilon_\pi$ est donc injective. On veut montrer que son image est exactement l'ensemble des couple $\mathcal{J}, \epsilon$ où $\mathcal{J}$ est un ensemble de couples (ρ, a) sans multiplicité avec la condition sur la parité de a (a est pair si et seulement si $L(\rho, Sym^2, s)$ a un pôle en $s = 0$) et la condition de dimension

$$2n = \sum_{(\rho,a)} a\, d_\rho$$

et ϵ est une application de $Jord(\pi)$ dans $\{\pm 1\}$.

On fixe $\mathcal{J}$ un ensemble de couples (ρ, a) comme ci-dessus et ϵ une application de $\mathcal{J}$ dans $\{\pm 1\}$. On traite d'abord le cas cuspidal c'est-à-dire le cas où $\mathcal{J}$ est sans trou et où ϵ est alterné et on distingue encore 2 cas. On suppose d'abord que $\mathcal{J}$ ne contient aucun élément (ρ, a) avec ρ la représentation triviale. On note alors

$$\zeta = \times_{(\rho,a)\in\mathcal{J}} \epsilon(\rho, a).$$

On sait que l'on peut associer à $\mathcal{J}, \epsilon$ une représentation cuspidale de $SO(m, F)_\zeta$, où $m = 2n + 1$, notée $\tilde{\pi}_0$ telle que $Jord(\tilde{\pi}_0) = \mathcal{J}$ et $\epsilon_{\tilde{\pi}_0} = \epsilon$. On note $\tilde{\pi}_i$ pour $i = 1, 2$ les extensions de $\tilde{\pi}_0$ à $O(m, F)_\zeta$. Et pour $i = 1, 2$ et N grand, on note $\pi_{i,N}$ l'image de $\tilde{\pi}_i$ dans la correspondance de Howe $Mp(2n + N, F), O(m, F)_\zeta$. En utilisant la filtration de Kudla et le fait que $\tilde{\pi}_i$ est cuspidale, on voit que la première occurrence de $\tilde{\pi}_i$ est pour $N = 2$ ou $N = 0$. Montrons d'abord par l'absurde qu'au moins pour une valeur de i, $N = 0$. Supposons donc que pour $i = 1, 2$, la première occurrence de $\tilde{\pi}_i$ est pour $N = 2$; ceci est équivalent à dire que $Jac_{|\,|^{-1/2}} \pi_{i,2} = 0$. Ainsi $\pi_{i,2}$ est une représentation cuspidale. On calcule l'image de $\pi_{i,2}$ dans la correspondance $Mp(2n + 2, F), O(m + 2N', F)_\zeta$ pour N' grand. On vérifie que l'image est un sous-module irréductible, $\tilde{\pi}_{i,N'}$ de l'induite :

$$\langle |\,|^{-N'+1/2}, \cdots, |\,|^{1/2} \rangle \times \pi_i.$$

De plus, le fait que $Jac_{|\,|^{-1/2}} \pi_{i,2} = 0$ entraîne la même assertion pour $\tilde{\pi}_{i,N'}$ à cause de la filtration de Kudla. Ainsi $\tilde{\pi}_{1,N'} = \tilde{\pi}_{2,N'} \otimes sign$ et cela contredit l'inégalité fondamentale de 2.2. Ainsi pour au moins une valeur de i, $N = 0$. Pour vérifier que $\pi_i := \pi_{i,0}$, pour cette valeur de i convient, on calcule encore l'image de π_i dans la correspondance $Mp(2n,F), O(m+2N',F)_\zeta$ et on trouve un sous-module irréductible de l'induite

$$\langle |\,|^{-N'-1/2}, \cdots, |\,|^{-1/2}\rangle \times \tilde{\pi}_i.$$

Et il résulte des définitions que ϵ est l'application associée à π_i.

On suppose maintenant que $\mathcal{J}$ contient des blocs du type $(1,a)$ et on note a_1, l'entier a maximum avec cette propriété. On note alors $\mathcal{J}'$ l'ensemble $\mathcal{J}$ auquel on a enlevé $(1,a_1)$ ainsi $\mathcal{J}'$ contient $(1,b)$ pour b pair strictement inférieur à a_1. On pose ici $\zeta := \times_{(\rho,a)\in\mathcal{J}'}\epsilon(\rho,a)$ et on note pour $i=1,2$, $\tilde{\pi}_i$ les représentations cuspidales de $O(m',F)_\zeta$ correspondant à $\mathcal{J}'$ et à ϵ' la restriction de ϵ à $\mathcal{J}'$; ici $m' = 2n+1-2a_1$. On a déjà calculé la première occurrence des représentations $\tilde{\pi}_i$; d'après l'inégalité fondamentale de 2.2, on sait que pour au moins une valeur de i cette première occurrence se fait pour $2n_i > m'$. Fixons un tel i et notons π_i l'image de $\tilde{\pi}_i$; on sait que c'est une représentation cuspidale tel que $Jord(\pi_i)$ s'obtient en ajoutant un couple $(1,b)$ avec $b = a_1 - 2 + 2 = a_1$. Ainsi π_i est une représentation de $Mp(2n,F)$ et on vérifie comme ci-dessus qu'elle convient.

On ne fait plus d'hypothèse sur $\mathcal{J}$ et ϵ et on démontre maintenant l'existence de la représentation π associée à $\mathcal{J},\epsilon$ par récurrence sur n. On suppose d'abord que ϵ n'est pas alterné ; on fixe $(\rho,a),(\rho,a')$ consecutifs tels que $\epsilon(\rho,a)=\epsilon(\rho,a')$. On note $\mathcal{J}'$ l'ensemble $\mathcal{J}$ auquel on a enlevé (ρ,a) et (ρ,a'). On note π' la série discrète de $Mp(2n',F)$ (n' convenable) correspondant à $\mathcal{J}'$ et à la restriction de ϵ à $\mathcal{J}'$. On suppose que $a > a'$. D'après les propriétés de $Jord(\pi')$, on sait que l'induite $St(\rho,a')\times\pi'$ est réductible et elle a donc 2 sous-modules irréductibles non isomorphes (calcul de modules de Jacquet et réciprocité de Frobenius, argument déjà employé). Puis on vérifie que l'induite

$$\langle \rho|\,|^{(a-1)/2}, \cdots, \rho|\,|^{-(a'-1)/2}\rangle \times \pi'$$

a 2 sous-représentations irréductibles que l'on note π_i pour $i=1,2$; ces représentations sont non isomorphes car elles ont des modules de Jacquet non isomorphes. Le point est de démontrer que pour $i=1,2$, π_i est une série discrète et qu'elle correspond à $\mathcal{J}$ et une application ϵ_i ayant même restriction à $\mathcal{J}'$ que ϵ et vérifiant $\epsilon_i(\rho,a)=\epsilon_i(\rho,a')$.

On utilise $\tilde{\pi}_{i,N}$ l'image de π_i dans la correspondance avec $O(2n+1+2N,F)$; on a montré que $\tilde{\pi}_{i,N}$ est dans $\mathcal{E}_{\mathcal{J},N}$ et les propriétés des modules de Jacquet de π_i se lisent sur ceux de $\tilde{\pi}_{i,N}$. On obtient alors l'assertion.

Il reste à voir le cas où $\mathcal{J}$ a des trous, c'est-à-dire où il existe $(\rho,a)\in\mathcal{J}$ avec $a>2$ et $(\rho,a-2)\notin\mathcal{J}$. On appelle ici $\mathcal{J}'$ l'ensemble qui se déduit de $\mathcal{J}$ en remplaçant (ρ,a) par $(\rho,a-2)$ et on note ϵ' l'application qui se déduit naturellement de ϵ. On admet l'existence de π' une série discrète de $Mp(2n-2d_\rho,F)$ correspondant à $\mathcal{J}'$ et ϵ' et on considère l'induite $\rho|\,|^{(a-1)/2}\times\pi'$. On vérifie qu'elle a un unique sous-module irréductible que l'on note π et il faut vérifier que π est une série discrète et que π est associé à $\mathcal{J}$ et ϵ ; cela se fait comme ci-dessus.

Cela termine la preuve.

4. Construction générale

Onon On pose $\zeta_{G^*}=+1$ si G^* est un groupe (de similitudes) orthogonal et -1 sinon. On fixe un morphisme ψ de $W_F\times SL(2,\mathbb{C})\times SL(2,\mathbb{C})$ dans G^* ; on décompose ψ en sous-représentations irréductibles, $\psi = \oplus_{(\rho,a,b)\in Jord(\psi)}\rho\otimes[a]\otimes[b]$. On tient évidemment compte des multiplicités éventuelles de $Jord(\psi)$, ce qui suppose que $\rho \otimes [a] \otimes [b]$ est considéré avec le sous-espace dans lequel cette représentation est réalisée. Supposons que θ_{G^*} laisse stable la classe d'isomorphie de $\rho \otimes [a] \otimes [b]$; on peut alors construire une forme bilinéaire symétrique ou orthogonale telle que la représentation de $W_F \otimes SL(2,\mathbb{C}) \otimes SL(2,\mathbb{C})$ laisse stable cette forme éventuellement à homothétie près ; on note $\zeta_{\rho,a,b} = +1$ si cette forme est orthogonale et -1 sinon. Pour G non un groupe de type $GSpin$, on dit que (ρ, a, b) a bonne parité si la classe d'isomorphie de cette représentation est stable par θ_{G^*} et si $\zeta_{\rho,a,b} = \zeta_{G^*}$; le cas des groupes $GSpin$ est un peu différent et expliqué dans [**5**] ; il faut fixé ν_ψ, c'est-à-dire que l'on fixe un caractère de W_F et que l'on regarde les morphismes ψ à valeurs dans G^* qui dans l'inclusion dans $GL(m_G^*, \mathbb{C}) \times GL(1,\mathbb{C})$ se projette sur ν_ψ dans le facteur $GL(1,\mathbb{C})$.

On décompose $\psi = \psi_{mp} \oplus \psi_{bp}$ en la somme de ses sous représentations ayant bonne parité, ce qui donne ψ_{bp} et la somme des autres représentations, c'est-à-dire la somme de celles qui ne sont pas θ_{G^*}-invariante (à isomorphisme près) et de celles qui ont mauvaise parité. De l'algèbre linéaire élémentaire montre que l'on peut trouver une sous-représentation, $\psi_{1/2,mp}$ de ψ_{mp} telle que $\psi_{mp} = \psi_{1/2,mp} \oplus \theta_{G^*}(\psi_{1/2,mp})$. De plus, l'ensemble des éléments $z_{\rho,a,b}$ définis ci-dessus engendre le groupe des composantes du centralisateur de ψ dans G^* ; on peut donc ainsi identifier le groupe des composantes du commutant de ψ_{bp} (dans un groupe convenable) à celui de ψ.

Supposons défini un ensemble de représentations $\pi \in \Pi(\psi_{bp})$ et pour chacune de ces représentations un caractère du groupe des composantes du centralisateur de ψ_{bp}, noté ϵ_π, tel que le caractère

$$\sum_{\pi \in \Pi(\psi_{bp})} \epsilon_\pi(\psi_{bp}) \operatorname{tr} \pi$$

soit stable et se transfère en la trace tordue de $\pi^{GL}(\psi_{bp})$ pour un bon choix d'action de θ_G, où $\epsilon_\pi(\psi_{bp})$ est le produit des $\epsilon_\pi(z_{\rho,a,b})^{b+1}$ quand (ρ, a, b) parcourt $Jord(\psi_{bp})$ en tenant compte des multiplicités ; ce signe est la valeur du caractètre ϵ_π sur l'image par ψ_{bp} de l'élément non trivial du centre de la 2e copie de $SL(2,\mathbb{C})$.

On sait que $\pi^{GL}(\psi)$ est l'induite $\pi^{GL}(\psi_{1/2,mp}) \times \pi^{GL}(\psi_{bp}) \times \theta_G(\pi^{GL}_{1/2,mp})$ et toute action de θ sur $\pi^{GL}(\psi_{bp})$ se prolonge donc canoniquement à $\pi^{GL}(\psi)$. De plus pour le prolongement de l'action utilisée pour calculer le transfert ci-dessus, on a aussi que $\sum_{\pi \in \Pi(\psi_{bp})} \epsilon_\pi(\psi_{bp}) \pi^{GL}(\psi_{1/2,mp}) \times \pi$ a pour transfert la trace tordue de $\pi^{GL}(\psi)$.

Il y a donc 2 points à démontrer : d'une part construire $\Pi(\psi_{bp})$ et d'autre part démontrer que pour tout $\pi \in \Pi(\psi_{bp})$ l'induite $\pi^{GL}(\psi_{1/2,mp}) \times \pi$ est irréductible.

4.1. Construction dans le cas de bonne parité . La construction donnée ici est une variante de celle donnée dans [**21**]. Son avantage et qu'il est plus facile de définir l'action de θ sur $\pi^{GL}(\psi)$ pour laquelle on a l'égalité de transfert.

4.1.1. *Le cas de restriction discrète à la diagonale.* On suppose ici que la restriction de ψ à W_F fois la diagonale de $SL(2,\mathbb{C}) \times SL(2,\mathbb{C})$ est sans multiplicité ; c'est ce que l'on appelle le cas de restriction discrète à la diagonale, ce cas a été traité en [**21**] et cela signifie en termes concrets que pour tout $(\rho, a, b), (\rho, a', b') \in Jord(\psi)$, les segments $[|a-b|+1, a+b-1]$ et $[|a'-b'|, a'+b'-1]$ sont disjoints.

On reprend alors les constructions de [**21**] ; on fixe un couple $(\underline{t}, \underline{\eta})$ d'applications de $Jord(\psi)$ dans $\mathbb{Z}_{\geq 0} \times \{\pm 1\}$ soumis aux conditions suivantes pour tout $(\rho, a, b) \in Jord(\psi)$:

$$\underline{t}(\rho, a, b) \in [0, inf(a,b)/2] \cap \mathbb{Z},$$

et si $inf(a,b)/2 \in \mathbb{Z}$ et que $\underline{t}(\rho, a, b) = inf(a,b)/2$ alors $\underline{\eta}(\rho, a, b) = +$.

A un tel couple on associe une application, $\epsilon_{\underline{t},\underline{\eta}}$, de $Jord(\psi)$ dans ± 1 en posant :

$$\forall (\rho, a, b) \in Jord(\psi), \epsilon_{\underline{t},\underline{\eta}}(\rho, a, b) = \underline{\eta}(\rho, a, b)^{inf(a,b)}(-1)^{[inf(a,b)/2]+\underline{t}(\rho,a,b)}.$$

On pose encore $\epsilon_{\underline{t},\underline{\eta}}(\psi) := \prod_{(\rho,a,b)\in Jord(\psi)} \epsilon_{\underline{t},\underline{\eta}}(\rho, a, b)^{b+1}$.

Le but des constructions est d'associer à un tel couple une représentation irréductible, $\pi(\psi, \underline{t}, \underline{\eta})$ de G de telle sorte que la distribution

$$\sum_{\underline{t},\underline{\eta}} \epsilon_{\underline{t},\underline{\eta}}(\psi)\pi(\psi, \underline{t}, \underline{\eta}),$$

soit stable et se transfère en la θ-trace de $\pi^{GL}(\psi)$ (pour un choix d'action de θ à préciser), où l'on ne somme que sur les couples $\underline{t}, \underline{\eta}$ tels que $\epsilon_{\underline{t},\underline{\eta}}$ ait sa restriction au centre de G^* déterminé par G.

Cette construction se fait par récurrence où la récurrence porte sur

$$\ell(\psi) := \sum_{(\rho,a,b)\in Jord(\psi)} inf((a,b)-1).$$

On traite donc d'abord le cas où $\ell(\psi) = 0$.

Soit ρ une représentation cuspidale unitaire et x un demi-entier ; on suppose que $\rho \simeq \theta_G \rho$. Pour P un sous-groupe parabolique de Levi M de G, on écrit M comme un produit de facteurs GL et un groupe G_M de même type que G mais de rang en général plus petit. Soit σ une représentation irréductible de P triviale sur le radical unipotent de P et vue comme une représentation de M ; on écrit σ comme produit tensoriel $\sigma_{GL} \times \sigma_0$ où σ_0 est une représentation de G_M et σ_{GL} une représentation des facteurs GL. On pose $proj_{\rho,<x}\sigma = 0$ si le support cuspidal de σ_{GL} contient des termes qui ne sont pas de la forme $\rho|\,|^y$ avec $|y| < x$ et $proj_{\rho,<x}\sigma = \sigma$ sinon. On prolonge linéairement cette application au groupe de Grothendieck des représentations lisses de longueur finie de P triviales sur le radical unipotent de P.

On a défini en [**20**] $inv_{\rho,<x}$, une application dans le groupe de Grothendieck de G, en posant :

$$inv_{\rho,<x}(\pi) := \sum_P (-1)^{rgP} ind_P^G res_P^G(\pi),$$

où P parcourt l'ensemble des sous-groupes paraboliques standard de G. On peut définir de façon identique $inv_{\rho,\leq x}$ en remplaçant l'inégalité stricte par une inégalité large.

On considère $\pi_{temp}(\psi, \underline{\eta})$ comme étant la série discrète associée à la restriction, ψ_{temp} de ψ à W_F fois la diagonale de $SL(2,\mathbb{C}) \times SL(2,\mathbb{C})$ et au caractère $\underline{\eta}$, puisque $Jord(\psi_{temp})$ s'identifie très facilement à $Jord(\psi)$: ces 2 morphismes ont même restriction à W_F fois la diagonale de $SL(2,\mathbb{C})$.

On pose

$$\sigma_{\underline{\eta}}\pi(\psi, \underline{\eta}) := \circ_{(\rho,a,b)\in Jord(\psi); b>a} inv_{\rho,<(b-1)/2} inv_{\rho,\leq(b-1)/2} \pi(\psi_{temp}, \underline{\eta}),$$

où $\sigma_{\underline{\eta}} = \pm 1$ est tel que $\pi(\psi, \underline{\eta})$ est une représentation et non son opposée. Bien sûr il faut démontrer l'existence d'un tel signe, c'est fait en [**20**] où l'on montre en plus qu'avec cette définition, $\pi(\psi, \underline{\eta})$ est irréductible. On reviendra sur la question du signe plus bas. On avait une description plus constructive en 2.6 et [**20**] montre que les 2 définitions sont équivalentes.

On suppose maintenant que $\ell(\psi) > 0$ et on construit $\pi(\psi, \underline{t}, \underline{\eta})$ par récurrence. On fixe (ρ, a, b) avec $inf(a, b) > 1$; il est en fait facile de vérifier que les constructions ci-dessous sont indépendantes de ce choix (et c'est expliqué en [**21**]). On pose encore ζ le signe de $a - b$ en prenant $\zeta = +$ si $a = b$.

1e cas : $\underline{t}(\rho, a, b) > 0$; on pose ψ' le morphisme qui se déduit de ψ en changeant (ρ, a, b) en $(\rho, a, b-2)$ si $b \leq a$ et $(\rho, a-2, b)$ si $b > a$; il correspond à un groupe, G' de même type que G mais de rang plus petit. On vérifie que ψ' est encore de restriction discrète à W_F fois la diagonale de $SL(2, \mathbb{C}) \times SL(2, \mathbb{C})$; on considère $\underline{t}', \underline{\eta}'$ qui se déduisent de façon naturelle de $\underline{t}, \underline{\eta}$ sauf que l'on pose $\underline{t}'(\rho, a, b-2)$ (ou $\underline{t}'(\rho, a-2, b)$) $= \underline{t}(\rho, a, b) - 1$. Il est facile de voir que $\epsilon_{\underline{t}', \underline{\eta}}$ a la bonne restriction au centre de l'analogue pour G' de G^* si c'est le cas pour $\epsilon_{\underline{t}, \underline{\eta}}$. On suppose donc défini, $\pi(\psi', \underline{t}', \underline{\eta}')$. On a besoin de savoir que l'induite : $\langle \rho| \,|^{(a-b)/2}, \cdots, \rho| \,|^{-\zeta((a+b)/2-1)} \rangle \times \pi(\psi', \underline{t}', \underline{\eta}')$ a un unique sous-module irréductible et on pose alors $\pi(\psi, \underline{t}, \underline{\eta})$ cet unique sous-module irréductible, c'est-à-dire :

$$\pi(\psi, \underline{t}, \underline{\eta}) \hookrightarrow \langle \rho| \,|^{(a-b)/2}, \cdots, \rho| \,|^{-\zeta((a+b)/2-1)} \rangle \times \pi(\psi', \underline{t}', \underline{\eta}').$$

L'unicité du sous-module irréductible se montre en démontrant par récurrence des propriétés des modules de Jacquet des représentations ainsi construites ; on renvoie à [**21**].

2e cas : $\underline{t}(\rho, a, b) = 0$. On note ψ' le morphisme qui se déduit de ψ en remplaçant (ρ, a, b) par la somme portant sur les entiers $c \in [|a - b| + 1, a + b - 1]$ de même parité que $a + b - 1$ des représentations associées à $(\rho, c, 1)$ si $\zeta = +$ et $(\rho, 1, c)$ si $\zeta = -$. On définit $\underline{t}', \underline{\eta}'$ sur $Jord(\psi) - \{(\rho, a, b)\}$ en restreignant $\underline{t}$ et $\underline{\eta}$ et on pose $\underline{t}'(\rho, c, 1)$ (ou $(\rho, 1, c)) = 0$, $\underline{\eta}'(\rho, c, 1)$ (ou $(\rho, 1, c)) = \underline{\eta}(\rho, a, b)(-1)^{(c-|a-b|-1)/2}$ pour tout c comme ci-dessus. On vérifie que $\prod_c \epsilon_{\underline{t}', \underline{\eta}'}(\rho, c, 1)$ (ou $(\rho, 1, c)$) vaut

$$\underline{\eta}^{inf(a,b)}(-1)^{[inf(a,b)/2]};$$

ainsi $\epsilon_{\underline{t}, \underline{\eta}}$ et $\epsilon_{\underline{t}', \underline{\eta}'}$ ont même restriction au centre de G^*. De plus $\ell(\psi') = \ell(\psi) - inf(a, b) + 1 < \ell(\psi)$ et on pose alors simplement

$$\pi(\psi, \underline{t}, \underline{\eta}) = \pi(\psi', \underline{t}', \underline{\eta}').$$

Et $\Pi(\psi)$ est exactement l'ensemble des représentations $\pi(\psi, \underline{t}, \underline{\eta})$. Ceci sera justifié en 5. Toutes ces représentations sont non isomorphes entre elles, une fois ψ fixé.

4.1.2. *Le cas général de bonne parité.* On suppose ici que $\psi = \psi_{bp}$; on se ramène au cas de restriction discrète à la diagonale de la façon suivante. Pour tout $(\rho, a, b) \in Jord(\psi)$, on défini $\zeta_{\rho,a,b} = +1$ si $a \geq b$ et -1 sinon. On dit qu'un morphisme $\psi_{\gg}$ pour un groupe de même type que G mais de rang éventuellement plus grand domine fortement ψ s'il existe une application de $Jord(\psi)$ vu comme ensemble avec multiplicité dans $\mathbb{Z}_{\geq 0}$ notée $\underline{T}$ tel que l'on ait l'égalité d'ensemble avec multiplicité $Jord(\psi_{\gg}) =$

$$\{(\rho, a + (1 + \zeta_{\rho,a,b})\, \underline{T}(\rho, a, b), b + (1 - \zeta_{\rho,a,b})\, \underline{T}(\rho, a, b)); (\rho, a, b) \in Jord(\psi)\}.$$

En clair on augmente de $2\underline{T}(\rho,a,b)$ le plus grand des 2 entiers a,b et on laisse l'autre inchangé. Précisons que l'écriture prête à confusion car on voit dans $Jord(\psi)$ plusieurs copies d'un même élément (ρ,a,b) (cela dépend des multiplicités) et que $\underline{T}$ est défini sur chacune de ces copies indépendamment. On demande de plus que $\underline{T}$ vérifie la condition suivante : soient $(\rho,a,b),(\rho,a',b')\in Jord(\psi)$ (la même représentation ρ) tels que $\zeta_{\rho,a,b}=\zeta_{\rho,a',b'}$. Si $|a-b|>|a'-b'|$ ou $|(a-b)|=|a'-b'|$ et $\zeta_{\rho,a,b}=-;\zeta_{\rho,a',b'}=+$ alors soit $\underline{T}(\rho,a,b)\gg\underline{T}(\rho,a',b')$.

Plus généralement, on met un ordre total sur $Jord(\psi)$ tel que $(\rho,a,b)>_{Jord(\psi)}(\rho,a',b')$ si $|(a-b)/2|>|(a'-b')/2|$ ou si l'on a égalité mais $\zeta_{\rho,a,b}=-=-\zeta_{\rho,a',b'}$; le reste n'a pas d'importance. On fixe $(\rho_0,a_0,b_0)\in Jord(\psi)$ et on dit que $\psi_{>,(\rho_0,a_0,b_0}$ domine ψ si l'on a $\underline{T}(\rho,a,b)\gg\underline{T}(\rho,a',b')$ pour tout $(\rho,a,b)>_{Jord(\psi)}(\rho,a',b')\geq_{Jord(\psi)}(\rho_0,a_0,b_0$ et $T_{\rho,a,b}=0$ pour tout $(\rho,a,b)\leq_{Jord(\psi)}(\rho,a_0,b_0)$.

Grâce à l'hypothèse que $\psi_{bp}=\psi$, il existe des morphismes $\psi_{\gg}$ dominant fortement ψ et de restriction discrète à la diagonale ; on en fixe un c'est-à-dire une fonction $\underline{T}$ de $Jord(\psi)$ dans les entiers avec les hypothèses écrites ci-dessus pour qu'il n'y ait pas de confusion, on suppose (ce qui est loisible) que $\underline{T}(\rho_1,a_1,b_1)=0$ pour (ρ_1,a_1,b_1) le plus petit élément de $Jord(\psi)$; on peut donc voir $\psi_{\gg}$ comme $\psi_{>,\rho_1,a_1,b_1}$. On sait donc définir $\Pi(\psi_{>,\rho_1,a_1,b_1})$ pour un tel morphisme et on va obtenir $\Pi(\psi)$ en prenant des modules de Jacquet convenables.

Pour cela, pour $(\rho,a,b)\in Jord(\psi)$ et $\underline{T}$ fixé comme ci-dessus, on pose

$$\Delta(\rho,a,b,\underline{T}(\rho,a,b):=$$

$$\begin{matrix} (a-b)/2+\zeta_{\rho,a,b}\underline{T}(\rho,a,b) & \cdots & \zeta(\rho,a,b)((a+b)/2-1+\underline{T}(\rho,a,b)) \\ \vdots & \vdots & \vdots \\ (a-b)/2+\zeta_{\rho,a,b} & \cdots & \zeta_{\rho,a,b}(a+b)/2 \end{matrix} .$$

C'est-à-dire que si $\zeta_{\rho,a,b}=+$, les lignes sont des segments croissants (de gauche à droite) et les colonnes des segments décroissants (de haut en bas) et que si $\zeta_{\rho,a,b}=-$ les croissances sont inverses. On voit $\Delta(\rho,a,b,\underline{T}(\rho,a,b)$ comme un ensemble totalement ordonné (on commence en haut à gauche puis on prend l'ordre de la lecture française). A une telle collection de segments, Zelevinsky a associé une représentation irréductible, notons la $Z(\rho,a,b,\underline{T}_{\rho,a,b})$ comme unique sous-module d'une certaine induite associé au segment soit formés par les lignes soit pas les colonnes le résultat est le même.

On fixe $(\rho_0,a_0,b_0)\in Jord(\psi)$ et on suppose défini $\Pi(\psi_{>,\rho_0,a_0,b_0})$. On note (ρ,a,b) l'élément de $Jord(\psi)$, s'il existe, minimal parmi ceux qui sont plus grands que (ρ_0,a_0,b_0) ; s'il n'existe pas, on a fini et sinon on définit $\Pi(\psi_{>,\rho,a,b})$ de la façon suivante : on pose $B=|(a-b)/2|$, $A=(a+b)/2-1$ et $\Pi(\psi_{>,\rho,a,b})$ est par définition l'ensemble des représentations (a priori virtuelle)

$$\pi':=\circ_{i\in[1,T_{\rho,a,b}]}Jac_{\zeta(B+i),\cdots,\zeta(A+i)}\pi$$

quand π parcourt $\Pi(\psi_{>,\rho_0,a_0,b_0}$. On va montrer que pour π comme précédemment π' est nulle ou irréductible et que π est uniquement déterminé par π' si π' est non nul. Pour cela on montre les lemmes suivants :

LEMME 4.1.1. *Soit ψ' un morphisme tel que $\Pi(\psi')$ ait déjà été défini avec les propriétés ci-dessus. Soit $x\in\mathbb{R}$ et soit $\pi'\in\Pi(\psi')$. Soit m un entier positif. On a $Jac_{x,\cdots,x}\pi=0$ où x intervient m fois sauf s'il existe au moins m élément de $Jord(\psi')$ de la forme (ρ,a',b') avec $x=(a'-b')/2$.*

Ce lemme a été démontré en [**21**] 5.2 et est donc vrai pour les morphismes $\psi_\gg$ et les représentations qui leur sont associées. On admet donc le éléments de $\Pi(\psi_{>,\rho_0,a_0,b_0})$ et on le démontre pour les éléments de $\Pi(\psi_{>,\rho,a,b})$ avec les notations qui précédent l'énoncé ; on démontrera au passage que π' tel que défini ci-dessus est nul ou irréductible. On pose $\rho=\rho_0$; on démontre la propriété par récurrence décroissante sur $i\in[1,T_{\rho,a,b}]$, cela revient à remplacer $\psi_{>,\rho,a,b}$ par un morphisme qui s'en déduit en remplaçant (ρ,a,b) par $(\rho,a+2i-2,b)$ ou $(\rho,a,b+2i-2)$ suivant que $a\geq b$ ou $a<b$. On peut donc supposer que $T_{\rho,a,b}=1$ en ayant perdu l'hypothèse que $T_{\rho,a,b}$ est très grand. On note $\psi_>$ le morphisme relatif à π ; on remplace (ρ,a,b) par (ρ,A,B,ζ) en posant $B=|(a-b)|$, $A=(a+b)/2-1$ et ζ le signe de $(a-b)$ si $a-b\neq 0$ et $+$ sinon. On sait donc que l'on obtient ψ' en remplaçant dans $Jord(\psi_>)$ l'élément $(\rho,A+1,B+1,\zeta)$ par (ρ,A,B,ζ). Etant donné l'ordre mis sur $Jord(\psi)$ on sait encore que pour tout $(\rho,a',b',\zeta')\in Jord(\psi_>)$, on a soit $|(a'-b')|\leq B$ soit $(a'-b')/2\gg A$. On a donc défini, a priori dans le groupe de Grothendieck, π' par

$$\pi'=Jac_{\zeta(B+1),\cdots,\zeta(A+1)}\pi.$$

On suppose que $\pi'\neq 0$ et on fixe une représentation irréductible σ tel que par réciprocité de Frobenius, on a une inclusion

$$\pi\hookrightarrow\rho|\,|^{\zeta(B+1)}\times\cdots\times\rho|\,|^{\zeta(A+1)}\times\sigma. \tag{1}$$

En appliquant le lemme à π, on sait que $Jac_x\pi=0$ pour tout

$$x\in]\zeta(B+1),\zeta(A+1)],$$

l'inclusion ci-dessus ce factorise donc nécessairement par l'unique sous-module irréductible pour le GL convenable de l'induite $\rho|\,|^{\zeta(B+1)}\times\cdots\times\rho|\,|^{\zeta(A+1)}$. Ce sous-module est noté $\langle\rho|\,|^{\zeta(B+1)},\cdots,\rho|\,|^{\zeta(A+1)}\rangle$. C'est une représentation de Steinberg tordue si $\zeta=-$ et une représentation de Speh (tordue) si $\zeta=+$. On montre que σ satisfait au lemme. On prend x et m comme dans l'énoncé et on suppose que $Jac_{x,\cdots,x}\sigma\neq 0$ où x intervient m fois. Par réciprocité de Frobenius, il existe une représentation σ' et une inclusion

$$\sigma\hookrightarrow\rho|\,|^{x}\times\cdots\times\rho|\,|^{x}\times\sigma'. \tag{2}$$

En reporte (2) dans (1). On considère d'abord le cas où $x\neq\zeta B$ et $x\neq\zeta(A+2)$. Dans ce cas l'induite

$$\langle\rho|\,|^{\zeta(B+1)},\cdots,\rho|\,|^{\zeta(A+1)}\rangle\times\rho|\,|^{x}\times\cdots\times\rho|\,|^{x}$$

est irréductible et on a aussi $Jac_{x,\cdots,x}\pi\neq 0$ où x intervient m fois si $x\neq\zeta(B+1)$ et $m+1$ fois si $x=\zeta(B+1)$; le résultat pour σ se déduit donc du résultat pour π. On suppose maintenant que $x=\zeta B$; dans ce cas on note τ un sous-quotient irréductible de l'induite $\langle\rho|\,|^{\zeta(B+1)},\cdots,\rho|\,|^{\zeta(A+1)}\rangle\times\rho|\,|^{x}$ et on vérifie que $\tau\times\rho|\,|^{x}$ est irréductible. On a donc $Jac_{x,\cdots,x}\pi\neq 0$ mais pour uniquement $m-1$ copies de x ; le lemme pour π, x et $m-1$ donne le lemme pour σ,x et m puisque (ρ,A,B,ζ) est dans $Jord(\psi')$ sans être dans $Jord(\psi_>)$. Pour $x=\zeta(A+2)$ on veut montrer que $Jac_x\sigma=0$; on suppose donc que $m=1$. On sait que $Jac_x\pi=0$ donc on sait que si $Jac_x\sigma\neq 0$, on sait que (1) et (2) vont nécessairement se factoriser par :

$$\pi\hookrightarrow\langle\rho|\,|^{\zeta(B+1)},\cdots,\rho|\,|^{\zeta(A+2)}\rangle\times\sigma'.$$

Si $B+1$ est grand par rapport aux $|(a'-b')/2|$ avec $(\rho,a',b')\in Jord(\psi_<)$ vérifiant $|(a'-b')/2|\leq B$, on déduit directement de la construction dans le cas discret que

$Jac_{\zeta(B+1),\cdots,\zeta(B+2)}\pi = 0$. Si on n'a pas cette hypothèse on sait que π a été construit par module de Jacquet par la procédure qui précède l'énonce. Plus précisément, on considère l'induite

$$\langle\rho|\,|^{\zeta(B+2)},\cdots,\rho|\,|^{\zeta(A+2)}\rangle \times \pi.$$

Elle contient un unique sous-module irréductible et c'est d'après la définition de π une représentation dans le paquet associé à un morphisme qui se déduit de $\psi_>$ en remplaçant $(\rho, A+1, B+1, \zeta)$ par $(\rho, A+2, B+2, \zeta)$. On note π'' cette représentation et on peut appliquer le lemme à π'' et on a donc

$$\pi'' \hookrightarrow \langle\rho|\,|^{\zeta(B+2)},\cdots,\rho|\,|^{\zeta(A+2)}\rangle \times \langle\rho|\,|^{\zeta(B+1)},\cdots,\rho|\,|^{\zeta(A+2)}\rangle \times \sigma'.$$

Mais on peut échanger les 2 premiers facteurs par irréductibilité et on obtient donc $Jac_{\zeta(B+1)}\pi'' = 0$; ceci contredit le lemme pour $\pi'', x = \zeta(B+1)$ et $m = 1$. D'où la nullité cherchée. On revient maintenant à (1) pour calculer $Jac_{\zeta(B+1),\cdots,\zeta(A+1)}\pi$. On applique ce module de Jacquet au terme de droite de (1) ; on sait maintenant que $Jac_x\sigma = 0$ pour tout $x \in [\zeta(B+1), \zeta(A+1)]$. Ainsi les formules de Bernstein et Zelevinsky montrent que le résultat est σ, d'où nécessairement $\pi' = \sigma$. On a donc démontré le lemme.

LEMME 4.1.2. *On note $\Pi(\psi_{>,\rho,a,b})$ l'ensemble des éléments de la forme*

$$\circ_{i\in[1,T_{\rho,a,b}]} Jac_{\zeta(B+i),\cdots,\zeta(A+i)}\pi$$

en supprimant ceux qui sont nuls. Cet ensemble est sans multiplicité

Les paquets sont sans multiplicités si la restriction de ψ à W_F fois la diagonale de $SL(2,\mathbb{C})$ est sans multiplicité d'après [**21**]. On démontre donc le lemme par la récurrence qui suit la construction. On reprend la preuve précédente et on fait une récurrence sur i comme dans cette preuve. On considère donc π_1, π_2 distincts dans le paquet associé à $\psi_>$ (avec les notations de cette preuve) et on pose pour $i = 1, 2$, $\pi'_i = Jac_{\zeta(B+1),\cdots,\zeta(A+1)}\pi_i$. On sait que ce sont des représentations irréductibles ou nulles et le point est de démontrer que si elles sont toutes 2 non nulles, elles sont distincts. Mais on a démontrer l'inclusion pour $i = 1, 2$ tel que $\pi'_i \neq 0$:

$$\pi_i \hookrightarrow \langle\rho|\,|^{\zeta(B+1)},\cdots,\rho|\,|^{\zeta(A+1)}\rangle \times \pi'_i,$$

et que le membre de droite n'a qu'un unique sous-module irréductible. L'assertion est alors claire.

On peut résumer les résultat dans un lemme : pour tout choix de $\underline{t}_\gg, \underline{\eta}_\gg$ définis sur $Jord(\psi_\gg$

LEMME 4.1.3. *La représentation*

$$\pi(\psi, \underline{t}_\gg, \underline{\eta}_\gg) := \circ_{(\rho,a,b)} Jac_{\Delta(\rho,a,b,\underline{T}(\rho,a,b)}\pi(\psi_\gg, \underline{t}_\gg, \underline{\eta}_\gg)$$

est soit nulle soit irréductible, où les (ρ, a, b) sont pris dans l'ordre croissant (le plus grand est le plus à gauche avec les inversions usuelles dans les compositions d'applications). On note $\Pi(\psi)$ l'ensemble des représentations $\pi(\psi, \underline{t}_\gg, \underline{\eta}_\gg)$ qui sont non nulles ; cet ensemble est sans multiplicité, c'est-à-dire que toutes les représentations ainsi définies sont non isomorphes.

On peut améliorer ce lemme en précisant certains cas où la représentation obtenue est nulle et paramétrer les représentations de $\Pi(\psi)$ uniquement avec les couples $\underline{t}, \underline{\eta}$ défini sur $Jord(\psi)$ vu comme ensemble sans mulitplicité. On renvoie à [**21**] pour lénoncé et [**22**] pour la preuve. Le plus intéressant dans ce raffinement est que $\epsilon_{\underline{t}_{\gg},\underline{\eta}_{\gg}}$ est une application de $Jord(\psi)$ dans $\{\pm 1\}$ où $Jord(\psi)$ est maintenant vu sans multiplicité si $\pi(\psi, \underline{t}_{\gg}, \underline{\eta}_{\gg})$ est non nul. On utilise cette propriété pour simplifier l'écriture. Mais pour le résultat de multiplicité 1 dans les paquets d'Arthur, ce raffinement n'est pas utile.

La paramétrisation qui vient de nos constructions dépend très fortement du choix que l'on a fait pour ordonner $Jord(\psi)$; en fait tout ordre total sur $Jord(\psi)$ vérifiant la simple condition

$$(\rho, a, b) >_{Jord(\psi)} (\rho, a', b') \text{ si } (a-b)(a'-b') \geq 0,$$
$$|a-b| > |a'-b'| \text{ et } a+b > a'+b'$$

convient pour obtenir le lemme ; on obtient heureusement globalement le même ensemble de représentations mais la paramétrisation n'est pas la même ; pour démontrer que l'on a le même paquet de représentation on utilise l'interprétation de $\Pi(\psi)$ en termes de transfert, c'est tout à fait non trivial et on reviendra sur ces questions dans un autre article.

4.2. Le cas général. Ici on suppose que $\psi = \psi_{mp} \oplus \psi_{bp}$ et que ψ_{mp} n'est pas trivial. On a alors défini $\pi_{1/2,mp}$ et on pose pour $\underline{t}$ et $\underline{\eta}$ comme ci-dessus

$$\pi(\psi, \underline{t}, \underline{\eta}) = \pi^{GL}(\psi_{1/2,mp}) \times \pi(\psi_{bp}, \underline{t}, \underline{\eta}).$$

Il est démontré en **[22]** que cette induite est irréductible. En loc. cit., on fait la preuve sans supposer que $\pi(\psi_{bp}, \underline{t}, \underline{\eta})$ est unitaire. On donne ici la démonstration de l'irréductibilité en supposant que $\pi(\psi_{bp}, \underline{t}, \underline{\eta})$ est unitaire. Cette hypothèse est tout à fait raisonable puisqu'il résulte des résultats d'Arthur qu'une telle représentation est nécessairement composante locale d'une forme automorphe de carré intégrable ; d'où l'unitarité. Pour simplifier l'écriture, on pose $\pi := \pi(\psi_{bp}, \underline{t}, \underline{\eta})$ et on note ν_π le caractère central de π si G est de la forme $GSpin$ et ν_π le caractère trivial sinon.

On doit donc démontrer que l'induite $\pi^{GL}(\psi_{1/2,mp}) \times \pi(\psi_{bp}, \underline{t}, \underline{\eta})$ a un unique sous-module irréductible. On rappelle que $\pi^{GL}(\psi_{1/2,mp})$ est une induite irréductible de la forme

$$\times_{(\rho,a,b)\in Jord(\psi_{1/2,mp})} Speh(St(\rho, a), b).$$

On démontre l'irréductibilité cherchée par récurrence sur $|Jord(\psi_{1/2,mp})|$. Supposons d'abord qu'il existe $(\rho, a, b) \in Jord(\psi_{1/2,mp})$ tel que $\rho \not\simeq \rho^* \otimes \nu_\pi$; on fixe un tel ρ et on écrit $\psi'_{1/2,mp}$ le morphisme qui se déduit de $\psi_{1/2,mp}$ en enlevant toutes les représentations correspondant aux triplets (ρ, a, b) pour la représentation ρ fixé ; on suppose aussi (ce qui est loisible) que $Jord(\psi_{1/2,mp})$ ne contient aucun terme de la forme $(\rho^* \otimes \nu_\pi, a', b')$. Ainsi on a

$$\pi^{GL}(\psi_{1/2,mp}) \times \pi \simeq \times_{(a,b);(\rho,a,b)\in Jord(\psi_{1/2,mp})} Speh(St(\rho, a), b) \times \pi^{GL}(\psi'_{1/2,mp}) \times \pi.$$

On montre que le module de Jacquet de cette induite contient le terme

$$\left(\pi^{GL}(\psi_{1/2,mp}) \times \pi \simeq \times_{(a,b);(\rho,a,b)\in Jord(\psi_{1/2,mp})} Speh(St(\rho, a), b)\right) \otimes$$
$$\left(\pi^{GL}(\psi'_{1/2,mp}) \times \pi\right) \qquad (*)$$

et que c'est le seul terme de la forme $\sigma \otimes \pi'$ pour une représentation σ du groupe

$$GL\left(\sum_{(a,b);(\rho,a,b)\in Jord(\psi_{1/2,mp})} abd_\rho, F\right)$$

dont le support cuspidal est formée de représentation de la forme $\rho|\,|^t$ avec

$$t-(a-b)/2 \in \mathbb{Z};$$

ce sont les formules combinatoires de Bernstein-Zelevinsky et évidemment le fait que $\rho \not\simeq \rho^* \otimes \nu_\pi$. Par récurrence, on sait que la représentation $\pi^{GL}(\psi'_{1/2,mp}) \times \pi$ est irréductible et la réciprocité de Frobenius dit que toute sous-représentation irréductible de l'induite $\pi^{GL}(\psi_{1/2,mp}) \times \pi$ contient dans son module de Jacquet la représentation (*) comme quotient. Comme le foncteur de Jacquet est un foncteur exact, il y a donc au plus une sous-représentation irréductible.

On suppose donc maintenant que pour tout $(\rho,a,b) \in Jord(\psi_{1/2,mp})$, on a $\rho \simeq \rho^* \otimes \nu_\pi$.

On se ramène aisément au cas où ψ_{bp} est de restriction discrète à la diagonale puisque les Jac définis pour passer du cas de restriction discrète à la diagonale au cas général commutent à l'induction par des représentations $Speh(St(\rho,a),b)$ si (ρ,a,b) n'est pas de bonne parité. On se ramène aussi aisément (en suivant les définitions) au cas où $\underline{t} \equiv 0$ ce qui permet de supposer que $\ell(\psi_{bp})=0$, c'est-à-dire que ψ_{bp} est un morphisme élémentaire. Puis on se ramène encore au cas où ψ_{bp} est tempéré : l'argument consiste à définir plus précisément $proj_{\rho,<x}$ pour x un demi-entier en ne gardant que les supports cuspidaux des représentations des GL de la forme $\rho|\,|^z$ avec $|z|<x$ et $z-x \in \mathbb{Z}$. Alors l'application que l'on a définie pour passer de $\psi_{temp,bp}$ à ψ_{bp} commute alors à l'induction par $\pi^{GL}(\psi_{1/2,mp})$ et elle conserve l'irréductibilité éventuelle. L'intérêt de ces réductions est de traiter le cas où $\pi^{GL}(1/2,mp)$ est tempéré, qui est en fait le cas le plus difficile.

On suppose donc d'abord que $\psi_{1/2,mp}$ est tempéré. On sait donc que $b=1$ pour tout $(\rho,a,b) \in Jord(\psi_{1/2,mp})$ (c'est l'hypothèse tempéré) et que π est une série discrète ; on sait aussi que $(\rho,a) \notin Jord(\pi)$ parce qu'il n'existe pas a' de même parité que a avec $St(\rho,a')\times\pi$ réductible mais on a donc que l'induite $St(\rho,a)\times\pi$ est irréductible. L'irréductibilité de l'induite $\times_{(\rho,a,b)\in Jord(\psi_{1/2,mp})} St(\rho,a)\times\pi$ se montre alors en utilisant la théorie d'Harish-Chandra ; l'irréductibilité se voit sur les pôles des opérateurs d'entrelacement standard ; or ces opérateurs se factorisent et font nécessairement intervenir un opérateur d'entrelacement standard

$$St(\rho,a)|\,|^s \times \pi \to St(\rho,a)|\,|^{-s} \times \pi$$

qui a un pôle en $s=0$. D'où l'irréductibilité annoncée.

On suppose maintenant que $\pi^{GL}(\psi_{1/2,mp})$ n'est pas une représentation tempérée. On introduit la notation suivante ; soit (ρ,a) comme précédemment et soit t,t' des demi-entiers tels que $t'-t+1 \in \mathbb{Z}_{>0}$; on note $Q(St(\rho,a),t',t)$ l'unique sous-module irréductible de l'induite $St(\rho,a)|\,|^{t'} \times \cdots \times St(\rho,a)|\,|^t$ induite pour un groupe GL convenable. Pour $(\rho,a,b) \in Jord(\psi_{1/2,mp})$, on pose $\delta_b=\delta'_b=-1/2$ si b est pair et $\delta_b=-1$, $\delta'_b=0$ si b est impair. On considère l'induite

$$\times_{(\rho,a,b)\in Jord(\psi_{1/2,mp});b>1}\Big(Q(St(\rho,a),-(b-1)/2,\delta_b) \times Q(St(\rho,a),-(b-1)/2,\delta'_b)\Big)$$

$$\times_{(\rho,a,b)\in Jord(\psi_{1/2,mp});b=1} St(\rho,a) \times \pi,$$

où on ordonne les (ρ,a,b) qui interviennent de telle sorte que si (ρ,a,b) et (ρ,a',b') interviennent alors si $b>b'$, (ρ,a,b) est plus à gauche. L'intérêt est alors que l'induite $Q(St(\rho,a),-(b-1)/2,\delta'_b) \times Q(St(\rho,a),-(b'-1)/2,\delta_{b'})$ est irréductible (d'après [**27**]1.6.3). On vérifie alors que l'induite est une sous-représentation de l'induite

$$\times_{(\rho,a,b)\in Jord(\psi_{1/2,mp});b>1}\Big(Q(St(\rho,a),-(b-1)/2,\delta_b)\times Q(St(\rho,a),-(b-1)/2,\delta_b)\Big)$$
$$\times_{(\rho,a,b)\in Jord(\psi_{1/2,mp});b\equiv 1[2]}St(\rho,a)\times\pi. \qquad (2)$$

Ici il n'y a plus besoin de mettre d'ordre sur $Jord(\psi_{1/2,mp})$ grâce aux propriétés d'irréductibilité prouvées en loc. cit..

On vérifie que cette induite a un unique sous-module irréductible ; c'est un calcul de module de Jacquet, disons que l'on peut utiliser la théorie du quotient de Langlands si π est tempérée ce que l'on a le droit de supposer.

On note $Lang(\psi_{1/2,mp},\pi)$ ce sous-module irréductible ; il intervient en plus avec multiplicité 1 comme sous-quotient de l'induite. Avant de continuer, on remarque que $Lang(\psi_{1/2,mp},\pi)$ est certainement un sous-quotient de l'induite $\pi^{GL}(\psi_{1/2,mp})\times\pi$; c'est un calcul de module de Jacquet qui le prouve. Comme la représentation $\pi^{GL}(\psi_{1/2,mp})\times\pi$ est unitaire, l'irréductibilité cherchée est donc équivalente à l'existence d'une inclusion de $\pi^{GL}(\psi_{1/2,mp})\times\pi$ dans l'induite (2).

Il y a malheureusement quelques difficultés techniques pour démontrer ce résultat dans le cas où il existe $(\rho,a,b)\in Jord(\psi_{1/2,mp})$ avec b impair. On traitera ci-dessous le cas où $Jord(\psi_{1/2,mp})$ contient exactement un élément (ρ,a,b) avec $b>1$ et ici on se ramène par récurrence à ce cas ; la récurrence porte donc sur le cardinal de l'ensemble $(\rho',a',b')\in Jord(\psi_{1/2,mp})$ tel que $b'>1$. Fixons (ρ,a,b) avec $b>1$; on note $\psi'_{1/2,mp}$ le morphisme qui se déduit de $\psi_{1/2,mp}$ en enlevant la représentation associée à (ρ,a,b) et on suppose que $Jord(\psi'_{1/2,mp})$ contient aussi un élément (ρ',a',b') avec $b'>1$. Quitte à échanger ces 2 triplets, on suppose que $b\geq b'$. Par hypothèse de récurrence, on sait que $\pi^{GL}(\psi'_{1/2,mp})\times\pi$ est irréductible donc réduit à $Lang(\psi'_{1/2,mp},\pi)$; on note $\psi''_{1/2,mp}$ le morphisme qui se déduit de $\psi'_{1/2,mp}$ en enlevant (ρ',a',b') mais en rajoutant si b' est impair $(\rho',a',1)$. Donc on a des inclusions :

$$\pi^{GL}(\psi_{1/2,mp})\times\pi\hookrightarrow Q(St(\rho,a),-(b-1)/2,(b-1)/2)\times Q(St(\rho',a'),-(b'-1)/2,\delta_{b'})$$
$$\times Q(St(\rho',a'),-(b'-1)/2,\delta_{b'})\times\pi^{GL}(\psi''_{1/2,mp})\times\pi$$
$$\simeq Q(St(\rho',a'),-(b'-1)/2,\delta_{b'})\times Q(St(\rho',a'),-(b'-1)/2,\delta_{b'})\times$$
$$Q(St(\rho,a),-(b-1)/2,(b-1)/2)\times\pi^{GL}(\psi''_{1/2,mp})\times\pi.$$

Par récurrence, on sait que la représentation induite

$$Q(St(\rho,a),-(b-1)/2,(b-1)/2)\times\pi^{GL}(\psi''_{1/2,mp})\times\pi$$

est irréductible réduite au "bon" sous-quotient de Langlands. On peut donc prolonger les morphismes ci-dessus en une inclusion

$$\hookrightarrow Q(St(\rho',a'),-(b'-1)/2,\delta_{b'})\times Q(St(\rho',a'),-(b'-1)/2,\delta_{b'})$$
$$\times Q(St(\rho,a),-(b-1)/2,\delta_b)\times Q(St(\rho,a),-(b-1)/2,\delta_b)$$
$$\times\pi^{GL}(\psi''_{1/2,mp})\times St(\rho,a)\times\pi,$$

où le facteur $St(\rho,a)$ n'intervient que si b est impair. Il est alors facile de continuer pour trouver une inclusion de $\pi^{GL}(\psi_{1/2,mp})\times\pi$ dans l'induite (2). Cela réduit donc la preuve de l'irréductibilité de l'induite $\pi^{GL}(\psi_{1/2,mp})\times\pi$ au cas où $Jord(\psi_{1/2,mp})$ ne contient qu'un élément (ρ,a,b) avec $b>1$. On a le droit d'utiliser l'involution d'Iwahori-Matsumoto généralisée ([**9**], [**29**]) ; cette involution change π mais π reste dans un paquet associé à un morphisme ψ'_{bp} qui s'obtient simplement en échangeant les 2 copies de $SL(2,\mathbb{C})$. Et elle change $\psi_{1/2,mp}$ en échangeant les triplets (ρ,a,b) en (ρ,b,a) puisqu'ici c'est l'involution de Zelevinsky. On se ramène ainsi à des cas très particuliers de $\psi_{1/2,mp}$: pour tout $(\rho,a,b)\in Jord(\psi_{1/2,mp})$ on a $a=b=1$ sauf soit pour exactement 1 triplet de la forme (ρ,a,b) avec $inf(a,b)>1$ soit pour exactement 2 triplets l'un étant de la forme $(\rho,1,b)$ et l'autre de la forme $(\rho',a',1)$; dans le 2e cas, on écrit $a'=a$ et on suppose dans les 2 cas que $a\geq b$ (on s'y ramène éventuellement avec l'involution d'Iwahori-Matsumoto généralisée).

Maintenant, on fait en plus une récurrence sur $|Jord(\psi_{1/2,mp})|$. On suppose donc pour initialiser la récurrence que $Jord(\psi_{1/2,mp})$ n'a qu'un élément (ρ,a,b) avec $a\geq b>1$. On démontre d'abord que pour tout demi-entier x tel que $x-(b-1)/2$ soit dans $\mathbb{Z}$ l'induite :

$$St(\rho,a)|\,|^x\times\pi \tag{3}$$

est irréductible. Le cas $x=0$ (qui n'est possible que si b est impair) a déjà été traité. On suppose donc que $x>0$ et on remarque que l'induite (3) a un unique quotient irréductible qui est auss l'unique sous-module irréductible de l'induite $St(\rho,a)|\,|^{-x}\times\pi$. De plus cette représentation irréductible intervient avec multiplicité 1 comme sous-quotient de l'induite. Il suffit donc de démontrer que l'opérateur d'entrelacement standard :

$$St(\rho,a)|\,|^x\times\pi\rightarrow St(\rho,a)|\,|^{-x}\times\pi$$

est un isomorphisme. Montrons cela : on inclut π dans une induite de la forme $\times_{(\rho',z')}\rho|\,|^{z'}\times\pi_{cusp}$ où (ρ',z') parcourt un ensemble avec multiplicité formé d'une représentation cuspidale unitaire ρ' et d'un demi-entier z' et où π_{cusp} est une représentation cuspidale. Le point est que si (ρ',z') apparaît et si $\rho'\simeq\rho$ ou ρ^*, $z'-t\neq\pm1$ pour tout $t\in[(a-1)/2,-(a-1)/2]+x$ pour des questions de parité. De plus $St(\rho,a)|\,|^x\times\pi_{cusp}$ est aussi irréductible : les points de réductibilité des induites de la forme $\rho|\,|^s\times\pi_{cusp}$ pour s demi-entier sont aussi tels que $s-t\notin\mathbb{Z}$ pour tout t comme ci-dessus. On a alors facilement l'isomorphisme annoncée puisque l'opérateur d'entrelacement standard est la restriction d'un produit d'opérateur du même type qui correspondent aux induites que l'on vient de décrire. On écrit maintenant (ici on n'utilise pas le fait que $a\geq b$)

$$Q(St(\rho,a),-(b-1)/2,(b-1)/2)\times\pi\hookrightarrow$$

$$Q(St(\rho,a),-(b-1)/2,(b-3)/2)\times St(\rho,a)|\,|^{(b-1)/2}\times\pi$$

$$\simeq Q(St(\rho,a),-(b-1)/2,(b-3)/2)\times St(\rho,a)|\,|^{-(b-1)/2}\times\pi$$

$$\simeq St(\rho,a)|\,|^{-(b-1)/2}\times Q(St(\rho,a),-(b-1)/2,(b-3)/2)\times\pi.$$

Si $b=2$ ou 3 on a terminé. Sinon, on recommence en utilisant l'inclusion

$$Q(St(\rho,a),-(b-1)/2,(b-3)/2)\hookrightarrow$$

$$Q(St(\rho,a),-(b-1)/2,(b-5)/2)\times St(\rho,a)|\,|^{(b-3)/2}.$$

Finalement on trouve l'inclusion $Q(St(\rho,a),-(b-1)/2,(b-1)/2)\times\pi\hookrightarrow$

$$\times_{x\in[-(b-1)/2,\delta_b]}St(\rho,a)|\,|^x\times Q(St(\rho,a),-(b-1)/2,\delta_b)\times St(\rho,a)\times\pi,$$

où le dernier $St(\rho,a)$ n'intervient que si b est impair. On vérifie encore que la dernière induite à un unique sous-module irréductible et on conclut ainsi.

On peut alors traiter l'un des cas restants : on suppose que $|Jord(\psi_{1/2,mp})|$ a plus d'un élément et contient (ρ, a, b) avec $a \geq b > 1$. Ainsi il existe des triplets de la forme $(\rho', 1, 1)$ dnas $Jord(\psi_{1/2,mp})$ et on en fixe 1. On note $\pi'_{1/2,mp}$ le morphisme qui se déduit de $\psi_{1/2,mp}$ en enlevant $(\rho', 1, 1)$ et (ρ, a, b). On a

$$\pi^{GL}(\psi_{1/2,mp}) \times \pi = \rho' \times Q(St(\rho, a), -(b-1)/2, (b-1)/2) \times \pi(\psi'_{1/2,mp}) \times \pi.$$

Par récurrence, on sait que l'induite $Q(St(\rho, a), -(b-1)/2, (b-1)/2) \times \pi(\psi'_{1/2,mp}) \times \pi$ est irréductible, on peut donc continuer les égalités par une inclusion dans l'induite

$$\rho' \times Q(St(\rho, a), -(b-1)/2, \delta_b) \times Q(St(\rho, a), -(b-1)/2, \delta_b) \times \pi(\psi'_{1/2,mp}) \times St(\rho, a) \times \pi,$$

où le terme $St(\rho, a)$ n'intervient que si b est impair.

Comme $a \geq b$, l'induite $\rho' \times Q(St(\rho, a), -(b-1)/2, \delta_b)$ est irréductible (c'est très facile, toute induite $\rho' \times St(\rho, a)|\,|^x$ est irréductible si $x \in [-(b-1)/2, \delta_b]$ car le segment $[(a-1)/2 + x, -(a-1)/2 + x]$ contient 0). On obtient alors l'inclusion :

$$\pi^{GL}(\psi_{1/2,mp}) \times \pi \hookrightarrow$$

$$Q(St(\rho, a), -(b-1)/2, \delta_b) \times Q(St(\rho, a), -(b-1)/2, \delta_b) \times \rho' \times \pi(\psi'_{1/2,mp}) \times St(\rho, a) \times \pi,$$

et cette induite n'a qu'un unique sous-module irréductible $Lang(\psi_{1/2,mp}, \pi)$. Cela termine ce cas.

On suppose maintenant que $Jord(\psi_{1/2,mp})$ contient un terme $(\rho, 1, b)$ et un terme $(\rho', a, 1$ avec $a \geq b$ et éventuellement d'autres termes de la forme $(\rho'', 1, 1)$ que l'on regroupe en un morphisme noté $\psi'_{1/2,mp}$. On procède exactement comme ci-dessus :

$$\pi^{GL}(\psi_{1/2,mp}) \times \pi = St(\rho', a) \times Q(\rho, -(b-1)/2, (b-1)/2) \times \pi^{GL}(\psi'_{1/2,mp}) \times \pi$$

$$\hookrightarrow St(\rho', a) \times Q(\rho, -(b-1)/2, \delta_b) \times Q(\rho, -(b-1)/2, \delta_b) \times \pi^{GL}(\psi'_{1/2,mp}) \times \rho \times \pi,$$

où ρ n'intervient que si b est impair. Ici encore le fait que $a \geq b$ assure que pour tout $x \in [-(b-1)/2, \delta_b]$ l'induite $St(\rho', a) \times \rho|\,|^x$ est irréductible et on continue

$$\simeq Q(\rho, -(b-1)/2, \delta_b) \times Q(\rho, -(b-1)/2, \delta_b) \times St(\rho', a) \times \pi^{GL}(\psi'_{1/2,mp}) \times \rho \times \pi.$$

On conclut comme ci-dessus.

5. Transfert

Fixons ψ ; on considère la représentation $\pi^{GL}(\psi)$. Cette représentation a sa classe d'isomorphie qui est stable sous l'action de θ_G. On définit alors une action de θ_G sur cette représentation de la façon suivante : on écrit $\pi^{GL}(\psi)$ comme unique quotient irréductible d'une représentation induite de tempérée modulo le centre, c'est-à-dire comme quotient de Langlands. On vérifie alors qu'il existe une représentation σ d'un produit de groupe linéaire et une représentation tempérée σ_0 d'un seul groupe linéaire telle que $\pi^{GL}(\psi)$ soit l'unique quotient irréductible de l'induite $\sigma \times \sigma_0 \times \theta_G(\sigma)$ et que σ_0 a sa classe d'isomorphie qui est θ_G invariante (on donne ci-dessous σ et σ_0 ; ce n'est pas exactement la situation du quotient de Langlands mais on peut s'y ramener et l'action de θ ne dépend pas de cette variante). Comme σ_0 est une représentation tempérée d'un groupe linéaire, elle a un modèle de Whittaker par rapport à une fonctionnelle θ_G invariante ; on normalise l'action de θ_G sur σ_0 en demandant que θ_G agisse trivialement sur l'espace de Whittaker. Ensuite on prolonge canoniquement cette action de θ_G à l'induite $\sigma \times \sigma_0 \times \theta_G\sigma$ et par passage au quotient, on en déduit une action sur $\pi^{GL}(\psi)$. On écrit explicitement σ et σ_0 sous la forme :

$$\sigma_0 = \times_{(\rho,a,b)\in Jord(\psi_{bp});b\equiv 1[2]} St(\rho,a)$$

$$\sigma = \pi^{GL}(\psi_{1/2,mp}) \times \times_{(\rho,a,b)\in Jord(\psi_{bp};b\equiv 0[2]} \times_{c\in[(b-1)/2,1/2]} St(\rho,a)|\,|^c$$
$$\times_{(\rho,a,b)\in Jord(\psi_{bp});b\equiv 1[2]} \times_{c\in[(b-1)/2,1]} St(\rho,a)|\,|^c.$$

Cette action de θ_G est la normalisation de Whittaker (cf. [**28**] paragraphe 5)

On fixe ψ et on reprend les notations des paragraphes précédents qui donnent une paramétrisation de $\Pi(\psi)$. En particulier, pour tout choix de $\underline{t},\underline{\eta}$, on a le caractère du centralisateur de ψ défini par :

$$\epsilon_{\underline{t},\underline{\eta}}(\rho,a,b) = \underline{\eta}(\rho,a,b)^{inf(a,b)}(-1)^{[inf(a,b)/2]+\underline{t}(\rho,a,b)}.$$

Pour tout $(\rho,a,b)\in Jord(\psi)$, on pose, en suivant [**28**] 5.6 (où il y a une légère faute de frappe rectifié ici "ou $a-b=b'-a'$ et non $a-b=a'-b'$ comme écrit malencontreusement dans la 1e définition mais rectifié au bas de la page de loc. cit. dans la 2e définition),

$$\mathcal{Z}_{W,(\rho,a,b)} := \{(\rho,a',b'); inf(a,a')\equiv inf(b,b')\equiv 1[2]; sup(a,a')\equiv sup(b,b')\equiv 0[2];$$
$$a+a'<b+b' \text{ ou } a-b=b'-a'\neq 0 \text{ et } (a-b)(a+b-a'-b')<0\}.$$

On définit alors le caractère $\epsilon_{W,\psi}$ en posant pour tout $(\rho,a,b)\in Jord(\psi)$:

$$\epsilon_{W,\psi}(\rho,a,b) = (-1)^{|\mathcal{Z}_{W,\rho,a,b}|}.$$

On note z_ψ l'image par ψ de l'élément non trivial du centre de la 2e copie de $SL(2,\mathbb{C})$; c'est un élément du centralisateur de ψ et on a :

THÉORÈME 5.0.1. *La distribution*

$$\epsilon_{W,\psi}(z_\psi)\sum_{\underline{t}_\gg,\underline{\eta}_\gg} \epsilon_{\underline{t},\underline{\eta}}(z_\psi)\mathrm{tr}\,\pi(\psi,\underline{t}_\gg,\underline{\eta}_\gg)$$

est stable et a pour transfert la trace tordue de $\pi^{GL}(\psi)$ pour l'action de θ que l'on vient de définir.

Ce théorème est prouvé dans [**28**] 4.7.1, et 5.7.1.

On va redonner les très grandes lignes de la preuve du transfert "au signe près", c'est-à-dire sans préciser l'action de θ sur $\pi^{GL}(\psi)$; le calcul du signe est un calcul de facteur de normalisation d'opérateurs d'entrelacement que l'on ne refait pas ici.

Si le transfert est prouvé pour ψ_{bp}, il est immédiat qu'il est vrai pour ψ. On considère maintenant le cas où ψ est de restriction discrète à la diagonale

On fixe une application ϵ de $Jord(\psi)$ dans ± 1 de restriction à Z_{G^*} fixée par G et on pose

$$\pi(\psi,\epsilon) = \sum_{\underline{t},\underline{\eta};\epsilon=\epsilon_{\underline{t},\underline{\eta}}} \pi(\psi,\underline{t},\underline{\eta}).$$

C'est cette représentation $\pi(\psi,\epsilon)$ qui se calcule bien dans le groupe de Grothendieck des représentations de G permettant de prouver le théorème.

Supposons d'abord que ψ est élémentaire, dans ce cas ϵ s'identifie à $\underline{\eta}$ et il n'y a qu'un terme dans la somme. La définition même de $\epsilon_{\underline{t},\underline{\eta}}(\psi)\pi(\psi,\epsilon)$ est alors une formule dans le groupe de Grothendieck qui met en jeu des induites de restriction de la série discrète $\pi_{temp}(\psi,\underline{\eta})$. On a montré que la même formule valait pour $\pi^{GL}(\psi)$

dans ce cas en [**28**] 3.1.2 et 3.2.2 et on en a déduit en loc. cit. paragraphe 4 le transfert cherché.

Supposons maintenant que ψ n'est pas élémentaire; on fixe (ρ, a, b) avec l'hypothèse que $inf(a, b) > 1$. Pour simplifier l'écriture, on suppose que $a > b$; si $a \leq b$, il faut échanger les rôles de a et b. On pose $\epsilon_0 := \epsilon(\rho, a, b)$ et on note ψ' le morphisme qui se déduit de ψ en enlevant le bloc (ρ, a, b). On note aussi ϵ' la restriction de ϵ à $Jord(\psi')$. En [**21**], on a montré la formule : $\pi(\psi, \epsilon) =$

$$\sum_{c \in](a-b)/2,(a+b)/2[} \langle \rho| \,|^{-c}, \cdots, \rho| \,|^{-(a+b)/2+1} > \times Jac_{\rho| \,|^{(a-b)/2+2}, \cdots, \rho| \,|^{c}} \pi\big((\psi', \epsilon')$$

$$\oplus((\rho, a+2, b-2), \epsilon_0))\big)$$

$$\oplus_{\eta=\pm 1}(-1)^{[b/2]}\eta^b \epsilon_0^{b+1} \pi\big((\psi', \epsilon') \oplus ((\rho, a+1, b-1), \eta) \oplus ((\rho, a-b+1, 1), \epsilon_0\eta)\big),$$

où la première somme n'intervient pas si $b = 2$ et $\epsilon_0 = +$ et où, par exemple $(\psi', \epsilon') \oplus ((\rho, a+1, b-1), \eta) \oplus ((\rho, a-b+1, 1), \epsilon_0\eta)$ signifie que l'on regarde le morphisme, $\tilde{\psi}$, qui se déduit de ψ' en ajoutant les représentations associées à $(\rho, a+1, b-1)$ et à $(\rho, a-b+1, 1)$ et que l'on considère l'application de $Jord(\tilde{\psi})$ dans ± 1 qui vaut ϵ' sur $Jord(\psi')$ et vaut η sur $(\rho, a+1, b-1)$ et $\epsilon_0\eta$ sur $(\rho, a-b+1, 1)$. Cette formule n'est pas simple à montrer; la preuve se fait en utilisant les modules de Jacquet.

On pose $\epsilon(\psi) = \times_{(\rho,a,b) \in Jord(\psi)} \epsilon(\rho, a, b)^{b+1}$. On remarque que cela vaut $\epsilon_{\underline{t},\underline{\eta}}(z_\psi)$ avec les notations précédentes. Toujours avec ces notations, on a clairement $\epsilon(\psi) = \epsilon'(\psi')\epsilon_0^{b-2+1}$. On pose $\psi'' := \psi' \oplus (\rho, a+2, b-2)$ et ϵ'' le prolongement de ϵ' qui vaut ϵ_0 sur $(\rho, a+2, b-2)$; ici le groupe, G'', correspondant est de rang plus petit que celui de G (la différence est $2d_\rho(a-b+2)$) mais de même type et la restriction de ϵ'' au centre du groupe dual de G'' est la bonne. On note ψ'''_η le morphisme $\psi' \oplus (\rho, a+1, b-1) \oplus (\rho, a-b+1, 1)$; il est relatif au groupe G. Si $a < b$ le morphisme est $\psi' \oplus (\rho, a-1, b+1) \oplus (\rho, 1, |a-b|+1)$ et pour $\eta = \pm$, on note ϵ'''_η le prolongement de ϵ' qui vaut η sur $(\rho, a \pm 1, b \mp 1)$ et $\eta\epsilon_0$ sur le 2e bloc ajouté. On calcule $\epsilon'''_\eta(\psi''')$ en séparant les cas.

1e cas : $a \geq b$:

$$\eta^b \epsilon_0^{b+1} \epsilon(\psi) = \eta^b \epsilon_0^{b+1} \epsilon'(\psi')\epsilon_0^{b+1} = \eta^b \epsilon'(\psi') = \epsilon'''_\eta(\psi''').$$

2e cas : $a < b$; le signe dans la somme est $(-1)^{[a/2]}\eta^a \epsilon_0^{a+1}$ et on a :

$$\eta^a \epsilon_0^{a+1} \epsilon(\psi) = \eta^a \epsilon_0^{a+1} \epsilon'(\psi')\epsilon_0^{b+1}.$$

$$\epsilon'''_\eta = \epsilon'(\psi')\eta^b(\epsilon_0\eta)^{b-a} = \epsilon'(\psi')\eta^a \epsilon_0^{b+a}$$

et donc encore l'égalité de ces 2 expressions. Quand on va sommer sur η, on remplace ϵ'''_η par ϵ''' en imposant la restriction à $Jord(\psi')$ mais en fait on sommera aussi sur cette restriction; la restriction de ϵ''' au centre de G^* est bien celle fixé par G. Ainsi on obtient, en posant $\zeta_{\rho,a,b} = +$ si $a \geq b$ et $-$ sinon : $\sum_\epsilon \epsilon(\psi)\pi(\psi, \epsilon) =$

$$\sum_{\epsilon''} \sum_{c \in](b-a)/2, -\zeta_{\rho,a,b}(a+b)/2[} \epsilon''(\psi'')$$

$$\langle \rho| \,|^{(a-b)/2}, \cdots, \rho| \,|^{c} \rangle \times Jac_{\zeta_{\rho,a,b}(\rho| \,|^{(a-b)/2+2}, \cdots, \rho| \,|^{c}} \pi(\psi'', \epsilon'')$$

$$\oplus(-1)^{[inf(a,b)/2]} \sum_{\epsilon'''} \epsilon'''(\psi''')\pi(\psi''', \epsilon''').$$

En [**28**] 2.3.1, on a montré une formule du même type pour $\pi^{GL}(\psi)$ en suivant l'action de θ et on a montré comment le théorème se déduit de la comparaison de ces formules en [**28**] paragraphe 4.

On passe du cas de restriction discrète à la diagonale au cas général, en prenant des modules de Jacquet partiel et ceci n'altère pas le théorème ; on remarque d'ailleurs que le signe qui vient naturellement devant $\pi(\psi, \underline{t}_{\gg}, \underline{\eta}_{\gg})$ est précisément $\epsilon_{\underline{t}_{\gg}, \underline{\eta}_{\gg}}(\psi_{\gg}$. Et si on ne s'intéresse qu'à la multiplicité 1 (la question du signe étant réglé par d'autres méthodes) on n'a pas besoin de savoir que $\epsilon_{\underline{t}_{\gg}, \underline{\eta}_{\gg}}$ s'identifie à un caractère du centralisateur de ψ.

Dans tout ce qui précède, on a utiliser uniquement le caractère $\epsilon_{\underline{t}, \underline{\eta}}$; la raison en est que localement la normalisation de l'action de θ sur $\pi^{GL}(\psi)$ n'est pas la plus naturelle et on a utilisé une normalisation ad hoc qui donne un transfert avec ces signes. Pour passer à la normalisation de Whittaker, on a comparé les normalisations en [**28**] paragraphe 5.

6. Paquets d'Arthur pour les groupes métaplectiques

On ne veut pas utiliser le fait que les représentations dans les paquets d'Arthur sont unitaires ce qui compliquent les démonstrations. On va admettre la propriété suivante : le caractère d'une représentation irréductible de $Mp(2n, F)$ est une fonction localement L^1.

Dans le cas des groupes métaplectiques, on n'a pas de transfert disponible pour le moment. Pour ψ un morphisme de $W_F \times SL(2, \mathbb{C}) \times SL(2, \mathbb{C})$ à valeurs dans $Sp(2n, \mathbb{C})$ ayant les propriétés de tout ce travail, on peut associer un paquet de représentations de $Mp(2n, F)$ en généralisant les formules de 4.1 et 4.2. Pour tout N grand, on note ψ_N le morphisme qui se déduit de ψ en ajoutant la représentation $1 \times 1 \times [2N]$ de $W_F \times SL(2, \mathbb{C}) \times SL(2, \mathbb{C})$ trivial sur les 2 premiers facteurs et étant la représentation irréductible de dimension $2N$ sur la 2e copie de $SL(2, \mathbb{C})$ (on reprend évidemment ici une idée d'Adams). On a alors, sous l'hypothèse faite sur le caractère d'une représentation irréductible de $Mp(2n, F)$:

THÉORÈME 6.0.2. *Les représentations de $\Pi(\psi)$ sont irréductibles et ce sont exactement les représentations irréductibles de $Mp(2n, F)$ qui pour tout N grand ont une image par la correspondance de Howe appartenant au paquet de représentations de $O(2n + 2N + 1, F)$ dont la restriction à $SO(2n + 2N + 1, F)$ est un élément du paquet associé à ψ_N.*

On commence par traiter le cas où $\psi = \psi_{bp}$. On utilise la filtration de Kudla pour montrer que le théorème est vrai s'il est vrai dans le cas où la restriction de ψ à W_F fois la diagonale de $SL(2, \mathbb{C})$ est sans multiplicité. Ensuite on a une description explicite des représentations et on montre le théorème par récurrence en utilisant la filtration de Kudla comme pour la classification des séries discrètes.

On passe ensuite au cas où ψ_{mp} n'est pas trivial ; la difficulté ici est qu'il n'est pas raisonable d'admettre l'unitarité des représentations dans le paquet associé à ψ_{bp}. Le point clé est de démontrer l'irréductibilité de l'induite $\pi^{GL}(\psi_{1/2,mp}) \times \pi$ pour toute représentation π dans le paquet associé à ψ_{bp}. On se ramène au cas où π est tempérée : passer du cas élémentaire au cas tempéré se fait avec une application qui bien que n'étant défini que sur le groupe de Grothendieck conserve l'irréductibilité des représentations que l'on considère (au signe près). Passer du cas de restriction discrète à la diagonale au cas élémentaire se fait par récurrence sur $\sum_{(\rho,a,b)\in Jord(\psi_{bp})}(inf(a,b)-1)$; la représentation π est associé à des données $\underline{t}$ et $\underline{\eta}$. Si $\underline{t}$ est identiquement nul, on est directement ramené au cas élémentaire et sinon on sait qu'il existe une représentation π' de même nature que π tel que π soit l'unique sous-module irréductible de l'induite $St(\rho,a)|\,|^{-(b-1)/2}\times\pi'$ si $a\geq b$ et de l'induite $Speh(\rho,b)|\,|^{(a-1)/2}\times\pi'$ si $b>a$. On considère le cas où $b>a$ pour alléger l'écriture ; l'induite $Speh(\rho,b)|\,|^{(a-1)/2}\times\pi^{GL}(\psi_{1/2,mp})$ pour le GL convenable est irréductible. Ainsi on a une inclusion :

$$\pi^{GL}(\psi_{1/2,mp})\times\pi\hookrightarrow Sp(\rho,b)|\,|^{(a-1)/2}\times\pi^{GL}(\psi_{1/2,mp})\times\pi'.$$

Soit σ un sous module irréductible du membre de gauche. Par récurrence, on sait que l'induite $\pi^{GL}(\psi_{1/2,mp})\times\pi'$ est irréductible. Par réciprocité de Frobenius, on voit que le module de Jacquet de σ contient la représentation $Sp(\rho,b)|\,|^{(a-1)/2}\otimes\pi^{GL}(\psi_{1/2,mp})\times\pi'$ comme quotient irréductible ; donc en particulier

$$Jac_{\rho|\,|^{(a-b)/2},\cdots,\rho|\,|^{(a+b)/2-1}}\sigma\neq 0$$

et coïncide avec $\pi^{GL}(\psi_{1/2,mp})\times\pi'$. Mais par les formules standard, ceci vaut encore $Jac_{\rho|\,|^{(a-b)/2},\cdots,\rho|\,|^{(a+b)/2-1}}\pi^{GL}(\psi_{1/2,mp})\times\pi$; en particulier l'induite a un unique sous-module irréductible et ce sous-module irréductible intervient avec multiplicité 1 comme sous-quotient de l'induite. Par irréductibilité $\pi^{GL}(\psi_{1/2,mp})\times\pi'\simeq\pi^{GL}(\psi_{1/2,mp})^*\times\pi'$. Ainsi σ est aussi l'unique sous-module irréductible de l'induite $\pi^{GL}(\psi_{1/2,mp})^*\times\pi$; on a donc fini si π est unitaire. Sinon, soit τ un quotient irréductible de l'induite $\pi^{GL}(\psi_{1/2,mp})\times\pi$. On introduit δ un élément de $Gsp(2n,F)$ de norme symplectique -1 ; on suppose qu'il est dans le sous-groupe de Levi de $Sp(2n,F)$ image de celui que l'on utilise dans $Mp(2n,F)$ pour induire. On sait avec [**26**] 4.II.2 que τ^* est isomorphe à l'image de τ sous l'action de δ. Par dualité τ^* est un sous-module irréductible de l'induite $\pi^{GL}(\psi_{1/2,mp})^*\times\pi^*$. Ainsi

$$\tau^\delta\hookrightarrow\pi^{GL}(\psi_{1/2,mp})\times\pi^\delta.$$

En faisant agir δ, on obtient l'inclusion de τ dans l'induite $\pi^{GL}(\psi_{1/2,mp})^*\times\pi$. Ainsi τ est isomorphe à σ et comme σ intervient avec multiplicité 1 comme sous-quotient de l'induite $\pi^{GL}(\psi_{1/2,mp})\times\pi$, cette induite est nécessairement irréductible. On est donc ramené au cas où π est tempérée et ici on a déjà admis qu'une représentation tempérée de $Mp(2n,F)$ est unitaire. Ainsi l'induite $\pi^{GL}(\psi_{1/2,mp})\times\pi$ est semi-simple. Pour N grand, on note $\tilde{\pi}_N$ l'image de π pour la correspondance de Howe, $Mp(2n_\pi,F),O(2n_\pi+2N+1,F)$, où n_π est défini par π. Et en utilisant la filtration de Kudla comme en 2.2, on montre que la longueur de la représentation $\pi^{GL}(\psi_{1/2,mp})\times\pi$ est inférieure ou égale à celle de l'induite $\pi^{GL}(\psi_{1/2,mp})\times\tilde{\pi}_N$, c'est-à-dire est irréductible d'après 4.2. Cela termine la preuve.

7. Paquet de Langlands à l'intérieur d'un paquet d'Arthur

On fixe ψ. On note i_2 le morphisme de W_F dans le tore de la 2e copie de $SL(2,\mathbb{C})$, $w\in W_F\mapsto\begin{pmatrix}|w|^{1/2} & 0\\ 0 & |w|^{-1/2}\end{pmatrix}$. D'après Arthur, le paquet associé à ψ doit contenir tout le paquet de Langlands associé au morphisme ψ_L obtenue en restreignant ψ à $W_F\times SL(2,\mathbb{C})$ où W_F s'envoie dans $W_F\times SL(2,\mathbb{C})\times SL(2,\mathbb{C})$ par l'identité fois i_2 et $SL(2,\mathbb{C})$ s'identifie à la première copie de $SL(2,\mathbb{C})$. Décrivons les représentations dans ce paquet de Langlands, cela sera utilisé dans la preuve et c'est utile pour d'autres démonstrations. On note $\psi_{0,L}$ la sous-représentation de

$W_F \times SL(2,\mathbb{C})$ dans ψ_L correspondant au poids 0 de W_F c'est-à-dire celle où W_F agit de façon unitaire ; en d'autres termes la décomposition en sous-représentations irréductibles de $\psi_{0,L}$ correspond à l'ensemble des $(\rho,a,1)$ pour lesquels il existe b impair avec $(\rho,a,b) \in Jord(\psi)$ et la multiplicité d'un $(\rho,a,1)$ est exactement le nombre d'entiers b impairs avec $(\rho,a,b) \in Jord(\psi)$. On reprend des notations déjà introduites, pour tout entier b, on pose $\delta_b = -1$ si b est impair et $\delta_b = -1/2$ si b est pair et pour $(\rho,a,b) \in Jord(\psi)$, pour $(\rho,a,b) \in Jord(\psi)$, on note $Q(St(\rho,a),-(b-1)/2,\delta_b)$ l'unique sous-module irréductible pour le GL convenable de l'induite $St(\rho,a)|\,|^{-(b-1)/2} \times \cdots St(\rho,a)|\,|^{\delta_b}$. On vérifie que pour toute représentation irréductible π_0 appartenant au paquet tempéré $\Pi(\psi_{0,L})$ l'induite

$$\times_{(\rho,a,b)\in Jord(\psi)} Q(St(\rho,a),-(b-1)/2,\delta_b) \times \pi_0$$

a un unique sous-module irréductible. On le note $\pi_{0,L}$ et l'ensemble des $\pi_{0,L}$ quand π_0 varie est exactement l'ensemble des représentations dans le paquet de Langlands défini ci-dessus.

Proposition 7.0.3. *Le paquet de Langlands associé à ψ_L est inclus dans le paquet d'Arthur.*

On fixe ψ et on raisonne par récurrence sur $b[\psi] := \sum_{(\rho,a,b)\in Jord(\psi)}(b-1)$. Si $b[\psi]=0$, le paquet associé à ψ est tempéré et coïncide avec le paquet de Langlands. Le résultat est alors évident. On suppose donc que $b[\psi]>0$. En utilisant 4.2, on se ramène aisément au cas où $\psi=\psi_{bp}$: le paquet d'Arthur associé à ψ est formé des représentations irréductibles $\pi^{GL}(\psi_{1/2,mp}) \times \pi'$ où π' parcourt le paquet associé à ψ_{bp}. Et ceci est aussi vrai pour le paquet de Langlands par la même référence.

On suppose donc que $\psi=\psi_{bp}$ et on suppose d'abord qu'il existe $(\rho,a,b) \in Jord(\psi)$ tel que $a<b$. On fixe un tel triplet $(\rho,a,b) \in Jord(\psi)$ tel que $b-a$ est maximum et si ce maximum est atteint par plusieurs triplets on suppose alors que $a+b$ est maximum parmi ces triplets et on note m_0 la multiplicité dans $Jord(\psi)$ d'un triplet atteignant ces maximums. On calcule alors

$$Jac^{\theta}_{\rho|\,|^{(a-b)/2,\cdots,-(a+b)/2+1}} \circ \cdots \circ Jac^{\theta}_{\rho|\,|^{(a-b)/2,\cdots,-(a+b)/2+1}} \pi^{GL}(\psi) \tag{1}$$

où on effectue m_0 fois cette opération. On note ψ' le morphisme qui se déduit de ψ en remplaçant (ρ,a,b) par $(\rho,a,b-2)$ pour toutes les occurrences de (ρ,a,b). Un calcul qui utilise uniquement le lemme combinatoire de Bernstein Zelevinsky pour calculer les modules de Jacquet d'une induite montre que le résultat vaut exactement $m_0!$ fois $\pi^{GL}(\psi')$. De plus ce résultat est le transfert de

$$Jac_{\rho|\,|^{(a-b)/2,\cdots,-(a+b)/2+1}} \circ \cdots \circ Jac^{\theta}_{\rho|\,|^{(a-b)/2,\cdots,-(a+b)/2+1}} \sum_{\pi\in\Pi(\psi)} \epsilon_\pi \pi$$

où ϵ_π est un signe convenable. Ainsi pour tout $\pi' \in \Pi(\psi')$, il existe au moins une représentation $\pi \in \Pi(\psi)$ tel que π' soit une composante de

$$Jac_{\rho|\,|^{(a-b)/2,\cdots,-(a+b)/2+1}} \circ \cdots \circ Jac^{\theta}_{\rho|\,|^{(a-b)/2,\cdots,-(a+b)/2+1}} \pi$$

On rappelle que $Jac_{\rho|\,|^x}\pi = 0$ si $x<(a-b)/2$; ainsi on montre qu'il existe nécessairement une représentation irréductible σ d'un groupe de même type que G mais de rang plus petit et une inclusion

$$\pi \hookrightarrow St(\rho,a)|\,|^{-(b-1)/2} \times \cdots \times St(\rho,a)|\,|^{-(b-1)/2} \times \sigma, \tag{2}$$

où il y a m_0 copies de $St(\rho,a)|\,|^{-(b-1)/2}$. Et en calculant les modules de Jacquet on vérifie que nécessairement $\sigma \simeq \pi'$. Ainsi on a une inclusion

$$\pi \hookrightarrow St(\rho,a)|\,|^{-(b-1)/2} \times \cdots \times St(\rho,a)|\,|^{-(b-1)/2} \times \pi'. \qquad (3)$$

Par récurrence on sait que $\Pi(\psi')$ contient le paquet de Langlands associé à ψ'_L. Supposons maintenant que π' soit un des éléments du paquet de Langlands associé à ψ'_L. Soit (ρ',a',b') un triplet de $Jord(\psi')$, on reprend les notations de 4.2, c'est-à-dire que l'on note $\delta_{b'} = -1/2$ ou -1 suivant que b' est pair ou impair et $Q(\rho',a',b')$ est le sous-module de Langlands de l'induite $St(\rho',a')|\,|^{-(b'-1)/2} \times \cdots \times St(\rho',a')|\,|^{\delta_{b'}}$; la notation est différente de celle introduite avant l'énoncé mais plus simple. On note aussi ψ_{imp} qui se déduit de ψ' (ou ψ, cela revient au même) en changeant les triplets (ρ',a',b') avec b' impair en $(\rho',a',1)$ et en supprimant les triplets (ρ',a',b') avec b' pair. Donc par hypothèse, il existe un caractère η du centralisateur de ψ'_{imp} dans le groupe convenable tel que π' soit l'unique sous-module irréductible de l'induite :

$$\times_{(\rho',a',b')\in Jord(\psi')} Q(\rho',a',b') \times \pi(\psi'_{imp},\eta).$$

L'ordre dans lequel on prend les éléments de $Jord(\psi')$ est indifférent, on suppose donc que le plus à gauche est $(\rho,a,b-2)$. Ainsi, on a l'inclusion

$$\pi \hookrightarrow \left(St(\rho,a)|\,|^{-(b-1)/2} \times Q(\rho,a,b-2) \right) \times_{(\rho',a',b')\in Jord(\psi')-\{(\rho,a,b)\}} (\psi'_{imp},\eta).$$

Par irréductibilité de π, on peut remplacer la parenthèse par un sous-quotient irréductible de l'induite écrite à l'intérieur ; mais comme $Jac_{\rho|\,|^x}\pi = 0$ si $x \in](a-b)/2, -(a+b)/2+1]$ par minimalité de $(a-b)/2$, le seul sous-quotient possible est $Q(\rho,a,b)$. On a ainsi démontré que π est dans le paquet de Langlands associé à ψ. On a donc démontré la proposition dans ce cas.

Supposons maintenant que pour tout $(\rho,a,b) \in Jord(\psi)$, on ait $a \geq b$. On fixe encore (ρ,a,b) tel que $b > 1$ et $a-b$ est minimum avec cette propriété ; si plusieurs triplets de $Jord(\psi)$ vérifient ces conditions on fixe encore $a+b$ maximum et on note m_0 la multiplicité dans $Jord(\psi)$ d'un triplet satisfaisant ces conditions d'extrêmums. On calcule encore (1) ; ici il peut y avoir $(\rho,a',b') \in Jord(\psi)$ avec $b'=1$ et $(a'-1)/2 \leq (a-b)/2$ mais on a alors sûrement

$$-(b+a)/2+1 = -b+1+(b-a)/2 < (b-a)/2 \leq -(a'-1)/2$$

et cela assure encore que (1) vaut $\pi^{GL}(\psi')$ avec la même définition de ψ' que ci-dessus. Ensuite, on prouve encore (2) ; ici il faut utiliser le fait que l'on ne peut avoir $Jac_{\rho|\,|^x,\cdots,\rho|\,|^{-(a+b)/2+1}}\pi \neq 0$ que si $x \geq (a-b)/2$. Ensuite, on a (3) facilement et on conclut comme dans le cas précédent mais en utilisant ici que $Jac_{\rho|\,|^x,\cdots,\rho|\,|^{-(a+b)/2+1}}\pi \neq 0$ est impossible si $x < (a-b)/2$.

Cette démonstration ne calcule pas les paramètres des représentations dans le paquet de Langlands.

Références

[1] Arthur J., *Unipotent automorphic representations, conjectures* in Orbites unipotentes et représentations II, Astérisque 171-172, 1989, pp. 13-72

[2] Arthur J., *On local character relations* Selecta Math. 2, No 4 (1996), 501-579

[3] Arthur J., *An introduction to the trace formula* Clay Mathematics Proceedings, volume 4, 2005, pp. 1-253

[4] Arthur J., *A note on L-packets* Pure and applied math. quarterly, 2 (special issue : in honor of J. H. Coates, part 1 of 2), 2006, pp. 199-217

[5] Arthur J., *Automorphic Representations of GSp(4)* In Contributions to automorphic forms, geometry, and number theory (Shalika volume) edited by H. Hida, D. Ramakr- ishnan, and F. Shahidi, pages 65-81. Johns Hopkins Univ. Press, Baltimore, MD, 2004.

[6] Arthur J., *Automorphic Representations of Classical Groups* en préparation

[7] Asgari M., *Local L-functions for split spinor groups* Canad. J. Math., 54, 2002, pp. 673-693

[8] Asgari M., Shahidi F., *Generic transfer for general Spin groups* Duke Math. J., 132, 2006, pp. 137-190

[9] Aubert A.-M., *Dualité dans le groupe de Grothendieck de la catégorie des représentations lisses de longueur finie d'un groupe réductif p-adique* TAMS, 347, 1995, pp. 2179-2189 avec l'erratum publié dans TAMS, 348, 1996, pp. 4687-4690

[10] Bernstein I. N., Zelevinsky A. V., *Induced Representations of Reductive p-adic groups* 1 Ann de l'ENS, 10, 1977, pp. 147-185

[11] Goldberg D., *Some results on reducibility for unitary groups and local Asai L-functions* Crelle J., 448, 1994, pp. 65-95

[12] Harris, M. ; Taylor, R., *The geometry and cohomology of some simple Shimura varieties* Annals of Math Studies, 151, Princeton Univ. Press, 2001

[13] Henniart, G., *Une preuve simple des conjectures de Langlands pour GL_n sur un corps p-adique* Invent. Math., 139, 2000, pp. 439-455

[14] Kim H., *On local L-function and normalized intertwining operators* Canad. J. Math, 57, 2005, pp. 535-597

[15] Kudla S., *On the local theta correspondence* Inv. Math 83, 1986, pp. 229-255

[16] Kudla S., Rallis S., *On First Occurrence in the Local Theta Correspondence* in Automorphic Representations, L-functions and Applications, Volume en l'honneur de S. Rallis, De Gruyter, 2005

[17] Moeglin C., *Representations quadratiques unipotentes des groupes classiques p-adiques* Duke, 84, 1996, pp. 267-332

[18] Moeglin C., *Points de réductibilité pour les induites de cuspidales* Journal of Algebra 268, 2003, pp. 81-117

[19] Moeglin C., *Sur la classification des séries discrètes des groupes classiques p-adiques ; paramètre de Langlands et exhaustivité* JEMS, 4, 2002, pp. 143-200

[20] Mœglin C., *Sur certains paquets d'Arthur et involution d'Aubert-Schneider-Stuhler généralisée* ERT, volume 10, 2006, pp. 86-129

[21] Mœglin C., *Paquets d'Arthur discrets pour un groupe classique p-adique* Contemp. Math. 489, 2009, pp-179-257

[22] Mœglin C., *Paquets d'Arthur pour les groupes classiques p-adiques ; point de vue combinatoire* prépublication 2006

[23] Mœglin C., *Classification des séries discrètes pour certains groupes classiques p-adiques* à paraître en l'honneur de R. Howe, Lecture Notes Series, Institute for Mathematical Sciences, National University of Singapore, Nov 2007

[24] Mœglin C., *Classification et Changement de base pour les séries discrètes des groupes unitaires p-adiques* Pacific J. of Math., 233, 2007, pp. 159-204

[25] Moeglin C., Tadic M., *Construction of discrete series for classical p-adic groups* JAMS, 15, 2002, pp. 715-786

[26] Mœglin C., Vignéras M.-F., Waldspurger J.-L., *Correspondance de Howe sur un corps p-adique* LN 1291, Springer Verlag

[27] Mœglin C., Waldspurger J.-L., *Le spectre résiduel de GL(n)* Ann. de l'ENS, 22, 1989, pp. 605-674

[28] Mœglin C., Waldspurger J.-L., *Sur le transfert des traces tordues d'un groupe linéaire à un groupe classique p-adique* Selecta mathematica, 2006, vol 12 (3), pp. 433-516

[29] Schneider M., Stuhler U., *Representation theory and sheaves on the Bruhat-Tits building* Publ. Math. IHES 85, 1997, pp. 97-191

[30] Shahidi F., *Local coefficients and intertwining operators for GL(n)* Compositio Math, 48, 1983, pp. 271-295

[31] Shahidi. F. *On the Ramanujan conjecture and finiteness of poles for certain L-functions* Ann. of Math., 127, 1988, pp. 547-584

[32] Shahidi. F., *A proof of Langlands' conjecture on Plancherel measures; complementary series for p-adic groups* Ann. of Math. , 132, 1990, pp. 273-330

[33] Silberger, A., *Special representations of reductive p-adic groups are not integrable* Ann. of Math., 111, 1980, pp. 571-587

[34] Waldspurger, J.-L., *Représentation métaplectique et conjectures de Howe* Séminaire Bourbaki, 674, 1986-1987, Nov 86

[35] Waldspurger, J.-L., *La formule de Plancherel pour les groupes p-adiques, d'après Harish-Chandra* JIMJ, 2, 2003, pp. 235-333

[36] Waldspurger J.-L., *Le groupe GL_N tordu, sur un corps p-adique ,1e partie* Duke Math. Journal, 137, 2007, pp. 185-234

[37] Waldspurger J.-L., *Le groupe GL_N tordu, sur un corps p-adique ,2e partie* Duke Math. Journal, 137, 2007, pp. 235-336

[38] Zelevinsky A. V., *Induced Representations of Reductive p-adic groups II* Ann de l'ENS, 13, 1980, pp. 165-210

INSTITUT DE MATHÉMATIQUES DE JUSSIEU, CNRS
E-mail address: `moeglin@math.jussieu.fr`

Clay Mathematics Proceedings
Volume 13, 2011

Unramified Unitary Duals for Split Classical p–adic Groups; The Topology and Isolated Representations

Goran Muić and Marko Tadić

For Freydoon Shahidi, with our admiration and appreciation

Introduction

This paper is our attempt to understand the work of Barbasch and Moy [**BbMo2**] on unramified unitary dual for split classical p–adic groups from a different point of view.

The classification of irreducible unitary representations of reductive groups over local fields is a fundamental problem of harmonic analysis with various possible applications, like those in number theory and the theory of automorphic forms. The class of unramified unitary representations is especially important in the aforementioned applications. These representations occur in the following set-up that we fix in this paper. Let F be a non–Archimedean local field of arbitrary characteristic and $\mathcal{O}$ is its ring of integers. When we work with the classical groups we are obliged to require that the characteristic of F is different from 2. Let G be the group of F–points of an F–split reductive group $\mathbf{G}$. An irreducible (complex) representation π of G is unramified if it has a vector fixed under $\mathbf{G}(\mathcal{O})$. The set of equivalence classes $\mathrm{Irr}^{unr}(G)$ of unramified irreducible representations of G is usually described by the Satake classification (see [**Cr**]). This classification is essentially the Langlands classification for those representations. We write $\mathrm{Irr}^{u,unr}(G) \subset \mathrm{Irr}^{unr}(G)$ for the subset consisting of unramified unitarizable representations. We equip that set with the topology of uniform convergence of matrix coefficients on compact subsets ([**F**], [**Di**]; see also [**T1**], [**T6**]). A good understanding of unramified unitarizable representations is fundamental for the theory of automorphic forms since almost all components of cuspidal and residual automorphic representations are unramified and unitarizable. By a good understanding we mean the following:

(1) To have an explicit classification of unramified unitary duals with explicit parameters and with Satake parameters easily computed from them.

2010 *Mathematics Subject Classification.* Primary 22E50. Secondary 22E35.
Both authors were partly supported by Croatian Ministry of Science, Education and Sports grant # 037-0372794-2804.

The converse to (1) is not trivial and it is important. More precisely, from the point of view of the theory of automorphic forms, it would be important to have a way to decide from Satake parameters if a representation is unitarizable. Even for general linear groups we need a simple algorithm. This leads to the following:

(2) To have an effective way (an algorithm) for testing unitarity of an arbitrary unramified representation in $\mathrm{Irr}^{unr}(G)$ given by its Satake parameter.
(3) To understand the topology in terms of the classification, especially of isolated points in $\mathrm{Irr}^{u,unr}(G)$, which are exactly the isolated points in the whole unitary dual which are unramified representations. This would be particularly interesting to understand from the point of view of automorphic spectra.

If $\mathbf{G} = \mathrm{GL}(n)$, then those tasks and much more were accomplished in the works of the second–named author ([**T4**], [**T5**]; see also [**T2**], [**T3**]) more than twenty years ago. In Section 4 of the present paper we give a simple solution due to the second named author to those problems based entirely on [**Ze**], without use of a result of Bernstein on irreducibility of unitary parabolic induction proved in [**Be2**] and used in the earlier proof of the classification (see Theorem 4-1.)

In this paper we give the solutions to problems (1)–(3) in the case of the split classical groups $G = S_n$ where S_n is one of the groups $\mathrm{Sp}(2n, F)$, $\mathrm{SO}(2n+1, F)$, $\mathrm{O}(2n, F)$ (see Section 1 for the precise description of the groups). Regarding (3), we have an explicit description of isolated points, and (2) gives an algorithm for getting limit points for a given sequence in the dual. Further, the algorithm from (2) gives parameters in (1) in the case of unitarizability (the other direction is obvious).

This paper is the end of a long effort ([**M4**], [**M6**]). The approach to problem (1) is motivated and inspired by the earlier work [**LMT**], in creation of which ideas of E. Lapid played an important role (see also [**T12**] which is a special case of [**LMT**]). On a formal level, the formulation of the solution (see Theorem 0-8) to problem (1) is "dual" to that of the one in [**LMT**], but in our unramified case it has a much more satisfying formulation. This is not surprising since we are dealing with very explicit representations. On the other hand, the proofs are more involved. For example, the proof of the unitarity of the basic "building blocks" (see Theorem 0-4 below) requires complicated arguments with the poles of degenerate Eisenstein series (see [**M6**]). The problems (2) and (3) have not yet been considered for split classical groups. A characteristic of our approach is that at no point in the proofs does the explicit internal structure of representations play a role. This is the reason that this can be considered as an external approach to the unramified unitary duals (of classical groups), which is a kind of a continuation of such approaches in [**T3**], [**T2**], [**LMT**], etc.

We expect that our approach has a natural Archimedean version similar to the way that [**LMT**] covers both the non-Archimedean and Archimedean cases, or the way the earlier paper [**T3**] has a corresponding Archimedean version [**T2**], with the same description of unitary duals for general linear groups and proofs along the same lines.

Now we describe our results. They are stated in Section 5 in more detail than here. After becoming acquainted with the basic notation in Section 1, the reader may proceed directly to read Section 5.

In the introduction, we use classical notation for induced representations. In the rest of the paper we shall use notation adapted to the case of general linear and classical groups, which very often substantially simplifies arguments in proofs. A part of the exposition below makes perfect sense also for Archimedean fields (we shall comment on this later).

We fix the absolute value $|\ |$ of F which satisfies $d(ax) = |a|dx$. Let χ be an unramified character of $F^\times$ and $l \in \mathbb{Z}_{>0}$. Then we consider the following induced representation of $\mathrm{GL}(l, F)$:

$$\mathrm{Ind}_{P_\emptyset}^{\mathrm{GL}(l,F)}(|\ |^{\frac{l-1}{2}}\chi \otimes |\ |^{\frac{l-1}{2}-1}\chi \otimes \cdots \otimes |\ |^{-\frac{l-1}{2}}\chi)$$

which has a character $\chi \circ \det$ as the unique irreducible quotient. We denote this character by:

$$\langle [-\frac{l-1}{2}, \frac{l-1}{2}]^{(\chi)} \rangle.$$

We introduce Langlands dual groups as follows:

$$\begin{array}{ll} G = S_n = \mathrm{SO}(2n+1, F) & \hat{G}(\mathbb{C}) = \mathrm{Sp}(2n, \mathbb{C}) \subset \mathrm{GL}(N, \mathbb{C}); N = 2n \\ G = S_n = \mathrm{O}(2n, F) & \hat{G}(\mathbb{C}) = \mathrm{O}(2n, \mathbb{C}) \subset \mathrm{GL}(N, \mathbb{C}); N = 2n \\ G = S_n = \mathrm{Sp}(2n, F) & \hat{G}(\mathbb{C}) = \mathrm{SO}(2n+1, \mathbb{C}) \subset \mathrm{GL}(N, \mathbb{C}); N = 2n+1. \end{array}$$

The local functorial lift $\sigma^{\mathrm{GL}(N,F)}$ of $\sigma \in \mathrm{Irr}^{unr}(S_n)$ to $\mathrm{GL}(N, F)$ is always defined and it is an unramified representation. (See (10-1) in Section 10 for the precise description.) It is an easy exercise to check that the map $\sigma \mapsto \sigma^{\mathrm{GL}(N,F)}$ is injective. This lift plays the key role in the solutions to problems (2) and (3).

In order to describe $\mathrm{Irr}^{u,unr}(S_n)$ we need to introduce more notation. Let $\mathbf{sgn}_u$ be the unique unramified character of order two of $F^\times$ and let $\mathbf{1}_{F^\times}$ be the trivial character of $F^\times$. Let $\chi \in \{\mathbf{1}_{F^\times}, \mathbf{sgn}_u\}$. Then we define α_χ as follows:

$$\begin{array}{l} \text{if } S_n = \mathrm{O}(2n, F), \text{ then } \alpha_\chi = 0 \\ \text{if } S_n = \mathrm{SO}(2n+1, F), \text{ then } \alpha_\chi = \dfrac{1}{2} \\ \text{if } S_n = \mathrm{Sp}(2n, F), \text{ then } \alpha_{\mathbf{sgn}_u} = 0 \text{ and } \alpha_{\mathbf{1}_{F^\times}} = 1. \end{array}$$

We refer to Remark 5-3 for an explanation of this definition in terms of rank–one reducibility.

A pair (m, χ), where $m \in \mathbb{Z}_{>0}$ and χ is an unramified unitary character of $F^\times$, is called a Jordan block. The following definition can be found in [**M4**] (see also Definition 5-4 in Section 5):

DEFINITION 0-1. *Let $n > 0$. We denote by $\mathrm{Jord}_{sn}(n)$ the collection of all the sets* Jord, *which consist of Jordan blocks, such that the following hold:*

$$\chi \in \{\mathbf{1}_{F^\times}, \mathbf{sgn}_u\} \text{ and } m - (2\alpha_\chi + 1) \in 2\mathbb{Z} \text{ for all } (m,\chi) \in \mathrm{Jord}$$
$$\sum_{(m,\chi)\in \mathrm{Jord}} m = \begin{cases} 2n & \text{if } S_n = \mathrm{SO}(2n+1,F) \text{ or } S_n = \mathrm{O}(2n,F); \\ 2n+1 & \text{if } S_n = \mathrm{Sp}(2n,F), \end{cases}$$
and, additionally, if $\alpha_\chi = 0$, *then* card $\{k; (k,\chi) \in \mathrm{Jord}\} \in 2\mathbb{Z}$.

Let $\mathrm{Jord} \in \mathrm{Jord}_{sn}(n)$. Then, for $\chi \in \{\mathbf{1}_{F^\times}, \mathbf{sgn}_u\}$, we let

$$\mathrm{Jord}_\chi = \{k; (k,\chi) \in \mathrm{Jord}\}.$$

We let

$$\mathrm{Jord}'_\chi = \begin{cases} \mathrm{Jord}(\chi); \ \mathrm{card}(\mathrm{Jord}_\chi) \text{ is even}; \\ \mathrm{Jord}(\chi) \cup \{-2\alpha_\chi + 1\}; \ \mathrm{card}(\mathrm{Jord}_\chi) \text{ is odd}. \end{cases}$$

We write elements of Jord'_χ in the following way (the case $l_{1_{F^\times}} = 0$ or $l_{sgn_u} = 0$ is not excluded):

$$\begin{cases} \text{for } \chi = 1_{F^\times} \quad \text{as} \quad a_1 < a_2 < \cdots < a_{2l_{1_{F^\times}}} \\ \text{for } \chi = sgn_u \quad \text{as } b_1 < b_2 < \cdots < b_{2l_{sgn_u}}. \end{cases}$$

Next, we associate to $\mathrm{Jord} \in \mathrm{Jord}_{sn}(n)$ the unramified representation $\sigma(\mathrm{Jord})$ of S_n defined as the unique irreducible unramified subquotient of the representation parabolically induced from the representation

$$\left(\otimes_{i=1}^{l_{1_{F^\times}}} \langle [-\frac{a_{2i}-1}{2}, \frac{a_{2i-1}-1}{2}]^{(\mathbf{1}_{F^\times})}\rangle\right) \otimes \left(\otimes_{j=1}^{l_{\mathbf{sgn}_u}} \langle [-\frac{b_{2j}-1}{2}, \frac{b_{2j-1}-1}{2}]^{(\mathbf{sgn}_u)}\rangle\right).$$

Recall that irreducible tempered (resp., square integrable) representations of a reductive group can be characterized as satisfying certain inequalities (resp., strict inequalities). In [**M4**], the first author defines negative (resp., strongly negative) irreducible representations as those which satisfy the reverse inequalities (resp., strict inequalities). An unramified representation is strongly negative if its Aubert dual is in the discrete series. See [**M4**] for more details. We have the following result (see [**M4**]; Theorem 5-8 in Section 5 of this paper):

THEOREM 0-2. *Let* $n \in \mathbb{Z}_{>0}$. *The map* $\mathrm{Jord} \mapsto \sigma(\mathrm{Jord})$ *defines a one-to-one correspondence between the set* $\mathrm{Jord}_{sn}(n)$ *and the set of all strongly negative unramified representations of* S_n.

The inverse mapping to $\mathrm{Jord} \mapsto \sigma(\mathrm{Jord})$ will be denoted by $\sigma \mapsto \mathrm{Jord}(\sigma)$. Let us note that the set $\mathrm{Jord}_{sn}(n)$ also parameterizes the generic irreducible square integrable representations with Iwahori fixed vector.

An unramified representation is negative if its Aubert dual is tempered. Negative representations are classified in terms of strongly negative as follows ([**M4**]; Theorem 5-10 in Section 5 of this paper):

THEOREM 0-3. *Let* $\sigma_{neg} \in \mathrm{Irr}^{unr}(S_n)$ *be a negative representation. Then there exists a sequence of pairs* $(l_1, \chi_1), \ldots, (l_k, \chi_k)$ $(l_i \in \mathbb{Z}_{\geq 1}$, χ_i *is an unramified unitary character of* $F^\times)$, *unique up to a permutation and taking inverses of characters, and* *a unique strongly negative representation σ_{sn} such that σ_{neg} is a subrepresentation of the parabolically induced representation*

$$\mathrm{Ind}^{S_n}\left(\langle[-\frac{l_1-1}{2},\frac{l_1-1}{2}]^{(\chi_1)}\rangle\otimes\cdots\otimes\langle[-\frac{l_k-1}{2},\frac{l_k-1}{2}]^{(\chi_k)}\rangle\otimes\sigma_{sn}\right).$$

Conversely, for a sequence of pairs $(l_1,\chi_1),\ldots,(l_k,\chi_k)$ ($l_i\in\mathbb{Z}_{>0}$, χ_i an unramified unitary character of $F^\times$) and a strongly negative representation σ_{sn}, the unique irreducible unramified subquotient of

$$\mathrm{Ind}^{S_n}\left(\langle[-\frac{l_1-1}{2},\frac{l_1-1}{2}]^{(\chi_1)}\rangle\otimes\cdots\otimes\langle[-\frac{l_k-1}{2},\frac{l_k-1}{2}]^{(\chi_k)}\rangle\otimes\sigma_{sn}\right)$$

is negative and it is a subrepresentation.

We let $\mathrm{Jord}(\sigma_{neg})$ be the multiset

$$\mathrm{Jord}(\sigma_{sn})+\sum_{i=1}^{k}\{(l_i,\chi_i),(l_i,\chi_i^{-1})\}$$

(multisets are sets where multiplicities are allowed).

The proofs of Theorems 0-2 and 0-3 given in [**M4**] are obtained with Jacquet modules techniques enabling the results to hold for F of any characteristic different from two.

The key result for this paper is the following result of the first author (see [**M6**]; see Theorem 5-11):

THEOREM 0-4. *Every negative representation is unitarizable. Every strongly negative representation is a local component of a global representation appearing in the residual spectrum of a split classical group defined over a global field.*

The unitarizability of negative representations was obtained earlier by D. Barbasch and A. Moy. It follows from their unitarity criterion in [**BbMo**] and [**BbMo1**], which says that unitarizability can already be detected on Iwahori fixed vectors.

The following theorem is a consequence of the above results:

THEOREM 0-5. *Let $\sigma\in\mathrm{Irr}^{unr}(S_n)$ be a negative representation. Then its lift to $\mathrm{GL}(N,F)$ is given by*

$$\sigma^{\mathrm{GL}(N,F)}\simeq\times_{(l,\chi)\in\mathrm{Jord}(\sigma)}\ \langle[-\frac{l-1}{2},\frac{l-1}{2}]^{(\chi)}\rangle.$$

Moreover, its Arthur parameter $W_F\times\mathrm{SL}(2,\mathbb{C})\times\mathrm{SL}(2,\mathbb{C})\to\hat{G}(\mathbb{C})\subset\mathrm{GL}(N,\mathbb{C})$ is given by:

$$\oplus_{(l,\chi)\in\mathrm{Jord}(\sigma)}\ \chi\otimes V_1\otimes V_l,$$

where V_l is the unique algebraic representation of $\mathrm{SL}(2,\mathbb{C})$ of dimension l.

In order to describe the whole of $\mathrm{Irr}^{u,unr}(S_n)$, we need to introduce more notation. We write $\mathcal{M}^{unr}(S_n)$ for the set of pairs $(\mathbf{e},\sigma_{neg})$, where:

- $\mathbf{e}$ is a (perhaps empty) multiset consisting of a finite number of triples (l, χ, α) where $l \in \mathbb{Z}_{>0}$, χ is an unramified unitary character of $F^\times$, and $\alpha \in \mathbb{R}_{>0}$.
- $\sigma_{neg} \in \text{Irr } S_{n_{neg}}$ (this defines n_{neg}) is negative satisfying

$$n = \sum_{(l,\chi)} l \cdot \text{card } \mathbf{e}(l, \chi) + n_{neg}.$$

For $l \in \mathbb{Z}_{>0}$ and an unramified unitary character χ of $F^\times$, we denote by $\mathbf{e}(l, \chi)$ the submultiset of $\mathbf{e}$ consisting of all positive real numbers α (counted with multiplicity) such that $(l, \chi, \alpha) \in \mathbf{e}$.

We attach $\sigma \in \text{Irr}^{unr}(S_n)$ to $(\mathbf{e}, \sigma_{neg})$ in a canonical way. By definition, σ is the unique irreducible unramified subquotient of the following induced representation:

$$\text{(0-6)} \qquad \text{Ind}^{S_n}\left(\left(\otimes_{(l,\chi,\alpha)\in\mathbf{e}} \ \langle[-\frac{l-1}{2}, \frac{l-1}{2}]^{(|\ |^\alpha\chi)}\rangle\right) \otimes \sigma_{neg}\right).$$

We remark that the definition of σ does not depend on the choice of ordering of elements in $\mathbf{e}$.

In order to obtain unitary representations, we impose further conditions on $\mathbf{e}$ in the following definition (see Definition 5-13):

DEFINITION 0-7. *Let $\mathcal{M}^{u,unr}(S_n)$ be the subset of $\mathcal{M}^{unr}(S_n)$ consisting of the pairs $(\mathbf{e}, \sigma_{neg})$ satisfying the following conditions:*

(1) *If $\chi \notin \{\mathbf{1}_{F^\times}, \mathbf{sgn}_u\}$, then $\mathbf{e}(l, \chi) = \mathbf{e}(l, \chi^{-1})$ and $0 < \alpha < \frac{1}{2}$ for all $\alpha \in \mathbf{e}(l, \chi)$.*

(2) *If $\chi \in \{\mathbf{1}_{F^\times}, \mathbf{sgn}_u\}$ and $l - (2\alpha_\chi + 1) \notin 2\mathbb{Z}$, then $0 < \alpha < \frac{1}{2}$ for all $\alpha \in \mathbf{e}(l, \chi)$.*

(3) *If $\chi \in \{\mathbf{1}_{F^\times}, \mathbf{sgn}_u\}$ and $l - (2\alpha_\chi + 1) \in 2\mathbb{Z}$, then $0 < \alpha < 1$ for all $\alpha \in \mathbf{e}(l, \chi)$. Moreover, if we write the exponents that belong to $\mathbf{e}(l, \chi)$ as follows:*

$$0 < \alpha_1 \leq \cdots \leq \alpha_u \leq \frac{1}{2} < \beta_1 \leq \cdots \leq \beta_v < 1.$$

(We allow $u = 0$ or $v = 0$.) Then we must have the following:

(a) *If $(l, \chi) \notin \text{Jord}(\sigma_{neg})$, then $u + v$ is even.*

(b) *If $u > 1$, then $\alpha_{u-1} \neq \frac{1}{2}$.*

(c) *If $v \geq 2$, then $\beta_1 < \cdots < \beta_v$.*

(d) *$\alpha_i \notin \{1 - \beta_1, \ldots, 1 - \beta_v\}$ for all i.*

(e) *If $v \geq 1$, then the number of indices i such that $\alpha_i \in]1 - \beta_1, \frac{1}{2}]$ is even.*

(f) *If $v \geq 2$, then the number of indices i such that $\alpha_i \in]1 - \beta_{j+1}, 1 - \beta_j[$ is odd.*

The main result of the paper is the following explicit description of $\text{Irr}^{u,unr}(S_n)$ (see Theorem 5-14):

THEOREM 0-8. *Let* $(\mathbf{e}, \sigma_{neg}) \in \mathcal{M}^{u,unr}(S_n)$. *Then*

$$Ind^{S_n}\left(\left(\otimes_{(l,\chi,\alpha)\in\mathbf{e}} \ \langle[-\frac{l-1}{2}, \frac{l-1}{2}]^{(|\ |^{\alpha}\chi)}\rangle\right) \otimes \sigma_{neg}\right)$$

is irreducible. Moreover, the map

$$(\mathbf{e}, \sigma_{neg}) \longmapsto \mathrm{Ind}^{S_n}\left(\left(\otimes_{(l,\chi,\alpha)\in\mathbf{e}} \ \langle[-\frac{l-1}{2}, \frac{l-1}{2}]^{(|\ |^{\alpha}\chi)}\rangle\right) \otimes \sigma_{neg}\right)$$

is a one–to–one correspondence between $\mathcal{M}^{u,unr}(S_n)$ *and* $\mathrm{Irr}^{u,unr}(S_n)$.

This result is proved in Sections 7, 8, and 9. The preparation for the proof is done in the first two of these three sections. In Section 2, where we recall (from [**T8**]) some general principles for proving unitarity and non–unitarity, we also prove a new criterion for non–unitarity (see (RP) in Section 2). In Section 6 we describe all necessary reducibility facts explicitly (most of them are already established in [**M4**]).

Theorem 0-8 clearly solves problem (1) for the split classical groups. In Section 10 we describe a simple algorithm that has:
INPUT: an arbitrary irreducible unramified representation $\sigma \in \mathrm{Irr}^{unr}(S_n)$ given by its Satake parameter.
OUTPUT: tests the unitarity of σ and at the same time constructs the corresponding pair $(\mathbf{e}, \sigma_{neg})$ if σ is unitary.

The algorithm is based on the observation that, for a unitarizable σ, the Zelevinsky data (see [**Ze**]; or Theorem 1-7 here) of the lift $\sigma^{\mathrm{GL}(N,F)}$ is easy to describe from the datum $(\mathbf{e}, \sigma_{neg})$ of σ. (See Lemma 10-7.) This solves problem (2) stated above. We observe that this problem is almost trivial for $\mathrm{GL}(n, F)$. (See Theorem 4-1.)

The algorithm is very simple, and one can go almost directly to the algorithm in Section 10, to check if some irreducible unramified representation given in terms of Satake parameters is unitarizable (Definition 5-13 is relevant for the algorithm). In Section 12 we give examples of the use of this algorithm. The algorithm has ten steps, some of them quite easy, but usually only a few of them enter the test (see Section 12). It would be fairly easy to write a computer program, possible to handle classical groups of ranks exceeding tens of thousands, for determining unitarizability in terms of Satake parameters.

Finally, we come to problem (3). In Section 3 we show that $\mathrm{Irr}^{u,unr}(S_n)$ is naturally homeomorphic to a compact subset of the complex manifold consisting of all Satake parameters for S_n (see Theorem 3-5 and Theorem 3-7). The results of this section almost directly follow from [**T6**]. In Section 11 we determine the isolated points in $\mathrm{Irr}^{u,unr}(S_n)$. To describe the result, we introduce more notation. Let $\sigma \in \mathrm{Irr}^{u,unr}(S_n)$ be a strongly negative representation. Let $\chi \in \{\mathbf{1}_{F^\times}, \mathbf{sgn}_u\}$. Then we write $\mathrm{Jord}(\sigma)_\chi$ for the set of l such that $(l, \chi) \in \mathrm{Jord}(\sigma)$. If $a \in \mathrm{Jord}(\sigma)_\chi$ is not the minimum, then we write a_- for the greatest $b \in \mathrm{Jord}(\sigma)_\chi$ such that $b < a$. We have the following:

$$a - a_- \text{ is even (whenever } a_- \text{ is defined).}$$

Now, we are ready to state the main result of Section 11. It is the following theorem (see Theorem 11-3):

THEOREM 0-9. *A representation* $\sigma \in \mathrm{Irr}^{u,unr}(S_n)$ *is isolated if and only if* σ *is strongly negative, and for every* $\chi \in \{\mathbf{1}_{F^\times}, \mathbf{sgn}_u\}$ *such that* $\mathrm{Jord}(\sigma)_\chi \neq \emptyset$ *the following holds:*

(1) $a - a_- \geq 4$, *for all* $a \in \mathrm{Jord}(\sigma)_\chi$ *whenever* a_- *is defined.*
(2) *If* $\mathrm{Jord}_\chi \neq \{1\}$, *then* $\min \mathrm{Jord}_\chi \setminus \{1\} \geq 4$.

(We do not claim that $1 \in \mathrm{Jord}_\chi$ in (2). If $1 \notin \mathrm{Jord}_\chi$, then (2) claims that $\min \mathrm{Jord}_\chi \geq 4$.)

Since $\mathrm{Irr}^{u,unr}(S_n)$ is an open subset of $\mathrm{Irr}^{u}(S_n)$, the above theorem also classifies the isolated representations in the whole unitary dual $\mathrm{Irr}^{u}(S_n)$, which are unramified.

As an example, let $S_1 = \mathrm{Sp}(2,F) = \mathrm{SL}(2,F)$. Then the trivial representation $\mathbf{1}_{SL(2,F)}$ is strongly negative and $\mathrm{Jord}(\mathbf{1}_{\mathrm{SL}(2,F)}) = \{(3, \mathbf{1}_{F^\times})\}$. As is well–known, it is not isolated and this theorem confirms that. Let $n \geq 2$ and let $S_n = \mathrm{Sp}(2n, F)$. Then the trivial representation $\mathbf{1}_{\mathrm{Sp}(2n,F)}$ is strongly negative and $\mathrm{Jord}(\mathbf{1}_{\mathrm{Sp}(2n,F)}) = \{(2n+1, \mathbf{1}_{F^\times})\}$. Clearly, $\mathbf{1}_{\mathrm{Sp}(2n,F)}$ is isolated (as is well–known from [**K**]). We may consider the degenerate case $n = 0$. Then $\mathrm{Sp}(0, F)$ is the trivial group and $\mathbf{1}_{Sp(0,F)}$ is its trivial representation. It is reasonable to call such representation strongly negative and let $\mathrm{Jord}(\mathbf{1}_{\mathrm{Sp}(0,F)}) = \{(1, \mathbf{1}_{F^\times})\}$. Apart from that case, one always has $\mathrm{Jord}_\chi \neq \{1\}$.

Similarly, if we let $S_n = \mathrm{SO}(2n+1, F)$ $(n > 0)$, then $\mathbf{1}_{\mathrm{SO}(2n+1,F)}$ is strongly negative. We have $\mathrm{Jord}(\mathbf{1}_{\mathrm{SO}(2n+1,F)}) = \{(2n, \mathbf{1}_{F^\times})\}$. As is well–known, it is not isolated for $n = 1$ and this theorem confirms that. It is isolated for $n \geq 2$ (as is well–known from [**K**]).

We close this introduction with several comments. First recall that in [**BbMo2**], D. Barbasch and A. Moy address the first of the three problems that we consider in our paper. Their related paper [**BbMo**] contains some very deep fundamental results on unitarizability, like the fact that the Iwahori-Matsumoto involution preserves unitarity in the Iwahori fixed vector case. Their Hecke algebra methods are opposite to our methods. Their approach is based on a careful study of the internal structure of representations on Iwahori fixed vectors, based on the Kazhdan-Lusztig theory [**KLu**]. Their main result –Theorem A on page 23 of [**BbMo2**]– states that a parameter of any irreducible unitarizable unramified representation of a classical group is a "complementary series from an induced from a tempered representation tensored with a GL-complementary series". In other words, that it comes from a complementary series starting with a representation induced by a negative representation (from Theorem 0-4) tensored with a GL-complementary series. They do not determine parameters explicitly (they observe that "the parameters are hard to describe explicitly"; see page 23 of their paper). They get the unitarizability of negative representations by local (Hecke algebra) methods, but do not relate them to the automorphic spectrum like Theorem 0-4 does. Summing up without going into the details, the description in [**BbMo2**] partially covers Theorem 0-8.

There are complementary series in a number of cases, but their paper does not give the full picture: the explicit parameterization of the unramified unitary duals. In [**Bb**] there is also a description of the unramified unitary dual (see the very beginning of that paper for more precise description of the contents of that paper). On http:/www.liegroups.org there is an implementation of an algorithm for unitarity based on an earlier version of Barbasch's paper [**Bb**] (one can find more information regarding this on that site).

Our quite different approach gives explicit parameters of different type, and these explicit parameters have a relatively simple combinatorial description. We observe that all that we need for describing unitary duals are one-dimensional unramified unitary characters of general linear groups, parabolic induction, and taking irreducible unramified subquotients.

Let us note that except for motivations coming directly from the theory of automorphic forms, like applications to analytic properties of L-functions, etc., a motivation for us to get explicit classification of unramified unitary duals was to be able to answer question (2) (which would definitely be important also for the study of automorphic forms), and to classify isolated points in (3) (which can be significant in the study of automorphic spectra).

Recall that in two important cases where the unitarizability is understood (the case of general linear groups and the case of generic representations of quasi-split classical groups), the classification theorem is uniform for the Archimedean case as well as for the non-Archimedean case (see [**T3**], [**T2**] and [**LMT**]). Moreover, the proofs are essentially the same (not only analogous). Therefore, it is natural to expect this to be the case for unramified unitarizable representations of classical split groups. Having this in mind, we shall comment briefly on the Archimedean case. Assume $F = \mathbb{R}$ or $F = \mathbb{C}$. Let $|\ |$ be the ordinary absolute value (resp., square of it) if $F = \mathbb{R}$ (resp., $F = \mathbb{C}$), i.e., the modulus character of F (like in the non-Archimedean case). If we fix some suitable maximal compact subgroup K of S_n, then we may consider K–spherical representations, and call them unramified. Then the above constructions and statements make sense. More precisely, define $\mathrm{Jord}_{sn}(n)$ using only $\mathbf{1}_{F^\times}$ (i.e., all unramified characters χ of $F^\times$ which satisfy $\chi = \chi^{-1}$). Call representations from $\{\sigma(\mathrm{Jord});\ \mathrm{Jord} \in \mathrm{Jord}_{sn}\}$ strongly negative (define $\sigma(\mathrm{Jord})$ in the same way as in the non-Archimedean case). Define negative representations as those which arise as irreducible unramified subrepresentations of representations displayed in Theorem 0-3. Then Theorem 0-4 follows from [**M6**] (where is found the uniform proof for the non-Archimedean and Archimedean cases). Now, it is natural to ask if Theorem 0-8 is also true in that set–up. We have not been able to check that using the results of [**Bb**]. But there are a number of facts which suggest this. The first is Theorem 0-4. Second is that a number of arguments in the proof of Theorem 0-8 make sense in the Archimedean case (as in [**LMT**]). The complex case shows a particular similarity (consult [**T12**]). We expect that the approach of this paper will be extended to the Archimedean case. We also expect that Theorem 0-9 describing isolated representations holds in the Archimedean case, with a similar proof. We plan to address the Archimedean case in the future.

At the end, let us note that one possible strategy to get the answer to (1) (but not to (2) and (3)) would be to try to get Theorem 0-8 from the classification in

[**LMT**], using the Barbasch-Moy fundamental result that the Iwahori-Matsumoto involution preserves unitarity in the Iwahori fixed vector case (the proof of which is based on the Kazhdan-Lusztig theory [**KLu**]). This much less direct approach would still leave a number of questions to be solved. Furthermore, we expect that the approach that we present in our paper has a much greater chance for generalization than the one that we just discussed above as was the case for general linear groups, where the classification of unramified irreducible unitary representations was first obtained using the Zelevinsky classification, which very soon led to the classification of general irreducible unitary representations in terms of the Zelevinsky, as well as the Langlands, classification.

The first named author would like to thank D. Ban for explaining to him the results of [**Bn**]. The second named author is thankful to D. Renard and A. Moy for discussions on the topics related (directly or indirectly) to this paper. We would like to thank J. Schwermer for several invitations to the Schroedinger Institute for Mathematical Physics in Vienna, where various versions of the paper were written. We would also like to thank the referee, who read the paper very carefully, corrected a number of errors, and helped improve the style of presentation.

1. Preliminary Results

Let $\mathbb{Z}$, $\mathbb{R}$, and $\mathbb{C}$ be the ring of rational integers, the field of real numbers, and the field of complex numbers, respectively.

Let F be a non–Archimedean field of characteristic different from 2. We write $\mathcal{O}$ for the maximal compact subring of F. Let $\mathfrak{p}$ be the unique maximal ideal in $\mathcal{O}$. Let ϖ be a fixed generator of $\mathfrak{p}$ and let q be the number of elements in the corresponding residue field of $\mathcal{O}$. We write ν for the normalized absolute value of F. Let χ be a character of $F^\times$. We can uniquely write $\chi = \nu^{e(\chi)}\chi^u$ where χ^u is a unitary character and $e(\chi) \in \mathbb{R}$.

Let G be an l–group (see [**BeZ**]). We will consider smooth representations of G on complex vector spaces. We simply call them representations. If σ is a representation of G, then we write V_σ for its space. Its contragredient representation is denoted by $\widetilde{\sigma}$ and the corresponding non–degenerate canonical pairing by $\langle\ ,\ \rangle : V_{\widetilde{\sigma}} \times V_\sigma \to \mathbb{C}$. If σ_1 and σ_2 are representations of G, then we write $\mathrm{Hom}_G(\sigma_1, \sigma_2)$ for the space of all G–intertwining maps $\sigma_1 \to \sigma_2$. We say that σ_1 and σ_2 are equivalent, $\sigma_1 \simeq \sigma_2$, if there is a bijective $\varphi \in \mathrm{Hom}_G(\sigma_1, \sigma_2)$. Let $\mathrm{Irr}(G)$ be the set of equivalence classes of irreducible admissible representations of G. Let $\mathrm{R}(G)$ be the Grothendieck group of the category $\mathcal{M}_{adm.fin.leng.}(G)$ of all admissible representations of finite length of G. If σ is an object of $\mathcal{M}_{adm.fin.leng.}(G)$, then we write $s.s.(\sigma)$ for its semi–simplification in $\mathrm{R}(G)$. Frequently, in computations we simply write σ instead of $s.s.(\sigma)$. If G is the trivial group, then we write its unique irreducible representation as $\mathbf{1}$.

Next, we fix the notation for the general linear group $\mathrm{GL}(n, F)$. Let I_n be the identity matrix in $\mathrm{GL}(n, F)$. Let ${}^t g$ be the transposed matrix of $g \in \mathrm{GL}(n, F)$. The transposed matrix of $g \in \mathrm{GL}(n, F)$ with respect to the second diagonal will be denoted by ${}^\tau g$. If χ is a character of $F^\times$ and π is a representation of $\mathrm{GL}(n, F)$, then the representation $(\chi \circ \det) \otimes \pi$ of $\mathrm{GL}(n, F)$ will be written as $\chi\pi$.

We fix the minimal parabolic subgroup $P_{min}^{\mathrm{GL}_n}$ of $\mathrm{GL}(n, F)$ consisting of all upper triangular matrices in $\mathrm{GL}(n, F)$. A standard parabolic subgroup P of $\mathrm{GL}(n, F)$

is a parabolic subgroup containing $P_{min}^{\mathrm{GL}_n}$. There is a one–to–one correspondence between the set of all ordered partitions α of n, $\alpha = (n_1, \ldots, n_k)$ ($n_i \in \mathbb{Z}_{>0}$), and the set of standard parabolic subgroups of $\mathrm{GL}(n, F)$, attaching to a partition α the parabolic subgroup P_α consisting of all block–upper triangular matrices:

$$p = (p_{ij})_{1\le i,j\le k},\ p_{ij} \text{ is an } n_i \times n_j \text{ matrix},\ p_{ij} = 0\ (i > j).$$

The parabolic subgroup P_α admits a Levi decomposition $P_\alpha = M_\alpha N_\alpha$, where

$$M_\alpha = \{\mathrm{diag}(g_1, \ldots, g_k);\ g_i \in \mathrm{GL}(n_i, F)\ (1 \le i \le k)\}$$
$$N_\alpha = \{p \in P_\alpha;\ p_{ii} = I_{n_i}\ (1 \le i \le k)\}.$$

Let π_i be a representation of $\mathrm{GL}(n_i, F)$ $(1 \le i \le k)$. Then we consider $\pi_1 \otimes \cdots \otimes \pi_k$ as a representation of M_α as usual:

$$\pi_1 \otimes \cdots \otimes \pi_k(\mathrm{diag}(g_1, \ldots, g_k)) = \pi_1(g_1) \otimes \cdots \otimes \pi_k(g_k),$$

and extend it trivially across N_α to the representation of P_α denoted by the same letter. Then we form (normalized) induction, written as follows (see [**BeZ1**], [**Ze**]):

$$\pi_1 \times \cdots \times \pi_k = i_{n,\alpha}(\pi_1 \otimes \cdots \otimes \pi_k) := \mathrm{Ind}_{P_\alpha}^{\mathrm{GL}(n,F)}(\pi_1 \otimes \cdots \otimes \pi_k)$$

In this way we obtain the functor

$$\mathcal{M}_{adm.fin.leng.}(M_\alpha) \xrightarrow{i_{n,\alpha}} \mathcal{M}_{adm.fin.leng.}(\mathrm{GL}(n, F))$$

and a group homomorphism $\mathrm{R}(M_\alpha) \xrightarrow{i_{n,\alpha}} \mathrm{R}(\mathrm{GL}(n, F))$. Next, if π is a representation of $\mathrm{GL}(n, F)$, then we form the normalized Jacquet module $r_{\alpha,n}(\pi)$ of π (see [**BeZ1**]). It is a representation of M_α. In this way we obtain a functor $\mathcal{M}_{adm.fin.leng.}(\mathrm{GL}(n, F)) \xrightarrow{r_{\alpha,n}} \mathcal{M}_{adm.fin.leng.}(M_\alpha)$ and a group homomorphism $\mathrm{R}(\mathrm{GL}(n, F)) \xrightarrow{r_{\alpha,n}} \mathrm{R}(M_\alpha)$. The functors $i_{n,\alpha}$ and $r_{\alpha,n}$ are related by Frobenius reciprocity:

$$\mathrm{Hom}_{\mathrm{GL}(n,F)}(\pi,\ i_{n,\alpha}(\pi_1 \otimes \cdots \otimes \pi_k)) \simeq \mathrm{Hom}_{M_\alpha}(r_{\alpha,n}(\pi),\ \pi_1 \otimes \cdots \otimes \pi_k).$$

We list some additional basic properties of induction:

$$\pi_1 \times (\pi_2 \times \pi_3) \simeq (\pi_1 \times \pi_2) \times \pi_3,$$
$$\pi_1 \times \pi_2 \text{ and } \pi_2 \times \pi_1 \text{ have the same composition series},$$
$$\text{if } \pi_1 \times \pi_2 \text{ is irreducible, then } \pi_1 \times \pi_2 \simeq \pi_2 \times \pi_1,$$
$$\chi(\pi_1 \times \pi_2) \simeq (\chi\pi_1) \times (\chi\pi_2), \qquad \text{for a character } \chi \text{ of } F^\times,$$
$$\widetilde{\pi_1 \times \pi_2} \simeq \tilde{\pi}_1 \times \tilde{\pi}_2.$$

We take $\mathrm{GL}(0, F)$ to be the trivial group (we consider formally the unique element of this group as a 0×0 matrix and the determinant map $\mathrm{GL}(0, F) \to F^\times$). We extend $\times$ formally as follows: $\pi \times \mathbf{1} = \mathbf{1} \times \pi := \pi$ for every representation π of $\mathrm{GL}(n, F)$. The listed properties hold in this extended setting. We also let $r_{(0),0}(\mathbf{1}) = \mathbf{1}$.

Now, we fix the basic notation for the split classical groups. Let

$$J_n = \begin{bmatrix} 00 & \ldots & 01 \\ 00 & \ldots & 10 \\ \vdots & & \\ 10 & \ldots & 0 \end{bmatrix} \in \mathrm{GL}(n, F).$$

The symplectic group (of rank $n \geq 1$) is defined as follows:

$$\mathrm{Sp}(2n,F) = \left\{ g \in \mathrm{GL}(2n,F);\ g \cdot \begin{bmatrix} 0 & J_n \\ -J_n & 0 \end{bmatrix} \cdot {}^t g = \begin{bmatrix} 0 & J_n \\ -J_n & 0 \end{bmatrix} \right\}.$$

Next, the split orthogonal groups special odd-orthogonal groups (both of rank $n \geq 1$) are defined by

$$\mathrm{SO}(n,F) = \{ g \in \mathrm{SL}(n,F);\ g \cdot J_n \cdot {}^t g = J_n \}$$
$$\mathrm{O}(n,F) = \{ g \in \mathrm{GL}(n,F);\ g \cdot J_n \cdot {}^t g = J_n \}$$

We take $\mathrm{Sp}(0,F), \mathrm{SO}(0,F), \mathrm{O}(0,F)$ to be the trivial groups (we consider their unique element formally as a 0×0 matrix). In the sequel, we fix one of the following three series of groups:

$$S_n = \mathrm{Sp}(2n,F),\ n \geq 0$$
$$S_n = \mathrm{O}(2n,F),\ n \geq 0$$
$$S_n = \mathrm{SO}(2n+1,F),\ n \geq 0.$$

Let $n > 0$. Then the minimal parabolic subgroup $P^{S_n}_{\min}$ of S_n consisting of all upper triangular matrices is fixed. A standard parabolic subgroup P of S_n is a parabolic subgroup containing $P^{S_n}_{min}$. There is a one–to–one correspondence between the set of all finite sequences of positive integers of total mass $\leq n$ and the set of standard parabolic subgroups of S_n defined as follows. For $\alpha = (m_1, \ldots, m_k)$ of total mass $m := \sum_{i=1}^{l} m_i \leq n$, we let

$$P^{S_n}_\alpha := \begin{cases} P_{(m_1,\ldots,m_k,\ 2(n-m),\ m_k,m_{k-1},\ldots,m_1)} \cap S_n;\ S_n = \mathrm{Sp}(2n,F),\ \mathrm{O}(2n,F) \\ P_{(m_1,\ldots,m_k,\ 2(n-m)+1,\ m_k,m_{k-1},\ldots,m_1)} \cap S_n;\ S_n = \mathrm{SO}(2n+1,F). \end{cases}$$

(The middle term $2(n-m)$ is omitted if $m = n$.) The parabolic subgroup $P^{S_n}_\alpha$ admits a Levi decomposition $P_\alpha = M^{S_n}_\alpha N^{S_n}_\alpha$, where

$$M^{S_n}_\alpha = \{ \mathrm{diag}(g_1, \ldots, g_k, g,\ {}^\tau g_k^{-1}, \ldots, {}^\tau g_1^{-1});\ g_i \in \mathrm{GL}(m_i,F)\ (1 \leq i \leq k),\ g \in S_{n-m} \}$$
$$N^{S_n}_\alpha = \{ p \in^{S_n}_\alpha;\ p_{ii} = I_{n_i}\ \forall i \}$$

Let π_i be a representation of $\mathrm{GL}(n_i,F)$ $(1 \leq i \leq k)$. Let σ be a representation of S_{n-m}. Then we consider $\pi_1 \otimes \cdots \otimes \pi_k \otimes \sigma$ as a representation of $M^{S_n}_\alpha$ as usual:

$$\pi_1 \otimes \cdots \otimes \pi_k \otimes \sigma(\mathrm{diag}(g_1, \ldots, g_k, g,\ {}^\tau g_k^{-1}, \ldots, {}^\tau g_1^{-1})) = \pi_1(g_1) \otimes \cdots \otimes \pi_k(g_k) \otimes \sigma(g),$$

and extend it trivially across $N^{S_n}_\alpha$ to the representation of $P^{S_n}_\alpha$ denoted by the same letter. Then we form (normalized) induction written as follows (see [**T9**]):

$$\pi_1 \times \cdots \times \pi_k \rtimes \sigma = \mathrm{I}_{n,\alpha}(\pi_1 \otimes \cdots \otimes \pi_k \otimes \sigma) := \mathrm{Ind}^{S_n}_{P^{S_n}_\alpha}(\pi_1 \otimes \cdots \otimes \pi_k \otimes \sigma)$$

In this way we obtain a functor $\mathcal{M}_{adm.fin.leng.}(M^{S_n}_\alpha) \xrightarrow{I_{n,\alpha}} \mathcal{M}_{adm.fin.leng.}(S_n)$ and a group homomorphism $\mathrm{R}(M^{S_n}_\alpha) \xrightarrow{I_{n,\alpha}} \mathrm{R}(S_n)$. Next, if π is a representation of S_n, then we form the normalized Jacquet module $\mathrm{Jacq}_{\alpha,n}(\pi)$ of π. It is a representation of $M^{S_n}_\alpha$. In this way obtain a functor $\mathcal{M}_{adm.fin.leng.}(S_n) \xrightarrow{\mathrm{Jacq}_{\alpha,n}} \mathcal{M}_{adm.fin.leng.}(M^{S_n}_\alpha)$ and a group homomorphism $\mathrm{R}(S_n) \xrightarrow{\mathrm{Jacq}_{\alpha,n}} \mathrm{R}(M^{S_n}_\alpha)$. Here Frobenius reciprocity implies

$$\mathrm{Hom}_{S_n}(\pi,\ \mathrm{I}_{n,\alpha}(\pi_1 \otimes \cdots \otimes \pi_k \otimes \sigma)) \simeq \mathrm{Hom}_{M^{S_n}_\alpha}(\mathrm{Jacq}_{\alpha,n}(\pi),\ \pi_1 \otimes \cdots \otimes \pi_k \otimes \sigma).$$

Further

$$\pi_1 \rtimes (\pi_2 \rtimes \sigma) \simeq (\pi_1 \times \pi_2) \rtimes \sigma,$$
$$\widetilde{\pi \rtimes \sigma} \simeq \widetilde{\pi} \rtimes \widetilde{\sigma}$$
$\pi \rtimes \sigma$ and $\tilde{\pi} \rtimes \sigma$ have the same composition series,
if $\pi \rtimes \sigma$ is irreducible, then $\pi \rtimes \sigma \simeq \tilde{\pi} \rtimes \sigma$.

We remark that the third listed property follows, for example, from the general result ([**BeDeK**], Lemma 5.4), but in our case there is a proof that is simpler and based on the following result of Waldspurger (see [**MœViW**]):

$$\widetilde{\sigma} \simeq \sigma, \qquad S_n = \mathrm{SO}(2n+1, F),\ \mathrm{O}(2n, F)$$
$$\widetilde{\sigma} \simeq \sigma^x, \qquad S_n = \mathrm{Sp}(2n, F),$$

where $x \in \mathrm{GL}(2n, F)$ satisfies $x \cdot \begin{bmatrix} 0 & J_n \\ -J_n & 0 \end{bmatrix} \cdot {}^t x = (-1) \begin{bmatrix} 0 & J_n \\ -J_n & 0 \end{bmatrix}$, and $\sigma^x(g) = \sigma(x^{-1}gx)$, $g \in S_n$. Now, the stated property is obvious for $S_n = \mathrm{SO}(2n+1, F)$ and $O(2n, F)$. Let $S_n = \mathrm{Sp}(2n, F)$. Then

$$(\pi \rtimes \sigma)^y \simeq \pi \rtimes \sigma^x,$$

where $y = \mathrm{diag}(I_m, x, I_m)$ (π is a representation of $\mathrm{GL}(m, F)$). If $\widetilde{\pi \rtimes \widetilde{\sigma}} = \sum m_\rho \rho$ is a decomposition into irreducible representations in $\mathrm{R}(S_n)$, then $\pi \rtimes \widetilde{\sigma} = \sum m_\rho \widetilde{\rho}$, and, by a result of Waldspurger, we have the following:

$$\pi \rtimes \sigma = \pi \rtimes \widetilde{\sigma}^x = (\pi \rtimes \widetilde{\sigma})^y = \sum m_\rho \rho^y = \sum m_\rho \widetilde{\rho} = \widetilde{\pi \rtimes \widetilde{\sigma}} = \widetilde{\pi} \rtimes \sigma. \tag{1-1}$$

In this paper we work mostly with unramified representations. Let $n \geq 1$. If G is one of the groups $\mathrm{GL}(n, F)$, $\mathrm{Sp}(2n, F)$, $\mathrm{O}(2n, F)$, or $\mathrm{SO}(2n+1, F)$, then we let K be its maximal compact subgroup of the form $\mathrm{GL}(n, \mathcal{O})$, $\mathrm{Sp}(2n, \mathcal{O})$, $\mathrm{O}(2n, \mathcal{O})$, or $\mathrm{SO}(2n+1, \mathcal{O})$, respectively. We say that $\sigma \in \mathrm{Irr}(G)$ is unramified if it has a non–zero vector invariant under K. Unramified representations of G are classified using the Satake classification.

To explain the Satake classification, we let $P_{\min} = M_{\min} N_{\min}$ be the minimal parabolic subgroup of G as described above:

$$M_{\min} = \{\mathrm{diag}(x_1, \ldots, x_n)\}; \qquad G = \mathrm{GL}(n, F)$$
$$M_{\min} = \{\mathrm{diag}(x_1, \ldots, x_n, x_n^{-1}, \ldots, x_1^{-1})\}; \qquad G = \mathrm{Sp}(2n, F), \mathrm{O}(2n, F)$$
$$M_{\min} = \{\mathrm{diag}(x_1, \ldots, x_n, 1, x_n^{-1}, \ldots, x_1^{-1})\}; \quad G = \mathrm{SO}(2n+1, F).$$

Let $W = N_G(M_{\min})/M_{\min}$ be the Weyl group of G. It acts on $M_{\min}$ by conjugation: $w \cdot m = wmw^{-1}$, $w \in W$, $m \in M_{\min}$. This action extends to an action on the characters χ of $M_{\min}$ in the usual way: $w(\chi)(m) = \chi(w^{-1}mw)$, $w \in W$, $m \in M_{\min}$.

Explicitly, using the above description of $M_{\min}$, we fix the isomorphism $M_{\min} \simeq (F^\times)^n$ (considering only the first n coordinates). If $G = \mathrm{GL}(n, F)$, then W acts on $M_{\min}$ as the group of permutations of n letters. If G is one of the groups $\mathrm{Sp}(2n, F)$, $\mathrm{O}(2n, F)$, or $\mathrm{SO}(2n+1, F)$, then W acts on $M_{\min}$ as a group generated by the group of permutations of n letters and the following transformation:

$$(x_1, x_2, \ldots, x_n) \mapsto (x_1^{-1}, x_2, \ldots, x_n).$$

We have the following classification result (see [**Cr**]; [**R**] for $\mathrm{O}(2n, F)$):

THEOREM 1-2. *(i) Let $\chi_1, \dots, \chi_n$ be a sequence of unramified characters of $F^\times$. Then the induced representation $\mathrm{Ind}_{P_{\min}}^G(\chi_1 \otimes \cdots \otimes \chi_n)$ contains a unique unramified irreducible subquotient, denoted by $\sigma^G(\chi_1, \dots, \chi_n)$.*
(ii) Assume that $\chi_1, \dots, \chi_n$ and $\chi_1', \dots, \chi_n'$ are two sequences of unramified characters of $F^\times$. Then $\sigma^G(\chi_1, \dots, \chi_n) \simeq \sigma^G(\chi_1', \dots, \chi_n')$ if and only if there is $w \in W$ such that $\chi_1' \otimes \cdots \otimes \chi_n' = w(\chi_1 \otimes \cdots \otimes \chi_n)$. In other words, if and only if there is a permutation α of $\{1, 2, \dots, n\}$ letters and a sequence $\epsilon_1, \dots, \epsilon_n \in \{\pm 1\}$ such that $\chi_i' = \chi_{\alpha(i)}^{\epsilon_i}$, $i = 1, \dots, n$. ($\epsilon_1 = 1, \dots, \epsilon_n = 1$ for $G = \mathrm{GL}(n, F)$.)
(iii) Assume that $\sigma \in \mathrm{Irr}(G)$ is an unramified representation. Then there exists a sequence $(\chi_1, \dots, \chi_n)$ of unramified characters of $F^\times$ such that $\sigma \simeq \sigma^G(\chi_1, \dots, \chi_n)$. Every such sequence we call a supercuspidal support of σ.

We let $\mathrm{Irr}^{unr}(G)$ be the set of equivalence classes of irreducible unramified representations of G. We consider the trivial representation of the trivial group to be unramified. We let

$$\begin{cases} \mathrm{Irr}^{unr}(\mathrm{GL}) = \cup_{n \geq 0} \mathrm{Irr}^{unr}(\mathrm{GL}(n, F)) \\ \mathrm{Irr}^{unr}(S) = \cup_{n \geq 0} \mathrm{Irr}^{unr}(S_n). \end{cases} \tag{1-3}$$

There is another more precise classification of the elements of $\mathrm{Irr}^{unr}(\mathrm{GL})$ that we describe (see [**Ze**]).

In order to write down the Zelevinsky classification we introduce some notation. Let χ be an unramified character of $F^\times$, and let $n_1, n_2 \in \mathbb{R}$, $n_2 - n_1 \in \mathbb{Z}_{\geq 0}$. We denote by

$$[\nu^{n_1}\chi, \nu^{n_2}\chi] \qquad \text{or} \qquad [n_1, n_2]^{(\chi)}$$

the set $\{\nu^{n_1}\chi, \nu^{n_1+1}\chi, \dots, \nu^{n_2}\chi\}$, and call it a segment of unramified characters. To such a segment $\Delta = [\nu^{n_1}\chi, \nu^{n_2}\chi]$, Zelevinsky has attached a representation which is the unique irreducible subrepresentation of $\nu^{n_1}\chi \times \nu^{n_1+1}\chi \times \cdots \times \nu^{n_2}\chi$. This representation is the character

$$\nu^{(n_1+n_2)/2}\chi \; \mathbf{1}_{\mathrm{GL}(n_2-n_1+1,F)}. \tag{1-4}$$

We find it convenient to write it as follows:

$$\langle \Delta \rangle \qquad \text{or} \qquad \langle [n_1, n_2]^{(\chi)} \rangle. \tag{1-5}$$

We let

$$e(\Delta) = e([n_1, n_2]^{(\chi)}) = (n_1 + n_2)/2 + e(\chi). \tag{1-6}$$

Related to Theorem 1-2, we see

$$\langle \Delta \rangle = \langle [n_1, n_2]^{(\chi)} \rangle = \sigma^{\mathrm{GL}(n_1+n_2+1,F)}(\nu^{n_1}\chi, \nu^{n_1+1}\chi, \dots, \nu^{n_2}\chi).$$

The segments Δ_1 and Δ_2 of unramified characters are called linked if and only $\Delta_1 \cup \Delta_2$ is a segment but $\Delta_1 \not\subset \Delta_2$ and $\Delta_2 \not\subset \Delta_1$. We consider the empty set as a segment of unramified characters. It is not linked to any other segment. We let

$$\langle \emptyset \rangle = \mathbf{1} \in \mathrm{Irr}\ \mathrm{GL}(0, F).$$

Now, we give the Zelevinsky classification.

THEOREM 1-7. *(i) Let $\Delta_1, \dots, \Delta_k$ be a sequence of segments of unramified characters. Then $\langle \Delta_1 \rangle \times \cdots \times \langle \Delta_k \rangle$ is reducible if and only there are indices i, j, such that the segments Δ_i and Δ_j are linked. Moreover, if $\langle \Delta_1 \rangle \times \cdots \times \langle \Delta_k \rangle$ is irreducible, then it belongs to $\mathrm{Irr}^{unr}(\mathrm{GL})$.*

(ii) Conversely, if $\sigma \in \mathrm{Irr}^{unr}(\mathrm{GL})$, *then there is, up to a permutation, a unique sequence of segments of unramified characters* $\Delta_1, \ldots, \Delta_k$ *such that* $\sigma \simeq \langle \Delta_1 \rangle \times \cdots \times \langle \Delta_k \rangle$.

A similar classification exists in the case of the classical groups (see [**M4**]). We recall it in Sections 5 and 6. We end this section with the following remark:

REMARK 1-8. *It follows from Theorem 1-2 (ii) that every unramified representation* $\sigma \in \mathrm{Irr}^{unr}(S)$ *is self–dual. Also, if* $\pi \in \mathrm{Irr}^{unr}(\mathrm{GL})$, *then there exists a unique unramified irreducible subquotient, say* σ_1 *of* $\pi \rtimes \sigma$. *The representation* σ_1 *is self–dual, and it is also a subquotient of* $\widetilde{\pi} \rtimes \sigma$. *(See the basic properties for the induction for the split classical groups listed above.)*

2. Some General Results on Unitarizability

Let G be a connected reductive p–adic group or $\mathrm{O}(2n, F)$ $(n \geq 0)$. We recall that the contragredient representation π of G is denoted by $\widetilde{\pi}$. We write $\bar{\pi}$ for the complex conjugate representation of the representation π. We remind the reader that this means the following: In the representation space V_π we change the multiplication to $\alpha ._{new} v := \bar{\alpha} ._{old} v$, $\alpha \in \mathbb{C}$, $v \in V_\pi$. In this way we obtain $V_{\bar{\pi}}$. We let $\bar{\pi}(g)v = \pi(g)v$, $g \in G$, $v \in V_{\bar{\pi}} = V_\pi$. It is easy to see the following:

$$\bar{\widetilde{\pi}} \simeq \widetilde{\bar{\pi}}.$$

The Hermitian contragredient of the representation of π is defined as follows:

$$\pi^+ := \bar{\widetilde{\pi}}.$$

Let $P = MN$ be a parabolic subgroup of G. We have the following:

(H-IC) $\mathrm{Ind}_P^G(\sigma)^+ \simeq \mathrm{Ind}_P^G(\sigma^+)$.

A representation $\pi \in \mathrm{Irr}(G)$ is said to be Hermitian if there is a non–degenerate G–invariant Hermitian form $\langle\ ,\ \rangle$ on V_π. This means the following:

$$\langle \alpha v + \beta w, u \rangle = \alpha \langle v, u \rangle + \beta \langle w, u \rangle \tag{2-1}$$

$$\overline{\langle v, w \rangle} = \langle w, v \rangle \tag{2-2}$$

$$\langle \pi(g)v, \pi(g)v \rangle = \langle v, w \rangle, \tag{2-3}$$

$$\langle v, w \rangle = 0,\ \forall w \in V_\pi,\ \text{implies } v = 0. \tag{2-4}$$

Since π is irreducible, the Hermitian form $\langle\ ,\ \rangle$ is unique up to a non–zero real scalar. Let $\mathrm{Irr}^+(G)$ be the set of equivalence classes of irreducible Hermitian representations of G. Since we work with unramified representations, we let

$$\begin{cases} \mathrm{Irr}^{+,unr}(\mathrm{GL}) = \cup_{n \geq 0} \mathrm{Irr}^{+,unr}(\mathrm{GL}(n, F)) \\ \mathrm{Irr}^{+,unr}(S) = \cup_{n \geq 0} \mathrm{Irr}^{+,unr}(S_n). \end{cases} \tag{2-5}$$

We list the following basic properties of Hermitian representations:

(H-Irr) If $\pi \in \mathrm{Irr}(G)$, then $\pi \in \mathrm{Irr}^+(G)$ if and only if $\pi \simeq \pi^+$

(H-Ind) Let $P = MN$ be a parabolic subgroup of G. Let $\sigma \in \mathrm{Irr}^+(M)$. Then there is a non–trivial G–invariant Hermitian form on $\mathrm{Ind}_P^G(\sigma)$. In particular, if $\mathrm{Ind}_P^G(\sigma)$ is irreducible, then it is Hermitian.

In addition, a Hermitian representation $\pi \in \mathrm{Irr}(G)$ is said to be unitarizable if the form $\langle\ ,\ \rangle$ is definite. Let $\mathrm{Irr}^u(G)$ be the set of equivalence classes of irreducible unitarizable representations of G. We have the following:

$$\mathrm{Irr}^u(G) \subset \mathrm{Irr}^+(G) \subset \mathrm{Irr}(G).$$

In this paper we classify unramified unitarizable representations (see (1-3))

$$\begin{cases} \mathrm{Irr}^{u,unr}(\mathrm{GL}) = \cup_{n\geq 0}\mathrm{Irr}^{u,unr}(\mathrm{GL}(n,F)) \\ \mathrm{Irr}^{u,unr}(S) = \cup_{n\geq 0}\mathrm{Irr}^{u,unr}(S_n). \end{cases} \tag{2-6}$$

Now, we recall some principles used in the construction and classification of unitarizable unramified representations. Some of them are already well–known (see [**T8**]), but some of them are new (see (NU-RP)). Below, $P = MN$ denotes a parabolic subgroup of G and σ an irreducible representation of M.

(UI) **Unitary parabolic induction**: the unitarizability of σ implies that the parabolically induced representation $\mathrm{Ind}_P^G(\sigma)$ is unitarizable.

(UR) **Unitary parabolic reduction**: if σ is a Hermitian representation such that the parabolically induced representation $\mathrm{Ind}_P^G(\sigma)$ is irreducible and unitarizable, then σ is an (irreducible) unitarizable representation.

(D) **Deformation (or complementary series)**: let X be a connected set of characters of M such that each representation $\mathrm{Ind}_P^G(\chi\sigma)$, $\chi \in X$, is Hermitian and irreducible. Now, if $\mathrm{Ind}_P^G(\chi_0\sigma)$ is unitarizable for some $\chi_0 \in X$, then the whole family $\mathrm{Ind}_P^G(\chi\sigma)$, $\chi \in X$, consists of unitarizable representations.

(ED) **Ends of deformations**: suppose that Y is a set of characters of M, and X a dense subset of Y satisfying (D); then each irreducible subquotient of any $\mathrm{Ind}_P^G(\chi\sigma)$, $\chi \in Y$, is unitarizable.

Sometimes we get important irreducible unitarizable representations in the following way. Let Z denote the center of G. Let k be a global field, P_k the set of places of k, k_v the completion of k at the place v, $\mathbb{A}_k$ the ring of adèles of k, ω a unitary character of $Z(\mathbb{A}_k)$ and $L^2(\omega, G(k)\backslash G(\mathbb{A}_k))$ the representation of $G(\mathbb{A}_k) \simeq \otimes_{v\in P_k} G(k_v)$ by right translations on the space of square integrable functions on $G(\mathbb{A}_k)$ which transform under the action of $Z(\mathbb{A}_k)$ according to ω. Suppose that π is an irreducible representation of $G = G(F)$.

(RS) Residual automorphic spectrum factors: if $F \simeq k_v$ for some global field k and $v \in P_k$, and there exists an irreducible (non-cuspidal) subrepresentation Π of $L^2(\omega, G(\mathbb{A}_k))$ such that π is isomorphic to a (corresponding) tensor factor of Π, then π is unitarizable.

It is evident that π as above is unitarizable. But this construction is technically much more complicated than the above four. It requires computation of residues of Eisenstein series.

The last principle is not necessary to use for the classification of $\mathrm{Irr}^{u,unr}(\mathrm{GL})$, but we use it in [**M6**] in order to prove the unitarity of "basic building blocks" of $\mathrm{Irr}^{u,unr}(S)$. (See Theorem 5-11 in Section 5.)

In addition, the following simple remark is useful for proving non–unitarity. Obviously, the Cauchy-Schwarz inequality implies that matrix coefficients of unitarizable representations are bounded. Now, (D) directly implies the following:

REMARK 2-7. ***(unbounded matrix coefficients)*** *Let X be a connected set of characters of M such that each representation $\mathrm{Ind}_P^G(\chi\sigma)$, $\chi \in X$, is Hermitian and irreducible and that $\mathrm{Ind}_P^G(\chi_0\sigma)$ has an unbounded matrix coefficient for some $\chi_0 \in X$. Then all $\mathrm{Ind}_P^G(\chi\sigma)$, $\chi \in X$, are non–unitarizable.*

In addition, we use the following two criteria for proving non–unitarity. The criteria are very technical. In a special case they were already applied in [**LMT**]. We present them here in a more general form. Let $P = MN$ be a maximal parabolic subgroup of G. Assume that the Weyl group $W(M) = N_G(M)/M$ has two elements. (It always has one or two elements.) We write w_0 for a representative of the nontrivial element in $W(M)$. Assume that $\sigma \in \mathrm{Irr}(M)$ is an irreducible unitarizable representation such that $w_0(\sigma) \simeq \sigma$. Then there is a standard normalized intertwining operator (at least in the cases that we need) $N(\delta_P^s\sigma) : \mathrm{Ind}_P^G(\delta_P^s\sigma) \to \mathrm{Ind}_P^G(\delta_P^{-s}\sigma)$. We have the following:

(N-1) $N(\delta_P^s\sigma)N(\delta_P^{-s}\sigma) = id$

(N-2) $N(\delta_P^s\sigma)$ is Hermitian, and therefore holomorphic, for $s \in \sqrt{-1}\mathbb{R}$.

Let $\langle\ ,\ \rangle_\sigma$ be an M–invariant definite Hermitian form on V_σ. Then

$$\langle f_1, f_2\rangle_s = \int_K \langle f_1(k),\ N(\delta_P^s\sigma)f_2(k)\rangle_\sigma dk \tag{2-8}$$

is a Hermitian form on $\mathrm{Ind}_P^G(\delta_P^s\pi)$. It is non–degenerate whenever $\mathrm{Ind}_P^G(\delta_P^s\sigma)$ is irreducible and $N(\delta_P^s\sigma)$ is holomorphic. Now, we make the following two assumptions:

(A-1) If $\mathrm{Ind}_P^G(\delta_P^s\sigma)$ is reducible at $s = 0$, then $N(\sigma)$ is non–trivial.

(A-2) If $\mathrm{Ind}_P^G(\delta_P^s\sigma)$ is irreducible at $s = 0$, let $s_1 > 0$ be the first point of reducibility (this must exist because of Remark 2-7). We assume that $N(\delta_P^s\sigma)$ is holomorphic and non–trivial for $s \in]0,\ s_1]$. (Then (N-1) implies that $N(\delta_P^{-s}\sigma)$ is holomorphic for $s \in]0,\ s_1[$). We assume that $N(\delta_P^{-s}\sigma)$ has a pole at $s = s_1$ of odd order.

If (A-1) holds, then (N-1) implies that $\mathrm{Ind}_P^G(\sigma)$ is a direct sum of two non–trivial (perhaps reducible) representations on which $N(\sigma)$ acts as $-$id and id, respectively. Now, since $\mathrm{Ind}_P^G(\delta_P^s\sigma)$ is irreducible and $N(\delta_P^s\sigma)$ is holomorphic for $s > 0$, s close to 0, we conclude that $\langle\ ,\ \rangle_s$ is not definite. Hence $\mathrm{Ind}_P^G(\delta_P^s\sigma)$ is not unitarizable for $s > 0$, s close to 0.

If (A-2) holds, then we write k for the order of the pole of $N(\delta_P^{-s}\sigma)$ at $s = s_1$. We realize the family of representations $\mathrm{Ind}_P^G(\delta_P^s\sigma)$ ($s \in \mathbb{C}$) in the compact picture, say with space X. Let $f \in X$ be such that

$$F_s := (s - s_1)^k N(\delta_P^{-s}\sigma)f \text{ is holomorphic and non–zero at } s = s_1.$$

We see that F_s is real analytic near s_1. Let $h \in X$. Using (N-1), we compute:

$$\begin{aligned}\langle h, F_s\rangle_s &= \int_K \langle h(k),\ N(\delta_P^s\pi)F_s(k)\rangle_\sigma dk \\ &= (s - s')^k \int_K \langle h(k),\ f(k)\rangle_\sigma dk.\end{aligned} \tag{2-9}$$

Now, we apply some elementary results from linear algebra (see ([**Vo**], Theorem 3.2, Proposition 3.3)) to our situation. First, we may assume that f belongs to some fixed K–isotypic component, say E, of X. Since X is an admissible representation of K, we see that $\dim E < \infty$. We consider the restriction of the family of Hermitian forms $\langle\ ,\ \rangle_s$ to E. We write this restriction as $(\ ,\ )_t$, where $t = s - s_1$. Let

$$E = E^0 \supset E^1 \supset \cdots \supset E^N = \{0\}$$

be the filtration of E defined as follows. The space E^n is the space of vectors $e \in E$ for which there is a neighborhood U of 0 and a (real) analytic function $f_e : U \longrightarrow E$ satisfying

(i) $f_e(0) = e$
(ii) $\forall e' \in E$ the function $s \longmapsto (f_e(s), e')_t$ vanishes at 0 to order at least n.

Let $F = F_{s_1}$. Since $t \longmapsto F_{t+s_1}$ is a real analytic function from a neighborhood U of 0 into E, (2-9) implies that $F \in E^k$. Moreover, since also $f \in E$, we see that (2-9) applied to $f = h$ implies that $F \notin E^{k+1}$. We conclude

$$E^k/E^{k+1} \neq 0. \tag{2-10}$$

Next, we define a Hermitian form $(\ ,\ )^n$ on E^n by the formula

$$(e, e')^n = \lim_{t \longrightarrow 0} \frac{1}{t^n}(f_e(s), f_{e'}(s))_s.$$

(It is easy to see that this definition is independent of the choices of f_e and $f_{e'}$.) The radical of the form $(\ ,\ )^n$ is exactly E^{n+1}. We write (p_n, q_n) for the signature on E^n/E^{n+1}. It is proved in ([**Vo**], Proposition 3.3) that for t small positive, $(\ ,\ )_t$ has signature

$$(\sum_n p_n, \sum_n q_n)$$

and for t small negative

$$(\sum_{n \text{ even}} p_n + \sum_{n \text{ odd}} q_n, \sum_{n \text{ odd}} p_n + \sum_{n \text{ even}} q_n).$$

Now, we are ready to show the the non–unitarity of $\mathrm{Ind}_P^G(\delta_P^s\sigma)$ for $s - s_1$ small positive. It is enough to show that the Hermitian form $\langle\ ,\ \rangle_s$ is not definite.

Without loss of generality we may assume that $\langle\ ,\ \rangle_s$ ($s \in]0, s_1[$) is positive definite. Then it is positive definite on $\mathrm{Ind}_P^G(\delta_P^{s_1}\sigma)/\ker N(\delta_P^{(s_1)}\sigma)$. Thus, if there is unitarity immediately after s_1, then the form $\langle\ ,\ \rangle_s$ is positive definite for $s > s_1$ close to s_1. In particular, $(\ ,\ )_t$ is positive definite for $t > 0$ close to 0. Hence

$$\sum_n q_n = \sum_{n \text{ odd}} p_n + \sum_{n \text{ even}} q_n = 0.$$

Since k is odd, we see that

$$p_k = q_k = 0.$$

This contradicts (2-10). We have proved the following non–unitarity criteria:

(RP) Let $P = MN$ be a self-dual maximal parabolic subgroup of G. We write w_0 for the representative of the nontrivial element in $W(M)$. Assume that $\sigma \in \mathrm{Irr}(M)$ is an irreducible unitarizable representation such that $w_0(\sigma) \simeq \sigma$. Then we have the following:

(i) If (A-1) holds (i.e., $\mathrm{Ind}_P^G(\sigma)$ is reducible and $N(\sigma)$ is non–trivial), then $\mathrm{Ind}_P^G(\delta_P^s\sigma)$ is not unitarizable for $s > 0$, s close to 0.

(ii) If (A-2) holds (i.e., $\mathrm{Ind}_P^G(\delta_P^s\sigma)$ is irreducible at $s=0$, $s_1>0$ is the first reducibility point, $N(\delta_P^s\sigma)$ is holomorphic and non–trivial for $s\in]0,\ s_1]$ and $N(\delta_P^{-s}\sigma)$ has a pole at $s=s_1$ of odd order), then $\mathrm{Ind}_P^G(\delta_P^s\sigma)$ is not unitarizable for $s>s_1$, s close to s_1.

3. The Topology of the Unramified Dual

Let G be a connected reductive p–adic group or $\mathrm{O}(2n,F)$ $(n\geq 0)$. The topology on the non–unitary dual $\mathrm{Irr}(G)$ is given by the uniform convergence of matrix coefficients on compact sets ([**T1**], [**T6**]; see also [**F**], [**Di**]). Then $\mathrm{Irr}^u(G)$ is closed subset in $\mathrm{Irr}(G)$. We supply $\mathrm{Irr}^u(G)$ with the relative topology.

Now, we assume that G is one of the groups $\mathrm{GL}(n,F)$, $\mathrm{O}(2n,F)$, $\mathrm{SO}(2n+1,F)$ or $\mathrm{Sp}(2n,F)$. Let K be the maximal compact subgroup introduced in the paragraph before Theorem 1-2. The Weyl group W of G acts naturally on the analytic manifold $D_n=(\mathbb{C}^\times)^n$. The space of W–orbits D_n^W has the structure of an analytic manifold. The manifold D_n parameterizes unramified principal series of G as follows:

$$\mathrm{Ind}_{P_{\min}}^G(\chi_1\otimes\cdots\otimes\chi_n)\rightarrow(\chi_1(\varpi),\ldots,\chi_n(\varpi)).$$

The manifold D_n^W parameterizes unramified principal series of G. Let $\mathrm{Irr}^I(G)$ be the set of equivalence classes of irreducible representations σ of G for which there exists a representation in unramified principal series, say $\mathrm{Ind}_{P_{\min}}^G(\chi_1\otimes\cdots\otimes\chi_n)$, such that σ is an irreducible subquotient of $\mathrm{Ind}_{P_{\min}}^G(\chi_1\otimes\cdots\otimes\chi_n)$. The principal series is determined, up to association, uniquely by this condition. We have a well–defined map

$$\varphi_G:\mathrm{Irr}^I(G)\rightarrow D_n^W \tag{3-1}$$

defined by

$$\varphi_G(\sigma)=W\text{–orbit of a } n\text{–tuple } (\chi_1(\varpi),\ldots,\chi_n(\varpi)).$$

We call $\varphi_G(\sigma)$ the infinitesimal character of σ. The fibers of φ_G are finite. Its restriction to $\mathrm{Irr}^{unr}(G)$ induces a bijection $\varphi_G:\mathrm{Irr}^{unr}(G)\rightarrow D_n^W$,

$$\varphi_G(\sigma^G(\chi_1,\ldots,\chi_n))=W\text{–orbit of the } n\text{–tuple } (\chi_1(\varpi),\ldots,\chi_n(\varpi)). \tag{3-2}$$

Now, we recall some results from [**T6**].

LEMMA 3-3. *Suppose that G is connected (later we discuss the case of $\mathrm{O}(2n,F)$). Then the set $\mathrm{Irr}^I(G)$ is a connected component of $\mathrm{Irr}(G)$. Therefore it is open and closed there. The map φ_G given by (3-1) is continuous and closed.*

Next, ([**T6**], Lemma 5.8) implies the following:

LEMMA 3-4. *Suppose that G is connected. Then $\mathrm{Irr}^{unr}(G)$ is an open subset of $\mathrm{Irr}^I(G)$.*

We have the following description of the topology on $\mathrm{Irr}^{unr}(G)$:

THEOREM 3-5. *Suppose that G is connected. Then the map (3-2) is a homeomorphism.*

PROOF. As it is continuous and bijective, it is enough to show that it is closed. So, let Z be a closed set in $\mathrm{Irr}^{unr}(G)$. We must show that $\varphi_G(Z)$ is closed. In order to prove that, let $Cl(\varphi_G(Z))$ be its closure. We must show that $Cl(\varphi_G(Z))=$

$\varphi_G(Z)$. Let $x \in Cl(\varphi_G(Z))$. Then there exists a sequence $(x_m)_{m\geq 1}$ in $\varphi_G(Z)$ such that $\lim_m x_n = x$. (We remark that D_n^W is a complex analytic manifold.) We write

$$x_m = W\text{–orbit of the } n\text{–tuple } (s_{m,1}, \ldots, s_{m,n}) \in D_n \quad (m \geq 1);$$
$$x = W\text{–orbit of the } n\text{–tuple } (s_1, \ldots, s_n) \in D_n.$$

After passing to a subsequence and making the appropriate identification, we may assume $\lim_m s_{m,i} = s_i$ for $i = 1, \ldots, n$. We may take unramified characters $\chi_{m,i}$ and $\chi_1, \ldots, \chi_n$, $m \geq 1, i = 1, \ldots, n$, such that $\chi_{m,i}(\varpi) = s_{m,i}$ and $\chi_i(\varpi) = s_i$. Clearly, we have the following:

$$\varphi_G(\sigma^G(\chi_{m,1}, \ldots, \chi_{m,n})) = x_m \quad (m \geq 1);$$
$$\varphi_G(\sigma^G(\chi_1, \ldots, \chi_n)) = x.$$

Now, Proposition 5.2 of [**T6**] tells us that, passing to a subsequence, we may assume that the characters of $\sigma^G(\chi_{m,1}, \ldots, \chi_{m,n})$ converge pointwise, and that there are irreducible subquotients $\sigma_1, \ldots, \sigma_l$ of $\mathrm{Ind}_{P_{\min}}^G(\chi_1 \otimes \cdots \otimes \chi_n)$ and $k_1, \ldots, k_l \in \mathbb{Z}_{>0}$ such that the pointwise limit is a character of $\sum_{i=1}^l k_i\sigma_i$. Since the representations $\sigma^G(\chi_{m,1}, \ldots, \chi_{m,n})$ are unramified, among the representations $\sigma_1, \ldots, \sigma_l$ is $\sigma^G(\chi_1, \ldots, \chi_n)$. Therefore, one of the equivalent descriptions of the topology in [**T6**] implies $\sigma^G(\chi_1, \ldots, \chi_n) \in Z$. Hence $x = \varphi_G(\sigma^G(\chi_1, \ldots, \chi_n)) \in \varphi_G(Z)$. This shows that $\varphi_G(Z)$ is closed. □

REMARK 3-6. *Suppose that G is connected. Then the set $\mathrm{Irr}^{u,unr}(G)$ is a closed subset of $\mathrm{Irr}^{unr}(G)$ (see [**T6**]). Therefore, it can be identified via φ_G with a closed subset of D_n^W.*

In this paper we shall need only the topology of the unitary dual. The following theorem describes it.

THEOREM 3-7. *Let G be one of the groups $\mathrm{GL}(n, F)$, $\mathrm{O}(2n, F)$, $\mathrm{SO}(2n+1, F)$ or $\mathrm{Sp}(2n, F)$. Then map (3-2) restricts to a homeomorphism*

$$\varphi_G : \mathrm{Irr}^{u,unr}(G) \to D_n^W \tag{3-8}$$

of $\mathrm{Irr}^{u,unr}(G)$ onto a compact (closed) subset of D_n^W.

PROOF. If G is connected, then φ_G is a homeomorphism onto the image by Theorem 3-5. The image is compact by Theorem 3.1 of [**T1**] (this is also Theorem 2.5 of [**T6**]).

Now we briefly explain the proof in the case of $G = \mathrm{O}(2n, F)$ (below, sometimes we do not distinguish between elements in D_n^G and the W-orbits that they determine; one can easily complete the details). The compactness for the case of $\mathrm{SO}(2n, F)$ implies that the image of φ_G has compact closure. Further, the topology can be described by characters (see [**Mi**]). Suppose that we have a convergent sequence $\psi_m \to \psi$ in D_n^W, such that the sequence ψ_m is contained in the image of φ_G. Suppose that ψ_m corresponds to unramified characters ψ_m', and ψ to ψ'. Let π_m be such that $\varphi_G(\pi_m) = \psi_m$. Now, ([**T6**], Proposition 5.2) says that we can pass to a subsequence such that characters of π_m converge pointwise to the character of a subquotient π of the representation induced by ψ'. It is obvious that π has an irreducible unramified subquotient, say π'. Clearly, $\varphi_G(\pi') = \psi$. Now, [**T7**] implies that all irreducible subquotients are unitarizable. So, π' is unitarizable.

This implies that ψ is in the image of φ_G. Thus, the image is closed. Denote the image by X.

Let $Y \subset X$, and let ψ be a point in the closure of Y. Take a sequence ψ_m in Y converging to ψ. Let the ψ_m correspond to unramified characters ψ'_m, and ψ to ψ'. Take π_m such that $\varphi_G(\pi_m) = \psi_m$. This means that $\pi_m \in \varphi_G^{-1}(Y)$. As above we can pass to a subsequence such that characters of π_m converge pointwise to the character of a subquotient π of the representation induced by ψ'. Further, π has an irreducible unramified subquotient π' with $\varphi_G(\pi') = \psi$ and π' unitarizable. The description of the topology by characters implies that π' is a limit of the sequence π_m. This implies that π' is in the closure of $\varphi_G^{-1}(Y)$. This implies that $\varphi_G^{-1} : X \to \mathrm{Irr}^{u,unr}(G)$ is continuous.

Now, let $S \subset \mathrm{Irr}^{unr,u}(G)$. Take $\pi \in \mathrm{Irr}^{unr,u}(G)$ from the closure of S. Then we can find a sequence $\pi_m \in S$ converging to π. Let π_m and π be subquotients of representations induced by unramified characters ψ_m and ψ, respectively. Since X is compact, we can pass to a subsequence such that ψ_m converges (to some ψ_0). Next, arguing as above, we can pass to a subsequence of π_m such that all limits are subquotients of the representation induced by ψ_0. Now, the linear independence of characters of irreducible representations implies that $\psi = \psi_0$. Let $\psi'_m, \psi' \in D_n^W$ correspond to ψ_m, ψ, respectively. Observe that $\psi'_m = \varphi_G(\pi_m) \in \varphi_G(S)$, $\varphi_G(\pi) = \psi'$. Therefore, $\varphi_G(\pi)$ is in the closure of $\varphi_G(S)$. This ends the proof of continuity of φ_G. The proof of the theorem is now complete. □

4. The Unramified Unitary Dual of $\mathrm{GL}(n, F)$

The second named author classified unramified unitarizable representations $\mathrm{Irr}^{u,unr}(\mathrm{GL})$ in [**T4**]. The proof was based on Theorem 1-7 and a result of Bernstein on irreducibility of unitary parabolic induction proved in [**Be2**]. In this section we give the classification of $\mathrm{Irr}^{u,unr}(\mathrm{GL})$ without using the result of Bernstein. The main result of this section is the following theorem:

THEOREM 4-1. *(i) Let $\phi_1, \ldots, \phi_a, \psi_1, \ldots, \psi_b \in \mathrm{Irr}^{unr}(\mathrm{GL})$ be a sequence of unramified unitary characters (one–dimensional unramified representations). Let $\alpha_1, \ldots, \alpha_b \in \,]0, \frac{1}{2}[$ be a sequence of real numbers. (The possibility $a = 0$ or $b = 0$ is not excluded here.) Then*

$$(4\text{-}2)\quad \phi_1 \times \cdots \times \phi_a \times (\nu^{\alpha_1}\psi_1 \times \nu^{-\alpha_1}\psi_1) \times \cdots \times (\nu^{\alpha_b}\psi_b \times \nu^{-\alpha_b}\psi_b) \in \mathrm{Irr}^{u,unr}(\mathrm{GL}).$$

(ii) Let $\pi \in \mathrm{Irr}^{u,unr}(\mathrm{GL})$. Then there exist $\phi_1, \ldots, \phi_a$, $\psi_1, \ldots, \psi_b$, $\alpha_1, \ldots, \alpha_b$ as in (i) such that π is isomorphic to the induced representation given by (4-2). Each sequence $\phi_1, \ldots, \phi_a$ and $(\psi_1, \alpha_1), \ldots, (\psi_b, \alpha_b)$ is uniquely determined by π up to a permutation.

PROOF. Applying (H-IC) and (H-Irr), we see that a representation given by (4-2) is Hermitian. Next, fixing $\phi_1, \ldots, \phi_a, \psi_1, \ldots, \psi_b$ and letting $0 \le \alpha_1, \ldots, \alpha_b < 1/2$ vary, the representations in (4-2) form a continuous family of irreducible Hermitian representations with a unitarizable representation in it (namely, the one attached to $\alpha_1 = \cdots = \alpha_b = 0$). Thus, by (D), they are all unitarizable. The uniqueness in (ii) follows from the Zelevinsky classification (see Theorem 1-7).

Let $\pi \in \mathrm{Irr}^{u,unr}(\mathrm{GL})$. It remains to prove that π can be written in the form of (4-2). First, being unramified, the Zelevinsky classification (see Theorem 1-7) implies that π is fully–induced from (not necessarily unitary) characters in $\mathrm{Irr}^{unr}(\mathrm{GL})$.

Now, since $\pi \in \mathrm{Irr}^{+,unr}(\mathrm{GL})$, using (H-IC), (H-Irr), and (H-Ind), we obtain

$$\pi \simeq \phi_1 \times \cdots \times \phi_a \times (\nu^{\alpha_1}\psi_1 \times \nu^{-\alpha_1}\psi_1) \times \cdots \times (\nu^{\alpha_b}\psi_b \times \nu^{-\alpha_b}\psi_b), \tag{4-3}$$

where everything is as in (i) except that we have only $\alpha_1, \ldots, \alpha_b > 0$. To prove the theorem, we need to prove that $\alpha_i < 1/2$, $i = 1, \ldots, b$.

First, as all representations $\phi_1, \ldots, \phi_a, \nu^{\alpha_1}\psi_1 \times \nu^{-\alpha_1}\psi_1, \ldots, \nu^{\alpha_b}\psi_b \times \nu^{-\alpha_b}\psi_b$ are Hermitian and π is unitarizable, we conclude that $\phi_1 \otimes \cdots \otimes \phi_a \otimes (\nu^{\alpha_1}\psi_1 \times \nu^{-\alpha_1}\psi_1) \otimes \cdots \otimes (\nu^{\alpha_b}\psi_b \times \nu^{-\alpha_b}\psi_b)$ is unitarizable (see (UR)). In particular,

$$\nu^{\alpha_1}\psi_1 \times \nu^{-\alpha_1}\psi_1, \ \ldots, \ \nu^{\alpha_b}\psi_b \times \nu^{-\alpha_b}\psi_b \ \text{ are unitarizable representations.} \tag{4-4}$$

Suppose that some $\alpha_i \geq 1/2$ for some i. Let $\alpha = \alpha_i$ and $\psi = \psi_i$. We can write $\psi = \langle [-x, \ x]^{(\chi)} \rangle$, where χ is a unitary unramified character of $F^\times$ and $x \in \mathbb{Z}_{\geq 0}$ (see (1-4) and (1-5)). We let

$$\pi_{\beta,x} = \nu^\beta\psi \times \nu^{-\beta}\psi = \langle [-x+\beta, \ x+\beta]^{(\chi)} \rangle \times \langle [-x-\beta, \ x-\beta]^{(\chi)} \rangle, \text{ where } \beta \in \mathbb{R}. \tag{4-5}$$

Note that

$$\text{if } \pi_{\beta,x} \ \text{ is irreducible, then } \pi_{\beta,x} \in \mathrm{Irr}^+(\mathrm{GL}) \tag{4-6}$$

$$\pi_{\beta,x} \text{ is reducible if and only if} [-x+\beta, \ x+\beta]^{(\chi)}, [-x-\beta, \ x-\beta]^{(\chi)} \text{ are linked.} \tag{4-7}$$

Now, we consider the two cases.

First, we assume that $\alpha - x > 1/2$. Then (4-4), (4-6) and (4-7) imply that the continuous family of representations $\pi_{\beta,x}$ $(\beta \geq \alpha)$ is irreducible, Hermitian and, at $\beta = \alpha$, unitarizable. Therefore it is unitarizable everywhere. But this contradicts Remark 2-7 since for large enough β, $\pi_{\beta,x}$ has unbounded matrix coefficients. (See [**T1**], [**T6**].)

Therefore $\alpha - x \leq 1/2$. Now, using the definition (4-6) and (4-7), the irreducibility of $\pi_{\alpha,x}$ implies

$$\alpha \notin (1/2)\,\mathbb{Z}. \tag{4-8}$$

Next, there exists $k \in \mathbb{Z}_{>0}$ such that

$$\left| \frac{(-x+\alpha-k)+(x+\alpha-1)}{2} \right| = |\alpha - k/2 - 1/2| < 1/2$$

(there are exactly two such k's). Now, the representation

$$\pi_{\alpha-(k+1)/2, x+(k-1)/2} = \langle [-x+\alpha-k, x+\alpha-1]^{(\chi)} \rangle \times \langle [-x-\alpha+1, x-\alpha+k]^{(\chi)} \rangle$$

is irreducible and unitarizable by (i). Hence

$$\pi := \pi_{\alpha,x} \times \pi_{\alpha-(k+1)/2, x+(k-1)/2} \tag{4-9}$$

is a unitarizable representation. Next, (4-8) implies that

(4-10)

$a - b \notin \mathbb{Z}$, where a (resp., b) belongs to the first (resp., the last) two sequences:

$$\begin{cases} -x+\alpha, \ldots, x+\alpha, \\ -x+\alpha-k, \ldots, x+\alpha-1, \\ -x-\alpha, \ldots, x-\alpha, \\ -x-\alpha+1, \ldots, x-\alpha+k. \end{cases}$$

In particular, this and [**Ze**] imply

$$\langle[-x-\alpha,\ x-\alpha]^{(\chi)}\rangle \times \langle[-x+\alpha-k, x+\alpha-1]^{(\chi)}\rangle \simeq \langle[-x+\alpha-k, x+\alpha-1]^{(\chi)}\rangle \times \langle[-x-\alpha,\ x-\alpha]^{(\chi)}\rangle.$$

Hence, $\pi = \pi_{\alpha,x} \times \pi_{\alpha-(k+1)/2, x+(k-1)/2}$ is isomorphic to

$$\text{(4-11)}\quad \left(\langle[-x+\alpha,\ x+\alpha]^{(\chi)}\rangle \times \langle[-x+\alpha-k, x+\alpha-1]^{(\chi)}\rangle\right) \times \left(\langle[-x-\alpha,\ x-\alpha]^{(\chi)}\rangle \times \langle[-x-\alpha+1, x-\alpha+k]^{(\chi)}\rangle\right).$$

Since, by the Zelevinsky classification, the induced representations in both parentheses in (4-11) reduce, we conclude that π has at least four irreducible subrepresentations. Since, by definition,

$$\text{(4-12)}\quad \pi \simeq \langle[-x+\alpha,\ x+\alpha]^{(\chi)}\rangle \times \langle[-x-\alpha,\ x-\alpha]^{(\chi)}\rangle \times \langle[-x+\alpha-k, x+\alpha-1]^{(\chi)}\rangle \times \langle[-x-\alpha+1, x-\alpha+k]^{(\chi)}\rangle,$$

Frobenius reciprocity implies that the multiplicity of

$$\tau := \langle[-x+\alpha,\ x+\alpha]^{(\chi)}\rangle \times \langle[-x-\alpha,\ x-\alpha]^{(\chi)}\rangle \otimes \langle[-x+\alpha-k, x+\alpha-1]^{(\chi)}\rangle \times \langle[-x-\alpha+1, x-\alpha+k]^{(\chi)}\rangle,$$

in the Jacquet module

$$r_{(4x+2, 4x+2k),\ 8x+2k+2}(\pi)$$

must be at least four. This contradicts (the following) Lemma 4-13, and proves the theorem. □

It remains to prove the following lemma:

LEMMA 4-13. *The multiplicity of* τ *in* $r_{(4x+2,4x+2k),\ 8x+2k+2}(\pi)$ *is exactly two.*

PROOF. We begin by introducing some notation. If ρ is an admissible representation of $\mathrm{GL}(n, F)$, then we let

$$m^*(\rho) = \mathbf{1} \otimes \rho + \sum_{i=1}^{n-1} r_{(i,n-i),\ n}(\pi) + \rho \otimes \mathbf{1}$$

in $(\oplus_{n\geq 0}\mathrm{R}(\mathrm{GL}(n,F))) \otimes (\oplus_{n\geq 0}\mathrm{R}(\mathrm{GL}(n,F)))$. By [**Ze**], m^* is multiplicative:

$$m^*(\rho_1 \times \rho_2) = m^*(\rho_1) \times m^*(\rho_2).$$

Also, we recall (see [**Ze**])

$$m^*(\langle[a,b]^{(\chi)}\rangle) = \sum_{k=a-1}^{b} \langle[a,k]^{(\chi)}\rangle \otimes \langle[k+1,b]^{(\chi)}\rangle.$$

Combining this with the expression for π given by (4-12), we compute $m^*(\pi)$ as follows:

$$\sum \langle[-x+\alpha, k_1]^{(\chi)}\rangle \times \langle[-x-\alpha, k_2]^{(\chi)}\rangle \times \langle[-x+\alpha-k, k_3]^{(\chi)}\rangle \times \langle[-x-\alpha+1, k_4]^{(\chi)}\rangle \otimes$$
$$\langle[k_1+1\ x+\alpha]^{(\chi)}\rangle \times \langle[k_2+1,\ x-\alpha]^{(\chi)}\rangle \times \langle[k_3+1, x+\alpha-1]^{(\chi)}\rangle \times \langle[k_4+1, x-\alpha+k]^{(\chi)}\rangle$$

where the summation runs over

$$\begin{cases} -x+\alpha-1 \leq k_1 \leq x+\alpha \\ -x-\alpha-1 \leq k_2 \leq x-\alpha \\ -x+\alpha-k-1 \leq k_3 \leq x+\alpha-1 \\ -x-\alpha \leq k_4 \leq x-\alpha+k. \end{cases}$$

Now, we determine the multiplicity of τ in that expression. First, we find all possible terms where it occurs. Applying (4-10), we see that $k_1 = x+\alpha$ and $k_3 = -x+\alpha-k-1$. The expression for τ shows that $k_3 \geq -x-\alpha$. There are two cases. First, if $k_3 = -x-\alpha$, then the expression for τ shows that $k_4 = x-\alpha$. The term contains τ with multiplicity one since it is the tensor product of two induced representations where τ is the unique unramified irreducible subquotient. If $k_3 > -x-\alpha$, then $k_4 = -x-\alpha$. Hence, the expression for τ shows $k_3 = x-\alpha$. The resulting term is τ itself. □

Now, we turn our attention to the topological structure of $\mathrm{Irr}^{u,unr}(\mathrm{GL}(n,F))$. The topology of the unitary dual of $\mathrm{GL}(n,F)$ is described in [**T5**]. Here we recall a simple description in the unramified case which follows directly from the general and simple Theorem 3-5.

Let X be a subset of the unramified unitary dual of $\mathrm{GL}(n,F)$. We describe its closure $Cl(X)$. We consider all sequences in X of the form:

$$\pi^{(k)} \simeq \phi_1^{(k)} \times \cdots \times \phi_a^{(k)} \times (\nu^{\alpha_1^{(k)}}\psi_1^{(k)} \times \nu^{-\alpha_1^{(k)}}\psi_1^{(k)}) \times \cdots \times (\nu^{\alpha_b^{(k)}}\psi_b^{(k)} \times \nu^{-\alpha_b^{(k)}}\psi_b^{(k)})$$

where $\phi_i^{(k)}$ (resp., $\psi_j^{(k)}$) is a convergent sequence (in the obvious natural topology) of unramified unitary characters of a fixed general linear group, converging to some ϕ_i (resp., ψ_j), and $0 < \alpha_j^{(k)} < 1/2$ converges to $0 \leq \alpha_j \leq 1/2$ (the possibility $a=0$ or $b=0$ is not excluded). Let

$$\pi \simeq \phi_1 \times \cdots \times \phi_a \times (\nu^{\alpha_1}\psi_1 \times \nu^{-\alpha_1}\psi_1) \times \cdots \times (\nu^{\alpha_b}\psi_b \times \nu^{-\alpha_b}\psi_b). \tag{4-14}$$

This representation might be reducible, but its unique irreducible unramified subquotient $\pi^{\#}$ is unitarizable. Then $Cl(X)$ is exactly the set all possible such $\pi^{\#}$. The representation $\pi^{\#}$ can be described in the form given by Theorem 4-1 (i) as follows. If $\alpha_j = 1/2$, for some j in (4-14), we write $\psi_j = \chi_j 1_{\mathrm{GL}(h_j,F)}$, where χ_j is an unramified unitary character of $F^\times$, and in (4-14) change $\nu^{\alpha_j}\psi_j \times \nu^{-\alpha_j}\psi_j = \nu^{1/2}\chi_j 1_{\mathrm{GL}(h_j,F)} \times \nu^{-1/2}\chi_j 1_{\mathrm{GL}(h_j,F)}$ to $\chi_j 1_{\mathrm{GL}(h_j+1,F)} \times \chi_j 1_{\mathrm{GL}(h_j-1,F)}$.

5. The Unramified Unitary Dual $\mathrm{Irr}^{u,unr}(S)$

In this section we state the result on the classification of the unitary unramified dual $\mathrm{Irr}^{u,unr}(S)$. We begin by recalling some results of [**M4**].

DEFINITION 5-1. *Let* $\mathbf{sgn}_u$ *be the unique unramified character of order two of* $F^\times$. *Let* $\mathbf{1}_{F^\times}$ *be the trivial character of* $F^\times$.

We remark that $\mathbf{sgn}_u(\varpi) = -1$.

The following definition is crucial for us:

DEFINITION 5-2. *Let $\chi \in \{\mathbf{1}_{F^\times}, \mathbf{sgn}_u\}$. Then we define α_χ as follows:*

if $S_n = \mathrm{O}(2n, F)$ $(n \geq 0)$, then $\alpha_\chi = 0$

if $S_n = \mathrm{SO}(2n+1, F)$ $(n \geq 0)$, then $\alpha_\chi = \dfrac{1}{2}$

if $S_n = \mathrm{Sp}(2n, F)$ $(n \geq 0)$, then $\alpha_{\mathbf{sgn}_u} = 0$ and $\alpha_{\mathbf{1}_{F^\times}} = 1$.

Next, we recall the following well–known result that explains Definition 5-2:

REMARK 5-3. *For an unramified unitary character χ of $F^\times$ and $s \in \mathbb{R}$, we have the following:*

(i) *$\nu^s\chi \rtimes \mathbf{1}$ (a representation of S_1; see Section 1 for the notation), and $\nu^{-s}\chi^{-1} \rtimes \mathbf{1}$ have the same composition series (and therefore $\nu^s\chi \rtimes \mathbf{1}$ reduces if and only if $\nu^{-s}\chi^{-1} \rtimes \mathbf{1}$ reduces).*

(ii) *$\nu^s\chi \rtimes \mathbf{1}$ is irreducible if $\chi \notin \{\mathbf{1}_{F^\times}, \mathbf{sgn}_u\}$.*

(iii) *Suppose $\chi \in \{\mathbf{1}_{F^\times}, \mathbf{sgn}_u\}$. Then $\nu^s\chi \rtimes \mathbf{1}$ reduces if and only if $s = \pm\alpha_\chi$.*

A pair (m, χ), where $m \in \mathbb{Z}_{>0}$ and χ is an unramified unitary character of $F^\times$ is called a Jordan block.

DEFINITION 5-4. *Let $n > 0$. We write $\mathrm{Jord}_{sn}(n)$ for the collection of all sets* Jord *of Jordan blocks such that the following holds:*

$$\chi \in \{\mathbf{1}_{F^\times}, \mathbf{sgn}_u\} \text{ and } m - (2\alpha_\chi + 1) \in 2\mathbb{Z} \text{ for all } (m, \chi) \in \mathrm{Jord}$$

$$\sum_{(m,\chi)\in \mathrm{Jord}} m = \begin{cases} 2n & \text{if } S_n = \mathrm{SO}(2n+1, F) \text{ or } S_n = \mathrm{O}(2n, F); \\ 2n+1 & \text{if } S_n = \mathrm{Sp}(2n, F), \end{cases}$$

and, additionally, if $\alpha_\chi = 0$, then $\mathrm{card}\,\{k; (k, \chi) \in \mathrm{Jord}\} \in 2\mathbb{Z}$.

REMARK 5-5. *Let $(m, \chi) \in \mathrm{Jord} \in \mathrm{Jord}_{sn}(n)$ be a Jordan block. Then m is even if we are dealing with odd-orthogonal groups, and odd otherwise (i.e., if we are dealing with even-orthogonal or symplectic groups).*

Let $\mathrm{Jord} \in \mathrm{Jord}_{sn}(n)$. Then, for $\chi \in \{\mathbf{1}_{F^\times}, \mathbf{sgn}_u\}$, we let

$$\mathrm{Jord}_\chi = \{k; (k, \chi) \in \mathrm{Jord}\}.$$

We let

$$\mathrm{Jord}'_\chi = \begin{cases} \mathrm{Jord}(\chi); \ \mathrm{card}(\mathrm{Jord}_\chi) \text{ is even;} \\ \mathrm{Jord}(\chi) \cup \{-2\alpha_\chi + 1\}; \ \mathrm{card}(\mathrm{Jord}_\chi) \text{ is odd.} \end{cases}$$

We write Jord'_χ according to the character χ (the case $l_{\mathbf{1}_{F^\times}} = 0$ or $l_{\mathbf{sgn}_u} = 0$ is not excluded):

$$\begin{cases} \chi = \mathbf{1}_{F^\times}: & a_1 < a_2 < \cdots < a_{2l_{\mathbf{1}_{F^\times}}} \\ \chi = \mathbf{sgn}_u: & b_1 < b_2 < \cdots < b_{2l_{\mathbf{sgn}_u}} \end{cases} \tag{5-6}$$

(here $a_i, b_j \in 1 + 2\mathbb{Z}_{\geq 0}$ if $S_n = \mathrm{Sp}(2n, F)$ or $S_n = \mathrm{O}(2n, F)$, and $a_i, b_j \in 2\mathbb{Z}_{>0}$ if $S_n = \mathrm{SO}(2n+1, F)$).

Next, we associate to $\mathrm{Jord} \in \mathrm{Jord}_{sn}(n)$, the unramified representation $\sigma(\mathrm{Jord})$ of S_n defined as the unique irreducible unramified subquotient of the induced representation

(5-7)

$$\left(\times_{i=1}^{l_{\mathbf{1}_{F^\times}}} \langle [-\frac{a_{2i}-1}{2}, \frac{a_{2i-1}-1}{2}]^{(\mathbf{1}_{F^\times})}\rangle\right) \times \left(\times_{j=1}^{l_{\mathbf{sgn}_u}} \langle [-\frac{b_{2j}-1}{2}, \frac{b_{2j-1}-1}{2}]^{(\mathbf{sgn}_u)}\rangle\right) \rtimes \mathbf{1}.$$

In fact, $\sigma(\text{Jord})$ is a subrepresentation of the induced representation in (5-7).

We have the following result (see [**M4**]):

THEOREM 5-8. *Let $n \in \mathbb{Z}_{>0}$. The map* $\text{Jord} \mapsto \sigma(\text{Jord})$ *defines a one-to-one correspondence between the set* $\text{Jord}_{sn}(n)$ *and the set of all strongly negative unramified representations of S_n. (An unramified representation is strongly negative if its Aubert dual is in the discrete series.)*

The inverse mapping to $\text{Jord} \mapsto \sigma(\text{Jord})$ is denoted by $\sigma \mapsto \text{Jord}(\sigma)$.

For technical reasons, we consider the trivial representation of the trivial group S_0 to be strongly negative. We associate the set of Jordan blocks (depending on the tower S_n $(n \geq 0)$) as follows

$$\text{Jord}(\mathbf{1}) = \begin{cases} \{(1, \mathbf{1}_{F^\times})\}; & \text{if } S_n = \text{Sp}(2n, F) \ (n \geq 0), \\ \emptyset; & \text{otherwise.} \end{cases} \tag{5-9}$$

If we let $\text{Jord}_{sn}(0) = \{\text{Jord}(\mathbf{1})\}$ and $\mathbf{1} = \sigma(\text{Jord})$, $\text{Jord} \in \text{Jord}_{sn}(0)$, then Theorem 5-8 holds for $n = 0$. (We remark that Definition 5-4 and (5-7) hold for $\text{Jord} \in \text{Jord}_{sn}(0)$.)

An unramified representation is negative if its Aubert dual is tempered. Negative representations are classified in terms of strongly negative ones as follows:

THEOREM 5-10. *Let $\sigma_{neg} \in \text{Irr}^{unr}(S)$ be a negative representation. Then there exists a sequence of pairs $(l_1, \chi_1), \ldots, (l_k, \chi_k)$ ($l_i \in \mathbb{Z}_{\geq 1}$, χ_i an unramified unitary character of $F^\times$), unique up to a permutation and taking inverses of characters, and unique strongly negative representation σ_{sn} such that*

$$\sigma_{neg} \hookrightarrow \quad \langle[-\frac{l_1-1}{2}, \frac{l_1-1}{2}]^{(\chi_1)}\rangle \times \cdots \times \langle[-\frac{l_k-1}{2}, \frac{l_k-1}{2}]^{(\chi_k)}\rangle \rtimes \sigma_{sn}.$$

Conversely, for a sequence of the pairs $(l_1, \chi_1), \ldots, (l_k, \chi_k)$ ($l_i \in \mathbb{Z}_{>0}$, χ_i is an unramified unitary character of $F^\times$) and a strongly negative representation σ_{sn}, the unique irreducible unramified subquotient of

$$\langle[-\frac{l_1-1}{2}, \frac{l_1-1}{2}]^{(\chi_1)}\rangle \times \cdots \times \langle[-\frac{l_k-1}{2}, \frac{l_k-1}{2}]^{(\chi_k)}\rangle \rtimes \sigma_{sn}$$

is negative and it is a subrepresentation.

For the irreducible negative unramified representation $\sigma_{neg} \in \text{Irr}^{unr}(S)$ given by the above Theorem 5-10, one defines $\text{Jord}(\sigma_{neg})$ to be the multiset

$$\text{Jord}(\sigma_{sn}) + \sum_{i=1}^{k} \{(l_i, \chi_i), (l_i, \chi_i^{-1})\}$$

(multisets are sets where multiplicities are allowed). For a unitary unramified character χ of $F^\times$, we let $\text{Jord}(\sigma_{neg})_\chi$ be the multiset consisting of all l (counted with multiplicity) such that $(l, \chi) \in \text{Jord}(\sigma_{neg})$.

Now, we turn our attention to $\text{Irr}^{u,unr}(S)$. First, we have the following particular case of ([**M6**]):

THEOREM 5-11. *Let $\sigma \in \text{Irr}^{unr}(S)$ be a negative representation. Then σ is unitarizable.*

In order to describe the whole $\mathrm{Irr}^{u,unr}(S)$ we need to introduce more notation. We write $\mathcal{M}^{unr}(S)$ for the set of pairs $(\mathbf{e}, \sigma_{neg})$, where $\mathbf{e}$ is a (perhaps empty) multiset consisting of a finite number of triples (l, χ, α) where $l \in \mathbb{Z}_{>0}$, χ is an unramified unitary character of $F^{\times}$, and $\alpha \in \mathbb{R}_{>0}$. For $l \in \mathbb{Z}_{>0}$ and an unramified unitary character χ of $F^{\times}$, we let $\mathbf{e}(l, \chi)$ to be the submultiset of $\mathbf{e}$ consisting of all positive real numbers α (counted with multiplicity) such that $(l, \chi, \alpha) \in \mathbf{e}$. We have the following:

$$\mathbf{e} = \sum_{(l,\chi)} \sum_{\alpha \in \mathbf{e}(l,\chi)} \{(l, \chi, \alpha)\}.$$

We define the map $n : \mathcal{M}^{unr}(S) \to \mathbb{Z}$ as follows:

$$n(\mathbf{e}, \sigma_{neg}) = \sum_{(l,\chi)} l \cdot \mathrm{card}\ \mathbf{e}(l, \chi) + n_{neg}$$

where n_{neg} is defined by $\sigma_{neg} \in \mathrm{Irr}\ S_{n_{neg}}$.

We attach $\sigma \in \mathrm{Irr}^{unr}(S)$ to $(\mathbf{e}, \sigma_{neg})$ in a canonical way. By definition, σ is the unique irreducible unramified subquotient of the following induced representation:

$$\left(\times_{(l,\chi,\alpha) \in \mathbf{e}} \ \langle [-\frac{l-1}{2}, \frac{l-1}{2}]^{(\nu^{\alpha}\chi)} \rangle \right) \rtimes \sigma_{neg}. \tag{5-12}$$

It is a representation of $S_{n(\mathbf{e}, \sigma_{neg})}$.

We remark that the definition of σ does not depend on the choice of ordering of characters in (5-12). Next, the results of [**M4**] (see Lemma 6-2 in Section 6) imply that the constructed map $\mathcal{M}^{unr}(S) \to \mathrm{Irr}^{unr}(S)$ is surjective but not injective.

In order to obtain unitary representations, we impose further conditions on $\mathbf{e}$ in the following definition:

DEFINITION 5-13. *Let $\mathcal{M}^{u,unr}(S)$ be the subset of $\mathcal{M}^{unr}(S)$ consisting of the pairs $(\mathbf{e}, \sigma_{neg})$ satisfying the following conditions:*

(1) *If $\chi \notin \{\mathbf{1}_{F^{\times}}, \mathbf{sgn}_u\}$, then $\mathbf{e}(l, \chi) = \mathbf{e}(l, \chi^{-1})$ and $0 < \alpha < \frac{1}{2}$ for all $\alpha \in \mathbf{e}(l, \chi)$.*
(2) *If $\chi \in \{\mathbf{1}_{F^{\times}}, \mathbf{sgn}_u\}$ and $l - (2\alpha_{\chi} + 1) \notin 2\mathbb{Z}$, then $0 < \alpha < \frac{1}{2}$ for all $\alpha \in \mathbf{e}(l, \chi)$.*
(3) *If $\chi \in \{\mathbf{1}_{F^{\times}}, \mathbf{sgn}_u\}$ and $l - (2\alpha_{\chi} + 1) \in 2\mathbb{Z}$, then $0 < \alpha < 1$ for all $\alpha \in \mathbf{e}(l, \chi)$. Moreover, if we write the exponents that belong to $\mathbf{e}(l, \chi)$ as follows:*

$$0 < \alpha_1 \leq \cdots \leq \alpha_u \leq \frac{1}{2} < \beta_1 \leq \cdots \leq \beta_v < 1.$$

(We allow $u = 0$ or $v = 0$.) Then we also require the following:
(a) *If $(l, \chi) \notin \mathrm{Jord}(\sigma_{neg})$, then $u + v$ is even.*
(b) *If $u > 1$, then $\alpha_{u-1} \neq \frac{1}{2}$.*
(c) *If $v \geq 2$, then $\beta_1 < \cdots < \beta_v$.*
(d) *$\alpha_i \notin \{1 - \beta_1, \ldots, 1 - \beta_v\}$ for all i.*
(e) *If $v \geq 1$, then the number of indices i such that $\alpha_i \in]1 - \beta_1, \frac{1}{2}]$ is even.*
(f) *If $v \geq 2$, then the number of indices i such that $\alpha_i \in]1 - \beta_{j+1}, 1 - \beta_j[$ is odd.*

We advise the reader to construct some pairs $(\mathbf{e}, \sigma_{neg}) \in \mathcal{M}^{u,unr}(S)$. This should be done in the following way. One first chooses an arbitrary σ_{neg}. Then one adds multisets $\mathbf{e}(l, \chi)$ to $\mathbf{e}$ following (1)–(3) and (a)–(g), in that order.

The following theorem gives an explicit classification (with explicit parameters) of unramified unitary duals of classical groups S_n, i.e., of $\mathrm{Irr}^{u,unr}(S)$. The proof of the classification theorem is in Sections 7, 8, and 9. At no point in the proof does the explicit internal structure of representations play a role. This is the reason that this can be considered as an external approach to the unramified unitary duals (of classical groups), along the lines of such approaches in [**T3**], [**T2**], [**LMT**], etc.

THEOREM 5-14. *Let $n \in \mathbb{Z}_{\geq 0}$. We write $\mathcal{M}^{u,unr}(S_n)$ for the set of all $(\mathbf{e}, \sigma_{neg}) \in \mathcal{M}^{u,unr}(S)$ such that $n(\mathbf{e}, \sigma_{neg}) = n$. Then, for $(\mathbf{e}, \sigma_{neg}) \in \mathcal{M}^{u,unr}(S_n)$, the induced representation (5-12) is an irreducible unramified representation of S_n. Moreover, the map $(\mathbf{e}, \sigma_{neg}) \longmapsto \times_{(l,\chi,\alpha)\in \mathbf{e}} \langle [-\frac{l-1}{2}, \frac{l-1}{2}]^{(\nu^\alpha \chi)} \rangle \rtimes \sigma_{neg}$ is a one–to–one correspondence between $\mathcal{M}^{u,unr}(S_n)$ and $\mathrm{Irr}^{u,unr}(S_n)$.*

This result, along with Theorem 5-11, was partly obtained by Barbasch and Moy (see [**Bb**], [**BbMo**], [**BbMo1**], [**BbMo2**]). See the introduction for more explanation.

REMARK 5-15. *We separate the conditions of Definition 5-13 into three groups:*
Irreducibility conditions: (b), (d) in (3).
Hermiticity condition: (1).
Unitarizability conditions: (2) and (a), (c), (e), (f) in (3), and also the condition $0 < \alpha < 1$ in (3).

6. Some Technical Results

In this section we recall some results from [**M4**] and prove some results about reducibility and subquotients of certain induced representations needed in the proof of Theorem 5-14. The reader should skip this section at the first reading. We begin with the following lemma:

LEMMA 6-1. *Let $(\mathbf{e}, \sigma_{neg}) \in \mathcal{M}^{unr}(S)$. Then the induced representation (5-12) is reducible if and only if one of the following holds:*

(1) *there exist $(l, \chi, \alpha), (l', \chi', \alpha') \in \mathbf{e}$ such that the segments $[-\frac{l-1}{2}, \frac{l-1}{2}]^{(\nu^\alpha \chi)}$ and $[-\frac{l'-1}{2}, \frac{l'-1}{2}]^{(\nu^{\alpha'} \chi')}$ are linked*
(2) *there exist $(l, \chi, \alpha), (l', \chi', \alpha') \in \mathbf{e}$ such that the segments $[-\frac{l-1}{2}, \frac{l-1}{2}]^{(\nu^\alpha \chi)}$ and $[-\frac{l'-1}{2}, \frac{l'-1}{2}]^{(\nu^{-\alpha'} \chi')}$ are linked*
(3) *there exist $(l, \chi, \alpha) \in \mathbf{e}$ such that $\langle [-\frac{l-1}{2}, \frac{l-1}{2}]^{(\nu^\alpha \chi)} \rangle \rtimes \sigma_{neg}$ reduces*

Further, let $(l, \chi, \alpha) \in \mathbf{e}$ and consider the following statements:

(4) *there exists $(l', \chi') \in \mathrm{Jord}(\sigma_{neg})$ such that the segments $\langle [-\frac{l-1}{2}, \frac{l-1}{2}]^{(\nu^\alpha \chi)} \rangle$ and $\langle [-\frac{l'-1}{2}, \frac{l'-1}{2}]^{(\chi')} \rangle$ are linked*
(5) *$\chi \in \{\mathbf{1}_{F^\times}, \mathbf{sgn}_u\}$ and $l - (2|\alpha - \alpha_\chi| + 1) \in 2\mathbb{Z}_{\geq 0}$*

Then we have the following:

- *If $\chi = \mathbf{1}_{F^\times}$, $\alpha_\chi = 1$, $\mathrm{card}\ \mathrm{Jord}(\sigma_{neg})_{\mathbf{1}_{F^\times}}$ is odd, and $-\frac{l-1}{2} + \alpha = 1$, then the induced representation in (3) is reducible if and only if (4) holds.*

- *Otherwise, the induced representation in (3) is reducible if and only if (4) or (5) holds.*

PROOF. First, ([**M4**], Lemma 4.8) implies that the induced representation (5-12) is reducible if and only if (1) or (2) holds or $\langle[-\frac{l-1}{2}, \frac{l-1}{2}]^{(\nu^\alpha\chi)}\rangle \rtimes \sigma_{neg}$ is reducible. We will describe the reducibility of $\langle[-\frac{l-1}{2}, \frac{l-1}{2}]^{(\nu^\alpha\chi)}\rangle \rtimes \sigma_{neg}$, and this will conclude the proof. We write σ_{neg} as in Theorem 5-10. Then ([**M4**], Corollary 4.2) implies that $\langle[-\frac{l-1}{2}, \frac{l-1}{2}]^{(\nu^\alpha\chi)}\rangle \rtimes \sigma_{neg}$ reduces if and only if one of the following holds:

(a) $[-\frac{l-1}{2}, \frac{l-1}{2}]^{(\nu^\alpha\chi)}$ is linked with $[-\frac{l_i-1}{2}, \frac{l_i-1}{2}]^{(\chi_i)}$ for some i
(b) $[-\frac{l-1}{2}, \frac{l-1}{2}]^{(\nu^\alpha\chi)}$ is linked with $[-\frac{l_i-1}{2}, \frac{l_i-1}{2}]^{(\chi_i^{-1})}$ for some i
(c) $\langle[-\frac{l-1}{2}, \frac{l-1}{2}]^{(\nu^\alpha\chi)}\rangle \rtimes \sigma_{sn}$ reduces.

Using ([**M4**], Lemma 5.6), we see that (c) holds if and only if $\chi \in \{\mathbf{1}_{F^\times}, \mathbf{sgn}_u\}$, $l + 2\alpha - (2\alpha_\chi + 1) \in 2\mathbb{Z}$ and one of the following holds:

(d) card $\mathrm{Jord}(\sigma_{sn})_\chi$ is even.
 (d-1) there exists $(l', \chi') \in \mathrm{Jord}(\sigma_{sn})$ such that $\langle[-\frac{l-1}{2}, \frac{l-1}{2}]^{(\nu^\alpha\chi)}\rangle$ and $\langle[-\frac{l'-1}{2}, \frac{l'-1}{2}]^{(\chi')}\rangle$ are linked
 (d-2) $\langle[-\frac{l-1}{2}, \frac{l-1}{2}]^{(\nu^\alpha\chi)}\rangle \rtimes \mathbf{1}$ reduces (which is equivalent to $\chi \in \{\mathbf{1}_{F^\times}, \mathbf{sgn}_u\}$ and $l - (2|\alpha - \alpha_\chi| + 1) \in 2\mathbb{Z}_{\geq 0}$ by ([**M4**], Lemma 5.6 (i)).
(e) card $\mathrm{Jord}(\sigma_{sn})_\chi$ is odd; $\alpha_\chi = 1/2$, or $\alpha_\chi = 1$ and $-\frac{l-1}{2} + \alpha \neq 1$. Let $l_{min} = \min \mathrm{Jord}(\sigma_{sn})_\chi$.
 (e-1) there exists $l' \in \mathrm{Jord}(\sigma_{sn}) - \{l_{min}\}$ such that $\langle[-\frac{l-1}{2}, \frac{e-1}{2}]^{(\nu^\alpha\chi)}\rangle$ and $\langle[-\frac{l'-1}{2}, \frac{l'-1}{2}]^{(\chi)}\rangle$ are linked
 (e-2) $\langle[-\frac{l-1}{2}, \frac{l-1}{2}]^{(\nu^\alpha\chi)}\rangle$ and $\langle[\alpha_\chi, \frac{l_{min}-1}{2}]^{(\chi)}\rangle$ are linked
 (e-3) $\langle[-\frac{l-1}{2}, \frac{l-1}{2}]^{(\nu^\alpha\chi)}\rangle$ and $\langle[-\frac{l_{min}-1}{2}, -\alpha_\chi]^{(\chi)}\rangle$ are linked
 (e-4) $\langle[-\frac{l-1}{2}, \frac{l-1}{2}]^{(\nu^\alpha\chi)}\rangle \rtimes \mathbf{1}$ reduces (which is equivalent to $\chi \in \{\mathbf{1}_{F^\times}, \mathbf{sgn}_u\}$ and $l - (2|\alpha - \alpha_\chi| + 1) \in 2\mathbb{Z}_{\geq 0}$ by ([**M4**], Lemma 5.6 (i))
(f) card $\mathrm{Jord}(\sigma_{sn})_\chi$ is odd; $\alpha_\chi = 1$ and $-\frac{l-1}{2} + \alpha = 1$. Then $\chi = \mathbf{1}_{F^\times}$. Let $l_{min} = \min \mathrm{Jord}(\sigma_{sn})_{\mathbf{1}_{F^\times}}$.
 (f-1) there exists $l' \in \mathrm{Jord}(\sigma_{sn})_{\mathbf{1}_{F^\times}} - \{l_{min}\}$ such that $\langle[-\frac{l-1}{2}, \frac{l-1}{2}]^{(\nu^\alpha\mathbf{1}_{F^\times})}\rangle$ and $\langle[-\frac{l'-1}{2}, \frac{l'-1}{2}]^{(\mathbf{1}_{F^\times})}\rangle$ are linked
 (f-2) $\langle[-\frac{l-1}{2}, \frac{l-1}{2}]^{(\nu^\alpha\mathbf{1}_{F^\times})}\rangle$ and $\langle[1, \frac{l_{min}-1}{2}]^{(\mathbf{1}_{F^\times})}\rangle$ are linked
 (f-3) $\langle[-\frac{l-1}{2}, \frac{l-1}{2}]^{(\nu^\alpha\mathbf{1}_{F^\times})}\rangle$ and $\langle[-\frac{l_{min}-1}{2}, -1]^{(\mathbf{1}_{F^\times})}\rangle$ are linked
 (f-4) $l > l_{min}$.

It is easy to check that (e-2), (e-3) or (e-4) holds if and only if one of the following holds:

(e'-2) $\langle[-\frac{l-1}{2}, \frac{l-1}{2}]^{(\nu^\alpha\chi)}\rangle$ and $\langle[-\frac{l_{min}-1}{2}, \frac{l_{min}-1}{2}]^{(\chi)}\rangle$ are linked
(e'-4) $\langle[-\frac{l-1}{2}, \frac{l-1}{2}]^{(\nu^\alpha\chi)}\rangle \rtimes \mathbf{1}$ reduces.

It is easy to check that (f-2), (f-3) or (f-4) holds if and only if the following holds:

(f'-2) $\langle[-\frac{l-1}{2}, \frac{l-1}{2}]^{(\nu^\alpha\mathbf{1}_{F^\times})}\rangle$ and $\langle[-\frac{l_{min}-1}{2}, \frac{l_{min}-1}{2}]^{(\mathbf{1}_{F^\times})}\rangle$ are linked.

Clearly, this analysis completes the proof of the lemma. □

The next lemma will play a crucial role in determining surjectivity of the map in Theorem 5-14 (see [**M4**], Theorem 4.3):

LEMMA 6-2. *Let $\sigma \in \mathrm{Irr}^{unr}(S)$. Then there exists a unique $(\mathbf{e}, \sigma_{neg}) \in \mathcal{M}^{unr}(S)$ such that σ is isomorphic to the induced representation given by (5-12).*

Next, we determine when the representation $\sigma \in \mathrm{Irr}^{unr}(S)$, given by Lemma 6-2, is Hermitian. We compute using (H-IC) and Theorem 5-11:

$$\begin{aligned}
\sigma^+ &\simeq \left(\times_{(l,\chi,\alpha)\in\mathbf{e}} \langle[-\frac{l-1}{2}, \frac{l-1}{2}]^{(\nu^\alpha\chi)}\rangle \rtimes \sigma_{neg}\right)^+ \\
&\simeq \times_{(l,\chi,\alpha)\in\mathbf{e}} \left(\langle[-\frac{l-1}{2}, \frac{l-1}{2}]^{(\nu^\alpha\chi)}\rangle\right)^+ \rtimes (\sigma_{neg})^+ \\
&\simeq \times_{(l,\chi)} \times_{\alpha\in\mathbf{e}(l,\chi)} \langle[-\frac{l-1}{2}, \frac{l-1}{2}]^{(\nu^{-\alpha}\chi)}\rangle \rtimes \sigma_{neg} \\
&\simeq \times_{(l,\chi)} \times_{\alpha\in\mathbf{e}(l,\chi)} \langle[-\frac{l-1}{2}, \frac{l-1}{2}]^{(\nu^{\alpha}\chi^{-1})}\rangle \rtimes \sigma_{neg}.
\end{aligned} \tag{6-3}$$

(The last isomorphism follows from the fact that every representation in $\mathrm{Irr}^{unr}(S)$ is self-dual. See Remark 1-8.) Therefore, (H-Irr), (6-3), and

$$\sigma \simeq \times_{(l,\chi)} \times_{\alpha\in\mathbf{e}(\chi,l)} \langle[-\frac{l-1}{2}, \frac{l-1}{2}]^{(\nu^\alpha\chi)}\rangle \rtimes \sigma_{neg}$$

imply the following result:

LEMMA 6-4. *Let $\sigma \in \mathrm{Irr}^{unr}(S)$ be given by Lemma 6-2. Then $\sigma \in \mathrm{Irr}^{+,unr}(S)$ if and only if $\mathbf{e}(l,\chi) = \mathbf{e}(l,\chi^{-1})$ for all $l \in \mathbb{Z}_{\geq 1}$ and all unitary unramified characters χ of $F^\times$.*

LEMMA 6-5. *Let $\sigma_{neg} \in \mathrm{Irr}^{unr}(S)$ be a negative representation. Let χ be a unitary unramified character of $F^\times$ and $l \in \mathbb{Z}_{\geq 1}$. Then $\langle[-\frac{l-1}{2}, \frac{l-1}{2}]^{(\chi)}\rangle \rtimes \sigma_{neg}$ is reducible if and only if $\chi \in \{\mathbf{1}_{F^\times}, \mathbf{sgn}_u\}$, $l-(2\alpha_\chi+1) \in 2\mathbb{Z}$ and $(l,\chi) \notin \mathrm{Jord}(\sigma_{neg})$. If $\langle[-\frac{l-1}{2}, \frac{l-1}{2}]^{(\chi)}\rangle \rtimes \sigma_{neg}$ is reducible, then it is the direct sum of two non–equivalent representations (one of them is negative).*

PROOF. The proof of this result is standard and in the dual picture well–known (see [**MœT**]). We indicate the steps to explain why the result holds for local fields F of all characteristics. First, we apply Aubert's involution, extended to orthogonal groups by C. Jantzen [**Jn**], to reduce to the tempered case. Then we use the results of Goldberg [**G**] [1], extended to orthogonal groups using simple Mackey machinery (see [**LMT**], Section 2), and some general algebraic considerations based on them (see [**LMT**], Lemma 2.2 and Corollary 2.3), to reduce the claim to the case when the image $\hat{\sigma}_{neg}$ of σ_{neg} under Aubert's involution is in the discrete series. As σ_{neg} and $\hat{\sigma}_{neg}$ have the same supercuspidal support which is explicitly known by Theorem 5-8, we can easily compute the Plancherel measure attached to the induced representation $\chi\mathrm{Steinberg}_{\mathrm{GL}(l,F)} \rtimes \hat{\sigma}_{neg}$ (which has the same reducibility as $\langle[-\frac{l-1}{2}, \frac{l-1}{2}]^{(\chi)}\rangle \rtimes \sigma_{neg}$). The computation of the Plancherel measure is done by using the factorization (see [**W**]) and reduction to the split rank–one case. Now, we use the usual theory developed by Harish–Chandra to decompose $\chi\mathrm{Steinberg}_{\mathrm{GL}(l,F)} \rtimes \hat{\sigma}_{neg}$.

[1] Goldberg stated his results in the characteristic zero, but this assumption is not necessary. In fact, all fundamental results of Harish–Chandra used there follow from [W2] as was explained to the first named author by V. Heiermann.

Finally, assume that $\langle[-\frac{l-1}{2},\frac{l-1}{2}]^{(\chi)}\rangle\rtimes\sigma_{neg}$ is reducible. Then $\langle[-\frac{l-1}{2},\frac{l-1}{2}]^{(\chi)}\rangle\rtimes\hat{\sigma}_{neg}$ is a direct sum of two non–equivalent tempered representations (see [**LMT**], Lemma 2.2 and Corollary 2.3). Hence the composition series of $\langle[-\frac{l-1}{2},\frac{l-1}{2}]^{(\chi)}\rangle\rtimes\sigma_{neg}$ of two non–equivalent irreducible representations. Hence the last claim of the lemma follows from Theorems 5-10 and 5-11. □

Next, we prove the following lemma:

LEMMA 6-6. *Let $\chi\in\{\mathbf{1}_{F^\times},\mathbf{sgn}_u\}$ and let $l\in\mathbb{Z}_{\geq1}$ such that $l-(2\alpha_\chi+1)$ is not an even integer. Let σ_{neg} be a negative representation. Then the induced representation $\langle[-\frac{l-1}{2},\frac{l-1}{2}]^{(\nu^\alpha\chi)}\rangle\rtimes\sigma_{neg}$, where $\alpha\in\mathbb{R}_{>0}$, reduces at $\alpha=1/2$ and its unique unramified irreducible subquotient σ'_{neg} is a negative representation. Further, we have the following:*

$$\mathrm{Jord}(\sigma'_{neg})=\mathrm{Jord}(\sigma_{neg})+\{(l-1,\chi),(l+1,\chi)\}$$

(If $l=1$, then we omit $(l-1,\chi)$.)

PROOF. First, Lemma 6-1 implies that $\langle[-\frac{l-1}{2},\frac{l-1}{2}]^{(\nu^\alpha\chi)}\rangle\rtimes\sigma_{neg}$ is reducible at $\alpha=1/2$. We need to show that the unique unramified irreducible subquotient there is negative. First, applying Theorem 5-10, we find a sequence of pairs $(l_1,\chi_1),\ldots,(l_k,\chi_k)$ ($l_i\in\mathbb{Z}_{>0}$, χ_i is an unramified unitary character of $F^\times$), unique up to a permutation and taking inverses of characters, and the unique strongly negative representation σ_{sn} such that

$$\sigma_{neg}\hookrightarrow\langle[-\frac{l_1-1}{2},\frac{l_1-1}{2}]^{(\chi_1)}\rangle\times\cdots\times\langle[-\frac{l_k-1}{2},\frac{l_k-1}{2}]^{(\chi_k)}\rangle\rtimes\sigma_{sn}\tag{6-7}$$

and

$$\mathrm{Jord}(\sigma_{neg})=\mathrm{Jord}(\sigma_{sn})+\sum_{i=1}^{k}\{(l_i,\chi_i),(l_i,\chi_i^{-1})\}.\tag{6-8}$$

Now, using Theorem 5-8 and the explicit description of strongly negative representations (see (5-6) and (5-7)), it is easy to check the following:

- If $(l-1,\chi),(l+1,\chi)\not\in\mathrm{Jord}(\sigma_{sn})$, then there is a strongly negative representation σ'_{sn} such that $\mathrm{Jord}(\sigma'_{sn})=\mathrm{Jord}(\sigma_{sn})+\{(l-1,\chi),(l+1,\chi)\}$. We let $\sigma''_{neg}=\sigma'_{sn}$.
- If $(l-1,\chi)\in\mathrm{Jord}(\sigma_{sn})$, $(l+1,\chi)\not\in\mathrm{Jord}(\sigma_{sn})$, then there is a unique strongly negative representation σ'_{sn} such that
$$\mathrm{Jord}(\sigma'_{sn})=\mathrm{Jord}(\sigma_{sn})-\{(l-1,\chi)\}+\{(l+1,\chi)\}.$$
Let σ''_{neg} be the unique irreducible unramified subrepresentation of $\langle[-\frac{l-2}{2},\frac{l-2}{2}]^{(\chi)}\rangle\rtimes\sigma_{sn}$. Then
$$\mathrm{Jord}(\sigma''_{neg})=\mathrm{Jord}(\sigma'_{sn})+2\cdot\{(l-1,\chi)\}.$$
- If $(l-1,\chi)\not\in\mathrm{Jord}(\sigma_{sn})$, $(l+1,\chi)\in\mathrm{Jord}(\sigma_{sn})$, then there is a unique strongly negative representation σ'_{sn} such that
$$\mathrm{Jord}(\sigma'_{sn})=\mathrm{Jord}(\sigma_{sn})+\{(l-1,\chi)\}-\{(l+1,\chi)\}.$$
Let σ''_{neg} be the unique irreducible unramified subrepresentation of $\langle[-\frac{l}{2},\frac{l}{2}]^{(\chi)}\rangle\rtimes\sigma'_{sn}$. Then
$$\mathrm{Jord}(\sigma''_{neg})=\mathrm{Jord}(\sigma'_{sn})+2\cdot\{(l+1,\chi)\}.$$

- If $(l-1,\chi),(l+1,\chi)\in \mathrm{Jord}(\sigma_{sn})$, then there is a unique strongly negative representation σ'_{sn} such that
$$\mathrm{Jord}(\sigma'_{sn})=\mathrm{Jord}(\sigma_{sn})-\{(l-1,\chi),(l+1,\chi)\}.$$
Let σ''_{neg} be the unique irreducible unramified subrepresentation of $\langle[-\frac{l-2}{2},\frac{l-2}{2}]^{(\chi)}\rangle\times\langle[-\frac{l}{2},\frac{l}{2}]^{(\chi)}\rangle\rtimes\sigma'_{sn}$. Then
$$\mathrm{Jord}(\sigma''_{neg})=\mathrm{Jord}(\sigma'_{sn})+2\cdot\{(l-1,\chi),(l+1,\chi)\}.$$

Now, the unique irreducible unramified subquotient of $\langle[-\frac{l-1}{2},\frac{l-1}{2}]^{(\nu^{\frac{1}{2}}\chi)}\rangle\rtimes\sigma_{sn}$ is σ''_{neg}, which is described above. Combining this with (6-7) and (6-8), we find that the unique irreducible unramified subquotient σ'_{neg} of $\langle[-\frac{l-1}{2},\frac{l-1}{2}]^{(\nu^{\frac{1}{2}}\chi)}\rangle\rtimes\sigma_{neg}$ is a subrepresentation of $\langle[-\frac{l_1-1}{2},\frac{l_1-1}{2}]^{(\chi_1)}\rangle\times\cdots\times\langle[-\frac{l_k-1}{2},\frac{l_k-1}{2}]^{(\chi_k)}\rangle\rtimes\sigma''_{neg}$. Clearly, it is negative and $\mathrm{Jord}(\sigma'_{neg})=\mathrm{Jord}(\sigma_{neg})+\{(l-1,\chi),(l+1,\chi)\}$. □

We end this section by proving the following lemma:

LEMMA 6-9. *Assume that χ,χ' are unitary unramified characters of $F^\times$, $l,l'\in\mathbb{Z}_{\geq 1}$, and $\alpha,\alpha'\in\mathbb{R}_{>0}$. Then we have the following:*

(i) *If $\alpha\in]0,1[$ and $\alpha'\in]0,\frac{1}{2}]$, then the segments $\langle[-\frac{l-1}{2},\frac{l-1}{2}]^{(\nu^\alpha\chi)}\rangle$ and $\langle[-\frac{l'-1}{2},\frac{l'-1}{2}]^{(\nu^{-\alpha'}\chi')}\rangle$ are linked if and only if $\alpha+\alpha'=1$, $\chi'=\chi$, $l'=l$.*

(ii) *If $\alpha,\alpha'\in]0,1[$, then $\langle[-\frac{l-1}{2},\frac{l-1}{2}]^{(\nu^\alpha\chi)}\rangle$ and $\langle[-\frac{l'-1}{2},\frac{l'-1}{2}]^{(\nu^{\alpha'}\chi')}\rangle$ are not linked.*

(iii) *If $\alpha,\alpha'\in]\frac{1}{2},1[$, then $\langle[-\frac{l-1}{2},\frac{l-1}{2}]^{(\nu^\alpha\chi)}\rangle$ and $\langle[-\frac{l'-1}{2},\frac{l'-1}{2}]^{(\nu^{-\alpha'}\chi')}\rangle$ are linked if and only if $\alpha+\alpha'=3/2$, $\chi'=\chi$, and $l'=l\pm 1$.*

(iv) *If $\alpha\in]1,\frac{3}{2}[$ and $\alpha'\in]0,\frac{1}{2}[$, then the segments $\langle[-\frac{l-1}{2},\frac{l-1}{2}]^{(\nu^\alpha\chi)}\rangle$ and $\langle[-\frac{l'-1}{2},\frac{l'-1}{2}]^{(\nu^{-\alpha'}\chi')}\rangle$ are linked if and only if $\alpha+\alpha'=3/2$, $\chi'=\chi$, and $l'=l\pm 1$.*

(v) *If $\alpha\in]1,\frac{3}{2}[$ and $\alpha'\in]0,\frac{1}{2}[$, then the segments $\langle[-\frac{l-1}{2},\frac{l-1}{2}]^{(\nu^\alpha\chi)}\rangle$ and $\langle[-\frac{l'-1}{2},\frac{l'-1}{2}]^{(\nu^{\alpha'}\chi')}\rangle$ are linked if and only if $\alpha=1+\alpha'$, $\chi'=\chi$, and $l'=l$.*

PROOF. We prove (i). The segments are linked if and only if $\chi'=\chi$, and there exist $m,m'\in\mathbb{Z}_{\geq 1}$ such that

$$\begin{cases}\frac{l'-1}{2}-\alpha'=-m+\frac{l-1}{2}+\alpha\geq-\frac{l-1}{2}+\alpha-1\\ -\frac{l'-1}{2}-\alpha'=-m'-\frac{l-1}{2}+\alpha.\end{cases}\tag{6-10}$$

Adding the equalities, we obtain $(m+m')/2=\alpha+\alpha'$. Since $\alpha+\alpha'\in]0,\ \frac{3}{2}[$, we obtain $\alpha-\alpha'=1$ and $m=m'=1$. This proves one direction in (i). The opposite direction is obvious. We prove (ii). We may assume $\alpha'\leq\alpha$. If the segments are linked, then $\chi'=\chi$, and there exist $m,m'\in\mathbb{Z}_{\geq 1}$ such that

$$\begin{cases}\frac{l'-1}{2}+\alpha'=-m+\frac{l-1}{2}+\alpha\geq-\frac{l-1}{2}+\alpha-1\\ -\frac{l'-1}{2}+\alpha'=-m'-\frac{l-1}{2}+\alpha.\end{cases}\tag{6-11}$$

Adding the equalities, we obtain $\alpha=(m+m')/2+\alpha'\geq 1$. This is a contradiction. We prove (iii). If the segments are linked, then $\chi'=\chi$, and there exist $m,m'\in\mathbb{Z}_{\geq 1}$ such that (6-10) holds. Adding the equalities in (6-10), we obtain $(m+m')/2=$

$\alpha + \alpha'$. Since $\alpha + \alpha' \in]1,\ 2[$, we obtain $\alpha + \alpha' = \frac{3}{2}$, and $m = 1, m' = 2$ or $m = 2, m' = 1$. If $m = 1, m' = 2$, then $l' = l + 1$. Otherwise, $l' = l - 1$. The converse is obvious. The proof of (iv) is similar to that of (iii). We prove (v). Adding the equalities in (6-11), we obtain $\alpha = (m + m')/2 + \alpha'$. Since $\alpha \in]1,\ \frac{3}{2}[$, $\alpha' \in]0,\ \frac{1}{2}[$, and $(m + m')/2 \in \frac{1}{2}\mathbb{Z}$, we find $m = m' = 1$ and $\alpha = 1 + \alpha'$.

□

7. A Result on Non–Unitarity

In this section we use analytic techniques from [**M6**] to prove the non–unitarity of certain representations. The non–trivial part is an application of (RP) (see Section 2). The proof of the surjectivity of the map from Theorem 5-14 given in Section 9 depends critically on that result. We advise the reader to skip this section on the first reading.

The main result is the following theorem:

THEOREM 7-1. *Let $\chi \in \{\mathbf{1}_{F^\times}, \mathbf{sgn}_u\}$ and let $l \in \mathbb{Z}_{\geq 1}$. Let σ_{neg} be a negative representation. Then the induced representation $\langle[-\frac{l-1}{2}, \frac{l-1}{2}]^{(\nu^\alpha\chi)}\rangle \rtimes \sigma_{neg}$ is not unitarizable in the following two cases:*

- *for $\alpha \in]0,\ 1[$ if $l - (2\alpha_\chi + 1) \in 2\mathbb{Z}$ and $(l, \chi) \notin \mathrm{Jord}(\sigma_{neg})$*
- *for $\alpha \in]\frac{1}{2},\ 1[$ if $l - (2\alpha_\chi + 1) \notin 2\mathbb{Z}$*

The remainder of this section is devoted to the proof of Theorem 7-1. We freely use the notation and results of [**M6**]. We consider a continuous family of representations:

$$\sigma_s = \langle[-\frac{l-1}{2}, \frac{l-1}{2}]^{(\nu^s\chi)}\rangle \rtimes \sigma_{neg} \quad (s \in \mathbb{C})$$

of S_n. Let K be the maximal compact subgroup of S_n fixed in Section 1. Restricting to K, we may realize all representations σ_s on the same space X.

Let w_0 be the non–trivial element of the Weyl group $W(M)$, where M is the Levi subgroup of the standard maximal parabolic subgroup $P = MN$ of S_n such that $\langle[-\frac{l-1}{2}, \frac{l-1}{2}]^{(\nu^s\chi)}\rangle \otimes \sigma_{neg}$ is a representation of M. Hence $\sigma_s = \mathrm{Ind}_P^{S_n}(\langle[-\frac{l-1}{2}, \frac{l-1}{2}]^{(\nu^s\chi)}\rangle \otimes \sigma_{neg})$. We fix (and denote by the same letter) the representative of w_0 as explained in ([**M6**], Section 2). Next, let $N(s, w_0)$ be the standard normalized intertwining operator

$$N(s, w_0) : \sigma_s \to \sigma_{-s}$$

as explained in ([**M6**], Section 2) (see also [**Sh2**]). The geometric construction is given in [**M7**]. We consider it realized in the compact picture. We list its basic properties.

(norm-1) $N(s, w_0) \neq 0$ since it takes a suitable normalized K–invariant vector $0 \neq f_0 \in X$ onto itself.
(norm-2) $N(s, w_0)N(-s, w_0) = N(-s, w_0)N(s, w_0) = id_X$
(norm-3) $N(s, w_0)$ is Hermitian for $s \in \sqrt{-1}\mathbb{R}$, and therefore holomorphic there.

Now, we begin the proof of Theorem 7-1. We consider the family of Hermitian forms introduced in (2-8). We remark that σ_0 reduces if and only if $l-(2\alpha_\chi+1) \in 2\mathbb{Z}$ and $(l, \chi) \notin \mathrm{Jord}(\sigma_{neg})$ (see Lemma 6-5) while σ_s is irreducible for $s \in]0,\ 1[-\{\frac{1}{2}\}$ by Lemma 6-1. Next, $\sigma_{1/2}$ is reducible if and only if $l - (2\alpha_\chi + 1) \notin 2\mathbb{Z}$.

Assume that $l-(2\alpha_\chi+1)\in 2\mathbb{Z}$ and $(l,\chi)\notin \mathrm{Jord}(\sigma_{neg})$. Then, by Lemma 6-5, σ_0 is a direct sum of two irreducible representations. Then, using standard properties of normalized intertwining operators, we reduce the proof of non-triviality of $N(0,w_0)$ to the case when σ_{neg} is strongly negative. Then we may apply [**Bn**]. Therefore, $N(0,w_0)$ acts on one of the representations as $+\mathrm{id}$ while on the other it acts as $-\mathrm{id}$. Therefore the Hermitian form defined by (2-8) is not definite for $s\in]0,\ 1[$, proving the first claim.

Assume that $l-(2\alpha_\chi+1)\notin 2\mathbb{Z}$. We apply our general principle (RP) (see Section 2) to prove the second claim. We must check its assumption. We prove the following result which completes the proof of Theorem 7-1:

LEMMA 7-2. *Maintaining the above assumptions, $N(s,w_0)$ is holomorphic and $N(-s,w_0)$ has a simple pole at $s=1/2$.*

First, let π be the unique irreducible subrepresentation of $\sigma_{1/2}$ (which is equivalent to the unique irreducible quotient of $\sigma_{-1/2}$) (see [**M6**]). By the classification of irreducible unramified representations [**M6**] and Lemma 6-6, π is not unramified. Therefore, we have the following:

$$N(-s,w_0^{-1}) \text{ has a pole at } s=1/2. \tag{7-3}$$

From this point, the argument is standard and it follows the lines of the proof of ([**M6**], Lemma 3.5). First, we reduce to the case where σ_{neg} is strongly negative. Applying Theorem 5-10, we can find $l'\in\mathbb{Z}_{\geq 1}$, a unitary unramified character χ', and a negative representation σ'_{neg} such that

$$\sigma_{neg}\hookrightarrow \langle[-\frac{l'-1}{2},\frac{l'-1}{2}]^{(\chi')}\rangle\rtimes\sigma'_{neg}.$$

This implies the following commutative diagram (all involved intertwining operators are standard normalized operators; i_s is an embedding depending holomorphically on s):

$$\begin{array}{ccc}
\sigma_s & \xrightarrow{i_s} & \langle[-\frac{l-1}{2},\frac{l-1}{2}]^{(\nu^s\chi)}\rangle\times\langle[-\frac{l'-1}{2},\frac{l'-1}{2}]^{(\chi')}\rangle\rtimes\sigma'_{neg}\\
 & & \Big\downarrow N_1(s)\\
 & & \langle[-\frac{l'-1}{2},\frac{l'-1}{2}]^{(\chi')}\rangle\times\langle[-\frac{l-1}{2},\frac{l-1}{2}]^{(\nu^s\chi)}\rangle\rtimes\sigma'_{neg}\\
N(s,w_0)\Big\downarrow & & \Big\downarrow N_2(s)\\
 & & \langle[-\frac{l'-1}{2},\frac{l'-1}{2}]^{(\chi')}\rangle\times\langle[-\frac{l-1}{2},\frac{l-1}{2}]^{(\nu^{-s}\chi)}\rangle\rtimes\sigma'_{neg}\\
 & & \Big\downarrow N_3(s)\\
\sigma_{-s} & \xrightarrow{i_{-s}} & \langle[-\frac{l-1}{2},\frac{l-1}{2}]^{(\nu^{-s}\chi)}\rangle\times\langle[-\frac{l'-1}{2},\frac{l'-1}{2}]^{(\chi')}\rangle\rtimes\sigma'_{neg}.
\end{array}$$

At $s=\pm 1/2$, $\langle[-\frac{l-1}{2},\frac{l-1}{2}]^{(\nu^{\pm s}\chi)}\rangle\times\langle[-\frac{l'-1}{2},\frac{l'-1}{2}]^{(\chi')}$ is irreducible, and therefore $N_1(s)$ and $N_3(s)$ are holomorphic (by the clear analogies of (norm-1) and (norm-2) for them). This proves the first step of the reduction; we assume that σ_{neg} is strongly negative. To avoid any confusion we write σ_{sn} instead of σ_{neg}.

Next, we consider the following diagram:

$$
\text{(7-4)}\qquad
\begin{array}{ccc}
\sigma_s & \xrightarrow{\;i_s\;} & \langle[-\frac{l-2}{2},\frac{l-2}{2}]^{(\nu^{s-\frac{1}{2}}\chi)}\rangle \times \nu^{\frac{l-1}{2}+s}\chi \rtimes \sigma_{sn} \\
 & & \Big\downarrow {\scriptstyle N_1(s)} \\
 & & \langle[-\frac{l-2}{2},\frac{l-2}{2}]^{(\nu^{s-\frac{1}{2}}\chi)}\rangle \times \nu^{-\frac{l-1}{2}-s}\chi \rtimes \sigma_{sn} \\
{\scriptstyle N(s,w_0)}\Big\downarrow & & \Big\downarrow {\scriptstyle N_2(s)} \\
 & & \nu^{-\frac{l-1}{2}-s}\chi \times \langle[-\frac{l-2}{2},\frac{l-2}{2}]^{(\nu^{s-\frac{1}{2}}\chi)}\rangle \rtimes \sigma_{sn} \\
 & & \Big\downarrow {\scriptstyle N_3(s)} \\
\sigma_{-s} & \xrightarrow{\;i_{-s}\;} & \nu^{-\frac{l-1}{2}-s}\chi \times \langle[-\frac{l-2}{2},\frac{l-2}{2}]^{(\nu^{-s+\frac{1}{2}}\chi)}\rangle \rtimes \sigma_{sn}.
\end{array}
$$

(If $l = 1$, then $N_2(s)$ and $N_3(s)$ are not present.)

Now, as in the proof of ([**M6**], Lemma 3.5) (or adapting the argument for the normalized intertwining operator (7-8) below), it follows that $N_1(s)$ is holomorphic at $s = 1/2$. Next, $N_2(s)$ is holomorphic at $s = 1/2$ since we have the following diagram (j_s is an embedding depending holomorphically on s):

$$
\text{(7-5)}\qquad
\begin{array}{ccc}
\langle[-\frac{l-2}{2},\frac{l-2}{2}]^{(\nu^{s-\frac{1}{2}}\chi)}\rangle \times \nu^{-\frac{l-1}{2}-s}\chi & \xrightarrow{\;j_s\;} & \mathrm{Ind}(s) \\
\Big\downarrow {\scriptstyle N_2(s)} & & \Big\downarrow {\scriptstyle N_4(s)} \\
\nu^{-\frac{l-1}{2}-s}\chi \times \langle[-\frac{l-2}{2},\frac{l-2}{2}]^{(\nu^{s-\frac{1}{2}}\chi)}\rangle & \xrightarrow{\;j_{-s}\;} & \mathrm{Ind}_1(s),
\end{array}
$$

where

$$
\begin{cases}
\mathrm{Ind}(s) = \nu^{-\frac{l-1}{2}+s}\chi \times \cdots \times \nu^{\frac{l-1}{2}-1+s}\chi \times \nu^{-\frac{l-1}{2}-s}\chi \\
\mathrm{Ind}_1(s) = \nu^{-\frac{l-1}{2}-s}\chi \times \nu^{-\frac{l-1}{2}+s}\chi \times \cdots \times \nu^{\frac{l-1}{2}-1+s}\chi,
\end{cases}
$$

and the normalized operator $N_4(s)$ is a composition of normalized operators induced from the rank–one operators:

$$
Q_i(s) : \nu^{-\frac{l-1}{2}+s+i}\chi \times \nu^{-\frac{l-1}{2}-s}\chi \to \nu^{-\frac{l-1}{2}-s}\chi \times \nu^{-\frac{l-1}{2}+s+i}\chi,
$$

where $i = 0, \ldots l-2$. The normalized intertwining operators $Q_i(s)$ are holomorphic at $s = 1/2$ by the basic property of the normalization (see for example ([**M6**], Theorem 2.1)). Finally, by the analogue of (norm-3), $N_3(s)$ is holomorphic. This proves that $N(s, w_0)$ is holomorphic at $s = 1/2$ (see (7-4)). It remains to prove that $N(-s, w_0)$ has a simple pole at $s = 1/2$. To accomplish this we reverse the vertical arrows in (7-4) and change s into $-s$ in the arguments of all $N_i(\cdot)$ and $N(\cdot, w_0)$. We remind the reader that $N_2(-s)$ and $N_3(-s)$ are present if and only if $l > 1$. We assume $l > 1$. Now, arguing as above, we see that $N_3(-s)$ is holomorphic. Similarly, arguing as in (7-5), we see that $N_2(-s)$ has at most a simple pole at $s = 1/2$. The pole must be present since, by [**Ze**], the unique irreducible quotient of $\nu^{-\frac{l}{2}}\chi \times \langle[-\frac{l-2}{2},\frac{l-2}{2}]^{(\chi)}\rangle$ is the unique irreducible subrepresentation of $\langle[-\frac{l-2}{2},\frac{l-2}{2}]^{(\chi)}\rangle \times \nu^{-\frac{l}{2}}\chi$ and it is different than $\langle[-\frac{l}{2},\frac{l}{2}-1]^{(\chi)}\rangle$. Thus, we obtain

$$
\text{(7-6)}\qquad N_2(-s) \text{ has a simple pole at } s = 1/2.
$$

We investigate the influence of that pole on the image of $N_3(-1/2)$. Then the discussion above shows that $N_3(-1/2)$ is not an isomorphism if and only if $\langle[-\frac{l-2}{2},\frac{l-2}{2}]^{(\chi)}\rangle \rtimes \sigma_{sn}$ reduces. Applying Lemma 6-5 and our assumption $l-(2\alpha_\chi+1)$

is not an even integer, we see that if $l > 1$, then $N_3(-1/2)$ is not an isomorphism if and only if $(l-1, \chi) \notin \mathrm{Jord}(\sigma_{sn})$. In particular, if $N_3(-1/2)$ is an isomorphism (hence, a scalar multiple of the identity), then the image of $\sigma_{-1/2}$ under $i_{-1/2}$ is the same as its image under $N_3(-1/2)i_{-1/2}$; $N_2(-1/2)$ is holomorphic on that image. Thus, we summarize the discussion as follows:

(7-7) If $l > 1$ and $(l-1, \chi) \in \mathrm{Jord}(\sigma_{sn})$, then $N_2(-s)N_3(-s)i_{-s}$ is holomorphic at $s = 1/2$.

Next, we consider

$$N_1(-s) : \nu^{-\frac{l-1}{2}-s}\chi \rtimes \sigma_{sn} \to \nu^{\frac{l-1}{2}+s}\chi \rtimes \sigma_{sn} \tag{7-8}$$

at $s = 1/2$. First, we have the following:

(7-9) if $\nu^{\frac{l}{2}}\chi \rtimes \sigma_{sn}$ is irreducible, then $N_1(-s)$ is holomorphic at $s = 1/2$.

We describe the reducibility of $\nu^{\frac{l}{2}}\chi \rtimes \sigma_{sn}$ using Lemma 6-1:

(red-1) Assume $l > 1$. Then $\nu^{\frac{l}{2}}\chi \rtimes \sigma_{sn}$ is reducible if and only if $(l-1, \chi) \in \mathrm{Jord}(\sigma_{sn})$.

(red-2) Assume $l = 1$. Then the assumption $l - (2\alpha_\chi + 1) \notin 2\mathbb{Z}$ implies $\alpha_\chi = 1/2$. In this case we have the reducibility.

We analyze $N_1(-s)$ at $s = 1/2$. First, we apply Theorem 5-8 and (5-7) to obtain:

$$\sigma_{sn} \hookrightarrow \times_{i=1}^{l_{1_{F^\times}}} \langle [-\frac{a_{2i}-1}{2}, \frac{a_{2i-1}-1}{2}]^{(\mathbf{1}_{F^\times})} \rangle \times_{j=1}^{l_{\mathbf{sgn}_u}} \langle [-\frac{b_{2j}-1}{2}, \frac{b_{2j-1}-1}{2}]^{(\mathbf{sgn}_u)} \rangle \rtimes \mathbf{1}. \tag{7-10}$$

Hence we can write the following commutative diagram (where the vertical arrows are normalized intertwining operators; j_s is an embedding depending holomorphically on s):

$$\begin{array}{ccc} \nu^{-\frac{l-1}{2}-s}\chi \rtimes \sigma_{sn} & \xrightarrow{j_{-s}} & \nu^{-\frac{l-1}{2}-s}\chi \rtimes \mathrm{Ind} \\ {\scriptstyle N_1(-s)}\downarrow & & \downarrow{\scriptstyle N_1'(s)} \\ \nu^{\frac{l-1}{2}+s}\chi \rtimes \sigma_{sn} & \xrightarrow{j_s} & \nu^{\frac{l-1}{2}+s}\chi \rtimes \mathrm{Ind}, \end{array}$$

where

$$\mathrm{Ind} = \times_{i=1}^{l_{1_{F^\times}}} \langle [-\frac{a_{2i}-1}{2}, \frac{a_{2i-1}-1}{2}]^{(\mathbf{1}_{F^\times})} \rangle \times_{j=1}^{l_{\mathbf{sgn}_u}} \langle [-\frac{b_{2j}-1}{2}, \frac{b_{2j-1}-1}{2}]^{(\mathbf{sgn}_u)} \rangle \rtimes \mathbf{1}.$$

Next, the normalized operator $N_1'(s)$ can be factorized into the product of the following normalized operators:

$$\nu^{-\frac{l-1}{2}-s}\chi \times \langle [-\frac{a_{2i}-1}{2}, \frac{a_{2i-1}-1}{2}]^{(\mathbf{1}_{F^\times})} \rangle \xrightarrow{Q_{i,\mathbf{1}_{F^\times}}} \langle [-\frac{a_{2i}-1}{2}, \frac{a_{2i-1}-1}{2}]^{(\mathbf{1}_{F^\times})} \rangle \times \nu^{-\frac{l-1}{2}-s}\chi,$$

$$\nu^{-\frac{l-1}{2}-s}\chi \times \langle [-\frac{b_{2i}-1}{2}, \frac{b_{2i-1}-1}{2}]^{(\mathbf{sgn}_u)} \rangle \xrightarrow{Q_{i,\mathbf{sgn}_u}} \langle [-\frac{b_{2i}-1}{2}, \frac{b_{2i-1}-1}{2}]^{(\mathbf{sgn}_u)} \rangle \times \nu^{-\frac{l-1}{2}-s}\chi,$$

$$\nu^{-\frac{l-1}{2}-s}\chi \rtimes \mathbf{1} \xrightarrow{P(s)} \nu^{\frac{l-1}{2}+s}\chi \rtimes \mathbf{1},$$

$$\langle[-\frac{a_{2i}-1}{2}, \frac{a_{2i-1}-1}{2}]^{(\mathbf{1}_{F^\times})}\rangle \times \nu^{\frac{l-1}{2}+s}\chi \xrightarrow{R_{i,\mathbf{1}_{F^\times}}} \nu^{\frac{l-1}{2}+s}\chi \times \langle[-\frac{a_{2i}-1}{2}, \frac{a_{2i-1}-1}{2}]^{(\mathbf{1}_{F^\times})}\rangle,$$

and

$$\langle[-\frac{b_{2i}-1}{2}, \frac{b_{2i-1}-1}{2}]^{(\mathbf{sgn}_u)}\rangle \times \nu^{\frac{l-1}{2}+s}\chi \xrightarrow{R_{i,\mathbf{sgn}_u}} \nu^{\frac{l-1}{2}+s}\chi \times \langle[-\frac{b_{2i}-1}{2}, \frac{b_{2i-1}-1}{2}]^{(\mathbf{sgn}_u)}\rangle$$

We analyze them at $s=-1/2$. First, if $l=1$, then all of them except $P(s)$ are holomorphic. By the rank–one theory $P(s)$ has a simple pole at $s=-1/2$. This proves that $N(-s,w_0)$ has a simple pole in this case. We assume $l>1$. Now, if $(l-1,\chi)\not\in \mathrm{Jord}(\sigma_{sn})$, then (red-1) and (7-9) imply that $N_1(-s)$ is holomorphic at $s=-1/2$; combining (7-3) and (7-6), $N(-s,w_0)$ has a simple pole at $s=-1/2$. ¿From now on, we assume $(l-1,\chi)\in \mathrm{Jord}(\sigma_{sn})$. Then we need to prove that $N_1(-s)$ has at most a simple pole at $s=1/2$ since then (7-3) and (7-7) imply $N(-s,w_0)$ has a simple pole at $s=-1/2$.

Assume that $l>2$ or $l=2$ and $\alpha_\chi\neq 1$. Then $Q_{i,\mathbf{1}_{F^\times}}(s)$ (resp., $Q_{i,\mathbf{sgn}_u}(s)$) has at most a simple pole at $s=-1/2$ if and only if $\chi=\mathbf{1}_{F^\times}$ (resp., $\chi=\mathbf{sgn}_u$) (noting $(l-1,\chi)\in \mathrm{Jord}(\sigma_{sn})$). If this is so, $R_{i,\mathbf{1}_{F^\times}}(s)$ and $R_{i,\mathbf{sgn}_u}(s)$ are ismorphisms (hence holomorphic) at $s=-1/2$. The converse statement is also true. Thus, the contribution of all the normalized operators $Q_{i,\mathbf{1}_{F^\times}}(s), Q_{i,\mathbf{sgn}_u}(s), R_{i,\mathbf{1}_{F^\times}}(s), R_{i,\mathbf{1}_{F^\times}}(s)$ is just at a simple pole at $s=-1/2$. Further, since $l>2$ or $l=2$ and $\alpha_\chi\neq 1$, $R(s)$ is holomorphic at $s=-1/2$. This proves that $N(-s,w_0)$ has a simple pole at $s=-1/2$ in this case. Finally, we assume $l=2$ and $\alpha_\chi=1$. Then $S_n=\mathrm{Sp}(2n,F)$ and card $\mathrm{Jord}(\sigma_{sn})_{\mathbf{1}_{F^\times}}$ is odd (see Definition 5-4). Applying (5-6), we obtain $a_1=-1<0<a_2<\cdots$. Since $(1,\mathbf{1}_{F^\times})\in \mathrm{Jord}(\sigma_{sn})$, $a_2=1$. Thus, $\langle[-\frac{a_2-1}{2},\frac{a_1-1}{2}]^{(\mathbf{1}_{F^\times})}\rangle = \langle[0,-1]^{(\mathbf{1}_{F^\times})}\rangle$ is empty. In particular, $Q_{1,\mathbf{1}_{F^\times}}(s)$ and $R_{1,\mathbf{1}_{F^\times}}(s)$ are not present. Thus, all normalized operators $Q_{i,\mathbf{1}_{F^\times}}(s), Q_{i,\mathbf{sgn}_u}(s), R_{i,\mathbf{1}_{F^\times}}(s), R_{i,\mathbf{sgn}_u}(s)$ are holomorphic at $s=-1/2$. Since $R(s)$ has a simple pole at $s=-1/2$, the proof is complete.

8. The Injectivity of $\mathcal{M}^{u,unr}(S_n)\to \mathrm{Irr}^{u,unr}(S_n)$

In this section we show the map $\mathcal{M}^{u,unr}(S_n)\to \mathrm{Irr}^{u,unr}(S_n)$ given by

$$(\mathbf{e},\sigma_{neg}) \longmapsto \times_{(l,\chi,\alpha)\in\mathbf{e}} \ \langle[-\frac{l-1}{2},\frac{l-1}{2}]^{(\nu^\alpha\chi)}\rangle \rtimes \sigma_{neg}$$

(see Theorem 5-14) is well–defined and injective. Lemmas 8-1 and 8-2 show that the map is well-defined; injectivity then follows.

LEMMA 8-1. *Let* $(\mathbf{e},\sigma_{neg})\in \mathcal{M}^{u,unr}(S)$. *Then the induced representation* $\left(\times_{(l,\chi,\alpha)\in\mathbf{e}} \ \langle[-\frac{l-1}{2},\frac{l-1}{2}]^{(\nu^\alpha\chi)}\rangle\right)\rtimes\sigma_{neg}$ *is irreducible.*

PROOF. This follows from Lemma 6-1 and Lemma 6-9 (i), (ii) and (iii). □

LEMMA 8-2. *Let* $(\mathbf{e}, \sigma_{neg}) \in \mathcal{M}^{u,unr}(S)$. *Then the induced representation* $(\times_{(l,\chi,\alpha)\in\mathbf{e}} \ \langle[-\frac{l-1}{2}, \frac{l-1}{2}]^{(\nu^\alpha\chi)}\rangle) \rtimes \sigma_{neg}$ *is unitarizable (and irreducible).*

PROOF. We let σ be that induced (irreducible) representation. We prove the unitarity of σ by induction on $m :=$ card $\mathbf{e}$. If $m = 0$, then $\mathbf{e} = \emptyset$. Therefore $\sigma = \sigma_{neg}$ is unitarizable by Theorem 5-11. Assume that the claim is true for all $(\mathbf{e}', \sigma'_{neg}) \in \mathcal{M}^{u,unr}(S)$ with card $\mathbf{e}' < m$. Now, we proceed according to Definition 5-13 (1)–(3) as follows.

$(Def-1)$ Assume that there exists $\chi \notin \{\mathbf{1}_{F^\times}, \mathbf{sgn}_u\}$ and $l \in \mathbb{Z}_{\geq 1}$ such that $\mathbf{e}(l,\chi) \neq \emptyset$. Then we pick some $\alpha \in \mathbf{e}(l,\chi)$. Applying Definition 5-13 (1), $\alpha \in \mathbf{e}(l,\chi^{-1})$. We let $\mathbf{e}' = \mathbf{e} - \{(l,\chi,\alpha),(l,\chi^{-1},\alpha)\}$. Then it is easy to see that $(\mathbf{e}', \sigma_{neg}) \in \mathcal{M}^{u,unr}(S)$. Since card $\mathbf{e}' <$ card $\mathbf{e} =: m$, we apply the inductive assumption to obtain the unitarity of σ' defined by

$$\sigma' = \times_{(l',\chi',\alpha')\in\mathbf{e}'} \ \langle[-\frac{l'-1}{2}, \frac{l'-1}{2}]^{(\nu^{\alpha'}\chi')}\rangle \rtimes \sigma_{neg}.$$

Next, we have

$$\sigma \simeq \langle[-\frac{l-1}{2}, \frac{l-1}{2}]^{(\nu^\alpha\chi)}\rangle \times \langle[-\frac{l-1}{2}, \frac{l-1}{2}]^{(\nu^\alpha\chi^{-1})} \rtimes \sigma'$$
$$\simeq \left(\langle[-\frac{l-1}{2}, \frac{l-1}{2}]^{(\nu^\alpha\chi)}\rangle \times \langle[-\frac{l-1}{2}, \frac{l-1}{2}]^{(\nu^{-\alpha}\chi)}\rangle\right) \rtimes \sigma'.$$

Since, by Definition 5-13 (1), $\alpha \in]0, \frac{1}{2}[$, the unitarity of σ follows from Theorem 4-1 and (UI).

$(Def-2)$ Assume that there exists $\chi \in \{\mathbf{1}_{F^\times}, \mathbf{sgn}_u\}$ and $l \in \mathbb{Z}_{\geq 1}$, such that $l - (2\alpha_\chi + 1) \notin 2\mathbb{Z}$ and $\mathbf{e}(l,\chi) \neq \emptyset$. Then we pick some $\alpha \in \mathbf{e}(l,\chi)$. We let $\mathbf{e}' = \mathbf{e} - \{(l,\chi,\alpha)\}$. Then it is obvious that $(\mathbf{e}', \sigma_{neg}) \in \mathcal{M}^{u,unr}(S)$. Since card $\mathbf{e}' <$ card $\mathbf{e} =: m$, we apply the inductive assumption to obtain the unitarity of σ' defined by $\sigma' = \times_{(l',\chi',\alpha')\in\mathbf{e}'} \ \langle[-\frac{l'-1}{2}, \frac{l'-1}{2}]^{(\nu^{\alpha'}\chi')}\rangle \rtimes \sigma_{neg}$. We claim that $\langle[-\frac{l-1}{2}, \frac{l-1}{2}]^{(\chi)}\rangle \rtimes \sigma'$ is irreducible. Namely, since $l - (2\alpha_\chi + 1) \notin 2\mathbb{Z}$, Lemma 6-5 implies that $\langle[-\frac{l-1}{2}, \frac{l-1}{2}]^{(\chi)}\rangle \rtimes \sigma_{neg}$ is irreducible. Clearly, this representation is negative and we denote it by σ'_{neg}. Using this it is easy to show $(\mathbf{e}', \sigma'_{neg}) \in \mathcal{M}^{u,unr}(S)$. Next, the attached induced representation

$$\times_{(l',\chi',\alpha')\in\mathbf{e}'} \ \langle[-\frac{l'-1}{2}, \frac{l'-1}{2}]^{(\nu^{\alpha'}\chi')}\rangle \rtimes \sigma'_{neg} \simeq$$
$$\times_{(l',\chi',\alpha')\in\mathbf{e}'} \ \langle[-\frac{l'-1}{2}, \frac{l'-1}{2}]^{(\nu^{\alpha'}\chi')}\rangle \rtimes \left(\langle[-\frac{l-1}{2}, \frac{l-1}{2}]^{(\chi)}\rangle \rtimes \sigma_{neg}\right) \simeq$$
$$\langle[-\frac{l-1}{2}, \frac{l-1}{2}]^{(\chi)}\rangle \times \left(\times_{(l',\chi',\alpha')\in\mathbf{e}'} \ \langle[-\frac{l'-1}{2}, \frac{l'-1}{2}]^{(\nu^{\alpha'}\chi')}\rangle \rtimes \sigma_{neg}\right) \simeq$$
$$\langle[-\frac{l-1}{2}, \frac{l-1}{2}]^{(\chi)}\rangle \rtimes \sigma'$$

is irreducible by Lemma 8-1. Similarly, using induction in stages, Lemma 6-1 implies the irreducibility of $\sigma_s = \langle[-\frac{l-1}{2}, \frac{l-1}{2}]^{(\nu^s\chi)}\rangle \rtimes \sigma'$ for $s \in]0, \frac{1}{2}[$. Now, (D) implies the unitarity of σ_s. Since $\sigma \simeq \sigma_\alpha$, we have proved its unitarity.

$(Def-3)$ Assume that there exists $\chi \in \{\mathbf{1}_{F^\times}, \mathbf{sgn}_u\}$ and $l \in \mathbb{Z}_{\geq 1}$, such that $l - (2\alpha_\chi + 1) \in 2\mathbb{Z}$ and $\mathbf{e}(l,\chi) \neq \emptyset$. We use the notation introduced in Definition 5-13 (3). If there are indices $i_1 \neq i_2$ such that α_{i_1} and α_{i_2} both belong to one of the segments $]1-\beta_1, \frac{1}{2}]$, $]0, 1-\beta_v[$, or $]1-\beta_{j+1}, 1-\beta_j[$ (for some j) or simply if $v = 0$

but $u \geq 2$ (so we can arbitrarily pick the two different indices $i_1 \neq i_2$), then we let $\mathbf{e}' = \mathbf{e} - \{(l, \chi, \alpha_{i_1}), (l, \chi, \alpha_{i_2})\}$. Then it is obvious that $(\mathbf{e}', \sigma_{neg}) \in \mathcal{M}^{u,unr}(S)$. Since card $\mathbf{e}'$ < card $\mathbf{e}$ =: m, we apply the inductive assumption to obtain the unitarity of σ' defined by $\sigma' = \times_{(l',\chi',\alpha')\in\mathbf{e}'} \langle[-\frac{l'-1}{2}, \frac{l'-1}{2}]^{(\nu^{\alpha'}\chi')}\rangle \rtimes \sigma_{neg}$. It is easy to see that we can write σ as follows:

$$\sigma \simeq \langle[-\frac{l-1}{2}, \frac{l-1}{2}]^{(\nu^{\alpha_{i_1}}\chi)}\rangle \times \langle[-\frac{l-1}{2}, \frac{l-1}{2}]^{(\nu^{\alpha_{i_2}}\chi)} \rtimes \sigma'$$
$$\simeq \left(\langle[-\frac{l-1}{2}, \frac{l-1}{2}]^{(\nu^{\alpha_{i_1}}\chi)}\rangle \times \langle[-\frac{l-1}{2}, \frac{l-1}{2}]^{(\nu^{-\alpha_{i_2}}\chi)}\rangle\right) \rtimes \sigma'.$$

The inductive assumption applied to σ' and Lemma 6-4 implies that σ is Hermitian. Now, since

$$\left(\langle[-\frac{l-1}{2}, \frac{l-1}{2}]^{(\nu^{\alpha_{i_2}}\chi)}\rangle \times \langle[-\frac{l-1}{2}, \frac{l-1}{2}]^{(\nu^{-\alpha_{i_2}}\chi)}\rangle\right) \rtimes \sigma' \tag{8-3}$$

is unitarizable by (UI) (applying $\alpha_{i_2} \in]0, \frac{1}{2}[$ and Theorem 4-1), the way we have chosen α_{i_1} and α_{i_2} enables us to deform the first exponent α_{i_2} (see (8-3)) to α_{i_1} proving the unitarity of σ by (D). Thus, if $v = 0$, we may assume that $u \in \{0, 1\}$. If $v = 0, u = 1$, then $u + v = 1$ is odd. Hence $(l, \chi) \in \text{Jord}(\sigma_{neg})$ (see Definition 5-13 (3) (a)). Therefore, by Lemma 6-1, $\langle[-\frac{l-1}{2}, \frac{l-1}{2}]^{(\chi)}\rangle \rtimes \sigma_{neg}$ is irreducible, and we proceed as in the case (Def-2) with $\alpha = \alpha_1$.

Now, we assume $v = 1$. Then, by our reduction, we may assume that $]1-\beta_1, \frac{1}{2}]$ does not contain any α_i, while $]0, 1-\beta_1[$ contains all. Therefore we may assume $u \in \{0, 1\}$. If $u = 0$, then $u+v = 1$ is odd. Hence $(l, \chi) \in \text{Jord}(\sigma_{neg})$ (see Definition 5-13 (3) (a)). Therefore, by Lemma 6-1, $\langle[-\frac{l-1}{2}, \frac{l-1}{2}]^{(\chi)}\rangle \rtimes \sigma_{neg}$ is irreducible, and we proceed as in the case $(Def-2)$ with $\alpha = \beta_1$. If $u = v = 1$, then we need to prove the unitarity of

$$\sigma \simeq \langle[-\frac{l-1}{2}, \frac{l-1}{2}]^{(\nu^{\alpha_1}\chi)}\rangle \times \langle[-\frac{l-1}{2}, \frac{l-1}{2}]^{(\nu^{\beta_1}\chi)}\rangle \rtimes \sigma',$$

where σ' is attached to $(\mathbf{e}', \sigma_{neg})$ with

$$\mathbf{e}' = \mathbf{e} - \{(l, \chi, \alpha_1), (l, \chi, \beta_1)\}.$$

(Clearly, by induction, σ' is unitarizable.) We start from the following family of induced representations:

$$\sigma_s := \langle[-\frac{l-1}{2}, \frac{l-1}{2}]^{(\nu^{\alpha_1}\chi)}\rangle \times \langle[-\frac{l-1}{2}, \frac{l-1}{2}]^{(\nu^{s}\chi)}\rangle \rtimes \sigma' \simeq$$
$$\langle[-\frac{l-1}{2}, \frac{l-1}{2}]^{(\nu^{\alpha_1}\chi)}\rangle \times \langle[-\frac{l-1}{2}, \frac{l-1}{2}]^{(\nu^{-s}\chi)}\rangle \rtimes \sigma',$$

where $s \in [\alpha_1, 1-\alpha_1[$. Lemma 8-1 implies that every representation σ_s (for $s \in [\alpha_1, 1-\alpha_1[$) is irreducible. Since it is unitarizable for $s = \alpha_1$ by the above isomorphism and Theorem 4-1, (D) implies the unitarizability of σ_s for every $s \in [\alpha_1, 1-\alpha_1[$. Since $\beta_1 \in [\alpha_1, 1-\alpha_1[$, we see that $\sigma = \sigma_{\beta_1}$ is unitarizable.

Finally, we assume $v \geq 2$. Then, by our reduction, we may assume that the interval $]1-\beta_1, \frac{1}{2}]$ does not contain any α_i while $]1-\beta_2, 1-\beta_1[$ must contain a unique α_i. Hence $u \geq 1$ and $\alpha_u \in]1-\beta_2, 1-\beta_1[$. We need to prove the unitarity of

$$\sigma \simeq \langle[-\frac{l-1}{2}, \frac{l-1}{2}]^{(\nu^{\alpha_u}\chi)}\rangle \times \langle[-\frac{l-1}{2}, \frac{l-1}{2}]^{(\nu^{\beta_1}\chi)}\rangle \rtimes \sigma',$$

where σ' is attached to $(\mathbf{e}', \sigma_{neg})$ with

$$\mathbf{e}' = \mathbf{e} - \{(l, \chi, \alpha_u), (l, \chi, \beta_1)\}.$$

(Clearly, by induction, σ' is unitarizable.) We start from the following family of induced representations:

$$\sigma_s := \langle[-\frac{l-1}{2}, \frac{l-1}{2}]^{(\nu^{\alpha_1}\chi)}\rangle \times \langle[-\frac{l-1}{2}, \frac{l-1}{2}]^{(\nu^{s}\chi)}\rangle \rtimes \sigma' \simeq$$
$$\langle[-\frac{l-1}{2}, \frac{l-1}{2}]^{(\nu^{\alpha_u}\chi)}\rangle \times \langle[-\frac{l-1}{2}, \frac{l-1}{2}]^{(\nu^{-s}\chi)}\rangle \rtimes \sigma',$$

where $s \in [\alpha_u, 1-\alpha_u[$. Lemma 8-1 implies that every representation σ_s (for $s \in [\alpha_u, 1-\alpha_u[$) is irreducible. Since it is unitarizable for $s = \alpha_u$ by the above isomorphism and Theorem 4-1, (D) implies the unitarizability of σ_s for every $s \in [\alpha_u, 1-\alpha_u[$. Since $\beta_1 \in [\alpha_u, 1-\alpha_u[$, we see that $\sigma = \sigma_{\beta_1}$ is unitarizable. This completes the proof of the lemma. □

9. The Surjectivity of $\mathcal{M}^{u,unr}(S_n) \to \mathrm{Irr}^{u,unr}(S_n)$

In this section we prove the surjectivity of the map $\mathcal{M}^{u,unr}(S_n) \to \mathrm{Irr}^{u,unr}(S_n)$ given by $(\mathbf{e}, \sigma_{neg}) \longmapsto \times_{(l,\chi,\alpha)\in\mathbf{e}} \ \langle[-\frac{l-1}{2}, \frac{l-1}{2}]^{(\nu^{\alpha}\chi)}\rangle \rtimes \sigma_{neg}$. This completes the proof of Theorem 5-14.

The proof will be done by induction on n. If $n = 0$ then $\mathcal{M}^{u,unr}(S_n) = \{(\emptyset, \mathbf{1})\}$, $\mathrm{Irr}^{u,unr}(S_n) = \{\mathbf{1}\}$, and above map is just $(\emptyset, \mathbf{1}) \longmapsto \mathbf{1}$. Therefore, the theorem is obvious in this case. Assume the surjectivity of the maps for all non–negative integers $< n$. Then we prove the surjectivity of the map for n. More precisely, for

(9-1) $$\sigma \in \mathrm{Irr}^{u,unr}(S_n)$$

we need to produce the datum $(\mathbf{e}, \sigma_{neg}) \in \mathcal{M}^{u,unr}(S_n)$ such that

(9-2) $$\sigma \simeq \times_{(l,\chi,\alpha)\in\mathbf{e}} \ \langle[-\frac{l-1}{2}, \frac{l-1}{2}]^{(\nu^{\alpha}\chi)}\rangle \rtimes \sigma_{neg}.$$

First, by Lemma 6-2, there is a unique

$$(\mathbf{e}, \sigma_{neg}) \in \mathcal{M}^{unr}(S_n)$$

such that (9-2) holds. Therefore it remains to prove the following theorem:

THEOREM 9-3. $(\mathbf{e}, \sigma_{neg}) \in \mathcal{M}^{u,unr}(S_n)$ *(that is,* $(\mathbf{e}, \sigma_{neg})$ *satisfies Definition 5-13).*

The proof of this result (that is, the proof of the inductive step) will occupy the remainder of this section. It is done by (another) induction on $m = \mathrm{card}\ \mathbf{e}$, $(\mathbf{e}, \sigma_{neg}) \in \mathcal{M}^{u,unr}(S_n)$. If $m = 0$, then the representation is $\sigma \simeq \sigma_{neg}$, and, clearly, $(\emptyset, \sigma_{neg})$ satisfies Definition 5-13. Next, we state the following useful observation that will be used several times in the proof below:

REMARK 9-4. *Lemma 6-1 and (D) imply that "being in complementary series" is an "open condition". This means, for every* $(l, \chi, \alpha) \in \mathbf{e}$ *we may choose* ϵ *having small absolute value such that* $\times_{(l,\chi,\alpha)\in\mathbf{e}}\langle[-\frac{l-1}{2}, \frac{l-1}{2}]^{(\nu^{\alpha+\epsilon}\chi)}\rangle \rtimes \sigma_{neg}$ *is irreducible and unitarizable, and* $\alpha + \epsilon \notin (1/2)\mathbb{Z}$, $(\alpha+\epsilon) \pm (\alpha'+\epsilon') \notin (1/2)\mathbb{Z}$ *for all* $(l, \chi, \alpha) \neq (l', \chi', \alpha') \in \mathbf{e}$. *We refer to this perturbation of exponents as bringing* σ *into general position.*

The appropriate definition is the following:

DEFINITION 9-5. *We say that* σ *is in general position if* $\alpha \notin (1/2)\mathbb{Z}$, $\alpha \pm \alpha' \notin (1/2)\mathbb{Z}$ *for all* $(l, \chi, \alpha) \neq (l', \chi', \alpha') \in \mathbf{e}$.

The first step in the proof is easy:

LEMMA 9-6. *If there exist* $\chi \notin \{\mathbf{1}_{F^\times}, \mathbf{sgn}_u\}$ *and* $l \in \mathbb{Z}_{\geq 1}$ *such that* $\mathbf{e}(l, \chi) \neq \emptyset$, *then* $(\mathbf{e}, \sigma_{neg}) \in \mathcal{M}^{u,unr}(S_n)$.

PROOF. By our assumption, σ is unitarizable. Therefore, σ is Hermitian. Now, Lemma 6-4 implies $\mathbf{e}(l', \chi') = \mathbf{e}(l', (\chi')^{-1})$ for $\chi' \notin \{\mathbf{1}_{F^\times}, \mathbf{sgn}_u\}$ and $l' \in \mathbb{Z}_{\geq 1}$. If $\mathbf{e}(l, \chi) \neq \emptyset$, then let $\alpha \in \mathbf{e}(l, \chi)$. Then $\alpha \in \mathbf{e}(l, \chi^{-1})$. We let $\mathbf{e}' = \mathbf{e} - \{(l, \chi, \alpha), (l, \chi^{-1}, \alpha)\}$. Then $(\mathbf{e}', \sigma_{neg}) \in \mathcal{M}^{unr}(S_n)$. Let σ' be the irreducible unramified representation attached to $(\mathbf{e}', \sigma_{neg})$. By definition, it is an irreducible subquotient of $\times_{(l',\chi',\alpha')\in\mathbf{e}'} \langle[-\frac{l'-1}{2}, \frac{l'-1}{2}]^{(\nu^{\alpha'}\chi')}\rangle \rtimes \sigma_{neg}$. As we can permute the characters in (9-2), we may write

$$\sigma \simeq \langle[-\frac{l-1}{2}, \frac{l-1}{2}]^{(\nu^{\alpha}\chi)}\rangle \times \langle[-\frac{l-1}{2}, \frac{l-1}{2}]^{(\nu^{\alpha}\chi^{-1})}\rangle$$
$$\times_{(l',\chi',\alpha')\in\mathbf{e}'} \quad \langle[-\frac{l'-1}{2}, \frac{l'-1}{2}]^{(\nu^{\alpha'}\chi')}\rangle \rtimes \sigma_{neg}.$$

This shows that

$$\sigma' \simeq \times_{(l',\chi',\alpha')\in\mathbf{e}'} \quad \langle[-\frac{l'-1}{2}, \frac{l'-1}{2}]^{(\nu^{\alpha'}\chi')}\rangle \rtimes \sigma_{neg} \tag{9-7}$$

and

$$\sigma \simeq \langle[-\frac{l-1}{2}, \frac{l-1}{2}]^{(\nu^{\alpha}\chi)}\rangle \times \langle[-\frac{l-1}{2}, \frac{l-1}{2}]^{(\nu^{\alpha}\chi^{-1})}\rangle \rtimes \sigma'. \tag{9-8}$$

Since σ is Hermitian, Lemma 6-4, the definition of $\mathbf{e}'$ and (9-7) imply that σ' is also Hermitian. Next, the isomorphism (9-8) implies

$$\sigma \simeq \langle[-\frac{l-1}{2}, \frac{l-1}{2}]^{(\nu^{\alpha}\chi)}\rangle \times \langle[-\frac{l-1}{2}, \frac{l-1}{2}]^{(\nu^{-\alpha}\chi)}\rangle \rtimes \sigma'.$$

Therefore, σ is fully–induced from the tensor product of two irreducible Hermitian representations: $\langle[-\frac{l-1}{2}, \frac{l-1}{2}]^{(\nu^{\alpha}\chi)}\rangle \times \langle[-\frac{l-1}{2}, \frac{l-1}{2}]^{(\nu^{-\alpha}\chi)}\rangle$ and σ'. Since σ is unitarizable, (UR) implies that $\langle[-\frac{l-1}{2}, \frac{l-1}{2}]^{(\nu^{\alpha}\chi)}\rangle \times \langle[-\frac{l-1}{2}, \frac{l-1}{2}]^{(\nu^{-\alpha}\chi)}\rangle$ and σ' are unitarizable. By induction, this means that $(\mathbf{e}', \sigma_{neg}) \in \mathcal{M}^{u,unr}(S_{n'})$, where $n' < n$ is defined by $\sigma' \in \mathrm{Irr}^{unr}(S_{n'})$. Now, by induction, we have the following:

$$(\mathbf{e}', \sigma_{neg}) \in \mathcal{M}^{u,unr}(S_{n'}). \tag{9-9}$$

Also, since $\langle[-\frac{l-1}{2}, \frac{l-1}{2}]^{(\nu^{\alpha}\chi)}\rangle \times \langle[-\frac{l-1}{2}, \frac{l-1}{2}]^{(\nu^{-\alpha}\chi)}\rangle$ is irreducible and unitarizable, Theorem 4-1 implies that $\alpha \in]\frac{-1}{2}, \frac{1}{2}[$. Since by definition of $(\mathbf{e}, \sigma_{neg})$ we have $\alpha > 0$, we obtain $0 < \alpha < \frac{1}{2}$. Now, since $\mathbf{e}' = \mathbf{e} - \{(l, \chi, \alpha), (l, \chi^{-1}, \alpha)\}$ and (9-9) holds, it is easy to check that $(\mathbf{e}, \sigma_{neg}) \in \mathcal{M}^{u,unr}(S_n)$ (see Definition 5-13). □

In the remainder of the proof of Theorem 9-3, Lemma 9-6 enables us to assume that $\chi \in \{\mathbf{1}_{F^\times}, \mathbf{sgn}_u\}$ whenever $e(l, \chi) \neq \emptyset$ for some l. (See Definition 5-13 (1).) Next, we prove the following lemma:

LEMMA 9-10. *Assume that there exist* $\chi \in \{\mathbf{1}_{F^\times}, \mathbf{sgn}_u\}$ *and* $l \in \mathbb{Z}_{\geq 1}$ *such that* $\mathbf{e}(l,\chi)$ *contains* α *and* β, $\alpha \leq \beta$ *(if* $\alpha = \beta$*, then we assume* α *is contained with multiplicity at least two) such that the following hold:*

(1) $]\alpha, \beta[\, \cap \frac{1}{2}\mathbb{Z} = \emptyset$

(2) *there is no* $\gamma \in]\alpha, \beta[$ *such that* $[-\frac{l-1}{2}, \frac{l-1}{2}]^{(\nu^{\pm\gamma}\chi)}$ *is linked with a segment* $[-\frac{l'-1}{2}, \frac{l'-1}{2}]^{(\nu^{\alpha'}\chi')}$ *for* $(l', \chi', \alpha') \in \mathbf{e}$.

Then $\alpha, \beta \in]0, \frac{1}{2}[$, *and* $(\mathbf{e}, \sigma_{neg}) \in \mathcal{M}^{u,unr}(S_n)$.

PROOF. We consider the following family of induced representations:

$$(9\text{-}11)\quad \langle[-\frac{l-1}{2}, \frac{l-1}{2}]^{(\nu^{\gamma}\chi)}\rangle \times \langle[-\frac{l-1}{2}, \frac{l-1}{2}]^{(\nu^{\beta}\chi)}\rangle \times \\ \times_{(l',\chi',\alpha') \in \mathbf{e} - \{(l,\chi,\alpha),\ (l,\chi,\beta)\}} \langle[-\frac{l'-1}{2}, \frac{l'-1}{2}]^{(\nu^{\alpha'}\chi')}\rangle \rtimes \sigma_{neg},$$

where $\gamma \in [\alpha, \beta]$. We prove the irreducibility of the induced representation in (9-11). First, applying Lemma 6-1 (see the beginning of the proof of Lemma 6-1) we list necessary and sufficient conditions for irreducibility:

(i) for $(l_1, \chi_1, \alpha_1), (l_2, \chi_2, \alpha_2) \in \{(l, \chi, \gamma)\} + \mathbf{e} - \{(l, \chi, \alpha)\}$, the segments $[-\frac{l_1-1}{2}, \frac{l_1-1}{2}]^{(\nu^{\pm\alpha_1}\chi_1)}$ and $[-\frac{l_2-1}{2}, \frac{l_2-1}{2}]^{(\nu^{\alpha_2}\chi_2)}$ are not linked

(ii) for $(l', \chi', \alpha') \in \{(l, \chi, \gamma)\} + \mathbf{e} - \{(l, \chi, \alpha)\}$, the induced representation $\langle[-\frac{l'-1}{2}, \frac{l'-1}{2}]^{(\nu^{\alpha'}\chi')}\rangle \rtimes \sigma_{neg}$ is irreducible.

Now, since for $\gamma = \alpha$ the induced representation is isomorphic to σ, it is irreducible. Thus, (i) and (ii) hold for $\gamma = \alpha$. Combining this with (2) shows that (i) holds for any $\gamma \in [\alpha, \beta]$. If $\gamma = \beta$, (ii) is obviously satisfied, proving the irreducibility for $\gamma = \beta$. Let $\gamma \in]\alpha, \beta[$. Then (1) implies $\gamma \notin \frac{1}{2}\mathbb{Z}$. Hence, Lemma 6-1 implies that $\langle[-\frac{l-1}{2}, \frac{l-1}{2}]^{(\nu^{\gamma}\chi)}\rangle \rtimes \sigma_{neg}$ is irreducible. Thus, (ii) always holds. Thus, the induced representation (9-11) is irreducible for all $\gamma \in [\alpha, \beta]$.

Since we assume $\chi' \in \{\mathbf{1}_{F^\times}, \mathbf{sgn}_u\}$ when $e(l', \chi') \neq \emptyset$ for some l', the family of representations (9-11) is Hermitian (see Lemma 6-4). Finally, it is unitarizable for $\gamma = \alpha$ (since it is isomorphic to σ), and therefore for all $\gamma \in [\alpha, \beta]$ (see (D)). In particular, it is irreducible and unitarizable for $\gamma = \beta$. Since in that case we can write (9-11) as follows:

$$\langle[-\frac{l-1}{2}, \frac{l-1}{2}]^{(\nu^{\beta}\chi)}\rangle \times \langle[-\frac{l-1}{2}, \frac{l-1}{2}]^{(\nu^{-\beta}\chi)}\rangle \\ \times_{(l',\chi',\alpha') \in \mathbf{e} - \{(l,\chi,\alpha),\ (l,\chi,\beta)\}} \langle[-\frac{l'-1}{2}, \frac{l'-1}{2}]^{(\nu^{\alpha'}\chi')}\rangle \rtimes \sigma_{neg},$$

we conclude that the following two induced representations are irreducible and Hermitian:

$$(9\text{-}12)\quad \begin{cases} \langle[-\frac{l-1}{2}, \frac{l-1}{2}]^{(\nu^{\beta}\chi)}\rangle \times \langle[-\frac{l-1}{2}, \frac{l-1}{2}]^{(\nu^{-\beta}\chi)}\rangle \\ \times_{(l',\chi',\alpha') \in \mathbf{e} - \{(l,\chi,\alpha),\ (l,\chi,\beta)\}} \ \langle[-\frac{l'-1}{2}, \frac{l'-1}{2}]^{(\nu^{\alpha'}\chi')}\rangle \rtimes \sigma_{neg}. \end{cases}$$

Therefore they are unitarizable by (UR). Now, Theorem 4-1 implies $\beta \in]0, \frac{1}{2}[$. Since $0 < \alpha \leq \beta$. We conclude $\alpha \in]0, \frac{1}{2}[$. Now, since $\alpha, \beta \in]0, \frac{1}{2}[$ and the other representation in 9-12 is unitarizable, we conclude by induction that $(\mathbf{e}, \sigma_{neg}) \in \mathcal{M}^{u,unr}(S_n)$. (Lemma 6-1 needs to be applied for the irreducibility conditions.) □

LEMMA 9-13. *Assume that there exist* $\chi \in \{\mathbf{1}_{F^\times}, \mathbf{sgn}_u\}$ *and* $l \in \mathbb{Z}_{\geq 1}$ *such that* $\mathbf{e}(l,\chi)$ *contains* $\alpha \in]0, \frac{1}{2}[$ *satisfying the following:*

there is no $\beta \in]0, \alpha[$ *such that* $[-\frac{l-1}{2}, \frac{l-1}{2}]^{(\nu^{\pm\beta}\chi)}$ *is linked with a segment* $[-\frac{l'-1}{2}, \frac{l'-1}{2}]^{(\nu^{\alpha'}\chi')}$ *for* $(l', \chi', \alpha') \in \mathbf{e}$.

If $\chi' \neq \chi$ *or* $l' \neq l$, *then* $\mathbf{e}(l', \chi')$ *satisfies Definition 5-13 (2) and (3).*

PROOF. We may assume that σ is in general position. We consider the following family of induced representations:

$$(9\text{-}14)\quad \langle [-\frac{l-1}{2}, \frac{l-1}{2}]^{(\nu^\beta\chi)}\rangle \times_{(l',\chi',\alpha')\in \mathbf{e}-\{(l,\chi,\alpha)\}} \langle [-\frac{l'-1}{2}, \frac{l'-1}{2}]^{(\nu^{\alpha'}\chi')}\rangle \rtimes \sigma_{neg},$$

where $\beta \in]0, \alpha]$. As in the proof of Lemma 9-10, we conclude the irreducibility of the induced representation given by (9-14).

Next, since we assume $\chi' \in \{\mathbf{1}_{F^\times}, \mathbf{sgn}_u\}$ when $e(l', \chi') \neq \emptyset$ for some l', the family of representations (9-14) is Hermitian (see Lemma 6-4). Finally, it is unitarizable for $\beta = \alpha$ (since it is isomorphic to σ), and therefore for all $\beta \in]0, \alpha]$ (see (D)). Applying (ED), we conclude that all irreducible subquotients of

$$(9\text{-}15)\quad \langle [-\frac{l-1}{2}, \frac{l-1}{2}]^{(\chi)}\rangle \times_{(l',\chi',\alpha')\in \mathbf{e}-\{(l,\chi,\alpha)\}} \langle [-\frac{l'-1}{2}, \frac{l'-1}{2}]^{(\nu^{\alpha'}\chi')}\rangle \rtimes \sigma_{neg}$$

are unitarizable. In particular, its unique irreducible unramified subquotient is unitarizable. We determine this subquotient. First, let σ'_{neg} be the unique irreducible unramified subquotient of $\langle [-\frac{l-1}{2}, \frac{l-1}{2}]^{(\chi)}\rangle \rtimes \sigma_{neg}$. Since σ_{neg} is unitarizable (see Theorem 5-11), we see that

$$(9\text{-}16)\qquad \sigma'_{neg} \hookrightarrow \langle [-\frac{l-1}{2}, \frac{l-1}{2}]^{(\chi)}\rangle \rtimes \sigma_{neg}.$$

Now, the classification of negative representations (see Theorem 5-10) implies that σ'_{neg} is negative. and

$$(9\text{-}17)\qquad \mathrm{Jord}(\sigma'_{neg}) = \mathrm{Jord}(\sigma_{neg}) + \{2 \cdot (l, \chi)\}.$$

Next, since σ is in general position, we easily see that the induced representation

$$\times_{(l',\chi',\alpha')\in \mathbf{e}-\{(l,\chi,\alpha)\}} \langle [-\frac{l'-1}{2}, \frac{l'-1}{2}]^{(\nu^{\alpha'}\chi')}\rangle \rtimes \sigma'_{neg}$$

is irreducible; it is the unique irreducible unramified subquotient of (9-15). Since it is unitarizable and $\mathrm{card}(\mathbf{e} - \{(l, \chi, \alpha)\}) < \mathrm{card}\, \mathbf{e}$, by induction we conclude that

$$(9\text{-}18)\qquad (\mathbf{e} - \{(l, \chi, \alpha)\}, \sigma'_{neg}) \in \mathcal{M}^{u,unr}(S_n)$$

Since $\chi' \neq \chi$ or $l' \neq l$, we see $\mathbf{e}(l', \chi') \subset \mathbf{e} - \{(l, \chi, \alpha)\}$. Thus, (9-18) implies that $\mathbf{e}(l', \chi')$ satisfies Definition 5-13 (2) and (3). □

We record the following corollary to the proof of Lemma 9-13:

COROLLARY 9-19. *Assume that there exist* $\chi \in \{\mathbf{1}_{F^\times}, \mathbf{sgn}_u\}$ *and* $l \in \mathbb{Z}_{\geq 1}$ *such that* $\mathbf{e}(l, \chi)$ *contains* $\alpha \in]0, \frac{1}{2}[$ *satisfying the following:*

there is no $\beta \in]0, \alpha[$ *such that* $[-\frac{l-1}{2}, \frac{l-1}{2}]^{(\nu^{\pm\beta}\chi)}$ *is linked with a segment* $[-\frac{l'-1}{2}, \frac{l'-1}{2}]^{(\nu^{\alpha'}\chi')}$ *for* $(l', \chi', \alpha') \in \mathbf{e}$.

We define σ'_{neg} *using (9-16) (or, equivalently, using (9-17)). Then*

$$(\mathbf{e} - \{(l,\chi,\alpha)\}, \sigma'_{neg}) \in \mathcal{M}^{u,unr}(S_n).$$

PROOF. The claim follows from (9-18). □

Similarly we prove the following result:

LEMMA 9-20. *Assume that there exist* $\chi \in \{\mathbf{1}_{F^\times}, \mathbf{sgn}_u\}$ *and* $l \in \mathbb{Z}_{\geq 1}$ *such that* $l - (2\alpha_\chi + 1) \notin 2\mathbb{Z}$ *and* $\mathbf{e}(l,\chi)$ *contains* $\alpha \in]0, \frac{1}{2}[$ *satisfying the following:*

> *there is no* $\beta \in]\alpha, \frac{1}{2}[$ *such that* $[-\frac{l-1}{2}, \frac{l-1}{2}]^{(\nu^{\pm\beta}\chi)}$ *is linked with a segment* $[-\frac{l'-1}{2}, \frac{l'-1}{2}]^{(\nu^{\alpha'}\chi')}$ *for* $(l',\chi',\alpha') \in \mathbf{e}$.

We define σ'_{neg} *using Lemma 6-6. Then* $(\mathbf{e} - \{(l,\chi,\alpha)\}, \sigma'_{neg}) \in \mathcal{M}^{u,unr}(S_n)$.

The next step is a technical result used several times in the proof below.

LEMMA 9-21. *Assume that* σ *is in general position. Then, for every submultiset* $\mathbf{e}_0$ *of* $\mathbf{e}$, *there exists a multiset* $\mathbf{e}_1$, *consisting of triples of the form* (l,χ,α) $(\chi \in \{\mathbf{1}_{F^\times}, \mathbf{sgn}_u\}$, $l \in \mathbb{Z}_{\geq 1}$, $\alpha \in]0, \frac{1}{2}[)$, *such that the induced representation* $\times_{(l,\chi,\alpha)\in \mathbf{e}_0+\mathbf{e}_1} \langle[-\frac{l-1}{2}, \frac{l-1}{2}]^{(\nu^\alpha\chi)}\rangle \rtimes \sigma_{neg}$ *is an irreducible unitarizable representation of* S_n. *Moreover, we may choose* $\mathbf{e}_1$ *such that* card$(\mathbf{e} - \mathbf{e}_0) \leq$ card $\mathbf{e}_1$, *and if all* $(l,\chi,\alpha) \in \mathbf{e} - \mathbf{e}_0$ *have the same* $\chi = \chi_0$, *then all* $(l,\chi,\alpha) \in \mathbf{e}_1$ *satisfy* $\chi = \chi_0$.

PROOF. If $\mathbf{e}_0 = \mathbf{e}$, then there is nothing to be proved; we may take $\mathbf{e}_1 = \emptyset$. Therefore, we may assume $\mathbf{e}_0 \neq \mathbf{e}$. If for all $(l,\chi,\alpha) \in \mathbf{e} - \mathbf{e}_0$ we have $\alpha \in]0, \frac{1}{2}[$, we are done; we may take $\mathbf{e}_1 = \mathbf{e} - \mathbf{e}_0$. Therefore, let $(l,\chi,\alpha) \in \mathbf{e} - \mathbf{e}_0$ such that $\alpha \geq \frac{1}{2}$. Since σ is in general position, we must have $\alpha \notin \frac{1}{2}\mathbb{Z}$. Then there is a unique $k \in \mathbb{Z}_{\geq 1}$ such that $\alpha \in]\frac{k}{2}, \frac{k+1}{2}[$. Then $\frac{k+1}{2} - \alpha \in]0, \frac{1}{2}[$, and the following induced representation is in a GL–complementary series (see Theorem 4-1):

(9-22)
$$\pi = \langle[-\frac{l+k-2}{2}, \frac{l+k-2}{2}]^{(\nu^{-\alpha+\frac{k+1}{2}}\chi)}\rangle \times \langle[-\frac{l+k-2}{2}, \frac{l+k-2}{2}]^{(\nu^{\alpha-\frac{k+1}{2}}\chi)}\rangle.$$

Therefore, the representation $\pi \rtimes \sigma$ is unitarizable, but reducible. We determine its unique irreducible unramified subquotient. Since

$$\begin{aligned}\langle[-\frac{l+k-2}{2}, \frac{l+k-2}{2}]^{(\nu^{\alpha-\frac{k+1}{2}}\chi)}\rangle \times \langle[-\frac{l-1}{2}, \frac{l-1}{2}]^{(\nu^\alpha\chi)}\rangle \\ = \langle[\alpha - \frac{l-1}{2} - k, \alpha + \frac{l-1}{2} - 1]^{(\chi)}\rangle \times \langle[\alpha - \frac{l-1}{2}, \alpha + \frac{l-1}{2}]^{(\chi)}\rangle,\end{aligned}$$

and $\alpha \notin \frac{1}{2}\mathbb{Z}$, Zelevinsky theory implies that the unique irreducible unramified subquotient of

$$\begin{aligned}\langle[-\frac{l+k-2}{2}, \frac{l+k-2}{2}]^{(\nu^{-\alpha+\frac{k+1}{2}}\chi)}\rangle \times \langle[-\frac{l+k-2}{2}, \frac{l+k-2}{2}]^{(\nu^{\alpha-\frac{k+1}{2}}\chi)}\rangle \times \\ \langle[-\frac{l-1}{2}, \frac{l-1}{2}]^{(\nu^\alpha\chi)}\rangle\end{aligned}$$

is exactly

$$\langle[-\frac{l+k-2}{2}, \frac{l+k-2}{2}]^{(\nu^{-\alpha+\frac{k+1}{2}}\chi)}\rangle \times \langle[\alpha - \frac{l-1}{2} - k, \alpha + \frac{l-1}{2}]^{(\chi)}\rangle \times \langle[\alpha - \frac{l-1}{2}, \alpha + \frac{l-1}{2} - 1]^{(\chi)}\rangle,$$

or written differently,

$$\langle[-\frac{l+k-2}{2}, \frac{l+k-2}{2}]^{(\nu^{-\alpha+\frac{k+1}{2}}\chi)}\rangle \times \langle[-\frac{l+k-1}{2}, \frac{l+k-1}{2}]^{(\nu^{\alpha-\frac{k}{2}}\chi)}\rangle \times \langle[\frac{-(l-2)}{2}, \frac{(l-2)}{2}]^{(\nu^{\alpha-\frac{1}{2}}\chi)}\rangle.$$

(We remark that the segment $[\frac{-(l-2)}{2}, \frac{(l-2)}{2}]^{(\nu^{\alpha-\frac{1}{2}}\chi)}$ is empty if $l=1$, and should be omitted.) Since σ is in general position, and the segments

$$\begin{cases} [-\frac{l+k-2}{2}, \frac{l+k-2}{2}]^{(\nu^{\alpha-\frac{k+1}{2}}\chi)}, \ [-\frac{l+k-1}{2}, \frac{l+k-1}{2}]^{(\nu^{\alpha-\frac{k}{2}}\chi)} \\ [-\frac{l+k-2}{2}, \frac{l+k-2}{2}]^{(\nu^{\alpha-\frac{k+1}{2}}\chi)}, \ [\frac{-(l-2)}{2}, \frac{(l-2)}{2}]^{(\nu^{\alpha-\frac{1}{2}}\chi)} \end{cases}$$

are not linked, Lemma 6-1 implies that the unique irreducible unramified subquotient of $\pi \rtimes \sigma$ is

$$\langle[-\frac{l+k-2}{2}, \frac{l+k-2}{2}]^{(\nu^{-\alpha+\frac{k+1}{2}}\chi)}\rangle \times \langle[-\frac{l+k-1}{2}, \frac{l+k-1}{2}]^{(\nu^{\alpha-\frac{k}{2}}\chi)}\rangle \times \langle[\frac{-(l-2)}{2}, \frac{(l-2)}{2}]^{(\nu^{\alpha-\frac{1}{2}}\chi)}\rangle \times_{(l',\chi',\alpha')\in \mathbf{e}-\{(l,\chi,\alpha)\}} \langle[-\frac{l'-1}{2}, \frac{l'-1}{2}]^{(\nu^{\alpha'}\chi')}\rangle \rtimes \sigma_{neg}. \tag{9-23}$$

Now, since $-\alpha + \frac{k+1}{2}, \alpha - \frac{k}{2} \in]0, \frac{1}{2}[$, and $\alpha - \frac{1}{2} < \alpha$, we may iterate this procedure until we obtain what we want. □

Next, we prove the following lemma:

LEMMA 9-24. *Assume that σ is in general position. Assume that there exist $l_1, k_1 \in \mathbb{Z}_{\geq 1}$, such that $\mathbf{e}(l_1, \mathbf{1}_{F^\times}) \neq \emptyset$ and $\mathbf{e}(k_1, \mathbf{sgn}_u) \neq \emptyset$. Then $(\mathbf{e}, \sigma_{neg}) \in \mathcal{M}^{u,unr}(S_n)$.*

PROOF. Let $\chi_0 \in \{\mathbf{1}_{F^\times}, \mathbf{sgn}_u\}$ and $l_0 \in \mathbb{Z}_{\geq 1}$. We need to show that the exponents from $\mathbf{e}(l_0, \chi_0)$ satisfy Definition 5-13 (2) or (3). By the assumption of the lemma, we may find $\chi' \in \{\mathbf{1}_{F^\times}, \mathbf{sgn}_u\}$ and $l' \in \mathbb{Z}_{\geq 1}$ such that $\chi' \neq \chi_0$ and $\mathbf{e}(l', \chi') \neq \emptyset$. Letting $\mathbf{e}_0 = \sum_{l'_0} \mathbf{e}(l'_0, \chi_0)$ in Lemma 9-21, we may assume that $\alpha < 1/2$ for all $(l, \chi, \alpha) \in \mathbf{e} - \mathbf{e}_0$. Now, we take some (l, χ), $\chi \neq \chi_0$, such that $\mathbf{e}(l, \chi) \neq \emptyset$. If card $\mathbf{e}(l, \chi) > 1$, then we apply Lemma 9-10 to complete the proof of the lemma. Otherwise, we use Lemma 9-13. □

In the remainder of the proof we assume that there is a unique $\chi \in \{\mathbf{1}_{F^\times}, \mathbf{sgn}_u\}$ such that if $\mathbf{e}(l', \chi') \neq \emptyset$, then $\chi' = \chi$. We prove the following lemma:

LEMMA 9-25. *Assume that σ is in general position. Assume that $\mathbf{e}(1, \chi) \neq \emptyset$. Then, if $\alpha > 1$, for some $\alpha \in \mathbf{e}(1, \chi)$, then*

$$k - (2\alpha_\chi + 1) \in 2\mathbb{Z}, \tag{9-26}$$

where $k \in \mathbb{Z}_{\geq 2}$ is defined by $\alpha \in]\frac{k}{2}, \frac{k+1}{2}[$.

PROOF. Applying Lemma 9-21, we may assume that every $(l', \chi, \alpha') \in \mathbf{e} - \{(1, \chi, \alpha)\}$ satisfies $\alpha' \in]0, \frac{1}{2}[$. We let $(l', \chi, \alpha') \in \mathbf{e} - \{(1, \chi, \alpha)\}$. Then the segment $[-\frac{l'-1}{2}, \frac{l'-1}{2}]^{(\nu^{\alpha'}\chi)}$ is linked with the segment $\{\nu^\alpha\chi\}$ if and only if $\alpha = \frac{l'-1}{2} + \alpha' + 1$. Since $\alpha \in]\frac{k}{2}, \frac{k+1}{2}[$ and $\alpha' \in]0, \frac{1}{2}[$, we see that this is equivalent to $l' = k-1$ and $\alpha' = \alpha - k/2$. Similarly, $[-\frac{l'-1}{2}, \frac{l'-1}{2}]^{(\nu^{-\alpha'}\chi)}$ is linked with the segment $\{\nu^\alpha\chi\}$ if and only if $l' = k$ and $\alpha' = \frac{k+1}{2} - \alpha$. Therefore, since the induced representation (9-2) is irreducible, we see that for $(l', \chi, \alpha') \in \mathbf{e} - \{(1, \chi, \alpha)\}$ we must have

$$(l', \alpha') \neq (k-1, \alpha - \frac{k}{2}),\ (k, \frac{k+1}{2} - \alpha).$$

We remark that the segments $[-\frac{l'-1}{2}, \frac{l'-1}{2}]^{(\nu^{x'}\chi)}$ and $[-\frac{l''-1}{2}, \frac{l''-1}{2}]^{(\nu^{\pm x''}\chi)}$, where $x', x'' \in]0, \frac{1}{2}[$, are never linked.

Those observations enable us to assume that there are no triples $(l', \chi, \alpha') \in \mathbf{e} - \{(1, \chi, \alpha)\}$ such that one of the following holds:

$$\begin{cases} l' \notin \{k-1, k\} \\ l' = k-1 \text{ and } \alpha' < \alpha - \frac{k}{2} \\ l' = k \text{ and } \alpha' < \frac{k+1}{2} - \alpha \end{cases}$$

applying Corollary 9-19 several times.

Next, applying Lemma 9-10, we may assume that $\mathbf{e} - \{(1, \chi, \alpha)\}$ contains at most two elements (each with multiplicity at most one) which are necessarily of the form

$$\begin{cases} (k-1, \chi, \beta), \text{ where } \beta \in]\alpha - \frac{k}{2},\ \frac{1}{2}[; \\ (k, \chi, \gamma), \text{ where } \gamma \in]\frac{k+1}{2} - \alpha,\ \frac{1}{2}[. \end{cases}$$

Thus, we may assume the following:

$$\mathbf{e} = \{(1, \chi, \alpha),\ n_\beta \cdot (k-1, \chi, \beta),\ n_\gamma \cdot (k, \chi, \gamma)\}. \tag{9-27}$$

(Here $n_\beta, n_\gamma \in \{0, 1\}$ are the multiplicities.)

Now, proceed as follows: We use the complementary series ($l = 1$ in our case)

$$\pi = \langle[-\frac{l+k-3}{2}, \frac{l+k-3}{2}]^{(\nu^{-\alpha+\frac{k}{2}}\chi)}\rangle \times \langle[-\frac{l+k-3}{2}, \frac{l+k-3}{2}]^{(\nu^{\alpha-\frac{k}{2}}\chi)}\rangle, \tag{9-28}$$

and repeat the steps of the proof of Lemma 9-21 from the point (9-22) up to (9-23) where instead of (9-23) we obtain a new irreducible unitarizable unramified representation σ' which is isomorphic to ($l = 1$ in our case)

$$\begin{aligned}&\langle[-\frac{l+k-3}{2}, \frac{l+k-3}{2}]^{(\nu^{\alpha-\frac{k}{2}}\chi)}\rangle \times \langle[-\frac{l+k-2}{2}, \frac{l+k-2}{2}]^{(\nu^{\alpha-\frac{k-1}{2}}\chi)}\rangle \times \\ &\langle[\frac{-(l-2)}{2}, \frac{(l-2)}{2}]^{(\nu^{\alpha-\frac{1}{2}}\chi)}\rangle \times_{(l',\chi',\alpha')\in\mathbf{e}-\{(l,\chi,\alpha)\}} \langle[-\frac{l'-1}{2}, \frac{l'-1}{2}]^{(\nu^{\alpha'}\chi')}\rangle \rtimes \sigma_{neg}.\end{aligned} \tag{9-29}$$

Thus, σ' is attached to $(\mathbf{e}', \sigma_{neg})$ where

$$\mathbf{e}' = \{(k-1, \chi, \alpha - \frac{k}{2}),\ (k, \chi, \alpha - \frac{k-1}{2}),\ n_\beta \cdot (k-1, \chi, \beta),\ n_\gamma \cdot (k, \chi, \gamma)\}.$$

We remark that $\alpha - \frac{k-1}{2} \in]\frac{1}{2}, 1[$.

Now, using Corollary 9-19 and Lemma 6-9 (i), (ii), we obtain a new unitarizable unramified representation σ'' attached to $(\mathbf{e}'', \sigma'_{neg})$ where

$$\mathbf{e}'' = \{(k, \chi, \alpha - \frac{k-1}{2}),\ n_\beta \cdot (k-1, \chi, \beta),\ n_\gamma \cdot (k, \chi, \gamma)\}$$

and

$$\mathrm{Jord}(\sigma'_{neg}) = \mathrm{Jord}(\sigma_{neg}) + 2 \cdot \{(k-1, \chi)\}.$$

Therefore, either by the inductive assumption (that is, in the case $n_\beta > 0$ or $n_\gamma > 0$) or by Theorem 7-1 (if $n_\beta = n_\gamma = 0$) we obtain (9-26). □

LEMMA 9-30. *Assume that σ is in general position. Assume that $\mathbf{e}(1, \chi) \neq \emptyset$. Then $\alpha < 3/2$ for $\alpha \in \mathbf{e}(1, \chi)$.*

PROOF. Assume to the contrary that there exists $\alpha \in \mathbf{e}(1, \chi)$ such that $\alpha \geq \frac{3}{2}$. Then since σ is in general position, there exists $k \in \mathbb{Z}_{\geq 3}$ such that $\alpha \in]\frac{k}{2}, \frac{k+1}{2}[$. Then Lemma 9-25 implies that

$$k - (2\alpha_\chi + 1) \in 2\mathbb{Z}. \tag{9-31}$$

Next, applying Lemma 9-21, we may assume that every $(l', \chi, \alpha') \in \mathbf{e} - \{(1, \chi, \alpha)\}$ satisfies $\alpha' \in]0, \frac{1}{2}[$. Equipped with this, we may assume the reduction (9-27).

We would like to "move" α from $]\frac{k}{2}, \frac{k+1}{2}[$ into $]\frac{k-1}{2}, \frac{k}{2}[$. We have two cases. First, we assume that there exists $\epsilon > 0$ such that the induced representation

$$\nu^x\chi \times \langle[-\frac{k-2}{2}, \frac{k-2}{2}]^{(\nu^y\chi)}\rangle \times \langle[-\frac{k-1}{2}, \frac{k-1}{2}]^{(\nu^z\chi)}\rangle \rtimes \sigma_{neg}$$

is irreducible and unitarizable for $x \in [\frac{k}{2} - \epsilon, \alpha]$, $y \in]\alpha - \frac{k}{2}, \frac{1}{2} + \epsilon[$, and $z \in]\frac{k+1}{2} - \alpha, \frac{1}{2} + \epsilon[$. We explain this assumption. The irreducibility is an easy consequence of Lemma 6-9, and the unitarity follows from (D) since at $(x, y, z) = (\alpha, \beta, \gamma)$ is unitarizable, except that reducibility might occur for $x = \frac{k}{2}$ (y, z are arbitrary). Now, the induced representation $\nu^\alpha\chi \times \langle[-\frac{k-2}{2}, \frac{k-2}{2}]^{(\nu^\beta\chi)}\rangle \times \langle[-\frac{k-1}{2}, \frac{k-1}{2}]^{(\nu^\gamma\chi)}\rangle \rtimes \sigma_{neg}$ is irreducible and unitarizable for some $\alpha \in]\frac{k-1}{2}, \frac{k}{2}[$. Since $k - 1 \geq 2$, then Lemma 9-25 shows that $k - 1 - (2\alpha_\chi + 1) \in 2\mathbb{Z}$. This is a contradiction since we already have $k - (2\alpha_\chi + 1) \in 2\mathbb{Z}$. (See (9-31).)

If we have reducibility at $x = \frac{k}{2}$ for some y and z, then Lemma 6-1 and $k - (2\alpha_\chi + 1) \in 2\mathbb{Z}$ would imply that $(k-1, \chi) \in \mathrm{Jord}(\sigma_{neg})$. Then, applying Lemma 6-5, since $k - 1 - (2\alpha_\chi + 1) \notin 2\mathbb{Z}$, $(k-1, \chi)$ appears at least twice in $\mathrm{Jord}(\sigma_{neg})$, and there exists a negative representation σ'_{neg} such that

$$\sigma_{neg} \simeq \langle[-\frac{k-2}{2}, \frac{k-2}{2}]^{(\chi)}\rangle \rtimes \sigma'_{neg}.$$

Then σ is of the form

$$\begin{aligned}\nu^\alpha\chi \times \langle[-\frac{k-2}{2}, \frac{k-2}{2}]^{(\nu^\beta\chi)}\rangle \times \langle[-\frac{k-1}{2}, \frac{k-1}{2}]^{(\nu^\gamma\chi)}\rangle \rtimes \sigma_{neg} \simeq \\ \simeq \langle[-\frac{k-2}{2}, \frac{k-2}{2}]^{(\chi)}\rangle \times \\ \left(\nu^\alpha\chi \times \langle[-\frac{k-2}{2}, \frac{k-2}{2}]^{(\nu^\beta\chi)}\rangle \times \langle[-\frac{k-1}{2}, \frac{k-1}{2}]^{(\nu^\gamma\chi)}\rangle \rtimes \sigma'_{neg}\right)\end{aligned}$$

Now, applying (UR) we obtain the unitarity of

$$\nu^{\alpha}\chi \times \langle[-\frac{k-2}{2}, \frac{k-2}{2}]^{(\nu^{\beta}\chi)}\rangle \times \langle[-\frac{k-1}{2}, \frac{k-1}{2}]^{(\nu^{\gamma}\chi)}\rangle \rtimes \sigma'_{neg}$$

and this contradicts the general inductive assumption. □

LEMMA 9-32. *Let $l \in \mathbb{Z}_{\geq 1}$ such that $\mathbf{e}(l,\chi) \neq \emptyset$. Then, if $l-(2\alpha_\chi+1) \notin 2\mathbb{Z}$ (resp., $l-(2\alpha_\chi+1) \in 2\mathbb{Z}$), then $\alpha < 1/2$ (resp., $\alpha < 1$) for $\alpha \in \mathbf{e}(l,\chi)$.*

PROOF. We may assume that σ is in general position. (See Remark 9-4.) Assume that the claim is not true for $\alpha \in \mathbf{e}(l,\chi)$. Applying Lemma 9-21, we may assume that every $(l',\chi,\alpha') \in \mathbf{e} - \{(l,\chi,\alpha)\}$ satisfies $\alpha' \in]0, \frac{1}{2}[$.

Now, we proceed as in the proof of Lemma 9-21 (that is, we imitate that proof "multiplying" σ by π given by (9-22) and repeating the steps done there to obtain (9-23)), keep replacing (l,χ,α) by $(l-1,\chi,\alpha-1/2)$ while $l \geq 2$. (This keeps all other exponents in $]0, \frac{1}{2}[$.) As a result, we may assume that one of the following holds:

(a) $l-(2\alpha_\chi+1) \notin 2\mathbb{Z}$ and $\alpha \in]\frac{1}{2}, 1[$
(b) $l-(2\alpha_\chi+1) \notin 2\mathbb{Z}$ and $\alpha \in]1, \frac{3}{2}[$
(c) $l=1$ and there exists $k \in \mathbb{Z}_{\geq 2}$ such that $\alpha \in]\frac{k}{2}, \frac{k+1}{2}[$.

Next, Lemma 9-30 implies that $k=2$ in (c). Now, we reduce that case to the previous two. Since $k-(2\alpha_\chi+1) \in 2\mathbb{Z}$ for $k=2$ (see (9-26)), we see that $l-(2\alpha_\chi+1) \notin 2\mathbb{Z}$ for $l=1$. Hence this is just the case (b). It remains to consider the cases (a) and (b). The case (a) is easy. We apply Corollary 9-19 several times in order to reduce to the case $\mathbf{e} = \{(l,\chi,\alpha)\}$. Then we apply Theorem 7-1 to obtain a contradiction. We consider the case (b). Arguing as in the proof of Lemma 9-25, we may assume the following:

$$\mathbf{e} = \{(l,\chi,\alpha),\ n_\beta \cdot (l-1,\chi,\beta),\ n_\gamma \cdot (l+1,\chi,\gamma)\}.$$

where $\beta \in]\alpha-1, \frac{1}{2}[$ and $\gamma \in]\frac{3}{2}-\alpha, \frac{1}{2}[$. (Here $n_\beta, n_\gamma \in \{0,1\}$ are the multiplicities.)

Now, we "move" α into $]\frac{1}{2}, 1[$ arguing as in the last part of the proof of Lemma 9-30 reducing (b) to (a). In more detail, there exists $\epsilon > 0$ such that the induced representation

$$\langle[-\frac{l-1}{2}, \frac{l-1}{2}]^{(\nu^{x}\chi)}\rangle \times \langle[-\frac{l-2}{2}, \frac{l-2}{2}]^{(\nu^{y}\chi)}\rangle \times \langle[-\frac{l}{2}, \frac{l}{2}]^{(\nu^{z}\chi)}\rangle \rtimes \sigma_{neg}$$

is irreducible and unitarizable for $x \in [1-\epsilon, \alpha]$, $y \in]\alpha-1, \frac{1}{2}+\epsilon[$, and $z \in]\frac{3}{2}-\alpha, \frac{1}{2}+\epsilon[$. Hence $\langle[-\frac{l-1}{2}, \frac{l-1}{2}]^{(\nu^{\alpha}\chi)}\rangle \times \langle[-\frac{l-2}{2}, \frac{l-2}{2}]^{(\nu^{\beta}\chi)}\rangle \times \langle[-\frac{l}{2}, \frac{l}{2}]^{(\nu^{\gamma}\chi)}\rangle \rtimes \sigma_{neg}$ is unitarizable for some (new) $\alpha \in]\frac{1}{2}, 1[$ where β and γ are less than but close to $\frac{1}{2}$. We are now in case (a). □

Let us summarize what we have done so far. We have reduced the proof that $(\mathbf{e}, \sigma_{neg}) \in \mathcal{M}^{unr}(S_n)$ attached to $\sigma \in \mathrm{Irr}^{u,unr}(S_n)$ (see (9-1)) satisfies Definition 5-13 to the following. We may assume that there is a unique $\chi \in \{\mathbf{1}_{F^\times}, \mathbf{sgn}_u\}$ such that if $\mathbf{e}(l',\chi') \neq \emptyset$, then $\chi' = \chi$. Looking at Definition 5-13 and since Lemma 9-32 holds, we may apply Lemma 6-9 and (several times) Corollary 9-19 to assume that there is also a unique l such that if $\mathbf{e}(l',\chi) \neq \emptyset$, then $l = l'$. Now, if $l-(2\alpha_\chi+1) \notin 2\mathbb{Z}$, we see that $(\mathbf{e}, \sigma_{neg}) \in \mathcal{M}^{unr}(S_n)$ satisfies Definition 5-13 (applying Lemma 9-32), and Theorem 9-3 is proved. Thus, it remains to consider the case $l-(2\alpha_\chi+1) \in 2\mathbb{Z}$. We need to check Definition 5-13 (3). First, Lemma 9-32 implies $0<\alpha<1$ for all $\alpha \in \mathbf{e}(l,\chi)$. Then we write this multiset as in Definition 5-13 (3). Hence (9-2) can be written as follows:

$$\text{(9-33)} \qquad \sigma \simeq \times_{i=1}^{u} \langle [-\frac{l-1}{2}, \frac{l-1}{2}]^{(\nu^{\alpha_i}\chi)} \times_{j=1}^{v} \langle [-\frac{l-1}{2}, \frac{l-1}{2}]^{(\nu^{\beta_j}\chi)} \rangle \rtimes \sigma_{neg}.$$

Now, we check (a)–(f) in Definition 5-13 (3). First, since the induced representation (9-33) is irreducible, (b) and (d) hold. (See Lemma 6-9.) Next, Lemma 9-10 implies (c). It remains to prove (a), (e), and (f).

Now, if there exist indices $i_1 \neq i_2$ such that $\alpha_{i_1}, \alpha_{i_1} \in]1-\beta_1, \frac{1}{2}]$ or $\alpha_{i_1}, \alpha_{i_1} \in]1-\beta_{j+1}, 1-\beta_j[$ for some j, then we apply Lemma 9-10 and we are done. Otherwise, we may assume that the number of indices i such that $\alpha_i \in]1-\beta_1, \frac{1}{2}]$ (resp., $\alpha_i \in]1-\beta_{j+1}, 1-\beta_j[$ for some fixed j) is 0 or 1. We must show that this number is one for $]1-\beta_{j+1}, 1-\beta_j[$ for every j, and zero for $]1-\beta_1, \frac{1}{2}]$.

If we can find j such that the number is zero for $]1-\beta_{j+1}, 1-\beta_j[$, then we apply Lemma 9-10 to deform β_j into β_{j+1}, and obtain $\beta_j, \beta_{j+1} < \frac{1}{2}$, which is a contradiction. If on the other hand $]1-\beta_1, \frac{1}{2}]$ contains a unique α_i, then we may deform it to β_1 and Lemma 9-10 would imply $\beta_1 < \frac{1}{2}$, which is a contradiction. This proves (e) and (f).

In the same reduction (that is, no α_i's in $]1-\beta_{j+1}, 1-\beta_j[$ for every j, and there is a unique α_i in $]1-\beta_1, \frac{1}{2}]$) we prove (a).

If $v>0$, we may also assume that the number of indices i such that $\alpha_i \in]0,\ 1-\beta_v[$ is either 0 or 1. We must show if $(l,\chi) \notin \mathrm{Jord}(\sigma_{neg})$, then $u+v$ is even. We accomplish this as follows.

First, if $v=0$ (that is, no β_i's), then then there is no i such that $\alpha_i \in]0,\ 1-\beta_v[$ and the claim follows from from Theorem 7-1 ($u=v=0$ here). Next, we assume $v \geq 1$.

We reduce this case to the case $v=0$ as follows. We apply the complementary series (9-22) with $\alpha=\beta_1$ and $k=1$. We obtain a new unitary representation σ_1 attached to $(\mathbf{e}_1, \sigma_{neg})$, where

$$\mathbf{e}_1 = \mathbf{e} - \{(l,\chi,\beta_1)\} + \{(l,\chi,1-\beta_1), (l+1,\chi,\beta_1-\frac{1}{2}), (l-1,\chi,\beta_1-\frac{1}{2})\}.$$

We apply Lemma 6-9 and Corollary 9-19 to obtain a new unitary representation σ' attached to $(\mathbf{e}', \sigma'_{neg})$, where

$$\begin{cases} \mathbf{e}' = \mathbf{e}_1 - \{(l+1,\chi,\beta_1-\frac{1}{2}), (l-1,\chi,\beta_1-\frac{1}{2})\} \\ \mathrm{Jord}(\sigma'_{neg}) = \mathrm{Jord}(\sigma_{neg}) + \{(l-1,\chi), (l+1,\chi\}. \end{cases}$$

Clearly, $(l,\chi) \notin \mathrm{Jord}(\sigma'_{neg})$. Then by induction, we have $u'+v'$ is even. Since $v'=v-1$ and $u'=u-1$, we obtain the claim. This proves that all conditions (a)–(f) hold. This completes the proof of Theorem 5-14, and therefore the proof of the surjectivity of the map in Theorem 5-14.

10. Functoriality, Satake Parameters and an Algorithm for Testing Unitarity

In this section we present an algorithm that describes an effective and easy way of testing unitarity of an unramified representation given by its Satake parameter (see Theorem 1-2). We introduce the Langlands dual groups as follows:

$$G = S_n = \mathrm{SO}(2n+1,F) \quad \hat{G}(\mathbb{C}) = \mathrm{Sp}(2n,\mathbb{C}) \subset \mathrm{GL}(N,\mathbb{C}); N = 2n$$
$$G = S_n = \mathrm{O}(2n,F) \quad \hat{G}(\mathbb{C}) = \mathrm{O}(2n,\mathbb{C}) \subset \mathrm{GL}(N,\mathbb{C}); N = 2n$$
$$G = S_n = \mathrm{Sp}(2n,F) \quad \hat{G}(\mathbb{C}) = \mathrm{SO}(2n+1,\mathbb{C}) \subset \mathrm{GL}(N,\mathbb{C}); N = 2n+1.$$

Assume that $(\chi_1, \dots, \chi_n)$ is a sequence of unramified characters of $F^\times$. Then the induced representation

$$\chi_1 \times \cdots \times \chi_n \rtimes \mathbf{1} = \mathrm{Ind}_{P_{\min}}^G (\chi_1 \otimes \cdots \otimes \chi_n)$$

contains the unique unramified irreducible subquotient,

$$\sigma^G := \sigma^G(\chi_1, \dots, \chi_n).$$

(See Theorem 1-2.) Its Langlands lift to $\mathrm{GL}(N,F)$ is an unramified representation given by

$$\sigma^{\mathrm{GL}(N,F)} := \sigma^{\mathrm{GL}(N,F)}(\chi_1, \dots, \chi_n, \chi_1^{-1}, \dots, \chi_n^{-1}, \star), \tag{10-1}$$

where

$$\star = \begin{cases} \text{should be omitted if } G = \mathrm{O}(2n,F),\ \mathrm{SO}(2n+1,F) \\ \mathbf{1}_{F^\times} \text{ if } G = \mathrm{Sp}(2n,F). \end{cases}$$

In particular, the lift is an irreducible subquotient of

$$\chi_1 \times \cdots \times \chi_n \times \chi_1^{-1} \times \cdots \times \chi_n^{-1} \times \star. \tag{10-2}$$

Obviously, we have the following:

$$\sigma^{\mathrm{GL}(N,F)} \text{ is self–dual: } \widetilde{\sigma}^{\mathrm{GL}(N,F)} \simeq \sigma^{\mathrm{GL}(N,F)}, \tag{10-3}$$

and

$$\sigma^{\mathrm{GL}(N,F)} \text{ has a trivial central character.} \tag{10-4}$$

Since $\sigma^{\mathrm{GL}(N,F)}$ is an irreducible subquotient of (10-2), its description in the Zelevinsky classification can be obtain by the well–known process of "linking" (see [**Ze**]):

$$\sigma^{\mathrm{GL}(N,F)} \simeq \langle \Delta_1 \rangle \times \cdots \times \langle \Delta_k \rangle, \tag{10-5}$$

where $\Delta_1, \dots, \Delta_k$ is up to a permutation, the unique sequence of segments of unramified characters characterized by the following two conditions:

- There is an equality of the multisets:
$$\Delta_1 + \cdots + \Delta_k = \{\chi_1, \dots, \chi_n, \chi_1^{-1}, \dots, \chi_n^{-1}, \star\}.$$
- There are no indices i, j, such that the segments Δ_i and Δ_j are linked.

The expression (10-5) is easy to find for an unitary representation σ^G, and this is the basis for our algorithm. Assume that σ^G is unitarizable. Then we apply Theorem 5-14 to find $(\mathbf{e}, \sigma_{neg}) \in \mathcal{M}^{u,unr}(S)$ such that

$$\sigma^G \simeq \left(\times_{(l,\chi,\alpha)\in \mathbf{e}} \ \langle [-\frac{l-1}{2}, \frac{l-1}{2}]^{(\nu^\alpha \chi)} \rangle \right) \rtimes \sigma_{neg}. \tag{10-6}$$

We have the following lemma:

LEMMA 10-7. *Assume that σ^G is unitarizable and given by (10-6). Then the representation $\sigma^{\mathrm{GL}(N,F)}$ is isomorphic to the following induced representation:*

$$\times_{(l,\chi,\alpha)\in\mathbf{e};\ \alpha\neq\frac{1}{2}} \langle[-\frac{l-1}{2},\frac{l-1}{2}]^{(\nu^\alpha\chi)}\rangle \times \langle[-\frac{l-1}{2},\frac{l-1}{2}]^{(\nu^{-\alpha}\chi^{-1})}\rangle$$
$$\times_{(l,\chi,\alpha)\in\mathbf{e};\ \alpha=\frac{1}{2}} \langle[-\frac{l-2}{2},\frac{l-2}{2}]^{(\chi)}\rangle \quad \textit{(we omit the segment if } l=1)$$
$$\times_{(l,\chi,\alpha)\in\mathbf{e};\ \alpha=\frac{1}{2}} \langle[-\frac{l}{2},\frac{l}{2}]^{(\chi)}\rangle$$
$$\times_{(l,\chi)\in\mathrm{Jord}(\sigma_{neg})} \langle[-\frac{l-1}{2},\frac{l-1}{2}]^{(\chi)}\rangle$$

PROOF. Since $(\mathbf{e},\sigma_{neg})$ satisfies Definition 5-13, the claim easily follows from Lemma 6-1. (To compute the lift of σ_{neg} one applies Theorems 5-8 and 5-10; if $\sigma_{neg}=\mathbf{1}\in\mathrm{Irr}\ S_0$ we apply the definition (5-9).) We leave the simple verification to the reader. The reader should realize that the lemma does not hold if σ^G is not unitarizable. □

Now, we present the following:

Algorithm for testing the unitarity of $\sigma^G(\chi_1,\ldots,\chi_n)$.

It has the following steps:

(1) Introduce the multiset $\{\chi_1,\ldots,\chi_n,\chi_1^{-1},\ldots,\chi_n^{-1},\star\}$.
(2) Among the characters in (1), perform the (maximal) linking, to get the multisegment $\{\Delta_1,\ldots,\Delta_k\}$ which satisfies:
   - There is an equality of the multisets:
   $$\Delta_1+\cdots+\Delta_k=\{\chi_1,\ldots,\chi_n,\chi_1^{-1},\ldots,\chi_n^{-1},\star\}.$$
   - There are no indices i,j, such that the segments Δ_i and Δ_j are linked. [2]

We begin our second stage of the algorithm (where we apply Lemma 10-7). We recursively construct the multisets Jord and $\mathbf{e}$ that must be $\mathrm{Jord}(\sigma_{neg})$ and $\mathbf{e}$ for $\sigma^G(\chi_1,\ldots,\chi_n)$ if this representation is unitarizable. We start with $\mathrm{Jord}=\emptyset$, $\mathbf{e}=\emptyset$ and the multiset $\eta=\{\Delta_1,\ldots,\Delta_k\}$, and modify them recursively. We execute the algorithm until $\eta=\emptyset$.

It is easy to show that $\tilde{\eta}=\eta$.

(3) Denote by $\eta_{nsd,unit}$ the multiset of all $\Delta\in\eta$; $\Delta=[-\frac{l-1}{2},\frac{l-1}{2}]^{(\nu^\alpha\chi)}$ (where as usual $l\in\mathbb{Z}_{\geq 1}$, χ is an unitary unramified character of $F^\times$, and $\alpha\in\mathbb{R}$) such that $\chi\notin\{\mathbf{1}_{F^\times},\mathbf{sgn}_u\}$ and $\alpha=0$. Add to Jord the multiset of all (l,χ) when Δ runs over $\eta_{nsd,unit}$, replace η by $\eta-\eta_{nsd,unit}$, and keep $\mathbf{e}$ unchanged. It is easy to see that for Δ from $\eta_{nsd,unit}$, the segments Δ and $\widetilde{\Delta}$ shows up in $\eta_{nsd,unit}$ (and η)) with the same multiplicity.

[2]The multisegment $\{\Delta_1,\ldots,\Delta_k\}$ is the result of the following simple algorithm: Let $\Delta_1=\{\chi_1\},\ldots,\Delta_n=\{\chi_n\},\Delta_{n+1}=\{\chi_1^{-1}\},\ldots,\Delta_{2n}=\{\chi_n^{-1}\},\Delta_N=\{\star\}$ and $k=N$ be the starting sequence $\Delta_1,\ldots,\Delta_k$ of the segments. Repeat the following recursive step until it is not possible: find two indices $i<j$ such that Δ_i and Δ_j are linked and replace

$$\begin{cases}\Delta_i\leftrightarrow\Delta_i\cup\Delta_j\\ \Delta_j\leftrightarrow\Delta_i\cap\Delta_j \text{ (omit this segment if the intersection is empty)}\end{cases}$$

(4) Denote by $\eta_{nsd,+}$ the multiset of all $\Delta \in \eta$; $\Delta = [-\frac{l-1}{2}, \frac{l-1}{2}]^{(\nu^\alpha \chi)}$ such that $\chi \notin \{\mathbf{1}_{F^\times}, \mathbf{sgn}_u\}$ and $\alpha > 0$. **If some** $\alpha \geq 1/2$, then the **algorithm stops** and the representation is not unitarizable. Further, **if** $\eta_{nsd,+} \neq \tilde{\eta}_{nsd,+}$ (this is a Hermitian condition, which is equivalent to $\mathbf{e}(l,\chi) \neq \mathbf{e}(l,\chi^{-1})$ for some χ and l as above), then the **algorithm stops** and the representation is not unitarizable. If neither of these happen, then add to $\mathbf{e}$ the multiset of all (l,χ,α) when Δ runs over $\eta_{nsd,+}$, replace η by $\eta - \eta_{nsd,+} - \tilde{\eta}_{nsd,+}$, and keep Jord unchanged.

(5) Denote by $\eta_{sd,\{\frac{1}{2}\},+}$ the multiset of all $\Delta \in \eta$; $\Delta = [-\frac{l-1}{2}, \frac{l-1}{2}]^{(\nu^\alpha \chi)}$ such that $\chi \in \{\mathbf{1}_{F^\times}, \mathbf{sgn}_u\}$, $l-(2\alpha_\chi+1) \notin 2\mathbb{Z}$ and $\alpha > 0$. **If some** $\alpha \geq 1/2$ (in which case $\alpha > 1/2$), then **the algorithm stops** and the representation is not unitarizable. If not, then add to $\mathbf{e}$ the multiset of all (l,χ,α) when Δ runs over $\eta_{sd,\{\frac{1}{2}\},+}$, replace η by $\eta-\eta_{sd,\{\frac{1}{2}\},+}-\tilde{\eta}_{sd,\{\frac{1}{2}\},+}$, and keep Jord unchanged.

(6) Denote by $\eta_{sd,\{0,1\},+}$ the multiset of all $\Delta \in \eta$; $\Delta = [-\frac{l-1}{2}, \frac{l-1}{2}]^{(\nu^\alpha \chi)}$ such that $\chi \in \{\mathbf{1}_{F^\times}, \mathbf{sgn}_u\}$, $l-(2\alpha_\chi+1) \in 2\mathbb{Z}$ and $\alpha > 0$. Observe that we cannot have $\alpha = 1/2$ (in η we do not have linked segments). **If some** $\alpha \geq 1$, then **the algorithm stops** and the representation is not unitarizable. If not (i.e., if all $\alpha < 1$), then for all (l,χ) coming from Δ's in $\eta_{sd,\{0,1\},+}$, check if the multiset $\mathbf{e}(l,\chi)$ of all α such that $\langle[-\frac{l-1}{2}, \frac{l-1}{2}]^{(\nu^\alpha \chi)}\rangle$ is in $\eta_{sd,\{0,1\},+}$ satisfies condition (c) of (3) in Definition 5-13 (observe that condition (d) of (3) in Definition 5-13 is satisfied, since $\tilde{\eta} = \eta$ and in η we do not have linked segments). **If all these conditions are not satisfied**, then **the algorithm stops** and the representation is not unitarizable. If not, then add to $\mathbf{e}$ the multiset of all (l,χ,α) when Δ runs over $\eta_{sd,\{0,1\},+}$, replace η by $\eta - \eta_{sd,\{0,1\},+} - \tilde{\eta}_{sd,\{0,1\},+}$, and keep Jord unchanged.

(7) Denote by $\eta_{sd,unit,red}$ the multiset of all $\Delta \in \eta$; $\Delta = [-\frac{l-1}{2}, \frac{l-1}{2}]^{(\nu^\alpha \chi)}$ such that $\chi \in \{\mathbf{1}_{F^\times}, \mathbf{sgn}_u\}$, $\alpha = 0$ and $l-(2\alpha_\chi+1) \in 2\mathbb{Z}$. [3] Add to Jord the multiset of all (l,χ) when Δ runs over $\eta_{sd,unit,red}$, replace η by $\eta - \eta_{sd,unit,red}$, and keep $\mathbf{e}$ unchanged.

(8) Take $\Delta \in \eta$; $\Delta = [-\frac{l-1}{2}, \frac{l-1}{2}]^{(\nu^\alpha \chi)}$ with the largest possible l (as usual, $l \in \mathbb{Z}_{\geq 1}$, χ is an unitary unramified character of $F^\times$, and $\alpha \in \mathbb{R}$). Then $\alpha = 0$, $\chi \in \{\mathbf{1}_{F^\times}, \mathbf{sgn}_u\}$ and $l-(2\alpha_\chi+1) \notin 2\mathbb{Z}$. Form the multiset η_Δ consisting of all $\Psi \in \eta$ such that $\Psi = \Delta$.
   (i) If card η_Δ is even, say $2m$, we replace Jord by Jord $+ 2m\{(l,\chi)\}$, remove η_Δ from η and keep $\mathbf{e}$ unchanged.
   (ii) If card η_Δ is odd, say $2m+1$, then we perform the following steps (see the second line in the displayed formula in Lemma (10-7)):
      (a) **If** $l = 1$, then **the algorithm stops** and the representation $\sigma^G(\chi_1, \ldots, \chi_n)$ is not unitarizable.
      (b) If $l = 2$, then we replace $\mathbf{e}$ by $\mathbf{e} + \{(1,\chi,\frac{1}{2})\}$, Jord by Jord $+ 2m\{(l,\chi)\}$ and η by $\eta - \eta_\Delta$.
      (c) If $l \geq 3$, then we let $\eta_{[-\frac{l-3}{2}, \frac{l-3}{2}]^{(\chi)}}$ to be the sub–multiset of η corresponding to $\Psi = [-\frac{l-3}{2}, \frac{l-3}{2}]^{(\chi)}$. **If** card $\eta_{[-\frac{l-3}{2}, \frac{l-3}{2}]^{(\chi)}}$ **is even**, then the **algorithm stops** and the representation

[3] Fix $\chi \in \{\mathbf{1}_{F^\times}, \mathbf{sgn}_u\}$. One sees directly using (10-4) that the sum of multiplicities of all $\Delta = [-\frac{l-1}{2}, \frac{l-1}{2}]^{(\chi)}$ in $\eta_{sd,unit,red}$ is even.

$\sigma^G(\chi_1, \ldots, \chi_n)$ is not unitary. If it is odd, say $2m' + 1$, we replace $\mathbf{e}$ by $\mathbf{e} + \{(l-1, \chi, \frac{1}{2})\}$, Jord by Jord $+ 2m\{(l, \chi)\} + 2m'\{(l-2)\}$ and η by $\eta - \eta_\Delta - \eta_\Psi$.

(9) If $\eta = \emptyset$, we go to the following step. Otherwise, we go back to the Step 8. The above procedure provides that (b) in (3) of Definition 5-13 is satisfied.

(10) One easily sees that there exists σ_{neg} such that $\mathrm{Jord}(\sigma_{neg}) = \mathrm{Jord}$ (we can construct the representation σ_{neg} attached to Jord following the steps described in Section 5 (see Theorems 5-8 and 5-10), but we can finish the algorithm without constructing σ_{neg}). Check if, for all (l, χ) from steps (6) and (8), the corresponding multiset $\mathbf{e}(l, \chi)$ satisfies conditions (a), (e) and (f) of (3) in Definition 5-13 with respect to the Jord that we have obtained. **If not, $\sigma^G(\chi_1, \ldots, \chi_n)$ is not unitarizable**. Otherwise, $\sigma^G(\chi_1, \ldots, \chi_n)$ is unitarizable.

This terminates the algorithm.

Observe that in the case of unitarizability of $\sigma^G(\chi_1, \ldots, \chi_n)$, the multisets $\mathbf{e}$ and Jord that we have obtained at the end of algorithm determine the parameters $(\mathbf{e}, \sigma(\mathrm{Jord}))$ of $\sigma^G(\chi_1, \ldots, \chi_n)$ from Theorem 5-14.

11. Isolated Points in $\mathrm{Irr}^{u,unr}(S_n)$

We equip $\mathrm{Irr}^{u,unr}(S_n)$ with the topology described in Section 3. In this section we determine all isolated representations in $\mathrm{Irr}^{u,unr}(S_n)$. It is based on our classification result Theorem 5-14 as well as the description of topology given by Theorem 3-7. In more detail, since $\mathrm{Irr}^{u,unr}(S_n)$ is a closed subset of $\mathrm{Irr}^{unr}(S_n)$, it is homeomorphic (via φ_{S_n}) to a closed subset of a complex manifold having a countable base of the topology. Therefore, we have the following trivially: $\sigma \in \mathrm{Irr}^{u,unr}(S_n)$ is not an isolated point if and only if there is a sequence $(\sigma_m)_{m \in \mathbb{Z}_{>0}}$ in $\mathrm{Irr}^{unr}(S_n) \setminus \{\sigma\}$ such that

$$\lim_m \sigma_m = \sigma \quad (\text{equivalently, } \lim_m \varphi_{S_n}(\sigma_m) = \varphi_{S_n}(\sigma)). \tag{11-1}$$

We begin with the following lemma:

LEMMA 11-2. *Assume that $\sigma \in \mathrm{Irr}^{u,unr}(S_n)$ is isolated. Then it must be strongly negative.*

PROOF. Let $(\mathbf{e}, \sigma_{neg}) \in \mathcal{M}^{u,unr}(S_n)$ such that

$$\sigma \simeq \times_{(l,\chi,\alpha)\in\mathbf{e}} \ \langle[-\frac{l-1}{2}, \frac{l-1}{2}]^{(\nu^\alpha \chi)}\rangle \rtimes \sigma_{neg}.$$

If σ is not strongly negative, then either $\mathbf{e} \neq \emptyset$, or $\mathbf{e} = \emptyset$ and $\sigma \simeq \sigma_{neg}$ is negative but not strongly negative.

If $\mathbf{e} \neq \emptyset$, then pick some $(l_0, \chi_0, \alpha_0) \in \mathbf{e}$. Applying Theorem 5-14 and Definition 5-13, we choose $\epsilon > 0$ small enough and a sequence $(\alpha_m)_{m \in \mathbb{Z}_{>0}}$ contained in $]\alpha_0 - \epsilon, \ \alpha_0 + \epsilon[\setminus\{\alpha_0\}$ converging to α_0 such that $(\mathbf{e}^{(m)}, \sigma_{neg}) \in \mathcal{M}^{u,unr}(S_n)$, where

$$\mathbf{e}^{(m)} = \mathbf{e} - \{(l_0, \chi_0, \alpha_0)\} + \{(l_0, \chi_0, \alpha_m)\} \ \text{ for all } m \in \mathbb{Z}_{>0}.$$

Now, we define a sequence of unramified unitary representations $(\sigma_m)_{m \in \mathbb{Z}_{>0}} \in \mathrm{Irr}^{u,unr}(S_n)$ by $\sigma_m \simeq \times_{(l^{(m)},\chi^{(m)},\alpha^{(m)})\in\mathbf{e}^{(m)}} \langle[-\frac{l^{(m)}-1}{2}, \frac{l^{(m)}-1}{2}]^{(\nu^{\alpha^{(m)}}\chi^{(m)})}\rangle \rtimes \sigma_{neg}$. Obviously, $\sigma_m \not\simeq \sigma$ for all m and $\lim_m \varphi_{S_n}(\sigma_m) = \varphi_{S_n}(\sigma)$. Hence σ is not isolated.

If σ is a negative but not strongly negative representation, then there exists $l \in \mathbb{Z}_{>0}$, an unramified unitary character χ of $F^\times$, and a negative representation σ'_{neg} such that $\sigma \hookrightarrow \langle[-\frac{l-1}{2}, \frac{l-1}{2}]^{(\chi)}\rangle \rtimes \sigma'_{neg}$. Now, $\chi(\varpi)$ is a complex number of absolute value one. We choose a sequence $(\alpha_m)_{m\in\mathbb{Z}_{>0}}$ of complex numbers of absolute value one converging to $\chi(\varpi)$ such that $\alpha_m \neq \chi(\varpi)$ for all m. Then we define a sequence $(\chi_m)_{m\in\mathbb{Z}_{>0}}$ of unramified unitary characters of $F^\times$ by $\chi_m(\varpi) = \alpha_m$ and a sequence of unramified (unitary) negative representations $(\sigma_m)_{m\in\mathbb{Z}_{>0}}$ in $\mathrm{Irr}^{u,unr}(S_n)$ by $\sigma_m \hookrightarrow \langle[-\frac{l-1}{2}, \frac{l-1}{2}]^{(\chi_m)}\rangle \rtimes \sigma'_{neg}$. Obviously, $\sigma_m \not\cong \sigma$ for all m and $\lim_m \varphi_{S_n}(\sigma_m) = \varphi_{S_n}(\sigma)$. Hence σ is not isolated. □

Now, we assume that σ is strongly negative. We write $\mathrm{Jord} = \mathrm{Jord}(\sigma)$ for the set of its Jordan blocks (See Theorem 5-8 and the notation introduced before that theorem.) If $\chi \in \{\mathbf{1}_{F^\times}, \mathbf{sgn}_u\}$ and $a \in \mathrm{Jord}_\chi$ which is not the minimum, then we write a_- for the greatest $b \in \mathrm{Jord}_\chi$ such that $b < a$. We have

$$a - a_- \text{ is even (whenever } a_- \text{ is defined).}$$

This follows from the fact that $a-(2\alpha_\chi+1) \in 2\mathbb{Z}$ for all $a \in \mathrm{Jord}_\chi$. (See Definition 5-4.)

The main result of this section is the following theorem:

THEOREM 11-3. *Let $n > 0$. Then $\sigma \in \mathrm{Irr}^{u,unr}(S_n)$ is isolated if and only if σ is strongly negative, and for every $\chi \in \{\mathbf{1}_{F^\times}, \mathbf{sgn}_u\}$ such that $\mathrm{Jord}_\chi \neq \emptyset$, the following holds:*

(1) *$a - a_- \geq 4$, for all $a \in \mathrm{Jord}_\chi$ whenever a_- is defined.*
(2) *If $\mathrm{Jord}_\chi \neq \{1\}$, then $\min \mathrm{Jord}_\chi \setminus \{1\} \geq 4$.*

We do not claim that $1 \in \mathrm{Jord}_\chi$ in (2). If $1 \notin \mathrm{Jord}_\chi$, then (2) claims that $\min \mathrm{Jord}_\chi \geq 4$.

We start the proof of Theorem 11-3 with the following lemma:

LEMMA 11-4. *Assume that $\sigma \in \mathrm{Irr}^{u,unr}(S_n)$ is strongly negative and isolated. Assume that $\mathrm{Jord}_\chi \neq \emptyset$ for some $\chi \in \{\mathbf{1}_{F^\times}, \mathbf{sgn}_u\}$. Then $a - a_- \geq 4$ for all $a \in \mathrm{Jord}_\chi$ whenever a_- is defined.*

PROOF. Assume that there exists $\chi \in \{\mathbf{1}_{F^\times}, \mathbf{sgn}_u\}$ such that $\mathrm{Jord}_\chi \neq \emptyset$ and there is a gap in Jord_χ of 2, say $a-a_- = 2$ for $a, a_- \in \mathrm{Jord}_\chi$. Then the construction of strongly negative representations (see the text before Theorem 5-8) shows that

$$\mathrm{Jord}' := \mathrm{Jord} - \{(a_-, \chi), (a, \chi)\}$$

is a set of Jordan blocks for some strongly negative representation $\sigma' \in \mathrm{Irr}^{u,unr}(S_{n'})$ (See Definition 5-4.) Moreover, Theorem 5-8 and Remark 1-8 imply that σ is a subquotient of

$$\langle[-\frac{a-1}{2}, \frac{a_- - 1}{2}]^{(\chi)}\rangle \rtimes \sigma' = \langle[-\frac{l-1}{2}, \frac{l-1}{2}]^{(\nu^{\frac{1}{2}}\chi)}\rangle \rtimes \sigma', \quad l = a_- + 1. \tag{11-5}$$

Now, we look at the family of induced representations $\langle[-\frac{l-1}{2}, \frac{l-1}{2}]^{(\nu^s\chi)} \rtimes \sigma'$ ($s \in [0, \frac{1}{2}]$). Since $l-(2\alpha_\chi+1) = a_-+1-(2\alpha_\chi+1) \notin 2\mathbb{Z}$, Lemma 6-5 implies reducibility at $s = 0$. Therefore we have unitarity and irreducibility for $s \in [0, \frac{1}{2}[$. At $s = 1/2$ we have reducibility, and σ appears as a subquotient (see (11-5)). Hence σ cannot be isolated, arguing as in the proof of Lemma 11-2. □

LEMMA 11-6. *Assume that* $\sigma \in \mathrm{Irr}^{u,unr}(S_n)$ *is strongly negative and isolated. Assume that* $\mathrm{Jord}_\chi \setminus \{1\} \neq \emptyset$ *for* $\chi \in \{\mathbf{1}_{F^\times}, \mathbf{sgn}_u\}$. *Then* $\min \mathrm{Jord}_\chi \setminus \{1\} \geq 4$.

PROOF. We consider several cases.

First, we assume that $\alpha_\chi = 1$ (hence, $\chi = \mathbf{1}_{F^\times}$ and $S_n = \mathrm{Sp}(2n, F)$; see Lemma 5-2) and $1 \in \mathrm{Jord}_\chi$. Then the claim follows from the previous lemma. Assume that $\alpha_\chi = 1$ and $1 \notin \mathrm{Jord}_\chi$. Then the elements of Jord_χ are odd integers (since $a - (2\alpha_\chi + 1) \in 2\mathbb{Z}$). We need to show $3 \notin \mathrm{Jord}_\chi$. Assume that $3 \in \mathrm{Jord}_\chi$. Then we define a strongly negative representation σ' using

$$\mathrm{Jord}' := \mathrm{Jord} - \{(3, \chi)\} + \{(1, \chi)\}.$$

Since in this case $S_n = \mathrm{Sp}(2n, F)$, Definition 5-4 implies that card Jord_χ is odd. Hence, by Theorem 5-8 and Remark 1-8, we obtain that σ is an irreducible subquotient of $\nu\chi \rtimes \sigma'$. Now, we consider the family of representations $\nu^s\chi \rtimes \sigma'$ ($s \in [0, 1]$). Lemma 6-5 implies the irreducibility at $s = 0$ (since $(1, \chi) \in \mathrm{Jord}'$). Lemma 6-1 implies its irreducibility for $s \in]0, 1[$. Hence σ is not isolated.

Assume $\alpha_\chi = 0$. Then card Jord_χ is even by the very last property stated in Definition 5-4. Now, if $1 \in \mathrm{Jord}_\chi$, then $3 \notin \mathrm{Jord}_\chi$ by the previous lemma. If $1 \notin \mathrm{Jord}_\chi$, then we need to show that $3 \notin \mathrm{Jord}_\chi$. To prove this, assume to the contrary that $1 \notin \mathrm{Jord}_\chi$ and $3 \in \mathrm{Jord}_\chi$. Then we can arrive at the contradiction as in the case $\alpha_\chi = 1$ and $1 \notin \mathrm{Jord}_\chi$ above.

Assume $\alpha_\chi = \frac{1}{2}$. Then, by Definition 5-4, $S_n = \mathrm{SO}(2n+1, F)$ and Jord_χ consists of even integers. In this case we need to show $2 \notin \mathrm{Jord}_\chi$. Assume that $2 \in \mathrm{Jord}_\chi$. Then we define a strongly negative representation σ' using

$$\mathrm{Jord}' := \mathrm{Jord} - \{(2, \chi)\}.$$

Then, by Theorem 5-8 and Remark 1-8, we obtain that σ is an irreducible subquotient of $\nu^{1/2}\chi \rtimes \sigma'$. Now, we consider the family of representations $\nu^s\chi \rtimes \sigma'$ ($s \in [0, 1]$). Lemma 6-5 implies the irreducibility at $s = 0$. Lemma 6-1 implies its irreducibility for $s \in]0, \frac{1}{2}[$. Hence σ is not isolated. □

Lemmas 11-4 and 11-6 prove that the conditions imposed upon σ in Theorem 11-3 are necessary. We need to show that they are sufficient. We start by constructing for an arbitrary sequence in $\mathrm{Irr}^{u,unr}(S_n)$ a convergent subsequence. Let $(\sigma_m)_{m \in \mathbb{Z}_{>0}}$ be a sequence in $\mathrm{Irr}^{u,unr}(S_n)$. The classification of unramified unitarizable representations (see Theorem 5-14) implies that there exists a unique sequence $(\mathbf{e}^{(m)}, \sigma_{neg}^{(m)}) \in \mathcal{M}^{u,unr}(S_n)$, $m \in \mathbb{Z}_{>0}$, such that

$$\sigma_m \simeq \times_{(l^{(m)}, \chi^{(m)}, \alpha^{(m)}) \in \mathbf{e}^{(m)}} \langle [-\frac{l^{(m)} - 1}{2}, \frac{l^{(m)} - 1}{2}]^{(\nu^{\alpha^{(m)}}\chi^{(m)})} \rangle \rtimes \sigma_{neg}^{(m)}.$$

Since σ_m is a representation of S_n, we have

$$\sum_{(l^{(m)}, \chi^{(m)}, \alpha^{(m)}) \in \mathbf{e}^{(m)}} l^{(m)} \leq n$$

for all $m \in \mathbb{Z}_{>0}$. Therefore, if we choose some enumeration writing elements of every $\mathbf{e}^{(m)}$ as a sequence:

$$\mathbf{e}^{(m)} \ldots (l_1^{(m)}, \chi_1^{(m)}, \alpha_1^{(m)}), \ldots, (l_{a^{(m)}}^{(m)}, \chi_{a^{(m)}}^{(m)}, \alpha_{a^{(m)}}^{(m)}),$$

then, passing to a subsequence, we may assume that the following is independent of m:

$$\begin{cases} a^{(m)} = a \\ l_i^{(m)} = l_i, \ i = 1, \dots, a. \end{cases}$$

Next, the complex absolute value of $\chi_i^{(m)}(\varpi)$ is equal to 1. Hence, passing to subsequences we may assume that every sequence $(\chi_i^{(m)}(\varpi))_{m \in \mathbb{Z}_{>0}}$, $i = 1, \dots, a$, converges. We define a sequence of unramified unitary characters $\chi_1, \dots, \chi_a$ of $F^\times$ by

$$\lim_m \chi_i^{(m)}(\varpi) = \chi_i(\varpi).$$

Since every sequence $(\alpha_i^{(m)})$ is bounded (see Definition 5-13), we see that we may assume that it converges:

$$\lim_m \alpha_i^{(m)} = \alpha_i.$$

Next, we apply Theorem 5-10 to the sequence $(\sigma_{neg}^{(m)})_{m \in \mathbb{Z}_{>0}}$. For every m, we find a sequence of the pairs $(k_1^{(m)}, \mu_1^{(m)}), \dots, (k_{b(m)}^{(m)}, \mu_{b(m)}^{(m)})$ ($k_i \in \mathbb{Z}_{\geq 1}$, $\mu_i^{(m)}$ is an unramified unitary character of $F^\times$), and a strongly negative representation $\sigma_{sn}^{(m)}$ such that

$$\sigma_{neg}^{(m)} \hookrightarrow \langle [-\frac{k_1^{(m)} - 1}{2}, \frac{k_1^{(m)} - 1}{2}]^{(\mu_1^{(m)})} \rangle \times \cdots \times \langle [-\frac{k_{b(m)}^{(m)} - 1}{2}, \frac{k_{b(m)}^{(m)} - 1}{2}]^{(\mu_{b(m)}^{(m)})} \rangle \rtimes \sigma_{sn}^{(m)}.$$

As above, passing to a subsequence, we may assume that the following is independent of m:

$$\begin{cases} b^{(m)} = b \\ k_i^{(m)} = k_i, \ i = 1, \dots, a. \end{cases}$$

Hence, we may define a sequence of unramified unitary characters $\mu_1, \dots, \mu_b$ of $F^\times$ by

$$\lim_m \mu_i^{(m)}(\varpi) = \mu_i(\varpi).$$

Next, since there are only finitely many strictly negative representations in

$$\cup_{0 \leq m \leq n} \mathrm{Irr}^{u,unr}(S_m),$$

we may assume that

$$\sigma_{sn}^{(m)} = \sigma_{sn}$$

is independent of m. We write σ_{sn} in the form $\sigma^{S_b}(\lambda_1, \dots, \lambda_b)$ (see Theorem 1-2). Now, we have that $\varphi_{S_n}(\sigma_m)$ is the W–orbit of the n tuple

$$\begin{aligned} (q^{-\frac{l_1-1}{2} - \alpha_1^{(m)}} \chi_1^{(m)}(\varpi), q^{-\frac{l_1-1}{2}+1-\alpha_1^{(m)}} \chi_1^{(m)}(\varpi), \dots, q^{\frac{l_1-1}{2} - \alpha_1^{(m)}} \chi_1^{(m)}(\varpi), \\ \dots, \\ q^{-\frac{l_a-1}{2} - \alpha_a^{(m)}} \chi_a^{(m)}(\varpi), q^{-\frac{l_a-1}{2}+1-\alpha_a^{(m)}} \chi_a^{(m)}(\varpi), \dots, q^{\frac{l_a-1}{2} - \alpha_a^{(m)}} \chi_a^{(m)}(\varpi), \\ q^{-\frac{k_1-1}{2}} \mu_1^{(m)}(\varpi), q^{-\frac{k_1-1}{2}+1} \mu_1^{(m)}(\varpi), \dots, q^{\frac{k_1-1}{2}} \mu_1^{(m)}(\varpi), \\ \dots, \\ q^{-\frac{k_b-1}{2}} \mu_b^{(m)}(\varpi), q^{-\frac{k_b-1}{2}+1} \mu_b^{(m)}(\varpi), \dots, q^{\frac{k_b-1}{2}} \mu_b^{(m)}(\varpi), \\ \lambda_1(\varpi), \dots, \lambda_b(\varpi)). \end{aligned}$$

Clearly, the sequence $(\varphi_{S_n}(\sigma_m))$ converges to the W–orbit of the n tuple

$$\begin{aligned}(q^{-\frac{l_1-1}{2}-\alpha_1}\chi_1(\varpi), q^{-\frac{l_1-1}{2}+1-\alpha_1}\chi_1(\varpi), \ldots, q^{\frac{l_1-1}{2}-\alpha_1}\chi_1(\varpi),\\ \ldots,\\ q^{-\frac{l_a-1}{2}-\alpha_a}\chi_a(\varpi), q^{-\frac{l_a-1}{2}+1-\alpha_a}\chi_a(\varpi), \ldots, q^{\frac{l_a-1}{2}-\alpha_a}\chi_a(\varpi),\\ q^{-\frac{k_1-1}{2}}\mu_1(\varpi), q^{-\frac{k_1-1}{2}+1}\mu_1(\varpi), \ldots, q^{\frac{k_1-1}{2}}\mu_1(\varpi),\\ \ldots,\\ q^{-\frac{k_b-1}{2}}\mu_b(\varpi), q^{-\frac{k_b-1}{2}+1}\mu_b(\varpi), \ldots, q^{\frac{k_b-1}{2}}\mu_b(\varpi),\\ \lambda_1(\varpi), \ldots, \lambda_b(\varpi)).\end{aligned}$$

The corresponding representation $\sigma \in \mathrm{Irr}^{unr}(S_n)$ is unitary (since $\varphi_{S_n}(\mathrm{Irr}^{u,unr}(S_n))$ is a closed subset of D_n^W; see Theorem 3-7), and clearly the unique irreducible unramified subquotient of

$$(11\text{-}7)\qquad \times_{i=1}^{a}\ \langle[-\frac{l_i-1}{2}, \frac{l_i-1}{2}]^{(\nu^{\alpha_i}\chi_i)}\rangle \times_{i=1}^{b}\ \langle[-\frac{k_i-1}{2}, \frac{k_i-1}{2}]^{(\mu_i)}\rangle \rtimes \sigma_{sn}.$$

We summarize the assumptions on the sequence:

$$(11\text{-}8)\qquad \begin{aligned}&\sigma_m \simeq \times_{i=1}^{a}\ \langle[-\frac{l_i-1}{2}, \frac{l_i-1}{2}]^{(\nu^{\alpha_i^{(m)}}\chi_i^{(m)})}\rangle \rtimes \sigma_{neg}^{(m)},\\ &\mathbf{e}^{(m)} = \{(l_i, \chi_i^{(m)}, \alpha_i^{(m)}),;\ i = 1, \ldots, a\},\ \ (\mathbf{e}^{(m)},\ \sigma_{neg}^{(m)}) \in \mathcal{M}^{u,unr}(S_n)\\ &\sigma_{neg}^{(m)} \hookrightarrow \times_{i=1}^{b}\ \langle[-\frac{k_i-1}{2}, \frac{k_i-1}{2}]^{(\mu_i^{(m)})}\rangle \rtimes \sigma_{sn}.\\ &\lim_m \chi_i^{(m)}(\varpi) = \chi_i(\varpi),\ \ \lim_m \mu_i^{(m)}(\varpi) = \mu_i(\varpi),\ \ \lim_m \alpha_i^{(m)} = \alpha_i.\end{aligned}$$

Now, in order to complete the proof of Theorem 11-3, we need to prove the following lemma:

LEMMA 11-9. *Assume that $\sigma \in \mathrm{Irr}^{u,unr}(S_n)$ is strongly negative such that (1) and (2) of Theorem 11-3 hold. Then for every sequence $(\sigma_m)_{m\in\mathbb{Z}_{>0}}$, satisfying (11-8), such that $\varphi_{S_n}(\sigma_m) \to \varphi_{S_n}(\sigma)$, there exists m_0 such that $\sigma_m = \sigma_{sn}$ for $m \geq m_0$.*

PROOF. Assume that $(\sigma_m)_{m\in\mathbb{Z}_{>0}}$ is a sequence that satisfies (11-8). Assume that $\varphi_{S_n}(\sigma_m) \to \varphi_{S_n}(\sigma)$ but there is no m_0 such that $\sigma_m = \sigma_{sn}$ for $m \geq m_0$. Then passing to a subsequence we may assume that $a + b > 0$ for all $m > 0$. (a and b are defined in (11-8).) We show that this is not possible.

Put $G = S_n$. We begin by computing the Langlands lift $\tau := \sigma^{\mathrm{GL}(N,F)}$ of σ to $\mathrm{GL}(N, F)$. (See (10-1) for the definition of the lift and the first displayed formula in Section 10 for the definition of the number N.) We can compute the lift in two ways. First, since by our assumption σ is strongly negative, we have the following:

$$(11\text{-}10)\qquad \tau \simeq \times_{(l,\chi)\in\mathrm{Jord}(\sigma)}\langle[-\frac{l-1}{2}, \frac{l-1}{2}]^{(\chi)}\rangle.$$

Also, since σ is the limit of the sequence $(\sigma_m)_{m\in\mathbb{Z}_{>0}}$, it is an irreducible subquotient of the induced representation given by (11-7). Therefore, we obtain the following:

$$\tau \text{ is the unique unramified irreducible subquotient of}$$
$$(11\text{-}11)\quad \begin{aligned} &\times_{i=1}^{a}\langle[-\frac{l_i-1}{2},\frac{l_i-1}{2}]^{(\nu^{\alpha_i}\chi_i)}\rangle \times_{i=1}^{b}\langle[-\frac{k_i-1}{2},\frac{k_i-1}{2}]^{(\mu_i)}\rangle\times\\ &\qquad\times_{(l',\chi')\in\mathrm{Jord}(\sigma_{sn})}\langle[-\frac{l'-1}{2},\frac{l'-1}{2}]^{(\chi')}\rangle\times\\ &\times_{i=1}^{a}\langle[-\frac{l_i-1}{2},\frac{l_i-1}{2}]^{(\nu^{-\alpha_i}\chi_i^{-1})}\rangle \times_{j=1}^{b}\langle[-\frac{k_i-1}{2},\frac{k_i-1}{2}]^{(\mu_i^{-1})}\rangle.\end{aligned}$$

Now, since only $\chi\in\{\mathbf{1}_{F^\times},\mathbf{sgn}_u\}$ appears in (11-10), we see that

$$\chi_1,\dots,\chi_a,\mu_1,\dots,\mu_b\in\{\mathbf{1}_{F^\times},\mathbf{sgn}_u\}.$$

Likewise, we have $\alpha_1,\dots,\alpha_a\in\frac{1}{2}\mathbb{Z}$. Since $\lim_m\alpha_i^{(m)}=\alpha_i$ and $\alpha_i^{(m)}\in]0,1[$ (see Definition 5-13), we see that

$$(11\text{-}12)\quad \begin{aligned}&\text{if } a>0,\text{ then }\ \alpha_1,\dots,\alpha_a\in\{0,\frac{1}{2},1\},\ \chi_1,\dots,\chi_a\in\{\mathbf{1}_{F^\times},\mathbf{sgn}_u\}\\ &\text{if } b>0,\text{ then }\ \mu_1,\dots,\mu_b\in\{\mathbf{1}_{F^\times},\mathbf{sgn}_u\}.\end{aligned}$$

Using this we analyse (11-11). First, we observe that the unique irreducible unramified subquotient of

$$\langle[-\frac{l_i-1}{2},\frac{l_i-1}{2}]^{(\nu^{\alpha_i}\chi_i)}\rangle\times\langle[-\frac{l_i-1}{2},\frac{l_i-1}{2}]^{(\nu^{-\alpha_i}\chi_i^{-1})}\rangle,$$

for $\alpha_i=\frac{1}{2}$ is

$$\langle[-\frac{l_i}{2},\frac{l_i}{2}]^{(\chi_i)}\rangle\times\langle[-\frac{l_i-2}{2},\frac{l_i-2}{2}]^{(\chi_i)}\rangle.$$

Next, we observe that the unique irreducible unramified subquotient of

$$\langle[-\frac{l_i-1}{2},\frac{l_i-1}{2}]^{(\nu^{\alpha_i}\chi_i)}\rangle\times\langle[-\frac{l_i-1}{2},\frac{l_i-1}{2}]^{(\nu^{-\alpha_i}\chi_i^{-1})}\rangle,$$

for $\alpha_i=1$ is

$$\begin{cases}\langle[-\frac{l_i+1}{2},\frac{l_i+1}{2}]^{(\chi_i)}\rangle\times\langle[-\frac{l_i-3}{2},\frac{l_i-3}{2}]^{(\chi_i)}\rangle; & l_i\ge 2\\ \nu\chi_i\times\nu^{-1}\chi_i; & l_i=1.\end{cases}$$

Therefore, (11-11) implies that τ is an irreducible subquotient of

$$(11\text{-}13)\quad \begin{aligned}&\times_{i,\ \alpha_i=0}\langle[-\frac{l_i-1}{2},\frac{l_i-1}{2}]^{(\chi_i)}\rangle\times\langle[-\frac{l_i-1}{2},\frac{l_i-1}{2}]^{(\chi_i)}\rangle\times\\ &\times_{i,\ \alpha_i=\frac{1}{2}}\langle[-\frac{l_i}{2},\frac{l_i}{2}]^{(\chi_i)}\rangle\times\langle[-\frac{l_i-2}{2},\frac{l_i-2}{2}]^{(\chi_i)}\rangle\times\\ &\times_{i,\ \alpha_i=1,\ l_i\ge2}\langle[-\frac{l_i+1}{2},\frac{l_i+1}{2}]^{(\chi_i)}\rangle\times\langle[-\frac{l_i-3}{2},\frac{l_i-3}{2}]^{(\chi_i)}\rangle\times\\ &\times_{i=1}^{b}\langle[-\frac{k_i-1}{2},\frac{k_i-1}{2}]^{(\mu_i)}\rangle\times\langle[-\frac{k_i-1}{2},\frac{k_i-1}{2}]^{(\mu_i)}\rangle\\ &\times_{(l',\chi')\in\mathrm{Jord}(\sigma_{sn})}\langle[-\frac{l'-1}{2},\frac{l'-1}{2}]^{(\chi')}\rangle\times\\ &\times_{i,\ \alpha_i=1,\ l_i=1}\nu\chi_i\times\nu^{-1}\chi_i.\end{aligned}$$

We show that there is no i such that $\alpha_i = 1$ and $l_i = 1$. If this is not the case, then (11-10) and (11-13) imply that $(3, \chi_i) \in \mathrm{Jord}(\sigma)$ for some i such that $\alpha_i = 1$ and $l_i = 1$. This contradicts (2) of Theorem 11-3. Now, we have that τ is isomorphic to

$$\begin{aligned}
&\times_{i,\ \alpha_i=0} \langle [-\frac{l_i-1}{2}, \frac{l_i-1}{2}]^{(\chi_i)}\rangle \times \langle [-\frac{l_i-1}{2}, \frac{l_i-1}{2}]^{(\chi_i)}\rangle \times \\
&\times_{i,\ \alpha_i=\frac{1}{2}} \langle [-\frac{l_i}{2}, \frac{l_i}{2}]^{(\chi_i)}\rangle \times \langle [-\frac{l_i-2}{2}, \frac{l_i-2}{2}]^{(\chi_i)}\rangle \times \\
&\times_{i,\ \alpha_i=1,\ l_i\geq 2} \langle [-\frac{l_i+1}{2}, \frac{l_i+1}{2}]^{(\chi_i)}\rangle \times \langle [-\frac{l_i-3}{2}, \frac{l_i-3}{2}]^{(\chi_i)}\rangle \times \\
&\times_{i=1}^{b} \langle [-\frac{k_i-1}{2}, \frac{k_i-1}{2}]^{(\mu_i)}\rangle \times \langle [-\frac{k_i-1}{2}, \frac{k_i-1}{2}]^{(\mu_i)}\rangle \\
&\times_{(l',\chi')\in \mathrm{Jord}(\sigma_{sn})} \langle [-\frac{l'-1}{2}, \frac{l'-1}{2}]^{(\chi')}\rangle .
\end{aligned} \tag{11-14}$$

Since $\mathrm{Jord}(\sigma)$ is a set (see Theorem 5-8), we see that (11-10) and (11-14) imply that $b = 0$ and there is no i such that $\alpha_i = 0$. Thus, we see that τ is isomorphic to

$$\begin{aligned}
&\times_{i,\ \alpha_i=\frac{1}{2}} \langle [-\frac{l_i}{2}, \frac{l_i}{2}]^{(\chi_i)}\rangle \times \langle [-\frac{l_i-2}{2}, \frac{l_i-2}{2}]^{(\chi_i)}\rangle \times \\
&\times_{i,\ \alpha_i=1,\ l_i\geq 2} \langle [-\frac{l_i+1}{2}, \frac{l_i+1}{2}]^{(\chi_i)}\rangle \times \langle [-\frac{l_i-3}{2}, \frac{l_i-3}{2}]^{(\chi_i)}\rangle \times \\
&\times_{(l',\chi')\in \mathrm{Jord}(\sigma_{sn})} \langle [-\frac{l'-1}{2}, \frac{l'-1}{2}]^{(\chi')}\rangle .
\end{aligned}$$

If there is an i such that $\alpha_i = \frac{1}{2}$, then $(l_i+1, \chi_i) \in \mathrm{Jord}(\sigma)$ and $(l_i-1, \chi_i) \in \mathrm{Jord}(\sigma)$ $(l_i \geq 2)$. Now, if $l_i = 1$, then $(2, \chi_i) \in \mathrm{Jord}(\sigma)$. This contradicts (2) of Theorem 11-3. On the other hand, if $l_i \geq 2$, then $(l_i \pm 1, \chi_i) \in \mathrm{Jord}(\sigma)$. Clearly, if we put $a = l_i + 1$, then $a_- = l_i - 1$. Hence $a - a_- = 2$. This contradicts (1) of Theorem 11-3. Thus, we see that τ is isomorphic to

$$\times_{i,\ \alpha_i=1,\ l_i\geq 2} \langle [-\frac{l_i+1}{2}, \frac{l_i+1}{2}]^{(\chi_i)}\rangle \times \langle [-\frac{l_i-3}{2}, \frac{l_i-3}{2}]^{(\chi_i)}\rangle \times \times_{(l',\chi')\in \mathrm{Jord}(\sigma_{sn})} \langle [-\frac{l'-1}{2}, \frac{l'-1}{2}]^{(\chi')}\rangle . \tag{11-15}$$

To complete the proof we need to show that there is no i such that $\alpha_i = 1$ and $l_i \geq 2$. Assume that this is not the case. Let us fix some i_0 such that $\alpha_{i_0} = 1$ and $l_{i_0} \geq 2$. Then, (11-15) implies that

$$(l_{i_0} + 2, \chi_{i_0}) \in \mathrm{Jord}(\sigma) \tag{11-16}$$

and

$$\sigma_m \simeq \times_{i,\ \alpha_i=1,\ l_i\geq 2} \langle [-\frac{l_i-1}{2}, \frac{l_i-1}{2}]^{(\nu^{\alpha_i^{(m)}} \chi_i^{(m)})}\rangle \rtimes \sigma_{sn}, \text{ for all } m > 0.$$

Since $\lim_m \alpha_i^{(m)} = \alpha_i = 1$, we may assume that $\alpha_i^{(m)} \in]\frac{1}{2}, 1[$. Then, since σ_m is unitary, Theorem 5-14 implies that $\mathbf{e}^{(m)}(l_{i_0}, \chi_{i_0}^{(m)})$ satisfies Definition 5-13 (3). In particular, $\chi_{i_0}^{(m)}(\varpi) = -1$. Now, $\lim_m \chi_{i_0}^{(m)}(\varpi) = \chi_{i_0}(\varpi)$ implies that we have $\chi_{i_0}^{(m)} = \chi_{i_0}$ for all $m > 0$. Next, according to Definition 5-13 (3) (a) (applied to any σ_m) we have the two cases.

Assume $(l_{i_0}, \chi_{i_0}^{(m)}) \in \mathrm{Jord}(\sigma_{sn})$. Then (11-15) implies that $(l_{i_0}, \chi_{i_0}) \in \mathrm{Jord}(\sigma)$. If we put $a = l_{i_0} + 2$ and apply (11-16), then we obtain that $a_- = l_{i_0}$. This contradicts (1) of Theorem 11-3.

Assume $(l_{i_0}, \chi_{i_0}^{(m)}) \notin \mathrm{Jord}(\sigma_{sn})$. Then, according to Definition 5-13 (3) (a) (applied to any σ_m), there must exist $i \neq i_0$ such that $l_i = l_{i_0}$, $\chi_i = \chi_{i_0} = \chi_{i_0}^{(m)} = \chi_i^{(m)}$, and $\alpha_i^{(m)} \in \mathbf{e}^{(m)}(l_{i_0}, \chi_{i_0}^{(m)})$. Hence $l_i = l_{i_0}$, $\chi_i = \chi_{i_0}$, and $\alpha_i = \alpha_{i_0} = 1$. Then (11-15) and (11-16) imply that $(l_{i_0} + 2, \chi_{i_0})$ appears twice in $\mathrm{Jord}(\sigma)$. This contradicts the fact that $\mathrm{Jord}(\sigma)$ is a set. (See Theorem 5-8.) □

12. Examples

In this section we give examples of our algorithm presented in Section 10. We use the notation introduced there. In particular, when we speak about steps we mean the steps of the algorithm in Section 10. We begin by the following remark:

REMARK 12-1. *Suppose we consider a representation*

$$\sigma = \sigma^G(\chi_1, \ldots, \chi_n). \tag{12-2}$$

Let χ be a unitary (unramified) character such that $\chi = \chi_i^u$ for some index i. Consider the subsequence $\varphi_1, \ldots, \varphi_m$ of $\chi_1, \ldots, \chi_n$ formed by χ_i for which $\chi_i^u \in \{\chi, \chi^{-1}\}$, and the representation

$$\sigma^G(\varphi_1, \ldots, \varphi_m). \tag{12-3}$$

From the classification theorem (see Theorem 5-14 and Definition 5-13) it is clear that if (12-2) is unitarizable, then (12-3) is unitarizable. The converse also holds: if (12-3) is unitarizable for all χ as above, then (12-2) is unitarizable.

Therefore, it is enough to understand how the algorithm works in the case that all χ_i^u belong to one $\{\chi, \chi^{-1}\}$. We consider below only such examples.

A. First consider the easy case: $\chi \neq \chi^{-1}$, i.e., $\chi \notin \{\mathbf{1}_{F^\times}, \mathbf{sgn}_u\}$. In this group of examples we always assume that G is not a symplectic group. (If one adds the segment $\{\mathbf{1}_{F^\times}\}$ in the multiset η, then one would obtain examples for symplectic groups.)

12.1. Example. Look at $\sigma = \sigma^G(\chi, \nu\chi)$. Now, steps 1 and 2 give

$$\eta = \{[0,1]^{(\chi)}, [-1,0]^{(\chi^{-1})}\} = \{[-\frac{1}{2}, \frac{1}{2}]^{(\frac{1}{2}\chi)}, [-\frac{1}{2}, \frac{1}{2}]^{(-\frac{1}{2}\chi^{-1})}\}.$$

Step 3 is not executed here. In step 4, $\eta_{nsd,+} = \{[-\frac{1}{2}, \frac{1}{2}]^{(\frac{1}{2}\chi)}\}$. Since $1/2 \geq 1/2$, σ is not unitarizable.

12.2. Example. Look at $\sigma = \sigma^G(\chi, \nu\chi, \nu\chi^{-1})$. Now, steps 1 and 2 give

$$\eta = \{[-1,1]^{(\chi)}, [-1,1]^{(\chi^{-1})}\}.$$

In step 3 we have $\mathrm{Jord} = \{(3,\chi)\,(3,\chi^{-1})\}$, $\eta_{nsd,unit} = \{[-1,1]^{(\chi)}, [-1,1]^{(\chi^{-1})}\}$, and the new η is $\eta - \eta_{nsd,unit}$. The steps 4–8 are not executed for the new η (since it is empty). Step 9 sends us directly to step 10. Step 10 implies that σ is negative (therefore unitarizable) with $\mathrm{Jord}(\sigma) = \{(3,\chi)\,(3,\chi^{-1})\}$ (and $\mathbf{e} = \emptyset$).

12.3. Example. Look at $\sigma = \sigma^G(\nu^{-1/4}\chi, \nu^{3/4}\chi)$. Now, steps 1 and 2 give

$$\eta = \{[-1/4, 3/4]^{(\chi)}, [-3/4, 1/4]^{(\chi^{-1})}\} = \{[-\frac{1}{2}, \frac{1}{2}]^{(\frac{1}{4}\chi)}, [-\frac{1}{2}, \frac{1}{2}]^{(-\frac{1}{4}\chi^{-1})}\}.$$

Step 3 is not executed. In step 4 we have $\eta_{nsd,+} = \{[-\frac{1}{2}, \frac{1}{2}]^{(\frac{1}{4}\chi)}\}$ Now, we have a non-unitarizability since the Hermitian condition is not satisfied: $\eta_{nsd,+} \neq \bar{\eta}_{nsd,+}$. (The exponents are $< 1/2$.)

B. Now, consider the case $\chi = \chi^{-1}$. This means $\chi \in \{\mathbf{1}_{F^\times}, \mathbf{sgn}_u\}$. We shall deal with χ through the constant $\alpha_\chi \in \{0, \frac{1}{2}, 1\}$ defined in Definition 5-2. We write every χ_i (uniquely) in the form $\nu^{e(\chi_i)}\chi$ where $e(\chi_i) \in \mathbb{R}$, and instead of assigning $\sigma = \sigma^G(\chi_1, \dots, \chi_n)$, we assign the sequence of exponents $e(\chi_1), \dots, e(\chi_n)$. (Repeated exponents as in the sequence $1, 1, 1, 2, 2, 3$ shall be written as follows: $1^{(3)}, 2^{(2)}, 3$.) In what follows $[a, b]$ means $[a, b]^{(\chi)} = [-\frac{b-a}{2}, \frac{b-a}{2}]^{(\nu^{\frac{a+b}{2}}\chi)}$. We shall start with simple examples.

12.4. Example. We consider the exponents $1, 2$. Let $\alpha_\chi = 1/2$. Then σ is a representation of $\mathrm{SO}(5, F)$. Now, step 2 gives $\eta = \{[-2, -1], [1, 2]\}$. Now, step 6 is relevant and it implies $[1, 2] \in \eta_{sd,\{0,1\},+}$. Since $\frac{1+2}{2} \geq 1$, σ is not unitarizable.

12.5. Example. We consider the exponents $1, 2$. Let $\alpha_\chi = 0$. Now, step 2 gives $\eta = \{[-2, -1], [1, 2], \dots\}$. (In addition to the displayed segments, one needs to include $\{\mathbf{1}_{F^\times}\}$ if $G = \mathrm{Sp}(4, F)$; in which case $\chi = \mathbf{sgn}_u$.) Now, step 5 is relevant and it implies $[1, 2] \in \eta_{sd,\frac{1}{2},+}$. Since $\frac{1+2}{2} \geq 1/2$, σ is not unitarizable.

12.6. Example. We consider the exponents $1, 2$. Consider $\alpha_\chi = 1$. (Then σ is a representation of $\mathrm{Sp}(4, F)$ and $\chi = \mathbf{1}_{F^\times}$; see Definition 5-2). Step 2 gives $\eta = \{[-2, 2]\}$. Since $(5 - 1)/2 = 2$ and $5 - (2\alpha_\chi + 1) = 2 \in 2\mathbb{Z}$, step 7 is relevant. It implies $\mathrm{Jord} = \{(5, \mathbf{1}_{F^\times})\}$. We remove $[-2, 2]$ from η and proceed further to step 9. Step 9 takes us to step 10. Step 10 shows that σ is (strongly) negative.

12.7. Example. We consider the exponents $1/2, 3/2$. We assume that σ is a representation of $\mathrm{O}(4, F)$. Then $\alpha_\chi = 0$. Now step 2 gives $\eta = \{[-3/2, 3/2]\}$. Since $(4 - 1) = 3/2$ and $4 - (2\alpha_\chi + 1) = 3 \notin 2\mathbb{Z}$, we proceed to step 8. Since $4 \geq 3$, we see that σ is not unitarizable by (c) of (ii) in step 8.

12.8. Example. We consider the exponents $1/2, 3/2$. Consider $\alpha_\chi = 1/2$. Then σ is a representation of $\mathrm{SO}(5, F)$. Again step 2 gives $\eta = \{[-3/2, 3/2]\}$. Since $(4 - 1) = 3/2$ and $4 - (2\alpha_\chi + 1) = 2 \in 2\mathbb{Z}$, we proceed to step 7. We remove $[-2, 2]$ from η and proceed further to step 9. Step 9 takes us to step 10. Step 10 shows that σ is (strongly) negative.

12.9. Example. We consider the exponents $1/2, 3/2$. Consider $\alpha_\chi = 1$. Then σ is a representation of $\mathrm{Sp}(4, F)$ and $\chi = \mathbf{1}_{F^\times}$. Step 2 gives $\eta = \{[-3/2, 3/2], [0, 0]\}$. Step 7 will put $(1, \mathbf{1}_{F^\times})$ in Jord, but we get non-unitarizability from (c) of (ii) in step 8 (applied to $\Delta = [-3/2, 3/2]$).

Now come some slightly more complicated examples.

12.10. Example. We consider the exponents $0^{(4)}, 1^{(5)}, 2^{(3)}, 3$. Then $n = 4 + 5 + 3 + 1 = 13$. We assume that σ is a representation of $\mathrm{O}(26, F)$ or $\mathrm{SO}(27, F)$. We perform step 2

characters of σ^{GL}	η
$-3, -2^{(3)}, -1^{(5)}, 0^{(8)}, 1^{(5)}, 2^{(3)}, 3$	$\emptyset$
$-2^{(2)}, -1^{(4)}, 0^{(7)}, 1^{(4)}, 2^{(2)}$	$[-3,3]$
$-2, -1^{(3)}, 0^{(6)}, 1^{(3)}, 2$	$[-3,3], [-2,2]$
$-1^{(2)}, 0^{(5)}, 1^{(2)}$	$[-3,3], 2 \cdot [-2,2]$
$-1, 0^{(4)}, 1$	$[-3,3], 2 \cdot [-2,2], [-1,1]$
$0^{(3)}$	$[-3,3], 2 \cdot [-2,2], 2 \cdot [-1,1]$
$\emptyset$	$[-3,3], 2 \cdot [-2,2], 2 \cdot [-1,1], 3 \cdot [0,0]$.

Consider first $\alpha_\chi = 0$. Then σ is a representation of $\mathrm{O}(26, F)$. Now, step 7 gives $\mathrm{Jord} = \{(7, \chi), 2 \cdot (5, \chi), 2 \cdot (3, \chi), (1, \chi)\}$ and $\mathbf{e} = \emptyset$. We proceed to step 10. The representation σ is a negative representation attached to Jord.

Now consider reducibility at $\alpha_\chi = 1/2$. Then σ is a representation of $\mathrm{SO}(27, F)$. Now, step 8 is relevant. In the first iteration of step 8 the largest possible l is $l = 2 \cdot 3 + 1 = 7$ and the corresponding Δ is $\Delta = [-3, 3]$. We apply (ii) (c), to see that σ is not unitarizable.

12.11. Example. Consider the exponents $0^{(6)}, 1^{(8)}, 2^{(3)}, 3^{(2)}, 4$. Let $n = 6 + 8 + 3 + 2 + 1 = 20$. Assume that σ is a representation of $\mathrm{SO}(41, F)$. Then $\alpha_\chi = 1/2$. As in the previous example, step 2 gives

$$\eta = \{[-4,4], [-3,3], [-2,2], 5 \cdot [-1,1], 4 \cdot [0,0]\}.$$

Applying step 8 (ii) (c) twice and step 8 (i) twice, we obtain

$$\mathbf{e} = \{(8, \chi, 1/2), (4, \chi, 1/2)\}, \quad \mathrm{Jord} = \{4 \cdot [-1,1], 4 \cdot [0,0]\}.$$

We proceed directly to step 10. We obtain unitarizability.

12.12. Example. Consider the exponents $1/4$, $4/6$ and $5/6$. Let $n = 1 + 1 + 1 = 3$. Assume that σ is a representation of $\mathrm{Sp}(6, F)$. Let $\chi = \mathbf{1}_{F^\times}$. Then $\alpha_\chi = 1$. Step 2 gives the multiset

$$\eta = \{[-1/4, -1/4], [-4/6, -4/6], [-5/6, -5/6], [0,0], [1/4, 1/4], [4/6, 4/6], [5/6, 5/6]\}.$$

Now, step 6 implies

$$\mathbf{e} = \{(1, \chi, 1/4), (1, \chi, 4/6), (1, \chi, 5/6)\}.$$

Next, step 7 implies $\mathrm{Jord} = \{(1, \chi)\}$. Step 10 implies unitarizability.

References

[A] A. M. Aubert, *Dualité dans le groupe de Grothendieck de la catégorie des représentations lisses de longueur finie d'un groupe réductif p-adique*, Trans. Amer. Math. Soc. **347** (1995), 2179-2189. (and *Erratum*, Trans. Amer. Math. Soc **348** (1996), 4687-4690.)

[BdRe] A. I. Badulescu, D. A. Renard, *Sur une conjecture de Tadić*, Glasnik Mat., **39** no. 1 (2004), 49-54.

[Bn] D. Ban, *Linear indepedence of intertwining operators*, J. Algebra **271** (2004), 749-767.

[Bb] D. Barbasch, *The unitary spherical spectrum for split classical groups*, J. Inst. Math. Jussieu **9** (2010), 265–356.

[BbMo] D. Barbasch, A. Moy, *A unitary criterion for p-adic groups*, Invent. Math. **98** (1989), 19-37.

[BbMo1] D. Barbasch, A. Moy, *Reduction to real infinitensimal character in affine Hecke algebras*, J. Amer. Math. Soc. **6**, (1993), 611-635.

[BbMo2] D. Barbasch, A. Moy, *Unitary spherical spectrum for p-adic classical groups. Representations of Lie groups, Lie algebras and their quantum analogues,* Acta Appl. Math. **44** (1996), no. 1-2, 3–37.

[Bu] D. Baruch, *A proof of Kirillov's conjecture,* Ann. of Math. (2) **158** (2003), no. 1, 207–252.

[BeDeK] J. Bernstein, P. Deligne, D. Kazhdan, *Trace Paley-Wiener theorem for reductive p-adic groups*, J. Analyse Math. **47** (1986), 180-192.

[Be] J. Bernstein, rédigé par P. Deligne, *Le 'centre' de Bernstein*, Représentations des groupes réductif sur un corps local, Herman, Paris, 1984.

[Be1] J. Bernstein, *Draft of: Representations of p-adic groups* (lectures at Harvard University, 1992, written by Karl E. Rumelhart).

[Be2] J. Bernstein, *P-invariant distributions on* GL(N) *and the classification of unitary representations of* GL(N) *(non-archimedean case),* Lie Group Representations II, Lecture Notes in Math. **1041**, Springer-Verlag, Berlin, 1984, 50-102.

[BeZ] J. Bernstein, A. V. Zelevinsky, *Representations of the group* GL(n, F), *where F is a local non-Archimedean field* (Russian), Uspehi Mat. Nauk **31** (1976), no. 3(189), 5–70.

[BeZ1] J. Bernstein, A. V. Zelevinsky, *Induced representations of reductive p-adic groups I,* Ann. Sci. École Norm. Sup., **10** (1977), 441-472.

[Cr] P. Cartier, *Representations of p-adic groups: a survey. Automorphic forms, representations and L-functions (Proc. Sympos. Pure Math., Oregon State Univ., Corvallis, Ore., 1977), Part 1,* Proc. Sympos. Pure Math. **XXXIII**, Amer. Math. Soc., Providence, R.I (1979), 111–155.

[Cs] W. Casselman, *Introduction to the theory of admissible representations of p-adic reductive groups*, preprint (1995).

[Di] J. Dixmier, *Les C^*-algebras et leurs Représentations,* Gauthiers-Villars, Paris, 1969.

[F] J. M. G. Fell, *Non-unitary dual space of groups,* Acta Math. **114**, 1965, 267-310.

[G] D. Goldberg, *Reducibility of induced representations for* Sp($2n$) *and* SO(n), Amer. J. Math. **116** (1994), 1101–1151.

[Jn] C. Jantzen, *Duality and supports of induced representations for orthogonal groups,* Canad. J. Math. **57** (2005), no. 1, 159–179.

[Jc] H. Jacquet, *On the residual spectrum of* GL(n), Lie Group Representations II, Lecture Notes in Math. **1041**, Springer-Verlag, Berlin, 1984, 185-208.

[K] D. Kazhdan, *Connection of the dual space of a group with the structure of its closed subgroups,* Functional Anal. Appl. **1** (1967), 63-65.

[KLu] D. Kazhdan, G. Lusztig, *Proof of the Deligne–Langlands conjecture for Hecke algebras,* Invent. Math. **87** (1987), 153-215.

[LMT] E. Lapid, G. Muić, M. Tadić, *On the generic unitary dual of quasisplit classical groups,* Int. Math. Res. Not. (2004), **no. 26**, 1335–1354.

[Mi] D. Miličić, *On C^*-algebras with bounded trace,* Glasnik Mat. **8(28)** (1973), 7-21.

[Mœ] C. Mœglin, *Sur la classification des séries discrètes des groupes classiques p-adiques: paramètres de Langlands et exhaustivité,* J. Eur. Math. Soc. (JEMS) **4** (2002), 143-200.

[MœT] C. Mœglin, M. Tadić, *Construction of discrete series for classical p-adic groups,* J. Amer. Math. Soc. **15** (2002), 715-786.

[MœViW] C. Mœglin, M.-F. Vignéras, J.L. Waldspurger, *Correspondence de Howe sur un corps p–adique,* Lecture Notes in Math. **1291**, 1987.

[M] G. Muić, *Some results on square integrable representations; Irreducibility of standard representations,* Internat. Math. Res. Notices **14** (1998), 705-726.

[M1] G. Muić, *On generic irreducible representation for $Sp(n, F)$ and $SO(2n+1, F)$,* Glas. Mat. Ser III **33(53)** (1988), 19–31.

[M2] G. Muić, *A proof of Casselman-Shahidi's conjecture for quasi-split classical groups,* Canad. Math. Bull., **44** (2001), 298-312.

[M3] G. Muić, *Composition Series of Generalized Principal Series; The Case of Strongly Positive Discrete Series,* Israel J. Math. **149** (2004), 157-202.

[M4] G. Muić, *On the Non-Unitary Unramified Dual for Classical p-adic Groups,* Trans. Amer. Math. Soc. **358** (2006), 4653–4687.

[M5] G. Muić, *Construction of Steinberg type representations for reductive p-adic groups,* Math. Z. **253** (2006), no. 3, 635–652.

[M6] G. Muić, *On Certain Classes of Unitary Representations for Split Classical Groups,* Canad. J. Math. **59** (2007), 148–185.

[M7] G. Muić, *A Geometric Construction of Intertwining Operators for Reductive p–adic Groups,* Manuscripta Math. **125 (2)** (2008), 241–272.

[MSh] G. Muić, F. Shahidi, *Irreducibility of standard representations for Iwahori-spherical representations,* Math. Ann. **312**(1998), 151–165.

[R] S. Rallis, *Langlands' functoriality and the Weil representation,* Amer. J. Math. **104** (1982), 469–515.

[Sh1] F. Shahidi, *On certain L-functions*, Amer. J. Math. **103** (1981), 297-356.

[Sh2] F. Shahidi, *A proof of Langlands' conjecture on Plancherel measures; complementary series for p-adic groups*, Ann. of Math. **132** (1990), 273-330.

[Sp] B. Speh, *Unitary representations of* $\mathrm{GL}(n,\mathbb{R})$ *with non-trivial* (g,K)*- cohomology,* Invent. Math. **71** (1983), 443-465.

[T1] M. Tadić, *The topology of the dual space of a reductive group over a local field,* Glas. Mat. Ser. III **18(38)** (1983), no. 2, 259–279.

[T2] M. Tadić, *Unitary representations of general linear group over real and complex field,* preprint MPI/SFB 85-22 Bonn (1985).

[T3] M. Tadić, *Classification of unitary representations in irreducible representations of general linear group (non-Archimedean case)*, Ann. Sci. Ecole Norm. Sup. **19** (1986), 335-382.

[T4] M. Tadić, *Spherical unitary dual of general linear group over non-Archimedean local field,* Ann. Inst. Fourier (Grenoble) **36** (1986), no. 2, 47–55.

[T5] M. Tadić, *Topology of unitary dual of non–Archimedean* $\mathrm{GL}(n)$, Duke Math. J. **55** (1987), no. 2, 385–422.

[T6] M. Tadić, *Geometry of dual spaces of reductive groups (non–archimedean case),* J. Analyse Math. Vol. **51** (1988), 139-181.

[T7] M. Tadić, *On limits of characters of irreducible unitary representations,* Glas. Mat. Ser. III **23(43)** (1988), no. 1, 15–25.

[T8] M. Tadić, *An external approach to unitary representations,* Bull. Amer. Math. Soc. (N.S.) **28** (1993), no. 2, 215–252.

[T9] M. Tadić, *Representations of p-adic symplectic groups*, Compositio Math. **90** (1994), 123-181.

[T10] M. Tadić, *On reducibility of parabolic induction,* Israel J. Math. **107** (1998), 29–91.

[T11] M. Tadić, *A family of square integrable representations of classical p-adic groups in the case of general half-integral reducibilities,* Glas. Mat. Ser. III **37(57)** (2002), no. 1, 21–57.

[T12] M. Tadić, *An exercise on unitary representations in the case of complex classical groups*, Functional analysis IX (Dubrovnik, 2005), 91-102., Various Publ. Ser. 48, Univ. Aarhus, Aarhus, 2007.

[Vo] D. A. Vogan, *Unitarizability of certain series of representations*, Ann. of Math. **120** (1984), 141-187.

[W] J. L. Waldspurger, *La formule de Plancherel pour les groupes p-adiques, d'après Harish-Chandra*, Journal of the Inst. of Math. Jussieu **2(2)** (2003), 235-333.

[Ze] A. V. Zelevinsky, *Induced representations of reductive p–adic groups. On irreducible representations of* $\mathrm{GL}(n)$, Ann. Sci. Ecole Norm. Sup. **13** (1980), 165-210.

Dept. of Mathematics, University of Zagreb, Bijenička 30, 10000 Zagreb, Croatia
E-mail address: `gmuic@math.hr`

Dept. of Mathematics, University of Zagreb, Bijenička 30, 10000 Zagreb, Croatia
E-mail address: `tadic@math.hr`

Clay Mathematics Proceedings
Volume 13, 2011

Parametrization of Tame Supercuspidal Representations

Fiona Murnaghan

This paper is dedicated to Freydoon Shahidi on the occasion of his 60th birthday.

ABSTRACT. Let G be a connected reductive p-adic group that splits over a tamely ramified extension. We describe a parametrization of the equivalence classes of tame supercuspidal representations of G which is valid subject to mild hypotheses. Certain of these equivalence classes are parametrized by G-conjugates of quasicharacters of elliptic maximal tori in G. We adapt the parametrization to obtain a parametrization of the equivalence classes of self-contragredient tame supercuspidal representations. Finally, we discuss how some features of the parametrization are reflected in properties of the characters of the representations.

1. Introduction

Let F be a nonarchimedean local field, and let $\mathbf{G}$ be a connected reductive F-group. Throughout this paper, we assume that $\mathbf{G}$ splits over a tamely ramified extension of F. Under this assumption, the results of [**Y**] give a construction of irreducible supercuspidal representations of the group $G = \mathbf{G}(F)$ of F-rational points of $\mathbf{G}$. We will refer to these representations as tame supercuspidal representations. In general, there exist supercuspidal representations of G that are not tame (see, for example, [**BK**] and [**S**]). However, as shown in [**K**], if G satisfies some additional tameness hypotheses, then all irreducible supercuspidal representations of G are tame.

The construction of [**Y**] begins with the definition of collections of data which in this paper we refer to as (reduced) cuspidal G-data (see Section 8). Yu associates a family of irreducible supercuspidal representations of G to each cuspidal G-datum. He proves that all of the representations in each such family belong to the same equivalence class, thereby obtaining a map from the set of cuspidal G-data to the set of equivalence classes of supercuspidal representations of G. This map is not one-to-one, and the problem of determining criteria that detect when two distinct

2010 *Mathematics Subject Classification.* Primary: 22E50; Secondary: 20G05, 20G25.
Key words and phrases. Admissible representation, supercuspidal, K-type, reductive p-adic group.
The author's research is supported by an NSERC Discovery Grant.

G-data give rise to the same equivalence class of representations is not addressed in [**Y**].

In [**HM**], Hakim and the author define an equivalence relation on the set of cuspidal G-data and, subject to certain hypotheses (concerning the behaviour of quasicharacters of twisted Levi subgroups of G), we prove that two cuspidal G-data are equivalent if and only if the corresponding equivalence classes of tame supercuspidal representations coincide. In this paper, we describe (see Theorem 9.1) a parametrization of the (equivalence classes of) tame supercuspidal representations which is obtained from the equivalence relation on cuspidal G-data.

A cuspidal G-datum comprises several parts. One part is an irreducible depth zero supercuspidal representation π' of $G' = \mathbf{G}'(F)$, where $\mathbf{G}'$ is an elliptic tamely ramified twisted Levi subgroup of $\mathbf{G}$. Other parts of the G-datum include a tamely ramified twisted Levi sequence $\vec{\mathbf{G}}$ in $\mathbf{G}$, having the property that the first group in the sequence is $\mathbf{G}'$, along with a sequence $\vec{\phi}$ of quasicharacters of the F-rational points of the various groups occurring in $\vec{\mathbf{G}}$. These quasicharacters are required to satisfy several conditions which we do not specify here. The reader may refer to Definition 8.2 for more information about the conditions on the quasicharacters in $\vec{\phi}$ and other ingredients of a cuspidal G-datum. We may define a quasicharacter ϕ of G' by taking the product of the restrictions to G' of the quasicharacters in $\vec{\phi}$. Thus, to each cuspidal G-datum, we associate a triple of the form (G', π', ϕ), where G' and π' are as above, and ϕ is a quasicharacter of G'. As discussed below, when $G' \neq G$, some quasicharacters of G' do not occur in triples that are associated to cuspidal G-data.

We define an equivalence relation on the set of triples (G', π', ϕ), where ϕ is an arbitrary quasicharacter of G', and $\mathbf{G}'$ is a (not necessarily elliptic) twisted Levi subgroup of G, by declaring that if $\dot{G} = g^{-1}G'g$ for some $g \in G$, and the representations $\pi'\phi$ and ${}^g(\dot{\pi}\dot{\phi})$ are equivalent representations of G', then $(\dot{G}, \dot{\pi}, \dot{\phi})$ is G-equivalent to (G', π', ϕ). (Here the element g is conjugating the representation $\dot{\pi}\dot{\phi}$ of $\dot{G}$ to a representation ${}^g(\dot{\pi}\dot{\phi})$ of G' - see Section 2.) The equivalence relation on cuspidal G-data is defined in such a way that the triples associated to the elements of an equivalence class of cuspidal G-data comprise one G-equivalence class of triples. Therefore, results of [**HM**] (stated in Theorem 8.7) tell us that the equivalence classes of tame supercuspidal representations of G correspond to the G-equivalence classes of triples that are associated to cuspidal G-data.

Because of the properties of cuspidal G-data, the quasicharacters ϕ that occur in triples which are associated to cuspidal G-data must have specific properties. In particular, each such quasicharacter ϕ must be G-factorizable in the sense of Definitions 5.1 and 5.3 and G-regular on G'_{0+} in the sense of Definition 6.1. In this case, ϕ can be factorized in different ways. However, as shown in Proposition 5.4, any two factorizations are related in a specific manner. As indicated in Lemma 8.9, each of these factorizations can be used, along with π', to produce a cuspidal G-datum that maps to the triple (G', π', ϕ).

If the equivalence class of a tame supercuspidal representation corresponds to the equivalence class of a triple where $G' = T$ is a maximal torus in G, we say that the representation is toral. In this case, T is elliptic, that is, T is compact modulo the centre of G. As stated in Corollary 9.4, the toral supercuspidal representations of G, up to equivalence, G can be parametrized by pairs (T, ϕ), up to G-conjugacy, where T runs over the conjugacy classes of tamely ramified elliptic maximal tori in G, and ϕ varies over the set of G-factorizable quasicharacters of T that are G-regular on T_{0+}.

In the first part of Section 10 (see Proposition 10.8 and Corollary 10.10), using the theory of Deligne-Lusztig representations of finite groups of Lie type, we describe a parametrization of some classes of depth zero supercuspidal representations in terms of orbits of certain depth zero quasicharacters of elliptic maximal tori. In the second part of the section (see Theorem 10.13), we apply Corollary 10.10 and Theorem 9.1 to parametrize some more equivalence classes of tame supercuspidals in terms of orbits of certain quasicharacters of tamely ramified elliptic maximal tori. Except when G is a general linear group, these classes of tame supercuspidal representations do not exhaust the classes of tame supercuspidal representations.

When F has characteristic zero, Howe ([**H**]) parametrized the equivalence classes of tame supercuspidal representations of general linear groups in terms of equivalence classes of F-admissible quasicharacters of multiplicative groups of tamely ramified extensions of F. Howe's parametrization may be viewed as a correspondence between the equivalence classes of tame supercuspidal representations of general linear groups and the conjugacy classes of particular kinds of quasicharacters of elliptic maximal tori. In Section 11, we describe the relation between Howe's parametrization and the parametrization of Theorem 9.1 (as it applies to general linear groups).

As shown in Section 12, the equivalence classes of self-contragredient tame supercuspidal representations are parametrized by equivalence classes that contain triples of a particular form. If π is self-contragredient, then either π has depth zero, or π corresponds to a triple (G', π', ϕ) having the following property: there exists $g \in G$ such that g normalizes G', $g \notin G'$, $g^2 \in G'$, ${}^g\phi = \phi^{-1}$, and ${}^g\pi'$ is equivalent to the contragredient of π'. If π is a self-contragredient toral supercuspidal representation, then the pair (T, ϕ) of Corollary 9.4 may be chosen so that ${}^g\phi = \phi^{-1}$ for some g such that g normalizes T and $g^2 \in T$.

Suppose that (G', π', ϕ) is a triple whose equivalence class parametrizes the equivalence class of a tame supercuspidal representation π. In Section 13, we discuss results that use properties of ϕ to describe aspects of the behaviour of the character Θ_π of π. We also discuss some cases in which properties of Θ_π may be obtained from properties of π'.

The author thanks the referee for detailed comments on the paper.

2. Basic notation

Let F be a nonarchimedean local field, with residue field $\mathfrak{f}$. In order to be able to apply some results from [**HM**] and [**Y**], we assume that the characteristic of $\mathfrak{f}$ is odd. We normalize the valuation ν on F so that $\nu(F^\times) = \mathbb{Z}$. Let $\mathbf{G}$ be a connected reductive group defined over F, and let $\mathfrak{g}$ be the Lie algebra of $\mathbf{G}$. The notation $G = \mathbf{G}(F)$ and $\mathfrak{g} = \mathfrak{g}(F)$ will be used to denote the F-rational points of $\mathbf{G}$ and $\mathfrak{g}$, respectively.

Let $\mathcal{B}(G) = \mathcal{B}(\mathbf{G}, F)$ be the (extended) Bruhat-Tits building of G. As in [**MP1**], we can associate to any point x in $\mathcal{B}(G)$ a parahoric subgroup $G_{x,0}$ of G and a filtration $\{ G_{x,r} \mid r \in \mathbb{R},\ r \geq 0 \}$ of the parahoric, together with a filtration $\{ \mathfrak{g}_{x,r} \mid r \in \mathbb{R} \}$ of the Lie algebra $\mathfrak{g}$. We remark that the indexing of these filtrations depends on the choice of valuation ν. With our choice of valuation ν, if ϖ is a prime element in F, we have $\varpi \mathfrak{g}_{x,r} = \mathfrak{g}_{x,r+1}$ for all $x \in \mathcal{B}(G)$, and all $r \in \mathbb{R}$. If $r \in \mathbb{R}$ and

$x \in \mathcal{B}(G)$, set $\mathfrak{g}_{x,r+} = \bigcup_{s>r} \mathfrak{g}_{x,s}$ and, if $r \geq 0$, $G_{x,r+} = \bigcup_{s>r} G_{x,s}$. If $r \geq 0$, let $G_r = \bigcup_{x \in \mathcal{B}(G)} G_{x,r}$ and $G_{r+} = \bigcup_{x \in \mathcal{B}(G)} G_{x,r+}$.

Moy and Prasad also define lattices $\{\, \mathfrak{g}^*_{x,r} \mid r \in \mathbb{R} \,\}$ in the dual $\mathfrak{g}^*$ of $\mathfrak{g}$, as follows:

$$\mathfrak{g}^*_{x,r} = \{\, X^* \in \mathfrak{g}^* \mid X^*(\mathfrak{g}_{x,(-r)+}) \subset \mathfrak{p} \,\},$$

where $\mathfrak{p}$ is the maximal ideal in the ring of integers of F. Set $\mathfrak{g}^*_{x,r+} = \bigcup_{s>r} \mathfrak{g}^*_{x,s}$.

From this point onward, we assume that $\mathbf{G}$ splits over a tamely ramified extension of F. The centre of $\mathbf{G}$ will be denoted by $\mathbf{Z}$ and its Lie algebra by $\mathfrak{z}$. Whenever convenient, we identify the dual $\mathfrak{z}^*$ of $\mathfrak{z} = \mathfrak{z}(F)$ with the subspace of $\mathfrak{g}^*$ consisting of those elements that are invariant under the co-adjoint action Ad^* of G on $\mathfrak{g}^*$. There are filtrations on $\mathfrak{z}$ and $\mathfrak{z}^*$ which have the property that, if $r \in \mathbb{R}$, then $\mathfrak{z}_r = \mathfrak{z} \cap \mathfrak{g}_{x,r}$, and $\mathfrak{z}^*_r = \mathfrak{z}^* \cap \mathfrak{g}^*_{x,r}$, for any point $x \in \mathcal{B}(G)$. Set $\mathfrak{z}_{r+} = \mathfrak{z} \cap \mathfrak{g}_{x,r+}$ and $\mathfrak{z}^*_{r+} = \mathfrak{z}^* \cap \mathfrak{g}^*_{x,r+}$.

If $x \in \mathcal{B}(G)$, $r \in \mathbb{R}$, $r > 0$, and $s = r/2$, we have an isomorphism

$$e = e_{x,r} : \mathfrak{g}_{x,s+}/\mathfrak{g}_{x,r+} \to G_{x,s+}/G_{x,r+}$$

of abelian groups (see Lemma 1.3 and Corollary 2.4 of [**Y**]).

Let Λ be a character of F that is nontrivial on the ring of integers of F and trivial on the maximal ideal $\mathfrak{p}$.

DEFINITION 2.1. If $r > 0$, $s = r/2$, $x \in \mathcal{B}(G)$, and S is a subgroup of $G_{x,s+}$ that contains $G_{x,r+}$, let $\mathfrak{s}$ be the lattice in $\mathfrak{g}_{x,s+}$ such that $\mathfrak{s} \supset \mathfrak{g}_{x,r+}$ and $e(\mathfrak{s}/\mathfrak{g}_{x,r+}) = S/G_{x,r+}$. An element $X^* \in \mathfrak{g}^*_{x,-r}$ defines a character of $S/G_{x,r+}$, hence of S, as follows:

$$e(Y + \mathfrak{g}_{x,r+}) \mapsto \Lambda(X^*(Y)), \qquad Y \in \mathfrak{s}.$$

We say that the element X^* of $\mathfrak{g}^*_{x,-r}$ *realizes* this character of S.

If H is a subgroup of G and $g \in G$, let ${}^gH = gHg^{-1}$. If τ is a representation of H, we define a representation ${}^g\tau$ of ${}^gH$ by setting ${}^g\tau(g_0) = \tau(g^{-1}g_0g)$, $g_0 \in {}^gH$. We denote the normalizer of H in G by $N_G(H)$.

We will use the notation $\tau_1 \simeq \tau_2$ to indicate equivalence of two representations τ_1 and τ_2 of H. If τ and χ are representations of H and χ is one-dimensional, we will use the notation $\tau\chi$ for the twist of τ by χ: $(\tau\chi)(h) = \chi(h)\tau(h)$, $h \in H$.

If K is an open subgroup of G that contains Z, the quotient K/Z is compact, and ρ is an irreducible smooth representation of K, the representation of G obtained via compact induction from ρ will be denoted by $\mathrm{Ind}_K^G \rho$.

3. Twisted Levi subgroups and Levi sequences

DEFINITION 3.1. Suppose that $\mathbf{G}'$ is an E-Levi F-subgroup of $\mathbf{G}$ for some finite extension E of F. Such a group will be called a *twisted Levi subgroup* of $\mathbf{G}$.

If we can choose E to be tamely ramified over F, then we say that $\mathbf{G}'$ is *tamely ramified*. If the centre $\mathbf{Z}'$ of $\mathbf{G}'$ has the property that $\mathbf{Z}'/\mathbf{Z}$ is F-anistropic, we say that $\mathbf{G}'$ is *elliptic*.

If $\mathbf{G}'$ is a twisted Levi subgroup of $\mathbf{G}$, we will refer to the group $G' = \mathbf{G}'(F)$ as a twisted Levi subgroup of G. We will refer to G' as elliptic whenever $\mathbf{G}'$ is elliptic.

Note that the elliptic twisted Levi subgroups of $\mathbf{G}$ are the twisted Levi subgroups that do not lie in proper F-Levi subgroups of $\mathbf{G}$. Suppose that E is a finite tamely ramified extension of F and $\mathbf{G}'$ is the stabilizer in $\mathbf{G}$ of a semisimple element in $\mathfrak{g}$ that lies in the F-rational points of the Lie algebra of an F-torus that splits over E. As shown in Lemma 2.4 of [**K**], $\mathbf{G}'$ is a tamely ramified twisted Levi subgroup of $\mathbf{G}$.

Let $\mathbf{G}'$ be a tamely ramified twisted Levi subgroup of $\mathbf{G}$. Let $\mathfrak{g}'$ and $\mathfrak{z}'$ be the Lie algebras of $\mathbf{G}'$ and $\mathbf{Z}'$, respectively. Following the same notational conventions as for $\mathbf{G}$ and $\mathfrak{g}$, let $\mathfrak{g}' = \mathfrak{g}'(F)$, $\mathfrak{z}' = \mathfrak{z}'(F)$, and $Z' = \mathbf{Z}'(F)$. There is a natural family of embeddings of the building $\mathcal{B}(G')$ of G' into $\mathcal{B}(G)$. Although there is not a canonical way to distinguish one member of the family, all of the embeddings have the same image. Identifying $x \in \mathcal{B}(G')$ with its image under any of these embeddings, we have $G'_{x,r} = G_{x,r} \cap G'$ for $r > 0$ and $\mathfrak{g}'_{x,r} = \mathfrak{g}_{x,r} \cap \mathfrak{g}'$ for $r \in \mathbb{R}$, with analogous equalities when r is replaced by r^+. If we identify $\mathfrak{g}'^*$ with the $\mathrm{Ad}^*\, Z'$-fixed elements in $\mathfrak{g}^*$, we have $\mathfrak{g}'^*_{x,r} = \mathfrak{g}_{x,r} \cap \mathfrak{g}'^*$, and similarly when r^+ replaces r.

DEFINITION 3.2. A sequence $\vec{\mathbf{G}} = (\mathbf{G}^0, \ldots, \mathbf{G}^d)$ of connected reductive F-subgroups of $\mathbf{G}$ is a *twisted Levi sequence* in $\mathbf{G}$ if

$$\mathbf{G}^0 \subsetneq \mathbf{G}^1 \subsetneq \cdots \subsetneq \mathbf{G}^d = \mathbf{G}$$

and there exists a finite extension E of F such that $\mathbf{G}^0 \otimes E$ splits over E and $\mathbf{G}^i \otimes E$ is a Levi subgroup (that is, an E-Levi subgroup) of $\mathbf{G} \otimes E$ for $0 \leq i \leq d$. If E can be chosen to be chosen to be tamely ramified over F, we say that $\vec{\mathbf{G}}$ is *tamely ramified.*

4. Quasicharacters

We refer to a smooth one-dimensional representation of a totally disconnected group as a quasicharacter of the group. The notation $\mathbf{1}$ will be used for the trivial character.

Recall from Section 2.6 of [**HM**] that the depth of a quasicharacter ϕ of G (that is, the Moy-Prasad depth of ϕ as a smooth irreducible representation of G) can be characterized as follows:

DEFINITION 4.1. The depth of a quasicharacter ϕ of G is the smallest nonnegative real number r that satisfies any of the following equivalent conditions:

(1) $\phi \,|\, G_{x,r^+} \equiv 1$ for some $x \in \mathcal{B}(G)$,
(2) $\phi \,|\, G_{x,r^+} \equiv 1$ for all $x \in \mathcal{B}(G)$,
(3) $\phi \,|\, G_{r^+} \equiv 1$.

HYPOTHESIS C(G). *Let ϕ be a quasicharacter of G of positive depth r. If $x \in \mathcal{B}(G)$, then $\phi \,|\, G_{x,(r/2)^+}$ is realized by an element of $\mathfrak{z}^*_{-r}$.*

We are not assuming Hypothesis C(G) at this point. Later in the paper, in order to apply results of [**HM**], we will need to assume Hypothesis C(G') for various tamely ramified twisted Levi subgroups $\mathbf{G}'$ of $\mathbf{G}$. We remark that Hypothesis C(G) holds when G is a general linear group (see Lemma 2.50 of [**HM**]).

For some results, we will only need the following weaker hypothesis.

HYPOTHESIS C$(G)_w$. *Let ϕ be a quasicharacter of G of positive depth r. If $x \in \mathcal{B}(G)$, then $\phi \,|\, G_{x,r}$ is realized by an element of $\mathfrak{z}^*_{-r}$.*

LEMMA 4.2. (Lemma 2.51 of [**HM**]) *Let ϕ be a quasicharacter of G of positive depth r. Let x, $y \in \mathcal{B}(G)$. Suppose that Γ_x, $\Gamma_y \in \mathfrak{z}^*_{-r}$ realize $\phi \,|\, G_{x,r}$ and $\phi \,|\, G_{y,r}$, respectively. Then $\Gamma_x - \Gamma_y \in \mathfrak{z}^*_{(-r)^+}$.*

Throughout the rest of this section, we assume that $\mathbf{G}'$ is a tamely ramified twisted Levi subgroup of $\mathbf{G}$. In some situations, we will need to assume the above hypothesis for twisted Levi subgroups of $\mathbf{G}$ that contain $\mathbf{G}'$.

HYPOTHESIS $\mathrm{C}(G')_w^+$. *Hypothesis* $\mathrm{C}(\dot{G})_w$ *holds for all twisted Levi subgroups* $\dot{\mathbf{G}}$ *of* $\mathbf{G}$ *that contain* $\mathbf{G}'$.

Let $\mathfrak{z}'$ be the Lie algebra of the centre of G', and let $\mathfrak{z}'^*$ be the dual of $\mathfrak{z}'$.

DEFINITION 4.3. Let $\Gamma \in \mathfrak{z}'^*_{-r}$. We say that Γ is *G-generic of depth $-r$* if it is G-generic of depth r in the sense of [**Y**].

The reason for our convention regarding depth is as follows. After the publication of [**Y**], the notion of depth for nonnilpotent elements of Lie algebras (which has an obvious analogue for elements of the duals of Lie algebras) was defined in [**AD1**]. Relative to the latter notion of depth, an element Γ as in Definition 4.3 has depth $-r$, because $\Gamma \in \mathfrak{z}'^*_{-r} \setminus \mathfrak{z}'^*_{(-r)+}$.

The key property of G-generic elements that we will use is stated in Lemma 4.4. This is Lemma 5.16 of [**HM**]. The proof of the lemma uses Lemma 8.3 of [**Y**] and does not require the hypotheses assumed in [**HM**]. Under weak hypotheses, as in [**AR**], the lemma can be obtained from Proposition 7.5 of [**AR**] (see Lemmas 2.2.4 and 2.3.6 of [**KM1**]).

LEMMA 4.4. *Let $t \in \mathbb{R}$, and let $\Gamma \in \mathfrak{z}'^*_t$ be a G-generic element of depth t. Then*

$$G' = \{\, g \in G \mid \mathrm{Ad}^*(g)(\Gamma + \mathfrak{g}'^*_{t+}) \cap (\Gamma + \mathfrak{g}'^*_{t+}) \neq \emptyset \,\}.$$

DEFINITION 4.5. Let $x \in \mathcal{B}(G')$. A quasicharacter ϕ of G' which is of positive depth r is said to be *G-generic relative to x* if $\phi \,|\, G'_{x,r}$ is realized by a G-generic element $\Gamma \in \mathfrak{z}'^*$ of depth $-r$.

As shown in Lemma 4.7, whenever Hypothesis $\mathrm{C}(G')_w$ is satisfied, then genericity is independent of the point x. (See also Remark 9.1 of [**Y**].)

LEMMA 4.6. (Lemma 5.18 of [**HM**]) *Let $(\mathbf{G}', \mathbf{G}^\natural, \mathbf{G})$ and $(\mathbf{G}', \mathbf{G}^\flat, \mathbf{G})$ be tamely ramified twisted Levi sequences in $\mathbf{G}$. Let $\phi^\natural$ and $\phi^\flat$ be quasicharacters of $G^\natural$ and $G^\flat$, respectively. Let $x^\natural$, $x^\flat \in \mathcal{B}(G')$. Suppose that $\phi^\natural$ and $\phi^\flat$ are G-generic of depth r relative to $x^\natural$ and $x^\flat$, respectively. If $\phi^\natural$ and $\phi^\flat$ agree on $G'_{x^\natural,r} \cap G'_{x^\flat,r}$, then $G^\natural = G^\flat$.*

LEMMA 4.7. *Suppose that Hypothesis $\mathrm{C}(G')_w$ holds. Let ϕ be a quasicharacter of G' of positive depth r. Let x, $y \in \mathcal{B}(G')$.*

(1) *An element $\Gamma \in \mathfrak{z}'^*_{-r}$ realizes $\phi \,|\, G'_{x,r}$ if and only if Γ also realizes $\phi \,|\, G'_{y,r}$.*
(2) *The quasicharacter ϕ is G-generic relative to x if and only if ϕ is G-generic relative to y.*

PROOF. First, let $\Gamma \in \mathfrak{z}'^*_{-r}$, and assume that Γ realizes $\phi \,|\, G'_{x,r}$. According to Hypothesis $\mathrm{C}(G')_w$, there exists $\Gamma' \in \mathfrak{z}'^*_{-r}$ that realizes $\phi \,|\, G'_{y,r}$. Applying Lemma 4.2, we see that $\Gamma - \Gamma' \in \mathfrak{z}'^*_{(-r)+}$. Since $\mathfrak{z}'^*_{(-r)+} \subset \mathfrak{g}'^*_{y,(-r)+}$, we have $\Gamma \in \Gamma' + \mathfrak{g}'^*_{y,(-r)+}$. Thus Γ realizes $\phi \,|\, G'_{y,r}$.

Part (2) is a consequence of part (1). □

DEFINITION 4.8. Suppose that Hypothesis $\mathrm{C}(G')_w$ holds. We say that a quasicharacter ϕ of G' of positive depth is *G-generic* if ϕ is G-generic relative to some (equivalently every) $x \in \mathcal{B}(G')$. If $\Gamma \in \mathfrak{z}'^{*}_{-r}$ realizes $\phi \,|\, G'_{x,r}$ for all $x \in \mathcal{B}(G')$, we say that Γ *realizes* ϕ *on* G'_r.

LEMMA 4.9. *Assume that Hypothesis* $\mathrm{C}(G)_w$ *holds. Suppose that* $G' \neq G$. *Let* χ *be a quasicharacter of* G *and let* $r \in \mathbb{R}$ *be such that* $r > 0$.

(1) *The depth of* $\chi \,|\, G'$ *is equal to the depth of* χ.
(2) *If* $\Gamma \in \mathfrak{z}'^{*}_{-r}$ *is* G*-generic of depth* $-r$, *then* $\Gamma + \mathfrak{z}^{*}_{-r}$ *consists of* G*-generic elements of depth* $-r$.
(3) *Suppose that* ϕ *is a quasicharacter of* G' *of depth* r *and* ϕ *is* G*-generic relative to some* $x \in \mathcal{B}(G')$. *If* χ *has depth at most* r, *then* $\phi(\chi \,|\, G')$ *has depth* r *and is* G*-generic relative to* x.
(4) *Suppose that* ϕ *and* x *are as above. Then* $\phi \,|\, G'_{x,r}$ *cannot be of the form* $\dot{\chi} \,|\, G'_{x,r}$ *for any quasicharacter* $\dot{\chi}$ *of* G.

PROOF. Let s be the depth of χ. Let $x \in \mathcal{B}(G')$. If $s > 0$, since Hypothesis $\mathrm{C}(G)_w$ holds, there exists $\Gamma_\chi \in \mathfrak{z}^{*}_{-s}$ that realizes $\chi \,|\, G_{x,s}$. It follows from Lemma 2.52 of [**HM**] that $\chi \,|\, G'$ has depth s. If $s = 0$, then $\chi \,|\, G_{x,0+} \equiv 1$ and $G'_{x,0+} = G' \cap G_{x,0+}$, so we have $\chi \,|\, G'_{x,0+} \equiv 1$. This implies that the depth of $\chi \,|\, G'$ is 0.

The second part is Lemma 4.21 of [**HM**].

For the third part, since ϕ is G-generic relative to $x \in \mathcal{B}(G')$, there exists an element $\Gamma \in \mathfrak{z}'^{*}_{-r}$ that is G-generic of depth $-r$ and realizes $\phi \,|\, G'_{x,r}$. If $s < r$, then $\phi\chi \,|\, G'_{x,r} = \phi \,|\, G'_{x,r}$ and part (3) follows in this case because the property of being G-generic of depth r relative to x depends only on the values of the quasicharacter on $G'_{x,r}$. If $r = s$, let Γ_χ be as above. Then $\Gamma + \Gamma_\chi$ belongs to $\Gamma + \mathfrak{z}^{*}_{-r}$ and realizes $\phi\chi \,|\, G'_{x,r}$. According to part (2), $\Gamma + \Gamma_\chi$ is G-generic of depth $-r$. Thus $\phi(\chi \,|\, G')$ is G-generic relative to x (and has depth r).

For the last part, suppose that $\dot{\chi}$ is a quasicharacter of G such that $\dot{\chi} \,|\, G'_{x,r} = \phi \,|\, G'_{x,r}$. Note that, according to part (1), $\dot{\chi}$ has depth r. Applying part (3) to $\phi(\dot{\chi}^{-1} \,|\, G')$, we find that $\phi(\dot{\chi}^{-1} \,|\, G')$ has depth r and therefore cannot be trivial on $G'_{x,r}$. This contradicts our starting assumption. $\square$

5. Factorizations of quasicharacters

As in the previous section, we assume that $\mathbf{G}'$ is a tamely ramified twisted Levi subgroup of $\mathbf{G}$.

DEFINITION 5.1. Let $x \in \mathcal{B}(G')$. A quasicharacter ϕ of G' is *G-factorizable relative to x* if there exist a tamely ramified twisted Levi sequence $\vec{\mathbf{G}} = (\mathbf{G}^0, \ldots, \mathbf{G}^d = \mathbf{G})$ and quasicharacters $\phi_0, \ldots, \phi_d$ of $G^0, \ldots, G^d$, respectively, such that

(F1) $\mathbf{G}'$ is a tamely ramified twisted Levi subgroup of $\mathbf{G}^0$.
(F2) $\phi \,|\, G'_{0+} = \prod_{i=0}^{d} (\phi_i \,|\, G'_{0+})$.
(F3) If $d \geq 1$, then $0 < r_0 < \cdots < r_{d-1}$, where r_i is the depth of ϕ_i, $0 \leq i \leq d-1$.
(F4) The quasicharacter ϕ_i of G^i is G^{i+1}-generic relative to x for $0 \leq i \leq d-1$.
(F5) If $d \geq 1$ and ϕ_d is nontrivial, then the depth r_d of ϕ_d satisfies $r_d > r_{d-1}$.

If ϕ is G-factorizable relative to x, we set $\vec{\phi} = (\phi_0, \ldots, \phi_d)$ and we refer to $(\vec{\mathbf{G}}, \vec{\phi})$ as a *G-factorization of ϕ relative to x*.

REMARK 5.2. Suppose that $x \in \mathcal{B}(G')$ and there exists a factorization $(\vec{\mathbf{G}}, \vec{\phi})$ of ϕ relative to x having the property that $\mathbf{G}^0 = \mathbf{G}'$. Let $\chi = \phi\,(\prod_{i=0}^{d} \phi_i^{-1} \,|\, G^0)$. Condition (F2) says that χ has depth zero. Hence $\chi\phi_0 \,|\, G^0_{r_0} = \phi_0 \,|\, G^0_{r_0}$. If $d > 0$, then, because ϕ_0 is G^1-generic of depth r_0 relative to x and this property is determined by the restriction $\phi_0 \,|\, G^0_{r_0} = \chi\phi_0 \,|\, G^0_{r_0}$, we have that $\chi\phi_0 \,|\, G^0$ is G^1-generic of depth r_0 relative to x. Therefore, we may replace ϕ_0 with $\chi\phi_0$ to produce a factorization of ϕ for which the equality of Condition (F2) extends to $G' = G^0$.

If Hypothesis $\mathrm{C}(G')^+_w$ holds, and x and y belong to $\mathcal{B}(G')$, then Lemma 4.7(2) tells us that a G-factorization $(\vec{\mathbf{G}}, \vec{\phi})$ of ϕ relative to x is a G-factorization of ϕ relative to y, so the property of being G-factorizable is independent of the choice of element in $\mathcal{B}(G')$.

DEFINITION 5.3. Assume that Hypothesis $\mathrm{C}(G')^+_w$ holds. We say that a quasicharacter ϕ of G' is *G-factorizable* or *factorizable* if ϕ is G-factorizable relative to some (equivalently every) $x \in \mathcal{B}(G')$. A G-factorization $(\vec{\mathbf{G}}, \vec{\phi})$ of ϕ relative to a point $x \in \mathcal{B}(G')$ may also just be called a *G-factorization* or *factorization* of ϕ.

Suppose that Hypothesis $\mathrm{C}(G')^+_w$ holds and $(\vec{\mathbf{G}}, \vec{\phi})$ is a factorization of a quasicharacter ϕ of G'. Then, as described below, we can modify the quasicharacters that appear in $\vec{\phi}$ to produce another G-factorization of ϕ. Lemma 5.4 shows that all G-factorizations of ϕ can be obtained from $(\vec{\mathbf{G}}, \vec{\phi})$ in this way.

If $d = 0$, choose a quasicharacter $\dot{\phi}_0$ of G such that $\dot{\phi}_0 \,|\, G'_{0+} = \phi_0 \,|\, G'_{0+}$. If $d > 0$ and ϕ_d is trivial, set $\dot{\phi}_d = \phi_d$. If $d > 0$ and ϕ_d is nontrivial, choose a quasicharacter $\dot{\phi}_d$ of G that agrees with ϕ_d on $G_{r^+_{d-1}}$. If $d-1 > 0$, choose a quasicharacter $\dot{\phi}_{d-1}$ of G^{d-1} such that

$$\dot{\phi}_{d-1} \,|\, G^{d-1}_{r^+_{d-2}} = \phi_{d-1}\phi_d\dot{\phi}_d^{-1} \,|\, G^{d-1}_{r^+_{d-2}}.$$

Continuing in this manner, choose quasicharacters $\dot{\phi}_{d-2}, \dot{\phi}_{d-3}, \ldots, \dot{\phi}_1$ of G^{d-2}, $G^{d-3}, \ldots, G^1$, respectively, such that

$$\dot{\phi}_i \,|\, G^i_{r^+_{i-1}} = \phi_i \prod_{j=i+1}^{d} \phi_j\dot{\phi}_j^{-1} \,|\, G^i_{r^+_{i-1}} \qquad \text{for } 1 \le i \le d-1.$$

Finally, choose a quasicharacter $\dot{\phi}_0$ of G^0 such that

$$\dot{\phi}_0 \,|\, G^0_{0+} = \phi_0 \prod_{j=1}^{d} \phi_j\dot{\phi}_j^{-1} \,|\, G^0_{0+}.$$

For $0 \le i \le d-1$, set

$$\chi_{i+1} = \prod_{j=i+1}^{d} \phi_j\dot{\phi}_j^{-1} \,|\, G^{i+1}.$$

By construction, χ_{i+1} has depth at most r_i. For convenience of notation, set $r_{-1} = 0$. Note that $\dot{\phi}_i \,|\, G^i_{r^+_{i-1}} = \phi_i\chi_{i+1} \,|\, G^i_{r^+_{i-1}}$. An application of Lemma 4.9(3) shows that $\dot{\phi}_i$ is G^{i+1}-generic of depth r_i. Let $\vec{\dot{\phi}} = (\dot{\phi}_0, \ldots, \dot{\phi}_d)$. Then $(\vec{\mathbf{G}}, \vec{\dot{\phi}})$ is a factorization of ϕ.

PROPOSITION 5.4. *Assume that Hypothesis* $\mathrm{C}(G')^+_w$ *holds. Suppose that* $(\vec{\mathbf{G}}, \vec{\phi})$ *is a factorization of a quasicharacter* ϕ *of* G'. *If* $(\vec{\dot{\mathbf{G}}}, \vec{\dot{\phi}})$ *is another factorization of* ϕ, *where* $\vec{\dot{\mathbf{G}}} = (\dot{\mathbf{G}}^0, \dots, \dot{\mathbf{G}}^\ell = \mathbf{G})$ *and* $\vec{\dot{\phi}} = (\dot{\phi}_0, \dots, \dot{\phi}_\ell)$, *then*

(1) $d = \ell$, *and* $\dot{\mathbf{G}}^i = \mathbf{G}^i$, $0 \le i \le d$.

(2) *If* $\dot{r}_i$ *is the depth of* $\dot{\phi}_i$, *and* $0 \le i \le d-1$, *then* $\dot{r}_i = r_i$. *The quasicharacter* $\dot{\phi}_d$ *is nontrivial if and only if* ϕ_d *is nontrivial, and in that case, the depth* $\dot{r}_d$ *of* $\dot{\phi}_d$ *equals* r_d.

(3) *Set* $r_{-1} = 0$. *Then* $\prod_{j=i}^d \phi_j \,|\, G^i_{r^+_{i-1}} = \prod_{j=i}^d \dot{\phi}_j \,|\, G^i_{r^+_{i-1}}$ *for* $0 \le i \le d$.

PROOF. If $\phi \,|\, G'_{0+} \equiv 1$, then, by Definition 5.1(F5), each factorization of ϕ has the form $\vec{\mathbf{G}} = (\mathbf{G})$ and $\vec{\phi} = (\phi_0)$, where ϕ_0 is a quasicharacter of G of depth zero, so clearly the proposition holds in this case. Therefore we assume that $\phi \,|\, G'_{0+}$ is nontrivial.

Case 1: Suppose that $d = 0$. Then $\vec{\mathbf{G}} = (\mathbf{G}^0) = (\mathbf{G})$ and $\vec{\phi} = (\phi_0)$, where ϕ_0 is a quasicharacter of G such that $\phi_0 \,|\, G'_{0+} = \phi \,|\, G'_{0+}$. Suppose that $\ell > 0$. Then $\dot{\phi}_{\ell-1}$ is a G-generic quasicharacter of $\dot{G}^{\ell-1}$ of depth $\dot{r}_{\ell-1}$. Set $s = \dot{r}_{\ell-1}$. It follows from Condition (F2) that $\phi_0 \,|\, G'_s = \dot{\phi}_{\ell-1}\dot{\phi}_\ell \,|\, G'_s$. Thus $\phi_0^{-1}\dot{\phi}_{\ell-1}\dot{\phi}_\ell \,|\, G'_s$ is trivial. Applying Lemma 4.9(1), we have that the quasicharacter $\dot{\phi}_{\ell-1}(\dot{\phi}_\ell\phi_0^{-1} \,|\, \dot{G}^{\ell-1})$ of $\dot{G}^{\ell-1}$ has depth less than s. But this implies $\dot{\phi}_{\ell-1} \,|\, \dot{G}^{\ell-1}_s = \phi_0^{-1}\dot{\phi}_\ell \,|\, \dot{G}^{\ell-1}_s$. According to Lemma 4.9(4), this cannot happen, because $\dot{\phi}_{\ell-1}$ has depth s and is G-generic. Thus our assumption that $\ell > 0$ is false. Upon reversing the roles of d and ℓ, we conclude that $d = 0$ if and only if $\ell = 0$. In this case, $\ell = d = 0$ and $\dot{\mathbf{G}}^0 = \mathbf{G} = \mathbf{G}^0$, and $\dot{\phi}_0$ is a quasicharacter of G that agrees with ϕ on G'_{0+}. Thus the depths of the quasicharacters $\phi_0 \,|\, G'$, $\dot{\phi}_0 \,|\, G'$ and ϕ are equal (as the restrictions of these quasicharacters to G'_{0+} coincide). Now Lemma 4.9(1) implies that ϕ_0 and $\dot{\phi}_0$ have the same depth. Thus the proposition holds when $d = 0$.

Case 2: Assume that $d > 0$. The results above tell us that $\ell > 0$. Let u be the depth of ϕ_{d-1} and let s be the depth of $\dot{\phi}_{\ell-1}$.

Suppose that $u > s$. From Definition 5.1(F3) and Lemma 4.9(1), we have that u is greater than the depth of $\phi_i \,|\, G'$ for $0 \le i \le d-2$. Similarly, as $u > s$, we find that u is greater than the depth of $\dot{\phi}_i \,|\, G'$ for $0 \le i \le \ell - 1$. This implies that

$$\dot{\phi}_\ell \,|\, G'_u = \phi \,|\, G'_u = \phi_{d-1}\phi_d \,|\, G'_u.$$

It now follows from Lemma 4.9(1) that the depth of $\phi_{d-1}(\phi_d\dot{\phi}_\ell^{-1} \,|\, G^{d-1})$ is less than u. Hence $\phi_{d-1} \,|\, G^{d-1}_u = \phi_d\dot{\phi}_\ell^{-1} \,|\, G^{d-1}_u$. However, an application of Lemma 4.9(4) leads to a contradiction, as ϕ_{d-1} has depth u and is G-generic. Thus $u > s$ is impossible. Reversing the roles of u and s, we find that $s > u$ is also impossible.

We have shown that $s = u$. Note that

$$\phi \,|\, G'_{s+} = \phi_d \,|\, G'_{s+} = \dot{\phi}_\ell \,|\, G'_{s+}.$$

Thus the depth of $\phi_d\dot{\phi}_\ell^{-1} \,|\, G'$ is at most s. An application of Lemma 4.9(1) gives

$$\phi_d \,|\, G_{s+} = \dot{\phi}_\ell \,|\, G_{s+}.$$

Note that this implies that the depth of ϕ_d is equal to the depth of $\dot{\phi}_\ell$. Since the depth of $\phi_d^{-1}\dot{\phi}_\ell$ is at most s, we may apply Lemma 4.9(3) to conclude that $\dot{\phi}_{\ell-1}(\dot{\phi}_\ell\phi_d^{-1} \,|\, \dot{G}^{\ell-1})$ is G-generic of depth s.

Next, observe that $\phi_{d-1} \,|\, G'_s = \dot\phi_{\ell-1}(\dot\phi_\ell \phi_d^{-1}) \,|\, G'_s$. If $G' \neq \dot G^{\ell-1}$ and $G' \neq G^{d-1}$, then Lemma 4.6 implies that $\dot G^{\ell-1} = G^{d-1}$. If $G' = \dot G^{\ell-1}$ and $G' \neq G^{d-1}$, then we have that the G-generic quasicharacter $\dot\phi_{\ell-1}$ of G', of depth s, agrees on G'_s with the quasicharacter $\phi_{d-1}(\phi_d \dot\phi_\ell^{-1} \,|\, G^{d-1})$ of G^{d-1}. By Lemma 4.9(4), this is impossible. A similar argument gives a contradiction if $G' \neq \dot G^{\ell-1}$ and $G' = G^{d-1}$. Thus we conclude that $G^{d-1} = \dot G^{\ell-1}$.

We have shown that (when $d > 0$) the depths of ϕ_d and $\dot\phi_\ell$ are equal. Furthermore, $\dot G^{\ell-1} = G^{d-1}$ and the depths of $\dot\phi_{\ell-1}$, $\dot\phi_{\ell-1}(\dot\phi_\ell\phi_d^{-1} \,|\, G^{d-1})$ and ϕ_{d-1} are equal. As we have proved the proposition in the case $d = 0$, we may argue by induction on d. The sequence $(\mathbf{G}^0, \ldots, \mathbf{G}^{d-1})$ (in $\mathbf{G}^{d-1}$) and the sequence of quasicharacters $(\phi_0, \ldots, \phi_{d-1})$ constitute a G^{d-1}-factorization of the quasicharacter $\phi(\phi_d^{-1} \,|\, G')$. We obtain another G^{d-1}-factorization of $\phi(\phi_d^{-1} \,|\, G')$ from the twisted Levi sequence $(\dot{\mathbf{G}}^0, \ldots, \dot{\mathbf{G}}^{\ell-1} = \mathbf{G}^{d-1})$ together with the sequence $(\dot\phi_0, \ldots, \dot\phi_{\ell-2}, \dot\phi_{\ell-1}(\dot\phi_\ell\phi_d^{-1} \,|\, G^{d-1}))$. By induction, we conclude that $d-1 = \ell-1$, $\dot{\mathbf{G}}^i = \mathbf{G}^i$ for $0 \le i \le d-1$, the depth $\dot r_i$ of $\dot\phi_i$ equals r_i for $0 \le i \le d-2$, and

$$\prod_{j=i}^{d-1} \dot\phi_j \,|\, G^i_{r^+_{i-1}} = \left(\prod_{j=i}^{d-1} \dot\phi_j \,|\, G^i_{r^+_{i-1}}\right) \dot\phi_\ell \phi_d^{-1} \,|\, G^i_{r^+_{i-1}}, \qquad 0 \le i \le d-1.$$

This is clearly equivalent to the last part of the statement of the proposition, except for the case $i = d$. The latter case is implied by the equality of depths of $\dot\phi_{d-1}$, $\dot\phi_{\ell-1}(\dot\phi_\ell\phi_d^{-1} \,|\, G^{d-1})$ and $\dot\phi_{\ell-1}$ (which was shown above). □

LEMMA 5.5. *Assume that Hypothesis* $\mathrm{C}(G')^+_w$ *holds. Let* ϕ *be a* G*-factorizable quasicharacter of* G' *and let* $(\vec{\mathbf{G}}, \vec\phi)$ *be a factorization of* ϕ *(with notation as in Definition 5.1). Let* $\dot\phi$ *be a quasicharacter of a twisted Levi subgroup* $\dot G$ *of* G *such that* $G' \subset \dot G$*. If* $\dot\phi \,|\, G'_{r_0} = \phi \,|\, G'_{r_0}$*, then* $\dot G \subset G^0$*.*

PROOF. If $d = 0$, then $G^0 = G$, so there is nothing to show. Assume that $d \ge 1$. Suppose that ϕ_d is nontrivial. Then $(\vec{\mathbf{G}}, (\phi_0, \ldots, \phi_{d-1}, \mathbf{1}))$ is a G-factorization of the quasicharacter $\phi(\phi_d^{-1} \,|\, G')$, and $\phi\phi_d^{-1} \,|\, G'_{r_0} = \dot\phi\phi_d^{-1} \,|\, G'_{r_0}$. Therefore, after replacing ϕ by $\phi(\phi_d^{-1} \,|\, G')$ and $\dot\phi$ by $\dot\phi(\phi_d^{-1} \,|\, \dot G)$, we may (and do) assume that ϕ_d is trivial.

Note that $\phi \,|\, G'_{r_{d-1}} = \phi_{d-1} \,|\, G'_{r_{d-1}} = \dot\phi \,|\, G'_{r_{d-1}}$. According to Lemma 4.9(1), $\phi_{d-1} \,|\, G'$ has depth r_{d-1}. Thus $\dot\phi \,|\, G'$ also has depth r_{d-1}. Applying Lemma 4.9(1) again, we have that the depth of $\dot\phi$ is r_{d-1}. Let $\dot{\mathfrak{z}}$ be the Lie algebra of the centre of $\dot G$. As we have assumed Hypothesis $\mathrm{C}(\dot G)_w$, there exists $\dot\Gamma \in \dot{\mathfrak{z}}^*_{-r_{d-1}}$ that realizes $\dot\phi \,|\, \dot G_{r_{d-1}}$. Let $\Gamma_{d-1} \in \mathfrak{z}^{d-1\,*}_{-r_{d-1}}$ be a G-generic element of depth $-r_{d-1}$ that realizes $\phi_{d-1} \,|\, G^{d-1}_{r_{d-1}}$. Because ϕ_{d-1} and $\dot\phi$ agree on $G'_{r_{d-1}}$, we have by Lemma 4.2 that $\dot\Gamma \in \Gamma_{d-1} + \mathfrak{z}'^*_{(-r_{d-1})^+}$. Since $\mathfrak{z}'^*_{(-r_{d-1})^+} \subset \mathfrak{g}^{d-1\,*}_{(-r_{d-1})^+}$ and $\mathrm{Ad}^*(g)\dot\Gamma = \dot\Gamma$ for all $g \in \dot G$, it follows that

$$\mathrm{Ad}^*(g)(\Gamma_{d-1} + \mathfrak{g}^{d-1\,*}_{(-r_{d-1})^+}) \cap (\Gamma_{d-1} + \mathfrak{g}^{d-1\,*}_{(-r_{d-1})^+}) \neq \emptyset \ \ \forall\, g \in \dot G.$$

According to Lemma 4.4, $\dot G \subset G^{d-1}$. If $d = 1$, this completes the proof.

Suppose that $d > 1$. Let $\phi' = \phi(\phi_{d-1}^{-1} \,|\, G')$. Note that the twisted Levi sequence $(\mathbf{G}^0, \ldots, \mathbf{G}^{d-1})$, together with the sequence $(\phi_0, \ldots, \phi_{d-2}, \mathbf{1})$, is a G^{d-1}-factorization of ϕ'. Let $\dot\phi' = \dot\phi(\phi_{d-1}^{-1} \,|\, G')$. Then $\phi' \,|\, G'_{r_0} = \dot\phi' \,|\, G'_{r_0}$. By induction, this equality implies that $\dot G \subset G^0$. □

The following lemma will be used in the proof of Theorem 12.1.

LEMMA 5.6. *Assume that Hypothesis* $\mathrm{C}(G')^+_w$ *holds. Let* ϕ *be a* G*-factorizable quasicharacter of* G'. *Suppose that there exists* $g \in N_G(G')$ *such that* $\phi \,|\, G'_{0+} = {}^g\phi \,|\, G'_{0+}$. *Let* $(\vec{\mathbf{G}}, \vec{\phi})$ *be a* G*-factorization of* ϕ, *with* $\vec{\mathbf{G}} = (\mathbf{G}^0, \ldots, \mathbf{G}^d)$. *Then* $g \in G^0$.

REMARK 5.7. Note that it follows from Proposition 5.4 that G^0 is independent of the choice of factorization of ϕ.

PROOF. As there is nothing to show if $d = 0$, we assume that $d \geq 1$. Let $\vec{\phi} = (\phi_0, \ldots, \phi_d)$. Observe that $(({}^g\mathbf{G}^0, \ldots, {}^g\mathbf{G}^d = \mathbf{G}), ({}^g\phi_0, \ldots, {}^g\phi_d = \phi_d))$ is also a G-factorization of ϕ. Applying Proposition 5.4, we have ${}^g\mathbf{G}^j = \mathbf{G}^j$, $0 \leq j \leq d$, and

$$\prod_{j=i}^{d} \phi_j \,|\, G^i_{r^+_{i-1}} = \prod_{j=i}^{d} {}^g\phi_j \,|\, G^i_{r^+_{i-1}}, \quad 0 \leq i \leq d.$$

(Recall that $r_{-1} = 0$.) In particular, as $\phi_d = {}^g\phi_d$ and $r_{d-1} > r_i$ for $0 \leq i < d-1$, we have $\phi_{d-1} \,|\, G^{d-1}_{r_{d-1}} = {}^g\phi_{d-1} \,|\, G^{d-1}_{r_{d-1}}$. Let $\Gamma_{d-1} \in \mathfrak{z}^{d-1\,*}_{-r_{d-1}}$ be a G-generic element that realizes $\phi_{d-1} \,|\, G^{d-1}_{r_{d-1}}$. Then the equality above implies (by Lemma 4.2) that $\Gamma_{d-1} - \mathrm{Ad}^*(g^{-1})\Gamma_{d-1} \in \mathfrak{z}^{d-1\,*}_{(-r_{d-1})^+}$. In particular, $\mathrm{Ad}^*(g^{-1})\Gamma_{d-1} \in \Gamma_{d-1} + \mathfrak{g}^{d-1\,*}_{(-r_{d-1})^+}$. According to Lemma 4.4, this implies that $g \in G^{d-1}$. If $d = 1$ this completes the proof.

Suppose that $d > 1$. Let $\phi' = \phi(\phi_d^{-1} \,|\, G')$. The sequence $(\mathbf{G}^0, \ldots, \mathbf{G}^{d-1})$, along with the sequence $(\phi_0, \ldots, \phi_{d-2}, \phi_{d-1})$, is a G^{d-1}-factorization of ϕ'. Using ${}^g\phi_d = \phi_d$ we find that $\phi' \,|\, G'_{0+} = {}^g\phi' \,|\, G'_{0+}$. By induction, we have $g \in G^0$. □

The next lemma shows that under certain conditions all quasicharacters of a group are factorizable.

LEMMA 5.8. *Let* G' *be a proper tamely ramified twisted Levi subgroup of* G. *Assume that Hypothesis* $\mathrm{C}(G')^+_w$ *holds. Suppose that the following conditions hold for every twisted Levi subgroup* H *of* G *that contains* G'.

(1) *Suppose that* $r > 0$ *and* $\Gamma \in \mathfrak{z}^{H\,*}_{-r} \setminus \mathfrak{z}^{H\,*}_{(-r)^+}$, *where* $\mathfrak{z}^H$ *is the centre of* $\mathfrak{h}$. *For* $x \in \mathcal{B}(H)$, *the character of* $H_{x,r}$ *realized by* Γ *is trivial on* $H_{x,r} \cap H_{\mathrm{der}}$, *where* H_{der} *is the derived group of the topological group* H.
(2) *If* $r > 0$, $\Gamma' \in \mathfrak{z}'^{\,*}_{-r} \setminus \mathfrak{z}'^{\,*}_{(-r)+}$ *and* $(\Gamma' + \mathfrak{z}'^{\,*}_{(-r)+}) \cap \mathfrak{z}^{H\,*} = \emptyset$, *then there exist a twisted Levi subgroup* $\dot{G}$ *of* H, *containing* G', *and an element* $\dot{\Gamma} \in \dot{\mathfrak{z}}^* \cap (\Gamma' + \mathfrak{z}'_{(-r)+})$ *that is* H*-generic of depth* $-r$, *where* $\dot{\mathfrak{z}}$ *is the Lie algebra of the centre of* $\dot{G}$.

Then every quasicharacter of G' *is* G*-factorizable.*

REMARK 5.9. As shown in Section 4 of [**AR**], in some cases, it is possible to identify Lie algebras and their duals in a way that is compatible with the Moy-Prasad filtrations. In those cases, Proposition 5.4 of [**AR**] gives information about groups for which conditions as in part (2) of the lemma are known to hold. We also remark that the hypotheses assumed in Section 3.4 of [**K**] are sufficient to guarantee factorizability of quasicharacters.

PROOF. Let ϕ be a quasicharacter of G'. If ϕ has depth zero, then $((\mathbf{G}), (\phi))$ is a factorization of ϕ (see the beginning of the proof of Proposition 5.4). Therefore we assume that ϕ has positive depth r. Since we have assumed that Hypothesis $\mathrm{C}(G')_w$ holds, there exists $\Gamma \in \mathfrak{z}'^*_{-r} \setminus \mathfrak{z}'^*_{(-r)+}$ that realizes ϕ on G'_r. If $(\Gamma + \mathfrak{z}'^*_{(-r)+}) \cap \mathfrak{z}^*$ is nonempty, then, after replacing Γ by a $\mathfrak{z}'^*_{(-r)+}$-translate, if necessary, we may assume that $\Gamma \in \mathfrak{z}^*_{-r}$. Let $x \in \mathcal{B}(G)$. By assumption (1), the character ϕ_Γ of $G_{x,r}$ that is realized by Γ is trivial on $G_{x,r} \cap G_{\mathrm{der}}$. This allows us to transfer ϕ_Γ to a character of the group $G_{x,r}G_{\mathrm{der}}/G_{\mathrm{der}} \simeq G_{x,r}/(G_{x,r} \cap G_{\mathrm{der}})$. Since $G_{x,r}$ is compact, we have that $G_{x,r}G_{\mathrm{der}}/G_{\mathrm{der}}$ is a closed subgroup of the locally compact abelian group G/G_{der}, so there exists an extension of ϕ_Γ to G/G_{der}. Pulling this extension back to G, we obtain a quasicharacter χ_1 of G that agrees with ϕ_Γ on $G_{x,r}$ (hence has depth r). Now $\phi\chi_1^{-1} \,|\, G'$ has depth less than r (according to Lemma 4.9(1)). If $\phi\chi_1^{-1} \,|\, G'$ has positive depth and can be realized by an element of $\mathfrak{z}^*$ (of the appropriate depth), we may repeat the above process to obtain a quasicharacter χ_2 of G such that $\phi((\chi_1\chi_2)^{-1} \,|\, G')$ has depth less than the depth of $\phi\chi_1^{-1} \,|\, G'$.

The set of depths of quasicharacters is discrete, so it follows from the discussion above that there exists a quasicharacter χ of G (which is trivial if the original coset $\Gamma + \mathfrak{z}'^*_{(-r)+}$ does not intersect $\mathfrak{z}^*$) such that either $\phi(\chi^{-1} \,|\, G')$ has depth zero, or $\phi(\chi^{-1} \,|\, G')$ has positive depth $s \leq r$, and, for $x \in \mathcal{B}(G')$, its restriction to $G'_{x,s}$ is realized by an element $\Gamma' \in \mathfrak{z}'^*_{-s}$ such that $\Gamma' + \mathfrak{z}'^*_{(-s)+}$ does not intersect $\mathfrak{z}^*$. In the first case, we set $\vec{\mathbf{G}} = (\mathbf{G})$ and $\vec{\phi} = (\chi)$ to obtain a G-factorization of ϕ.

In the second case, according to the second condition of the lemma (taking $H = G$ and replacing r by s), there exist a twisted Levi subgroup $\dot{G}$ of G, containing G', and an element $\dot{\Gamma} \in \dot{\mathfrak{z}}^* \cap (\Gamma' + \mathfrak{z}'^*_{(-s)+})$ that is G-generic of depth $-s$. Let $y \in \mathcal{B}(\dot{G})$. Condition (1) guarantees that the character of $\dot{G}_{y,s}$ realized by $\dot{\Gamma}$ is trivial on $\dot{G}_{y,s} \cap \dot{G}_{\mathrm{der}}$. Arguing as in the first paragraph of the proof, we find that there exists an extension $\dot{\chi}_1$ of this character of $\dot{G}_{y,s}$ to $\dot{G}$. Note that the depth of the quasicharacter $\phi(\chi^{-1} \,|\, G')(\dot{\chi}_1^{-1} \,|\, G')$ is less than s. Now proceed recursively with $\dot{G}$ in place of G. □

6. Regular quasicharacters

Let $\mathbf{G}'$ be a proper twisted Levi subgroup of $\mathbf{G}$.

DEFINITION 6.1. Suppose that ϕ is a quasicharacter of G' and S is a subset of G' containing an open neighbourhood of some point in G. We say that ϕ is *G-regular on S* whenever $\phi \,|\, S$ is not the restriction to S of a quasicharacter of a twisted Levi subgroup $\dot{G}$ of G such that $G' \subsetneq \dot{G}$.

REMARK 6.2. If $r > 0$, G-regularity of ϕ on G'_r is equivalent to G-regularity of ϕ on $G'_{x,r}$, for any $x \in \mathcal{B}(G')$. Indeed, if ϕ is not regular on $G'_{x,r}$, then $\phi \,|\, G'_{x,r} = \dot{\phi} \,|\, G'_{x,r}$ for a quasicharacter $\dot{\phi}$ of a twisted Levi subgroup $\dot{G}$ of G such that $G' \subsetneq \dot{G}$. This implies that $\phi\dot{\phi}^{-1} \,|\, G'$ has depth strictly less than r, and thus $\phi\dot{\phi}^{-1} \,|\, G'_r \equiv 1$ (see Definition 4.1) so ϕ is not regular on G'_r.

LEMMA 6.3. *Assume that Hypothesis $\mathrm{C}(G')^+_w$ holds. Let ϕ be a G-factorizable quasicharacter of G'. Assume that the depth of ϕ is positive. Let $(\vec{\mathbf{G}}, \vec{\phi})$ be a G-factorization of ϕ with $\vec{\mathbf{G}} = (\mathbf{G}^0, \ldots, \mathbf{G})$ and $\vec{\phi} = (\phi_0, \ldots, \phi_d)$. Let r_0 be the depth of ϕ_0. Then ϕ is G-regular on G'_{r_0} if and only if $G^0 = G'$.*

PROOF. Because the depth of ϕ is positive and ϕ is G-factorizable, it follows from Definition 5.1(F2,F3) that $r_0 > 0$. If $G^0 \neq G'$, then Condition (F2) in

Definition 5.1 implies that ϕ is is not G-regular on G'_{0+}, as $\prod_{j=0}^{d}(\phi_j \,|\, G^0)$ is a quasicharacter of G^0 that agrees with ϕ on G'_{0+}. Since $G_{0+} \supset G'_{r_0}$, it follows that ϕ is not G-regular on G'_{r_0}.

Suppose that $G^0 = G'$. It follows from Lemma 5.5 that if $\dot{\phi}$ is a quasicharacter of a twisted Levi subgroup $\dot{G}$ of G such that $G' \subset \dot{G}$ and $\phi \,|\, G'_{r_0} = \dot{\phi} \,|\, G'_{r_0}$, then $\dot{G} = G'$. $\square$

REMARK 6.4. Suppose that Hypotheses $\mathrm{C}(G)_w$ and $\mathrm{C}(G')_w$ hold. Let ϕ be a G-generic quasicharacter of G' of positive depth r. Then ϕ is G-regular on G'_r. To see this, set $\vec{\mathbf{G}} = (\mathbf{G}', \mathbf{G})$ and $\vec{\phi} = (\phi, \mathbf{1})$. Since ϕ is G-generic, we have that $(\vec{\mathbf{G}}, \vec{\phi})$ is a G-factorization of ϕ. Applying Lemma 6.3, we see that ϕ is G-regular on G'_r.

COROLLARY 6.5. *Let notation and assumptions be as in Lemma 6.3. Then ϕ is G-regular on G'_{0+} if and only if $G^0 = G'$.*

PROOF. If $G^0 \neq G'$, then, as observed in the first part of the proof of Lemma 6.3, condition Condition (F2) in Definition 5.1 implies that ϕ is is not G-regular on G'_{0+}.

If $G^0 = G'$, then an application of Lemma 6.3 gives that ϕ is G-regular on G'_{r_0}. It then follows from the definition of G-regular and the fact that $G'_{r_0} \subset G'_{0+}$ that ϕ is G-regular on G'_{0+}. $\square$

7. Depth zero supercuspidal representations

In this section, we state key results concerning depth zero irreducible supercuspidal representations of connected reductive p-adic groups. Theorem 7.1, Proposition 7.3 and Proposition 7.4 are Proposition 6.8, Proposition 6.6 and Theorem 6.11(2), respectively, of [**MP2**]. Some of these results were also proved by Morris, in Proposition 1.4, Proposition 2.1 and Corollary 3.5 of [**Mo4**].

THEOREM 7.1. *Let π be a depth zero irreducible supercuspidal representation of G. Then π is equal to the representation $\mathrm{Ind}_K^G \rho$ obtained via (compact) induction from an irreducible smooth representation ρ of K, where*

(1) *There exists a maximal parahoric subgroup $G_{x,0}$ of G such that $K = N_G(G_{x,0})$,*
(2) *$\rho \,|\, G_{x,0^+}$ is a multiple of the trivial representation of $G_{x,0^+}$,*
(3) *$\rho \,|\, G_{x,0}$ contains a representation of $G_{x,0}$ that factors to an irreducible cuspidal representation of $G_{x,0}/G_{x,0^+}$.*

REMARK 7.2. Let $\mathfrak{f}$ be the residue class field of F. The group $G_{x,0}/G_{x,0^+}$ is the $\mathfrak{f}$-rational points of a connected reductive $\mathfrak{f}$-group, so the notion of cuspidal representation makes sense. Because $G_{x,0}$ is a parahoric subgroup of G, the group $N_G(G_{x,0})$ is compact modulo the centre of G. Hence ρ is finite-dimensional.

PROPOSITION 7.3. *Let K be the normalizer of a maximal parhahoric subgroup $G_{x,0}$ of G. Let ρ be an irreducible representation of K that satisfies conditions (2) and (3) of Theorem 7.1 and let $\pi = \mathrm{Ind}_K^G \rho$. Then π is a depth zero irreducible supercuspidal representation of G.*

PROPOSITION 7.4. *Let $K = N_G(G_{x,0})$ and ρ be as in Theorem 7.1. Suppose that $\dot{K} = N_G(G_{y,0})$ for some $y \in \mathcal{B}(G)$, $\dot{\rho}$ is an irreducible smooth representation of $\dot{K}$ such that $\dot{\rho} \,|\, G_{y,0^+}$ is a multiple of the trivial representation of $G_{y,0^+}$, and $\mathrm{Ind}_{\dot{K}}^G \dot{\rho} \simeq \mathrm{Ind}_K^G \rho$. Then there exists $g \in G$ such that $G_{x,0} = {}^g G_{y,0}$ and $\rho \simeq {}^g \dot{\rho}$.*

8. G-equivalence and cuspidal generic G-data

We begin this section with the definition of an equivalence relation that will be used in the parametrization of tame supercuspidal representations. Next we recall the definition of a cuspidal G-datum, and state the theorem from [**HM**] which gives criteria for equivalence of tame supercuspidal representations in terms of conditions on the cuspidal G-data used to produce the representations. Finally, we verify some properties of quasicharacters that occur in various triples which are associated to cuspidal G-data.

Suppose that G' and $\dot{G}$ are twisted Levi subgroups of G, π' and $\dot{\pi}$ are depth zero irreducible supercuspidal representations of G' and $\dot{G}$, and ϕ and $\dot{\phi}$ are quasicharacters of G' and $\dot{G}$, respectively.

DEFINITION 8.1. We say that the triples (G', π', ϕ) and $(\dot{G}, \dot{\pi}, \dot{\phi})$ are *G-equivalent* whenever there exists a $g \in G$ such that

$$G' = {}^g\dot{G} \quad \text{and} \quad \pi'\phi \simeq {}^g(\dot{\pi}\dot{\phi}).$$

It is clear that G-equivalence is an equivalence relation. Note that the definition does not assume that G' is tamely ramified or elliptic, nor that ϕ is G-regular or G-factorizable. However, if (G', π', ϕ) and $(\dot{G}, \dot{\pi}, \dot{\phi})$ are G-equivalent, then G' is tamely ramified (respectively, elliptic) if and only if $\dot{G}$ is tamely ramified (respectively, elliptic). Furthermore, when G' and $\dot{G}$ are tamely ramified, then ϕ is G-regular on G'_{0+} (respectively, G-factorizable) if and only if $\dot{\phi}$ is G-regular on $\dot{G}_{0+}$ (respectively, G-factorizable).

We now recall the definition of a cuspidal G-datum. The definition is due to Yu, although the adjective cuspidal is not used in [**Y**].

DEFINITION 8.2. A *(reduced) cuspidal G-datum* is a triple $\Psi = (\vec{\mathbf{G}}, \pi', \vec{\phi})$ that satifies the following conditions:

(1) $\vec{\mathbf{G}} = (\mathbf{G}^0, \dots, \mathbf{G}^d = \mathbf{G})$ is a tamely ramified twisted Levi sequence in $\mathbf{G}$, and $\mathbf{G}^0$ is elliptic in $\mathbf{G}$.
(2) π' is an irreducible supercuspidal representation of G^0 of depth zero.
(3) $\vec{\phi} = (\phi_0, \dots, \phi_d)$, where ϕ_i is a quasicharacter of G^i, $0 \leq i \leq d$. Let $x \in \mathcal{B}(G^0)$ be such that $G^0_{x,0}$ is a maximal parahoric subgroup of G^0 and π' is induced from a representation ρ of $N_{G^0}(G^0_{x,0})$ whose restriction to $G^0_{x,0+}$ is a multiple of the trivial representation (see Theorem 7.1). If $d > 0$, then, for $0 \leq i \leq d-1$, ϕ_i is G^{i+1}-generic of depth r_i relative to x, with $0 < r_0 < \cdots < r_{d-1}$. If ϕ_d is nontrivial, then the depth r_d of ϕ_d satisfies $r_d > r_{d-1}$ if $d > 0$, and $r_d \geq 0$ otherwise.

REMARK 8.3. Let $\Psi = (\vec{\mathbf{G}}, \pi', \vec{\phi})$ be a cuspidal G-datum. Let x and ρ be as in item (3) above. In [**HM**], we referred to the triple Ψ as a reduced cuspidal G-datum, and called the 4-tuple $(\vec{\mathbf{G}}, x, \rho, \vec{\phi})$ an extended cuspidal G-datum. Yu's construction associates an irreducible supercuspidal representation of G to each extended cuspidal G-datum. Yu shows that all of the extended cuspidal G-data arising from a given reduced cuspidal G-datum give rise to equivalent supercuspidal representations. Thus we have a map from the set of reduced cuspidal G-data onto the set of equivalence classes of tame supercuspidal representations of G.

DEFINITION 8.4. If $\Psi = (\vec{\mathbf{G}}, \pi', \vec{\phi})$ is a cuspidal G-datum, and $\phi = \prod_{i=0}^{d} \phi_i \,|\, G^0$, we say that the triple (G^0, π', ϕ) is *associated* to Ψ.

REMARK 8.5. In [**A2**], Adler constructed some supercuspidal representations of connected reductive p-adic groups that split over tamely ramified extensions. These representations correspond to cuspidal G-data of the form $\Psi = ((\mathbf{G}'), \mathbf{1}, (\phi))$, where G' is a tamely ramified twisted Levi subgroup that is compact modulo the centre of G, and ϕ is a G-generic quasicharacter of G' of positive depth. Note that the triple associated to Ψ by Definition 8.4 is $(G', \mathbf{1}, \phi)$.

REMARK 8.6. Let π be a tame supercuspidal representation associated, via Yu's construction, to a cuspidal G-datum $(\vec{\mathbf{G}}, \pi', \vec{\phi})$. Let notation be as in Definition 8.2. Let $\phi = \prod_{i=0}^{d} \phi_i \,|\, G^0$. If $d = 0$, then the depth of π is equal to the depth r_d of $\phi_d = \phi$. If $d > 0$, that is, if $G^0 \neq G$, then the depth of π is equal to $\max(r_{d-1}, r_d)$, which, by Definition 8.2(3), is positive. Assuming that Hypothesis $\mathrm{C}(G)_w$ holds, it follows from Lemma 4.9(1) that the depth of ϕ is also equal to $\max(r_{d-1}, r_d)$.

The following theorem (from Chapter 6 of [**HM**]) gives criteria that determine explicitly when two cuspidal G-data give rise to the same equivalence class of tame supercuspidal representations.

THEOREM 8.7. *Suppose that* $\Psi = (\vec{\mathbf{G}}, \pi', \vec{\phi})$ *and* $\dot{\Psi} = (\vec{\dot{\mathbf{G}}}, \dot{\pi}', \vec{\dot{\phi}})$ *are cuspidal G-data such that Hypothesis* $\mathrm{C}(G')$ *holds for all twisted Levi subgroups* $\mathbf{G}'$ *occurring in* $\vec{\mathbf{G}}$ *and* $\vec{\dot{\mathbf{G}}}$. *Then the supercuspidal representations of G obtained (via Yu's construction) from* Ψ *and* $\dot{\Psi}$ *belong to the same equivalence class if and only if the triples associated to* Ψ *and* $\dot{\Psi}$ *in the sense of Definition 8.4 are equivalent in the sense of Definition 8.1.*

Next we verify that whenever Hypothesis $\mathrm{C}(G')_w$ holds for all elliptic tamely ramified twisted Levi subgroups G' of G, we can identify all of the triples of the form (G', π', ϕ) that are associated to cuspidal G-data (and therefore to tame supercuspidal representations) in terms of properties of ϕ. As shown below, if a quasicharacter ϕ of G' is G-factorizable, and G-regular on G'_{0+}, and π' is an irreducible supercuspidal representation of G' of depth zero, then each factorization of ϕ gives rise to a cuspidal G-datum to which the triple (G', π', ϕ) is associated.

LEMMA 8.8. *Let* $\Psi = (\vec{\mathbf{G}}, \pi', \vec{\phi})$ *be a cuspidal G-datum, with* $\vec{\mathbf{G}} = (\mathbf{G}^0, \dots, \mathbf{G}^d = \mathbf{G})$ *and* $\vec{\phi} = (\phi_0, \dots, \phi_d)$. *Let* $\phi = \prod_{i=0}^{d} \phi_i \,|\, G^0$. *Let* $x \in \mathcal{B}(G^0)$ *be as in Definition 8.2(3). Then* $(\vec{\mathbf{G}}, \vec{\phi})$ *is a G-factorization of ϕ relative to x. If Hypothesis* $\mathrm{C}(G^0)_w^+$ *holds, then ϕ is G-regular on* G^0_{0+}.

PROOF. It is immediate from Definitions 8.2(3) and 5.1 that $(\vec{\mathbf{G}}, \vec{\phi})$ is a G-factorization of ϕ relative to x.

Assume that Hypothesis $\mathrm{C}(G^0)_w^+$ holds. Then Lemma 4.7 tells us that the genericity condition on ϕ_i, $0 \leq i \leq d-1$, is independent of the choice of x in Definition 8.2(3). Thus $(\vec{\mathbf{G}}, \vec{\phi})$ is a factorization of ϕ. We may now apply Corollary 6.5 to conclude that ϕ is G-regular on G^0_{0+}. □

LEMMA 8.9. *Suppose that G' is an elliptic tamely ramified twisted Levi subgroup of G. Assume that Hypothesis* $\mathrm{C}(G')_w^+$ *holds. Let ϕ be a G-factorizable quasicharacter of G' that is G-regular on* G'_{0+}. *Let* $(\vec{\mathbf{G}}, \vec{\phi})$ *be a G-factorization of ϕ. Then* $G' = G^0$. *If π' is an irreducible depth zero supercuspidal representation of G', then* $\Psi = (\vec{\mathbf{G}}, \pi', \vec{\phi})$ *is a cuspidal G-datum to which* (G', π', ϕ) *is associated.*

PROOF. Because we are assuming Hypothesis $C(G')_w^+$, we may apply Corollary 6.5 to conclude that, because $(\vec{\mathbf{G}}, \vec{\phi})$ is a factorization of ϕ, regularity of ϕ on G'_{0+} forces $G' = G^0$.

According to Theorem 7.1, there exists $x \in \mathcal{B}(G^0)$ such that $G^0_{x,0}$ is a maximal parahoric subgroup of G^0 and π' is induced from a representation of $N_{G^0}(G^0_{x,0})$ whose restriction to $G^0_{x,0+}$ is a multiple of the trivial representation. Since $G' = G^0 \subset G^i$ for $0 \leq i \leq d-1$ and Hypothesis $C(G^i)_w$ holds, ϕ_i is G^{i+1}-generic (in particular, relative to x). It is now clear that the conditions in Definition 8.2 hold. □

9. The parametrization

Throughout this section, we assume that Hypothesis $C(G')$ holds for all elliptic tamely ramified twisted Levi subgroups G' of G. Note that this assumption guarantees that Hypothesis $C(G')_w^+$ also holds for all elliptic tamely ramified twisted Levi subgroups G' of G.

Recall that in Definition 8.1, we defined an equivalence relation on the set of triples of the form (G', π', ϕ), where G' is a tamely ramified twisted Levi subgroup of G, π' is a depth zero irreducible supercuspidal representation of G', and ϕ is a quasicharacter of G'.

THEOREM 9.1. *The equivalence classes of irreducible tame supercuspidal representations correspond bijectively to the equivalence classes of triples (G', π', ϕ) as above such that G' is elliptic, and ϕ is G-factorizable and G-regular on G'_{0+}.*

PROOF. Let π and $\dot{\pi}$ be tame supercuspidal representations of G. Let Ψ and $\dot{\Psi}$ be cuspidal G-data that give rise, via Yu's construction, to (the equivalence classes of) π and $\dot{\pi}$, respectively. Let $\Sigma := (G^0, \pi', \phi)$ and $\dot{\Sigma} := (\dot{G}^0, \dot{\pi}', \dot{\phi})$ be the triples associated to Ψ and $\dot{\Psi}$ in the sense of Definition 8.4. According to Lemma 8.8, ϕ and $\dot{\phi}$ are G-factorizable, and G-regular on G^0_{0+} and $\dot{G}^0_{0+}$, respectively. Note that Theorem 8.7 tells us that π and $\dot{\pi}$ are equivalent if and only if the two triples Σ and $\dot{\Sigma}$ are equivalent in the sense of Definition 8.1.

To complete the proof, we observe that Lemma 8.9 guarantees that any triple (G', π', ϕ) having the properties indicated in the statement of the theorem is associated to some cuspidal G-datum. □

Suppose that $\mathbf{T}$ is an elliptic maximal F-torus in $\mathbf{G}$ that splits over a tamely ramified extension of F. Then $\mathbf{T}$ is an elliptic tamely ramified twisted Levi subgroup of $\mathbf{G}$. We will refer to $T = \mathbf{T}(F)$ as a tamely ramified elliptic maximal torus in G.

DEFINITION 9.2. If the equivalence class of a tame supercuspidal representation π of G corresponds to the equivalence class of a triple (G', π', ϕ), where G' is a tamely ramified elliptic maximal torus in G, we say that π is *toral.*

Suppose that T is a tamely ramified elliptic maximal torus in G. The set of depth zero supercuspidal representations of T is just the set of depth zero quasicharacters of T. If π' is a depth zero quasicharacter of T and ϕ is a quasicharacter of T, then the G-equivalence class of (T, π', ϕ) consists of triples $(\dot{T}, \dot{\pi}', \dot{\phi})$ such that $T = {}^g\dot{T}$ for some $g \in G$, and $\pi'\phi = {}^g(\dot{\pi}'\dot{\phi})$. Furthermore, since π' has depth zero, ϕ is G-regular on T_{0+} (respectively, G-factorizable) if and only if $\pi'\phi$ is G-regular on T_{0+} (respectively, G-factorizable).

DEFINITION 9.3. Suppose that $\mathbf{T}$ and $\dot{\mathbf{T}}$ are maximal F-tori in $\mathbf{G}$, ϕ is a quasicharacter of $T = \mathbf{T}(F)$, and $\dot{\phi}$ is a quasicharacter of $\dot{T} = \dot{\mathbf{T}}(F)$. We say that (T, ϕ) and $(\dot{T}, \dot{\phi})$ are *G-equivalent* if there exists $g \in G$ such that $T = {}^g\dot{T}$ and $\phi = {}^g\dot{\phi}$.

Sometimes we will refer to the G-equivalence class of a pair (T, ϕ) as the G-orbit of (T, ϕ). The toral supercuspidal representations of G can be parametrized as follows:

COROLLARY 9.4. *The equivalence classes of toral supercuspidal representations of G correspond bijectively to the equivalence classes of pairs (T, ϕ), where T is a tamely ramified elliptic maximal torus in G and ϕ is a G-factorizable quasicharacter of T that is G-regular on T_{0^+}.*

PROOF. In view of Theorem 8.7 and the comments preceding Definition 9.3, it suffices to show that if π and $\dot{\pi}$ are toral supercuspidal representations whose equivalence classes correspond to triples (T, π', ϕ) and $(\dot{T}, \dot{\pi}', \dot{\phi})$, then these triples are G-equivalent if and only if the pairs $(T, \pi'\phi)$ and $(\dot{T}, \dot{\pi}'\dot{\phi})$ are G-equivalent in the sense of Definition 9.3. This is immediate from the definitions. □

10. Supercuspidal representations and quasicharacters of elliptic maximal tori

In Definition 9.3, we defined a notion of G-equivalence on the set of pairs (T, ϕ), where $\mathbf{T}$ is a maximal F-torus in $\mathbf{G}$ and ϕ is a quasicharacter of $T = \mathbf{T}(F)$. The equivalence classes of toral supercuspidal representations are parametrized by G-orbits of pairs (T, ϕ), where T is a tamely ramified elliptic maximal torus in G, and ϕ is a quasicharacter of T that has the properties indicated in Corollary 9.4. In this section, we show that some other equivalence classes of tame supercuspidal representations can be parametrized by G-equivalence classes of pairs (T, ϕ) where T is a tamely ramified elliptic maximal torus in G and ϕ is a quasicharacter of T that has certain properties.

In the first part of the section (see Definition 10.1), we define a set $\mathcal{S}_0^G$ consisting of pairs (T, ϕ) where T is a particular kind of tamely ramified elliptic maximal torus in G and ϕ is a depth zero quasicharacter of T that satisfies some conditions. To each pair (T, ϕ) we associate a depth zero irreducible supercuspidal representation $\pi_{(T,\phi)}$ that is induced from (an extension of) the inflation of a cuspidal Deligne-Lusztig representation of a reductive group over the residue field of F. Then (see Proposition 10.8 and Corollary 10.10), we verify that the equivalence classes of the representations $\pi_{(T,\phi)}$ are parametrized by the G-equivalence classes of the pairs (T, ϕ) in $\mathcal{S}_0^G$.

In the second part of the section, we describe some tame supercuspidal representations whose construction involves both a depth zero part and a positive depth part and we show, in Theorem 10.13, that the equivalence classes of such representations are parametrized by G-equivalence classes of certain kinds of pairs (T, ϕ).

In Section 11, we discuss tame supercuspidal representations of general linear groups. The Howe parametrization of the equivalence classes of these representations is a parametrization in terms of certain G-equivalence classes of pairs (T, ϕ).

Recall that $\mathfrak{f}$ is the residue field of F. Let T be a maximal $\mathfrak{f}$-torus in a connected reductive $\mathfrak{f}$-group G, and let θ be a character of $\mathsf{T}(\mathfrak{f})$. In [**DL**], Deligne and Lusztig defined a class function $R^{\mathbf{G}}_{\mathbf{T},\theta}$ on $\mathbf{G}(\mathfrak{f})$. When θ is in general position, $R^{\mathbf{G}}_{\mathbf{T},\theta}$ is, up to sign, the character of an irreducible representation of $\mathbf{G}(\mathfrak{f})$, which we will refer to as a Deligne-Lusztig representation. This representation is cuspidal whenever $\mathbf{T}$ is elliptic. Note that, according to Lemma 2.4.1 of [**D3**], $\mathbf{G}$ always contains an elliptic maximal $\mathfrak{f}$-torus.

If $x \in \mathcal{B}(G)$, let $\mathbf{G}_x$ be the connected reductive $\mathfrak{f}$-group denoted by $\overline{G}_F^{\text{red}}$ in Section 3.5 of [**T**] for which $\mathbf{G}_x(\mathfrak{f}) \simeq G_{x,0}/G_{x,0^+}$. Suppose that $\mathbf{T}$ is an elliptic maximal torus in $\mathbf{G}_x$. As shown in §2.4 of [**D3**]. there exists an elliptic maximal torus T in G such that $T \subset N_G(G_{x,0})$ and the image of $T \cap G_{x,0}$ in $\mathbf{G}_x(\mathfrak{f})$ is equal to $\mathbf{T}(\mathfrak{f})$. Although T is not unique, it follows from Lemma 2.2.2 of [**D3**] that any two choices of T are conjugate by an element of $G_{x,0^+}$.

Recall that Z denotes the centre of G.

DEFINITION 10.1. Let $\mathcal{S}_0^G$ be the set of pairs (T,ϕ) such that

(1) T is a tamely ramified elliptic maximal torus in G that normalizes a maximal parahoric $G_{x,0}$ of G, and the image of $T \cap G_{x,0}$ in $\mathbf{G}_x(\mathfrak{f})$ is the $\mathfrak{f}$-rational points of an elliptic maximal $\mathfrak{f}$-torus $\mathbf{T}$ in $\mathbf{G}_x$.

(2) ϕ is a depth zero quasicharacter of T such that $\phi \mid T \cap G_{x,0}$ factors to a character θ of $\mathbf{T}(\mathfrak{f})$ that is in general position.

(3) The representation ρ of $K := N_G(G_{x,0})$ constructed as follows is irreducible. Let ρ_0 be the inflation to $G_{x,0}$ of the representation of $\mathbf{G}_x(\mathfrak{f})$ whose character is equal to $\pm R^{\mathbf{G}_x}_{\mathbf{T},\theta}$ and let ρ_1 be the (unique) extension of ρ_0 to $ZG_{x,0}$ such that if $z \in Z \cap G_{x,0}$, $\rho_1(z)$ is equal to $\phi(z)$ times the identity operator on the space of ρ_0. Let $\rho = \operatorname{Ind}^K_{ZG_{x,0}} \rho_1$.

REMARK 10.2. Let $\mathbf{Z}_x(\mathfrak{f})$ be the image of $Z \cap G_{x,0}$ in $\mathbf{G}_x(\mathfrak{f})$. Since $\mathbf{Z}_x(\mathfrak{f})$ is a subgroup of the centre of $\mathbf{G}_x(\mathfrak{f})$, $\mathbf{Z}_x(\mathfrak{f})$ is a subgroup of $\mathbf{T}(\mathfrak{f})$. The restricition of the class function $\pm R^{\mathbf{G}_x}_{\mathbf{T},\theta}$ to $\mathbf{Z}_x(\mathfrak{f})$ is equal to the restriction of θ to $\mathbf{Z}_x(\mathfrak{f})$ multiplied by the degree of the representation ρ_0. Therefore it follows from Definition 10.1(2) that, for $z \in Z \cap G_{x,0}$, $\rho_0(z)$ is equal to $\phi(z)$ times the identity operator on the space of ρ_0. Hence the extension ρ_1 of Definition 10.1(3) is well-defined.

REMARK 10.3. In general, there can exist pairs (T,ϕ) for which the first two conditions of Definition 10.1 hold, but the third condition is not satisfied. For example, if $G = \mathbf{GSp}_4(F)$, there exist pairs (T,ϕ) that satisfy Definition 10.1(1) and (2), with $G_{x,0}$ a nonspecial maximal parahoric subgroup of G (that is, the image of x in the reduced building of G is not a special vertex). Note that the index of $ZG_{x,0}$ in K is equal to 2 in this example. It is easy to see that there are some choices of (T,ϕ) for which the representation $\rho = \operatorname{Ind}^K_{ZG_{x,0}} \rho_1$ is reducible. The details are left to the reader.

DEFINITION 10.4. Let $(T,\phi) \in \mathcal{S}_0^G$. Set $\pi_{(T,\phi)} := \operatorname{Ind}^G_K \rho$.

According to Proposition 7.3, $\pi_{(T,\phi)}$ is a depth zero irreducible supercuspidal representation of G. Note that if we replace ρ_0 by an equivalent representation ρ_0' of $G_{x,0}$ and carry out the same construction as in Definition 10.1(3), using ρ_0' in place of ρ_0 and using the quasicharacter $\phi \mid Z$ to extend ρ_0' to a representation ρ_1' of $ZG_{x,0}$, we have that ρ_1' is equivalent to ρ_1. It follows that the representation $\operatorname{Ind}^G_{ZG_{x,0}} \rho_1'$ is equivalent to $\operatorname{Ind}^G_{ZG_{x,0}} \rho_1 = \pi_{(T,\phi)}$.

LEMMA 10.5. *Let* $(T,\phi) \in \mathcal{S}_0^G$. *Then* $T = Z(T \cap G_{x,0})$.

PROOF. Let $\gamma \in T$. Since T is elliptic, γ fixes the image of x in the reduced building of G, so conjugation by γ^{-1} induces an $\mathfrak{f}$-automorphism $c_{\gamma^{-1}}$ of $\mathbf{G}_x$. It follows from properties of the Deligne-Lusztig construction that

$$R^{\mathbf{G}_x}_{c_{\gamma^{-1}}(\mathbf{T}),\theta\circ c_\gamma} \circ c_{\gamma^{-1}} = R^{\mathbf{G}_x}_{\mathbf{T},\theta}.$$

Combining this with the fact that the restriction of c_γ to $\mathbf{T}$ is the identity, we see that $R^{\mathbf{G}_x}_{\mathbf{T},\theta} \circ c_{\gamma^{-1}} = R^{\mathbf{G}_x}_{\mathbf{T},\theta}$. It follows that ${}^\gamma\rho_0$ is equivalent to ρ_0. Because ${}^\gamma\rho_1(zk_1) = \phi_Z(z){}^\gamma\rho_0(k_1)$ for all $z \in Z$ and $k_1 \in G_{x,0}$, we see that ${}^\gamma\rho_1$ is equivalent to ρ_1. Irreduciblity of ρ then implies that $\gamma \in ZG_{x,0}$, because (by Mackey theory) ${}^k\rho_1$ cannot be equivalent to ρ_1 if $k \in K$ and $k \notin ZG_{x,0}$. We conclude that $T \subset ZG_{x,0}$. That is, $T = Z(T \cap G_{x,0})$. □

The next lemma tells us that if $(T,\phi) \in \mathcal{S}_0^G$, then the equivalence class of $\pi_{(T,\phi)}$ is determined by $G_{x,0}$, $\mathbf{T}$, θ and $\phi\,|\,Z$.

LEMMA 10.6. *Let (T,ϕ), $(\dot{T},\dot{\phi}) \in \mathcal{S}_0^G$. Let x, $\mathbf{T}$, θ, and $\dot{x}$, $\dot{\mathbf{T}}$, $\dot{\theta}$, respectively, be as in Definition 10.1, relative to the pair (T,ϕ) and the pair $(\dot{T},\dot{\phi})$. If $G_{x,0} = G_{\dot{x},0}$, $\dot{\mathbf{T}} = \mathbf{T}$, $\dot{\theta} = \theta$, and $\phi\,|\,Z = \dot{\phi}\,|\,Z$, then*

(1) $\pi_{(\dot{T},\dot{\phi})} \simeq \pi_{(T,\phi)}$.

(2) *There exists $k \in G_{x,0+}$ such that $\dot{T} = {}^kT$ and $\dot{\phi} = {}^k\phi$.*

(3) *$(\dot{T},\dot{\phi})$ is G-equivalent to (T,ϕ) in the sense of Definition 9.3.*

PROOF. Let ρ and $\dot{\rho}$ be the representations of $K = N_G(G_{x,0}) = N_G(G_{\dot{x},0})$ that are associated, as in Definition 10.1(3), to (T,ϕ) and $(\dot{T},\dot{\phi})$, respectively. For part (1), note that the assumptions of the lemma guarantee that $\rho \simeq \dot{\rho}$.

As mentioned in comments preceding Definition 10.1, since $G_{x,0} = G_{\dot{x},0}$ and $\mathbf{T} = \dot{\mathbf{T}}$, according to a result of DeBacker(Lemma 2.2.2 of [**D3**]), there exists $k \in G_{x,0+}$ such that $\dot{T} = {}^kT$. Note that, because $k \in G_{x,0+}$ ${}^k\phi$ is the inflation of θ to $\dot{T} \cap G_{x,0}$ and ${}^k\phi\, Z \cap G_{x,0} = \phi\,|\,G_{x,0}$. Combining this information, Lemma 10.5 (applied to $\dot{T}$), and the assumptions $\theta = \dot{\theta}$ and $\phi\,|\,Z = \dot{\phi}\,|\,Z$, we have that $\dot{\phi} = {}^k\phi$. This proves part (2) of the lemma. Part (3) is an immediate consequence of part (2). □

LEMMA 10.7. *Let $g \in G$ and let $(T,\phi) \in \mathcal{S}_0^G$. Set $\dot{T} = {}^gT$ and $\dot{\phi} = {}^g\phi$. Then $(\dot{T},\dot{\phi}) \in \mathcal{S}_0^G$ and $\pi_{(\dot{T},\dot{\phi})} \simeq \pi_{(T,\phi)}$.*

PROOF. Because $\dot{T}/Z$ is compact and $x \in \mathcal{B}(T)$, the torus $\dot{T}$ determines the maximal parahoric subgroup ${}^gG_{x,0} = G_{gx,0}$ in the sense that $G_{y,0} = G_{gx,0}$ for all $y \in \mathcal{B}(\dot{T})$. There exists an elliptic maximal $\mathfrak{f}$-torus $\dot{\mathbf{T}}$ in $\mathbf{G}_{gx}$ having the property that $\dot{\mathbf{T}}(\mathfrak{f})$ is the image of $\dot{T} \cap G_{gx,0}$ in $\mathbf{G}_{gx}(\mathfrak{f})$ (see Section 3.5 of [**T**]). Thus condition (1) of Definition 10.1 is satisfied for $(\dot{T},\dot{\phi})$.

The restriction $\dot{\phi}\,|\,\dot{T} \cap G_{gx,0}$ factors to a character $\dot{\theta}$ of $\dot{\mathbf{T}}(\mathfrak{f})$. Conjugation by g factors to an $\mathfrak{f}$-isomorphism c_g between $\mathbf{G}_x$ and $\mathbf{G}_{gx}$. Because θ is in general position, $\dot{\mathbf{T}} = c_g(\mathbf{T})$ and $\dot{\theta} = \theta \circ c_g^{-1}$, we have that the character $\dot{\theta}$ is in general position. Hence condition (2) of Definition 10.1 is satisfied for $(\dot{T},\dot{\phi})$.

Let $\dot{\rho}_0$ be the representation of $G_{gx,0}$ that is the inflation of the representation of $\mathbf{G}_{gx}(\mathfrak{f})$ whose character is $\pm R^{\mathbf{G}_{gx}}_{\dot{\mathbf{T}},\dot{\theta}}$. The relation $R^{\mathbf{G}_{gx}}_{\dot{\mathbf{T}},\dot{\theta}} \circ c_g = R^{\mathbf{G}_x}_{\mathbf{T},\theta}$ implies that $\dot{\rho}_0 \simeq {}^g\rho_0$. Note that $\dot{\phi}\,|\,Z = {}^g\phi\,|\,Z = \phi\,|\,Z$. Since the same quasicharacter is used to extend the equivalent representations $\dot\rho_0$ and ${}^g\rho_0$ to representations $\dot\rho_1$ and ${}^g\rho_1$ of $ZG_{gx,0}$, we have $\dot\rho_1 \simeq {}^g\rho_1$. Set $\dot\rho = \mathrm{Ind}_{ZG_{gx,0}}^{{}^gK}\dot\rho_1$. Then $\dot\rho \simeq {}^g\rho$. Thus the irreducibility of ρ implies that $\dot\rho$ is irreducible. This completes the verification that condition (3) of Definition 10.1 is satisfied by $\dot\rho$. Let $\dot\pi = \pi_{(\dot T,\dot\phi)} = \mathrm{Ind}_{{}^gK}^G\dot\rho$. The equivalence of $\dot\rho$ and ${}^g\rho$ implies equivalence of $\dot\pi$ and $\mathrm{Ind}_{{}^gK}^G{}^g\rho$, and the latter representation is equivalent to $\pi_{(T,\phi)} = \mathrm{Ind}_K^G\rho$. □

PROPOSITION 10.8. *Let (T,ϕ), $(\dot T,\dot\phi) \in \mathcal{S}_0^G$. Then $\pi_{(T,\phi)} \simeq \pi_{(\dot T,\dot\phi)}$ if and only if (T,ϕ) and $(\dot T,\dot\phi)$ are G-equivalent in the sense of Definition 9.3.*

PROOF. One direction of the proposition is proved in Lemma 10.7. For the other direction, assume that $\pi_{(T,\phi)}$ and $\pi_{(\dot T,\dot\phi)}$ are equivalent representations. We use the notations x, $\mathbf{T}$, θ, K, ρ_0, etc., and $\dot x$, $\dot K$, $\dot\rho$, etc., for the objects appearing in Definition 10.1 as it applies to (T,ϕ) and $(\dot T,\dot\phi)$, respectively. According to Proposition 7.4, there exists $g \in G$ such that ${}^g\dot K = K$ and ${}^g\dot\rho \simeq \rho$. After replacing $(\dot T,\dot\phi)$ by the G-equivalent pair $({}^g\dot T, {}^g\dot\phi)$, we may assume that $\dot T = T$, $\dot K = K$, $\dot\rho \simeq \rho$, and $G_{\dot x,0} = G_{x,0}$. Note that we now also have $\mathbf{T} = \dot{\mathbf{T}}$. It follows from the equivalence of ρ and $\dot\rho$ that

$$\dot\phi \,|\, Z = \phi \,|\, Z \quad\text{and}\quad \pm R_{\mathbf{T},\dot\theta}^{\mathbf{G}_x} = \pm R_{\mathbf{T},\theta}^{\mathbf{G}_x}.$$

By results of [**DL**] concerning equivalence of Deligne-Lusztig representations, the second equality above implies that there exists $h \in \mathbf{G}_x(\mathfrak{f})$ that normalizes $\mathbf{T}(\mathfrak{f})$ and satisfies $\theta = {}^h\dot\theta$. Choose $k \in G_{x,0}$ whose image in $\mathbf{G}_x(\mathfrak{f})$ is equal to h. Then the image of ${}^k\dot T = {}^kT$ in $\mathbf{G}_x$ is equal to $\mathbf{T}(\mathfrak{f})$ and ${}^k\dot\phi \,|\, {}^kT \cap G_{x,0}$ is the inflation of ${}^h\dot\theta = \theta$. Also, ${}^k\dot\phi \,|\, Z = \phi \,|\, Z$. We may now apply Lemma 10.6(3) to the pairs (T,ϕ) and $({}^k\dot T, {}^k\dot\phi)$ to conclude that they are G-equivalent in the sense of Definition 9.3. □

DEFINITION 10.9. Let $\mathcal{A}_{0,\mathrm{cusp}}^G := \{\, \pi_{(T,\phi)} \mid (T,\phi) \in \mathcal{S}_0^G \,\}$.

The following is an immediate consequence of Definition 10.9 and Proposition 10.8.

COROLLARY 10.10. *The equivalence classes of representations in $\mathcal{A}_{0,\mathrm{cusp}}^G$ correspond bijectively to the G-equivalence classes of pairs (T,ϕ) in $\mathcal{S}_0^G$.*

As discussed in Section 11, if $G = \mathbf{GL}_n(F)$, the set $\mathcal{A}_{0,\mathrm{cusp}}^G$ contains all of the depth zero supercuspidal representations of G.

The paper [**DR**] of DeBacker and Reeder gives an explicit construction of some depth zero supercuspidal L-packets for pure inner forms of unramified p-adic groups. The representations in these L-packets belong to $\mathcal{A}_{0,\mathrm{cusp}}^G$. However, there can exist representations in $\mathcal{A}_{0,\mathrm{cusp}}^G$ which do not lie in these L-packets. For example, using notation from earlier in the section, if $G = \mathbf{Sp}_4(F)$ and $G_{x,0}$ is a nonspecial maximal parahoric subgroup of G, then $K = G_{x,0}$ and there are some choices of pairs $(T,\phi) \in \mathcal{S}_0^G$ such that $\pi_{(T,\phi)}$ does not lie in any of the L-packets constructed in [**DR**]. Roughly speaking, this happens because the regularity conditions imposed on the parameters that determine the L-packets in [**DR**] can correspond to a condition on the character θ that determines ρ_0 which is stronger than the general position condition. The notation here is as in Definition 10.1.

Let G' be a proper elliptic tamely ramified twisted Levi subgroup of G.

DEFINITION 10.11. Let $\mathcal{R}^G_{G'}$ be the set of equivalence classes of tame supercuspidal representations of G that are parametrized, via Theorem 9.1, by the G-equivalence classes of triples of the form (G', π', χ), where $\pi' \in \mathcal{A}^{G'}_{0,\mathrm{cusp}}$ and χ is a G-factorizable quasicharacter of G' that is G-regular on G'_{0+}.

DEFINITION 10.12. Let $\mathcal{T}^G_{G'}$ be the set of G-equivalence classes (in the sense of Definition 9.3) of pairs of the form $(T, \phi'\chi_T)$, where $(T, \phi') \in \mathcal{S}^{G'}_0$, χ is a G-factorizable quasicharacter of G' that is G-regular on G'_{0+}, and $\chi_T = \chi \,|\, T$.

THEOREM 10.13. *Suppose that an equivalence class π of representations in $\mathcal{R}^G_{G'}$ is parametrized by the G-equivalence class of a triple (G', π', χ). Let $(T, \phi') \in \mathcal{S}^{G'}_0$ be a pair whose G'-equivalence class, in the sense of Definition 9.3, parametrizes the equivalence class of π', via Corollary 10.10. Let $\chi_T = \chi \,|\, T$ and let $\mathcal{F}(\pi)$ be the G-equivalence class of the pair $(T, \phi'\chi_T)$. Then $\mathcal{F}$ is a well defined bijection between $\mathcal{R}^G_{G'}$ and $\mathcal{T}^G_{G'}$.*

PROOF. Let $(G', \dot\pi', \dot\chi)$ be another triple whose G-equivalence class parametrizes an equivalence class in $\mathcal{R}^G_{G'}$. Let $(\dot T, \dot\phi') \in \mathcal{S}^{G'}_0$ be a pair whose G'-equivalence class parametrizes the equivalence class of $\dot\pi'$. It follows from Theorem 9.1 that to prove the theorem, it suffices to show that the G-equivalence classes of the triples (G', π', χ) and $(G', \dot\pi', \dot\chi)$ are the same if and only if the G-equivalence classes of the pairs $(T, \phi'\chi_T)$ and $(\dot T, \dot\phi'\dot\chi_{\dot T})$ are the same.

Suppose that the G-equivalence classes of (G', π', χ) and $(G', \dot\pi', \dot\chi)$ are the same. Then, according to Definition 8.1, there exists $g \in N_G(G')$ such that ${}^g(\pi'\chi) \simeq \dot\pi'\dot\chi$. Observe that ${}^g\pi'$, and hence also the equivalent representation $\dot\pi'\dot\chi\, {}^g\chi^{-1}$, belongs to $\mathcal{A}^{G'}_{0,\mathrm{cusp}}$. The associated equivalence class of representations of G' is parametrized, via Corollary 10.10, by the G'-equivalence class of the pair $({}^gT, {}^g\phi')$ and also by the G'-equivalence class of the pair $(\dot T, \dot\phi(\dot\chi\, {}^g\chi^{-1})_{\dot T})$. This implies (see Definition 9.3) that the G'-equivalence classes of $({}^gT, {}^g(\phi'\chi_T))$ and $(\dot T, \dot\phi'\dot\chi_{\dot T})$ are the same. Hence the G-equivalence classes of the pairs $(T, \phi'\chi_T)$ and $(\dot T, \dot\phi\dot\chi_{\dot T})$ are the same.

Next, assume that the G-equivalence classes of $({}^gT, {}^g(\phi'\chi_T))$ and $(\dot T, \dot\phi'\dot\chi_{\dot T})$ are the same. Then there exists $g \in G$ such that ${}^gT = \dot T$ and ${}^g(\phi'\chi_T) = \dot\phi'\dot\chi_{\dot T}$. Because both ϕ' and $\dot\phi'$ have depth zero, we have that ${}^g\chi \,|\, \dot T_{0+} = \dot\chi \,|\, \dot T_{0+}$. If $T = G'$, then $\dot T = {}^gT = {}^gG'$, so, since $\dot T \subset G'$, we have that $T = G'$ if and only if $\dot T = G'$, and in this case $g \in N_G(G')$.

In this paragraph, we assume that $\dot T \neq G'$. Because $\dot T$ is a tamely ramified twisted Levi subgroup of G', we may apply Lemma 4.9(1) to conclude that the depths of ${}^g\chi$ and $\dot\chi$ are equal. Since $\dot\chi$ is G-regular on G'_{0+}, the depth of $\dot\chi$ is positive. Note that $(\dot{\mathbf{T}} = {}^g\mathbf{T}, {}^g\mathbf{G}', \mathbf{G})$ and $(\dot{\mathbf{T}}, \mathbf{G}', \mathbf{G})$ are tamely ramified twisted Levi sequences in $\mathbf{G}$. If $\dot\chi$ and χ (hence ${}^g\chi$) are G-generic, then we may apply Lemma 4.6 to conclude that the equality ${}^g\chi \,|\, \dot T_{0+} = \dot\chi \,|\, \dot T_{0+}$ implies that ${}^gG' = G$. In general, ${}^g\chi$ and $\dot\chi$ are not G-generic. However, we may use the fact that χ and $\dot\chi$ are G-factorizable and G-regular on G'_{0+}, together with an inductive argument, applying Lemma 4.6 at various stages, to conclude that the equality ${}^g\chi \,|\, \dot T_{0+} = \dot\chi \,|\, \dot T_{0+}$ implies ${}^gG' = G'$, that is $g \in N_G(G')$. The details are omitted.

We have shown that whenever $(T, \phi'\chi_T)$ and $(\dot T, \dot\phi'\dot\chi_{\dot T})$ are G-equivalent in the sense of Definition 9.3, there exists $g \in N_G(G')$ such that ${}^gT = \dot T$ and ${}^g(\phi'\chi_T) = \dot\phi'\dot\chi_{\dot T}$. We may rewrite this as ${}^g\phi' = \dot\phi'(\dot\chi\, {}^g\chi^{-1})_{\dot T}$. The representation $\dot\pi'(\dot\chi\, {}^g\chi^{-1})$

belongs to the equivalence class of representations in $\mathcal{A}^{G'}_{0,\mathrm{cusp}}$ that corresponds, via Corollary 10.10, to the G'-equivalence class of the pair $(\dot{T}, \dot{\phi}'(\dot{\chi}\, {}^g\chi^{-1})_{\dot{T}})$. Because this is the same as the G'-equivalence class of the pair $({}^gT, {}^g\phi')$, which parametrizes the equivalence class of ${}^g\pi'$, we have $\dot{\pi}'(\dot{\chi}\, {}^g\chi^{-1}) \simeq {}^g\pi'$. That is, $\dot{\pi}'\dot{\chi} \simeq {}^g(\pi'\chi)$. Hence the triples (G', π', χ) and $(G', \dot{\pi}', \dot{\chi})$ are G-equivalent in the sense of Definition 8.1. □

11. Tame supercuspidal representations of general linear groups

Throughout this section only, we assume that n is an integer such that $n \geq 2$ and $\mathbf{G} = \mathbf{GL}_n$ is the general linear group of rank n. We begin with a discussion of depth zero supercuspidal representations of general linear groups. Following that, we make some comments about the connections between the Howe parametrization of tame supercuspidal representations of general linear groups, Yu's construction as it applies to general linear groups, and the parametrization of Theorem 9.1.

Let $\mathfrak{o}$ be the ring of integers in F. The group $\mathbf{GL}_n(\mathfrak{o})$ is a maximal parahoric subgroup of G and its pro-unipotent radical $\mathbf{GL}_n(\mathfrak{o})^u$ consists of the elements $g \in G$ such that the entries of $g - 1$ lie in the maximal ideal $\mathfrak{p}$. The quotient $\mathbf{GL}_n(\mathfrak{o})/\mathbf{GL}_n(\mathfrak{o})^u$ is isomorphic to $\mathbf{GL}_n(\mathfrak{f})$. Choose an elliptic maximal torus $\mathbf{T}$ in $\mathbf{G}$ such that $T = \mathbf{T}(F)$ normalizes $\mathbf{GL}_n(\mathfrak{o})$ and the image of $T \cap \mathbf{GL}_n(\mathfrak{o})$ in $\mathbf{GL}_n(\mathfrak{f})$ is the $\mathfrak{f}$-points of an elliptic maximal torus T in GL_n. The normalizer of $\mathbf{GL}_n(\mathfrak{o})$ in G is equal to $Z\mathbf{GL}_n(\mathfrak{o})$, and $T = Z(T \cap \mathbf{GL}_n(\mathfrak{o}))$, where Z is the centre of G. Let ϕ be a quasicharacter of T such that $\phi \,|\, T \cap \mathbf{GL}_n(\mathfrak{o})$ is the inflation of a character θ of $\mathsf{T}(\mathfrak{f})$ that is in general position. Let ρ be a representation of $Z\mathbf{GL}_n(\mathfrak{o})$ such that $\rho \,|\, \mathbf{GL}_n(\mathfrak{o})$ is the inflation of a representation of $\mathbf{GL}_n(\mathfrak{f})$ whose character is $\pm R^{\mathsf{G}}_{\mathsf{T},\theta}$ and $\rho(z)$ is equal to $\phi(z)$ times the identity operator for $z \in Z$. Then $(T, \phi) \in \mathcal{S}^G_0$, where $\mathcal{S}^G_0$ is as in Definition 10.1, and the compactly induced representation $\pi_{(T,\phi)} := \mathrm{Ind}^G_{Z\mathbf{GL}_n(\mathfrak{o})}\rho$ is an irreducible depth zero supercuspidal representation of G. Up to conjugacy, $\mathbf{GL}_n(\mathfrak{o})$ is the only maximal parahoric subgroup of G. Furthermore, $\mathbf{GL}_n(\mathfrak{f})$ contains one conjugacy class of elliptic maximal tori, and all irreducible cuspidal representations of $\mathbf{GL}_n(\mathfrak{f})$ are Deligne-Lusztig representations. Consequently, it follows from Theorem 7.1 that a depth zero irreducible supercuspidal representation of G is equivalent to $\pi_{(T,\phi)}$, where T is as above and ϕ is a depth zero quasicharacter of T such that $\phi \,|\, T \cap \mathbf{GL}_n(\mathfrak{o})$ factors to a character in general position. Applying Corollary 10.10, we see that the equivalence classes of irreducible depth zero supercuspidal representations correspond bijectively to the set of G-equivalence classes that contain such a pair (T, ϕ). We remark that T is isomorphic to the multiplicative group $E^\times$ of an unramified degree n extension of F, and a depth zero quasicharacter ϕ of T factors to a character in general position if and only if the character of $E^\times$ corresponding to ϕ under any isomorphism of T and $E^\times$ is not fixed by any nontrivial element of the Galois group of E over F.

As shown by Moy ([**M**]), when F has characteristic zero and the residual characteristic p of F is prime to n, the equivalence classes of supercuspidal representations of $\mathbf{GL}_n(F)$ are parametrized, via Howe's construction ([**H**]), by equivalence classes of F-admissible quasicharacters of multiplicative groups of tamely ramified degree n extensions of F. For the rest of this section, we assume that the characteristic of F is zero. When p divides n, Howe's construction parametrizes only the equivalence classes of tame supercuspidal representations of $\mathbf{GL}_n(F)$. We remark that there is a bijective correspondence between the conjugacy classes of tamely ramified elliptic maximal tori in $\mathbf{GL}_n(F)$ and the isomorphism classes of degree n tamely ramified extensions of F. The Howe correspondence is actually a bijective correspondence between the equivalence classes of tame supercuspidal representations of $\mathbf{GL}_n(F)$ and the $\mathbf{GL}_n(F)$-orbits of pairs (T, ϕ) where T is a tamely ramified elliptic maximal torus in $\mathbf{GL}_n(F)$ and ϕ is a quasicharacter of T that is F-admissible in the sense of Howe.

The relation between cuspidal G-data and triples (G', π', ϕ), in our general setting, is analogous to the relation between Howe factorizations and F-admissible quasicharacters, in the setting of general linear groups and division algebras. In Section 3.5 of [**HM**], Hakim and the author describe how to pass back and forth between cuspidal G-data and Howe factorizations of F-admissible quasicharacters. This is done in such a way that the equivalence class of tame supercuspidal representations arising from a cuspidal G-datum coincides with the equivalence class arising from the F-admissible quasicharacter whose Howe factorization is associated to the cuspidal G-datum. Lemmas 8.8 and 8.9 of this paper show how to relate cuspidal G-data to the triples whose G-orbits parametrize the equivalence classes of tame supercuspidal representations (as in Theorem 9.1). Combining these lemmas with information from Section 3.5 of [**HM**] yields a bijective correspondence between the equivalence classes of F-admissible quasicharacters of multiplicative groups of tamely ramified degree n extensions of F and the G-orbits of triples that parametrize tame supercuspidal representations of G. This correspondence is described below.

Let (G', π', ϕ) be a triple whose G-equivalence class parametrizes an equivalence class of tame supercuspidal representations of G, via the correspondence of Theorem 9.1. First we consider the case when G' is a torus. In this case, G' is isomorphic to $E^\times$ for some degree n tamely ramified extension E of F, and, via this isomorphism, we identify π' and ϕ with quasicharacters of $E^\times$. Let $\varphi = \pi'\phi$. Then φ is an F-admissible quasicharacter of $E^\times$ whose equivalence class corresponds to the G-orbit of the triple (G', π', ϕ). The associated equivalence class of tame supercuspidal representations is toral and is also parametrized by the G-orbit of (G', φ), via Corollary 9.4.

Suppose that G' is not a torus. Then G' is isomorphic to $\mathbf{GL}_m(E')$, where m is a proper divisor of n and E' is a tamely ramified extension of F of degree n/m. According to the earlier discussion of depth zero supercuspidal representations of general linear groups, the equivalence class of π' corresponds to the G'-orbit of a pair (T, ϕ'), where T is an elliptic maximal torus in G' that is isomorphic to the multiplicative group $E^\times$ of an unramified degree m extension E of E' and ϕ' is a depth zero quasicharacter of T such that ϕ' is not fixed by any nontrivial element in the Galois group of E over E'. There exists a quasicharacter χ of $E^\times$ such that $\phi = \chi \circ \det_{G'}$, where $\det_{G'}$ is the determinant on G'. Let $N_{E/E'}$ be the norm map from $E^\times$ to $E'^\times$. Then $\phi \,|\, E^\times = \chi \circ N_{E/E'}$. Let $\varphi = \phi'(\chi \circ N_{E/E'})$. Then φ is F-admissible and the equivalence class of φ corresponds, via the results of Howe and Moy and Theorem 9.1, to the G-equivalence class of (G', π', ϕ).

To see how the correspondence works in the reverse direction, let E be a tamely ramified degree n extension of F and let φ be an F-admissible quasicharacter of $E^\times$. Let $\mathfrak{p}_E$ be the maximal ideal in the ring of integers of E.

Suppose that there exist an extension E' of F and a quasicharacter φ' of $E'^{\times}$ such that $E' \subsetneq E$ and

$$\varphi \,|\, 1+\mathfrak{p}_E = \varphi' \circ N_{E/E'} \,|\, 1+\mathfrak{p}_E.$$

It follows from admissibility of φ that E is unramified over E'. We choose E' and φ' so that the degree of E over E' is as large as possible. Choose an elliptic maximal torus T of G such that $T \simeq E^{\times}$. Let G' be the centralizer in G of the subgroup of T corresponding to $E'^{\times}$ and let m be the degree of E over E'. Then G' is isomorphic to $\mathbf{GL}_m(E')$ and G' is an elliptic tamely ramified twisted Levi subgroup of G. Define $\phi(g) = \varphi'(\det_{G'}(g))$ for $g \in G'$. It is not difficult to show that ϕ is G-factorizable and ϕ is G-regular on G'_{0+}. Also, the quasicharacter $\varphi(\varphi'^{-1} \circ N_{E/E'})$ of $E^{\times}$ has depth zero and is not fixed by any nontrivial element of the Galois group of E over E'. By definition of G', the image T of $E^{\times}$ in G lies in G' and T is a tamely ramified elliptic maximal torus in G'. The G'-orbit of $(T, \varphi(\varphi'^{-1} \circ N_{E/E'}))$ parametrizes an equivalence class of depth zero supercuspidal representations of G'. Let π' be a representation in this class. The G-orbit of the triple (G', π', ϕ) corresponds to the equivalence class of the F-admissible quasicharacter φ.

Next, we assume that there do not exist E' and φ' as in the previous paragraph. As above, we choose an elliptic maximal torus T of G such that $T \simeq E^{\times}$. We can check that φ is G-factorizable and is G-regular on $T_{0+} \simeq 1+\mathfrak{p}_E$. The trivial representation $\mathbf{1}$ of T is a depth zero supercuspidal representation of T. The G-orbit of the triple $(T, \mathbf{1}, \varphi)$ corresponds to the equivalence class of the F-admissible quasicharacter φ.

12. Self-contragredient tame supercuspidal representations

If π is a smooth representation of G, we use the notation $\widetilde{\pi}$ for the contragredient representation. We say that π is *self-contragredient* if π is equivalent to $\widetilde{\pi}$. If θ is an involution of G, that is, θ is an automorphism of $\mathbf{G}$ of order two that is defined over F, let G^{θ} be the subgroup of G consisting of those points in G that are fixed by θ. We say that a smooth representation π of G is *θ-distinguished* if there exists a nonzero G^{θ}-invariant linear form on the space of π. For certain G and θ, the irreducible smooth θ-distinguished representations π of G have the property that π and $\widetilde{\pi} \circ \theta$ are equivalent. (For a discussion of this kind of phenomenon, see [**Ha**], particularly Lemma 3.) Note that if θ is inner, that is, θ is given by conjugation by an element of G whose square lies in the centre of G, this property amounts to self-contragredience of π. For example, let n be an integer such that $n \geq 2$, and let k be an integer such that $1 \leq k \leq n-1$. Consider the involution θ of $\mathbf{GL}_n(F)$ given by conjugation by a diagonal matrix whose first k diagonal entries are equal to 1 and whose remaining diagonal entries are equal to -1. As shown by Jacquet and Rallis in [**JR**], any irreducible smooth representation of $\mathbf{GL}_n(F)$ that is θ-distinguished must be self-contragredient.

The results of [**HM**], as they apply to the case where θ is an inner involution of G and π is a θ-distinguished tame supercuspidal representation of G, can be used to show that π and $\widetilde{\pi}$ share many properties. For certain θ-distinguished toral supercuspidal representations, this implies self-contragredience.

The following theorem gives a parametrization of the self-contragredient tame supercuspidal representations of G. The proof of the theorem appears later in this section. We continue to assume that Hypothesis C(G') holds for all tamely ramified twisted Levi subgroups G' of G.

THEOREM 12.1. *The correspondence of Theorem 9.1 restricts to a bijective correspondence between the equivalence classes of self-contragredient tame supercuspidal representations of G and the G-equivalence classes that contain triples of the form (G', π', ϕ) where one of the following holds:*

(1) *$G' = G$, ϕ is trivial, and π' is self-contragredient.*
(2) *$G' \neq G$ and there exists $g \in N_G(G')$ such that $g^2 \in G'$, ${}^g\phi = \phi^{-1}$ and $\pi' \simeq {}^g\widetilde{\pi}'$.*

When (2) holds, $g \notin G'$. Furthermore, if K' is the normalizer of a maximal parahoric subgroup of G' and ρ is an irreducible smooth representation of K' such that $\pi' = \mathrm{Ind}_{K'}^{G'}\rho$ (as in Theorem 7.1), the element g may be chosen so that $g \in N_G(K')$ and $\rho \simeq {}^g\widetilde{\rho}$.

REMARK 12.2. Let π be a self-contragredient tame supercuspidal representation whose equivalence class is parametrized by the G-orbit of a triple (G', π', ϕ) as in Theorem 12.1. Recall (see Remark 8.6) that the depth of π is equal to the depth of ϕ. It follows that

(1) The G-equivalence classes containing triples as in Theorem 12.1(1) parametrize the equivalence classes of self-contragredient depth zero irreducible supercuspidal representations of G.
(2) The G-equivalence classes containing triples as in Theorem 12.1(2) parametrize the equivalence classes of self-contragredient positive depth tame supercuspidal representations of G.

In the toral case, the above theorem reduces (using Corollary 9.4) to the following (which is analogous to results obtained by Adler ([**A1**]), which describe self-contragredient supercuspidal representations of general linear groups):

COROLLARY 12.3. *The correspondence of Corollary 9.4 restricts to a bijective correspondence between the equivalence classes of self-contragredient toral supercuspidal representations of G and the equivalence classes which contain a pair (T, ϕ) having the property that there exists $g \in N_G(T)$ such that $g^2 \in T$ and ${}^g\phi = \phi^{-1}$.*

The following lemma will be used in the proof of Theorem 12.1.

LEMMA 12.4. *Let ϕ be a quasicharacter of a twisted Levi subgroup G' of G. Suppose that $g \in N_G(G')$ satisfies $g^2 \in G'$ and $\phi \,|\, G'_{0+} = {}^g\phi^{-1} \,|\, G'_{0+}$. Then there exists a quasicharacter $\dot\phi$ of G' such that $\dot\phi = {}^g\dot\phi^{-1}$ and $\dot\phi \,|\, G'_{0+} = \phi \,|\, G'_{0+}$.*

PROOF. Let G'_{der} be the derived group of the topological group G'. Let $\mathcal{G} = G'/G'_{\mathrm{der}}$. Let $\mathcal{H}$ be the image of the set $\{\, hg^{-1}hg \mid h \in G' \,\}$ in $\mathcal{G}$. Because g normalizes G', $\mathcal{H}$ is a subgroup of $\mathcal{G}$. Let $\mathcal{G}_+$ be the image of G'_{0+} in $\mathcal{G}$. It follows from results of [**D2**] that $G'_{0+} \subset G'_{x,0+}G'_{\mathrm{der}}$ for any $x \in \mathcal{B}(G')$. Thus $\mathcal{G}_+$ is also the image of $G'_{x,0+}$ in $\mathcal{G}$ for any $x \in \mathcal{B}(G')$. Because $\mathcal{H}$ is a closed subgroup of $\mathcal{G}$ and $\mathcal{G}_+$ is a compact subgroup of $\mathcal{G}$, it follows that $\mathcal{H}\mathcal{G}_+$ is a closed subgroup of the locally compact abelian group $\mathcal{G}$.

Let $x \in \mathcal{B}(G')$. The restriction $\phi \,|\, G'_{x,0+}$ factors to a character χ of $\mathcal{G}_+$. Suppose that $k \in G'_{x,0+}$ and $k \in hg^{-1}hgG'_{\mathrm{der}}$ for some $h \in G'$. Then, using that $g^2 \in G'$ and ϕ is a quasicharacter of G' (so is trivial on G'_{der}), we have

$$\phi(g^{-1}kg) = \phi(g^{-1}hg \cdot g^{-2}hg^2) = \phi(h)\phi(g^{-1}hg) = \phi(k).$$

However, since k and $g^{-1}kg$ both belong to G'_{0+}, and (by assumption) ϕ and ${}^g\phi^{-1}$ agree on G'_{0+}, we have $\phi(k) = \phi(g^{-1}kg)^{-1}$. Therefore $\phi(k)^2 = 1$. Since $\phi(k)^{p^n} = \phi(k^{p^n})$ tends to 1 as n tends to infinity and we have assumed that the residual characteristic of F is odd, this implies $\phi(k) = 1$. Thus $\chi \,|\, \mathcal{H} \cap \mathcal{G}_+$ is trivial. Therefore we may extend χ to a character of $\mathcal{H}\mathcal{G}_+$ by making it trivial on the subgroup $\mathcal{H}$. Next, we use the fact that $\mathcal{H}\mathcal{G}_+$ is a closed subgroup of the locally compact abelian group $\mathcal{G}$. This guarantees the existence of an extension of χ to a quasicharacter of $\mathcal{G}$.

Let $\dot\phi$ be a quasicharacter of G' that is inflated from an extension of χ to $\mathcal{G}$. The fact that χ is trivial on $\mathcal{H}$ means that $\dot\phi = {}^g\dot\phi^{-1}$. The fact that, by construction $\phi \,|\, G'_{x,0+}$ and $\dot\phi \,|\, G'_{x,0+}$ both factor to $\chi \,|\, \mathcal{G}_+$ guarantees that they agree on $G'_{x,0+}$, which implies, as a result of properties of depths of quasicharacters (see Definition 4.1), that $\dot\phi$ and ϕ agree on G'_{0+}. □

Proof. (Theorem 12.1) Let (G', π', ϕ) be a triple as in Theorem 9.1, and let π be an element of the corresponding equivalence class of supercuspidal representations of G. It follows from Theorem 4.25 of [**HM**] that the equivalence class of $\widetilde{\pi}$ corresponds to the equivalence class of the triple $(G', \widetilde{\pi}', \phi^{-1})$. Assume that π is self-contragredient. By Theorem 8.7, the triples (G', π', ϕ) and $(G', \widetilde{\pi}', \phi^{-1})$ are equivalent in the sense of Definition 8.1. Therefore there exists $g \in N_G(G')$ such that $\pi'\phi \simeq {}^g(\widetilde{\pi}'\phi^{-1})$.

Suppose that $G' = G$. Note that the triple (G, π', ϕ) is associated, via Definition 8.4, to the cuspidal G-datum $\Psi := ((\mathbf{G}), \pi', (\phi))$, which gives rise, via Yu's construction, to the supercuspidal representation $\pi = \pi'\phi$. The relations at the end of the previous paragraph imply that $\pi'\phi$ is self-contragredient. In particular, since the depth of π' is zero, we have $\phi \,|\, G_{0+} = \phi^{-1} \,|\, G_{0+}$. That is, $\phi^2 \,|\, G_{0+} \equiv 1$. Since we have assumed that the residual characteristic of F is odd, this forces $\phi \,|\, G_{0+} \equiv 1$. Thus $\pi = \pi'\phi$ is a depth zero self-contragredient irreducible supercuspidal representation of G. The triples (G, π', ϕ) and $(G, \pi'\phi, \mathbf{1})$ are G-equivalent, and the latter triple has the required form.

Next, suppose that $G' \neq G$. Let g be as above. Since π' has depth zero, we see that $\phi \,|\, G'_{0+} = {}^g\phi^{-1} \,|\, G'_{0+}$. After conjugating by g once more, we find that $\phi \,|\, G'_{0+} = {}^{g^2}\phi \,|\, G'_{0+}$. Because $\phi \,|\, G'_{0+}$ is G-regular and G-factorizable, we may apply Lemma 5.6 and Corollary 6.5 to deduce that $g^2 \in G'$. According to Lemma 12.4, there exists a quasicharacter $\dot\phi$ of G' such that ϕ and $\dot\phi$ agree on G'_{0+} and $\dot\phi = {}^g\dot\phi^{-1}$.

Let $\dot\pi = \pi'(\phi\dot\phi^{-1})$. Note that $\dot\pi$ has depth zero and

$$\dot\pi \simeq {}^g(\widetilde{\pi}'\phi^{-1})\dot\phi^{-1} = {}^g(\widetilde{\pi}'\phi^{-1}\dot\phi) = {}^g\widetilde{\dot\pi}.$$

Therefore the triple $(G', \dot\pi, \dot\phi)$, which is G-equivalent to (G', π', ϕ), and hence, by Theorem 9.1, corresponds to the equivalence class of π, has the properties described in part (2) of the statement of the theorem. Note that if $g \in G'$, then $\dot\phi = \dot\phi^{-1}$. Since p is assumed to be odd, this implies that $\dot\phi \,|\, G'_{0+}$ is trivial. This contradicts the fact that ϕ, hence also $\dot\phi$, is G-regular on G'_{0+}. Therefore $g \notin G'$.

Let K' and ρ and g be as in the statement of the theorem, with $\mathrm{Ind}_{K'}^{G'}\rho$ equivalent to ${}^g\mathrm{Ind}_{K'}^{G'}\widetilde{\rho}$. As the latter representation is equivalent to $\mathrm{Ind}_{{}^gK'}^{G'}{}^g\widetilde{\rho}$, an application of Proposition 7.4 yields the existence of $h \in G'$ such that $hg \in N_G(K')$ and $\rho \simeq {}^{hg}\widetilde{\rho}$. Upon replacing g by hg, we obtain the desired result. □

REMARK 12.5. Let G', g, and ϕ be as in statement (2) of Theorem 12.1. Let $(\vec{\mathbf{G}}, \vec{\phi})$ be a G-factorization of ϕ. Set ${}^g\vec{\mathbf{G}} = ({}^g\mathbf{G}^0, \dots, {}^g\mathbf{G}^d)$ and ${}^g\vec{\phi}^{-1} = ({}^g\phi_1^{-1}, \dots, {}^g\phi_d^{-1})$. Because ${}^g\phi^{-1} = \phi$, the pair $({}^g\vec{\mathbf{G}}, {}^g\vec{\phi}^{-1})$ is another factorization of ϕ. Applying Proposition 5.4, we find that ${}^g\mathbf{G}^i = \mathbf{G}^i$ for $0 \leq i \leq d$. It is possible to show, using arguments similar to those used in the proof of Lemma 12.4, that there exists some factorization $(\vec{\mathbf{G}}, \vec{\phi})$ that satisfies $\phi_i = {}^g\phi_i^{-1}$ for $0 \leq i \leq d$.

13. Characters of tame supercuspidal representations

Suppose that π is a tame supercuspidal representation of G whose equivalence class is parametrized, via Theorem 9.1, by a triple (G', π', ϕ). In this section, we discuss results that indicate how properties of such a triple influence the asymptotic behaviour of the character of π.

Before turning to a discussion of characters, we make some comments about Fourier transforms of orbital integrals. Let $C_c^\infty(\mathfrak{g})$ and $C_c^\infty(\mathfrak{g}^*)$ be the spaces of complex-valued, locally constant, compactly supported functions on $\mathfrak{g}$ and $\mathfrak{g}^*$, respectively. Recall that we have fixed a character Λ of F that is nontrivial on the ring of integers of F and trivial on the maximal ideal $\mathfrak{p}$. If $f \in C_c^\infty(\mathfrak{g})$, the Fourier transform $\widehat{f}$ of f belongs to $C_c^\infty(\mathfrak{g}^*)$ and is defined by:

$$\widehat{f}(X^*) = \int_{\mathfrak{g}} \Lambda(X^*(Y))\, f(Y)\, dY, \qquad X^* \in \mathfrak{g}^*,$$

where dY is a Haar measure on $\mathfrak{g}$. If $\mathcal{O}$ is a co-adjoint orbit in $\mathfrak{g}^*$, and the orbital integral $\mu_{\mathcal{O}}$ converges, then the Fourier transform of the orbital integral $\mu_{\mathcal{O}}$ is the distribution $\widehat{\mu}_{\mathcal{O}}$ on $\mathfrak{g}$ defined by

$$\widehat{\mu}_{\mathcal{O}}(f) = \mu_{\mathcal{O}}(\widehat{f}), \qquad f \in C_c^\infty(\mathfrak{g}).$$

This distribution is represented by a locally integrable function, also denoted by $\widehat{\mu}_{\mathcal{O}}$, on $\mathfrak{g}$. This function is locally constant on the set $\mathfrak{g}^{reg}$ of regular elements in $\mathfrak{g}$.

Let B be an F-valued, nondegenerate, G-invariant, symmetric bilinear form on $\mathfrak{g}$. We may use B to identify $\mathfrak{g}^*$ with $\mathfrak{g}$. The linear functional that is identified with a point $X \in \mathfrak{g}$ is defined by $Y \mapsto B(X, Y)$, $Y \in \mathfrak{g}$. Because this identification is G-equivariant, it can be used to identify co-adjoint orbits in $\mathfrak{g}^*$ with adjoint orbits in $\mathfrak{g}$. In addition, we can identify Fourier transforms of functions in $C_c^\infty(\mathfrak{g})$ with functions in $C_c^\infty(\mathfrak{g})$. When orbital integrals converge, these identifications allow us to talk about Fourier transforms of orbital integrals relative to adjoint orbits in $\mathfrak{g}$, rather than co-adjoint orbits in $\mathfrak{g}^*$. Such identifications are made in [**KM1**], [**KM2**], [**M1**] and [**M2**], but not in [**AD2**].

As shown in Proposition 4.1 of [**AR**], under some explicitly stated mild conditions the form B can be chosen in such a way that, under the associated identification of $\mathfrak{g}$ with $\mathfrak{g}^*$, the lattices $\mathfrak{g}_{x,r}$ and $\mathfrak{g}^*_{x,r}$ correspond, for all $x \in \mathcal{B}(G)$ and $r \in \mathbb{R}$. We remark that the hypotheses assumed in [**KM1**], [**KM2**], [**M1**] and [**M2**] guarantee that the conditions in Proposition 4.1 of [**AR**] are satisfied.

If F has characteristic zero, or, as shown in Section 3 of [**Mo1**], if F has characteristic p and p is sufficiently good for $\mathbf{G}$, then the set of nilpotent orbits in $\mathfrak{g}$ (or in $\mathfrak{g}^*$) is finite. If Hypothesis 3.4.3 of [**D1**] is satisfied for $\mathbf{G}$, then nilpotent orbital integrals in $\mathfrak{g}$ (or in $\mathfrak{g}^*$) converge. Throughout this section, we will assume that, for all subgroups $\mathbf{H}$ of $\mathbf{G}$ that arise as identity components of stabilizers of semisimple elements of $\mathfrak{g}$ or of $\mathfrak{g}^*$, F and $\mathbf{H}$ satisfy hypotheses which guarantee that the set of nilpotent H-orbits in $\mathfrak{h}$ (or in $\mathfrak{h}^*$) is finite, and also that the associated nilpotent orbital integrals converge. With these assumptions, the set of G-orbits in $\mathfrak{g}$ (or in $\mathfrak{g}^*$) whose closures contain the orbit of a fixed semisimple element is finite. In addition, all orbital integrals converge.

In what follows, if $\mathcal{O}$ is a coadjoint orbit or an adjoint orbit that is identified with a coadjoint orbit (as discussed above), the notation $\widehat{\mu}_{\mathcal{O}}$ will be used for the function representing the Fourier transform of the orbital integral $\mu_{\mathcal{O}}$.

We say that an element of $\mathfrak{g}$ (or of $\mathfrak{g}^*$) is elliptic if its stabilizer in G is compact modulo Z. The adjoint (or co-adjoint) orbit of an elliptic element will be referred to as an elliptic orbit.

Let Θ_π be the character of an irreducible supercuspidal representation π, and let $d(\pi)$ be the formal degree of π. Certain supercuspidal representations have the property that on some neighbourhood of zero in $\mathfrak{g}$ intersected with the regular set $\mathfrak{g}^{reg}$, $\Theta_\pi \circ \exp$ is equal to $d(\pi)$ times the Fourier transform of an elliptic orbital integral. That is, there exists an elliptic orbit $\mathcal{O}_\pi$ such that

$$\Theta_\pi(\exp X) = d(\pi)\,\widehat{\mu}_{\mathcal{O}_\pi}(X), \qquad X \in \mathcal{V}_\pi \cap \mathfrak{g}^{reg}, \tag{13.1}$$

where $\mathcal{V}_\pi$ is an open neighbourhood of zero in $\mathfrak{g}$, and exp is the exponential map, or some reasonable substitute, such as a truncated exponential map. As shown in [**M1**], when F has characteristic zero and the residual characteristic p of F is greater than n, all supercuspidal representations of $\mathbf{GL}_n(F)$ exhibit this property. The associated orbits consist of regular elements: this is a property of elliptic $\mathbf{GL}_n(F)$-orbits.

The results of [**M2**] show that characters of some of the supercuspidal representations of classical groups constructed by Morris ([**Mo2**], [**Mo3**]) also display this kind of behaviour.

More recently, Adler and DeBacker ([**AD2**]) have shown that, for supercuspidal representations π of $\mathbf{GL}_n(F)$, when $p > n$, the relation (13.1) holds for $\mathcal{V}_\pi = \mathfrak{g}_{\rho(\pi)^+}$, where $\rho(\pi)$ is the depth of π. They also show that (13.1) holds on $\mathcal{V}_\pi = \mathfrak{g}_{\rho(\pi)^+}$ for the supercuspidal representations reductive p-adic groups constructed in Adler ([**A2**]), but here the orbit $\mathcal{O}_\pi$ does not necessarily consist of regular elements.

The results of [**KM1**] and [**KM2**] describe the asymptotic behaviour of characters of irreducible admissible (not necessarily supercuspidal) representations that contain particular kinds of K-types. Here we discuss the results of [**KM2**] as they apply to tame supercuspidal representations. These results are proved subject to some hypotheses (see Section 3.2 of [**KM2**]), including the assumption that F has characteristic zero, which guarantee that the conditions in Proposition 4.1 of [**AR**] hold. They also imply that Hypothesis C(G') holds for every tamely ramified twisted Levi subgroup G' of G. Therefore, under the hypotheses of [**KM2**], the equivalence classes of tame supercuspidal representations are parametrized as in Theorem 9.1. Let π be a tame supercuspidal representation of G whose equivalence class is parametrized by the G-equivalence class of a triple (G', π', ϕ), where G' is a proper tamely ramified elliptic twisted Levi subgroup of G. Let $(\vec{\mathbf{G}}, \vec{\phi})$ be a G-factorization of ϕ. Let $x \in \mathcal{B}(G')$. For $0 \le i \le d-1$, let $\Gamma_i \in \mathfrak{z}^{i\,*}$ be a G^{i+1}-generic element of depth $-r_i$ that realizes ϕ_i on G^i_{x,s_i^+}, where $s_i = r_i/2$. In this situation, we are identifying $\mathfrak{g}^i$ and $\mathfrak{g}^{i\,*}$, so we identify Γ_i with an element of $\mathfrak{z}^i$, which we will also denote by Γ_i. If ϕ_d is trivial, set $\Gamma_d = 0$. Otherwise, let $\Gamma_d \in \mathfrak{z}$ be an element that realizes $\phi_d \mid G_{x,(r_d/2)^+}$. Set $\Gamma = \sum_{i=0}^{d} \Gamma_d$. Note that, since $G^0 = G'$, we have $\Gamma \in \mathfrak{z}'$ and $C_G(\Gamma) = G'$ ([**KM2**]).

Note that if π is toral then G' is an elliptic maximal torus, so G'/Z is always compact in the toral case. For the moment, suppose that G'/Z is compact. Then the element Γ is elliptic. Also, $\mathfrak{g}'$ does not contain any nontrivial nilpotent elements. This means that $\mathcal{O}_\Gamma$ is the only orbit whose closure contains Γ. According to Theorem 4.4.1 of [**KM2**], the relation (13.1) holds, up to a positive scalar multiple, with $\mathcal{O}_\pi = \mathcal{O}_\Gamma$ and $\mathcal{V}_\pi = \mathfrak{g}_{s_{d-1}^+}$. (We remark that, once we have this sort of relation, it is a simple matter to verify that the scalar is 1, as long as the the measure on $\mathcal{O}_\Gamma$ is compatible with the one used to compute $d(\pi)$.)

As shown in [**KM2**], in cases where where G'/Z is not compact, the stabilizer of Γ in G is still equal to G'. By contrast with the case where G'/Z is compact, in this case $\mathcal{O}_\Gamma$ lies in the closure of other orbits. Let $\mathcal{O}(\Gamma)$ be the (finite) set of $\operatorname{Ad} G$-orbits in $\mathfrak{g}$ whose closures contain $\mathcal{O}_\Gamma$. According to Theorem 4.4.1 of [**KM2**], there exist complex numbers $c_{\mathcal{O}}(\pi)$, indexed by the orbits $\mathcal{O} \in \mathcal{O}(\Gamma)$, such that

$$\Theta_\pi(\exp X) = \sum_{\mathcal{O} \in \mathcal{O}(\Gamma)} c_{\mathcal{O}}(\pi) \widehat{\mu_{\mathcal{O}}}(X), \qquad X \in \mathfrak{g}_{s_{d-1}^+} \cap \mathfrak{g}^{reg}.$$

We expect that, under weaker hypotheses than those assumed in [**KM2**], the above equality would still hold, but only on the smaller domain $\mathfrak{g}_{\rho(\pi)^+}$.

When G'/Z is noncompact, in some special cases, for example, cases such as the ones discussed in the following paragraph, a relation like (13.1) holds, but we do not expect that in general. The methods and results of [**KM2**] (as they apply to a tame supercuspidal representation whose equivalence class is parametrized by a triple (G', π', ϕ)) do not depend in any way on properties of the depth zero supercuspidal representation π' of G'.

Let T be a tamely ramified elliptic maximal torus in G' that normalizes a maximal parahoric subgroup $G'_{x,0}$ of G' and is such that the image of $T \cap G'_{x,0}$ in $G'_{x,0}/G'_{x,0^+}$ is the group of $\mathfrak{f}$-rational points of a maximal elliptic torus. Let ϕ' be a depth zero quasicharacter of T whose restriction to $T \cap G_{x,0}$ factors to a character θ of $(T \cap G_{x,0})/(T \cap G_{x,0^+})$ that is in general position. Suppose that π' is associated to $\pm R_{\mathbf{T},\theta}^{\mathbf{G}'}$, as in Sections 7 and 10. Then it may be possible to use properties of cuspidal Deligne-Lusztig representations to produce an element Γ' in the Lie algebra of T such that a relation like (13.1) holds for $\Theta_{\pi'}$, with $\mathcal{O}_{\pi'}$ equal to the G'-orbit of Γ'. In this case, we may have the relation (13.1) for Θ_π, with $\mathcal{O}_\pi = \mathcal{O}_{\Gamma'+\Gamma}$ and, possibly, $\mathcal{V}_\pi = \mathfrak{g}_{\rho(\pi)^+}$. This is what happens for general linear groups and for some of the representations studied in [**M2**] (where $\mathcal{V}_\pi$ is not specified explicitly).

When G'/Z is noncompact and π' does not arise via the Deligne-Lusztig construction, then we do not expect a relation like (13.1) to hold for Θ_π. In some cases, it is possible to obtain a more general relation that involves finitely many Fourier transforms of orbital integrals (where the set of orbits appearing does not coincide with the set of orbits whose closures contain a fixed semisimple orbit). In some cases, the orbits appearing in such relations are regular and semisimple. The results of Cunningham ([**C**]) describe this kind of relation for depth zero supercuspidal representations of 4×4 symplectic groups. The aim of ongoing work of DeBacker and Kazhdan, beginning with the preprint [**DK**], is to obtain such relations for depth zero supercuspidal representations.

Some recent results show that data that parametrize tame supercuspidal representations can be used to describe properties of their characters on domains that are not necessarily close to the identity.

For example, DeBacker and Reeder show that expressions for characters of cuspidal Deligne-Lusztig representations of reductive groups over finite fields in terms of combinations of Green functions can be used to produce expressions for the characters of the depth zero supercuspidal representations studied in [**DR**]. Near the identity, the expression looks like (13.1). However, on some domains away from the identity element, expressions for characters as linear combinations (usually involving more than one term) of Fourier transforms of orbital integrals are obtained.

A recent paper of Adler and Spice ([**AS**]) computes character values for those tame supercuspidal representations that are parametrized by triples (G', π', ϕ) having the property that factorizations of ϕ involve twisted Levi sequences $\vec{\mathbf{G}}$ for which G^i/Z is compact for $0 \leq i \leq d-1$. The formulas obtained in [**AS**] involve expressions that are linear combinations of Fourier transforms of orbital integrals, where the orbits that appear are defined in terms of the data that parametrize the representations.

References

[A1] J.D. Adler, *Self-contragredient supercuspidal representations of GL_n*, Proc. Amer. Math. Soc. **125** (1997), 2471–2479.

[A2] J.D. Adler, *Refined anisotropic K-types and supercuspidal representations*, Pacific J. Math. **185** (1998), 1–32.

[AD1] J.D. Adler and S. DeBacker, *Some applications of Bruhat-Tits theory to harmonic analysis on the Lie algebra of a reductive p-adic group*, Michigan Math. J. **50** (2002), 263–285.

[AD2] J.D. Adler and S. DeBacker, *Murnaghan-Kirillov theory for supercuspidal representations of tame general linear groups*, J. reine angew. Math. **575** (2004), 1–35.

[AR] J.D. Adler and A. Roche, *An intertwining result for p-adic groups*, Canad. J. Math. **52** (2000), 449–467.

[AS] J.D. Adler and L. Spice, *Supercuspidal characters of reductive p-adic groups*, Amer. J. Math. **131** (2009), 1137–1210.

[BK] C.J. Bushnell and P.C. Kutzko, *The Admissible Dual of GL(N) via Compact Open Subgroups*, Annals of Mathematics Studies, Number 119, Princeton University Press (1993).

[C] C. Cunningham, *Characters of depth-zero supercuspidal representations of the rank-2 symplectic group*, Canad. J. Math. **52** (2000), 306–331.

[D1] S. DeBacker, *Homogeneity results for invariant distributions of a reductive p-adic group*, Ann. Sci. École Norm. Sup. **35** (2002), 391–422.

[D2] S. DeBacker, *Some applications of Moy-Prasad filtrations to harmonic analysis on a reductive p-adic group*, Michigan Math. J. **50** (2002), 241–261.

[D3] S. DeBacker, *Parameterizing conjugacy classes of maximal unramified tori*, Michigan Math. J. **54** (2006), 157–178.

[DK] S. DeBacker and D. Kazhdan, *Murnaghan-Kirillov theory for depth zero supercuspidal representations: reduction to Lusztig functions*, preprint.

[DR] S. DeBacker and M. Reeder, *Depth-zero supercuspidal L-packets and their stability*, Ann. of Math. **169** (2009), 1137–1210.

[DL] P. Deligne and G. Lusztig, *Representations of reductive groups over finite fields*, Ann. of Math., **103** (1976), 103–161.

[Ha] J. Hakim, *Supercuspidal Gelfand pairs*, J. Number Theory **100** (2003), 251–269.

[HM] J. Hakim and F. Murnaghan, *Distinguished tame supercuspidal representations*, Internat. Math. Res. Papers, 2008, no. 2, 166pp.

[H] R. Howe, *Tamely ramified supercuspidal representations of GL_n*, Pacific J. Math. **73** (1977), 437–460.

[K] J.-L. Kim, *Supercuspidal representations: an exhaustion theorem*, J. Amer. Math. Soc. **20** (2007), 273–320.

[KM1] J.-L. Kim and F. Murnaghan, *Character expansions and unrefined minimal K-types*, Amer. J. Math. **125** (2003), 1199-1234.

[KM2] J.-L. Kim and F. Murnaghan, *K-types and Γ-aysmptotic expansions*, J. reine angew. Math. **592** (2006), 189–236.

[JR] H. Jacquet and S. Rallis, *Uniqueness of linear periods*, Compositio Math. **102** (1996), 65–123.

[Mo1] L.E. Morris, *Rational conjugacy classes of unipotent elements and maximal tori, and some axioms of Shalika*, J. London Math. Soc. **38** (1988), 112-124.

[Mo2] L.E. Morris, *Tamely ramified supercuspidal representations of classical groups I. Filtrations*, Ann. scient. Ec. Norm. Sup., **24** (1991), 705–738.

[Mo3] L.E. Morris, *Tamely ramified supercuspidal representations of classical groups II: Representation theory*, Ann. scient. Ec. Norm. Sup., **25** (1992), 233–274.

[Mo4] L.E. Morris, *Tamely ramified supercusidal representations*, Ann. scient. Ec. Norm. Sup., **29** (1996), 639–667.

[M] A. Moy, *Local constants and the tame Langlands correspondence*, Amer. J. Math. **64** (1991), 863–930.

[MP1] A. Moy and G. Prasad, *Unrefined minimal K-types for p-adic groups*, Invent. Math. **116** (1994), 393–408.

[MP2] A. Moy and G. Prasad, *Jacquet functors and unrefined minimal K-types*, Comment. Math. Helvetici **71** (1996), 98–121.

[M1] F. Murnaghan, *Local character expansions and Shalika germs for $GL(n)$*, **304** (1996), 423–455.

[M2] F. Murnaghan, *Characters of supercuspidal representations of classical group*, Ann. Sci. Ec. Norm. Sup. **29** (1996), 49–105.

[S] S. Stevens, *The supercuspidal representations of p-adic classical groups*, Invent. Math. **172** (2008), 289–352.

[T] J. Tits, *Reductive groups over local fields*, in "Automorphic Forms, Representations, and L-Functions", Proceedings of Symposia in Pure Mathematics **33**, part 1, Amer. Math. Soc., Providence RI, 1979, pp. 29–69.

[Y] J.-K. Yu, *Construction of tame supercuspidal representations*, J. Amer. Math. Soc. **14** (2001), 579–622.

Department of Mathematics, University of Toronto, Toronto, Canada M5S 2E4
E-mail address: fiona@math.toronto.edu

Clay Mathematics Proceedings
Volume 13, 2011

On the Sato-Tate Conjecture, II

V. Kumar Murty

To Freydoon Shahidi on his 60th birthday

ABSTRACT. We formulate and investigate a weak form of the Sato-Tate conjecture. By replacing the normal use of the Ikehara-Wiener Tauberian theorem with a result of Heilbronn and Landau, we show that if all symmetric power L-functions are analytic at $s = 1$, then the weak Sato-Tate conjecture holds.

1. Introduction

The Sato-Tate conjecture describes the distribution of the "angles of Frobenius". We consider the L-function of an elliptic curve over $\mathbb{Q}$

$$L(s) \ = \ P(s)\prod \left(1 - \pi_p p^{-s}\right)^{-1} \left(1 - \overline{\pi}_p p^{-s}\right)^{-1}.$$

Here, the product is over primes p that do not divide the conductor of the curve and $P(s)$ is an Euler product supported only at the primes dividing the conductor. Also,

$$\pi_p \ = \ p^{1/2} e^{i\theta_p}$$

with $0 \leq \theta_p \leq \pi$. Let us write

$$\alpha_p \ = \ e^{i\theta_p}$$

and

$$\beta_p \ = \ e^{-i\theta_p}.$$

Let I be an interval in $[0, \pi]$. The conjecture of Sato-Tate [**11**] asserts that if the elliptic curve does not have complex multiplication, then

$$\#\{p \leq x : \theta_p \in I\} \ \sim \ \left(\int_I \frac{2}{\pi}(\sin^2\theta) d\theta\right) \pi(x) \tag{1}$$

where $\pi(x)$ denotes the number of primes $p \leq x$. (In the case that the curve has complex multiplication, the θ_p are almost equidistributed in $[0, \pi]$ apart from a 'skewing' that occurs at $\pi/2$. See [**4**] for a precise statement.)

2010 *Mathematics Subject Classification.* Primary 11G40 ; Secondary 11G05, 11F66, 11M45.
Key words and phrases. Sato-Tate conjecture, Tauberian theorems, symmetric power L-functions.
Research partially supported by a grant from NSERC.

The conjecture implies the following apparently more general statement: let f be an integrable function on the unit circle. Then

$$\lim_{x\to\infty}\frac{1}{\pi(x)}\sum_{p\le x}f(\theta_p) = \int_0^{\pi}\frac{2}{\pi}(\sin^2\theta)f(\theta)d\theta. \tag{2}$$

Important progress on (1) has been made in recent work of Taylor, Clozel, Harris and Shepherd-Barron. (See the exposé of Carayol [**1**].) In particular, they are able to prove the conjecture if the elliptic curve has a prime of semistable reduction.

The conjecture can be formulated more generally for any normalized holomorphic cusp form for a congruence subgroup that is an eigenfunction for the Hecke operators and which does not have complex multiplication (in the sense of Ribet). This general version of the conjecture is still open.

In our discussion below, we shall be working with a normalized eigenfunction f of the Hecke operators for the group $\Gamma_0(N)$ of weight $k\ge 2$. We assume that f does not have complex multiplication. In this case, for a prime p that does not divide N, we have $\pi_p = p^{(k-1)/2}e^{i\theta_p}$ with $0\le\theta_p\le\pi$ and the Sato-Tate conjecture asserts that (1) holds.

In this paper, we discuss the following weaker statement.

CONJECTURE 1.1. *For I an interval in $[0,\pi]$, we have*

$$\lim_{x\to\infty}\frac{1}{\log x}\sum_{\substack{p\le x\\ \theta_p\in I}}\frac{\log p}{p} = \int_I\frac{2}{\pi}(\sin^2\theta)d\theta. \tag{3}$$

Let us call this the *weak Sato-Tate conjecture.*

It is clear that the Sato-Tate conjecture implies the weak Sato-Tate conjecture. Indeed, applying (1) and using partial summation, we see that the weak Sato-Tate conjecture follows. We do not expect the reverse implication as the situation is a little similar to the case of the distribution of prime numbers. The estimate

$$\sum_{p\le x}\frac{\log p}{p} = (1+\mathbf{o}(1))\log x$$

can be derived in a few lines from the Chebyshev bound $\pi(x) = \mathbf{O}(x/\log x)$, while the proof of the prime number theorem is somewhat more complicated.

We note that the weak Sato-Tate conjecture implies the following estimate. Given an interval I of positive measure, and any $\epsilon>0$, we have

$$\#\{p\le x : \theta_p\in I\} \gg_I x^{1-\epsilon} \tag{4}$$

where the implied constant depends on I. Indeed, applying (3) to x and $x^{1-\epsilon}$, we find that

$$\sum_{\substack{x^{1-\epsilon}\le p\le x\\ \theta_p\in I}}\frac{\log p}{p} \gg \log x.$$

Since each term in the sum is $\mathbf{O}((\log x)/x^{1-\epsilon})$, (4) follows. If instead of (3), we have the stronger statement

$$\frac{1}{\log x}\sum_{\substack{p\le x\\ \theta_p\in I}}\frac{\log p}{p} = \int_I \frac{2}{\pi}(\sin^2\theta)d\theta + \mathbf{O}\left(\frac{1}{\log x}\right) \tag{5}$$

then we would be able to deduce the bound

$$\#\{p\le x : \theta_p\in I\} \gg_I \pi(x). \tag{6}$$

The main result of this paper is to show that the weak Sato-Tate conjecture is a consequence of the holomorphy at $s=1$ of a collection of L-functions. Denote by $M_r(s)$ the r-th symmetric power L-function. It is given by an Euler product whose factor at a prime that does not divide N is

$$\left(1-\frac{\alpha_p^r}{p^s}\right)^{-1}\left(1-\frac{\alpha_p^{r-1}\beta_p}{p^s}\right)^{-1}\cdots\left(1-\frac{\beta_p^r}{p^s}\right)^{-1}.$$

The indexing is such that $M_0(s)=\zeta(s)$, and $M_1(s)=L(s-\frac{k-1}{2})$. In [**4**], it was shown that if all M_r have an analytic continuation to $\Re(s)\ge 1$, then the Sato-Tate conjecture holds. In this note, we shall show that if all $M_r(s)$ are analytic at the point $s=1$, then the weak Sato-Tate conjecture holds.

In [**5**], R. Murty used information on the analytic continuation of a finite number of the M_r to deduce oscillation theorems for the numbers $\{a_n\}$, where

$$L(s) = \sum_{n=1}^{\infty}\frac{a_n}{n^s}.$$

We show that the hypotheses of some of those theorems can be replaced by weaker ones, and in particular, analytic continuation of the M_r can be replaced with holomorphy at $s=1$.

The analytic continuation of the symmetric power L-functions has been, and is being, studied by many authors. The work of Shimura and Gelbart-Jacquet gave the analytic properties of $M_2(s)$. The work of Shahidi [**10**] established the analytic continuation of $M_3(s)$ under a certain non-vanishing hypothesis. More recent work of Kim and Shahidi [**3**] showed that $M_r(s)$ is regular for $\Re(s)\ge 1$ and $m\le 9$ with the possible exception that $M_9(s)$ may have a pole at $s=1$. (See the article by R. Murty [**6**] for a discussion of this and other related work.)

Acknowledgements. It is a pleasure to thank Sanoli Gun, Ram Murty and the referee for helpful comments on an earlier version. It is also a pleasure to dedicate this paper in friendship to Freydoon Shahidi on the occasion of his sixtieth birthday.

2. A lemma

We prove the following result. It is a variant of [**4**], Lemma 3.2.

LEMMA 2.1. *Let $f(s)$ be a function satisfying the following hypotheses:*
(a) f is holomorphic in $\sigma>1$ and non-zero there;
(b) f is holomorphic in the disc $|s-1|<c$ except for a pole of order $e\ge 0$ at $s=1$;

(c) *$\log f(s)$ can be written as a Dirichlet series $\sum_{n=1}^{\infty} b_n/n^s$ with $b_n \geq 0$ for $\sigma > 1$.*

Then in the disc $|s-1| < c/2(e+2)$, any zero of f with $\sigma = 1$ has order $\leq e/2$.

Proof. Consider the disc $|s-1| < c/2(e+2)$. Let $1+it_0$ be a zero of f of order k and in this disc. Suppose that $k > e/2$. We wish to derive a contradiction. Set $b = [e/2]+1$, that is, the smallest integer larger than $e/2$. In particular, $k \geq b$.

Consider the function

$$g(s) \;=\; f(s)^{2b+1} \prod_{j=1}^{2b} f(s+ijt_0)^{2(2b+1-j)}.$$

Clearly g is holomorphic for $\sigma > 1$. Moreover, as

$$|2bt_0| \;<\; 2bc/2(e+2) \;\leq\; c/2$$

we see that g is holomorphic in $|s-1| < c/2$. It vanishes to at least first order at $s = 1$, as

$$4b^2 - (2b+1)e \;\geq\; 4b^2 - (2b+1)(2b-1) \;=\; 1.$$

But for $\sigma > 1$,

$$\log g(s) \;=\; \sum_{n\geq 1} b_n n^{-s} \left(2b+1+\sum_{j=1}^{2b} 2(2b+1-j)n^{-ijto}\right).$$

Let $\phi_n = t_0 \log n$. Then for $\sigma > 1$,

$$\Re \log g(\sigma) \;=\; \log|g(\sigma)| \;=\; \sum_{n\geq 1} b_n n^{-\sigma} \left(2b+1+\sum_{j=1}^{2b} 2(2b+1-j)\cos(j\phi_n)\right).$$

Now, we have the identity

$$(7) \qquad 2b+1+\sum_{j=1}^{2b} 2(2b+1-j)\cos(j\theta) \;=\; \left(1+2\sum_{j=1}^{b} \cos j\theta\right)^2 \geq 0.$$

Hence, $\log|g(\sigma)| \geq 0$ for $\sigma > 1$ and so $|g(\sigma)| \geq 1$. This contradicts g having a zero at $s = 1$. This proves the lemma.

Remark. We have not attempted to find the largest disc in which one can assert the conclusion of the Lemma.

3. Non-vanishing of the Symmetric power L-functions

Suppose we know that all M_r for $r > 0$ are analytic at $s = 1$. We will use the Lemma of the previous section to deduce that there is a small neighbourhood of $s = 1$ in which no M_r vanishes.

THEOREM 3.1. *Suppose that each M_r (for $0 < r \leq 2m$) can be analytically continued to a disc $|s-1| < c$. Then, no M_r ($r \leq m$) vanishes at a point $1+it$ with $|t| < c/8$.*

Proof. For $r \le m$, consider the function

$$G_r = M_0 M_2 \cdots M_{2r}.$$

Then G_r has at most a simple pole at $s = 1$ and no other pole in the disc $|s-1| < c$. Moreover, we have

$$\log M_j(s) = \sum_{n,p} \left(\frac{\sin(j+1)n\theta_p}{\sin n\theta_p} \right) \frac{1}{np^{ns}}$$

and using the identity (see [**5**], p. 435)

$$\sum_{j=0}^{r} \frac{\sin(2j+1)\theta}{\sin\theta} = \left(\frac{\sin(r+1)\theta}{\sin\theta} \right)^2,$$

we see that $\log G_r(s)$ is a Dirichlet series with non-negative coefficients. Thus, Lemma 2.1 implies that it does not vanish at any point $1 + it$ with $|t| < c/6$. This proves that no M_r, with $r \le m$ and even, can have a zero at a point $1 + it$ with $|t| < c/6$ unless $t = 0$.

Next, consider

$$H_r = (M_0 M_1 \cdots M_{2r-1})^2 M_{2r}.$$

Again, H_r is analytic in $|s - 1| < c$ except possibly for a pole of order ≤ 2 at $s = 1$. Moreover, $\log H_r(s)$ is a Dirichlet series with non-negative coefficients. This is because of (7) and

$$\log H_r(s) = \sum_{n,p} \left((2r+1) + \sum_{j=1}^{2r} 2(2r+1-j)\cos jn\theta_p \right) \frac{1}{np^{ns}}.$$

Thus, Lemma 2.1 implies that any zero $1 + it$ of H_r with $|t| < c/8$ is at most of order 1. This means that none of $M_1, \ldots, M_{2r-1}$ can have a zero at $s = 1+it$, with $0 < |t| < c/8$.

Finally, we have to deal with possible zeros at $s = 1$. If both M_j and M_k (with $j \ne k$) have zeros at $s = 1$, then H_r would have a double zero there (for any $r > j, k$) contradicting the result established in the previous paragraph. For the same reason, no M_j can have a double zero at $s = 1$. This means that there is at most one j_0 such that M_{j_0} vanishes at $s = 1$. If such a j_0 exists, then M_{j_0} has a simple zero at $s = 1$.

It is well-known that M_1 and M_2 are analytic at $s = 1$ and do not vanish there. Thus, we may suppose that $j_0 \ge 3$. Choose any even ℓ with $0 < \ell < j_0$ and consider the quantity

$$\left(1 + \frac{\sin(\ell+1)\theta}{\sin\theta} + \frac{\sin(j_0+1)\theta}{\sin\theta} \right)^2.$$

We expand this and use the fact that for $a \ge b$, we have

$$\frac{\sin(a+1)\theta}{\sin\theta} \cdot \frac{\sin(b+1)\theta}{\sin\theta} = \sum_{k=0}^{b} \frac{\sin(a+b-2k+1)\theta}{\sin\theta}$$

to express the entire quantity as a linear combination of terms of the form $\sin j\theta/\sin\theta$ for various values of j. Hence, if we set

$$\mathcal{G}(s) \;=\; M_0 M_\ell^2 M_{j_0}^2 \left(\prod_{a=0}^{\ell} M_{2a}\right)\left(\prod_{b=0}^{j_0} M_{2b}\right)\left(\prod_{k=0}^{\ell} M_{j_0+\ell-2k}^2\right)$$

then

$$\log g(s) \;=\; \sum \frac{1}{np^{ns}}\left(1+\frac{\sin(\ell+1)(n\theta_p)}{\sin(n\theta_p)} + \frac{\sin(j_0+1)(n\theta_p)}{\sin(n\theta_p)}\right)^2$$

and so has non-negative coefficients. As it converges for $\Re(s) > 1$, it follows that for $\sigma > 1$, we have $\log|g(\sigma)| \geq 0$ and in particular, g cannot vanish at $s = 1$. But examining the product defining g, we see that the only terms that can contribute either a zero or pole are M_0 and M_{j_0} and that M_0 occurs to the third power. As $j_0 \geq 3$, and we have chosen ℓ to be even, it is easy to see that M_{j_0} occurs to at least the fourth power. Thus g vanishes at $s = 1$ and this is a contradiction. This completes the proof of the theorem.

4. Application of a result of Heilbronn and Landau

Heilbronn and Landau prove the following result [**2**]. Suppose that the Dirichlet series

$$f(s) \;=\; \sum_{n=1}^{\infty} a_n n^{-s}$$

converges for $\sigma > 1$ and that $a_n \geq 0$. Assume that

$$f(s) - \frac{1}{s-1}$$

is regular at $s = 1$. Then

$$A(x) \;=\; \sum_{n\leq x} \frac{a_n}{n} \;=\; \log x \;+\; \mathbf{O}(1).$$

We shall use the above result to prove the following bound.

THEOREM 4.1. *Suppose that all M_r $(r > 0)$ are holomorphic at $s = 1$. Then, for $r > 0$, we have the estimate*

$$\sum_{p\leq x} \frac{\sin(r+1)\theta_p}{\sin\theta_p}\cdot\frac{\log p}{p} \;=\; \mathbf{O}_r(1) \tag{8}$$

where the notation indicates that the implied constant may depend on r.

Remark. Notice that we do not need to assume holomorphy in a fixed disc.

Proof. We shall prove it for even values of r first. For this purpose, we again use the function

$$G(s) \;=\; G_r(s) \;=\; M_0 M_2 \cdots M_{2r}.$$

As observed in the proof of Theorem 3.1, this is a Dirichlet series with the property that $\log G(s)$ has non-negative coefficients. Moreover, $G(s)$ has a simple pole at $s = 1$, and the hypothesis that each M_i is holomorphic at $s = 1$ means that there is some disc $|s - 1| < c_r$ on which all the $M_2, M_4, \cdots, M_{2r}$ are holomorphic. By

Theorem 3.1, there is a smaller disc $|s-1| < c'_r$ (say) in which none of the M_i vanish. This implies that $-G'/G(s)$ has a simple pole at $s=1$ and that

$$-\frac{G'}{G}(s) - \frac{1}{s-1}$$

is analytic in this disc. Noting that

$$-\frac{G'}{G}(s) = -\frac{M'_0}{M_0}(s) - \sum_{j=1}^{r} \frac{M'_{2j}}{M_{2j}}(s)$$

and applying the theorem of Heilbronn and Landau, we see that

$$\sum_{j=0}^{r} \sum_{p^n \leq x} \frac{\sin(2j+1)\theta_p}{\sin\theta_p} \cdot \frac{\log p}{p^n} = \log x + \mathbf{O}_r(1).$$

Noting that the contribution of prime powers to the left hand side is bounded and that

$$\sum_{p \leq x} \frac{\log p}{p} = \log x + \mathbf{O}(1),$$

we get inductively that (8) holds.

For odd values of r, we use the function

$$G(s) = (M_0 M_1 \cdots M_{2r-1})^2 M_{2r}.$$

Again, we apply the Heilbronn-Landau theorem to $-G'/G(s)$ and proceed by induction. This proves that (8) holds for all values of $r > 0$.

COROLLARY 4.1. *With the hypothesis as in Theorem (4.1), we have for $r \neq 0, 2$,*

$$\sum_{p \leq x} \exp(ir\theta_p) \cdot \frac{\log p}{p} = \mathbf{O}_r(1).$$

For $r = 0, 2$, we have

$$\sum_{p \leq x} \exp(ir\theta_p) \cdot \frac{\log p}{p} = (-1)^{r/2} \log x + \mathbf{O}_r(1).$$

This is proved by induction on r using Theorem (4.1).

Let us set

$$c_m = \lim_{x\to\infty} \frac{1}{\log x} \sum_{p \leq x} \exp(im\theta_p) \cdot \frac{\log p}{p}.$$

Thus, Corollary 4.1 asserts that

(9) $$c_0 = 1,\ c_2 = -1,\ c_m = 0\ (m \neq 0, 2).$$

5. The Weak Sato-Tate conjecture

To deduce the weak Sato-Tate conjecture, we use a variant of a result ([**7**], Theorem 8) of R. Murty and Sinha. By Weyl's criterion, a sequence of points $x_n \in [0, \pi]$ is equidistributed (with respect to the usual measure) provided for every $m > 0$, we have

$$\kappa_m = \lim_{V\to\infty} \frac{1}{V} \sum_{n\le V} \exp(imx_n) = 0.$$

When this condition holds, we have for any interval $I \subset [0, \pi]$,

$$\lim_{V\to\infty} \frac{1}{V} \#\{n \le V : x_n \in I\} = \ell(I)$$

where $\ell(I)$ denotes the length of I (equivalently, the measure of I). A classical result of Erdös and Turán gives an effective version of this in the sense that an estimate for

$$\left| \frac{1}{V} \#\{n \le V : x_n \in I\} - \ell(I) \right|$$

is derived in terms of finitely many of the sums

$$\sum_{n\le V} \exp(imx_n).$$

In [**7**], Theorem 8, this result is generalized to the case that the x_n are equidistributed with respect to a measure other than the usual one. In this case, the κ_m may be non-zero and the density function is given by

$$F(x) = \sum_{m=-\infty}^{\infty} \kappa_m \exp\{-imx\}.$$

Thus,

$$\lim_{V\to\infty} \frac{1}{V} \#\{n \le V : x_n \in I\} = \int_I F(x)dx.$$

The condition

$$\sum_{m=-\infty}^{\infty} |\kappa_m| < \infty$$

is imposed to ensure absolute convergence of the series defining F. We need a slightly generalized version of their result to take into account the weight $(\log p)/p$ in the sums.

PROPOSITION 5.1. *Let $I = [a, b] \subset [0, \pi]$ be an interval. Consider the quantity*

$$E(x) = \left| \frac{1}{\log x} \sum_{\substack{p\le x \\ \theta_p \in I}} \frac{\log p}{p} - \int_I \frac{2}{\pi}(\sin^2\theta)d\theta \right|.$$

Then, for any positive integer M, we have

$$E(x) \ll \frac{1}{M} + \sum_{1\le|m|\le M} \left(\tfrac{1}{M+1} + \min(b-a, \tfrac{1}{|m|}) \right) \times \left| \tfrac{1}{\log x} \textstyle\sum_{p\le x} \exp(im\theta_p) \tfrac{\log p}{p} - c_m \right|$$

with the c_m *defined by* (9).

Proof. This essentially follows along the lines of a very similar result in [**7**] and so we shall outline the argument. In [**7**], Section 3, the M-th order Beurling-Selberg trigonometric polynomials $S_M^{\pm}$ relative to the interval $I = [a, b]$ are introduced. They have the property that

$$S_M^- \leq \chi_I \leq S_M^+$$

where χ_I denotes the characteristic function of the interval I. Moreover, their Fourier coefficients satisfy

$$|\hat{S}_M^{\pm}(n) - \hat{\chi}_I(n)| \leq \frac{1}{M+1}. \tag{10}$$

Moreover, the Fourier coefficients of χ_I itself satisfy

$$|\hat{\chi}_I(n)| \leq \min\left(b-a, \frac{1}{|n|}\right). \tag{11}$$

Finally, we have

$$||S_M^{\pm} - \chi_I||_{L^1} \leq \frac{1}{M+1}. \tag{12}$$

Now, let x_n be a sequence of points in $[0, \pi]$ and let $g \geq 0$ be a piecewise continuous function on $\mathbb{R}$ and define

$$d_m = \lim_{V\to\infty} \frac{\sum_{n\leq V} \exp(imx_n)g(n)}{\sum_{n\leq V} g(n)}.$$

Define the density function

$$F(t) = \sum_m d_m \exp(-imt)$$

and denote by μ the measure $F(t)dt$. We want to estimate

$$\sum_{n\leq x} \chi_I(x_n)g(n)$$

and we will do so by estimating

$$\sum_{n\leq x} S_M^{\pm}(x_n)g(n).$$

The following estimate enables us to separate $S_M^{\pm}$ and the function g.

LEMMA 5.1. *We have*

$$\left|\sum_{n\leq V} S_M^{\pm}(x_n)g(n) - \left(\int_0^{\pi} S_M^{\pm}(t)d\mu\right)\left(\sum_{n\leq V} g(n)\right)\right|$$

$$\leq \sum_{1\leq|m|\leq M} \left|\hat{S}_M^{\pm}(m)\right| \left|\sum_{n\leq V} \exp(imx_n)g(n) - d_m \sum_{n\leq V} g(n)\right|.$$

Proof. Since the Fourier coefficients $S_M^{\pm}(m)$ vanish for $|m| > M$, we have

$$\sum_{n\le V} S_M^{\pm}(x_n)g(n) = \sum_{n\le V}\sum_{|m|\le M} \hat{S}_M^{\pm}(m)\exp(imx_n)g(n). \tag{13}$$

Rearranging the double sum, we see that this is equal to

$$\begin{aligned}
&= \sum_{|m|\le M} \hat{S}_M^{\pm}(m)\left\{\sum_{n\le V}\exp(imx_n)g(n)\right\}\\
&= \sum_{|m|\le M} \hat{S}_M^{\pm}(m)\left\{\sum_{n\le V}\exp(imx_n)g(n) - d_m\sum_{n\le V} g(n)\right\}\\
&\quad + \left(\sum_{n\le V} g(n)\right)\left(\sum_{|m|\le V}\hat{S}_M^{\pm}(m)d_m\right).
\end{aligned}$$

Now,

$$\begin{aligned}
\sum_{|m|\le M}\hat{S}_M^{\pm}(m)d_m &= \sum_m d_m\int_0^{\pi} S_M^{\pm}(t)\exp(-imt)dt\\
&= \int_0^{\pi} S_M^{\pm}(t)\left(\sum_m d_m\exp(-imt)\right)dt\\
&= \int_0^{\pi} S_M^{\pm}(t)d\mu.
\end{aligned}$$

Putting these together, we deduce that

$$\begin{aligned}
&\sum_{n\le V} S_M^{\pm}(x_n)g(n) - \left(\int_0^{\pi} S_M^{\pm}(t)d\mu\right)\left(\sum_{n\le V} g(n)\right)\\
&= \sum_{1\le|m|\le M}\hat{S}_M^{\pm}(m)\left(\sum_{n\le V}\exp(imx_n)g(n) - d_m\sum_{n\le V} g(n)\right).
\end{aligned}$$

The Lemma follows from this.

Using the estimate (12), we see that

$$\int_0^{\pi} S_M^{\pm}(t)d\mu = \mu(I) + \mathbf{O}\left(\frac{1}{M}\right). \tag{14}$$

Now using (10) and (11), we deduce that

$$|\hat{S}_M^{\pm}(m)| \le \frac{1}{M+1} + \min\left(b-a, \frac{1}{|m|}\right). \tag{15}$$

Now, we apply Lemma 5.1 with the function

$$g(n) = \begin{cases}(\log p)/p & \text{if } n = p \text{ is prime}\\ 0 & \text{if } n \text{ is not prime.}\end{cases}$$

Using the above estimates (14) and (15), dividing through by

$$\sum_{n\leq V} g(n) \sim \log V$$

and using (9), the Proposition 5.1 follows.

Now we are ready to prove the main result.

THEOREM 5.1. *Suppose that all M_r $(r > 0)$ are holomorphic at $s = 1$. Then the weak Sato-Tate conjecture holds.*

Proof. Let $\epsilon > 0$ and choose $M = [1/\epsilon]$. Using Corollary 4.1, and Proposition 5.1, we see that

$$E(x) \ll \epsilon + \frac{c(\epsilon)\log(1/\epsilon)}{\log x}.$$

Now taking $x \to \infty$, the result follows.

6. Variants and Consequences

We can ask what can be deduced if we only know that a finite number of the M_r are holomorphic in a disc about $s = 1$. The following variant of [**5**], Lemma 2 follows by the same method as in [**5**] by using Theorem 5.1.

THEOREM 6.1. *Suppose that M_r is analytic in a disc $|s-1| < c$ for all $r \leq 2m+2$. Then, as $x \to \infty$, and for $r \leq m+1$,*

$$\sum_{p\leq x}(2\cos\theta_p)^{2r}\frac{\log p}{p} = \frac{1}{r+1}\binom{2r}{r}(1+\mathbf{o}(1))\log x.$$

For $r \leq m$,

$$\sum_{p\leq x}(2\cos\theta_p)^{2r+1}\frac{\log p}{p} = \mathbf{o}(\log x).$$

Using this, we can deduce the following result which is a strengthening of [**5**], Theorem 4.

THEOREM 6.2. *Suppose that M_r is analytic at $s = 1$ for all $r \leq 2m+2$. Then each of the following statements holds for a set of primes p of positive density:*

(i) For any $\delta > 0$,

$$-\delta < 2\cos\theta_p < \frac{2}{\delta(m+2)}$$

(ii) for any $\epsilon > 0$,

$$|2\cos\theta_p| > \sqrt{\frac{4m+2}{m+2}} - \epsilon$$

(iii) for any $\epsilon > 0$,

$$2\cos\theta_p > \beta_m - \epsilon$$

where

$$\beta_m = \left\{\frac{1}{4(m+2)}\binom{2m+2}{m+1}\right\}^{\frac{1}{2m+1}}.$$

We omit the proof as it follows along the same lines as in [**5**].

References

[1] H. Carayol, La conjecture de Sato-Tate (d'aprés Clozel, Harris, Shepherd-Barron, Taylor), *Sem. Bourbaki*, **59**(2006-2007), no. 977.

[2] H. Heilbronn and E. Landau, Anwendungen der N. Wienerschen Methode, *Math. Zeitschrift*, **37**(1933), 18-21.(See also *The Collected Papers of Hans Arnold Heilbronn*, pp. 168-172, eds. E. Kani and R. A. Smith, Canadian Mathematical Society, Wiley, Toronto, 1988.)

[3] H. Kim and F. Shahidi, Functorial products for $GL_2 \times GL_3$ and the symmetric cube for GL_2, *Annals of Math.*, **155**(2002), 837-893.

[4] V. Kumar Murty, On the Sato-Tate conjecture, in: Number Theory related to Fermat's Last Theorem, pp. 195-205, (ed. N. Koblitz), *Progress in Mathematics*, Volume **26**, Birkhauser, Boston, 1982.

[5] M. Ram Murty, Oscillations of Fourier coefficients of modular forms, *Math. Annalen*, **262**(1983), 431-446.

[6] M. Ram Murty, Applications of Symmetric Power L-functions, in: *Lectures on automorphic L-functions*, pp. 204-283, eds. J. Cogdell, H. Kim, R. Murty, Fields Institute Lecture Notes, AMS, 2006.

[7] M. Ram Murty and K. Sinha, Effective equidistribution of eigenvalues of Hecke operators, *J. Number Theory*, **129**(2009), 681-714.

[8] M. Ram Murty and V. Kumar Murty, The Sato-Tate conjecture and generalizations, *Proc. Indian Acad. Sci.*, to appear.

[9] J.-P. Serre, Abelian ℓ-adic representations and elliptic curves, Benjamin, New York, 1968.

[10] F. Shahidi, Third symmetric power L-functions for $GL(2)$, *Compositio Math.*, **70**(1989), 24-273.

[11] J. Tate, Algebraic cycles and poles of zeta functions, in: Arithmetical Algebraic Geometry, pp. 93-110, ed. O. F. G. Schilling, Harper and Row, New York, 1965.

Department of Mathematics, University of Toronto, 40 St. George Street, Toronto, ON M5S 2E4. CANADA

E-mail address: murty@math.toronto.edu

Clay Mathematics Proceedings
Volume 13, 2011

Icosahedral Fibres of the Symmetric Cube and Algebraicity

Dinakar Ramakrishnan

To Freydoon Shahidi, on the occasion of his sixtieth birthday
Tavalodet Mubarak!

ABSTRACT. For any number field F, call a cusp form $\pi = \pi_\infty \otimes \pi_f$ on $\mathrm{GL}(2)/F$ *special icosahedral*, or just *s-icosahedral* for short, if π is not solvable polyhedral, and for a suitable "conjugate" cusp form π' on $\mathrm{GL}(2)/F$, $\mathrm{sym}^3(\pi)$ is isomorphic to $\mathrm{sym}^3(\pi')$, and the symmetric fifth power L-series of π equals the Rankin-Selberg L-function $L(s, \mathrm{sym}^2(\pi') \times \pi)$ (up to a finite number of Euler factors). Then the point of this Note is to obtain the following result:

Let π be s-icosahedral (of trivial central character). Then π_f is algebraic without local components of Steinberg type, π_∞ is of Galois type, and π_v is tempered everywhere. Moreover, if π' is also of trivial central character, it is s-icosahedral, and the field of rationality $\mathbb{Q}(\pi_f)$ (of π_f) is $K := \mathbb{Q}[\sqrt{5}]$, with π'_f being the Galois conjugate of π_f under the non-trivial automorphism of K.

There is an analogue in the case of non-trivial central character ω, with the conclusion that π_f is algebraic when ω is, and when ω has finite order, $\mathbb{Q}(\pi_f)$ is contained in a cyclotomic field.

1. Introduction

Let us begin with some motivation and consider a continuous irreducible representation ρ with trivial determinant of the absolute Galois group $\mathcal{G}_F$ (of a number field F) into $\mathrm{GL}(2, \mathbb{C})$ which is *icosahedral*, i.e., whose image in $\mathrm{PGL}(2, \mathbb{C})$ is the alternating group A_5. Then it is well known that ρ is rational over $K = \mathbb{Q}[\sqrt{5}]$, but not equivalent to its Galois conjugate ρ' under the non-trivial automorphism of K. Moreover, one has (see **[Kim2]**, **[Wan]**, for example): (i) $\mathrm{sym}^3(\rho) \simeq \mathrm{sym}^3(\rho')$, and (ii) $\mathrm{sym}^5(\rho) \simeq \mathrm{sym}^2(\rho') \otimes \rho$. These two are the signature properties of such representations, and they lend themselves to natural automorphic analogues.

Let π be a cuspidal automorphic representation of $\mathrm{GL}(2, \mathbb{A}_F)$ with central character ω. For every $m \geq 1$ one has its *symmetric m-th power L-function* $L(s, \pi; \mathrm{sym}^m)$, which is an Euler product over the places v of F, with the v-factors

2010 *Mathematics Subject Classification.* Primary: 11F70; Secondary: 11F80, 22E55.

Partly supported by the NSF grant DMS-0701089.

(for finite unramified v of norm q_v) being given by

$$L_v(s, \pi; \mathrm{sym}^m) = \prod_{j=0}^{m} (1 - \alpha_v^j \beta_v^{m-j} {q_v}^{-s})^{-1},$$

where the unordered pair $\{\alpha_v, \beta_v\}$ defines the diagonal conjugacy class in $\mathrm{GL}_2(\mathbb{C})$ attached to π_v. Even at a ramified (resp. archimedean) place v, one has by the local Langlands correspondence a 2-dimensional representation σ_v of the extended Weil group $W_{F_v} \times \mathrm{SL}(2, \mathbb{C})$ (resp. of the Weil group W_{F_v}), and the v-factor of the symmetric m-th power L-function is associated to $\mathrm{sym}^m(\sigma_v)$. A special case of the *principle of functoriality* of Langlands asserts that there is, for each m, an (isobaric) automorphic representation $\mathrm{sym}^m(\pi)$ of $\mathrm{GL}(m+1, \mathbb{A})$ whose standard (degree $m+1$) L-function $L(s, \mathrm{sym}^m(\pi))$ agrees, at least at the primes not dividing $\mathcal{N}$, with $L(s, \pi; \mathrm{sym}^m)$. This was known to be true for $m = 2$ long ago by the work of Gelbart and Jacquet ([**GJ**]), and more recently for $m = 3, 4$ by the deep works of Shahidi and Kim ([**KS2, Kim1, KS1**]). Write $A^{2j}(\pi)$ for $\mathrm{sym}^{2j}(\pi) \otimes \omega^{-j}$ when the latter is automorphic; it is also customary to denote $A^2(\pi)$ by $\mathrm{Ad}(\pi)$. We will say that π is *solvable polyhedral* if $\mathrm{sym}^m(\pi)$ is Eisensteinian for some $m \leq 4$.

Suppose π' is another cusp form on $\mathrm{GL}(2)/F$, say of the same central character as π, such that $\mathrm{sym}^2(\pi) \simeq \mathrm{sym}^2(\pi')$. Then one knows ([**Ram1**]) that π and π' must be abelian twists of each other. One could ask if the same conclusion holds if $\mathrm{sym}^3(\pi)$ is isomorphic to $\mathrm{sym}^3(\pi')$. The answer is in the negative in that case, and a counterexample would be furnished by a π associated to a 2-dimensional icosahedral Galois representation ρ (of trivial determinant), i.e., with its image in $\mathrm{PGL}(2, \mathbb{C})$ being isomorphic to the alternating group A_5. Indeed, as remarked above, ρ would be defined over $\mathbb{Q}[\sqrt{5}]$ and π' would be associated to the Galois conjugate ρ' of ρ (under $a + b\sqrt{5} \mapsto a - b\sqrt{5}$). However, the *even* Galois representations are not (at all) known to be modular. Nevertheless, cusp forms of such icosahedral type are of great interest to study, for themselves and for understanding the fibres of the symmetric cube transfer. What we do is give a definition of an s-icosahedral cusp form which does not depend on any conjecture, and which is robust enough to furnish consequences which one would usually know only when there is an associated Galois representation.

Let us call a cusp form π on $\mathrm{GL}(2)/F$ (of central character ω) *s-icosahedral* if we have

(sI-1) π is not solvable polyhedral;
(sI-2) $\mathrm{sym}^3(\pi) \simeq \mathrm{sym}^3(\pi')$, for a cusp form π' on $\mathrm{GL}(2)/F$; and
(sI-3) the following identity of L-functions holds (outside a finite set S of places of F containing the archimedean and ramified places):

$$L^S(s, \pi; \mathrm{sym}^5) = L^S(s, \mathrm{Ad}(\pi') \times \pi \otimes \omega^2),$$

with π' as in (sI-2).

Observe that if (π, π') are associated as above, then $(\pi, \pi' \otimes \nu)$ will also be associated if ν is a cubic character, but this ambiguity can be eliminated by requiring that the ratio of the central characters ω, ω', of π, π' respectively, is not cubic. Note also that the property of being s-icosahedral is invariant under twisting $\pi \mapsto \pi \otimes \chi$ by idèle class characters χ, with $\pi' \mapsto \pi' \otimes \chi$, in particular under taking contragredients $\pi \mapsto \pi^\vee = \pi \otimes \omega^{-1}$.

Given cusp forms π_1, π_2 on $\mathrm{GL}(n_1)/F$, $\mathrm{GL}(n_2)/F$ respectively, one has good analytic properties ([**Sha2, Sha1, JPSS, JS**]) of the associated Rankin-Selberg L-function $L(s, \pi_1 \times \pi_2)$. When $(n_1, n_2) = (2, 2)$ ([**Ram1**]) and $(n_1, n_2) = (3, 2)$ ([**KS2**]), one also knows the existence of an isobaric automorphic form $\pi_1 \boxtimes \pi_2$ on $\mathrm{GL}(n_1 n_2)$. Thus the hypothesis (sI-3) implies that $\mathrm{sym}^5(\pi)$ is modular; in fact, $\mathrm{Ad}(\pi') \boxtimes \pi \otimes \omega^2$ represents it at every finite place (and at infinity), as seen by the standard stability results due to Shahidi and others (see section 2 below).

What we want to do in this Note is to look at the situation where one does not know of the existence of a corresponding Galois representation ρ, like when π is a suitable Maass form on $\mathrm{GL}(2)/\mathbb{Q}$, and see what arithmetic properties one can still deduce *unconditionally*. However, it is a *Catch 22* situation because it is not easy, either by the trace formula or by other means, to construct such Maass forms of Laplacian eigenvalue 1/4 except by starting with a Galois representation. Nevertheless, here is what we prove:

Theorem A *Let F be a number field and $\pi = \pi_\infty \otimes \pi_f$ a cuspidal automorphic representation of $GL_2(\mathbb{A}_F)$ with algebraic central character ω. Suppose π is s-icosahedral, i.e., satisfies (sI-1), (sI-2) and (sI-3) relative to a cusp form π' on $GL(2)/F$ of central character ω'. Then we have the following:*

(a) *π_f is algebraic with no local components of Steinberg type*
(b) *If ω is of finite order, π_f is rational over a cyclotomic extension of $\mathbb{Q}$, and π_∞ is of Galois type*
(c) *π is tempered, i.e., satisfies the Ramanujan hypothesis.*
(d) *If $\omega = \omega' = 1$, we have:*
  (1) *π_f is rational over $\mathbb{Q}[\sqrt{5}]$, but not over $\mathbb{Q}$;*
  (ii) *π' is unique, with π'_f being the Galois conjugate of π_f under the non-trivial automorphism τ of $\mathbb{Q}[\sqrt{5}]$; and*
  (iii) *π' is also s-icosahedral and of Galois type.*

Concerning (b), let us recall that π_∞ is said to be of *Galois type* if at every archimedean place v, the associated 2-dimensional representation σ_v of the Weil group W_{F_v} is trivial upon restriction to $\mathbb{C}^*$.

One writes $\mathbb{Q}(\pi_f)$ for the field of rationality of π_f ([**Clo**], [**Wal**]). We will use the algebraic parametrization at any v which is equivariant for the action of Aut $\mathbb{C}$ on the two sides of the local Langlands correspondence ([**Hen2**]. In particular, for v unramified, it corresponds to the *Tate parametrization* of [**Clo**], which was introduced so as not to worry about spurious square-roots of $q_v = Nv$; the normalization is unitary in [**Wal**].

In [**Ram2**], we introduced a notion of quasi-icosahedrality, based on a conditional assumption that certain symmetric powers of π are automorphic. More precisely, an irreducible cuspidal automorphic representation π of $\mathrm{GL}(2, \mathbb{A}_F)$ is called *quasi-icosahedral* iff we have

(i) $\mathrm{sym}^m(\pi)$ is automorphic for every $m \leq 6$;
(ii) $\mathrm{sym}^m(\pi)$ is cuspidal for every $m \leq 4$; and
(iii) $\mathrm{sym}^6(\pi)$ is not cuspidal.

The key result of [**Ram2**] (see part (b) of Theorem A′ of section 2) is that, for every such quasi-icosahedral π, there exists another cusp form π' of $\mathrm{GL}(2)/F$ such that the symmetric fifth power of such a quasi-icosahedral cusp form π is necessarily a character twist of the functorial product $\mathrm{Ad}(\pi') \boxtimes \pi$. From this we obtain

Proposition B *Let π be an s-icosahedral cusp form on $GL(2)/F$ of central character ω. Assume that $sym^6(\pi)$ is automorphic. Then π is quasi-icosahedral.*

It will be left to the astute reader to figure out the conditions on a quasi-icosahedral π which will make it s-icosahedral. We do not pursue this any further here because we want to stay in the unconditional realm here.

Sometimes in Mathematics, when one makes a right prediction and puts it in a workable framework, the proof is not hard to find; Theorem A here is an instance of that and the proof mostly requires just bookkeeping. Nevertheless, we hope that the conclusion is of some interest. It should be mentioned that given the definition, it is not surprising that an s-icosahedral π is algebraic, but what is nice is that when $\omega = 1$, π is even rational over $\mathbb{Q}(\sqrt{5})$.

Finally, one can easily construct non-singleton fibres of the symmetric cube transfer by taking, for any cusp form π on GL(2), the collection $\{\pi \otimes \chi \,|\, \chi^3 = 1\}$. (By [**KS1**], $\mathrm{sym}^3(\pi)$ will be cuspidal iff π is not dihedral or tetrahedral.) These fibres are not terribly interesting, however, especially compared to the icosahedral ones.

We thank the referee for helpful comments, and the NSF for continued support through the grant DMS-0701089.

This article is dedicated to Freydoon Shahidi, from whom I have learnt a lot about automorphic L-functions and the Langlands-Shahidi method, and who has been a longtime friend and in effect a *baradar e bozorgtar.*

2. The symmetric cube constraint

Let F be a number field with adèle ring $\mathbb{A}_F$. For each place v, denote by F_v the corresponding local completion of F, and for v finite, by $\mathfrak{O}_v$ the ring of integers of F_v with uniformizer ϖ_v of norm q_v. Throughout this article, ω (resp. ω') denote the central character of a cusp form π (resp. π' on $\mathrm{GL}(2)/F$.

We will use, without mention, the notations and conventions of [**Ram2**], especially section 1 therein.

The symmetric cube condition (sI-2) (see Introduction) imposes strong constraints on how the local components π_v and π'_v are related at all v. We spell these out at the unramified v in the second part of the Lemma below, and show that, not surprisingly, when the central character ω is algebraic, π_v and π'_v are algebraically related. The first part show the effect of π satisfying both (sI-1) and (sI-2) on its "mirror" π'.

Lemma 2.1 *Let π be a cusp form on $GL(2)/F$ which is s-icosahedral relative to π'. Then*

(A) *π' satisfies (sI-j) for $j = 1, 2$, and moreover,*

$$(\omega')^3 = \omega^3.$$

In particular, ω' is algebraic iff ω is.

(B) *Let v be any finite place where π, π' are unramified, and denote by $\{a, b\}$ with $b = wa^{-1}$ (resp. $\{c, d\}$ with $d = w'c^{-1}$) the unordered pair of complex numbers associated to π_v (resp. π'_v), where $w = \omega_v(\varpi), w' = \omega'_v(\varpi)$. Write*

$$w = zw', \text{ with } z^3 = 1.$$

Then one of the following cases occurs (up to interchanging a and b):

(1) $\{c, d\} = \{za, zb\}$; $\mathrm{Ad}(\pi_v) \simeq \mathrm{Ad}(\pi'_v)$

(2) $\{c, d\} = \{\mu za, \mu^{-1}zb\}$, $\mu^4 = 1$, $a^2 = \mu w$; $\mathrm{Ad}(\pi_v) \simeq \mathrm{Ad}(\pi'_v)$

(3) $\{c, d\} = \{\zeta za, \zeta^{-1}zb\}$, $\zeta^5 = 1$, $a^2 = \zeta w$

In particular, when ω is algebraic, π_v and π'_v are both algebraic in cases (2) and (3). Moreover, if ω has finite order, $\mathbb{Q}(\pi_v)$ is contained, in these cases, in a finite abelian extension K of $\mathbb{Q}$, independent of v, containing $\mathbb{Q}(\omega)$.

Remark 2.2 In section 3 we will show, by also appealing to (sI-3), which is not used in Lemma 2.1, that (i) π_v is algebraic in *all* cases, and (ii) $a^4 = 1$ in case (2) of part (B) above, if $w = 1$. It is perhaps useful to note that if (sI-3) is not satisfied by π, one may take $F = \mathbb{Q}$ and π' to be a cubic twist of π, and such a π is expected to be transcendental if it is generated, for example, by a Maass form φ of weight 0 relative to a congruence subgroup of $\mathrm{SL}(2, \mathbb{Z})$ acting on the upper half plane with Laplacian eigenvalue $\lambda > 1/4$.

Proof of Lemma 2.1 (A) The fact that π and π' have isomorphic symmetric cubes implies immediately that ${\omega'}^6 = \omega^6$, which is not sufficient for us. We will first show that π' is not solvable polyhedral. First, since π is not solvable polyhedral, $\mathrm{sym}^j(\pi)$ is cuspidal for $j \leq 4$ (cf. [**KS1**]). By (sI-2), $\mathrm{sym}^3(\pi')$ is also cuspidal. Suppose $\mathrm{sym}^4(\pi')$ is not cuspidal. Then, by the criterion of Kim and Shahidi ([**KS1**]), $\mathrm{sym}^3(\pi')$ must be monomial, which forces it to admit a non-trivial self-twist by a quadratic character ν, say. Then by (sI-2), $\mathrm{sym}^3(\pi)$ also admits a self-twist by ν, implying that $\mathrm{sym}^4(\pi)$ is not cuspidal, contradicting (sI-1). Hence π' satisfies both (sI-j) for $j \leq 2$.

Next we appeal to Kim's theorem ([**Kim1**]) giving the automorphy in $\mathrm{GL}(6)/F$ of the exterior square $\Lambda^2(\Pi)$ of any cusp form Π on $\mathrm{GL}(4)/F$. Applying this with $\Pi = \mathrm{sym}^3(\eta)$ for a cusp form η on $\mathrm{GL}(2)/F$ with central character ν, we obtain the (well known) isobaric decomposition

$$\Lambda^2(\mathrm{sym}^3(\eta)) \;=\; \big(\mathrm{sym}^4(\eta) \otimes \nu\big) \boxplus \nu^3. \tag{2.3}$$

To prove this, we first note that by [**Kim1**], the L-functions agree at almost all places, and then appeal to the strong multiplicity one theorem for global isobaric representations, due to Jacquet and Shalika ([**JS**]). Now applying this to $\eta = \pi, \pi'$, we get by (sI-2), the following equivalence of isobaric sums:

$$\big(\mathrm{sym}^4(\pi) \otimes \omega\big) \boxplus \omega^3 \;=\; \big(\mathrm{sym}^4(\pi') \otimes \omega'\big) \boxplus {\omega'}^3. \tag{2.4}$$

Since $\mathrm{sym}^4(\pi)$ and $\mathrm{sym}^4(\pi')$ are both cuspidal, we are forced to have

$$(\omega')^3 \;=\; \omega^3,$$

as claimed.

(B) Preserving the notations used in part (B) of the Lemma, and noting that

$$b = wa^{-1},\ d = w'c^{-1},$$

we get by (sI-2), the equality of sets:

$$\{a^3, a^2b, ab^2, b^3\} = \{c^3, c^2d, cd^2, d^3\}. \tag{2.5}$$

Clearly,
(2.6)

$$a^2b = wa,\ ab^2 = w^2a^{-1}, b^3 = w^3a^{-3} \text{ and } c^2d = w'c,\ cd^2 = w'^2c^{-1},\ d^3 = w'^3c^{-3}.$$

Note that, since $w' = z^2w$, we have

$$c = \alpha a \implies d = w'\alpha^{-1}a^{-1} = z^2\alpha^{-1}b, \tag{2.7}$$

and

$$c = \beta b \implies d = w'\beta^{-1}b^{-1} = z^2\beta^{-1}a. \tag{2.8}$$

A priori, a^3 has four possibilities to satisfy (2.5). However, by interchanging c and d, we are reduced to considering only *two main cases*:

Case **I**: $\mathbf{a^3 = c^3}$

In this case, since $\omega'^3 = \omega^3$ and $w' = z^2w$, (2.5) and (2.6) yield, after dividing by w,

$$\{a, wa^{-1}\} = \{-z^2c, zwc^{-1}\}. \tag{2.9}$$

If $a = -z^2c$, then $c = za$ and $d = zb$ (by (2.7)), putting us in the situation (1). Since we have

$$\{c, d\} = \{a, b\} \cdot z,$$

π_v is (at most) a cubic twist of π'_v; it follows that $\mathrm{Ad}(\pi_v)$ and $\mathrm{Ad}(\pi'_v)$ are isomorphic.

So we may assume $a = zwc^{-1}$. Then we get (using (2.8))

$$c = zwa^{-1} = zb, \text{ and } d = za, \tag{2.10}$$

again landing us in (1).

Case **II**: $\mathbf{a^3 = w'c}$

In this case, $c = w'^{-1}a^3$, and we obtain from (2.5) and (2.6),

$$\{wa, w^2a^{-1}\} = \{w^{-3}a^9, w^6a^{-9}\}. \tag{2.11}$$

If $wa = w^6a^{-9}$, then $a^{10} = w^5$, so that

$$a^2 = \zeta w, \text{ with } \zeta^5 = 1. \tag{2.12}$$

Hence $a^3 = \zeta wa$, and so

$$c = w'^{-1}\zeta wa = \zeta za. \tag{2.13}$$

It follows that

$$d = w'c^{-1} = (z^2w)\zeta^{-1}z^{-1}a^{-1} = \zeta zb. \tag{2.14}$$

Thus we are in situation (3).

It is left to consider when $wa = w^{-3}a^9$. In this case, $a^8 = w^4$, so that

$$a^2 = \mu w, \text{ with } \mu^4 = 1. \tag{2.15}$$

Hence $a^3 = \mu w a$, and arguing as above, we deduce

$$c = \mu z a, \quad \text{and} \quad d = \mu^{-1} z b. \tag{2.16}$$

This puts us in situation (2).

We still need to show that $\mathrm{Ad}(\pi_v) \simeq \mathrm{Ad}(\pi'_v)$ in this case as well. To this end, note that since $a^2 = \mu w$ and $c = \mu z a$,

$$c = (a^2 w^{-1}) z a = w^{-1} z a^3. \tag{2.17}$$

On the other hand, raising $c = \mu z a$ to the fourth power, we get, since $\mu^4 = 1$,

$$c^4 = z a^4 = z a (a^3) = z a (w z^{-1} c), \tag{2.18}$$

which in turn yields

$$c^3 = w z a. \tag{2.19}$$

Furthermore,

$$c^2 = w^{-2} z^2 a^6 = w^{-2} z (z a^4) a^2 = w^{-2} z c^4 a^2,$$

implying

$$c^{-2} = w^{-2} z a^2, \quad \text{and} \quad c^2 = w^2 z^2 a^{-2}. \tag{2.20}$$

The unramified representation $\mathrm{Ad}(\pi_v)$ is described by the unordered triple

$$\{ab^{-1}, 1, a^{-1}b\} = \{w^{-1}a^2, 1, wa^{-2}\}. \tag{2.21}$$

Similarly, $\mathrm{Ad}(\pi'_v)$ is given by the unordered triple

$$\{cd^{-1}, 1, c^{-1}d\} = \{zw^{-1}c^2, 1, z^{-1}wc^{-2}\},$$

which, by (2.20), is the same as

$$\{wa^{-2}, 1, w^{-1}a^2\}. \tag{2.22}$$

The assertion follows. □

3. The nicety of $\mathrm{sym}^5(\pi)$ for π s-icosahedral

As mentioned in the Introduction, the condition (s-I3) implies that for an s-icosahedral π of central character ω, the automorphic representation $\mathrm{Ad}(\pi') \boxtimes \pi \otimes \omega^2$ of $\mathrm{GL}_6(\mathbb{A}_F)$, whose existence is given by [**KS2**], represents $\mathrm{sym}^5(\pi)$ at all places outside a finite set S of places containing the archimedean and ramified places. In fact we have the following strengthening:

Proposition 3.1 *Let π be an s-icosahedral cusp form on $GL(2)/F$ with central character ω.*

(a) *At every finite place v,*

$$L(s, \pi_v; \mathrm{sym}^5) = L(s, \mathrm{Ad}(\pi'_v) \boxtimes \pi_v \otimes \omega_v^2) \tag{3.2}$$

and

$$\varepsilon(s, \pi_v; \mathrm{sym}^5) = \varepsilon(s, \mathrm{Ad}(\pi'_v) \boxtimes \pi_v \otimes \omega_v^2) \tag{3.3}$$

(b) *If Σ_∞ denotes the set of archimedean places of F, we have*

$$L(s, \pi_\infty; \mathrm{sym}^5) = L(s, \mathrm{Ad}(\pi'_\infty) \boxtimes \pi_\infty \otimes \omega_\infty^2), \tag{3.4}$$

where π_∞ (resp. π'_∞) denotes $\otimes_{v \in \Sigma_\infty} \pi_v$ (resp. $\otimes_{v \in \Sigma_\infty} \pi'_v$).

(c) *If π, π' are not dihedral twists of each other, i.e., if their symmetric squares are not twist equivalent, then sym*$^5(\pi)$*, defined to be* $\mathrm{Ad}(\pi') \boxtimes \pi \otimes \omega^2$*, is cuspidal, and hence* $L(s, \pi; \mathrm{sym}^5)$ *is entire in this case.*

Remark Of course we are dealing with a very special class of cusp forms π here. For the status of results in the general case (which is much more complicated), see [**Sha2**], [**Sha3**] and [**KS1**].

Proof. At any place v, we have the well known factorization (using Clebsch-Gordan):

$$L(s, \mathrm{sym}^4(\pi_v) \times \pi_v) = L(s, \pi; \mathrm{sym}^5) L(s, \mathrm{sym}^3(\pi_v) \otimes \omega_v). \tag{3.5}$$

Similarly for the ε-factors, making use of the local Langlands conjecture for $\mathrm{GL}(n)$, now known by the works of Harris-Taylor [**HT**] and Henniart [**Hen1**]. So it holds for the γ-factors as well. We claim that one has, for a sufficiently ramified character ν_v of F_v^*,

$$L(s, \pi; \mathrm{sym}^5 \otimes \nu_v) = 1. \tag{3.6}$$

Indeed, by the standard stability results for the Rankin-Selberg L-functions, we may choose a highly ramified character ν_v such that both $L(s, \mathrm{sym}^4(\pi_v) \times \pi_v \otimes \nu_v)$ and $L(s, \mathrm{sym}^3(\pi_v) \otimes \omega_v \nu_v)$ are both 1. One way to see this is to use the fact that the local Langlands correspondence preserves the L-functions of pairs, and make use of the well known result on the Galois side. The claim now follows, thanks to (3.5). We can deduce analogously the same statement for the γ-factors and ε-factors. For a representation theoretic proof of the stability of γ-factors (which yields what we want for the ε-factors because stability holds for the L-factors), see for example [**CPSS**], where a general situation is treated.

Consider the global L-functions $L_1(s) := L(s, \mathrm{sym}^4(\pi) \times \pi)$ and

$$L_2(s) := L(s, \mathrm{Ad}(\pi') \boxtimes \pi \otimes \omega^2) L(s, \mathrm{sym}^3(\pi) \otimes \omega). \tag{3.7}$$

They both have functional equations and analytic continuations, and moreover, thanks to (s-I3) and (3.5), they are the same local factors out side a finite set of places containing Σ_∞ and the ramified finite places. First choose a global character ν such that at each finite place in S, ν_v is sufficiently ramified so that the ν_v twists of $L_{1,v}(s)$ and $L_{2,v}(s)$ are both 1; we can do this by the (standard) argument above. Comparing functional equations, and using the fact that the archimedean factors of $L_j(s), j = 1, 2$, have no poles to the right of $\Re(s) > \frac{1}{2}$, we get

$$L_{1,\infty}(s) = L_{2,\infty}(s),$$

which furnishes (3.4).

Now pick any finite place v in S and choose a ν such that for every $u \in S, u \neq v$, ν_u is sufficiently ramified. Comparing the functional equations again, we get (3.2); here we use the invertibility of the ε-factors and the shape of the local factors, as well as the very weak Ramanujan at v which separates the poles of $L_{j,v}(s)$ from its dual. After that we may, in the same way, deduce (3.3) as well. We have now proved parts (a) and (b) of the Proposition.

Part (c) of the Proposition follows easily from the general cuspidality criterion ([**RW**]) for the Kim-Shahidi functorial transfer from $\mathrm{GL}(2) \times \mathrm{GL}(3)$ to $\mathrm{GL}(6)$. In fact, already by Theorem 1.2 of [**Wan**], given that π is not solvable polyhedral, which is our hypothesis, the only way $\mathrm{Ad}(\pi') \boxtimes \pi \otimes \omega^2$ can fail to be cuspidal is for

$\mathrm{sym}^2(\pi)$ and $\mathrm{sym}^2(\pi')$ to be twists of each other by a character. But this has been ruled out by the hypothesis, and we are done. □

4. Proof of part (a) of Theorem A

Let π be as in Theorem A, attached to (π', ν), with ω algebraic. We will show the following:

(4.1)

- (i) At each place v, π_v is algebraic.
- (ii) The global representation π_f is rational over a finite extension of $\mathbb{Q}$ (containing $\mathbb{Q}(\omega)$).

To deduce (ii) from (i), we need to show in addition that $\mathbb{Q}(\pi_v)$ has degree bounded independent of the place v.

4.1. When π_v and π'_v are unramified. Let v be a finite place where π_v and π'_v are unramified. Preserving the notations of section 1, let us recall that the only case where we do not yet know the algebraicity of π_v is case (1) of Lemma 2.1, part (B). We treat this case now, making use of (sI-3).

Lemma 4.1.1 *Let v be a finite place where π_v is unramified. Suppose we have $\mathrm{Ad}(\pi_v)) \simeq \mathrm{Ad}(\pi'_v)$, which is satisfied in cases (1), (2) of Lemma 2.1, part (B). Then (sI-3) implies that one of the following identities holds:*

$$a^2 = w,\ a^4 = w^2,\ a^6 = w^3.$$

In particular, π_v is algebraic if ω_v is algebraic. Moreover, if $w = 1$, then $a^m = 1$ with $m \in \{4, 6\}$.

Remark 4.1.2 The reason for also including here the case (2) of Lemma 2.1, part (B), is the following. We knew earlier that in this case, $a^2 = \mu w$, with $\mu^4 = 1$, so that when $w = 1$, $a^4 = \pm 1$. But now, using Lemma 4.1.1 in addition, we rule out the (potentially troublesome) possibility $a^4 = -1$ (when $w = 1$), which will be important to us later.

Proof of Lemma 4.1.1. Since $\mathrm{Ad}(\pi'_v)$ is by assumption (in this Lemma) isomorphic to $\mathrm{Ad}(\pi_v)$, it is described by the triple $\{w^{-1}a^2, 1, wa^{-2}\}$. So $\mathrm{Ad}(\pi'_v)\otimes\omega_v^2$ is associated to $\{wa^2, w^2, w^3a^{-2}\}$. Thus $\mathrm{Ad}(\pi'_v) \otimes \pi_v \otimes \omega_v^2$ is given by the sextuple

$$\{wa^3, w^2a, w^2a, w^3a^{-1}, w^3a^{-1}, w^4a^{-3}\}.$$

On the other hand, $\mathrm{sym}^5(\pi_v)$ is attached to the sextuple

$$\{a^5, wa^3, w^2a, w^3a^{-1}, w^4a^{-3}, w^5a^{-5}\}.$$

Comparing these two tuples, we get

$$\{wa^3, w^2a, w^3a^{-1}, w^4a^{-3}\} = \{a^5, wa^3, w^4a^{-3}, w^5a^{-5}\}.$$

Looking at the possibilities for w^2a, we see that *either* $a^4 = w^2$ (which subsumes both the possibilities $w^2a = a^5$ and $w^2a = w^4a^{-3}$), *or* $a^2 = w$ (when $w^2a = wa^3$), *or* $a^6 = w^3$ (when $w^2a = w^5a^{-5}$).

□

Thanks to the Lemma, we see that at all the unramified places v, $\mathbb{Q}(\pi_v)$ is contained in a finite solvable extension of $\mathbb{Q}(\omega)$, with its degree over $\mathbb{Q}(\omega)$ bounded independent of v (see Lemma 2.1 for cases (2) through (4). And since ω is by hypothesis (of Theorem A) algebraic, i.e., of type A_0, $\mathbb{Q}(\omega)$ is a number field; in fact it is a CM field or totally real. Consequently, we have proved the following:

Lemma 4.1.3 *Let π be as in Theorem A, and let S be the union of the archimedean places and the finite places where π_v is ramified. Let π^S denote (as usual) the restricted tensor product of the local components π_v as v runs over all the places outside S. Then π^S is algebraic, rational over a finite extension K of $\mathbb{Q}(\omega)$. In fact, K is contained in a compositum of cyclotomic and Kummer extensions over $\mathbb{Q}(\omega)$.*

4.2. Proof of non-occurrence of π_v of Steinberg type. Suppose π_v is a special representation. Then there is a character λ of F_v^* such that the associated 2-dimensional representation σ_v of the extended Weil group $W'_{F_v} = W_{F_v} \times \mathrm{SL}(2,\mathbb{C})$ is of the form $\lambda \otimes st$ where st denotes the standard representation of $\mathrm{SL}(2,\mathbb{C})$. From (sI-2), it follows that π'_v is also necessarily special, with its associated 2-dimensional representation σ'_v of W'_{F_v} being of the form $\lambda' \otimes st$. Then $\mathrm{Ad}(\pi'_v)$ has parameter $\mathrm{Ad}(\sigma'_v) \simeq 1 \otimes \mathrm{sym}^2(st)$, so that $\mathrm{Ad}(\pi'_v) \otimes \pi_v \otimes \omega^2$ has the parameter

$$\lambda^3 \otimes \left(\mathrm{sym}^3(st) \oplus st\right). \tag{4.2.1}$$

On the other hand, $\mathrm{sym}^5(\pi_v)$ has the parameter

$$\lambda^5 \otimes \mathrm{sym}^5(st). \tag{4.2.2}$$

But the representations (4.2.1) and (4.2.2) cannot be isomorphic, contradicting (sI-3). Thus π_v cannot be of Steinberg type. □

4.3. When π_v is unramified, but π'_v is ramified. Thanks to (sI-2) and (sI-3), $\mathrm{sym}^3(\pi'_v)$ and $\mathrm{Ad}(\pi'_v)$ must both be unramified in this case. Suppose π'_v is a principal series representation. So we may write

$$\pi'_v \simeq \mu_1 \boxplus \mu_2, \, \omega' = \mu_1\mu_2,$$

with $\mu_1^3, \mu_2^3, \mu_1^2\mu_2, \mu_1\mu_2^2, \mu_1\mu_2^{-1}$ unramified. Then μ_1 and μ_2 must themselves be unramified, forcing π'_v to be in the unramified principal series.

The only remaining possibility is for π'_v to be a supercuspidal representation with (irreducible) parameter $\sigma_v = \mathrm{Ind}_E^{F_v}(\chi)$, for a character χ of the multiplicative group of a quadratic extension E of F_v. If θ denotes the non-trivial automorphism of E/F_v, then recall that the irreducibility of σ_v implies that $\chi \circ \theta \neq \chi$. We have

$$\mathrm{sym}^2(\sigma_v) \simeq \mathrm{Ind}_E^{F_v}(\chi^2) \oplus \chi_0,$$

where χ_0 is the restriction of χ to F_v^*. The determinant of σ_v, which corresponds to ω'_v, is then $\chi_0\nu$, where ν is the quadratic character of W_{F_v} corresponding to E/F_v. Since $\mathrm{Ad}(\pi'_v) = \mathrm{sym}^2(\sigma_v) \otimes (\chi_0\nu)^{-1}$ is unramified, we must have ν, i.e., E/F_v, unramified, and

$$\chi^2 \circ \theta = \chi^2,$$

which implies

$$\chi = \lambda(\mu \circ N_{E/F_v}), \quad \text{with} \quad \lambda^2 = 1, (\lambda \circ \theta) \neq \lambda,$$

with μ a character of F_v^*. Then $\chi_0 = \lambda_0\mu^2$ and

$$\mathrm{Ad}(\sigma_v) \simeq \lambda_0 \oplus \lambda_0\nu \oplus \nu.$$

Moreover, λ_0 must be unramified as well. If $\lambda_0 = 1$, then $\mathrm{Ad}(\sigma_v)$ contains the trivial representation $\underline{1}$, and therefore

$$\dim_{\mathbb{C}}\left(\mathrm{Hom}_{W_{F_v}}(\underline{1}, \mathrm{End}(\sigma_v))\right) = 2,$$

since $\mathrm{End}(\sigma_v) \simeq \mathrm{Ad}(\sigma_v) \oplus \underline{1}$. This contradicts, by Schur's Lemma, the irreducibility of σ_v. So we must have $\lambda_0 \neq 1$. But then, as λ_0 is an unramified quadratic character, it must coincide with ν, making $\lambda_0\nu = 1$. Again, we get $\underline{1} \subset \mathrm{Ad}(\sigma_v)$, leading to a contradiction.

So we conclude that π'_v must be unramified when π_v is.

4.4. When π_v is ramified. Suppose π_v is a ramified principal series representation attached to the characters μ_1, μ_2 of F_v^*, with $\omega_v = \mu_1\mu_2$. Then π'_v is also necessarily a ramified principal series representation, attached to characters μ'_1, μ'_2, with $\omega'_v = \mu'_1\mu'_2$. The criteria (sI-1) and (sI-2) give conditions relating various powers of these characters and of ν_v. The situation is similar to the unramified case, and we conclude algebraicity as before. Again, $\mathbb{Q}(\pi_v)$ is a finite extension of Kummer type over $\mathbb{Q}(\omega)$.

It remains to consider the case when π_v is supercuspidal, in which case the constraints force π'_v to also be supercuspidal. It is well known that a supercuspidal representation with algebraic central character ω_v is algebraic, in fact rational over $\mathbb{Q}(\omega_v)$. Indeed, as mentioned in the Introduction, we are using the algebraic parametrization ([**Hen2**]), which is equivariant for the action of $\mathrm{Aut}(\mathbb{C})$. So it suffices to verify this for the parameter σ_v of π_v. Since we are over a local field, the image G_v of σ_v is necessarily solvable, and in particular, the image of G_v in $\mathrm{PGL}(2, \mathbb{C})$ must be dihedral, tetrahedral, or octahedral. The assertion about rationality follows from the known results on the irreducible representations of coverings of D_{2n}, A_4 and S_4.

So we have now proved part (a) of Theorem A. □

5. Proof of part (b) of Theorem A

Here we are assuming that ω is of finite order. Then so is ω', and the arguments of sections 1 and 2 imply immediately that $\mathbb{Q}(\pi_f)$ is a cyclotomic field. To be precise, this is clear outside places v where π_v is not square-integrable, and the assertion in the supercuspidal case holds because it is rational over $\mathbb{Q}(\omega)$, which is cyclotomic.

We need to show that at any archimedean place v, π_v is of Galois type. Let σ_v, resp. σ'_v, denote the 2-dimensional representation of the Weil group W_{F_v} associated to π_v, resp. π'_v. First suppose that the restriction of σ_v to $\mathbb{C}^*$ is a one-dimensional twist of $(z/|z|)^m \oplus (z/|z|)^{-m}$ with $m > 0$; this happens when either v is real and π_v is a discrete series representation (of lowest weight m), or v is complex and π_v is the base change of a discrete series representation of $\mathrm{GL}(2, F_{\mathbb{R}})$. In either case, (sI-2) implies that π_v is also of the same form. But then we see that the restriction to $\mathbb{C}^*$ of the parameter of $\mathrm{Ad}(\pi'_v) \otimes \pi_v$ is a one-dimensional twist of

$$\left((z/|z|)^{2m} \oplus 1 \oplus (z/|z|)^{-2m}\right) \otimes \left((z/|z|)^m \oplus (z/|z|)^{-m}\right),$$

which is

$$(z/|z|)^{3m} \oplus (z/|z|)^m \oplus ((z/|z|)^m \oplus (z/|z|)^{-m} \oplus (z/|z|)^{-m} \oplus (z/|z|)^{-3m}.$$

Since $m \neq 0$, this representation cannot possibly be a one-dimensional twist of the restriction to $\mathbb{C}^*$ of $\mathrm{sym}^5(\sigma_v)$, which is evidently multiplicity-free. Note that we would have no contradiction if we had allowed $m = 0$, which corresponds to the Galois type situation.

Thus we may assume, by the classification of unitary representations, that π_v is a unitary character twist of a spherical representation $\sigma_v^0 = \mu_1 \otimes \mu_2$, where μ_1, μ_2 are unramified characters of F_v^*. Then, since ω_v has finite order by assumption, the restriction of σ_v^0 to $\mathbb{C}^*$ is necessarily of the form $|\cdot|^s \oplus |\cdot|^{-s}$. Applying (sI-2), we see that the restriction of σ'_v to $\mathbb{C}^*$ will also need to be a unitary character twist of $\sigma_v^0|_{\mathbb{C}^*}$. Then $(\mathrm{Ad}(\sigma'_v) \otimes \sigma_v)|_{\mathbb{C}^*}$ cannot, unlike $\mathrm{sym}^5(\sigma_v)|_{\mathbb{C}^*}$, contain a unitary character twist of $|\cdot|^{5s}$, unless $s = 0$. So we get a contradiction to (sI-3) if $s \neq 0$. Hence s must be zero, and since ω_v has finite order, σ_v is a finite order twist of a σ_v^0 whose restriction to $\mathbb{C}^*$ is $1 \oplus 1$. In other words, π_v is of Galois type.

□

6. Temperedness of π

It suffices to prove temperedness at each place v. When v is archimedean, we have already shown that π_v is even of Galois type, so we may assume that v is finite. Suppose π_v is a principal series representation with parameter σ_v. If π_v is non-tempered, i.e., violates the Ramanujan hypothesis, we must have, since $\pi_v^\vee \simeq \overline{\pi}_v$,

$$\sigma_v \simeq \lambda_v \otimes \left(|\cdot|^t \oplus |\cdot|^{-s}\right), \quad \text{with} \quad \lambda_v^2 = \omega_v,\, t \in \mathbb{R}_+^*.$$

As ω_v has finite order, λ_v does as well. As in the archimedean spherical case, (sI-2) implies that the parameter of π'_v is necessarily of the form

$$\sigma'_v \simeq \lambda'_v \otimes \left(|\cdot|^t \oplus |\cdot|^{-s}\right), \quad \text{with} \quad {\lambda'_v}^2 = \omega'_v,\, t \in \mathbb{R}_+^*,$$

for the same t; here we have used the fact that ν_v has finite order. It follows (since $t > 0$) that

$$\mathrm{Hom}\left(|\cdot|^{5t}, \mathrm{Ad}(\sigma'_v) \otimes \sigma_v \otimes \omega_v^2\right) = 0,$$

while

$$\mathrm{Hom}\left(|\cdot|^{5t}, \mathrm{sym}^5(\sigma_v)\right) \neq 0.$$

This contradicts (sI-3) and so t must be zero. In other words, π_v must be tempered.

Finally, as is well known, if π_v is a discrete series representation with unitary central character, it is necessarily tempered.

This proves part (c) of Theorem A.

□

7. Proof of part (d) of Theorem A

Proposition 7.1 *Let π, π' be as in Theorem A, with $\omega = \omega' = 1$. Then at any finite place v, π_v is either a unitary principal series or a supercuspidal representation, with π'_v of the same type, and furthermore,*

$$\mathbb{Q}(\pi_v) \subset \mathbb{Q}[\sqrt{5}].$$

More precisely, we have

(a) *When π_v, π'_v are in the unitary principal series,*

(ai) $\mathbb{Q}(\pi_v) = \mathbb{Q}$, *if* $\mathrm{Ad}(\pi_v) \simeq \mathrm{Ad}(\pi'_v)$*;*
(aii) $\mathbb{Q}(\pi_v) = \mathbb{Q}[\sqrt{5}]$, *if* $\mathrm{Ad}(\pi_v) \not\simeq \mathrm{Ad}(\pi'_v)$.
(b) *When* π_v *is supercuspidal,* $\mathbb{Q}(\pi_v) = \mathbb{Q}$.

Corollary 7.2 *Let* π, π' *be as above (in Proposition 7.1). Then we have* $\mathbb{Q}(\pi_f) \subset \mathbb{Q}[\sqrt{5}]$*, and moreover,*

$$\mathbb{Q}(\pi_f) = \mathbb{Q}[\sqrt{5}] \Leftrightarrow \pi \not\simeq \pi'.$$

Proposition 7.1 $\implies$ **Corollary 7.2**:

Since $\mathbb{Q}(\pi_f)$ is the compositum of all the $\mathbb{Q}(\pi_v)$ as v runs over finite places (cf. [**Clo**], Prop. 3.1, for example), we have

$$\mathbb{Q}(\pi_f) \subset \mathbb{Q}[\sqrt{5}]. \tag{7.2}$$

First suppose $\mathrm{Ad}(\pi_f) \not\simeq \mathrm{Ad}(\pi'_f)$. Then at some finite place u, say, $\mathrm{Ad}(\pi_u)$ and $\mathrm{Ad}(\pi'_u)$ need to be non-isomorphic, forcing, by (aii) of Prop. 7.1,

$$\mathbb{Q}(\pi_f) = \mathbb{Q}(\pi_u) = \mathbb{Q}[\sqrt{5}].$$

On the other hand, if $\mathrm{Ad}(\pi_f)$ and $\mathrm{Ad}(\pi'_f)$ are isomorphic, their local components are isomorphic as well, and (ai) of Prop. 7.1 then implies that

$$\mathbb{Q}(\pi_f) = \mathbb{Q}.$$

Now we *claim* that for our π, π',

$$\mathrm{Ad}(\pi_f) \simeq \mathrm{Ad}(\pi'_f) \implies \pi_f \simeq \pi'_f. \tag{7.3}$$

Indeed, by the *multiplicity one theorem for* $SL(2)$ (cf. [**Ram1**]), we have

$$\pi'_f \simeq \pi_f \otimes \nu_f, \tag{7.4}$$

for some idèle class character ν of F. Since $\omega = \omega' = 1$, we must have

$$\nu^2 = 1.$$

Anyhow, (7.4) implies

$$\mathrm{sym}^3(\pi'_f) \simeq \mathrm{sym}^3(\pi_f) \otimes \nu_f^3,$$

Since π and π' have, by (sI-2), isomorphic symmetric cubes, we must then have

$$\nu^3 = 1,$$

or else, $\mathrm{sym}^3(\pi)$ will need to admit a non-trivial self-twist, which is not possible as π is not solvable polyhedral. So it follows that $\nu = 1$, *proving the claim.*

This also finishes the proof of the Corollary assuming Prop. 7.1.

□

Proof of Proposition 7.1

Since we are assuming here that $\omega = \omega' = 1$, we first claim that at the places v where π_v is supercuspidal, $\mathbb{Q}(\pi_v)$ is just $\mathbb{Q}$. Indeed, as we are using the algebraic parametrization, we can transfer the problem to the field of rationality of the associated irreducible 2-dimensional representation σ_v of W_{F_v}. This is a known fact, and we briefly sketch an argument. Let θ be an automorphism of $\mathbb{C}$. It suffices to check that $\varepsilon(s, \pi_v^\theta \otimes \chi^\theta)$ equals $\varepsilon(s, \sigma_v^\theta \otimes \chi^\theta)$, for all characters χ; we are identifying, by class field theory, the characters χ of F_v^* with the corresponding ones of W_{F_v}. (We will also identify the central character ω_v with $\det(\sigma_v)$.) If we look at the root number $W(\sigma_v \otimes \chi) = \varepsilon(1/2, \sigma_v \otimes \chi)$, then the ratio $W(\sigma_v^\theta \otimes \chi^\theta)/W(\omega_v^\tau \otimes \chi^\theta)$

equals $(W(\sigma_v \otimes \chi)/W(\omega_v \otimes \chi))^\tau$. An analogous property holds on the GL(2) side, as seen by the integral representation of the local zeta functions, using the Whittaker model, which is compatible with the θ-action ([**Wal**]. The conductors also correspond, and one gets the Galois equivariance of $\pi_v \mapsto \sigma_v$. Next, note that we may write σ_v as an unramified abelian twist of an irreducible 2-dimensional representation of $\mathrm{Gal}(\overline{F}_v/F_v)$ which will have solvable, finite image. As we saw at the end of section 3, such a σ_v is rational over $\mathbb{Q}(\omega_v)$. Hence the claim.

Moreover, we know that π_v will not be of Steinberg type, nor non-tempered. So let us focus on the finite places v where π_v is in the unitary principal series.

First let v be an unramified place for π. Then $w = w' = z = \delta = 1$ by hypothesis, and combining Lemma 2.1 with Proposition 4.1.1, we get the following:

Lemma 7.5 *When π_v, π'_v are unramified with $\omega_v = \omega'_v = 1$, one of the following happens (up to exchanging a with $b = a^{-1}$ and c with $d = c^{-1}$):*

(i) $a^m = 1$, *for* $m \in \{4, 6\}$, *and* $c = \pm a$; $\mathrm{Ad}(\pi_v) \simeq \mathrm{Ad}(\pi'_v)$
(ii) $c = a^3, a = c^{-3}, a^{10} = 1$; $\mathrm{Ad}(\pi_v) \not\simeq \mathrm{Ad}(\pi'_v)$

In the first case, $\mathrm{Ad}(\pi'_v)$ and $\mathrm{Ad}(\pi_v)$ are isomorphic, and $\mathbb{Q}(\pi_v)$ is contained in either $\mathbb{Q}(\zeta_4)$ or $\mathbb{Q}(\zeta_6)$, where by ζ_n we mean a primitive n-th root of unity in $\mathbb{C}$. In case (ii), it is clear that $\mathbb{Q}(\pi_v)$ is contained in $\mathbb{Q}(\zeta_{10})$. On the other hand, since π_v is selfdual, the trace and determinant of the conjugacy class of π_v are both real, and so π_f is rational over the real subfield of $\mathbb{Q}(\zeta_n)$ for appropriate n. In other words, when π_v, π'_v are unramified with trivial central character,

$$\mathbb{Q}(\pi_v) = \mathbb{Q} \quad \text{when} \quad \mathrm{Ad}(\pi_v) \simeq \mathrm{Ad}(\pi'_v),$$

and

$$\mathbb{Q}(\pi_v) = \mathbb{Q}(\sqrt{5}) \quad \text{when} \quad \mathrm{Ad}(\pi_v) \not\simeq \mathrm{Ad}(\pi'_v),$$

It remains to consider the case when π_v is a ramified, tempered principal series representation with $\omega_v = 1$. Then its parameter decomposes as

$$\sigma_v \simeq \mu \oplus \mu^{-1},$$

for a ramified character μ of F_v^*. We have, thanks to the algebraic normalization of parameters,

$$\mathbb{Q}(\pi_v) = \mathbb{Q}(\mu)^+,$$

where the right hand side is the totally real subfield of the cyclotomic field $\mathbb{Q}(\mu)$. Again, the reason is that $\mathbb{Q}(\pi_v)$ is *a priori* contained in $\mathbb{Q}(\mu)$, and the selfduality of π_v makes it rational over the real subfield.

When π_v is such a ramified principal series representation, π'_v is also forced to be of similar type, thanks to (sI-2), with parameter $\sigma'_v = \mu' \oplus \mu'^{-1}$. Furthermore, we need to have

$$\mu^3 \oplus \mu \oplus \mu^{-1} \oplus \mu^{-3} \simeq \mu'^3 \oplus \mu' \oplus \mu'^{-1} \oplus \mu'^{-3}$$

This implies that either $\pi_v \simeq \pi'_v$ or, up to interchanging the roles of μ' and μ'^{-1}, $\mu' = \mu^3$. In the former case, arguing as in the unramified case above, we deduce that either $\mu^4 = 1$ or $\mu^6 = 1$. So $\mathbb{Q}(\pi_v) = \mathbb{Q}$ in this case. So we may assume that $\mu' = \mu^3$. Then (sI-3) yields

$$\mu^5 \oplus \mu^3 \oplus \mu \oplus \mu^{-1} \oplus \mu^{-3} \oplus \mu^{-5} \simeq \left(\mu^6 \oplus 1 \oplus \mu^{-6}\right) \otimes \left(\mu \oplus \mu^{-1}\right).$$

Hence

$$\mu^3 \oplus \mu^{-3} \simeq \mu^7 \oplus \mu^{-7}.$$

In other words, we have

$$\mu^4 = 1 \quad \text{or} \quad \mu^{10} = 1.$$

Thus $\mathbb{Q}(\mu)$ is generated over $\mathbb{Q}$ by an m-th root of unity ζ_m with $m \in \{4, 10\}$. Consequently, since $\mathbb{Q}(\zeta_4)^+ = \mathbb{Q}$ and $\mathbb{Q}(\zeta_{10})^+ = \mathbb{Q}(\sqrt{5})$,

$$\mathbb{Q}(\pi_v) = \mathbb{Q}(\zeta_4)^+ = \mathbb{Q} \quad \text{or} \quad \mathbb{Q}(\pi_v) = \mathbb{Q}(\sqrt{5}).$$

This finishes the proof of Proposition 7.1.

□

Let τ be the non-trivial automorphism of $\mathbb{Q}(\sqrt{5})$. At each finite v, let π_v^τ denote the τ-conjugate representation of π_v, which makes sense because $\mathbb{Q}(\pi_v) \subset \mathbb{Q}(\sqrt{5})$. Similarly, let π_f^τ denote the τ-conjugate of π_f. Then π_f^τ is an admissible irreducible, generic representation because π_f has those properties. However, it is not at all clear if π_f^τ is automorphic. In the present case, we can deduce the following:

Proposition 7.6 *Let π be an s-icosahedral cusp form (with trivial central character) on $GL(2)/F$ relative to π', also of trivial central character. Then we have*

(i) *π' is unique, satisfying*

$$\pi_f^\tau \simeq \pi_f'.$$

(ii) *π_f is not rational over $\mathbb{Q}$.*

Proof. We will first prove (ii) by contradiction. Suppose π_f is rational over $\mathbb{Q}$. Then by Corollary 7.2, we must have $\pi_f' \simeq \pi_f$. Hence

$$\mathrm{sym}^2(\pi_f') \otimes \pi_f \simeq \mathrm{sym}^3(\pi_f) \boxplus \pi_f.$$

This means that (by (sI-3),

$$L(s, \pi_f; \mathrm{sym}^5) = L(s, \mathrm{sym}^3(\pi_f))L(s, \pi_f)$$

On the other hand, by Clebsch-Gordan,

$$L(s, \mathrm{sym}^4(\pi_f) \times \pi_f) = L(s, \pi_f; \mathrm{sym}^5)L(s, \mathrm{sym}^3(\pi_f))$$

So

$$L(s, \mathrm{sym}^4(\pi_f) \times \pi_f) = L(s, \mathrm{sym}^3(\pi_f))^2 L(s, \pi_f).$$

This leads to the identity

$$L(s, \mathrm{sym}^4(\pi_f) \times (\pi_f \boxtimes \pi_f)) = L(s, \mathrm{sym}^3(\pi_f) \times \pi_f)L(s, \pi_f \boxtimes \pi_f).$$

The rightmost L-function has a pole at $s = 1$ since π is selfdual, and one knows by Shahidi that $L(s, \mathrm{sym}^3(\pi_f) \times \pi_f)$ does not vanish at $s = 1$. It then forces the left hand side L-function to have a pole at $s = 1$. But this can't be, because $\mathrm{sym}^4(\pi)$ is, thanks to the condition (sI-1), a cusp form on $\mathrm{GL}(5)/F$ (see [**KS1**]) and $\pi \boxtimes \pi$ is an automorphic form on $\mathrm{GL}(4)/F$. This contradiction proves that π_f cannot be rational over $\mathbb{Q}$.

To deduce (i), it suffices to prove, locally at each finite place v, that

$$\pi_v' \simeq \pi_v^\tau.$$

Let v be a finite place where π_v is in the principal series with parameter $\sigma_v = \mu \oplus \mu^{-1}$, where μ may or may not be ramified. Then π_v' will also be of the same form, say with parameter $\sigma'_v = \mu' \oplus \mu'^{-1}$. Thanks to (sI-2), we must have, up to permuting μ' and μ'^{-1},

$$\mu' = \mu^m, \quad \text{where} \quad m \in \{1, 3\}.$$

When $\mu' = \mu$, $\pi'_v \simeq \pi_v$, and as we have seen above, $\mu^m = 1$ with $m \in \{4, 6\}$, implying that $\mathbb{Q}(\pi_v) = \mathbb{Q}$. So let us assume that $\mu' = \mu^3$. Then (again as above) $\mu^4 = 1$ or $\mu^1 0 = 1$. In the former case, we get $\pi'_v \simeq \pi_v$, while in the latter case, either $\mu^2 = 1$, again implying $\pi'_v \simeq \pi_v$, or else, $\mu^{10} = 1$, but $\mu^2 \neq 1$. Let $\tilde{\tau}$ be the automorphism of $M := \mathbb{Q}(\zeta_{10})$ given by $\zeta_{10} \mapsto \zeta_{10}^3$. The totally real subfield of M is $K = \mathbb{Q}(\sqrt{5})$ and $\tilde{\tau}$ restricts to the non-trivial automorphism τ of K. It follows that

$$\pi'_v \simeq \pi_v^\tau,$$

which holds at every principal series place v. In fact it holds at every finite place because at the supercuspidal places, $\mathbb{Q}(\pi_v) = \mathbb{Q}$ due to π_v having trivial central character. Consequently, $\pi'_v \simeq \pi_v^\tau$ at each finite place v. Moreover, π' is unique by the strong multiplicity one theorem.

This proves Proposition 7.6.

□

We claim that π' is s-icosahedral with $(\pi')' \simeq \pi$. Indeed, (sI-2) is a reflexive condition, and it also proves that π' is not dihedral or tetrahedral. We get (sI-3) (for π') by applying τ (to the (sI-3) for π) and making use of π'_f being π_f^τ. Furthermore, suppose $\mathrm{sym}^4(\pi')$ is Eisensteinian. Then by [**KS1**], $\mathrm{sym}^3(\pi')$ must admit a quadratic self-twist. Then so must $\mathrm{sym}^3(\pi)$ by (sI-2), resulting in the non-cuspidality of $\mathrm{sym}^4(\pi)$. But this contradicts (sI-1), and the claim follows.

Finally, since π' is s-icosahedral, it also follows, by the reasoning we employed for π, that π' is also of Galois type at infinity.

This proves part (d) of Theorem A.

□

Remark 7.7 It should be noted that as π is of Galois type at infinity with trivial central character, it cannot correspond, for F totally real, to a holomorphic Hilbert modular form. So there is no reason at all, given the state of our current knowledge, to assert that its Galois conjugate π^τ should be modular (though it is expected). However, in our special case, it is automorphic because it happens to be isomorphic to the shadow cusp form π'.

References

[Clo] L. Clozel. Motifs et formes automorphes: applications du principe de fonctorialité. In *Automorphic forms, Shimura varieties, and L-functions, Vol. I (Ann Arbor, MI, 1988)*, volume 10 of *Perspect. Math.*, pages 77–159. Academic Press, Boston, MA, 1990.

[CPSS] J. W. Cogdell, I. I. Piatetski-Shapiro, and F. Shahidi. Stability of γ-factors for quasi-split groups. *J. Inst. Math. Jussieu* **7** (2008), 27–66.

[GJ] S. Gelbart and H. Jacquet. A relation between automorphic representations of GL(2) and GL(3). *Ann. Sci. École Norm. Sup. (4)* **11** (1978), 471–542.

[HT] M. Harris and R. Taylor. *The geometry and cohomology of some simple Shimura varieties*, volume 151 of *Annals of Mathematics Studies*. Princeton University Press, Princeton, NJ, 2001. With an appendix by Vladimir G. Berkovich.

[Hen1] G. Henniart. Une preuve simple des conjectures de Langlands pour GL(n) sur un corps p-adique. *Invent. Math.* **139** (2000), 439–455.

[Hen2] G. Henniart. Une caractérisation de la correspondance de Langlands locale pour $GL(n)$. *Bull. Soc. Math. France* **130** (2002), 587–602.

[JPSS] H. Jacquet, I. I. Piatetskii-Shapiro, and J. A. Shalika. Rankin-Selberg convolutions. *Amer. J. Math.* **105** (1983), 367–464.

[JS] H. Jacquet and J. A. Shalika. On Euler products and the classification of automorphic forms. II. *Amer. J. Math.* **103** (1981), 777–815.

[Kim1] H. H. Kim. Functoriality for the exterior square of GL_4 and the symmetric fourth of GL_2. *J. Amer. Math. Soc.* **16** (2003), 139–183 (electronic). With Appendix 1 by Dinakar Ramakrishnan and Appendix 2 by Kim and Peter Sarnak.

[Kim2] H. H. Kim. An example of non-normal quintic automorphic induction and modularity of symmetric powers of cusp forms of icosahedral type. *Invent. Math.* **156** (2004), 495–502.

[KS1] H. H. Kim and F. Shahidi. Cuspidality of symmetric powers with applications. *Duke Math. J.* **112** (2002), 177–197.

[KS2] H. H. Kim and F. Shahidi. Functorial products for $GL_2 \times GL_3$ and the symmetric cube for GL_2. *Ann. of Math. (2)* **155** (2002), 837–893. With an appendix by Colin J. Bushnell and Guy Henniart.

[Ram1] D. Ramakrishnan. Modularity of the Rankin-Selberg L-series, and multiplicity one for $SL(2)$. *Ann. of Math. (2)* **152** (2000), 45–111.

[Ram2] D. Ramakrishnan. Remarks on the symmetric powers of cusp forms on $GL(2)$. In *Automorphic Forms and L-functions I: Global Aspects, Proceedings of a Workshop in honor of Steve Gelbert*, pages 237–256. IMCP/CONM, AMS Series, 2009.

[RW] D. Ramakrishnan and S. Wang. A cuspidality criterion for the functorial product on $GL(2) \times GL(3)$ with a cohomological application. *Int. Math. Res. Not.* (2004), 1355–1394.

[Sha1] F. Shahidi. On certain L-functions. *Amer. J. Math.* **103** (1981), 297–355.

[Sha2] F. Shahidi. On the Ramanujan conjecture and finiteness of poles for certain L-functions. *Ann. of Math. (2)* **127** (1988), 547–584.

[Sha3] F. Shahidi. Symmetric power L-functions for $GL(2)$. In *Elliptic curves and related topics*, volume 4 of *CRM Proc. Lecture Notes*, pages 159–182. Amer. Math. Soc., Providence, RI, 1994.

[Wal] J.-L. Waldspurger. Quelques propriétés arithmétiques de certaines formes automorphes sur $GL(2)$. *Compositio Math.* **54** (1985), 121–171.

[Wan] S. Wang. On the symmetric powers of cusp forms on $GL(2)$ of icosahedral type. *Int. Math. Res. Not.* (2003), 2373–2390.

253-37 Caltech, Pasadena, CA 91125, USA
E-mail address: dinakar@caltech.edu

Clay Mathematics Proceedings
Volume 13, 2011

Pseudo Eisenstein Forms and the Cohomology of Arithmetic Groups III: Residual Cohomology Classes

Jürgen Rohlfs and Birgit Speh†

Dedicated to Freydoon Shahidi

ABSTRACT. Let $\mathbb{G}$ be a reductive algebraic group defined over $\mathbb{Q}$ and fix a maximal compact subgroup K of $\mathbb{G}(\mathbb{R})$. A compact subgroup $K_A = K\,K_f \subset \mathbb{G}(\mathbf{A})$ defines a locally symmetric space $S(K_f) = \mathbb{G}(\mathbb{Q})\backslash\mathbb{G}(\mathbf{A})/K\,K_f A_G$. We consider the "residual Eisenstein cohomology" of $S(K_f)$. Its classes can be represented by harmonic differential forms on $S(K_f)$, which are the residues of Eisenstein forms. Then we construct pseudo Eisenstein forms representing nontrivial cohomology classes with compact support in the Poincaré dual of the "residual Eisenstein cohomology". We use these classes to prove that Sp_4 defines a nontrivial modular symbol in $S(K_f)$ for $\mathbb{G} = \mathrm{GL}_4$. We also sketch the connection with the formula for the volume of a locally symmetric space.

I. Introduction

Let $\mathbb{G}$ be a semisimple algebraic group defined over $\mathbb{Q}$ and let Γ be a torsion-free arithmetic subgroup of $G = \mathbb{G}(\mathbb{R})$, K a maximal compact subgroup of G, and $X_\infty = G/K$ the symmetric space. Denote by $\Gamma\backslash X_\infty$ the associated locally symmetric space.

The de Rham cohomology $H^*(\Gamma\backslash X_\infty, \mathbb{C})$ is isomorphic to the relative Lie algebra cohomology

$$H^*(\mathfrak{g}, K, \mathcal{A}(\Gamma\backslash G)),$$

where $\mathcal{A}(\Gamma\backslash G)$ is the space of automorphic forms [**9**] and $\mathfrak{g} = \mathrm{Lie}(G)$. Consider the $(\mathfrak{g}, K)$–module $\mathcal{A}_{cusp}(\Gamma\backslash G)$ of cusp forms. By a result of A.Borel [**4**] there is an injection

$$H^*(\mathfrak{g}, K, \mathcal{A}_{cusp}(\Gamma\backslash G)) \hookrightarrow H^*(\mathfrak{g}, K, \mathcal{A}(\Gamma\backslash G))$$

and its image is called the cuspidal cohomology of Γ.

In this paper we assume that $\Gamma\backslash G$ is not compact (mod center) and consider the "residual cohomology"

$$H^*_{res}(\mathfrak{g}, K, \mathcal{A}(\Gamma\backslash G))$$

2010 *Mathematics Subject Classification.* Primary 11F75; Secondary 11F67, 11E55.
†partially supported by NSF grant DMS-0070561.

which is defined as follows: Consider the $(\mathfrak{g}, K)$–module of the residual automorphic functions $\mathcal{A}_{res}(\Gamma\backslash G)$. The inclusion

$$J_{res} : \mathcal{A}_{res}(\Gamma\backslash G) \hookrightarrow \mathcal{A}(\Gamma\backslash G)$$

defines a map

$$J^*_{res} : H^*(\mathfrak{g}, K, \mathcal{A}_{res}(\Gamma\backslash G)) \to H^*(\mathfrak{g}, K, \mathcal{A}(\Gamma\backslash G)),$$

which is not injective. Its image is denoted by $H^*_{res}(\mathfrak{g}, K, \mathcal{A}(\Gamma\backslash G))$.

The residual cohomology $H^0_{res}(\mathfrak{g}, K, \mathcal{A}(\Gamma\backslash G))$ is always nontrivial since the constant function representing the trivial representation is in the residual spectrum and represents a cohomology class in degree 0. Determining $J^*_{res}(H^*(\mathfrak{g}, K, \mathbb{C}))$ is equivalent to determining the nontrivial cohomology classes which are represented by invariant differential forms. This was solved by J.Franke in [**10**].

In general for irreducible unitary representations $A_\mathfrak{q}$ in the residual spectrum $J^*_{res}(H^*(\mathfrak{g}, K, A_\mathfrak{q}))$ is unknown. In this paper we show

THEOREM I.1. *Suppose that $A_\mathfrak{q}$ is a representation in the residual spectrum and let $r_\mathfrak{q}$ be the lowest degree so that*

$$H^{r_\mathfrak{q}}(\mathfrak{g}, K, A_\mathfrak{q}) \neq 0.$$

Suppose that $P = MAN$ is a parabolic subgroup of G, π a tempered irreducible representation of M and that $I(P, \pi, \nu)$ is a principal series representation with Langlands subquotient $A_\mathfrak{q}$. Suppose that we have a residual Eisenstein intertwining operator

$$E_{res} : I(P, \pi, \nu) \to A_\mathfrak{q} \subset \mathcal{A}_{res}(\Gamma\backslash G).$$

Then

$$J^{r_\mathfrak{q}}_{res}(H^{r_\mathfrak{q}}(\mathfrak{g}, K, A_\mathfrak{q}))$$

is a nontrivial class in

$$H^{r_\mathfrak{q}}_{res}(\mathfrak{g}, K, \mathcal{A}(\Gamma\backslash G)).$$

For all presently known subrepresentations of the residual spectrum the assumptions of the theorem are satisfied, but since not all representations π in the cuspidal spectrum are tempered, these assumptions might not be always satisfied.

Denoting the cohomology with compact support by $H^*_c(\Gamma\backslash X_\infty, \mathbb{C})$ and using Poincaré duality and the ideas and techniques introduced in [**25**] we deduce

THEOREM I.2. *Under the assumptions of the theorem*

$$H_c^{\dim X_\infty - r_\mathfrak{q}}(\Gamma\backslash X_\infty, \mathbb{C}) \neq 0.$$

The nontrivial classes in $H_c^{\dim X_\infty - r_\mathfrak{q}}(\Gamma\backslash X_\infty, \mathbb{C})$ are represented by differential forms $E(\tilde{\omega}_{\mathfrak{q},\mu_o})$ with compact support given as pseudo Eisenstein forms in section IV.

We use these results to describe some residual cohomology classes of $S(K_f) = GL_n(\mathbb{Q})\backslash GL_n(\mathbf{A})/K\,K_f A_G$, (see also 5.6 in [**11**]) . Here $\mathbf{A}$ are the adéles of $\mathbb{Q}$, A_G the connected component of the identity of the scalar matrices and K_f is an open compact subgroup of the finite adéles $\mathbf{A}_f$.

THEOREM I.3. *Suppose that $\mathbb{G} = GL_n$ and that $n = r\,m$. Then for K_f small enough*

$$H^j(S(K_f), \mathbb{C}) \neq 0$$

if

(1) *r and m even and* $j = \frac{r(r+1)m}{4} + \frac{r^2 m(m+2)}{2}$

(2) *r even and m odd and* $j = \frac{r(r+1)(m-1)}{4} + \frac{r^2(m-1)(1+m)}{2}$,

(3) *m=2 and* $j = r(r+1)/2$.

The second part of the paper is concerned with applications of theorem I.2 to modular symbols and to a formula for the volume of a symmetric space.

Suppose that $\mathbb{H} \subset \mathbb{G}$ is a $\mathbb{Q}$–rational reductive subgroup so that $K \cap H$ is maximal compact in $H := \mathbb{H}(\mathbb{R})$. Then the inclusion

$$H \to G$$

induces an inclusion

$$X_{H,\infty} := H/H \cap K \to X_\infty = G/K$$

and hence it induces a proper map

$$j : \Gamma \cap X_{H,\infty} \backslash X_{H,\infty} \to \Gamma \backslash X_\infty.$$

Assume that $\Gamma \backslash X_\infty$ is not compact and that $\Gamma \cap H \backslash X_{H,\infty}$ is an oriented noncompact manifold. If a closed d–form ω represents

$$[\omega] \in H_c^d(\Gamma \backslash X_\infty, \mathbb{C}), \quad d = \dim X_{H,\infty}$$

then

$$\int_{\Gamma \cap H \backslash X_{H,\infty}} j^*\omega$$

is defined. This means that the integral over $\Gamma \cap H \backslash X_{H,\infty}$ determines a map

$$[X_{\Gamma \cap H \backslash H}] : H_c^d(\Gamma \backslash X_\infty, \mathbb{C}) \to \mathbb{C},$$

which is called the *modular symbol* attached to $\mathbb{H}$. Using Poincaré duality we identify the modular symbol $[X_{\Gamma \cap H \backslash H}]$ with an element in $H^*(\Gamma \backslash X_\infty, \mathbb{C})$. If we can find a class $[\omega] \in H_c^*(\Gamma \backslash X_\infty, \mathbb{C})$ such that $[X_{\Gamma \cap H \backslash H}]([\omega]) \neq 0$ then $[X_{\Gamma \cap H \backslash H}]$ is a non-trivial modular symbol.

Ash, Ginzburg and Rallis list 6 families of pairs $(\mathbb{G}, \mathbb{H})$ and show that the restriction of $[X_H]$ to the cuspidal cohomology is zero. One of these pairs is $\mathbb{G} = GL_{2n}$ and $\mathbb{H} = Sp_{2n}$, $n \leq 2$.

We consider generalized modular symbols corresponding to $\mathbb{H} = Sp_4$ in GL_4 in theorem VI.3. and detect a non-trivial modular symbol by use of a pseudo Eisenstein series.

THEOREM I.4. *Suppose that $\mathbb{G} = GL_4$ and $\mathbb{H}$ is a symplectic group compatible with the choice of the maximal compact subgroup $K \subset GL_4(\mathbb{R})$. For K_f small enough*

$$[\mathbb{H}(\mathbb{Q}) \backslash \mathbb{H}(\mathbf{A})/(K \cap \mathbb{H}(\mathbb{R}))(K_f \cap \mathbb{H}(\mathbf{A}_f))]$$

is a nontrivial modular symbol in $H^3(\mathbb{G}(\mathbb{Q}) \backslash \mathbb{G}(\mathbf{A})/A_G K K_f, \mathbb{C})$.

In the arithmetic of modular curves and automorphic forms modular symbols provide a link between geometry and arithmetic. "Period integrals" of Eisenstein classes or cuspidal cohomology classes over compact modular symbols have been used by G.Harder to obtain information about special values of L-functions [**15**], [**16**]. We conjecture that the value of the modular symbol on this residual cohomology class is related to special values of Rankin convolutions of cusp forms.

In the last section we sketch the connection of our techniques with the formula for the volume of a locally symmetric space due to Langlands [**21**]. In the proof of theorem I.2 we have used a formula for the integral over a product of a pseudo Eisenstein form with an Eisenstein form. We show how this formula is related to the computation of the Tamagawa number $\tau(\mathbb{G})$ of $\mathbb{G}$. We sketch this only in the simplest case, i.e. if $\mathbb{G}/\mathbb{Q}$ is split and simply connected.

The results of this article will be used in a sequel to this paper to construct other nontrivial modular symbols.

The outline of the paper is as follows. We first introduce the notation and then we relate the $(\mathfrak{g}, K)$-cohomology of principal series representations to the $(\mathfrak{g}, K)$–cohomology of their unitary Langlands subrepresentations. We exhibit in section III. nontrivial "residual" cohomology classes of $S(K_f)$ and their harmonic representatives. In the next section we discuss the example GL_n. In section V. we construct a family of differential forms with compact support which represent classes in the Poincaré dual of the residual cohomology classes constructed in the previous section. In section VI. we apply these techniques to show that Sp_4 defines a nontrivial modular symbol in $S(K_f)$ for $\mathbb{G} = GL_4$. We sketch in the last section the connection of our techniques with the well known formula for the volume of $S(K_f)$ for a maximal compact subgroup K_f.

II. Representations with nontrivial $(\mathfrak{g}_0, K)$–cohomology

In this section we first introduce the notation. Then we prove some results relating the $(\mathfrak{g}, K)$-cohomology of principal series representations and their unitary Langlands subrepresentations. The results and techniques of this section are purely representation theoretic. They will be used in the later sections in the context of representations in the residual spectrum.

2.1 Let $\mathbb{G}$ be a reductive algebraic group over $\mathbb{Q}$, $G = \mathbb{G}(\mathbb{R})$, K a maximal compact subgroup of G, $\mathbb{A}_G$ be a maximal $\mathbb{Q}$–split torus in the center of $\mathbb{G}$ and let A_G be the connected component of $\mathbb{A}_G(\mathbb{R})$. This defines the symmetric space

$$X_\infty = G/K \cdot A_G.$$

For a rational subgroup $\mathbb{H}$ of $\mathbb{G}$ the Lie algebra of H will be denoted by $\mathfrak{h}$ and we put $\mathfrak{h}_{\mathbb{C}} = \mathfrak{h} \otimes \mathbb{C}$. Let $\mathfrak{g}_0$ be the Lie algebra of the real points of the intersection of the kernels of all rational characters of $\mathbb{G}$. Then $\mathfrak{g}/\mathfrak{a}_G \simeq \mathfrak{g}_0$. The Cartan decomposition is denoted by $\mathfrak{g}_0 = \mathfrak{k} \oplus \mathbf{p}$. Let θ be the corresponding Cartan involution.

We fix a minimal parabolic subgroup P_0 of G. Parabolic subgroups and their decompositions are denoted by $P = LN = MA_PN$. We denote the positive roots of $(\mathfrak{a}_P/\mathfrak{a}_G, \mathfrak{g}_0)$ defined by the parabolic subgroup P by Σ_P^+ and define

$$\rho_P := 1/2 \sum_{\beta \in \Sigma_P^+} \beta.$$

For a Cartan subalgebra $\mathfrak{h}_\mathbb{C}$ of $\mathfrak{g}_\mathbb{C}$ and positive roots compatible with the choice of P_0 we denote by ρ_G half the sum of the positive roots of $(\mathfrak{h}_\mathbb{C}, \mathfrak{g}_\mathbb{C})$.

2.2 We consider a standard representation

$$I(P, \pi_L, \mu_0) = \mathrm{ind}_{MA_PN}^G \pi_L \otimes e^{\mu_0 + \rho_P} \otimes 1$$

with nontrivial $(\mathfrak{g}, K)$–cohomology

$$H^*(\mathfrak{g}_0, K, I(P, \pi_L, \mu_0)) \neq 0.$$

Recall that μ_0 is the differential of a character of A and that π_L is tempered. We assume that $P = MA_PN$ is a standard parabolic subgroup and that $\Re(\mu_0) \in \mathfrak{a}_P^*$ is in the interior of the dominant Weyl chamber with respect to Σ_P^+, so that there is a unique maximal nontrivial subrepresentation $U(P, \pi_L, \mu_0)$ and Langlands subquotient

$$L(P, \pi_L, \mu_0) := I(P, \pi_L, \mu_0)/U(P, \pi_L, \mu_0)$$

(section IV in [**7**]). Furthermore we assume in this section that the Langlands subquotient $L(P, \pi_L, \mu_0))$ is unitary and that A_G^0 acts trivially.

This implies that the hermitian dual $I(P, \bar{\pi}_L^*, -\mu_0)$ of the representation $I(P, \pi_L, \mu_0))$ is isomorphic to $I(P, \pi_L^{w_0}, w_0(\mu_0))$ for an element $w_0 \neq 1$ with $w_0^2 = 1$ of the Weyl group $W_P = \mathrm{Norm}(A_P)/\mathrm{Cent}(A_P)$ of the parabolic subgroup P [**30**]. Here π^* is the contragredient representation of π. Note that π_L unitary implies that $\bar{\pi}_L^*$ is isomorphic to π_L.

Since $I(P, \pi_L, \mu_0)$ has nontrivial $(\mathfrak{g}, K)$– cohomology the infinitesimal characters of $I(P, \pi_L, \mu_0)$ and $I(P, \bar{\pi}_L^*, -\mu_0)$ are equal to ρ_G. Using the formulas in Borel-Wallach [3.3 in [**7**]] we conclude that there exists s in the Weyl group W of $(\mathfrak{h}_\mathbb{C}, \mathfrak{g}_\mathbb{C})$ so that

$$\mu_0 = -(s\,\rho_G)_{|\mathfrak{a}_P}$$

and there exists a finite dimensional irreducible representation E_s of M so that

$$\begin{aligned} &H^{l(s)+q}(\mathfrak{g}_0, K, I(P, \pi_L, \mu_0)) = \\ &= (H^*(\mathfrak{m}_0, K_M, \pi_L \otimes E_s) \otimes \wedge^*(\mathfrak{a}_P/\mathfrak{a}_G)^*)^q \\ &= \oplus_{r+k=q} H^r(\mathfrak{m}_0, K_M, \pi_L \otimes E_s) \otimes \wedge^k(\mathfrak{a}_P/\mathfrak{a}_G)^* \end{aligned}$$

where $l(s)$ is the length of s and $K_M = K \cap M$. In particular the highest weight of E_s is $(s(\rho_G) - \rho_G)_{|\mathbf{m}}$.

Notation: *For later reference we denote the lowest degree i for which*

$$H^i(\mathfrak{g}_0, K, I(P, \pi_L, \mu_0)) \neq 0$$

by $e(P, \pi_L, \mu_0)$. *If no confusion is possible we drop the subscript L.*

Poincaré duality implies that the $(\mathfrak{g}, K)$–cohomology of $I(P, \bar{\pi}_L^*, -(\mu_0))$ is nonzero and isomorphic as a vector space to

$$H^*(\mathfrak{m}_0, K_M, \bar{\pi}_L^* \otimes \bar{E}_s^*) \otimes \wedge(\mathfrak{a}_P/\mathfrak{a}_G)^*.$$

where E_s^* is the contragredient representation to E_s.

2.3 We obtain a representative of the one-dimensional space $H^{e(P,\pi,\mu)}(\mathfrak{g}_0, K, I(P,\pi,\mu))$ as follows:

Recall that $\mathbf{p}$ is the -1 eigenspace of the Cartan involution θ. Let ω_N be a harmonic form representing the lowest weight of the $\mathfrak{m}$-module E_s, let ω_M be the highest weight vector of a representation in $\bigwedge^*(\mathbf{p}\cap\mathfrak{m})$ and v a highest weight vector in the lowest K_M–type of π_L. Then $v \otimes \omega_N \otimes \omega_M^*$ represents a nontrivial map in

$$\mathrm{Hom}_{M\cap K}(\wedge^*(\mathbf{p}\cap\mathfrak{m}), \pi_L \otimes E_s) = H^*(\mathfrak{m}_0, K_M, \pi_L \otimes E_s).$$

Let

$$\psi : \mathfrak{n} \to \mathbf{p} \quad \text{defined by} \quad X \to X - \theta(X).$$

Then $\omega_M^* \wedge \psi(\omega_N)$ is the lowest weight vector of a representation of K. This representation of K is a K-type of $I(P,\pi,\mu)$ and thus it defines a nontrivial element in $\mathrm{Hom}_K(\bigwedge^{e(P,\pi,\mu_0)}\mathbf{p}, I(P,\pi,\mu))$. It represents the nontrivial $(\mathfrak{g}, K)$–cohomology class in degree $e(P,\pi,\mu_0)$.

For more details see [**27**] or [**13**]

2.4 For a connected semisimple Lie group with a maximal compact subgroup K the unitary $(\mathfrak{g}, K)$–modules with nontrivial $(\mathfrak{g}, K)$–cohomology have been constructed and classified by Parthasarathy, Vogan and Zuckerman [**31**]. They are parametrized by equivalence classes of θ–stable parabolic subalgebras $\mathfrak{q}$ of $\mathfrak{g}_{\mathbb{C}}$ and are denoted by $A_{\mathfrak{q}}$. For a more general description for disconnected groups G see [**20**].

By 2.2 there exists an intertwining operator

$$M(P,\pi_L,\mu_0,w) : I(P,\pi_L,\mu_0) \to I(P,\bar{\pi}_L,-\mu_0).$$

Since all unitary representations with infinitesimal character ρ_G have non-trivial $(\mathfrak{g}, K)$–cohomology there exists a θ–stable parabolic subalgebra $\mathfrak{q}$ and the unique irreducible subrepresentation $L(P,\bar{\pi}_L,\mu_0)$ of $I(P,\bar{\pi}_L,-\mu_0)$ is isomorphic to a representation $A_{\mathfrak{q}}$ for a θ–stable parabolic $\mathfrak{q}$ of $\mathfrak{g}_0 \otimes \mathbb{C}$ [**26**]. Here we extend the $(\mathfrak{g}_0, K)$–module $A_{\mathfrak{q}}$ to a $(\mathfrak{g}, K)$–module by $\mathfrak{a}_G$ acting trivially on $A_{\mathfrak{q}}$.

If $\mathfrak{q} = \mathfrak{l}_{\mathfrak{q}} \oplus \mathfrak{n}_{\mathfrak{q}}$ is the Levi decomposition of $\mathfrak{q}$ we define $r_{\mathfrak{q}} = \dim \mathfrak{n}_{\mathfrak{q}} \cap \mathbf{p}$. Then

$$\begin{aligned} H^i(\mathfrak{g}_0, K, A_{\mathfrak{q}}) &= 0 \quad \text{if } i < r_{\mathfrak{q}} \\ &= \mathbb{C} \quad \text{if } i = r_{\mathfrak{q}} \end{aligned}$$

Furthermore, since $A_{\mathfrak{q}}$ is unitary and irreducible

$$H^*(\mathfrak{g}_0, K, A_{\mathfrak{q}}) = \mathrm{Hom}_K(\wedge^*\mathbf{p}, A_{\mathfrak{q}}).$$

Denote by $\mathfrak{t}_K$ the intersection of a fundamental θ–stable Cartan subalgebra of $\mathfrak{g}$ contained in $\mathfrak{l}$ with $\mathfrak{k}$.. The minimal K–type $F_{\mathfrak{q}}$ of $A_{\mathfrak{q}}$ has highest weight $\lambda_{\mathfrak{q}}$, the sum of the weights of $\mathfrak{t}_K$ on $\mathfrak{n}_{\mathfrak{q}} \cap \mathbf{p}$. See [**31**].

PROPOSITION II.1. *Suppose that $A_{\mathfrak{q}}$ is the Langlands subrepresentation of $I(P,\bar{\pi}_L,-\mu_0)$. Then*

$$H^{e(P,\pi,-\mu_0)}(\mathfrak{g}_0, K, A_{\mathfrak{q}}) \neq 0.$$

Furthermore the inclusion of $A_{\mathfrak{q}}$ into $I(P,\bar{\pi}_L,-\mu_0)$ defines an injective map of $H^{e(P,\pi,-\mu_0)}(\mathfrak{g}_0, K, A_{\mathfrak{q}})$ into $H^{e(P,\pi,-\mu_0)}(\mathfrak{g}_0, K, I(P,\bar{\pi}_L,-\mu_0))$.

Proof: The formulas on page 82 of [**31**] show that a K–type of $\wedge^{e(P,\pi,-\mu_0)}\mathbf{p}$ which represents the nontrivial class in

$$\mathrm{Hom}_K(\wedge^{e(P,\pi,-\mu_0)}\mathbf{p}, I(P,\bar{\pi}_L,-\mu_0))$$

has extremal weight $\lambda_{\mathfrak{q}}$ and hence is equal to the K–type $F_{\mathfrak{q}}$, which also occurs in the subrepresentation $A_{\mathfrak{q}}$. This K–type has multiplicity one in $I(P,\pi_L,-\mu_0)$ and hence the map $\mathrm{Hom}_K(F_{\mathfrak{q}}, I(P,\bar{\pi}_L,-\mu_0))$ factors over $A_{\mathfrak{q}}$. □

To prove that

$$\begin{aligned} H^i(\mathfrak{g}_0, K, A_{\mathfrak{q}}) &= 0 \quad \text{if } i < e(P,\pi,-\mu_0) \\ &= \mathbb{C} \quad \text{if } i = e(P,\pi,-\mu_0) \end{aligned}$$

i.e. that $r_{\mathfrak{q}} = e(P,\pi,-\mu_0)$ we review some results from exterior algebra. See Chapter I in [**8**].

Let V be a finite dimensional vector space. A subring I of $\wedge^*(V^*)$ is an ideal if
a.) $\alpha \in I$ implies $\alpha \wedge \beta \in I$ for all $\beta \in \wedge^*(V^*)$
b.) $\alpha \in I$ implies that all its homogeneous components in $\wedge^*(V^*)$ are contained in I.

Given an ideal I of $\wedge^*(V^*)$ its retracting space is the smallest subspace $W^* \subset V^*$ so that I is generated as an ideal by a set S of elements of $\wedge^*(W^*)$, i.e. an element of I is a sum of elements of the form $\sigma \wedge \alpha$ with $\sigma \in S$ and $\alpha \in \wedge^*(V^*)$.

Let $W^* \subset V^*$ be a subspace of dimension r and let I_W be the ideal generated by $\wedge^*(W^*)$. Then by 1.3 of [**8**] W^* is also the retracting space of I_W.

THEOREM II.2. *Suppose that $A_{\mathfrak{q}}$ is the Langlands subrepresentation of the standard representation $I(P,\bar{\pi}_L,-\mu_0)$. Then*

$$r_{\mathfrak{q}} = e(P,\pi,-\mu_0).$$

Proof: Let P^0 be the kernel of the absolute values of the rational characters of P. Since the symmetric space X_∞ has a covering $A_P \times P^0/K_M$ we have a vector space decomposition

$$\mathbf{p} = \mathfrak{a}_P \oplus \mathbf{p}_{\mathfrak{m}} \oplus \psi(\mathfrak{n}),$$

where $\mathfrak{m} = \mathfrak{k} \cap \mathfrak{m} \oplus \mathbf{p}_{\mathfrak{m}}$ is the corresponding Cartan decomposition of $\mathfrak{m}$. Consider the subspace $\mathfrak{a}_P \subset \mathbf{p}$ and the ideal $I(\mathfrak{a}_P)$ of $\wedge^*(\mathbf{p})$ generated by $\mathfrak{a}_P$. Note that the ideal $I(\mathfrak{a}_P)$ is a $U(\mathfrak{m})$-invariant subspace of $\wedge^*(\mathbf{p})$.

The differential forms representing the nontrivial $(\mathfrak{g}_0, K)$–cohomology classes of $I(P,\bar{\pi}_L,-\mu_0)$ have representatives in $\mathrm{Hom}_K(\wedge^*\mathbf{p}, I(P,\bar{\pi}_L,-\mu_0))$ and are determined by their values on the K–isotypic components of $\wedge^*(\mathbf{p})$. Consider the subrepresentation $F_{\mathfrak{q}}$ of $\wedge^*(\mathbf{p})$ in degree $e(P,\pi,-\mu_0)$. We showed that this representation determines an nontrivial $(\mathfrak{g}_0, K)$–cohomology class of $I(P,\bar{\pi}_L,-\mu_0)$ in degree $e(P,\pi,-\mu_0)$.

Considering a realization of $I(P,\bar{\pi}_L,-\mu_0)$ in the functions on G we consider $I(P,\bar{\pi}_L,-\mu_0) \otimes \wedge^*\mathbf{p}^*$ as differential forms on X. The formula in 2.3 shows together with the results in [**13**], [**27**] that we can find a harmonic representative of the nontrivial cohomology class of $I(P,\bar{\pi}_L,-\mu_0)$ in degree $e(P,\pi,-\mu_0)$. This representative is not in $I(P,\bar{\pi}_L,-\mu_0) \otimes I(\mathfrak{a}_P)^*$ and none of its homogeneous components are in $I(P,\bar{\pi}_L,-\mu_0) \otimes I(\mathfrak{a}_P)^*$.

Since $A_{\mathfrak{q}}$ is unitary and irreducible

$$H^*(\mathfrak{g}_0, K, A_{\mathfrak{q}}) = \mathrm{Hom}_K(\wedge^* \mathbf{p}, A_{\mathfrak{q}}).$$

Now $F_{\mathfrak{q}}$ is the only subrepresentation of $\wedge^* \mathbf{p}$ which is also a K–type of $A_{\mathfrak{q}}$ [**31**]. Write $\mathfrak{q} = \mathfrak{l}_{\mathfrak{q}} \oplus \mathfrak{n}_{\mathfrak{q}}$ for the Levi decomposition. D.Vogan and G.Zuckerman show in [**31**] that the highest weight vectors of the subrepresentations $F_{\mathfrak{q}} \subset \wedge^*(\mathbf{p})$ are of the form $\omega_L \wedge V_{r_{\mathfrak{q}}}$ where ω_L is an L–invariant differential form in $\wedge^*(\mathbf{p} \cap \mathfrak{l}_{\mathfrak{q}})$ and $V_{r_{\mathfrak{q}}}$ is the representation with highest weight $2\rho(\mathfrak{n}_{\mathfrak{q}} \cap \mathbf{p})$. These forms are harmonic representatives of the cohomology classes.

Using that $A_{\mathfrak{q}} \hookrightarrow I(P, \bar{\pi}_L, -\mu_0)$ we can conclude that these forms are also harmonic forms on X and in degree $e(P, \pi, -\mu_0)$ there is also a harmonic form representing a nontrivial class of $I(P, \bar{\pi}_L, -\mu_0)$.

Now A_P is the maximal $\mathbb{R}$–split torus in $L_{\mathfrak{q}}$ [**31**], and so the only harmonic representative for the nontrivial $(\mathfrak{g}_0, K)$–cohomology classes of $A_{\mathfrak{q}}$ which may not have a homogeneous component in $I(\mathfrak{a}_P^*)$ is in degree $r_{\mathfrak{q}}$.

Thus the previous proposition implies that $r_{\mathfrak{q}} = e(P, \pi, -\mu_0)$. □

III. Residual Eisenstein classes

In this section residual Eisenstein forms and classes are introduced. The main reference for Eisenstein series and their residues is the book by C.Mœglin and Waldspurger [23]. We also freely use their terminology.

3.1 Let $\mathbf{A}_f \subset \mathbf{A}$ be the finite adéles of the adéle ring $\mathbf{A}$ over $\mathbb{Q}$. We give $\mathbb{G}(\mathbf{A}_f)$ the topology induced by the topology of $\mathbf{A}_f$ and define the global symmetric space $X = X_\infty \times \mathbb{G}(A_f)$. Then $\mathbb{G}(\mathbb{Q})$ acts on X and we obtain the adélic locally symmetric space $S = \mathbb{G}(\mathbb{Q}) \backslash X$. For a compact open subgroup K_f of $\mathbb{G}(\mathbf{A}_f)$ we consider the locally symmetric space

$$S(K_f) = \mathbb{G}(\mathbb{Q}) \backslash X / K_f.$$

We fix a minimal parabolic subgroup $\mathbb{P}_0$ of $\mathbb{G}$ defined over $\mathbb{Q}$ and a Levi subgroup $\mathbb{L}_0$ of $\mathbb{P}_0$ also defined over $\mathbb{Q}$.

3.2 Let $\mathbb{P}$ be a standard parabolic subgroup of $\mathbb{G}$ defined over $\mathbb{Q}$ with Levi decomposition $\mathbb{P} = \mathbb{U}_P \mathbb{L}_P$ and let A_P be the connected component of the maximally split torus in the center of $L_P := \mathbb{L}_P(\mathbb{R})$. Following Arthur we define the height function

$$H_P : \mathbb{L}_P(\mathbf{A}) \to A_P.$$

The kernel of H_P is denoted by $\mathbb{L}^1(\mathbf{A})$. We write $\mathbb{P}^1(\mathbf{A}) = \mathbb{U}_P(\mathbf{A}) \mathbb{L}^1_P(\mathbf{A})$.

A parameter $\mu \in \mathfrak{a}_P^*$ defines character χ_μ of $\mathbb{L}_P(\mathbf{A})$ by

$$\chi_\mu(l_{\mathbf{A}}) = e^{\mu(log\ H_P(l_{\mathbf{A}}))}, \quad l_{\mathbf{A}} \in \mathbb{L}_P(\mathbf{A}).$$

Let Σ_P^+ be the roots of A_P on $U_P = \mathbb{U}_P(\mathbb{R})$ and ρ_P half the sum of the positive roots. If no confusion is possible we omit the subscript P.

Suppose that $\mathbb{G}^\flat$ is a reductive subgroup of $\mathbb{G}$. Recall that an irreducible unitary representation $\pi_{\mathbf{A}}$ of a reductive group $\mathbb{G}^\flat(\mathbf{A})$ is called automorphic if it occurs discretely in $L^2(\mathbb{G}^\flat(\mathbb{Q})^\flat(\mathbf{A}), \xi)$ for a character ξ which is trivial on $\mathbb{A}_{G^\flat}(\mathbb{Q})$ and $A_{G^\flat}$.

We now fix a unitary character ξ of $\mathbb{A}_G(\mathbf{A})$ which is trivial on $\mathbb{A}_G(\mathbb{Q})$ and A_G.

Let $\pi_{\mathbb{L}(\mathbf{A})} = \prod_v \pi_v$ be an irreducible unitary automorphic representation of $\mathbb{L}(\mathbf{A})$ on V_π which transforms under $\mathbb{A}_G(\mathbf{A})$ by ξ. We define using normalized induction a representation

$$\begin{aligned} I(\mathbb{P}, \pi_{\mathbb{L}(\mathbf{A})}, \mu) &= \operatorname{ind}_{\mathbb{P}(\mathbf{A})}^{\mathbb{G}(\mathbf{A})} \chi_{\mu+\rho_P} \otimes \pi_{\mathbb{L}(\mathbf{A})} = \\ &= \prod_\nu \operatorname{ind}_{\mathbb{P}(\mathbb{Q}_\nu)}^{\mathbb{G}(\mathbb{Q}_\nu)} (\chi_{\mu+\rho_P})_\nu \otimes \pi_{\mathbb{L}(\mathbb{Q}_\nu)} \end{aligned}$$

This representation acts on the space of functions f on $\mathbb{G}(\mathbf{A})$ with values in V_π which satisfy

$$f(z_\mathbf{A} u_\mathbf{A} l_\mathbf{A} g_\mathbf{A}) = \xi(z_\mathbf{A}) \pi_{\mathbb{L}(\mathbf{A})}(l_\mathbf{A}) \chi_{\mu+\rho_P}(l_\mathbf{A}) f(g_\mathbf{A}).$$

Here $l_\mathbf{A} \in \mathbb{L}(\mathbf{A}),\ u_\mathbf{A} \in \mathbb{U}(\mathbf{A}),\ g_\mathbf{A} \in \mathbb{G}(\mathbf{A}), z_\mathbf{A} \in \mathbb{A}_G(\mathbf{A})$.

The automorphic representation

$$\pi_{\mathbb{L}(\mathbf{A})} = \prod_\nu \pi_{\mathbb{L}(\mathbb{Q}_\nu)} = \pi_L \prod_p \pi_{\mathbb{L}(\mathbb{Q}_p)}$$

is in the cuspidal spectrum of

$$\mathbb{L}(\mathbb{Q}) A_P \backslash \mathbb{L}(\mathbf{A}) / K K_f \cap \mathbb{L}(\mathbf{A})$$

if its factor π_L at the infinite places is tempered [**32**]. We will therefore assume this for the rest of the article.

3.3 Assume now as in the previous section that $\Re(\mu)$ is in the dominant Weyl chamber with respect to Σ_P^+. Let $W(L)$ be the set of elements w in the Weyl group of $\mathbb{G}$ of minimal length modulo the Weyl group of $\mathbb{L}$, and such that $w\mathbb{L}w^{-1}$ is also a standard Levi of $\mathbb{G}$ [23]. For every w in $W(L)$ there exists an intertwining operator

$$\mathbf{M}(\pi_{\mathbb{L}(\mathbf{A})}, \mu, w) : I(\mathbb{P}, \pi_{\mathbb{L}(\mathbf{A})}, \mu) \to I(\mathbb{P}, \pi^w_{\mathbb{L}(\mathbf{A})}, w(\mu)).$$

Choosing an isomorphism

$$A_w : V_\pi \to V_{\pi^w} = V_\pi$$

we identify it with an intertwining operator $M_\mathbf{A}(\pi_{\mathbb{L}(\mathbf{A})}, \mu, w)$, which has for $\Re(\mu)$ large an Euler product

$$M_\mathbf{A}(\pi_{\mathbb{L}(\mathbf{A})}, \mu, w) = \prod_\nu M(\pi_{\mathbb{L}(\mathbf{Q}_\nu)}, \mu, w).$$

(see II.1.9 in [**23**] for details).

There exists a meromorphic function $m(\pi_{\mathbb{L}(\mathbb{Q}_\nu)}, \mu, w)$ so that the local intertwining operator

$$\mathcal{M}(\pi_{\mathbb{L}(\mathbb{Q}_\nu)}, \mu, w) = m(\pi_{\mathbb{L}(\mathbb{Q}_\nu)}, \mu, w) M(\pi_{\mathbb{L}(\mathbb{Q}_\nu)}, \mu, w)$$

is holomorphic and nonzero for μ in the dominant Weyl chamber. See [**1**]. There is also a meromorphic function $m(\pi_{\mathbb{L}(\mathbf{A})}, \mu, w)$, which for large dominant μ is the product of the local factors, so that

$$\mathcal{M}(\pi_{\mathbb{L}(\mathbf{A})}, \mu, w) = m(\pi_{\mathbb{L}(\mathbf{A})}, \mu, w) M_\mathbf{A}(\pi_{\mathbb{L}(\mathbf{A})}, \mu, w)$$

is holomorphic and nonzero for μ in the dominant Weyl chamber.

We say that the operator $M(\pi_{\mathbb{L}(\mathbf{A})}, \mu, w)$ has a pole of order r at μ_0 if the function $m(\pi_{\mathbb{L}(\mathbf{A})}, \mu, w_0)$ has a zero at $\mu = \mu_0$ of order r.

We fix the level K_f and are interested in automorphic forms on $\mathbb{G}(\mathbb{Q})\backslash\mathbb{G}(\mathbf{A})/A_G K_f$. Thus from now on we consider representations $I(\mathbb{P}, \pi_{\mathbb{L}(\mathbf{A}_f)}, \mu)$ with a K_f-invariant vector, i.e. $I(\mathbb{P}, \pi_{\mathbb{L}(\mathbf{A}_f)}, \mu)^{K_f} \neq 0$.

3.4 By our assumptions the representation $\pi_{\mathbb{L}(\mathbf{A})}$ is a subrepresentation of the cuspidal spectrum of $\mathbb{L}(\mathbb{Q})A_L\backslash\mathbb{L}(\mathbf{A})$; hence we consider $I(\mathbb{P}, \pi_{\mathbb{L}(\mathbf{A})}, \mu)$ as realized in the functions on $\mathbb{P}(\mathbb{Q})\mathbb{U}(\mathbf{A})\backslash\mathbb{G}(\mathbf{A})$. A form

$$\eta_{\pi_{\mathbb{L}(\mathbf{A})},\mu} \in \wedge^i \mathbf{p}^* \otimes_K I(\mathbb{P}, \pi_{\mathbb{L}(\mathbf{A})}, \mu)^{K_f} \cong \mathrm{Hom}_K(\wedge^i \mathbf{p}, I(\mathbb{P}, \pi_{\mathbb{L}(\mathbf{A})}, \mu))^{K_f}$$

defines a differential form on $\mathbb{P}(\mathbb{Q})\mathbb{U}(\mathbf{A})\backslash\mathbb{G}(\mathbf{A})/KK_f$. For $\Re(\mu)$ large and dominant we define following [**13**] the Eisenstein differential form

$$E(\eta_{\pi_{\mathbb{L}(\mathbf{A})},\mu}) = \sum_{g^{-1}\in\mathbb{P}(\mathbb{Q})\backslash\mathbb{G}(\mathbb{Q})} g^* \eta_{\pi_{\mathbb{L}(\mathbf{A})},\mu}$$

on $S(K_f)$.

The constant term $E(\eta_{\pi_{\mathbb{L}(\mathbf{A})},\mu})^P$ of the form $E(\eta_{\pi_{\mathbb{L}(\mathbf{A})},\mu})$ with respect to the parabolic subgroup $\mathbb{P} = \mathbb{L}\mathbb{U}$ is equal to

$$\begin{aligned} E(\eta_{\pi_{\mathbb{L}(\mathbf{A})},\mu})^P &= \sum_{w\in W(L)} M(\pi_{\mathbb{L}(\mathbf{A})}, \mu, w)\eta_{\pi_{\mathbb{L}(\mathbf{A})},\mu} \\ &= \sum_{w\in W(L)} m(\pi_{\mathbb{L}(\mathbf{A})}, \mu, w)^{-1}\mathcal{M}(\pi_{\mathbb{L}(\mathbf{A})}, \mu, w)\eta_{\pi_{\mathbb{L}(\mathbf{A})},\mu}. \end{aligned}$$

Recall that $W(L)$ is the set of elements w in the Weyl group of G of minimal length modulo the Weyl group of L, and such that wLw^{-1} is also a standard Levi of G [23]. Furthermore

$$M(\pi_{\mathbb{L}(\mathbf{A})}, \mu, w)\eta_{\pi_{\mathbb{L}(\mathbf{A})},\mu} \in \wedge^i \mathbf{p}^* \otimes_K M(\pi_{\mathbb{L}(\mathbf{A})}, \mu, w) I(\mathbb{P}, \pi_{\mathbb{L}(\mathbf{A})}, \mu)^{K_f}.$$

If the constant term $E(\eta_{\pi_{\mathbb{L}(\mathbf{A})},\mu})^P$ has a pole of order $\dim(\mathfrak{a}_P)$ for $\mu = \mu_0$ dominant, then so does the Eisenstein form $E(\eta_{\pi_{\mathbb{L}(\mathbf{A})},\mu})$ [**23**]. We write $E(\eta^{res}_{\pi_{\mathbb{L}(\mathbf{A})},\mu_0})$ for the residue at $\mu = \mu_0$ and call this a residual Eisenstein form. The form

$$E(\eta^{res}_{\pi_{\mathbb{L}(\mathbf{A})},\mu_0})$$

is square integrable.

Let w_0 be the Weyl group element considered in 2.2. If the constant term $E(\eta_{\pi_{\mathbb{L}(\mathbf{A})},\mu})^P$ has a pole of order $\dim(\mathfrak{a}_P)$ for $\mu = \mu_0$ dominant, then $M(\pi_{L(\mathbf{A})}, \mu, w_0)$ has a pole at $\mu = \mu_0$ of order $\dim(\mathfrak{a}_P)$ [**23**]. The image of $\mathcal{M}(\pi_{L(\mathbf{Q_A})}, \mu_0, w_0)$ is a unitary representation which is isomorphic to a representation in the residual spectrum.

The constant term of the residual Eisenstein form $E(\eta^{res}_{\pi_{\mathbb{L}(\mathbf{A})},\mu_0})$ determines its growth at infinity. Since a residual Eisenstein form is square integrable, it is decaying at infinity and its constant term $E(\eta^{res}_{\pi_{\mathbb{L}(\mathbf{A})},\mu_0})^P$ has only the term

$$m(\pi_{\mathbb{L}(\mathbf{A})}, \mu_0, w_0)^{-1}\mathcal{M}(\pi_{\mathbb{L}(\mathbf{A})}, \mu_0, w_0)\eta_{\pi_{\mathbb{L}(\mathbf{A})},\mu_0}$$

for $w_0 \in W(L)$ as introduced in 2.2.

3.5 If in addition

$$H^*(\mathfrak{g}_0, K, I(\mathbb{P}, \pi_{\mathbb{L}(\mathbf{A})}, \mu_0)) \neq 0$$

then the factor $I(P,\pi_L,\mu_0)$ of $I(\mathbb{P},\pi_{\mathbb{L}(\mathbf{A})},\mu_0)$ at the infinite place satisfies the assumptions of the previous section and we conclude that in degree $r_{\mathfrak{q}} = e(P,\pi,-\mu_0)$ the form

$$E(\eta^{res}_{\pi_{\mathbb{L}(\mathbf{A})},\mu_0}) \in \mathrm{Hom}_K(\wedge^{r_{\mathfrak{q}}}\mathbf{p}, L^2_{res}(\mathbb{G}(\mathbb{Q})A_G\backslash\mathbb{G}(\mathbf{A})))_{A_{\mathfrak{q}}}$$

where $L^2_{res}(\mathbb{G}(\mathbb{Q})A_G\backslash\mathbb{G}(\mathbf{A}))_{A_{\mathfrak{q}}}$ is the isotypic component of type $A_{\mathfrak{q}}$, i.e.

$$L^2_{res}(\mathbb{G}(\mathbb{Q})A_G\backslash\mathbb{G}(\mathbf{A}))_{A_{\mathfrak{q}}} = \mathrm{Hom}_{(\mathfrak{g},K)}(A_{\mathfrak{q}}, L^2_{res}(\mathbb{G}(\mathbb{Q})A_G\backslash\mathbb{G}(\mathbf{A}))\otimes A_{\mathfrak{q}}).$$

The standard principal series representations $I(P,\pi_L,\mu_0)$ and $I(P,\bar\pi_L,-\mu_0)$ have $A_{\mathfrak{q}}$ as Langlands subquotient, respectively subrepresentation, and they have both the same K–types. Thus we have a nonzero closed form

$$\eta_{\mathfrak{q}} \in \wedge^{r_{\mathfrak{q}}}\mathbf{p}\otimes_K I(\mathbb{P},\pi_{\mathbb{L}(\mathbf{A})},\mu_0)$$

defined by the subrepresentation of $\wedge^{r_{\mathfrak{q}}}\mathbf{p}$ with highest weight

$$2\rho(\mathfrak{n}_{\mathfrak{q}}\cap\mathbf{p}_{\mathbb{C}}) = \sum_{\beta\in\Sigma(\mathfrak{n}_{\mathfrak{q}}\cap\mathbf{p}_{\mathbb{C}})}\beta,$$

where $\mathfrak{n}_{\mathfrak{q}}$ is the nilradical of the θ–stable parabolic subalgebra $\mathfrak{q}$.

The constant Fourier coefficient $E(\eta^{res}_{\mathfrak{q}})^P$is a nonzero multiple of $\mathcal{M}(\pi_{\mathbb{L}(\mathbf{A})},\mu_0,w_0)\eta_{\mathfrak{q}}$. We have the exact sequence

$$\begin{aligned}0 \to\ & \mathrm{Im}(\mathcal{M}(\pi_{\mathbb{L}(\mathbf{A})},\mu_0,w_0)) \to I(\mathbb{P},\pi_{\mathbb{L}(\mathbf{A})},-\mu_0)\\ &\to\ I(\mathbb{P},\pi_{\mathbb{L}(\mathbf{A})},-\mu_0)/\mathrm{Im}(\mathcal{M}(\pi_{\mathbb{L}(\mathbf{A})},\mu_0,w_0)) \to 0\end{aligned}$$

and by 2.2 it induces in degree $r_{\mathfrak{q}}$ the isomorphism

$$H^{r_{\mathfrak{q}}}(\mathfrak{g},K,\ \mathrm{Im}(\mathcal{M}(\pi_{\mathbb{L}(\mathbf{A})},\mu_0,w_0))) \quad = \quad H^{r_{\mathfrak{q}}}(\mathfrak{g},K,I(\mathbb{P},\pi_{\mathbb{L}(\mathbf{A})},-\mu_0)).$$

Hence

$$[\mathcal{M}(\pi_{\mathbb{L}(\mathbf{A})},\mu,w_0)\eta_{\mathfrak{q}}]$$

defines a nontrivial class in $H^{e(P,\pi,-\mu_0)}(\mathfrak{g}_0,K,I(\mathbb{P},\pi_{\mathbb{L}(\mathbf{A})},-\mu_0))$ and so does $[E(\eta^{res}_{\mathfrak{q}})^P]$.

3.6 To prove that there are nontrivial residual Eisenstein classes $[E(\eta^{res}_{\Pi_{L(\mathbf{A})},\mu_0})]$ in degree $r_{\mathfrak{q}}$ we follow the ideas of [**13**] and [**27**] and consider the restriction of a form to a face

$$e(\mathbb{P},K_f) = \mathbb{P}(\mathbb{Q})\backslash\mathbb{G}(\mathbf{A})/K\ K_fA_P$$

of the boundary of the Borel-Serre compactification. See [**6**] for details on the Borel-Serre compactification.

The cohomology of the face $e(\mathbb{P},K_f)$ is isomorphic to

$$\mathrm{ind}^{\mathbb{G}(\mathbf{A}_f)}_{\mathbb{P}(\mathbf{A}_f)} H^*(\mathbb{L}(\mathbb{Q})\backslash\mathbb{L}(\mathbf{A})/K_{\mathbb{L}(\mathbf{A})}A_P, H^*(\mathbf{n},\mathbb{C}))$$

where $K_{\mathbb{L}(\mathbf{A})} = (K\cap L)(K_{\mathbf{A}_f}\cap\mathbb{L}(\mathbf{A}_f)) = K_L\,K_{\mathbb{L}(\mathbf{A}_f)}$. We have

$$\begin{aligned}&H^*_{\mathrm{cusp}}(\mathbb{L}(\mathbb{Q})\backslash\mathbb{L}(\mathbf{A})/K_{\mathbb{L}(\mathbf{A})}A_P, H^*(\mathbf{n},\mathbb{C}))\\ &\qquad\subset H^*(\mathbb{L}(\mathbb{Q})\backslash\mathbb{L}(\mathbf{A})/K_{\mathbb{L}(\mathbf{A})}A_P, H^*(\mathbf{n},\mathbb{C}))\end{aligned}$$

and

$$H^*_{\mathrm{cusp}}(\mathbb{L}(\mathbb{Q})\backslash\mathbb{L}(\mathbf{A})/K_{\mathbb{L}(\mathbf{A})}A_P, H^*(\mathbf{n},\mathbb{C}))$$

is equal to the direct sum of

$$\mathrm{Hom}_{\mathbb{L}(\mathbf{A})}(\Pi_{\mathbb{L}(\mathbf{A})}, L^2_{\mathrm{cusp}}(\mathbb{L}(\mathbb{Q})\backslash\mathbb{L}(\mathbf{A})/K_{\mathbb{L}(\mathbf{A})}A_P)\otimes\mathrm{Hom}_{K_L}(\wedge^*\mathbf{p}_{\mathfrak{m}}, \Pi_{\mathbb{L}(\mathbf{A})}\otimes H^*(\mathfrak{n},\mathbb{C}))),$$

where $\mathfrak{p}_{\mathfrak{m}} = \mathfrak{p} \cap \mathfrak{m}$.

Since $\pi_{\mathbb{L}(\mathbf{A})}$ is a cuspidal representation, $[E(\eta_{\mathfrak{q}}^{res})^P]$ defines by section II a nontrivial class in

$$\operatorname{ind}_{\mathbb{P}(\mathbf{A}_f)}^{\mathbb{G}(\mathbf{A}_f)} H^*(\mathfrak{m}, K_M, \pi_{\mathbb{L}(\mathbf{A})} \otimes H^*(\mathfrak{n}, \mathbb{C})) \otimes \wedge^0 \mathfrak{a}_P.$$

But $H^*(\mathfrak{m}, K_M, \pi_{\mathbb{L}(\mathbf{A})} \otimes H^*(\mathfrak{n}, \mathbb{C})) \otimes \wedge^0 \mathfrak{a}_P$ can be considered as a subspace of

$$H^*_{\text{cusp}}(\mathbb{L}(\mathbb{Q})\backslash\mathbb{L}(\mathbf{A})/K_{\mathbb{L}(\mathbf{A})}A_P, H^*(\mathfrak{n}, \mathbb{C})).$$

Hence $[E(\eta_{\mathfrak{q}}^{res})^P]$ can be considered as a class in $H^*_{cusp}(e(\mathbb{P}, K_f))$.

In [**13**] it is proved that computing the restriction of an Eisenstein form to the face $e(\mathbb{P}, K_f)$ corresponds to taking the constant Fourier coefficient. Thus we proved

THEOREM III.1. *Under our assumptions $[E(\eta_{\mathfrak{q}}^{res})]$ is a nontrivial cohomology class of $S(K_f)$ in degree $e(P, \pi, -\mu_0) = r_{\mathfrak{q}}$.*

IV. An example: Residual cohomology classes for $\mathbf{GL}_n$

Let $\mathbf{A}$ be the adéles of $\mathbb{Q}$. We use the description of the residual spectrum of the general linear group by C. Mœglin and J.L.Waldspurger and the nonvanishing results of the cuspidal cohomology of [**6**] to prove nonvanishing results for the residual cohomology of the general linear group.

4.1 In [**3**] it is proved that there are cuspidal representations $\pi_{\mathbf{A}}$ of $GL_n(\mathbf{A})$ which are invariant under the Cartan involution, i.e. are self-dual and which satisfy

$$H^*(\mathfrak{g}_0, K, \pi_{\mathbf{A}} \otimes F_\lambda) \neq 0$$

for a finite dimensional representation F_λ with highest weight λ provided that $\langle \lambda, \alpha \rangle \in 2\mathbb{N}^+$. If $n = 2$ it is known from the classical theory of automorphic forms that for all finite dimensional representations F_λ of $GL_2(\mathbb{R})$ there exist cuspidal automorphic representations with

$$H^*(\mathfrak{g}_0, K, \pi_{\mathbf{A}} \otimes F_\lambda) \neq 0.$$

C. Mœglin and J.L. Waldspurger proved that the representations in the residual spectrum of $GL_n(\mathbb{Q})A_G\backslash GL_n(\mathbf{A})$ are obtained as follows [**24**]: A partition $(m, m, \ldots, m)$ of $n = r\,m$ defines a parabolic subgroup $\mathbb{P}_m$. Then

$$\mathbb{L}_m(\mathbf{A}) = \prod GL_m(\mathbf{A}).$$

Let $\pi_{\mathbf{A},m}$ be a representation in the cuspidal spectrum of $GL_m(\mathbb{Q})A_{G_m}\backslash GL_m(\mathbf{A})$. Consider the standard representation

$$I(\mathbb{P}_m, \prod \pi_{\mathbf{A},m}, \mu) = \operatorname{ind}_{\mathbb{P}_m(\mathbf{A})}^{GL_n(\mathbf{A})} \chi_{\mu+\rho_P} \otimes \prod \pi_{\mathbf{A},m}$$

for μ dominant with respect to Σ_{P_m}. If $\langle \alpha, \mu_0 \rangle = 1$ for all simple roots of in Σ_{P_0} the representation $I(\mathbb{P}_m, \prod \pi_{\mathbf{A},m}, \mu_0)$ is reducible and its Langlands quotient $L(\mathbb{P}_m, \prod \pi_{\mathbf{A},m}, \mu_0)$ is in the residual spectrum. Then $L(\mathbb{P}_1, \prod \pi_{\mathbf{A},1}, \mu_0)$ is one-dimensional and $L(\mathbb{P}_2, \prod \pi_{\mathbf{A},2}, \mu_0)$ is a Speh representation discussed in [**28**].

4.2 Now suppose that $I(\mathbb{P}_m, \prod \pi_{\infty,m}, \mu_0)$ has nontrivial $(\mathfrak{g}, K)$-cohomology. Then $I(\mathbb{P}_m, \prod \pi_{\infty,m}, \mu_0)$ has infinitesimal character ρ. This determines the infinitesimal character of $\pi_{\infty,m}$ [**7**]. Since $\pi_{\infty,m}$ is tempered there exists exactly one representation of $GL(m,\mathbb{R})$ with this property. A computation shows that if r is even then the infinitesimal character of $\pi_{\infty,m}$ satisfies the condition of [**3**] and hence there exists a cuspidal representation $\pi_{\mathbf{A},m}$ with $\pi_{\infty,m}$ as factor at the infinite place. If $m=2$ then such representations exist for all r.

THEOREM IV.1. *Suppose that r or m is even. Then there are representations $L(\mathbb{P}_m,\prod\pi_{\mathbf{A},m},\mu_0)$ with nontrivial $(\mathfrak{g},K)$-cohomology in the residual spectrum.*

4.3 Since representations are tempered and hence induced, a straightforward computation shows

LEMMA IV.2. *Suppose that*

$$L(\mathbb{P}_m,\prod\pi_{\mathbf{A},m},\mu_0)=L(\mathbb{P}_m,\prod\pi_{\infty,m},\mu_0)L(\mathbb{P}_m,\prod\pi_{\mathbf{A}_f,m},\mu_0))$$

is in the residual spectrum and that

$$H^*(\mathfrak{g}_0,K,L(\mathbb{P}_m,\prod\pi_{\infty,m},\mu_0))\neq 0.$$

If m is even, the Langlands subquotient $L(\mathbb{P}_m,\prod\pi_m,\mu_0)$ is unitarily induced from a parabolic subgroup $\mathbb{P}_{2r}$ and Speh representations on each factor of $\mathbb{L}(\mathbf{A})$. If m is odd then it is unitarily induced from a parabolic for a partition $(2r,2r,\dots,2r,r)$ On each of the factors of the Levi subgroup isomorphic to GL_{2r} we induce from a Speh representation and from the trivial representation on the factor isomorphic to GL_r.

Using the formula 4.2.1 in [**28**] we conclude

LEMMA IV.3. *We keep the assumptions of the previous lemma. If m is even then*

$$\begin{aligned}H^i(\mathfrak{g}_0,K,L(\mathbb{P}_m,\prod\pi_{\infty,m},\mu_0)) &=\mathbb{C} \quad \textit{if } i=\frac{r(r-1)m}{4}+\frac{r^2m(m+1)}{2}\\ &=0 \quad \textit{if } i<\frac{r(r-1)m}{4}+\frac{r^2m(m+1)}{2}.\end{aligned}$$

If m is odd then

$$\begin{aligned}&H^i(\mathfrak{g}_0,K,L(\mathbb{P}_m,\prod\pi_{\infty,m},\mu_0))=\\ &\quad=\mathbb{C} \quad \textit{if } i=\frac{r(r+1)(m-1)}{4}+\frac{r^2(m-1)(1+m)}{2}\\ &\quad=0 \quad \textit{if } i<\frac{r(r+1)(m-1)}{4}+\frac{r^2(m-1)(1+m)}{2}.\end{aligned}$$

Now II.2 implies

THEOREM IV.4. *Suppose that $\mathbb{G}=GL_n$ and that $n=r\,m$. Then for K_f small enough*

$$H^j(S(K_f),\mathbb{C})\neq 0$$

if

(1) *r and m even and* $j = \frac{r(r+1)m}{4} + \frac{r^2 m(m+2)}{2}$

(2) *r even and m odd and* $j = \frac{r(r+1)(m-1)}{4} + \frac{r^2(m-1)(1+m)}{2}$,

(3) $m=2$ *and* $j = r(r+1)/2$.

Remark: All the residual cohomology classes in this theorem are in degrees below $(1/2) \dim X_\infty$.

V. Some classes with compact support

If K_f is small enough $S(K_f)$ is orientable. We fix an orientation and have Poincaré duality, i.e. a nondegenerate pairing of $H^*(S(K_f), \mathbb{C})$ and $H^*_c(S(K_f), \mathbb{C})$. In this section we first recall the construction of closed pseudo Eisenstein forms with compact support [**25**]. Then we compute the integral of their cup product with residual Eisenstein forms to show, that the residual Eisenstein forms represent non-trivial classes in $H^*_c(S(K_f), \mathbb{C})$.

5.1 Suppose that $\rho : G \to \mathrm{End}(V)$ is a finite dimensional representation of G. Choose an admissible scalar product on $\mathfrak{g}$. The corresponding volume form is denoted by

$$dx \in \Omega^d(X, \mathbb{C}), \quad d = \dim X_\infty.$$

Fix an admissible inner product on V (II.2 in [**7**].) Then there is a hermitian scalar product $\langle\ \rangle$ on $\wedge^i(\mathfrak{g}_0/\mathfrak{k})^* \otimes V$ where the superscript indicates the dual space, There is an induced pointwise scalar product on $\Omega^i(X, \tilde{V})$ and on $\Omega^i(S(K_f), \tilde{V})$ and we obtain a map

$$A : \Omega^i(X, \tilde{V}) \times \Omega^{d-i}(X, \tilde{V}^*) \to \Omega^d(X, \mathbb{C})$$

where V^* is the contragredient representation of V. Let $\bar{V}^*$ be the complex conjugate. Then there is a map

$$* : \Omega^i(X, \tilde{V}) \to \Omega^i(X, \tilde{\bar{V}}^*)$$

characterized by the equation

$$A(\omega_1, *\omega_2) = \langle \omega_1, \omega_2 \rangle dx$$

for $\omega_1, \omega_2 \in \Omega^i(X, \tilde{V})$ and similarly for forms on $S(K_f)$ and on $\mathbb{P}(\mathbb{Q})\backslash X/KK_f$.

5.2 Suppose that $\pi_{\mathbb{L}(\mathbf{A})}$ is a cuspidal representation with $\pi_{\mathbb{L}(\mathbf{A})}^{K_f \cap \mathbb{L}(\mathbf{A}_f)} \neq 0$ and that

$$H^*(\mathfrak{m}_0, K \cap L, \pi_{\mathbb{L}(\mathbf{A})} \otimes F_L) \neq 0$$

for an irreducible finite dimensional representation F_L. Let r_π be the lowest degree in which the $(\mathfrak{m}, K \cap L)$–cohomology of $\pi_{\mathbb{L}(\mathbf{A})}$ is non-trivial. Consider a non-trivial cohomology class $[\omega_\pi]$ in degree r_{π_L}. Since $\pi_{\mathbb{L}(\mathbf{A})}$ is a cuspidal representation with a $K_f \cap \mathbb{L}(\mathbf{A}_f)$–fixed vector we consider ω_π as a differential form representing a nontrivial class in

$$\mathrm{ind}_{\mathbb{P}(\mathbf{A}_f)}^{\mathbb{G}(\mathbf{A}_f)} H^{r_\pi}_{cusp}(\mathbb{L}(\mathbb{Q})\backslash \mathbb{L}(\mathbf{A})/A_L\,(K\,K_f \cap \mathbb{L}(\mathbf{A})),\ F_L).$$

We fix an admissible inner product on F_L and a volume form dx_L. Then $*\omega_\pi$ is a differential form with compact support representing a class in

$$\mathrm{ind}_{\mathbb{P}(\mathbf{A}_f)}^{\mathbb{G}(\mathbf{A}_f)} H_{cusp}^{d_L-r_\pi}(\mathbb{L}(\mathbb{Q})\backslash\mathbb{L}(\mathbf{A})/A_L\,(K\,K_f\cap\mathbb{L}(\mathbf{A})),\bar{F_L}^*)$$

dual to it with respect to Poincaré duality. Here d_L is the dimension of the symmetric space of L.

On the other hand, following I.1.4 and I.7.1 in [**7**] we can use representation theory to define the Poincaré dual

$$*_{rep}\omega_\pi \in \mathrm{ind}_{\mathbb{P}(\mathbf{A}_f)}^{\mathbb{G}(\mathbf{A}_f)} H^{d_L-r_\pi}(\mathfrak{m}_0, K\cap L, \bar{\pi}^*_{L(\mathbf{A})}\otimes \bar{F}_L^*)$$

of the form

$$\omega_\pi \in \mathrm{ind}_{\mathbb{P}(\mathbf{A}_f)}^{\mathbb{G}(\mathbf{A}_f)} H^{r_\pi}(\mathfrak{m}_0, K_\infty\cap L, \pi_{L(\mathbf{A})}\otimes F_L).$$

Since $\pi_{L(\mathbf{A})}$ is a unitary representation, $\bar{\pi}^*_{L(\mathbf{A})}\cong \pi_{L(\mathbf{A})}$ and thus we have $\bar{F_L}^*\cong F_L$. The Poincaré dual of a cuspidal class is again a cuspidal class and the results in section II of [**7**] show that $[*_{rep}\omega_\pi]$ and $[*\omega_\pi]$ represent the same cohomology class in

$$\mathrm{ind}_{\mathbb{P}(\mathbf{A}_f)}^{\mathbb{G}(\mathbf{A}_f)} H_{cusp}^{d_L-r_\pi}(\mathbb{L}(\mathbb{Q})\backslash\mathbb{L}(\mathbf{A})/A_L(K\,K_f\cap\mathbb{L}(\mathbf{A})),F_L).$$

5.3 We write $A_P = A_G A_L$ where $\mathfrak{a}_L\subset\mathfrak{g}_0$. Let ω_A be a differential form of compact support on A_L in the top degree so that

$$0\neq[\omega_A]\in H_c^{\dim\ A_L}(A_L)$$

and normalized so that $\int_{A_L}\omega_A = 1$.

We use the notation and assumptions of section III. Suppose that $[E(\eta_{\mathfrak{q}}^{\mathrm{res}})]$ is a nontrivial cohomology class of $S(K_f)$ in degree

$$e(P,\pi,-\mu_0) = r_{\mathfrak{q}}\,.$$

Then $r_{\mathfrak{q}} = r_\pi + r_N$ and there exists an irreducible cuspidal representation $\pi_{\mathbb{L}(\mathbf{A})}$ and a finite dimensional subrepresentation $F_L\subset H^{r_N}(\mathfrak{n},\mathbb{C})$ of L so that

$$H^{r_\pi}(\mathfrak{m}_0, K\cap L, \pi_{\mathbb{L}(\mathbf{A})}\otimes F_L)\neq 0.$$

We write ω_π for a closed and coclosed form so that

$$[\omega_\pi]\in \mathrm{ind}_{\mathbb{P}(\mathbf{A}_f)}^{\mathbb{G}(\mathbf{A}_f)} H^{r_\pi}(\mathfrak{m}_0, K_\infty\cap L, \pi_{\mathbb{L}(\mathbf{A})}\otimes F_L),$$

represents the same class as $E(\eta_{\mathfrak{q}}^{res})^P$.

Note that we have Poincaré duality on $H^*(\mathfrak{n},\mathbb{C})$. So we consider $F\xrightarrow{\sim}\bar{F}^*\subset H^{\dim\ (\mathfrak{n})-r_N}(\mathfrak{n},\mathbb{C})$. Then

$$[*_{rep}\omega_\pi]\in \mathrm{ind}_{\mathbb{P}(\mathbf{A}_f)}^{\mathbb{G}(\mathbf{A}_f)} H^{d_L-r_\pi}(\mathfrak{m}_0, K\cap L, \pi_{\mathbb{L}(\mathbf{A})}\otimes H^{\dim\ \mathfrak{n}-r_N}(\mathfrak{n},\mathbb{C}))$$

can be considered as a form in degree $d_L - r_\pi + \dim\mathfrak{n} - r_N$.

We consider $*\omega_\pi$ as a form on the face $e(\mathbb{P},K_f)$ of the Borel–Serre boundary on $S(K_f)$ and use the natural identification

$$\mathbb{P}(\mathbb{Q})\backslash X/KK_f \xrightarrow{\sim} e(\mathbb{P},K_f)\times A_P\,.$$

Then

$$\tilde{\omega}_{\pi,\mu} := *\omega_\pi\wedge\chi_{\mu-\rho_P}\omega_A$$

is a closed form with compact support on $\mathbb{P}(\mathbb{Q})\backslash X/KK_f$ and we get a compactly supported Eisenstein form

$$E(\tilde{\omega}_{\pi,\mu}) = \sum_{\gamma\in G(\mathbb{Q})/\mathbb{P}(\mathbb{Q})} \gamma^*\tilde{\omega}_{\pi,\mu}.$$

For more detail of this construction see [25].

THEOREM V.1. *The pseudo Eisenstein $E(\tilde{\omega}_{\pi,\mu_0})$ series represents a non-trivial cohomology class with compact support such that*

$$\int_{\mathbb{G}(\mathbb{Q})\backslash(\mathbb{G}(\mathbf{A})} E(\tilde{\omega}_{\pi,\mu_0}) \wedge \operatorname{res}_{\mu=\mu_0} E(\eta_q,\mu) \neq 0\,.$$

Proof: We have

$$\begin{aligned}
&\int_{\mathbb{G}(\mathbb{Q})A_G\backslash\mathbb{G}(\mathbf{A})} E(\tilde{\omega}_{\pi,\mu_0}) \wedge\ \operatorname{res}_{\mu=\mu_0} E(\eta_q,\mu)\\
&= \int_{\mathbb{P}(\mathbb{Q})\mathbf{A}_G\backslash G(\mathbf{A})} \tilde{\omega}_{\pi,\mu_0} \wedge\ \operatorname{res}_{\mu=\mu_0} E(\eta_q,\mu)\\
&= \int_{\mathbb{P}(\mathbb{Q})\mathbb{N}(\mathbf{A})A_G\backslash G(\mathbf{A})} \tilde{\omega}_{\pi,\mu_0} \wedge\ \operatorname{res}_{\mu=\mu_0} E(\eta_\rho,\mu_0)^P\\
&= \operatorname{vol}(KK_f) \int_{A_P/A_G} \int_{\mathbb{L}(\mathbb{Q})\mathbb{L}(\mathbf{A})/A_P} *\omega_\pi \wedge \omega_\pi \chi_{\mu_0-\rho_P}\chi_{\rho_P-\mu_0}\omega_A\\
&= vol(KK_f) \int_{\mathbb{L}(\mathbf{A})\cap KK_f} ||\omega_\pi||^2(x)dx \neq 0\,.
\end{aligned}$$

Hence the second claim holds and by Stokes' theorem $E(\tilde{\omega}_{\pi,\mu_0})$ cannot be the boundary of a compactly supported form. q.e.d.

COROLLARY V.2. *$E(\tilde{\omega}_{\pi,\mu_0})$ represents a nontrivial cohomology class in the L^2-cohomology of $S(K_f)$.*

Proof: The form $E(\tilde{\omega}_{\mathfrak{q},\mu_0})$ is square integrable. Therefore we can interpret our formulas in L^2-cohomology. □

The pseudo Eisenstein form $E(\omega_{\mathfrak{q},\mu_0})$ is a differential form with compact support. So we can use the methods of Langlands (as written up in the Mœglin-Waldspurger) to compute its "Fourier " expansion in the L^2– spectrum and in particular its projection on the residual cohomology $\operatorname{Hom}_K(\wedge^{n-r_\mathfrak{q}}\mathbf{p}, L^2_{\text{res}}(\mathbb{G}(\mathbb{Q})A_G\backslash\mathbb{G}(\mathbf{A})))$. This idea will be pursued in the following sections.

VI. A symplectic modular symbol

Suppose that $\mathbb{G} = GL_4$. We prove in the section that for K_f small enough the symplectic group defines a non-trivial modular symbol for $S(K_f)$.

6.1 Let $\mathbb{G} = GL_4$ over $\mathbb{Q}$ and let $\theta : g \to (g^t)^{-1}$. Define the skew-symmetric matrix

$$J = \begin{pmatrix} 0 & I \\ -I & 0 \end{pmatrix}$$

and let $\mathbb{H}$ be the symplectic subgroup of $\mathbb{G}$ defined as the fixed points of the involution

$$\sigma : g \to -J\,(g^t)^{-1}\,J.$$

Since $\sigma\theta = \theta\sigma$ the involution θ also defines a Cartan involution on H for the maximal compact subgroup $H \cap K$.

We fix a Borel subgroup $\mathbb{B}$ consisting of the upper triangular matrices.

6.2 Let $\mathfrak{q} = \mathfrak{l} \oplus \mathfrak{n}$ be the θ-stable parabolic subalgebra in $gl_4(\mathbb{C})$ defined by J. Then the centralizer L of J in $GL_4(\mathbb{R})$ is isomorphic to $GL(2,\mathbb{C})$. Furthermore $\mathbf{p}_{\mathbb{C}} = (\mathfrak{n} \cap \mathbf{p}_{\mathbb{C}}) \oplus (\mathbf{p}_{\mathbb{C}} \cap \mathfrak{h}_{\mathbb{C}})$. [**31**]

We also denote by $A_\mathfrak{q}$ the unitary representation of the cohomologically induced Harish-Chandra module $A_\mathfrak{q}$. It is a subrepresentation of $L^2(H\backslash G)$. [**29**]

The abelian subgroup

$$A_P = \left\{ \begin{pmatrix} a_1 & 0 & 0 & 0 \\ 0 & a_2 & 0 & 0 \\ 0 & 0 & a_1^{-1} & 0 \\ 0 & 0 & 0 & a_2^{-1} \end{pmatrix}, a_1, a_2 \in \mathbb{R}^* \right\}$$

is a maximally split torus in H. Let $\mathbb{P}$ be a rational standard maximal parabolic subgroup of $\mathbb{G}$ with Levi factor $\mathbb{L}$ of $GL_2 \times GL_2$. For K_f small enough there exists a representation $\Pi_{\mathbb{L}(\mathbf{A})}$ in the cuspidal spectrum of $L^2(\mathbb{L}(\mathbb{Q})A_P\backslash\mathbb{L}(\mathbf{A})/\mathbb{L}(\mathbf{A}_f) \cap K_f)$ which has a discrete series representation at the infinite place with infinitesimal character 2ρ. The Eisenstein intertwining operator

$$E(\mathbb{P}, \pi_{\mathbb{L}(\mathbf{A})}, \mu) : I(\mathbb{P}, \pi_{\mathbb{L}(\mathbf{A})}, \mu) \to C^\infty(\mathbb{G}(\mathbb{Q})\backslash\mathbb{G}(\mathbf{A})/A_G\,(K\,K_f))$$

has a pole for $\mu_0 = \frac{1}{2}\rho_P$. The image of the residual intertwining operator for $\mu_0 = \frac{1}{2}\rho_P$ is a representation

$$\Pi_{\mathbf{A}} = \Pi_{\mathbf{A}}(\pi_{\mathbb{L}(\mathbf{A})}, \mu_0) = \prod_v \Pi_v = \Pi_\infty \Pi_{\mathbf{A}_f}$$

with $\Pi_\infty = A_\mathfrak{q}$ with $r_\mathfrak{q} = 3$. The representation $\Pi_{\mathbf{A}}$ is isomorphic to a subrepresentation of the residual spectrum of $L^2(\mathbb{G}(\mathbb{Q})\backslash\mathbb{G}(\mathbf{A})/A_G K_f)$. For details see [**28**].

By theorem III.1 the representation $\Pi_{\mathbf{A}}$ defines for K_f small enough a residual cohomology class $[E(\eta_\mathfrak{q}^{res})]$ in degree $r_\mathfrak{q}$= 3. We have (see V.2)

$$0 \neq [E(\tilde{\omega}_{\mathfrak{q},\mu_0})] \in H_c^6(S(K_f), \mathbb{C}).$$

Following [**29**] we let $0 \neq \omega_H \in \wedge^{\dim \mathfrak{h}\cap\mathbf{p}}(\mathfrak{h} \cap \mathbf{p})^*$. Then $*\omega_H \in \wedge^{\dim \mathfrak{n}\cap\mathbf{p}}\mathbf{p}^*$ defines a differential form on $S(K_f)$. Assuming that K_f is small enough and we have a compatible orientation on $S_H(K_f \cap \mathbb{H}(\mathbf{A}_f)$ we identify

$$E(\tilde{\omega}_{\mathfrak{q},\mu_0}) \wedge *\omega_H$$

with a function $\mathcal{E}_\mathfrak{q} \neq 0$ on $S(K_f)$. There exists a measure $dh_{\mathbf{A}}$ on $S_H(K_f \cap \mathbb{H}(\mathbf{A}_f))$ induced by a left invariant measure on $\mathbb{H}(\mathbf{A})/(K\,K_f \cap \mathbb{H}(\mathbf{A}))$ so that

LEMMA VI.1. *Under our assumptions*

$$\int_{S_H(K_f\cap\mathbb{H}(\mathbf{A}_f))} *E(\eta_{\mathfrak{q}}^{res}) = \int_{S_H(K_f\cap\mathbb{H}(\mathbf{A}_f))} \mathcal{E}_{\mathfrak{q}}(h_{\mathbf{A}})dh_{\mathbf{A}}.$$

Proof: This follows from the arguments on page 5.6 in [**29**]. □

6.3 The representations in the discrete spectrum of $L^2(\mathbb{G}(\mathbb{Q})\backslash\mathbb{G}(\mathbf{A})/A_G)$ with nontrivial $(\mathfrak{g}, K)$–cohomology are [**28**] :

(1) tempered and in the cuspidal spectrum; they have nontrivial $(\mathfrak{g}_0, K)$–cohomology in degrees 4 and 5,
(2) the trivial representation $\mathbf{1}$ in the residual spectrum; it has non-trivial $(\mathfrak{g}, K)$–cohomology in degree 1, 4, 5 and 9,
(3) the representations $\Pi_{\mathbf{A}}$ in the residual spectrum described in 6.2; they have nontrivial $(\mathfrak{g}, K)$–cohomology in degree 3 and 6.

The L^2–cohomology $H^*_{L^2}(S(K_F), \mathbb{C})$ is infinite in degree 4 and 5 [**5**], but finite in degree 1, 3, 6 and 9. In degree 3 it is represented by residual harmonic forms $E(\eta_{\mathfrak{q}}^{res}) \in \mathrm{Hom}_K(\wedge^3\mathbf{p}, \Pi_{\mathbf{A}})$ for representations $\Pi_{\mathbf{A}}$.

By the previous lemma and the results of Jacquet and Rallis for K_f small enough [**18**]

$$I_{S_H(K_f\cap\mathbb{H}(\mathbf{A}_f))} : \omega \to \int_{S_H(K_f\cap\mathbb{H}(\mathbf{A}_f))} \omega$$

defines a map

$$I_{S_H(K_f\cap\mathbb{H}(\mathbf{A}_f))} : \mathrm{Hom}_K(\wedge^6\mathbf{p}, L^2_{dis}(\mathbb{G}(\mathbb{Q})\backslash\mathbb{G}(\mathbf{A})/A_GK_f)) \to \mathbb{C}.$$

Let $\Omega_{\pi_{\mathbf{A}}}$ be a smooth form with compact support representing the same L^2–cohomology class as $*E(\eta_{\mathfrak{q}}^{res})$. Then

$$*E(\eta_{\mathfrak{q}}^{res}) - \Omega_{\pi_{\mathbf{A}}} = d\eta.$$

The integral $\int_{S_H(K_f\cap\mathbb{H}(\mathbf{A}_f))} d\eta$ is finite.

Let d_H be the differential on $S_H(K_f \cap \mathbb{H}(\mathbf{A}_f))$. An easy check shows that for $h \in S_H(K_f \cap \mathbb{H}(\mathbf{A}_f))$

$$d\eta(h) \wedge *\omega_H(h) = d_H\eta(h) \wedge *\omega_H(h).$$

Thus by the L^2–Stokes theorem for complete manifolds [**12**]

$$\int_{S_H(K_f\cap\mathbb{H}(\mathbf{A}_f))} d\omega = 0.$$

Thus we proved

PROPOSITION VI.2. *For K_f small enough the map*

$$I_{S_H(K_f\cap\mathbb{H}(\mathbf{A}_f))} : \omega \to \int_{S_H(K_f\cap\mathbb{H}(\mathbf{A}_f))} \omega$$

defines a modular symbol

$$[I_{S_H(K_f\cap\mathbb{H}(\mathbf{A}_f))}] \in H^3(S(K_f), \mathbb{C}).$$

THEOREM VI.3. *Suppose that $\mathbb{G} = GL(4)$ and $\mathbb{H}$ is a symplectic group compatible with the choice of the maximal compact subgroup $K \subset GL(4,\mathbb{R})$. There exists a K_f so that*

$$[I_{S_H(K_f \cap \mathbb{H}(\mathbf{A}_f))}]$$

is not zero.

Proof: The linear functional

$$f \to \int_{S_H(K_f \cap \mathbb{H}(\mathbf{A}_f))} f(h_{\mathbf{A}})dh_{\mathbf{A}}$$

is not zero on the K–finite functions f by the work of Jacquet and Rallis in the residual spectrum [**18**]. Let $\mathcal{E}_{\mathfrak{q}}$ be the function defined in VI.1. By the previous considerations we may assume that this function is in $L^2_{res}(\mathbb{G}(\mathbb{Q})\backslash\mathbb{G}(\mathbf{A}/A_G K_f)$. There exists a $g_0 \in \mathbb{G}(\mathbb{Q})$ near the identity so that

$$\int_{S_H(K_f \cap \mathbb{H}(\mathbf{A}_f))} \mathcal{E}_{\mathfrak{q}}(g_0 h_{\mathbf{A}})dh_{\mathbf{A}} \neq 0.$$

Here we integrate over the orbit of g_0 under $\mathbb{H}(\mathbf{A})$. Since the rational elements are dense in G we may assume that g_0 is rational. There is a subgroup $K_{g_0,f}$ of finite index in both K_f and its g_0 conjugate so that $[S_H(K_{g_0,f} \cap \mathbb{H}(\mathbf{A}_f))](\mathcal{E}_{\mathfrak{q}}) \neq 0$ □

VII. On the volume of $S(K_f)$

In the proof of theorem V.1 we have used a formula for the integral of a wedge product of a pseudo Eisenstein form with an Eisenstein form. We show how this formula is related to the computation of the Tamagawa number $\tau(\mathbb{G})$ of $\mathbb{G}$. We sketch this only in the simplest case, i.e. if $\mathbb{G}/\mathbb{Q}$ is split and simply connected.

7.1. Preliminaries. Let $\mathbb{G}/\mathbb{Q}$ be a split semisimple connected and simply connected algebraic group with $\mathbb{Q}$–Lie algebra $\mathfrak{g}$. We fix a system R of roots, a system $R^+ \subset R$ of positive roots and a Chevalley basis $\{X_\alpha, H_i\}, i = 1,\dots,l, \alpha \in R$, of $\mathfrak{g}$ in standard notation. Then the H_i are a $\mathbb{Q}$–basis of the $\mathbb{Q}$–Lie algebra of a split torus $\mathbb{T}$ of $\mathbb{G}$ which is contained in a Borel group $\mathbb{B}$ of $\mathbb{G}$ and $\mathbb{B}$ has unipotent radical $\mathbb{N}$, where the $\mathbb{Q}$–Lie algebra of $\mathbb{N}$ has basis $\{X_\alpha\}_{\alpha\in R^+}$. The Chevalley basis determines an invariant differential form of highest degree and hence invariant measures ω_ν^G on $\mathbb{G}(\mathbb{Q}_\nu)$ for all places v of $\mathbb{Q}$. Then the restricted tensor product

$$\omega^G := \widehat{\bigotimes_v} \omega_v$$

is an invariant measure on $\mathbb{G}(\mathbf{A})$. The measure ω^G induces a measure dg on $\mathbb{G}(\mathbb{Q})\backslash\mathbb{G}(\mathbf{A})$ and by definition the Tamagawa number $\tau(\mathbb{G})$ is defined as

$$\tau(\mathbb{G}) = \int_{G(\mathbb{Q})\backslash G(\mathbf{A})} dg.$$

For all this see [21].

Let p be a prime. Since $\mathbb{G}/\mathbb{Q}$ is a Chevalley group $\mathbb{G}(\mathbb{Z}_p) =: K_p$ makes sense. Put $K_\infty = K$ and let $K \subset \mathbb{G}(\mathbb{R})$ be a maximal compact subgroup of $G(\mathbb{R})$. For all places v of $\mathbb{Q}$ put $N_v := \mathbb{N}(\mathbb{Q}_v), T_v = \mathbb{T}(\mathbb{Q}_v)$, and $G_v = \mathbb{G}(\mathbb{Q}_v)$. Then we have the Iwasawa decomposition $G_v = N_v T_v K_v$.

Put

$$\rho_v(t_v) = |\det \ \mathrm{Ad}(t_v)_{|\mathfrak{n}_v}|_v^{1/2}.$$

Here $|\ |_v$ is the normalized absolute value on $\mathbb{Q}_v$, $\mathfrak{n}_v$ the $\mathbb{Q}_v$–Lie-algebra of N_v and $\mathrm{Ad}(t_v)$ is the adjoint action of $t_v \in T_v$ on $\mathfrak{n}_v$. Put $|\rho|(t) = \prod_v \rho_v(t_v)$ for $t \in \mathbb{T}(\mathbf{A})$. For $v \neq \infty$ let ω_v^T be the measure on T_v given by the $\{H_i\}$ with normalization $\mathrm{vol}_{\omega_v}(\mathbb{T}(\mathbb{Z}_v)) = 1$. If $v = \infty$ let ω_v^T be the measure determined by the $\{H_i\}$. Then the restricted product

$$\omega^T = \widehat{\bigotimes_v} \, \omega_v^T$$

is a measure on $\mathbb{T}(\mathbf{A})$. We choose on $\mathbb{N}(\mathbf{A})$ the measure $\omega^N = \otimes_v \omega_v^N$, where the measures ω_v^N are determined by the $\{X_\alpha\}_{\alpha\in\mathbb{R}^+}$. Then we have for the finite places

$$\omega_v^G = \rho_v^{-2}\omega_v^N \otimes \omega_v^T \otimes \omega_v^K ,$$

where ω_v^K is the measure on K_v given by restriction of ω_v^G . We choose a measure ω_∞^K on K_∞ such that this formula also holds for $v = \infty$. To simplify the notation we write for $\varphi \in C_c(\mathbb{G}(\mathbf{A})$ as usual

$$\begin{aligned}\int_{\mathbb{G}(\mathbf{A})} \varphi(g) d\omega^G(g) \quad &=: \quad \int_{\mathbb{G}(\mathbf{A})} \varphi(g)dg \\ &= \quad \int_{\mathbb{N}(\mathbf{A})} \int_{\mathbb{T}(\mathbf{A})} \int_{KK_f} \varphi(ntk)|\rho|^{-2}(t) dndtdk.\end{aligned}$$

The measure dn induces a measure of mass 1, also denoted by dn, on the quotient $\mathbb{N}(\mathbb{Q})\backslash\mathbb{N}(\mathbf{A})$. The measure dt induces a measure, again denoted by dt, on the quotient $\mathbb{T}(\mathbb{Q})\backslash\mathbb{T}(\mathbf{A})$.

7.2. Let $\varphi : \mathbb{T}(\mathbb{Q})\backslash\mathbb{T}(\mathbf{A}) \longrightarrow \mathbb{R}$ be a continuous compactly supported function which is right $(KK_f) \cap \mathbb{T}(\mathbf{A})$–invariant such that

$$\int_{\mathbb{T}(\mathbb{Q})\backslash\mathbb{T}(\mathbf{A})} \varphi(t)|\rho|^{-2}(t)dt = 1.$$

Since

$$\mathbb{T}(\mathbb{Q})\backslash\mathbb{T}(\mathbf{A})/(KK_f) \cap \mathbb{T}(\mathbf{A}) \xrightarrow{\sim} \mathbb{N}(\mathbf{A})\mathbb{B}(\mathbb{Q})\backslash\mathbb{G}(\mathbf{A})/KK_f$$

we view φ as left $\mathbb{N}(\mathbf{A})\mathbb{B}(\mathbb{Q})$–invariant and right KK_f–invariant function on $\mathbb{G}(\mathbf{A})$. Then the pseudo-Eisenstein series

$$E(\varphi)(g) := \sum_{\gamma\in\mathbb{B}(\mathbb{Q})\backslash\mathbb{G}(\mathbb{Q})} \varphi(\gamma g), \quad g \in \mathbb{G}(\mathbf{A}) ,$$

is a compactly supported function on $\mathbb{G}(\mathbb{Q})\backslash\mathbb{G}(\mathbf{A})$. We use 7.1 and get

$$\int_{\mathbb{G}(\mathbb{Q})\backslash\mathbb{G}(\mathbf{A})} E(\varphi)(g)dg = \mathrm{vol}(KK_f).$$

where dg also denotes the induced measure on $\mathbb{G}(\mathbb{Q})\backslash\mathbb{G}(\mathbf{A})$. So if $\mathbf{1}$ denotes the constant function,

$$(\mathrm{vol}(KK_f)/\tau(\mathbb{G})) \cdot \mathbf{1}$$

is the projection of $E(\varphi)$ on the constant function in the L^2-spectral decomposition.

We can also interpret this integral as the integral of a wedge product of a residual Eisenstein form in degree 0 with a pseudo Eisenstein form in degree $d = \dim X_\infty$.

7.3 To determine another formula for projection of the function $E(\varphi)$ onto the trivial representation we use Fourier analysis on the torus. Let $\lambda : \mathbb{T}(\mathbb{Q})\backslash\mathbb{T}(\mathbf{A}) \longrightarrow \mathbb{C}^*$ be a continuous character and put

$$\hat{\varphi}_\lambda(g) = \int_{\mathbb{T}(\mathbb{Q})\backslash\mathbb{T}(\mathbf{A})} |\rho|^{-1}(t)\lambda^{-1}(t)\varphi(tg)dt\,.$$

The function $\hat{\varphi}_\lambda$ is $\mathbb{N}(\mathbf{A})\mathbb{B}(\mathbb{Q})$–left invariant and KK_f–right invariant. Moreover

$$\hat{\varphi}_\lambda(t) = \lambda(t)|\rho|(t)\hat{\varphi}_\lambda(1)$$

and

$$\hat{\varphi}_{|\rho|}(1) = 1$$

by our choice of φ.

As usual we put

$$E(\varphi,\lambda)(g) := \sum_{\gamma\in\mathbb{B}(\mathbb{Q})\backslash\mathbb{G}(\mathbb{Q})} \hat{\varphi}_\lambda(\gamma g).$$

It is well known that this Eisenstein series is convergent if $\mathrm{Re}(\lambda)$ is sufficiently big. This holds in particular for $\lambda = |\rho|^s$ and $1 < s \in \mathbb{R}$. We have

$$\int_{\mathbb{N}(\mathbb{Q})\backslash\mathbb{N}(\mathbf{A})} E(\varphi,\lambda)(ng)dn = \sum_{w\in W} M(w,\lambda)\hat{\varphi}_\lambda(g),$$

where

$$M(\omega,\lambda)\hat{\varphi}_\lambda(g) = \int_{\mathbb{N}(\mathbb{Q})\backslash\mathbb{N}(\mathbf{A})} \hat{\varphi}_\lambda(wng)dn.$$

It is well known that $E(\varphi,\lambda)$ has a meromorphic continuation in λ with a residue of order ℓ at $\lambda = |\rho|$, which is denoted by $\mathrm{res}_{\lambda=|\rho|}E(\varphi,\lambda)$. Moreover this residue is a multiple $\mathbf{c}$ of the constant function $\mathbf{1}$ on $\mathbb{G}(\mathbb{Q})\backslash\mathbb{G}(\mathbf{A})$. To obtain an exact formula for $\mathbf{c}$ and to calculate the residue it suffices to consider the one-dimensional problem $\lambda_s = |\rho|^s, 1 < s \in \mathbb{R}$. We use

$$(M(w,|\rho|^s)\hat{\varphi}_{|\rho|^s})(t) = |\rho|(t)|w\rho|^s(t)(M(w,|\rho|^s)\hat{\varphi}_{|\rho|^s})(1), \quad t \in \mathbb{T}(\mathbf{A}).$$

It is well known that that for $s > 1, (M(w,|\rho|^s)\hat{\varphi}_{|\rho|^s})(1)$ can be calculated as a product of local contributions using the results of Gindikin-Karpelevič for all places v of $\mathbb{Q}$. We get

$$\mathrm{res}_{s=1}(M(w_0,|\rho|)^s)\hat{\varphi}_{|\rho|^s})(1) = \mathrm{vol}(KK_f).$$

Hence

$$\mathrm{res}_{s=1}E(\varphi,|\rho|^s) = \mathrm{vol}(KK_f)\,\mathbf{1},$$

where $\mathbf{1}$ is the constant function on $\mathbb{G}(\mathbb{Q})\backslash\mathbb{G}(\mathbf{A})$.

We use the notation given in the proof of V.1 and deduce that

$$\mathcal{M}(\pi_{\mathbb{L}(\mathbf{A})},\mu_0,\omega_0)\eta_q = \mathrm{vol}(KK_f)\mathbf{1}.$$

So integrating the projection of $E(\varphi)$ on the constant function we get $\mathrm{vol}(KK_f)\tau(\mathbb{G})$. On the other hand in 7.2 we showed that this integral is $\mathrm{vol}(KK_f)$. Hence

$$\tau(\mathbb{G}) = 1.$$

This is due to Langlands [21].

In a sequel to this paper we will use similar techniques to compute the integrals defined by other residual modular symbols.

References

[1] J. Arthur, Intertwining operators and residues I. Weighted characters, J. of Functional Analysis 84, (1989) 19 - 84.

[2] A. Ash, D. Ginzburg, S. Rallis, Vanishing periods of cusp forms over modular symbols. Math. Ann. 296 (1993), no. 4, 709–723.

[3] D. Barbasch and B. Speh, Cuspidal representations of reductive groups, Preprint 2008.

[4] A. Borel, Stable real cohomology of arithmetic groups. II. Manifolds and Lie groups (Notre Dame, Ind., 1980), pp. 21–55, Progr. Math., 14, Birkhäuser, Boston, Mass., 1981.

[5] A. Borel and W. Casselman, L^2-cohomology of locally symmetric manifolds of finite volume. Duke Math. J. 50 (1983), no. 3, 625–647.

[6] A. Borel and J.-P. Serre, Corners and arithmetic groups. Avec un appendice: Arrondissement des variétés à coins, par A. Douady et L. Hérault. Comment. Math. Helv. 48 (1973), 436–491.

[7] A. Borel and N. Wallach, *Continuous Cohomology, Discrete Subgroups, and Representations of Reductive Groups*, Annals of Mathematics Studies 94, Princeton University Press (1980).

[8] R. Bryant, S. S. Chern, R. B. Gardner, H. L. Goldschmid, and P. Griffiths, *Exterior Differential Systems* , Mathematical Sciences Research Institute Publications, Springer Verlag (1991).

[9] J. Franke, Harmonic analysis in weighted L^2-spaces. Ann. Sci. École Norm. Sup. (4) 31 (1998), no. 2, 181–279.

[10] J. Franke, A topological model for some summand of the Eisenstein cohomology of congruence subgroups, Progr. Math 2008, vol.258, 27-85.

[11] J. Franke and J. Schwermer, A decomposition of spaces of automorphic forms, and the Eisenstein cohomology of arithmetic groups. Math. Ann. 311 (1998), no. 4, 765–790.

[12] M.P. Gaffney and P. Matthew, A special Stokes's theorem for complete Riemannian manifolds. Ann. of Math. (2) 60, (1954) 140–145.

[13] G. Harder, On the cohomology of discrete arithmetically defined groups. *Discrete subgroups of Lie groups and applications to moduli* (Internat. Colloq., Bombay, 1973), pp. 129–160. Oxford Univ. Press, Bombay, 1975.

[14] G. Harder, On the cohomology of $SL(2, O)$. *Lie groups and their representations* (Proc. Summer School on Group Representations of the Bolyai János Math. Soc., Budapest, 1971), pp. 139–150. Halsted, New York, 1975.

[15] G. Harder, Eisenstein cohomology of arithmetic groups and its applications to number theory. Proceedings of the International Congress of Mathematicians, Vol. I, II (Kyoto, 1990), 779–790, Math. Soc. Japan, Tokyo, 1991.

[16] G. Harder, Eisenstein cohomology of arithmetic groups. The case GL_2. Invent. Math. 89 (1987), no. 1, 37–118.

[17] Harish-Chandra, *Automorphic forms on semi-simple Lie groups*, Springer Lecture Notes 62, Springer Verlag (1968).

[18] H. Jacquet and S. Rallis, Symplectic periods. J. Reine Angew. Math. 423 (1992), 175–197.

[19] A.W. Knapp, *Representation theory of semisimple groups. An overview based on examples.* Princeton Mathematical Series, 36. Princeton University Press, Princeton, NJ, 1986.

[20] A.W. Knapp and D. Vogan, Jr. *Cohomological induction and unitary representations.* Princeton Mathematical Series, 45. Princeton University Press, Princeton, NJ, 1995.

[21] R.P. Langlands, The volume of the fundamental domain for some arithmetical subgroups of Chevalley groups. 1966 Algebraic Groups and Discontinuous Subgroups (Proc. Sympos. Pure Math., Boulder, Colo., 1965) pp. 143–148.

[22] R.P. Langlands, *On the functional equations satisfied by Eisenstein series.* Lecture Notes in Mathematics, Vol. 544. Springer-Verlag, Berlin-New York.

[23] C. Mœglin et J.-L. Waldspurger, *Spectral decomposition and Eisenstein series*, Cambridge Tract in Mathematics 113, Cambridge University Press (1995).

[24] C. Mœglin et J.-L. Waldspurger, Le spectre résiduel de $GL(n)$. Ann. Sci. École Norm. Sup. (4) 22 (1989), no. 4, 605–674.
[25] J. Rohlfs and B. Speh, Pseudo Eisenstein forms and the cohomology of arithmetic groups I, Manuscripta Math. 106 (2001), no. 4, 505–518.
[26] S. Salamanca-Riba, On the unitary dual of real reductive Lie groups and the $A_q(\lambda)$ modules: the strongly regular case. Duke Math. J. 96 (1999), no. 3, 521–546.
[27] J. Schwermer, *Kohomologie arithmetisch definierter Gruppen und Eisensteinreihen*, Lecture Notes in Mathematics 988, Springer–Verlag (1983).
[28] B. Speh, Representation of $\mathrm{Gl}(\mathrm{n},\mathbb{R})$ with nontrivial $(\mathfrak{g}, K)$–cohomology, Invent. Math 71, 443–465 (1983).
[29] Y. Tong and S.P. Wang, Geometric realisation of discrete series for semisimple symmetric spaces, Invent. math. 96, 425-458 (1989).
[30] D. Vogan Jr., *Unitary representations of reductive Lie groups*.e
Annals of Mathematics Studies, 118. Princeton University Press, Princeton, NJ, 1987.
[31] D. Vogan Jr. and G. Zuckerman. Unitary representations with nonzero cohomology. Compositio Math. 53 (1984), no. 1, 51–90.
[32] N. Wallach, On the constant term of a square integrable automorphic form. Operator algebras and group representations, Vol. II (Neptun, 1980), 227–237, Monogr. Stud. Math., 18, Pitman, Boston, MA, 1984.

Jürgen Rohlfs, Katholische Universität Eichstätt, Ostenstr.26 - 28, 85072 Eichstätt, Germany
E-mail address: juergen.rohlfs@ku-eichstaett.de

Birgit Speh, Department of Mathematics, 310 Malott Hall, Cornell University, Ithaca, NY 14853-4201, U.S.A.
E-mail address: speh@math.cornell.edu

Clay Mathematics Proceedings
Volume 13, 2011

A Functor from Smooth o-Torsion Representations to (φ, Γ)-Modules

Peter Schneider and Marie-France Vigneras

To Freydoon Shahidi

Contents

Introduction		525
0.	Notations and conventions	527
1.	$\Lambda(P_+)$-modules	528
2.	P_+-subrepresentations	530
3.	The case of a compactly induced representation	532
4.	The basic cohomological functor	538
5.	Towards (φ, Γ)-modules	540
6.	A functor in the reverse direction	542
7.	Dependence on N_0 and ℓ	545
8.	Some topological localizations	551
9.	Generalized (φ, Γ)-modules	578
10.	(φ, Γ)-modules	588
11.	The case $GL_2(\mathbb{Q}_p)$	593
12.	Subquotients of principal series	599
References		600

Introduction

Let $\mathcal{G}_{\mathbb{Q}_p} := \mathrm{Gal}(\overline{\mathbb{Q}}_p/\mathbb{Q}_p)$ denote the absolute Galois group of the field $\mathbb{Q}_p$ of p-adic numbers. If ℓ is any prime number different from p then the local Langlands philosophy predicts a close relation between finite dimensional ℓ-adic representations of $\mathcal{G}_{\mathbb{Q}_p}$ and the (infinite dimensional) smooth representation theory in characteristic zero of reductive groups over $\mathbb{Q}_p$. In recent years it has become increasingly clear that most probably some kind of extension of this correspondence to the case $\ell = p$ exists. On the Galois side the theory of p-adic representations of $\mathcal{G}_{\mathbb{Q}_p}$ has reached maturity through the work of Colmez, Faltings, Fontaine, and

2010 *Mathematics Subject Classification.* Primary: 11S37. Secondary: 22E50.

others. On the reductive group side the foundations of a theory of continuous representations in p-adic locally convex vector spaces have been laid in the work of Schneider/Teitelbaum and others. Nevertheless a precise conjectural framework how these two sides correspond to each other is still missing.

It is all the more remarkable that despite a missing general picture Colmez recently has managed to establish such a correspondence for the group $GL_2(\mathbb{Q}_p)$. His starting point is Fontaine's theorem ([**Fon**]) that the category of p-adic Galois representations is naturally equivalent to the category of (finitely generated) etale (φ, Γ)-modules. Moreover, this equivalence arises from an analogous equivalence over any of the finite rings $\mathbb{Z}/p^m\mathbb{Z}$. The concept of a (φ, Γ)-module is a purely (semi)linear algebra notion. It is best viewed as being part of the module theory of the multiplicative monoid of p-adic integers $\mathbb{Z}_p^\bullet := \mathbb{Z}_p \setminus \{0\}$ over a certain big coefficient ring $\Lambda_F(\mathbb{Z}_p)$. This monoid can be identified with the submonoid in $GL_2(\mathbb{Q}_p)$ of dominant diagonal matrices (modulo the center). In addition, the coefficient ring $\Lambda_F(\mathbb{Z}_p)$ can be understood in terms of the unipotent radical of the standard Borel subgroup P in $GL_2(\mathbb{Q}_p)$. In fact, Colmez establishes in [**Co1**] and [**Co2**] a functorial relationship between smooth torsion P-representations and etale (φ, Γ)-modules.

This paper constitutes an attempt to understand the smooth representation theory with $\mathbb{Z}_p$-torsion coefficients of a Borel subgroup in a general reductive group in terms of new objects which we call generalized (φ, Γ)-modules. We place ourselves in the context of a $\mathbb{Q}_p$-split connected reductive group over $\mathbb{Q}_p$ (whose center we usually assume, for technical simplicity, to be connected), and we fix a Borel subgroup $P = TN$ of its group G of $\mathbb{Q}_p$-rational points with split torus T and unipotent radical N. We also fix an appropriate compact open subgroup $N_0 \subseteq N$ which in turn gives rise to the "dominant" submonoid $T_+ := \{t \in T : tN_0t^{-1} \subseteq N_0\}$ in T. On the one side we consider the abelian category $\mathcal{M}_{o-tor}(P)$ of all smooth P-representations in o-torsion modules where o is the ring of integers in a fixed finite extension $K/\mathbb{Q}_p$. On the other side we note that the monoid T_+ acts, by functoriality, on the completed group ring $\Lambda(N_0) := o[[N_0]]$. A first step towards generalizing (φ, Γ)-modules would be to consider $\Lambda(N_0)$-modules with an additional semilinear T_+-action. But technically it is preferable to introduce the monoid $P_+ := N_0T_+$, a corresponding monoid ring $\Lambda(P_+)$ which is an overring of $\Lambda(N_0)$, and to work with the category $\mathcal{M}(\Lambda(P_+))$ of all (left unital) $\Lambda(P_+)$-modules instead. Such a module M will be called etale if every $t \in T_+$ acts, informally speaking, with slope zero on M. We show in section 1 that the etale $\Lambda(P_+)$-modules form an abelian category $\mathcal{M}_{et}(\Lambda(P_+))$.

In sections 2 to 4 we construct a universal δ-functor $V \longmapsto D^i(V)$ for $i \geq 0$ from the category $\mathcal{M}_{o-tor}(P)$ into the category $\mathcal{M}_{et}(\Lambda(P_+))$. The idea is to consider inside a representation V in $\mathcal{M}_{o-tor}(P)$ all P_+-subrepresentations which still generate V, to pass to their Pontrjagin duals as modules over $\Lambda(N_0)$, and to form the inductive limit denoted by $D(V)$. But the result is a $\Lambda(N_0)$-module which carries an additional action of the inverse monoid T_+^{-1} and not the monoid T_+ which we want. Somewhat as a miracle it turns out (section 3) that for any compactly induced V this T_+^{-1}-action has a natural right inverse T_+-action which does lead to a $\Lambda(P_+)$-module structure on $D(V)$. In section 4 we then use a functorial resolution by compactly induced representations to produce our δ-functor $D^i(V)$.

With the help of a Whittaker type functional ℓ on N we pass in section 5 from the category $\mathcal{M}_{et}(\Lambda(P_+))$ to the category $\mathcal{M}_{et}(\Lambda(S_\star))$ for the "standard" monoid $S_\star$ in $GL_2(\mathbb{Q}_p)$. We point out that our standard monoid $S_\star$ corresponds to Fontaine's original definition of (φ, Γ)-modules in which $\Gamma \cong \mathbb{Z}_p$. Colmez instead works mostly with a $\Gamma \cong \mathbb{Z}_p^\times$. Both points of view are equivalent; but the former, in our context, has less technical complications.

In section 6 we construct, in a totally elementary way, a functor from the category $\mathcal{M}_{et}(\Lambda(P_+))$ back into the category of all P-representations. Its precise relationship to the functor $D^0(V)$ remains unknown at this point.

In section 7 we show that our δ-functors are independent, up to a natural isomorphism, of the choices made for N_0 and ℓ.

An object in the category $\mathcal{M}_{et}(\Lambda(S_\star))$, even if finitely generated, is not yet a (φ, Γ)-module in the sense of Fontaine. The reason is that the base ring for the latter is not $\Lambda(\mathbb{Z}_p)$ but a certain p-adically completed localization $\Lambda_F(\mathbb{Z}_p)$ of $\Lambda(\mathbb{Z}_p)$. In section 8 we therefore take up the technically rather involved task to construct in a general framework a topological localization which when applied in section 9 to the ring $\Lambda(P_+)$ will lead to a ring $\Lambda_\ell(P_\star)$ and the corresponding abelian category $\mathcal{M}_{et}(\Lambda_\ell(P_\star))$ of etale $\Lambda_\ell(P_\star)$-modules which we view as a generalization of Fontaine's etale (φ, Γ)-modules. This construction relies in a crucial way on the interpretation in [**SV2**] of certain microlocalized completed group rings as skew Laurent series rings. By base extension our δ-functor $D^i(V)$ gives rise to a δ-functor $D^i_\ell(V)$ into the category $\mathcal{M}_{et}(\Lambda_\ell(P_\star))$. In section 10 we show that our specialization technique along ℓ extends to the ring $\Lambda_\ell(P_\star)$ finally leading to a δ-functor $D^i_{\Lambda_F(S_\star)}(V)$ into the category of not necessarily finitely generated (φ, Γ)-modules à la Fontaine. The question of finite generation remains the fundamental open question in this paper. We give an initial sufficient criterion, though, in Remark 11.4. We expect that for V in a suitable category of smooth o-torsion representations of G the etale (φ, Γ)-modules $D^i_{\Lambda_F(S_\star)}(V)$ indeed are finitely generated and therefore correspond to p-adic Galois representations.

As explained in section 11 Colmez' functor for the group $G = GL_2(\mathbb{Q}_p)$ (originally defined on the smooth o-torsion representations of G which are admissible, of finite length, and have a central character) coincides with our functor $D^0_{\Lambda_F(S_\star)}(V)$.

In the final section 12 we discuss the example of principal series representations.

The first author thanks the IHES as well as Columbia University, where parts of this paper were written, for their hospitality and financial support. The second author, for the same reasons, thanks the Radcliffe Institute and the Mathematics Department at Harvard University.

It is a pleasure to dedicate this article to Freydoon Shahidi. His influence on the theory of automorphic representations has been fundamental. We admire his generosity in sharing mathematical ideas, and we feel extremely lucky to have him as a wonderful friend for almost 30 years.

0. Notations and conventions

We denote by $| \ |$ the absolute value of the field $\overline{\mathbb{Q}}_p$. Let G be the group of $\mathbb{Q}_p$-rational points of a $\mathbb{Q}_p$-split connected reductive group over $\mathbb{Q}_p$. We fix a Borel subgroup $P = TN$ in G with maximal split torus T and unipotent radical N. Let Φ^+ denote, as usual, the set of roots of T positive with respect to P and let $\Delta \subseteq \Phi^+$ be the subset of simple roots. For any $\alpha \in \Phi^+$ we have the root subgroup $N_\alpha \subseteq N$.

We recall that $N = \prod_{\alpha \in \Phi^+} N_\alpha$ (set-theoretically) for any total ordering of Φ^+. Let $T_0 \subseteq T$ be the maximal compact subgroup. We fix a compact open subgroup $N_0 \subseteq N$ which is totally decomposed, i.e., such that $N_0 = \prod_\alpha (N_0 \cap N_\alpha)$ for any total ordering of Φ^+. Then $P_0 := T_0 N_0$ is a group. We introduce the submonoid $T_+ \subseteq T$ of all $t \in T$ such that $tN_0t^{-1} \subseteq N_0$, or equivalently, such that $|\alpha(t)| \leq 1$ for any $\alpha \in \Delta$. Obviously

$$P_+ := N_0 T_+ = P_0 T_+ P_0$$

then is a submonoid of P.

We also fix a finite extension $K/\mathbb{Q}_p$ with ring of integers o, prime element π, and residue class field k. By a representation we always will mean a linear action of the respective group (or monoid) in a torsion o-module V. It is called smooth if the stabilizer of each element in V is open in the group. It is called finitely generated if there are finitely many elements in V such that the smallest submodule of V which contains all translates by the group (or monoid) of these elements is V itself.

We recall that Pontrjagin duality $V \longmapsto V^* := \mathrm{Hom}_o(V, K/o)$ sets up an anti-equivalence between the category of all torsion o-modules and the category of all compact linear-topological o-modules. In particular, the functor $V \longmapsto V^*$ is exact on torsion o-modules.

For any compact open subgroup $G_0 \subseteq G$ let $\Lambda(G_0) := o[[G_0]]$, resp. $\Omega(G_0) := k[[G_0]] = \Lambda(G_0)/\pi\Lambda(G_0)$, denote the completed group ring of the profinite group G_0 over o, resp. over k. Any smooth G_0-representation V is the filtered union of its finite subrepresentations. Its Pontrjagin dual V^* therefore is a (compact) module over $\Lambda(G_0)$ (always considered as a left $\Lambda(G_0)$-module through the inversion map on G_0).

1. $\Lambda(P_+)$-modules

The monoid P_+ acts by conjugation upon itself. Let $\mathcal{N}(P_+)$ denote the set of all "open left-normal subgroups" of P_+, i. e., all open normal subgroups $Q \subseteq P_0$ which satisfy $bQb^{-1} \subseteq Q$ for any $b \in P_+$. This is a fundamental system of open neighbourhoods of the unit element in P_+. For each $Q \in \mathcal{N}(P_+)$ we may form the factor monoid $Q \backslash P_+$ as well as the corresponding monoid ring $o[Q\backslash P_+]$. The conjugation action of P_+ passes to $Q\backslash P_+$ and hence to an action on the ring $o[Q\backslash P_+]$. In the projective limit we obtain the unital o-algebra

$$\widetilde{\Lambda}(P_+) := \varprojlim_Q o[Q\backslash P_+]$$

together with an action of P_+ on it. The obvious map

$$\Lambda(P_0) \otimes_{o[P_0]} o[P_+] \longrightarrow \widetilde{\Lambda}(P_+)$$

is injective and its image is a subring $\Lambda(P_+)$ of $\widetilde{\Lambda}(P_+)$ which is invariant under the P_+-action. We will write $\phi_b : \Lambda(P_+) \longrightarrow \Lambda(P_+)$ for the ring endomorphism given by the action of the element $b \in P_+$. If $b \in P_0$ then we obviously have $\phi_b(\lambda) = b\lambda b^{-1}$ for any $\lambda \in \Lambda(P_+)$. For any ϕ_b-invariant subring $\Lambda \subseteq \Lambda(P_+)$ we denote by $\Lambda \otimes_{\Lambda, b} -$ the base change functor for left Λ-modules along the ring homomorphism ϕ_b. Let $\Theta \subseteq T_+$ be a subset of representatives for the cosets in T_+/T_0. Then, as a left $\Lambda(P_0)$-module, we have

$$\Lambda(P_+) = \bigoplus_{t' \in \Theta} \Lambda(P_0)t' . \tag{1}$$

LEMMA 1.1. *For any $b = n_0t \in P_+ = N_0T_+$ the ring endomorphism ϕ_b is injective and makes $\Lambda(P_+)$ a free right module of rank $[N_0 : bN_0b^{-1}] = [N_0 : tN_0t^{-1}]$ over itself; the map*

$$\begin{aligned} \Lambda(N_0) \otimes_{\Lambda(N_0),b} \Lambda(P_+) &\xrightarrow{\cong} \Lambda(P_+) \\ \nu \otimes \lambda &\longmapsto \nu\phi_b(\lambda) \end{aligned}$$

is an isomorphism.

PROOF. We have $tP_0t^{-1} = tN_0t^{-1}T_0$ and hence the disjoint decomposition

$$P_0 = \bigcup_{n \in N_0/tN_0t^{-1}} n(tP_0t^{-1}) \ .$$

Conjugation by n_0 gives

$$P_0 = n_0P_0n_0^{-1} = \bigcup_{n \in N_0/tN_0t^{-1}} n_0nn_0^{-1}(bP_0b^{-1}) = \bigcup_{n \in N_0/bN_0b^{-1}} n(bP_0b^{-1}) \ .$$

It follows that

$$\begin{aligned} \Lambda(P_+) = n_0\Lambda(P_+)n_0^{-1} &= \bigoplus_{t' \in \Theta} \Lambda(P_0)n_0t'n_0^{-1} \\ &= \bigoplus_{t' \in \Theta} \bigoplus_{n \in N_0/bN_0b^{-1}} n\Lambda(bP_0b^{-1})bt'b^{-1} \\ &= \bigoplus_{n \in N_0/bN_0b^{-1}} n \operatorname{im}(\phi_b) \ . \end{aligned}$$

□

Let $\mathcal{M}(\Lambda(P_+))$ be the abelian category of all left (unital) $\Lambda(P_+)$-modules and $D(\Lambda(P_+))$ the corresponding derived category.

DEFINITION 1.2. *A left unital $\Lambda(P_+)$-module M is called etale if, for any $b \in P_+$, the $\Lambda(P_+)$-linear map*

$$\begin{aligned} \Lambda(P_+) \otimes_{\Lambda(P_+),b} M &\xrightarrow{\cong} M \\ \lambda \otimes x &\longmapsto \lambda bx \end{aligned}$$

is bijective.

Obviously the condition in the above definition is automatically satisfied for any $b \in P_0$ and therefore needs to be checked only for every $t' \in \Theta$. In fact, because of Lemma 1.1 a $\Lambda(P_+)$-module M is etale if and only if

$$\begin{aligned} \Lambda(N_0) \otimes_{\Lambda(N_0),t'} M &\xrightarrow{\cong} M \\ \lambda \otimes x &\longmapsto \lambda t'x \end{aligned}$$

is bijective for any $t' \in \Theta$. Let $\mathcal{M}_{et}(\Lambda(P_+))$ denote the full subcategory of all etale $\Lambda(P_+)$-modules in $\mathcal{M}(\Lambda(P_+))$.

PROPOSITION 1.3. *The subcategory $\mathcal{M}_{et}(\Lambda(P_+))$ of $\mathcal{M}(\Lambda(P_+))$ is closed under the formation of kernels, cokernels, extensions and arbitrary inductive and projective limits; in particular, $\mathcal{M}_{et}(\Lambda(P_+))$ is an abelian category.*

PROOF. This is a straightforward consequence of Lemma 1.1. □

In general the subcategory $\mathcal{M}_{et}(\Lambda(P_+))$ of $\mathcal{M}(\Lambda(P_+))$ is not closed under the passage to submodules.

It also follows that the full subcategory $D_{et}(\Lambda(P_+))$ of all those complexes in $D(\Lambda(P_+))$ whose cohomology modules are etale is a triangulated subcategory.

In a completely symmetric way we have the ring $\widetilde{\Lambda}(P_+^{-1})$ for the monoid P_+^{-1}. The map $b \longmapsto b^{-1}$ induces an anti-isomorphism of rings $\widetilde{\Lambda}(P_+) \xrightarrow{\cong} \widetilde{\Lambda}(P_+^{-1})$.

We also will need the following straightforward variants of everything above. Let $T_\star \subseteq T_+$ be any submonoid and put $P_\star := N_0 T_\star$. We then have the subring

$$\Lambda(P_\star) := \Lambda(P_\star \cap P_0) \otimes_{o[P_\star \cap P_0]} o[P_\star] \subseteq \Lambda(P_+)$$

as well as the abelian categories $\mathcal{M}(\Lambda(P_\star))$ and $\mathcal{M}_{et}(\Lambda(P_\star))$ together with the forgetful functors

$$\mathcal{M}(\Lambda(P_+)) \longrightarrow \mathcal{M}(\Lambda(P_\star)) \qquad \text{and} \qquad \mathcal{M}_{et}(\Lambda(P_+)) \longrightarrow \mathcal{M}_{et}(\Lambda(P_\star)) \ .$$

For the latter observe that by Lemma 1.1 the map

$$\begin{aligned} \Lambda(P_\star) \otimes_{\Lambda(P_\star),b} \Lambda(P_+) &\xrightarrow{\cong} \Lambda(P_+) \\ \mu \otimes \lambda &\longmapsto \mu\phi_b(\lambda) \ , \end{aligned}$$

for any $b \in P_\star$, is an isomorphism.

Of particular interest is the case of the group $GL_2(\mathbb{Q}_p)$ and its Borel subgroup of lower triangular matrices $P_2(\mathbb{Q}_p)$. The "standard monoid" in $P_2(\mathbb{Q}_p)_+$ is

$$S_\star := \{ \begin{pmatrix} 1 & 0 \\ a & b \end{pmatrix} : a \in \mathbb{Z}_p, b \in (1 + p^{\epsilon(p)}\mathbb{Z}_p)p^{\mathbb{N}_0} \}.$$

with $\epsilon(2) := 2$ and $\epsilon(p) := 1$ for odd p. In $S_\star$ we have the subgroups

$$S_0 := \{ \begin{pmatrix} 1 & 0 \\ a & 1 \end{pmatrix} : a \in \mathbb{Z}_p \} \qquad \text{and} \qquad \Gamma := \{ \begin{pmatrix} 1 & 0 \\ 0 & b \end{pmatrix} : b \in 1 + p^{\epsilon(p)}\mathbb{Z}_p \} \cong \mathbb{Z}_p$$

and the element $\varphi := \left(\begin{smallmatrix} 1 & 0 \\ 0 & p \end{smallmatrix}\right)$. The ring $\Lambda(S_\star) = \Lambda(S_0\Gamma)[\varphi; \phi_\varphi]$ is a skew polynomial ring over $\Lambda(S_0\Gamma)$. Later on we will see that these data are very closely related to Fontaine's notion of a (φ, Γ)-module.

2. P_+-subrepresentations

Fix a smooth P-representation V. An o-submodule $M \subseteq V$ will be called *generating* if $V = PM$.

LEMMA 2.1. *For any P_+-subrepresentation M of V the following assertions are equivalent:*

i. *M is generating;*
ii. *for any $v \in V$ there is a $t \in T_+$ such that $tv \in M$.*

PROOF. It is trivial that ii. implies i. Let M therefore be generating and let $v \in V$ be any vector. We then find elements $n_1, \ldots, n_r \in N$, $t_1, \ldots, t_r \in T$, and $v_1, \ldots, v_r \in M$ such that

$$v = \sum_{i=1}^{r} n_i t_i v_i \ .$$

We choose a $t \in T_+$ such that

$$t n_i t_i = (t n_i t^{-1})(t t_i) \in N_0 T_+$$

for any $1 \leq i \leq r$. We obtain $tv \in M$. □

LEMMA 2.2. *Let $M_0, M_1 \subseteq V$ be two generating P_+-subrepresentations; then $M_0 \cap M_1$ is a generating P_+-subrepresentation as well.*

PROOF. Clearly $M_0 \cap M_1$ is P_+-invariant. We check the assertion ii. in Lemma 2.1. Let $v \in V$ be any vector. Since M_i is generating we find a $t_i \in T_+$ such that $t_i v \in M_i$. Then $t_1 t_2 v = t_2 t_1 v \in M_1 \cap M_2$. □

We let $\mathcal{P}_+(V)$ denote the set of all generating P_+-subrepresentations of V. The above lemma says that the set $\mathcal{P}_+(V)$ is decreasingly filtered with respect to the partial order given by inclusion.

LEMMA 2.3. *For any map $f : V_1 \longrightarrow V_2$ of smooth P-representations and any $M \in \mathcal{P}_+(V_2)$ we have $f^{-1}(M) \in \mathcal{P}_+(V_1)$.*

PROOF. Obviously $f^{-1}(M)$ is P_+-invariant. Using again Lemma 2.1 let $v \in V_1$ be any vector. Since M is generating we find a $t \in T_+$ such that $tf(v) = f(tv) \in M$. Hence $tv \in f^{-1}(M)$. □

From $\mathcal{P}_+(V)$ we obtain by dualizing the filtered inductive system $\{M^*\}_{M\in\mathcal{P}_+(V)}$ of (left) $\widetilde{\Lambda}(P_+^{-1})$-modules. We define

$$D(V) := \varinjlim_{M\in\mathcal{P}_+(V)} M^*$$

as a $\widetilde{\Lambda}(P_+^{-1})$-module. We note that $D(V)$ actually is a quotient of V^* because the restriction maps in the filtered inductive system are surjective:

$$D(V) = V^*/\{x \in V^* : x|M = 0 \text{ for some } M \in \mathcal{P}_+(V)\}.$$

It follows from Lemma 2.3 that $D(V)$ is contravariantly functorial in V. Moreover, since $g^{-1}(M) \cap f^{-1}(M) \subseteq (g+f)^{-1}(M)$ for any two maps $g, f : V_1 \longrightarrow V_2$ this functor D is additive.

REMARK 2.4. *Let $f : V_1 \longrightarrow V_2$ be a map of smooth P-representations; we have:*

i. *If f is injective then $D(f)$ is surjective;*
ii. *if f is surjective then $D(f)$ is injective.*

PROOF. i. In fact in the commutative diagram of restriction maps

$$\begin{array}{ccc} V_2^* & \xrightarrow{f^*} & V_1^* \\ \downarrow & & \downarrow \\ D(V_2) & \xrightarrow{D(f)} & D(V_1) \end{array}$$

all four maps are surjective.

ii. This is immediate from the fact that for any $M \in \mathcal{P}_+(V_1)$ we have, by the surjectivity of the map f, that $f(M) \in \mathcal{P}_+(V_2)$. □

LEMMA 2.5. *Let $0 \longrightarrow V_1 \longrightarrow V_2 \longrightarrow V_3 \longrightarrow 0$ be a short exact sequence of smooth P-representations. Suppose that the following holds:*

(2) *For any $M_1 \in \mathcal{P}_+(V_1)$ we find $M_2 \in \mathcal{P}_+(V_2)$ such that $M_2 \cap V_1 \subseteq M_1$.*

Then the sequence $0 \longrightarrow D(V_3) \longrightarrow D(V_2) \longrightarrow D(V_1) \longrightarrow 0$ is exact.

PROOF. Straightforward. □

Example: (The Steinberg representation)
The group $P = NT$ acts on N by $(nt)(n') := ntn't^{-1}$. Consider the induced action of P on the vector space $V_{St} := C_c^\infty(N)$ of k-valued locally constant functions with compact support on N. It is straightforward to see that the subspace $C^\infty(N_0)$ of locally constant functions on N_0 is generating and P_+-invariant (cf. [**Vi2**] Lemme 4).

LEMMA 2.6. *Any generating P_+-subrepresentation $M \subseteq V_{St}$ contains $C^\infty(N_0)$.*

PROOF. By Lemma 2.1 we find a $t \in T_+$ such that $t\,\mathrm{char}_{N_0} = \mathrm{char}_{tN_0t^{-1}} \in M$ where $\mathrm{char}_?$ denotes the characteristic function of a compact open subgroup $? \subseteq N$. But $tN_0t^{-1} \subseteq N_0$ easily implies that $C^\infty(N_0) = N_0T_+ \cdot k\,\mathrm{char}_{tN_0t^{-1}}$. □

It follows that $D(V_{St}) = \Lambda(N_0)/\pi\Lambda(N_0)$.

3. The case of a compactly induced representation

We now analyze in detail the following kind of P-representations. Let V_0 be a smooth P_0-representation and form the compact induction $V := \mathrm{ind}_{P_0}^P(V_0)$. This is the o-module

$$\mathrm{ind}_{P_0}^P(V_0) := \text{all compactly supported functions } \psi : P \longrightarrow V_0 \text{ such that}$$
$$\psi(bb_0) = b_0^{-1}\psi(b) \quad \text{for any } b \in P, b_0 \in P_0$$

with the group P acting by left translations. As a piece of notation we denote, for any right P_0-invariant subset $X \subseteq P$, by $\mathrm{ind}_{P_0}^X(V_0)$ the o-submodule in V of all those functions with support in X. Clearly the map

$$\begin{aligned} \mathrm{ind}_{P_0}^{P_0}(V_0) &\xrightarrow{\cong} V_0 \\ \psi &\longmapsto \psi(1) \end{aligned}$$

is a P_0-equivariant isomorphism. By abuse of notation, we let any $v \in V_0$ denote at the same time the function in the left hand side corresponding to it. For any $s \in T_+$ we have the subset $P_+s = N_0T_+sP_0$ in P so that we may introduce the P_+-subrepresentation

$$M_s := M_s(V_0) := \mathrm{ind}_{P_0}^{P_+s}(V_0)$$

of V. Containing sV_0 any M_s generates V as a P-representation. Hence $M_s \in \mathcal{P}_+(V)$. If V_0 is finite then for the same reason M_s is finitely generated as a P_+-representation. We have $M_{s'} \subseteq M_s$ whenever $s' \in T_+s$. Hence these subrepresentations are decreasingly filtered.

We also define, for each $t \in T_+$, the P_0-subrepresentation

$$M(t) := M(t)(V_0) := \mathrm{ind}_{P_0}^{N_0tP_0}(V_0)\ .$$

Since P_+ is the disjoint union

$$P_+ = N_0T_+P_0 = \bigcup_{t \in T_+/T_0} N_0tP_0\ .$$

we have the P_0-invariant decomposition

$$M_s = \bigoplus_{t \in (T_+s)/T_0} M(t)\ .$$

LEMMA 3.1. *Suppose that V_0 is finite; for any generating P_+-subrepresentation $M \subseteq V$ there is an $s \in T_+$ such that $M_s \subseteq M$.*

PROOF. Since M is generating we find finitely many elements $b_1, \ldots, b_r \in P$ such that

$$V_0 \subseteq \sum_{i=1}^{r} b_i M \ .$$

Choose an $s \in T_+$ such that $sb_i \in P_+$ for any $1 \leq i \leq r$. Then

$$M_s = P_+ s V_0 \subseteq \sum_{i=1}^{r} P_+ s b_i M \subseteq M \ .$$

□

For general V_0 we introduce the set $\mathrm{Sub}(V_0)$, resp. $\mathrm{Fin}(V_0)$, of all, resp. all finite, P_0-subrepresentations of V_0. Both are partially ordered by inclusion. On the other hand, the monoid T_+/T_0 is preordered by $t'T_0 \leq tT_0$ if $t' \in T_+ t$. Let $\sigma : T_+/T_0 \longrightarrow \mathrm{Sub}(V_0)$ be any order reversing map which satisfies

$$\bigcup_{t \in T_+/T_0} \sigma(t) = V_0 \ . \tag{3}$$

We form the subspace

$$M_\sigma := M_\sigma(V_0) := \bigoplus_{t \in T_+/T_0} M(t)(\sigma(t))$$

of $V = \mathrm{ind}_{P_0}^{P}(V_0)$.

LEMMA 3.2. i. *$M_\sigma \in \mathcal{P}_+(V)$.*
ii. *Given any $M \in \mathcal{P}_+(V)$ there is a σ as above such that $M_\sigma \subseteq M$.*

PROOF. i. Given any U_0 in $\mathrm{Sub}(V_0)$ the subspace $M(t)(U_0)$ is P_0-invariant and satisfies

$$t'M(t)(U_0) \subseteq M(t't)(U_0) \qquad \text{for any } t' \in T_+ \ .$$

It follows that

$$t'\big(M(t)(\sigma(t))\big) \subseteq M(t't)(\sigma(t)) \subseteq M(t't)(\sigma(t't))$$

since σ is order reversing. Hence M_σ is P_+-invariant. Again since σ is order reversing we have

$$M_\sigma \supseteq M_s(\sigma(s)) \qquad \text{for any } s \in T_+ \ .$$

It follows that $PM_\sigma \supseteq \mathrm{ind}_{P_0}^{P}(\sigma(s))$ for any $s \in T_+$ and hence, as a consequence of the condition (3), that

$$PM_\sigma \supseteq \sum_{s \in T_+} \mathrm{ind}_{P_0}^{P}(\sigma(s)) = \mathrm{ind}_{P_0}^{P}\big(\sum_{s \in T_+} \sigma(s)\big) = \mathrm{ind}_{P_0}^{P}(V_0) \ .$$

ii. It follows from Lemma 2.3 that, for any $U_0 \in \mathrm{Fin}(V_0)$, the P_+-subrepresentation $M \cap \mathrm{ind}_{P_0}^{P}(U_0)$ is generating in $\mathrm{ind}_{P_0}^{P}(U_0)$. Using Lemma 3.1 we find an $s(U_0) \in T_+$

such that $M_{s(U_0)}(U_0) \subseteq M$. Hence M contains

$$\begin{aligned}\sum_{U_0 \in \mathrm{Fin}(V_0)} M_{s(U_0)}(U_0) &= \sum_{U_0 \in \mathrm{Fin}(V_0)} \bigoplus_{t \in T_+ s(U_0)/T_0} M(t)(U_0) \\ &= \bigoplus_{t \in T_+/T_0} \sum_{\substack{U_0 \in \mathrm{Fin}(V_0) \\ t \in T_+ s(U_0)}} M(t)(U_0) \\ &= \bigoplus_{t \in T_+/T_0} M(t)\Big(\sum_{\substack{U_0 \in \mathrm{Fin}(V_0) \\ t \in T_+ s(U_0)}} U_0 \Big) .\end{aligned}$$

We therefore set

$$\sigma(t) := \sum_{U_0 \in \mathrm{Fin}(V_0), t \in T_+ s(U_0)} U_0 .$$

The union of all $\sigma(t)$ contains the union of all $U_0 \in \mathrm{Fin}(V_0)$ and consequently is equal to V_0. □

The dual of $\mathrm{ind}_{P_0}^P(V_0)$ can be described explicitly as follows. Let

$$\begin{aligned}\mathrm{Ind}_{P_0}^P(V_0^*) := \text{all functions } \Phi : P \longrightarrow V_0^* \text{ such that}\\ \Phi(bb_0) = b_0^{-1}\Phi(b)\\ \text{for any } b \in P, b_0 \in P_0\end{aligned}$$

with the group P acting by left translations. The map

$$\begin{aligned}\mathrm{Ind}_{P_0}^P(V_0^*) &\xrightarrow{\cong} \mathrm{ind}_{P_0}^P(V_0)^* \\ \Phi &\longmapsto \big[\psi \mapsto \sum_{b \in P/P_0} \Phi(b)(\psi(b))\big]\end{aligned}$$

is a P-equivariant o-linear isomorphism. Under this isomorphism the orthogonal complement $M_\sigma^\perp$ of M_σ in $\mathrm{ind}_{P_0}^P(V_0)^*$ corresponds to the P_+^{-1}-invariant subspace

$$J_\sigma := J_\sigma(V_0) := \{\Phi \in \mathrm{Ind}_{P_0}^P(V_0^*) : \Phi(N_0 t) \subseteq \sigma(t)^\perp \text{ for any } t \in T_+\}$$

of $\mathrm{Ind}_{P_0}^P(V_0^*)$. We therefore obtain a natural isomorphism

$$D(\mathrm{ind}_{P_0}^P(V_0)) \cong \mathrm{Ind}_{P_0}^P(V_0^*)/\bigcup_\sigma J_\sigma$$

of $\widetilde{\Lambda}(P_+^{-1})$-modules. On the other hand there is the P_0-invariant decomposition

$$\mathrm{Ind}_{P_0}^P(V_0^*) = J^+ \oplus J^-$$

with

$$J^\pm := J^\pm(V_0) := \{\Phi \in \mathrm{Ind}_{P_0}^P(V_0^*) : \Phi|\Big(\begin{cases} P \setminus P_+ \\ P_+ \end{cases}\Big) = 0\} .$$

We write $\Phi = \Phi^+ + \Phi^-$ for the corresponding decomposition of any function $\Phi \in \mathrm{Ind}_{P_0}^P(V_0^*)$ into functions $\Phi^\pm \in J^\pm$. The subspace J^+ in $\mathrm{Ind}_{P_0}^P(V_0^*)$ is P_+-invariant.

Since obviously $J^- \subseteq J_\sigma$ for any σ the natural map

$$J^+ \longrightarrow \mathrm{Ind}_{P_0}^P(V_0^*)/\bigcup_\sigma J_\sigma \tag{4}$$

is surjective. For any order reversing map $\sigma : T_+/T_0 \longrightarrow \mathrm{Sub}(V_0)$ satisfying (3) and any $s \in T_+$ we define the new map

$$\sigma_s : T_+/T_0 \longrightarrow \mathrm{Sub}(V_0)$$
$$t \longmapsto \begin{cases} \sigma(s^{-1}t) & \text{if } t \in T_+s, \\ \{0\} & \text{otherwise} \end{cases}$$

which again is order reversing and satisfies (3). We have $J_\sigma \subseteq J_{\sigma_s}$.

LEMMA 3.3. i. *We have* $s(J^+ \cap J_\sigma) \subseteq J^+ \cap J_{\sigma_s}$ *for any* σ *satisfying* (3) *and any* $s \in T_+$.
ii. $J^+ \cap \bigcup_\sigma J_\sigma$ *is* P_+*-invariant.*

PROOF. i. Let $\Phi \in J^+ \cap J_\sigma$. We have to show that $s\Phi$ lies in J_{σ_s}. Let $n_0 \in N_0$ and $t \in T_+$. We have

$$(s\Phi)(n_0t) = \Phi(s^{-1}n_0t) = \Phi((s^{-1}n_0s)(s^{-1}t)) \ .$$

If $s^{-1}n_0t \notin P_+$ then

$$(s\Phi)(n_0t) = \Phi(s^{-1}n_0t) = 0 \in \sigma_s(t)^\perp \ .$$

If $s^{-1}n_0t \in P_+$ then $s^{-1}n_0s \in N_0$ and $s^{-1}t \in T_+$ so that

$$(s\Phi)(n_0t) \in \Phi(N_0(s^{-1}t)) \subseteq \sigma(s^{-1}t)^\perp = \sigma_s(t)^\perp \ .$$

ii. This is an immediate consequence of i. □

By (4) we have the $\Lambda(P_0)$-equivariant isomorphism

$$J^+/J^+ \cap \bigcup_\sigma J_\sigma \xrightarrow{\cong} \mathrm{Ind}_{P_0}^P(V_0^*)/\bigcup_\sigma J_\sigma \cong D(\mathrm{ind}_{P_0}^P(V_0)) \ .$$

The right hand side in fact carries a $\widetilde{\Lambda}(P_+^{-1})$-action whereas the left hand side, by the lemma, has a $\Lambda(P_+)$-action. By transport of structure we view this latter action also on $D(\mathrm{ind}_{P_0}^P(V_0))$. If we represent an element $x \in D(\mathrm{ind}_{P_0}^P(V_0))$ by a function $\Phi \in J^+$ then $t\Phi \in J^+$ represents tx for any $t \in T_+$. Since $t^{-1}(t\Phi) = \Phi$ we see that, for any $t \in T_+$, the operator t^{-1} on $D(\mathrm{ind}_{P_0}^P(V_0))$ is a distinguished left inverse of the operator t.

PROPOSITION 3.4. *For any map* $f : \mathrm{ind}_{P_0}^P(U_0) \longrightarrow \mathrm{ind}_{P_0}^P(V_0)$ *of compactly induced smooth* P*-representations the map* $D(f) : D(\mathrm{ind}_{P_0}^P(V_0)) \longrightarrow D(\mathrm{ind}_{P_0}^P(U_0))$ *is a map of* $\Lambda(P_+)$*-modules.*

PROOF. By Lemma 2.3 and Lemma 3.2.ii we find an order reversing map $\sigma : T_+/T_0 \longrightarrow \mathrm{Sub}(U_0)$ satisfying (3) such that

$$f(M_\sigma(U_0)) \subseteq M_1(V_0) \ . \tag{5}$$

Dually we then have

$$f^*(J^-(V_0)) \subseteq J_\sigma(U_0) \ .$$

Therefore, if $s \in T_+$ is any element, all maps in the diagram

$$\begin{array}{ccc}
J^+(V_0) & \xrightarrow{\quad s \quad} & J^+(V_0) \\
\cong \downarrow & & \downarrow \cong \\
\mathrm{Ind}_{P_0}^P(V_0^*)/J^-(V_0) & & \mathrm{Ind}_{P_0}^P(V_0^*)/J^-(V_0) \\
f^* \downarrow & & \downarrow f^* \\
\mathrm{Ind}_{P_0}^P(U_0^*)/J_\sigma(U_0) & & \mathrm{Ind}_{P_0}^P(U_0^*)/J_{\sigma_s}(U_0) \\
\cong \uparrow & & \uparrow \cong \\
J^+(U_0)/J^+(U_0) \cap J_\sigma(U_0) & \xrightarrow{s} & J^+(U_0)/J^+(U_0) \cap J_{\sigma_s}(U_0)
\end{array}$$

are well defined. It suffices to show that this diagram is commutative. But to do this we first have to observe another consequence of (5). Suppose that a function $\psi \in \mathrm{ind}_{P_0}^P(U_0)$ is supported on $NT_+ \setminus P_+$ and such that $\psi(Nt') \subseteq \sigma(t')$ for any $t' \in T_+$. Let R be a set of representatives for the cosets in N/N_0 which contains the unit element 1. Because of the disjoint decomposition

$$NT_+ = \bigcup_{n \in R} nN_0T_+$$

we may write ψ as a finite sum

$$\psi = n_1\psi_{n_1} + \ldots + n_r\psi_{n_r} \qquad \text{with } n_1, \ldots, n_r \in R \setminus \{1\} \text{ and } \psi_{n_i} \in M_\sigma(U_0).$$

It follows from (5) that

$$f(\psi) = \sum_i n_i f(\psi_{n_i}) \in \sum_i n_i f(M_\sigma(U_0)) \subseteq \sum_i n_i M_1(V_0)$$

and hence is supported on $P \setminus P_+$. Passing to orthogonal complements we obtain that

$$(6) \qquad f^*(\Phi)((N \setminus N_0)t') \subseteq \sigma(t')^\perp \qquad \text{for any } t' \in T_+ \text{ and any } \Phi \in J^+(V_0).$$

To now check the commutativity of the diagram we let $\Phi \in J^+(V_0)$. Since $J^- \subseteq J_\sigma$ its image under the composed map in the left column is $f^*(\Phi)^+$. Hence we have to prove that

$$s(f^*(\Phi)^+) - f^*(s\Phi)^+ = s(f^*(\Phi)^+) - (sf^*(\Phi))^+ \in J_{\sigma_s}(U_0)$$

or, equivalently, that

$$s(f^*(\Phi)^+)(n_0t) - (sf^*(\Phi))^+(n_0t) \in \sigma(s^{-1}t)^\perp \qquad \text{for any } n_0 \in N_0 \text{ and } t \in T_+s$$

holds true. If $t = t's$ we have

$$\begin{aligned} s(f^*(\Phi)^+)(n_0t) - (sf^*(\Phi))^+(n_0t) &= f^*(\Phi)^+(s^{-1}n_0st') - f^*(\Phi)(s^{-1}n_0st') \\ &= \begin{cases} -f^*(\Phi)(s^{-1}n_0st') & \text{if } s^{-1}n_0s \notin N_0, \\ 0 & \text{otherwise.} \end{cases} \end{aligned}$$

This reduces us to showing that

$$f^*(\Phi)(s^{-1}n_0st') \in \sigma(t')^\perp \qquad \text{for any } n_0 \in N_0 \setminus sN_0s^{-1} \text{ and } t' \in T_+$$

which we already did in (6). □

PROPOSITION 3.5. *The $\Lambda(P_+)$-module $D(\mathrm{ind}_{P_0}^{P}(V_0))$ is etale.*

PROOF. Similarly as with $\mathrm{ind}_{P_0}^{X}$ we will use $\mathrm{Ind}_{P_0}^{X}(.)$, for any right P_0-invariant subset $X \subseteq P$, to denote the submodule of all functions $\Phi \in \mathrm{Ind}_{P_0}^{P}(.)$ with support in X.

We have to show that, for any $s \in T_+$, the map

$$\Lambda(P_+) \otimes_{\Lambda(P_+),s} (J^+/J^+ \cap \bigcup_\sigma J_\sigma) \xrightarrow{\cong} J^+/J^+ \cap \bigcup_\sigma J_\sigma$$
$$\lambda \otimes x \longmapsto \lambda s x$$

is bijective. Let $n_1, \ldots, n_r \in N_0$ be representatives for the cosets in N_0/sN_0s^{-1}. By Lemma 1.1 the above map may be viewed as the map

$$(J^+/J^+ \cap \bigcup_\sigma J_\sigma)^r \longrightarrow J^+/J^+ \cap \bigcup_\sigma J_\sigma$$
$$(x_1, \ldots, x_r) \longmapsto \sum_{i=1}^{r} n_i s x_i \ .$$

Lemma 3.3.i then reduces us to showing that, for each σ, the map

$$(J^+/J^+ \cap J_\sigma)^r \longrightarrow J^+/J^+ \cap J_{\sigma_s}$$
$$(x_1, \ldots, x_r) \longmapsto \sum_{i=1}^{r} n_i s x_i$$

is bijective. But we have the N_0-equivariant decompositions

$$J^+/J^+ \cap J_\sigma = \prod_{t \in T_+/T_0} \mathrm{Ind}_{P_0}^{N_0 t P_0}((V_0/\sigma(t))^*)$$

and

$$J^+/J^+ \cap J_{\sigma_s} = \prod_{t \in T_+/T_0} \mathrm{Ind}_{P_0}^{N_0 s t P_0}((V_0/\sigma(t))^*)$$

which are respected by the action of s. It finally comes down therefore to the bijectivity of the map

$$(\mathrm{Ind}_{P_0}^{N_0 t P_0}(.))^r \longrightarrow \mathrm{Ind}_{P_0}^{N_0 s t P_0}(.)$$
$$(\Phi_1, \ldots, \Phi_r) \longmapsto \sum_{i=1}^{r} n_i s \Phi_i$$

which is straightforward from the disjoint union

$$N_0 s t P_0 = \bigcup_{i=1}^{r} n_i (s N_0 s^{-1}) s t P_0 = \bigcup_{i=1}^{r} n_i s N_0 t P_0 \ .$$

□

Sometimes it is technically useful to notice that $M_1(V_0)^*$ is through transport of structure along the isomorphism

$$J^+(V_0) \xrightarrow{\cong} \mathrm{Ind}_{P_0}^{P}(V_0^*)/J^-(V_0) \cong M_1(V_0)^*$$

a $\Lambda(P_+)$-module. It is not etale, though, but of course also carries a $\widetilde{\Lambda}(P_+^{-1})$-action providing distinguished left inverses. The natural surjection $M_1(V_0)^* \twoheadrightarrow D(\mathrm{ind}_{P_0}^{P}(V_0))$ respects all these structures.

4. The basic cohomological functor

Let $\mathcal{M}_{o-tor}(P)$ denote the abelian category of all smooth P-representations. Any V in $\mathcal{M}_{o-tor}(P)$ comes with the canonical epimorphism

$$\begin{aligned} \rho_V : \operatorname{ind}_{P_0}^{P}(V) &\longrightarrow V \\ \psi &\longmapsto \sum_{b\in P/P_0} b\psi(b) \end{aligned}$$

in $\mathcal{M}_{o-tor}(P)$. This leads to the canonical resolution

$$\mathcal{I}_\bullet(V) : \ldots \longrightarrow \mathcal{I}_{n+1}(V) \xrightarrow{\rho_n} \mathcal{I}_n(V) \xrightarrow{\rho_{n-1}} \ldots \xrightarrow{\rho_0} \mathcal{I}_0(V)$$

in $\mathcal{M}_{o-tor}(P)$ inductively defined by $\mathcal{I}_0(V) := \operatorname{ind}_{P_0}^{P}(V)$, $\rho_{-1} := \rho_V$,

$$\mathcal{I}_{n+1}(V) := \operatorname{ind}_{P_0}^{P}(\ker \rho_{n-1}), \text{ and } \rho_n := \rho_{\ker \rho_{n-1}} \text{ for } n \geq 0$$

such that

$$\mathcal{I}_\bullet(V) \xrightarrow[\rho_V]{\simeq} V$$

is a quasi-isomorphism.

As usual, let $C^{\geq 0}(\mathcal{M}_{et}(\Lambda(P_+)))$ denote the category of cohomological complexes in nonnegative degrees in $\mathcal{M}_{et}(\Lambda(P_+))$. We have the contravariant functor

$$\begin{aligned} RD : \mathcal{M}_{o-tor}(P) &\longrightarrow C^{\geq 0}(\mathcal{M}_{et}(\Lambda(P_+))) \\ V &\longmapsto D(\mathcal{I}_\bullet(V)) \ . \end{aligned}$$

We put

$$D^i(V) := h^i(D(\mathcal{I}_\bullet(V)))$$

for $i \geq 0$. By Remark 2.4.ii there is a natural injection $D(V) \hookrightarrow D^0(V)$ but which in general is not bijective.

LEMMA 4.1. *The functor* $V_0 \longmapsto D(\operatorname{ind}_{P_0}^{P}(V_0))$ *on* $\mathcal{M}_{o-tor}(P_0)$ *is exact.*

PROOF. Clearly the functor $V_0 \longmapsto \operatorname{ind}_{P_0}^{P}(V_0)$ is exact. It therefore is sufficient to check the hypothesis (2) in Lemma 2.5 for any inclusion $\operatorname{ind}_{P_0}^{P}(V_0') \subseteq \operatorname{ind}_{P_0}^{P}(V_0)$ coming from a pair $V_0' \subseteq V_0$ of smooth P_0-representations. In view of Lemma 3.2 we give ourselves a order reversing map $\sigma' : T_+/T_0 \longrightarrow \operatorname{Sub}(V_0')$ satisfying $\bigcup_{t\in T_+/T_0} \sigma'(t) = V_0'$. We also pick a strictly dominant element $s_0 \in T_+$, which means that $|\alpha(s_0)| < 1$ for any $\alpha \in \Delta$. We note that the subset $\{s_0^m : m \geq 0\}$ is cofinal in the preordered set T_+. Using Zorn's lemma we find inductively, for any $m \geq 0$, a subrepresentation $\sigma(s_0^m) \in \operatorname{Sub}(V_0)$ which is maximal with respect to the properties that

$$\sigma(s_0^m) \cap V_0' = \sigma'(s_0^m) \quad \text{and} \quad \sigma(s_0^m) \supseteq \sigma(s_0^{m-1})$$

(where $\sigma(s_0^{-1}) := \{0\}$). We now define the order reversing function $\sigma : T_+/T_0 \longrightarrow \operatorname{Sub}(V_0)$ by

$$\sigma(t) := \sigma(s_0^m) \quad \text{if } \ t \in s_0^m T_+ \setminus s_0^{m+1} T_+ \ .$$

In order to check that

$$\bigcup_{t\in T_+/T_0} \sigma(t) = \bigcup_{m\geq 0} \sigma(s_0^m) = V_0 \tag{7}$$

holds true we consider any $U_0 \in \mathrm{Fin}(V_0)$. For any $n \geq m$ we have the obvious commutative diagram

$$\begin{array}{ccccc} [\sigma(s_0^m)+U_0]\cap V_0'/\sigma'(s_0^m) & \hookrightarrow & \sigma(s_0^m)+U_0/\sigma(s_0^m) & \xleftarrow{\cong} & U_0/U_0\cap\sigma(s_0^m) \\ \downarrow & & \downarrow & & \downarrow \\ [\sigma(s_0^n)+U_0]\cap V_0'/\sigma'(s_0^n) & \hookrightarrow & \sigma(s_0^n)+U_0/\sigma(s_0^n) & \xleftarrow{\cong} & U_0/U_0\cap\sigma(s_0^n) \end{array}$$

in which the left horizontal arrows are injective. In particular all members of the diagram are finite. By the finiteness of U_0 the increasing sequence of subspaces $U_0\cap\sigma(s_0^m)$ has to stabilize. This means that the right vertical arrow in the diagram is bijective for sufficiently big m. On the other hand, since $\bigcup_{m\geq 0}\sigma'(s_0^m)=V_0'$, the left vertical arrow in the diagram is the zero map whenever the difference $n-m$ is sufficiently big. It follows that $[\sigma(s_0^m)+U_0]\cap V_0'=\sigma'(s_0^m)$ for big m which, by the maximality property of $\sigma(s_0^m)$, means that $U_0\subseteq\sigma(s_0^m)$. This establishes (7). By construction we have

$$M_\sigma(V_0)\cap \mathrm{ind}_{P_0}^P(V_0') = M_{\sigma'}(V_0') .$$

□

LEMMA 4.2. *For any short exact sequence* $0\longrightarrow V_1\longrightarrow V_2\longrightarrow V_3\longrightarrow 0$ *in* $\mathcal{M}_{o-tor}(P)$ *we have the long exact sequence*

$$\begin{aligned} 0 \longrightarrow D^0(V_3) \longrightarrow D^0(V_2) \longrightarrow D^0(V_1) \longrightarrow D^1(V_3) \longrightarrow \dots \\ \longrightarrow D^i(V_3) \longrightarrow D^i(V_2) \longrightarrow D^i(V_1) \longrightarrow \dots \end{aligned}$$

in $\mathcal{M}_{et}(\Lambda(P_+))$.

PROOF. We apply the functor D to the short exact sequence of resolutions

$$0\longrightarrow \mathcal{I}_\bullet(V_1)\longrightarrow \mathcal{I}_\bullet(V_2)\longrightarrow \mathcal{I}_\bullet(V_3)\longrightarrow 0$$

and obtain the sequence of complexes

$$0\longrightarrow D(\mathcal{I}_\bullet(V_3))\longrightarrow D(\mathcal{I}_\bullet(V_2))\longrightarrow D(\mathcal{I}_\bullet(V_1))\longrightarrow 0$$

in $C^{\geq 0}(\mathcal{M}_{et}(\Lambda(P_+)))$. It is exact by Lemma 4.1. The associated long exact cohomology sequence is the asserted sequence. □

LEMMA 4.3. *Any compactly induced smooth* P*-representation* $V=\mathrm{ind}_{P_0}^P(V_0)$ *satisfies* $D^0(V)=D(V)$ *and* $D^i(V)=0$ *for* $i\geq 1$.

PROOF. In this case we have two more P-equivariant maps besides ρ_V. These are

$$\begin{aligned} \epsilon_V : \mathrm{ind}_{P_0}^P(V) &\longrightarrow \mathrm{ind}_{P_0}^P(V_0)=V \\ \psi &\longmapsto \epsilon_V(\psi)(b):=\psi(b)(1) \end{aligned}$$

and

$$\begin{aligned} \sigma_V : V=\mathrm{ind}_{P_0}^P(V_0) &\longrightarrow \mathrm{ind}_{P_0}^P(V) \\ \psi &\longmapsto \sigma_V(\psi)(b)(c):=\begin{cases}\psi(bc) & \text{if } c\in P_0,\\ 0 & \text{otherwise.}\end{cases} \end{aligned}$$

It is straightforward to check that

$$\rho_V\circ\sigma_V=\mathrm{id}_V=\epsilon_V\circ\sigma_V .$$

It follows that

$$\mathrm{ind}_{P_0}^{P}(V) = \mathrm{im}(\sigma_V) \oplus \ker(\epsilon_V) = \mathrm{im}(\sigma_V) \oplus \ker(\rho_V) \ .$$

In particular, we have $\ker(\rho_V) \cong \ker(\epsilon_V)$. But $\ker(\epsilon_V) = \mathrm{ind}_{P_0}^{P}(\mathrm{ind}_{P_0}^{P \backslash P_o}(V_0))$ is compactly induced. Proceeding inductively we obtain that each $\ker \rho_{n-1}$ is compactly induced and is a direct summand of $\mathcal{I}_n(V)$. It easily follows that $D(\mathcal{I}_\bullet(V))$ is an exact resolution of $D(V)$. □

We see that the δ-functor $V \longmapsto D^i(V)$ is coeffacable and hence universal.

COROLLARY 4.4. *Let* $\ldots \longrightarrow \mathrm{ind}_{P_0}^{P}(V_n) \longrightarrow \ldots \longrightarrow \mathrm{ind}_{P_0}^{P}(V_0) \longrightarrow V \longrightarrow 0$ *be any exact sequence in* $\mathcal{M}_{o-tor}(P)$*; we then have* $D^i(V) \cong h^i(D(\mathrm{ind}_{P_0}^{P}(V_\bullet)))$ *for any* $i \geq 0$.

By the usual procedure the functor RD extends to an exact contravariant functor

$$\begin{aligned} RD : D^-(\mathcal{M}_{o-tor}(P)) &\longrightarrow D^+(\mathcal{M}_{et}(\Lambda(P_+))) \\ V^\bullet &\longmapsto \mathrm{Tot} D(\mathcal{I}_\bullet(V^\bullet)) \ . \end{aligned}$$

from the bounded above derived category $D^-(\mathcal{M}_{o-tor}(P))$ of the abelian category $\mathcal{M}_{o-tor}(P)$ into the bounded below derived category $D^+(\mathcal{M}_{et}(\Lambda(P_+)))$ of the abelian category $\mathcal{M}_{et}(\Lambda(P_+))$.

5. Towards (φ, Γ)-modules

At this point we fix once and for all, as part of an épinglage for the group G, isomorphisms of algebraic groups $\iota_\alpha : N_\alpha \xrightarrow{\cong} \mathbb{Q}_p$, for $\alpha \in \Delta$, such that

$$\iota_\alpha(tnt^{-1}) = \alpha(t)\iota_\alpha(n) \qquad \text{for any } n \in N_\alpha, t \in T.$$

Using that $\prod_{\alpha \in \Delta} N_\alpha$ naturally is a quotient of $N/[N,N]$ we then introduce the homomorphism of groups

$$\ell := \sum_{\alpha \in \Delta} \iota_\alpha : N \longrightarrow \mathbb{Q}_p \ .$$

It induces a (continuous) epimorphism of (compact) rings $\Lambda(N_0) \longrightarrow \Lambda(\ell(N_0)) \cong \Lambda(\mathbb{Z}_p)$, also denoted by ℓ. It is convenient to normalize the ι_α in such a way that

$$\iota_\alpha(N_0 \cap N_\alpha) = \mathbb{Z}_p \qquad \text{for any } \alpha \in \Delta$$

holds true. We then, in particular, have $\ell(N_0) = \mathbb{Z}_p$.

To avoid technicalities we assume from now on that the center of the group G is connected. Then the quotient $X^*(T)/\oplus_{\alpha \in \Delta} \mathbb{Z}\alpha$ is free. Hence we find a cocharacter $\xi \in X_*(T)$ such that $\alpha \circ \xi = \mathrm{id}_{\mathbb{G}_m}$ for any $\alpha \in \Delta$. It is injective and uniquely determined up to a central cocharacter. We once and for all fix such a ξ. It satisfies

$$\xi(\mathbb{Z}_p \setminus \{0\}) \subseteq T_+$$

and

$$\ell(\xi(a) n \xi(a^{-1})) = a\ell(n) \qquad \text{for any } a \in \mathbb{Q}_p^\times, n \in N \ . \tag{8}$$

Put

$$T_\star := \xi((1 + p^{\epsilon(p)}\mathbb{Z}_p)p^{\mathbb{N}_0}) \cong (1 + p^{\epsilon(p)}\mathbb{Z}_p)p^{\mathbb{N}_0} \qquad \text{and} \qquad P_\star := N_0 T_\star \ .$$

It follows from (8) that

$$P_\star = N_0 T_\star \longrightarrow S_\star$$
$$n_0 t \longmapsto \begin{pmatrix} 1 & 0 \\ \ell(n_0) & \xi^{-1}(t) \end{pmatrix}$$

is an epimorphism of monoids. If we define

$$N_1 := \ker(N_0 \xrightarrow{\ell} \mathbb{Z}_p)$$

then, in fact, we have the isomorphism

$$N_1 \backslash P_\star \xrightarrow{\cong} S_\star \ .$$

It follows that

$$\Lambda(S_\star) = o \otimes_{\Lambda(N_1)} \Lambda(P_\star)$$

holds true at least as bimodules. Hence we have a natural isomorphism

$$\Lambda(S_\star) \otimes_{\Lambda(P_\star)} M \cong o \otimes_{\Lambda(N_1)} M$$

for any M in $\mathcal{M}(\Lambda(P_\star))$. But $\Lambda(P_\star)$ is free as a left $\Lambda(N_0\xi(1+p^{\epsilon(p)}\mathbb{Z}_p))$-module by (1). Furthermore, $\Lambda(N_0\xi(1+p^{\epsilon(p)}\mathbb{Z}_p))$ is flat as a left $\Lambda(N_1)$-module (cf. the proof of Lemma 5.5 in [**OV**]). This implies that any flat and, in particular, any projective left $\Lambda(P_\star)$-module is flat as a $\Lambda(N_1)$-module. By using $\Lambda(P_\star)$-projective resolutions the above natural isomorphism therefore extends to natural isomorphisms

$$\mathrm{Tor}_i^{\Lambda(P_\star)}(\Lambda(S_\star), M) \cong \mathrm{Tor}_i^{\Lambda(N_1)}(o, M)$$

for any $i \geq 0$ and any M in $\mathcal{M}(\Lambda(P_\star))$. Since N_1 is pro-p and torsionfree the ring $\Lambda(N_1)$ is noetherian of finite global dimension by [**Neu**]. It follows that there is an integer $d(N_1) \geq 0$ such that

$$\mathrm{Tor}_i^{\Lambda(P_\star)}(\Lambda(S_\star), M) = 0 \qquad \text{for any } i > d(N_1)$$

and any M in $\mathcal{M}(\Lambda(P_\star))$. From this we conclude by a standard argument (cf. [**Har**] I.5.3) that the total derived tensor product

$$\Lambda(S_\star) \otimes^{\mathbb{L}}_{\Lambda(P_\star)} : D(\Lambda(P_\star)) \longrightarrow D(\Lambda(S_\star))$$

is well defined on the whole derived category and respects the bounded below derived categories

$$\Lambda(S_\star) \otimes^{\mathbb{L}}_{\Lambda(P_\star)} : D^+(\Lambda(P_\star)) \longrightarrow D^+(\Lambda(S_\star)) \ .$$

Moreover, suppose that M is an etale $\Lambda(P_\star)$-module. We observe that any projective $\Lambda(P_\star)$-module necessarily is etale. Let $F_\bullet \longrightarrow M$ be a projective resolution of M. Then $F_\bullet$ is a complex of etale $\Lambda(P_\star)$-modules and, similarly, $\Lambda(S_\star) \otimes_{\Lambda(P_\star)} F_\bullet$ is a complex of etale $\Lambda(S_\star)$-modules. Hence, as a consequence of Prop. 1.3,

$$\mathrm{Tor}_i^{\Lambda(P_\star)}(\Lambda(S_\star), M) = h_i(\Lambda(S_\star) \otimes_{\Lambda(P_\star)} F_\bullet) \qquad \text{for any } i \geq 0$$

is an etale $\Lambda(S_\star)$-module. In other words, the total derived tensor product restricts to a functor

$$\Lambda(S_\star) \otimes^{\mathbb{L}}_{\Lambda(P_\star)} : D^+_{et}(\Lambda(P_\star)) \longrightarrow D^+_{et}(\Lambda(S_\star)) \ .$$

between the bounded below derived categories with etale cohomology modules.

Altogether we obtain the composed functor

$$RD_{\Lambda(S_\star)} : D^-(\mathcal{M}_{o-tor}(P)) \xrightarrow{RD} D^+_{et}(\Lambda(P_+))$$
$$\xrightarrow{\text{forget}} D^+_{et}(\Lambda(P_\star)) \xrightarrow{\Lambda(S_\star)\otimes^{\mathbb{L}}_{\Lambda(P_\star)}} D^+_{et}(\Lambda(S_\star)) .$$

It gives rise on $\mathcal{M}_{o-tor}(P)$ to the δ-functor

$$D^i_{\Lambda(S_\star)}(V) := h^i(RD_{\Lambda(S_\star)}(V)) \qquad \text{for } i \geq 0$$

into etale $\Lambda(S_\star)$-modules.

6. A functor in the reverse direction

As noted in section 1 the conjugation action of T_+ on P_+ induces an action of T_+ by (continuous injective) ring endomorphisms ϕ_t on $\Lambda(P_+)$.

On the other hand it is not difficult to see that, for each $t \in T_+$, there is a unique surjective continuous linear (but not multiplicative) map $\psi_t : \Lambda(P_+) \longrightarrow \Lambda(P_+)$ which on group elements $b \in P_+$ satisfies

$$\psi_t(b) = \begin{cases} t^{-1}bt & \text{if } b \in tP_+t^{-1}, \\ 0 & \text{otherwise.} \end{cases}$$

By continuity it suffices to check the following formulas on group elements where they are straightforward:

i. $\psi_t \circ \phi_t = \mathrm{id}_{\Lambda(P_+)}$ for any $t \in T_+$.
ii. $\psi_t = \phi_t^{-1}$ for $t \in T_0$.
iii. $\psi_{t_1} \circ \psi_{t_2} = \psi_{t_1 t_2}$ for any $t_1, t_2 \in T_+$.
iv. $\psi_t(\lambda\phi_t(\mu)) = \psi_t(\lambda)\mu$ and $\psi_t(\phi_t(\lambda)\mu) = \lambda\psi_t(\mu)$ for any $t \in T_+$ and $\lambda, \mu \in \Lambda(P_+)$.

Let D be any etale $\Lambda(P_+)$-module. Using the identity iv. above we may introduce, for any $t \in T_+$, the composed map

$$\begin{array}{rcccccc} \psi_t : & D & \xleftarrow{\cong} & \Lambda \otimes_{\Lambda,t} D & \longrightarrow & D \\ & \lambda t x & \longleftrightarrow & \lambda \otimes x & \longmapsto & \psi_t(\lambda)x . \end{array}$$

It satisfies:

v. $\psi_t \circ t = \mathrm{id}_D$ for any $t \in T_+$.
vi. $\psi_t = t^{-1}$ for $t \in T_0$.
vii. $\psi_{t_1} \circ \psi_{t_2} = \psi_{t_1 t_2}$ for any $t_1, t_2 \in T_+$.
viii. $\psi_t(\phi_t(\lambda)x) = \lambda\psi_t(x)$ for any $\lambda \in \Lambda(P_+)$ and $x \in D$.

The last is a consequence of the identity iv. above. Moreover, D carries the P_+^{-1}-action defined by

$$bx := \psi_t(n_0 x) \qquad \text{for } b = t^{-1}n_0 \in P_+^{-1} = T_+^{-1}N_0 \text{ and } x \in D.$$

REMARK 6.1. *If $D = D^i(V)$ for some representation V in $\mathcal{M}_{o-tor}(P)$ then the operator ψ_t coincides with the action of t^{-1} on D given by the construction of the δ-functor D^i.*

PROOF. By the construction of the δ-functor it suffices to establish the assertion for any $D = D(\mathrm{ind}_{P_0}^{P}(V_0))$. We have to show that

$$t^{-1}(\lambda t y) = \psi_t(\lambda)y$$

holds true for any $\lambda \in \Lambda(P_+)$ and any $y \in D$. But D is a quotient of $M_1(V_0)^*$ where $M_1(V_0) = \mathrm{ind}_{P_0}^{P_+}(V_0)$. So we are further reduced to checking the above relation for $y \in M_1(V_0)^*$. But $M_1(V_0)^*$ is a compact $\Lambda(P_+)$-module. By continuity we therefore need to establish only the relation

$$t^{-1}(nty) = \psi_t(n)y$$

for any $n \in N_0$ and any $y \in M_1(V_0)^*$. If $n = tn't^{-1}$ for some $n' \in N_0$ then the right hand side is $n'y$. Using that t^{-1} is a left inverse of t as an operator on $M_1(V_0)^*$ we compute $t^{-1}(nty) = t^{-1}(tn't^{-1}ty) = t^{-1}(tn'y) = n'y$. If $n \notin tN_0t^{-1}$ then the right hand side is equal to zero. Evaluating the left hand side in a function $\psi \in M_1$ is the same as evaluating y in the function $\tilde{\psi}$ where

$$\tilde{\psi}(b) := \begin{cases} \psi(t^{-1}ntb) & \text{if } b \in P_+, \\ 0 & \text{if } b \in P \setminus P_+. \end{cases} \tag{9}$$

Since $t^{-1}nt \notin N_0$ the set $t^{-1}ntP_+$ is disjoint from P_+ on which ψ is supported. Hence $\tilde{\psi} = 0$. □

Generalizing notation introduced by Colmez we now define the o-module

$$\psi^{-\infty}(D) := \{(x_t)_t \in \prod_{t \in T_+} D : \psi_{t_1}(x_{t_1 t_2}) = x_{t_2} \text{ for any } t_1, t_2 \in T_+\} .$$

The monoid T_+ acts o-linearly on $\psi^{-\infty}(D)$ by

$$t' \cdot (x_t)_t := (x_{t't})_t \qquad \text{for } t' \in T_+.$$

In fact, acting by t' has the inverse

$$(x_t)_t \longmapsto (\psi_{t'}(x_t))_t .$$

Hence T_+ acts by automorphisms. Observing that any element in T is a quotient of two elements in T_+ it therefore follows that this T_+-action extends uniquely to an action of the full torus T on $\psi^{-\infty}(D)$.

On the other hand N_0 acts o-linearly on $\psi^{-\infty}(D)$ by

$$n \cdot (x_t)_t := (\phi_t(n)x_t)_t \qquad \text{for } n \in N_0.$$

To see this we compute

$$\begin{aligned} \psi_{t_1}(\phi_{t_1 t_2}(n)x_{t_1 t_2}) &= \psi_{t_1}(\phi_{t_1}(\phi_{t_2}(n))x_{t_1 t_2}) \\ &= \phi_{t_2}(n)\psi_{t_1}(x_{t_1 t_2}) \\ &= \phi_{t_2}(n)x_{t_2} \end{aligned}$$

and

$$\begin{aligned} n' \cdot (n \cdot (x_t)_t) = n' \cdot (\phi_t(n)x_t)_t &= (\phi_t(n')\phi_t(n)x_t)_t \\ &= (\phi_t(n'n)x_t)_t \\ &= (n'n) \cdot (x_t)_t . \end{aligned}$$

These two actions are compatible in the sense that, for $t' \in T_+$ and $n \in N_0$, we have

$$\begin{aligned} t' \cdot (n \cdot (x_t)_t) &= t' \cdot (\phi_t(n)x_t)_t = (\phi_{t't}(n)x_{t't})_t \\ &= (\phi_t(t'nt'^{-1})x_{t't})_t = (t'nt'^{-1}) \cdot (x_{t't})_t \\ &= (t'nt'^{-1}) \cdot (t' \cdot (x_t)_t) . \end{aligned}$$

In a next intermediate step we claim that for arbitrary $t' \in T$ and $n \in N_0$ but such that $t'nt'^{-1} \in N_0$ we have

$$t' \cdot n \cdot (t')^{-1} \cdot \xi = (t'nt'^{-1}) \cdot \xi \qquad \text{for any } \xi \in \psi^{-\infty}(D).$$

Write $t' = t_1 t_2^{-1}$ with $t_1, t_2 \in T_+$. Using the previous formula we compute

$$\begin{aligned}(t'nt'^{-1}) \cdot \xi &= (t_2)^{-1} \cdot t_2 \cdot (t_2^{-1} t_1 n t_1^{-1} t_2) \cdot \xi \\ &= (t_2)^{-1} \cdot (t_1 n t_1^{-1}) \cdot t_2 \cdot \xi \\ &= (t_2)^{-1} \cdot (t_1 n t_1^{-1}) \cdot t_1 \cdot (t_1)^{-1} \cdot t_2 \cdot \xi \\ &= (t_2)^{-1} \cdot t_1 \cdot n \cdot (t_1)^{-1} \cdot t_2 \cdot \xi \\ &= t' \cdot n \cdot (t')^{-1} \cdot \xi \,.\end{aligned}$$

Let now $n \in N$ be completely arbitrary. We choose a $t' \in T_+$ such that $t'nt'^{-1} \in N_0$ and define

$$n \cdot \xi := (t')^{-1} \cdot (t'nt'^{-1}) \cdot t' \cdot \xi \qquad \text{for any } \xi \in \psi^{-\infty}(D).$$

By the intermediate computation this definition is independent of the choice of t' and hence extends the N_0-action to an o-linear action of N on $\psi^{-\infty}(D)$. In fact, by construction, this N-action and the previous T-action combine into a P-action on $\psi^{-\infty}(D)$.

Everything above is natural. Hence we have constructed a covariant functor $\psi^{-\infty}$ from the category of etale $\Lambda(P_+)$-modules into the category of P-representations (in arbitrary o-modules). This functor can be viewed as a form of induction. Let

$$\begin{aligned}\mathrm{Ind}_{P_+}^{P}(D) := \text{all functions } &F : P \longrightarrow D \text{ such that} \\ &F(bb_+) = b_+^{-1} F(b) \\ &\text{for any } b \in P, b_+ \in P_+\end{aligned}$$

with the group P acting by left translations. Using, as above, that $P = T_+^{-1} P_+$ one checks that

$$\begin{aligned}\mathrm{Ind}_{P_+}^{P}(D) &\xrightarrow{\cong} \psi^{-\infty}(D) \\ F &\longmapsto (F(t^{-1}))_t\end{aligned}$$

is a P-equivariant isomorphism.

LEMMA 6.2. *The functor $\psi^{-\infty}$ is exact.*

PROOF. We pick an element $s \in T_+$ which is strictly dominant, i. e., which satisfies $|\alpha(s)| < 1$ for any $\alpha \in \Phi^+$. The subset $\{s^m\}_{m \in \mathbb{N}}$ is cofinal in the preordered set T_+. Hence $\psi^{-\infty}(D)$, for any D in $\mathcal{M}_{et}(\Lambda(P_+))$, is the projective limit of the sequence

$$\ldots \xrightarrow{\psi_s} D \xrightarrow{\psi_s} \ldots \xrightarrow{\psi_s} D \,.$$

Since ψ_s is surjective the exactness follows immediately. □

Let V be any representation in $\mathcal{M}_{o-tor}(P)$. There is the natural P_+^{-1}-equivariant map

$$\tilde{a}_V : V^* \twoheadrightarrow D(V) \hookrightarrow D^0(V) \,.$$

By Remark 6.1 we have

$$\psi^{-\infty}(D^0(V)) := \{(x_t)_t \in \prod_{t\in T_+} D^0(V) : t_1^{-1}x_{t_1t_2} = x_{t_2} \text{ for any } t_1, t_2 \in T_+\} .$$

Hence the map $\tilde{a}_V$ lifts to a natural transformation

$$\begin{aligned} a_V : V^* &\longrightarrow \psi^{-\infty}(D^0(V)) \\ x &\longmapsto (\tilde{a}_V(tx))_t . \end{aligned}$$

of functors on $\mathcal{M}_{o-tor}(P)$.

LEMMA 6.3. *The map a_V is P-equivariant.*

PROOF. The T-equivariance only needs to be checked on elements $t' \in T_+^{-1}$. Using Remark 6.1 we compute

$$a_V(t'x) = (\tilde{a}_V(tt'x))_t = (t'\tilde{a}_V(tx))_t = t' \cdot (\tilde{a}_V(tx))_t = t' \cdot a_V(x) .$$

Knowing T-equivariance already we need to check N-equivariance only for $n \in N_0$. We compute

$$\begin{aligned} a_V(nx) &= (\tilde{a}_V(tnx))_t = (\tilde{a}_V(tnt^{-1}tx))_t = (\phi_t(n)\tilde{a}_V(tx))_t = n \cdot (\tilde{a}_V(tx))_t \\ &= n \cdot a_V(x) . \end{aligned}$$

□

REMARK 6.4. *Suppose that, for some $M \in \mathcal{P}_+(V)$, the composed natural map $M^* \to D(V) \to D^0(V)$ is bijective; then the map a_V is an isomorphism.*

PROOF. Because of Remark 6.1 we have to show that the map

$$\begin{aligned} V^* &\longrightarrow \{(x_t)_t \in \prod_{t\in T_+} M^* : t_1^{-1}x_{t_1t_2} = x_{t_2} \text{ for any } t_1, t_2 \in T_+\} \\ x &\longmapsto ((tx)|M)_t \end{aligned}$$

is bijective. This is equivalent to the map

$$\begin{aligned} V^* &\longrightarrow \{(y_t)_t \in \prod_{t\in T_+} (t^{-1}M)^* : y_{t_1t_2}|t_2^{-1}M = y_{t_2} \text{ for any } t_1, t_2 \in T_+\} \\ x &\longmapsto (x|t^{-1}M)_t \end{aligned}$$

being bijective which is obvious since we have $V = \bigcup_{t\in T_+} t^{-1}M$. □

7. Dependence on N_0 and ℓ

In this section we will investigate the question in which way our δ-functor D^i depends on the initial choice of the compact open subgroup N_0. We therefore make our notation more precise and write $D(N_0; V)$ and $D^i(N_0; V)$ instead of $D(V)$ and $D^i(V)$, respectively. Let $N_0' \subseteq N$ be another choice of a totally decomposed compact open subgroup. Since then $N_0 \cap N_0'$ is totally decomposed and compact open as well it suffices to treat the case

$$N_0 \subseteq N_0'$$

which we assume from now on throughout this section. Let $P_0' := T_0 N_0'$ and $P_+' := N_0' T_+$. The natural embedding of rings $\Lambda(P_+) \hookrightarrow \Lambda(P_+')$ makes $\Lambda(P_+')$ a right $\Lambda(P_+)$-module which, as a consequence of the decomposition

$$P_+' = \bigcup_{n \in N_0'/N_0} nP_+ \ ,$$

is free of rank equal to the index $[N_0' : N_0]$. We therefore have the exact base extension functor

$$\Lambda(P_+') \otimes_{\Lambda(P_+)} . : \mathcal{M}(\Lambda(P_+)) \longrightarrow \mathcal{M}(\Lambda(P_+')) \ .$$

It is straightforward to check that this functor respects etale modules. Our goal is to establish the following result.

PROPOSITION 7.1. *There is a natural isomorphism of δ-functors*

$$\Lambda(P_+') \otimes_{\Lambda(P_+)} D^i(N_0; .) \cong D^i(N_0'; .) \ .$$

on $\mathcal{M}_{o-tor}(P)$.

PROOF. The argument is formally similar to the proof of Prop. 3.5. Let V_0 be any smooth P_0-representation and $V := \mathrm{ind}_{P_0}^P(V_0)$ the compactly induced P-representation. Then $V_0' := \mathrm{ind}_{P_0}^{P_0'}(V_0)$ is a smooth P_0'-representation and, by the transitivity of induction, we have

$$V = \mathrm{ind}_{P_0}^P(V_0) = \mathrm{ind}_{P_0'}^P(V_0') \ .$$

On the dual side we correspondingly (and more precisely) have the isomorphisms

$$\begin{array}{ccc} & & \mathrm{Ind}_{P_0'}^P(\mathrm{ind}_{P_0}^{P_0'}(V_0^*)) \\ & \overset{\mathcal{I}}{\nearrow} & \big\downarrow \\ \mathrm{Ind}_{P_0}^P(V_0^*) & & \\ & \underset{\mathcal{J}}{\searrow} & \\ & & \mathrm{Ind}_{P_0'}^P((V_0')^*) \end{array}$$

where

$$\mathcal{I}(\Phi)(b)(b_0') := \Phi(bb_0')$$

for $b \in P$, $b_0' \in P_0'$ and

$$\mathcal{J}(\Phi)(b)(\psi) := \sum_{b_0' \in P_0'/P_0} \Phi(bb_0')(\psi(b_0')) = \sum_{n' \in N_0'/N_0} \Phi(bn')(\psi(n'))$$

for $b \in P$, $\psi \in V_0'$. Clearly, for any right P_0'-invariant subset $X \subseteq P$ the map $\mathcal{J}$ restricts to a bijection

$$\mathrm{Ind}_{P_0}^X(V_0^*) \xrightarrow{\cong} \mathrm{Ind}_{P_0'}^X((V_0')^*) \ . \tag{10}$$

In particular,

$$J^+(V_0) = \mathrm{Ind}_{P_0}^{P_+}(V_0^*) \subseteq \mathrm{Ind}_{P_0}^{P_+'}(V_0^*) \underset{\mathcal{J}}{\xrightarrow{\cong}} \mathrm{Ind}_{P_0'}^{P_+'}((V_0')^*) = J^+(V_0') \ .$$

Notations like $J^{\pm}(V_0)$, $\mathrm{Sub}(V_0)$, and $J_\sigma(V_0)$ depend on V_0 as a representation for a specific subgroup (here P_0) of P. In order not to make the notation too heavy we agree in this proof to the abuse that when writing $J^{\pm}(V_0')$ etc. we refer to the subgroup P_0'. We now consider any order reversing map $\sigma : T_+/T_0 \longrightarrow \mathrm{Sub}(V_0)$ satisfying (3) and define the map

$$\begin{aligned} \mathrm{ind}(\sigma) : T_+/T_0 &\longrightarrow \mathrm{Sub}(V_0') \\ t &\longrightarrow \mathrm{ind}_{P_0}^{P_0'}(\sigma(t)) \end{aligned}$$

which again is order reversing and satisfies (3). Using that any $\Phi \in J^+(V_0) \cap J_\sigma(V_0)$ satisfies $\Phi(N_0't) \subseteq \sigma(t)^\perp$ for any $t \in T_+$ one easily checks that

$$\mathcal{J}\big(J^+(V_0) \cap J_\sigma(V_0)\big) \subseteq J^+(V_0') \cap J_{\mathrm{ind}(\sigma)}(V_0')$$

holds true. Hence the map

(11)

$$\begin{aligned} \Lambda(P_+') \underset{\Lambda(P_+)}{\otimes} \big(J^+(V_0)/(J^+(V_0) \cap \bigcup_\sigma J_\sigma(V_0))\big) &\to J^+(V_0')/(J^+(V_0') \cap \bigcup_\sigma J_{\mathrm{ind}(\sigma)}(V_0')) \\ \lambda \otimes x &\mapsto \lambda \mathcal{J}(x) \end{aligned}$$

is well defined. To check that it is bijective we let $n_1, \ldots, n_r \in N_0'$ be representatives for the cosets in N_0'/N_0. Then the above map may be viewed as the map

$$\begin{aligned} \big(J^+(V_0)/(J^+(V_0) \cap \bigcup_\sigma J_\sigma(V_0))\big)^r &\longrightarrow J^+(V_0')/(J^+(V_0') \cap \bigcup_\sigma J_{\mathrm{ind}(\sigma)}(V_0')) \\ (x_1, \ldots, x_r) &\longmapsto \sum_{i=1}^r n_i \mathcal{J}(x_i) \ . \end{aligned}$$

Of course it suffices to show that, for each σ, the map

$$\begin{aligned} \big(J^+(V_0)/(J^+(V_0) \cap J_\sigma(V_0))\big)^r &\longrightarrow J^+(V_0')/(J^+(V_0') \cap J_{\mathrm{ind}(\sigma)}(V_0')) \\ (x_1, \ldots, x_r) &\longmapsto \sum_{i=1}^r n_i \mathcal{J}(x_i) \end{aligned}$$

is bijective. But we have the decompositions

$$J^+(V_0)/(J^+(V_0) \cap J_\sigma(V_0)) = \prod_{t \in T_+/T_0} \mathrm{Ind}_{P_0}^{N_0 t P_0}((V_0/\sigma(t))^*)$$

and

$$J^+(V_0')/(J^+(V_0') \cap J_{\mathrm{ind}(\sigma)}(V_0')) = \prod_{t \in T_+/T_0} \mathrm{Ind}_{P_0'}^{N_0' t P_0'}((V_0'/\mathrm{ind}(\sigma)(t))^*)$$

which are respected by $\mathcal{J}$. Using (10) it therefore comes down to the bijectivity of

$$\begin{aligned} (\mathrm{Ind}_{P_0}^{N_0 t P_0}(.))^r &\longrightarrow \mathrm{Ind}_{P_0}^{N_0' t P_0'}(.) \\ (\Phi_1, \ldots, \Phi_r) &\longmapsto \sum_{i=1}^r n_i \Phi_i \end{aligned}$$

which is straightforward from the disjoint union

$$N_0' t P_0' = \bigcup_{i=1}^r n_i N_0 t P_0 \ .$$

This establishes the bijectivity of (11). In order to read (11) as an isomorphism

(12) $$\Lambda(P_+') \otimes_{\Lambda(P_+)} D(N_0; V) \xrightarrow{\cong} D(N_0'; V)$$

it remains to check that the $M_{\mathrm{ind}(\sigma)}(V_0')$ with varying σ are cofinal among all generating P_+'-subrepresentations M of V. But M then a fortiori is a generating P_+-subrepresentation of V. By Lemma 3.2.ii we therefore find a σ such that $M_\sigma(V_0) \subseteq M$. It follows that $M_{\mathrm{ind}(\sigma)}(V_0') = P_+' M_\sigma(V_0) \subseteq M$.

Next we have to convince ourselves that the isomorphism (12) is natural in maps $f : \mathrm{ind}_{P_0}^P(U_0) \longrightarrow \mathrm{ind}_{P_0}^P(V_0)$ of compactly induced representations. Put $U_0' := \mathrm{ind}_{P_0}^{P_0'}(U_0)$. Viewing f as a map $\mathrm{ind}_{P_0}^P(U_0') \longrightarrow \mathrm{ind}_{P_0}^P(V_0')$ means to consider $j_{V_0} \circ f \circ j_{U_0}^{-1}$ where

$$j_{V_0} : \mathrm{ind}_{P_0}^P(V_0) \xrightarrow{\cong} \mathrm{ind}_{P_0}^P(V_0')$$

is the transitivity isomorphism such that $(j_{V_0}^*)^{-1} = \mathcal{J}$ (and correspondingly for j_{U_0}). We therefore have to check the commutativity of the diagram

$$\begin{array}{ccc}
J^+(V_0)/(J^+(V_0) \cap \bigcup_\sigma J_\sigma(V_0)) & \xrightarrow{\mathcal{J}} & J^+(V_0')/(J^+(V_0') \cap \bigcup_\sigma J_{\mathrm{ind}(\sigma)}(V_0')) \\
\cong \downarrow & & \downarrow \cong \\
\mathrm{Ind}_{P_0}^P(V_0^*)/\bigcup_\sigma J_\sigma(V_0) & & \mathrm{Ind}_{P_0'}^P((V_0')^*)/\bigcup_\sigma J_{\mathrm{ind}(\sigma)}(V_0') \\
D(N_0;f) \downarrow & & \downarrow D(N_0';j\circ f\circ j^{-1}) \\
\mathrm{Ind}_{P_0}^P(U_0^*)/\bigcup_\tau J_\tau(U_0) & & \mathrm{Ind}_{P_0'}^P((U_0')^*)/\bigcup_\tau J_{\mathrm{ind}(\tau)}(U_0') \\
\cong \uparrow & & \uparrow \cong \\
J^+(U_0)/(J^+(U_0) \cap \bigcup_\tau J_\tau(U_0)) & \xrightarrow{\mathcal{J}} & J^+(U_0')/(J^+(U_0') \cap \bigcup_\tau J_{\mathrm{ind}(\tau)}(U_0'))
\end{array}$$

or equivalently that

$$\mathcal{J}((f^*(\Phi))^+) - ((j \circ f \circ j^{-1})^*(\mathcal{J}(\Phi)))^+ = \mathcal{J}((f^*(\Phi))^+) - (\mathcal{J}(f^*(\Phi)))^+$$

lies in $\bigcup_\tau J_{\mathrm{ind}(\tau)}(U_0')$ for any $\Phi \in J^+(V_0)$. But we know from (6) that we find in fact a single map $\tau : T_+/T_0 \longrightarrow \mathrm{Sub}(U_0)$ such that

$$f^*(\Phi)((N \setminus N_0)t) \subseteq \tau(t)^\perp \tag{13}$$

for any $t \in T_+$ and any $\Phi \in J^+(V_0)$. We claim that $\mathcal{J}((f^*(\Phi))^+) - (\mathcal{J}(f^*(\Phi)))^+ \subseteq J_{\mathrm{ind}(\tau)}(U_0')$ which amounts to

$$\mathcal{J}\big((f^*(\Phi))^+ - f^*(\Phi)\big)(n_0't) \in \mathrm{ind}_{P_0}^{P_0'}(\tau(t))^\perp$$

for any $n_0' \in N_0'$ and $t \in T_+$. Let $\psi \in \mathrm{ind}_{P_0}^{P_0'}(\tau(t))$. We compute

$$\mathcal{J}\big((f^*(\Phi))^+ - f^*(\Phi)\big)(n_0't)(\psi) = \sum_{n' \in N_0'/N_0} \big((f^*(\Phi))^+ - f^*(\Phi)\big)(n_0'tn')(\psi(n'))$$

We always have $n_0'tn' \in N_0't$. If even $n_0'tn' \in N_0t$ then in the corresponding summand the two terms coincide so that their difference is zero. Otherwise we have $(f^*(\Phi))^+(n_0'tn') = 0$ and $f^*(\Phi)(n_0'tn') \in \tau(t)^\perp$ by (13). So the value of the corresponding difference on $\psi(n')$ again is zero. This establishes the claim and hence the naturality of the isomorphism (12).

Finally we consider a general representation V in $\mathcal{M}_{o-tor}(P)$ and the corresponding resolution $\mathcal{I}_\bullet(P_0; V)$ from section 4 by representations compactly induced from P_0. There is an obvious natural homomorphism of resolutions

$$\begin{array}{ccc} \mathcal{I}_\bullet(P_0;V) & \longrightarrow & \mathcal{I}_\bullet(P_0';V) \\ \downarrow & & \downarrow \\ V & \xrightarrow{\mathrm{id}} & V. \end{array}$$

Since by the above discussion $\mathcal{I}_\bullet(P_0;V)$ can also be viewed as a resolution of V by representations compactly induced from P_0' Cor. 4.4 implies that this homomorphism induces isomorphisms

$$h^i(D(N_0';\mathcal{I}_\bullet(P_0;V))) \xrightarrow{\cong} D^i(N_0';V) .$$

On the other hand, by the above discussion the natural map

$$\Lambda(P_+') \otimes_{\Lambda(P_+)} D(N_0;\mathcal{I}_\bullet(P_0;V)) \xrightarrow{\sim} D(N_0';\mathcal{I}_\bullet(P_0;V))$$

is an isomorphism. Together with the exactness of the functor $\Lambda(P_+')\otimes_{\Lambda(P_+)}$ this implies our assertion. □

COROLLARY 7.2. *For any $t \in T$ there is a natural isomorphism of δ-functors*

$$\Lambda(tP_+t^{-1}) \otimes_{\Lambda(P_+),t} D^i(N_0;.) \cong D^i(tN_0t^{-1};.)$$

on $\mathcal{M}_{o-tor}(P)$.

PROOF. First we assume that $t \in T_+$. We then have

$$\begin{aligned} &\Lambda(tP_+t^{-1}) \otimes_{\Lambda(P_+),t} D^i(N_0;.) \\ &\quad \cong \Lambda(tP_+t^{-1}) \otimes_{\Lambda(P_+),t} \left(\Lambda(P_+) \otimes_{\Lambda(tP_+t^{-1})} D^i(tN_0t^{-1};.)\right) \\ &\quad = \Lambda(tP_+t^{-1}) \otimes_{\Lambda(tP_+t^{-1}),t} D^i(tN_0t^{-1};.) \\ &\quad \cong D^i(tN_0t^{-1};.) . \end{aligned}$$

Here the first isomorphism comes from Prop. 7.1 and the last one is the fact that the $\Lambda(tP_+t^{-1})$-modules $D^i(tN_0t^{-1};.)$ are etale. Now let $t \in T$ be arbitrary and write $t = t_1t_2^{-1}$ with $t_i \in T_+$. The above isomorphism for the group $t_2^{-1}N_0t_2$ and the elements t_1 and t_2 gives

$$\Lambda(tP_+t^{-1}) \otimes_{\Lambda(t_2^{-1}P_+t_2),t_1} D^i(t_2^{-1}N_0t_2;.) \cong D^i(tN_0t^{-1};.)$$

and

$$\Lambda(P_+) \otimes_{\Lambda(t_2^{-1}P_+t_2),t_2} D^i(t_2^{-1}N_0t_2;.) \cong D^i(N_0;.) ,$$

respectively. In combination we obtain

$$\begin{aligned} D^i(tN_0t^{-1};.) &\cong \Lambda(tP_+t^{-1}) \otimes_{\Lambda(t_2^{-1}P_+t_2),t_1} D^i(t_2^{-1}N_0t_2;.) \\ &= \Lambda(tP_+t^{-1}) \otimes_{\Lambda(P_+),t} \Lambda(P_+) \otimes_{\Lambda(t_2^{-1}P_+t_2),t_2} D^i(t_2^{-1}N_0t_2;.) \\ &\cong \Lambda(tP_+t^{-1}) \otimes_{\Lambda(P_+),t} D^i(N_0;.) . \end{aligned}$$

□

As another consequence of the above results we can justify our specific choice of the homomorphism $\ell : N \longrightarrow \mathbb{Q}_p$ in section 5. The unipotent factor group $N/[N,N]$ is naturally a $\mathbb{Q}_p$-vector space. A homomorphism $\ell' : N \longrightarrow \mathbb{Q}_p$ is called *generic* if it induces a linear map $N/[N,N] \longrightarrow \mathbb{Q}_p$ satisfying $\ell'|N_\alpha \neq 0$ for any $\alpha \in \Delta$. We have $N/[N,N] = \prod_{\alpha\in\Delta} N_\alpha$ (cf. [**BT**] Prop. 4.7.(iii) and Remark 4.11). Hence the épinglage $(\iota_\alpha)_\alpha$ provides an isomorphism between $N/[N,N]$ and the standard vector space $\mathbb{Q}_p^{|\Delta|}$. A generic homomorphism then corresponds to an element in the dual vector space $(\mathbb{Q}_p^{|\Delta|})^*$ all of whose (standard) coordinates are nonzero. On the other hand, by our assumption that the center C of G is connected, the simple roots $\alpha \in \Delta$ form a basis of the character group $X^*(T/C)$. This implies that the homomorphism $\prod_{\alpha\in\Delta} \alpha : T \longrightarrow (\mathbb{Q}_p^*)^{|\Delta|}$ is surjective. It follows that the action of T on $N/[N,N]$ corresponds to the standard action of $(\mathbb{Q}_p^*)^{|\Delta|}$ on $\mathbb{Q}_p^{|\Delta|}$. The subset of vectors with nonzero coordinates is a single orbit for this action. This proves that for any generic homomorphism $\ell' : N \longrightarrow \mathbb{Q}_p$ we find a $t \in T$ such that $\ell'(.) = \ell(t.t^{-1})$.

Since $\ell'(N_0)$ will not be equal to $\mathbb{Z}_p$ in general we introduce the monoid

$$S_{\ell'} := \{ \begin{pmatrix} 1 & 0 \\ a & b \end{pmatrix} : a \in \ell'(N_0), b \in (1 + p^{\epsilon(p)}\mathbb{Z}_p)p^{\mathbb{N}_0} \}.$$

We then have the epimorphism of monoids

$$\begin{aligned} \ell' : P_\star = N_0 T_\star &\longrightarrow S_{\ell'} \\ n_0 t &\longmapsto \begin{pmatrix} 1 & 0 \\ \ell'(n_0) & \xi^{-1}(t) \end{pmatrix} \end{aligned}$$

as well as the corresponding ring homomorphism $\ell' : \Lambda(P_\star) \longrightarrow \Lambda(S_{\ell'})$, which, for simplicity, we denote all by the same symbol ℓ'.

COROLLARY 7.3. *Let $\ell', \ell'' : N \longrightarrow \mathbb{Q}_p$ be any two generic homomorphisms such that $\ell'(N_0) \subseteq \ell''(N_0)$; then there are isomorphisms*

$$\mathrm{Tor}_i^{\Lambda(P_\star)}(\Lambda(S_{\ell''}), D^j(N_0;V)) \cong \Lambda(S_{\ell''}) \otimes_{\Lambda(S_{\ell'})} \mathrm{Tor}_i^{\Lambda(P_\star)}(\Lambda(S_{\ell'}), D^j(N_0;V))$$

for $i, j \geq 0$ which are natural in V in $\mathcal{M}_{o-tor}(P)$.

PROOF. By our above discussion we find a $t \in T$ such that $\ell' = \ell''(t.t^{-1})$. Suppose first that $t \in T_+$. Having in mind the commutative diagram of rings

$$\begin{array}{ccccc} \Lambda(P_\star) & \xrightarrow[\cong]{\phi_t} & \Lambda(tP_\star t^{-1}) & \xrightarrow{\subseteq} & \Lambda(P_\star) \\ \downarrow{\scriptstyle \ell'} & & \downarrow{\scriptstyle \ell''} & \swarrow{\scriptstyle \ell''} & \\ \Lambda(S_{\ell'}) & \xrightarrow{\subseteq} & \Lambda(S_{\ell''}) & & \end{array}$$

we then compute

$$\begin{aligned} &\Lambda(S_{\ell''}) \otimes_{\Lambda(S_{\ell'})} \mathrm{Tor}_i^{\Lambda(P_\star),\ell'}(\Lambda(S_{\ell'}), D^j(N_0;V)) \\ &= \mathrm{Tor}_i^{\Lambda(P_\star),\ell'}(\Lambda(S_{\ell''}), D^j(N_0;V)) \\ &= \mathrm{Tor}_i^{\Lambda(tP_\star t^{-1}),\ell''}(\Lambda(S_{\ell''}), \Lambda(tP_\star t^{-1}) \otimes_{\Lambda(P_\star),t} D^j(N_0;V)) \\ &\cong \mathrm{Tor}_i^{\Lambda(tP_\star t^{-1}),\ell''}(\Lambda(S_{\ell''}), D^j(tN_0t^{-1};V)) \\ &= \mathrm{Tor}_i^{\Lambda(P_\star),\ell''}(\Lambda(S_{\ell''}), \Lambda(P_\star) \otimes_{\Lambda(tP_\star t^{-1})} D^j(tN_0t^{-1};V)) \\ &= \mathrm{Tor}_i^{\Lambda(P_\star),\ell''}(\Lambda(S_{\ell''}), \Lambda(P_+) \otimes_{\Lambda(tP_+t^{-1})} D^j(tN_0t^{-1};V)) \\ &\cong \mathrm{Tor}_i^{\Lambda(P_\star),\ell''}(\Lambda(S_{\ell''}), D^j(N_0;V)) \ . \end{aligned}$$

Here the two isomorphisms come from Cor. 7.2 and Prop. 7.1, respectively. Also, for greater clarity we have inserted superscripts ℓ' and ℓ'' to indicate that the respective

Tor-functor is formed with respect to the corresponding ring homomorphism. A general element $t \in T$ can be written $t = t_1 t_2^{-1}$ with $t_i \in T_+$. Our claim then follows easily by using the above isomorphisms consecutively for the pairs $(\ell''(t_1.t_1^{-1}), \ell') = (\ell'(t_2.t_2^{-1}), \ell')$ and $(\ell''(t_1.t_1^{-1}), \ell'')$. $\square$

8. Some topological localizations

This section is devoted to the construction of certain topological localizations of completed group rings which are needed later on. It is entirely technical and should be skipped at first reading.

Let H_0 be a compact p-adic Lie group. In this case the ring $\Lambda(H_0)$ is well known to be noetherian. In addition we suppose given a closed normal subgroup $H_1 \subseteq H_0$ such that the factor group H_0/H_1 is isomorphic to the additive group of p-adic integers $\mathbb{Z}_p$. First of all we recall from [**SV1**] that $\Lambda(H_0)$ can be viewed as a skew power series ring $\Lambda(H_1)[[t_0; \sigma_0, \delta_0]]$ over $\Lambda(H_1)$. For this one picks a topological generator γ_0 of a subgroup of H_0 which maps isomorphically onto H_0/H_1. One defines $t_0 := \gamma_0 - 1$, the ring automorphism σ_0 of $\Lambda(H_1)$ by $\sigma_0(\lambda) := \gamma_0 \lambda \gamma_0^{-1}$, and the σ_0-derivation $\delta_0 := \sigma_0 - \mathrm{id}$. As a consequence of [**SV1**] Lemma 1.6 the σ_0-derivation δ_0 is topologically nilpotent (and hence σ_0-nilpotent). Of course, $\Lambda(H_0/H_1)$ is a commutative formal power series ring in one variable over o.

From now on we also assume that H_0 is a pro-p-group without element of order p. Then $\Lambda(H_0)$ and $\Lambda(H_1)$ are integral domains ([**Neu**]) which are strict-local[1] with residue field k. Let $\mathfrak{m}(H_0)$ and $\mathfrak{m}(H_1)$ denote the respective maximal ideals. The ideal

$$J := J(H_0, H_1) := \ker\big(\Lambda(H_0) \longrightarrow \Lambda(H_0/H_1) \longrightarrow \Lambda(H_0/H_1)/\pi\Lambda(H_0/H_1)\big)$$

is equal to $J = \mathfrak{m}(H_1)\Lambda(H_0) = \Lambda(H_0)\mathfrak{m}(H_1)$. According to [**CFKSV**] Thm. 2.4 and Prop. 2.6 (and bottom of p. 203) the multiplicatively closed subset $S := S(H_0, H_1) := \Lambda(H_0) \setminus J$ of $\Lambda(H_0)$ satisfies the (left and right) Ore condition. Hence the localization $\Lambda(H_0)_S$ of $\Lambda(H_0)$ with respect to S exists. It is a strict-local integral domain with maximal ideal $J\Lambda(H_0)_S = \mathfrak{m}(H_1)\Lambda(H_0)_S$ and residue field equal to the field of fractions of $\Omega(H_0/H_1)$ (which is isomorphic to a Laurent series field in one variable over k). We now define $\Lambda_{H_1}(H_0)$ to be the $\mathfrak{m}(H_1)$-adic completion of $\Lambda(H_0)_S$. Of course, this again is a strict-local ring whose maximal ideal we denote by $\mathfrak{m}_{H_1}(H_0)$. We have $\mathfrak{m}_{H_1}(H_0) = J\Lambda_{H_1}(H_0) = \mathfrak{m}(H_1)\Lambda_{H_1}(H_0)$. In [**SV2**] Thm. 4.7 (and Lemma 4.2(ii)) it is shown that $\Lambda_{H_1}(H_0)$ is a noetherian pseudocompact ring which can be viewed as an (infinite) skew Laurent series ring in the variable t_0 over $\Lambda(H_1)$ and which is flat over $\Lambda(H_0)_S$ (and hence $\Lambda(H_0)$). Later on we need the following technical fact.

LEMMA 8.1. *For any $m \geq 0$ and any $s \in S(H_0, H_1)$ there is an $l \geq 0$ such that*

$$\mathfrak{m}(H_0)^l \subseteq \big(\Lambda(H_0)\mathfrak{m}(H_1)^m + \mathfrak{m}(H_0)^m s\big) \cap \big(\mathfrak{m}(H_1)^m \Lambda(H_0) + s\mathfrak{m}(H_0)^m\big)\ .$$

PROOF. Since the topology of the noetherian ring $\Lambda(H_0)$ is the $\mathfrak{m}(H_0)$-adic one it suffices to show that the left ideal $\Lambda(H_0)\mathfrak{m}(H_1)^m + \mathfrak{m}(H_0)^m s$ and the right ideal $\mathfrak{m}(H_1)^m\Lambda(H_0) + s\mathfrak{m}(H_0)^m$ both are open in $\Lambda(H_0)$. By symmetry we only discuss the former. By [**CFKSV**] Prop. 2.6 the $\Lambda(H_1)$-module $\Lambda(H_0)/\Lambda(H_0)s$ is finitely generated. Then also $\Lambda(H_0)/\mathfrak{m}(H_0)^m s$ is finitely generated over $\Lambda(H_1)$. It follows that $\Lambda(H_0)/\big(\Lambda(H_0)\mathfrak{m}(H_1)^m + \mathfrak{m}(H_0)^m s\big)$ is finitely generated over $\Lambda(H_1)/\mathfrak{m}(H_1)^m$

[1] A local ring is called strict-local if its quotient by the maximal ideal is a skew field.

and hence is finite. The closed left ideal $\Lambda(H_0)\mathfrak{m}(H_1)^m + \mathfrak{m}(H_0)^m s$ therefore must be open. □

In this section we will consider a triple $H_1 \subseteq H_0 \subseteq H$ of pro-p p-adic Lie groups without elements of order p where H_1 and H_0 are closed normal subgroups of H and both factor groups H_0/H_1 and H/H_0 are isomorphic to $\mathbb{Z}_p$. We then have the inclusions of rings

$$\begin{array}{ccccc} & & & & \Lambda(H) \\ & & & & \uparrow \subseteq \\ \Lambda_{H_1}(H_0) & \xleftarrow{\supseteq} & \Lambda(H_0)_{S(H_0,H_1)} & \xleftarrow{\supseteq} & \Lambda(H_0). \end{array}$$

We do not know whether $S(H_0, H_1)$ also is an Ore set in the bigger ring $\Lambda(H)$. Our goal is to construct, under the additional assumption that H is a semidirect product $H \cong H_1 \rtimes (H/H_1)$, a topological ring which contains the bimodule $\Lambda_{H_1}(H_0) \otimes_{\Lambda(H_0)} \Lambda(H)$ as a dense subgroup.

We pick a topological generator γ of a subgroup of H which maps isomorphically onto H/H_0, and we define $t := \gamma - 1$, $\sigma(\lambda) := \gamma\lambda\gamma^{-1}$ for $\lambda \in \Lambda(H_0)$, and $\delta := \sigma - \mathrm{id}_{\Lambda(H_0)}$. Then $\Lambda(H) = \Lambda(H_0)[[t;\sigma,\delta]]$. Since H_1 is normal in H the ring automorphism σ of $\Lambda(H_0)$ respects the ideals $\mathfrak{m}(H_1)$ and $J(H_0, H_1)$ and the Ore set $S(H_0, H_1)$. Hence σ extends first to a ring automorphism of $\Lambda(H_0)_{S(H_0,H_1)}$ and then further to a ring automorphism of the completion $\Lambda_{H_1}(H_0)$ which we also denote by σ. Correspondingly δ extends to the σ-derivation $\delta := \sigma - \mathrm{id}_{\Lambda_{H_1}(H_0)}$. Recall from the above that $\Lambda_{H_1}(H_0)$ is pseudocompact and noetherian. In particular, the pseudocompact topology is the $\mathfrak{m}_{H_1}(H_0)$-adic one. Hence σ necessarily is a topological automorphism of $\Lambda_{H_1}(H_0)$ and δ is continuous. But unfortunately δ is not topologically nilpotent on $\Lambda_{H_1}(H_0)$ for the pseudocompact topology. So the formalism of [**SV1**] §1 (in particular Remark 1.1.i) to construct the skew power series ring $\Lambda_{H_1}(H_0)[[t;\sigma,\delta]]$ does not apply formally.

But there is the following coarser topology on $\Lambda_{H_1}(H_0)$, introduced and called the weak topology in [**SV2**] §1.2. It is given by the fundamental system of open zero neighbourhoods

$$B_m := \mathfrak{m}_{H_1}(H_0)^m + \mathfrak{m}(H_0)^m$$

for $m \geq 0$. Since each B_m is a $(\Lambda(H_0), \Lambda(H_0))$-sub-bimodule of $\Lambda_{H_1}(H_0)$ this certainly makes $\Lambda_{H_1}(H_0)$ an additive topological group. We obviously have

$$B_k B_l \subseteq B_{\min(k,l)} \ .$$

LEMMA 8.2. i. *The weak topology is a ring topology.*
 ii. *The weak topology on $\Lambda_{H_1}(H_0)$ induces the compact topology on the subring $\Lambda(H_0)$.*
 iii. $\bigcap_m B_m = \{0\}$.
 iv. *The weak topology is complete.*
 v. *Each $\mathfrak{m}_{H_1}(H_0)^m$, for $m \geq 0$, is closed for the weak topology.*
 vi. *σ is a topological automorphism for the weak topology.*

PROOF. i. It remains to show that for any $m \geq 0$ and any $\mu \in \Lambda_{H_1}(H_0)$ there is an $l \geq 0$ such that $\mu B_l \cup B_l \mu \subseteq B_m$. The other one being analogous we only show the inclusion $\mu B_l \subseteq B_m$. We may write

$$\mu = \mu' + s^{-1}\lambda \qquad \text{with } \mu' \in \mathfrak{m}_{H_1}(H_0)^m, s \in S(H_0, H_1), \text{ and } \lambda \in \Lambda(H_0).$$

According to Lemma 8.1 we find an $l \geq m$ such that $\mathfrak{m}(H_0)^l \subseteq \mathfrak{m}(H_1)^m\Lambda(H_0) + s\mathfrak{m}(H_0)^m$. It follows that

$$s^{-1}\mathfrak{m}(H_0)^l \subseteq \mathfrak{m}_{H_1}(H_0)^m + \mathfrak{m}(H_0)^m .$$

Altogether we obtain

$$\mu B_l \subseteq \mathfrak{m}_{H_1}(H_0)^l + \mu\mathfrak{m}(H_0)^l \subseteq \mathfrak{m}_{H_1}(H_0)^m + s^{-1}\mathfrak{m}(H_0)^l \subseteq B_m .$$

ii., iii., and iv. According to [**SV2**] Lemma 4.3.i we may identify $\Lambda_{H_1}(H_0)$ as a (left) $\Lambda(H_1)$-module with

$$\{(\lambda_j)_j \in \prod_{j\in\mathbb{Z}} \Lambda(H_1) : \lim_{j\to-\infty} \lambda_j = 0\}$$

in such a way that

$$\mathfrak{m}_{H_1}(H_0)^m = \{(\lambda_j)_j \in \prod_{j\in\mathbb{Z}} \mathfrak{m}(H_1)^m : \lim_{j\to-\infty} \lambda_j = 0\},$$

$$\Lambda(H_0) = \prod_{j\geq 0} \Lambda(H_1), \text{ and}$$

$$\mathfrak{m}(H_1)^m\Lambda(H_0) = \prod_{j\geq 0} \mathfrak{m}(H_1)^m$$

(loc. cit. Lemma 1.12 and Prop. 2.26.i). We easily read off from this that

$$\mathfrak{m}_{H_1}(H_0)^m \cap \Lambda(H_0) = \mathfrak{m}(H_1)^m\Lambda(H_0) \subseteq \mathfrak{m}(H_0)^m \tag{14}$$

and that

$$\bigcap_{m\geq 0} (\mathfrak{m}_{H_1}(H_0)^m + \Lambda(H_0)) = \Lambda(H_0) .$$

We deduce from the former equation that

$$B_m \cap \Lambda(H_0) = \big(\mathfrak{m}_{H_1}(H_0)^m \cap \Lambda(H_0)\big) + \mathfrak{m}(H_0)^m = \mathfrak{m}(H_0)^m$$

which is ii. Together with the latter equation we get

$$\bigcap_{m\geq 0} B_m = \bigcap_{m\geq 0} B_m \cap \Lambda(H_0) = \bigcap_{m\geq 0} \mathfrak{m}(H_0)^m = \{0\}$$

hence iii. As a $\Lambda(H_1)$-module we have

$$\Lambda_{H_1}(H_0) = \Lambda^-_{H_1}(H_0) \oplus \Lambda(H_0)$$

where, with the above identification,

$$\Lambda^-_{H_1}(H_0) := \{(\lambda_j)_j \in \prod_{j<0} \Lambda(H_1) : \lim_{j\to-\infty} \lambda_j = 0\}$$

and

$$B_m = \big(\mathfrak{m}_{H_1}(H_0)^m \cap \Lambda^-_{H_1}(H_0)\big) \oplus \mathfrak{m}(H_0)^m .$$

This means that the weak topology is the direct sum topology of the subspace topology on $\Lambda^-_{H_1}(H_0)$ induced by the pseudocompact topology on $\Lambda_{H_1}(H_0)$ on the one hand and the compact topology on $\Lambda(H_0)$ on the other hand. The latter clearly is complete and the former as well once we show that $\Lambda^-_{H_1}(H_0)$ is closed in $\Lambda_{H_1}(H_0)$ with respect to the pseudocompact topology. But one easily checks that the maps

$$\begin{aligned} \Lambda_{H_1}(H_0) &\longrightarrow \Lambda(H_1) \\ (\lambda_j)_j &\longmapsto \lambda_{j_0} , \end{aligned}$$

for any $j_0 \in \mathbb{Z}$, are continuous for the pseudocompact topologies, and $\Lambda^-_{H_1}(H_0)$ is the simultaneous kernel of these maps for $j_0 \geq 0$.

v. Using the above descriptions we have

$$\begin{aligned}\bigcap_{l\geq 0}(\mathfrak{m}_{H_1}(H_0)^m + B_l) &= \bigcap_{l\geq m}(\mathfrak{m}_{H_1}(H_0)^m + \mathfrak{m}(H_0)^l)\\ &= \big(\mathfrak{m}_{H_1}(H_0)^m \cap \Lambda^-_{H_1}(H_0)\big) \oplus \bigcap_{l\geq m}\big((\mathfrak{m}_{H_1}(H_0)^m \cap \Lambda(H_0)) + \mathfrak{m}(H_0)^l\big)\\ &= \big(\mathfrak{m}_{H_1}(H_0)^m \cap \Lambda^-_{H_1}(H_0)\big) \oplus \bigcap_{l\geq m}\big(\mathfrak{m}(H_1)^m\Lambda(H_0) + \mathfrak{m}(H_0)^l\big)\ .\end{aligned}$$

But as a finitely generated ideal $\mathfrak{m}(H_1)^m\Lambda(H_0)$ is closed in the compact ring $\Lambda(H_0)$. This implies that

$$\bigcap_{l\geq m}\big(\mathfrak{m}(H_1)^m\Lambda(H_0) + \mathfrak{m}(H_0)^l\big) = \mathfrak{m}(H_1)^m\Lambda(H_0)$$

and hence that

$$\bigcap_{l\geq 0}(\mathfrak{m}_{H_1}(H_0)^m + B_l) = \mathfrak{m}_{H_1}(H_0)^m\ .$$

vi. This is obvious. □

With σ, of course, also δ is continuous for the weak topology. We are able to say more when the group H is a semidirect product $H \cong H_1 \rtimes (H/H_1)$. In this case we may pick the above elements $\gamma \in H$ and $\gamma_0 \in H_0$ in such a way that we have

$$\gamma\gamma_0\gamma^{-1} = \gamma_0^x \qquad \text{for some } x \in 1 + p\mathbb{Z}_p.$$

LEMMA 8.3. *Suppose that $H \cong H_1 \rtimes (H/H_1)$ is a semidirect product; for any $i \geq 1$ we have:*

i. $\delta^{m+i}(t_0^{-i}) \in B_m$ *for any* $m \geq 1$;

ii. $t^i t_0 \in t_0\mathfrak{m}(H)^i$.

PROOF. Inside $\Lambda(H_0)$ we have the subring $o[[t_0]]$. Let $\mathfrak{n}$ denote the maximal ideal of this latter ring. It is generated by π and t_0. Since $o[[t_0]]/\mathfrak{n} = k$ we must have $\delta(o[[t_0]]) \subseteq \mathfrak{n}$. We compute

$$\sigma(t_0) = \gamma(\gamma_0 - 1)\gamma^{-1} = \gamma_0^x - 1 = (t_0+1)^x - 1 = \sum_{j\geq 1}\binom{x}{j}t_0^j$$

and see that

$$\sigma(t_0) \in xt_0 + o[[t_0]]t_0^2 = xt_0(1 + o[[t_0]]t_0)\ .$$

This implies first of all that

$$\delta(t_0) \in (x-1)t_0 + o[[t_0]]t_0^2 \subseteq \big(\pi o + o[[t_0]]t_0\big)t_0 = \mathfrak{n}t_0 \subseteq \mathfrak{n}^2\ ,$$

hence $\delta(\mathfrak{n}) \subseteq \mathfrak{n}^2$, and then inductively

$$(15) \qquad \delta(\mathfrak{n}^j) \subseteq \mathfrak{n}^{j+1} \qquad \text{for any } j \geq 0.$$

Because of $1 + o[[t_0]]t_0 \subseteq o[[t_0]]^\times$ it also implies that

$$\sigma(t_0^{-1}) \in x^{-1}t_0^{-1}(1 + o[[t_0]]t_0) = x^{-1}t_0^{-1} + o[[t_0]]\ .$$

It follows inductively that

$$\sigma(t_0^{-i}) \in x^{-i}t_0^{-i} + o[[t_0]]t_0^{-(i-1)}$$

and hence that

$$\delta(t_0^{-i}) \in \pi o t_0^{-i} + o[[t_0]] t_0^{-(i-1)} = \mathfrak{n} t_0^{-i} .$$

Using (15) we deduce by another induction that

$$\delta^j(t_0^{-i}) \in \mathfrak{n}^j t_0^{-i} = \sum_{l=0}^{j} \pi^l t_0^{j-l-i} o[[t_0]] \qquad \text{for any } j \geq 1.$$

In this last sum the summands for $l \geq m$ lie in $\mathfrak{m}_{H_1}(H_0)^m$. On the other hand, for $l < m$ and $j \geq m + i$ we have $j - l - i \geq 0$ and $l + (j - l - i) \geq m$. Hence the corresponding summand in this case lies in $\mathfrak{m}(H_0)^m$. This proves the first assertion.

For the second assertion it suffices, by induction, to consider the case $i = 1$. We compute

$$t t_0 = \sigma(t_0) t + \delta(t_0) \in t_0 o[[t_0]]^\times t + t_0 \mathfrak{n} \subseteq t_0 \mathfrak{m}(H) .$$

□

PROPOSITION 8.4. *Suppose that $H \cong H_1 \rtimes (H/H_1)$ is a semidirect product; then δ is locally topologically nilpotent, i. e., for any $\mu \in \Lambda_{H_1}(H_0)$ and any $m \geq 1$ there is a $k \geq 1$ such that $\delta^l(\mu) \in B_m$ for any $l \geq k$.*

PROOF. We fix m. We know from [**SV2**] Prop. 1.2 and Remark 1.11 that $\Theta := \{t_0^j\}_{j \geq 0}$ is an Ore set in $\Lambda(H_0)/\mathfrak{m}(H_1)^m \Lambda(H_0)$ and that

$$\Lambda_{H_1}(H_0)/\mathfrak{m}_{H_1}(H_0)^m = \big(\Lambda(H_0)/\mathfrak{m}(H_1)^m \Lambda(H_0)\big)_\Theta$$

is the corresponding localization. Hence any element $\mu \in \Lambda_{H_1}(H_0)$ can be written, modulo $\mathfrak{m}_{H_1}(H_0)^m$, in the form

$$\mu = \mu_i t_0^{-i} + \mu_{i-1} t_0^{-(i-1)} + \ldots + \mu_1 t_0^{-1} + \mu_0$$

for some $i \geq 0$, $\mu_1, \ldots, \mu_i \in \Lambda(H_1)$, and $\mu_0 \in \Lambda(H_0)$. Since $\mathfrak{m}_{H_1}(H_0)^m$ is δ-invariant and since δ is topologically nilpotent on $\Lambda(H_0)$ ([**SV1**] Lemma 1.6) it therefore suffices to consider elements μ of the form

$$\mu = \mu_1 t_0^{-i} \qquad \text{for some } i \geq 1 \text{ and } \mu_1 \in \Lambda(H_1).$$

As δ is a σ-derivation and $\Lambda(H_1)$ is σ-invariant one easily verifies by induction that

$$\delta^l(\Lambda(H_1) t_0^{-i}) \subseteq \sum_{j=0}^{l} \delta^{l-j}(\Lambda(H_1)) \delta^j(t_0^{-i}) \qquad \text{for any } l \geq 1.$$

Again since δ is topologically nilpotent on $\Lambda(H_0)$ and hence on $\Lambda(H_1)$ we have $\delta^{l-j}(\Lambda(H_1)) \subseteq \mathfrak{m}(H_1)^m$ and hence $\delta^{l-j}(\Lambda(H_1))\delta^j(t_0^{-i}) \subseteq B_m$ whenever $l - j$ is sufficiently big. This reduces us finally to the case $\mu = t_0^{-i}$ for any $i \geq 1$ which we have dealt with in Lemma 8.3.i. □

It will be convenient for us to write elements in the countable direct product $\prod_{i \geq 0} \Lambda_{H_1}(H_0)$, viewed as a left $\Lambda_{H_1}(H_0)$-module, as formal power series

$$\sum_{i \geq 0} \mu_i t^i \qquad \text{with } \mu_i \in \Lambda_{H_1}(H_0).$$

The ring we want to construct will be a certain submodule of this direct product. For its definition we first have to recall the notion of boundedness in topological rings as well as some of its elementary properties.

DEFINITION 8.5. *Let R be a Hausdorff topological ring; a subset $A \subseteq R$ is called bounded if for any neighbourhood of zero $U' \subseteq R$ there is another neighbourhood of zero $U \subseteq R$ such that $U \cdot A \cup A \cdot U \subseteq U'$ (where $X \cdot Y := \{xy : x \in X, y \in Y\}$).*

REMARK 8.6. *Let A, A_1, and A_2 be subsets of a Hausdorff topological ring R; we then have:*

i. *If A is bounded then any subset $A_1 \subseteq A$ also is bounded;*
ii. *with A_1 and A_2 also $A_1 \cup A_2$, $A_1 + A_2$, and $A_1 \cdot A_2$ are bounded;*
iii. *any compact A (in particular, any finite A) is bounded;*
iv. *if A is bounded then also its closure $\overline{A}$ is bounded;*
v. *suppose that R has a fundamental system of neighbourhoods of zero consisting of additive subgroups; then with A also the additive subgroup generated by A is bounded;*
vi. *any convergent sequence in R forms a bounded subset.*

PROOF. i. and v. are obvious. vi. is easy. See [**War**] Thm. 12.3 for iii. and Cor. 12.5 for ii. and iv. □

LEMMA 8.7. *Let $A \subseteq \Lambda_{H_1}(H_0)$ be a bounded subset (for the weak topology); then the smallest σ-invariant additive subgroup containing A is bounded as well (and is δ-invariant).*

PROOF. Using the fact that $\sigma(B_m) = B_m$ one easily checks that $\bigcup_{j\geq 0} \sigma^j(A)$ is bounded. Now apply Remark 8.6.v. □

LEMMA 8.8. *For any subset $A \subseteq \Lambda_{H_1}(H_0)$ the following conditions are equivalent:*

i. *A is bounded;*
ii. *for any $m \geq 1$ there is an $l \geq 1$ such that $t_0^l A \subseteq \mathfrak{m}_{H_1}(H_0)^m + \Lambda(H_0)$;*
iii. *for any $m \geq 1$ there is an $l \geq 1$ such that $At_0^l \subseteq \mathfrak{m}_{H_1}(H_0)^m + \Lambda(H_0)$.*

PROOF. By definition the set A is bounded if and only if for any $m \geq 1$ there is an $l \geq m$ such that

$$B_l \cdot A \cup A \cdot B_l \subseteq B_m \; .$$

Since $\mathfrak{m}_{H_1}(H_0)^l$ is a (two-sided) ideal in $\Lambda_{H_1}(H_0)$ this inclusion is equivalent to the inclusion

$$\mathfrak{m}(H_0)^l \cdot A \cup A \cdot \mathfrak{m}(H_0)^l \subseteq B_m$$

and then also to the inclusion

$$\mathfrak{m}(H_0)^l \cdot A \cup A \cdot \mathfrak{m}(H_0)^l \subseteq \mathfrak{m}_{H_1}(H_0)^m + \Lambda(H_0)$$

(for a possibly different l). The latter trivially implies that

$$t_0^l A \cup At_0^l \subseteq \mathfrak{m}_{H_1}(H_0)^m + \Lambda(H_0) \; . \tag{16}$$

Suppose, vice versa, that (16) holds true for some l (depending on m) which we then may assume to satisfy $l \geq m$. By Lemma 8.1 applied to $s := t_0^l$ we find an $l' \geq 0$ such that

$$\mathfrak{m}(H_0)^{l'} \subseteq \big(J(H_0,H_1)^l + \Lambda(H_0)t_0^l\big) \cap \big(J(H_0,H_1)^l + t_0^l\Lambda(H_0)\big) \; .$$

It easily follows that

$$\mathfrak{m}(H_0)^{l'} \cdot A \cup A \cdot \mathfrak{m}(H_0)^{l'} \subseteq \mathfrak{m}_{H_1}(H_0)^m + \Lambda(H_0) \; .$$

It remains to show that the conditions ii. and iii. are equivalent. In fact, we fix m. As recalled earlier we know from [**SV2**] Remark 1.11 that

$$\Lambda_{H_1}(H_0)/\mathfrak{m}_{H_1}(H_0)^m = \big(\Lambda(H_1)/\mathfrak{m}(H_1)^m\big)((t_0;\overline{\sigma}_0,\overline{\delta}_0))$$

is a skew Laurent series (with finite negative parts) ring where $\overline{\sigma}_0$ and $\overline{\delta}_0$ denote the maps on the coefficients induced by σ_0 and δ_0, respectively. The condition $t_0^l A \subseteq \mathfrak{m}_{H_1}(H_0)^m + \Lambda(H_0)$ therefore means that the image in $\Lambda_{H_1}(H_0)/\mathfrak{m}_{H_1}(H_0)^m$ of each element in A can be written in the form

$$\sum_{i\geq -l} t_0^i r_i \qquad \text{with } r_i \in \Lambda(H_1)/\mathfrak{m}(H_1)^m \ .$$

But $\overline{\delta}_0$ is nilpotent on $\Lambda(H_1)/\mathfrak{m}(H_1)^m$, say $\overline{\delta}_0^{N+1} = 0$. The formula (1.4) in [**SV2**] then shows an identity of the form

$$\sum_{i\geq -l} t_0^i r_i = \sum_{j\geq -l-N} r'_j t_0^j$$

whatever the coefficients r_i are. It therefore follows that $At_0^{l+N} \subseteq \mathfrak{m}_{H_1}(H_0)^m + \Lambda(H_0)$. In the other direction the argument works in the same way (cf. (1.3) in [**SV2**]). □

REMARK 8.9. *The proof of Lemma 8.8 in particular shows that a subset $A \subseteq \Lambda_{H_1}(H_0)$ is bounded if and only if it is left bounded if and only if it is right bounded.*

COROLLARY 8.10. *Suppose that $H \cong H_1 \rtimes (H/H_1)$ is a semidirect product; for any bounded subset $A \subseteq \Lambda_{H_1}(H_0)$ and any $m \geq 1$ there is a $k \geq 1$ such that $\delta^k(A) \subseteq B_m$.*

PROOF. By Lemma 8.8 we find, for any $m \geq 1$, an $i \geq 1$ such that

$$A \subseteq \Lambda(H_1)t_0^{-i} + \ldots + \Lambda(H_1)t_0^{-1} + \Lambda(H_0) + \mathfrak{m}_{H_1}(H_0)^m \ .$$

From here on the argument proceeds as in the proof of Prop. 8.4. □

We now define

$$\Lambda_{H_0,H_1}(H) := \{\sum_{i\geq 0} \mu_i t^i : \{\mu_i\}_i \text{ is bounded in } \Lambda_{H_1}(H_0)\} \ .$$

By Remark 8.6.ii/iii this is a $\Lambda_{H_1}(H_0)$-submodule of the direct product. We view $\Lambda_{H_1}(H_0)$ as being contained in $\Lambda_{H_0,H_1}(H)$ through $\mu \mapsto \mu + 0t + 0t^2 + \ldots$ On the other hand viewing $\Lambda(H) = \Lambda(H_0)[[t;\sigma,\delta]]$ as a skew power series ring in t and noting that $\Lambda(H_0)$ is compact and therefore, by Lemma 8.2.ii and Remark 8.6.iii, bounded in $\Lambda_{H_1}(H_0)$ we see that the ring $\Lambda(H)$ in an obvious way is contained in $\Lambda_{H_0,H_1}(H)$ as well. More generally, by Remark 8.6.ii the map

$$\begin{aligned} \Lambda_{H_1}(H_0) \otimes_{\Lambda(H_0)} \Lambda(H) &\longrightarrow \Lambda_{H_0,H_1}(H) \\ \lambda \otimes (\sum_{i\geq 0} \mu_i t^i) &\longmapsto \sum_{i\geq 0} \lambda\mu_i t^i \end{aligned} \tag{17}$$

is well defined.

REMARK 8.11. *The above map (17) is injective.*

PROOF. Suppose that the element $\sum_{k=0}^{r} \lambda_k \otimes (\sum_{i\geq 0} \mu_{k,i} t^i)$ lies in the kernel of (17). This means that

$$\sum_{k=0}^{r} \lambda_k \mu_{k,i} = 0 \qquad \text{for any } i \geq 0.$$

Let X denote the right $\Lambda(H_0)$-submodule of $\Lambda(H_0)^r$ generated by the vectors $(\mu_{k,i})_k$ for $i \geq 0$. Since $\Lambda(H_0)$ is noetherian this module X is finitely generated. Hence we find vectors $(\nu_{1,k})_k, \ldots, (\nu_{s,k})_k \in X$ and elements $c_{1,i}, \ldots, c_{s,i} \in \Lambda(H_0)$ for $i \geq 0$ such that

$$(\mu_{k,i})_k = (\nu_{1,k})_k c_{1,i} + \ldots + (\nu_{s,k})_k c_{s,i} \qquad \text{for any } i \geq 0.$$

Of course we have

$$\sum_{k=0}^{r} \lambda_k \nu_{1,k} = \ldots = \sum_{k=0}^{r} \lambda_k \nu_{s,k} = 0 \ .$$

We now compute

$$\begin{aligned}
\sum_{k=0}^{r} \lambda_k \otimes (\sum_{i\geq 0} \mu_{k,i} t^i) &= \sum_{k=0}^{r} \lambda_k \otimes \sum_{i\geq 0} (\nu_{1,k} c_{1,i} + \ldots + \nu_{s,k} c_{s,i}) t^i \\
&= \sum_{k=0}^{r} \sum_{l=0}^{s} \lambda_k \otimes \nu_{l,k} (\sum_{i\geq 0} c_{l,i} t^i) \\
&= \sum_{l=0}^{s} (\sum_{k=0}^{r} \lambda_k \nu_{l,k}) \otimes (\sum_{i\geq 0} c_{l,i} t^i) \\
&= 0 \ .
\end{aligned}$$

□

On $\Lambda_{H_0,H_1}(H)$ we have the descending filtration

$$F^m \Lambda_{H_0,H_1}(H) := \{\sum_{i\geq 0} \mu_i t^i \in \Lambda_{H_0,H_1}(H) : \{\mu_i\}_i \subseteq \mathfrak{m}_{H_1}(H_0)^m\} \qquad \text{for } m \geq 0$$

by $\Lambda_{H_1}(H_0)$-submodules. By taking this filtration as a fundamental system of open zero neighbourhoods we obtain the *strong* topology on $\Lambda_{H_0,H_1}(H)$. The fact that $\mathfrak{m}_{H_1}(H_0)^m \Lambda_{H_0,H_1}(H) \subseteq F^m \Lambda_{H_0,H_1}(H)$ implies that $\Lambda_{H_0,H_1}(H)$ with the strong topology is a (left) topological module over $\Lambda_{H_1}(H_0)$ with its pseudocompact topology.

LEMMA 8.12.
i. *$\Lambda_{H_0,H_1}(H)$ is Hausdorff and complete in the strong topology.*
ii. *$\mathfrak{m}_{H_1}(H_0)^m \Lambda_{H_0,H_1}(H)$, for any $m \geq 0$, is dense in $F^m \Lambda_{H_0,H_1}(H)$ for the strong topology.*
iii. *$\Lambda_{H_1}(H_0) \otimes_{\Lambda(H_0)} \Lambda(H)$ is dense in $\Lambda_{H_0,H_1}(H)$ for the strong topology.*
iv. *For any $m \geq 0$ the natural map*

$$\big(\Lambda_{H_1}(H_0)/\mathfrak{m}_{H_1}(H_0)^m\big) \otimes_{\Lambda(H_0)} \Lambda(H) \xrightarrow{\cong} \Lambda_{H_0,H_1}(H)/F^m \Lambda_{H_0,H_1}(H)$$

is an isomorphism.

PROOF. i. We have

$$\begin{aligned}\Lambda_{H_0,H_1}(H) &\subseteq \varprojlim_m \left(\Lambda_{H_0,H_1}(H)/F^m\Lambda_{H_0,H_1}(H)\right)\\ &\subseteq \varprojlim_m \Big(\prod_{i\geq 0}\Lambda_{H_1}(H_0)/\prod_{i\geq 0}\mathfrak{m}_{H_1}(H_0)^m\Big)\\ &= \prod_{i\geq 0}\varprojlim_m \left(\Lambda_{H_1}(H_0)/\mathfrak{m}_{H_1}(H_0)^m\right)\\ &= \prod_{i\geq 0}\Lambda_{H_1}(H_0)\end{aligned}$$

Therefore it suffices to show that

$$\bigcap_{m\geq 0}\Big(\Lambda_{H_0,H_1}(H)+\prod_{i\geq 0}\mathfrak{m}_{H_1}(H_0)^m\Big)=\Lambda_{H_0,H_1}(H)\ .$$

Let $\sum_{i\geq 0}\mu_i t^i$ be a power series contained in the left hand side, and let $k\geq 0$. By assumption we find a power series $\sum_{i\geq 0}\nu_i t^i\in\Lambda_{H_0,H_1}(H)$ such that $\mu_i-\nu_i\in\mathfrak{m}_{H_1}(H_0)^k\subseteq B_k$ for any $i\geq 0$. There is an $l\geq k$ such that $B_l\cdot\{\nu_i\}_i\subseteq B_k$. It follows that

$$B_l\cdot\{\mu_i\}_i\subseteq B_l\cdot\{\mu_i-\nu_i\}_i+B_l\cdot\{\nu_i\}_i\subseteq B_k\ .$$

Hence $\{\mu_i\}_i$ is left bounded and therefore bounded by Remark 8.9.

ii. and iii. Let $\lambda=\sum_{i\geq 0}\lambda_i t^i\in F^m\Lambda_{H_0,H_1}(H)$, and let $m'\geq m$. By Lemma 8.8.ii we find an $l\geq 1$ and elements $\mu_i\in\Lambda(H_0)$ such that

$$\lambda_i-t_0^{-l}\mu_i\in\mathfrak{m}_{H_1}(H_0)^{m'}\qquad\text{for any } i\geq 0.$$

We put $\mu:=\sum_{i\geq 0}\mu_i t^i\in\Lambda(H)$ and obtain $\lambda-t_0^{-l}\mu\in F^{m'}\Lambda_{H_0,H_1}(H)$. In particular, we have $t_0^{-l}\mu\in F^m\Lambda_{H_0,H_1}(H)$ and hence

$$\mu\in F^m\Lambda_{H_0,H_1}(H)\cap\Lambda(H)=\mathfrak{m}(H_1)^m\Lambda(H)$$

where the right hand identity comes from (14) in the proof of Lemma 8.2. We see that λ, modulo $F^{m'}\Lambda_{H_0,H_1}(H)$, is congruent to $t_0^{-l}\mu\in\mathfrak{m}_{H_1}(H_0)^m\Lambda(H)$. This proves both assertions.

iv. Surjectivity is immediate from iii., and the injectivity is a computation totally analogous to the one in the proof of Remark 8.11. □

Obviously, the skew polynomial ring $\Lambda_{H_1}(H_0)[t;\sigma,\delta]$ is contained as a $\Lambda_{H_1}(H_0)$-submodule in $\Lambda_{H_0,H_1}(H)$. The multiplication in this skew polynomial ring is given by the formula

$$(18)\qquad \Big(\sum_{i\geq 0}\lambda_i t^i\Big)\Big(\sum_{j\geq 0}\mu_j t^j\Big)=\sum_{l\geq 0}\Big(\sum_{k=0}^{l}\sum_{i\geq k}\binom{i}{k}\lambda_i\delta^{i-k}(\sigma^k(\mu_{l-k}))\Big)t^l\ .$$

PROPOSITION 8.13. *Suppose that $H\cong H_1\rtimes(H/H_1)$ is a semidirect product; then the formula (18) defines (via convergence of the sums on the right hand side for the weak topology on $\Lambda_{H_1}(H_0)$) a multiplication map on $\Lambda_{H_0,H_1}(H)$ making the latter into a ring in which the $F^m\Lambda_{H_0,H_1}(H)$ are two-sided ideals.*

PROOF. We begin by checking that the coefficients on the right hand side of (18) are well defined in $\Lambda_{H_1}(H_0)$ whenever the two factors on the left hand side lie in $\Lambda_{H_0,H_1}(H)$. Because of Lemma 8.2.iv it suffices to show that, for any bounded sequence $\{\nu_1, \nu_2, \ldots\} \subseteq \Lambda_{H_1}(H_0)$ and any singleton $\nu \in \Lambda_{H_1}(H_0)$, the sequence $\nu_1\delta(\nu), \nu_2\delta^2(\nu), \nu_3\delta^3(\nu), \ldots$ tends to zero with respect to the weak topology. Given any neighbourhood of zero B_m we find, by boundedness, some $m' \geq 0$ such that $\bigcup_{k\geq 0} \nu_k B_{m'} \subseteq B_m$. But according to Prop. 8.4 we have $\delta^k(\nu) \in B_{m'}$ for any sufficiently big k. It follows that $\nu_k\delta^k(\nu) \in B_m$ for any sufficiently big k.

That with $\{\lambda_i\}_i$ and $\{\mu_j\}_j$ also the sequence of coefficients on the right hand side of (18) is bounded is a straightforward consequence of Lemma 8.7 and Remark 8.6.ii/iv/v.

The resulting multiplication map

$$\cdot : \Lambda_{H_0,H_1}(H) \times \Lambda_{H_0,H_1}(H) \longrightarrow \Lambda_{H_0,H_1}(H)$$

clearly is bi-additive. Since the $\mathfrak{m}_{H_1}(H_0)^m$ are σ- and hence δ-invariant (two-sided) ideals in $\Lambda_{H_1}(H_0)$ which are closed with respect to the weak topology by Lemma 8.2.v we easily see that this multiplication satisfies

$$F^m\Lambda_{H_0,H_1}(H) \cdot \Lambda_{H_0,H_1}(H) \cup \Lambda_{H_0,H_1}(H) \cdot F^m\Lambda_{H_0,H_1}(H) \subseteq F^m\Lambda_{H_0,H_1}(H)$$

for any $m \geq 0$. It remains to establish the associativity of this multiplication. Let $\lambda = \sum_i \lambda_i t^i, \mu = \sum_j \mu_j t^j$, and $\nu = \sum_k \nu_k t^k$ be three elements in $\Lambda_{H_0,H_1}(H)$. We have

$$\alpha := \lambda \cdot \mu = \sum_m \alpha_m t^m \qquad \text{with} \quad \alpha_m := \sum_{a=0}^{m} \sum_b \binom{b}{a} \lambda_b \delta^{b-a}(\sigma^a(\mu_{m-a}))$$

$$\beta := \mu \cdot \nu = \sum_n \beta_n t^n \qquad \text{with} \quad \beta_n := \sum_{c=0}^{n} \sum_f \binom{f}{c} \mu_f \delta^{f-c}(\sigma^c(\nu_{n-c}))$$

and

$$(\lambda \cdot \mu) \cdot \nu = \alpha \cdot \nu = \sum_l \Big(\sum_{e=0}^{l} \sum_f \binom{f}{e} \alpha_f \delta^{f-e}(\sigma^e(\nu_{l-e}))\Big)t^l$$

$$\lambda \cdot (\mu \cdot \nu) = \lambda \cdot \beta = \sum_l \Big(\sum_{g=0}^{l} \sum_b \binom{b}{g} \lambda_b \delta^{b-g}(\sigma^g(\beta_{l-g}))\Big)t^l.$$

Hence we have to show the identity

$$\sum_{e=0}^{l} \sum_f \binom{f}{e} \alpha_f \delta^{f-e}(\sigma^e(\nu_{l-e})) = \sum_{g=0}^{l} \sum_b \binom{b}{g} \lambda_b \delta^{b-g}(\sigma^g(\beta_{l-g})) \tag{19}$$

for any $l \geq 0$. We first compute the right hand side. By inserting the definition of β_n and using that σ and hence also δ are continuous for the weak topology (Lemma 8.2.vi) we obtain

$$\begin{aligned}
&\sum_{g=0}^{l} \sum_b \binom{b}{g} \lambda_b \sum_{c=0}^{l-g} \sum_f \binom{f}{c} \delta^{b-g}(\sigma^g(\mu_f \delta^{f-c}(\sigma^c(\nu_{l-g-c})))) \\
&= \sum_b \lambda_b \sum_f \sum_{g=0}^{l} \sum_{c=0}^{l-g} \binom{b}{g} \binom{f}{c} \delta^{b-g}(\sigma^g(\mu_f)\delta^{f-c}(\sigma^{g+c}(\nu_{l-(g+c)}))) \\
&= \sum_b \lambda_b \sum_f \sum_{g=0}^{l} \sum_{e=g}^{l} \binom{b}{g} \binom{f}{e-g} \delta^{b-g}(\sigma^g(\mu_f)\delta^{f-e+g}(\sigma^e(\nu_{l-e}))).
\end{aligned}$$

where in the last identity we have substituted e for $g+c$. By applying the general Leibniz type rule for the σ-derivation δ which, since σ and δ commute, reads

$$\delta^r(ab) = \sum_{s=0}^{r} \binom{r}{s} \delta^{r-s}(\sigma^s(a))\delta^s(b)$$

we continue computing

$$= \sum_b \lambda_b \sum_f \sum_{g=0}^{l} \sum_{e=g}^{l} \binom{b}{g}\binom{f}{e-g} \sum_{s=0}^{b-g} \binom{b-g}{s} \delta^{b-g-s}(\sigma^{g+s}(\mu_f))\delta^{f-e+g+s}(\sigma^e(\nu_{l-e}))$$

$$= \sum_b \lambda_b \sum_f \sum_{g=0}^{l} \sum_{a=g}^{b} \sum_{e=g}^{l} \binom{b}{g}\binom{f}{e-g}\binom{b-g}{a-g} \delta^{b-a}(\sigma^a(\mu_f))\delta^{f-e+a}(\sigma^e(\nu_{l-e}))$$

$$= \sum_b \lambda_b \sum_f \sum_{a=0}^{b} \sum_{e=0}^{l} [\sum_{g=0}^{\min(a,e)} \binom{b}{g}\binom{f}{e-g}\binom{b-g}{a-g}] \delta^{b-a}(\sigma^a(\mu_f))\delta^{f-e+a}(\sigma^e(\nu_{l-e}))$$

where in the middle identity we have substituted a for $g+s$. One easily checks that

$$\sum_{g=0}^{\min(a,e)} \binom{b}{g}\binom{f}{e-g}\binom{b-g}{a-g} = \binom{b}{a} \sum_{g=0}^{e} \binom{a}{g}\binom{f}{e-g} = \binom{b}{a}\binom{f+a}{e} .$$

Hence the right side of (19) is equal to

$$\sum_b \lambda_b \sum_f \sum_{a=0}^{b} \sum_{e=0}^{l} \binom{b}{a}\binom{f+a}{e} \delta^{b-a}(\sigma^a(\mu_f))\delta^{f+a-e}(\sigma^e(\nu_{l-e})) .$$

Next we argue that in this latter multi-sum the summations over b and over f can be interchanged. For this we first check that, for any given $m \geq 0$, all but finitely many of the elements

$$x_{b,f} := \sum_{a=0}^{b} \sum_{e=0}^{l} \binom{b}{a}\binom{f+a}{e} \delta^{b-a}(\sigma^a(\mu_f))\delta^{f+a-e}(\sigma^e(\nu_{l-e}))$$

(recall that l is arbitrary but fixed) lie in B_m. By Lemma 8.7 the set

$$A := \{\delta^{b-a}(\sigma^a(\mu_f))\}_{a,b,f} \cup \{\delta^{f+a-e}(\sigma^e(\nu_{l-e}))\}_{a,e,f} .$$

is bounded. Hence we find an $m' \geq 0$ such that

$$A \cdot B_{m'} \cup B_{m'} \cdot A \subseteq B_m .$$

By Prop. 8.4 there is an $N_1 \geq 0$ such that $\delta^{f+a-e}(\sigma^e(\nu_{l-e})) \in B_{m'}$ whenever $f+a \geq N_1$. By Cor. 8.10 there is an $N_2 \geq 0$ such that $\delta^{b-a}(\sigma^a(\mu_f)) \in B_{m'}$ whenever $b-a \geq N_2$. We conclude that

$$\delta^{b-a}(\sigma^a(\mu_f))\delta^{f+a-e}(\sigma^e(\nu_{l-e})) \in B_m$$

provided $f+a \geq N_1$ or $b-a \geq N_2$. Since $f+a < N_1$ and $b-a < N_2$ together imply $f+b < N_1+N_2$ we finally see that $x_{b,f} \in B_m$ for $b+f \geq N_1+N_2$.

Since $\{\lambda_b\}_b$ is bounded the family $\{\lambda_b x_{b,f}\}_{b,f}$ also has the property that, for any given $m \geq 0$, all but finitely many of its elements lie in B_m. It follows that the right hand side of (19) is equal to

$$\sum_f \sum_b \sum_{a=0}^{b} \sum_{e=0}^{l} \binom{b}{a}\binom{f+a}{e} \lambda_b \delta^{b-a}(\sigma^a(\mu_f))\delta^{f+a-e}(\sigma^e(\nu_{l-e}))$$

$$= \sum_f \sum_b \sum_{a=0}^{f} \sum_{e=0}^{l} \binom{b}{a}\binom{f}{e} \lambda_b \delta^{b-a}(\sigma^a(\mu_{f-a}))\delta^{f-e}(\sigma^e(\nu_{l-e}))$$

$$= \sum_{e=0}^{l} \sum_f \binom{f}{e} (\sum_{a=0}^{f} \sum_b \binom{b}{a} \lambda_b \delta^{b-a}(\sigma^a(\mu_{f-a})))\delta^{f-e}(\sigma^e(\nu_{l-e}))$$

$$= \sum_{e=0}^{l} \sum_f \binom{f}{e} \alpha_f \delta^{f-e}(\sigma^e(\nu_{l-e}))$$

which is the left hand side of (19). □

For the remainder of this section we *assume that* $H \cong H_1 \rtimes (H/H_1)$ *is a semidirect product.* In the proof of Prop. 8.4 we had recalled already from **[SV2]** that $\Theta = \{t_0^j\}_{j\geq 0}$ is an Ore set in $\Lambda(H_0)/\mathfrak{m}(H_1)^m\Lambda(H_0)$, for any $m \geq 1$ with

$$\Lambda_{H_1}(H_0)/\mathfrak{m}_{H_1}(H_0)^m = \left(\Lambda(H_0)/\mathfrak{m}(H_1)^m\Lambda(H_0)\right)_\Theta .$$

LEMMA 8.14. *The set* Θ *consists of regular elements and satisfies the (left and right) Ore condition in* $\Lambda(H)/\mathfrak{m}(H_1)^m\Lambda(H)$ *for any* $m \geq 1$.

PROOF. According to Lemma 8.3.ii we have $t^i t_0 = t_0 \nu_i$ with $\nu_i \in \mathfrak{m}(H)^i$. We fix an $m \geq 1$, and we let $\lambda = \sum_{i\geq 0} t^i \lambda_i \in \Lambda(H)$ with $\lambda_i \in \Lambda(H_0)$ be an arbitrary element. By the "right" version of **[SV2]** Lemma 1.1.ii there exist an $M \geq 0$ and $\mu_i \in \Lambda(H_0)$ such that

$$\lambda_i t_0^M \equiv t_0 \mu_i \mod \mathfrak{m}(H_1)^m\Lambda(H_0)$$

for any $i \geq 0$. We obtain

$$\lambda t_0^M = \sum_{i\geq 0} t^i \lambda_i t_0^M \equiv \sum_{i\geq 0} t^i t_0 \mu_i = t_0(\sum_{i\geq 0} \nu_i \mu_i) \mod \mathfrak{m}(H_1)^m\Lambda(H)$$

where $\mu := \sum_{i\geq 0} \nu_i\mu_i \in \Lambda(H)$ is well defined because of $\nu_i \in \mathfrak{m}(H)^i$; note that $\mathfrak{m}(H_1)^m\Lambda(H) = \Lambda(H)\mathfrak{m}(H_1)^m$. By a straightforward induction we deduce from this that for any $j \geq 1$ and any $\lambda \in \Lambda(H)$ we find a $\mu \in \Lambda(H)$ such that

$$\lambda t_0^{Mj} \equiv t_0^j \mu \mod \mathfrak{m}(H_1)^m\Lambda(H) .$$

This in particular gives the asserted right Ore condition.

For right regularity we write an arbitrary element $\lambda \in \Lambda(H)$ as $\lambda = \sum_{i\geq 0} \lambda_i t^i$ with $\lambda_i = \sum_{k\geq 0} t_0^k \mu_{i,k} \in \Lambda(H_0)$ and $\mu_{i,k} \in \Lambda(H_1)$. If $t_0^j \lambda \in \mathfrak{m}(H_1)^m\Lambda(H)$ then $t_0^j\lambda_i = \sum_{k\geq 0} t_0^{j+k}\mu_{i,k} \in \mathfrak{m}(H_1)^m\Lambda(H_0) = \Lambda(H_0)\mathfrak{m}(H_1)^m$ for any $i \geq 0$, hence $\mu_{i,k} \in \mathfrak{m}(H_1)^m$ for any $i, k \geq 0$, and therefore $\lambda \in \mathfrak{m}(H_1)^m\Lambda(H)$.

The left Ore condition and left regularity follow by analogous arguments. □

We have the obvious injective ring homomorphisms

$$\Lambda(H_0)/\mathfrak{m}(H_1)^m\Lambda(H_0) \hookrightarrow \Lambda(H)/\mathfrak{m}(H_1)^m\Lambda(H) \hookrightarrow \Lambda_{H_0,H_1}(H)/F^m\Lambda_{H_0,H_1}(H)$$

where the injectivity of the right hand map is a consequence of (14). By the universal property of localization they extend to injective ring homomorphisms

$$\big(\Lambda(H_0)/\mathfrak{m}(H_1)^m\Lambda(H_0)\big)_\Theta \hookrightarrow \big(\Lambda(H)/\mathfrak{m}(H_1)^m\Lambda(H)\big)_\Theta \hookrightarrow \Lambda_{H_0,H_1}(H)/F^m\Lambda_{H_0,H_1}(H)\ .$$

PROPOSITION 8.15. *We have as rings:*

i. $\big(\Lambda(H)/\mathfrak{m}(H_1)^m\Lambda(H)\big)_\Theta = \Lambda_{H_0,H_1}(H)/F^m\Lambda_{H_0,H_1}(H)$;

ii. $\Lambda_{H_0,H_1}(H) = \varprojlim_m \big(\Lambda(H)/\mathfrak{m}(H_1)^m\Lambda(H)\big)_\Theta$;

iii. *The ring $\Lambda_{H_0,H_1}(H)$ is, up to isomorphism, independent of the choice of the variables t_0 and t.*

PROOF. i. Because of

$$\begin{aligned}
&\big(\Lambda(H)/\mathfrak{m}(H_1)^m\Lambda(H)\big)_\Theta\\
&= \big(\Lambda(H_0)/\mathfrak{m}(H_1)^m\Lambda(H_0)\otimes_{\Lambda(H_0)}\Lambda(H)\big)_\Theta\\
&= \big(\Lambda(H_0)/\mathfrak{m}(H_1)^m\Lambda(H_0)\big)_\Theta \underset{\Lambda(H_0)/\mathfrak{m}(H_1)^m\Lambda(H_0)}{\otimes} \big(\Lambda(H_0)/\mathfrak{m}(H_1)^m\Lambda(H_0) \underset{\Lambda(H_0)}{\otimes}\Lambda(H)\big)\\
&= \big(\Lambda(H_0)/\mathfrak{m}(H_1)^m\Lambda(H_0)\big)_\Theta\otimes_{\Lambda(H_0)}\Lambda(H)\\
&= \Lambda_{H_1}(H_0)/\mathfrak{m}_{H_1}(H_0)^m\otimes_{\Lambda(H_0)}\Lambda(H)
\end{aligned}$$

this follows from Lemma 8.12.iv.

ii. As $\Lambda_{H_0,H_1}(H) = \varprojlim_m \Lambda_{H_0,H_1}(H)/F^m\Lambda_{H_0,H_1}(H)$ by Lemma 8.12.i this is a consequence of i.

iii. Because of ii. it suffices to show that $\big(\Lambda(H)/\mathfrak{m}(H_1)^m\Lambda(H)\big)_\Theta$ is independent of the choice of t_0. Let $\widetilde{\Theta} := \{\widetilde{t_0}^j\}_{j\geq 0}$ for some other choice. We have

$$\big(\Lambda(H_0)/\mathfrak{m}(H_1)^m\Lambda(H_0)\big)_\Theta = \Lambda_{H_1}(H_0)/\mathfrak{m}_{H_1}(H_0)^m = \big(\Lambda(H_0)/\mathfrak{m}(H_1)^m\Lambda(H_0)\big)_{\widetilde{\Theta}}\ .$$

Hence $\widetilde{t_0}$ and t_0 are units in $\big(\Lambda(H)/\mathfrak{m}(H_1)^m\Lambda(H)\big)_\Theta$ and $\big(\Lambda(H)/\mathfrak{m}(H_1)^m\Lambda(H)\big)_{\widetilde{\Theta}}$, respectively. This implies

$$\big(\Lambda(H)/\mathfrak{m}(H_1)^m\Lambda(H)\big)_\Theta = \big(\Lambda(H)/\mathfrak{m}(H_1)^m\Lambda(H)\big)_{\widetilde{\Theta}}\ .$$

□

It follows from Prop. 8.15 that any element $\mu\in\Lambda_{H_0,H_1}(H)$ can be written in the form $\mu = \sum_{i\geq 0} t^i\mu_i$ with $\{\mu_i\}_i\subseteq\Lambda_{H_1}(H_0)$ a bounded subset.

We want to define and investigate a "weak" topology on the ring $\Lambda_{H_0,H_1}(H)$ which actually will be more important than the strong topology. To motivate the definition we point out that as a consequence of Lemma 8.8 we have

$$\Lambda_{H_0,H_1}(H) = \bigcup_{j\geq 0} F^m\Lambda_{H_0,H_1}(H) + t_0^{-j}\Lambda(H) \qquad \text{for any } m\geq 0. \tag{20}$$

An additive subgroup $C\subseteq\Lambda_{H_0,H_1}(H)$ will be called open for the weak topology if

– $F^m\Lambda_{H_0,H_1}(H)\subseteq C$ for some $m\geq 0$ and

– for any $j\geq 0$ there is an $\ell(j)\geq 0$ such that $C\supseteq t_0^{-j}\mathfrak{m}(H)^{\ell(j)}$.

Correspondingly the *weak* topology on $\Lambda_{H_0,H_1}(H)$ is defined to be the topology for which the additive subgroups

$$C_{m,\ell} := F^m\Lambda_{H_0,H_1}(H) + \sum_{j\geq 0} t_0^{-j}\mathfrak{m}(H)^{\ell(j)}\ ,$$

with $m \geq 0$ and $\ell : \mathbb{N}_0 \longrightarrow \mathbb{N}_0$ any function, form a fundamental system of open zero neighbourhoods. This certainly makes $\Lambda_{H_0,H_1}(H)$ into an additive topological group. But one easily checks that multiplication by t_0^k, for any $k \in \mathbb{Z}$, is a topological automorphism.

REMARK 8.16. *The weak topology on $\Lambda_{H_0,H_1}(H)$ is independent of the choice of the variable t_0.*

PROOF. For the purposes of this proof we write $C_{m,\ell}(t_0) := C_{m,\ell}$. Let $\widetilde{t_0}$ be another choice of variable. For any $j \geq 0$ there is an $i(j,m) \geq 0$ such that

$$\widetilde{t_0}^{-j} \in \mathfrak{m}_{H_1}(H_0)^m + t_0^{-i(j,m)}\Lambda(H_0) .$$

We define a new function $\widetilde{\ell} : \mathbb{N}_0 \longrightarrow \mathbb{N}_0$ by $\widetilde{\ell}(j) := \ell(i(j,m))$ and obtain $C_{m,\widetilde{\ell}}(\widetilde{t_0}) \subseteq C_{m,\ell}(t_0)$. Hence, by symmetry, the two neighbourhood bases $\{C_{m,\ell}(\widetilde{t_0})\}_{m,\ell}$ and $\{C_{m,\ell}(t_0)\}_{m,\ell}$ define the same topology. □

LEMMA 8.17. *Given any $m' \geq m \geq 0$ there is a function ℓ such that for any $\lambda = \sum_{i\geq 0} \lambda_i t^i \in C_{m,\ell}$ we have*

$$\lambda_i \in \mathfrak{m}_{H_1}(H_0)^m + \mathfrak{m}(H_0)^{m'} \subseteq B_m \qquad \text{for any } 0 \leq i \leq m'.$$

PROOF. Applying Lemma 8.1 to m' and the element $s = t_0^j$, for any $j \geq 0$, we find an $\ell'(j) \geq 0$ such that

$$\mathfrak{m}(H_0)^{\ell'(j)} \subseteq \mathfrak{m}(H_1)^{m'}\Lambda(H_0) + t_0^j\mathfrak{m}(H_0)^{m'} \subseteq \mathfrak{m}_{H_1}(H_0)^{m'} + t_0^j\mathfrak{m}(H_0)^{m'}$$

and hence

$$t_0^{-j}\mathfrak{m}(H_0)^{\ell'(j)} \subseteq \mathfrak{m}_{H_1}(H_0)^{m'} + \mathfrak{m}(H_0)^{m'} \subseteq \mathfrak{m}_{H_1}(H_0)^m + \mathfrak{m}(H_0)^{m'} .$$

Since the $\mathfrak{m}(H)$-adic topology on $\Lambda(H) = \Lambda(H_0)[[t;\sigma,\delta]] \cong \prod_{i\geq 0}\Lambda(H_0)$ coincides with the direct product topology (cf. [**SV1**] §1) we then may pick $\ell(j) \geq 0$ in such a way that

$$\mathfrak{m}(H)^{\ell(j)} \subseteq (\sum_{i=0}^{m'} \mathfrak{m}(H_0)^{\ell'(j)}t^i) + \Lambda(H)t^{m'+1} .$$

Suppose now that $\lambda = \sum_{i\geq 0}\lambda_i t^i \in C_{m,\ell}$. Then λ_i, for any $0 \leq i \leq m'$, lies in

$$\mathfrak{m}_{H_1}(H_0)^m + \sum_{j\geq 0} t_0^{-j}\mathfrak{m}(H_0)^{\ell'(j)} \subseteq \mathfrak{m}_{H_1}(H_0)^m + \mathfrak{m}(H_0)^{m'} .$$

□

PROPOSITION 8.18. i. *The weak topology of $\Lambda_{H_0,H_1}(H)$ induces on the subrings $\Lambda_{H_1}(H_0)$ and $\Lambda(H)$ the weak and compact topology, respectively.*

ii. *$F^m\Lambda_{H_0,H_1}(H)$, for any $m \geq 0$, is closed in $\Lambda_{H_0,H_1}(H)$ for the weak topology.*

iii. *$\Lambda_{H_0,H_1}(H)$ is Hausdorff and complete in the weak topology.*

PROOF. i. We obviously have $B_{\max(m,\ell(0))} \subseteq C_{m,\ell}$. On the other hand, if for a given $m \geq 0$ we choose the function ℓ as in the above Lemma 8.17 then we obtain

$$C_{m,\ell} \cap \Lambda_{H_1}(H_0) \subseteq B_m .$$

To determine the topology induced on $\Lambda(H)$ we note that obviously $\mathfrak{m}(H)^{\ell(0)} \subseteq C_{m,\ell}$. On the other hand, given any $m \geq 0$, we find, by the direct product topology argument, an $m' \geq 0$ such that

$$(\sum_{i=0}^{m'} \mathfrak{m}(H_0)^{m'} t^i) + \Lambda(H) t^{m'+1} \subseteq \mathfrak{m}(H)^m .$$

We consider now any $\lambda = \sum_{i \geq 0} \lambda_i t^i \in C_{m',\ell} \cap \Lambda(H)$ where the function ℓ is chosen as in Lemma 8.17 for the pair (m', m'). Then λ_i, for any $0 \leq i \leq m'$, lies in

$$\big(\mathfrak{m}_{H_1}(H_0)^{m'} + \mathfrak{m}(H_0)^{m'}\big) \cap \Lambda(H_0) = \big(\mathfrak{m}_{H_1}(H_0)^{m'} \cap \Lambda(H_0)\big) + \mathfrak{m}(H_0)^{m'} = \mathfrak{m}(H_0)^{m'}$$

where the last identity uses the fact (cf. formula (14) in the proof of Lemma 8.2) that

$$\mathfrak{m}_{H_1}(H_0)^{m'} \cap \Lambda(H_0) = \mathfrak{m}(H_1)^{m'} \Lambda(H_0) \subseteq \mathfrak{m}(H_0)^{m'} .$$

By the choice of m' this means that $\lambda \in \mathfrak{m}(H)^m$.

ii. We choose for any $m' \geq m$ a function $\ell_{m',m}$ as in Lemma 8.17. Any $\lambda = \sum_{i \geq 0} \lambda_i t^i \in \bigcap_{m' \geq m} C_{m,\ell_{m',m}}$ satisfies

$$\lambda_i \in \bigcap_{m' \geq 0} \big(\mathfrak{m}_{H_1}(H_0)^m + \mathfrak{m}(H_0)^{m'}\big) \qquad \text{for any } i \geq 0.$$

But we know from Lemma 8.2.v that this intersection is equal to $\mathfrak{m}_{H_1}(H_0)^m$. It follows that

$$\bigcap_{\ell} C_{m,\ell} = F^m \Lambda_{H_0,H_1}(H) .$$

iii. That the weak topology is Hausdorff follows from ii. and Lemma 8.12.i. To establish the completeness let $(\lambda^{(\alpha)})_{\alpha \in \Xi}$ with $\lambda^{(\alpha)} = \sum_{i \geq 0} \lambda_i^{(\alpha)} t^i$ be a Cauchy net in $\Lambda_{H_0,H_1}(H)$ for the weak topology. Lemma 8.17 immediately implies that each $(\lambda_i^{(\alpha)})_{\alpha \in \Xi}$, for $i \geq 0$, is a Cauchy net for the weak topology in $\Lambda_{H_1}(H_0)$. The latter is complete by Lemma 8.2.iv. Hence each of these Cauchy nets has a limit $\lambda_i \in \Lambda_{H_1}(H_0)$. We will show that $\lambda := \sum_{i \geq 0} \lambda_i t^i$ lies in $\Lambda_{H_0,H_1}(H)$ and is the limit, for the weak topology, of the original Cauchy net $(\lambda^{(\alpha)})_{\alpha \in \Xi}$.

As a piece of notation we let $\mathcal{L}$ denote the set of all functions $\ell : \mathbb{N}_0 \longrightarrow \mathbb{N}_0$. It is partially ordered by $\ell_1 \leq \ell_2$ if $\ell_1(j) \leq \ell_2(j)$ for any $j \geq 0$.

In a first step we will establish, for each $m \geq 0$, the existence of a $j(m) \geq 0$ such that for any function ℓ the subset

$$\Xi(m,\ell) := \{\alpha \in \Xi : \lambda^{(\alpha)} \in C_{m,\ell} + t_0^{-j(m)} \Lambda(H)\}$$

is cofinal in Ξ. Fixing m there otherwise is, for any $k \geq 0$, a function ℓ_k and an index $\alpha_k \in \Xi$ such that

$$\lambda^{(\alpha)} \notin C_{m,\ell_k} + t_0^{-k} \Lambda(H) \qquad \text{for all } \alpha \geq \alpha_k.$$

We certainly may assume that $\ell_0 \leq \ell_1 \leq \ell_2 \leq \ldots$ We now define a new function ℓ by $\ell(j) := \ell_j(j)$. For $k \leq j$ we have $\ell(j) = \ell_j(j) \geq \ell_k(j)$ and hence

$$t_0^{-j} \mathfrak{m}(H)^{\ell(j)} \subseteq t_0^{-j} \mathfrak{m}(H)^{\ell_k(j)} \subseteq C_{m,\ell_k} .$$

For $k \geq j$ we have

$$t_0^{-j} \mathfrak{m}(H)^{\ell(j)} \subseteq t_0^{-j} \Lambda(H) \subseteq t_0^{-k} \Lambda(H) .$$

It follows that

$$C_{m,\ell} + t_0^{-k}\Lambda(H) \subseteq C_{m,\ell_k} + t_0^{-k}\Lambda(H) \qquad \text{for any } k \geq 0.$$

In particular, for any $k \geq 0$, we obtain

$$\lambda^{(\alpha)} \notin C_{m,\ell} + t_0^{-k}\Lambda(H) \qquad \text{for all } \alpha \geq \alpha_k.$$

Now we choose an index $\beta \in \Xi$ such that

$$\lambda^{(\beta_1)} - \lambda^{(\beta_2)} \in C_{m,\ell} \qquad \text{for any } \beta_1, \beta_2 \geq \beta.$$

We also choose, by (20), the integer k large enough so that $\lambda^{(\beta)} \in C_{m,\ell} + t_0^{-k}\Lambda(H)$. Then

$$\lambda^{(\alpha)} \in C_{m,\ell} + t_0^{-k}\Lambda(H) \qquad \text{for all } \alpha \geq \beta.$$

Since Ξ is directed this is a contradiction.

In the next step we show that $\lambda \in \Lambda_{H_0,H_1}(H)$. We have

$$\lambda_0^{(\alpha)}, \ldots, \lambda_{m'}^{(\alpha)} \in \mathfrak{m}_{H_1}(H_0)^m + t_0^{-j(m)}\Lambda(H_0) \qquad \text{for any } \alpha \in \Xi(m, \ell_{m',m}).$$

Since $\mathfrak{m}_{H_1}(H_0)^m + t_0^{-j(m)}\Lambda(H_0)$ is closed in $\Lambda_{H_1}(H_0)$ for the weak topology by Lemma 8.2 the cofinality of $\Xi(m, \ell_{m',m})$ implies that

$$\lambda_0, \ldots, \lambda_{m'} \in \mathfrak{m}_{H_1}(H_0)^m + t_0^{-j(m)}\Lambda(H_0)\ .$$

But m' was arbitrary. We therefore obtain

$$\{\lambda_i\}_{i\geq 0} \subseteq \mathfrak{m}_{H_1}(H_0)^m + t_0^{-j(m)}\Lambda(H_0) \qquad \text{for any } m \geq 0.$$

This means, by Lemma 8.8, that $\{\lambda_i\}_{i\geq 0}$ is bounded and hence that $\lambda \in \Lambda_{H_0,H_1}(H)$.

For the time being we fix an $m \geq 0$. The product set $\Xi \times \mathcal{L}$ is a directed partially ordered set by $(\alpha, \ell) \geq (\beta, \ell')$ if $\alpha \geq \beta$ and $\ell \geq \ell'$. We construct a net $\{\nu^{(\alpha,\ell)}\}_{(\alpha,\ell)\in\Xi\times\mathcal{L}}$ in $t_0^{-j(m)}\Lambda(H)$ in the following way. By cofinality we may pick an index $\alpha' \geq \alpha$ in $\Xi(m,\ell)$. We then find a $\nu^{(\alpha,\ell)} = \sum_{i\geq 0} \nu_i^{(\alpha,\ell)} t^i \in t_0^{-j(m)}\Lambda(H)$ such that $\lambda^{(\alpha')} - \nu^{(\alpha,\ell)} \in C_{m,\ell}$. Let us check that the net $\{\nu_i^{(\alpha,\ell)}\}_{(\alpha,\ell)\in\Xi\times\mathcal{L}}$, for any $i \geq 0$, converges to λ_i in the quotient $\Lambda_{H_1}(H_0)/\mathfrak{m}_{H_1}(H_0)^m$. Given any $m' \geq \max(m, i)$ we choose an $\alpha \in \Xi$ such that

$$\lambda_i^{(\beta)} - \lambda_i \in B_{m'} \subseteq \mathfrak{m}_{H_1}(H_0)^m + \mathfrak{m}(H_0)^{m'} \qquad \text{for any } \beta \geq \alpha.$$

By construction we find, for any $(\beta, \ell) \geq (\alpha, \ell_{m',m})$, an index $\beta' \geq \beta$ such that

$$\lambda^{(\beta')} - \nu^{(\beta,\ell)} \in C_{m,\ell} \subseteq C_{m,\ell_{m',m}}\ .$$

In particular, by Lemma 8.17, we have

$$\lambda_i^{(\beta')} - \nu_i^{(\beta,\ell)} \in \mathfrak{m}_{H_1}(H_0)^m + \mathfrak{m}(H_0)^{m'}\ .$$

It follows that

$$\nu_i^{(\beta,\ell)} - \lambda_i = (\lambda_i^{(\beta')} - \lambda_i) - (\lambda_i^{(\beta')} - \nu_i^{(\beta,\ell)}) \in \mathfrak{m}_{H_1}(H_0)^m + \mathfrak{m}(H_0)^{m'}$$

for any $(\beta, \ell) \geq (\alpha, \ell_{m',m})$.

Now we observe that the weak topology induces, by i., on the quotient

$$\left(t_0^{-j(m)}\Lambda(H) + F^m\Lambda_{H_0,H_1}(H)\right)/F^m\Lambda_{H_0,H_1}(H) \cong \prod_{i\geq 0} t_0^{-j(m)}\Lambda(H_0)/\mathfrak{m}(H_1)^m\Lambda(H_0)$$

the (compact) direct product topology. It follows that the net $\{\nu^{(\alpha,\ell)}\}_{(\alpha,\ell)\in\Xi\times\mathfrak{L}}$ converges to λ for the weak topology in the quotient $\Lambda_{H_0,H_1}(H)/F^m\Lambda_{H_0,H_1}(H)$.

Next we claim that with $\{\nu^{(\alpha,\ell)}\}_{(\alpha,\ell)\in\Xi\times\mathfrak{L}}$ also $(\lambda^{(\alpha)})_{\alpha\in\Xi}$ converges to λ in the quotient $\Lambda_{H_0,H_1}(H)/F^m\Lambda_{H_0,H_1}(H)$. Given any function ℓ_0 we choose an $(\alpha,\ell_1)\in\Xi\times\mathcal{L}$ such that

$$\nu^{(\beta,\ell)}-\lambda\in C_{m,\ell_0}\qquad\text{for any } (\beta,\ell)\geq(\alpha,\ell_1)$$

and

$$\lambda^{(\beta_1)}-\lambda^{(\beta_2)}\in C_{m,\ell_0}\qquad\text{for any } \beta_1,\beta_2\geq\alpha.$$

We put $\ell_2:=\max(\ell_0,\ell_1)$ and pick the $\alpha'\geq\alpha$ such that

$$\lambda^{(\alpha')}-\nu^{(\alpha,\ell_2)}\in C_{m,\ell_2}\ .$$

Then

$$\begin{aligned}\lambda^{(\beta)}-\lambda&=(\lambda^{(\beta)}-\lambda^{(\alpha')})+(\lambda^{(\alpha')}-\nu^{(\alpha,\ell_2)})+(\nu^{(\alpha,\ell_2)}-\lambda)\\&\in C_{m,\ell_0}+C_{m,\ell_2}+C_{m,\ell_0}\subseteq C_{m,\ell_0}\end{aligned}$$

for any $\beta\geq\alpha$.

We now have shown that our original Cauchy net $(\lambda^{(\alpha)})_{\alpha\in\Xi}$ converges to λ for the weak topology in the quotient $\Lambda_{H_0,H_1}(H)/F^m\Lambda_{H_0,H_1}(H)$ for any $m\geq 0$. It is clear from the explicit definition of the weak topology that this means that $(\lambda^{(\alpha)})_{\alpha\in\Xi}$ converges to λ in $\Lambda_{H_0,H_1}(H)$. □

Behind the above proof is the general principle that a (countable) strict inductive limit of complete topological abelian groups again is complete. But the notion of an inductive limit for topological algebraic structures is not entirely straightforward in the sense that it has the tendency to depend on the precise category one is working in. Since we did not want to get into a discussion of these questions we preferred to explicitly work out the argument in our case.

LEMMA 8.19. *Suppose that $H\cong H_1\rtimes(H/H_1)$ is a semidirect product; for any $m,k\geq 0$ there are $N(m)\geq 0$ and $l(k),l'(k)\geq 0$ such that we have:*

i. $\mathfrak{m}(H_0)^{l(k)}t_0^{-1}\subseteq\mathfrak{m}_{H_1}(H_0)^m+t_0^{-1-N(m)}\mathfrak{m}(H_0)^k$;
ii. $\mathfrak{m}(H)^{l'(k)}\cdot t_0^{-1}\subseteq F^m\Lambda_{H_0,H_1}(H)+t_0^{-1-N(m)}\mathfrak{m}(H)^k$.

PROOF. Since δ_0 is topologically nilpotent we find an $N(m)\geq 0$ such that $\delta_0^{N(m)+1}(\Lambda(H_1))\subseteq\mathfrak{m}(H_1)^m$. i. Let $k\geq 1$. The $\mathfrak{m}(H_0)$-adic topology on $\Lambda(H_0)=\Lambda(H_1)[[t_0;\sigma_0,\delta_0]]\cong\prod_{i\geq 0}\Lambda(H_1)$ coincides with the direct product topology. Hence $\mathfrak{m}(H_1)^k+\mathfrak{m}(H_0)^{k-1}t_0$ is open which means that it contains some $\mathfrak{m}(H_0)^{l(k)}$. We in fact consider any $\lambda=\nu+\mu t_0$ with $\nu\in\mathfrak{m}(H_1)^k$ and $\mu\in\mathfrak{m}(H_0)^{k-1}$. Using formula (1.5) in [**SV2**] we obtain

$$\lambda t_0^{-1}=\nu t_0^{-1}+\mu=\Big(\sum_{i\leq -1}t_0^i\sigma_0\delta_0^{-i-1}(\nu)\Big)+\mu\ .$$

Due to the choice of $N(m)$ the right hand side is contained in

$$\begin{aligned}
&\Big(\sum_{i=-1-N(m)}^{-1} t_0^i\sigma_0\delta_0^{-i-1}(\nu)\Big) + \mu + \mathfrak{m}_{H_1}(H_0)^m \\
&\subseteq \Big(\sum_{i=-1-N(m)}^{-1} t_0^i\mathfrak{m}(H_1)^k\Big) + \mathfrak{m}(H_0)^{k-1} + \mathfrak{m}_{H_1}(H_0)^m \\
&= t_0^{-1-N(m)}\Big(\Big(\sum_{i=0}^{N(m)} t_0^i\mathfrak{m}(H_1)^k\Big) + t_0^{1+N(m)}\mathfrak{m}(H_0)^{k-1}\Big) + \mathfrak{m}_{H_1}(H_0)^m \\
&\subseteq t_0^{-1-N(m)}\mathfrak{m}(H_0)^k + \mathfrak{m}_{H_1}(H_0)^m .
\end{aligned}$$

In the case $k = 0$ the same computation actually gives

$$\Lambda(H_0)t_0^{-1} \subseteq t_0^{-1-N(m)}\Lambda(H_0) + \mathfrak{m}_{H_1}(H_0)^m .$$

ii. We now have to use, for $\Lambda(H)$, the direct product topology argument twice. First we observe that there is a $k' \geq 0$ such that

$$\sum_{l=0}^{k'} \mathfrak{m}(H_0)^{k'}t^l + \Lambda(H)t^{k'+1} \subseteq \mathfrak{m}(H)^k .$$

We then have available the integer $l(k')$ form the first assertion. Secondly we note that

$$\sum_{i=0}^{l(k')+k'} \mathfrak{m}(H_0)^{l(k')+k'-i}t^i + \Lambda(H)t^{l(k')+k'+1}$$

is open in $\Lambda(H)$ and hence contains some $\mathfrak{m}(H)^{l'(k)}$. We actually will show that

$$\Big(\sum_{i=0}^{l(k')+k'} \mathfrak{m}(H_0)^{l(k')+k'-i}t^i + \Lambda(H)t^{l(k')+k'+1}\Big)\cdot t_0^{-1}$$

$$\subseteq F^m\Lambda_{H_0,H_1}(H) + t_0^{-1-N(m)}\Big(\sum_{l=0}^{k'} \mathfrak{m}(H_0)^{k'}t^l + \Lambda(H)t^{k'+1}\Big)$$

holds true. We therefore consider any

$$\lambda = \sum_{i\geq 0} \lambda_i t^i \qquad \text{with } \lambda_i \in \mathfrak{m}(H_0)^{l(k')+k'-i} \text{ for } 0 \leq i \leq l(k') + k'.$$

By construction we have

$$\lambda \cdot t_0^{-1} = \sum_{l\geq 0}\Big(\sum_{i\geq l}\binom{i}{l}\lambda_i\delta^{i-l}(\sigma^l(t_0^{-1}))\Big)t^l ,$$

and we claim that the coefficients on the right hand side lie in $\mathfrak{m}(H_0)^{l(k')}t_0^{-1}$, resp. $\Lambda(H_0)t_0^{-1}$, for $0 \leq l \leq l(k')$, resp. $l > l(k')$. We know from the proof of Lemma 8.3 that

$$\delta^{i-l}(t_0^{-1}) \in \mathfrak{n}^{i-l}t_0^{-1} ,$$

where $\mathfrak{n}$ denotes the maximal ideal in $o[[t_0]]$, and that $\sigma(t_0^{-1}) \subseteq o[[t_0]]t_0^{-1}$. We deduce that

$$\delta^{i-l}(\sigma^l(t_0^{-1})) = \sigma^l(\delta^{i-l}(t_0^{-1})) \subseteq \sigma^l(\mathfrak{n}^{i-l}t_0^{-1}) \subseteq \mathfrak{n}^{i-l}t_0^{-1} .$$

For any $l \le i$ we therefore always have $\lambda_i \delta^{i-l}(\sigma^l(t_0^{-1})) \in \Lambda(H_0)t_0^{-1}$. If in addition $l \le k'$ then

$$\lambda_i \delta^{i-l}(\sigma^l(t_0^{-1})) \in \mathfrak{m}(H_0)^{l(k')+k'-i}\mathfrak{n}^{i-l}t_0^{-1} \subseteq \mathfrak{m}(H_0)^{l(k')+k'-l}t_0^{-1} \subseteq \mathfrak{m}(H_0)^{l(k')}t_0^{-1}$$

for $l \le i \le l(k') + k'$ and

$$\lambda_i \delta^{i-l}(\sigma^l(t_0^{-1})) \in \mathfrak{m}(H_0)^{i-l}t_0^{-1} \subseteq \mathfrak{m}(H_0)^{l(k')+k'-l}t_0^{-1} \subseteq \mathfrak{m}(H_0)^{l(k')}t_0^{-1} .$$

for $i > l(k') + k'$. Since any $\mathfrak{m}(H_0)^j t_0^{-1}$ is compact and hence closed for the weak topology this establishes our claim, i. e., we have

$$\lambda \cdot t_0^{-1} = \sum_{l \ge 0} \mu_l t_0^{-1} t^l \quad \text{with } \mu_l \in \mathfrak{m}(H_0)^{l(k')} \text{ for } 0 \le l \le k' \text{ and } \in \Lambda(H_0) \text{ for } l > k'.$$

Applying now the first assertion we obtain that

$$\mu_l t_0^{-1} \in t_0^{-1-N(m)} \begin{Bmatrix} \mathfrak{m}(H_0)^{k'} \\ \Lambda(H_0) \end{Bmatrix} + \mathfrak{m}_{H_1}(H_0)^m \qquad \begin{cases} \text{if } 0 \le l \le k', \\ \text{if } l > k'. \end{cases}$$

This means that

$$\lambda \cdot t_0^{-1} \in F^m \Lambda_{H_0,H_1}(H) + t_0^{-1-N(m)} \Big(\sum_{l=0}^{k'} \mathfrak{m}(H_0)^{k'} t^l + \Lambda(H) t^{k'+1} \Big) .$$

□

PROPOSITION 8.20. *Suppose that $H \cong H_1 \rtimes (H/H_1)$ is a semidirect product; the multiplication map in the ring $\Lambda_{H_0,H_1}(H)$ is separately continuous for the weak topology.*

PROOF. We first consider the left multiplication by some $\lambda \in \Lambda_{H_0,H_1}(H)$. For any $m \ge 0$ and $j \ge 0$ we find, by applying Lemma 8.8 to the set A of coefficients of $\lambda \cdot t_0^{-j}$, a $k(\lambda, m, j) \ge 0$ such that

$$\lambda \cdot t_0^{-j} \subseteq F^m \Lambda_{H_0,H_1}(H) + t_0^{-k(\lambda,m,j)} \Lambda(H) .$$

Hence

$$\begin{aligned} \lambda \cdot (F^m \Lambda_{H_0,H_1}(H) + t_0^{-j}\mathfrak{m}(H)^k) &\subseteq F^m \Lambda_{H_0,H_1}(H) + \lambda \cdot t_0^{-j}\mathfrak{m}(H)^k \\ &\subseteq F^m \Lambda_{H_0,H_1}(H) + t_0^{-k(\lambda,m,j)}\mathfrak{m}(H)^k \end{aligned}$$

for any $k \ge 0$. Suppose given now any open $C_{m,\ell}$. If we define a new function ℓ' by $\ell'(j) := \ell(k(\lambda, m, j))$ then

$$\lambda \cdot C_{m,\ell'} \subseteq C_{m,\ell} .$$

The argument for the right multiplication by λ is similar but in addition is crucially based on Lemma 8.19.ii. Let $l'' := l' \circ \ldots \circ l'$ denote the $k(\lambda, m, 0)$-fold iteration of the function l' in that lemma. By a correspondingly iterated application of that lemma we obtain

$$\begin{aligned} &(F^m \Lambda_{H_0,H_1}(H) + t_0^{-j}\mathfrak{m}(H)^{l''(k)}) \cdot \lambda \\ &\quad \subseteq \big(F^m \Lambda_{H_0,H_1}(H) + t_0^{-j}\mathfrak{m}(H)^{l''(k)}\big) \cdot \big(F^m \Lambda_{H_0,H_1}(H) + t_0^{-k(\lambda,m,0)} \Lambda(H)\big) \\ &\quad \subseteq F^m \Lambda_{H_0,H_1}(H) + t_0^{-j}\mathfrak{m}(H)^{l''(k)} \cdot t_0^{-k(\lambda,m,0)} \Lambda(H) \\ &\quad \subseteq F^m \Lambda_{H_0,H_1}(H) + t_0^{-j-k(\lambda,m,0)(1+N(m))}\mathfrak{m}(H)^k \end{aligned}$$

for any $k \geq 0$. If we this time, given any function ℓ, define a new function ℓ' by

$$\ell'(j) := l''(\ell(j + k(\lambda, m, 0)(1 + N(m))))$$

then we have

$$C_{m,\ell'} \cdot \lambda \subseteq C_{m,\ell} \ .$$

□

Under our standing assumption that $H \cong H_1 \rtimes (H/H_1)$ we now have the commutative diagram of rings

$$\begin{array}{ccc} \Lambda_{H_0,H_1}(H) & \xleftarrow{\ \supseteq\ } & \Lambda(H) \\ \uparrow{\scriptstyle\subseteq} & & \uparrow{\scriptstyle\subseteq} \\ \Lambda_{H_1}(H_0) & \xleftarrow{\ \supseteq\ } & \Lambda(H_0) \end{array} \tag{21}$$

where, in addition, all maps are topological inclusions for the weak, resp. compact, topologies on the rings in the left, resp. right, column.

PROPOSITION 8.21. *Suppose that $H \cong H_1 \rtimes (H/H_1)$ is a semidirect product; we then have:*

i. *$\Lambda_{H_0,H_1}(H)$ is (left and right) noetherian;*
ii. *$\Lambda_{H_0,H_1}(H)$ is flat as a left as well as right $\Lambda_{H_1}(H_0)$-module;*
iii. *$\Lambda_{H_0,H_1}(H)$ is flat as a left as well as right $\Lambda(H)$-module.*

PROOF. *Step 1:* Let $H_1' \subseteq H_1$ be an open subgroup which is normal in H. We put

$$H_0' := \overline{< H_1', \gamma_0 >} \quad \text{and} \quad H' := \overline{< H_1', \gamma_0, \gamma >} \ .$$

Then H' is an open subgroup of H such that $H' \cong H_1' \rtimes (H/H_1)$. Obviously $\Lambda(H_1)$ is free of rank $[H_1 : H_1']$ as a left or right $\Lambda(H_1')$-module. Each of the rings $A := \Lambda(H_0), \Lambda(H), \Lambda_{H_1}(H_0)$, or $\Lambda_{H_0,H_1}(H)$ contains the corresponding ring $A' := \Lambda(H_0'), \Lambda(H'), \Lambda_{H_1'}(H_0')$, or $\Lambda_{H_0',H_1'}(H')$. We claim that in each case

$$A = \Lambda(H_1) \otimes_{\Lambda(H_1')} A' = A' \otimes_{\Lambda(H_1')} \Lambda(H_1)$$

holds true. For $A = \Lambda(H_0)$ and $A = \Lambda(H)$ this follows immediately from their descriptions

$$\Lambda(H_0) = \{\sum_{i \geq 0} \mu_i t_0^i : \mu_i \in \Lambda(H_1)\} = \{\sum_{i \geq 0} t_0^i \mu_i : \mu_i \in \Lambda(H_1)\}$$

and

$$\Lambda(H) = \{\sum_{i \geq 0} \lambda_i t^i : \lambda_i \in \Lambda(H_0)\} = \{\sum_{i \geq 0} t^i \lambda_i : \lambda_i \in \Lambda(H_0)\}$$

in terms of skew power series. Similarly, using that

$$\begin{aligned} \Lambda_{H_1}(H_0) &= \{\sum_{i \in \mathbb{Z}} \mu_i t_0^i : \mu_i \in \Lambda(H_1), \lim_{i \to -\infty} \mu_i = 0\} \\ &= \{\sum_{i \in \mathbb{Z}} t_0^i \mu_i : \mu_i \in \Lambda(H_1), \lim_{i \to -\infty} \mu_i = 0\} \end{aligned}$$

together with the fact that on $\Lambda(H_1)$ the $\mathfrak{m}(H_1)$-adic and the $\mathfrak{m}(H_1')\Lambda(H_1)$-adic topology coincide the claim is clear for the ring $A = \Lambda_{H_1}(H_0)$ as well. Finally, for the ring $A = \Lambda_{H_0,H_1}(H)$ we use Prop. 8.15 and obtain

$$\begin{aligned}
\Lambda_{H_0',H_1'}(H') \otimes_{\Lambda(H_1')} \Lambda(H_1) &= \big(\varprojlim \big(\Lambda(H')/\mathfrak{m}(H_1')^m\Lambda(H')\big)_\Theta\big) \otimes_{\Lambda(H_1')} \Lambda(H_1) \\
&= \varprojlim \big(\big(\Lambda(H')/\mathfrak{m}(H_1')^m\Lambda(H')\big)_\Theta \otimes_{\Lambda(H_1')} \Lambda(H_1)\big) \\
&= \varprojlim \big(\Lambda(H')/\mathfrak{m}(H_1')^m\Lambda(H') \otimes_{\Lambda(H_1')} \Lambda(H_1)\big)_\Theta \\
&= \varprojlim \big(\Lambda(H)/\mathfrak{m}(H_1')^m\Lambda(H)\big)_\Theta \\
&= \varprojlim \big(\Lambda(H)/\mathfrak{m}(H_1)^m\Lambda(H)\big)_\Theta \\
&= \Lambda_{H_0,H_1}(H)
\end{aligned}$$

where the second, resp. the second last, identity is due to the fact that $\Lambda(H_1)$ is free over $\Lambda(H_1')$ of rank $[H_1 : H_1']$, resp. to the cofinality of $\{\mathfrak{m}(H_1)^m\}_{m\geq 1}$ and $\{\mathfrak{m}(H_1')^m\Lambda(H_1)\}_{m\geq 1}$. The symmetric identity follows in the same way. This establishes our claim and shows that in order to prove our assertion we may replace, whenever convenient, the triple $H_1 \subseteq H_0 \subseteq H$ by the triple $H_1' \subseteq H_0' \subseteq H'$.

Step 2: The above commutative diagram of rings (21) in fact is a diagram of filtered rings with complete and separated filtrations defined by the two-sided ideals

$$F^m\Lambda(H_0) := \mathfrak{m}(H_1)^m\Lambda(H_0), \ F^m\Lambda(H) := \mathfrak{m}(H_1)^m\Lambda(H),$$
$$F^m\Lambda_{H_1}(H_0) := \mathfrak{m}_{H_1}(H_0)^m = \mathfrak{m}(H_1)^m\Lambda_{H_1}(H_0),$$

and $F^m\Lambda_{H_0,H_1}(H)$ as before. We obtain a corresponding commutative diagram of graded rings

(22)
$$\begin{array}{ccc}
\mathrm{gr}^\bullet \Lambda_{H_0,H_1}(H) & \longleftarrow & \mathrm{gr}^\bullet \Lambda(H) \\
\uparrow & & \uparrow \\
\mathrm{gr}^\bullet \Lambda_{H_1}(H_0) & \longleftarrow & \mathrm{gr}^\bullet \Lambda(H_0).
\end{array}$$

By the way, all four maps in this diagram again are injective (cf. (14) and [**SV2**] Lemma 1.12.i). By [**LvO**] Prop.s II.1.2.1 and II.1.2.3 our assertions follow from:

iv. All four graded rings in the diagram (22) are left and right noetherian;
v. $\mathrm{gr}^\bullet \Lambda_{H_0,H_1}(H)$ is flat as a left and as a right $\mathrm{gr}^\bullet \Lambda_{H_1}(H_0)$-module;
vi. $\mathrm{gr}^\bullet \Lambda_{H_0,H_1}(H)$ is flat as a left and as a right $\mathrm{gr}^\bullet \Lambda(H)$-module.

Ad v.: As a consequence of Lemma 8.12.iv and of the flatness of $\Lambda(H)$ over $\Lambda(H_0)$ we have

$$\mathrm{gr}^\bullet \Lambda_{H_1}(H_0) \otimes_{\Lambda(H_0)/\mathfrak{m}(H_1)\Lambda(H_0)} \Lambda(H)/\mathfrak{m}(H_1)\Lambda(H) \xrightarrow{\cong} \mathrm{gr}^\bullet \Lambda_{H_0,H_1}(H) \ .$$

The same flatness then implies, by base extension, that $\mathrm{gr}^\bullet \Lambda_{H_0,H_1}(H)$ is flat as a left $\mathrm{gr}^\bullet \Lambda_{H_1}(H_0)$-module. Using Prop. 8.15.i one sees that one also has

$$\Lambda(H)/\mathfrak{m}(H_1)\Lambda(H) \otimes_{\Lambda(H_0)/\mathfrak{m}(H_1)\Lambda(H_0)} \mathrm{gr}^\bullet \Lambda_{H_1}(H_0) \xrightarrow{\cong} \mathrm{gr}^\bullet \Lambda_{H_0,H_1}(H)$$

which implies the asserted flatness as a right module.
Ad vi.: Because of [**SV2**] lemma 1.12.ii we have

$$F^m\Lambda(H) = \{\sum_{i\geq 0} t^i\nu_i \in \Lambda(H) : \nu_i \in \mathfrak{m}(H_1)^m\Lambda(H_0)\}$$

and

$$\mathfrak{m}(H_1)^m\Lambda(H_0) = \{\sum_{i\geq 0}\alpha_i t_0^i \in \Lambda(H_0) : \alpha_i \in \mathfrak{m}(H_1)^m\}.$$

It follows first of all that the set $\Theta = \{t_0^j\}_{j\geq 0}$ is mapped, by the symbol map, injectively into $\mathrm{gr}^0\,\Lambda(H)$. We therefore will not distinguish, in the notation, between the elements in Θ and their symbols. Secondly, let $\nu = \sum_{i\geq 0} t^i\nu_i \in \Lambda(H)$ and assume that $\nu t_0^j \in F^m\Lambda(H)$ for some $j \geq 0$. Then $\nu_i t_0^j \in \mathfrak{m}(H_1)^m\Lambda(H_0)$ and hence $\nu_i \in \mathfrak{m}(H_1)^m\Lambda(H_0)$ for any $i \geq 0$. We conclude that $\nu \in F^m\Lambda(H)$. It follows that the elements in Θ are left regular in $\mathrm{gr}^0\,\Lambda(H)$. In particular, Θ as a subset of $\mathrm{gr}^\bullet\,\Lambda(H)$ is multiplicatively closed. Right regularity follows by a symmetric argument. We consider now any element $a = \overline{\lambda_0} + \ldots + \overline{\lambda_r} \in \mathrm{gr}^\bullet\,\Lambda(H)$ with $\overline{\lambda_j} = \lambda_j + F^{j+1}\Lambda(H) \in \mathrm{gr}^j\,\Lambda(H)$. By Lemma 8.14 we find, after choosing some $m > r$, an integer $M > 0$ and elements $\mu_0, \ldots, \mu_r \in \Lambda(H)$ such that

$$t_0^M\lambda_j \equiv \mu_j t_0 \mod F^m\Lambda(H) \qquad \text{for any } 0 \leq j \leq r.$$

The regularity of t_0 then implies that $\mu_j \in F^j\Lambda(H)$ and that

$$t_0^M\overline{\lambda_j} = \overline{\mu_j}t_0 \qquad \text{with } \overline{\mu_j} := \mu_j + F^{j+1}\Lambda(H) \in \mathrm{gr}^j\,\Lambda(H).$$

By setting $b := \overline{\mu_0} + \ldots + \overline{\mu_r}$ we obtain

$$t_0^M a = bt_0 \qquad \text{in } \mathrm{gr}^\bullet\,\Lambda(H).$$

This means that $\Theta \subseteq \mathrm{gr}^\bullet\,\Lambda(H)$ satisfies the left Ore condition. Again the right Ore condition holds as well by a symmetric argument. Since t_0 is invertible in $\mathrm{gr}^\bullet\,\Lambda_{H_0,H_1}(H)$ the injective homomorphism $\mathrm{gr}^\bullet\,\Lambda(H) \longrightarrow \mathrm{gr}^\bullet\,\Lambda_{H_0,H_1}(H)$ extends to an injective ring homomorphism

$$\left(\mathrm{gr}^\bullet\,\Lambda(H)\right)_\Theta \longrightarrow \mathrm{gr}^\bullet\,\Lambda_{H_0,H_1}(H)\ .$$

As a straightforward consequence of Prop. 8.15.i it also is surjective and hence is an isomorphism. As a localization in Θ the ring $\mathrm{gr}^\bullet\,\Lambda_{H_0,H_1}(H)$ of course is (left and right) flat over $\mathrm{gr}^\bullet\,\Lambda(H)$.

Step 3: For this proof we do not need to establish the assertion iv. in full generality. Because of Step 1 it in fact suffices to do this after replacing the given triple $H_1 \subseteq H_0 \subseteq H$ by an appropriate "smaller" one. By [**Wil**] Prop. 8.5.2 there is an open normal subgroup $\widetilde{H} \subseteq H$ which is extra-powerful. Then $H_1' := H_1 \cap \widetilde{H}$ is extra-powerful, too, since $\widetilde{H}/H_1'$ is torsionfree. Hence it suffices to prove iv. under the additional assumption that H_1 is extra-powerful.

We have seen that $\mathrm{gr}^\bullet\,\Lambda_{H_0,H_1}(H)$ is a localization of $\mathrm{gr}^\bullet\,\Lambda(H)$. By exactly analogous arguments $\mathrm{gr}^\bullet\,\Lambda_{H_1}(H_0)$ is a localization of $\mathrm{gr}^\bullet\,\Lambda(H_0)$. Since being noetherian is preserved by localization (cf. [**MCR**] Prop. 2.1.16.iii) we therefore need only to consider the rings $\mathrm{gr}^\bullet\,\Lambda(H)$ and $\mathrm{gr}^\bullet\,\Lambda(H_0)$. They contain the graded ring $\mathrm{gr}^\bullet\,\Lambda(H_1)$ for the filtration $F^m\Lambda(H_1) := \mathfrak{m}(H_1)^m$. Using [**SV2**] Lemma 1.12.i one checks that

$$\mathrm{gr}^\bullet\,\Lambda(H) = \mathrm{gr}^\bullet\,\Lambda(H_1) \otimes_k \Omega(H/H_1) = \Omega(H/H_1) \otimes_k \mathrm{gr}^\bullet\,\Lambda(H_1)$$

and

$$\mathrm{gr}^\bullet\,\Lambda(H_0) = \mathrm{gr}^\bullet\,\Lambda(H_1) \otimes_k \Omega(H_0/H_1) = \Omega(H_0/H_1) \otimes_k \mathrm{gr}^\bullet\,\Lambda(H_1)$$

hold true (at least as bimodules). We see, first of all, that if $H' \subseteq H$ is an open subgroup containing H_1 then $\mathrm{gr}^\bullet\,\Lambda(H)$ and $\mathrm{gr}^\bullet\,\Lambda(H_0)$ are finitely generated free modules over $\mathrm{gr}^\bullet\,\Lambda(H')$ and $\mathrm{gr}^\bullet\,\Lambda(H_0 \cap H')$, respectively. Due to [**DDMS**] Lemma

3.4 and Cor. 8.34 there is an open normal subgroup $N \subseteq H_1$ such that each element of N is a p-th power in H_1. Since, by the above observation, we may replace H by any appropriate open $H' \supseteq H_1$ it suffices to consider the special case where conjugation by γ_0 and by γ both induce the identity on H_1/N. This implies that the commutators $[\gamma_0, h]$ and $[\gamma, h]$, for any $h \in H_1$, are p-th powers in H_1. The computation in the proof of [**SV2**] Lemma 4.3.ii then shows that the two factors $\mathrm{gr}^\bullet \Lambda(H_1)$ and $\Omega(H/H_1)$ in the above tensor product representation of $\mathrm{gr}^\bullet \Lambda(H)$ centralize each other.

At this point we make use of our additional assumption that H_1 is extra-powerful. Then $\mathrm{gr}^\bullet \Lambda(H_1)$ is a finitely generated commutative k-algebra by [**SV2**] Lemma 4.3.iii. It follows that $\mathrm{gr}^\bullet \Lambda(H)$, resp. $\mathrm{gr}^\bullet \Lambda(H_0)$, is an almost normalizing extension of the noetherian ring $\Omega(H/H_1)$, resp. $\Omega(H_0/H_1)$, and hence is noetherian by [**MCR**] Thm. 1.6.14. □

Consider any finitely generated $\Lambda_{H_1}(H_0)$-module M. We choose a presentation of M as a quotient $\Lambda_{H_1}(H_0)^n \twoheadrightarrow M$ of a finitely generated free $\Lambda_{H_1}(H_0)$-module. On $\Lambda_{H_1}(H_0)^n$ we have the product topology of the weak topology on each factor $\Lambda_{H_1}(H_0)$, and then on M we consider the corresponding quotient topology. The latter is easily shown to be independent of the particular presentation of M used and to make M into a topological $\Lambda_{H_1}(H_0)$-module. It will be called the *weak* topology on M.

LEMMA 8.22. *For any finitely generated $\Lambda_{H_1}(H_0)$-module M we have:*

i. *Every submodule $L \subseteq M$ is closed for the weak topology;*
ii. *M is is complete and Hausdorff in its weak topology.*

PROOF. i. By the definition of the weak topology we only need to consider the case $M = \Lambda_{H_1}(H_0)^n$. Note that in this case M is a $\Lambda_{H_1}(H_0)$-bimodule so that we may multiply by any ring element from the right. We recall that the ring $\Lambda_{H_1}(H_0)$ is noetherian and pseudocompact. Hence M is a finitely generated pseudocompact $\Lambda_{H_1}(H_0)$-module. The general theory of pseudocompact rings then tells us that L is (finitely generated and hence) closed for the pseudocompact topology. As a consequence we have by [**Gab**] IV.3 Prop. 11 that

$$L = \bigcap_{m \geq 0} (L + \mathfrak{m}_{H_1}(H_0)^m M) .$$

Note that $\mathfrak{m}_{H_1}(H_0)^m M$ is closed in M by Lemma 8.2.v. Let now $\{x_i\}_{i \in \mathbb{N}}$ be a sequence in L which converges to some x in M. It suffices to show that $x \in L + \mathfrak{m}_{H_1}(H_0)^m M$ for any $m \geq 0$. We fix some $m \geq 0$. By Remark 8.6.vi and Lemma 8.8 we find an $l \geq 0$ such that all $x_i t_0^l$ as well as $x t_0^l$ lie in $\Lambda(H_0)^n + \mathfrak{m}_{H_1}(H_0)^m M$. Hence modulo $\mathfrak{m}_{H_1}(H_0)^m M$ the sequence $\{x_i t_0^l\}_i$ lies in

$$\big((L t_0^l + \mathfrak{m}_{H_1}(H_0)^m M)/\mathfrak{m}_{H_1}(H_0)^m M\big) \cap \big((\Lambda(H_0)^n + \mathfrak{m}_{H_1}(H_0)^m M)/\mathfrak{m}_{H_1}(H_0)^m M\big)$$

and converges to $x t_0^l$ in $(\Lambda(H_0)^n + \mathfrak{m}_{H_1}(H_0)^m M)/\mathfrak{m}_{H_1}(H_0)^m M$ with respect to the topology induced by the weak topology. By Lemma 8.2.ii this topology coincides with the natural compact topology on the latter as a finitely generated module over the noetherian compact ring $\Lambda(H_0)$. The former is a (necessarily finitely generated) $\Lambda(H_0)$-submodule of the latter and as such has to be closed. This shows that $x t_0^l$ lies in $L t_0^l + \mathfrak{m}_{H_1}(H_0)^m M$ and hence that $x \in L + \mathfrak{m}_{H_1}(H_0)^m M$.

ii. The assertion i. immediately implies that M is Hausdorff for the weak topology. By construction the weak topology on $\Lambda_{H_1}(H_0)^n$ has a countable fundamental system of open neighbourhoods of zero. Hence it is metrizable (cf. [**War**] Thm. 6.12). It also is complete by Lemma 8.2.iv. In this situation any factor group of $\Lambda_{H_1}(H_0)^n$ by a closed subgroup, so in particular M by i., is complete as well (cf. [**War**] Thm. 6.12). □

REMARK 8.23. *Let M be a complete Hausdorff linear-topological o-module; any continuous (left) H-action on M extends uniquely to a continuous (left) $\Lambda(H)$-module structure on M.*

PROOF. [**Laz**] Thm. II.2.2.6. □

Let $\Gamma \subseteq H$ denote the closed subgroup topologically generated by our choice of γ; in particular, $\Gamma \xrightarrow{\cong} H/H_0 \cong \mathbb{Z}_p$. Since H_1 is normal in H the conjugation action of Γ on H_0 induces an action of Γ on the ring $\Lambda_{H_1}(H_0)$. We let $\sigma_{\gamma'}$ denote the ring automorphism corresponding to $\gamma' \in \Gamma$.

REMARK 8.24. i. *The $\sigma_{.}$-action of Γ on $\Lambda_{H_1}(H_0)$ is continuous for the weak topology.*

ii. *We have $\sigma_{\gamma'}(\lambda) = \gamma' \cdot \lambda \cdot \gamma'^{-1}$ for any $\gamma' \in \Gamma$ and $\lambda \in \Lambda_{H_1}(H_0)$ (where $\cdot$ denotes the multiplication in the ring $\Lambda_{H_0,H_1}(H)$.*

PROOF. i. The Γ-action respects the rings $\Lambda(H_0) \supseteq \Lambda(H_1)$ and hence their unique maximal ideals. It follows immediately that

$$\sigma_{\gamma'}(B_m) = B_m \qquad \text{for any } \gamma' \in \Gamma \text{ and any } m \geq 0.$$

For the asserted continuity it therefore remains to show that for any $\mu \in \Lambda_{H_1}(H_0)$ and any $m \geq 0$ there is an open subgroup $\Gamma' \subseteq \Gamma$ such that

$$\sigma_{\gamma'}(\mu) \in \mu + B_m \qquad \text{for any } \gamma' \in \Gamma'.$$

Since this relation only depends on μ modulo B_m we may assume that μ is of the form $\mu = t_0^{-l}\nu$ for some $l \geq 0$ and some $\nu \in \Lambda(H_0)$. We fix an $m' \geq m$ such that $t_0^{-l}B_{m'} \subseteq B_m$. First of all, contemplating the diagram

$$\begin{array}{ccc} \Gamma \times \Lambda(H_0) & \xrightarrow{\quad \sigma_{.} \quad} & \Lambda(H_0) \\ {\scriptstyle (\gamma,\mu)\mapsto(\gamma,\mu,\gamma^{-1})}\big\downarrow & & \big\downarrow{\scriptstyle \subseteq} \\ \Lambda(H) \times \Lambda(H_0) \times \Lambda(H) & \xrightarrow{\text{product}} & \Lambda(H) \end{array}$$

we see that the $\sigma_{.}$-action on $\Lambda(H_0)$ is continuous. Hence there is an open subgroup $\Gamma_1 \subseteq \Gamma$ such that

$$\sigma_{\gamma'}(\nu) \in \nu + \mathfrak{m}(H_0)^{m'} \subseteq \nu + B_{m'} \qquad \text{for any } \gamma' \in \Gamma_1.$$

Secondly we have to revisit the computation in Lemma 8.3. We recall that $\mathfrak{n}$ denotes the maximal ideal of the subring $o[[t_0]]$ in $\Lambda(H_0)$. Define the continuous homomorphism $e : \Gamma \longrightarrow 1 + p\mathbb{Z}_p$ by

$$\gamma'\gamma_0\gamma'^{-1} = \gamma_0^{e(\gamma')}$$

and put $\Gamma_2 := e^{-1}(1 + p^{m+l+1}\mathbb{Z}_p)$. We have

$$\sigma_{\gamma'}(t_0) = e(\gamma')t_0 + \sum_{j\geq 2} \binom{e(\gamma')}{j} t_0^j \; .$$

If v_p denotes the p-adic valuation then, for $j \geq 2$ and $y \in \mathbb{Z}_p$, one has (cf. [**Sch**] Prop. 47.4)

$$v_p(\binom{y}{j}) \geq v_p(y-1) - j \ .$$

It follows that for $\gamma' \in \Gamma_2$ we have

$$\begin{aligned}\sigma_{\gamma'}(t_0) &= e(\gamma')t_0\big(1 + \sum_{j\geq 2} e(\gamma')^{-1}\binom{e(\gamma')}{j} t_0^{j-1}\big) \\ &\in e(\gamma')t_0\big(1 + \sum_{j\geq 2} (\pi o)^{m+l+1-j} t_0^{j-1}\big) \\ &\subseteq e(\gamma')t_0\big(1 + \sum_{j=2}^{l+1} (\pi o)^{m} t_0^{j-1} + \mathfrak{n}^m t_0^l\big) \\ &\subseteq e(\gamma')t_0\big(1 + (\pi o)^m o[[t_0]] + \mathfrak{n}^m t_0^l\big) \ ,\end{aligned}$$

hence

$$\sigma_{\gamma'}(t_0^l) \in e(\gamma')^l t_0^l\big(1 + (\pi o)^m o[[t_0]] + \mathfrak{n}^m t_0^l\big) \ ,$$

and therefore

$$\begin{aligned}\sigma_{\gamma'}(t_0^{-l}) &\in e(\gamma')^{-l} t_0^{-l}\big(1 + (\pi o)^m o[[t_0]] + \mathfrak{n}^m t_0^l\big) \\ &\subseteq e(\gamma')^{-l} t_0^{-l} + \mathfrak{m}_{H_1}(H_0)^m + \mathfrak{m}(H_0)^m \\ &\subseteq e(\gamma')^{-l} t_0^{-l} + B_m = t_0^{-l} + (e(\gamma')^{-l} - 1)t_0^{-l} + B_m \\ &= t_0^{-l} + B_m \ .\end{aligned}$$

Together we obtain

$$\sigma_{\gamma'}(t_0^{-l}\nu) \in (t_0^{-l} + B_m)(\nu + B_{m'}) \subseteq t_0^{-l}\nu + t_0^{-l}B_{m'} + B_m + B_m B_{m'} \subseteq t_0^{-l}\nu + B_m$$

for any $\gamma' \in \Gamma' := \Gamma_1 \cap \Gamma_2$.

ii. Since

$$\Lambda_{H_0,H_1}(H) = \varprojlim_m \big(\Lambda_{H_0,H_1}(H)/F^m\Lambda_{H_0,H_1}(H)\big)$$

we may do the comparison modulo $F^m\Lambda_{H_0,H_1}(H)$. By (20) we therefore may assume that $\lambda = t_0^{-j}\mu$ for some $j \geq 0$ and some $\mu \in \Lambda(H)$. At this point we emphasize that, for any fixed γ', we assert the equality of two ring automorphisms. Hence we are reduced to showing that the assertion holds true in the two cases $\lambda = t_0$ and $\lambda = \mu$. The first case, of course, is subsumed by the second one. But for $\lambda \in \Lambda(H)$ our assertion is clear since the multiplication $\cdot$ restricts to the usual multiplication in $\Lambda(H)$. □

DEFINITION 8.25. *A $(\Lambda_{H_1}(H_0), \Gamma)$ -module is a finitely generated left $\Lambda_{H_1}(H_0)$-module with a $\sigma_.$-linear left Γ-action which is continuous for the weak topology.*

Let us consider a (left) $\Lambda_{H_0,H_1}(H)$-module M such that

- M is finitely generated over $\Lambda_{H_1}(H_0)$ and
- the module multiplication $\Lambda_{H_0,H_1}(H) \times M \longrightarrow M$ is separately continuous for the weak topologies.

By applying Prop. 8.18.i to a commutative diagram

$$\begin{array}{ccc} \Lambda_{H_0,H_1}(H)^n & & \\ \uparrow \subseteq & \searrow\!\!\!\!\searrow & \\ & & M \\ & \nearrow\!\!\!\!\nearrow & \\ \Lambda_{H_1}(H_0)^n & & \end{array}$$

of presentations of M the latter requirement implies that the weak topology on M coincides with the quotient topology derived from some presentation $\Lambda_{H_0,H_1}(H)^n \twoheadrightarrow M$. As another consequence of Prop. 8.18.i the compact ring $\Lambda(H)$ acts separately continuously on M. Since both, $\Lambda(H)$ and M, are complete metrizable abelian groups by Lemma 8.2 this action, in fact, has to be continuous ([**CF**] Thm. 2). In particular, by further restriction we obtain a continuous Γ-action on M. It is $\sigma_{.}$-linear by Remark 8.24.ii. We see that M is a $(\Lambda_{H_1}(H_0), \Gamma)$-module. By Lemma 8.12.ii the (left) ideal $\mathfrak{m}_{H_1}(H_0)^m \Lambda_{H_0,H_1}(H)$ is dense in $F^m \Lambda_{H_0,H_1}(H)$ for the strong and hence the weak topology. On the other hand $\mathfrak{m}_{H_1}(H_0)^m M$ is closed in M by Lemma 8.22.i. We point out that therefore M has the additional property that

$$F^m \Lambda_{H_0,H_1}(H) \cdot M = \mathfrak{m}_{H_1}(H_0)^m M \qquad \text{for any } m \geq 0.$$

Vice versa, let us start now with a $(\Lambda_{H_1}(H_0), \Gamma)$-module M. By Lemma 8.2.ii the compact ring $\Lambda(H_0)$ and hence the group H_0 act continuously on M. Therefore $H = H_0 \rtimes \Gamma$ acts continuously on M. Because of Lemma 8.22.ii we may apply Remark 8.23 and see that the $\Lambda(H_0)$-action extends to a continuous action of $\Lambda(H)$ on M. We want to see that the actions of $\Lambda_{H_1}(H_0)$ and $\Lambda(H)$ on M, which we have so far, combine and further extend to a separately continuous action of the ring $\Lambda_{H_0,H_1}(H)$. For this it is useful to first make the following observation. Being finitely generated over the noetherian pseudocompact ring $\Lambda_{H_1}(H_0)$ the module M is pseudocompact for the $\mathfrak{m}_{H_1}(H_0)$-adic topology. It therefore follows from [**Gab**] IV.3 Prop. 10 that the natural map

$$(23) \qquad M \xrightarrow{\cong} \varprojlim_m M/\mathfrak{m}_{H_1}(H_0)^m M$$

is an isomorphism of $\Lambda_{H_1}(H_0)$-modules. The $\sigma_{.}$-action, of course, respects the maximal ideal $\mathfrak{m}_{H_1}(H_0)$. Hence $\sigma_{.}$-linearity implies that Γ respects the submodules $\mathfrak{m}_{H_1}(H_0)^m M$. In particular, (23) is an isomorphism of $(\Lambda_{H_1}(H_0), \Gamma)$-modules.

In order to construct an action by $\Lambda_{H_0,H_1}(H)$ on M we therefore may assume, provided we do this in a functorial way, that

$$(24) \qquad \mathfrak{m}_{H_1}(H_0)^m M = 0 \qquad \text{for some } m \geq 0.$$

Let now $\lambda \in \Lambda_{H_0,H_1}(H)$. By (20) we may write $\lambda = \mu + t_0^{-j}\nu$ for appropriate $j \geq 0$, $\mu \in F^m \Lambda_{H_0,H_1}(H)$, and $\nu \in \Lambda(H)$. We define

$$\lambda \cdot x := t_0^{-j}(\nu x) \qquad \text{for any } x \in M.$$

In order to see that this is well defined let $\lambda = \mu' + t_0^{-j'}\nu'$ be another such decomposition. We may assume that $j \geq j'$. Then $t_0^{-j}(\nu - t_0^{j-j'}\nu') = \mu' - \mu \in F^m \Lambda_{H_0,H_1}(H)$ and hence $\nu - t_0^{j-j'}\nu' \in F^m \Lambda_{H_0,H_1}(H) \cap \Lambda(H) = \mathfrak{m}(H_1)^m \Lambda(H)$ where the last identity comes from (14) in the proof of Lemma 8.2. Because of (24) it follows that

$(\nu - t_0^{j-j'}\nu')x = 0$ and consequently that

$$t_0^{-j}(\nu x) = t_0^{-j}((t_0^{j-j'}\nu')x) = t_0^{-j}(t_0^{j-j'}(\nu' x)) = t_0^{-j'}(\nu' x) .$$

One easily deduces from this computation that our definition also is independent of the choice of a specific m in (24). It is straightforward to check that the resulting map

$$\cdot : \Lambda_{H_0,H_1}(H) \times M \longrightarrow M$$

is o-bilinear and functorial in M (at least as long as M satisfies (24)).

LEMMA 8.26. *Assuming* (24) *the map* $\cdot$ *is associative.*

PROOF. *Step 1:* Let $\lambda, \lambda' \in \Lambda_{H_0,H_1}(H)$ and $x \in M$ be any elements. We have to show that

$$\lambda \cdot (\lambda' \cdot x) = (\lambda \cdot \lambda') \cdot x$$

holds true. Choose $j \geq 0$ and $\nu, \nu' \in \Lambda(H)$ such that

$$\lambda - t_0^{-j}\nu, \lambda' - t_0^{-j}\nu' \in F^m\Lambda_{H_0,H_1}(H) .$$

Then

$$\lambda \cdot \lambda' - (t_0^{-j}\nu) \cdot (t_0^{-j}\nu') \in F^m\Lambda_{H_0,H_1}(H) .$$

Hence the above identity amounts to

$$t_0^{-j}(\nu(t_0^{-j}(\nu' x))) = ((t_0^{-j}\nu) \cdot (t_0^{-j}\nu')) \cdot x . \tag{25}$$

Step 2: We claim that both sides of (25) depend continuously on $\nu, \nu' \in \Lambda(H)$. For the left hand side this is an immediate consequence of the continuity of the $\Lambda_{H_1}(H_0)$- and $\Lambda(H)$-actions on M. To see this on the right hand side we must rewrite it. By Lemma 8.19.ii there are integers $N = N(m,j) \geq 0$ and $l = l(j) \geq 0$ such that

$$t_0^l\Lambda(H) \cdot t_0^{-j} \subseteq \mathfrak{m}(H)^l \cdot t_0^{-j} \subseteq F^m\Lambda_{H_0,H_1}(H) + t_0^{-N}\Lambda(H) .$$

Moreover, the same lemma says that the resulting map

$$\begin{aligned} \mathfrak{m}(H)^l &\longrightarrow \Lambda(H)/\mathfrak{m}(H_1)^m\Lambda(H) \\ \nu &\longmapsto \tilde{\nu} + \mathfrak{m}(H_1)^m\Lambda(H) \text{ where } \nu \cdot t_0^{-j} - t_0^{-N}\tilde{\nu} \in F^m\Lambda_{H_0,H_1}(H) \end{aligned}$$

is continuous. It follows that

$$((t_0^{-j}\nu) \cdot (t_0^{-j}\nu')) \cdot x = t_0^{-j-l-N}((\widetilde{(t_0^l\nu)}\nu')x) ,$$

and that the right hand side is continuous in ν and ν'.

Step 3: The elements in H span a dense o-submodule of $\Lambda(H)$. By the continuity property established in Step 2 it therefore suffices to prove the identity (25) for group elements $\nu = h$ and $\nu' = h'$. Write $h = h_0\gamma_1$ and $h' = h_0'\gamma_2$ with $h_0, h_0' \in H_0$ and $\gamma_1, \gamma_2 \in \Gamma$. Then (25) becomes a special case of the identity

$$\alpha(\gamma_1(\beta(\gamma_2 x))) = (\alpha \cdot \gamma_1 \cdot \beta \cdot \gamma_2) \cdot x \qquad \text{for any } \alpha, \beta \in \Lambda_{H_1}(H_0).$$

Using the σ.-linearity of the Γ-action the left hand side is equal to

$$\alpha(\sigma_{\gamma_1}(\beta)(\gamma_1(\gamma_2 x))) = \alpha\sigma_{\gamma_1}(\beta)((\gamma_1\gamma_2)x) = (\alpha\sigma_{\gamma_1}(\beta)) \cdot ((\gamma_1\gamma_2) \cdot x) .$$

Using the Remark 8.24.ii the right hand side is equal to

$$(\alpha \cdot \gamma_1 \cdot \beta \cdot \gamma_1^{-1} \cdot \gamma_1 \cdot \gamma_2) \cdot x = (\alpha\sigma_{\gamma_1}(\beta) \cdot \gamma_1\gamma_2) \cdot x .$$

This reduces us to the special case of associativity dealt with in the subsequent last step.

Step 4: For $\mu \in \Lambda_{H_1}(H_0)$ and $\nu \in \Lambda(H)$ we have

$$(\mu \cdot \nu) \cdot x = \mu \cdot (\nu \cdot x) .$$

Write $\mu = \mu' + t_0^{-l}\nu'$ for appropriate $l \geq 0$, $\mu' \in \mathfrak{m}_{H_1}(H_0)^m$, and $\nu' \in \Lambda(H_0)$. Then $\mu \cdot \nu = \mu' \cdot \nu + t_0^{-l}\nu'\nu$ with $\mu' \cdot \nu \in F^m\Lambda_{H_0,H_1}(H)$ and $\nu'\nu \in \Lambda(H)$. Hence

$$(\mu \cdot \nu) \cdot x = t_0^{-l}((\nu'\nu)x) = t_0^{-l}(\nu'(\nu x)) = (t_0^{-l}\nu')(\nu x) = \mu \cdot (\nu \cdot x) .$$

□

By using (23) our construction extends to arbitrary $(\Lambda_{H_1}(H_0), \Gamma)$-modules M in an obvious way. We leave it to the reader to check that this construction is functorial. We have achieved in this way a fully faithful embedding of the category of $(\Lambda_{H_1}(H_0), \Gamma)$-modules into the category of $\Lambda_{H_0,H_1}(H)$-modules. (Of course, we always keep supposing that $H \cong H_1 \rtimes (H/H_1)$ is a semidirect product.) The image of this embedding is characterized by the next proposition.

REMARK 8.27. *The map* (23) *is a topological isomorphism (with the right hand side given the projective limit topology of the weak topologies).*

PROOF. Since $\mathfrak{m}_{H_1}(H_0)$ is an ideal in $\Lambda_{H_1}(H_0)$ any open neighbourhood of zero in M contains some $\mathfrak{m}_{H_1}(H_0)^m M$. □

PROPOSITION 8.28. *For any $(\Lambda_{H_1}(H_0), \Gamma)$-module M the corresponding action of $\Lambda_{H_0,H_1}(H)$ on M is separately continuous for the weak topologies.*

PROOF. By the Remark 8.27 we again may assume that M satisfies (24). Then the multiplication by any $\lambda \in \Lambda_{H_0,H_1}(H)$ is the composite of the multiplication by some $\nu \in \Lambda(H)$ and the multiplication by some $t_0^{-j} \in \Lambda_{H_1}(H_0)$ both of which are already known to be continuous. On the other hand let $x \in M$ be a fixed element. The o-linear map

$$\begin{aligned} \rho_x : \Lambda_{H_0,H_1}(H) &\longrightarrow M \\ \lambda &\longmapsto \lambda \cdot x \end{aligned}$$

whose continuity remains to be seen, by construction, vanishes on $F^m\Lambda_{H_0,H_1}(H)$. Let $U \subseteq M$ be any open neighbourhood of zero which we may assume to be an additive subgroup. For any $j \geq 0$ there is an open neighbourhood of zero $U_j \subseteq M$ such that $t_0^{-j}U_j \subseteq U$. Moreover, since $\rho_x|\Lambda(H)$ is continuous we find an $\ell(j) \geq 0$ such that $\rho_x(\mathfrak{m}(H)^{\ell(j)}) \subseteq U_j$. It follows that

$$\rho_x(t_0^{-j}\mathfrak{m}(H)^{\ell(j)}) = t_0^{-j}\rho_x(\mathfrak{m}(H)^{\ell(j)}) \subseteq t_0^{-j}U_j \subseteq U$$

and hence that $\rho_x(C_{m,\ell}) \subseteq U$.

□

9. Generalized (φ, Γ)-modules

It cannot be expected that the modules $D^i(V)$ have good properties in general. To improve the situation we propose to pass to a specific topological localization. To do so we will apply the construction of the previous section to the situation introduced at the beginning of section 5. Specifically we put $H_1 := N_1 \subseteq H_0 := N_0$. As γ_0 we choose any element in $N_0 \cap N_\alpha$ for some $\alpha \in \Delta$ such that $\ell(\gamma_0) = 1$. We also put $\Gamma := \xi(1 + p^{\epsilon(p)}\mathbb{Z}_p)$, let γ be any topological generator of Γ, and define

$H := N_0\Gamma$. The semidirect product condition needed for most of the previous section is satisfied and we have available the diagram of rings

$$\begin{array}{ccc} \Lambda_\ell(N_0\Gamma) := \Lambda_{N_0,N_1}(N_0\Gamma) & \xleftarrow{\supseteq} & \Lambda(N_0\Gamma) \\ \uparrow \subseteq & & \uparrow \subseteq \\ \Lambda_\ell(N_0) := \Lambda_{N_1}(N_0) & \xleftarrow{\supseteq} & \Lambda(N_0). \end{array}$$

The $\sigma_.$-action of Γ on $\Lambda_\ell(N_0)$ extends the Γ-action on $\Lambda(N_0)$ denoted by $\phi_.$ in section 1. Remark 8.24.ii says that these Γ-actions are induced by the conjugation by Γ on $\Lambda_\ell(N_0\Gamma)$.

We want to go one step further. The group $N_0\Gamma$ is the group part of the monoid $P_\star = N_0\Gamma\varphi^{\mathbb{N}_0}$ where $\varphi := \xi(p)$. Correspondingly we have the inclusion of rings $\Lambda(N_0\Gamma) \subseteq \Lambda(P_\star)$. More precisely, if we let σ_φ denote the (injective but not surjective) continuous ring endomorphism of $\Lambda(N_0\Gamma)$ induced by the conjugation by φ on $N_0\Gamma$ then $\Lambda(P_\star) = \Lambda(N_0\Gamma)[\varphi;\sigma_\varphi]$ is the skew polynomial ring over $\Lambda(N_0\Gamma)$ with respect to σ_φ. We note that σ_φ fixes the subring $\Lambda(\Gamma)$ which means that in $\Lambda(P_\star)$ the two variables t and φ commute. The endomorphism σ_φ respects the subrings $\Lambda(N_1) \subseteq \Lambda(N_0)$ and their maximal ideals and, since it still is injective on $\Omega(N_0/N_1)$, also the Ore set $S := S(N_0, N_1)$. It therefore extends to a ring endomorphism first of the localization $\Lambda(N_0)_S$ and then of its $\mathfrak{m}(N_1)$-adic completion $\Lambda_\ell(N_0)$, still denoted by σ_φ. Since φ and γ commute in $T_\star$ the endomorphism σ_φ commutes with $\sigma = \sigma_\gamma$ and $\delta = \sigma - \mathrm{id}$. We visibly have $\sigma_\varphi(B_m) \subseteq B_m$ for any $m \geq 0$ which implies that σ_φ is continuous for the weak topology on $\Lambda_\ell(N_0)$.

LEMMA 9.1. i. $\sigma_\varphi(t_0) = (t_0+1)^p - 1$.
ii. *$t_0 = u\sigma_\varphi(t_0)$ for some unit u in $\Lambda_{\{1\}}(N_0 \cap N_\alpha) \subseteq \Lambda_\ell(N_0)$.*
iii. *σ_φ respects bounded subsets for the weak topology on $\Lambda_\ell(N_0)$.*

PROOF. i. This follows immediately from our choice of γ_0 and the fact that $\alpha \circ \xi = \mathrm{id}_{\mathbb{G}_m}$.

ii. The ring $\Lambda_{\{1\}}(N_0 \cap N_\alpha)$ is a commutative local ring with maximal ideal generated by π. By i. we have $\sigma_\varphi(t_0) = t_0 v$ where $v := \sum_{i=1}^p \binom{p}{i} t_0^{i-1} = p + \ldots + t_0^{p-1} \in o[t_0]$ does not lie in this maximal ideal. Hence its inverse $u := v^{-1}$ exists.

iii. Let $A \subseteq \Lambda_\ell(N_0)$ be any bounded subset. For a given $m \geq 0$ let $l \geq 0$ be such that $t_0^l A \subseteq \mathfrak{m}_{N_1}(N_0)^m + \Lambda(N_0)$ (cf. Lemma 8.8). Applying σ_φ and using ii. we obtain

$$t_0^l \sigma_\varphi(A) = u^l \sigma_\varphi(t_0^l A) \subseteq \mathfrak{m}_{N_1}(N_0)^m + u^l \Lambda(N_0)\ .$$

If we choose $k \geq 0$ such that $t_0^k u \in \pi^m \Lambda_{\{1\}}(N_0 \cap N_\alpha) + o[[t_0]]$ then $t_0^{(k+1)l}\sigma_\varphi(A) \subseteq \mathfrak{m}_{N_1}(N_0)^m + \Lambda(N_0)$. It now follows from Lemma 8.8 that $\sigma_\varphi(A)$ is bounded. □

We therefore may define the map

$$\begin{aligned} \sigma_\varphi : \Lambda_\ell(N_0\Gamma) &\longrightarrow \Lambda_\ell(N_0\Gamma) \\ \sum_{i\geq 0} \mu_i t^i &\longmapsto \sum_{i\geq 0} \sigma_\varphi(\mu_i) t^i\ . \end{aligned}$$

It is immediate from (18), the continuity of σ_φ on $\Lambda_\ell(N_0)$, and its commutation with σ and δ that this extended σ_φ in fact is an endomorphism of the ring $\Lambda_\ell(N_0\Gamma)$.

REMARK 9.2. *The endomorphism σ_φ is continuous for the strong as well as the weak topology on $\Lambda_\ell(N_0\Gamma)$.*

PROOF. The case of the strong topology is obvious since σ_φ respects the filtration $F^m\Lambda_\ell(N_0\Gamma)$. The case of the weak topology is a straightforward consequence of the following computation based on Lemma 9.1.ii. Again let $k \geq 0$ be such that $t_0^k u \in \pi^m \Lambda_{\{1\}}(N_0 \cap N_\alpha) + o[[t_0]]$. For any $j, l \geq 0$ we then have

$$\begin{aligned}
\sigma_\varphi(t_0^{-j}\mathfrak{m}(N_0\Gamma)^l) &\subseteq \sigma_\varphi(t_0)^{-j}\mathfrak{m}(N_0\Gamma)^l \\
&= t_0^{-j}u^j\mathfrak{m}(N_0\Gamma)^l \\
&= t_0^{-j(1+k)}t_0^{kj}u^j\mathfrak{m}(N_0\Gamma)^l \\
&\subseteq t_0^{-j(1+k)}(\pi^m\Lambda_{\{1\}}(N_0 \cap N_\alpha) + o[[t_0]])\mathfrak{m}(N_0\Gamma)^l \\
&\subseteq F^m\Lambda_\ell(N_0\Gamma) + t_0^{-j(1+k)}\mathfrak{m}(N_0\Gamma)^l
\end{aligned}$$

□

This allows us to form the skew polynomial ring

$$\Lambda_\ell(P_\star) := \Lambda_\ell(N_0\Gamma)[\varphi; \sigma_\varphi] \ .$$

As a bimodule it satisfies

$$\Lambda_\ell(P_\star) = \Lambda_\ell(N_0\Gamma) \otimes_{\Lambda(N_0\Gamma)} \Lambda(P_\star) \ . \tag{26}$$

Our basic diagram for the following now is

$$\begin{array}{ccc}
\Lambda_\ell(P_\star) & \xleftarrow{\supseteq} & \Lambda(P_\star) \\
\uparrow \subseteq & & \uparrow \subseteq \\
\Lambda_\ell(N_0) & \xleftarrow{\supseteq} & \Lambda(N_0).
\end{array}$$

DEFINITION 9.3. *A $(\Lambda_\ell(N_0), \Gamma, \varphi)$-module M is a $(\Lambda_\ell(N_0), \Gamma)$-module with an additional σ_φ-linear endomorphism φ_M which commutes with the Γ-action.*

We point out that, as σ_φ is continuous on $\Lambda_\ell(N_0)$, the σ_φ-linear endomorphism φ_M of a $(\Lambda_\ell(N_0), \Gamma, \varphi)$-module M, which by definition is finitely generated over $\Lambda_\ell(N_0)$, automatically is continuous for the weak topology on the module M.

In Lemma 8.26 we have seen that the $(\Lambda_\ell(N_0), \Gamma)$-modules form a full subcategory of all $\Lambda_\ell(N_0\Gamma)$-modules.

LEMMA 9.4. *Let M be a $(\Lambda_\ell(N_0), \Gamma, \varphi)$-module viewed as a $\Lambda_\ell(N_0\Gamma)$-module; the endomorphism φ_M is σ_φ-linear with respect to the $\Lambda_\ell(N_0\Gamma)$-action.*

PROOF. We may assume that $\mathfrak{m}_{N_1}(N_0)^m M = 0$ for some $m \geq 0$. By the definition of the $\Lambda_\ell(N_0\Gamma)$-module structure we then have to show the identity

$$\varphi_M((t_0^{-j}\nu) \cdot x) = \sigma_\varphi(t_0^{-j}\nu) \cdot \varphi_M(x)$$

for any $j \geq 0$, $\nu \in \Lambda(N_0\Gamma)$, and $x \in M$. As a consequence of Prop. 8.28, Remark 9.2, and the continuity of φ_M both sides of this identity depend continuously on ν. Hence it suffices to consider any ν of the form $\nu = \nu_0 + \nu_1 t + \ldots + \nu_k t^k$ with $\nu_i \in \Lambda(N_0)$. By assumption φ_M is σ_φ-linear with respect to scalars in $\Lambda_\ell(N_0)$. This, in fact, reduces us to the identity

$$\varphi_M(t \cdot x) = \sigma_\varphi(t) \cdot \varphi_M(x) \qquad \text{for any } x \in M.$$

On the left hand side φ_M commutes with the Γ-action by assumption. On the right hand side we have $\sigma_\varphi(t) = t$. Hence both sides are equal to $t \cdot \varphi_M(x)$. $\square$

This lemma implies that by letting $\varphi \in \Lambda(P_\star)$ act as φ_M on a $(\Lambda_\ell(N_0), \Gamma, \varphi)$-module M we obtain a $\Lambda_\ell(P_\star)$-module. In this way the $(\Lambda_\ell(N_0), \Gamma, \varphi)$-modules form a full subcategory of all $\Lambda_\ell(P_\star)$-modules.

We recall from section 1 that a $\Lambda(P_\star)$-module M is etale if the $\Lambda(P_\star)$-linear map

$$\begin{aligned} \Lambda(P_\star) \otimes_{\Lambda(P_\star),\sigma_\varphi} M &\xrightarrow{\cong} M \\ \mu \otimes x &\longmapsto \mu\varphi x \end{aligned}$$

is an isomorphism. We observe that the endomorphisms σ_φ of $\Lambda_\ell(N_0\Gamma)$ and of $\Lambda(P_\star)$ both come by restriction from the ring endomorphism

$$\begin{aligned} \sigma_\varphi : \Lambda_\ell(P_\star) = \Lambda_\ell(N_0\Gamma)[\varphi;\sigma_\varphi] &\longrightarrow \Lambda_\ell(P_\star) = \Lambda_\ell(N_0\Gamma)[\varphi;\sigma_\varphi] \\ \sum_{k\geq 0} \lambda_k \varphi^k &\longrightarrow \sum_{k\geq 0} \sigma_\varphi(\lambda_k)\varphi^k \ . \end{aligned}$$

This suggests the following definition.

DEFINITION 9.5. *A $\Lambda_\ell(P_\star)$-module M is called etale if the $\Lambda_\ell(P_\star)$-linear map*

$$\begin{aligned} \Lambda_\ell(P_\star) \otimes_{\Lambda_\ell(P_\star),\sigma_\varphi} M &\xrightarrow{\cong} M \\ \mu \otimes x &\longmapsto \mu\varphi x \end{aligned}$$

is an isomorphism.

PROPOSITION 9.6. *The endomorphism σ_φ of $\Lambda_\ell(P_\star)$ is injective and makes $\Lambda_\ell(P_\star)$ a free right module of rank $[N_0 : \varphi N_0 \varphi^{-1}]$ over itself; the map*

$$\begin{aligned} \Lambda_\ell(N_0) \otimes_{\Lambda_\ell(N_0),\sigma_\varphi} \Lambda_\ell(P_\star) &\xrightarrow{\cong} \Lambda_\ell(P_\star) \\ \nu \otimes \mu &\longmapsto \nu\sigma_\varphi(\mu) \end{aligned}$$

is an isomorphism.

PROOF. *Preliminary observation:* In the general situation of section 8 let $H_0' \subseteq H_0$ be an open subgroup and put $H_1' := H_1 \cap H_0'$. We then have of course $\Lambda(H_0') \subseteq \Lambda(H_0)$ and $J(H_0', H_1') \subseteq J(H_0, H_1)$. But since $\Omega(H_0'/H_1') \subseteq \Omega(H_0/H_1)$ we also have $S(H_0', H_1') \subseteq S(H_0, H_1)$. By localization and completion we therefore obtain a natural ring homomorphism $\Lambda_{H_1'}(H_0') \longrightarrow \Lambda_{H_1}(H_0)$ which gives rise to a natural homomorphism of bimodules

$$\Lambda(H_0) \otimes_{\Lambda(H_0')} \Lambda_{H_1'}(H_0') \longrightarrow \Lambda_{H_1}(H_0) \ .$$

Step 1: We claim that the natural map

$$\Lambda(N_0) \otimes_{\Lambda(\varphi N_0 \varphi^{-1})} \Lambda_{\varphi N_1 \varphi^{-1}}(\varphi N_0 \varphi^{-1}) \xrightarrow{\cong} \Lambda_{N_1}(N_0) = \Lambda_\ell(N_0)$$

is bijective. (Note that $\varphi N_1 \varphi^{-1} = N_1 \cap \varphi N_0 \varphi^{-1}$.) We choose an open subgroup $N_0' \subseteq \varphi N_0 \varphi^{-1} \subseteq N_0$ which is normal in N_0. We now apply (a symmetric version of) [**SV2**] Prop. 4.5 to the pairs $N_0' \trianglelefteq \varphi N_0 \varphi^{-1}$ and $N_0' \trianglelefteq N_0$ obtaining the isomorphisms

$$\Lambda(\varphi N_0 \varphi^{-1}) \otimes_{\Lambda(N_0')} \Lambda_{N_1 \cap N_0'}(N_0') \xrightarrow{\cong} \Lambda_{N_1 \cap \varphi N_0 \varphi^{-1}}(\varphi N_0 \varphi^{-1})$$

and

$$\Lambda(N_0) \otimes_{\Lambda(N_0')} \Lambda_{N_1 \cap N_0'}(N_0') \xrightarrow{\cong} \Lambda_{N_1}(N_0) \ .$$

The combination of the two gives our claim. In addition we know that the ring $\Lambda_{\varphi N_1 \varphi^{-1}}(\varphi N_0 \varphi^{-1})$ is flat as a (left) $\Lambda(\varphi N_0 \varphi^{-1})$-module. It follows that the natural ring homomorphism

$$\begin{aligned} \Lambda_{\varphi N_1 \varphi^{-1}}(\varphi N_0 \varphi^{-1}) = \Lambda(\varphi N_0 \varphi^{-1}) \otimes_{\Lambda(\varphi N_0 \varphi^{-1})} \Lambda_{\varphi N_1 \varphi^{-1}}(\varphi N_0 \varphi^{-1}) \\ \longrightarrow \Lambda(N_0) \otimes_{\Lambda(\varphi N_0 \varphi^{-1})} \Lambda_{\varphi N_1 \varphi^{-1}}(\varphi N_0 \varphi^{-1}) = \Lambda_\ell(N_0) \end{aligned}$$

is injective. In an obvious reformulation we have shown so far that the ring endomorphism σ_φ of $\Lambda_\ell(N_0)$ is injective and that the bimodule map

$$\begin{aligned} \Lambda(N_0) \otimes_{\Lambda(N_0), \sigma_\varphi} \Lambda_\ell(N_0) &\xrightarrow{\cong} \Lambda_\ell(N_0) \\ \nu \otimes \mu &\longmapsto \nu \sigma_\varphi(\mu) \end{aligned}$$

is bijective.

Step 2: We let $r := [N_0 : \varphi N_0 \varphi^{-1}]$ and we fix representatives $n_1, \ldots, n_r \in N_0$ for the cosets in $N_0/\varphi N_0 \varphi^{-1}$. In the previous step we have seen that the map

$$\begin{aligned} I: \qquad \Lambda_\ell(N_0)^r &\xrightarrow{\cong} \Lambda_\ell(N_0) \\ (\mu^{(1)}, \ldots \mu^{(r)}) &\longmapsto n_1 \sigma_\varphi(\mu^{(1)}) + \ldots + n_r \sigma_\varphi(\mu^{(r)}) \end{aligned}$$

is bijective. We claim that I also is a homeomorphism for the weak topology, resp. the direct product of the weak topologies, on the right, resp. left, hand side. To see this we pick an open subgroup $N' \subseteq \varphi N_1 \varphi^{-1}$ which is normal in N_0. In particular, N' is normal in $\varphi N_0 \varphi^{-1}$. It then follows from **[SV2]** Lemma 4.4 that in $\Lambda_\ell(N_0)$, resp. in $\Lambda_{\varphi N_1 \varphi^{-1}}(\varphi N_0 \varphi^{-1})$ the $\mathfrak{m}_{N_1}(N_0)$-adic and the $\mathfrak{m}(N')\Lambda_\ell(N_0)$-adic filtrations, resp. the $\mathfrak{m}_{\varphi N_1 \varphi^{-1}}(\varphi N_0 \varphi^{-1})$-adic and the $\mathfrak{m}(N')\Lambda_\ell(\varphi N_0 \varphi^{-1})$-adic filtrations, are equivalent. In fact, this means that in $\Lambda_\ell(N_0)$ all three filtrations, the $\mathfrak{m}_{N_1}(N_0)$-adic one, the $\mathfrak{m}(N')\Lambda_\ell(N_0)$-adic one, and the $\mathfrak{m}(\varphi^{-1}N'\varphi)\Lambda_\ell(N_0)$-adic one are equivalent. Visibly under the map I the product filtration $\big(\mathfrak{m}(\varphi^{-1}N'\varphi)^m \Lambda_\ell(N_0)\big)^r$ on the left hand side corresponds to the filtration

$$\begin{aligned} n_1 \mathfrak{m}(N')^m \sigma_\varphi(\Lambda_\ell(N_0)) + \ldots + n_r \mathfrak{m}(N')^m \sigma_\varphi(\Lambda_\ell(N_0)) = \\ \mathfrak{m}(N')^m \big(n_1 \sigma_\varphi(\Lambda_\ell(N_0)) + \ldots + n_r \sigma_\varphi(\Lambda_\ell(N_0))\big) = \mathfrak{m}(N')^m \Lambda_\ell(N_0) \ . \end{aligned}$$

on the right hand side. It remains to observe that the restriction of the map I to $\Lambda(N_0)^r \xrightarrow{\cong} \Lambda(N_0)$ is a homeomorphism by compactness.

Step 3: It follows immediately from the first step that both, the ring endomorphism σ_φ of $\Lambda_\ell(N_0\Gamma)$ as well as the bimodule map

$$\begin{aligned} \Lambda(N_0) \otimes_{\Lambda(N_0), \sigma_\varphi} \Lambda_\ell(N_0\Gamma) &\longrightarrow \Lambda_\ell(N_0\Gamma) \\ \nu \otimes \mu = \nu \otimes \sum_{i \geq 0} \mu_i t^i &\longmapsto \nu \sigma_\varphi(\mu) = \sum_{i \geq 0} \nu \sigma_\varphi(\mu_i) t^i \ , \end{aligned} \tag{27}$$

are injective. To establish the surjectivity of the latter map let $\lambda = \sum_{i \geq 0} \lambda_i t^i \in \Lambda_\ell(N_0\Gamma)$ be any element. According to the first step there are, for any $i \geq 0$, uniquely determined elements $\mu_i^{(1)}, \ldots, \mu_i^{(r)} \in \Lambda_\ell(N_0)$ such that

$$\lambda_i = n_1 \sigma_\varphi(\mu_i^{(1)}) + \ldots + n_r \sigma_\varphi(\mu_i^{(r)}) \ .$$

We claim that, for each $1 \leq k \leq r$, the subset $\{\mu_i^{(k)}\}_{i\geq 0} \subseteq \Lambda_\ell(N_0)$ is bounded for the weak topology. Let $m' \geq 0$. By the second step we find an $m'' \geq 0$ such that $I(B_{m'}^r) \supseteq B_{m''}$. The boundedness of the set $\{\lambda_i\}_{i\geq 0}$ implies the existence of an $m \geq 0$ such that $\lambda_i B_m \subseteq B_{m''}$ for any $i \geq 0$. For any $\nu \in B_m$ we then obtain $I((\mu_i^{(1)}\nu, \ldots, \mu_i^{(r)}\nu)) = \lambda_i \sigma_\varphi(\nu) \in \lambda_i \sigma_\varphi(B_m) \subseteq \lambda_i B_m \subseteq B_{m''}$, hence $\mu_i^{(1)}\nu, \ldots, \mu_i^{(r)}\nu \in B_{m'}$, and therefore

$$\mu_i^{(k)} B_m \subseteq B_{m'} \qquad \text{for any } i \geq 0 \text{ and } 1 \leq k \leq r.$$

It follows that the elements $\mu^{(k)} := \sum_{i\geq 0} \mu_i^{(k)} t^k$ are well defined in $\Lambda_\ell(N_0\Gamma)$ and that we have

$$\lambda = n_1\sigma_\varphi(\mu^{(1)}) + \ldots + n_r\sigma_\varphi(\mu^{(r)}) .$$

This proves the surjectivity and hence bijectivity of (27).

Step 4: The injectivity of σ_φ on $\Lambda_\ell(P_\star)$ is a trivial consequence of the injectivity of σ_φ on $\Lambda_\ell(N_0\Gamma)$ established in the third step. The isomorphisms (26) and (27) combine into the isomorphisms

$$\Lambda(N_0) \otimes_{\Lambda(N_0),\sigma_\varphi} \Lambda_\ell(P_\star) \xrightarrow{\cong} \Lambda_\ell(P_\star) \quad \text{and} \quad \Lambda_\ell(N_0) \otimes_{\Lambda_\ell(N_0),\sigma_\varphi} \Lambda_\ell(P_\star) \xrightarrow{\cong} \Lambda_\ell(P_\star) .$$

□

COROLLARY 9.7. *Let M be a $(\Lambda_\ell(N_0), \Gamma, \varphi)$-module; viewed as a $\Lambda_\ell(P_\star)$-module M is etale if and only if the map*

$$\begin{aligned} \Lambda_\ell(N_0) \otimes_{\Lambda_\ell(N_0),\sigma_\varphi} M &\xrightarrow{\cong} M \\ \nu \otimes x &\longmapsto \nu\varphi_M(x) \end{aligned}$$

is bijective.

PROOF. By Prop. 9.6 we have

$$\Lambda_\ell(P_\star) \otimes_{\Lambda_\ell(P_\star),\sigma_\varphi} M = \Lambda_\ell(N_0) \otimes_{\Lambda_\ell(N_0),\sigma_\varphi} \Lambda_\ell(P_\star) \otimes_{\Lambda_\ell(P_\star)} M = \Lambda_\ell(N_0) \otimes_{\Lambda_\ell(N_0),\sigma_\varphi} M .$$

□

Let $\mathcal{M}(\Lambda_\ell(P_\star))$ be the abelian category of (left unital) $\Lambda_\ell(P_\star)$-modules and $D^+(\Lambda_\ell(P_\star))$ the corresponding bounded below derived category. Let $\mathcal{M}_{et}(\Lambda_\ell(P_\star))$ denote the full subcategory in $\mathcal{M}(\Lambda_\ell(P_\star))$ of etale modules and $D^+_{et}(\Lambda_\ell(P_\star))$ the full subcategory in $D^+(\Lambda_\ell(P_\star))$ of all complexes whose cohomology modules are etale.

COROLLARY 9.8.
i. *The subcategory $\mathcal{M}_{et}(\Lambda_\ell(P_\star))$ of $\mathcal{M}(\Lambda_\ell(P_\star))$ is closed under the formation of kernels, cokernels, extensions and arbitrary inductive and projective limits; in particular, $\mathcal{M}_{et}(\Lambda_\ell(P_\star))$ is an abelian category.*
ii. *$D^+_{et}(\Lambda_\ell(P_\star))$ is a triangulated subcategory of $D^+(\Lambda_\ell(P_\star))$.*

The base change functor for modules, which by Prop. 8.21.iii and (26) is exact, obviously restricts to an exact functor

$$\Lambda_\ell(P_\star) \otimes_{\Lambda(P_\star)} : \mathcal{M}_{et}(\Lambda(P_\star)) \longrightarrow \mathcal{M}_{et}(\Lambda_\ell(P_\star)) .$$

and then extends to the functor

$$\Lambda_\ell(P_\star) \otimes_{\Lambda(P_\star)} : D^+_{et}(\Lambda(P_\star)) \longrightarrow D^+_{et}(\Lambda_\ell(P_\star)) .$$

between derived categories. We introduce the composed functor

$$RD_\ell : D^-(\mathcal{M}_{o-tor}(P)) \xrightarrow{RD} D^+_{et}(\Lambda(P_+))$$
$$\xrightarrow{\text{forget}} D^+_{et}(\Lambda(P_\star)) \xrightarrow{\Lambda_\ell(P_\star)\otimes_{\Lambda(P_\star)}} D^+_{et}(\Lambda_\ell(P_\star)) \ .$$

as well as the δ-functor

$$D^i_\ell(V) := h^i(RD_\ell(V)) = \Lambda_\ell(P_\star) \otimes_{\Lambda(P_\star)} D^i(V) \qquad \text{for } i \geq 0$$

from $\mathcal{M}_{o-tor}(P)$ into etale $\Lambda_\ell(P_\star)$-modules. The **fundamental open question** in this context is for which representations V in $\mathcal{M}_{o-tor}(P)$ the etale module $D^0_\ell(V)$ (or even any $D^i_\ell(V)$) is a $(\Lambda_\ell(N_0), \Gamma, \varphi)$-module.

A first result in this direction can be obtained by generalizing to our noncommutative setting the arguments in [**Eme**].

PROPOSITION 9.9. *Let V be an admissible representation in $\mathcal{M}_{o-tor}(G)$; suppose that for some $M \in \mathcal{P}_+(V)$ there is an exact sequence of P_+-representations of the form*

$$\ldots \longrightarrow \mathrm{ind}^{P_+}_{P_0}(V_n) \longrightarrow \ldots \longrightarrow \mathrm{ind}^{P_+}_{P_0}(V_0) \longrightarrow M \longrightarrow 0$$

with V_n finite for any $n \geq 0$; then M^* and hence $D(V)$ are finitely generated $\Lambda(N_0)$-modules, and $\mathcal{P}_+(V)$ contains a unique minimal element.

PROOF. *Step 1:* At first we let $M \in \mathcal{P}_+(V)$ be arbitrary. Since $D(V)$ is a quotient of M^* and since $\Lambda(N_0)$ is noetherian it suffices to show that the compact $\Lambda(N_0)$-module M^* is finitely generated. By the topological Nakayama lemma ([**BH**]) this reduces to the finiteness of $M^*/\mathfrak{m}(N_0)M^*$. The latter is the Pontrjagin dual of the group $H^0(N_0, M^{\pi=0})$ of N_0-invariants in the k-vector space $M^{\pi=0}$ consisting of all elements in M which are annihilated by π. We therefore will show that $H^0(N_0, M^{\pi=0})$ is finite.

The N_0-invariants $H^0(N_0, V)$ in V do not form a P_+-subrepresentation. But the monoid T_+ acts on $H^0(N_0, V)$ via the so called Hecke action which is defined by

$$t \cdot v := \sum_{n \in N_0/tN_0t^{-1}} ntv \qquad \text{for } t \in T_+ \text{ and } v \in V.$$

Since for T_0 the Hecke action coincides with the group action we see that the Hecke action extends to a $\Lambda(P_+)$-module structure. By [**Em**] Thm. 3.2.3(1) the admissibility of V implies that $H^0(N_0, V)$ is a union of Hecke invariant finitely generated o-submodules. It follows that $H^0(N_0, M^{\pi=0})$ is a union of Hecke invariant finite k-vector spaces. Hence $H^0(N_0, M^{\pi=0})$ is finite if and only if it is finitely generated as a $\Lambda(P_+)$-module (for the Hecke action).

Step 2: Next we apply duality to $H^0(N_0, M^{\pi=0})$. First of all we observe that

$$H^0(N_0, M^{\pi=0}) = \mathrm{Hom}_{\Lambda(N_0)}(k, M) \ .$$

Let d denote the dimension of the p-adic Lie group N_0. The ring $\Lambda(N_0)$ is a regular local noetherian integral domain of global dimension $d+1$ ([**Neu**]). We therefore have, for any finitely generated $\Lambda(N_0)$-module X, the natural duality isomorphism

$$\mathrm{Ext}^*_{\Lambda(N_0)}(X, .) = \mathrm{Tor}^{\Lambda(N_0)}_{d+1-*}(\mathcal{D}_{\Lambda(N_0)}(X), .)$$

between functors on the category of all (left) $\Lambda(N_0)$-modules where the dualizing complex

$$\mathcal{D}_{\Lambda(N_0)}(X) := \mathrm{RHom}_{\Lambda(N_0)}(X, \Lambda(N_0))$$

is placed in degrees $-(d+1)$ up to 0. On the other hand N_0 is a Poincaré group ([**Laz**] Thm. V.2.5.8). Therefore ([**NSW**] Cor. 5.4.15(ii)) the dualizing complex $\mathcal{D}_{\Lambda(N_0)}(k)$ in fact is quasi-isomorphic to the trivial module k placed in degree zero. (We note that the character $\chi : N_0 \longrightarrow \mathbb{Z}_p^\times$ which describes the action of N_0 on the dualizing module $I \cong \mathbb{Q}_p/\mathbb{Z}_p$ has values in $1 + p\mathbb{Z}_p$ so that $I^{p=0}$ is a trivial N_0-module.) It follows that the duality isomorphism specializes to

$$\mathrm{Ext}^*_{\Lambda(N_0)}(k,.) = \mathrm{Tor}^{\Lambda(N_0)}_{d+1-*}(k,.)\ ,$$

and we obtain in particular that

$$H^0(N_0, M^{\pi=0}) = \mathrm{Tor}^{\Lambda(N_0)}_{d+1}(k,M) \tag{28}$$

for any $M \in \mathcal{P}_+(V)$.

Step 3: In this step we identify the Hecke action on the right hand side of (28). We begin with a completely general observation. Let H_0 be any profinite group and $H_1 \subseteq H_0$ any open subgroup. Then

$$\begin{aligned} \mathrm{Hom}_{\Lambda(H_1)}(\Lambda(H_0),\Lambda(H_1)) &\xrightarrow{\cong} \Lambda(H_0) \\ f &\longmapsto \sum_{g\in H_0/H_1} gf(g^{-1}) \end{aligned}$$

is an isomorphism of left $\Lambda(H_0)$-modules. Hence

$$\mathrm{Hom}_{\Lambda(H_1)}(\Lambda(H_0),Y) = \mathrm{Hom}_{\Lambda(H_1)}(\Lambda(H_0),\Lambda(H_1)) \otimes_{\Lambda(H_1)} Y \cong \Lambda(H_0) \otimes_{\Lambda(H_1)} Y$$

for any left $\Lambda(H_1)$-module Y. It follows that the exact scalar extension functor $\Lambda(H_0) \otimes_{\Lambda(H_1)} .$ preserves injective modules and is right adjoint to the scalar restriction functor. We consequently have the adjunction isomorphism

$$\begin{aligned} \mathrm{Hom}_{\Lambda(H_1)}(X,Y) &\xrightarrow{\cong} \mathrm{Hom}_{\Lambda(H_0)}(X,\Lambda(H_0)\otimes_{\Lambda(H_1)} Y) \\ f &\longmapsto [x \longmapsto \sum_{g\in H_0/H_1} g\otimes f(g^{-1}x)] \end{aligned} \tag{29}$$

for any $\Lambda(H_0)$-module X and any $\Lambda(H_1)$-module Y. More generally, by using an injective resolution of Y, we obtain

$$\mathrm{RHom}_{\Lambda(H_1)}(X,Y) \cong \mathrm{RHom}_{\Lambda(H_0)}(X,\Lambda(H_0)\otimes_{\Lambda(H_1)} Y)\ .$$

For $X = \Lambda(H_0)$ one checks by a straightforward computation that the isomorphism (29) is compatible with the above duality isomorphism. For purely formal reasons the same then holds true for any X which is finitely generated and projective over $\Lambda(H_0)$ and hence over $\Lambda(H_1)$.

Let $t \in T_+$. We apply the above discussion to the groups $H_0 := N_0$ and $H_1 := tN_0t^{-1}$ and obtain the commutative diagram of isomorphisms

$$\begin{array}{ccc} \mathrm{Ext}^*_{\Lambda(tN_0t^{-1})}(X,Y) & \xrightarrow{\cong} & \mathrm{Tor}^{\Lambda(tN_0t^{-1})}_{d+1-*}(\mathcal{D}_{\Lambda(tN_0t^{-1})}(X),Y) \\ \downarrow{\cong} & & \downarrow{\cong} \\ \mathrm{Ext}^*_{\Lambda(N_0)}(X,\Lambda(N_0)\otimes_{\Lambda(tN_0t^{-1})} Y) & \xrightarrow{\cong} & \mathrm{Tor}^{\Lambda(N_0)}_{d+1-*}(\mathcal{D}_{\Lambda(N_0)}(X),\Lambda(N_0)\otimes_{\Lambda(tN_0t^{-1})} Y) \end{array}$$

for any finitely generated $\Lambda(N_0)$-module X and any $\Lambda(tN_0t^{-1})$-module Y where the horizontal, resp. perpendicular, arrows come from duality, resp. adjunction. In the special case $X = k$ we may rewrite this as a commutative diagram of isomorphisms

$$\begin{array}{ccc} \mathrm{Ext}^*_{\Lambda(N_0)}(k,Y) & \xrightarrow{\cong} & \mathrm{Tor}^{\Lambda(N_0)}_{d+1-*}(k,Y) \\ \cong\downarrow & & \downarrow\cong \\ \mathrm{Ext}^*_{\Lambda(N_0)}(k,\Lambda(N_0)\otimes_{\Lambda(N_0),t} Y) & \xrightarrow{\cong} & \mathrm{Tor}^{\Lambda(N_0)}_{d+1-*}(k,\Lambda(N_0)\otimes_{\Lambda(N_0),t} Y) \end{array}$$

for any $\Lambda(N_0)$-module Y.

We now suppose that Y is a $\Lambda(P_+)$-module. We then have the $\Lambda(N_0)$-equivariant map

$$(30) \qquad \begin{aligned} \Lambda(N_0)\otimes_{\Lambda(N_0),t} Y &\longrightarrow Y \\ \lambda\otimes y &\longmapsto \lambda t y\ . \end{aligned}$$

Hence the naturality of the duality isomorphism together with the above commutative diagram gives rise to the commutative diagram

$$\begin{array}{ccc} \mathrm{Ext}^*_{\Lambda(N_0)}(k,Y) & \xrightarrow{\cong} & \mathrm{Tor}^{\Lambda(N_0)}_{d+1-*}(k,Y) \\ \downarrow & & \downarrow \\ \mathrm{Ext}^*_{\Lambda(N_0)}(k,Y) & \xrightarrow{\cong} & \mathrm{Tor}^{\Lambda(N_0)}_{d+1-*}(k,Y) \end{array}$$

where the vertical arrows are induced by (30). We see that the duality isomorphism respects the natural $\Lambda(P_+)$-actions on both sides. Moreover, using (29) one easily checks that under the identification $H^0(N_0, M^{\pi=0}) = \mathrm{Hom}_{\Lambda(N_0)}(k,M)$ the Hecke action on the left hand side corresponds to the natural $\Lambda(P_+)$-action on the right hand side.

Step 4: It remains to show that under the assumption imposed on M in our assertion $\mathrm{Tor}^{\Lambda(N_0)}_{d+1}(k,M)$ is finitely generated over $\Lambda(P_+)$ (for the natural action). In fact, since N_0 of course acts trivially on $\mathrm{Tor}^{\Lambda(N_0)}_*(k,M)$, it is the factor ring $\Lambda(P_+)/\mathfrak{m}(N_0)\Lambda(P_+)$ by the two-sided ideal $\mathfrak{m}(N_0)\Lambda(P_+)$ which really acts. This factor ring is isomorphic to the ring $\Omega(T_+)$ which is formally constructed in exactly the same way as $\Lambda(P_+)$ but starting from the monoid rings over k of the factor monoids of T_+. In fact, in this situation we simply have

$$\Omega(T_+) = \Omega(T_0)[T_+/T_0]$$

which obviously is a commutative ring. Moreover, since the factor monoid T_+/T_0 is finitely generated the ring $\Omega(T_+)$ is a finitely generated algebra over the noetherian ring $\Omega(T_0)$ and therefore is noetherian.

We now compute $\mathrm{Tor}^{\Lambda(N_0)}_*(k,M)$ as a $\Lambda(T_+)$-module in the following way. Let

$$\ldots \longrightarrow \bigoplus_{I_n}\Lambda(P_+) \longrightarrow \ldots \longrightarrow \bigoplus_{I_0}\Lambda(P_+) \longrightarrow M \longrightarrow 0$$

be any resolution of M by free $\Lambda(P_+)$-modules. Since $\Lambda(P_+)$ is free as a left $\Lambda(P_0)$-module and $\Lambda(P_0)$ is flat over $\Lambda(N_0)$ this in particular is a resolution of M by flat

$\Lambda(N_0)$-modules. Hence

$$\begin{aligned}\mathrm{Tor}_*^{\Lambda(N_0)}(k,M) &= h_*\big(k\otimes_{\Lambda(N_0)}\big(\bigoplus_{I_\bullet}\Lambda(P_+)\big)\big) = h_*\big(\bigoplus_{I_\bullet}\Lambda(P_+)/\mathfrak{m}(N_0)\Lambda(P_+)\big)\\ &= h_*\big(\bigoplus_{I_\bullet}\Omega(T_+)\big).\end{aligned}$$

It follows that if the index set I_{d+1} is finite then, since $\Omega(T_+)$ is noetherian, $\mathrm{Tor}_{d+1}^{\Lambda(N_0)}(k,M)$ is a finitely generated $\Omega(T_+)$-module. Therefore it suffices to show that our assumption on M guarantees the existence of a resolution of M by finitely generated free $\Lambda(P_+)$-modules. A double complex argument further reduces us to showing that each representation $\mathrm{ind}_{P_0}^{P_+}(V)$ with finite V has a resolution by finitely generated free $\Lambda(P_+)$-modules. But, $\Lambda(P_0)$ being noetherian, V certainly has a resolution by finitely generated free $\Lambda(P_0)$-modules. We only have to apply the exact functor $\mathrm{ind}_{P_0}^{P_+}(.)=\Lambda(P_+)\otimes_{\Lambda(P_0)}-$ to this latter resolution. □

Having Lemma 8.12.iv in mind it also seems interesting to investigate the "completed" base change functor which sends M to

$$\begin{aligned}&\varprojlim_m\Big(\big(\Lambda_\ell(N_0\Gamma)/F^m\Lambda_\ell(N_0\Gamma)\big)\otimes_{\Lambda(N_0\Gamma)}M\Big)\\ &\quad=\varprojlim_m\Big(\big(\Lambda_\ell(N_0\Gamma)/F^m\Lambda_\ell(N_0\Gamma)\big)\otimes_{\Lambda(N_0\Gamma)}\big(M/\mathfrak{m}(N_1)^mM\big)\Big)\\ &\quad=\varprojlim_m\Big(\big(\Lambda_\ell(N_0)/\mathfrak{m}_{N_1}(N_0)^m\big)\otimes_{\Lambda(N_0)}M\Big)\,.\end{aligned}$$

Unfortunately it has no apparent exactness properties.

We finish this section by showing that by an appropriate choice of the subgroup N_0 one can improve the properties of the ring $\Lambda_\ell(N_0)$.

PROPOSITION 9.10. i. *If the pro-p group N_0 satisfies $[N_0,N_0]\subseteq N_0^{p^2}$ then $\Lambda_\ell(N_0)$ is an integral domain.*

ii. *Let*

$$\iota_\alpha(N_0\cap N_\alpha)=p^{n(\alpha)}\mathbb{Z}_p\qquad \text{for any } \alpha\in\Phi^+$$

and suppose that the function $n:\Phi^+\longrightarrow\mathbb{Z}$ satisfies $n(i\alpha+j\beta)<in(\alpha)+jn(\beta)-1$ for any $\alpha,\beta\in\Phi^+$ and $i,j>0$ such that $i\alpha+j\beta\in\Phi^+$; we then have $[N_0,N_0]\subseteq N_0^{p^2}$.

PROOF. i. Pro-p groups satisfying the commutator condition in our assertion are called extra-powerful in the literature. Since N_0/N_1 is torsionfree the subgroup N_1 is extra-powerful as well. Since even N_0 is torsionfree and since N_0 and N_1 are topologically finitely generated we know from [**DDMS**] Thm. 4.5 that N_0 and N_1 are uniform pro-p groups. It suffices to show that the graded ring $\mathrm{gr}^\bullet\Lambda_\ell(N_0)$ for the $\mathfrak{m}(N_1)$-adic filtration is an integral domain. It is known (cf. [**Ven**] Thm. 3.22) that for a uniform and extra-powerful N_1 the graded ring $\mathrm{gr}^\bullet\Lambda(N_1)$ for the $\mathfrak{m}(N_1)$-adic filtration is a polynomial ring in finitely many variables over k and hence is an integral domain. On the other hand in our situation [**SV2**] Lemma 4.3(ii) and (iii) hold and say that the assumptions in [**SV2**] Prop. 1.15 are satisfied the proof of which then tells us that $\mathrm{gr}^\bullet\Lambda_\ell(N_0)$ is a subring of the commutative Laurent series ring $(\mathrm{gr}^\bullet\Lambda(N_1))((t_0))$. With $\mathrm{gr}^\bullet\Lambda(N_1)$ also $(\mathrm{gr}^\bullet\Lambda(N_1))((t_0))$ is an integral domain.

ii. Since N_0 is topologically finitely generated it suffices, by [**DDMS**] §3.1, to show that $[N_0, N_0]$ is contained in the subgroup of N_0 generated by the p^2-powers (every element in this subgroup then in fact is a p^2-power). It further suffices to consider the commutators $[N_0 \cap N_\alpha, N_0 \cap N_\beta]$ for any $\alpha, \beta \in \Phi^+$. By the standard commutation rules in N together with our assumption on n we obtain

$$\begin{aligned}[N_0 \cap N_\alpha, N_0 \cap N_\beta] &= [\iota_\alpha^{-1}(p^{n(\alpha)}\mathbb{Z}_p), \iota_\beta^{-1}(p^{n(\beta)}\mathbb{Z}_p)] \\ &\subseteq \prod_{\substack{i,j>0 \\ i\alpha+j\beta\in\Phi}} \iota_{i\alpha+j\beta}^{-1}(p^{in(\alpha)+jn(\beta)}\mathbb{Z}_p) \\ &\subseteq \prod_{\substack{i,j>0 \\ i\alpha+j\beta\in\Phi}} \iota_{i\alpha+j\beta}^{-1}(p^{2+n(i\alpha+j\beta)}\mathbb{Z}_p) \\ &= \prod_{\substack{i,j>0 \\ i\alpha+j\beta\in\Phi}} (N_0 \cap N_{i\alpha+j\beta})^{p^2}\end{aligned}$$

(where some order in the product is fixed). □

10. (φ, Γ)-modules

Everything in the preceding two sections applies in particular to the standard monoid $S_\star$ from section 1. (To be precise apply it to $P = P_2(\mathbb{Q}_p)$ and

$$\ell(\begin{pmatrix} 1 & 0 \\ a & 1 \end{pmatrix}) := a \quad \text{and} \quad \xi(b) := \begin{pmatrix} 1 & 0 \\ 0 & b \end{pmatrix}).$$

It is more convenient to use an independent notation in this case. We put

$\Lambda_F(S_0) := \Lambda_{\{1\}}(S_0)$, $\Lambda_F(S_0\Gamma) := \Lambda_{S_0,\{1\}}(S_0\Gamma)$, and $\Lambda_F(S_\star) := \Lambda_F(S_0\Gamma)[[\varphi; \sigma_\varphi]]$.

We have $\Lambda(S_0) = o[[t_0]]$, and

$$\Lambda_F(S_0) = \{\sum_{j\in\mathbb{Z}} a_j t_0^j : a_j \in o \text{ and } \lim_{j\to-\infty} a_j = 0\}$$

is the π-adic completion of the localization of $o[[t_0]]$ in the multiplicative subset $o[[t_0]] \backslash \pi o[[t_0]]$; it is a complete discrete valuation ring with residue field $k((t_0))$. The element $\left(\begin{smallmatrix} 1 & 0 \\ 0 & b \end{smallmatrix}\right) \in \Gamma$ acts on $\Lambda_F(S_0)$ by sending t_0 to $(t_0+1)^b - 1$. The endomorphism σ_φ sends t_0 to $(t_0+1)^p - 1$. We see that an etale $(\Lambda_F(S_0), \Gamma, \varphi)$-module in our sense is exactly the same as an etale (φ, Γ)-module in the sense of Fontaine ([**Fon**]).

We have the exact base change functor

$$\Lambda_F(S_\star)\otimes_{\Lambda(S_\star)} : \mathcal{M}_{et}(\Lambda(S_\star)) \longrightarrow \mathcal{M}_{et}(\Lambda_F(S_\star)) \ .$$

as well as the functor

$$\Lambda_F(S_\star)\otimes_{\Lambda(S_\star)} : D^+_{et}(\Lambda(S_\star)) \longrightarrow D^+_{et}(\Lambda_F(S_\star)) \ .$$

between derived categories.

REMARK 10.1. *For any $\Lambda(S_\star)$-module M such that $\pi^m M = 0$ for some $m \geq 0$ we have*

$$\Lambda_F(S_\star) \otimes_{\Lambda(S_\star)} M = \Lambda_F(S_0) \otimes_{\Lambda(S_0)} M \ .$$

PROOF. This follows from (26) and Lemma 8.12.iv using the fact that in the present situation we have $F^m\Lambda_F(S_0\Gamma) = \pi^m\Lambda_F(S_0\Gamma)$. □

By composition we obtain the functor

$$\begin{aligned} RD_{\Lambda_F(S_\star)} &:= \Lambda_F(S_\star) \otimes^{\mathbb{L}}_{\Lambda(P_\star)} RD \\ &= \Lambda_F(S_\star) \otimes_{\Lambda(S_\star)} RD_{\Lambda(S_\star)} : D^-(\mathcal{M}_{o-tor}(P)) \longrightarrow D^+_{et}(\Lambda_F(S_\star)). \end{aligned}$$

The corresponding δ-functor on $\mathcal{M}_{o-tor}(P)$ is

$$D^i_{\Lambda_F(S_\star)}(V) := h^i(RD_{\Lambda_F(S_\star)}(V)) = \Lambda_F(S_\star) \otimes_{\Lambda(S_\star)} D^i_{\Lambda(S_\star)}(V) \qquad \text{for } i \geq 0.$$

By the universal properties of localization and adic completion the homomorphism $\ell : \Lambda(N_0) \longrightarrow \Lambda(S_0)$ from section 5 extends naturally to a surjective homomorphism of pseudocompact rings $\ell : \Lambda_\ell(N_0) \longrightarrow \Lambda_\ell(S_0)$ (in terms of Laurent series it is given by applying the augmentation map $\Lambda(N_1) \longrightarrow o$ to the coefficients). One easily checks that the weak topology on $\Lambda_\ell(S_0)$ is the quotient topology with respect to the map ℓ of the weak topology on $\Lambda_\ell(N_0)$. Using our boundedness criterion Lemma 8.8 one sees that ℓ further extends to the surjective map

$$\begin{aligned} \ell : \Lambda_\ell(N_0\Gamma) &\longrightarrow \Lambda_F(S_0\Gamma) \\ \sum_{i\geq 0} \mu_i t^i &\longmapsto \sum_{i \geq 0} \ell(\mu_i) t^i \ . \end{aligned}$$

Since, as a consequence of (8), the original map ℓ respects σ and δ this extended map ℓ in fact is a ring homomorphism. Again it is easy to check that this extension still is strict for the weak topologies. But ℓ also respects σ_φ (again by (8)). So we finally obtain the surjective ring homomorphism

$$\begin{aligned} \ell : \quad \Lambda_\ell(P_\star) &\longrightarrow \Lambda_F(S_\star) \\ \sum_{k\geq 0} \lambda_k \varphi^k &\longmapsto \sum_{k \geq 0} \ell(\lambda_k) \varphi^k \ . \end{aligned}$$

This allows us to introduce the right exact base change functor

$$\Lambda_F(S_\star) \otimes_{\Lambda_\ell(P_\star)} : \mathcal{M}(\Lambda_\ell(P_\star)) \longrightarrow \mathcal{M}(\Lambda_F(S_\star))$$

as well as its left derived functor

$$\Lambda_F(S_\star) \otimes^{\mathbb{L}}_{\Lambda_\ell(P_\star)} : D(\Lambda_\ell(P_\star)) \longrightarrow D(\Lambda_F(S_\star))$$

between the corresponding unbounded derived categories ([**KS**] Thm. 14.4.3). We recall that to compute the latter for a complex $M^\bullet$ in $D(\Lambda_\ell(P_\star))$ one chooses a homotopically projective resolution $P^\bullet \xrightarrow{\sim} M^\bullet$ and one has

$$\Lambda_F(S_\star) \otimes^{\mathbb{L}}_{\Lambda_\ell(P_\star)} M^\bullet \simeq \Lambda_F(S_\star) \otimes_{\Lambda_\ell(P_\star)} P^\bullet \ .$$

Obviously the former functor restricts to

$$\Lambda_F(S_\star) \otimes_{\Lambda_\ell(P_\star)} : \mathcal{M}_{et}(\Lambda_\ell(P_\star)) \longrightarrow \mathcal{M}_{et}(\Lambda_F(S_\star)) \ .$$

As far as the derived functor is concerned let $D_{et}(\Lambda_\ell(P_\star))$ and $D_{et}(\Lambda_F(S_\star))$ denote the respective full triangulated subcategories of complexes with etale cohomology modules. Let $M^\bullet$ be a complex in $D_{et}(\Lambda_\ell(P_\star))$. As a consequence of Prop. 9.6 the natural map

$$\Lambda_\ell(P_\star) \otimes_{\Lambda_\ell(P_\star), \sigma_\varphi} M^\bullet \xrightarrow{\simeq} M^\bullet$$

then is a quasi-isomorphism. By functoriality we obtain the isomorphisms

$$h^\bullet(\Lambda_F(S_\star) \otimes^{\mathbb{L}}_{\Lambda_\ell(P_\star)} (\Lambda_\ell(P_\star) \otimes_{\Lambda_\ell(P_\star), \sigma_\varphi} M^\bullet)) \cong h^\bullet(\Lambda_F(S_\star) \otimes^{\mathbb{L}}_{\Lambda_\ell(P_\star)} M^\bullet) \ .$$

To further compute the left hand term we fix a homotopically projective resolution $P^\bullet \xrightarrow{\sim} M^\bullet$. Since any base change has the restriction functor as an exact right adjoint

$$\Lambda_\ell(P_\star) \otimes_{\Lambda_\ell(P_\star),\sigma_\varphi} P^\bullet \xrightarrow{\sim} \Lambda_\ell(P_\star) \otimes_{\Lambda_\ell(P_\star),\sigma_\varphi} M^\bullet$$

is a homotopically projective resolution as well. We compute

$$\begin{aligned} &h^\bullet\big(\Lambda_F(S_\star) \otimes^{\mathbb{L}}_{\Lambda_\ell(P_\star)} \big(\Lambda_\ell(P_\star) \otimes_{\Lambda_\ell(P_\star),\sigma_\varphi} M^\bullet\big)\big) \\ &\quad = h^\bullet\big(\Lambda_F(S_\star) \otimes_{\Lambda_\ell(P_\star)} \big(\Lambda_\ell(P_\star) \otimes_{\Lambda_\ell(P_\star),\sigma_\varphi} P^\bullet\big)\big) \\ &\quad = h^\bullet\big(\Lambda_F(S_\star) \otimes_{\Lambda_\ell(P_\star),\sigma_\varphi} P^\bullet\big) \\ &\quad = h^\bullet\big(\Lambda_F(S_\star) \otimes_{\Lambda_F(S_\star),\sigma_\varphi} \big(\Lambda_F(S_\star) \otimes_{\Lambda_\ell(P_\star)} P^\bullet\big)\big) \\ &\quad = \Lambda_F(S_\star) \otimes_{\Lambda_F(S_\star),\sigma_\varphi} h^\bullet\big(\Lambda_F(S_\star) \otimes_{\Lambda_\ell(P_\star)} P^\bullet\big) \\ &\quad = \Lambda_F(S_\star) \otimes_{\Lambda_F(S_\star),\sigma_\varphi} h^\bullet\big(\Lambda_F(S_\star) \otimes^{\mathbb{L}}_{\Lambda_\ell(P_\star)} M^\bullet\big) \end{aligned}$$

where the second to last identity uses Prop. 9.6. It follows that $\Lambda_F(S_\star) \otimes^{\mathbb{L}}_{\Lambda_\ell(P_\star)} M^\bullet$ lies in $D_{et}(\Lambda_F(S_\star))$.

The following commutative diagram displays all the functors we have constructed:

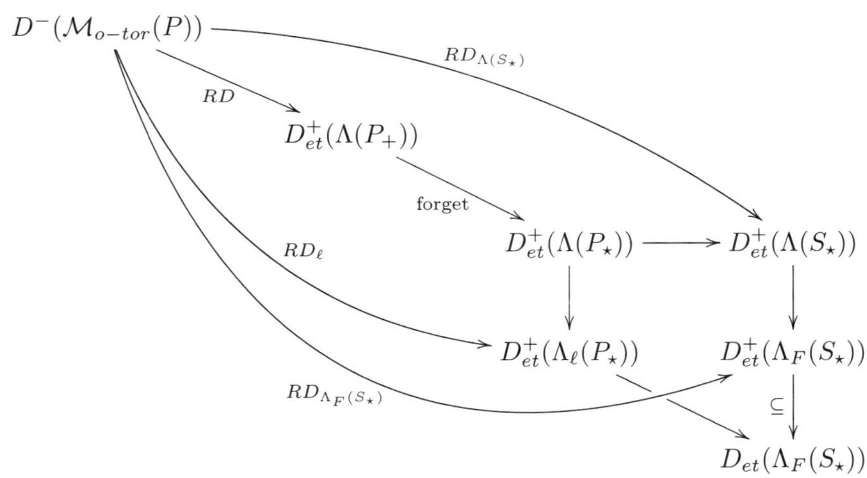

The unnamed arrows are the respective (derived) base change functors. For the commutativity of the corresponding box of base changes see [**KS**] Prop. 14.4.7 (and Ex. 13.3.9).

LEMMA 10.2. *For any $\Lambda(P_\star)$-module M such that $\pi^m M = 0$ for some $m \geq 0$ we have*

$$\mathrm{Tor}_i^{\Lambda(P_\star)}(\Lambda_F(S_\star), M) = \Lambda_F(S_0) \otimes_{\Lambda(S_0)} \mathrm{Tor}_i^{\Lambda(N_0)}(\Lambda(S_0), M) \qquad \textit{for any } i \geq 0.$$

PROOF. Since $\Lambda_F(S_0\Gamma)$ is flat over $\Lambda(S_0\Gamma)$, $\Lambda(P_\star)$ is flat over $\Lambda(N_0\Gamma)$, and

$$\Lambda_F(S_\star) = \Lambda_F(S_0\Gamma) \otimes_{\Lambda(S_0\Gamma)} \Lambda(S_\star) \quad \text{as well as} \quad \Lambda(S_\star) = \Lambda(S_0\Gamma) \otimes_{\Lambda(N_0\Gamma)} \Lambda(P_\star)$$

we have

$$\mathrm{Tor}_i^{\Lambda(P_\star)}(\Lambda_F(S_\star), M) = \Lambda_F(S_0\Gamma) \otimes_{\Lambda(S_0\Gamma)} \mathrm{Tor}_i^{\Lambda(N_0\Gamma)}(\Lambda(S_0\Gamma), M)\ .$$

With M each $\mathrm{Tor}_i^{\Lambda(N_0\Gamma)}(\Lambda(S_0\Gamma), M)$ is annihilated by π^m. Hence we may apply Lemma 8.12.iv (cf. also Remark 10.1) to the right hand side and obtain

$$\Lambda_F(S_0\Gamma) \otimes_{\Lambda(S_0\Gamma)} \mathrm{Tor}_i^{\Lambda(N_0\Gamma)}(\Lambda(S_0\Gamma), M) = \Lambda_F(S_0) \otimes_{\Lambda(S_0)} \mathrm{Tor}_i^{\Lambda(N_0\Gamma)}(\Lambda(S_0\Gamma), M) .$$

Finally we note that $\Lambda(N_0\Gamma)$ is flat over $\Lambda(N_0)$ and that

$$\Lambda(S_0\Gamma) = \Lambda(S_0) \otimes_{\Lambda(N_0)} \Lambda(N_0\Gamma)$$

which imply

$$\mathrm{Tor}_i^{\Lambda(N_0\Gamma)}(\Lambda(S_0\Gamma), M) = \mathrm{Tor}_i^{\Lambda(N_0)}(\Lambda(S_0), M) .$$

□

For finitely generated compactly induced representations the functor $RD_{\Lambda_F(S_\star)}$ has an interesting stability property. Let $V = \mathrm{ind}_{P_0}^P(V_0)$ with a finite V_0. Using the notations from section 3 we have

$$D(V) = \varinjlim_{s \in T_+} M_s^* ,$$

as $\widetilde{\Lambda}(P_+^{-1})$-modules, by Lemma 3.1. As noted before, via the identification

$$J^+(V_0) \xrightarrow{\cong} \mathrm{Ind}_{P_0}^P(V_0^*)/J^-(V_0) = M_1(V_0^*)$$

the Pontrjagin dual of $M_1(V_0) = \mathrm{ind}_{P_0}^{P_+}(V_0)$ is a $\Lambda(P_+)$-module, and the natural map

$$M_1(V_0)^* \longrightarrow D(V)$$

is a map of $\Lambda(P_+)$-modules both of which are annihilated by some power of π.

PROPOSITION 10.3. *For any representation compactly induced from a finite V_0 the map*

$$\Lambda_F(S_\star) \otimes^{\mathbb{L}}_{\Lambda(P_\star)} M_1(V_0)^* \xrightarrow{\simeq} \Lambda_F(S_\star) \otimes^{\mathbb{L}}_{\Lambda(P_\star)} D(\mathrm{ind}_{P_0}^P(V_0))$$

is a quasi-isomorphism.

PROOF. The assertion is that the maps

$$\mathrm{Tor}_i^{\Lambda(P_\star)}(\Lambda_F(S_\star), M_1(V_0)^*) \xrightarrow{\cong} \mathrm{Tor}_i^{\Lambda(P_\star)}(\Lambda_F(S_\star), D(V))$$

are isomorphisms. By Lemma 10.2 we are reduced to showing that

$$\Lambda_F(S_0) \otimes_{\Lambda(S_0)} \mathrm{Tor}_i^{\Lambda(N_0)}(\Lambda(S_0), M_1(V_0)^*) \xrightarrow{\cong} \Lambda_F(S_0) \otimes_{\Lambda(S_0)} \mathrm{Tor}_i^{\Lambda(N_0)}(\Lambda(S_0), D(V))$$

are isomorphisms. But Tor-functors commute with inductive limits. Hence it suffices to show that the maps

$$\Lambda_F(S_0) \otimes_{\Lambda(S_0)} \mathrm{Tor}_i^{\Lambda(N_0)}(\Lambda(S_0), M_1(V_0)^*) \xrightarrow{\cong} \Lambda_F(S_0) \otimes_{\Lambda(S_0)} \mathrm{Tor}_i^{\Lambda(N_0)}(\Lambda(S_0), M_s(V_0)^*)$$

are isomorphisms, or equivalently, that

$$\Lambda_F(S_0) \otimes_{\Lambda(S_0)} \mathrm{Tor}_i^{\Lambda(N_0)}(\Lambda(S_0), (M_1(V_0)/M_s(V_0))^*) = 0$$

for any $s \in T_+$ and any $i \geq 0$. But as an N_0-representation we have

$$M_1(V_0)/M_s(V_0) = \bigoplus_{t \in (T_+ - T_+ s)/T_0} M(t)(V_0) . \tag{31}$$

For each direct summand there is the N_0-equivariant isomorphism

$$M(t)(V_0) = \mathrm{ind}_{P_0}^{N_0 t P_0}(V_0) \xrightarrow{\cong} \mathrm{ind}_{tN_0t^{-1}}^{N_0}(t_* V_0)$$
$$\psi \longmapsto \phi(n_0) := \psi(n_0 t) \ ,$$

and so

$$M(t)(V_0)^* \cong \mathrm{ind}_{tN_0t^{-1}}^{N_0}(t_*V_0)^* = \Lambda(N_0)\otimes_{\Lambda(tN_0t^{-1})}(t_*V_0)^* = \Lambda(N_0)\otimes_{\Lambda(tN_0t^{-1})} t_*V_0^* \ .$$

Since the tensor product with a finitely generated module over a noetherian ring commutes with arbitrary direct products we obtain

$$\begin{aligned}&\mathrm{Tor}_i^{\Lambda(N_0)}(\Lambda(S_0), (M_1(V_0)/M_s(V_0))^*) \\ &\qquad \cong \prod_{t\in (T_+ - T_+ s)/T_0} \mathrm{Tor}_i^{\Lambda(N_0)}(\Lambda(S_0), \Lambda(N_0)\otimes_{\Lambda(tN_0t^{-1})} t_*V_0^*) \ .\end{aligned}$$

Since in the commutative diagram of rings

$$\begin{array}{ccc} \Lambda(tN_0t^{-1}) & \longrightarrow & \Lambda(tN_0t^{-1}N_1/N_1) \\ \subseteq\downarrow & & \downarrow\subseteq \\ \Lambda(N_0) & \longrightarrow & \Lambda(S_0) \end{array}$$

the vertical maps make the lower ring a free module of finite rank over the upper ring each term in the right hand direct product, as a $\Lambda(S_0)$-module, can be rewritten as

$$\begin{aligned}&\mathrm{Tor}_i^{\Lambda(N_0)}(\Lambda(S_0), \Lambda(N_0)\otimes_{\Lambda(tN_0t^{-1})} t_*V_0^*) \\ &\quad = \Lambda(S_0)\otimes_{\Lambda(tN_0t^{-1}N_1/N_1)} \mathrm{Tor}_i^{\Lambda(tN_0t^{-1})}(\Lambda(tN_0t^{-1}N_1/N_1), t_*V_0^*) \\ &\quad = \Lambda(S_0)\otimes_{\Lambda(tN_0t^{-1}N_1/N_1)} t_*\, \mathrm{Tor}_i^{\Lambda(N_0)}(\Lambda(N_0t^{-1}N_1t/t^{-1}N_1t), V_0^*) \\ &\quad = \Lambda(S_0)\otimes_{\Lambda(tN_0t^{-1}N_1/N_1)} t_*\, \mathrm{Tor}_i^{\Lambda(N_0)}(\Lambda(N_0/N_0\cap t^{-1}N_1t), V_0^*) \\ &\quad = t_*\big(\Lambda(t^{-1}N_0t/t^{-1}N_1t)\otimes_{\Lambda(N_0/N_0\cap t^{-1}N_1t)} \mathrm{Tor}_i^{\Lambda(N_0)}(\Lambda(N_0/N_0\cap t^{-1}N_1t), V_0^*)\big).\end{aligned}$$

To go further we divide up the index set $T_+ - T_+s$ into finitely many subsets $T_{s,\alpha}$ indexed by the simple roots $\alpha \in \Delta$ by defining

$$T_{s,\alpha} := \{t \in T_+ : |\alpha(t)| > |\alpha(s)|\}.$$

Since $t \notin T_+$ if and only if $|\alpha(t)| > 1$ for some $\alpha \in \Delta$ it is clear that

$$T_+ - T_+s = \bigcup_{\alpha\in\Delta} T_{s,\alpha} \ .$$

We claim that, for any given $\alpha \in \Delta$, we find an open subgroup $N''_\alpha \subseteq N_0 \cap N_\alpha$ which, through the injective projection map $N''_\alpha \hookrightarrow N_0/N_0\cap t^{-1}N_1t$, acts trivially on $\mathrm{Tor}_i^{\Lambda(N_0)}(\Lambda(N_0/N_0\cap t^{-1}N_1t), V_0^*)$ for any $t \in T_+$.

To establish this claim we let $N_0^c := \prod_{\beta\in\Phi^+\setminus\Delta} N_0 \cap N_\beta$. It follows from the standard commutator relations in N (cf. [**BT**] Prop. 4.7.(iii) and Remark 4.11) that N_0/N_0^c is commutative. Moreover, $N_0^c \subseteq N_0 \cap t^{-1}N_1t$ for any $t \in T_+$, and we have the spectral sequence

$$\mathrm{Tor}_i^{\Lambda(N_0/N_0^c)}\big(\Lambda(N_0/N_0\cap t^{-1}N_1t),\mathrm{Tor}_j^{\Lambda(N_0)}(\Lambda(N_0/N_0^c),V_0^*)\big) \Longrightarrow \mathrm{Tor}_{i+j}^{\Lambda(N_0)}(\Lambda(N_0/N_0\cap t^{-1}N_1t),V_0^*)\ .$$

We first investigate the o-modules $\mathrm{Tor}_j^{\Lambda(N_0)}(\Lambda(N_0/N_0^c),V_0^*)$ under the action of $N_{\alpha,0} := N_0\cap N_\alpha$. First of all, the ring $\Lambda(N_0)$ having finite global dimension ([**Neu**]) at most finitely many of them can be nonzero. Secondly, since V_0^* is finite they are killed by some power of π. Since $\Lambda(N_0/N_0^c) = o\otimes_{\Lambda(N_0^c)}\Lambda(N_0)$ and $\Lambda(N_0)$ is flat over $\Lambda(N_0^c)$ (cf. the proof of Lemma 5.5 in [**OV**]) we have

$$\mathrm{Tor}_j^{\Lambda(N_0)}(\Lambda(N_0/N_0^c),V_0^*)\cong \mathrm{Tor}_j^{\Lambda(N_0^c)}(o,V_0^*)$$

as o-modules. Using a resolution

$$\ldots\longrightarrow\Lambda(N_0^c)^{m_j}\longrightarrow\ldots\longrightarrow\Lambda(N_0^c)^{m_0}\longrightarrow V_0^*\longrightarrow 0$$

by finitely generated free modules over the noetherian ring $\Lambda(N_0^c)$ we compute

$$\begin{aligned}\mathrm{Tor}_j^{\Lambda(N_0^c)}(o,V_0^*) &= h_j(o\otimes_{\Lambda(N_0^c)}\Lambda(N_0^c)^{m_\bullet})\\ &= h_j(o^{m_\bullet})\end{aligned}$$

which shows that these groups are finitely generated over o. Altogether we see that the o-modules $\mathrm{Tor}_j^{\Lambda(N_0)}(\Lambda(N_0/N_0^c),V_0^*)$ in fact are finite and therefore are fixed by some open subgroup $N''_\alpha\subseteq N_{\alpha,0}$. The ring $\Lambda(N_0/N_0\cap t^{-1}N_1t)$ being commutative we conclude that all terms in the above spectral sequence are fixed by N''_α which, in particular, establishes our claim.

We deduce from this claim that

$$\begin{aligned}&\mathrm{Tor}_i^{\Lambda(N_0)}(\Lambda(S_0),\Lambda(N_0)\otimes_{\Lambda(tN_0t^{-1})}t_*V_0^*)\\ &= t_*\big(\Lambda(t^{-1}N_0t/t^{-1}N_1t)\otimes_{\Lambda(N_0/N_0\cap t^{-1}N_1t)}\mathrm{Tor}_i^{\Lambda(N_0)}(\Lambda(N_0/N_0\cap t^{-1}N_1t),V_0^*)\big)\ ,\end{aligned}$$

for any $t\in T_+$, is fixed by $tN''_\alpha t^{-1}$. But

$$N'_\alpha := \bigcap_{t\in T_{s,\alpha}} tN''_\alpha t^{-1}$$

still is open in $N_{\alpha,0}$. We pick an element $\gamma_\alpha\in S_0$ which is the image of a topological generator of N'_α, and we finally obtain that the $\Lambda(S_0)$-module

$$\prod_{t\in T_{s,\alpha}/T_0}\mathrm{Tor}_i^{\Lambda(N_0)}(\Lambda(S_0),\Lambda(N_0)\otimes_{\Lambda(tN_0t^{-1})}t_*V_0^*)$$

is killed by $\gamma_\alpha - 1$, which in particular implies that its base change to $\Lambda_F(S_0)$ vanishes. Forming the finite direct sum over the $\alpha\in\Delta$ then gives the vanishing of $\Lambda_F(S_0)\otimes_{\Lambda(S_0)}\mathrm{Tor}_i^{\Lambda(N_0)}(\Lambda(S_0),(M_1(V_0)/M_s(V_0))^*)$. $\square$

11. The case $GL_2(\mathbb{Q}_p)$

Throughout this section we let G be the group $GL_2(\mathbb{Q}_p)$, and we make our choices of $P = P_2(\mathbb{Q}_p),\ldots$ as at the end of section 1. This case is particularly simple since we obviously have

$$D^i_{\Lambda_F(S_\star)}(V) = \Lambda_F(S_\star)\otimes_{\Lambda(S_\star)}D^i(V)\qquad\text{for any } i\geq 0 \text{ and any } V \text{ in } \mathcal{M}_{o-tor}(P).$$

PROPOSITION 11.1. *If V in $\mathcal{M}_{o-tor}(P)$ is finitely generated then the map*

$$\Lambda_F(S_0) \otimes_{\Lambda(S_0)} M^* \xrightarrow{\cong} \Lambda_F(S_0) \otimes_{\Lambda(S_0)} D(V) = \Lambda_F(S_\star) \otimes_{\Lambda(S_\star)} D(V)$$

is an isomorphism for any sufficiently small M in $\mathcal{P}_+(V)$.

PROOF. We write V as a quotient

$$\mathrm{ind}_{P_0}^P(V_0) \xrightarrow{f} V \longrightarrow 0$$

of a representation compactly induced from a finite V_0. Let $M := f(M_1(V_0)) \in \mathcal{P}(V)$, and consider any $M' \in \mathcal{P}(V)$ which is contained in M. Then f induces an isomorphism

$$M_1(V_0)/f^{-1}(M') \cap M_1(V_0) \xrightarrow{\cong} M/M' .$$

We pick an $s \in T_+$ such that $M_s(V_0) \subseteq f^{-1}(M')$. By the proof of Prop. 10.3 we have $\Lambda_F(S_0) \otimes_{\Lambda(S_0)} \big(M_1(V_0)/M_s(V_0)\big)^* = 0$ and a fortiori $\Lambda_F(S_0) \otimes_{\Lambda(S_0)} (M/M')^* = \Lambda_F(S_0) \otimes_{\Lambda(S_0)} \big(M_1(V_0)/f^{-1}(M') \cap M_1(V_0)\big)^* = 0$ which implies that

$$\Lambda_F(S_0) \otimes_{\Lambda(S_0)} M^* \xrightarrow{\cong} \Lambda_F(S_0) \otimes_{\Lambda(S_0)} M'^*$$

is an isomorphism. Since the tensor product commutes with inductive limits the assertion follows. For the additional identity in the assertion we recall Remark 10.1. □

A representation V in $\mathcal{M}_{o-tor}(P)$ is called *finitely presented* if there is an exact sequence in $\mathcal{M}_{o-tor}(P)$ of the form

$$\mathrm{ind}_{P_0}^P(U_1) \xrightarrow{\rho} \mathrm{ind}_{P_0}^P(U_0) \longrightarrow V \longrightarrow 0$$

with finite U_0 and U_1. According to Cor. 4.4 we then have the exact sequence of (etale) $\Lambda(S_\star)$-modules

$$0 \longrightarrow D^0(V) \longrightarrow D(\mathrm{ind}_{P_0}^P(U_0)) \xrightarrow{D(\rho)} D(\mathrm{ind}_{P_0}^P(U_1)) .$$

Using Propositions 10.3 and 11.1 (rather their proofs) we see that $D^0_{\Lambda_F(S_\star)}(V)$ can be computed as the kernel

$$D^0_{\Lambda_F(S_\star)}(V) = \Lambda_F(S_0) \otimes_{\Lambda(S_0)} \ker\big(M_s(U_0)^* \xrightarrow{\rho^*} M_{s'}(U_1)^*\big) \tag{32}$$

for any $s, s' \in T_+$ such that $\rho(M_{s'}(U_1)) \subseteq M_s(U_0))$.

LEMMA 11.2. *For any finite subset $X \subseteq P$ and any sufficiently big $n \geq 0$ we have*

$$P_+\varphi^{2n}X \subseteq P_+\varphi^n \quad \text{and} \quad (P \setminus P_+)X \subseteq P \setminus P_+\varphi^n .$$

PROOF. We choose n big enough so that $\varphi^n X \cap \varphi^n X^{-1} \subseteq P_+$ holds true. Then, of course, $P_+\varphi^{2n}X \subseteq P_+\varphi^n P_+ = P_+\varphi^n$. Moreover, if $b \in P \setminus P_+$ satisfies $bx = b_+\varphi^n$ for some $x \in X$ and $b_+ \in P_+$ then $b = b_+\varphi^n x^{-1} \in P_+$ which is a contradiction. □

PROPOSITION 11.3. *If V in $\mathcal{M}_{o-tor}(P)$ is finitely presented then the map*

$$\Lambda_F(S_0) \otimes_{\Lambda(S_0)} D(V) \xrightarrow{\cong} \Lambda_F(S_0) \otimes_{\Lambda(S_0)} D^0(V) = D^0_{\Lambda_F(S_\star)}(V)$$

is an isomorphism.

PROOF. Consider any finite presentation

$$\mathrm{ind}_{P_0}^{P}(U_1) \xrightarrow{\rho} \mathrm{ind}_{P_0}^{P}(U_0) \xrightarrow{f} V \longrightarrow 0$$

of V. By (32) we have

$$\Lambda_F(S_0) \otimes_{\Lambda(S_0)} D^0(V) = \Lambda_F(S_0) \otimes_{\Lambda(S_0)} \ker\big(M_{\varphi^n}(U_0)^* \xrightarrow{\rho^*} M_{\varphi^m}(U_1)^*\big)$$

for any $m, n \geq 0$ such that $\rho(M_{\varphi^m}(U_1)) \subseteq M_{\varphi^n}(U_0))$. We now choose n big enough so that $M := f(M_{\varphi^n}(U_0)) \subseteq V$ satisfies Prop. 11.1 so that we obtain (also using the flatness of $\Lambda_F(S_0)$ over $\Lambda(S_0)$) the exact sequence

$$0 \to \Lambda_F(S_0) \underset{\Lambda(S_0)}{\otimes} D(V) \to \Lambda_F(S_0) \underset{\Lambda(S_0)}{\otimes} M_{\varphi^n}(U_0)^* \to \Lambda_F(S_0) \underset{\Lambda(S_0)}{\otimes} \rho^{-1}(M_{\varphi^n}(U_0))^*.$$

We see that to prove our assertion it is sufficient to establish that

$$\Lambda_F(S_0) \otimes_{\Lambda(S_0)} \big(\rho^{-1}(M_{\varphi^n}(U_0))/M_{\varphi^m}(U_1)\big)^* = 0$$

vanishes for appropriate choices of m, n. There is a finite subset $X \subseteq P$ such that $\rho(U_1) \subseteq XU_0$. Hence by possibly enlarging m, n we may, according to Lemma 11.2, assume that

$$P_+\varphi^m X \subseteq P_+\varphi^n \quad \text{and} \quad (P \setminus P_+)X \subseteq P \setminus P_+\varphi^n$$

holds true. It follows that in the decompositions

$$\mathrm{ind}_{P_0}^{P}(U_1) = \mathrm{ind}_{P_0}^{P\setminus P_+}(U_1) \oplus \mathrm{ind}_{P_0}^{P_+\setminus P_+\varphi^m}(U_1) \oplus M_{\varphi^m}(U_1)$$

and

$$\mathrm{ind}_{P_0}^{P}(U_0) = \mathrm{ind}_{P_0}^{P\setminus P_+\varphi^n}(U_0) \oplus M_{\varphi^n}(U_0)$$

the homomorphism ρ respects the first and the last summands. We deduce from this that $\rho^{-1}(M_{\varphi^n}(U_0))/M_{\varphi^m}(U_1)$ is isomorphic to a subrepresentation of the P_+-representation $\mathrm{ind}_{P_0}^{P_+\setminus P_+\varphi^m}(U_1)$. Hence we are further reduced to proving that

$$\Lambda_F(S_0) \otimes_{\Lambda(S_0)} \mathrm{ind}_{P_0}^{P_+\setminus P_+\varphi^m}(U_1)^* = 0 \ .$$

But this we have done already in the proof of Prop. 10.3 (cf. (31)). □

REMARK 11.4. *For a general split group G arguments as above show that for any finitely presented V in $\mathcal{M}_{o-tor}(P)$ the map*

$$\Lambda_F(S_0) \otimes_{\Lambda(N_0)} D(V) \longrightarrow \Lambda_F(S_0) \otimes_{\Lambda(N_0)} D^0(V)$$

at least is surjective. It therefore follows from Prop. 9.9 that for general G and any admissible V in $\mathcal{M}_{o-tor}(G)$ the following holds: If for some $M \in \mathcal{P}_+(V)$ there is an exact sequence of P_+-representations of the form

$$\ldots \longrightarrow \mathrm{ind}_{P_0}^{P_+}(V_n) \longrightarrow \ldots \longrightarrow \mathrm{ind}_{P_0}^{P_+}(V_0) \longrightarrow M \longrightarrow 0$$

with V_n finite for any $n \geq 0$ then V is finitely presented for P and the $\Lambda_F(S_0)$-module $\Lambda_F(S_0) \otimes_{\Lambda(N_0)} D^0(V)$ is finitely generated.

COROLLARY 11.5. *Let $(0 \longrightarrow)V_1 \longrightarrow V_2 \longrightarrow V_3 \longrightarrow 0$ be an exact sequence in $\mathcal{M}_{o-tor}(P)$; if V_3 is finitely presented then the sequence*

$$0 \longrightarrow \Lambda_F(S_0) \underset{\Lambda(S_0)}{\otimes} D(V_3) \longrightarrow \Lambda_F(S_0) \underset{\Lambda(S_0)}{\otimes} D(V_2) \longrightarrow \Lambda_F(S_0) \underset{\Lambda(S_0)}{\otimes} D(V_1)(\longrightarrow 0)$$

is exact.

PROOF. We contemplate the commutative diagram

$$\begin{array}{ccccccc} & & 0 & & 0 & & 0 \\ & & \downarrow & & \downarrow & & \downarrow \\ 0 & \longrightarrow & \Lambda_F(S_0)\otimes_{\Lambda(S_0)} D(V_3) & \longrightarrow & \Lambda_F(S_0)\otimes_{\Lambda(S_0)} D(V_2) & \longrightarrow & \Lambda_F(S_0)\otimes_{\Lambda(S_0)} D(V_1) \\ & & \downarrow{\cong} & & \downarrow & & \downarrow \\ 0 & \longrightarrow & \Lambda_F(S_0)\otimes_{\Lambda(S_0)} D^0(V_3) & \longrightarrow & \Lambda_F(S_0)\otimes_{\Lambda(S_0)} D^0(V_2) & \longrightarrow & \Lambda_F(S_0)\otimes_{\Lambda(S_0)} D^0(V_1) \end{array}$$

in which the lower row is exact by construction, the columns are exact by Remark 2.4.ii, and the upper row is a complex which is exact at the left spot again by Remark 2.4.ii (always using in addition the flatness of $\Lambda_F(S_0)$ over $\Lambda(S_0)$). But by our assumption and Prop. 11.3 the left vertical map is an isomorphism. In this situation an easy diagram chase shows that the upper row has to be exact as well. For the additional zero we use Remark 2.4.i. □

COROLLARY 11.6. *Let $V_1 \longrightarrow V_2 \longrightarrow V_3$ be an exact sequence in $\mathcal{M}_{o-tor}(P)$; we have:*

i. *If V_1 is finitely generated and V_2 is finitely presented then the sequence*

$$\Lambda_F(S_0)\otimes_{\Lambda(S_0)} D(V_3) \longrightarrow \Lambda_F(S_0)\otimes_{\Lambda(S_0)} D(V_2) \longrightarrow \Lambda_F(S_0)\otimes_{\Lambda(S_0)} D(V_1)$$

is exact;

ii. *if all three representations V_1, V_2, and V_3 are finitely presented then the sequence*

$$D^0_{\Lambda_F(S_\star)}(V_3) \longrightarrow D^0_{\Lambda_F(S_\star)}(V_2) \longrightarrow D^0_{\Lambda_F(S_\star)}(V_1)$$

is exact.

PROOF. i. Let $V_3' := \mathrm{im}(V_2 \longrightarrow V_3)$. We leave it to the reader to check that the quotient of a finitely presented representation by a finitely generated subrepresentation is finitely presented. Hence we may apply the previous corollary to the exact sequence $V_1 \longrightarrow V_2 \longrightarrow V_3' \longrightarrow 0$. In addition we observe the Remark 2.4.i.

ii. This is an immediate consequence of i. and Prop. 11.3. □

Colmez in [**Col**] investigates particularly nice finite presentations of a modified form. To review his result let, at first, V be an arbitrary finitely generated smooth G-representation which has a central character. We note right away that, as a consequence of the Iwasawa decomposition, V then also is finitely generated as a P-representation. Let Z denote the center of G, put $G_0 := GL_2(\mathbb{Z}_p)$, and recall the notation $\varphi = \left(\begin{smallmatrix} 1 & 0 \\ 0 & p \end{smallmatrix}\right)$. Let G_1 denote the normalizer of $G_0 \cap \varphi^{-1}G_0\varphi$ in G. It contains $(G_0\cap\varphi^{-1}G_0\varphi)Z$ with index two and with $\left(\begin{smallmatrix} 0 & p \\ 1 & 0 \end{smallmatrix}\right)$ representing the nontrivial coset. We pick a finite G_0Z-subrepresentation $U_0 \subseteq V$ which generates V so that we obtain a surjection

$$\mathrm{ind}^G_{G_0Z}(U_0) \xrightarrow{f} V \longrightarrow 0 .$$

One checks that $U_0 \cap \varphi U_0$ is $\varphi G_1 \varphi^{-1}$-invariant ([**Col**] Lemma 2.6). To avoid confusion we denote, in this section, the natural inclusion $U_0 \hookrightarrow \mathrm{ind}^G_{G_0Z}(U_0)$ by $v \mapsto \tilde{v}$. The finite o-submodule

$$U_1 := \{\widetilde{\varphi^{-1}v} - \varphi^{-1}\tilde{v} : v \in U_0 \cap \varphi U_0\}$$

of $\mathrm{ind}_{G_0Z}^G(U_0)$ lies in the kernel of f and is G_1-invariant. Hence we obtain a complex

$$\mathrm{ind}_{G_1}^G(U_1) \xrightarrow{\rho} \mathrm{ind}_{G_0Z}^G(U_0) \xrightarrow{f} V \longrightarrow 0 \tag{33}$$

in $\mathcal{M}_{o-tor}(G)$. Colmez calls (33) a standard presentation of V provided that it is exact. We emphasize that the center Z acts on all three terms of this sequence through the same central character.

THEOREM 11.7. *(Colmez) For any smooth G-representation V which is admissible and of finite length and which has a central character U_0 can be chosen in such a way that*

$$0 \longrightarrow \mathrm{ind}_{G_1}^G(U_1) \xrightarrow{\rho} \mathrm{ind}_{G_0Z}^G(U_0) \xrightarrow{f} V \longrightarrow 0 \tag{34}$$

is a short exact sequence.

PROOF. [**Co1**] Thm. 3.1 and Lemma 2.8 (compare also [**Oll**] and [**Vi3**] for modulo p representations). □

By using the Iwasawa decompositions $G = PG_0 = PG_1$ as well as $G_0Z \cap P = G_1Z \cap P = P_0Z$ we may rewrite (34) as

$$0 \longrightarrow \mathrm{ind}_{P_0Z}^P(U_1) \xrightarrow{\rho} \mathrm{ind}_{P_0Z}^P(U_0) \xrightarrow{f} V \longrightarrow 0\ . \tag{35}$$

We have

$$P_0Z = P_0 \times \zeta^{\mathbb{Z}} \qquad \text{with } \zeta := \left(\begin{smallmatrix} p & 0 \\ 0 & p \end{smallmatrix}\right).$$

Let $\mathcal{M}_{G-good}(P)$ denote the full subcategory of $\mathcal{M}_{o-tor}(P)$ whose objects are all V which arise by restriction from an admissible finite length smooth G-representation which has a central character.

LEMMA 11.8. *For any smooth P_0Z-representation U_0 we have*

i. *$D^i(\mathrm{ind}_{P_0Z}^P(U_0)) = 0$ for any $i \geq 1$;*

ii. *if ζ acts as a scalar on U_0 then $D^0(\mathrm{ind}_{P_0Z}^P(U_0)) = D(\mathrm{ind}_{P_0Z}^P(U_0))$.*

PROOF. We view

$$0 \longrightarrow o[\zeta^{\mathbb{Z}}] \xrightarrow{(\zeta-1)\cdot} o[\zeta^{\mathbb{Z}}] \xrightarrow{\zeta\mapsto 1} o \longrightarrow 0$$

as a short exact sequence of smooth (but not o-torsion) P_0Z-representations with P_0Z acting through its projection onto $\zeta^{\mathbb{Z}}$. It splits as a sequence of o-modules. With respect to the diagonal action

$$0 \longrightarrow U_0 \otimes_o o[\zeta^{\mathbb{Z}}] \xrightarrow{\rho_0} U_0 \otimes_o o[\zeta^{\mathbb{Z}}] \longrightarrow U_0 \longrightarrow 0\ ,$$

where $\rho_0 := \mathrm{id} \otimes (\zeta - 1)$, therefore is a short exact sequence of smooth (and o-torsion) P_0Z-representations. It gives rise to the short exact sequence of smooth P-representations

$$0 \longrightarrow \mathrm{ind}_{P_0Z}^P(U_0 \otimes_o o[\zeta^{\mathbb{Z}}]) \xrightarrow{\mathrm{ind}(\rho_0)} \mathrm{ind}_{P_0Z}^P(U_0 \otimes_o o[\zeta^{\mathbb{Z}}]) \longrightarrow \mathrm{ind}_{P_0Z}^P(U_0) \longrightarrow 0\ . \tag{36}$$

But the map

$$\begin{aligned} \mathrm{ind}_{P_0}^{P_0Z}(U_0) &\xrightarrow{\cong} U_0 \otimes_o o[\zeta^{\mathbb{Z}}] \\ \psi &\longmapsto \sum_{n\in\mathbb{Z}} \zeta^n(\psi(\zeta^n)) \otimes \zeta^n \end{aligned}$$

is a P_0Z-equivariant isomorphism. Inserting this into (36) and using the transitivity of compact induction we arrive at an exact sequence of smooth P-representations of the form

$$0 \longrightarrow \mathrm{ind}_{P_0}^{P}(U_0) \longrightarrow \mathrm{ind}_{P_0}^{P}(U_0) \longrightarrow \mathrm{ind}_{P_0Z}^{P}(U_0) \longrightarrow 0 \ .$$

Applying Cor. 4.4 and Lemma 2.4.i to it gives the first assertion.

To establish the second assertion it suffices to check the condition (2) in Lemma 2.5 for the short exact sequence (36). We first make some general observations. Let V_0 be any smooth P_0Z-representation on which ζ acts as a scalar. Since $P_+ \supseteq P_0Z$ the P_+-subrepresentations of $\mathrm{ind}_{P_0Z}^{P}(V_0)$ defined by

$$M'_s(V_0) := \mathrm{ind}_{P_0Z}^{P_+s}(V_0) \qquad \text{for any } s \in T_+$$

and

$$M'_\sigma(V_0) := \oplus_{t \in T_+/T_0Z} \mathrm{ind}_{P_0Z}^{N_0tP_0Z}(\sigma(t))$$

for any order reversing map $\sigma : T_+/T_0Z \longrightarrow \mathrm{Sub}(V_0)$ (note that by our assumption any element in $\mathrm{Sub}(V_0)$ automatically is a P_0Z-subrepresentation) satisfying (3) make perfect sense. It is easily checked that analogs of Lemma 3.1 and Lemma 3.2 hold true. Given this background the verification of (2) for (36) proceeds in exactly the same way as the proof of Lemma 4.1. □

REMARK 11.9. *Any V in an exact sequence $\mathrm{ind}_{P_0Z}^{P}(U_1) \longrightarrow \mathrm{ind}_{P_0Z}^{P}(U_0) \longrightarrow V \longrightarrow 0$ with finite U_0 and U_1 is of finite presentation.*

PROOF. In the proof of Lemma 11.8 we have seen that $\mathrm{ind}_{P_0Z}^{P}(U_0)$ is of finite presentation. Hence V being the quotient of a finitely presented representation by a finitely generated subrepresentation is finitely presented as well. □

PROPOSITION 11.10. i. *For any V in $\mathcal{M}_{G-good}(P)$ we have $D^i(V) = 0$ for any $i \geq 1$.*
ii. *The functor D^0 restricted to $\mathcal{M}_{G-good}(P)$ is exact.*

PROOF. ii. is an immediate consequence of i. Lemma 11.8 says that we can use (35) to compute the δ-functor D^i on V. Hence

$$D^i(V) = h^i(D(\mathrm{ind}_{P_0Z}^{P}(U_0)) \xrightarrow{D(\rho)} D(\mathrm{ind}_{P_0Z}^{P}(U_1)) \longrightarrow 0 \longrightarrow \ldots) \ .$$

By Remark 2.4.i the map $D(\rho)$ is surjective. It follows that $D^i(V) = 0$ for $i \geq 1$. □

In view of Remark 10.1 and our discussion of etale $(\Lambda_F(S_0), \Gamma, \varphi)$-modules in section 9 our functor $D^0_{\Lambda_F(S_\star)} = \Lambda_F(S_\star) \otimes_{\Lambda(S_\star)} D^0$ restricted to $\mathcal{M}_{G-good}(P)$ coincides with the functor constructed by Colmez in [**Co1**].

PROPOSITION 11.11. *For every representation V in $\mathcal{M}_{G-good}(P)$ the set $\mathcal{P}_+(V)$ has a (unique) minimal element M_0; in particular, we have $D(V) = M_0^*$.*

PROOF. According to [**Co1**] Lemma 4.25 there is an $M \in \mathcal{P}_+(V)$ such that the map $M^* \longrightarrow \Lambda_F(S_0) \otimes_{\Lambda(S_0)} D(V)$ and a fortiori the map $M^* \longrightarrow D(V)$ have a finite kernel. Let now $M' \subseteq M$ be some other element in $\mathcal{P}_+(V)$. Then

$$\ker(M^* \longrightarrow M'^*) \subseteq \ker(M^* \longrightarrow D(V)) \ .$$

Hence the finite groups $\ker(M^* \longrightarrow M'^*)$ for decreasing M' must stabilize. It follows that M'^* and M' stabilize. □

12. Subquotients of principal series

Let $\chi : P \longrightarrow k^\times$ be a locally constant character and

$$\mathrm{Ind}_P^G(\chi) := \{F \in C^\infty(G) : F(gb) = \chi(b)^{-1}F(g) \text{ for any } g \in G, b \in P\}$$

the corresponding principal series representation of G (by left translation). We recall that, for any topological space X we let $C^\infty(X)$, resp. $C_c^\infty(X)$, denote the k-vector space of all k-valued locally constant, resp. locally constant with compact support, functions on X. As a matter of further notation we write

$$\mathrm{Ind}_P^X(\chi) := \{F \in \mathrm{Ind}_P^G(\chi) : F|(G \setminus X) = 0\}$$

for any open right P-invariant subset $X \subseteq G$. Furthermore, we fix a representative $\dot{w}$ in the normalizer $N(T)$ of T in G of every element w in the Weyl group $W := N(T)/T$.

As a P-representation $\mathrm{Ind}_P^G(\chi)$ is best understood by using the Bruhat decomposition $G = \bigcup_{w \in W} PwP$. Choosing once and for all a total order on W refining the Bruhat order we obtain the P-invariant filtration $\{\mathrm{Ind}_P^{PwP}(\chi)\}_{w \in W}$ of $\mathrm{Ind}_P^G(\chi)$. Its bottom term is $\mathrm{Ind}_P^{Pw_0P}(\chi)$, corresponding to the Steinberg representation of G, where w_0 denotes the longest element in W. Each filtration step is, via $F \longmapsto F(\cdot\dot{w})$, isomorphic to $V(w, \chi) := C_c^\infty(N/N_w)$, where $N_w := N \cap \dot{w}N\dot{w}^{-1}$, with N acting by left translation and T acting by

$$(t\phi)(n) := \chi(w^{-1}tw)\phi(t^{-1}nt) \ .$$

In particular, $V(w, \chi)$ is a character twist of $V(w, 1)$. In $V(w, \chi)$ we have the generating P_+-subrepresentation $M(w, \chi) := C^\infty(N_0/N_{w,0})$ where $N_{w,0} := N_0 \cap \dot{w}N\dot{w}^{-1}$.

PROPOSITION 12.1. *$M(w, \chi)$ is the (unique) minimal element in $\mathcal{P}_+(V(w, \chi))$; in particular, we have $D(V(w, \chi)) = M(w, \chi)^* = \Lambda(N_0) \otimes_{\Lambda(N_{w,0})} k$.*

PROOF. This is the same argument as for Lemma 2.6 (cf. [**Vi2**] Lemma 4). □

Since $N_{w_0w,0} \xrightarrow{\cong} N_0/N_{w,0}$ one can write $D(V(w, \chi)) = \Omega(N_{w_0w,0})$; but this does not reflect the $\Lambda(N_0)$-action very well.

PROPOSITION 12.2. *$\Lambda_F(S_0) \otimes_{\Lambda(N_0)} D(V(w, \chi)) = 0$ when $w \neq w_0$ and is equal to $\Lambda_F(S_0) \otimes_o k$ when $w = w_0$.*

PROOF. The case $w = w_0$ being clear let $w \neq w_0$. Then (cf. [**MS**] Lemma 4.1) there is a simple root α such that $N_\alpha \subseteq N \cap \dot{w}N\dot{w}^{-1}$. It follows that S_0, being the image of $N_{\alpha,0}$, acts trivially on

$$\Lambda(S_0) \otimes_{\Lambda(N_0)} D(V(w, \chi)) = \Lambda(S_0) \otimes_{\Lambda(N_{w,0})} k$$

(compare the end of the proof of Prop. 10.3). □

Both propositions remain true (with the same proofs) when the coefficient ring is $o/\pi^m o$, for some $m \geq 1$, instead of k.

References

[BH] Balister P.N., Howson S.: *Note on Nakayama's Lemma for Compact Λ-modules.* Asian J. Math. 1, 224-229 (1997)

[BT] Borel A., Tits J.: *Homomorphismes "Abstraits" de Groupes Algebriques Simples.* Ann. Math. 97, 499-571 (1973)

[CF] Christensen J.P.R., Fischer P.: *Joint Continuity of Measurable Biadditive Mappings.* Proc. AMS 103, 1125-1128 (1988)

[CFKSV] Coates J., Fukaya T., Kato K., Sujatha R., Venjakob O.: *The GL_2 main conjecture for elliptic curves without complex multiplication.* Publ. Math. IHES. 101, 163-208 (2005)

[Co1] Colmez P.: *Représentations de $GL_2(\mathbb{Q}_p)$ et (φ,Γ)-modules.* Astérisque 330, 281-509 (2010)

[Co2] Colmez P.: *(φ,Γ)-modules et représentations du mirabolique de $GL_2(\mathbb{Q}_p)$.* Astérisque 330, 61-153 (2010)

[DDMS] Dixon J.D., du Sautoy M.P.F., Mann A., Segal D.: *Analytic Pro-p-Groups.* Cambridge Univ. Press 1999

[Em] Emerton M.: *Ordinary parts of admissible representations of p-adic reductive groups II. Derived functors.* Astérisque 331, 403-459 (2010)

[Eme] Emerton M.: *On a class of coherent rings, with applications to the smooth representation theory of $GL_2(\mathbb{Q}_p)$ in characteristic p.* Preprint 2008

[Fon] Fontaine J.-M.: *Représentations p-adiques des corps locaux.* In The Grothendieck Festschrift (eds. Cartier, Illusie, Katz, Laumon, Manin, Ribet), vol. II, pp. 249-309. Progress in Math. 87, Birkhäuser: Boston 1990

[Gab] Gabriel P.: *Des Catégories Abéliennes.* Bull. Soc. math. France 90, 323-448 (1962)

[Har] Hartshorne R.: *Residues and Duality.* Springer Lect. Notes Math. 20 (1966)

[KS] Kashiwara M., Schapira P.: *Categories and Sheaves.* Springer: Berlin - Heidelberg - New York 2006

[Laz] Lazard M.: *Groupes analytiques p-adiques.* Publ. Math. IHES 26, 389-603 (1965)

[LvO] Li H., van Oystaeyen F.: *Zariskian filtrations.* Kluwer: Dordrecht 1996

[MCR] McConnell J.C., Robson J.C.: *Noncommutative Noetherian rings.* American Math. Soc.: Providence 2001

[MS] Miličić D., Soergel W.: *Twisted Harish-Chandra sheaves and Whittaker modules: the non-degenerate case.* Preprint

[NSW] Neukirch J., Schmidt A., Wingberg K.: *Cohomology of Number Fields.* 2nd edition. Springer: Berlin - Heidelberg - New York 2008

[Neu] Neumann A.: *Completed group algebras without zero divisors.* Archiv Math. 51, 496-499 (1988)

[OV] Ochi Y., Venjakob O.: *On the structure of Selmer groups over p-adic Lie extensions.* J. Alg. Geometry 11, 547-580 (2002)

[Oll] Ollivier R.: *Mod p representations for p-adic GL_2 and coefficient systems on the tree.* Preprint

[Sch] Schikhof W.: *Ultrametric calculus.* Cambridge Univ. Press 1984

[SV1] Schneider P., Venjakob O.: *On the codimension of modules over skew power series rings with applications to Iwasawa algebras.* J. Pure Appl. Algebra 204, 349-367 (2005)

[SV2] Schneider P., Venjakob O.: *Localisations and completions of skew power series rings.* American J. Math. 132, 1-36 (2010)

[Ven] Venjakob O.: *On the structure theory of the Iwasawa algebra of a p-adic Lie group.* J. Eur. Math. Soc. 4, 271-311 (2002)

[Vi1] Vigneras M.-F.: *Représentations ℓ-modulaires d'un groupe réductif p-adique avec $\ell \neq p$.* Progress in Math. 137, Birkhäuser: Boston 1996

[Vi2] Vigneras M.-F.: *Série principale modulo p de groupes réductifs p-adiques.* GAFA 17, 2090-2112 (2008)

[Vi3] Vigneras M.-F.: *An integrality criterion for the homology of an equivariant coefficient system on the tree.* Pure Appl. Math. Quart., special issue dedicated to Prof. Serre (2008)

[War] Warner S.: *Topological Rings.* Elsevier: Amsterdam 1993

[Wil] Wilson J.S.: *Profinite Groups.* Oxford Univ. Press 1988

Universität Münster, Mathematisches Institut, Einsteinstr. 62, 48291 Münster, Germany

E-mail address: pschnei@math.uni-muenster.de

Université de Paris 7, Institut de Mathematiques de Jussieu, 175 rue du Chevaleret, Paris 75013, France

E-mail address: vigneras@math.jussieu.fr

Clay Mathematics Proceedings
Volume 13, 2011

Motivic Galois Groups and L-Groups

Hiroyuki Yoshida

Dedicated to Professor Freydoon Shahidi on the occasion of his 60th birthday

ABSTRACT. Given a pure motive, we formulate a conjecture which predicts the existence of an automorphic representation of a group related to the Hodge group, corresponding to the motive. To describe an expected automorphic representation on a small group, we discuss the existence of a splitting field of the canonical cohomology class attached to a motivic λ-adic representation.

Introduction

Let E and F be number fields. Let M be a pure motive of weight w over F with coefficients in E. Take an embedding $E \hookrightarrow \mathbf{C}$. The problem to be considered in this paper is:

PROBLEM. Find a connected reductive algebraic group G over F, an irreducible automorphic representation $\pi = \otimes_v \pi_v$ of $G(F_{\mathbf{A}})$ and a representation r of the L-group ${}^L G$ such that $L(M,s) = L(s,\pi,r)$.

Around 1964, Shimura conjectured that every elliptic curve over $\mathbf{Q}$ is modular. This conjecture was striking at that time and the beginning of the entire business related to the problem (cf. Lang [La], Shimura [Sh4], note on [64e]). In the general case, Langlands made deep and extensive studies in the 1970's culminating in the introduction of the automorphic Langlands group ([Lan1]). The conjecture was formulated as a relation between the motivic Galois group and the automorphic Langlands group. As a recent important paper on this formalism, we should mention Arthur [A2].

The answer (G,π,r) to our problem is not unique in general. In the late 1970's, there was a consensus of specialists that π exists on $\mathrm{GL}(d)$, where d is the rank of M. Clozel ([Cl]]) discussed this GL-conjecture in great detail. In this paper, we will concentrate on finding G as small as possible, for which we can formulate a natural solution of the problem. Then the other answers could be derived by the functoriality principle.

2010 *Mathematics Subject Classification.* Primary 11F70, Secondary 11S37.
Key words and phrases. Motive, Automorphic representation.

The importance of finding a small G can be seen by the following classical example. Let $\mathcal{E}$ be an elliptic curve defined over $\mathbf{Q}$ with complex multiplication by an imaginary quadratic field K. Then there exists a Hecke character ψ of $K_{\mathbf{A}}^{\times}$ such that $L(s,\psi)$ is equal to the zeta function of $\mathcal{E}$. Let $T = \mathrm{R}_{K/\mathbf{Q}}(\mathbf{G}_m)$ be a non-split torus over $\mathbf{Q}$. We have $T(\mathbf{Q}_{\mathbf{A}}) = K_{\mathbf{A}}^{\times}$ and ψ can be regarded as an automorphic representation of $T(\mathbf{Q}_{\mathbf{A}}) \subset \mathrm{GL}(2,\mathbf{Q}_{\mathbf{A}})$.

Now we are going to explain our ideas. Let λ be a finite place of E. Then we have a λ-adic representation ρ_λ attached to M:

$$\rho_\lambda : \mathrm{Gal}(\overline{F}/F) \longrightarrow \mathrm{GL}(d, E_\lambda).$$

Let H be the Zariski closure of the image and let H^0 be its identity component. A basic conjecture (Conjecture 4.1), which we assume throughout the paper, is:

The algebraic group H is defined over E and does not depend on λ. Moreover $H^0 = \mathrm{Hg}(M)$ where $\mathrm{Hg}(M)$ is the Hodge group of M (cf. §1, §4).

This conjecture implies that H^0 is reductive. Let K be the finite Galois extension of F such that

$$(1) \qquad H^0 \cap \rho_\lambda(\mathrm{Gal}(\overline{F}/F)) = \rho_\lambda(\mathrm{Gal}(\overline{F}/K)), \qquad \mathrm{Gal}(\overline{F}/K) \supset \mathrm{Ker}(\rho_\lambda).$$

Take an embedding $E_\lambda \hookrightarrow \mathbf{C}$. Then we have an exact sequence

$$(2) \qquad 1 \longrightarrow H^0(\mathbf{C}) \longrightarrow H(\mathbf{C}) \longrightarrow \mathrm{Gal}(K/F) \longrightarrow 1.$$

From (2), we obtain a homomorphism

$$\mu_{H^0} : \mathrm{Gal}(K/F) \longrightarrow \mathrm{Aut}(\mathcal{R}_0(H^0)) \cong \mathrm{Aut}(H^0)/\mathrm{Inn}(H^0).$$

Here $\mathcal{R}_0(H^0)$ denotes the based root datum of H^0. Now we can find a connected reductive algebraic group G defined over F such that (i) G is quasi-split over F, (ii) ${}^LG^0 = H^0(\mathbf{C})$, (iii) $\mu_G = \mu_{H^0}$. Here ${}^LG^0$ is the connected L-group of G; we regard μ_{H^0} as a homomorphism of $\mathrm{Gal}(\overline{F}/F)$ into $\mathrm{Aut}(\mathcal{R}_0(H^0))$ and μ_G is the homomorphism of $\mathrm{Gal}(\overline{F}/F)$ into $\mathrm{Aut}(\mathcal{R}_0({}^LG^0))$ (cf. §3).

We expect that an automorphic representation π which corresponds to M exists on $G(F_{\mathbf{A}})$, except for anomalous cases which can occur only when $w = 0$. To examine this problem, we naturally compare $H^0(\mathbf{C})$ with the L-group of G and will find an interesting arithmetical problem. The L-group is defined as the semi-direct product of ${}^LG^0 = H^0(\mathbf{C})$ and $\mathrm{Gal}(\overline{F}/F)$ with respect to μ_G. However there are examples for which (2) does not split. Looking more closely, we find that in the equivalence class of factor sets obtained from (2), there exists a canonical factor set f_Z taking values in the center $Z(H^0(\mathbf{C}))$ of $H^0(\mathbf{C})$. In other words, f_Z defines a canonical cohomology class in $H^2(\mathrm{Gal}(K/F), Z(H^0(\mathbf{C})))$ (cf. §5). If there exists a finite extension L of K which is normal over F such that f_Z splits after the inflation to $\mathrm{Gal}(L/F)$, we call L a *splitting field.* If such an L exists, we define a finite version of the L-group by ${}^LG = {}^LG^0 \rtimes \mathrm{Gal}(L/F)$. Then we have a homomorphism ${}^LG \longrightarrow H(\mathbf{C})$. This homomorphism is canonical up to the action of $H^1(\mathrm{Gal}(L/F), Z(H^0(\mathbf{C})))$. We define $r : {}^LG \longrightarrow \mathrm{GL}(d,\mathbf{C})$ as the composite of this homomorphism with the inclusion map $H(\mathbf{C}) \hookrightarrow \mathrm{GL}(d,\mathbf{C})$. For a place v of F, there exists a local parameter $\psi_v : W'_{F_v}(\mathbf{C}) \longrightarrow H(\mathbf{C})$, where W'_{F_v} is the Weil–Deligne group scheme. We can show that ψ_v lifts to the Langlands parameter $\phi_v : W'_{F_v}(\mathbf{C}) \longrightarrow {}^LG$, which should give the L-packet to which the local component π_v belongs (§5, Main Conjecture).

In the text, we take a reductive algebraic group $\widetilde{H}$ defined over E such that $H \subset \widetilde{H} \subset \mathrm{GL}(d)$. We call the case $\widetilde{H} = H$ minimal. Similarly to the minimal case, we obtain a quasi-split group $\widetilde{G}$ defined over F and a canonical cohomology class in $H^2(\mathrm{Gal}(\widetilde{K}/F), Z(\widetilde{H}^0(\mathbf{C})))$. (Here $\widetilde{K}$ is defined by (1) with $\widetilde{H}^0$ in place of H^0.) Assuming the existence of a splitting field of this cohomology class, we formulate the Main Conjecture so that it predicts the existence of an automorphic representation of $\widetilde{G}(F_{\mathbf{A}})$ corresponding to M. The minimal case and the general case are connected by the functoriality principle. [1]

In the latter half of this paper, we discuss the problem in the minimal case. Since it is rather technical, we summarize the main features. We assume that ρ_λ is absolutely irreducible and that $w \neq 0$.

1. By Clifford's theorem, either $\rho_\lambda|\mathrm{Gal}(\overline{F}/K)$ is isotypic or there exists a field $F \subsetneq K' \subset K$ and a λ-adic representation τ_λ of $\mathrm{Gal}(\overline{F}/K')$ such that $\rho_\lambda \cong \mathrm{Ind}_{\mathrm{Gal}(\overline{F}/K')}^{\mathrm{Gal}(\overline{F}/F)} \tau_\lambda$. In the isotypic case, $Z(H^0(\mathbf{C}))$ is connected and the splitting field exists by Tate's theorem. In the alternative case, we can reduce the problem to τ_λ. By this procedure, we can predict an automorphic representation corresponding to M on a slightly "bigger"group than $G_{\mathbf{A}}$. (See the discussion at the end of section 5.)

2. The existence of the splitting field can be reduced to the case where $\rho_\lambda \cong \mathrm{Ind}_{\mathrm{Gal}(\overline{F}/K)}^{\mathrm{Gal}(\overline{F}/F)} \tau_\lambda$ with K normal over F (cf. §6). For every $\sigma \in \mathrm{Gal}(K/F)$, a symmetry operator $P(\sigma)$ acts on $H^0(\mathbf{C})$ as an automorphism. For this assertion we need the Hodge conjecture. A main result of this paper (Theorem 5.6) asserts that a splitting field exists under the hypothesis (Hypothesis 8.2) that $P(\sigma)$, $\sigma \in \mathrm{Gal}(K/F)$ stabilize a splitting datum for H^0. We note that the existence of a splitting field of the factor set defined by (2) is proved under the Hodge conjecture.

3. The hypothesis explained above may not be completely convincing. In section 9, we show that assuming the local splitting of f_Z, we can still predict an automorphic representation on $G_{\mathbf{A}}$ corresponding to M. We show the local splitting at unramified places v of F assuming the semisimplicity of $\rho_\lambda(\Phi_v)$, where Φ_v is a Frobenius element.

Now let us explain the organization of this paper. In section 1, we will discuss the Hodge group of an E-rational Hodge structure. Since the coefficient field E is not taken into account in the available literature, we will prove the relevant results in appendix I. In section 2, we will review a result of Deligne [D1] on λ-adic representations of local Galois groups. In section 3, we will review L-groups and L-packets briefly. In section 4, we will explain the basic conjecture on ρ_λ and the local parameter ψ_v. In section 5, we will formulate the main conjecture assuming the existence of a splitting field. The existence of a splitting field in the minimal case will be proved under a hypothesis in section 6 to section 8. Section 6 is preliminary. In section 7, we will prove the existence of a splitting field when H^0 is a torus. We will treat the general case in section 8. In section 9, we will discuss the local splitting of f_Z. In section 10, we will briefly discuss the use of a z-extension to our problem, which was suggested by the referee. Appendix II contains three results related to the descent of the field of definition. First we will show the existence of

[1] We think that though the minimal case is naturally interesting, the cases where the $\widetilde{H}$'s are connected are more basic for practical applications and as the targets for the attempts to give proofs.

a quasi-split G as above. The second result is the following. Let A be an abelian variety defined over F with sufficiently many complex multiplications. I showed ([Y2]) that there exists a finite Galois extension K of F and a representation ρ of the Weil group $W_{F,K}$ such that the zeta function of A is equal to $L(s,\rho,W_{F,K})$. Then I asked whether there exists an A for which ρ is not equivalent to a direct sum of monomial representations. We will give an affirmative answer. The third result is an example for which (2) does not split.

The author would like to thank Professor D. Blasius for interesting discussions concerning the topics dealt with in this paper. He thanks Dr. K. Hiraga for useful discussions on L-packets and on A-packets.

It is my pleasure to dedicate this paper to my respected friend Freydoon Shahidi.

Notation. For a group scheme G over a commutative ring R and an R-algebra S, $G(S)$ denotes the group of S-valued points of G. In particular, if G is an algebraic group defined over a field F and K is a field containing F, $G(K)$ denotes the group of K-rational points of G. The identity component of G is denoted by G^0 and the center of G is denoted by $Z(G)$. For a group G and $g \in G$, $i(g)$ denotes the inner automorphism $i(g)(x) = gxg^{-1}$ of G. By $\mathrm{Aut}(G)$ and $\mathrm{Inn}(G)$, we denote the automorphism group and the inner automorphism group of G respectively. When G is an algebraic group, $\mathrm{Aut}(G)$ and $\mathrm{Inn}(G)$ denote the corresponding groups in the category of algebraic groups. For a field F and its separable extension K of finite degree, $\mathrm{R}_{K/F}$ denotes the restriction of scalars functor of Weil. For an algebraic number field F of finite degree, $F_{\mathbf{A}}$ and $F_{\mathbf{A}}^{\times}$ denote the adele ring and the idele group of F respectively. For $a \in F_{\mathbf{A}}^{\times}$, a_∞ denotes the infinite part of a. We denote by F_{ab} the maximal abelian extension of F in an algebraic closure of F. Let K be a Galois extension of F of finite degree. The relative Weil group for K/F is denoted by $W_{K/F}$ or by $W_{F,K}$.

1. The Hodge group

Let V be a finite dimensional vector space over $\mathbf{Q}$. Let w be an integer. A decomposition

$$V \otimes_{\mathbf{Q}} \mathbf{C} = \oplus_{p+q=w} V^{p,q}, \qquad \overline{V^{p,q}} = V^{q,p}, \tag{1.1}$$

$V^{p,q}$ being a $\mathbf{C}$-subspace of $V \otimes_{\mathbf{Q}} \mathbf{C}$, is called a $\mathbf{Q}$-rational Hodge structure of weight w on V. Let $S = \mathrm{R}_{\mathbf{C}/\mathbf{R}}(\mathbf{G}_m)$; S is a torus defined over $\mathbf{R}$ and we have $S(K) = (\mathbf{C} \otimes_{\mathbf{R}} K)^{\times}$ for an $\mathbf{R}$-algebra K. There exists a morphism $h : S \longrightarrow \mathrm{GL}(V)$ of algebraic groups defined over $\mathbf{R}$ such that

$$h(z)v^{p,q} = z^{-p}\bar{z}^{-q}v^{p,q}, \qquad v^{p,q} \in V^{p,q}, \qquad z \in \mathbf{C}^{\times} = S(\mathbf{R}). \tag{1.2}$$

Definition 1.1. The Hodge group $\mathrm{Hg}(V)$ is the smallest algebraic subgroup defined over $\mathbf{Q}$ of $\mathrm{GL}(V)$ which contains the image of h. [2]

Proposition 1.2. ([D3], Proposition 3.6.) *$\mathrm{Hg}(V)$ is connected. If the Hodge structure is polarizable, then $\mathrm{Hg}(V)$ is reductive.*

[2]Strictly speaking, $\mathrm{Hg}(V,h)$ is the proper notation. We use $\mathrm{Hg}(V)$ for simplicity. The condition means that $(*)$ $\mathrm{Hg}(V)(\mathbf{C}) \supset h(S(\mathbf{C}))$. We can show easily that if $\mathrm{Hg}(V)(\mathbf{C}) \supset h(S(\mathbf{R}))$, then $(*)$ holds.

Let E be an algebraic number field of finite degree. We will generalize the notion of the Hodge group for an E-rational Hodge structure. Let V be a finite dimensional vector space over E. When we regard V as a vector space over $\mathbf{Q}$, we denote it by $\underline{V}$. A decomposition

$$(1.3) \qquad V \otimes_{\mathbf{Q}} \mathbf{C} = \oplus_{p+q=w} V^{p,q}, \qquad \overline{V^{p,q}} = V^{q,p},$$

$V^{p,q}$ being an $E\otimes_{\mathbf{Q}}\mathbf{C}$-submodule of $V\otimes_{\mathbf{Q}}\mathbf{C}$, is called an E-rational Hodge structure of weight w.

Define $h : S \longrightarrow \mathrm{GL}(\underline{V})$ using the underlying $\mathbf{Q}$-rational Hodge structure $\underline{V}$. Since h is E-linear, we have

$$\mathrm{Im}(h) \subset \mathrm{GL}(V)(E \otimes_{\mathbf{Q}} \mathbf{C}) \subset \mathrm{GL}(\underline{V})(\mathbf{C}).$$

DEFINITION 1.3. The Hodge group $\mathrm{Hg}(V)$ is the smallest algebraic subgroup defined over E of $\mathrm{GL}(V)$ such that the group of $(E \otimes_{\mathbf{Q}} \mathbf{C})$-valued points contains $\mathrm{Im}(h)$.

Let $\mathrm{R}_{E/\mathbf{Q}}$ be the restriction of scalars functor of Weil. Let

$$H \subset \mathrm{R}_{E/\mathbf{Q}}(\mathrm{GL}(V)) \subset \mathrm{GL}(\underline{V})$$

be an algebraic subgroup defined over $\mathbf{Q}$ of $\mathrm{R}_{E/\mathbf{Q}}(\mathrm{GL}(V))$. We call the smallest algebraic subgroup G defined over E of $\mathrm{GL}(V)$ such that $\mathrm{R}_{E/\mathbf{Q}}(G) \supset H$ the *E-envelope* of H. We see easily that if H is connected, then its E-envelope is connected.

PROPOSITION 1.4. $\mathrm{Hg}(V)$ *is the E-envelope of* $\mathrm{Hg}(\underline{V})$.

PROOF. We have

$$\mathrm{Im}(h) = h(S(\mathbf{C})) \subset \mathrm{Hg}(\underline{V})(\mathbf{C}) \subset \mathrm{R}_{E/\mathbf{Q}}(\mathrm{GL}(V))(\mathbf{C}).$$

Since

$$\mathrm{R}_{E/\mathbf{Q}}(G)(\mathbf{C}) = G(E \otimes_{\mathbf{Q}} \mathbf{C}),$$

for an algebraic group defined over E, the assertion follows. □

PROPOSITION 1.5. $\mathrm{Hg}(V)$ *is connected. If V is polarizable, then it is reductive.*

The proof will be given in Appendix I. In fact, we will prove that if H is reductive, then its E-envelope is reductive.

REMARK 1.6. We have

$$h(S(\mathbf{C})) \subset \mathrm{GL}(V)(E \otimes_{\mathbf{Q}} \mathbf{C}) = \prod_{\sigma \in J_E} (\sigma\mathrm{GL}(V))(\mathbf{C}).$$

Here J_E denotes the set of all isomorphisms of E into $\mathbf{C}$; $\sigma\mathrm{GL}(V)$ denotes the conjugate of $\mathrm{GL}(V)$ by σ, which is an algebraic group defined over $\sigma(E)$. Regard E as a subfield of $\mathbf{C}$ and take the embedding id : $E \hookrightarrow \mathbf{C}$. Then we see that $\mathrm{Hg}(V)$ is the closure of the projection of $h(S(\mathbf{R}))$ to the id-component with respect to the E-Zariski topology.

We are going to consider the Mumford-Tate group for an E-rational Hodge structure V. Let $E(1)$ be the E-rational Hodge structure of weight -2 of rank 1. Define $h' : S \longrightarrow \mathrm{GL}(\underline{E(1)})$ as above and put $\widetilde{h} = (h, h') : S \longrightarrow \mathrm{GL}(\underline{V}) \times \mathrm{GL}(\underline{E(1)})$. We have

$$\widetilde{h}(S(\mathbf{C})) \subset \mathrm{GL}(E \otimes_{\mathbf{Q}} \mathbf{C}) \times \mathbf{G}_m(E \otimes_{\mathbf{Q}} \mathbf{C}).$$

Here we consider $\mathbf{G}_m$ as an algebraic group defined over E.

DEFINITION 1.7. The Mumford-Tate group $\mathrm{MT}(V)$ is the smallest algebraic subgroup defined over E of $\mathrm{GL}(V) \times \mathbf{G}_m$ such that the group of $(E \otimes_{\mathbf{Q}} \mathbf{C})$-valued points contains $\mathrm{Im}(\widetilde{h})$.

By this definition, it is clear that $\mathrm{Hg}(V)$ is the projection of $\mathrm{MT}(V)$ to the first component $\mathrm{GL}(V)$. Now we consider a tensor space

$$T = V^{\otimes l} \otimes_E \check{V}^{\otimes m} \otimes_E E(p)$$

for $0 \le l, m \in \mathbf{Z}$, $p \in \mathbf{Z}$. Here $\check{V}$ denotes the dual Hodge structure to V. We have an E-rational Hodge structure on T. We define the action of $\mathrm{GL}(V) \times \mathbf{G}_m$ on T as follows: $\mathrm{GL}(V)$ acts on $V^{\otimes l} \otimes_E \check{V}^{\otimes m}$ canonically and on $E(p)$ trivially; $\mathbf{G}_m$ acts on $V^{\otimes l} \otimes_E \check{V}^{\otimes m}$ trivially and on $E(p)$ by $u \cdot v = u^{-p}v$, $u \in \mathbf{G}_m$, $v \in E(p)$.

PROPOSITION 1.8. *Let G be the subgroup of $\mathrm{GL}(V) \times \mathbf{G}_m$ which fixes all tensors of type $(0,0)$ for every tensor space T as above. We assume that V is polarizable. Then we have $\mathrm{MT}(V) = G$.*

PROOF. It is clear that: $t \in T$ is of type $(0,0) \iff t$ is fixed by $\widetilde{h}(S(\mathbf{R}))$. Since $\mathrm{MT}(V)(E \otimes_{\mathbf{Q}} \mathbf{C}) \supset \widetilde{h}(S(\mathbf{R}))$, if t is fixed by $\mathrm{MT}(V)$, then it is of type $(0,0)$. The converse assertion holds since the similar fact to Remark 1.6 holds for $\mathrm{MT}(V)$.

Now G is the group which fixes all tensors fixed by $\mathrm{MT}(V)$. Since $\mathrm{MT}(V)$ is the E-envelope of $\mathrm{MT}(\underline{V})$, it is reductive (cf. Appendix I). Applying the criterion [D3], Proposition 3.1, (c), we conclude that $G = \mathrm{MT}(V)$. $\square$

EXAMPLE 1.9. Let A be an n-dimensional abelian variety defined over $\mathbf{C}$. Then $H^1(A, \mathbf{Q})$ defines a $\mathbf{Q}$-rational Hodge structure of weight 1 and $H_1(A, \mathbf{Q})$ defines a $\mathbf{Q}$-rational Hodge structure of weight -1. They are dual each other and the Hodge groups attached to them are isomorphic. We put $\mathrm{Hg}(A) = \mathrm{Hg}(H_1(A, \mathbf{Q}))$. If ψ is a polarization of A, then we have $\mathrm{Hg}(A) \subset GSp(V, \psi)$.

Assume that $\mathrm{End}(A) = \mathbf{Z}$. If $n \le 2$ or n is odd, we have $\mathrm{Hg}(A) = GSp(V, \psi)$. When $n = 4$, there is an example due to Mumford [Mu2] such that $\mathrm{Hg}(A) \subsetneqq GSp(V, \psi)$.

Assume that A has sufficiently many complex multiplications and is of type (K, Φ), where K is a CM-field of degree $2n$ and Φ is a CM-type of K (cf. Shimura [S2], §17, §18). Let (K', Φ') be the reflex of (K, Φ). There is a morphism between algebraic tori: $\det \Phi' : \mathrm{R}_{K'/\mathbf{Q}}(\mathbf{G}_m) \longrightarrow \mathrm{R}_{K/\mathbf{Q}}(\mathbf{G}_m)$. Then

$$\mathrm{Hg}(A) \cong \mathrm{Image}(\det \Phi') \subset \mathrm{R}_{K/\mathbf{Q}}(\mathbf{G}_m).$$

If K is an imaginary quadratic field, we have $\mathrm{Hg}(A) \cong \mathrm{R}_{K/\mathbf{Q}}(\mathbf{G}_m)$. If we consider $H_1(A, \mathbf{Q})$ as a K-rational Hodge structure, its Hodge group is isomorphic to $\mathbf{G}_m$.

EXAMPLE 1.10. Let $E_1, \ldots, E_n$ be elliptic curves defined over a number field F without complex multiplication. We assume that E_i and E_j are not isogenous over $\overline{\mathbf{Q}}$ for $i \ne j$. Then it is known that (Serre [Se2], p. 183)

$$\mathrm{Hg}(E_1 \times \cdots \times E_n) \cong \{(g_1, \ldots, g_n) \in \mathrm{GL}(2)^n \mid \det g_1 = \cdots = \det g_n\}.$$

2. λ-adic representations

A homomorphism of the Weil-Deligne group naturally corresponds to a λ-adic representation of the local Galois group. In this section, we will review briefly this correspondence. For details, we refer the reader to [D1].

Let F be a non-archimedean local field of characteristic 0. Let p be the residual characteristic of F and q be the cardinality of the residue field. Let $I \subset \mathrm{Gal}(\overline{F}/F)$ be the inertia group. We take a geometric Frobenius element $\Phi \in \mathrm{Gal}(\overline{F}/F)$. The local Weil group W_F is

$$W_F = I \rtimes \langle \Phi \rangle \subset \mathrm{Gal}(\overline{F}/F).$$

The topology of W_F is defined by taking a fundamental system of neighbourhoods of the identity of I as that of W_F. For $g \in W_F$, we write $g = \Phi^n i(g)$ with $n \in \mathbf{Z}$, $i(g) \in I$ and put $\|g\| = q^{-n}$. Let

$$W_F' = \mathbf{G}_a \rtimes W_F$$

be the Weil–Deligne group scheme. For a commutative ring R in which p is invertible, the group of R-valued points is

$$W_F'(R) = R \rtimes W_F$$

and the group law is given by

$$(x_1, g_1)(x_2, g_2) = (x_1 + \|g_1\| x_2, g_1 g_2), \qquad x_1, x_2 \in R,\ g_1, g_2 \in W_F.$$

Let $P \subset I$ be the wild ramification group. We have $I/P \cong \prod_{l \neq p} \mathbf{Z}_l$. Let $\mathbf{t}_l : I \to \mathbf{Z}_l$ be the canonical projection.

Now let $\ell \neq p$ be a prime number and let E_λ be a field such that $[E_\lambda : \mathbf{Q}_\ell] < \infty$. Let V be a finite dimensional vector space over E_λ. Let

$$\rho : W_F \longrightarrow \mathrm{GL}(V)$$

be a continuous representation. Here the topology of $\mathrm{GL}(V)$ is the λ-adic topology. A theorem of Grothendieck states that there exists an open subgroup I_0 of I and a nilpotent element $N \in \mathrm{End}(V)$ such that

$$\rho(h) = \exp(\mathbf{t}_l(h) N), \qquad h \in I_0.$$

Since $\mathbf{t}_\ell(ghg^{-1}) = \|g\| \mathbf{t}_\ell(h)$, $g \in W_F$, $h \in I$, we have

$$\rho(g) N \rho(g)^{-1} = \|g\| N, \qquad g \in W_F. \tag{2.1}$$

Put

$$\rho'(g) = \rho(g) \exp(-\mathbf{t}_l(i(g)) N), \qquad g \in W_F.$$

Then we can show that ρ' is a homomorphism of W_F into $\mathrm{GL}(V)$. Since ρ' is trivial on I_0, ρ' is continuous with respect to the discrete topology of $\mathrm{GL}(V)$. Then we can define a representation σ' of $W_F'(E_\lambda)$ into $\mathrm{GL}(V)$ by

$$\sigma'((x, g)) = \exp(xN) \rho(g) \exp(-\mathbf{t}_l(i(g)) N), \qquad (x, g) \in W_F'(E_\lambda).$$

Let

$$\rho(\Phi) = \rho(\Phi)^{ss} u$$

be the Jordan decomposition of $\rho(\Phi)$, with $\rho(\Phi)^{ss}$ semisimple, u unipotent. Then u commutes with N and with all $\rho(g)$, $g \in W_F$. Put

$$\rho'^{ss}(\Phi^n h) = \rho'(\Phi^n h) u^{-n}, \qquad n \in \mathbf{Z}, \quad h \in I.$$

We see that ρ'^{ss} is a representation of W_F into $\mathrm{GL}(V)$. We define a representation σ'^{ss} of $W_F'(E_\lambda)$ into $\mathrm{GL}(V)$ by

$$\sigma'^{ss}((x, \Phi^n h)) = \exp(xN) \rho'(\Phi^n h) u^{-n}, \qquad (x, \Phi^n h) \in W_F'(E_\lambda),\ n \in \mathbf{Z},\ h \in I.$$

We call σ'^{ss} the Φ-semisimplification of σ'. Its equivalence class does not depend on the choice of Φ. An element $(x, \Phi^n h) \in W_F'(E_\lambda)$ is called semisimple if $n \neq 0$ or if $n = 0$, $x = 0$. If $(x, \Phi^n h)$ is semisimple, we see that $\sigma'^{ss}((x, \Phi^n h))$ is semisimple.

Thus we have the procedure

$$\rho \longrightarrow \rho' \longrightarrow \rho'^{ss} \longrightarrow \sigma'^{ss}$$

to construct a Φ-semisimple representation σ'^{ss} of $W_F'(E_\lambda)$ from a given λ-adic representation $\rho : W_F \to \mathrm{GL}(V)$. We note the following. Let H be a (possibly non-connected) algebraic group defined over E_λ. Let $\rho : W_F \to H(E_\lambda)$ be a continuous homomorphism. The same procedure yields the continuous homomorphism $\sigma'^{ss} : W_F'(E_\lambda) \to H(E_\lambda)$ with the discrete topology on $H(E_\lambda)$, since the Jordan decomposition can be done inside $H(E_\lambda)$. Let $\iota : E_\lambda \hookrightarrow \mathbf{C}$ be an isomorphism. Then we have a representation $\phi : W_F'(\mathbf{C}) \to H(\mathbf{C})$. This is the local parameter attached to the λ-adic representation ρ, which will be used in §5.

3. L-groups and L-packets

In this section, we will review briefly the notion of L-groups and L-packets. We refer the reader to Borel [Bo2]. See Arthur [A3], Vogan [V] for advanced treatments of L-packets.

Let F be a field of characteristic 0. Let G be a connected reductive algebraic group defined over F. Consider G as a group defined over $\overline{F}$ and take a maximal torus T and a Borel subgroup B which contains T. Then we obtain a based root datum:

$$\mathcal{R}_0(G) = (X^*(T), \Delta, X_*(T), \check{\Delta}).$$

Here $X^*(T)$ is the character group of T and Δ is the set of simple roots; $X_*(T)$ is the group of cocharacters of T and $\check{\Delta}$ is the set of simple coroots. For $\alpha \in \Delta$, take an element $u_\alpha \neq 1$ from the root subgroup U_α corresponding to α. We call $(B, T, \{u_\alpha\}_{\alpha \in \Delta})$ a *splitting datum* for G. Then we have

$$\mathrm{Aut}(\mathcal{R}_0(G)) \cong \mathrm{Aut}(G, B, T, \{u_\alpha\}_{\alpha \in \Delta}).$$

Here the right-hand side denotes the group of automorphisms of G which leave B, T and the set of u_α, $\alpha \in \Delta$ stable; by the action of $\mathrm{Gal}(\overline{F}/F)$ on B and T, we obtain a homomorphism

$$\mu_G : \mathrm{Gal}(\overline{F}/F) \longrightarrow \mathrm{Aut}(\mathcal{R}_0(G)).$$

Take a connected reductive algebraic group ${}^L G^0$ over $\mathbf{C}$ such that

$$\mathcal{R}_0({}^L G^0) = (X_*(T), \check{\Delta}, X^*(T), \Delta)$$

and form a semi-direct product:

$${}^L G = {}^L G^0 \rtimes \mathrm{Gal}(\overline{F}/F).$$

Here, choosing a splitting datum $({}^L B^0, {}^L T^0, \{\check{u}_\alpha\}_{\alpha \in \check{\Delta}})$ for ${}^L G^0$, we have

$$\mathrm{Aut}({}^L G^0, {}^L B^0, {}^L T^0, \{\check{u}_\alpha\}_{\alpha \in \check{\Delta}}) \cong \mathrm{Aut}(\mathcal{R}_0({}^L G^0)) \cong \mathrm{Aut}(\mathcal{R}_0(G)),$$

and $\mathrm{Gal}(\overline{F}/F)$ acts on ${}^L G^0$ through μ_G. [3]

[3] In [Bo2], the L-group defined by specifying the choice of $\{\check{u}_\alpha\}_{\alpha \in \Delta}$ is called admissible. In general, the L-group is defined without specifying the choice of $\{\check{u}_\alpha\}_{\alpha \in \Delta}$. Then μ_G is well defined up to an inner automorphism by an element of ${}^L T^0$.

Now assume that F is a local field. If F is non-archimedean, put $W'_F = W'_F(\mathbf{C})$. If F is archimedean, put $W'_F = W_F$ and regard every element as semisimple.

DEFINITION 3.1. A homomorphism $\phi : W'_F \longrightarrow {}^LG$ is called a Langlands parameter if the diagram

$$\begin{array}{ccc} W'_F & \xrightarrow{\phi} & {}^LG \\ \downarrow & & \downarrow \\ \mathrm{Gal}(\overline{F}/F) & = & \mathrm{Gal}(\overline{F}/F) \end{array}$$

is commutative and the following conditions hold:

(i) ϕ is continuous, $\mathbf{G}_a$ is mapped to unipotent elements of ${}^LG^0$, and ϕ sends semisimple elements to semisimple elements.

(ii) If the image of ϕ is contained in a parabolic subgroup P of ${}^LG$, P is relevant.

If G is quasi-split, then the condition (ii) is automatically satisfied. (For the definition of relevance, see [Bo1], p. 32.) We say that two Langlands parameters are equivalent if one is tranformed to the other by an inner automorphism of ${}^LG^0$. Let $\Phi(G/F) = \Phi(G)$ be the set of equivalence classes of the Langlands parameters. Let $\Pi(G(F))$ be the set of equivalence classes of irreducible admissible representations of $G(F)$.

CONJECTURE 3.2. (Langlands) *For every $\phi \in \Phi(G)$, there exists a finite subset $\Pi_\phi = \Pi_\phi(G(F))$ of $\Pi(G(F))$ which partitions $\Pi(G(F))$:*

$$\Pi(G(F)) = \sqcup_{\phi\in\Phi(G)}\Pi_\phi.$$

We call Π_ϕ the L-packet associated to ϕ. The representations in Π_ϕ are expected to have the same property with respect to the L-function and the ϵ-factor (L-indistinguishable). Moreover Π_ϕ must satisfy a number of properties which are included in the conjecture. We list some of them.

(A) Π_ϕ contains a discrete series representation $\Longleftrightarrow$ Π_ϕ consists of discrete series representations $\Longleftrightarrow$ $\phi(W'_F)$ is not contained in any proper Levi subgroup of ${}^LG$.

(B) Assume that $\phi(\mathbf{G}_a) = \{1\}$. Then Π_ϕ contains a tempered representation $\Longleftrightarrow$ Π_ϕ consists of tempered representations $\Longleftrightarrow$ $\phi(W_F)$ is bounded.

We also note that there is a law which gives the central character of representations in Π_ϕ in terms of ϕ.

When $F = \mathbf{C}, \mathbf{R}$, Langlands [Lan3] proved the conjecture. When $G = \mathrm{GL}(n)$ and F is non-archimedean, the result of Harris-Taylor-Henniart establishes Conjecture 3.2. For $G = \mathrm{GL}(n)$, Π_ϕ is a singleton. In general, the structure of Π_ϕ can be very complicated. See [A3]. The existence of Π_ϕ remains conjectural in general.

4. Local parameters attached to a motive

Let E and F be number fields. Let M be a motive over F with coefficients in E. Let d be the rank of M. We assume that M is of pure weight w and is polarizable. Take an isomorphism $F \hookrightarrow \mathbf{C}$. On the Betti realization $H_B(M) \cong E^d$, we have an E-rational Hodge structure of weight w. Put $V = H_B(M)$, $\mathrm{Hg}(M) = \mathrm{Hg}(V)$. By Proposition 1.5, $\mathrm{Hg}(M)$ is a connected reductive algebraic group defined over E.

For a finite place λ of E, we have the λ-adic realization $H_\lambda(M)$ and the λ-adic representation

$$\rho_\lambda : \mathrm{Gal}(\overline{F}/F) \longrightarrow \mathrm{GL}(H_\lambda(M)) \cong \mathrm{GL}(d, E_\lambda).$$

Let ℓ be a prime number. When we forget the coefficient field E, we have the ℓ-adic realization $H_\ell(M)$ and the ℓ-adic representation

$$\sigma_\ell : \mathrm{Gal}(\overline{F}/F) \longrightarrow \mathrm{GL}(H_\ell(M)) \cong \mathrm{GL}(gd, \mathbf{Q}_\ell),$$

where $g = [E : \mathbf{Q}]$. There is the relation

$$H_\ell(M) = \oplus_{\lambda|\ell}\, H_\lambda(M), \qquad \sigma_\ell \cong \oplus_{\lambda|\ell}\, \rho_\lambda. \tag{4.1}$$

By the comparison isomorphism, we have

$$\underline{V} \otimes_{\mathbf{Q}} \mathbf{Q}_\ell \cong H_\ell(M), \qquad V \otimes_E E_\lambda \cong H_\lambda(M).$$

Conjecture 4.1. *There exists an algebraic group H defined over E, which does not depend on λ, such that $H^0 = \mathrm{Hg}(M)$ and such that $\mathrm{Im}(\rho_\lambda)$ is Zariski dense in H. If $E = \mathbf{Q}$, then $\mathrm{Im}(\rho_\lambda)$ is open in $H(E_\lambda)$.*

Remark 4.2. Conjecture 4.1 tacitly assumes that $\mathrm{Hg}(M)$ does not depend on the choice of an embedding $F \hookrightarrow \mathbf{C}$. As far as the author knows, this fact is established for abelian varieties (by the theory of absolute Hodge cycles, [D3]) but not in general. We also note that semisimplicity of ρ_λ follows from Conjecture 4.1.

Remark 4.3. When $E = \mathbf{Q}$, a weaker form of Conjecture 4.1 is given in [Mu1] together with the definition of the special Mumford–Tate group. In [Se1], Conjecture 4.1 is stated also assuming $E = \mathbf{Q}$. The author could not find a reference for general E.

Let us clarify the relation of the conjecture for the case $E = \mathbf{Q}$ and for the case of general E. Let $\mathfrak{H}$ be the Zariski closure of $\mathrm{Im}(\sigma_l)$ and let $\widetilde{\mathfrak{H}}$ be the E-envelope of $\mathfrak{H}$. Then we have

$$\mathrm{Im}(\sigma_l) = \mathrm{Im}(\oplus_{\lambda|l}\, \rho_\lambda) \subset \mathfrak{H}(\mathbf{Q}_l) \subset \mathrm{R}_{E/\mathbf{Q}}(\widetilde{\mathfrak{H}})(\mathbf{Q}_l) = \prod_{\lambda|l} \widetilde{\mathfrak{H}}(E_\lambda).$$

Hence we obtain

$$\mathrm{Im}(\rho_\lambda) \subset \widetilde{\mathfrak{H}}(E_\lambda). \tag{4.2}$$

On the other hand, let H be the closure of $\mathrm{Im}(\rho_\lambda)$ ($\subset \mathrm{GL}(V)(E_\lambda)$) with respect to the E-Zariski topology. In other words, H is the smallest algebraic subgroup defined over E of $\mathrm{GL}(V)$ such that $H(E_\lambda) \supset \mathrm{Im}(\rho_\lambda)$. By (4.2), we have $H \subset \widetilde{\mathfrak{H}}$. Assume that H is independent of λ. Then we have

$$\mathrm{Im}(\sigma_l) = \mathrm{Im}(\oplus_{\lambda|l}\, \rho_\lambda) \subset \prod_{\lambda|\ell} H(E_\lambda) = \mathrm{R}_{E/\mathbf{Q}}(H)(\mathbf{Q}_\ell).$$

Hence we have $\mathfrak{H} \subset \mathrm{R}_{E/\mathbf{Q}}(H)$. By the minimality of $\widetilde{\mathfrak{H}}$, we have $\widetilde{\mathfrak{H}} \subset H$. Therefore we obtain $\widetilde{\mathfrak{H}} = H$. Thus we have proved the following proposition.

Proposition 4.4. *Assume Conjecture 4.1 for the case $E = \mathbf{Q}$ and let $\mathfrak{H}$ be the Zariski closure of $\mathrm{Im}(\sigma_l)$. Let H be the closure of $\mathrm{Im}(\rho_\lambda)$ with respect to the E-Zariski topology. Assume that H is independent of λ. Then H is equal to the E-envelope of $\mathfrak{H}$.*

In Proposition 4.4, we note that H^0 is the E-envelope of $\mathfrak{H}^0$. In view of Proposition 1.4, Proposition 4.4 states that Conjecture 4.1 for general E follows from the case $E = \mathbf{Q}$ under the independence assumption of H.

Assuming Conjecture 4.1, we are going to construct a homomorphism

$$\psi_v : W'_{F_v} \longrightarrow H(\mathbf{C})$$

for every place v of F.

Let v be non-archimedean. Take a finite place λ of E which is prime to v. By restriction, we have a λ-adic representation

$$\rho_{\lambda,v} : \mathrm{Gal}(\overline{F_v}/F_v) \longrightarrow H(E_\lambda).$$

Take an embedding $E_\lambda \hookrightarrow \mathbf{C}$. By the procedure of §2, we obtain a representation

$$\psi_v = \sigma'^{ss}_{\lambda,v} : W'_{F_v}(\mathbf{C}) \longrightarrow H(\mathbf{C}),$$

which sends semisimple elements to semisimple elements.

REMARK 4.5. The equivalence class of ψ_v may depend on the choice of λ. If M is attached to the cohomology of a projective smooth algebraic variety X defined over F, the independence is known when either X is an abelian variety or X has good reduction at v.

Let v be archimedean. Take an embedding $E \hookrightarrow \mathbf{C}$. We have a homomorphism

$$h : S(\mathbf{R}) = \mathbf{C}^\times \longrightarrow H(\mathbf{C})$$

attached to the Hodge structure on $H_B(M)$.

If v is imaginary, then $W_\mathbf{C} = \mathbf{C}^\times$. We take $\psi_v = h$.

Suppose that v is real. We have the non-split exact sequence

$$1 \longrightarrow \mathbf{C}^\times \longrightarrow W_\mathbf{R} \longrightarrow \mathrm{Gal}(\mathbf{C}/\mathbf{R}) \longrightarrow 1.$$

We can realize $W_\mathbf{R}$ inside $\mathbf{H}^\times$, the group of nonzero Hamilton quaternions, as

$$W_\mathbf{R} = \langle \mathbf{C}^\times, j \rangle, \qquad j^2 = -1, \quad jzj^{-1} = \bar{z}, \ z \in \mathbf{C}^\times.$$

Put $\tau' = \rho_\lambda(c) \in H(E_\lambda) \subset H(\mathbf{C})$, where $c \in \mathrm{Gal}(\overline{F}_v/F_v) \cong \mathrm{Gal}(\mathbf{C}/\mathbf{R})$ is the complex conjugation. Then we can show that

$$\tau' h(z) = h(\bar{z})\tau', \qquad z \in S(\mathbf{R}) = \mathbf{C}^\times.$$

If $w = 0$, we put $\tau = \tau'$. If $w \neq 0$, then the scalars $\mathbf{C}^\times$ is contained in the center Z of $H(\mathbf{C})$. Taking $t \in Z$ such that $t^2 = h(-1) = (-1)^w$, we put $\tau = \tau' t$. Now τ satisfies

$$\tau^2 = h(-1), \qquad \tau h(z) = h(\bar{z})\tau, \qquad z \in S(\mathbf{R}).$$

Therefore we can define a representation ψ_v of $W_\mathbf{R}$ into $H(\mathbf{C})$ by

$$\psi_v|\mathbf{C}^\times = h, \qquad \psi_v(j) = \tau.$$

5. Main Conjecture

As in the previous section, we consider the λ-adic representation

$$\rho_\lambda : \mathrm{Gal}(\overline{F}/F) \longrightarrow \mathrm{GL}(H_\lambda(M))$$

attached to a motive M. We assume Conjecture 4.1. Put $V = H_B(M)$. We fix an embedding $E_\lambda \hookrightarrow \mathbf{C}$. There exists the unique finite Galois extension K of F such that the diagram

(R)

$$\begin{array}{ccccccccc} 1 & \longrightarrow & \rho_\lambda(\mathrm{Gal}(\overline{F}/K)) & \longrightarrow & \rho_\lambda(\mathrm{Gal}(\overline{F}/F)) & \longrightarrow & \mathrm{Gal}(K/F) & \longrightarrow & 1 \\ & & \downarrow & & \downarrow & & \| & & \\ 1 & \longrightarrow & H^0(\mathbf{C}) & \longrightarrow & H(\mathbf{C}) & \longrightarrow & \mathrm{Gal}(K/F) & \longrightarrow & 1 \end{array}$$

is commutative. Here $H^0 = \mathrm{Hg}(M)$, the vertical arrows are inclusion maps and both rows are exact. In fact, K is uniquely determined by the condition

$$(5.1) \qquad H^0(\mathbf{C}) \cap \rho_\lambda(\mathrm{Gal}(\overline{F}/F)) = \rho_\lambda(\mathrm{Gal}(\overline{F}/K)), \qquad \mathrm{Gal}(\overline{F}/K) \supset \mathrm{Ker}(\rho_\lambda).$$

More generally let $\widetilde{H}$ be a reductive algebraic group defined over E such that $H \subset \widetilde{H} \subset \mathrm{GL}(V)$. We have $H^0 \subset \widetilde{H}^0$. Let $\widetilde{K}$ be the finite Galois extension of F determined by

$$(5.1') \qquad \widetilde{H}^0(\mathbf{C}) \cap \rho_\lambda(\mathrm{Gal}(\overline{F}/F)) = \rho_\lambda(\mathrm{Gal}(\overline{F}/\widetilde{K})), \qquad \mathrm{Gal}(\overline{F}/\widetilde{K}) \supset \mathrm{Ker}(\rho_\lambda).$$

We have $\widetilde{K} \subset K$. Similarly to (R), we have a commutative diagram

(R$'$)

$$\begin{array}{ccccccccc} 1 & \longrightarrow & \rho_\lambda(\mathrm{Gal}(\overline{F}/\widetilde{K})) & \longrightarrow & \rho_\lambda(\mathrm{Gal}(\overline{F}/F)) & \longrightarrow & \mathrm{Gal}(\widetilde{K}/F) & \longrightarrow & 1 \\ & & \downarrow & & \downarrow & & \| & & \\ 1 & \longrightarrow & \widetilde{H}^0(\mathbf{C}) & \longrightarrow & \widetilde{H}(\mathbf{C}) & \longrightarrow & \mathrm{Gal}(\widetilde{K}/F) & \longrightarrow & 1 \end{array}$$

We call the situation where $\widetilde{H} = H$ the *minimal case.*

The exact sequence in the second row of (R$'$) defines a homomorphism

$$\mathrm{Gal}(\widetilde{K}/F) \longrightarrow \mathrm{Aut}(\widetilde{H}^0)/\mathrm{Inn}(\widetilde{H}^0).$$

We have a split exact sequence

$$1 \longrightarrow \mathrm{Inn}(\widetilde{H}^0) \longrightarrow \mathrm{Aut}(\widetilde{H}^0) \longrightarrow \mathrm{Aut}(\mathcal{R}_0(\widetilde{H}^0)) \longrightarrow 1.$$

Here

$$\mathcal{R}_0(\widetilde{H}^0) = (X^*(T), \Delta, X_*(T), \check{\Delta}).$$

is a based root datum. Thus we have a homomorphism

$$\mu_{\widetilde{H}^0} : \mathrm{Gal}(\widetilde{K}/F) \longrightarrow \mathrm{Aut}(\mathcal{R}_0(\widetilde{H}^0)).$$

PROPOSITION 5.1. *Let M be a connected reductive algebraic group defined over $\mathbf{C}$. Let μ be a homomorphism of* $\mathrm{Gal}(\overline{F}/F)$ *into* $\mathrm{Aut}(\mathcal{R}_0(M))$. *Then there exists a connected reductive algebraic group G defined over F such that*
(i) *G is quasi-split over F.* (ii) ${}^LG^0 = M(\mathbf{C})$. (iii) $\mu_G = \mu$.

A proof will be given in Appendix II. Let $\widetilde{G}$ be the quasi-split group defined over F given by Proposition 5.1 by taking $M = \widetilde{H}^0$, $\mu = \mu_{\widetilde{H}^0}$. We are going to compare ${}^L\widetilde{G}$ with the group extension

$$(5.2)\qquad 1 \longrightarrow \widetilde{H}^0(\mathbf{C}) \longrightarrow \widetilde{H}(\mathbf{C}) \longrightarrow \mathrm{Gal}(\widetilde{K}/F) \longrightarrow 1$$

in the second row of (R$'$). For this purpose, let us recall a few basic facts on factor sets.

Let

$$1 \longrightarrow N \longrightarrow G \xrightarrow{\ \pi\ } F \longrightarrow 1$$

be an exact sequence of groups. For $\sigma \in F$, we choose $\tilde{\sigma} \in G$ so that $\pi(\tilde{\sigma}) = \sigma$. Then, for σ, $\tau \in F$, we have

$$f(\sigma,\tau)\widetilde{\sigma\tau} = \tilde{\sigma}\tilde{\tau}$$

with $f(\sigma,\tau) \in N$. We define $a(\sigma) \in \mathrm{Aut}(N)$ by

$$a(\sigma)n = \tilde{\sigma}n\tilde{\sigma}^{-1}, \qquad n \in N.$$

Then we have

$$(5.3)\qquad i(f(\sigma,\tau))a(\sigma\tau) = a(\sigma)a(\tau),$$

$$(5.4)\qquad f(\sigma,\tau)f(\sigma\tau,\rho) = (a(\sigma)f(\tau,\rho))f(\sigma,\tau\rho).$$

Here $i(f(\sigma,\tau))$ is the inner automorphism of N defined by $f(\sigma,\tau)$. The datum $\{a(\sigma) \in \mathrm{Aut}(N)\}$, $\{f(\sigma,\tau) \in N\}$, or simply $\{a(\sigma), f(\sigma,\tau)\}$, is called a *factor set* of F taking values in N.

We say two factor sets are *equivalent* if there exists $\{\alpha_\sigma \in N\}_{\sigma\in F}$ such that

$$(5.5)\qquad a'(\sigma) = i(\alpha_\sigma)a(\sigma),$$

$$(5.6)\qquad f'(\sigma,\tau) = \alpha_\sigma(a(\sigma)\alpha_\tau)f(\sigma,\tau)\alpha_{\sigma\tau}^{-1}.$$

The transformation $\{a,f\} \longrightarrow \{a',f'\}$ is caused when we change $\widetilde{\sigma}$ to $\alpha_\sigma\widetilde{\sigma}$. When a factor set is given, we can define a group law on $N \times F$ by

$$(5.7)\qquad (n_1,\sigma)(n_2,\tau) = (n_1(a(\sigma)n_2)f(\sigma,\tau),\sigma\tau).$$

This group structure determines factor sets in the equivalence class of the given one.

Let $\varphi : \widetilde{F} \longrightarrow F$ be a group homomorphism. Then $\{a(\varphi(\sigma)),\ f(\varphi(\sigma),\varphi(\tau))\}$ is a factor set of $\widetilde{F}$ taking values in N. We call this process the *inflation* by φ. We say a factor set *splits* if it is equivalent to a factor set with $f(\sigma,\tau) = 1$ for all σ, τ. This is the case if and only if G is isomorphic to the semi-direct product of N with F: $G = N \rtimes F$.

When N is abelian, N is a left F-module by the action a of F on N, and (5.4) is the 2-cocycle condition on f. Two factor sets are equivalent if and only if they are cohomologous.

The following lemma will be used to lift the local parameter ψ_v defined in §4 to the Langlands parameter.

LEMMA 5.2. *Let N and F be groups. Let $\{a(\sigma) \in \mathrm{Aut}(N)\}$, $\{f(\sigma,\tau) \in N\}$ be a factor set of F taking values in N. Let G be the group whose underlying set is $N \times F$ and whose group law is defined by (5.7). Let $\widetilde{F}$ be a group and $\varphi : \widetilde{F} \longrightarrow F$ be a homomorphism. Let $\widetilde{G}$ be the group whose underlying set is $N \times \widetilde{F}$ and whose group law is defined by (5.7) using the inflated factor set. Define a homomorphism*

$p : \widetilde{G} \longrightarrow G$ by $p((n,\sigma)) = (n, \varphi(\sigma))$, and a homomorphism $\pi : G \longrightarrow F$ by $\pi((n,\sigma)) = \sigma$. Define a homomorphism $\widetilde{\pi} : \widetilde{G} \longrightarrow \widetilde{F}$ similarly. Let $\mathfrak{G}$ be a group and $\psi : \mathfrak{G} \longrightarrow G$ be a homomorphism. Put $q = \pi \circ \psi$. If $\widetilde{q} : \mathfrak{G} \longrightarrow \widetilde{F}$ is a homomorphism satisfying $\varphi \circ \widetilde{q} = q$, then there exists a homomorphism $\widetilde{\psi} : \mathfrak{G} \longrightarrow \widetilde{G}$ such that $p \circ \widetilde{\psi} = \psi$. Moreover such a $\widetilde{\psi}$ is unique if we impose the condition $\widetilde{\pi} \circ \widetilde{\psi} = \widetilde{q}$.

PROOF. Write

$$\psi(g) = (n(g), q(g)), \qquad n(g) \in N,\ q(g) \in F, \quad g \in \mathfrak{G}.$$

The condition that ψ is a homomorphism is given by

$$(n(g_1)(a(q(g_1))n(g_2))f(q(g_1), q(g_2)), q(g_1)q(g_2)) = (n(g_1g_2), q(g_1g_2)) \tag{5.8}$$

for g_1, $g_2 \in \mathfrak{G}$. Put

$$\widetilde{\psi}(g) = (n(g), \widetilde{q}(g)), \qquad g \in \mathfrak{G}.$$

Then, using (5.8), we can verify immediately that $\widetilde{\psi}$ is a homomorphism. Since

$$p(\widetilde{\psi}(g)) = (n(g), \varphi(\widetilde{q}(g))) = (n(g), q(g)),$$

we have $p \circ \widetilde{\psi} = \psi$. It is clear that $\widetilde{\pi} \circ \widetilde{\psi} = \widetilde{q}$. The uniqueness assertion can be verified easily. This completes the proof. □

Now let us compare ${}^L\widetilde{G}$ with $\widetilde{H}(\mathbf{C})$. Let $\{\widetilde{a}(\sigma), \widetilde{f}(\sigma,\tau)\}$ be a factor set of $\mathrm{Gal}(\widetilde{K}/F)$ taking values in $\widetilde{H}^0(\mathbf{C})$ defined by the exact sequence (5.2). Explicitly it is given as follows. For $\sigma \in \mathrm{Gal}(\widetilde{K}/F)$, take $\widetilde{\sigma} \in \mathrm{Gal}(\overline{F}/F)$ so that $\widetilde{\sigma}|\widetilde{K} = \sigma$. Then we have

$$\widetilde{a}(\sigma)h = \rho_\lambda(\widetilde{\sigma})h\rho_\lambda(\widetilde{\sigma})^{-1}, \qquad h \in \widetilde{H}^0(\mathbf{C}), \tag{5.9}$$

$$\widetilde{f}(\sigma,\tau) = \rho_\lambda(\widetilde{\sigma}\widetilde{\tau}(\widetilde{\sigma\tau})^{-1}). \tag{5.10}$$

The exact sequence

$$1 \longrightarrow \mathrm{Inn}(\widetilde{H}^0) \longrightarrow \mathrm{Aut}(\widetilde{H}^0) \xrightarrow{\ \pi\ } \mathrm{Out}(\widetilde{H}^0) \longrightarrow 1 \tag{5.11}$$

splits. Let T be a maximal torus and B be a Borel subgroup containing T. The splitting of (5.11) is given by (cf. §3)

$$\mathrm{Out}(\widetilde{H}^0) \cong \mathrm{Aut}(\mathcal{R}_0(\widetilde{H}^0)) \cong \mathrm{Aut}(\widetilde{H}^0, B, T, \{u_\alpha\}_{\alpha\in\Delta}).$$

Let

$$s : \mathrm{Out}(\widetilde{H}^0) \longrightarrow \mathrm{Aut}(\widetilde{H}^0)$$

be a homomorphism such that $\pi \circ s = \mathrm{id}$. Take $\sigma \in \mathrm{Gal}(\widetilde{K}/F)$. Since $\pi(s(\pi(\widetilde{a}(\sigma)))) = \pi(\widetilde{a}(\sigma))$, there exists $\alpha_\sigma \in \widetilde{H}^0(\mathbf{C})$ such that

$$s(\pi(\widetilde{a}(\sigma))) = i(\alpha_\sigma)\widetilde{a}(\sigma). \tag{5.12}$$

Now we consider an equivalent factor set to $\{\widetilde{a}(\sigma), \widetilde{f}(\sigma,\tau)\}$.

$$\widetilde{a}_Z(\sigma) = i(\alpha_\sigma)\widetilde{a}(\sigma), \tag{5.13}$$

$$\widetilde{f}_Z(\sigma,\tau) = \alpha_\sigma(\widetilde{a}(\sigma)\alpha_\tau)\widetilde{f}(\sigma,\tau)\alpha_{\sigma\tau}^{-1}. \tag{5.14}$$

As (5.12) shows, the mapping $\mathrm{Gal}(\widetilde{K}/F) \ni \sigma \longrightarrow \widetilde{a}_Z(\sigma) \in \mathrm{Aut}(\widetilde{H}^0(\mathbf{C}))$ is a homomorphism. By (5.3), we see that $i(\widetilde{f}_Z(\sigma,\tau)) = 1$. This implies that $\widetilde{f}_Z(\sigma,\tau) \in Z(\widetilde{H}^0(\mathbf{C}))$. Thus we obtain a cohomology class of $\widetilde{f}_Z$ in $H^2(\mathrm{Gal}(\widetilde{K}/F), Z(\widetilde{H}^0(\mathbf{C})))$

from (5.2). It does not depend on the choices of $\widetilde{\sigma}$, α_σ and the section s, which we will verify below.

1°. Suppose that we change $\widetilde{\sigma}$ to $u_\sigma\widetilde{\sigma}$ with $u_\sigma \in \mathrm{Gal}(\overline{F}/\widetilde{K})$. Then we see easily that $\{\widetilde{a}_Z(\sigma)|Z(\widetilde{H}^0(\mathbf{C})), \widetilde{f}_Z(\sigma,\tau)\}$ does not change.

2°. Suppose that we change α_σ to $z_\sigma\alpha_\sigma$ with $z_\sigma \in Z(\widetilde{H}^0(\mathbf{C}))$. Then $\widetilde{a}_Z(\sigma)$ does not change and $\widetilde{f}_Z(\sigma,\tau)$ changes to

$$\widetilde{f}_Z(\sigma,\tau)z_\sigma(\widetilde{a}_Z(\sigma)z_\tau)z_{\sigma\tau}^{-1}.$$

Thus the cohomology class of $\widetilde{f}_Z$ in $H^2(\mathrm{Gal}(\widetilde{K}/F), Z(\widetilde{H}^0(\mathbf{C})))$ does not change.

3°. Suppose that we change s to another section s' which comes from another splitting datum:

$$\mathrm{Out}(\widetilde{H}^0) \cong \mathrm{Aut}(\mathcal{R}_0(\widetilde{H}^0)) \cong \mathrm{Aut}(\widetilde{H}^0, B', T', \{u'_\alpha\}_{\alpha\in\Delta}).$$

Since $B' = i(h)B$, $T' = i(h)T$, $u'_\alpha = i(h)u_\alpha$ with $h \in \widetilde{H}^0(\mathbf{C})$, we have $s'(x) = i(h)s(x)i(h)^{-1}$, $x \in \mathrm{Out}(\widetilde{H}^0)$. Put $i(h)i(\alpha_\sigma)\widetilde{a}(\sigma)i(h)^{-1} = i(\alpha'_\sigma)\widetilde{a}(\sigma)$ with $\alpha'_\sigma \in \widetilde{H}^0(\mathbf{C})$. Then we have $\alpha'_\sigma\rho_\lambda(\widetilde{\sigma}) = z_\sigma h\alpha_\sigma\rho_\lambda(\widetilde{\sigma})h^{-1}$ with $z_\sigma \in Z(\widetilde{H}^0(\mathbf{C}))$. An easy calculation shows that $\widetilde{a}_Z(\sigma)$ changes to $i(h)\widetilde{a}_Z(\sigma)i(h)^{-1}$ and $\widetilde{f}_Z$ changes as in 2°. Thus the action of $\widetilde{a}_Z(\sigma)$ on $Z(\widetilde{H}^0(\mathbf{C}))$ and the cohomology class of $\widetilde{f}_Z(\sigma,\tau)$ do not change.

Therefore the cohomology class of $\widetilde{f}_Z$ in $H^2(\mathrm{Gal}(\widetilde{K}/F), Z(\widetilde{H}^0(\mathbf{C})))$ is uniquely determined by (5.2). In the minimal case, it depends only on ρ_λ. We fix a splitting datum for $\widetilde{H}^0$ and a section s. Since we have

$$\mu_{\widetilde{G}}(\sigma) = \mu_{\widetilde{H}^0}(\sigma) = \pi(\widetilde{a}(\sigma)), \qquad \sigma \in \mathrm{Gal}(\widetilde{K}/F),$$

$\mathrm{Gal}(\widetilde{K}/F)$ acts on ${}^L\widetilde{G}^0 = \widetilde{H}^0(\mathbf{C})$ by $s(\pi(\widetilde{a}(\sigma))) = \widetilde{a}_Z(\sigma)$.
Suppose that $\widetilde{f}_Z$ splits, i.e., the cohomology class of $\widetilde{f}_Z$ in $H^2(\mathrm{Gal}(\widetilde{K}/F), Z(\widetilde{H}^0(\mathbf{C})))$ is trivial. Then, for ${}^L\widetilde{G} = {}^L\widetilde{G}^0 \rtimes \mathrm{Gal}(\widetilde{K}/F)$, we have ${}^L\widetilde{G} \cong \widetilde{H}(\mathbf{C})$ by (ii) and (iii) of Proposition 5.1. In this case, the local parameter ψ_v gives the Langlands parameter for ${}^L\widetilde{G}$ and it is not difficult to formulate the main conjecture. However $\widetilde{f}_Z$ does not split in general, as will be shown in Appendix II.

DEFINITION 5.3. Let $\{\widetilde{a}_Z(\sigma), \widetilde{f}_Z(\sigma,\tau)\}$ be as above. A finite Galois extension L of F containing $\widetilde{K}$ is called a splitting field for $\widetilde{f}_Z$ if the cohomology class of $\widetilde{f}_Z$ in $H^2(\mathrm{Gal}(L/F), Z(\widetilde{H}^0(\mathbf{C})))$ becomes trivial after inflation to $\mathrm{Gal}(L/F)$ by the canonical map $\mathrm{Gal}(L/F) \longrightarrow \mathrm{Gal}(\widetilde{K}/F)$. In the minimal case $\widetilde{H} = H$, a splitting field for $\widetilde{f}_Z$ is called a splitting field for ρ_λ.

THEOREM 5.4. *Assume that $Z(\widetilde{H}^0(\mathbf{C}))$ is connected and that the action of $\widetilde{a}_Z(\sigma)$ on it is trivial for every $\sigma \in \mathrm{Gal}(\widetilde{K}/F)$. Then a splitting field for $\widetilde{f}_Z$ exists.*

PROOF. Since $\widetilde{H}^0$ is reductive, the identity component of the center is a torus. By the assumption $Z(\widetilde{H}^0(\mathbf{C}))$ is isomorphic to $(\mathbf{C}^\times)^r$. Hence we may regard $\widetilde{f}_Z(\sigma,\tau)$ as a 2-cocycle of $\mathrm{Gal}(\widetilde{K}/F)$ taking values in $(\mathbf{C}^\times)^r$. By a theorem of

Tate ([Se2], Theorem 4), we obtain [4]

$$H^2(\mathrm{Gal}(\overline{F}/F), (\mathbf{C}^\times)^r) = 1. \tag{5.15}$$

Therefore a splitting field exists. This completes the proof. □

It is worth recording the following theorem, though we will not use it in this paper. The proof is identical to the above consideration. (In view of Langlands' result, this theorem also holds for non-archimedean local fields and function fields of one variable over a finite field, without assuming the triviality of the action of $a(\sigma)$ on the center.)

THEOREM 5.5. *Let F be a number field and K be a finite Galois extension. Let G be a connected reductive group over $\mathbf{C}$. Let $\{a(\sigma), f(\sigma,\tau)\}$ be a factor set of $\mathrm{Gal}(K/F)$ taking values in $G(\mathbf{C})$. We assume that the center of G is connected and that $a(\sigma)$, $\sigma \in \mathrm{Gal}(K/F)$ are automorphisms of G as an algebraic group acting trivially on the center. Then there exists a finite Galois extension $L \supset K$ of F such that the factor set inflated to $\mathrm{Gal}(L/F)$ by the canonical map $\mathrm{Gal}(L/F) \longrightarrow \mathrm{Gal}(K/F)$ splits.*

We note that the center of H^0 is not necessarily connected: For the case of Example 1.10, the center of the Hodge group has 2^{n-1} connected components. A main result of this paper, which concerns the minimal case, is:

THEOREM 5.6. *We assume Hypothesis 8.2 in section 8. Then a splitting field for ρ_λ exists if ρ_λ is absolutely irreducible and $w \neq 0$.*

The proof will be completed in §8. Here we note the following Lemma.

LEMMA 5.7. *Let K be a number field and $\tau_\lambda : \mathrm{Gal}(\overline{K}/K) \longrightarrow \mathrm{GL}(m, E_\lambda)$ be a λ-adic representation. Let C be the Zariski closure of $\mathrm{Im}(\tau_\lambda)$. Assume that C contains infinitely many scalar matrices. If τ_λ is absolutely irreducible, then the center of C is isomorphic to $\mathbf{G}_m$.*

PROOF. The center of C commutes with every element of $\mathrm{Im}(\tau_\lambda)$. By Schur's lemma, the center of C consists of scalar matrices. Therefore, by the assumption, the center of C must be isomorphic to $\mathbf{G}_m$. □

REMARK 5.8. Assume that τ_λ is obtained from a motive M over K with coefficients in E. If $w \neq 0$, Conjecture 4.1 implies that the center of C contains infinitely many scalar matrices. (cf. [Se1], 2.3 for an unconditional result of Deligne.)

Let L be a splitting field for $\widetilde{f}_Z$. Put

$${}^L\widetilde{G} = {}^L\widetilde{G}^0 \rtimes \mathrm{Gal}(L/F).$$

Then (5.2) is embedded into a commutative diagram

$$\begin{array}{ccccccccc} 1 & \longrightarrow & {}^L\widetilde{G}^0 & \longrightarrow & {}^L\widetilde{G} & \longrightarrow & \mathrm{Gal}(L/F) & \longrightarrow & 1 \\ & & \| & & \downarrow & & \downarrow & & \\ 1 & \longrightarrow & \widetilde{H}^0(\mathbf{C}) & \longrightarrow & \widetilde{H}(\mathbf{C}) & \longrightarrow & \mathrm{Gal}(\widetilde{K}/F) & \longrightarrow & 1. \end{array} \tag{5.16}$$

[4]Langlands ([Lan2], Lemma 4) proved that a 2-cocycle of $\mathrm{Gal}(K/F)$ taking values in a torus over $\mathbf{C}$ splits after the inflation to the relative Weil group $W_{L/F}$, where L is a finite Galois extension of F containing K. His result holds also for local fields and function fields of one variable over a finite field. It is not difficult to derive Tate's theorem, which concerns the case with trivial action, from Langlands' result.

We define a homomorphism r of ${}^L\widetilde{G}$ into $\mathrm{GL}(H_B(M))(\mathbf{C})$ as the composite of the homomorphism ${}^L\widetilde{G} \longrightarrow \widetilde{H}(\mathbf{C})$ in (5.16) and the canonical injection. By Lemma 5.2, the mapping $\psi_v : W'_{F_v} \longrightarrow \widetilde{H}(\mathbf{C})$ can be lifted to the mapping $\phi_v : W'_{F_v} \longrightarrow {}^L\widetilde{G}$ so that the following diagram is commutative:

$$\begin{array}{ccc} W'_{F_v} & \xrightarrow{\phi_v} & {}^L\widetilde{G} \\ \| & & \downarrow \\ W'_{F_v} & \xrightarrow{\psi_v} & \widetilde{H}(\mathbf{C}). \end{array} \tag{5.17}$$

We see that ϕ_v sends semisimple elements to semisimple elements and that $\mathbf{G}_a$ is mapped to unipotent elements contained in ${}^L\widetilde{G}^0$. In view of the latter part of Lemma 5.2, we see that it is a Langlands parameter. Now we can formulate the main conjecture.

MAIN CONJECTURE. *Assume that ρ_λ is absolutely irreducible and that a splitting field L for $\widetilde{f}_Z$ exists. Take a quasi-split connected reductive algebraic group $\widetilde{G}$ defined over F and define r and ϕ_v as above. Then there exists an irreducible automorphic representation $\pi = \otimes_v \pi_v$ of $\widetilde{G}(F_{\mathbf{A}})$ such that $L(s,\pi,r) = L(M,s)$. Moreover*
(i) *π_v belongs to the L-packet $\Pi_{\phi_v}(\widetilde{G}/F_v)$.*
(ii) *π is cuspidal.*
(iii) *π is essentially unitary and tempered. In other words, there exists a morphism $\nu : \widetilde{G} \longrightarrow \mathbf{G}_m$ defined over F and a quasicharacter χ of $F_{\mathbf{A}}^\times$ such that $\pi \otimes (\chi \circ \nu)$ is unitary and tempered.*

REMARK 5.9. The homomorphism ${}^L\widetilde{G} \longrightarrow \widetilde{H}(\mathbf{C})$ depends on the splitting of the 2-cocycle inflated from $\widetilde{f}_Z$. In other words, $H^1(\mathrm{Gal}(L/F), Z(\widetilde{H}^0(\mathbf{C})))$ acts on the equivalence class of r. If we change the splitting and r, then ϕ_v will change; however $L(s,\pi_v,r_v)$ does not change.

REMARK 5.10. The π on $\widetilde{G}(F_{\mathbf{A}})$ corresponding to M is not unique in general. To know which π will appear in the tempered spectrum, we need the (conjectural) multiplicity formula due to Labesse-Langlands-Kottwitz ([LL], [K]). To deduce more precise information, we need to know the structure of the L-packet $\Pi_{\phi_v}(\widetilde{G}(F_v))$. Non-tempered π will appear if we consider a mixed motive.

REMARK 5.11. The main conjectures for H (the minimal case) and $\widetilde{H}$ are related by the functoriality principle. More precisely an L-homomorphism f : ${}^LG \longrightarrow {}^L\widetilde{G}$ is given as follows. The cocycle $\widetilde{f}_Z$ for H is denoted by f_Z. Let L be a common splitting field for f_Z and $\widetilde{f}_Z$. Set ${}^LG = {}^LG^0 \rtimes \mathrm{Gal}(L/F)$, ${}^L\widetilde{G} = {}^L\widetilde{G}^0 \rtimes \mathrm{Gal}(L/F)$. Let φ (resp. $\widetilde{\varphi}$) be the canonical map of $\mathrm{Gal}(L/F)$ onto $\mathrm{Gal}(K/F)$ (resp. $\mathrm{Gal}(\widetilde{K}/F)$). Let π be the canonical map of $\mathrm{Gal}(K/F)$ onto $\mathrm{Gal}(\widetilde{K}/F)$. Let

$$f_Z(\varphi(\sigma),\varphi(\tau)) = \beta_\sigma(a_Z(\varphi(\sigma))\beta_\tau)\beta_{\sigma\tau}^{-1}, \qquad \sigma,\tau \in \mathrm{Gal}(L/F), \quad \beta_* \in Z(H^0(\mathbf{C})),$$

$$\widetilde{f}_Z(\widetilde{\varphi}(\sigma),\widetilde{\varphi}(\tau)) = \widetilde{\beta}_\sigma(\widetilde{a}_Z(\widetilde{\varphi}(\sigma))\widetilde{\beta}_\tau)\widetilde{\beta}_{\sigma\tau}^{-1}, \qquad \sigma,\tau \in \mathrm{Gal}(L/F), \quad \widetilde{\beta}_* \in Z(\widetilde{H}^0(\mathbf{C}))$$

be splittings. The element α_σ defined by (5.12) for $\widetilde{H}$ is denoted by $\widetilde{\alpha}_\sigma$. Then

$$f((h,\sigma)) = (h\beta_\sigma^{-1}\alpha_{\varphi(\sigma)}\rho_\lambda(\widetilde{\varphi(\sigma)})\rho_\lambda(\widetilde{\pi(\sigma)})^{-1}\widetilde{\alpha}_{\pi(\sigma)}^{-1}\widetilde{\beta}_\sigma,\ \sigma), \qquad h \in H^0(\mathbf{C}), \quad \sigma \in \mathrm{Gal}(L/F). \tag{5.18}$$

The predicted automorphic representations of $G(F_\mathbf{A})$ and of $\widetilde{G}(F_\mathbf{A})$ are in functorial correspondence under the L-homomorphism f. The same remark applies for the relation between H_1 and $\widetilde{H}$, where $H \subset H_1 \subset \widetilde{H}$.

REMARK 5.12. Let $W_{L/F}$ be the relative Weil group, where L is a finite Galois extension of F which contains $\widetilde{K}$. By the canonical map $W_{L/F} \longrightarrow \mathrm{Gal}(\widetilde{K}/F)$, we can inflate $\widetilde{f}_Z$ to the 2-cocycle of $W_{L/F}$ taking values in $Z(\widetilde{H}^0(\mathbf{C}))$. By the result of Langlands cited above, this cocycle splits for a sufficiently large L if $Z(\widetilde{H}^0(\mathbf{C}))$ is connected. Therefore, for the Weil form of the L-group ${}^L\widetilde{G} = \widetilde{H}^0(\mathbf{C}) \rtimes W_{L/F}$, we can formulate the main conjecture in the similar manner, whenever $Z(\widetilde{H}^0(\mathbf{C}))$ is connected.

We can interpret Remark 5.11 as showing that the minimal case is most challenging. After the next section, we will discuss the existence of a splitting field in the minimal case. We end this section by the observation that a slightly weaker formulation than the main conjecture is possible for the minimal case using only Theorem 5.4 (and not using the existence of the splitting field).

Assume that ρ_λ is absolutely irreducible and $w \neq 0$. Suppose that $\rho_\lambda|\mathrm{Gal}(\overline{F}/K)$ is isotypic. By Lemma 5.7 and Remark 5.8, the center of H^0 is connected. (We clearly see that $a(\sigma)$ acts on the center trivially.) Hence the splitting field exists by Theorem 5.4.

Next assume that $\rho_\lambda|\mathrm{Gal}(\overline{F}/K)$ is not isotypic. By Clifford's theorem, there exists a field $F \subsetneqq K' \subset K$ and a λ-adic representation

$$\tau_\lambda : \mathrm{Gal}(\overline{F}/K') \longrightarrow \mathrm{GL}(W)$$

such that $\rho_\lambda \cong \mathrm{Ind}_{\mathrm{Gal}(\overline{F}/K')}^{\mathrm{Gal}(\overline{F}/F)}\tau_\lambda$. Here W denotes a finite dimensional vector space over E_λ. (In terms of motives, this case corresponds to the following situation: There exists a motive $\widetilde{M}$ over K' with coefficients in E such that $M = R_{K'/F}(\widetilde{M})$ and τ_λ is the λ-adic representation associated to $\widetilde{M}$.) Let $\mathfrak{H}$ be the Zariski closure of $\mathrm{Im}(\tau_\lambda)$. Since the Zariski closure of $\tau_\lambda(\mathrm{Gal}(\overline{F}/K))$ is a homomorphic image of H^0, we have $\tau_\lambda(\mathrm{Gal}(\overline{F}/K)) \subset \mathfrak{H}^0(\mathbf{C})$. By the proof of Clifford's theorem, we see that $\tau_\lambda|\mathrm{Gal}(\overline{F}/K)$ is isotypic. Therefore the center of $\mathfrak{H}^0$ is isomorphic to $\mathbf{G}_m$, and a splitting field for τ_λ exists by Theorem 5.4. By the Main Conjecture, there exist a connected reductive algebraic group $\widetilde{G}$ defined over K', an irreducible automorphic representation $\widetilde{\pi}$ of $\widetilde{G}(K'_\mathbf{A})$ and a representation $\widetilde{r}$ of ${}^L\widetilde{G}$ such that $L(\widetilde{M},s) = L(M,s) = L(s,\widetilde{\pi},\widetilde{r})$. Put $G' = \mathrm{R}_{K'/F}(\widetilde{G})$. The L-group of G' is the induced group of ${}^L\widetilde{G}$ (cf. [Bo2], p. 35). Since $G'(F_\mathbf{A}) = \widetilde{G}(K'_\mathbf{A})$, we can find an irreducible automorphic representation π' of $G'(F_\mathbf{A})$ and a representation r' of ${}^LG'$ such that $L(M,s) = L(s,\pi',r')$. The Langlands parameter to which π' corresponds can be described explicitly, though we omit the details. In this way, the problem is "almost solved". However the group G' is slightly "bigger" than G. In this sense, this construction is unsatisfactory.

Let us consider an example. Let A be an elliptic curve defined over K' without complex multiplication. We assume that $\alpha(A)$ is not isogenous to $\beta(A)$ for two different embeddings α and β of K' into $\mathbf{C}$. We consider the motive $H_1(R_{K'/F}(A))$. Then we have $G' = \mathrm{R}_{K'/F}(\mathrm{GL}(2))$. On the other hand, by Example 1.10, we have $H^0(\mathbf{C}) = \{(g_1, \ldots, g_n) \in \mathrm{GL}(2, \mathbf{C})^n \mid \det g_1 = \cdots = \det g_n\}$, where $n = [K' : F]$. Put $G_1 = \mathrm{GL}(2)^n/Z$, $Z = \{z_1 \cdot 1_2, \ldots, z_n \cdot 1_2 \mid z_1 \cdots z_n = 1\}$ and consider G_1 as an algebraic group defined over K'. Since the dual group to H^0 is G_1 over $\mathbf{C}$, we see that $G = \mathrm{R}_{K'/F}(G_1)$ by considering the action of $\mathrm{Gal}(\overline{F}/F)$. There is a canonical surjective central homomorphism $G' \longrightarrow G$. An automorphic representation corresponding to M exists on $\mathrm{GL}(2, K'_{\mathbf{A}}) = G'(F_{\mathbf{A}})$ by a standard conjecture. Since this representation has trivial central character, the corresponding representation exists on $G_1(K'_{\mathbf{A}}) = G(F_{\mathbf{A}})$. We see that a splitting field for f_Z (in the minimal case) exists because Lemma 8.1 holds unconditionally and (8.10) holds in this case.

6. Preliminary considerations on splitting fields

Let ρ_λ be the λ-adic representation attached to a motive M. To show the existence of a splitting field for ρ_λ, we may enlarge the coefficient field E. Hereafter we assume that H^0 splits over E. It is convenient to consider $\{a(\sigma), f(\sigma, \tau)\}$ as a factor set taking values in $H^0(\overline{E_\lambda})$.

Take the finite Galois extension K of F so that

(6.1) $$H^0(\overline{E_\lambda}) \cap \rho_\lambda(\mathrm{Gal}(\overline{F}/F)) = \rho_\lambda(\mathrm{Gal}(\overline{F}/K)), \qquad \mathrm{Gal}(\overline{F}/K) \supset \mathrm{Ker}(\rho_\lambda).$$

The field K is the same as that defined by (5.1). We have the exact sequence

(6.2) $$1 \longrightarrow H^0(\overline{E_\lambda}) \longrightarrow H(\overline{E_\lambda}) \longrightarrow \mathrm{Gal}(K/F) \longrightarrow 1.$$

For $\sigma \in \mathrm{Gal}(K/F)$, take $\widetilde{\sigma} \in \mathrm{Gal}(\overline{F}/F)$ so that $\widetilde{\sigma}|K = \sigma$. Then a factor set of $\mathrm{Gal}(K/F)$ attached to the exact sequence (6.2) is

(6.3) $$a(\sigma)h = \rho_\lambda(\widetilde{\sigma})h\rho_\lambda(\widetilde{\sigma})^{-1}, \qquad h \in H^0(\overline{E_\lambda}),$$

(6.4) $$f(\sigma, \tau) = \rho_\lambda(\widetilde{\sigma}\widetilde{\tau}(\widetilde{\sigma\tau})^{-1}).$$

The exact sequence

$$1 \longrightarrow \mathrm{Inn}(H^0) \longrightarrow \mathrm{Aut}(H^0) \xrightarrow{\ \pi\ } \mathrm{Out}(H^0) \longrightarrow 1$$

splits. Take a splitting datum $(B, T, \{u_\alpha\}_{\alpha \in \Delta})$ for H^0 and let

$$s : \mathrm{Out}(H^0) \longrightarrow \mathrm{Aut}(H^0)$$

be the corresponding section. Take $\alpha_\sigma \in H^0(\overline{E_\lambda})$ so that

(6.5) $$s(\pi(a(\sigma))) = i(\alpha_\sigma)a(\sigma).$$

Define an equivalent factor set to $\{a(\sigma), f(\sigma, \tau\}$ by

(6.6) $$a_Z(\sigma) = i(\alpha_\sigma)a(\sigma),$$

(6.7) $$f_Z(\sigma, \tau) = \alpha_\sigma(a(\sigma)\alpha_\tau)f(\sigma, \tau)\alpha_{\sigma\tau}^{-1}.$$

Then $f_Z(\sigma, \tau)$ is a 2-cocycle of $\mathrm{Gal}(K/F)$ taking values in $Z(H^0(\overline{E_\lambda}))$.

LEMMA 6.1. *Let L be a finite Galois extension of F which contains K. Let $\varphi : \mathrm{Gal}(L/F) \longrightarrow \mathrm{Gal}(K/F)$ be the canonical map. For $\sigma \in \mathrm{Gal}(L/F)$, take $\bar{\sigma} \in \mathrm{Gal}(\overline{F}/F)$ so that $\bar{\sigma}|L = \sigma$. Put*

$$\bar{a}(\sigma)h = \rho_\lambda(\bar{\sigma})h\rho_\lambda(\bar{\sigma})^{-1}, \qquad h \in H^0(\overline{E_\lambda}), \tag{6.8}$$

$$\bar{f}(\sigma, \tau) = \rho_\lambda(\bar{\sigma}\bar{\tau}(\overline{\sigma\tau})^{-1}). \tag{6.9}$$

Then $\{\bar{a}(\sigma), \bar{f}(\sigma, \tau)\}$ is a factor set of $\mathrm{Gal}(L/F)$ taking values in $H^0(\overline{E_\lambda})$ which is equivalent to the factor set obtained from $\{a(\sigma), f(\sigma, \tau)\}$ by inflation by φ. For $\sigma \in \mathrm{Gal}(L/F)$, take $\beta_\sigma \in H^0(\overline{E_\lambda})$ so that

$$s(\pi(\bar{a}(\sigma))) = i(\beta_\sigma)\bar{a}(\sigma)$$

and put

$$\bar{a}_Z(\sigma) = i(\beta_\sigma)\bar{a}(\sigma), \tag{6.10}$$

$$\bar{f}_Z(\sigma, \tau) = \beta_\sigma(a(\sigma)\beta_\tau)\bar{f}(\sigma, \tau)\beta_{\sigma\tau}^{-1}. \tag{6.11}$$

Then the cohomology class of $\bar{f}_Z$ in $H^2(\mathrm{Gal}(L/F), Z(H^0(E_\lambda)))$ coincides with the class obtained from f_Z by inflation by φ.

PROOF. For $\sigma \in \mathrm{Gal}(L/F)$, put

$$\widetilde{\varphi(\sigma)} = u_\sigma\bar{\sigma}, \quad u_\sigma \in \mathrm{Gal}(\overline{F}/K), \qquad \beta_\sigma = \alpha_{\varphi(\sigma)}\rho_\lambda(u_\sigma).$$

With this choice of β_σ, define a_Z and f_Z by (6.10) and (6.11). It suffices to show

$$\bar{a}_Z(\sigma) = a_Z(\varphi(\sigma)), \qquad \sigma \in \mathrm{Gal}(L/F),$$

$$\bar{f}_Z(\sigma, \tau) = f_Z(\varphi(\sigma), \varphi(\tau)), \qquad \sigma, \tau \in \mathrm{Gal}(L/F).$$

We have

$$\bar{a}_Z(\sigma) = i(\alpha_{\varphi(\sigma)}\rho_\lambda(u_\sigma))\bar{a}(\sigma) = i(\alpha_{\varphi(\sigma)})a(\varphi(\sigma)) = a_Z(\varphi(\sigma)),$$

$$f_Z(\sigma, \tau) = \alpha_{\varphi(\sigma)}\rho_\lambda(u_\sigma)\rho_\lambda(\bar{\sigma})\alpha_{\varphi(\tau)}\rho_\lambda(u_\tau)\rho_\lambda(\bar{\sigma})^{-1}\rho_\lambda(\bar{\sigma}\bar{\tau}(\overline{\sigma\tau})^{-1})\rho_\lambda(u_{\sigma\tau})^{-1}\alpha_{\varphi(\sigma\tau)}^{-1}$$

$$= \alpha_{\varphi(\sigma)}\rho_\lambda(\widetilde{\varphi(\sigma)})\alpha_{\varphi(\tau)}\rho_\lambda(\widetilde{\varphi(\tau)})\rho_\lambda(\widetilde{\varphi(\sigma\tau)})^{-1}\alpha_{\varphi(\sigma\tau)}^{-1}$$

$$= \alpha_{\varphi(\sigma)}(\rho_\lambda(\widetilde{\varphi(\sigma)})\alpha_{\varphi(\tau)}\rho_\lambda(\widetilde{\varphi(\sigma)})^{-1})\rho_\lambda(\widetilde{\varphi(\sigma)}\widetilde{\varphi(\tau)}\widetilde{\varphi(\sigma\tau)}^{-1})\alpha_{\varphi(\sigma\tau)}^{-1} = f_Z(\varphi(\sigma), \varphi(\tau)).$$

Hence the assertion follows. □

LEMMA 6.2. *Let η_λ^i, $1 \le i \le m$ be λ-adic representations of $\mathrm{Gal}(\overline{F}/F)$ attached to motives M_i over F with coefficients in E. Put $\tau_\lambda = \oplus_{i=1}^m \eta_\lambda^i$. If τ_λ has a splitting field, then every η_λ^i, $1 \le i \le m$ has a splitting field.*

PROOF. Let H_i be the Zariski closure of $\mathrm{Im}(\eta_\lambda^i)$, $1 \le i \le m$ and let H be the Zariski closure of τ_λ. Clearly we see that

$$H \subset H_1 \times H_2 \times \cdots \times H_m, \qquad H^0 \subset H_1^0 \times H_2^0 \times \cdots \times H_m^0$$

and that the projection map $p_i : H \longrightarrow H_i$ is a surjective homomorphism satisfying $p_i(H^0) = H_i^0$, $p_i(Z(H^0)) \subset Z(H_i^0)$. Let K_i be the finite Galois extension of F determined by

$$H_i^0(\overline{E_\lambda}) \cap \eta_\lambda^i(\mathrm{Gal}(\overline{F}/F)) = \eta_\lambda^i(\mathrm{Gal}(\overline{F}/K_i)), \qquad \mathrm{Gal}(\overline{F}/K_i) \supset \mathrm{Ker}(\eta_\lambda^i).$$

Let L be a splitting field for τ_λ. We may assume that L contains K_i, $1 \le i \le m$. For $\sigma \in \mathrm{Gal}(L/F)$, take $\bar\sigma \in \mathrm{Gal}(\overline{F}/F)$ so that $\bar\sigma|L = \sigma$. We define a factor set $\{a_Z(\sigma), f_Z(\sigma,\tau)\}$ taking values in $Z(H^0(\overline{E_\lambda}))$ as in Lemma 6.1:

$$a(\sigma)h = \tau_\lambda(\bar\sigma)h\tau_\lambda(\bar\sigma)^{-1}, \quad h \in H^0(\overline{E_\lambda}), \qquad f(\sigma,\tau) = \tau_\lambda(\bar\sigma\bar\tau(\overline{\sigma\tau})^{-1}),$$

$$s(\pi(a(\sigma))) = i(\beta_\sigma)a(\sigma),$$

$$a_Z(\sigma) = i(\beta_\sigma)a(\sigma), \qquad f_Z(\sigma,\tau) = \beta_\sigma(a(\sigma)\beta_\tau)f(\sigma,\tau)\beta_{\sigma\tau}^{-1}.$$

We also define a factor set $\{a_{Z,i}(\sigma), f_{Z,i}(\sigma,\tau)\}$ taking values in $Z(H_i^0(\overline{E_\lambda}))$ by

$$a_i(\sigma)h = \eta_\lambda^i(\bar\sigma)h\eta_\lambda^i(\bar\sigma)^{-1}, \quad h \in H_i^0(\overline{E_\lambda}), \qquad f_i(\sigma,\tau) = \eta_\lambda^i(\bar\sigma\bar\tau(\overline{\sigma\tau})^{-1}),$$

$$s_i(\pi_i(a_i(\sigma))) = i(\beta_{\sigma,i})a_i(\sigma),$$

$$a_{Z,i}(\sigma) = i(\beta_{\sigma,i})a_i(\sigma), \qquad f_{Z,i}(\sigma,\tau) = \beta_{\sigma,i}(a_i(\sigma)\beta_{\tau,i})f_i(\sigma,\tau)\beta_{\sigma\tau,i}^{-1}.$$

Let $(B, T, \{u_\alpha\}_{\alpha\in\Delta})$ be a splitting datum for H^0. Let U be the unipotent radical of B. The homomorphic images $B_i = p_i(B)$, $U_i = p_i(U)$, $T_i = p_i(T)$ are a Borel subgroup, the unipotent radical of B_i and a maximal torus of H_i. Let Σ^+ (resp. Σ_i^+) be the set of positive roots defined by (B,T) (resp. (B_i,T_i)). Let $\beta \in \Sigma^+$ and let $U_\beta \subset U$ be the corresponding root subgroup. Assume that $p_i(U_\beta) \ne \{1\}$. Since $p_i(U_\beta) \cong \mathbf{G}_a$ and is normalized by T_i, we have $p_i(U_\beta) = U_\alpha^i$ for some $\alpha \in \Sigma_i^+$, where U_α^i is the root subgroup corresponding to α. Then we find easily that $\beta = \alpha \circ p_i$. Conversely, given $\alpha \in \Sigma_i^+$, examining root subgroups contained in $p_i^{-1}(U_\alpha^i) \cap U$, we see that there exists $\beta \in \Sigma^+$ such that $p_i(U_\beta) = U_\alpha^i$. Thus we have shown that $p_i(U_\beta) \ne \{1\}$ if and only if $\beta = p_i \circ \alpha$ for some $\alpha \in \Sigma_i^+$. In this case, $p_i(U_\beta) = U_\alpha^i$; moreover β is simple if α is simple. Let $\Delta_i \subset \Sigma_i^+$ be the set of simple roots. For $\alpha \in \Delta_i$, we put $u_\alpha^i = p_i(u_\beta)$, $\beta = \alpha \circ p_i$. Then $i(p_i(\beta_\sigma))a_i(\sigma)$ stabilizes the splitting datum $(B_i, T_i, \{u_\alpha^i\}_{\alpha\in\Delta_i})$. Therefore we may take $\beta_{\sigma,i} = p_i(\beta_\sigma)$. Then it is immediate to see that

$$a_{Z,i}(\sigma)(p_i(h)) = p_i(a_Z(\sigma)h), \quad h \in H^0, \qquad f_{Z,i}(\sigma,\tau) = p_i(f_Z(\sigma,\tau)).$$

This completes the proof. □

REMARK 6.3. The converse to Lemma 6.2 could well be true but seems to be difficult to prove.

Let K be a finite Galois extension of F such that

$$(6.12) \qquad \rho_\lambda(\mathrm{Gal}(\overline{F}/K)) \subset H^0(\overline{E_\lambda}), \qquad \mathrm{Gal}(\overline{F}/K) \supset \mathrm{Ker}(\rho_\lambda).$$

By Lemma 6.1, we see that to show the existence of a splitting field, it is enough to consider the 2-cocycle $f_Z(\sigma,\tau)$ defined by (6.3)–(6.7). Hereafter, we denote by K a field satisfying (6.12) without assuming (6.1).

Let $\tau_\lambda : \mathrm{Gal}(\overline{F}/K) \longrightarrow \mathrm{GL}(W)$ be a λ-adic representation, where W is a finite dimensional vector space over E_λ. Put $\rho_\lambda = \mathrm{Ind}_{\mathrm{Gal}(\overline{F}/K)}^{\mathrm{Gal}(\overline{F}/F)}\tau_\lambda$. We assume that τ_λ is associated to a motive $\widetilde{M}$ over K with coefficients in E. Let $M = \mathrm{R}_{K/F}(\widetilde{M})$. Then ρ_λ is associated to the motive M. By Lemma 6.2, the problem of the existence of a splitting field can be reduced to this situation. Also by Lemma 6.2, enlarging K if necessary, we may assume that K is normal over F, the Zariski closure of $\mathrm{Im}(\tau_\lambda)$ is connected and that the condition (6.12) is satisfied. Let H be the Zariski closure of $\mathrm{Im}(\rho_\lambda)$ as before and let $\mathfrak{H}$ be the Zariski closure of $\mathrm{Im}(\tau_\lambda)$. By Conjecture 4.1, we have $H = \mathrm{Hg}(M)$, $\mathfrak{H} = \mathrm{Hg}(\widetilde{M})$. In §7, we will deal with the case where τ_λ is one dimensional. The general case will be treated in §8.

7. The existence of splitting fields in abelian case

Let K be an algebraic number field. Let χ be a Hecke character of $K_{\mathbf{A}}^\times$ of conductor $\mathfrak{f}$. Let $I_{\mathfrak{f}}$ denote the group consisting of all fractional ideals prime to $\mathfrak{f}$. Let χ_* be a homomorphism of $I_{\mathfrak{f}}$ into $\mathbf{C}^\times$ associated to χ. We assume that χ_* satisfies

$$\chi_*((a)) = \prod_{\sigma \in J_K} \sigma(a)^{l(\sigma)}, \qquad a \equiv 1 \mod {}^\times\mathfrak{f}, \quad a \gg 0. \tag{7.1}$$

Here J_K denotes the set of all isomorphisms of K into $\mathbf{C}$ and $l(\sigma) \in \mathbf{Z}$; $\mod {}^\times\mathfrak{f}$ denotes the multiplicative congruence modulo $\mathfrak{f}$; $a \gg 0$ means that a is totally positive. Let E be the finite algebraic number field generated by $\chi_*(\mathfrak{a})$, $\mathfrak{a} \in I_{\mathfrak{f}}$. Let λ be a finite place of E and let ℓ be the prime number lying below λ. As was shown by Weil, there exists a λ-adic representation χ_λ of $\mathrm{Gal}(K_{\mathrm{ab}}/K)$ into $E_\lambda^\times$ associated to χ. They are related by

(7.2)

$$\chi(a) = \chi_\lambda([a]) \quad \text{if } a \in K_{\mathbf{A}}^\times \text{ satisfies } a_\infty = 1 \text{ and } a_v = 1 \text{ whenever } v \text{ divides } \ell.$$

Here $[a] \in \mathrm{Gal}(K_{\mathrm{ab}}/K)$ denotes the canonical image of a. We may consider χ_λ as a λ-adic representation of $\mathrm{Gal}(\overline{K}/K)$. Now assume that K is normal over an algebraic number field F and consider the induced representation

$$\rho_\lambda = \mathrm{Ind}_{\mathrm{Gal}(\overline{F}/K)}^{\mathrm{Gal}(\overline{F}/F)} \chi_\lambda.$$

In the notation of §6, $\tau_\lambda = \chi_\lambda$. We are going to prove that ρ_λ has a splitting field using the ℓ-adic method.

Put

$$\mathfrak{G} = \mathrm{Gal}(K/F), \qquad n = [K : F] = |\mathfrak{G}|.$$

Since we are assuming that the weight w of M is not zero, χ is of infinite order (cf. Remark 7.3). Then we have

$$\mathfrak{H} = \mathbf{G}_m, \qquad H^0 \subset (\mathbf{G}_m)^n.$$

Here we consider $\mathbf{G}_m$ as an algebraic group defined over E. Put $T = H^0$. Let $X^*((\mathbf{G}_m)^n) \cong \mathbf{Z}^n$ be the character group of $(\mathbf{G}_m)^n$. We write an element of $X^*((\mathbf{G}_m)^n)$ as $(x_\alpha)_{\alpha \in \mathfrak{G}}$, $x_\alpha \in \mathbf{Z}$. Let Y be the subgroup of $X^*((\mathbf{G}_m)^n)$ which annihilates T. For $\alpha \in \mathfrak{G}$, we choose $\widetilde{\alpha} \in \mathrm{Gal}(\overline{F}/F)$ so that $\widetilde{\alpha}|K = \alpha$. We put

$$\chi_\lambda^\alpha(x) = \chi_\lambda((\widetilde{\alpha})^{-1} x \widetilde{\alpha}), \quad x \in \mathrm{Gal}(\overline{F}/K), \qquad \chi^\alpha(x) = \chi(\alpha^{-1}(x)), \quad x \in K_{\mathbf{A}}^\times.$$

Then χ^α and χ_λ^α are related by (7.2). Since

$$\rho_\lambda | \mathrm{Gal}(\overline{F}/K) \cong \oplus_{\alpha \in \mathfrak{G}} \chi_\lambda^\alpha,$$

$(y_\alpha)_{\alpha \in \mathfrak{G}} \in \mathbf{Z}^n$ belongs to Y if and only if

$$\prod_{\alpha \in \mathfrak{G}} (\chi_*^\alpha)^{y_\alpha} = 1. \tag{7.3}$$

By

$$\chi_*^\alpha((a)) = \chi_*((\alpha^{-1}(a))) = \prod_{\sigma \in J_K} \sigma(a)^{l(\sigma\alpha)}, \qquad a \equiv 1 \mod {}^\times\mathfrak{f}, \quad a \gg 0,$$

we obtain

$$\sum_{\alpha \in \mathfrak{G}} y_\alpha l(\sigma\alpha) = 0 \quad \text{for all } \sigma \in J_K \tag{7.4}$$

from (7.3). Since H^0 is the Zariski closure of $\mathrm{Gal}(\overline{F}/L)$ for any finite extension L of K, we see that (7.4) is the necessary and sufficient condition for $(y_\alpha)_{\alpha \in \mathfrak{G}} \in Y$. As $X^*((\mathbf{G}_m)^n)/Y$ is torsion-free, we see that T is a split torus defined over E, whose character group is isomorphic to $\mathbf{Z}^n/Y$.

Now we regard ρ_λ as a λ-adic representation of $\mathrm{Gal}(K_{\mathrm{ab}}/F)$. For $\sigma \in \mathfrak{G}$, we take $\widetilde{\sigma} \in \mathrm{Gal}(K_{\mathrm{ab}}/F)$ so that $\widetilde{\sigma}|K = \sigma$. The factor set attached to ρ_λ is given by

$$a(\sigma)h = \rho_\lambda(\widetilde{\sigma})h\rho_\lambda(\widetilde{\sigma})^{-1}, \qquad h \in T(E_\lambda),$$

$$f(\sigma,\tau) = \rho_\lambda(\widetilde{\sigma}\widetilde{\tau}(\widetilde{\sigma\tau})^{-1}) = \oplus_{\alpha \in \mathfrak{G}} \chi_\lambda^\alpha(\widetilde{\sigma}\widetilde{\tau}(\widetilde{\sigma\tau})^{-1}).$$

Since H^0 is commutative, we have $a_Z(\sigma) = a(\sigma)$, $f_Z(\sigma,\tau) = f(\sigma,\tau)$. We put

$$\xi(\sigma,\tau) = \widetilde{\sigma}\widetilde{\tau}(\widetilde{\sigma\tau})^{-1}.$$

Then ξ is a 2-cocycle and the cohomology class of ξ is in $H^2(\mathfrak{G}, \mathrm{Gal}(K_{\mathrm{ab}}/K))$. We see easily that

$$\oplus_{\alpha \in \mathfrak{G}} \chi_\lambda^\alpha : \mathrm{Gal}(K_{\mathrm{ab}}/K) \longrightarrow T(E_\lambda)$$

is a $\mathfrak{G}$-homomorphism and that f is the image of ξ under this mapping. Let $C_K = K_{\mathbf{A}}^\times/K^\times$ be the idele class group of K and let D_K be the identity component of C_K. We have $\mathrm{Gal}(K_{\mathrm{ab}}/K) \cong C_K/D_K$. Since $H^{2r+1}(\mathfrak{G}, D_K) = 1$ (cf. Artin-Tate [AT], p. 91–92), the canonical map $H^2(\mathfrak{G}, C_K) \longrightarrow H^2(\mathfrak{G}, C_K/D_K)$ is surjective. Let $\eta \in H^2(\mathfrak{G}, C_K)$ denote the fundamental class, i.e., the canonical generator of $H^2(\mathfrak{G}, C_K) \cong \mathbf{Z}/n\mathbf{Z}$. The cohomology class of ξ is the image of η under the canonical map.

Let $\mathcal{T}$ be the maximal compact subgroup of $T(E_\lambda)$. Since

$$f(\sigma,\tau) = \oplus_{\alpha \in \mathfrak{G}} \chi_\lambda^\alpha(\xi(\sigma,\tau)), \tag{7.5}$$

the order of the cohomology class of f in $H^2(\mathfrak{G}, \mathcal{T})$ divides n. Now take λ so that l does not divide n. We replace K by its finite extension L so that $\rho_\lambda(\mathrm{Gal}(\overline{F}/L))$ is contained in the maximal pro-ℓ-subgroup $\mathcal{L}$ of $\mathcal{T}$. Since this is the inflation process by Lemma 6.1, the order of the cohomology class of f in $H^2(\mathrm{Gal}(L/F), \mathcal{T})$ still divides n. [5] Now the triviality of the cohomology class of f follows from the next Lemma.

LEMMA 7.1. *Let G be a finite group. Let $\mathcal{T}$ be a G-module and $\mathcal{L}$ be a G-submodule of $\mathcal{T}$. We assume that $\mathcal{T}$ is a profinite group with continuous action of G, $\mathcal{L}$ is a pro-ℓ-group and that $\mathcal{L}$ is open in $\mathcal{T}$. Suppose that $c \in H^2(G, \mathcal{T})$ is given. Let m be the order of c. If c is in the image of $H^2(G, \mathcal{L})$ under the canonical map $H^2(G, \mathcal{L}) \longrightarrow H^2(G, \mathcal{T})$ and m is not divisible by ℓ, then $c = 1$.*

PROOF. Take a decreasing sequence $\{\mathcal{T}_i\}_{i=1}^\infty$ of open subgroups of $\mathcal{T}$ so that $\cap_{i=1}^\infty \mathcal{T}_i = \{1\}$ and such that $\mathcal{T} = \varprojlim \mathcal{T}/\mathcal{T}_i$. Replacing $\mathcal{T}_i$ by $\cap_{\sigma \in G} \sigma\mathcal{T}_i$, we may assume that $\mathcal{T}_i$ is a G-submodule of $\mathcal{T}$. Put $\mathcal{L}_i = \mathcal{L} \cap \mathcal{T}_i$. Then $\mathcal{L}_i$ is an open subgroup of $\mathcal{L}$ and $\mathcal{L}/\mathcal{L}_i$ is a finite ℓ-group. Let c_i be the image of c under the

[5] More concretely we see the following. Let $\mathfrak{a}_K : K_{\mathbf{A}}^\times \longrightarrow \mathrm{Gal}(K_{\mathrm{ab}}/K)$ be the canonical morphism. Then (7.5) can be written as $f(\sigma,\tau) = \oplus_{\alpha \in \mathfrak{G}} \chi_\lambda^\alpha(\mathfrak{a}_K(\eta(\sigma,\tau)))$. If we replace K by L, f changes to $\oplus_{\alpha \in \mathfrak{G}} \chi_\lambda^\alpha(\mathfrak{a}_K(N_{L/K}(\widetilde{\eta}(\sigma,\tau))))$, where $\widetilde{\eta}$ is the fundamental class of $H^2(\mathrm{Gal}(L/F), C_L)$.

canonical map $H^2(G, \mathcal{T}) \longrightarrow H^2(G, \mathcal{T}/\mathcal{T}_i)$. Let f be a 2-cocycle which represents c.

We assert that there exists a positive integer N such that c_i is of order m if $i \geq N$. If this assertion is false, there exists $m' < m$ such that c_i is of order m' for infinitely many i. For such an i, there exists $a_i(\sigma) \in \mathcal{T}$, $\sigma \in G$ such that

$$f(\sigma, \tau)^{m'} \equiv a_i(\sigma)(\sigma a_i(\tau)) a_i(\sigma\tau)^{-1} \mod \mathcal{T}_i, \qquad \sigma, \tau \in G.$$

Take simultaneously convergent subsequences of $a_i(\sigma)$ for all $\sigma \in G$ and let $a(\sigma) \in \mathcal{T}$ be the limit. Then we have

$$f(\sigma, \tau)^{m'} = a(\sigma)(\sigma a(\tau)) a(\sigma\tau)^{-1} \qquad \sigma, \tau \in G.$$

This is a contradiction and the existence of N is established.

Let $i \geq N$. By the assumption, we see that c_i is in the image of the canonical map $H^2(G, \mathcal{L}/\mathcal{L}_i) \longrightarrow H^2(G, \mathcal{T}/\mathcal{T}_i)$. Since $H^2(G, \mathcal{L}/\mathcal{L}_i)$ is an ℓ-group, we must have $m = 1$. This completes the proof. $\square$

REMARK 7.2. Assume that K is a CM-field. Then there exists a motive $M(\chi)$ over K with coefficients in E attached to χ. The rank of $M(\chi)$ is 1 and the weight w of $M(\chi)$ is $l(\sigma) + l(\sigma c)$ which is independent of $\sigma \in J_K$, where c is the complex conjugation (cf. Blasius [Bl]). For ρ_λ, the associated motive is $M = \mathrm{R}_{K/F}(M(\chi))$. Regard l as a $\mathbf{Z}$-valued function on $\mathrm{Gal}(\overline{\mathbf{Q}}/\mathbf{Q})$ which is left invariant under $\mathrm{Gal}(\overline{\mathbf{Q}}/E)$ and right invariant under $\mathrm{Gal}(\overline{\mathbf{Q}}/K)$. We regard $\overline{\mathbf{Q}}$ as a subfield of $\mathbf{C}$. Take $\tau : K \hookrightarrow \overline{\mathbf{Q}}$ and $t \in \mathrm{Gal}(\overline{\mathbf{Q}}/\mathbf{Q})$ such that $t|K = \tau$. Consider E as a subfield of $\mathbf{C}$. Then for the rank 1 E-module $H_B(M(\chi^\alpha))$, we have

$$H_B(M(\chi^\alpha)) \otimes_{E,\mathrm{id}} \mathbf{C} = H^{l(t\alpha), w-l(t\alpha)}, \qquad H_B(M) \otimes_{E,\mathrm{id}} \mathbf{C} = \oplus_{\alpha \in \mathfrak{G}} H^{l(t\alpha), w-l(t\alpha)}.$$

Here we use the same letter α also for its extension to $\mathrm{Gal}(\overline{\mathbf{Q}}/\mathbf{Q})$. Using (7.4), we see that the Hodge group $\mathrm{Hg}(M)$ is independent of τ and is equal to T.

Let K be an arbitrary algebraic number field and let χ be an algebraic Hecke character satisfying (7.1). Then there exists a CM-subfield K_0 of K, an algebraic Hecke character χ_0 of K_0 and a character of finite order ψ of $K_{\mathbf{A}}^\times$ such that $\chi = \chi_0 \circ N_{K/K_0} \times \psi$. The motive attached to χ is given by $M(\chi_0)_K \otimes M(\psi)$, where $M(\chi_0)_K$ is the motive $M(\chi_0)$ considered as a motive over K and $M(\psi)$ denotes the Artin motive attached to ψ.

8. The existence of splitting fields in general case

We consider the situation in the last paragraph of §6. Thus $\rho_\lambda = \mathrm{Ind}_{\mathrm{Gal}(\overline{F}/K)}^{\mathrm{Gal}(\overline{F}/F)} \tau_\lambda$ for a λ-adic representation $\tau_\lambda : \mathrm{Gal}(\overline{F}/K) \longrightarrow \mathrm{GL}(W)$. We put

$$\mathfrak{G} = \mathrm{Gal}(K/F), \qquad n = [K : F].$$

We realize ρ_λ as follows. For $\alpha \in \mathfrak{G}$, we prepare a vector space αW, which is isomorphic to W, and put $V = \oplus_{\alpha \in \mathfrak{G}} \alpha W$. For $\alpha \in \mathfrak{G}$, we choose $\widetilde{\alpha} \in \mathrm{Gal}(\overline{F}/F)$ so that $\widetilde{\alpha}|K = \alpha$. Then we have

$$\mathrm{Gal}(\overline{F}/F) = \sqcup_{\alpha \in \mathfrak{G}} \widetilde{\alpha}\, \mathrm{Gal}(\overline{F}/K).$$

For $\sigma \in \mathrm{Gal}(\overline{F}/F)$ and $\alpha \in \mathfrak{G}$, we put

$$\sigma\widetilde{\alpha} = \widetilde{\beta}_\alpha h_\alpha, \qquad \beta_\alpha \in \mathfrak{G},\ h_\alpha \in \mathrm{Gal}(\overline{F}/K).$$

Then we define $\rho_\lambda(\sigma) \in \mathrm{GL}(V)$ by

$$\rho_\lambda(\sigma)(\oplus_{\alpha\in\mathfrak{G}}\, \alpha\cdot v_\alpha) = \oplus_{\alpha\in\mathfrak{G}}\, \beta_\alpha \cdot \tau_\lambda((\widetilde{\beta_\alpha})^{-1}\sigma\widetilde{\alpha})v_\alpha. \tag{8.1}$$

Here we label a vector of αW as $\alpha\cdot v_\alpha$, $v_\alpha \in W$. The right-hand side means the vector whose β_α-component is $\tau_\lambda((\widetilde{\beta_\alpha})^{-1}\sigma\widetilde{\alpha})v_\alpha$. If $\sigma \in \mathrm{Gal}(\overline{F}/K)$, we have

$$\rho_\lambda(\sigma)(\oplus_{\alpha\in\mathfrak{G}}\, \alpha\cdot v_\alpha) = \oplus_{\alpha\in\mathfrak{G}}\, \alpha \cdot \tau_\lambda((\widetilde{\alpha})^{-1}\sigma\widetilde{\alpha})v_\alpha, \qquad \sigma \in \mathrm{Gal}(\overline{F}/K). \tag{8.2}$$

For $\alpha \in \mathfrak{G}$, we define a λ-adic representation τ_λ^α of $\mathrm{Gal}(\overline{F}/K)$ by

$$\tau_\lambda^\alpha(\sigma) = \tau_\lambda((\widetilde{\alpha})^{-1}\sigma\widetilde{\alpha}), \qquad \sigma \in \mathrm{Gal}(\overline{F}/K).$$

By (8.1), we have $\rho_\lambda|\mathrm{Gal}(\overline{F}/K) \cong \oplus_{\alpha\in\mathfrak{G}}\, \tau_\lambda^\alpha$. Note that K is normal over F as in §6.

As was explained in §6, we may and do assume that $\mathfrak{H} = \mathfrak{H}^0$. Then we clearly see that

$$H^0 \subset \mathfrak{H}^{\mathfrak{G}} = \mathfrak{H}^n \tag{8.3}$$

and that the projection of H^0 to every direct factor of $\mathfrak{H}^n$ is surjective. Hence we have

$$Z(H^0) = Z(\mathfrak{H})^n \cap H^0. \tag{8.4}$$

Note that H^0 as a subgroup of $\mathfrak{H}^n$ may depend on the choice of $\{\widetilde{\alpha}\}$.

We write an element $d \in \prod_{\alpha\in\mathfrak{G}} \mathfrak{H} = \mathfrak{H}^{\mathfrak{G}}$ as $d = [d_\alpha]_{\alpha\in\mathfrak{G}}$, $d_\alpha \in \mathfrak{H}$. This notation applies more generally for an element of $\mathrm{GL}(W)^{\mathfrak{G}}$. For $\sigma \in \mathfrak{G}$, we define a permutation matrix $P(\sigma) \in \mathrm{GL}(V)$ by

$$P(\sigma)(\oplus_{\alpha\in\mathfrak{G}}\, \alpha\cdot v_\alpha) = \oplus_{\alpha\in\mathfrak{G}}\, \sigma\alpha\cdot v_\alpha.$$

Then we can verify easily that

$$P(\sigma)dP(\sigma)^{-1} = [d_{\sigma^{-1}\alpha}]_{\alpha\in\mathfrak{G}}, \qquad d = [d_\alpha]_{\alpha\in\mathfrak{G}} \in \mathrm{GL}(W)(\mathbf{C})^{\mathfrak{G}}. \tag{8.5}$$

By (8.1), we have

$$\rho_\lambda(\widetilde{\sigma})^{-1}(\oplus_{\alpha\in\mathfrak{G}}\, \alpha\cdot v_\alpha) = \oplus_{\alpha\in\mathfrak{G}}\, \sigma^{-1}\alpha \cdot \tau_\lambda((\widetilde{\sigma^{-1}\alpha})^{-1}\widetilde{\sigma}^{-1}\widetilde{\alpha})v_\alpha, \qquad \sigma \in \mathfrak{G}.$$

Put

$$D_\sigma = [\tau_\lambda((\widetilde{\sigma^{-1}\alpha})^{-1}\widetilde{\sigma}^{-1}\widetilde{\alpha})]_{\alpha\in\mathfrak{G}} \in \mathfrak{H}^{\mathfrak{G}} \subset \mathrm{GL}(W)(E_\lambda)^{\mathfrak{G}}. \tag{8.6}$$

Then we have

$$P(\sigma) = D_\sigma \rho_\lambda(\widetilde{\sigma}), \qquad \sigma \in \mathfrak{G}. \tag{8.7}$$

By abuse of notation, we denote the map $x \longrightarrow P(\sigma)xP(\sigma)^{-1}$ (resp. $x \longrightarrow D_\sigma x D_\sigma^{-1}$), $x \in GL(W)^{\mathfrak{G}}$ by $i(P(\sigma))$ (resp. $i(D_\sigma)$). Though the following Lemma is intuitively convincing, a proof is highly non-trivial.

LEMMA 8.1. *We assume the Hodge conjecture. Then* $i(P(\sigma)) \in \mathrm{Aut}(H^0)$ *for some choice of* $\{\widetilde{\alpha}\}$, $\alpha \in \mathfrak{G}$.

PROOF. If our lemma holds for a direct sum of representations, then it clearly holds for every direct summand. Hence we may assume that $E = \mathbf{Q}$ (cf. (4.1)). Furthermore, since every motive is a direct summand of $H^*(X)(q)$, we may assume that $\widetilde{M} = H^*(X)(q)$, $M = H^*(\mathrm{R}_{K/F}(X))(q)$. Here X is a projective smooth algebraic variety defined over K and (q) denotes the Tate twist. Let U be the E-rational Hodge structure attached to M. For $\alpha \in \mathfrak{G}$, let V_α be the E-rational Hodge structure attached to $\alpha(\widetilde{M})$. Since $M = \oplus_{\alpha\in\mathfrak{G}}\alpha(\widetilde{M})$ over K, we have $U = \oplus_{\alpha\in\mathfrak{G}}V_\alpha$. By the comparison isomorphism, we have

$$U \otimes_E E_\lambda \cong H_\lambda(M) = \oplus_{\alpha\in\mathfrak{G}}H_\lambda(\alpha(M)).$$

The representation ρ_λ of $\mathrm{Gal}(\overline{F}/F)$ is realized on $H_\lambda(M)$ as above for some choice of $\{\widetilde{\alpha}\}$. We have

$$U^{\otimes l} \otimes_E \check{U}^{\otimes m} \otimes_E E(p) = \left(\oplus_{\sum l_\alpha = l, \sum m_\alpha = m} \left(\otimes_{\alpha\in\mathfrak{G}} V_\alpha^{\otimes l_\alpha} \otimes_E \check{V}_\alpha^{\otimes m_\alpha}\right)\right) \otimes_E E(p).$$

A tensor of type $(0,0)$ of the left-hand side can be written as the sum of tensors of type $(0,0)$ of direct factors on the right-hand side. Hence $g \in \mathrm{MT}(U)$ if and only if g fixes every tensor of type $(0,0)$ for all tensor spaces of the form

$$\left(\otimes_{\alpha\in\mathfrak{G}} V_\alpha^{\otimes l_\alpha} \otimes_E \check{V}_\alpha^{\otimes m_\alpha}\right) \otimes_E E(p). \tag{8.8}$$

The cycle maps to the tensor space of (8.8) and to

$$(H_\lambda(\alpha(M))^{\otimes l_\alpha} \otimes H_\lambda(\alpha(\check{M}))^{\otimes m_\alpha})(p)$$

are compatible with the comparison isomorphism. Let t be a tensor of type $(0,0)$ in the tensor space (8.8). Let $\sigma \in \mathfrak{G}$. By the Hodge conjecture, t corresponds to an algebraic cycle c on $\otimes_{\alpha\in\mathfrak{G}}\alpha(\widetilde{M})^{\otimes l_\alpha} \otimes \alpha(\check{\widetilde{M}})^{\otimes m_\alpha}$. We may assume that c is rational over K. Then we have a tensor $\sigma(t)$ which corresponds to an algebraic cycle on $\otimes_{\alpha\in\mathfrak{G}}\sigma\alpha(\widetilde{M})^{\otimes l_\alpha} \otimes \sigma\alpha(\check{\widetilde{M}})^{\otimes m_\alpha}$. Write t in the form

$$t = \sum_i \otimes_{\alpha\in\mathfrak{G}} (\otimes_{j=1}^{l_\alpha} u_{\alpha,j,i}) \otimes (\otimes_{j=1}^{m_\alpha} v_{\alpha,j,i}) \otimes w, \qquad u_{\alpha,j,i} \in V_\alpha,\ v_{\alpha,j,i} \in \check{V}_\alpha,\ w \in E(p).$$

Then we have

$$\sigma(t) = \sum_i \otimes_{\alpha\in\mathfrak{G}} (\otimes_{j=1}^{l_{\sigma\alpha}} u_{\sigma\alpha,j,i}) \otimes (\otimes_{j=1}^{m_{\sigma\alpha}} v_{\sigma\alpha,j,i}) \otimes w.$$

In view of (8.3), we write an element of H^0 as $(x_\alpha)_{\alpha\in\mathfrak{G}}$. Put $x' = (x_{\sigma\alpha})_{\alpha\in\mathfrak{G}}$. There exists $u \in \mathbf{G}_m$ such that $(x,u) \in \mathrm{MT}(U)$, which is equivalent to $xt = u^{-p}t$. Then we have $x'\sigma(t) = u^{-p}\sigma(t)$. Since t is an arbitrary tensor of type $(0,0)$, we obtain $x' \in H^0$. This completes the proof. □

Hereafter we fix such a choice of $\{\widetilde{\alpha}\}$, $\alpha \in \mathfrak{G}$ assured by Lemma 8.1.

HYPOTHESIS 8.2. *The automorphisms $i(P(\sigma))$, $\sigma \in \mathfrak{G}$ stabilize a splitting datum of H^0.*

For symmetry reasons, this hypothesis seems plausible. However, the author is unable to prove it even assuming the Hodge conjecture. [6] We can prove it or dispense with it in several cases. To see this, let $(B, T, \{u_\delta\}_{\delta\in\Delta})$ be a splitting datum of H^0. Regard $H^0 \subset \mathfrak{H}^{\mathfrak{G}}$ and let $p_\alpha : H^0 \longrightarrow \mathfrak{H}$ be the projection map to the α component, $\alpha \in \mathfrak{G}$. Since p_α is surjective, we see that $(p_\alpha(B), p_\alpha(T), \{p_\alpha(u_\delta)\}_{\delta\in\Delta})$ is a splitting datum for $\mathfrak{H}$, in the same way as in Lemma 6.2. Here we take only such δ which satisfies $p_\alpha(u_\delta) \neq 1$. Put $S =$

[6] The subgroup of H^0 preserving the flag defined by the Hodge filtration can be parabolic. If this is so, it will provide a clue.

$(B, T, \{u_\delta\}_{\delta\in\Delta})$, $S_\alpha = (p_\alpha(B), p_\alpha(T), \{p_\alpha(u_\delta)\}_{\delta\in\Delta})$ for simplicity. By (8.5), we have

$$i(P(\sigma))(S_\alpha)_{\alpha\in\mathfrak{G}} = (S_{\sigma^{-1}\alpha})_{\alpha\in\mathfrak{G}}.$$

For $\alpha \in \mathfrak{G}$, take $\gamma_\alpha \in \mathfrak{H}$ so that $i(\gamma_\alpha)S_\alpha = S_1$. We see that γ_α is unique modulo $Z(\mathfrak{H})$. Put $\gamma = [\gamma_\alpha]_{\alpha\in\mathfrak{G}}$. Then we find $i(P(\sigma)\gamma P(\sigma)^{-1})(S_{\sigma^{-1}\alpha})_{\alpha\in\mathfrak{G}} = (S_1, S_1, \dots, S_1)$ and therefore

$$i(\gamma^{-1}(P(\sigma)\gamma P(\sigma)^{-1}))i(P(\sigma))(S_\alpha)_{\alpha\in\mathfrak{G}} = (S_\alpha)_{\alpha\in\mathfrak{G}}.$$

For $g \in \mathfrak{G}$, put $\sigma g = P(\sigma)gP(\sigma)^{-1}$. The above formula shows that

$$(8.9) \qquad i(\gamma^{-1}(\sigma\gamma))i(P(\sigma))S = S \qquad \text{for every} \quad \sigma \in \mathfrak{G}.$$

Suppose that

$$(8.10) \qquad \mathfrak{H}^n = \Delta(\mathfrak{H})Z(\mathfrak{H})^n H^0.$$

Here Δ denotes the diagonal embedding of $\mathfrak{H}$ into $\mathfrak{H}^n$. Then there exists a $\gamma' \in H^0$ which satisfies (8.9) with γ' in place of γ. In this case, we see that $i(P(\sigma))$ stabilizes the splitting datum $i(\gamma')S$ of H^0 and Hypothesis 8.2 holds true.

In general, let $s : \mathrm{Out}(H^0) \longrightarrow \mathrm{Aut}(H^0)$ be the section defined by a splitting datum S. Let $\pi : \mathrm{Aut}(H^0) \longrightarrow \mathrm{Out}(H^0)$ be the canonical homomorphism. We can find $\beta_\sigma \in H^0$ such that $i(\beta_\sigma)i(P(\sigma)) = s(\pi(i(P(\sigma))))$. This is equivalent to $i(\beta_\sigma)i(P(\sigma))S = S$. Hence by (8.9), we get $\beta_\sigma = z_\sigma\gamma^{-1}(\sigma\gamma)$ with $z_\sigma \in Z(\mathfrak{H})^n$. Then we see that $z_\sigma(P(\sigma)z_\tau P(\sigma)^{-1})z_{\sigma\tau}^{-1}$ defines a 2-cocycle of $\mathfrak{G}$ taking values in $Z(H^0)$. Hypothesis 8.2 implies that the cohomology class of this cocycle is trivial. Essentially needed in the last step of the argument given below is that this cohomology class has a splitting field.

Let $\mathfrak{K}$ be an algebraic extension of $\mathbf{Q}_\ell$ of finite degree. Let G be a connected semisimple algebraic group defined over $\mathfrak{K}$. We assume that G splits over $\mathfrak{K}$. Then we have the split exact sequence

$$(8.11) \qquad 1 \longrightarrow \mathrm{Inn}(G(\overline{\mathfrak{K}})) \cap \mathrm{Aut}_{\mathfrak{K}}(G) \longrightarrow \mathrm{Aut}_{\mathfrak{K}}(G) \longrightarrow \mathrm{Out}(G) \longrightarrow 1.$$

Here $\mathrm{Aut}_{\mathfrak{K}}(G)$ denotes the group of automorphisms of G as an algebraic group, the automorphisms being defined over $\mathfrak{K}$. Let $\mathrm{Aut}(G(\mathfrak{K}))$ be the automorphism group of the (ℓ-adic) topological group $G(\mathfrak{K})$. We consider $\mathrm{Aut}_{\mathfrak{K}}(G) \subset \mathrm{Aut}(G(\mathfrak{K}))$. The group of inner automorphisms by the elements in $G(\overline{\mathfrak{K}})$ is denoted by $\mathrm{Inn}(G(\overline{\mathfrak{K}}))$. For $\varphi \in \mathrm{Aut}(G(\mathfrak{K}))$, we define a continuous map ψ_φ from $G(\mathfrak{K})$ to $G(\mathfrak{K})$ by

$$\psi_\varphi(x) = x^{-1}\varphi(x), \qquad x \in G(\mathfrak{K}).$$

For a compact subset C of $G(\mathfrak{K})$ and an open subset U whose closure is compact, we put

$$W(C, U) = \{\varphi \in \mathrm{Aut}(G(\mathfrak{K})) \mid \psi_\varphi(C) \subset U\}.$$

PROPOSITION 8.3. *There exists a compact subset C and an open compact subgroup U of $G(\mathfrak{K})$ such that* $\mathrm{Aut}_{\mathfrak{K}}(G) \cap W(C, U) \subset \mathrm{Inn}(G(\overline{\mathfrak{K}})) \cap \mathrm{Aut}_{\mathfrak{K}}(G)$

PROOF. First we assume that G is of adjoint type. Then we have $\mathrm{Inn}(G(\overline{\mathfrak{K}})) \cap \mathrm{Aut}_{\mathfrak{K}}(G) = \mathrm{Inn}(G(\mathfrak{K}))$. Here $\mathrm{Inn}(G(\mathfrak{K}))$ denotes the group of inner automorphisms by the elements in $G(\mathfrak{K})$.

Let T be a maximal split torus and B be a Borel subgroup containing T. Let $\Psi_0(G)$ be the based root datum of G. We have

$$\mathrm{Out}(G) \cong \Psi_0(G) \cong \mathrm{Aut}(G, B, T, \{u_\alpha\}_{\alpha\in\Delta}).$$

For (8.11), we take the section $s : \mathrm{Out}(G) \longrightarrow \mathrm{Aut}_{\mathfrak{K}}(G)$, which is obtained from this isomorphism. We may assume that the elements in the images of the section are rational over $\mathfrak{K}$. Take $\sigma \in \mathrm{Out}(G)$ and put $\tau = s(\sigma)$.

Suppose that $i(g)\tau \in W(C, U)$. Then we have $y^{-1}g\tau(y)g^{-1} \in U$ for every $y \in C$. Put $x = \tau(y)$. Then, $i(g)\tau \in W(C, U)$ is equivalent to the fact that, for every $x \in \tau(C)$, there exists $u = u(x) \in U$ such that $gxg^{-1} = \tau^{-1}(x)u$. We write C for $\tau(C)$ and τ for τ^{-1}. Then the condition can be written as

$$(8.12) \qquad gxg^{-1} = \tau(x)u(x), \qquad u(x) \in U \quad \text{for every } x \in C.$$

Let $\mathcal{O}$ be the ring of integers of $\mathfrak{K}$. Let K be the subgroup of $G(\mathfrak{K})$ generated by $u_\alpha(t)$, $\alpha \in \Sigma$, $t \in \mathcal{O}$ and $\check{\alpha}(t)$, $t \in \mathcal{O}^\times$. Here Σ is the set of all roots and $\check{\alpha} \in X_*(T)$ is a coroot; $\{u_\alpha(t)\}$ denotes the root subgroup corresponding to α. Then K is a maximal compact subgroup of $G(\mathfrak{K})$. Since τ stabilizes $u_\alpha(t)$, $t \in \mathcal{O}$ and $\tau(\check{\alpha}(t)) = (\tau\check{\alpha})(t)$, we have

$$(8.13) \qquad \tau(K) = K.$$

We can find a decreasing sequence $\{U_n\}$ of open compact subgroups so that

$$U_0 = K, \qquad U_n \lhd K, \qquad \cap_{n=1}^{\infty} U_n = \{1\}.$$

First suppose that the condition (8.12) holds for $C = K$, $U = U_n$, $g = g_n$. Then we have $gKg^{-1} = K$. Put

$$T^+ = \{t \in T(\mathfrak{K}) \mid \alpha(t) \in \mathcal{O} \ \text{ for every positive root } \alpha\}.$$

Then we have a Cartan decomposition (cf. Steinberg [St1], Theorem 21)

$$G(\mathfrak{K}) = KT^+K.$$

If $t \in T(\mathfrak{K})$ normalizes K, we have $\alpha(t) \in \mathcal{O}$ for every root α, since $tu_\alpha(x)t^{-1} = u_\alpha(\alpha(t)x)$. This implies $\alpha(t) \in \mathcal{O}$ for every $\alpha \in X^*(T)$. By [St1], Lemma 49, we have $t \in K$. Thus we have shown that the normalizer of K in $G(\mathfrak{K})$ is K itself. In particular $g \in K$. The condition (8.12) can be written as

$$g_n x g_n^{-1} \equiv \tau(x) \mod U_n, \qquad x \in K.$$

Suppose that the assertion of our proposition is false. Then this equation must hold for infinitely many n for some $\tau \notin \mathrm{Inn}(G(\mathfrak{K}))$, since $\mathrm{Out}(G)$ is a finite group. Take a convergent subsequence of $\{g_n\}$ and let $g \in K$ be its limit. Then we have

$$gxg^{-1} = \tau(x), \qquad x \in K.$$

Put $\tau' = i(g^{-1})\tau$. We have $\tau'|K = \mathrm{id}$. Since $T \cap K$ is Zariski dense in T and τ' is an algebraic automorphism, we have $\tau'|T = \mathrm{id}$. By the Cartan decomposition, we obtain $\tau' = \mathrm{id}$. This is a contradiction and completes the proof in the case where G is of adjoint type.

In the general case, let $\psi : G \longrightarrow \overline{G}$, $\overline{G} = G/Z(G)$ be the central isogeny. Define a maximal compact subgroup K_G (resp. K) of $G(\mathfrak{K})$ (resp. $\overline{G}(\mathfrak{K})$) as above. Then $\psi(K_G) \subset K$. We see that the normalizer of $\psi(K_G)$ in $\overline{G}(\mathfrak{K})$ is K and that a Cartan decomposition with respect to $\psi(K_G)$ holds. Let $\tau \in \mathrm{Aut}_{\mathfrak{K}}(G)$. Then τ induces an automorphism $\bar{\tau} \in \mathrm{Aut}_{\mathfrak{K}}(\overline{G})$. Applying the argument above to $\overline{G}$ and $W = \psi(K_G)$, we may assume that there exists $\bar{g} \in \overline{G}(\mathfrak{K})$ such that $\bar{\tau}(x) = \bar{g}x\bar{g}^{-1}$, $x \in \overline{G}(\overline{\mathfrak{K}})$. Take $g \in G(\overline{\mathfrak{K}})$ so that $\psi(g) = \bar{g}$. Then we have $\tau(x) = z(x)gxg^{-1}$, $x \in G(\overline{\mathfrak{K}})$, $z(x) \in Z(G)$. Since $x \mapsto z(x) \in Z(G)$ is continuous with respect to the Zariski topology, we have $z(x) = 1$. This completes the proof. □

REMARK 8.4. We can make $\mathrm{Aut}(G(\mathfrak{K}))$ a Hausdorff topological group by assigning $W(C,U)\cap W(C,U)^{-1}$ as the fundamental system of open neighbourhoods of the identity. Then, by Proposition 8.3, we see that the projection map $\mathrm{Aut}_{\mathfrak{K}}(G) \longrightarrow \mathrm{Out}(G)$ in (8.11) is continuous with the discrete topology on $\mathrm{Out}(G)$ and the induced topology on $\mathrm{Aut}_{\mathfrak{K}}(G)$.

We consider the motive $\wedge^{\tilde{d}}\widetilde{M}$ where $\tilde{d}$ is the rank of $\widetilde{M}$. As a rank 1 motive over K, it is associated to an algebraic Hecke character χ of $K_{\mathbf{A}}^{\times}$. As in §7, we consider the λ-adic representation χ_λ attached to χ and the torus T associated to the induced representation $\mathrm{Ind}_{\mathrm{Gal}(\overline{F}/K)}^{\mathrm{Gal}(\overline{F}/F)}\chi_\lambda$. We have a morphism of algebraic groups defined over E:

$$\delta : H^0 \ni (x_\alpha)_{\alpha\in\mathfrak{G}} \longrightarrow (\det x_\alpha)_{\alpha\in\mathfrak{G}} \in T. \tag{8.14}$$

First we replace E_λ by its finite extension so that the 2-cocycle f_Z takes values in $Z(H^0(E_\lambda))$. We see that mapping the factor set f of (6.4) by δ, we obtain a 2-cocycle of $\mathfrak{G}$ taking values in $T(E_\lambda)$. When we change f to an equivalent factor set, the cohomology class of $\delta(f)$ in $H^2(\mathfrak{G}, T(E_\lambda))$ does not change. Let ℓ be the prime lying below λ. We choose λ so that ℓ is prime to $\tilde{d}[K:F]$ and to the order of the outer automorphism group of the derived group of H^0. As shown in §7, the cocycle $\delta(f)$ has a splitting field. We enlarge K so that $\delta(f)$ splits. This procedure corresponds to the following operation. We consider $\widetilde{M}$ as the motive over the enlarged field K and reset $M = R_{K/F}(\widetilde{M})$. Then enlarge K again so that $\tau_\lambda(\mathrm{Gal}(\overline{F}/K))$ is contained in a pro-ℓ-group.

Let $\mathcal{T}' \subset T(E_\lambda)$ be the maximal compact subgroup of $T(E_\lambda)$. If $\delta(f)$ takes values in $\mathcal{T}'$, then $\delta(f)$ splits in $H^2(\mathfrak{G}, \mathcal{T}')$ (cf. §7). Replacing E_λ by a finite extension, we may assume that for every $t' \in \mathcal{T}'$, there exists $t \in Z(H^0(E_\lambda))$ satisfying $\delta(t) = t'$. (Here $\mathcal{T}'$ denotes the group before the replacement of E_λ.) Let $\mathcal{T}$ be the maximal compact subgroup of $Z(H^0(E_\lambda))$. Then t as above can be taken from $\mathcal{T}$. Suppose that a factor set f takes values in $\mathcal{T}$. Then we see that the cohomology class of $f^{\tilde{d}}$ is trivial in $H^2(\mathfrak{G}, \mathcal{T})$.

LEMMA 8.5. *We have* $i(D_\sigma) \in \mathrm{Inn}(H^0)$.

PROOF. By Lemma 8.1 and (8.7), we have $i(D_\sigma) \in \mathrm{Aut}(H^0)$. Let $D(H^0)$ be the derived group and C be the connected center of H^0. We have $H^0 = C \cdot D(H^0)$ over $\overline{E_\lambda}$. Hence, by (8.4), it suffices to show that $i(D_\sigma)$ gives an inner automorphism of $D(H^0)$. We consider the canonical homomorphism $\pi : \mathrm{Aut}_{E_\lambda}(D(H^0)) \longrightarrow \mathrm{Out}(D(H^0))$ as in (8.11). Since D_σ belongs to a pro-ℓ-subgroup of $\mathfrak{H}(E_\lambda)^n$ (cf. (8.6)), $D_\sigma^{l^q}$ converges to the identity element when q tends to infinity, with respect to the λ-adic topology of $\mathfrak{H}(E_\lambda)^n$. By Proposition 8.3, there exists $0 < q \in \mathbf{Z}$ such that $\pi(D_\sigma^{l^q}) = 1$. Since ℓ does not divide the order of $\mathrm{Out}(D(H^0))$, we have $\pi(D_\sigma) = 1$. This completes the proof. □

LEMMA 8.6. *Replacing E_λ by a finite extension, we have $z_\sigma D_\sigma = \alpha_\sigma$ with $z_\sigma \in Z(\mathfrak{H}(E_\lambda))^n$, $\alpha_\sigma \in H^0(E_\lambda)$. Moreover we can take z_σ from a pro-ℓ-subgroup of $Z(\mathfrak{H}(E_\lambda))^n$.*

PROOF. By Lemma 8.5, there exists $\alpha_\sigma \in H^0(E_\lambda)$ such that $i(D_\sigma) = i(\alpha_\sigma)$, enlarging E_λ if necessary. Since the projection of H^0 to every ith factor (cf. (8.3)) is surjective, we see that there exists $z_\sigma \in Z(\mathfrak{H}(E_\lambda))^n$ such that $z_\sigma D_\sigma = \alpha_\sigma$. Our task is to show the latter assertion.

Put $D = D_\sigma$. We consider the condition that $xD \in H^0$ for $x = (y_\alpha \cdot 1_{\tilde{d}})_{\alpha \in \mathfrak{G}}$. This is the case if and only if $(xD, u) \in \mathrm{MT}(U)$ for some $u \in \mathbf{G}_m$. (Here $U = \oplus_{\alpha \in \mathfrak{G}} V_\alpha$ in the notation of the proof of Lemma 8.1.) For a tensor t of type $(0, 0)$ in the tensor space (8.8), we have

$$\prod_{\alpha \in \mathfrak{G}} y_\alpha^{(l_\alpha - m_\alpha)} u^{-p} Dt = t. \tag{8.15}$$

Since this equation holds for some $y_\alpha \in E_\lambda^\times$, we have $Dt = c(t)t$ with $c(t) \in E_\lambda^\times$. Then we see that $c(t)$ belongs to a pro-ℓ-subgroup of $\mathcal{O}_{E_\lambda}^\times$.

To verify $(xD, u) \in \mathrm{MT}(U)$, it suffices to check the condition (8.15) for finitely many tensors t. The equations (8.15) for these tensors can be written in the form

$$\prod_{j=1}^{N} y_j^{a_{ij}} = c_i, \qquad 1 \le i \le m. \tag{8.16}$$

Here $a_{ij} \in \mathbf{Z}$ and c_i belongs to a pro-ℓ-subgroup of $\mathcal{O}_{E_\lambda}^\times$. To complete the proof, we need the next sublemma.

SUBLEMMA 8.7. *Let $\mathcal{O}$ be the ring of integers of $\overline{E_\lambda}$. Regard (8.16) as simultaneous equations with respect to y_j, $1 \le j \le N$. If (8.16) has a solution $y_j \in \overline{E_\lambda}$, $1 \le j \le N$, then it has a solution $y_j \in \mathcal{O}^\times$, $1 \le j \le N$.*

PROOF. Put $A = (a_{ij}) \in M(m, N, \mathbf{Z})$. We can find $U = (u_{ij}) \in \mathrm{GL}(m, \mathbf{Z})$ and $V = (v_{ij}) \in \mathrm{GL}(N, \mathbf{Z})$ so that

$$UAV = \begin{pmatrix} e_1 & \cdots & 0 & 0 & \cdots & 0 \\ \vdots & \ddots & \vdots & \vdots & & \vdots \\ 0 & \cdots & e_r & 0 & \cdots & 0 \\ 0 & \cdots & 0 & 0 & \cdots & 0 \\ \vdots & & \vdots & \vdots & \ddots & \\ 0 & \cdots & 0 & 0 & \cdots & 0 \end{pmatrix}, \qquad 0 < e_i \in \mathbf{Z}.$$

Put

$$c_i' = \prod_{k=1}^{m} c_k^{u_{ik}}, \quad 1 \le i \le m \qquad y_j = \prod_{k=1}^{N} z_k^{v_{jk}}, \quad 1 \le j \le N.$$

Then (8.16) is transformed to equivalent equations:

$$z_i^{e_i} = c_i', \quad 1 \le i \le r, \qquad 1 = c_i', \quad r < i \le N. \tag{8.17}$$

By the assumption $c_i' = 1$, $r < i \le N$ holds. Hence (8.17) has a solution $z_i \in \mathcal{O}^\times$, $1 \le i \le r$, $z_i = 1$, $r < i \le N$, which gives a desired solution y_j, $1 \le j \le N$. □

By the sublemma, replacing E_λ by its finite extension, we can find a solution $y_j \in \mathcal{O}_{E_\lambda}^\times$, $1 \le j \le N$ of (8.16). Taking the projection to the pro-ℓ-part of $\mathcal{O}_{E_\lambda}^\times$, we obtain a solution y_j, $1 \le j \le N$ lying in the pro-ℓ-part of $\mathcal{O}_{E_\lambda}^\times$. This completes the proof of Lemma 8.6. □

Let $\mathcal{T}$ be the maximal compact subgroup of $Z(H^0(E_\lambda))$ and let $\mathcal{L}$ be the pro-ℓ-part of $\mathcal{T}$. Let $\mathcal{T}^*$ be the maximal compact subgroup of $Z(\mathfrak{H}(E_\lambda))^n$ and let $\mathcal{L}^*$ be the pro-ℓ-part of $\mathcal{T}^*$. We have $\mathcal{T} = \mathcal{T}^* \cap H^0(E_\lambda)$, $\mathcal{L} = \mathcal{L}^* \cap H^0(E_\lambda)$. By Hypothesis 8.2, we may assume $s(\pi(a(\sigma))) = i(P(\sigma))$, $\sigma \in \mathfrak{G}$ for a suitable section s. By (8.7), we have $a(\sigma) = i(D_\sigma)^{-1} i(P(\sigma))$. By Lemma 8.6, we can take $z_\sigma \in \mathcal{L}^*$ so that $z_\sigma D_\sigma \in H^0(E_\lambda)$. Then we have

$$s(\pi(a(\sigma))) = i(z_\sigma D_\sigma) a(\sigma),$$

$$f_Z(\sigma,\tau) = z_\sigma D_\sigma(a(\sigma)(z_\tau D_\tau)) f(\sigma,\tau)(z_{\sigma\tau} D_{\sigma\tau})^{-1}.$$

Using $z_\sigma D_\sigma = z_\sigma P(\sigma)\rho_\lambda(\widetilde{\sigma})^{-1}$ and (6.4), we obtain

$$a_Z(\sigma) = i(P(\sigma)),$$

$$f_Z(\sigma,\tau) = z_\sigma(P(\sigma) z_\tau P(\sigma)^{-1}) z_{\sigma\tau}^{-1}.$$

The last formula shows that f_Z is a 2-cocycle of $\mathfrak{G}$ taking values in $\mathcal{L}$. Now the splitting of f_Z in $H^2(\mathfrak{G}, \mathcal{T})$ follows from Lemma 7.1. This completes the proof of Theorem 5.6. □

EXAMPLE 8.8. Let us show that a splitting field does not necessarily exist for the case $w = 0$. Let $f \in S_k(\Gamma_0(N), \chi)$ be a primitive cusp form of weight k. Let E be the number field generated by the eigenvalues of Hecke operators for f. Associated to f, there exists a motive M_f over $\mathbf{Q}$ with coefficients in E; M_f is of weight $k-1$ and of rank 2. We assume that: (i) f is not a CM-form. (ii) k is odd. (iii) χ is primitive of conductor N and takes values in $\{\pm 1\}$. By a result of Ribet, $\mathrm{Im}(\rho_\lambda)$ is Zariski dense in $\mathrm{GL}(2)$. We put $M = M_f((k-1)/2)$. Then M is of weight 0 and we find that the exact sequence (5.2) takes the following form:

(8.18)
$$1 \longrightarrow H^0(\mathbf{C}) = \mathrm{SL}(2)(\mathbf{C}) \longrightarrow H(\mathbf{C}) \longrightarrow \mathrm{Gal}(K/\mathbf{Q}) \longrightarrow 1.$$

Here K is the imaginary quadratic field which corresponds to χ. This extension obviously splits. Let us calculate the cocycle f_Z following the procedure of §5, (5.12)–(5.14) with $\widetilde{H} = H$. Let σ be the generator of $\mathrm{Gal}(K/\mathbf{Q})$. We choose $\widetilde{\sigma} \in \mathrm{Gal}(\overline{\mathbf{Q}}/\mathbf{Q})$ so that $\widetilde{\sigma}|K = \sigma$ and take $\widetilde{1} = 1$. Then we have

$$a(1) = 1, \qquad a(\sigma) = i(\rho_\lambda(\widetilde{\sigma})),$$

$$f(1,1) = f(1,\sigma) = f(\sigma,1) = 1_2, \qquad f(\sigma,\sigma) = \rho_\lambda(\widetilde{\sigma})^2.$$

Since $\mathrm{Out}(\mathrm{SL}(2,\mathbf{C})) = \{1\}$, the section s is the trivial mapping. For $\tau \in \mathrm{Gal}(K/\mathbf{Q})$, we must take $\alpha_\tau \in \mathrm{SL}(2,\mathbf{C})$ so that $i(\alpha_\tau)a(\tau) = 1$. By (5.13), we have $a_Z(\tau) = 1$, $\tau \in \mathrm{Gal}(K/\mathbf{Q})$. We may take $\alpha_1 = 1$. Then we have

$$f_Z(1,1) = f_Z(1,\sigma) = f_Z(\sigma,1) = 1_2, \qquad f_Z(\sigma,\sigma) = (\alpha_\sigma \rho_\lambda(\widetilde{\sigma}))^2.$$

Since $i(\alpha_\sigma)a(\sigma) = 1$, we can write $\alpha_\sigma \rho_\lambda(\widetilde{\sigma}) = z_\sigma 1_2$ with $z_\sigma \in \mathbf{C}^\times$. By $\det \rho_\lambda(\widetilde{\sigma}) = -1$, we get $z_\sigma^2 = -1$. Thus we find that we may take $f_Z(\alpha,\beta) = f'(\alpha,\beta) \cdot 1_2$, where

$$f'(1,1) = f'(1,\sigma) = f'(\sigma,1) = 1, \qquad f'(\sigma,\sigma) = -1.$$

We can show easily that the cohomology class of f' in $H^2(\mathrm{Gal}(K/\mathbf{Q}), \{\pm 1\})$ does not have a splitting field, using the fact that there does not exist a quartic cyclic extension L of $\mathbf{Q}$ which contains K.

9. Local splittings

We use the notation of §6. Let f_Z be the 2-cocycle attached to ρ_λ defined by (6.3)–(6.7). In this section, we will show that assuming the local splitting of f_Z, we can still formulate an analogue of the main conjecture on $G_{\mathbf{A}}$. Here G is the quasi-split group over F defined in §5 (with $\widetilde{H} = H$).

We fix an embedding $E_\lambda \hookrightarrow \mathbf{C}$. Let V be the representation space of ρ_λ; V is considered as a vector space over $\mathbf{C}$. We regard $H(\mathbf{C})$ as the group whose underlying set is $H^0(\mathbf{C}) \times \mathrm{Gal}(K/F)$ and whose group law is defined by

$$(9.1)\qquad \begin{aligned}(h_1,\sigma)(h_2,\tau) =&(h_1(a_Z(\sigma)h_2)f_Z(\sigma,\tau),\sigma\tau),\\ &h_1,h_2 \in H^0(\mathbf{C}),\quad \sigma,\tau \in \mathrm{Gal}(K/F).\end{aligned}$$

Let $r_1 : H(\mathbf{C}) \hookrightarrow \mathrm{GL}(V)$ be the inclusion map. Then we can write

$$(9.2)\qquad r_1((h,\sigma)) = r_0(h)A_\sigma,$$

where $r_0 : H^0(\mathbf{C}) \hookrightarrow \mathrm{GL}(V)$ is the inclusion map and $A_\sigma = \alpha_\sigma\rho_\lambda(\widetilde{\sigma}) \in H(\mathbf{C})$. Let $\varphi : \mathrm{Gal}(\overline{F}/F) \longrightarrow \mathrm{Gal}(K/F)$ be the canonical map and put $\widetilde{a}_Z(\sigma) = a_Z(\varphi(\sigma))$, $\widetilde{f}_Z(\sigma,\tau) = f_Z(\varphi(\sigma),\varphi(\tau))$. Let $\widetilde{H}(\mathbf{C})$ be the group whose underlying set is $H^0(\mathbf{C})\times \mathrm{Gal}(\overline{F}/F)$ and whose group law is defined by

$$(h_1,\sigma)(h_2,\tau) = (h_1(\widetilde{a}_Z(\sigma)h_2)\widetilde{f}_Z(\sigma,\tau),\sigma\tau),\ h_1,h_2 \in H^0(\mathbf{C}),\sigma,\tau \in \mathrm{Gal}(\overline{F}/F).$$

We define ${}^LG$ as the semi-direct product $H^0(\mathbf{C}) \rtimes \mathrm{Gal}(\overline{F}/F)$. The group law is defined by

$$(9.3)\qquad (h_1,\sigma)(h_2,\tau) = (h_1(\widetilde{a}_Z(\sigma)h_2),\sigma\tau).$$

Let T be a torus contained in $\mathrm{GL}(V)$ which is stable under the action of $a_Z(\sigma)$, $\sigma \in \mathrm{Gal}(K/F)$. We assume that T contains $Z(H^0(\mathbf{C}))$, T centralizes $H^0(\mathbf{C})$ and that

$$(9.4)\qquad H^1(\mathrm{Gal}(\overline{F}/F),T(\mathbf{C})) = H^2(\mathrm{Gal}(\overline{F}/F),T(\mathbf{C})) = 1.$$

Here cohomology groups are defined using continuous cochains. We define a subgroup $\mathcal{H}(\mathbf{C})$ of $\mathrm{GL}(V)$ by

$$\mathcal{H}(\mathbf{C}) = T(\mathbf{C})H^0(\mathbf{C}) \rtimes \mathrm{Gal}(\overline{F}/F).$$

We have ${}^LG \subset \mathcal{H}(\mathbf{C})$. Let

$$(9.5)\qquad \widetilde{f}_Z(\sigma,\tau) = t_\sigma(\widetilde{a}_Z(\sigma)t_\tau)t_{\sigma\tau}^{-1},\qquad t_\sigma \in T(\mathbf{C})$$

be a splitting of $\widetilde{f}_Z$ in $T(\mathbf{C})$. Then we can define an injective homomorphism $\iota : \widetilde{H}(\mathbf{C}) \longrightarrow \mathcal{H}(\mathbf{C})$ by

$$(9.6)\qquad \iota((h,\sigma)) = (ht_\sigma,\sigma),\qquad h \in H^0(\mathbf{C}),\quad \sigma \in \mathrm{Gal}(\overline{F}/F).$$

We can define a homomorphism r^* of $\mathcal{H}(\mathbf{C})$ into $\mathrm{GL}(V)$ by the formula [7]

$$(9.7)\qquad r^*((h,\sigma)) = r_0(h)t_\sigma^{-1}A_\sigma,\qquad h \in T(\mathbf{C})H^0(\mathbf{C}),\quad \sigma \in \mathrm{Gal}(\overline{F}/F).$$

Here we use the same letter r_0 for the inclusion map $T(\mathbf{C})H^0(\mathbf{C}) \hookrightarrow \mathrm{GL}(V)$. Let $\iota^* : {}^LG \hookrightarrow \mathcal{H}(\mathbf{C})$ be the inclusion map. We define a homomorphism of ${}^LG$ into $\mathrm{GL}(V)$ by $r = r^* \circ \iota^*$.

[7] We normalize the cocycle f_Z so that $f_Z(1,\sigma) = f_Z(\sigma,1) = 1$, $\sigma \in \mathrm{Gal}(K/F)$.

Now let v be a place of F. We assume that $\widetilde{f}_Z$ splits locally at v, that is,

$$\widetilde{f}_Z(\sigma,\tau) = u_\sigma(\widetilde{a}_Z(\tau)u_\tau)u_{\sigma\tau}^{-1}, \qquad u_\sigma \in Z(H^0(\mathbf{C})), \quad \sigma,\tau \in \mathrm{Gal}(\overline{F_v}/F_v). \tag{9.8}$$

Here $\{u_\sigma\}$ is a continuous 1-cochain. This condition is equivalent to that $f_Z|\mathrm{Gal}(K_w/F_v)$ has a splitting field, where w is a place of K lying above v. We can construct a Langlands parameter $\phi_v : W'_{F_v} \longrightarrow {}^LG$ from ψ_v as follows. Let $\psi_v : W'_{F_v} \longrightarrow H(\mathbf{C})$ be the local parameter defined in §4 and put

$$\psi_v(g) = (\psi_v^0(g), \pi(g)), \qquad g \in W'_{F_v}.$$

Here $\pi(g)$ denotes the projection of g to $\mathrm{Gal}(K/F)$ and $\psi_v^0(g) \in H^0(\mathbf{C})$. By Lemma 5.2, ψ_v can be lifted to a homomorphism $\widetilde{\psi}_v : W'_{F_v} \longrightarrow \widetilde{H}(\mathbf{C})$ by the formula

$$\widetilde{\psi}_v(g) = (\psi_v^0(g), \bar{g}), \qquad g \in W'_{F_v}.$$

Here $\bar{g}$ denotes the projection of g to $\mathrm{Gal}(\overline{F_v}/F_v)$. Then we can define the Langlands parameter by the formula

$$\phi_v(g) = (\psi_v^0(g)u_{\bar{g}}, \bar{g}), \qquad g \in W'_{F_v}. \tag{9.9}$$

Calculating using definitions, we have, for $g \in W'_{F_v}$,

$$(r_1 \circ \psi_v)(g) = r_0(\psi_v^0(g))A_{\bar{g}},$$
$$(r \circ \phi_v)(g) = r_0(\psi_v(g))u_{\bar{g}}t_{\bar{g}}^{-1}A_{\bar{g}}.$$

By the condition (9.4), there exists $t \in T(\mathbf{C})$ such that

$$u_\sigma t_\sigma^{-1} = t^{-1}(\widetilde{a}_Z(\sigma)t), \qquad \sigma \in \mathrm{Gal}(\overline{F_v}/F_v).$$

Then we find

$$t(r_0(\psi_v(g))u_{\bar{g}}t_{\bar{g}}^{-1}A_{\bar{g}})t^{-1} = r_0(\psi_v^0(g))A_{\bar{g}}$$

and obtain

$$r_1 \circ \psi_v \cong r \circ \phi_v. \tag{9.10}$$

We devote the remaining part of this section to the proof of the local splitting in the simplest case. Let w be a place of K lying over v. We consider the restriction of f_Z to $\mathrm{Gal}(K_w/F_v)$. We assume that v is a finite place, K_w is unramified over F_v and that ρ_λ is unramified at v. Let $\Phi_v \in \mathrm{Gal}(\overline{F}_v/F_v)$ be a Frobenius element. We put $x = \rho_\lambda(\Phi_v) \in H$ and we assume that x is semisimple. (This semisimplicity is a widely believed conjecture.)

Put $f = f_v = [K_w : F_v]$ and let σ be the restriction of Φ_v to K_w. Then $\mathrm{Gal}(K_w/F_v)$ is the cyclic group of order f generated by σ. We have $x^f \in H^0$. Let $i(x)$ denote the automorphism of H^0 defined by $i(x)h = xhx^{-1}$. By a theorem of Steinberg ([St2], Theorem 7.5), there exists a Borel subgroup B of H^0 and a maximal torus $T \subset B$ which is stabilized by $i(x)$. Let D be the Zariski closure of $\{x^{fk} \mid k \in \mathbf{Z}\}$. Since D is diagonarizable, we have $D = T_1 \times \mathcal{F}$, where $T_1 = D^0$ is a torus and $\mathcal{F}$ is a finite group. Replacing K_w by an unramified extension, we may assume that $\mathcal{F} = \{1\}$. Then T_1 normalizes T. Since $N_{H^0}(T)/T$ is finite, we see that $T_1 \subset T$. Take $\{u_\alpha\}$ so that $(B, T, \{u_\alpha\}_{\alpha\in\Delta})$ is a splitting datum of H^0. Let $\pi : \mathrm{Aut}(H^0) \longrightarrow \mathrm{Out}(H^0)$ be the canonical homomorphism and s be the section defined by the splitting datum. Now take $\alpha_\sigma \in H^0(\mathbf{C})$ which satisfies (6.5), i.e.,

$$s(\pi(i(x))) = i(\alpha_\sigma)i(x). \tag{9.11}$$

Since $i(x)$ stabilizes (B,T), we have $\alpha_\sigma \in T(\mathbf{C})$. Put $t(\sigma)=\alpha_\sigma$. By (6.7), we have

$$f_Z(\sigma^i,\sigma^j)=t(\sigma^i)(i(x)^i t(\sigma^j))t(\sigma^{i+j})^{-1}f(\sigma^i,\sigma^j).$$

Taking $\widetilde{\sigma^i}=\Phi_v^i$, $0\le i<f$, we get, for $0\le i,j<f$,

$$f(\sigma^i,\sigma^j)=\begin{cases}1 & \text{if } i+j<f,\\ x^f & \text{if } i+j\ge f.\end{cases}$$

We may take

$$t(\sigma^i)=t(\sigma)(xt(\sigma)x^{-1})\cdots(x^{i-1}t(\sigma)x^{-(i-1)})=(t(\sigma)x)^i x^{-i}.$$

Then we find, for $0\le i,j<f$,

$$f_Z(\sigma^i,\sigma^j)=\begin{cases}1 & \text{if } i+j<f,\\ (t(\sigma)x)^f & \text{if } i+j\ge f.\end{cases}$$

We have $(t(\sigma)x)^f\in Z(H^0(\mathbf{C}))$. We can write $Z(H^0(\mathbf{C}))=C\times\mathcal{F}'$, where C is the connected center and $\mathcal{F}'$ is a finite group. Replacing K_w by a suitable unramified extension, we may assume that $(t(\sigma)x)^f\in C$. By a result of Langlands ([Lan2], Lemma 4), we conclude that $f_Z|\mathrm{Gal}(K_w/F_v)$ has a splitting field.

Let S be a set of places of F. We call ρ_λ has the *local splitting property outside* S if f_Z splits locally at $v\notin S$ and if there exists T satisfying (9.4). Then the obvious analogue of Lemma 6.2 holds and we are reduced to the case where ρ_λ is the induced representation as in the beginning of §8. In that case, we may take $T=Z(\mathfrak{H})^n$ since this group is the induced module from the trivial one and cohomologically trivial. (Thus (9.4) holds.) The discussion of this section can be summarized by the following theorem.

THEOREM 9.1. *Let S' be the set of finite places v of F such that ρ_λ is unramified at v, v is unramified in K and that the image of the Frobenius element is semisimple. Let S be the complement of S'. Then f_Z has the local splitting property outside S.*

10. Further discussions

To reduce problems appearing in fuctoriality conjectures to the case where the L-group has connected center, Langlands ([Lan2]) introduced the notion of a z-extension. Let us recall the situation briefly. Let G be a connected reductive group over F. For simplicity, we assume that G is quasi-split. Then Langlands constructed a connected reductive quasi-split group $\widetilde{G}$ defined over F such that $\widetilde{G}$ is a central extension of G over F and such that the center of ${}^L\widetilde{G}^0$ is connected. Moreover the maps $\widetilde{G}(F)\longrightarrow G(F)$, $\widetilde{G}(F_{\mathbf{A}})\longrightarrow G(F_{\mathbf{A}})$ are surjective, and the map $\widetilde{G}(F_v)\longrightarrow G(F_v)$ is surjective for every place v of F.

We can use this devise for an individual case in section 5, eliminating the problem of the splitting field at the cost of enlarging the group. If $\widetilde{G}$ happens to coincide with the group $\widetilde{G}$ obtained from $\widetilde{H}$ in section 5, it poses a subtle question.

Using the Weil form of the L-group, $\widetilde{f_Z}$ has a splitting field. Suppose that f_Z has a splitting field. Then the Main conjecture predicts automorphic representations π and $\widetilde{\pi}$ on $G(F_{\mathbf{A}})$ and on $\widetilde{G}(F_{\mathbf{A}})$ respectively. Here G is the quasi-split group in the minimal case. As in section 5, we have explicit Langlands parameters corresponding to π and $\widetilde{\pi}$. Let Z be the kernel of the central homomorphism $\widetilde{G}\longrightarrow G$. Then the central character of $\widetilde{\pi}$ must be trivial on $Z(F_{\mathbf{A}})$. The author has not verified that this is always the case. In the situation of Example 8.8, $H^0 = \mathrm{SL}(2)$, $\widetilde{H} = \mathrm{GL}(2)$, $G = \mathrm{PGL}(2)$, $\widetilde{G} = \mathrm{GL}(2)$. The central character of $\widetilde{\pi}$ is not trivial on $Z(\mathbf{Q}_{\mathbf{A}})$, which is consistent with the fact that f_Z does not have a splitting field. Another simple case is the example discussed at the end of section 5, in which the central character is trivial.

We include the following Proposition to supplement the discussions in section 5.

PROPOSITION 10.1. *Suppose that we are in the minimal case. If K is a cyclic extension of F, ρ_λ is absolutely irreducible and $w \neq 0$, then the cohomology group $H^2(\mathrm{Gal}(K/F), Z(H^0(\mathbf{C})))$ is trivial.*

PROOF. Since $\mathrm{Gal}(K/F)$ is cyclic, we have

$$H^2(\mathrm{Gal}(K/F), Z(H^0(\mathbf{C}))) \cong Z(H^0(\mathbf{C}))^{\mathrm{Gal}(K/F)}/N(Z(H^0(\mathbf{C}))).$$

Here $Z(H^0(\mathbf{C}))^{\mathrm{Gal}(K/F)}$ denotes the $\mathrm{Gal}(K/F)$-invariant submodule of $Z(H^0(\mathbf{C}))$ and N is the norm map. By the definition of the action of $\mathrm{Gal}(K/F)$, we see that $Z(H^0(\mathbf{C}))^{\mathrm{Gal}(K/F)} \subset Z(H(\mathbf{C}))$. By Lemma 5.7 and Remark 5.8, we have $Z(H(\mathbf{C})) \cong \mathbf{C}^\times$ consists of scalar matrices and $Z(H^0(\mathbf{C}))$ contains all scalar matrices. The triviality of the cohomology group in question follows immediately from these facts. □

Appendix I

Let k be a field of characteristic 0 and K be an algebraic extension of k of finite degree. Let V be a finite dimensional vector space over K. When we regard V as a vector space over k, we denote it by $\underline{V}$. Let $\mathrm{R}_{K/k}$ be the restriction of scalars functor of Weil. We have $\mathrm{R}_{K/k}(\mathrm{GL}(V)) \subset \mathrm{GL}(\underline{V})$.

DEFINITION AI.1. *Let H be an algebraic subgroup of $\mathrm{R}_{K/k}(\mathrm{GL}(V))$ defined over k. The K-envelope of H is the smallest algebraic subgroup G defined over K of $\mathrm{GL}(V)$ such that $\mathrm{R}_{K/k}(G) \supset H$.*

Since the identity component G^0 is defined over K and $\mathrm{R}_{K/k}(G^0)$ is the identity component of $\mathrm{R}_{K/k}(G)$, we see that if H is connected, then its K-envelope is connected.

For a Lie algebra $\mathfrak{g}$ over K, let $\mathrm{R}_{K/k}(\mathfrak{g})$ denote $\mathfrak{g}$ considered as a Lie algebra over k. Then we have the relation

$$\mathrm{Lie}(\mathrm{R}_{K/k}(G)) = \mathrm{R}_{K/k}(\mathrm{Lie}(G)) \tag{AI.1}$$

for an algebraic group G defined over K. We can prove this relation, for example, using the description of $\mathrm{R}_{K/k}$ given in [Y3], §1.10.

We can formally extend the concept of K-envelope to Lie algebras.

DEFINITION AI.2. *Let $\mathfrak{h}$ be a Lie subalgebra of $\mathrm{R}_{K/k}(\mathfrak{gl}(V))$ defined over k. The K-envelope of $\mathfrak{h}$ is the smallest Lie subalgebra $\mathfrak{g}$ defined over K of $\mathfrak{gl}(V)$ such that $\mathrm{R}_{K/k}(\mathfrak{g}) \supset \mathfrak{h}$.*

It is easy to see that the K-envelope of $\mathfrak{h}$ is the K-linear span of $\mathfrak{h}$. It is well known that $\mathfrak{h}$ is semisimple (resp. reductive) if and only if $\mathfrak{h} \otimes_k K$ is semisimple (resp. reductive) (cf. [Bou], §6, n°10). Since there exists a surjective homomorphism from $\mathfrak{h} \otimes_k K$ to the K-envelope of $\mathfrak{h}$, we obtain:

PROPOSITION AI.3. *Let* $\mathfrak{h}$ *be a Lie subalgebra defined over* k *of* $\mathrm{R}_{K/k}(\mathfrak{gl}(V))$. *If* $\mathfrak{h}$ *is semisimple (resp. reductive), then the* K*-envelope of* $\mathfrak{h}$ *is semisimple (resp. reductive).*

The purpose of this Appendix is to prove:

PROPOSITION AI. 4. *Let* H *be an algebraic subgroup defined over* k *of* $\mathrm{R}_{K/k}(\mathrm{GL}(V))$. *We assume that* H *is connected. If* H *is semisimple (resp. reductive), then the* K*-envelope of* H *is semisimple (resp. reductive).*

For the proof, we need some preparations. Let $\mathfrak{g}$ be a Lie subalgebra of $\mathfrak{gl}(V)$. We call $\mathfrak{g}$ *algebraic* if $\mathfrak{g} = \mathrm{Lie}(G)$ for a connected algebraic subgroup G defined over K of $\mathrm{GL}(V)$. The smallest algebraic Lie subalgebra of $\mathfrak{gl}(V)$ which contains $\mathfrak{g}$ is called the *algebraic envelope* of $\mathfrak{g}$ (cf. Chevalley, [Ch], Chapitre II, §14).

PROPOSITION AI.5. *Let* G *be a connected algebraic subgroup defined over* K *of* $\mathrm{GL}(V)$. *Let* $R(G)$ *be the radical of* G. *Then the radical of* $\mathrm{Lie}(G)$ *is equal to* $\mathrm{Lie}(R(G))$.

PROOF. Put $\mathfrak{g} = \mathrm{Lie}(G)$ and let $\mathfrak{r}$ be the radical of $\mathfrak{g}$. Since $\mathrm{Lie}(R(G))$ is a solvable ideal of $\mathfrak{g}$, we have $\mathrm{Lie}(R(G)) \subset \mathfrak{r}$. The point is to show the opposite inclusion. First let us show that $\mathfrak{r}$ is algebraic. Let $\mathfrak{r}'$ be the algebraic envelope of $\mathfrak{r}$. Take a connected algebraic group $R' \subset \mathrm{GL}(V)$ defined over K so that $\mathrm{Lie}(R') = \mathfrak{r}'$. We obviously have $\mathfrak{r}' \subset \mathfrak{g}$, $R' \subset G$. Let $g \in G$. Since $\mathrm{Ad}(g)\mathfrak{r}' = \mathrm{Lie}(gR'g^{-1})$ is algebraic and contains $\mathrm{Ad}(g)\mathfrak{r} = \mathfrak{r}$, we have

$$\mathrm{Ad}(g)\mathfrak{r}' = \mathfrak{r}', \qquad g \in G. \tag{AI.2}$$

By (AI.2), we see that R' is a normal subgroup of G. Hence $\mathfrak{r}'$ is an ideal of $\mathfrak{g}$. By [Ch], II, Théorème 13, $[\mathfrak{r}', \mathfrak{r}'] = [\mathfrak{r}, \mathfrak{r}]$. Hence $\mathfrak{r}'$ is solvable. Therefore $\mathfrak{r} = \mathfrak{r}'$ and $\mathfrak{r}$ is algebraic. Since $\mathfrak{r}$ is a solvable ideal, we see that R' is a connected solvable normal subgroup. Therefore we have $R' \subset R(G)$, $\mathfrak{r} \subset \mathrm{Lie}(R(G))$. This completes the proof. □

COROLLARY 1. If G is reductive, then $\mathrm{Lie}(G)$ is reductive.

In fact, $R(G)$ is a torus if G is reductive. By Proposition AI.5, the radical of $\mathrm{Lie}(G)$ is commutative. We note that the converse is false as the example $G = SL(2) \times \mathbf{G}_a$ shows.

COROLLARY 2. G is semisimple if and only if $\mathrm{Lie}(G)$ is semisimple.

In fact, we have

$$G \text{ is semisimple} \Longleftrightarrow R(G) = \{1\} \Longleftrightarrow \mathrm{Lie}(R(G)) = \{0\}$$
$$\Longleftrightarrow \text{The radical of } \mathrm{Lie}(G) = \{0\} \Longleftrightarrow \mathrm{Lie}(G) \text{ is semisimple.}$$

Remark. Proposition AI.5 and its Corollaries must be well known to specialists. In fact, Corollary 2 is given as Theorem 13.5 in Humphreys [H]. For the other statements, the author could not find appropriate references.

PROOF OF PROPOSITION AI.4. Let G be the K-envelope of H and put

$$\mathfrak{g} = \mathrm{Lie}(G), \qquad \mathfrak{h} = \mathrm{Lie}(H).$$

Let $\mathfrak{g}_0$ be the K-envelope of $\mathfrak{h}$. By (AI.1), we have $\mathfrak{g} \supset \mathfrak{g}_0$.

First assume that H is semisimple. By Corollary 2 to Proposition AI.5, $\mathfrak{h}$ is semisimple. Hence $\mathfrak{g}_0$ is semisimple by Proposition AI.3. By [Ch], II, Théorème 15,

$\mathfrak{g}_0$ is algebraic. Hence there exists a connected semisimple algebraic subgroup G_0 defined over K of G such that $\mathrm{Lie}(G_0) = \mathfrak{g}_0$. By (AI.1), we have

$$\mathrm{Lie}(\mathrm{R}_{K/k}(G_0)) = \mathrm{R}_{K/k}(\mathfrak{g}_0) \supset \mathfrak{h}.$$

Therefore $\mathrm{R}_{K/k}(G_0) \supset H$ and we obtain $G = G_0$. This proves our assertion in the semisimple case.

Next assume that H is reductive. By Corollary 1 to Proposition AI.5, $\mathfrak{h}$ is reductive. By Proposition AI.3, $\mathfrak{g}_0$ is reductive. Hence we have

$$\mathfrak{g}_0 = \mathfrak{s}_0 \times \mathfrak{c}_0.$$

Here $\mathfrak{s}_0$ is a simisimple subalgebra and $\mathfrak{c}_0$ is an abelian subalgebra. Let $\mathfrak{c} \subset \mathfrak{gl}(V)$ be the algebraic envelope of $\mathfrak{c}_0$. By [Ch], II, Théorème 13, $\mathfrak{c}$ is abelian. We have $\mathfrak{s}_0 \times \mathfrak{c} \subset \mathfrak{gl}(V)$. By [Ch], Théorème 14, $\mathfrak{s}_0 \times \mathfrak{c}$ is algebraic. There exists a connected algebraic subgroup G' defined over K of $\mathrm{GL}(V)$ such that $\mathrm{Lie}(G') = \mathfrak{s}_0 \times \mathfrak{c}$. By (AI.1), we have

$$\mathrm{Lie}(\mathrm{R}_{K/k}(G')) = \mathrm{R}_{K/k}(\mathfrak{s}_0 \times \mathfrak{c}) \supset \mathfrak{h}.$$

Hence $\mathrm{R}_{K/k}(G') \supset H$. This implies $G' \supset G$. Thus we obtain

$$\mathfrak{s}_0 \times \mathfrak{c} \supset \mathrm{Lie}(G) = \mathfrak{g} \supset \mathfrak{s}_0 \times \mathfrak{c}_0.$$

From this relation, we see that $\mathfrak{z}(\mathfrak{g}) = \mathfrak{c}$, $\mathrm{Lie}(G) = \mathrm{Lie}(G')$, $G = G'$. Here $\mathfrak{z}(\mathfrak{g})$ denotes the center of $\mathfrak{g}$. In particular, we have shown that $\mathfrak{g}$ is reductive.

Now write

$$\mathfrak{g} = \mathfrak{s} \times \mathfrak{c},$$

where $\mathfrak{s}$ is semisimple and $\mathfrak{c}$ is abelian. Let $R(G)$ be the radical of G. By Proposition AI.4, we have $\mathrm{Lie}(R(G)) = \mathfrak{c}$. Since $\mathfrak{c} = \mathrm{Lie}(Z(G)^0)$, we obtain $R(G) = Z(G)^0$. By [Bo1], p. 86, Theorem 4.7, we have

$$R(G) = R(G)_s \times R(G)_u.$$

Here $R(G)_s$ (resp. $R(G)_u$) denotes the algebraic subgroup of $R(G)$ consisting of semisimple (resp. unipotent) elements. Put $\mathfrak{c}_s = \mathrm{Lie}(R(G)_s)$. Since $Z(H)^0 \subset \mathrm{R}_{K/k}(Z(G)^0) = \mathrm{R}_{K/k}(R(G))$, we have $Z(H)^0 \subset \mathrm{R}_{K/k}(R(G)_s)$. Hence $\mathfrak{z}(\mathfrak{h}) \subset \mathrm{R}_{K/k}(\mathfrak{c}_s)$. Now take a connected algebraic subgroup G'' of G so that $\mathrm{Lie}(G'') = \mathfrak{s} \times \mathfrak{c}_s$. Then $R(G'') = R(G)_s$, so G'' is reductive. We have

$$\mathrm{Lie}(\mathrm{R}_{K/k}(G'')) = \mathrm{R}_{K/k}(\mathfrak{s} \times \mathfrak{c}_s) \supset \mathrm{R}_{K/k}(\mathfrak{s})\mathfrak{z}(\mathfrak{h}).$$

Since $\mathrm{R}_{K/k}(\mathfrak{s})$ contains the semisimple part of $\mathfrak{h}$, we get

$$\mathrm{Lie}(\mathrm{R}_{K/k}(G'')) \supset \mathfrak{h}, \qquad \mathrm{R}_{K/k}(G'') \supset H.$$

Therefore we obtain $G = G''$. This completes the proof. □

Appendix II

In this appendix, we will prove a few results related to the descent of the field of definition and a representation. In this appendix, we will use the right action of Galois groups.

Let F be a field and K be a Galois extension of F of finite degree. Let X be an algebraic variety defined over K. We assume that X is quasi-projective. Suppose that for every $\sigma \in \mathrm{Gal}(K/F)$, there is given an isomorphism $f_\sigma : X \longrightarrow X^\sigma$ defined over K such that the descent condition

(AII.1) $$f_{\sigma\tau} = f_\sigma^\tau f_\tau$$

is satisfied. Then a well known theorem of Weil ([W2]) tells that there exists an algebraic variety X_0 defined over F and an isomorphism $g : X_0 \longrightarrow X$ defined over K such that

$$f_\sigma = g^\sigma \circ g^{-1}. \tag{AII.2}$$

Take $x \in X_0$ and put $y = g(x)$. Then, for $\sigma \in \mathrm{Gal}(\overline{F}/F)$, we have

$$y^\sigma = g^\sigma(x^\sigma) = (f_{\sigma|K} \circ g)(x^\sigma).$$

Hence we obtain

$$x^\sigma = g^{-1} f_{\sigma|K}^{-1}(g(x)^\sigma). \tag{AII.3}$$

First we will give:

PROOF OF PROPOSITION 5.1. We take a connected reductive algebraic group G defined over F so that it splits over F and such that the based root datum

$$\mathcal{R}_0(G) = (X^*(T), \Delta, X_*(T), \check{\Delta}).$$

is the dual to $\mathcal{R}_0(M)$. Here T is the maximal F-split torus of G. Let $B \supset T$ be a Borel subgroup defined over F of G. Then we have

$$\mathrm{Aut}(\mathcal{R}_0(G)) \cong \mathrm{Aut}(G, B, T, \{u_\alpha\}_{\alpha\in\Delta}) \subset \mathrm{Aut}(G).$$

For $\nu \in \mathrm{Aut}(G, B, T, \{u_\alpha\}_{\alpha\in\Delta})$, we have

$$\nu(B) = B, \qquad \nu(T) = T, \qquad \alpha(\nu^{-1}(t)) = \nu(\alpha)(t), \quad t \in T.$$

Choose u_α so that $u_\alpha \in G(F)$. It is known that ν is defined over F (cf. [SGA3], Exposé XXIV, Théorème 1.3).

We may assume that μ is a homomorphism

$$\mu : \mathrm{Gal}(K/F) \longrightarrow \mathrm{Aut}(G, B, T, \{u_\alpha\}_{\alpha\in\Delta})$$

for a finite Galois extension K of F. For $\sigma \in \mathrm{Gal}(K/F)$, we put $f_\sigma = \mu(\sigma)^{-1}$. Since f_σ is defined over F, the relation $f_{\sigma\tau} = f_\sigma f_\tau$ is a descent condition. Hence there exists an algebraic group G_0 defined over F and an isomorphism $g : G_0 \longrightarrow G$ defined over K such that $f_\sigma = g^\sigma \circ g^{-1}$. We have ${}^LG_0^0 \cong {}^LG^0 \cong M(\mathbf{C})$. We define a Borel subgroup and a maximal torus of G_0 by

$$B_0 = g^{-1}(B), \qquad T_0 = g^{-1}(T).$$

For $\sigma \in \mathrm{Gal}(K/F)$, we have

$$B_0^\sigma = (g^\sigma)^{-1}(B^\sigma) = g^{-1}f_\sigma^{-1}(B) = g^{-1}\mu(\sigma)(B) = g^{-1}(B) = B_0.$$

Hence B_0 is defined over F, i.e., G_0 is quasi-split over F. Similarly we see that T_0 is defined over F. For $\alpha \in \Delta$, we define a character $\alpha_0 : T_0 \longrightarrow \mathbf{G}_m$ by the formula

$$\alpha_0(t) = \alpha(g(t)), \qquad t \in T_0.$$

Put $u^0_{\alpha_0} = g^{-1}(u_\alpha)$, $\alpha \in \Delta$. Then, for $\sigma \in \mathrm{Gal}(K/F)$, we have

$$\begin{aligned}
&\alpha_0^\sigma(t^\sigma) = \alpha_0(t)^\sigma = (\alpha(g(t)))^\sigma = \alpha(g^\sigma(t^\sigma)) = \alpha(\mu(\sigma)^{-1}g(t^\sigma)) \\
=&(\mu(\sigma)\alpha)(g(t^\sigma)) = (\mu(\sigma)\alpha_0)(t^\sigma), \\
&(u^0_{\alpha_0})^\sigma = (g^{-1}(u_\alpha))^\sigma = (g^\sigma)^{-1}(u_\alpha) = g^{-1}\mu(\sigma)(u_\alpha) = g^{-1}(u_{\mu(\sigma)\alpha}).
\end{aligned}$$

Therefore we obtain $\alpha_0^\sigma = \mu(\sigma)\alpha_0$, $\mu_{G_0} = \mu$. This completes the proof of Proposition 5.1.

Next we assume that X is an abelian variety of dimension d. We are going to describe the ℓ-adic representation of $\mathrm{Gal}(\overline{F}/F)$ attached to X_0 using the ℓ-adic representation of $\mathrm{Gal}(\overline{F}/K)$ attached to X and the descent data $\{f_\sigma\}$. We fix an isomorphism

$$\iota_0 : T_\ell(X_0) \cong \mathbf{Q}_\ell^{2d},$$

where $T_\ell(X_0)$ denotes the Tate module of X_0. We regard an element of $\mathbf{Q}_\ell^{2d}$ as a row vector. Let

$$g_* : T_\ell(X_0) \cong T_\ell(X)$$

be the isomorphism obtained from g. Put

$$\mathrm{Gal}(K/F) = \{\sigma_1, \sigma_2, \ldots, \sigma_n\}, \qquad \sigma_1 = \mathrm{id}.$$

We use the same letter σ_i for its extension to an element of $\mathrm{Gal}(\overline{F}/F)$. For $1 \le i \le n$, we choose an isomorphism $\iota_{\sigma_i} : T_\ell(X^{\sigma_i}) \cong \mathbf{Q}_\ell^{2d}$ so that

(AII.4) $$\iota_{\sigma_i}((g_*(x))^{\sigma_i}) = \iota_0(x), \qquad x \in T_\ell(X_0)$$

holds. We put

(AII.5) $$\iota_{\sigma_1}(x^\tau) = \iota_{\sigma_1}(x)\rho(\tau), \qquad x \in T_\ell(X), \quad \tau \in \mathrm{Gal}(\overline{F}/K)$$

with $\rho(\tau) \in \mathrm{GL}(2d, \mathbf{Q}_\ell)$; ρ is the ℓ-adic representation of $\mathrm{Gal}(\overline{F}/K)$ attached to X. Define $f^*_{\sigma_i} \in \mathrm{GL}(2d, \mathbf{Q}_\ell)$ by

(AII.6) $$\iota_{\sigma_1}(f_{\sigma_i}^{-1}(x)) = \iota_{\sigma_i}(x) f^*_{\sigma_i}, \qquad x \in T_\ell(X^{\sigma_i}).$$

Now take $\sigma \in \mathrm{Gal}(\overline{F}/F)$ and put $\sigma = \tau\sigma_i$, $\tau \in \mathrm{Gal}(\overline{F}/K)$. Then for $x \in T_\ell(X_0)$, we have, using (AII.3), (AII.4), (AII.6), (AII.4), (AII.5) in this order, that

$$\begin{aligned}\iota_0(x^\sigma) &= \iota_0(g_*^{-1} f_{\sigma_i}^{-1}(g_*(x)^\sigma)) = \iota_{\sigma_1}(f_{\sigma_i}^{-1}(g_*(x)^\sigma)) = \iota_{\sigma_i}(g_*(x)^{\tau\sigma_i}) f^*_{\sigma_i} \\ &= \iota_{\sigma_1}(g_*(x)^\tau) f^*_{\sigma_i} = \iota_{\sigma_1}(g_*(x))\rho(\tau) f^*_{\sigma_i}.\end{aligned}$$

Hence we obtain

(AII.7) $$\iota_0(x^\sigma) = \iota_0(x)\rho(\tau) f^*_{\sigma_i}, \qquad x \in T_\ell(X_0).$$

This formula shows that the ℓ-adic representation of $\mathrm{Gal}(\overline{F}/F)$ attached to X_0 is given by

$$\sigma = \tau\sigma_i \longrightarrow \rho(\tau) f^*_{\sigma_i}.$$

Now we assume that X is defined over F from the beginning. Thus X_0 is an F-form of X. Since ρ extends to an ℓ-adic representation of $\mathrm{Gal}(\overline{F}/F)$, (AII.5) can be written as

(AII.5′) $$\iota_{\sigma_1}(x^\tau) = \iota_{\sigma_1}(x)\rho(\tau), \qquad x \in T_\ell(X), \quad \tau \in \mathrm{Gal}(\overline{F}/F).$$

Since $X^{\sigma_i} = X$, we can write (AII.6) as

(AII.6′) $$\iota_{\sigma_1}(f_{\sigma_i}^{-1}(x)) = \iota_{\sigma_1}(x) f^*_{\sigma_1}, \qquad x \in T_\ell(X).$$

Now take $\sigma \in \mathrm{Gal}(\overline{F}/F)$ and $x \in T_\ell(X_0)$. We put $f^*_{\sigma|K} = f^*_{\sigma_i}$ if $\sigma|K = \sigma_i$. Calculating similarly to the above, we obtain

$$\begin{aligned}\iota_0(x^\sigma) &= \iota_0(g_*^{-1} f_{\sigma|K}^{-1}(g_*(x)^\sigma)) = \iota_{\sigma_1}(f_{\sigma|K}^{-1}(g_*(x)^\sigma)) = \iota_{\sigma_1}(g_*(x)^\sigma) f^*_{\sigma|K} \\ &= \iota_{\sigma_1}(g_*(x))\rho(\sigma) f^*_{\sigma|K},\end{aligned}$$

(AII.7′) $$\iota_0(x^\sigma) = \iota_0(x)\rho(\sigma) f^*_{\sigma|K}, \qquad x \in T_\ell(X_0).$$

This formula shows that the ℓ-adic representation of $\mathrm{Gal}(\overline{F}/F)$ attached to X_0 is given by

$$\sigma \longrightarrow \rho(\sigma) f^*_{\sigma|K}. \tag{AII.8}$$

From the descent condition $f_{\sigma\tau|K} = f^{\tau}_{\sigma|K} f_{\tau|K}$, we can derive the relatiom

$$f^*_{\sigma\tau|K} = \rho(\tau)^{-1} f^*_{\sigma|K} \rho(\tau) f^*_{\tau|K}, \qquad \sigma, \tau \in \mathrm{Gal}(\overline{F}/F). \tag{AII.9}$$

Using (AII.9), we can verify that (AII.8) defines a representation of $\mathrm{Gal}(\overline{F}/F)$.

Let A be an abelian variety defined over F and m be a positive integer. Put $X = A^m$, $d_0 = \dim A$, $d = md_0 = \dim X$. Let

$$\rho : \mathrm{Gal}(\overline{F}/F) \longrightarrow \mathrm{GL}(2d_0, \mathbf{Q}_\ell)$$

be the ℓ-adic representation attached to A. Let ρ_X be the ℓ-adic representation attached to X. Then we have

$$\rho_X = \rho \oplus \rho \oplus \cdots \oplus \rho \qquad (m \text{ times}).$$

Let K be a finite Galois extension of F and let $\eta : \mathrm{Gal}(K/F) \longrightarrow \mathrm{GL}(m, \mathbf{Z})$ be a representation. For $\sigma \in \mathrm{Gal}(K/F)$, we define an isomorphism $f_\sigma : X \longrightarrow X$ by

$$f_\sigma((x_1, \dots, x_m)) = (x_1, \dots, x_m)\eta(\sigma)^{-1}.$$

Then f_σ is defined over F and satisfies the descent condition $f_{\sigma\tau} = f_\sigma f_\tau$.

Proposition AII.1. *Let X_0 be the F-form of X defined by f_σ. Then the ℓ-adic representation attached to X_0 is equivalent to $\rho \otimes \eta$. Here we regard η as a representation of* $\mathrm{Gal}(\overline{F}/F)$ *by the canonical map* $\mathrm{Gal}(\overline{F}/F) \longrightarrow \mathrm{Gal}(K/F)$.

Proof. We take an isomorphism $\iota_A : T_\ell(A) \cong \mathbf{Q}_\ell^{2d_0}$ and put

$$\iota_{\sigma_1} = \iota_A \oplus \cdots \oplus \iota_A : T_\ell(X) = T_\ell(A)^m \cong \mathbf{Q}_\ell^{2d}.$$

Then, for $(x_1, \dots, x_m) \in T_\ell(A)$, we have

$$\begin{aligned} \iota_{\sigma_1}(f_{\sigma_i}^{-1}(x_1, \dots, x_m)) &= \iota_{\sigma_1}((x_1, \dots, x_m)) f^*_{\sigma_i} = (\iota_A(x_1), \dots, \iota_A(x_m)) f^*_{\sigma_i} \\ &= \iota_{\sigma_1}((x_1, \dots, x_m)\eta(\sigma_i)) = (\iota_A(x_1), \dots, \iota_A(x_m))(1_{2d_0} \otimes \eta(\sigma_i)). \end{aligned}$$

Hence we have $f^*_{\sigma|K} = 1_{2d_0} \otimes \eta(\sigma)$. The assertion follows from (AII.8). □

We say that an abelian variety A has sufficiently many complex multiplications if $\mathrm{End}(A) \otimes \mathbf{Q}$ contains a commutative semisimple algebra of dimension $2\dim A$. Assume furthermore that A is defined over a number field F. Let $\zeta(s, A/F)$ be the one dimensional part of the zeta function of A. Then the author showed ([Y2]) that there exists a representation $\rho : W_{F,K} \longrightarrow \mathrm{GL}(2n, \mathbf{C})$ such that $\zeta(s, A/F) = L(s, \rho, W_{F,K})$. Here $n = \dim A$, K is a finite Galois extension of F and $W_{F,K}$ is the relative Weil group. We are going to show that there exists an A for which ρ is not equivalent to a direct sum of monomial representations, solving a question raised in [Y2], p. 99. We note the following: Put $X = A^m$ and construct the F-form X_0 of X as above. Then X_0 is an abelian variety with sufficiently many complex multiplications. By Proposition AII.1, we have

$$\zeta(s, X_0/F) = L(s, \rho \otimes \eta, W_{F,K}). \tag{AII.10}$$

Let F be an imaginary quadratic field and E be an elliptic curve defined over F such that $\mathrm{End}(E) \otimes \mathbf{Q} \cong F$. Then we have ([Sh1], Theorem 7.43, [Sh2], 19.11)

$$\zeta(s, E/F) = L(s, \psi) L(s, \overline{\psi}).$$

Here ψ is a Hecke character of $F_{\mathbf{A}}^{\times}$. Now take a finite Galois extension K of F and a representation

$$\eta : \mathrm{Gal}(K/F) \longrightarrow \mathrm{GL}(m, \mathbf{Z})$$

so that η is irreducible and non-monomial when regarded as a representation to $\mathrm{GL}(m, \mathbf{C})$. (For example, take K so that $\mathrm{Gal}(K/F) \cong S_n$, $n \geq 5$. There exists an irreducible non-monomial representation η of S_n since it is not solvable; η is realizable over $\mathbf{Q}$. We see that η stabilizes a lattice in $\mathbf{Q}^m$, $m = \dim \eta$.) We regard ψ (resp. η) as a representation of $W_{F,K}$ by the transfer homomorphism (resp. the projection) $W_{F,K} \longrightarrow F_{\mathbf{A}}^{\times}$ (resp. $W_{F,K} \longrightarrow \mathrm{Gal}(K/F)$). Let X_0 be the F-form of E^m constructed using the descent data defined by η. By (AII.10), we have

$$\text{(AII.11)} \qquad \zeta(s, X_0/F) = L(s, (\psi \otimes \eta) \oplus (\overline{\psi} \otimes \eta), W_{F,K}).$$

PROPOSITION AII.2. *The representation $(\psi \otimes \eta) \oplus (\overline{\psi} \otimes \eta)$ of $W_{F,K}$ is not equivalent to a direct sum of monomial representations.*

PROOF. Assume that

$$(\psi \otimes \eta) \oplus (\overline{\psi} \otimes \eta) \cong \oplus_{i=1}^{n} \mathrm{Ind}_{W_{F_i,L}}^{W_{F,L}} \omega_i.$$

Here F_i is an extension of F of finite degree and ω_i is a Hecke character of $(F_i)_{\mathbf{A}}^{\times}$; L is a finite Galois extension of F which contains F_i, $1 \leq i \leq n$ and K; ω_i is regarded as a character of $W_{F_i,L}$ by the transfer homomorphism $W_{F_i,L} \longrightarrow (F_i)_{\mathbf{A}}^{\times}$; the representation on the left-hand side is regarded as the representation of $W_{F,L}$ by the canonical homomorphism $W_{F,L} \longrightarrow W_{F,K}$. By the irreducibility of $\psi \otimes \eta$ and $\overline{\psi} \otimes \eta$, we have

$$\mathrm{Ind}_{W_{F_1,L}}^{W_{F,L}} \omega_1 \cong \psi \otimes \eta \quad \text{or} \quad \overline{\psi} \otimes \eta \quad \text{or} \quad (\psi \otimes \eta) \oplus (\overline{\psi} \otimes \eta).$$

First we assume that

$$\mathrm{Ind}_{W_{F_1,L}}^{W_{F,L}} \omega_1 \cong \psi \otimes \eta.$$

Then we have

$$\text{(AII.12)} \qquad (\mathrm{Ind}_{W_{F_1,L}}^{W_{F,L}} \omega_1) \otimes \psi^{-1} = \mathrm{Ind}_{W_{F_1,L}}^{W_{F,L}} (\omega_1 \otimes (\psi \circ N_{F_1/F})^{-1}) \cong \eta.$$

Now we consider the commutative diagram

$$\text{(AII.13)} \qquad \begin{array}{ccccccccc} 1 & \longrightarrow & L_{\mathbf{A}}^{\times}/L^{\times} & \longrightarrow & W_{F_1,L} & \longrightarrow & \mathrm{Gal}(L/F_1) & \longrightarrow & 1 \\ & & \Vert & & \downarrow & & \downarrow & & \\ 1 & \longrightarrow & L_{\mathbf{A}}^{\times}/L^{\times} & \longrightarrow & W_{F,L} & \longrightarrow & \mathrm{Gal}(L/F) & \longrightarrow & 1. \end{array}$$

Here the vertical arrows are inclusion maps. Put $\pi = \mathrm{Ind}_{W_{F_1,L}}^{W_{F,L}} (\omega_1 \otimes (\psi \circ N_{F_1/F})^{-1})$. The restriction of π to $L_{\mathbf{A}}^{\times}/L^{\times}$ is trivial by (AII.12). By the Frobenius reciprocity, we have

$$\omega_1 \otimes (\psi \circ N_{F_1/F})^{-1} | (L_{\mathbf{A}}^{\times}/L^{\times}) = 1.$$

Let ω_1' denote the character of $\mathrm{Gal}(L/F_1)$ determined by $\omega_1 \otimes (\psi \circ N_{F_1/F})^{-1}$. By (AII.12), we have

$$\mathrm{Ind}_{\mathrm{Gal}(L/F_1)}^{\mathrm{Gal}(L/F)} \omega_1' \cong \eta.$$

Since the right-hand side factors through the map $\mathrm{Gal}(L/F) \longrightarrow \mathrm{Gal}(K/F)$, we see that $F_1 \subset K$ and that ω_1' is invariant under $\mathrm{Gal}(L/K)$. Hence it determines a character ω_1'' of $\mathrm{Gal}(K/F_1)$. Then we have $\mathrm{Ind}_{\mathrm{Gal}(K/F_1)}^{\mathrm{Gal}(K/F)} \omega_1'' \cong \eta$. This contradicts our assumption that η is not monomial. The case $\mathrm{Ind}_{W_{F_1,L}}^{W_{F,L}}\omega_1 \cong \overline{\psi}\otimes\eta$ can be treated similarly.

Next we consider the case

$$\mathrm{Ind}_{W_{F_1,L}}^{W_{F,L}}\omega_1 \cong (\psi\otimes\eta)\oplus(\overline{\psi}\otimes\eta).$$

We have

$$\text{(AII.14)} \qquad \mathrm{Ind}_{W_{F_1,L}}^{W_{F,L}}(\omega_1\otimes(\psi\circ N_{F_1/F})^{-1}) \cong \eta\oplus(\psi^{-1}\overline{\psi}\otimes\eta).$$

Let ω_1^* be the restriction of $\omega_1\otimes(\psi\circ N_{F_1/F})^{-1}$ to $L_{\mathbf{A}}^\times/L^\times$ in (AII.13). Since the transfer map $W_{F_1,L}\longrightarrow(F_1)_{\mathbf{A}}^\times$ induces the norm map N_{L/F_1} on $L_{\mathbf{A}}^\times/L^\times$, we have

$$\omega_1^* = (\omega_1\otimes(\psi\circ N_{F_1/F})^{-1})\circ N_{L/F_1}.$$

By the formula of induced characters, we see that the restriction of $\mathrm{Ind}_{W_{F_1,L}}^{W_{F,L}}(\omega_1\otimes(\psi\circ N_{F_1/F})^{-1})$ to $L_{\mathbf{A}}^\times/L^\times$ is equal to

$$\bigoplus_{x\in\mathrm{Gal}(L/F)/\mathrm{Gal}(L/F_1)} \omega_1^*(x^{-1}gx), \qquad g\in L_{\mathbf{A}}^\times/L^\times.$$

Since η is trivial on $L_{\mathbf{A}}^\times/L^\times$, $\omega_1^*(x^{-1}gx)$ is trivial for some $x\in\mathrm{Gal}(L/F)$. This implies that $\omega_1^*|L_{\mathbf{A}}^\times/L^\times$ is trivial. But on the right-hand side of (AII.14), the restriction of $\psi^{-1}\overline{\psi}\otimes\eta$ to $L_{\mathbf{A}}^\times/L^\times$ is equal to $(\psi^{-1}\overline{\psi})\circ N_{L/F}$, which is not trivial. This is a contradiction and completes the proof. □

Next we are going to construct an example in which the exact sequence

$$\text{(5.2)} \qquad 1 \longrightarrow H^0(\mathbf{C}) \longrightarrow H(\mathbf{C}) \longrightarrow \mathrm{Gal}(K/F) \longrightarrow 1$$

does not split ($\widetilde{H}=H$). Let F be a number field which contains a CM-field. Take an algebraic Hecke character χ of $F_{\mathbf{A}}^\times$ of infinite order as in §7. Take a finite Galois extension K of F. Then take a finite Galois extension L of K so that $L\subset K_{\mathrm{ab}}$ and that L is normal over F. Take an irreducible representation

$$\eta:\mathrm{Gal}(L/F)\longrightarrow \mathrm{GL}(m,\mathbf{C}).$$

We assume that η is faithful. We take an algebraic number field of finite degree E so that E contains $\chi(a)$ when $a\in F_{\mathbf{A}}^\times$ satisfies $a_\infty=1$ and such that η (changing to an equivalent one) is realized over E. Let λ be a finite place of E. Then, as explained in §7, we have a λ-adic representation

$$\chi_\lambda:\mathrm{Gal}(F_{\mathrm{ab}}/F)\longrightarrow E_\lambda^\times$$

associated to χ. We regard η as a representation of $\mathrm{Gal}(L/F)$ into $\mathrm{GL}(m,E_\lambda)$. We put

$$\rho=\chi\otimes\eta, \qquad \rho_\lambda=\chi_\lambda\otimes\eta.$$

We can regard ρ as a representation of $W_{F,K}$ into $\mathrm{GL}(m,\mathbf{C})$ and ρ_λ as a representation

$$\rho_\lambda:\mathrm{Gal}(K_{\mathrm{ab}}/F)\longrightarrow \mathrm{GL}(m,E_\lambda).$$

We regard ρ_λ also as a representation of $\mathrm{Gal}(\overline{F}/F)$ into $\mathrm{GL}(m,E_\lambda)$. For this ρ_λ, we have

$$H^0(\mathbf{C})=\mathbf{C}^\times\cdot 1_m\cong\mathbf{C}^\times,$$

since χ is of infinite order. We assume that:

(AII.15) The center of $\mathrm{Gal}(L/F)$ is $\mathrm{Gal}(L/K)$.

Then the condition (5.1) is satisfied since η is faithful. We have an exact sequence

$$\text{(AII.16)} \qquad 1 \longrightarrow \mathbf{C}^\times \longrightarrow H(\mathbf{C}) \longrightarrow \mathrm{Gal}(K/F) \longrightarrow 1.$$

Our task is to find an example for which (AII.16) does not split. As in §5, the factor set defining (AII.16) is given as follows. Fix an embedding $E_\lambda \hookrightarrow \mathbf{C}$. For $\sigma \in \mathrm{Gal}(K/F)$, take $\widetilde{\sigma} \in \mathrm{Gal}(K_{\mathrm{ab}}/F)$ so that $\widetilde{\sigma}|K = \sigma$. Put

$$\rho_\lambda(\widetilde{\sigma}\widetilde{\tau}(\widetilde{\sigma\tau})^{-1}) = f(\sigma,\tau)\cdot 1_m, \qquad \eta_\lambda(\widetilde{\sigma}\widetilde{\tau}(\widetilde{\sigma\tau})^{-1}) = f'(\sigma,\tau)\cdot 1_m.$$

Then $\{f(\sigma,\tau)\}$ is the factor set attached to (AII.16). We note that $\mathrm{Gal}(K/F)$ acts trivially on $\mathbf{C}^\times$. Thus $f(\sigma,\tau)$ defines a cohomology class $\xi \in H^2(\mathrm{Gal}(K/F), \mathbf{C}^\times)$. We have

$$f(\sigma,\tau) = f'(\sigma,\tau)\chi_\lambda(\widetilde{\sigma})\chi_\lambda(\widetilde{\tau})\chi_\lambda(\widetilde{\sigma\tau})^{-1}.$$

Hence $f'(\sigma,\tau)$ defines the same cohomology class ξ.

PROPOSITION AII.3. *In addition to (AII.15), assume that the commutator group* $[\mathrm{Gal}(K/F), \mathrm{Gal}(K/F)]$ *is equal to* $\mathrm{Gal}(K/F)$ *and that the exact sequence*

$$(*) \qquad 1 \longrightarrow \mathrm{Gal}(L/K) \longrightarrow \mathrm{Gal}(L/F) \longrightarrow \mathrm{Gal}(K/F) \longrightarrow 1$$

does not split. Then $\xi \neq 1$.

PROOF. By (AII.15), there exists a faithful character ω of $\mathrm{Gal}(L/K)$ such that $\eta(g) = \omega(g)\cdot 1_m$, $g \in \mathrm{Gal}(L/K)$. Then we have

$$f'(\sigma,\tau) = \omega(\widetilde{\sigma}\widetilde{\tau}(\widetilde{\sigma\tau})^{-1}).$$

Here we regard $\widetilde{\sigma}$ as an element of $\mathrm{Gal}(L/F)$ such that $\widetilde{\sigma}|K = \sigma$. Since ω is faithful, we have $\mathrm{Gal}(L/K) \cong \mathbf{Z}/n\mathbf{Z}$ for an integer $n \geq 2$ and $\omega(\mathrm{Gal}(L/K)) = \mu_n$. Here μ_n is the group consisting of all nth roots of unity. Let $\xi' \in H^2(\mathrm{Gal}(K/F), \mu_n)$ be the cohomology class defined by $f'(\sigma,\tau)$. Since $(*)$ does not split, we have $\xi' \neq 1$. The cohomology class ξ is the image of ξ' under the canonical map

$$\varphi : H^2(\mathrm{Gal}(K/F), \mu_n) \longrightarrow H^2(\mathrm{Gal}(K/F), \mathbf{C}^\times).$$

By our assumption, $H^1(\mathrm{Gal}(K/F), \mathbf{C}^\times/\mu_n) = 1$. Hence φ is injective. This completes the proof. □

As a concrete example, take F and L so that $\mathrm{Gal}(L/F) \cong \mathrm{SL}(2, \mathbf{Z}/5\mathbf{Z})$. (Take an elliptic curve defined over $\mathbf{Q}$ so that the field L generated by the 5-division points satisfies $\mathrm{Gal}(L/\mathbf{Q}) \cong \mathrm{GL}(2, \mathbf{Z}/5\mathbf{Z})$. Then let F be the subfield of L which is generated over $\mathbf{Q}$ by a primitive fifth root of unity.) Let K be the subfield of L corresponding to $\left\{\pm\begin{pmatrix}1 & 0\\ 0 & 1\end{pmatrix}\right\}$. We have $\mathrm{Gal}(K/F) \cong A_5$. We see that (AII.15) and the conditions of Proposition AII.3 are satisfied. Let η be an irreducible 2-dimensional representation of $\mathrm{SL}(2, \mathbf{Z}/5\mathbf{Z})$. Then η is faithful. This gives an explicit example for which (AII.16) does not split.

References

[A1] J. Arthur, Unipotent automorphic representations: Conjectures, Astérisque, 171-172 (1989), 13–71.

[A2] J. Arthur, A note on the automorphic Langlands group, Canad. Math. Bull. 45 (2002), 466–482.

[A3] J. Arthur, A note on L-packets, Pure and applied Mathematics Quarterly 2 (2006), 199–217.

[AT] E. Artin and J. Tate, Class field theory, Benjamin, New York, 1968.

[Bl] D. Blasius, On the critical values of Hecke L-series, Ann. of Math. 124 (1986), 23–63.

[Bo1] A. Borel, Linear algebraic groups, second enlarged edition, Graduate Texts in Mathematics 126, Springer Verlag, New York-Berlin-Heidelberg, 1991.

[Bo2] A. Borel, Automorphic L-functions, Proc. Sympos. Pure Math. 33 (1979), part 2, 27–61.

[Bou] N. Bourbaki, Groupes et algèbres de Lie, Chapitre I, Herman, Paris, 1971.

[BT] A. Borel and J. Tits, Groupes réductifs, Publ. Math. IHES 27 (1965), 55–150.

[Ch] C. Chevalley, Théorie des groupes de Lie, Tome II, Groupes algébriques, Herman, Paris, 1951.

[Cl] L. Clozel, Motifs et formes automorphes: applications du principe de fonctorialité, in "Automorphic forms, Shimura varieties and L-functions I", 77–159, Academic Press, Boston, 1990.

[CKPS] J. W. Cogdell, H. H. Kim, I. I. Piatetski–Shapiro and F. Shahidi, Functoriality for the classical groups, Publ. Math. IHES 99 (2004), 163–233.

[D1] P. Deligne, Les constantes des equations fonctionnelles des fonctions L, in modular functions of one variable II, Lecture notes in Math. 349 (1973), 501–597, Springer Verlag.

[D2] P. Deligne, Valeurs de fonctions L et périodes d'integrales, Proc. Symposia Pure Math. 33 (1979), part 2, 313–346.

[D3] P. Deligne, Hodge cycles on abelian varieties (Notes by J. S. Milne), Lecture notes in Math. 900 (1982), 9–100, Springer-Verlag.

[DM] P. Deligne and J. S. Milne, Tannakian categories, Lecture notes in Math. 900 (1982), 101–228, Springer-Verlag.

[H] J. E. Humphreys, Linear algebraic groups, Graduate texts in mathematics, Springer-Verlag, New York-Heidelberg-Berlin, 1975.

[J] U. Jannsen, Motives, numerical equivalence, and semi-simplicity, Invent. Math. 107 (1992), 447–452.

[JKS] U. Jannsen, S. Keiman and J. -P. Serre (eds.), Motives, Proc. Symposia Pure Math. 55 (1994), parts 1 and 2.

[K] R. Kottwitz, Stable trace formula: cuspidal tempered terms, Duke Math. J. 51 (1984), 611–650.

[Lab1] J. -P. Labesse, Cohomologie, L-groupes et fonctorialité, Comp. Math. 55 (1985), 163–184.

[Lab2] J. -P. Labesse (avec la collaboration de L. Breen et L. Clozel), Cohomologie, Stabilisation et changement de base, Astérisque 257 (1999).

[La] S. Lang, Some history of the Shimura-Taniyama conjecture, Notices of AMS. 42 (1995), 1301–1307.

[Lan1] R. P. Langlands, Automorphic forms, Shimura varieties, and motives. Ein Märchen, in Automorphic forms, Representations, and L-functions, Proc. Sympos. Pure Math. 33 (1979), 205–248.

[Lan2] R. P. Langlands, Stable conjugacy: Definitions and Lemmas, Can. J. Math. XXXI (1979), 700–725 .

[Lan3] R. P. Langlands, On the classification of representations of real algebraic groups, in Representation theory and harmonic analysis on semisimple Lie groups, Mathematical Surveys and Monographs 31, American Mathematical Society, 1989, 101–170.

[LL] J. -P. Labesse and R. P. Langlands, L-indistingushability for $SL(2)$, Can. J. Math. 31 (1979), 726–785.

[LS] R. P. Langlands and D. Shelstad, On the definition of transfer factors, Math. Ann. 278 (1987), 219–271.

[Mi] J. S. Milne, Arithmetic duality theorems, Perspectives in Mathematics 1, Academic Press, 1986.

[Mu1] D. Mumford, Families of abelian varieties, in Algebraic groups and discontinuous subgroups, Proc. Sympos. Pure Math. 9 (1966), 347–351.

[Mu2] D. Mumford, A note on Shimura's paper "Discontinuous groups and abelian varieties ", Math. Ann. 181 (1969), 345–351.

[R] D. Ramakrishnan, Pure motives and automorphic forms, Proc. Symposia Pure Math. 55 (1994), vol. II, 411–446.

[Se1] J. -P. Serre, Représentations ℓ-adiques, Kyoto Int. Symposium on Algebraic Number Theory, Japan Soc. for the Promotion of Science (1977), 177–193.

[Se2] J. -P. Serre, Modular forms of weight one and Galois representations, in Algebraic number fields (ed. A. Frölich), Academic Press (1977), 193–268.

[Se3] J. -P. Serre, Propriétés conjecturales des groupes de Galois motiviques et des représentations ℓ-adiques, Proc. Symp. Pure Math. 55 (1994), vol. I, 377–400.

[Se4] J. -P. Serre, Galois cohomology, Springer Verlag, 1997.

[SGA3] Schémas en groupes, Un séminaire dirigé par M. Demazure et A. Grothendieck, Lecture notes in Math. 151, 152, 153, Springer Verlag, 1970.

[Sh1] G. Shimura, Introduction to the Arithmetic Theory of Automorphic Functions, Iwanami Shoten and Princeton University Press, 1971.

[Sh2] G. Shimura, Abelian varieties with complex multiplication and modular functions, Princeton Mathematical Series 46, Princeton University Press, 1998.

[Sh3] G. Shimura, Elementary Dirichlet series and modular forms, Springer Monographs in Mathematics, Springer Verlag, 2007.

[Sh4] G. Shimura, Collected Papers I–IV, Springer Verlag, 2002.

[Sp] T. A. Springer, Reductive groups, Proc. of Symposia Pure Math., 33 (1979), 3–27.

[St1] R. Steinberg, Lectures on Chevalley groups, Yale University Lecture notes, 1967.

[St2] R. Steinberg, Endomorphisms of linear algebraic groups, Memoirs of American Mathematical Society 80, 1968.

[T1] J. Tilouine, Nearly ordinary rank four Galois representations and p-adic modular forms (with an appendix by D. Blasius), Comp. Math. 142 (2006), 1122–1156.

[T2] J. Tilouine, Siegel varieties and p-adic Siegel modular forms, Documenta Math. Extra Volume Coates (2006), 789–824.

[V] D. Vogan, The local Langlands conjecture, Contemp. Math. 145 (1993), 305–379.

[W1] A. Weil, Sur la théorie du corps de classes, J. Math. Soc. Japan 3 (1951), 1–35.

[W2] A. Weil, The field of definition of a variety, Amer. J. Math. 78 (1956), 509–524.

[Y1] H. Yoshida, Siegel's modular forms and the arithmetic of quadratic forms, Inv. Math. 60 (1980), 193–248.

[Y2] H. Yoshida, Abelian varieties with complex multiplication and representations of the Weil groups, Ann. of Math. 114 (1981), 87–102.

[Y3] H. Yoshida, On the zeta functions of Shimura varieties and periods of Hilbert modular forms, Duke Math. J. 75 (1994), 121–191.

[Y4] H. Yoshida, Motives and Siegel modular forms, Amer. J. Math. 123 (2001), 1171–1197.

Department of Mathematics, Kyoto University, Kyoto, Japan, 606-8502

E-mail address: `yoshida@math.kyoto-u.ac.jp`